Worldwide Guide to Equivalent Nonferrous Metals and Alloys

Paul M. Unterweiser
Staff Editor
Manager, Publications Development

Marilyn Penzenik
Project Coordinator

AMERICAN SOCIETY FOR METALS
METALS PARK, OHIO 44073

Nothing contained in this book is to be construed as a grant of any right of manufacture, sale, or use in connection with any method, process, apparatus, product or composition, whether or not covered by letters patent or registered trademark, nor as a defense against liability for the infringement of letters patent or registered trademark.

Library of Congress Cataloging in Publication Data

American Society for Metals.

Worldwide guide to equivalent nonferrous metals and alloys.

Includes index.

1. Nonferrous metals – Specifications. 2. Nonferrous alloys – Specifications. I. Unterweiser, Paul M. II. Penzenik, Marilyn. III. Title.

TA479.3.A46 1980 669 80-15791

ISBN 0-87170-101-4

PRINTED IN THE UNITED STATES OF AMERICA

Introduction

Worldwide Guide to Equivalent Nonferrous Metals and Alloys is the second volume in a new series of basic reference books, the Engineering Handbook series, compiled and published by American Society for Metals. An earlier volume, *Worldwide Guide to Equivalent Irons and Steels,* was published in 1979 and met with an understandably enthusiastic reception. At best, however, it marked completion of not more than half of a total assignment – a comprehensive and dependable compilation of equivalent metals and alloys, ferrous *and* nonferrous, produced in the leading industrial nations and used in virtually all nations throughout the world. With the publication of the present volume, this rather monumental assignment has been completed. Most important, the universal demand for such information – long believed impossible to satisfy – has at last been satisfied.

Both singly and as a companion set, these *Worldwide Guides* can, with justification, lay claim to several significant innovations. To the best of our knowledge, they stand alone in their exhaustive coverage of *both* ferrous and nonferrous metals and alloys. They are the first to encompass the specifications and designations of as many as 34 standards organizations, representing 18 nations, to which has been added the product of one international and three continental specification groups. They are also the first to present so thorough and comprehensive a selection of ferrous and nonferrous alloys and alloy products, wrought and cast. In both volumes, all of the common mill (wrought) and foundry (cast) products are represented.

In the present volume, which has been arranged in eleven sections, can be found eight major families of nonferrous metals and alloys, all in both wrought and cast product forms. Included are the following metals and their alloys – aluminum, copper, magnesium, tin, lead, zinc, nickel, and titanium. These were selected for the most obvious reason: they are the nonferrous metals and alloys of principal interest to industry throughout the world.

This volume readopts a concept introduced heretofore – a concept for the presentation of alloys and alloy products that facilitates the reader's selection on the bases of chemical composition, product form and nominal mechanical properties. This concept departs from the routine of citing equivalents and alternatives which, in many instances, prove to be little more than approximations and may fall short of the requirements established for fabrication and end use.

The initial justification given for this method of presentation is worth reiterating. "This approach to selection underscores a truism too often ignored, namely, that ultimate selection rests with the reader. Only he can satisfactorily equate a selection with the requirements that fulfill his needs. Bear in mind that individual requirements relating to product form, heat treatment or temper, cost, fabrication and end use may, and often do, vary considerably even when the identical alloy composition is selected. Accordingly, it is essential that the reader – before attempting to make full use of *Worldwide Guide* – first familiarize himself with the details and instructions set forth in the section on "Format and Guidelines for Use", beginning on the following page (page IV). The ground rules for using the *Guide* most efficiently and effectively are neither simplistic nor overly demanding. When observed, they lead to development of a skill. And in common with most useful and rewarding skills, performance tends to improve with practice."

Paul M. Unterweiser
Staff Editor
Manager, Publications Development
American Society for Metals

William H. Cubberly
Director of Reference Publications
American Society for Metals

Format and Guidelines for Use

General Description. This *Guide* is arranged in eleven sections according to major alloy families and product forms (wrought and/or cast). The types of alloys given in each of the sections and the respective product form(s) are set forth in the table of contents, on a title page preceding each section, and at the top of every page in each of the sections. Specification numbers and designation numbers are listed in full in *two separate indexes,* the first beginning on page 601 and the second on page 611; both indexes have been arranged in alphanumeric sequence. With the help of these indexes, it is easy to locate any alloy in the book when either the specification or designation number is known. In this connection, one characteristic of many of the current ASTM specification numbers is worthy of special mention. With adoption of these specifications by the American National Standards Institute, a standard such as ASTM B 618 now appears as ANSI/ASTM B 618.

Page Format. Data entries on individual pages follow a six-column arrangement which, reading from left to right, lists alloy specification number, designation, country of origin, product form(s), chemical composition, and mechanical properties and hardness values (sometimes expressed as a temper or condition). Of the six columns, only the last two – chemical composition and mechanical properties and hardness values – warrant further clarification.

> **Chemical composition** is given in conventional weight percent for each of the chemical elements (sometimes including allowable "tramp" elements) cited in the specification. Weight percent of each element, either a number or range, *precedes* the chemical symbol for the element to which it applies. Thus, 2.0-3.0 *Si*, 0.7-1.3 *Cu*, 0.10 *Mn*, 0.10 *Mg* describe the weight percentages of the four elements – silicon, copper, manganese and magnesium – listed in the specification for the alloy described. The alloy is an aluminum alloy, and thus the notation "rem *Al*" (for "remainder aluminum") also appears. "Tramp" or unspecified elements may be referred to in generalized terms, such as "0.30 others, total", or they may be cited specifically. Minor differences in styling, such as preferred sequence of elements or placement of "rem *Al*" at the beginning or end of a listing, generally reflect the styling followed in the original specification; they are not significant technically.

Mechanical Properties and Hardness Values. Property values are preceded by a *condition* and/or *temper* (e.g., annealed, as-cast, stress relieved, half-hard), sometimes related to a specific product form (e.g., aged [extruded bar]), for which the values, as cited in the specification, are applicable. If these values may be arrived at by mutual agreement of supplier and purchaser, the condition is described as "as agreed".

Condition and/or temper is followed by *product size*, that is, the sectional dimension (or range of dimensions) of the product to which the property values are applicable. This dimension most commonly refers to a diameter and is given in inches and/or millimeters, the metric equivalent. Some specifications do not cite a specific product size but rather indicate an upper or lower limit in size to which property values are applicable; others state that property values are common to all sizes. These stipulations are noted in the columnar listing.

Product size is followed by *mechanical properties and hardness values*, the former expressed in both English and metric units, where applicable. In sequence, the property values given apply to tensile strength, TS; yield strength, YS; and elongation, El. Elongation may be expressed as a percentage in 2 in. or 50 mm, or as a function of specimen diameter or section, e.g., elongation × 4D. In countries that use the metric system exclusively, only the metric values appear in specifications; this is reflected in the specific listing. Hardness values are expressed in the terms given in the original specification.

Certain specifications, most often those applicable to cast products, provide chemical composition but do not specify mechanical properties or hardness values.

Rationale. In terms of both number and quantity of alloying additives, most nonferrous alloys are far more complex than the average ferrous alloy. Thus, a relatively simple sorting system based on a convenient common denominator, such as ranking comparable alloys of a given class or family by descending carbon content – the system used successfully in the *Worldwide Guide to Equivalent Irons and Steels* – obviously could not be applied to the nonferrous alloys. It was necessary, therefore, to devise a workable sorting-ranking system for nonferrous alloys that would serve the same basic objective – to draw comparable or equivalent alloy compositions to within close proximity of each other on any given page or series of pages, thus facilitating identification of equivalent alloy compositions. In general, this was accomplished in two stages – a preliminary sorting into principal alloy sub-families, followed by a final arranging by rank accomplished with the aid of a computer program.

Without delving into the many intricacies and refinements involved, a simple example will serve to demonstrate how this two-stage, sorting-ranking system works. All wrought aluminum and aluminum alloys can be divided into seven sub-families (or series) based on principal alloying element or elements. This system, as adopted by the Aluminum Association and others, has evolved into the series designation system, beginning with the 1000 series and concluding with the 7000 series. The 1000 series comprise the aluminums, essentially unalloyed and containing a high of 99.75%, and a low of 99.00%, Al. These can be conveniently ranked by aluminum content, beginning with the highest (99.75%) and proceeding downward to the lowest (99.00%).

Similarly, the 2000 series alloys, which contain copper as the principal alloying addition, can be conveniently ranked by descending copper content from a high of 6.3% Cu to a low of 2.3% Cu. Except for the titanium alloys which were sorted and ranked on the basis of sub-family and principal phase constituent(s), most of the nonferrous metals and alloys in this volume were sorted into sub-families and ranked by modified versions of the so-called "high-to-low" system.

Thus, the objective of drawing comparable or equivalent alloys to within close proximity of each other was accomplished. Accordingly, when a type of alloy can be reasonably approximated on the basis of series number, classification, or major alloying constituent(s), equivalent alloy compositions can be found by careful review of a page or, at most, several pages of the *Guide*. Other factors such as country of origin, product form(s), and mechanical property and hardness values can also be readily identified and compared.

When the specification or designation number of the alloy is known, or if those of a similar alloy are known, then reference to either of the indexes is all that is required to locate the specific alloy (and accompanying data) in the *Guide*. Again, it will be found that this alloy is located within reasonably close proximity to other alloys that are comparable or equivalent.

Common Abbreviations and Symbols

diam	–	diameter or thickness of test specimen
TS	–	tensile strength
YS	–	yield strength
El	–	elongation
HB	–	Brinell hardness
HV	–	Vickers hardness
HR	–	Rockwell hardness, usually followed by a letter or a letter and number indicating the scale used
$<$	–	less than
$\leq$	–	up to and including
$>$	–	greater than
$\geq$	–	greater than and including
max	–	maximum
min	–	minimum
psi	–	pounds per square inch
MPa	–	megapascals
°C	–	degrees Celsius
°F	–	degrees Fahrenheit
h	–	hour, hours
in 4D	–	referring to elongation, a gage length 4 times the specimen diameter
OD	–	outer diameter

Contributor Standards Organizations

Nation or Organization	Standards Organization, Abbreviation
Australia	AS
Canada	CSA
Denmark	DS
Europe	AECMA
Finland	SFS
France	AIR, NFA
Germany	DIN
India	IS
International Organization for Standardization	ISO
Italy	UNI
Japan	JIS
Mexico	NOM, DGN
Norway	NS
Pan American	COPANT
South Africa	SABS
Sweden	SIS
Switzerland	VSM
United Kingdom	BS
United States	AA, AMS, ANSI/ASTM, ASTM, CDA, Federal, Military

Contents

WROUGHT ALUMINUM AND ALUMINUM ALLOYS

WROUGHT ALUMINUM AND ALUMINUM ALLOYS

specification number	designation	country	product forms	chemical composition	mechanical properties and hardness values
BS 1470	S1	UK	Plate, sheet, strip	99.99 **Al**, 0.01 **Cu+Si+Fe**	**Annealed**: 0.2–6.0 mm **diam**, 65 max MPa **TS**, 30% in 2 in. (50 mm) **El**; **Strain hardened to 1/2 hard**: 0.2–6.0 mm **diam**, 80 MPa **TS**, 7% in 2 in. (50 mm) **El**; **Strain hardened to hard**: 0.2–6.0 mm **diam**, 100 MPa **TS**, 3% in 2 in. (50 mm) **El**
IS 739	G1	India	Wire	99.99 min **Al**, 0.01 **Cu+Si+Fe**	**Annealed and cold worked to hard**: all **diam**, 93 MPa **TS**
JIS H 4170	IN99	Japan	Foil: high purity	99.99 min **Al**, 0.004 **Fe**, 0.010 **Cu+Si**	
CSA HA.7.1	.1100	Canada	Tube: drawn and extruded, pipe: drawn	99.90 min **Al**, 0.05–0.20 **Cu**, 0.05 **Mn**, 0.10 **Zn**, 1.0 **Si+Fe**, 0.05 each, 0.15 total, others	**Drawn and O**: 0.014–0.500 in. **diam**, 15.5 max ksi (107 max MPa) **TS**; **Drawn and H18**: 0.014–0.500 in. **diam**, 22.0 ksi (152 MPa) **TS**; **Extruded and O**: all **diam**, 11.0 ksi (76 MPa) **TS**, 3.0 ksi (21 MPa) **YS**, 25% in 2 in. (50 mm) **El**
CSA HA.7	.1100	Canada	Tube, pipe: seamless	99.90 min **Al**, 0.05–0.20 **Cu**, 0.05 **Mn**, 0.10 **Zn**, 1.0 **Si+Fe**, 0.05 each, 0.15 total, others	**Drawn and O**: 0.014–0.500 in. **diam**, 15.5 max ksi (107 max MPa) **TS**; **Drawn and H18**: 0.014–0.500 in. **diam**, 22.0 ksi (152 MPa) **TS**; **Extruded and O**: 11.0 ksi (76 MPa) **TS**, 3.0 ksi (21 MPa) **YS**, 25% in 2 in. (50 mm) **El**
JIS H 4170	IN90	Japan	Foil: high purity	99.90 min **Al**, 0.030 **Fe**, 0.080 **Cu+Si**	
AS 1866	1080A	Australia	Rod, bar, shape	99.80 min **Al**, 0.15 **Si**, 0.03 **Cu**, 0.02 **Mn**, 0.02 **Mg**, 0.06 **Zn**, 0.02 **Ti**, 0.15 **Fe**, 0.02 each, others	**Strain–hardened**: all **diam**, 53 MPa **TS**, 28% in 2 in. (50 mm) **El**
AS 1734	1080A	Australia	Sheet, plate	99.80 min **Al**, 0.15 **Si**, 0.03 **Cu**, 0.02 **Mn**, 0.02 **Mg**, 0.06 **Zn**, 0.02 **Ti**, 0.15 **Fe**, 0.02 each, others	**Annealed**: 0.15–0.50 mm **diam**, 78 max MPa **TS**, 15% in 2 in. (50 mm) **El**; **Strain–hardened to 1/2 hard**: 0.25–0.30 mm **diam**, 92 MPa **TS**, 1% in 2 in. (50 mm) **El**; **Strain–hardened to full hard**: 0.15–0.50 mm **diam**, 123 MPa **TS**, 1% in 2 in. (50 mm) **El**
BS 1475	G1A	UK	Wire	99.8 **Al**, 0.15 **Si**, 0.02 **Cu**, 0.03 **Mn**, 0.06 **Zn**, 0.15 **Fe**, 0.2 **Cu+Si+Fe+Mn+Zn**	**Annealed**: ≤10 mm **diam**, 90 max MPa **TS**; **As manufactured**: ≤10 mm **diam**; **Strain–hardened to hard**: ≤10 mm **diam**, 125 MPa **TS**

WROUGHT ALUMINUM AND ALUMINUM ALLOYS

specification number	designation	country	product forms	chemical composition	mechanical properties and hardness values
BS 1470	S1A	UK	Plate, sheet, strip	99.8 **Al**, 0.15 **Si**, 0.02 **Cu**, 0.03 **Mn**, 0.06 **Zn**, 0.15 **Fe**, 0.2 **Cu+Si+Fe+Mn+Zn**	**Annealed**: 0.2–6.0 mm **diam**, 90 max MPa **TS**, 29% in 2 in. (50 mm) **El**; **Strain hardened to hard**: 0.2–3.0 mm **diam**, 125 MPa **TS**, 3% in 2 in. (50 mm) **El**
COPANT 862	1080	COPANT	Wrought products	99.80 min **Al**, 0.15 **Si**, 0.03 **Cu**, 0.02 **Mn**, 0.02 **Mg**, 0.03 **Zn**, 0.03 **Ti**, 0.15 **Fe**	
SFS 2580	AL99.8	Finland	Shapes	99.8 min **Al**, 0.15 **Si**, 0.03 **Cu**, 0.03 **Mn**, 0.06 **Zn**, 0.15 **Fe**, 0.03 each, 0.15 total, others	**Hot finished**: all **diam**, 60 MPa **TS**, 25% in 2in. (50 mm) **El**
NF A 50 411	1080 A (A8)	France	Bar, wire, and tube	99.80 min **Al**, 0.15 **Si**, 0.03 **Cu**, 0.02 **Mn**, 0.02 **Mg**, 0.06 **Zn**, 0.02 **Ti**, 0.15 **Fe**, 0.03 **Ga**, 0.02 each, others	**1/2 hard (drawn bar)**: ≤50 mm **diam**, 100 MPa **TS**, 70 MPa **YS**, 6% in 2 in. (50 mm) **El**; **Hard (wire) ≤6 mm diam, 140 MPa TS, 110 MPa YS, 4% in 2 in. 5 mm El; 1/4 hard (tube)**: ≤50 mm **diam**, 85 MPa **TS**, 60 MPa **YS**, 7% in 2 in. (50 mm) **El**
NF A 50 451	1080 A (A8)	France	Sheet, plate, and strip	99.80 min **Al**, 0.15 **Si**, 0.03 **Cu**, 0.02 **Mn**, 0.02 **Mg**, 0.06 **Zn**, 0.02 **Ti**, 0.15 **Fe**, 0.03 **Ga**, 0.02 each, others	**Annealed**: 0.4–3.2 mm **diam**, 60 MPa **TS**, 38% in 2 in. (50 mm) **El**; **1/4 hard**: 0.4–3.2 mm **diam**, 80 MPa **TS**, 55 MPa **YS**, 12% in 2 in. (50 mm) **El**; **1/2 hard**: 0.4–1.6 mm **diam**, 100 MPa **TS**, 70 MPa **YS**, 7% in 2 in. (50 mm) **El**
IS 739	G1A	India	Wire	99.8 min **Al**, 0.15 **Si**, 0.02 **Cu**, 0.03 **Mn**, 0.06 **Zn**, 0.15 **Fe**, 0.20 **Cu+Si+Fe+Mn+Zn**	**Annealed**: all **diam**, 79 max MPa **TS**; **Annealed and cold worked to hard**: all **diam**, 123 MPa **TS**
IS 738	T1A	India	Tube: drawn	99.8 min **Al**, 0.15 **Si**, 0.02 **Cu**, 0.03 **Mn**, 0.06 **Zn**, 0.15 **Fe**, 0.20 **Cu+Si+Fe+Mn+Zn**	**Annealed**: all **diam**, 79 max MPa **TS**; **Annealed and cold worked to hard**: ≤80 mm **diam**, 98 MPa **TS**
IS 736	19800	India	Plate	99.8 min **Al**, 0.15 **Si**, 0.03 **Cu**, 0.03 **Mn**, 0.06 **Zn**, 0.15 **Fe**, 0.2 **Cu+Si+Fe+Mn+Zn**	**As manufactured**: 30 mm **diam**, 55 MPa **TS**, 30% in 2 in. (50 mm) **El**; **Annealed**: 34 mm **diam**, 90 max MPa **TS**, 34% in 2 in. (50 mm) **El**; **Half hard**: 8 mm **diam**, 85 MPa **TS**, 8% in 2 in. (50 mm) **El**
ISO R827	Al99.8	ISO	Extruded products	99.8 min **Al**, 0.15 **Si**, 0.03 **Cu**, 0.03 **Mn**, 0.06 **Zn**, 0.15 **Fe**, 0.2 **Cu+Si+Fe+Mn+Zn**, total	**As manufactured**: all **diam**, 49 MPa **TS**, 22% in 2 in. (50 mm) **El**
ISO R209	Al99.8	ISO	Wrought products	99.8 min **Al**, 0.15 **Si**, 0.03 **Cu**, 0.03 **Mn**, 0.06 **Zn**, 0.15 **Fe**, 0.2 **Cu+Si+Fe+Mn+Zn**, total	

WROUGHT ALUMINUM AND ALUMINUM ALLOYS

specification number	designation	country	product forms	chemical composition	mechanical properties and hardness values
ISO TR2778	Al99.8	ISO	Tube: drawn	99.8 min **Al**, 0.15 **Si**, 0.03 **Cu**, 0.03 **Mn**, 0.06 **Zn**, 0.15 **Fe**, 0.2 **Cu+Si+Fe+Mn+Zn**	**Annealed**: all **diam**, 50 MPa **TS**, 29% in 2 in. (50 mm) **El**; **Strain hardened**: all **diam**, 75 MPa **TS**, 4% in 2 in. (50 mm) **El**
ISO TR2136	Al99.8	ISO	Sheet, plate	99.8 min **Al**, 0.15 **Si**, 0.03 **Cu**, 0.03 **Mn**, 0.06 **Zn**, 0.15 **Fe**, 0.2 **Cu+Si+Fe+Mn+Zn**, total	**Annealed**: all **diam**, 90 max MPa **TS**, 25% in 2 in. (50 mm) **El**
JIS H 4000	1080	Japan	Sheet, plate, and strip	99.80 min **Al**, 0.15 **Si**, 0.03 **Cu**, 0.02 **Mn**, 0.02 **Mg**, 0.03 **Zn**, 0.03 **Ti**, 0.16 **Fe**	**Annealed**: <1.3 mm **diam**, 59 MPa **TS**, 20 MPa **YS**, 30% in 2 in. (50 mm) **El**; **Hard**: <0.3 mm **diam**, 118 MPa **TS**, 1% in 2 in. (50 mm) **El**; **Hot rolled**: <12 mm **diam**, 69 MPa **TS**, 39 MPa **YS**, 15% in 2 in. (50 mm) **El**
COPANT 862	1075	COPANT	Wrought products	99.75 min **Al**, 0.2 **Si**, 0.04 **Cu**, 0.03 **Mn**, 0.03 **Mg**, 0.04 **Zn**, 0.03 **Ti**, 0.2 **Fe**	
BS L.110		UK	Sheet and strip (cladding material)	99.7 **Al**, 0.15 **Si**, 0.02 **Cu**, 0.03 **Zn**, 0.20 **Fe**	**Solution heat treated, quenched and aged**: 0.4–0.8 mm **diam**, 390 MPa **TS**, 235 MPa **YS**, 12% in 2 in. (50 mm) **El**
BS L.109		UK	Sheet and strip (cladding material)	99.7 min **Al**, 0.15 **Si**, 0.02 **Cu**, 0.03 **Zn**, 0.20 **Fe**	**Solution heat treated, quenched and aged**: 0.4–0.8 mm **diam**, 405 MPa **TS**, 12% in 2 in. (50 mm) **El**
BS L.108		UK	Sheet, strip (cladding material)	99.7 min **Al**, 0.15 **Si**, 0.02 **Cu**, 0.03 **Zn**, 0.20 **Fe**	**Solution heat treated, quenched and aged**: 0.4–6.0 mm **diam**, 385 MPa **TS**, 240 MPa **YS**, 14% in 2 in. (50 mm) **El**
BS 2L.90		UK	Sheet and strip (cladding material)	99.7 min **Al**, 0.15 **Si**, 0.02 **Cu**, 0.03 **Zn**, 0.20 **Fe**	**Solution heat and precipitation treated and quenched**: 0.4–0.8 mm **diam**, 415 MPa **TS**, 350 MPa **YS**, 7% in 2 in. (50 mm) **El**
BS 2L.89		UK	Sheet and strip (cladding material)	99.7 min **Al**, 0.15 **Si**, 0.02 **Cu**, 0.03 **Zn**, 0.20 **Fe**	**Solution heat treated, quenched and naturally aged**: 0.4–0.8 mm **diam**, 385 MPa **TS**, 250 MPa **YS**, 13% in 2 in. (50 mm) **El**
BS 3L.72		UK	Sheet and strip (cladding material)	99.7 min **Al**, 0.15 **Si**, 0.02 **Cu**, 0.03 **Zn**, 0.20 **Fe**	**Solution heat treated, quenched and naturally aged**: 0.4–0.8 mm **diam**, 385 MPa **TS**, 250 MPa **YS**, 13% in 2 in. (50 mm) **El**
COPANT 862	1070	COPANT	Wrought products	99.70 min **Al**, 0.20 **Si**, 0.04 **Cu**, 0.03 **Mn**, 0.03 **Mg**, 0.04 **Zn**, 0.03 **Ti**, 0.25 **Fe**, 0.03 each, others	

WROUGHT ALUMINUM AND ALUMINUM ALLOYS

specification number	designation	country	product forms	chemical composition	mechanical properties and hardness values
DS 3012	1070	Denmark	Rolled, drawn, and extruded products	99.70 min **Al**, 0.20 **Si**, 0.04 **Cu**, 0.03 **Mn**, 0.03 **Mg**, 0.04 **Zn**, 0.03 **Ti**, 0.25 **Fe**, 0.03 each, others	**1/2 hard (rolled and drawn products)**: 0.5–6 mm **diam**, 95 MPa **TS**, 75 MPa **YS**, 6% in 2 in. (50 mm) **El**, 35 **HV**; **Hard (rolled and drawn products)**: 0.5–6 mm **diam**, 130 MPa **TS**, 120 MPa **YS**, 4% in 2 in. (50 mm) **El**, 40 **HV**; **As manufactured (extruded products)**: all **diam**, 65 MPa **TS**, 18% in 2 in. (50 mm) **El**
SFS 2581	A199.7	Finland	Shapes and sheet	99.7 min **Al**, 0.20 **Si**, 0.03 **Cu**, 0.03 **Mn**, 0.07 **Zn**, 0.25 **Fe**, 0.03 each, 0.15 total, others	**Hot finished (shapes)**: all **diam**, 60 MPa **TS**, 25% in 2 in. (50 mm) **El**; **Annealed (sheet)**: all **diam**, 100 MPa **TS**, 20 MPa **YS**, 35% in 2 in. (50 mm) **El**; **Hard (sheet)**: 3 mm **diam**, 130 MPa **TS**, 110 MPa **YS**, 4% in 2 in. (50 mm) **El**
NF A 50 451	1070 A (A7)	France	Sheet, plate, and strip	99.70 min **Al**, 0.20 **Si**, 0.03 **Cu**, 0.03 **Mn**, 0.03 **Mg**, 0.07 **Zn**, 0.03 **Ti**, 0.25 **Fe**, 0.03 each, others	**Annealed**: 0.4–3.2 mm **diam**, 60 MPa **TS**, 38% in 2 in. (50 mm) **El**; **1/4 hard**: 0.4–3.2 mm **diam**, 80 MPa **TS**, 55 MPa **YS**, 12% in 2 in. (50 mm) **El**; **1/2 hard**: 0.4–1.6 mm **diam**, 100 MPa **TS**, 70 MPa **YS**, 7% in 2 in. (50 mm) **El**
IS 736	19700	India	Plate	99.7 min **Al**, 0.2 **Si**, 0.03 **Cu**, 0.03 **Mn**, 0.06 **Zn**, 0.25 **Fe**, 0.3 **Cu+Si+Fe+Mn+Zn**	**As manufactured**: 30 mm **diam**, 60 MPa **TS**, 30% in 2 in. (50 mm) **El**; **Annealed**: 34 mm **diam**, 95 max MPa **TS**, 34% in 2 in. (50 mm) **El**; **Half hard**: 8 mm **diam**, 90 MPa **TS**, 8% in 2 in. (50 mm) **El**
ISO R827	Al99.7	ISO	Extruded products	99.7 min **Al**, 0.20 **Si**, 0.03 **Cu**, 0.03 **Mn**, 0.07 **Zn**, 0.25 **Fe**, 0.03 **Cu+Si+Fe+Mn+Zn**, total	**As manufactured**: all **diam**, 49 MPa **TS**, 22% in 2 in. (50 mm) **El**
ISO R209	Al99.7	ISO	Wrought products	99.7 min **Al**, 0.20 **Si**, 0.03 **Cu**, 0.03 **Mn**, 0.07 **Zn**, 0.25 **Fe**, 0.3 **Cu+Si+Fe+Mn+Zn**, total	
ISO TR2136	Al99.7	ISO	Sheet, plate	99.7 min **Al**, 0.20 **Si**, 0.03 **Cu**, 0.03 **Mn**, 0.07 **Zn**, 0.25 **Fe**, 0.3 **Cu+Si+Fe+Mn+Zn**, total	**Annealed**: all **diam**, 95 max MPa **TS**, 25% in 2 in. (50 mm) **El**
JIS Z 3232	1070	Japan	Rod and wire: welding and electrode	99.70 min **Al**, 0.20 **Si**, 0.04 **Cu**, 0.03 **Mn**, 0.03 **Mg**, 0.04 **Zn**, 0.03 **Ti**, 0.25 **Fe**, 0.0008 **Be**	**Annealed**: all **diam**, 59 MPa **TS**
JIS H 4160	1070	Japan	Foil	99.70 min **Al**, 0.20 **Si**, 0.04 **Cu**, 0.03 **Mn**, 0.03 **Mg**, 0.04 **Zn**, 0.03 **Ti**, 0.25 **Fe**	

WROUGHT ALUMINUM AND ALUMINUM ALLOYS

specification number	designation	country	product forms	chemical composition	mechanical properties and hardness values
JIS H 4120	1070	Japan	Wire and rod: rivet	99.70 min **Al**, 0.20 **Si**, 0.04 **Cu**, 0.03 **Mn**, 0.03 **Mg**, 0.04 **Zn**, 0.03 **Ti**, 0.25 **Fe**	**1/2 hard**: ≤25 mm **diam**, 78 MPa **TS**
JIS H 4080	1070	Japan	Tube: extruded	99.70 min **Al**, 0.20 **Si**, 0.04 **Cu**, 0.03 **Mn**, 0.03 **Mg**, 0.04 **Zn**, 0.03 **Ti**, 0.25 **Fe**	**As manufactured**: all **diam**, 49 MPa **TS**, 20 MPa **YS**
JIS H 4040	1070	Japan	Bar: extruded	99.70 min **Al**, 0.20 **Si**, 0.04 **Cu**, 0.03 **Mn**, 0.03 **Mg**, 0.04 **Zn**, 0.03 **Ti**, 0.25 **Fe**	**As manufactured**: all **diam**, 49 MPa **TS**, 20 MPa **YS**
JIS H 4000	1070	Japan	Sheet, plate, and strip	99.70 min **Al**, 0.20 **Si**, 0.04 **Cu**, 0.03 **Mn**, 0.03 **Mg**, 0.04 **Zn**, 0.03 **Ti**, 0.25 **Fe**	**Annealed**: <1.3 mm **diam**, 59 MPa **TS**, 30% in 2 in. (50 mm) **El**; **Hard**: <0.3 mm **diam**, 118 MPa **TS**, 1% in 2 in. (50 mm) **El**; **Hot rolled**: <12 mm **diam**, 69 MPa **TS**, 39 MPa **YS**, 15% in 2 in. (50 mm) **El**
AMS 4000E		US	Sheet and plate	99.60 min **Al**, 0.25 **Si**, 0.05 **Cu**, 0.03 **Mn**, 0.03 **Mg**, 0.05 **Zn**, 0.03 **Ti**, 0.35 **Fe**, 0.03 each, others	**Annealed**: 0.006–0.019 in. (0.15–0.48 mm) **diam**, 8 ksi (55 MPa) **TS**, 15% in 2 in. (50 mm) **El**
ANSI/ASTM B 210	1060	US	Tube: seamless	99.60 min **Al**, 0.25 **Si**, 0.05 **Cu**, 0.03 **Mn**, 0.03 **Mg**, 0.05 **Zn**, 0.03 **Ti**, 0.35 **Fe**, 0.05 **V**, 0.03 each, others	**Annealed**: 0.018–0.500 in. **diam**, 9 ksi (59 MPa) **TS**, 3 ksi (17 MPa) **YS**; **H12 (round tube)**: 0.018–0.500 in. **diam**, 10 ksi (69 MPa) **TS**, 4 ksi (28 MPa) **YS**; **H14 (round tube)**: 0.018–0.500 in. **diam**, 12 ksi (83 MPa) **TS**, 10 ksi (69 MPa) **YS**
ANSI/ASTM B 483	1060	US	Tube: for general purposes	99.60 min **Al**, 0.25 **Si**, 0.05 **Cu**, 0.03 **Mn**, 0.03 **Mg**, 0.05 **Zn**, 0.03 **Ti**, 0.35 **Fe**, 0.03 each, others	**Annealed**: 0.018–0.500 in. **diam**, 9 ksi (62 MPa) **TS**, 3 ksi (21 MPa) **YS**; **H12 (round tube)**: 0.018–0.500 in. **diam**, 10 ksi (69 MPa) **TS**, 4 ksi (28 MPa) **YS**; **H14 (round tube)**: 0.018–0.500 in. **diam**, 12 ksi (83 MPa) **TS**, 10 ksi (69 MPa) **YS**
ANSI/ASTM B 404	1060	US	Tube: seamless, for condensers and heat exchangers	99.60 min **Al**, 0.25 **Si**, 0.05 **Cu**, 0.03 **Mn**, 0.03 **Mg**, 0.05 **Zn**, 0.03 **Ti**, 0.35 **Fe**, 0.03 each, others	**Annealed**: 0.018–0.500 in. **diam**, 9 ksi (59 MPa) **TS**, 3 ksi (17 MPa) **YS**; **H14**: 0.018–0.500 in. **diam**, 12 ksi (83 MPa) **TS**, 10 ksi (69 MPa) **YS**
ANSI/ASTM B 345	1060	US	Tube: seamless	99.60 min **Al**, 0.25 **Si**, 0.05 **Cu**, 0.03 **Mn**, 0.03 **Mg**, 0.05 **Zn**, 0.03 **Ti**, 0.35 **Fe**, 0.03 each, others	**Annealed and H112**: all **diam**, 9 ksi (62 MPa) **TS**, 3 ksi (21 MPa) **YS**, 30% in 2 in. (50 mm) **El**

WROUGHT ALUMINUM AND ALUMINUM ALLOYS

specification number	designation	country	product forms	chemical composition	mechanical properties and hardness values
ANSI/ASTM B 241	1060	US	Tube: seamless	99.60 min **Al**, 0.25 **Si**, 0.05 **Cu**, 0.03 **Mn**, 0.03 **Mg**, 0.05 **Zn**, 0.03 **Ti**, 0.35 **Fe**, 0.03 each, others	**Annealed**: all **diam**, 9 ksi (59 MPa) **TS**, 3 ksi (17 MPa) **YS**, 25% in 2 in. (50 mm) **El**; **H112**: all **diam**, 9 ksi (59 MPa) **TS**, 3 ksi (17 MPa) **YS**
ANSI/ASTM B 234	1060	US	Tube: seamless, for condensers and heat exchangers	99.60 min **Al**, 0.25 **Si**, 0.05 **Cu**, 0.03 **Mn**, 0.03 **Mg**, 0.05 **Zn**, 0.03 **Ti**, 0.35 **Fe**, 0.03 each, others	**H14**: 0.010–0.200 in. **diam**, 12 ksi (83 MPa) **TS**, 10 ksi (69 MPa) **YS**
ANSI/ASTM B 221	1060	US	Bar, tube, rod, wire, and shapes	99.60 min **Al**, 0.25 **Si**, 0.05 **Cu**, 0.03 **Mn**, 0.03 **Mg**, 0.05 **Zn**, 0.03 **Ti**, 0.35 **Fe**, 0.03 each, others	**Annealed**: all **diam**, 9 ksi (59 MPa) **TS**, 3 ksi (17 MPa) **YS**, 25% in 2 in. (50 mm) **El**; **H112**: all **diam**, 9 ksi (59 MPa) **TS**, 3 ksi (17 MPa) **YS**
ANSI/ASTM B 211	1060	US	Bar, rod, and wire	99.60 min **Al**, 0.25 **Si**, 0.05 **Cu**, 0.03 **Mn**, 0.03 **Mg**, 0.05 **Zn**, 0.03 **Ti**, 0.35 **Fe**, 0.03 each, others	**Annealed**: ≥0.125 in. **diam**, 8 ksi (55 MPa) **TS**, 3 ksi (17 MPa) **YS**, 25% in 2 in. (50 mm) **El**; **H14**: ≤0.374 in. **diam**, 12 ksi (83 MPa) **TS**, 10 ksi (69 MPa) **YS**; **H18**: ≤0.374 in. **diam**, 16 ksi (110 MPa) **TS**, 13 ksi (90 MPa) **YS**
ANSI/ASTM B 209	1060	US	Sheet and plate	99.60 min **Al**, 0.25 **Si**, 0.05 **Cu**, 0.03 **Mn**, 0.03 **Mg**, 0.05 **Zn**, 0.03 **Ti**, 0.35 **Fe**, 0.05 **V**, 0.03 each, others	**Annealed**: 0.051–3.000 in. **diam**, 8 ksi (55 MPa) **TS**, 3 ksi (21 MPa) **YS**, 25% in 2 in. (50 mm) **El**; **H12**: 0.051–2.000 in. **diam**, 11 ksi (76 MPa) **TS**, 9 ksi (62 MPa) **YS**, 12% in. 2 in. (50 mm) **El**; **H14**: 0.051–1.000 in. **diam**, 12 ksi (83 MPa) **TS**, 10 ksi (69 MPa) **YS**, 10% in 2 in. (50 mm) **El**
COPANT 862	1060	COPANT	Wrought products	99.60 min **Al**, 0.25 **Si**, 0.05 **Cu**, 0.03 **Mn**, 0.03 **Mg**, 0.05 **Zn**, 0.03 **Ti**, 0.35 **Fe**	
JIS H 4180	1060	Japan	Plate	99.60 min **Al**, 0.25 **Si**, 0.05 **Cu**, 0.03 **Mn**, 0.03 **Mg**, 0.05 **Zn**, 0.03 **Ti**, 0.35 **Fe**	**1/2 hard**: 3–12 mm **diam**, 88 MPa **TS**, 59 MPa **YS**, 6% in 2 in. (50 mm) **El**; **Hot rolled**: <12 mm **diam**, 69 MPa **TS**, 39 MPa **YS**, 15% in 2 in. (50 mm) **El**; **As manufactured (extruded plate)**: 3–30 mm **diam**, 59 MPa **TS**, 29 MPa **YS**, 25% in 2 in. (50 mm) **El**
AECMA prEN2072	1050 A	AECMA	Sheet and strip	99.50 min **Al**, 0.25 **Si**, 0.05 **Cu**, 0.05 **Mn**, 0.05 **Mg**, 0.07 **Zn**, 0.05 **Ti**, 0.40 **Fe**, 0.03 each, other	**H14**: 0.4–2.0 mm **diam**, 100 MPa **TS**, 80 MPa **YS**, 5% in 2 in. (50 mm) **El**

WROUGHT ALUMINUM AND ALUMINUM ALLOYS

specification number	designation	country	product forms	chemical composition	mechanical properties and hardness values
AECMA prEN2073	Aluminum 1050 A	AECMA	Tube	99.50 min **Al**, 0.25 **Si**, 0.05 **Cu**, 0.05 **Mn**, 0.05 **Mg**, 0.07 **Zn**, 0.05 **Ti**, 0.040 **Fe**, 0.03 each, others	**H 14**: 0.4–2.0 mm **diam**, 100 MPa **TS**, 80 MPa **YS**, 6% in 2 in. (50 mm) **El**
ANSI/ASTM B 491	1050	US	Tube: coiled, for general purposes	99.50 min **Al**, 0.25 **Si**, 0.05 **Cu**, 0.05 **Mn**, 0.05 **Mg**, 0.05 **Zn**, 0.03 **Ti**, 0.40 **Fe**, 0.03 each, others	**H112**: 0.032–0.050 in **diam**, 9 ksi (62 MPa) **TS**, 3 ksi (21 MPa) **YS**, 25% in 2 in. (50 mm) **El**
ANSI/ASTM B 233	1350	US	Rod: redraw, for electrical purposes	99.50 min **Al**, 0.10 **Si**, 0.05 **Cu**, 0.01 **Mn**, 0.01 **Cr**, 0.05 **Zn**, 0.40 **Fe**, 0.05 **B**, 0.03 **Ga**, 0.02 **V+Ti**, 0.03 each, 0.10 total, others	**Annealed:** ≤0.500 in. (≤12.70 mm) **diam**, 9 ksi (59 MPa) **TS**; **H12 and H22:** ≤0.500 in. **diam**, 12 ksi (83 MPa) **TS**; **H14 and H12:** ≤0.500 in. **diam**, 15 ksi (103 MPa) **TS**
AS 1867	1050	Australia	Tube: drawn	99.50 min **Al**, 0.25 **Si**, 0.05 **Cu**, 0.05 **Mn**, 0.05 **Mg**, 0.05 **Zn**, 0.03 **Ti**, 0.40 **Fe**, 0.03 each, others	**Annealed**: all **diam**, 93 max MPa **TS**, 30% in 2 in. (50 mm) **El**; **Strain hardened to 1/4 hard**: all **diam**, 82 MPa **TS**; **Strain hardened to full hard**: all **diam**, 131 MPa **TS**
AS 1866	1050	Australia	Rod, bar, shape	99.50 min **Al**, 0.25 **Si**, 0.05 **Cu**, 0.05 **Mn**, 0.05 **Mg**, 0.05 **Zn**, 0.03 **Ti**, 0.40 **Fe**, 0.03 each, others	**Strain–hardened**: all **diam**, 62 MPa **TS**, 23% in 2 in. (50 mm) **El**
AS 1734	1050	Australia	Sheet, plate	99.50 min **Al**, 0.25 **Si**, 0.05 **Cu**, 0.05 **Mn**, 0.05 **Mg**, 0.05 **Zn**, 0.03 **Ti**, 0.40 **Fe**, 0.03 each, others	**Annealed**: 0.15–0.50 mm **diam**, 93 MPa **TS**, 15% in 2 in. (50 mm) **El**; **Strain–hardened to 1/2 hard**: 0.25–0.30 mm **diam**, 100 MPa **TS**, 2% in 2 in. (50 mm) **El**; **Strain–hardened to full hard**: 0.15–0.50 mm **diam**, 131 MPa **TS**, 1% in 2 in. (50 mm) **El**
AS 1734	1150	Australia	Sheet, plate	99.50 min **Al**, 0.05–0.20 **Cu**, 0.45 **Si+Fe**, 0.05 each, others	**Annealed**: 0.15–0.50 mm **diam**, 104 MPa **TS**, 15% in 2 in. (50 mm) **El**; **Strain–hardened to 1/2 hard**: 0.25–0.30 mm **diam**, 96 MPa **TS**, 1% in 2 in. (50 mm) **El**; **Strain–hardened to full hard**: 0.15–0.50 mm **diam**, 137 MPa **TS**, 1% in 2 in. (50 mm) **El**
BS 1474	E1B	UK	Bar, tube, section	99.5 **Al**, 0.3 **Si**, 0.05 **Cu**, 0.05 **Mn**, 0.10 **Zn**, 0.4 **Fe**, 0.5 **Cu+Si+Fe+Mn+Zn**	**As manufactured**: 60 MPa **TS**, 23% in 2 in. (50 mm) **El**
BS 1471	T1B	UK	Tube: drawn	99.5 **Al**, 0.3 **Si**, 0.05 **Cu**, 0.05 **Mn**, 0.10 **Zn**, 0.4 **Fe**, 0.5 **Cu+Si+Fe+Mn+Zn**	**Annealed:** ≤12.0 mm **diam**, 95 max MPa **TS**; **Strain–hardened to 1/2 hard:** ≤12.0 mm **diam**, 100 MPa **TS**; **Strain–hardened to hard:** ≤12.0 mm **diam**, 135 MPa **TS**

WROUGHT ALUMINUM AND ALUMINUM ALLOYS

specification number	designation	country	product forms	chemical composition	mechanical properties and hardness values
BS 2897		UK	Strip	99.5 **Al**, 0.05 **Cu**, 0.5 **Cu+Si+Fe**	**Annealed**: 93 max MPa **TS**, 25% in 2 in. (50 mm) **El**; **Strain–hardened to 1/2 hard**: 93 MPa **TS**, 8% in 2 in. (50 mm) **El**; **Strain–hardened to hard**: 157 MPa **TS**, 3% in 2 in. (50 mm) **El**
BS 1475	G1B	UK	Wire	99.5 **Al**, 0.3 **Si**, 0.05 **Cu**, 0.05 **Mn**, 0.10 **Zn**, 0.4 **Fe**, 0.5 **Cu+Si+Fe+Mn+Zn**	**Annealed**: ≤10 mm **diam**, 95 max MPa **TS**; **Strain–hardened to hard**: ≤10 mm **diam**, 135 MPa **TS**
BS 1473	RIB	UK	Rivet	99.5 **Al**, 0.3 **Si**, 0.05 **Mn**, 0.10 **Zn**, 0.4 **Fe**, 0.5 **Cu+Si+Fe+Mn+Zn**, total, others	**Strain hardened to 5/8 hard**: ≤12 mm **diam**, 110 MPa **TS**
BS 1470	S1B	UK	Plate, sheet, strip	99.5 **Al**, 0.3 **Si**, 0.05 **Cu**, 0.05 **Mn**, 0.10 **Zn**, 0.4 **Fe**, 0.5 **Cu+Si+Fe+Mn+Zn**	**Annealed**: 0.2–6.0 mm **diam**, 55 MPa **TS**, 22% in 2 in. (50 mm) **El** ; 22% in 2 in. (50 mm) **El**; **Strain hardened to 1/2 hard**: 0.2–12.5 mm **diam**, 100 MPa **TS**, 4% in 2 in. (50 mm) **El**; **Strain hardened to hard**: 0.2–3.0 mm **diam**, 135 MPa **TS**, 3% in 2 in. (50 mm) **El**
BS 1472	F1B	UK	Forging stock and forgings	99.5 **Al**, 0.3 **Si**, 0.05 **Cu**, 0.05 **Mn**, 0.10 **Zn**, 0.4 **Fe**, 0.5 **Cu+Si+Fe+Mn+Zn**	**As manufactured**: ≤150 mm **diam**, 60 MPa **TS**, 22% in 2 in. (50 mm) **El**
BS 2898	E1E	UK	Bar, tube, section	99.50 min **Al**, 0.05 **Cu**, 0.5 **Cu+Si+Fe**	**As manufactured**: all **diam**, 60 MPa **TS**, 23% in 2 in. (50 mm) **El**; **Strain–hardened to 1/4 hard**: all **diam**, 85 MPa **TS**, 13% in 2 in. (50 mm) **El**
BS L.153		UK	Sheet and strip (cladding material)	99.50 min **Al**, 0.25 **Si**, 0.05 **Cu**, 0.05 **Mn**, 0.05 **Mg**, 0.05 **Zn**, 0.03 **Ti**, 0.40 **Fe**, 0.03 each, others	**Solution heat treated, quenched and naturally aged**: 0.4–0.8 mm **diam**, 380 MPa **TS**, 235 MPa **YS**, 14% in 2 in. (50 mm) **El**
BS L.152		UK	Sheet, strip (cladding material)	99.50 min **Al**, 0.25 **Si**, 0.05 **Cu**, 0.05 **Mn**, 0.05 **Mg**, 0.05 **Zn**, 0.03 **Ti**, 0.40 **Fe**, 0.03 each, others	**Solution heat treated, quenched and artificially aged**: 0.4–0.8 mm **diam**, 415 MPa **TS**, 350 MPa **YS**, 7% in 2 in. (50 mm) **El**
BS L.151		UK	Sheet, strip (cladding material)	99.50 min **Al**, 0.25 **Si**, 0.05 **Cu**, 0.05 **Mn**, 0.05 **Mg**, 0.05 **Zn**, 0.03 **Ti**, 0.40 **Fe**, 0.03 each, others	**Solution heat treated, quenched and naturally aged**: 0.4–0.8 mm **diam**, 380 MPa **TS**, 235 MPa **YS**, 14% in 2 in. (50 mm) **El**
BS 4L.36		UK	Wire	99.5 min **Al**, 0.3 **Si**, 0.05 **Cu**, 0.05 **Mn**, 0.10 **Zn**, 0.4 **Fe**, 0.03 each, others	

WROUGHT ALUMINUM AND ALUMINUM ALLOYS

specification number	designation	country	product forms	chemical composition	mechanical properties and hardness values
CSA HA.7.1	.1050	Canada	Tube: drawn and extruded, pipe: drawn	99.50 min **Al**, 0.25 **Si**, 0.05 **Cu**, 0.05 **Mn**, 0.05 **Mg**, 0.05 **Zn**, 0.03 **Ti**, 0.40 **Fe**, 0.03 each, others	**Drawn and O**: 0.010–0.500 in. **diam**, 14.0 max ksi (97 max MPa) **TS**; **Drawn and H14**: 0.010–0.500 in. **diam**, 15.0 ksi (103 MPa) **TS**; **Extruded and O**: all **diam**, 14.0 max ksi (97 max MPa) **TS**
CSA HA.7	.1050	Canada	Tube, pipe: seamless	99.50 min **Al**, 0.25 **Si**, 0.05 **Cu**, 0.05 **Mn**, 0.05 **Mg**, 0.05 **Zn**, 0.03 **Ti**, 0.40 **Fe**, 0.03 each, others	**Drawn and O**: 0.010–0.500 in. **diam**, 14.0 max ksi (97 max MPa) **TS**; **Drawn and H18**: 0.010–0.500 in. **diam**, 19.0 ksi (131 MPa) **TS**; **Extruded and O**: all **diam**, 14.0 max ksi (97 max MPa) **TS**
CSA HA.4	.1050	Canada	Plate, sheet	99.50 min **Al**, 0.25 **Si**, 0.05 **Cu**, 0.05 **Mn**, 0.05 **Mg**, 0.05 **Zn**, 0.03 **Ti**, 0.40 **Fe**, 0.03 each, others	**O**: 0.006–0.012 in. **diam**, 8.5 ksi (59 MPa) **TS**, 3.0 ksi (21 MPa) **YS**, 12% in 2 in. (50 mm) **El**; **H14**: 0.009–0.019 in. **diam**, 15.0 ksi (103 MPa) **TS**, 13.0 ksi (90 MPa) **YS**, 2% in 2 in. (50 mm) **El**; **H18**: 0.009–0.019 in. **diam**, 19.0 ksi (131 MPa) **TS**, 15.0 ksi (103 MPa) **YS**, 1% in 2 in. (50 mm) **El**
COPANT 862	1050	COPANT	Wrought products	99.50 min **Al**, 0.25 **Si**, 0.05 **Cu**, 0.03 **Mn**, 0.03 **Mg**, 0.05 **Zn**, 0.05 **Ti**, 0.40 **Fe**, 0.03 each, 0.15 total, others	
DS 3012	1050	Denmark	Rolled, drawn, and extruded products	99.50 min **Al**, 0.05 **Cu**, 0.05 **Mg**, 0.05 **Zn**, 0.03 **Ti**, 0.40 **Fe**, 0.03 each, others	**1/2 hard (rolled and drawn products)**: 0.5–6 mm **diam**, 100 MPa **TS**, 80 MPa **YS**, 6% in 2 in. (50 mm) **El**, 35 **HV**; **Hard (rolled and drawn products)**: 0.5–6 mm **diam**, 135 MPa **TS**, 125 MPa **YS**, 4% in 2 in. (50 mm) **El**, 45 **HV**; **As manufactured (extruded products)**: all **diam**, 55 MPa **TS**, 22% in 2 in. (50 mm) **El**
SFS 2582	AL99.5	Finland	Shapes, sheet, tube, wire, and rod	99.5 min **Al**, 0.3 **Si**, 0.05 **Cu**, 0.05 **Mn**, 0.10 **Zn**, 0.4 **Fe**, 0.03 each, 0.15 total, others	**Hot finished (shapes)**: all **diam**, 70 MPa **TS**, 20 MPa **YS**, 23% in 2 in. (50 mm) **El**; **Annealed (sheet, tube, and wire)**: all **diam**, 100 max MPa **TS**, 20 MPa **YS**, 30% in 2 in. (50 mm) **El**; **Hard (tube and rod)**: $\leq$3 mm) **diam**, 140 MPa **TS**, 110 MPa **YS**, 4% in 2 in. (50 mm) **El**

WROUGHT ALUMINUM AND ALUMINUM ALLOYS

specification number	designation	country	product forms	chemical composition	mechanical properties and hardness values
SFS 2583	E–A199.5	Finland	Shapes, rod, tube, and wire	99.5 **Al**, 0.25 min **Si**, 0.05 min **Cu**, 0.05 **Mn**, 0.03 min **Zn**, 0.4 min **Fe**, 0.03 each, 0.15 total, others	**Hot finished (shapes)**: all **diam**, 70 MPa **TS**, 20 MPa **YS**, 23% in 2 in. (50 mm) **El**; **Annealed (except shapes)**: all **diam**, 100 max MPa **TS**, 30% in 2 in. (50 mm) **El**; **Hard (except shapes)**: all **diam**, 150 MPa **TS**, 4% in 2 in. (50 mm) **El**
NF A 50 411	1050 A (A5)	France	Bar, wire, tube, and shapes	99.50 min **Al**, 0.25 **Si**, 0.05 **Cu**, 0.05 **Mn**, 0.05 **Mg**, 0.07 **Zn**, 0.05 **Ti**, 0.40 **Fe**, 0.03 each, others	**1/2 hard (drawn bar)**: ≤50 mm **diam**, 100 MPa **TS**, 75 MPa **YS**, 5% in 2 in. (50 mm) **El**; **1/2 hard (drawn wire)**: ≤12 mm **diam**, 100 MPa **TS**, 75 MPa **YS**, 5% in 2 in. (50 mm) **El**; **1/4 hard (tube)**: 50 mm **diam**, 85 MPa **TS**, 65 MPa **YS**, 6% in 2 in. (50 mm) **El**
NF A 50 451	1050 A (A5)	France	Sheet, strip, and plate	99.50 min **Al**, 0.25 **Si**, 0.05 **Cu**, 0.05 **Mn**, 0.05 **Mg**, 0.07 **Zn**, 0.05 **Ti**, 0.40 **Fe**, 0.03 each, others	**Annealed**: 0.4–3.2 mm **diam**, 65 MPa **TS**, 35% in 2 in. (50 mm) **El**; **3/4 hard**: 0.4–3.2 mm **diam**, 120 MPa **TS**, 100 MPa **YS**, 5% in 2 in. (50 mm) **El**; **Hard**: 0.4–3.2 mm **diam**, 140 MPa **TS**, 125 MPa **YS**, 4% in 2 in. (50 mm) **El**
DIN 40501 part 4	E–AlF17/3.0257.32	Germany	Wire: electrical	99.5 min **Al**	**As manufactured (round)**: 1.5 mm **diam**, 170 MPa **TS**, 130 MPa **YS**; **As manufactured (flat)**: all **diam**, 170 MPa **TS**, 130 MPa **YS**
DIN 40501 part 4	E–AlF17/3.0257.32	Germany	Wire: electrical	99.5 min **Al**	**As manufactured (round)**: 3.1 mm **diam**, 160 MPa **TS**, 130 MPa **YS**; **As manufactured (flat)**: all **diam**, 160 MPa **TS**, 130 MPa **YS**
DIN 40501 part 4	E–Al F17/3.0257.32	Germany	Wire: electrical	99.5 min **Al**	**As manufactured (round)**: 0.2 mm **diam**, 180 MPa **TS**; **As manufactured (flat)**: all **diam**, 180 MPa **TS**
DIN 40501 part 4	E–AlF13/3.0257.30	Germany	Wire: electrical	99.5 min **Al**	**As manufactured (round)**: 1.5 mm **diam**, 130 MPa **TS**, 90 MPa **YS**, 2% in 4 in. (100 mm) **El**; **As manufactured (flat)**: 10 mm **diam**, 130 MPa **TS**, 90 MPa **YS**, 2% in 4 in. (100 mm) **El**
DIN 40501 part 4	E–AlF9/3.0257.26	Germany	Wire: electrical	99.5 min **Al**	**As manufactured (round)**: 1.5 mm **diam**, 90 MPa **TS**, 70 MPa **YS**, 3% in 4 in. (100 mm) **El**; **As manufactured (flat)**: 10 mm **diam**, 90 MPa **TS**, 70 MPa **YS**, 3% in 4 in. (100 mm) **El**

WROUGHT ALUMINUM AND ALUMINUM ALLOYS

specification number	designation	country	product forms	chemical composition	mechanical properties and hardness values
DIN 40501 part 4	E–AlF7/3.0257.10	Germany	Wire: electrical	99.5 min **Al**	**As manufactured (round):** 3.6 mm **diam**, 60 MPa **TS**, 60 max MPa **YS**, 25% in 4 in. (100 mm) **El**, 20–30 **HV**; **As manufactured (flat):** all **diam**, 60 MPa **TS**, 60 max MPa **YS**, 25% in 4 in. (100 mm) **El**, 20–30 **HV**
DIN 40501 part 3	E–AlMg Si 0.5 F22/3.3207.71	Germany	Section, bar	99.5 min **Al**	**As manufactured (round, square, hexagon):** all **diam**, 215 MPa **TS**, 160 MPa **YS**, 12% in 2 in. (50 mm) **El**, 65–90 **HB**; **As manufactured (flat):** 180 mm **diam**, 215 MPa **TS**, 160 MPa **YS**, 12% in 2 in. (50 mm) **El**, 65–90 **HB**
DIN 40501 part 3	E–AlMg Si 0.5 F17/3.3207.79	Germany	Section, bar	99.5 min **Al**	**As manufactured (round, square, hexagon):** all **diam**, 170 MPa **TS**, 120 MPa **YS**, 12% in 2 in. (50 mm) **El**, 45–65 **HB**; **As manufactured (flat):** 180 mm **diam**, 170 MPa **TS**, 120 MPa **YS**, 12% in 2 in. (50 mm) **El**, 45–65 **HB**
DIN 40501 part 3	E–AlF13/3.0257.30	Germany	Section, bar	99.5 min **Al**	**As manufactured (round, square, hexagon):** 10 mm **diam**, 130 MPa **TS**, 90 MPa **YS**, 5% in 2 in. (50 mm) **El**, 32–42 **HB**; **As manufactured (flat):** 12 mm **diam**, 130 MPa **TS**, 90 MPa **YS**, 5% in 2 in. (50 mm) **El**, 32–42 **HB**
DIN 40501 part 3	E–AlF10/3.0257.26	Germany	Section, bar	99.5 min **Al**	**As manufactured (round, square, hexagon):** 20 mm **diam**, 100 MPa **TS**, 70 MPa **YS**, 7% in 2 in. (50 mm) **El**, 28–38 **HB**; **As manufactured (flat):** 80 mm **diam**, 100 MPa **TS**, 70 MPa **YS**, 7% in 2 in. (50 mm) **El**, 28–38 **HB**
DIN 40501 part 3	E–AlF8/3.0257.09	Germany	Section, bar	99.5 min **Al**	**As manufactured (round, square, hexagon):** all **diam**, 80 MPa **TS**, 50 MPa **YS**, 15% in 2 in. (50 mm) **El**, 22–32 **HB**; **As manufactured (flat):** 120 mm **diam**, 80 MPa **TS**, 50 MPa **YS**, 15% in 2 in. (50 mm) **El**, 22–32 **HB**

WROUGHT ALUMINUM AND ALUMINUM ALLOYS

specification number	designation	country	product forms	chemical composition	mechanical properties and hardness values
DIN 40501 part 3	E-Al F6.5/3.0257.08	Germany	Section, bar	99.5 min **Al**	**As manufactured (round, square, hexagon)**: 63 mm **diam**, 65 MPa **TS**, 25 MPa **YS**, 23% in 2 in. (50 mm) **El**, 20–30 **HB**; **As manufactured (flat)**: 200 mm **diam**, 65 MPa **TS**, 25 MPa **YS**, 23% in 2 in. (50 mm) **El**, 20–30 **HB**
DIN 40501 part 3	E-Al p/3.0257.08	Germany	Section, bar	99.5 min **Al**	
DIN 40501 part 2	E-AlMgSi0.5F22/3.3 207.71	Germany	Tube	99.5 min **Al**	**As manufactured**: 250 mm **diam**, 215 MPa **TS**, 160 MPa **YS**, 12% in 2 in. (50 mm) **El**, 65–90 **HB**
DIN 40501 part 2	E-AlF10/3.0257.26	Germany	Tube	99.5 min **Al**	**As manufactured**: 120 mm **diam**, 100 MPa **TS**, 70 MPa **YS**, 6% in 2 in. (50 mm) **El**, 28–38 **HB**
DIN 40501 part 2	E-AlF7/3.0257.08	Germany	Tube	99.5 min **Al**	**As manufactured**: 250 mm **diam**, 70 MPa **TS**, 25 MPa **YS**, 20% in 2 in. (50 mm) **El**, 20–30 **HB**
DIN 40501 part 1	E-AlF16/3.0257.32	Germany	Sheet, strip	99.5 min **Al**	**Cold rolled**: 1.5 mm **diam**, 160 MPa **TS**, 140 MPa **YS**, 2% in 2 in. (50 mm) **El**, 37–52 **HB**
DIN 40501 part 1	E-AlF13/3.0257.30	Germany	Sheet, strip	99.5 min **Al**	**Cold rolled: sheet**: 4 mm **diam**, 130 MPa **TS**, 110 MPa **YS**, 3% in 2 in. (50 mm) **El**, 32–48 **HB**; **Cold rolled: strip**: 2.5 mm **diam**, 130 MPa **TS**, 110 MPa **YS**, 3% in 2 in. (50 mm) **El**, 32–48 **HB**
DIN 40501 part 1	E-AlF7/3.0257.10	Germany	Sheet, strip	99.5 min **Al**	**Cold rolled: sheet**: 20 mm **diam**, 70 MPa **TS**, 50 max MPa **YS**, 33% in 2 in. (50 mm) **El**, 18–30 **HB**; **Cold rolled: strip**: 3 mm **diam**, 70 MPa **TS**, 50 max MPa **YS**, 33% in 2 in. (50 mm) **El**, 18–30 **HB**
DIN 40501 part 1	E-AlF10/3.0257.26	Germany	Sheet, strip	99.5 min **Al**	**Cold rolled: sheet**: 6 mm **diam**, 100 MPa **TS**, 80 MPa **YS**, 7% in 2 in. (50 mm) **El**, 25–35 **HB**; **Cold rolled: strip**: 3 mm **diam**, 100 MPa **TS**, 80 MPa **YS**, 7% in 2 in. (50 mm) **El**, 25–35 **HB**
IS 739	G1B	India	Wire	99.5 min **Al**, 0.3 **Si**, 0.05 **Cu**, 0.05 **Mn**, 0.10 **Zn**, 0.4 **Fe**, 0.5 **Cu+Si+Fe+Mn+Zn**	**Annealed**: all **diam**, 93 max MPa **TS**, all **HV**; **Annealed and cold worked to hard**: 132 MPa **TS**

WROUGHT ALUMINUM AND ALUMINUM ALLOYS

specification number	designation	country	product forms	chemical composition	mechanical properties and hardness values
IS 738	T1B	India	Tube: drawn	99.5 min **Al**, 0.3 **Si**, 0.05 **Cu**, 0.05 **Mn**, 0.10 **Zn**, 0.4 **Fe**, 0.5 **Cu+Si+Fe+Mn+Zn**	**Annealed**: all **diam**, 93 max MPa **TS**; **Annealed and cold worked to hard**: ≤80.0 mm **diam**, 108 MPa **TS**
IS 736	19500	India	Plate	99.5 min **Al**, 0.3 **Si**, 0.05 **Cu**, 0.05 **Mn**, 0.1 **Zn**, 0.4 **Fe**, 0.5 **Cu+Si+Fe+Mn+Zn**	**As manufactured**: 28 mm **diam**, 65 MPa **TS**, 28% in 2 in. (50 mm) **El**; **Annealed**: 30 mm **diam**, 100 max MPa **TS**, 30% in 2 in. (50 mm) **El**; **Half hard**: 7 mm **diam**, 100 MPa **TS**, 7% in 2 in. (50 mm) **El**
IS 733	19500	India	Bar, rod, section	99.5 min **Al**, 0.3 **Si**, 0.05 **Cu**, 0.05 **Mn**, 0.1 **Zn**, 0.4 **Fe**, 0.5 **Cu+Si+Fe+Mn+Zn**	**As manufactured**: all **diam**, 65 MPa **TS**, 23% in 2 in. (50 mm) **El**
ISO R827	Al99.5	ISO	Extruded products	99.5 min **Al**, 0.3 **Si**, 0.05 **Cu**, 0.05 **Mn**, 0.10 **Zn**, 0.4 **Fe**, 0.5 **Cu+Si+Fe+Mn+Zn**, total	**As manufactured**: all **diam**, 49 MPa **TS**, 22% in 2 in. (50 mm) **El**
ISO R209	Al99.5	ISO	Wrought products	99.5 min **Al**, 0.3 **Si**, 0.05 **Cu**, 0.05 **Mn**, 0.10 **Zn**, 0.4 **Fe**, 0.5 **Cu+Si+Fe+Mn+Zn** total	
ISO TR2778	Al99.5	ISO	Tube: drawn	99.5 min **Al**, 0.3 **Si**, 0.05 **Cu**, 0.05 **Mn**, 0.10 **Zn**, 0.4 **Fe**, 0.5 **Cu+Si+Fe+Mn+Zn**, total	**Annealed**: all **diam**, 60 MPa **TS**, 22% in 2 in. (50 mm) **El**; **Strain hardened**: all **diam**, 95 MPa **TS**, 3% in 2 in. (50 mm) **El**
ISO TR2136	Al99.5	ISO	Sheet, plate	99.5 min **Al**, 0.3 **Si**, 0.05 **Cu**, 0.05 **Mn**, 0.10 **Zn**, 0.4 **Fe**, 0.5 **Cu+Si+Fe+Mn+Zn**, total	**Annealed**: all **diam**, 60 MPa **TS**, 20% in 2 in. (50 mm) **El**; **HB–Strain hardened**: all **diam**, 75 MPa **TS**, 6% in 2 in. (50 mm) **El**; **HD–Strain hardened**: all **diam**, 95 MPa **TS**, 4% in 2 in. (50 mm) **El**
JIS H 4120	1050	Japan	Wire and rod: rivet	99.50 min **Al**, 0.25 **Si**, 0.05 **Cu**, 0.05 **Mn**, 0.05 **Mg**, 0.05 **Zn**, 0.03 **Ti**, 0.40 **Fe**	**1/2 hard**: ≤25 mm **diam**, 88 MPa **TS**
JIS H 4090	1050	Japan	Pipe and tube: welded	99.50 min **Al**, 0.25 **Si**, 0.05 **Cu**, 0.05 **Mn**, 0.05 **Mg**, 0.05 **Zn**, 0.03 **Ti**, 0.40 **Fe**	**Annealed**: <1.3 mm **diam**, 69 MPa **TS**, 20 MPa **YS**, 25% in 2 in. (50 mm) **El**; **1/2 hard**: <1.3 mm **diam**, 98 MPa **TS**, 69 MPa **YS**, 4% in 2 in. (50 mm) **El**; **Hard**: <0.8 mm **diam**, 137 MPa **TS**, 2% in 2 in. (50 mm) **El**
JIS H 4080	1050	Japan	Tube: extruded	99.50 min **Al**, 0.25 **Si**, 0.05 **Cu**, 0.05 **Mn**, 0.05 **Mg**, 0.05 **Zn**, 0.03 **Ti**, 0.40 **Fe**	**As manufactured**: all **diam**, 59 MPa **TS**, 20 MPa **YS**
JIS H 4040	1050	Japan	Bar: extruded	99.50 min **Al**, 0.25 **Si**, 0.05 **Cu**, 0.05 **Mn**, 0.05 **Mg**, 0.05 **Zn**, 0.03 **Ti**, 0.40 **Fe**	**As manufactured**: all **diam**, 59 MPa **TS**, 20 MPa **YS**

WROUGHT ALUMINUM AND ALUMINUM ALLOYS

specification number	designation	country	product forms	chemical composition	mechanical properties and hardness values
JIS H 4000	1050	Japan	Sheet, plate, and strip	99.50 min **Al**, 0.25 **Si**, 0.05 **Cu**, 0.05 **Mn**, 0.05 **Mg**, 0.05 **Zn**, 0.03 **Ti**, 0.40 **Fe**	**Annealed**: <1.3 mm **diam**, 69 MPa **TS**, 20 MPa **YS**, 25% in 2 in. (50 mm) **El**; **Hard**: <0.3 mm **diam**, 137 MPa **TS**, 1% in 2 in. (50 mm) **El**; **Hot rolled**: <12 mm **diam**, 78 MPa **TS**, 49 MPa **YS**, 10% in 2 in. (50 mm) **El**
DGN–W–30	1S–H18	Mexico	Sheet	99.5 min **Al**	**Rolled**: 0.25–0.52 mm **diam**, 131 MPa **TS**, 1% in 2 in. (50 mm) **El**
DGN–W–30	1S–0	Mexico	Sheet	99.5 min **Al**	**Rolled**: 0.25–0.52 mm **diam**, 96 MPa **TS**, 15% in 2 in. (50 mm) **El**
DGN–W–30	1S–H14	Mexico	Sheet	99.5 min **Al**	**Rolled**: 0.53–0.80 mm **diam**, 103 MPa **TS**, 3% in 2 in. (50 mm) **El**
DGN–W–30	1S–H16	Mexico	Sheet	99.5 min **Al**	**Rolled**: 0.25–0.52 mm **diam**, 117 MPa **TS**, 1% in 2 in. (50 mm) **El**
NS 17 011	NS 17 011–02	Norway	Sheet, strip, and wire: for electrical purposes	99.5 min **Al**, 0.3 **Si**, 0.05 **Cu**, 0.10 **Zn**, 0.4 **Fe**, 0.5 **Cu+Fe+Mn+Si+Zn**, 0.03 **V+Mn+Cr+Ti**	**Soft annealed (sheet and strip)**: <0.5 mm **diam**, 100 MPa **TS**, 32% in 2 in. (50 mm) **El**; **Soft annealed (wire)**: 3.5–14 mm **diam**, 60 MPa **TS**, 60 MPa **YS**
NS 17 011	NS 17 011–14	Norway	Sheet, strip, bar, tube, wire, and profiles: for electrical purposes	99.5 min **Al**, 0.3 **Si**, 0.05 **Cu**, 0.10 **Zn**, 0.4 **Fe**, 0.5 **Cu+Fe+Mn+Si+Zn V+Mn+Cr+Ti**	**1/2 hard (sheet and strip)**: <0.5 mm **diam**, 100 MPa **TS**, 90 MPa **YS**, 5% in 2 in. (50 mm) **El**; **1/2 hard (bar)**: 5–15 mm **diam**, 98 MPa **TS**, 80 MPa **YS**, 6% in 2 in. (50 mm) **El**; **1/2 hard (tube)**: <6 mm **diam**, 93 MPa **TS**, 80 MPa **YS**, 6% in 2 in. (50 mm) **El**
NS 17 011	NS 17 011–18	Norway	Sheet, strip, bar, wire, and profiles: for electrical purposes	99.5 min **Al**, 0.3 **Si**, 0.05 **Cu**, 0.10 **Zn**, 0.4 **Fe**, 0.5 **Cu+Fe+Mn+Si+Zn**, 0.03 **V+Mn+Cr+Ti**	**Hard (sheet and strip)**: <0.5 mm **diam**, 140 MPa **TS**, 130 MPa **YS**, 3% in 2 in. (50 mm) **El**; **Hard (bar)**: 1–5 mm **diam**, 162 MPa **TS**, 150 MPa **YS**; **Hard (profiles)**
AMS 4011		US	Foil: for electronic applications	99.45 min **Al**, 0.05 **Cu**, 0.05 **Mn**, 0.55 **Fe+Si**, 0.03 each, others	
ANSI/ASTM B 373	1145	US	Foil: for capacitators	99.45 min **Al**, 0.05 **Cu**, 0.05 **Mn**, 0.55 **Si+Fe**, 0.03 each, others	
ANSI/ASTM B 236		US	Bar: for electrical purposes	99.45 min **Al**	**H12**: 0.125–1.000 in. **diam**, 8 ksi (55 MPa) **TS**; **H112**: 1.001–3.000 in. **diam**, 4 ksi (28 MPa) **TS**; **H111**: all **diam**, 4 ksi (28 MPa) **TS**

WROUGHT ALUMINUM AND ALUMINUM ALLOYS

specification number	designation	country	product forms	chemical composition	mechanical properties and hardness values
COPANT 862	1145	COPANT	Wrought products	99.45 min **Al**, 0.05 **Cu**, 0.05 **Mn**, 0.05 **Mg**, 0.05 **Zn**, 0.03 **Ti**, 0.55 **Si+Fe**, 0.03 each, others	
ANSI/ASTM B 491	1235	US	Tube: coiled, for general purposes	99.35 min **Al**, 0.05 **Cu**, 0.65 **Si+Fe**, 0.05 each, others	**H111**: 0.032–0.050 in. **diam**, 10 ksi (69 MPa) **TS**, 8 ksi (55 MPa) **YS**, 25% in 2 in. (50 mm) **El**; **H112**: 0.032–0.050 in. **diam**, 9 ksi (62 MPa) **TS**, 3 ksi (21 MPa) **YS**, 25% in 2 in. (50 mm) **El**
ANSI/ASTM B 483	1435	US	Tube: for general purposes	99.35 min **Al**, 0.15 **Si**, 0.02 **Cu**, 0.03 **Ti**, 0.30–0.50 **Fe**	**Annealed**: 0.018–0.500 in. **diam**, 10 ksi (69 MPa) **TS**, 3 ksi (21 MPa) **YS**; **H12 (round tube)**: 0.018–0.500 in. **diam**, 12 ksi (83 MPa) **TS**, 7 ksi (48 MPa) **YS**; **H14 (round tube)**: 0.018–0.500 in. **diam**, 14 ksi (97 MPa) **TS**, 12 ksi (83 MPa) **YS**
ANSI/ASTM B 345	1235	US	Foil: for capacitators	99.35 min **Al**, 0.05 **Cu**, 0.65 **Si+Fe** each, others	
COPANT 862	1235	COPANT	Wrought products	99.35 min **Al**, 0.05 **Cu**, 0.05 **Mn**, 0.05 **Mg**, 0.10 **Zn**, 0.03 **Ti**, 0.65 **Si+Fe**, 0.03 each, others	
ANSI/ASTM B 209	1230	US	Sheet and plate	99.30 min **Al**, 0.10 **Cu**, 0.05 **Mn**, 0.05 **Mg**, 0.10 **Zn**, 0.03 **Ti**, 0.7 **Si+Fe**, 0.03 each, others	
AS 1734	1230	Australia	Sheet, plate	99.30 min **Al**, 0.10 **Cu**, 0.05 **Mn**, 0.05 **Mg**, 0.10 **Zn**, 0.03 **Ti**, 0.7 **Si+Fe**, 0.05 each, others	
CSA HA.4	.1230	Canada	Plate, sheet (cladding for .2024 Alclad)	99.30 min **Al**, 0.10 **Cu**, 0.05 **Mn**, 0.10 **Zn**, 0.7 **Si+Fe**, 0.05 each, others	**O**: 0.010–0.032 in. **diam**, 30.0 max ksi (207 max MPa) **TS**, 14.0 max ksi (97 max MPa) **YS**, 12% in 2 in. (50 mm) **El**; **T3 (sheet)**: 0.010–0.020 in. **diam**, 59.0 ksi (407 MPa) **TS**, 39.0 ksi (269 MPa) **YS**, 12% in 2 in. (50 mm) **El**; **T42**: 0.010–0.020 in. **diam**, 57.0 ksi (393 MPa) **TS**, 34.0 ksi (234 MPa) **YS**, 12% in 2 in. (50 mm) **El**
JIS H 4160	IN30	Japan	Foil	99.30 min **Al**, 0.10 **Cu**, 0.05 **Mn**, 0.05 **Mg**, 0.05 **Zn**, 0.7 **Si+Fe**	
AMS 4062F		US	Tube: seamless	99.00 min **Al**, 0.05–0.20 **Cu**, 0.05 **Mn**, 0.10 **Zn**, 1.0 **Fe+Si**, 0.05 each, 0.15 total, others	**Strain hardened**: 0.014–0.500 in. (0.36–12.70 mm) **diam**, 16 ksi (110 MPa) **TS**

WROUGHT ALUMINUM AND ALUMINUM ALLOYS

specification number	designation	country	product forms	chemical composition	mechanical properties and hardness values
AMS 4001E		US	Sheet and plate	99.00 min **Al**, 0.05–0.20 **Cu**, 0.05 **Mn**, 0.10 **Zn**, 0.05–0.20 **Fe + Si**, 0.05 each, 0.15 total, others	**Annealed**: 0.006–0.019 in. (0.15–0.48 mm) **diam**, 11 ksi (76 MPa) **TS**, 15% in 2 in. (50 mm) **El**
AMS 4003E		US	Sheet and plate	99.00 min **Al**, 0.05–0.20 **Cu**, 0.05 **Mn**, 0.10 **Zn**, 1.0 **Fe + Si**, 0.05 each, 0.15 total, others	**Strain hardened to 1/2 hard**: 0.009–0.012 in. (0.23–0.30 mm) **diam**, 16 ksi (110 MPa) **TS**, 1% in 2 in. (50 mm) **El**
AMS 4180C		US	Wire	99.00 min **Al**, 0.05–0.20 **Cu**, 0.05 **Mn**, 0.10 **Zn**, 0.05 each, 0.15 total, others	**Drawn, hard**: all **diam**, 22 ksi (152 MPa) **TS**
AMS 4102D		US	Bar and rod: rolled, drawn, or cold finished	99.0 min **Al**, 0.05–0.20 **Cu**, 0.05 **Mn**, 0.10 **Zn**, 1.0 **Fe + Si**, 0.5 each, 0.15 total, others	
ASTM B 479		US	Foil: for flexible barrier applications	99.00 min **Al**, 0.01 **Pb**, 0.01 **As**, 0.01 **Cd**	
ANSI/ASTM B 547	1100	US	Tube: formed and arc-welded	99.00 min **Al**, 0.05–0.20 **Cu**, 0.05 **Mn**, 0.10 **Zn**, 1.0 **Si + Fe**, 0.05 each, 0.15 total, others	**Annealed**: 0.250–0.500 in. **diam**, 11 ksi (76 MPa) **TS**, 4 ksi (24 MPa) **YS**, 28% in 2 in. (50 mm) **El**; **H12**: 0.125–0.499 in. **diam**, 14 ksi (97 MPa) **TS**, 11 ksi (76 MPa) **YS**, 9% in 2 in. (50 mm) **El**; **H14**: 0.125–0.499 in. **diam**, 16 ksi (110 MPa) **TS**, 14 ksi (97 MPa) **YS**, 6% in 2 in. (50 mm) **El**
ANSI/ASTM B 491	1100	US	Tube: coiled, for general purposes	99.00 min **Al**, 0.05–0.20 **Cu**, 0.05 **Mn**, 0.10 **Zn**, 1.0 **Si + Fe**, 0.05 each, 0.15 total, others	**H112**: 0.032–0.050 in. **diam**, 11 ksi (76 MPa) **TS**, 3 ksi (21 MPa) **YS**, 25% in 2 in. (50 mm) **El**
ANSI/ASTM B 491	1200	US	Tube: coiled, for general purposes	99.00 min **Al**, 0.05 **Cu**, 0.05 **Mn**, 0.10 **Zn**, 1.0 **Si + Fe**, 0.05 each, 0.15 total, others	**H111**: 0.032–0.050 in. **diam**, 11 ksi (76 MPa) **TS**, 10 ksi (69 MPa) **YS**, 25% in 2 in. (50 mm) **El**; **H112**: 0.032–0.050 in. **diam**, 10 ksi (69 MPa) **TS**, 3 ksi (21 MPa) **YS**, 25% in 2 in. (50 mm) **El**
ANSI/ASTM B 483	1100	US	Tube: for general purposes	99.00 min **Al**, 0.05–0.20 **Cu**, 0.05 **Mn**, 0.10 **Zn**, 1.0 **Si + Fe**, 0.05 each, 0.15 total, others	**Annealed**: 0.018–0.500 in. **diam**, 11 ksi (76 MPa) **TS**, 4 ksi (28 MPa) **YS**; **H12 (round tube)**: 0.018–0.500 in. **diam**, 14 ksi (97 MPa) **TS**, 11 ksi (76 MPa) **YS**; **H14 (round tube)**: 0.018–0.500 in. **diam**, 16 ksi (110 MPa) **TS**, 14 ksi (97 MPa) **YS**
ANSI/ASTM B 316	1100	US	Wire and rod: rivet and cold-heading	99.00 min **Al**, 0.05–0.20 **Cu**, 0.05 **Mn**, 0.10 **Zn**, 1.0 **Si + Fe**, 0.05 each, 0.15 total, others	**Annealed**: $\leq$1.000 in. **diam**, 16 max ksi (107 max MPa) **TS**; **H12**: $\leq$1.000 in. **diam**, 16 ksi (110 MPa) **TS**

WROUGHT ALUMINUM AND ALUMINUM ALLOYS

specification number	designation	country	product forms	chemical composition	mechanical properties and hardness values
ANSI/ASTM B 247	1100	US	Die forgings	99.0 min **Al**, 0.05–0.20 **Cu**, 0.05 **Mn**, 0.10 **Zn**, 1.0 **Si+Fe**, 0.05 each, 0.15 total, others	**H112**: ≤4.000 in. **diam**, 11 ksi (76 MPa) **TS**, 4 ksi (28 MPa) **YS**, 18% in 2 in. (50 mm) **El**, 20 min **HB**
ANSI/ASTM B 241	1100	US	Tube: seamless	99.00 min **Al**, 0.05–0.20 **Cu**, 0.05 **Mn**, 0.10 **Zn**, 1.0 **Si+Fe**, 0.05 each, 0.15 total, others	**Annealed**: all **diam**, 11 ksi (76 MPa) **TS**, 3 ksi (21 MPa) **YS**, 25% in 2 in. (50 mm) **El**; **H112**: all **diam**, 11 ksi (76 MPa) **TS**, 3 ksi (21 MPa) **YS**
ANSI/ASTM B 221	1100	US	Bar, tube, rod, wire, and shapes	99.00 min **Al**, 0.05–0.20 **Cu**, 0.05 **Mn**, 0.10 **Zn**, 1.0 **Si+Fe**, 0.05 each, 0.15 total, others	**Annealed**: all **diam**, 11 ksi (76 MPa) **TS**, 3 ksi (21 MPa) **YS**, 25% in 2 in. (50 mm) **El**; **H112**: all **diam**, 9 ksi (59 MPa) **TS**, 3 ksi (21 MPa) **YS**
ANSI/ASTM B 211	1100	US	Bar, rod, and wire	99.00 min **Al**, 0.05–0.20 **Cu**, 0.05 **Mn**, 0.10 **Zn**, 1.0 **Si+Fe**, 0.05 each, 0.15 total, others	**Annealed**: ≥0.125 in. **diam**, 11 ksi (76 MPa) **TS**, 3 ksi (21 MPa) **YS**, 25% in 2 in. (50 mm) **El**; **H12**: ≤0.374 in. **diam**, 14 ksi (97 MPa) **TS**; **H112**: all **diam**, 11 ksi (76 MPa) **TS**, 3 ksi (21 MPa) **YS**
ANSI/ASTM B 210	1100	US	Tube: seamless	99.00 min **Al**, 0.05–0.20 **Cu**, 0.05 **Mn**, 0.10 **Zn**, 1.0 **Si+Fe**, 0.05 each, 0.15 total, others	**Annealed**: 0.018–0.500 in. **diam**, 11 ksi (76 MPa) **TS**, 4 ksi (24 MPa) **YS**; **H12**: 0.018–0.500 in. **diam**, 14 ksi (97 MPa) **TS**, 11 ksi (76 MPa) **YS**; **H14**: 0.018–0.500 in. **diam**, 16 ksi (110 MPa) **TS**, 14 ksi (97 MPa) **YS**
ANSI/ASTM B 209	1100	US	Sheet and plate	99.00 min **Al**, 0.05–0.20 **Cu**, 0.05 **Mn**, 0.10 **Zn**, 1.0 **Si+Fe**, 0.05 each, 0.15 total, others	**Annealed**: 0.250–3.000 in. **diam**, 11 ksi (76 MPa) **TS**, 4 ksi (28 MPa) **YS**, 28% in. 2 in. (50 mm) **El**; **H12 or H22**: 0.500–2.000 in. **diam**, 14 ksi **TS**, 11 ksi **YS**, 12% in 2 in. (50 mm) **El**; **H14 or H24**: 0.500–1.000 in. **diam**, 16 ksi (110 MPa) **TS**, 14 ksi (97 MPa) **YS**, 10% in 2 in. (50 mm) **El**
AS 1867	1200	Australia	Tube: drawn	99.00 min **Al**, 0.05 **Cu**, 0.05 **Mn**, 0.10 **Zn**, 0.05 **Ti**, 1.0 **Si+Fe**, 0.05 each, 0.15 total, others	**Annealed**: all **diam**, 100 max MPa **TS**; **Strain hardened to 1/4 hard**: all **diam**, 96 MPa **TS**; **Strain hardened to full hard**: all **diam**, 151 MPa **TS**
AS 1866	1100	Australia	Rod, bar, shape	99.00 min **Al**, 0.05–0.20 **Cu**, 0.05 **Mn**, 0.10 **Zn**, 0.05 each, 0.15 total, others	**Strain–hardened**: all **diam**, 75 MPa **TS**, 20 MPa **YS**, 18% in 2 in. (50 mm) **El**

WROUGHT ALUMINUM AND ALUMINUM ALLOYS

specification number	designation	country	product forms	chemical composition	mechanical properties and hardness values
AS 1866	1200	Australia	Rod, bar, shape	99.00 min **Al**, 0.05 **Cu**, 0.05 **Mn**, 0.10 **Zn**, 0.05 **Ti**, 0.05 each, 0.15 total, others	**Strain–hardened**: all **diam**, 75 MPa **TS**, 20 MPa **YS**, 18% in 2 in. (50 mm) **El**
AS 1734	1100	Australia	Sheet, plate	99.00 min **Al**, 0.05–0.20 **Cu**, 0.05 **Mn**, 0.10 **Zn**, 1.0 **Si + Fe**, 0.05 each, 0.15 total, others	**Annealed**: 0.15–0.50 mm **diam**, 107 MPa **TS**, 15% in 2 in. (50 mm) **El**; **Strain–hardened to 1/2 hard**: 0.25–0.30 mm **diam**, 110 MPa **TS**, 1% in 2 in. (50 mm) **El**; **Strain–hardened to full hard**: 0.15–0.50 mm **diam**, 151 MPa **TS**, 1% in 2 in. (50 mm) **El**
AS 1734	1200	Australia	Sheet, plate	99.00 min **Al**, 0.05 **Cu**, 0.05 **Mn**, 0.10 **Zn**, 0.05 **Ti**, 1.0 **Si + Fe**, 0.05 each, 0.15 total, others	**Annealed**: 0.15–0.50 mm **diam**, 107 MPa **TS**, 15% in 2 in. (50 mm) **El**; **Strain–hardened to 1/2 hard**: 0.25–0.30 mm **diam**, 110 MPa **TS**, 1% in 2 in. (50 mm) **El**; **Strain–hardened to full hard**: 0.15–0.50 mm **diam**, 151 MPa **TS**, 1% in 2 in.(50 mm) **El**
AS 1865	1200	Australia	Wire, rod, bar, strip	99.00 min **Al**, 0.05 **Cu**, 0.05 **Mn**, 0.10 **Zn**, 0.05 **Ti**, 1.0 **Si + Fe**, 0.05 each, 0.15 total, others	**Annealed**: all **diam**, 100 max MPa **TS**, 23% in 2 in. (50 mm) **El**; **Strain hardened to 1/4 hard**: 10 mm **diam**, 93 MPa **TS**; **Strain hardened to full hard**: 10 mm **diam**, 151 MPa **TS**
BS 1474	E1C	UK	Bar, tube, section	99.0 **Al**, 0.5 **Si**, 0.10 **Cu**, 0.1 **Mn**, 0.1 **Zn**, 0.7 **Fe**, 1.0 **Cu + Si + Fe + Mn + Zn**	**As manufactured**: 65 MPa **TS**, 18% in 2 in. (50 mm) **El**
BS 1471	T1C	UK	Tube: drawn	99.0 **Al**, 0.5 **Si**, 0.10 **Cu**, 0.1 **Mn**, 0.1 **Zn**, 0.7 **Fe**, 1.0 **Cu + Si + Fe + Mn + Zn**	**Annealed**: ≤12.0 mm **diam**, 105 MPa **TS**; **Strain–hardened to 1/2 hard**: ≤12.0 mm **diam**, 110 MPa **TS**; **Strain–hardened to hard**: ≤12.0 mm **diam**, 140 MPa **TS**
BS 1470	S1C	UK	Plate, sheet, strip	99.0 **Al**, 0.5 **Si**, 0.10 **Cu**, 0.1 **Mn**, 0.1 **Zn**, 0.7 **Fe**, 1.0 **Cu + Si + Fe + Mn + Zn**	**Annealed**: 0.2–6.0 mm **diam**, 70 MPa **TS**, 20% in 2 in. (50 mm) **El**; **Strain hardened to hard**: 0.2–3.0 mm **diam**, 140 MPa **TS**, 2% in 2 in. (50 mm) **El**
BS L.116		UK	Tube	99.0 min **Al**, 0.5 **Si**, 0.10 **Cu**, 0.10 **Mn**, 0.10 **Zn**, 0.7 **Fe**, 0.05 each, others	
BS 5. 17		UK	Sheet and strip	99.0 min **Al**, 0.5 **Si**, 0.10 **Cu**, 0.10 **Mn**, 0.10 **Zn**, 0.7 **Fe**, 0.05 each, others	**Cold rolled and annealed**: 0.4–0.8 mm **diam**, 70 MPa **TS**, 20% in 2 in. (50 mm) **El**
BS 5L.16		UK	Sheet and strip	99.0 min **Al**, 0.5 **Si**, 0.10 **Cu**, 0.10 **Mn**, 0.10 **Zn**, 0.7 **Fe**, 0.05 each, others	**Cold rolled or cold rolled and partially annealed**: 0.4–0.8 mm **diam**, 110 MPa **TS**, 3% in 2 in. (50 mm) **El**

WROUGHT ALUMINUM AND ALUMINUM ALLOYS

specification number	designation	country	product forms	chemical composition	mechanical properties and hardness values
BS 4L.34		UK	Forging stock, bar, section, forgings	99.0 min **Al**, 0.5 **Si**, 0.10 **Cu**, 0.10 **Mn**, 0.10 **Zn**, 0.7 **Fe**, 0.05 each, others	
BS 3L.54		UK	Tube	99.0 min **Al**, 0.5 **Si**, 0.10 **Cu**, 0.1 **Mn**, 0.1 **Mg**, 0.1 **Ni**, 0.10 **Zn**, 0.05 **Sn**, 0.7 **Fe**, 0.05 **Pb**, 0.20 **Zr+Ti**	**Cold drawn**: $\leq$75 mm **diam**, 125 MPa **TS**
CSA HA.6	.1100	Canada	Rod, wire: for welding and brazing	99.00 min **Al**, 0.05–0.20 **Cu**, 0.05 **Mn**, 0.10 **Zn**, 0.0008 **Be**, 1.0 **Si+Fe**, 0.05 each, 0.15 total, others	
CSA HA.6	.1200	Canada	Rivet, rod, wire	99.00 min **Al**, 0.05 **Cu**, 0.05 **Mn**, 0.10 **Zn**, 0.05 **Ti**, 1.0 **Si+Fe**, 0.05 each, 0.15 total, others	**O**: 0.051–1.000 in. **diam**, 15.5 max ksi (107 max MPa) **TS**; **H14**: 0.062–1.000 in **diam**, 16.0 ksi (157 MPa) **TS**
CSA HA.5	.1100	Canada	Bar, rod, wire, shapes	99.00 min **Al**, 0.05–0.20 **Cu**, 0.05 **Mn**, 0.10 **Zn**, 1.0 **Si+Fe**, 0.05 each, 0.15 total, others	**Drawn and O**: all **diam**, 11.0 ksi (76 MPa) **TS**, 3.0 ksi (21 MPa) **YS**, 25% in 2 in. (50 mm) **El**; **Drawn and H14**: <0.374 in. **diam**, 16.0 ksi (110 MPa) **TS**; **Extruded and O**: all in. **diam**, 22.0 ksi (76 MPa) **TS**, 21 MPa **YS**, 25% in 2 in. (50 mm) **El**
CSA HA.4	1100	Canada	Plate, sheet	99.00 min **Al**, 0.05–0.20 **Cu**, 0.05 **Mn**, 0.10 **Zn**, 1.0 **Si+Fe**, 0.05 each, 0.15 total, others	**O**: 0.006–0.019 in. **diam**, 11.0 ksi (76 MPa) **TS**, 3.5 ksi (24 MPa) **YS**, 15% in 2 in. (50 mm) **El**; **H12**: 0.032–0.050 in. **diam**, 14.0 ksi (97 MPa) **TS**, 11.0 ksi (76 MPa) **YS**, 6% in 2 in. (50 mm) **El**; **H14**: 0.009–0.012 in. **diam**, 16.0 ksi (110 MPa) **TS**, 14.0 ksi (97 MPa) **YS**, 1% in 2 in. (50 mm) **El**
COPANT 862	1200	COPANT	Wrought products	99.00 min **Al**, 0.05 **Cu**, 0.05 **Mn**, 0.10 **Zn**, 0.05 **Ti**, 1.0 **Si+Fe**, 0.05 each, 0.15 total, others	
COPANT 862	1100	COPANT	Wrought products	99.00 min **Al**, 0.05–0.20 **Cu**, 0.05 **Mn**, 0.10 **Zn**, 1.00 **Si+Fe**, 0.05 each, 0.15 total, others	
DS 3012	1200	Denmark	Rolled, drawn, and extruded products	99.00 min **Al**, 0.05 **Cu**, 0.05 **Mn**, 0.10 **Zn**, 0.01 **Fe**, 0.01 **Fe+Si** each, 0.15 total, others	**1/2 hard (rolled and drawn products)**: 0.5–6 mm **diam**, 110 MPa **TS**, 90 MPa **YS**, 5% in 2 in. (50 mm) **El**, 35 **HV**; **Hard (rolled and drawn products)**: 0.5–6 mm **diam**, 140 MPa **TS**, 130 MPa **YS**, 3% in 2 in. (50 mm) **El**, 45 **HV**; **As manufactured (extruded products)**: all **diam**, 65 MPa **TS**, 18% in in. (50 mm) **El**

WROUGHT ALUMINUM AND ALUMINUM ALLOYS

specification number	designation	country	product forms	chemical composition	mechanical properties and hardness values
WW-T-700/1D	1100	US	Tube	99.00 min **Al**, 0.20 **Cu**, 0.05 **Mn**, 0.10 **Zn**, 1.0 **Fe**, **Si** in **Fe** content; 0.05 each, 0.15 total, others	**Annealed**: 0.014–0.500 in. **diam**, 15.5 max ksi (107 max MPa) **TS**; **Strain hardened, one quarter hard temper**: 0.014–0.500 in **diam**, 14 ksi (97 MPa) **TS**; **Strain hardened, half hard temper**: 0.014–0.500 in. **diam**, 16 ksi(110 MPa) **TS**
QQ-A-1876	Grade B	US	Foil	99.00 min **Al**	
QQ-A-430B	1100	US	Rod, wire	99.00 min **Al**, 1.0 **Si**, 0.05–0.20 **Cu**, 0.05 **Mn**, 0.10 **Zn**, **Fe** in **Si** content; 0.05 each, 0.15 total, others	**Annealed**: 1.000 max in. **diam**, 15.5 max ksi 107 max MPa **TS**; **Strain hardened half hard temper**: 1.000 max in. **diam**, 16 ksi (110 MPa) **TS**
QQ-A-250/1E	1100	US	Plate, sheet	99.00 **Al**, 0.05–0.20 **Cu**, 0.05 **Mn**, 0.10 **Zn**, 1.0 **Fe**, **Si** in **Fe** content; 0.05 each, 0.15 total, others	**Annealed**: 0.006–0.019 in. **diam**, 11 ksi (76 MPa) **TS**, 3.5 ksi (24 MPa) **YS**, 15% in 2 in. (50 mm) **El**; **Strain hardened 1/4 hard temper**: 0.017–0.019 in. **diam**, 14 ksi (97 MPa) **TS**, 11 ksi (76 MPa) **YS**, 3% in 2 in. (50 mm) **El**; **Strain hardened and partially annealed to 1/4 hard temper**: 0.017–0.019 in. **diam**, 14 ksi (97 MPa) **TS**, 3% in 2 in (50 mm) **El**
QQ-A-225/1D	1100	US	Bar, rod, wire	99.00 min **Al**, 0.05–0.20 **Cu**, 0.05 **Mn**, 0.10 **Zn**, 1.0 **Fe**, **Si** in **Fe** content; 0.05 each, 0.15 total, others	**Annealed**: all **diam**, 15.5 ksi(107 MPa) **TS**, 25% in 2 in. (50 mm) **El**; **Strain hardened one quarter hard temper**: 0.0374 max in **diam**, 14 ksi (97 MPa) **TS**; **Strain hardened half hard temper**: 0.374 max in. **diam**, 16 ksi (110 MPa) **TS**
SFS 2584	AL99.0	Finland	Shapes, sheet, rod, tube, and wire	99.0 min **Al**, 0.5 **Si**, 0.10 **Cu**, 0.1 **Mn**, 0.1 **Zn**, 0.8 **Fe**, 0.05 each, 0.15 total, others	**Hot finished (shapes)**: all **diam**, 80 MPa **TS**, 30 MPa **YS**, 18% in 2 in. (50 mm) **El**; **Annealed (except shapes)**: all **diam**, 110 max MPa **TS**, 20 MPa **YS**, 30% in 2 in. (50 mm) **El**, 1/2 hard (except shapes): ≤6 **diam**, 110 MPa **TS**, 80 MPa **YS**, 5% in 2 in. (50 mm) **El**
NF A 50 411	1200 (A4)	France	Bar, wire, tube, and shapes	99.0 min **Al**, 0.05 **Cu**, 0.05 **Mn**, 0.10 **Zn**, 0.05 **Ti**, 1.0 **Fe+Si**, 0.05 each, 0.15 total, others	**1/2 hard (drawn bar)**: ≤50 mm **diam**, 110 MPa **TS**, 95 MPa **YS**, 4% in 2 in. (50 mm) **El**; **Drawn and annealed (wire)**: ≤10 mm **diam**, 75 MPa) **TS**, 110 MPa **YS**, 30% in 2 in. (50 mm) **El**; **1/4 hard (tube)**: ≤50 **diam**, 95 MPa **TS**, 75 MPa **YS**, 5% in 2 in. (50 mm) **El**

WROUGHT ALUMINUM AND ALUMINUM ALLOYS

specification number	designation	country	product forms	chemical composition	mechanical properties and hardness values
NF A 50 411	1100 (A45)	France	Bar, wire, and tube	99.00 min **Al**, 0.05–0.20 **Cu**, 0.05 **Mn**, 0.10 **Zn**, 1.0 **Fe+Si**	**1/2 hard (drawn bar):** ≤50 mm **diam**, 110 MPa **TS**, 95 MPa **YS**, 4% in 2 in. (50 mm) **El**; **Drawn and annealed (wire):** ≤10 mm **diam**, 75 MPa **TS**, 110 MPa **YS**, 30% in 2 in.(50 mm) **El**; **1/4 hard (tube):** ≤50 mm) **diam**, 95 MPa **TS**, 75 MPa **YS**, 5% in 2 in. (50 mm) **El**
NF A 50 451	1200 (A4)	France	Sheet, plate, and strip	99 min **Al**, 0.05 **Cu**, 0.05 **Mn**, 0.10 **Zn**, 0.05 **Ti**, 1.0 **Si+Fe**, 0.05 each, 0.15 total, others	**Annealed**: 0.4–3.2 mm **diam**, 75 MPa **TS**, 30% in 2 in.; **3/4 hard**: 0.4–3.2 mm **diam**, 130 MPa **TS**, 110 MPa **YS**, 4% in 2 in. (50 mm) **El**; **Hard**: 0.4–3.2 mm **diam**, 150 MPa **TS**, 130 MPa **YS**, 3% in 2 in. (50 mm) **El**
NF A 50 451	1100 (A 45)	France	Sheet, plate, and strip	99.00 min **Al**, 0.05–0.20 **Cu**, 0.05 **Mn**, 0.10 **Zn**, 1.0 **Si+Fe**, 0.05 each, 0.15 total, others	**Annealed**: 0.4–3.2 mm **diam**, 75 MPa **TS**, 30% in 2 in. 50 mm **El**; **3/4 hard**: 0.4–3.2 mm **diam**, 130 MPa **TS**, 110 MPa **YS**, 4% in 2 in. (50 mm) **El**; **Hard**: 0.4–3.2 mm **diam**, 150 MPa **TS**, 130 MPa **YS**, 3% in 2 in. (50 mm) **El**
IS 739	G1C	India	Wire	99 min **Al**, 0.5 **Si**, 0.10 **Cu**, 0.1 **Mn**, 0.7 **Fe**, 1.0 **Cu+Si+Fe+Mn+Zn**	**Annealed**: all **diam**, 98 max MPa **TS**; **Annealed and cold worked to hard**: all **diam**, 137 MPa **TS**
IS 5909	19000	India	Sheet, strip	99.0 min **Al**, 0.5 **Si**, 0.10 **Cu**, 0.1 **Mn**, 0.1 **Mg**, 0.1 **Ni**, 0.10 **Zn**, 0.15 **Ti**, 0.05 **Sn**, 0.7 **Fe**	**Annealed**: 30 mm **diam**, 79 MPa **TS**, 30% in 2 in. (50 mm) **El**; **Half hard**: 7 mm **diam**, 108 MPa **TS**, 7% in 2 in. (50 mm) **El**; **Hard**: 3 mm **diam**, 137 MPa **TS**, 3% in 2 in. (50 mm) **El**
IS 7882		India	Sheet, strip	99.0 **Al**, 0.5 **Si**, 0.10 **Cu**, 0.1 **Mn**, 0.1 **Mg**, 0.1 **Ni**, 0.10 **Zn**, 0.15 **Ti**, 0.05 **Sn** 0.7 **Fe**, 0.05 **Pb**	**Annealed**: 20 mm **diam**, 70–105 MPa **TS**; **Half hard**: 3 mm **diam**, 110–140 MPa **TS**; **Hard**: 1 mm **diam**, 150 MPa **TS**
IS 738	T1C	India	Tube: drawn	99 min **Al**, 0.5 **Si**, 0.10 **Cu**, 0.1 **Mn**, 0.7 **Fe**, 1.0 **Cu+Si+Fe+Mn+Zn**	**Annealed**: all **diam**, 98 max MPa **TS**; **Annealed and cold worked to hard**: ≤80.0 mm **diam**, 118 MPa **TS**
IS 736	19000	India	Plate	99.0 min **Al**, 0.5 **Si**, 0.1 **Cu**, 0.1 **Mn**, 0.2 **Mg**, 0.1 **Zn**, 0.7 **Fe**, 1.0 **Cu+Mg+Si+Fe+Mn+Zn**	**As manufactured**: 28 mm **diam**, 70 MPa **TS**, 28% in 2 in. (50 mm) **El**; **Annealed**: 28 mm **diam**, 70 MPa **TS**, 28% in 2 in. (50 mm) **El**; **Half hard**: 7 mm **diam**, 110 MPa **TS**, 7% in 2 in. (50 mm) **El**

WROUGHT ALUMINUM AND ALUMINUM ALLOYS

specification number	designation	country	product forms	chemical composition	mechanical properties and hardness values
IS 733	19000	India	Bar, rod, section	99.0 min **Al**, 0.5 **Si**, 0.1 **Cu**, 0.1 **Mn**, 0.2 **Mg**, 0.1 **Zn**, 0.7 **Fe**, 1.0 **Cu+Mg+Si+Fe+Mn+Zn**	**As manufactured**: all **diam**, 65 MPa **TS**, 18% in 2 in. (50 mm) **El**
IS 5902	19000	India	Wire, bar	99.0 min **Al**, 0.5 **Si**, 0.10 **Cu**, 0.1 **Mn**, 0.1 **Mg**, 0.1 **Ni**, 0.1 **Zn**, 0.1 **Ti**, 0.05 **Sn**, 0.7 **Fe**, 0.05 **Pb**	**As drawn**: 108 MPa **TS**
ISO R827	Al99.0	ISO	Extruded products	99.0 min **Al**, 0.5 **Si**, 0.10 **Cu**, 0.1 **Mn**, 0.1 **Zn**, 0.8 **Fe**, 1.0 **Cu+Si+Fe+Mn+Zn**, total	**As manufactured**: all **diam**, 59 MPa **TS**, 18% in 2 in. (50 mm) **El**
ISO R827	Al99.0Cu	ISO	Extruded products	99.0 min **Al**, 0.5 **Si**, 0.05–0.20 **Cu**, 0.1 **Mn**, 0.1 **Zn**, 0.8 **Fe**, 1.0 **Cu+Si+Fe+Mn+Zn**, total	**As manufactured**: all **diam**, 59 MPa **TS**, 18% in 2 in. (50 mm) **El**
ISO R209	Al99.0Cu	ISO	Wrought products	99.0 min **Al**, 0.5 **Si**, 0.05–0.20 **Cu**, 0.1 **Mn**, 0.1 **Zn**, 0.8 **Fe**, 1.0 **Cu+Si+Fe+Mn+Zn**, total	
ISO R209	Al99.0	ISO	Wrought products	99.0 min **Al**, 0.5 **Si**, 0.10 **Cu**, 0.1 **Mn**, 0.1 **Zn**, 0.8 **Fe**, 1.0 **Cu+Si+Fe+Mn+Zn**, total	
ISO TR2778	Al99.0	ISO	Tube: drawn	99.0 min **Al**, 0.5 **Si**, 0.10 **Cu**, 0.1 **Mn**, 0.1 **Zn**, 0.8 **Fe**, 1.0 **Cu+Fe+Si+Mn+Zn**, total	**Annealed**: all **diam**, 70 MPa **TS**, 20% in 2 in. (50 mm) **El**; **Strain hardened**: all **diam**, 105 MPa **TS**, 3% in 2 in. (50 mm) **El**
ISO TR2778	Al99.0Cu	ISO	Tube: drawn	99.0 min **Al**, 0.5 **Si**, 0.5–0.20 **Cu**, 0.1 **Mn**, 0.1 **Zn**, 0.8 **Fe**, 1.0 **Cu+Si+Fe+Mn+Zn**, total	**Annealed**: all **diam**, 70 MPa **TS**, 20% in 2 in. (50 mm) **El**; **Strain hardened**: all **diam**, 110 MPa **TS**, 3% in 2 in. (50 mm) **El**
ISO TR2136	Al99.0	ISO	Sheet, plate	99.0 min **Al**, 0.5 **Si**, 0.10 **Cu**, 0.1 **Mn**, 0.1 **Zn**, 0.8 **Fe**, 1.0 **Cu+Si+Fe+Mn+Zn**, total	**Annealed**: all **diam**, 70 MPa **TS**, 20% in 2 in. (50 mm) **El**; **HB–Strain hardened**: all **diam**, 90 MPa **TS**, 4% in 2 in. (50 mm) **El**; **HD–Strain hardened**: all **diam**, 105 MPa **TS**, 3% in 2 in. (50 mm) **El**
ISO TR2136	Al99.0Cu	ISO	Sheet, plate	99.0 min **Al**, 0.5 **Si**, 0.05–0.20 **Cu**, 0.1 **Mn**, 0.1 **Zn**, 0.8 **Fe**, 1.0 **Cu+Si+Fe+Mn+Zn**, total	**Annealed**: all **diam**, 75 MPa **TS**, 20% in 2 in. (50 mm) **El**; **HB–Strain hardened**: all **diam**, 95 MPa **TS**, 4% in 2 in. (50 mm) **El**; **HD–Strain hardened**: all **diam**, 110 MPa **TS**, 3% in 2 in. (50 mm) **El**
JIS Z 3232	1100	Japan	Rod and wire: welding and electrode	99.00 min **Al**, 0.05–0.20 **Cu**, 0.05 **Mn**, 0.10 **Zn**, 0.0008 **Be**, 1.0 **Si+Fe**	**Annealed**: all **diam**, 78 MPa **TS**
JIS Z 3232	1200	Japan	Rod and wire: welding and electrode	99.0 min **Al**, 0.05 **Cu**, 0.05 **Mn**, 0.10 **Zn**, 0.0008 **Be**, 1.0 **Si+Fe**	**Annealed**: all **diam**, 78 MPa **TS**

WROUGHT ALUMINUM AND ALUMINUM ALLOYS

specification number	designation	country	product forms	chemical composition	mechanical properties and hardness values
JIS H 4140	1100	Japan	Die forgings	99.00 min **Al**, 0.05–0.20 **Cu**, 0.05 **Mn**, 0.10 **Zn**, 1.0 **Si+Fe**, 0.15 total, others	**As manufactured**: ≤100 mm **diam**, 78 MPa **TS**, 29 MPa **YS**, 25% in 2 in. (50 mm) **El**
JIS H 4140	1200	Japan	Die forgings	99.00 min **Al**, 0.05 **Cu**, 0.05 **Mn**, 0.10 **Zn**, 1.0 **Si+Fe**, 0.15 total, others	**As manufactured**: ≤100 mm **diam**, 78 MPa **TS**, 29 MPa **YS**, 25% in 2 in. (50 mm) **El**
JIS H 4120	1100	Japan	Wire and rod: rivet	99.00 min **Al**, 0.05–0.20 **Cu**, 0.05 **Mn**, 0.10 **Zn**, 1.0 **Si+Fe**, 0.15, total, others	**1/2 hard**: ≤25 mm **diam**, 108 MPa **TS**
JIS H 4120	1200	Japan	Wire and rod: rivet	99.00 min **Al**, 0.05 **Cu**, 0.05 **Mn**, 0.10 **Zn**, 1.0 **Si+Fe**, 0.15 total, others	**1/2 hard**: ≤25 mm **diam**, 108 MPa **TS**
JIS H 4100	1100	Japan	Shapes: extruded	99.00 min **Al**, 0.05–0.20 **Cu**, 0.05 **Mn**, 0.10 **Zn**, 1.0 **Fe**, 1.0 **Si+Fe**, 0.15 total, others	**As manufactured**: all **diam**, 78 MPa **TS**, 20 MPa **YS**
JIS H 4100	1200	Japan	Shapes: extruded	99.00 min **Al**, 0.05 **Cu**, 0.05 **Mn**, 0.10 **Zn**, 1.0 **Si+Fe**, 0.15 total, others	**As manufactured**: all **diam**, 78 MPa **TS**, 20 MPa **YS**
JIS H 4090	1100	Japan	Pipe and tube: welded	99.00 min **Al**, 0.05–0.20 **Cu**, 0.05 **Mn**, 0.10 **Zn**, 1.0 **Si+Fe**, 0.15 total, others	**Annealed**: <1.3 mm **diam**, 78 MPa **TS**, 29 MPa **YS**, 25% in 2 in. (50 mm) **El**; **1/2 hard**: <1.3 mm **diam**, 118 MPa **TS**, 98 MPa **YS**, 4% in 2 in. (50 mm) **El**; **Hard**: <0.8 mm **diam**, 157 MPa **TS**, 2% in 2 in. (50 mm) **El**
JIS H 4090	1200	Japan	Pipe and tube: welded	99.00 min **Al**, 0.05 **Cu**, 0.05 **Mn**, 0.10 **Zn**, 1.00 **Si+Fe**, 0.15 total, others	**Annealed**: <1.3 mm **diam**, 78 MPa **TS**, 29 MPa **YS**, 25% in 2 in. (50 mm) **El**; **1/2 hard**: <1.3 mm **diam**, 118 MPa **TS**, 98 MPa **YS**, 4% in 2 in. (50 mm) **El**; **Hard**: <0.8 mm **diam**, 157 MPa **TS**, 2% in 2 in. (50 mm) **El**
JIS H 4080	1100	Japan	Tube: extruded	99.00 min **Al**, 0.05–0.20 **Cu**, 0.05 **Mn**, 0.10 **Zn**, 1.0 **Si+Fe**, 0.15 total, others	**As manufactured**: all **diam**, 78 MPa **TS**, 20 MPa **YS**, 25% in 2 in. (50 mm) **El**
JIS H 4080	1200	Japan	Tube: extruded	99.00 min **Al**, 0.05 **Cu**, 0.05 **Mn**, 0.10 **Zn**, 0.45 **Si+Fe**, 0.15 total, others	**As manufactured**: all **diam**, 78 MPa **TS**, 20 MPa **YS**, 25% in 2 in. (50 mm) **El**
JIS H 4040	1100	Japan	Bar: extruded	99.00 min **Al**, 0.05–0.20 **Cu**, 0.05 **Mn**, 0.10 **Zn**, 1.0 **Si+Fe**, 0.15 total, others	**As manufactured**: all **diam**, 78 MPa **TS**, 20 MPa **YS**
JIS H 4040	1200	Japan	Bar: extruded	99.00 min **Al**, 0.05 **Cu**, 0.05 **Mn**, 0.10 **Zn**, 1.0 **Si+Fe**, 0.15 total, others	**As manufactured**: all **diam**, 78 MPa **TS**, 20 MPa **YS**

WROUGHT ALUMINUM AND ALUMINUM ALLOYS

specification number	designation	country	product forms	chemical composition	mechanical properties and hardness values
JIS H 4000	1100	Japan	Sheet, plate, and strip	99.00 min Al, 0.05–0.20 Cu, 0.05 Mn, 0.10 Zn, 1.0 Si+Fe, 0.15 total, others	Annealed: <1.3 mm diam, 78 MPa TS, 29 MPa YS, 25% in 2 in. (50 mm) El; Hard: <0.3 mm diam, 157 MPa TS, 1% in 2 in. (50 mm) El; Hot rolled: <12 mm diam, 88 MPa TS, 49 MPa YS, 9% in 2 in. (50 mm) El
JIS H 4000	1200	Japan	Sheet, plate, and strip	99.00 min Al, 0.05 Cu, 0.05 Mn, 0.10 Zn, 1.0 Si+Fe, 0.15 total, others	Annealed: <1.3 mm diam, 78 MPa TS, 29 MPa YS, 25% in 2 in. (50 mm) El; Hard: <0.3 mm diam, 157 MPa TS, 1% in 2 in. (50 mm) El; Hot rolled: <12 mm diam, 88 MPa TS, 49 MPa YS, 9% in 2 in. (50 mm) El
DGN–W–30	2S–0	Mexico	Sheet	99 min Al	Rolled: 0.25–0.52 mm diam, 93 MPa TS, 38 MPa YS, 15% in 2 in. (50 mm) El
DGN–W–30	2S–H12	Mexico	Sheet	99 min Al	Rolled: 0.53–0.80 mm diam, 103 MPa TS, 90 MPa YS, 3% in 2 in. (50 mm) El
DGN–W–30	2S–H14	Mexico	Sheet	99 min Al	Rolled: 0.53–0.82 mm diam, 117 MPa TS, 110 MPa YS, 3% in 2 in. (50 mm) El
DGN–W–30	2S–H14	Mexico	Sheet	99 min Al	Rolled: 0.53–0.82 mm diam, 117 MPa TS, 110 MPa YS, 3% in 2 in. (50 mm) El
DGN–W–30	2S–H16	Mexico	Sheet	99 min Al	Rolled: 0.25–0.52 mm diam, 138 MPa TS, 131 MPa YS, 1% in 2 in. (50 mm) El
DGN–W–30	2S–H18	Mexico	Sheet	99 min Al	Rolled: 0.25–0.52 mm diam, 165 MPa TS, 159 MPa YS, 1% in 2 in. (50 mm) El
NS 17 010	NS 17 010–10	Norway	Rod	99.0 min Al, 0.3 Si, 0.05 Cu, 0.05 Mn, 0.10 Zn, 0.4 Fe, 0.5 Cu+Fe+Mn+Si+Zn	Precision drawn: 5–75 mm diam, 78 MPa TS, 49 MPa YS, 20% in 2 in. (50 mm) El
NS 17 005	NS 17 005–02	Norway	Sheet, strip, plate, tube, wire, and rod	99.0 min Al, 0.5 Si, 0.10 Cu, 0.1 Mn, 0.1 Zn, 0.8 Fe, 1.0 Cu+Fe+Mn+Si+Zn	Soft annealed (sheet, strip, plate): ≥0.5 mm diam, 69 MPa TS, 32% in 2 in. (50 mm) El; Soft annealed (tube and rod): 5–50 mm diam, 68 MPa TS, 30% in 2 in. (50 mm) El; Soft annealed (wire): all diam, 69 MPa TS
NS 17 005	NS 17 005–14	Norway	Sheet, strip, tube, and rod	99.0 min Al, 0.5 Si, 0.10 Cu, 0.1 Mn, 0.1 Zn, 0.8 Fe, 1.0 Cu+Fe+Mn+Si+Zn	1/2 hard (sheet and strip): <0.5 mm diam, 98 MPa TS, 88 MPa YS, 5% in 2 in. (50 mm) El; 1/2 hard (tube): <6 mm diam, 93 MPa TS, 78 MPa YS, 6% in 2 in. (50 mm) El; 1/2 hard (rod): 5–15 mm diam, 98 MPa TS, 78 MPa YS, 6% in 2 in. (50 mm) El

WROUGHT ALUMINUM AND ALUMINUM ALLOYS

specification number	designation	country	product forms	chemical composition	mechanical properties and hardness values
NS 17 005	NS 17 005–18	Norway	Sheet, strip, tube, and wire	99.0 min **Al**, 0.5 **Si**, 0.10 **Cu**, 0.1 **Mn**, 0.1 **Zn**, 0.8 **Fe**, 1.0 **Cu+Fe+Mn+Si+Zn**	**Hard (sheet and strip)**: <0.5 mm **diam**, 137 MPa **TS**, 117 MPa **YS**, 3% in 2 in. (50 mm) **El**; **Hard (tube)**: <1.5 mm **diam**, 123 MPa **TS**, 108 MPa **YS**, 4% in 2 in. (50 mm) **El**; **Hard (wire)**: 1–5 mm **diam**, 162 MPa **TS**, 147 MPa **YS**
NS 17 005	NS 17 005–08	Norway	Tube, wire, and rod	99.0 min **Al**, 0.5 **Si**, 0.10 **Cu**, 0.1 **Mn**, 0.1 **Zn**, 0.8 **Fe**, 1.0 **Cu+Fe+Mn+Si+Zn**	**Extruded**: all **diam**, 68 MPa **TS**, 30% in 2 in. (50 mm) **El**
NS 17 005	NS 17 005–10	Norway	Rod	99.0 min **Al**, 0.5 **Si**, 0.10 **Cu**, 0.1 **Mn**, 0.1 **Zn**, 0.8 **Fe**, 1.0 **Cu+Fe+Mn+Si+Zn**	**Precision drawn**: 5–75 mm **diam**, 78 MPa **TS**, 49 MPa **YS**, 20% in 2 in. (50 mm) **El**
NS 17 010	NS 17 010–02	Norway	Sheet, strip, plate, rod, tube, and wire	99.0 min **Al**, 0.3 **Si**, 0.05 **Cu**, 0.05 **Mn**, 0.10 **Zn**, 0.4 **Fe**, 0.5 **Cu+Fe+Mn+Si+Zn**	**Soft annealed (sheet, strip, plate)**: ≥0.5 mm **diam**, 98 MPa **TS**, 32% in 2 in. (50 mm) **El**; **Soft annealed (rod and wire)**: <5 mm **diam**, 64 MPa **TS**; **Soft annealed (tube)**: all **diam**, 30% in 2 in. (50 mm) **El**
NS 17 010	NS 17 010–18	Norway	Sheet, strip, tube, and wire	99.0 min **Al**, 0.3 **Si**, 0.05 **Cu**, 0.05 **Mn**, 0.10 **Zn**, 0.4 **Fe**, 0.5 **Cu+Fe+Mn+Si+Zn**	**Hard (sheet and strip)**: <0.5 mm **diam**, 127 MPa **TS**, 137 MPa **YS**, 3% in 2 in. (50 mm) **El**; **Hard (tube)**: <1.5 mm **diam**, 123 MPa **TS**, 108 MPa **YS**, 4% in 2 in. (50 mm) **El**; **Hard (wire)**: 1–5 mm **diam**, 162 MPa **TS**, 150 MPa **YS**
NS 17 010	NS 17 010–14	Norway	Sheet, strip, rod, and tube	99.0 min **Al**, 0.3 **Si**, 0.05 **Cu**, 0.05 **Mn**, 0.10 **Zn**, 0.4 **Fe**, 0.5 **Cu+Fe+Mn+Si+Zn**	**1/2 hard (sheet and strip)**: >0.5 mm **diam**, 98 MPa **TS**, 88 MPa **YS**, 5% in 2 in. (50 mm) **El**; **1/2 hard (rod)**: 5–15 mm **diam**, 98 MPa **TS**, 78 MPa **YS**, 6% in. (50 mm) **El**; **1/2 hard (tube)**: <6 mm **diam**, 93 MPa **TS**, 78 MPa **YS**, 6% in 2 in. (50 mm) **El**
NS 17 010	NS 17 010–08	Norway	Rod, tube, and wire	99.0 min **Al**, 0.3 **Si**, 0.05 **Cu**, 0.05 **Mn**, 0.10 **Zn**, 0.4 **Fe**, 0.5 **Cu+Fe+Mn+Si+Zn**	**Extruded**: all **diam**, 69 MPa **TS**, 30% in 2 in. (50 mm) **El**
QQ-A-1876	Grade A	US	Foil	98.6 min **Al**, 0.1 **Pb**, 0.1 **As**, 0.1 **Ca**	
AS 1734	8011	Australia	Sheet, plate	97.46 min **Al**, 0.50–0.90 **Si**, 0.10 **Cu**, 0.10 **Mn**, 0.05 **Mg**, 0.05 **Cr**, 0.10 **Zn**, 0.08 **Ti**, 0.6–1.0 **Fe**, 0.05 each, 0.15 total, others	

WROUGHT ALUMINUM AND ALUMINUM ALLOYS

specification number	designation	country	product forms	chemical composition	mechanical properties and hardness values
ANSI/ASTM B 313	1100	US	Tube: welded	0.05–0.20 **Cu**, 0.05 **Mn**, 0.10 **Zn**, 1.0 **Fe+Si**, 0.05 each, 0.15 total, others	**Annealed**: 0.051–0.125 in. **diam**, 11 ksi (76 MPa) **TS**, 4 ksi (28 MPa) **YS**, 30% in 2 in. (50 mm) **El**; **H12**: 0.051–0.113 in. **diam**, 14 ksi (97 MPa) **TS**, 11 ksi (76 MPa) **YS**, 8% in 2 in. (50 mm) **El**; **H14**: 0.051–0.113 in. **diam**, 16 ksi (110 MPa) **TS**, 14 ksi (97 MPa) **YS**, 5% in 2 in. (50 mm) **El**
AMS 4068		US	Tube: seamless	rem **Al**, 0.20 **Si**, 5.8–6.8 **Cu**, 0.20–0.40 **Mn**, 0.02 **Mg**, 0.10 **Zn**, 0.02–0.10 **Ti**, 0.30 **Fe**, 0.05–0.15 **V**, 0.10–0.25 **Zr**, 0.05 each, 0.15 total, others	**Solution heat treated and stress relieved**: 0.029–0.049 in. (0.74–1.24 mm) **diam**, 45 ksi (310 MPa) **TS**, 26 ksi (179 MPa) **YS**, 12% in 2 in. (50 mm) **El**; **Precipitation heat treated**: 0.029–0.049 in. (0.74–1.24 mm) **diam**, 60 ksi (414 MPa) **TS**, 42 ksi (290 MPa) **YS**, 6% in 2 in. (50 mm) **El**
AMS 4066		US	Tube: seamless	rem **Al**, 0.20 **Si**, 5.8–6.8 **Cu**, 0.20–0.40 **Mn**, 0.02 **Mg**, 0.10 **Zn**, 0.02–0.10 **Ti**, 0.30 **Fe**, 0.05–0.15 **V**, 0.10–0.25 **Zr**, 0.05 each, 0.15 total, others	**Solution and precipitation heat treated**: 0.029–0.049 in. (0.74–1.24 mm) **diam**, 60 ksi (414 MPa) **TS**, 42 ksi (290 MPa) **YS**, 6% in 2 in. (50 mm) **El**
AMS 4094		US	Sheet and plate: alclad	rem **Al**, 0.20 **Si**, 5.8–6.8 **Cu**, 0.20–0.40 **Mn**, 0.02 **Mg**, 0.10 **Zn**, 0.02–0.10 **Ti**, 0.30 **Fe**, 0.05–0.15 **V**, 0.10–0.25 **Zr**, 0.05 each, 0.15 total, others	**Solution and precipitation heat treated**: 0.020–0.039 in. **diam**, 49 ksi (338 MPa) **TS**, 37 ksi (255 MPa) **YS**, 6% in 2 in. (50 mm) **El**
AMS 4095		US	Sheet and plate: alclad	rem **Al**, 0.20 **Si**, 5.8–6.8 **Cu**, 0.20–0.40 **Mn**, 0.02 **Mg**, 0.10 **Zn**, 0.02–0.10 **Ti**, 0.30 **Fe**, 0.05–0.15 **V**, 0.10–0.25 **Zr**, 0.05 o each, 0.15 total, others	**Solution heat treated and cold worked**: 0.040–0.099 in. **diam**, 42 ksi (290 MPa) **TS**, 25 ksi (172 MPa) **YS**, 10% in 2 in. (50 mm) **El**; **Precipitation heat treated**: 0.020–0.039 in. **diam**, 49 ksi (338 MPa) **TS**, 37 ksi (255 MPa) **YS**, 6% in 2 in. (50 mm) **El**
AMS 4096		US	Sheet and plate: alclad	rem **Al**, 0.20 **Si**, 5.8–6.8 **Cu**, 0.20–0.40 **Mn**, 0.02 **Mg**, 0.10 **Zn**, 0.02–0.10 **Ti**, 0.30 **Fe**, 0.05–0.15 **V**, 0.10–0.25 **Zr**, 0.05 each, 0.15 total, others	**Annealed**: 0.020–2.000 in. **diam**, 32 ksi (221 MPa) **TS**, 16 ksi (110 MPa) **YS**, 12% in 2 in. (50 mm) **El**; **Solution and precipitation heat treated**: 0.020–0.039 in. **diam**, 44 ksi (303 MPa) **TS**, 29 ksi (200 MPa) **YS**, 6% in **El**
AMS 4031B		US	Sheet and plate	rem **Al**, 0.20 **Si**, 5.8–6.8 **Cu**, 0.20–0.40 **Mn**, 0.02 **Mg**, 0.10 **Zn**, 0.02–0.10 **Ti**, 0.30 **Fe**, 0.05–0.15 **V**, 0.10–0.25 **Zr**, 0.05 each, 0.15 total, others	**Annealed**: 0.020–2.000 in. **diam**, 32 ksi (221 MPa) **TS**, 16 ksi (110 MPa) **YS**, 12% in 2 in. (50 mm) **El**; **Solution and precipitation heat treated**: 0.020–0.039 in. **diam**, 54 ksi (372 MPa) **TS**, 36 ksi (248 MPa) **YS**, 6% in 2 in. (50 mm) **El**

WROUGHT ALUMINUM AND ALUMINUM ALLOYS

specification number	designation	country	product forms	chemical composition	mechanical properties and hardness values
AMS 4162A		US	Bar, tube, rod, wire, and shapes: extruded	rem **Al**, 0.20 **Si**, 5.8–6.8 **Cu**, 0.20–0.40 **Mn**, 0.02 **Mg**, 0.10 **Zn**, 0.02–0.10 **Ti**, 0.30 **Fe**, 0.05–0.15 **V**, 0.10–0.35 **Zr**, 0.05 each, 0.15 total, others	**Solution and precipitation heat treated:** <3.00 in. (<76.2 mm) **diam**, 58 ksi (400 MPa) **TS**, 42 ksi (290 MPa) **YS**, 6% in 2 in. (50 mm) **El**, 115 min **HB**
AMS 4163A		US	Bar, tube, rod, wire, and shapes: extruded	rem **Al**, 0.20 **Si**, 5.8–6.8 **Cu**, 0.20–0.40 **Mn**, 0.02 **Mg**, 0.10 **Zn**, 0.02–0.10 **Ti**, 0.30 **Fe**, 0.05–0.15 **V**, 0.10–0.25 **Zr**, 0.05 each, 0.15 total, others	**Solution heat treated:** ≤0.499 in. (≤12.67 mm) **diam**, 42 ksi (290 MPa) **TS**, 26 ksi (179 MPa) **YS**, 14% in 2 in. (50 mm) **El**, 77 min **HB**; **Precipitation heat treated:** 58 ksi (400 MPa) **TS**, 42 ksi (290 MPa) **YS**, 6% in 2 in. (50 mm) **El**
AMS 4313		US	Rings: rolled or forged	rem **Al**, 0.20 **Si**, 5.8–6.8 **Cu**, 0.20–0.40 **Mn**, 0.02 **Mg**, 0.10 **Zn**, 0.02–0.10 **Ti**, 0.30 **Fe**, 0.05–0.15 **V**, 0.10–0.25 **Zr**, 0.05 each, 0.15 total, others	**Solution heat treated:** all **diam**, 77 min **HB**; **Precipitation heat treated:** ≤3 in. (≤76 mm) **diam**, 60 ksi (414 MPa) **TS**, 48 ksi (331 MPa) **YS**, 6% in 2 in. (50 mm) **El**, 115 min **HB**
AMS 4143A		US	Die and hand forgings and rings	rem **Al**, 0.20 **Si**, 5.8–6.8 **Cu**, 0.20–0.40 **Mn**, 0.02 **Mg**, 0.10 **Zn**, 0.02–0.10 **Ti**, 0.30 **Fe**, 0.05–0.15 **V**, 0.10–0.25 **Zr**, 0.05 each, 0.15 total, others	**Solution and precipitation heat treated (die forgings):** ≤4 in. (≤102 mm) **diam**, 58 ksi (400 MPa) **TS**, 38 ksi (262 MPa) **YS**, 8% in 2 in. (50 mm) **El**, 110 min **HB**; **Solution and precipitation heat treated (rings):** ≤2.5 in. (64 mm) **diam**, 56 ksi (386 MPa) **TS**, 40 ksi (276 MPa) **YS**, 6% in 2 in. (50 mm) **El**, 110 min **HB**; **Solution and precipitation heat treated (hand forgings):** ≤4 in. (≤102 mm) **diam**, 58 ksi (400 MPa) **TS**, 40 ksi (276 MPa) **YS**, 6% in 2 in. (50 mm) **El**, 110 min **HB**
ANSI/ASTM B 316	2219	US	Wire and rod: rivet and cold-heading	rem **Al**, 0.20 **Si**, 5.8–6.8 **Cu**, 0.20–0.40 **Mn**, 0.02 **Mg**, 0.10 **Zn**, 0.02–0.10 **Ti**, 0.30 **Fe**, 0.05–0.15 **V**, 0.10–0.25 **Zr**, 0.05 each, 0.15 total, others	**Solution heat treated and naturally aged:** ≤1.000 in. **diam**, 45 ksi (310 MPa) **TS**; **Solution heat treated and artificially aged:** 0.061–1.000 in. **diam**, 55 ksi (379 MPa) **TS**, 35 ksi (241 MPa) **YS**, 6% in 2 in. (50 mm) **El**

WROUGHT ALUMINUM AND ALUMINUM ALLOYS

specification number	designation	country	product forms	chemical composition	mechanical properties and hardness values
ANSI/ASTM B 247	2219	US	Die and hand forgings	rem **Al**, 0.20 **Si**, 5.8–6.8 **Cu**, 0.20–0.40 **Mn**, 0.02 **Mg**, 0.10 **Zn**, 0.02–0.10 **Ti**, 0.30 **Fe**, 0.05–0.15 **V**, 0.10–0.25 **Zr**, 0.05 each, 0.15 total, others	**Solution heat treated and artificially aged (die forgings)**: ≤4.000 in. **diam**, 58 ksi (400 MPa) **TS**, 38 ksi (262 MPa) **YS**, 8% in 2 in. (50 mm) **El**, 100 min **HB**; **Solution heat treated and artificially aged (hand forgings)**: ≤4.000 in. **diam**, 58 ksi (400 MPa) **TS**, 40 ksi (276 MPa) **YS**, 6% in 2 in. (50 mm) **El**; **T852 (hand forgings)**: ≤4.000 in. **diam**, 62 ksi (427 MPa) **TS**, 50 ksi (345 MPa) **YS**, 6% in 2 in. (50 mm) **El**
ANSI/ASTM B 241	2219	US	Tube: seamless	rem **Al**, 0.20 **Si**, 5.8–6.8 **Cu**, 0.20–0.40 **Mn**, 0.02 **Mg**, 0.10 **Zn**, 0.02–0.10 **Ti**, 0.30 **Fe**, 0.05–0.15 **V**, 0.10–0.25 **Zr**, 0.05 each, 0.15 total, others	**Annealed**: all **diam**, 32 max ksi (221 max MPa) **TS**, 18 max ksi (124 max MPa) **YS**, 12% in 2 in. (50 mm) **El**; **T31**: ≤0.499 in. **diam**, 42 ksi (290 MPa) **TS**, 26 ksi (179 MPa) **YS**, 14% in 2 in. (50 mm) **El**; **Solution heat treated, cold worked, and artificially aged**: ≤2.999 in. **diam**, 58 ksi (400 MPa) **TS**, 42 ksi (290 MPa) **YS**, 6% in 2 in. (50 mm) **El**
ANSI/ASTM B 221	2219	US	Bar, tube, rod, wire, and shapes	rem **Al**, 0.20 **Si**, 5.8–6.8 **Cu**, 0.20–0.40 **Mn**, 0.02 **Mg**, 0.10 **Zn**, 0.02–0.10 **Ti**, 0.30 **Fe**, 0.05–0.15 **V**, 0.10–0.25 **Zr**, 0.05 each, 0.15 total, others	**Annealed**: all **diam**, 32 max ksi (221 max MPa) **TS**, 18 max ksi (124 max MPa) **YS**, 12% in 2 in. (50 mm) **El**; **T31**: 0.500–2.999 in. **diam**, 45 ksi (310 MPa) **TS**, 27 ksi (186 MPa) **YS**, 14% in 2 in. (50 mm) **El**; **T81**: ≤2.999 in. **diam**, 58 ksi (400 MPa) **TS**, 42 ksi (290 MPa) **YS**, 6% in 2 in. (50 mm) **El**
ANSI/ASTM B 211	2219	US	Bar, rod, and wire	rem **Al**, 0.20 **Si**, 5.8–6.8 **Cu**, 0.20–0.40 **Mn**, 0.02 **Mg**, 0.10 **Zn**, 0.02–0.10 **Ti**, 0.30 **Fe**, 0.05–0.15 **V**, 0.10–0.25 **Zr**	**T851**: 2.001–4.000 in. **diam**, 57 ksi (393 MPa) **TS**, 39 ksi (269 MPa) **YS**, 4% in 2 in. (50 mm) **El**
ANSI/ASTM B 209	Alclad 2219	US	Sheet and plate	rem **Al**, 0.20 **Si**, 5.8–6.8 **Cu**, 0.20–0.40 **Mn**, 0.02 **Mg**, 0.10 **Zn**, 0.02–0.10 **Ti**, 0.30 **Fe**, 0.05–0.15 **V**, 0.10–0.25 **Zr**, 7072 cladding material, 0.05 each, 0.15 total, others	**Annealed**: 0.020–2.000 in. **diam**, 32 max ksi (221 max MPa) **TS**, 16 max ksi (110 max MPa) **YS**, 12% in 2 in. (50 mm) **El**; **T81**: 0.100–0.249 in. **diam**, 58 ksi (400 MPa) **TS**, 43 ksi (296 MPa) **YS**, 7% in 2 in. (50 mm) **El**; **T87**: 0.250–0.499 in. **diam**, 60 ksi (414 MPa) **TS**, 48 ksi (331 MPa) **YS**, 7% in 2 in. (50 mm) **El**

WROUGHT ALUMINUM AND ALUMINUM ALLOYS

specification number	designation	country	product forms	chemical composition	mechanical properties and hardness values
ANSI/ASTM B 209	2219	US	Sheet and plate	rem **Al**, 0.20 **Si**, 5.8–6.8 **Cu**, 0.20–0.40 **Mn**, 0.02 **Mg**, 0.10 **Zn**, 0.02–0.10 **Ti**, 0.30 **Fe**, 0.05–0.15 **V**, 0.10–0.25 **Zr**, 0.05 each, 0.15 total, others	**Annealed**: 0.020–2.000 in. **diam**, 32 max ksi (221 max MPa) **TS**, 16 max ksi (110 max MPa) **YS**, 12% in 2 in. (50 mm) **El**; **T81 (sheet)**: 0.040–0.249 in. **diam**, 62 ksi (427 MPa) **TS**, 46 ksi (317 MPa) **YS**, 7% in 2 in. (50 mm) **El**; **T87**: 4.001–5.000 in. **diam**, 61 ksi (421 MPa) **TS**, 49 ksi (338 MPa) **YS**, 3% in 2 in. (50 mm) **El**
QQ-A-430B	2219	US	Rod, wire	0.20 **Si**, 5.8–6.8 **Cu**, 0.20–0.40 **Mn**, 0.02 **Mg**, 0.10 **Zn**, 0.02–0.10 **Ti**, 0.30 **Fe**, 0.05–0.15 **V**, 0.10–0.25 **Zr**, 0.05 each, 0.15 total, others (includes **V**+**Zr**)	**Solution heat treated and naturally aged**: 1.000 max in. **diam**, 45 ksi (310 MPa) **TS**; **Solution heat treated and artificially aged**: 0.063–1.000 in. **diam**, 55 ksi (379 MPa) **TS**, 35 ksi (242 MPa) **YS**, 6% in 2 in. (50 mm) **El**
QQ-A-250/30	2219	US	Plate, sheet	rem **Al**, 0.20 **Si**, 5.8–6.8 **Cu**, 0.20–0.40 **Mn**, 0.02 **Mg**, 0.10 **Zn**, 0.02–0.10 **Ti**, 0.30 **Fe**, 0.05–0.15 **V**, 0.10–0.25 **Zr**, 0.05 each, 0.15 total, others	**Annealed**: 0.020–2.000 in. **diam**, 32 max ksi (221 max MPa) **TS**, 16 max ksi (110 max MPa) **YS**, 12% in 2 in. (50 mm) **El**; **Re-solution heat treatment, artificial aging and temper T31**: 0.020–0.039 in. **diam**, 46 ksi (317 MPa) **TS**, 29 ksi (200 MPa) **YS**, 8% in 2 in. (50 mm) **El**; **Re-solution heat treatment, aging and temper T37**: 0.020–0.039 in. **diam**, 49 ksi (338 MPa) **TS**, 38 ksi (262 MPa) **YS**, 6% in 2 in. (50 mm) **El**
MIL-A-46118B		US	Plate and shapes	rem **Al**, 0.20 **Si**, 5.8–6.8 **Cu**, 0.20–0.40 **Mn**, 0.02 **Mg**, 0.10 **Zn**, 0.02–0.10 **Ti**, 0.30 **Fe**, 0.05–0.15 **V**, 0.10–0.25 **Zr**, 0.05 each, 0.15 total, others	**As rolled (plate)**: 0.250–1.000 in. **diam**, 427 MPa **TS**, 317 MPa **YS**, 8% in 2 in. (50 mm) **El**; **Die forged (shapes)**: <4.0 in. **diam**, 427 MPa **TS**, 345 MPa **YS**, 6% in 2 in. (50 mm) **El**
MIL-A-22771C	2219	US	Forgings	rem **Al**, 0.20 **Si**, 5.8–6.8 **Cu**, 0.20–0.40 **Mn**, 0.02 **Mg**, 0.10 **Zn**, 0.02–0.10 **Ti**, 0.30 **Fe**, 0.05–0.15 **V**, 0.10–0.25 **Zr**, 0.05 total, others	**T6**: ≤4 in. **diam**, 400 MPa **TS**, 262 MPa **YS**, 8% in 2 in. (50 mm) **El**, 100 **HB**
MIL-A-46083B	2219	US	Shapes	rem **Al**, 0.20 **Si**, 5.8–6.8 **Cu**, 0.20–0.40 **Mn**, 0.02 **Mg**, 0.10 **Zn**, 0.02–0.10 **Ti**, 0.30 **Fe**, 0.05–0.15 **V**, 0.10–0.25 **Zr**, 0.05 each, 0.15 total, others	**Class II**: ≤2.000 in. **diam**, 393 MPa **TS**, 331 MPa **YS**, 8% in 2 in. (50 mm) **El**

WROUGHT ALUMINUM AND ALUMINUM ALLOYS

specification number	designation	country	product forms	chemical composition	mechanical properties and hardness values
ANSI/ASTM B 211	2011	US	Bar, rod, and wire	rem **Al**, 0.40 **Si**, 5.0–6.0 **Cu**, 0.30 **Zn**, 0.7 **Fe**, 0.20–0.6 **Pb**, 0.20–0.6 **Bi**, 0.05 each, 0.15 total, others	**Solution heat treated and cold worked**: 0.125–1.500 in. **diam**, 45 ksi (310 MPa) **TS**, 38 ksi (262 MPa) **YS**, 10% in 2 in. (50 mm) **El**; **Solution heat treated and naturally aged**: 0.125–8.000 in. **diam**, 40 ksi (276 MPa) **TS**, 18 ksi (24 MPa) **YS**, 16% in 2 in. (50 mm) **El**; **Solution heat treated, cold worked, and artificially aged**: 0.125–3.250 in. **diam**, 54 ksi (372 MPa) **TS**, 40 ksi (276 MPa) **YS**, 10% in 2 in. (50 mm) **El**
ANSI/ASTM B 210	2011	US	Tube: seamless	rem **Al**, 0.40 **Si**, 5.0–6.0 **Cu**, 0.30 **Zn**, 0.7 **Fe**, 0.20–0.6 **Pb**, 0.20–0.6 **Bi**, 0.05 each, 0.15 total, others	**T3**: 0.050–0.500 in. **diam**, 47 ksi (324 MPa) **TS**, 40 ksi (276 MPa) **YS**, 8% in 2 in. (50 mm) **El**; **T4511**: 0.050–0.500 in. **diam**, 44 ksi (303 MPa) **TS**, 23 ksi (159 MPa) **YS**, 18% in 2 in. (50 mm) **El**
AS 1866	2011	Australia	Rod, bar, shape	rem **Al**, 0.40 **Si**, 5.0–6.0 **Cu**, 0.30 **Zn**, 0.7 **Fe**, 0.05 each, 0.15 total, others	**Solution heat treated and naturally aged**: all **diam**, 275 MPa **TS**, 124 MPa **YS**, 14% in 2 in. (50 mm) **El**; **Solution heat treated and artificially aged**: 25.0 mm **diam**, 351 MPa **TS**, 220 MPa **YS**, 8% in 2 in. (50 mm) **El**
AS 1865	2011	Australia	Wire, rod, bar, strip	rem **Al**, 4.0 **Si**, 5.0–6.0 **Cu**, 0.30 **Zn**, 0.7 **Fe**, 0.05 each, 0.15 total, others	**Solution heat treated and cold worked**: 3–40 mm **diam**, 310 MPa **TS**, 262 MPa **YS**, 8% in 2 in. (50 mm) **El**; **Solution heat treated, cold worked, and artificially aged**: 3–75 mm **diam**, 358 MPa **TS**, 275 MPa **YS**, 8% in 2 in. (50 mm) **El**
COPANT 862	2011	COPANT	Wrought products	rem **Al**, 0.4 **Si**, 5–6 **Cu**, 0.3 **Zn**, 0.7 **Fe**, 0.2–0.6 **Bi**, 0.05 each, 0.15 total, others	
CSA HA.5	.2011	Canada	Bar, rod, wire, shapes	rem **Al**, 0.40 **Si**, 5.0–6.0 **Cu**, 0.30 **Zn**, 0.7 **Fe**, 0.05 each, 0.15 total, others	**T3**: 0.125–1.500 in. **diam**, 45.0 ksi (310 MPa) **TS**, 38.0 ksi (262 MPa) **YS**, 10% in 2 in. (50 mm) **El**; **T8**: 0.125–2.500 in. **diam**, 54.0 ksi (372 MPa) **TS**, 40.0 ksi (276 MPa) **YS**, 10% in 2 in. (50 mm) **El**
DS 3012	2011	Denmark	Extruded products	93 **Al**, 0.40 **Si**, 5.0–6.0 **Cu**, 0.30 **Zn**, 0.7 **Fe**, 0.20–0.6 **Pb**, 0.20–0.6 **Bi**, 0.05 each, 0.15 total, others	**Solution annealed and cold worked**: 3–40 mm **diam**, 310 MPa **TS**, 260 MPa **YS**, 10% in 2 in. (50 mm) **El**, 90 **HV**; **Cold worked and artificially aged**: 3–75 mm **diam**, 370 MPa **TS**, 275 MPa **YS**, 10% in 2 in. (50 mm) **El**, 110 **HV**

WROUGHT ALUMINUM AND ALUMINUM ALLOYS

specification number	designation	country	product forms	chemical composition	mechanical properties and hardness values
QQ-A-225/3D	2011	US	Bar, rod, wire	rem **Al**, 0.40 **Si**, 5.0–6.0 **Cu**, 0.30 **Zn**, 0.7 **Fe**, 0.20–0.6 **Pb**, 0.20–0.6 **Bi**, 0.05 each, 0.15 total, others	**Solution heat treated and temper T3**: 0.125–1.500 in. **diam**, 45 ksi (310 MPa) **TS**, 38 ksi (262 MPa) **YS**, 10% in 2 in. (50 mm) **El**; **Solution heat treated and temper T8**: 0.125–3.250 in. **diam**, 54 ksi (372 MPa) **TS**, 40 ksi (276 MPa) **YS**, 10% in 2 in. (50 mm) **El**; **Solution heat treated and temper T4**: 8.000 max in. **diam**, 40 ksi (276 MPa) **TS**, 18 ksi (124 MPa) **YS**, 16% in 2 in. (50 mm) **El**
NF A 50 411	2011 (A-U5PbBi)	France	Bar and tube	rem **Al**, 0.40 **Si**, 5.0–6.0 **Cu**, 0.30 **Zn**, 0.70 **Fe**, 0.20–0.60 **Pb**, 0.20–0.60 **Bi**, 0.05 each, 0.15 total, others	**Cold finished (drawn bar)**: 310 MPa **TS**, 265 MPa **YS**, 9% in 2 in. (50 mm) **El**; **Cold finished (tube)**: ≤150 mm **diam**, 295 MPa **TS**, 235 MPa **YS**, 11% in 2 in. (50 mm) **El**
ISO 2779	Al-Cu6BiPb	ISO	Wrought products	rem **Al**, 0.40 **Si**, 5.0–6.0 **Cu**, 0.30 **Zn**, 0.20–0.6 **Pb**, 0.20–0.6 **Bi**, 0.05 each, 0.15 total, others	**Solution heat treated and naturally aged**: 0.125 in. (3 mm) **diam**, 275 MPa **TS**, 125 MPa **YS**, 16% in 2 in. (50 mm) **El**; **TD**: 0.125 in. (3 mm) **diam**, 310 MPa **TS**, 260 MPa **YS**, 10% in 2 in. (50 mm) **El**; **Solution heat treated and artificially aged**: 0.125 in. (3 mm) **diam**, 370 MPa **TS**, 275 MPa **YS**, 10% in 2 in. (50 mm) **El**
JIS H 4040	2011	Japan	Bar and wire: extruded	rem **Al**, 0.40 **Si**, 5.0–6.0 **Cu**, 0.30 **Zn**, 0.7 **Fe**, 0.20–0.6 **Pb**, 0.20–0.6 **Bi**, 0.15 total, others	**Solution heat treated and cold worked**: <6 mm **diam**, 314 MPa **TS**, 265 MPa **YS**; **Solution heat treated, cold worked, and artificially aged hardened**: <6 mm **diam**, 363 MPa **TS**, 275 MPa **YS**
SIS 14 43 55	SIS Al 43 55-03	Sweden	Bar	rem **Al**, 0.40 **Si**, 5.0–6.0 **Cu**, 0.30 **Zn**, 0.7 **Fe**, 0.20–0.6 **Pb**, 0.20–0.6 **Bi**, 0.05 each, 0.15 total, others	**Strain hardened and naturally aged**: 10–50 mm **diam**, 310 MPa **TS**, 260 MPa **YS**, 10% in 2 in. (50 mm) **El**, 95–140 **HV**
SIS 14 43 55	SIS Al 43 55-04	Sweden	Bar, section	rem **Al**, 0.40 **Si**, 5.0–6.0 **Cu**, 0.30 **Zn**, 0.7 **Fe**, 0.20–0.6 **Pb**, 0.20–0.6 **Bi**, 0.05 each, 0.15 total, others	**Artificially aged**: 10–130 mm **diam**, 275 MPa **TS**, 125 MPa **YS**, 14% in 2 in. (50 mm) **El**, 65–95 **HV**
SIS 14 43 55	SIS Al 43 55-06	Sweden	Bar, section	rem **Al**, 0.40 **Si**, 5.0–6.0 **Cu**, 0.30 **Zn**, 0.7 **Fe**, 0.20–0.6 **Pb**, 0.20–0.6 **Bi**, 0.05 each, 0.15 total, others	**Naturally aged**: 10–75 mm **diam**, 350 MPa **TS**, 230 MPa **YS**, 8% in 2 in. (50 mm) **El**, 110–150 **HV**
SIS 14 43 55	SIS Al 43 55-08	Sweden	Bar	rem **Al**, 0.40 **Si**, 5.0–6.0 **Cu**, 0.30 **Zn**, 0.7 **Fe**, 0.20–0.6 **Pb**, 0.20–0.6 **Bi**, 0.05 each, 0.15 total, others	**Strain hardened and artificially aged**: 10–50 mm **diam**, 370 MPa **TS**, 275 MPa **YS**, 8% in 2 in. (50 mm) **El**, 110–150 **HV**

WROUGHT ALUMINUM AND ALUMINUM ALLOYS

specification number	designation	country	product forms	chemical composition	mechanical properties and hardness values
BS L.119		UK	Ingot	rem **Al**, 0.25 **Si**, 4.7–5.5 **Cu**, 0.20–0.30 **Mn**, 0.10 **Mg**, 1.3–1.7 **Ni**, 0.10 **Zn**, 0.15–0.25 **Ti**, 0.40 **Fe**, 0.10–0.30 **Zr**, 0.50 **Ti+Zr**, 0.10–0.30 **Co**, 0.10–0.30 **Sb**, 0.05 each, 0.15 total, others	**Solution heat treated, quenched and artificially aged**: all **diam**, 215 MPa **TS**, 190 MPa **YS**, 1% in 2 in. (50 mm) **El**
BS L.119		UK	Castings	0.30 **Si**, 4.5–5.5 **Cu**, 0.20–0.30 **Mn**, 0.10 **Mg**, 1.3–1.7 **Ni**, 0.10 **Zn**, 0.15–0.25 **Ti**, 0.50 **Fe**, 0.10–0.30 **Zr**, 0.50 **Ti+Zr**, 0.10–0.30 **Co**, 0.10–0.30 **Sb**, 0.05 each, 0.15 total, others	**Solution heat treated, quenched and artificially aged**: all **diam**, 215 MPa **TS**, 190 MPa **YS**, 1% in 2 in. (50 mm) **El**
AECMA prEN2125	Al–P16–T6151	AECMA	Plate	rem **Al**, 0.50–1.20 **Si**, 3.9–5.0 **Cu**, 0.40–1.20 **Mn**, 0.20–0.80 **Mg**, 0.10 **Cr**, 0.25 **Zn**, 0.20 **Ti+Zr**, 0.15 each, 0.15 total, others	**Solution treated, quenched and artificially aged**: 6–10 mm **diam**, 460 MPa **TS**, 410 MPa **YS**, 8% in 2 in. (50 mm) **El**
AECMA prEN2124	Al–P16–T651	AECMA	Plate	rem **Al**, 0.50–1.20 **Si**, 3.9–5.0 **Cu**, 0.40–1.2 **Mn**, 0.20–0.8 **Mg**, 0.10 **Cr**, 0.25 **Zn**, 0.15 **Ti**, 0.20 **Ti+Zr**, 0.05 each, 0.15 total, others	**Solution treated, quenched and artificially aged**: 6–10 mm **diam**, 460 MPa **TS**, 410 MPa **YS**, 8% in 2 in. (50 mm) **El**
AMS 4028D		US	Sheet and plate	rem **Al**, 0.50–1.2 **Si**, 3.9–5.0 **Cu**, 0.40–1.2 **Mn**, 0.20–0.8 **Mg**, 0.10 **Cr**, 0.25 **Zn**, 0.15 **Ti**, 0.7 **Fe**, 0.20 **Zr+Ti**, 0.05 each, 0.15 total, others	**Annealed**: 0.020–0.499 in. (0.51–12.67 mm) **diam**, 32 ksi (221 MPa) **TS**, 16 ksi (110 MPa) **YS**, 16% in 2 in. (50 mm) **El**; **Solution and precipitation heat treated**: 0.020–0.039 in. (0.51–0.99 mm) **diam**, 64 ksi (441 MPa) **TS**, 57 ksi (393 MPa) **YS**
AMS 4029E		US	Sheet and plate	rem **Al**, 0.50–1.2 **Si**, 3.9–5.0 **Cu**, 0.40–1.2 **Mn**, 0.20–0.8 **Mg**, 0.10 **Cr**, 0.25 **Zn**, 0.15 **Ti**, 0.7 **Fe**, 0.05 each, 0.15 total, others	**Solution and precipitation heat treated**: 0.020–0.039 in. **diam**, 64 ksi (441 MPa) **TS**, 57 ksi (393 MPa) **YS**, 6% in 2 in. (50 mm) **El**
AMS 4153G		US	Bar, tube, rod, wire, and shapes: extruded	rem **Al**, 0.50–1.2 **Si**, 3.9–5.0 **Cu**, 0.40–1.2 **Mn**, 0.20–0.8 **Mg**, 0.10 **Cr**, 0.25 **Zn**, 0.15 **Ti**, 0.7 **Fe**, 0.20 **Zr+Ti** each, 0.15 total, others	**Solution and precipitation heat treated**: $\geq$0.499 in. (>gq>12.67 mm) **diam**, 60 ksi (414 MPa) **TS**, 53 ksi (365 MPa) **YS**, 7% in 2 in. (50 mm) **El**, 125 min **HB**
AMS 4314		US	Rings: rolled or forged	rem **Al**, 0.50–1.2 **Si**, 3.9–5.0 **Cu**, 0.40–1.2 **Mn**, 0.20–0.8 **Mg**, 0.10 **Cr**, 0.25 **Zn**, 0.15 **Ti**, 0.7 **Fe**, 0.20 **Zr+Ti**, 0.05 each, 0.15 total, others	**Solution and precipitation heat treated**: $\leq$2 in. ($\leq$51 mm) **diam**, 65 ksi (448 MPa) **TS**, 56 ksi (386 MPa) **YS**, 8% in 2 in. (50 mm) **El**, 120 min **HB**

specification number	designation	country	product forms	chemical composition	mechanical properties and hardness values
AMS 4133		US	Die and hand forgings and rings	rem **Al**, 0.50–1.2 **Si**, 3.9–5.0 **Cu**, 0.40–1.2 **Mn**, 0.20–0.8 **Mg**, 0.10 **Cr**, 0.25 **Zn**, 0.15 **Ti**, 0.7 **Fe**, 0.05 each, 0.15 total, others	**Solution and precipitation heat treated (die forgings)**: ≤2 in. (≤51 mm) **diam**, 65 ksi (448 MPa) **TS**, 56 ksi (386 MPa) **YS**, 6% in 2 in. (50 mm) **El**, 120 min **HB**; **Solution and precipitation heat treated (rings)**: ≤2.5 in. (≤64 mm) **diam**, 65 ksi (448 MPa) **TS**, 55 ksi (379 MPa) **YS**, 7% in 2 in. (50 mm) **El**, 120 min **HB**; **Solution and precipitation heat treated (hand forgings)**: ≤2 in. (≤51 mm) **diam**, 65 ksi (448 MPa) **TS**, 56 ksi (386 MPa) **YS**, 8% in 2 in. (50 mm) **El**, 120 min **HB**
AMS 4134B		US	Forgings	rem **Al**, 0.50–1.2 **Si**, 3.9–5.0 **Cu**, 0.40–1.2 **Mn**, 0.20–0.8 **Mg**, 0.10 **Cr**, 0.25 **Zn**, 0.50 **Ti**, 0.7 **Fe**, 0.20 **Zr+Ti**, 0.05 each, 0.15 total, others	**Solution heat treated**: ≤4 in. (102 mm) **diam**, 55 ksi (379 MPa) **TS**, 30 ksi (207 MPa) **YS**, 11% in 2 in. (50 mm) **El**, 100 min **HB**; **Precipitation heat treated**: all **diam**, 65 ksi (448 MPa) **TS**, 55 ksi (379 MPa) **YS**, 7% in 2 in. (50 mm) **El**, 120 min **HB**
AMS 4135K		US	Die and hand forgings and rings	rem **Al**, 0.50–1.2 **Si**, 3.9–5.0 **Cu**, 0.40–1.2 **Mn**, 0.20–0.8 **Mg**, 0.10 **Cr**, 0.25 **Zn**, 0.15 **Ti**, 0.7 **Fe**, 0.05 each, 0.15 total, others	**Solution and precipitation heat treated (die forgings and rings)**: ≤4 in. (≤102 mm) **diam**, 65 ksi (448 MPa) **TS**, 55 ksi (379 MPa) **YS**, 7% in 2 in. (50 mm) **El**, 120 min **HB**; **Solution and precipitation heat treated (hand forgings)**: ≤6 in. (152 mm) **diam**, 65 ksi (448 MPa) **TS**, 55 ksi (379 MPa) **YS**, 8% in 2 in. (50 mm) **El**, 120 min **HB**
AMS 4130H		US	Forgings	rem **Al**, 0.50–1.2 **Si**, 3.90–5.0 **Cu**, 0.40–1.2 **Mn**, 0.05 **Mg**, 0.10 **Cr**, 0.25 **Zn**, 0.15 **Ti**, 1.0 **Fe**, 0.05 each, 0.15 total, others	**Solution and precipitation heat treated**: ≤4 in. (102 mm) **diam**, 55 ksi (379 MPa) **TS**, 33 ksi (228 MPa) **YS**, 16% in 2 in. (50 mm) **El**, 100 min **HB**
AMS 4121D		US	Bar, rod, and wire	rem **Al**, 0.50–1.2 **Si**, 3.9–5.0 **Cu**, 0.40–1.2 **Mn**, 0.20–0.8 **Mg**, 0.10 **Cr**, 0.25 **Zn**, 0.15 **Ti**, 1.0 **Fe**, 0.05 each, 0.15 total, others	**Solution and precipitation heat treated**: all **diam**, 65 ksi (448 MPa) **TS**, 55 ksi (379 MPa) **YS**, 8% in 2 in. (50 mm) **El**, 125 min **HB**

WROUGHT ALUMINUM AND ALUMINUM ALLOYS

specification number	designation	country	product forms	chemical composition	mechanical properties and hardness values
ANSI/ASTM B 247	2014	US	Die and hand forgings	rem **Al**, 0.50–1.2 **Si**, 3.9–5.0 **Cu**, 0.40–1.2 **Mn**, 0.20–0.8 **Mg**, 0.10 **Cr**, 0.25 **Zn**, 0.15 **Ti**, 0.7 **Fe**, 0.20 **Zr+Ti**, 0.05 each, 0.15 total, others	**Solution heat treated and naturally aged (die forgings)**: ≤4.000 in. **diam**, 55 ksi (379 MPa) **TS**, 30 ksi (207 MPa) **YS**, 11% in 2 in. (50 mm) **El**, 100 min **HB**; **Solution heat treated and artificially aged (die forgings)**: ≤1.000 in. **diam**, 65 ksi (448 MPa) **TS**, 56 ksi (386 MPa) **YS**, 6% in 2 in. (50 mm) **El**, 125 min **HB**; **Solution heat treated and artificially aged (hand forgings)**: ≤2.000 in. **diam**, 65 ksi (448 MPa) **TS**, 56 ksi (386 MPa) **YS**, 8% in 2 in. (50 mm) **El**
ANSI/ASTM B 247	2025	US	Die forgings	rem **Al**, 0.50–1.2 **Si**, 3.9–5.0 **Cu**, 0.40–1.2 **Mn**, 0.05 **Mg**, 0.10 **Cr**, 0.25 **Zn**, 0.15 **Ti**, 1.0 **Fe**, 0.05 each, 0.15 total, others	**Solution heat treated and artificially aged**: ≤4.000 in. **diam**, 52 ksi (359 MPa) **TS**, 33 ksi (228 MPa) **YS**, 11% in 2 in. (50 mm) **El**, 100 min **HB**
ANSI/ASTM B 241	2014	US	Tube: seamless	rem **Al**, 0.50–1.2 **Si**, 3.9–5.0 **Cu**, 0.40–1.2 **Mn**, 0.20–0.8 **Mg**, 0.10 **Cr**, 0.25 **Zn**, 0.15 **Ti**, 0.7 **Fe**, 0.20 **Zr+Ti**, 0.05 each, 0.15 total, others	**Solution heat treated and artificially aged**: ≤0.499 in. **diam**, 60 ksi (414 MPa) **TS**, 53 ksi (365 MPa) **YS**, 7% in 2 in. (50 mm) **El**
ANSI/ASTM B 221	2014	US	Bar, tube, rod, wire, and shapes	rem **Al**, 0.50–1.2 **Si**, 3.9–5.0 **Cu**, 0.40–1.2 **Mn**, 0.20–0.8 **Mg**, 0.10 **Cr**, 0.25 **Zn**, 0.15 **Ti**, 0.7 **Fe**, 0.20 **Ti+Zr**, 0.05 each, 0.15 total, others	**Annealed**: all **diam**, 30 max ksi (207 max MPa) **TS**, 18 max ksi (124 max MPa) **YS**, 12% in 2 in. (50 mm) **El**; **Solution heat treated and naturally aged**: all **diam**, 50 ksi (345 MPa) **TS**, 35 ksi (241 MPa) **YS**, 12% in 2 in. (50 mm) **El**; **Solution heat treated and artificially aged**: ≥0.750 in. **diam**, 68 ksi (469 MPa) **TS**, 58 ksi (400 MPa) **YS**, 6% in 2 in. (50 mm) **El**
ANSI/ASTM B 211	2014	US	Bar, rod, and wire	rem **Al**, 0.50–1.2 **Si**, 3.9–5.0 **Cu**, 0.40–1.2 **Mn**, 0.20–0.8 **Mg**, 0.10 **Cr**, 0.25 **Zn**, 0.15 **Ti**, 0.7 **Fe**, 0.20 **Ti+Zr**, 0.05 each, 0.15 total, others	**Annealed**: 0.125–8.000 in. **diam**, 35 max ksi (241 max MPa) **TS**, 12% in 2 in. (50 mm) **El**; **Solution heat treated and naturally aged**: 0.125–4.000 in. **diam**, 55 ksi (379 MPa) **TS**, 32 ksi (221 MPa) **YS**, 16% in 2 in. (50 mm) **El**; **Solution heat treated and artificially aged**: 0.125–4.000 in. **diam**, 65 ksi (448 MPa) **TS**, 55 ksi (379 MPa) **YS**, 8% in 2 in. (50 mm) **El**

WROUGHT ALUMINUM AND ALUMINUM ALLOYS

specification number	designation	country	product forms	chemical composition	mechanical properties and hardness values
ANSI/ASTM B 209	Alclad 2014	US	Sheet and plate	rem **Al**, 0.50–1.2 **Si**, 3.9–5.0 **Cu**, 0.40–1.2 **Mn**, 0.20–0.8 **Mg**, 0.10 **Cr**, 0.25 **Zn**, 0.15 **Ti**, 0.7 **Fe**, 0.20 **Ti+Zr**, 0.05 each, 0.15 total, others, 6003 cladding material	**Annealed**: 0.020–0.499 in. **diam**, 30 max ksi (207 max MPa) **TS**, 14 max ksi (97 max MPa) **YS**, 16% in 2 in. (50 mm) **El**; **Solution heat treated and naturally aged (sheet)**: 0.040–0.249 in. **diam**, 57 ksi (393 MPa) **TS**, 34 ksi (234 MPa) **YS**, 15% in 2 in. (50 mm) **El**; **Solution heat treated and artificially aged**: 0.040–0.249 in. **diam**, 64 ksi (441 MPa) **TS**, 57 ksi (393 MPa) **YS**, 8% in 2 in. (50 mm) **El**
ANSI/ASTM B 210	2014	US	Tube: seamless	rem **Al**, 0.50–1.2 **Si**, 3.9–5.0 **Cu**, 0.40–1.2 **Mn**, 0.20–0.8 **Mg**, 0.10 **Cr**, 0.25 **Zn**, 0.15 **Ti**, 0.7 **Fe**, 0.20 **Ti+Zr**, 0.05 each, 0.15 total, others	**Annealed**: 0.018–0.500 in. **diam**, 32 max ksi (22 max MPa) **TS**, 16 max ksi (110 max MPa) **YS**; **Solution heat treated and naturally aged**: 0.260–0.500 in. **diam**, 54 ksi (372 MPa) **TS**, 30 ksi (207 MPa) **YS**, 12% in 2 in. (50mm) **El**; **Solution heat treated and artificially aged**: 0.260–0.500 in. **diam**, 65 ksi (448 MPa) **TS**, 55 ksi (379 MPa) **YS**, 8% in 2 in. (50 mm) **El**
ANSI/ASTM B 209	2014	US	Sheet and plate	rem **Al**, 0.50–1.2 **Si**, 3.9–5.0 **Cu**, 0.40–1.2 **Mn**, 0.20–0.8 **Mg**, 0.10 **Cr**, 0.25 **Zn**, 0.15 **Ti**, 0.7 **Fe**, 0.20 **Ti+Zr**, 0.05 each, 0.15 total, others	**Annealed**: 0.500–1.000 in. **diam**, 32 max ksi (220 max MPa) **TS**, 16 max ksi (110 max MPa) **YS**, 16% in 2 in. (50 mm) **El**; **Solution heat treated and naturally aged (sheet)**: 0.020–0.249 in. **diam**, 59 ksi (407 MPa) **TS**, 36 ksi (248 MPa) **YS**, 14% in 2 in. (50 mm) **El**; **Solution heat treated and artificially aged**: 0.040–0.249 in. **diam**, 66 ksi (455 MPa) **TS**, 58 ksi (400 MPa) **YS**, 7% in 2 in. (50 mm) **El**
AS 1866	2014	Australia	Rod, bar, shape	rem **Al**, 0.50–1.2 **Si**, 3.9–5.0 **Cu**, 0.40–1.2 **Mn**, 0.20–0.8 **Mg**, 0.10 **Cr**, 0.25 **Zn**, 0.15 **Ti**, 0.7 **Fe**, 0.05 each, 0.15 total, others	**Solution heat treated and naturally aged**: 10.0 mm **diam**, 372 MPa **TS**, 241 MPa **YS**, 15% in 2 in. (50 mm) **El**; **Solution heat treated and artificially aged**: 10.0 mm **diam**, 432 MPa **TS**, 386 MPa **YS**, 8% in 2 in. (50 mm) **El**
BS 1473	HR15	UK	Rivet	rem **Al**, 0.5–0.9 **Si**, 3.9–5.0 **Cu**, 0.4–1.2 **Mn**, 0.2–0.8 **Mg**, 0.10 **Cr**, 0.2 **Zn**, 0.2 **Ti**, 0.7 **Fe**	**Solution heat treated and naturally aged**: $\leq$12 mm **diam**, 385 MPa **TS**

WROUGHT ALUMINUM AND ALUMINUM ALLOYS

specification number	designation	country	product forms	chemical composition	mechanical properties and hardness values
BS 1470	HC15	UK	Plate, sheet, strip	rem **Al**, 0.5–0.9 **Si**, 3.9–5.0 **Cu**, 0.4–1.2 **Mn**, 0.2–0.8 **Mg**, 0.10 **Cr**, 0.2 **Zn**, 0.2 **Ti**, 0.7 **Fe**	**Solution heat treated and naturally aged**: 0.2–12.5 mm **diam**, 375 MPa **TS**, 230 MPa **YS**, 13% in 2 in. (50 mm) **El**; **Solution heat treated and precipitation treated**: 0.2–3.0 mm **diam**, 400 MPa **TS**, 325 MPa **YS**, 7% in 2 in. (50 mm) **El**
BS 1470	HS15	UK	Plate, sheet, strip	rem **Al**, 0.5–0.9 **Si**, 3.9–5.0 **Cu**, 0.4–1.2 **Mn**, 0.2–0.8 **Mg**, 0.10 **Cr**, 0.2 **Zn**, 0.2 **Ti**, 0.7 **Fe**	**Solution heat treated and naturally aged**: 0.2–25.0 mm **diam**, 385 MPa **TS**, 245 MPa **YS**, 13% in 2 in. (50 mm) **El**; **Solution heat treated and precipitation treated**: 0.2–3.0 mm **diam**, 430 MPa **TS**, 375 MPa **YS**, 6% in 2 in. (50 mm) **El**
BS 1472	HF15	UK	Forging stock and forgings	rem **Al**, 0.5–0.9 **Si**, 3.9–5.0 **Cu**, 0.4–1.2 **Mn**, 0.2–0.8 **Mg**, 0.10 **Cr**, 0.2 **Zn**, 0.2 **Ti**, 0.7 **Fe**	**Solution heat treated and naturally aged**: ≤150 mm **diam**, 370 MPa **TS**, 215 MPa **YS**, 13% in 2 in. (50 mm) **El**; **Solution heat treated and precipitation treated**: ≤150 mm **diam**, 450 MPa **TS**, 395 MPa **YS**, 6% in 2 in. (50 mm) **El**
BS 1474	HE15	UK	Bar, tube, section	rem **Al**, 0.5–0.9 **Si**, 3.9–5.0 **Cu**, 0.4–1.2 **Mn**, 0.2–0.8 **Mg**, 0.10 **Cr**, 0.2 **Zn**, 0.2 **Ti**, 0.7 **Fe**	**Solution heat treated and naturally aged**: ≤20 mm **diam**, 370 MPa **TS**, 230 MPa **YS**, 10% in 2 in. (50 mm) **El**; **Solution heat treated and precipitation treated**: ≤20 mm **diam**, 435 MPa **TS**, 370 MPa **YS**, 6% in 2 in. (50 mm) **El**
BS L.153		UK	Sheet and strip (core material)	rem **Al**, 0.50–1.1 **Si**, 3.9–5.0 **Cu**, 0.40–1.2 **Mn**, 0.20–0.8 **Mg**, 0.10 **Cr**, 0.05 **Ni**, 0.25 **Zn**, 0.15 **Ti**, 0.50 **Fe**, 0.20 **Zr+Ti**, 0.05 each, 0.15 total, others	**Solution heat treated, quenched and naturally aged**: 0.4–0.8 mm **diam**, 380 MPa **TS**, 235 MPa **YS**, 14% in 2 in. (50 mm) **El**
BS L.152		UK	Sheet, strip (core material)	rem **Al**, 0.50–1.1 **Si**, 3.9–5.0 **Cu**, 0.40–1.2 **Mn**, 0.20–0.8 **Mg**, 0.10 **Cr**, 0.05 **Ni**, 0.25 **Zn**, 0.15 **Ti**, 0.50 **Fe**, 0.20 **Zr+Ti**, 0.05 each, 0.15 total, others (clad with BS L 152 cladding material)	**Solution heat treated, quenched and artificially aged**: 0.4–0.8 mm) **diam**, 415 MPa **TS**, 350 MPa **YS**, 7% in 2 in. (50 mm) **El**
BS L.151		UK	Sheet, strip (core material)	rem **Al**, 0.50–1.1 **Si**, 3.9–5.0 **Cu**, 0.40–1.2 **Mn**, 0.20–0.8 **Mg**, 0.10 **Cr**, 0.05 **Ni**, 0.25 **Zn**, 0.15 **Ti**, 0.50 **Fe**, 0.20 **Zr+Ti**, 0.05 each, 0.15 total, others (clad with BS L 151 cladding material)	**Solution heat treated, quenched and naturally aged**: 0.4–0.8 mm **diam**, 380 MPa **TS**, 235 MPa **YS**, 14% in 2 in. (50 mm) **El**

WROUGHT ALUMINUM AND ALUMINUM ALLOYS

specification number	designation	country	product forms	chemical composition	mechanical properties and hardness values
BS L.150		UK	Sheet, strip	rem **Al**, 0.50–1.1 **Si**, 3.9–5.0 **Cu**, 0.40–1.2 **Mn**, 0.20–0.8 **Mg**, 0.10 **Cr**, 0.05 **Ni**, 0.25 **Zn**, 0.15 **Ti**, 0.50 **Fe**, 0.20 **Zr+Ti**, 0.05 each, 0.15 total, others	**Solution heat treated, quenched and artificially aged:** 0.4–0.8 mm **diam**, 430 MPa **TS**, 390 MPa **YS**, 6% in 2 in. (50 mm) **El**
BS L.106		UK	Sheet, strip	rem **Al**, 0.50–0.90 **Si**, 3.9–5.0 **Cu**, 0.40–1.2 **Mn**, 0.20–0.8 **Mg**, 0.10 **Cr**, 0.2 **Ni**, 0.2 **Zn**, 0.05 **Sn**, 0.5 **Fe**, 0.05 **Pb**, 0.2 **Zr+Ti**	**Solution heat treated, quenched and aged**: 0.4–6.0 mm **diam**, 400 MPa **TS**, 225 MPa **YS**, 14% in 2 in. (50 mm) **El**
BS L.105		UK	Tube	rem **Al**, 0.50–0.90 **Si**, 3.9–5.0 **Cu**, 0.40–1.2 **Mn**, 0.20–0.8 **Mg**, 0.10 **Cr**, 0.2 **Ni**, 0.2 **Zn**, 0.05 **Sn**, 0.5 **Fe**, 0.05 **Pb**, 0.2 **Zr+Ti**	**Solution heat treated, quenched and aged**: all **diam**, 400 MPa **TS**, 290 MPa **YS**, 10% in 2 in. (50 mm) **El**
BS L.103		UK	Forgings	rem **Al**, 0.50–0.90 **Si**, 3.9–5.0 **Cu**, 0.40–1.2 **Mn**, 0.20–0.8 **Mg**, 0.10 **Cr**, 0.2 **Ni**, 0.2 **Zn**, 0.05 **Sn**, 0.5 **Fe**, 0.05 **Pb**, 0.2 **Zr+Ti**	**Solution heat treated, quenched and aged**: all **diam**, 370 MPa **TS**, 225 MPa **YS**, 14% in 2 in. (50 mm) **El**
BS L.102		UK	Bar, section	rem **Al**, 0.50–0.90 **Si**, 3.9–5.0 **Cu**, 0.40–1.2 **Mn**, 0.20–0.8 **Mg**, 0.10 **Cr**, 0.2 **Ni**, 0.2 **Zn**, 0.05 **Sn**, 0.5 **Fe**, 0.05 **Pb**, 0.2 **Ti+Zr**	**Solution heat treated and aged:** $\leq$10 mm **diam**, 370 MPa **TS**, 235 MPa **YS**, 11% in 2 in. (50 mm) **El**
BS L.108		UK	Sheet, strip (core material)	rem **Al**, 0.50–0.09 **Si**, 3.9–5.0 **Cu**, 0.40–1.2 **Mn**, 0.20–0.8 **Mg**, 0.10 **Cr**, 0.2 **Ni**, 0.2 **Zn**, 0.05 **Sn**, 0.51 **Fe**, 0.05 **Pb**, 0.2 **Zr+Ti**, (clad with BS L.108 cladding material)	**Solution heat treated, quenched and aged**: 0.4–6.0 mm **diam**, 385 MPa **TS**, 240 MPa **YS**, 14% in 2 in. (50 mm) **El**
BS 2L.93		UK	Plate	rem **Al**, 0.50–0.90 **Si**, 3.9–5.0 **Cu**, 0.40–1.2 **Mn**, 0.20–0.8 **Mg**, 0.10 **Cr**, 0.2 **Ni**, 0.2 **Zn**, 0.05 **Sn**, 0.5 **Fe**, 0.05 **Pb**, 0.02 **Zr+Ti**	**Solution heat and precipitation treated and quenched:** 6–12.5 mm **diam**, 460 MPa **TS**, 410 MPa **YS**, 7% in 2 in. (50 mm) **El**
BS 2L.90		UK	Sheet and strip (core material)	rem **Al**, 0.50–0.90 **Si**, 3.9–5.0 **Cu**, 0.40–1.2 **Mn**, 0.20–0.8 **Mg**, 0.10 **Cr**, 0.2 **Ni**, 0.2 **Zn**, 0.05 **Sn**, 0.5 **Fe**, 0.05 **Pb**, 0.2 **Zr+Ti**, clad with BS 2L.90 cladding material)	**Solution heat and precipitation treated and quenched:** 0.4–0.8 mm **diam**, 415 MPa **TS**, 350 MPa **YS**, 7% in 2 in. (50 mm) **El**
BS 2L.89		UK	Sheet and strip (core material)	rem **Al**, 0.50–0.90 **Si**, 3.9–5.0 **Cu**, 0.40–1.2 **Mn**, 0.20–0.8 **Mg**, 0.10 **Cr**, 0.2 **Ni**, 0.2 **Zn**, 0.05 **Sn**, 0.5 **Fe**, 0.05 **Pb**, 0.2 **Zr+Ti**, clad with BS 2L.89 (cladding material)	**Solution heat treated, quenched and naturally aged:** 0.4–0.8 mm **diam**, 385 MPa **TS**, 250 MPa **YS**, 13% in 2 in. (50 mm) **El**
BS 2L.87		UK	Bar	rem **Al**, 0.50–0.90 **Si**, 3.9–5.0 **Cu**, 0.40–1.2 **Mn**, 0.20–0.8 **Mg**, 0.10 **Cr**, 0.2 **Ni**, 0.2 **Zn**, 0.05 **Sn**, 0.5 **Fe**, 0.05 **Pb**, 0.2 **Zr+Ti**	**Solution heat and precipitation treated and quenched**: all **diam**, 430 MPa **TS**, 390 MPa **YS**, 5% in 2 in. (50 mm) **El**
BS 6L.37		UK	Wire	rem **Al**, 0.50–0.90 **Si**, 3.9–5.0 **Cu**, 0.40–1.2 **Mn**, 0.20–0.8 **Mg**, 0.10 **Cr**, 0.2 **Ni**, 0.2 **Zn**, 0.05 **Sn**, 0.5 **Fe**, 0.05 **Pb**, 0.2 **Zr+Ti**	**Solution heat treated, quenched and naturally aged:** $\leq$6 mm **diam**, 400 MPa **TS**

WROUGHT ALUMINUM AND ALUMINUM ALLOYS

specification number	designation	country	product forms	chemical composition	mechanical properties and hardness values
BS 3L.72		UK	Sheet and strip (core material)	rem **Al**, 0.50–0.90 **Si**, 3.9–5.0 **Cu**, 0.40–1.2 **Mn**, 0.20–0.8 **Mg**, 0.10 **Cr**, 0.2 **Ni**, 0.2 **Zn**, 0.05 **Sn**, 0.5 **Fe**, 0.05 **Pb**, 0.2 **Zr+Ti**, (clad with BS 3L.72 cladding material)	**Solution heat treated, quenched and naturally aged:** 0.4–0.8 mm **diam**, 385 MPa **TS**, 250 MPa **YS**, 13% in 2 in. (50 mm) **El**
BS 3L.70		UK	Sheet and strip	rem **Al**, 0.50–0.90 **Si**, 3.9–5.0 **Cu**, 0.40–1.2 **Mn**, 0.20–0.8 **Mg**, 0.10 **Cr**, 0.2 **Ni**, 0.2 **Zn**, 0.05 **Sn**, 0.5 **Fe**, 0.05 **Pb**, 0.2 **Zr+Ti**	**Solution heat treated, quenched and naturally aged:** 0.4–0.8 mm **diam**, 400 MPa **TS**, 260 MPa **YS**, 13% in 2 in. (50 mm) **El**
BS 3L.65		UK	Bar, section	rem **Al**, 0.50–0.90 **Si**, 3.9–5.0 **Cu**, 0.40–1.2 **Mn**, 0.20–0.8 **Mg**, 0.10 **Cr**, 0.2 **Ni**, 0.2 **Zn**, 0.05 **Sn**, 0.5 **Fe**, 0.05 **Pb**, 0.2 **Zr+Ti**	**Solution heat and precipitation treated and quenched:** ≤10 mm **diam**, 435 MPa **TS**, 385 MPa **YS**, 7% in 2 in. (50 mm) **El**
BS 3L.63		UK	Tube	rem **Al**, 0.50–0.90 **Si**, 3.9–5.0 **Cu**, 0.40–1.2 **Mn**, 0.20–0.8 **Mg**, 0.10 **Cr**, 0.2 **Ni**, 0.2 **Zn**, 0.05 **Sn**, 0.5 **Fe**, 0.05 **Pb**, 0.2 **Zr+Ti**	**Solution heat and precipitation treated and quenched:** all **diam**, 450 MPa **TS**, 370 MPa **YS**, 7% in 2 in. (50 mm) **El**
COPANT 862	2014	COPANT	Wrought products	rem **Al**, 0.5–1.2 **Si**, 3.9–5.0 **Cu**, 0.4–1.20 **Mn**, 0.2–0.8 **Mg**, 0.10 **Cr**, 0.25 **Zn**, 0.15 **Ti** 0.70 **Fe**, 0.05 each, 0.15 total, others	
CSA HA.8	.2014	Canada	Forgings	rem **Al**, 0.50–1.2 **Si**, 3.9–5.0 **Cu**, 0.40–1.2 **Mn**, 0.20–0.8 **Mg**, 0.10 **Cr**, 0.25 **Zn**, 0.15 **Ti**, 0.7 **Fe**, 0.20 **Zr+Ti**, 0.05 each, 0.15 total, others	**T6 (die forgings)**: <1 in. **diam**, 65.0 ksi (448 MPa) **TS**, 56.0 ksi (386 MPa) **YS**, 6% in 2 in. (50 mm) **El**; **T6 (hand forgings**: <2.000 in. **diam**, 65.0 ksi (448 MPa) **TS**, 56.0 ksi (386 MPa) **YS**, 8% in 2 in. (50 mm) **El**
CSA HA.4	.2014 Alclad	Canada	Plate, sheet (core)	rem **Al**, 0.50–1.2 **Si**, 3.9–5.0 **Cu**, 0.40–1.2 **Mn**, 0.20–0.8 **Mg**, 0.10 **Cr**, 0.25 **Zn**, 0.15 **Ti**, 0.7 **Fe**, 0.05 each, 0.15 total, others, clad with alloy .6003	**T3 (flat sheet)**: 0.020–0.039 in. **diam**, 55.0 ksi (379 MPa) **TS**, 34.0 ksi (234 MPa) **YS**, 14% in 2 in. (50 mm) **El**; **T451 (plate)**: 0.250–0.499 in. **diam**, 57.0 ksi (393 MPa) **TS**, 36.0 ksi (248 MPa) **YS**, 15% in 2 in. (50 mm) **El**; **T6 (sheet)**: 0.020–0.039 in. **diam**, 63.0 ksi (434 MPa) **TS**, 55.0 ksi (379 MPa) **YS**, 7% in 2 in. (50 mm) **El**

WROUGHT ALUMINUM AND ALUMINUM ALLOYS

specification number	designation	country	product forms	chemical composition	mechanical properties and hardness values
CSA HA.5	.2014	Canada	Bar, rod, wire, shapes	rem **Al**, 0.50–1.2 **Si**, 3.9–5.0 **Cu**, 0.40–1.2 **Mn**, 0.20–0.8 **Mg**, 0.10 **Cr**, 0.25 **Zn**, 0.15 **Ti**, 0.7 **Fe**, 0.05 each, 0.15 total, others	**Extruded and O**: all **diam**, 30.0 max ksi (207 max MPa) **TS**, 18.0 max ksi (124 max MPa) **YS**, 12% in 2 in. (50 mm) **El**; **Extruded and T4**: ≥0.124 in. **diam**, 50.0 ksi (345 MPa) **TS**, 35.0 ksi (241 MPa) **YS**, 12% in 2 in. (50 mm) **El**; **Extruded and T6**: 0.125–0.499 in. **diam**, 60.0 ksi (414 MPa) **TS**, 53.0 ksi (365 MPa) **YS**, 7% in 2 in. (50 mm) **El**
QQ-A-250/3E	2014	US	Plate, sheet	rem **Al**, 0.50 **Si**, 3.9–5.0 **Cu**, 0.40–1.2 **Mn**, 0.20–0.8 **Mg**, 0.10 **Cr**, 0.25 **Zn**, 0.15 **Ti**, 0.7 **Fe**, 0.05 each, 0.15 total, others	**Annealed**: 0.020–0.499 in. **diam**, 30 max ksi (207 max MPa) **TS**, 14 max ksi (97 max MPa) **YS**, 16% in 2 in. (50 mm) **El**; **Solution heat treated and temper T451**: 0.250–0.499 in. **diam**, 57 ksi (393 MPa) **TS**, 36 ksi (248 MPa) **YS**, 15% in 2 in. (50 mm) **El**; **Solution treated, stress relieved and aged**: 0.250–0.499 in. **diam**, 64 ksi (441 MPa) **TS**, 57 ksi (393 MPa) **YS**, 8% in 2 in. (50 mm) **El**
QQ-A-225/4D	2014	US	Bar, rod, wire	rem **Al**, 0.50–1.2 **Si**, 3.9–5.0 **Cu**, 0.40–1.2 **Mn**, 0.20–0.8 **Mg**, 0.10 **Cr**, 0.25 **Zn**, 0.15 **Ti**, 0.7 **Fe**, 0.05 each, 0.15 total, others	**Annealed**: 8.000 max in. **diam**, 35 max ksi (241 max MPa) **TS**, 12% in 2 in. (50 mm) **El**; **Solution heat treated and naturally aged**: 8.000 max in. **diam**, 55 ksi (379 MPa) **TS**, 32 ksi (221 MPa) **YS**, 16% in 2 in. (50 mm) **El**; **Solution heat treated and artificially aged**: 8.000 max in. **diam**, 65 ksi (459 MPa) **TS**, 55 ksi (379 MPa) **YS**, 8% in 2 in. (50 mm) **El**
QQ-A-200/2D	2014	US	Bar, rod, shapes, tube, wire	rem **Al**, 0.50–1.2 **Si**, 3.9–5.0 **Cu**, 0.40–1.2 **Mn**, 0.20–0.8 **Mg**, 0.10 **Cr**, 0.15 **Ti**, 0.7 **Fe**, 0.5 each, 0.15 total, others	**Annealed**: all **diam**, 30 max ksi (207 max MPa) **TS**, 18 max ksi (124 max MPa) **YS**, 12% in 2 in. (50 mm) **El**; **Solution heated treated**: all **diam**, 50 ksi (345 MPa) **TS**, 35 ksi (241 MPa) **YS**, 12% in 2 in. (50 mm) **El**; **Solution heat treated and artificially aged**: 0.499 in. **diam**, 60 ksi (414 MPa) **TS**, 53 ksi (365 MPa) **YS**, 7% in 2 in. (50 mm) **El**

WROUGHT ALUMINUM AND ALUMINUM ALLOYS

specification number	designation	country	product forms	chemical composition	mechanical properties and hardness values
AIR 9051/A	A7–U4SG	France	Sheet, bar, billet, and plate: for forging	rem **Al**, 0.50–1.2 **Si**, 3.9–5.0 **Cu**, 0.40–1.2 **Mn**, 0.20–0.8 **Mg**, 0.10 **Cr**, 0.25 **Zn**, 0.15 **Ti**, 0.35 **Fe**, 0.20 **Ti+Zr**, 0.05 each, 0.15 total, others	**Rolled (sheet)**: ≤12 mm **diam**, 460 MPa **TS**, 410 MPa **YS**, 8% in 2 in. (50 mm) **El**; **Solution treated (bar)**: ≤250 mm **diam**, 450 MPa **TS**, 380 MPa **YS**, 6% in 2 in. (50 mm) **El**; **Water quenched and tempered (billet and plate)**: ≤500 mm **diam**, 450 MPa **TS**, 390 MPa **YS**, 8% in 2 in. (50 mm) **El**
NF A 50 411	2014–2014 A (A–U4SG)	France	Bar, tube, and shapes	rem **Al**, 0.50–1.2 **Si**, 3.9–5.0 **Cu**, 0.40–1.20 **Mn**, 0.20–0.80 **Mg**, 0.10 **Cr**, 0.25 **Zn**, 0.15 **Ti**, 0.70 **Fe**, 0.20 **Ti+Zr**, 0.05 each, 0.15 total, others	**Aged (extruded bar)**: 390 MPa **TS**, 255 MPa **YS**, 10% in 2 in. (50 mm) **El**; **Quenched and tempered (tube)**: ≤150 mm **diam**, 440 MPa **TS**, 390 MPa **YS**, 8% in 2 in. (50 mm) **El**; **Quenched and tempered (shapes)**: 410 MPa **TS**, 370 MPa **YS**, 8% in 2 in. (50 mm) **El**
NF A 50 451	2014 (A–U4SG)	France	Sheet, plate, strip	rem **Al**, 0.50–1.20 **Si**, 3.9–5.0 **Cu**, 0.40–1.20 **Mn**, 0.20–0.80 **Mg**, 0.10 **Cr**, 0.25 **Zn**, 0.15 **Ti**, 0.70 **Fe**, 0.20 **Ti+Zr**, 0.05 each, 0.15 total, others	**Annealed**: 0.4–1.6 mm **diam**, 140 max MPa **YS**, 13% in 2 in. (50 mm) **El**; **Aged**: 0.4–1.6 mm **diam**, 390 MPa **TS**, 240 MPa **YS**, 15% in 2 in. (50 mm) **El**; **Quenched and tempered**: 0.4–0.8 mm **diam**, 440 MPa **TS**, 390 MPa **YS**, 8% in 2 in. (50 mm) **El**
IS 5902	24435	India	Wire, bar	rem **Al**, 0.50–1.2 **Si**, 3.9–5.0 **Cu**, 0.40–1.2 **Mn**, 0.20–0.8 **Mg**, 0.10 **Cr**, 0.25 **Zn**, 0.15 **Ti**, 0.7 **Fe**	**Annealed and cold drawn**: 383 MPa **TS**
JIS H 4140	2014	Japan	Die forgings	rem **Al**, 0.50–1.2 **Si**, 3.9–5.0 **Cu**, 0.40–1.2 **Mn**, 0.20–0.8 **Mg**, 0.10 **Cr**, 0.25 **Zn**, 0.15 **Ti**, 0.7 **Fe**, 0.15 total, others	**Solution heat treated and age hardened**: ≤ 100 mm **diam**, 382 MPa **TS**, 206 MPa **YS**, 16% in 2 in. (50 mm) **El**; **Solution heat treated and artificially aged**: ≤ 100 mm **diam**, 451 MPa **TS**, 382 MPa **YS**, 8% in 2 in. (50 mm) **El**
JIS H 4140	2025	Japan	Die forgings	rem **Al**, 0.50–1.2 **Si**, 3.9–5.0 **Cu**, 0.40–1.2 **Mn**, 0.05 **Mg**, 0.10 **Cr**, 0.25 **Zn**, 0.15 **Ti**, 1.0 **Fe**, 0.15 total, others	**Solution heat treated and artificially age hardened**: ≤100 mm **diam**, 382 MPa **TS**, 226 MPa **YS**, 16% in 2 in. (50 mm) **El**

WROUGHT ALUMINUM AND ALUMINUM ALLOYS

specification number	designation	country	product forms	chemical composition	mechanical properties and hardness values
JIS H 4100	2014	Japan	Shapes: extruded	rem **Al**, 0.50–1.2 **Si**, 3.9–5.0 **Cu**, 0.40–1.2 **Mn**, 0.20–0.8 **Mg**, 0.10 **Cr**, 0.25 **Zn**, 0.15 **Ti**, 0.7 **Fe**, 0.15 total, others	**Annealed**: all **diam**, 245 MPa **TS**, 127 MPa **YS**, 12% in 2 in. (50 mm) **El**; **Solution heat treated and age hardened**: all **diam**, 343 MPa **TS**, 245 MPa **YS**, 12% in 2 in. (50 mm) **El**; **Solution heat treated and artificially age hardened**: < 13 mm **diam**, 412 MPa **TS**, 363 MPa **YS**, 7% in 2 in. (50 mm) **El**
JIS H 4080	2014	Japan	Tube: extruded	rem **Al**, 0.50–1.2 **Si**, 3.9–5.0 **Cu**, 0.40–1.2 **Mn**, 0.20–0.8 **Mg**, 0.10 **Cr**, 0.25 **Zn**, 0.15 **Ti**, 0.7 **Fe**, 0.15 total, others	**Annealed**: all **diam**, 245 MPa **TS**, 127 MPa **YS**, 12% in 2 in. (50 mm) **El**; **Solution heat treated and age hardened**: all **diam**, 343 MPa **TS**, 245 MPa **YS**, 12% in 2 in. **El**; **Solution heat treated and artificially age hardened**: <13 mm **diam**, 412 MPa **TS**, 363 MPa **YS**, 7% in 2 in. (50 mm) **El**
JIS H 4040	2014	Japan	Bar: extruded	rem **Al**, 0.50–1.2 **Si**, 3.9–5.0 **Cu**, 0.40–1.2 **Mn**, 0.20–0.8 **Mg**, 0.10 **Cr**, 0.25 **Zn**, 0.15 **Ti**, 0.7 **Fe**, 0.15 total, others	**Annealed**: all **diam**, 245 MPa **TS**, 127 MPa **YS**, 12% in 2 in. (50 mm) **El**; **Solution heat treated and age hardened**: all **diam**, 343 MPa **TS**, 245 MPa **YS**, 12% in 2 in. (50 mm) **El**; **Solution heat treated and artificially aged hardened**: <13 mm **diam**, 412 MPa **TS**, 363 MPa **YS**, 7% in 2 in. (50 mm) **El**
JIS H 4000	2014	Japan	Sheet, plate, and strip, and coil	rem **Al**, 0.50–1.2 **Si**, 3.9–5.0 **Cu**, 0.40–1.2 **Mn**, 0.20–0.8 **Mg**, 0.10 **Cr**, 0.25 **Zn**, 0.15 **Ti**, 0.7 **Fe**, 0.15 total, others	**Annealed**: <1.5 mm **diam**, 216 MPa **TS**, 108 MPa **YS**, 16% in 2 in. (50 mm) **El**; **Solution heat treated, age hardened, and cold worked**: all **diam**, 412 MPa **TS**, 14% in 2 in. (50 mm) **El**
MIL-T-15089C	2014	US	Tube: seamless	rem **Al**, 0.50–1.2 **Si**, 3.9–5.0 **Cu**, 0.40–1.2 **Mn**, 0.20–0.8 **Mg**, 0.10 **Cr**, 0.25 **Zn**, 0.15 **Ti**, 0.7 **Fe**, 0.20 **Zr+Ti**, 0.05 each, 0.15 total, others	**Heat treated**: all **diam**, 414 MPa **TS**, 365 MPa **YS**, 7% in 2 in. (50 mm) **El**
MIL-A-22771C	2014	US	Forgings	rem **Al**, 0.50–1.2 **Si**, 3.9–5.0 **Cu**, 0.40–1.2 **Mn**, 0.20–0.8 **Mg**, 0.10 **Cr**, 0.25 **Zn**, 0.15 **Ti**, 0.7 **Fe**, 0.05 total, others	**T6**: ≤1 in. **diam**, 448 MPa **TS**, 386 MPa **YS**, 6% in 2 in. (50 mm) **El**, 125 **HB**
SIS 14 43 38	SIS Al 43 38-00	Sweden	Bar	90.24 min **Al**, 0.50–1.2 **Si**, 3.9–5.0 **Cu**, 0.40–1.2 **Mn**, 0.20–0.8 **Mg**, 0.10 **Cr**, 0.25 **Zn**, 0.15 **Ti**, 0.7 **Fe**, 0.20 **Ti+Cr**, 0.05 each, 0.15 total, others	**Hot worked**: 10–150 mm **diam**, 300 MPa **TS**, 200 MPa **YS**, 8% in 2 in. (50 mm) **El**

WROUGHT ALUMINUM AND ALUMINUM ALLOYS

specification number	designation	country	product forms	chemical composition	mechanical properties and hardness values
SIS 14 43 38	SIS Al 43 38–02	Sweden	Plate, sheet, strip, bar, tube	90.24 min **Al**, 0.50–1.2 **Si**, 3.9–5.0 **Cu**, 0.40–1.2 **Mn**, 0.20–0.8 **Mg**, 0.10 **Cr**, 0.25 **Zn**, 0.15 **Ti**, 0.7 **Fe**, 0.20 **Ti+Cr**, 0.05 each, 0.15 total, others	**Annealed**: 5–30 mm **diam**, 220 MPa **TS**, 140 MPa **YS**, 14% in 2 in. (50 mm) **El**, 60–65 **HV**
SIS 14 43 38	SIS Al 43 38–04	Sweden	Plate, sheet, strip, bar, tube, section	90.24 min **Al**, 0.50–1.2 **Si**, 3.9–5.0 **Cu**, 0.40–1.2 **Mn**, 0.20–0.8 **Mg**, 0.10 **Cr**, 0.25 **Zn**, 0.15 **Ti**, 0.7 **Fe**, 0.20 **Ti+Cr**, 0.05 each, 0.15 total, others	**Naturally aged**: 5–30 mm **diam**, 385 MPa **TS**, 240 MPa **YS**, 14% in 2 in. (50 mm) **El**, 120–150 **HV**
SIS 14 43 38	SIS Al 43 38–06	Sweden	Plate, sheet, strip, bar, tube, section, forgings	90.24 **Al**, 0.50–1.2 **Si**, 3.9–5.0 **Cu**, 0.40–1.2 **Mn**, 0.20–0.8 **Mg**, 0.10 **Cr**, 0.25 **Zn**, 0.15 **Ti**, 0.7 **Fe**, 0.20 **Ti+Cr**; 0.05 each, 0.15 total, others	**Artificially aged**: 5–30 mm **diam**, 440 MPa **TS**, 380 MPa **YS**, 7% in 2 in. (50 mm) **El**, 130 **HV**
SIS 14 43 38	SIS Al 43 38–10	Sweden	Plate, sheet, strip	90.24 min **Al**, 0.50–1.2 **Si**, 3.9–5.0 **Cu**, 0.40–1.2 **Mn**, 0.20–0.8 **Mg**, 0.10 **Cr**, 0.25 **Zn**, 0.15 **Ti**, 0.7 **Fe**, 0.20 **Ti+Cr**, 0.05 each, 0.15 total, others	**Strain hardened**: 5–10 mm **diam**, 280 MPa **TS**, 260 MPa **YS**, 3% in 2 in. (50 mm) **El**, 80 **HV**
SFS 2595	AlCu4SiMg	Finland	Shapes and plate	rem **Al**, 0.5–1.2 **Si**, 3.8–5.0 **Cu**, 0.3–1.2 **Mn**, 0.2–0.8 **Mg**, 0.2 min **Ni**, 0.2 **Zn**, 0.7 **Fe**, 0.3 **Ti+Zr+Cr**, 0.05 each, 0.15 total, others	**Hot finished (shapes)**: all **diam**, 300 MPa **TS**, 200 MPa **YS**, 8% in 2 in. (50 mm) **El**; **Solution treated and naturally aged (plate)**: ≤12 mm **diam**, 380 MPa **TS**, 240 MPa **YS**, 14% in 2 in. (50 mm) **El**; **Solution treated and aged**: ≤12 mm **diam**, 430 MPa **TS**, 370 MPa **YS**, 7% in 2 in. (50 mm) **El**
IS 736	24345 Alclad	India	Plate	rem **Al**, 0.5–1.2 **Si**, 3.8–5.0 **Cu**, 0.3–1.2 **Mn**, 0.2–0.8 **Mg**, 0.3 **Cr**, 0.2 **Zn**, 0.3 **Ti**, 0.7 **Fe**	**Solution heat treated and naturally aged**: 12 mm **diam**, 375 MPa **TS**, 225 MPa **YS**, 12% in 2 in. (50 mm) **El**; **Solution and precipitation heat treated**: 7 mm **diam**, 420 MPa **TS**, 310 MPa **YS**, 7% in 2 in. (50 mm) **El**
IS 733	24345	India	Bar, rod, section	rem **Al**, 0.5–1.2 **Si**, 3.8–5.0 **Cu**, 0.3–1.2 **Mn**, 0.2–0.8 **Mg**, 0.3 **Cr**, 0.2 **Zn**, 0.3 **Ti**, 0.7 **Fe**	**Solution treated and naturally aged**: ≤10 mm **diam**, 375 MPa **TS**, 225 MPa **YS**, 10% in 2 in. (50 mm) **El**; **Solution and precipitation heat treated**: ≤10 mm **diam**, 430 MPa **TS**, 375 MPa **YS**, 6% in 2 in. (50 mm) **El**
ISO R827	Al–Cu4SiMg	ISO	Extruded products	rem **Al**, 0.5–1.2 **Si**, 3.8–5.0 **Cu**, 0.3–1.2 **Mn**, 0.2–0.8 **Mg**, 0.2 **Ni**, 0.2 **Zn**, 0.7 **Fe**, 0.03 **Ti+Zr+Cr**	**Solution treated and naturally aged**: 0.75 max in. (20 max mm) **diam**, 353 MPa **TS**, 226 MPa **YS**, 11% in 2 in. (50 mm) **El**; **Solution and precipitation treated**: 0.38 max in. (10 max mm) **diam**, 412 MPa **TS**, 363 MPa **YS**, 7% in 2 in. (50 mm) **El**
ISO R209	Al–CuSiMg	ISO	Wrought products	rem **Al**, 0.5–1.2 **Si**, 3.8–5.0 **Cu**, 0.3–1.2 **Mn**, 0.2–0.8 **Mg**, 0.2 **Ni**, 0.2 **Zn**, 0.7 **Fe**, 0.3 **Ti+Zr+Cr**	

WROUGHT ALUMINUM AND ALUMINUM ALLOYS

specification number	designation	country	product forms	chemical composition	mechanical properties and hardness values
ISO TR2778	Al–Cu4SiMg	ISO	Tube: drawn	rem **Al**, 0.5–1.2 **Si**, 3.8–5.0 **Cu**, 0.3–1.2 **Mn**, 0.2–0.8 **Mg**, 0.2 **Ni**, 0.2 **Zn**, 0.7 **Fe**, 0.3 **Ti+Zr+Cr**	**Solution heat treated and naturally aged**: all **diam**, 380 MPa **TS**, 235 MPa **YS**, 10% in 2 in. (50 mm) **El**; **Solution and precipitation treated**: all **diam**, 450 MPa **TS**, 380 MPa **YS**, 6% in 2 in. (50 mm) **El**
ISO R829	Al–Cu4SiMg	ISO	Forgings	rem **Al**, 0.5–1.2 **Si**, 3.8–5.0 **Cu**, 0.3–1.2 **Mn**, 0.2–0.8 **Mg**, 0.2 **Ni**, 0.2 **Zn**, 0.7 **Fe**, 0.3 **Ti+Zr+Cr**	**Solution treated and naturally aged**: all **diam**, 353 MPa **TS**, 206 MPa **YS**, 11% in 2 in. (50 mm) **El**; **Solution and precipitation treated**: all **diam**, 412 MPa **TS**, 363 MPa **YS**, 6% in 2 in. (50 mm) **El**
NS 17 105	NS 17 105–54	Norway	Sheet, strip, plate, and profiles	rem **Al**, 0.5–1.2 **Si**, 3.8–5.0 **Cu**, 0.3–1.2 **Mn**, 0.2–0.8 **Mg**, 0.2 **Ni**, 0.2 **Zn**, 0.7 **Fe**, 0.3 **Ti+Zr+Cr**	**Hardened and naturally aged (except profiles)**: 6.3–25 mm **diam**, 387 MPa **TS**, 240 MPa **YS**, 12% in 2 in. (50 mm) **El**; **Hardened and naturally aged (profiles)**: <20 mm **diam**, 353 MPa **TS**, 226 MPa **YS**, 11% in 2 in. (50 mm) **El**
NS 17 105	NS 17 105–56	Norway	Sheet, strip, plate, and profiles	rem **Al**, 0.5–1.2 **Si**, 3.8–5.0 **Cu**, 0.3–1.2 **Mn**, 0.2–0.8 **Mg**, 0.2 **Ni**, 0.2 **Zn**, 0.7 **Fe**, 0.3 **Ti+Zr+Cr**	**Hardened and artificially aged (except profiles)**: 12.5–25 mm **diam**, 441 MPa **TS**, 377 MPa **YS**, 6% in 2 in. (50 mm) **El**; **Hardened and artificially aged (profiles)**: <10 mm **diam**, 412 MPa **TS**, 363 MPa **YS**, 7% in 2 in. (50 mm) **El**
SABS 712	Al–Cu4SiMg	South Africa	Wrought products	rem **Al**, 0.5–1.2 **Si**, 3.9–4.9 **Cu**, 0.4–1.2 **Mn**, 0.2–0.8 **Mg**, 0.1 **Cr**, 0.25 **Zn**, 0.7 **Fe**, 0.2 **Ti+Zr**	
AECMA prEN2091	Al–P13 Pl–T4	AECMA	Sheet, strip	rem **Al**, 0.50 **Si**, 3.8–4.9 **Cu**, 0.30–0.9 **Mn**, 1.2–1.8 **Mg**, 0.10 **Cr**, 0.10 **Ni**, 0.25 **Zn**, 0.15 **Ti**, 0.20 **Ti+Zr**, 0.05 each, 0.15 total, others	**Solution treated, quenched and naturally aged**: 0.4–0.8 mm **diam**, 390 MPa **TS**, 235 MPa **YS**, 12% in 2 in. (50 mm) **El**
AECMA prEN2090	Al–P13 Pl–T3	AECMA	Sheet and strip	rem **Al**, 0.050 **Si**, 3.8–4.9 **Cu**, 0.30–0.9 **Mn**, 1.2–1.8 **Mg**, 0.10 **Cr**, 0.10 **Ni**, 0.25 **Zn**, 0.15 **Ti**, 0.20 **Ti+Zr**, 0.05 each, 0.15 total, others	**Solution treated, quenched and naturally aged**: 0.4–0.8 mm **diam**, 405 MPa **TS**, 270 MPa **YS**, 12% in 2 in. (50 mm) **El**
AMS 4060		US	Sheet and plate: alclad	rem **Al**, 0.50 **Si**, 3.8–4.9 **Cu**, 0.30–0.9 **Mn**, 1.2–1.8 **Mg**, 0.10 **Cr**, 0.25 **Zn**, 0.50 **Fe**, 0.05 each, 0.15 total, others	**Solution heat treated and cold reduced**: 0.020–0.062 in. **diam**, 61 ksi (421 MPa) **TS**, 47 ksi (324 MPa) **YS**, 8% in 2 in. (50 mm) **El**
AMS 4061		US	Sheet and plate: alclad	rem **Al**, 0.50 **Si**, 3.8–4.9 **Cu**, 0.30–0.9 **Mn**, 1.2–1.8 **Mg**, 0.10 **Cr**, 0.25 **Zn**, 0.50 **Fe**, 0.05 each, 0.15 total, others	**Solution heat treated and cold reduced**: 0.063–0.187 in. **diam**, 64 ksi (441 MPa) **TS**, 48 ksi (331 MPa) **YS**, 9% in 2 in. (50 mm) **El**

WROUGHT ALUMINUM AND ALUMINUM ALLOYS

specification number	designation	country	product forms	chemical composition	mechanical properties and hardness values
AMS 4072		US	Sheet and plate: alclad	rem **Al**, 0.50 **Si**, 3.8–4.9 **Cu**, 0.30–0.9 **Mn**, 1.2–1.8 **Mg**, 0.10 **Cr**, 0.25 **Zn**, 0.50 **Fe**, 0.05 each, 0.15 total, others	**Solution and precipitation heat treated**: 0.020–0.062 in. **diam**, 66 ksi (455 MPa) **TS**, 62 ksi (427 MPa) **YS**, 3% in 2 in. (50 mm) **El**
AMS 4073		US	Sheet and plate: alclad	rem **Al**, 0.50 **Si**, 3.8–4.9 **Cu**, 0.30–0.9 **Mn**, 1.2–1.8 **Mg**, 0.10 **Cr**, 0.25 **Zn**, 0.50 **Fe**, 0.05 each, 0.15 total, others	**Solution and precipitation heat treated**: 0.020–0.062 in. **diam**, 66 ksi (455 MPa) **TS**, 62 ksi (427 MPa) **YS**, 3% in 2 in. (50 mm) **El**
AMS 4074		US	Sheet and plate: alclad	rem **Al**, 0.50 **Si**, 3.8–4.9 **Cu**, 0.30–0.9 **Mn**, 1.2–1.8 **Mg**, 0.10 **Cr**, 0.25 **Zn**, 0.50 **Fe**, 0.05 each, 0.15 total, others	**Solution and precipitation heat treated**: 0.020–0.062 in. **diam**, 64 ksi (441 MPa) **TS**, 58 ksi (400 MPa) **YS**, 3% in 2 in. (50 mm) **El**
AMS 4075		US	Sheet and plate: alclad	rem **Al**, 0.50 **Si**, 3.8–4.9 **Cu**, 0.30–0.9 **Mn**, 1.2–1.8 **Mg**, 0.10 **Cr**, 0.25 **Zn**, 0.50 **Fe**, 0.05 each, 0.15 total, others	**Solution and precipitation heat treated**: 0.063–0.187 in. **diam**, 69 ksi (476 MPa) **TS**, 64 ksi (441 MPa) **YS**, 4% in 2 in. (50 mm) **El**
AMS 4077C		US	Sheet and plate: alclad one side	rem **Al**, 0.50 **Si**, 3.8–4.9 **Cu**, 0.30–0.9 **Mn**, 1.2–1.8 **Mg**, 0.10 **Cr**, 0.25 **Zn**, 0.15 **Ti**, 0.50 **Fe**, 0.05 each, 0.15 total, others	**Annealed**: 0.010–0.062 in. (0.25–1.57 mm) **diam**, 31 ksi (217 MPa) **TS**, 14 ksi (97 MPa) **YS**, 12% in 2 in. (50 mm) **El**; **Solution heat treated and aged**: 0.010–0.020 in. (0.25–0.51 mm) **diam**, 59 ksi (407 MPa) **TS**, 35 ksi (241 MPa) **YS**, 12% in 2 in. (50 mm) **El**
AMS 4086J		US	Tube: seamless	rem **Al**, 0.50 **Si**, 3.8–4.9 **Cu**, 0.30–0.9 **Mn**, 1.2–1.8 **Mg**, 0.10 **Cr**, 0.25 **Zn**, 0.15 **Ti**, 0.50 **Fe**, 0.20 **Zr+Ti**, 0.05 each, 0.15 total, others	**Solution heat treated and cold worked**: 0.018–0.024 in. (0.46–0.61 mm) **diam**, 64 ksi (441 MPa) **TS**, 42 ksi (290 MPa) **YS**, 10% in 2 in. (50 mm) **El**
AMS 4087D		US	Tube: seamless	rem **Al**, 0.50 **Si**, 3.8–4.9 **Cu**, 0.30–0.9 **Mn**, 1.2–1.8 **Mg**, 0.10 **Cr**, 0.25 **Zn**, 0.50 **Fe**, 0.05 each, 0.15 total, others	**Annealed**: 0.018–0.500 in. **diam**, 32 ksi (221 MPa) **TS**, 15 ksi (103 MPa) **YS**; **Solution heat treated and aged**: 0.018–0.024 in. **diam**, 64 ksi (441 MPa) **TS**, 40 ksi (276 MPa) **YS**, 10% in 2 in. (50 mm) **El**
AMS 4088F		US	Tube: seamless	rem **Al**, 0.50 **Si**, 3.8–4.9 **Cu**, 0.30–0.9 **Mn**, 1.2–1.8 **Mg**, 0.10 **Cr**, 0.25 **Zn**, 0.50 **Fe**, 0.05 each, 0.15 total, others	**Solution heat treated and cold worked**: 0.018–0.024 in. **diam**, 64 ksi (441 MPa) **TS**, 42 ksi (290 MPa) **YS**, 10% in 2 in.(50 mm) **El**
AMS 4097		US	Sheet and plate	rem **Al**, 0.50 **Si**, 3.8–4.9 **Cu**, 0.30–0.9 **Mn**, 1.2–1.8 **Mg**, 0.10 **Cr**, 0.25 **Zn**, 0.50 **Fe**, 0.05 each, 0.15 total, others	**Solution heat treated and cold reduced**: 0.020–0.062 in. **diam**, 69 ksi (476 MPa) **TS**, 52 ksi (359 MPa) **YS**, 8% in 2 in. (50 mm) **El**

WROUGHT ALUMINUM AND ALUMINUM ALLOYS

specification number	designation	country	product forms	chemical composition	mechanical properties and hardness values
AMS 4098		US	Sheet and plate	rem **Al**, 0.50 **Si**, 3.8–4.9 **Cu**, 0.30–0.9 **Mn**, 1.2–1.8 **Mg**, 0.10 **Cr**, 0.25 **Zn**, 0.50 **Fe**, 0.05 each, 0.15 total, others	**Solution heat treated and cold reduced**: 0.020–0.062 in. **diam**, 67 ksi (462 MPa) **TS**, 50 ksi (345 MPa) **YS**, 8% in 2 in. (50 mm) **El**
AMS 4099		US	Sheet and plate	rem **Al**, 0.50 **Si**, 3.8–4.9 **Cu**, 0.30–0.9 **Mn**, 1.2–1.8 **Mg**, 0.10 **Cr**, 0.25 **Zn**, 0.50 **Fe**, 0.05 each, 0.15 total, others	**Solution heat treated and cold reduced**: 0.063–0.249 in. **diam**, 67 ksi (462 MPa) **TS**, 50 ksi (345 MPa) **YS**, 8% in 2 in. (50 mm) **El**
AMS 4035F		US	Sheet and plate	rem **Al**, 0.50 **Si**, 3.8–4.9 **Cu**, 0.30–0.9 **Mn**, 1.2–1.8 **Mg**, 0.10 **Cr**, 0.25 **Zn**, 0.50 **Fe**, 0.05 each, 0.15 total, others	**Annealed**: 0.010–0.499 in. **diam**, 32 ksi (221 MPa) **TS**, 14 ksi (97 MPa) **YS**, 12% in 2 in. (50 mm) **El**; **Solution heat treated and aged**: 0.010–0.020 in. **diam**, 62 ksi (427 MPa) **TS**, 38 ksi (262 MPa) **YS**, 12% in 2 in. (50 mm) **El**
AMS 4037H		US	Sheet and plate	rem **Al**, 0.50 **Si**, 3.8–4.9 **Cu**, 0.30–0.9 **Mn**, 1.2–1.8 **Mg**, 0.10 **Cr**, 0.25 **Zn**, 0.50 **Fe**, 0.05 each, 0.15 total, others	**Solution heat treated and cold worked (sheet)**: 0.008–0.009 in. **diam**, 63 ksi (434 MPa) **TS**, 42 ksi (290 MPa) **YS**, 10% in 2 in. (50 mm) **El**
AMS 4040H		US	Sheet and plate: alclad	rem **Al**, 0.50 **Si**, 3.8–4.9 **Cu**, 0.30–0.9 **Mn**, 1.2–1.8 **Mg**, 0.10 **Cr**, 0.25 **Zn**, 0.15 **Ti**, 0.50 **Fe**, 0.05 each, 0.15 total, others	**Annealed**: 0.008–0.009 in. (0.20–0.30 mm) **diam**, 30 ksi (207 MPa) **TS**, 14 ksi (97 MPa) **YS**, 10% in 2 in. (50 mm) **El**; **Solution heat treated and aged**: 0.008–0.009 in. (0.20–0.23 mm) **diam**, 55 ksi (379 MPa) **TS**, 34 ksi (234 MPa) **YS**, 10% in 2 in. (50 mm) **El**
AMS 4041J		US	Sheet and plate: alclad	rem **Al**, 0.50 **Si**, 3.8–4.9 **Cu**, 0.30–0.9 **Mn**, 1.2–1.8 **Mg**, 0.10 **Cr**, 0.25 **Zn**, 0.50 **Fe**, 0.05 each, 0.15 total, others	**Solution heat treated and cold worked (sheet)**: 0.008–0.009 in. **diam**, 58 ksi (400 MPa) **TS**, 39 ksi (269 MPa) **YS**, 10% in 2 in. (50 mm) **El**
AMS 4192		US	Sheet and plate	rem **Al**, 0.50 **Si**, 3.8–4.9 **Cu**, 0.30–0.9 **Mn**, 1.2–1.8 **Mg**, 0.10 **Cr**, 0.25 **Zn**, 0.50 **Fe**, 0.05 each, 0.15 total, others	**Solution heat treated and cold reduced**: 0.020–0.062 in. **diam**, 67 ksi (462 MPa) **TS**, 50 ksi (345 MPa) **YS**, 8% in 2 in. (50 mm) **El**
AMS 4193		US	Sheet and plate	rem **Al**, 0.50 **Si**, 3.8–4.9 **Cu**, 0.30–0.9 **Mn**, 1.2–1.8 **Mg**, 0.10 **Cr**, 0.25 **Zn**, 0.50 **Fe**, 0.05 each, 0.15 total, others	**Solution and precipitation heat treated**: 0.020–0.062 in. **diam**, 70 ksi (483 MPa) **TS**, 62 ksi (427 MPa) **YS**, 3% in 2 in. (50 mm) **El**
AMS 4194		US	Sheet and plate: alclad	rem **Al**, 0.50 **Si**, 3.8–4.9 **Cu**, 0.30–0.9 **Mn**, 1.2–1.8 **Mg**, 0.10 **Cr**, 0.25 **Zn**, 0.50 **Fe**, 0.05 each, 0.15 total, others	**Solution heat treated and cold reduced**: 0.020–0.062 in. **diam**, 61 ksi (421 MPa) **TS**, 47 ksi (324 MPa) **YS**, 8% in 2 in. (50 mm) **El**

WROUGHT ALUMINUM AND ALUMINUM ALLOYS

specification number	designation	country	product forms	chemical composition	mechanical properties and hardness values
AMS 4195		US	Sheet and plate: alclad	rem **Al**, 0.50 **Si**, 3.8–4.9 **Cu**, 0.30–0.9 **Mn**, 1.2–1.8 **Mg**, 0.10 **Cr**, 0.25 **Zn**, 0.50 **Fe**, 0.05 each, 0.15 total, others	**Solution and precipitation heat treated**: 0.020–0.062 in. **diam**, 64 ksi (441 MPa) **TS**, 58 ksi (400 MPa) **YS**, 3% in 2 in. (50 mm) **El**
AMS 4164D		US	Bar, tube, rod, and shapes: extruded	rem **Al**, 0.50 **Si**, 3.8–4.9 **Cu**, 0.30–0.9 **Mn**, 1.2–1.8 **Mg**, 0.10 **Cr**, 0.25 **Zn**, 0.50 **Fe**, 0.05 each, 0.15 total, others	**Solution heat treated**: 0.749–1.499 in. **diam**, 65 ksi (448 MPa) **TS**, 46 ksi (317 MPa) **YS**, 10% in 2 in. (50 mm) **El**, 100 min **HB**
AMS 4165D		US	Bar, tube, rod, and shapes: extruded	rem **Al**, 0.50 **Si**, 3.8–4.9 **Cu**, 0.30–0.9 **Mn**, 1.2–1.8 **Mg**, 0.10 **Cr**, 0.25 **Zn**, 0.50 **Fe**, 0.05 each, 0.15 total, others	**Solution heat treated**: 0.749–1.499 in. **diam**, 65 ksi (448 MPa) **TS**, 46 ksi (317 MPa) **YS**, 10% in 2 in. (50 mm) **El**, 100 min **HB**
AMS 4152J		US	Bar, tube, rod, wire, and shapes: extruded	rem **Al**, 0.50 **Si**, 3.8–4.9 **Cu**, 0.30–0.9 **Mn**, 1.2–1.8 **Mg**, 0.10 **Cr**, 0.25 **Zn**, 0.15 **Ti**, 0.50 **Fe**, 0.20 **Zr+Ti**, 0.05 each, 0.15 total, others	**Solution heat treated (except tube)**: ≤0.249 in. (≤6.35 mm) **diam**, 57 ksi (393 MPa) **TS**, 42 ksi (290 MPa) **YS**, 12% in 2 in. (50 mm) **El**, 100 min **HB**; **Solution heat treated (tube)**: ≤0.249 in. (≤6.35 mm) **diam**, 57 ksi (393 MPa) **TS**, 42 ksi (290 MPa) **YS**, 10% in 2 in. (50 mm) **El**, 100 min **HB**
AMS 4042F		US	Sheet and plate: alclad	rem **Al**, 0.50 **Si**, 3.8–4.9 **Cu**, 0.30–0.9 **Mn**, 1.2–1.8 **Mg**, 0.10 **Cr**, 0.25 **Zn**, 0.50 **Fe**, 0.05 each, 0.15 total, others	**Solution heat treated and cold reduced**: 0.020–0.062 in. **diam**, 62 ksi (427 MPa) **TS**, 48 ksi (331 MPa) **YS**, 8% in 2 in. (50 mm) **El**
AMS 4007		US	Foil	rem **Al**, 0.50 **Si**, 3.8–4.9 **Cu**, 0.30–0.9 **Mn**, 1.2–1.8 **Mg**, 0.10 **Cr**, 0.25 **Zn**, 0.50 **Fe**, 0.05 each, 0.15 total, others	**Annealed**: all **diam**, 32 ksi (221 MPa) **TS**; **Solution heat treated**: all **diam**, 62 ksi (427 MPa) **TS**
AMS 4120H		US	Bar, rod, and wire	rem **Al**, 0.50 **Si**, 3.8–4.9 **Cu**, 0.30–0.9 **Mn**, 1.2–1.8 **Mg**, 0.10 **Cr**, 0.25 **Zn**, 0.15 **Ti**, 0.50 **Fe**, 0.20 **Zr+Ti**, 0.05 each, 0.15 total, others	**Solution heat treated**: 0.125–0.499 in. (3.18–12.67 mm) **diam**, 62 ksi (427 MPa) **TS**, 45 ksi (310 MPa) **YS**, 10% in 2 in. (50 mm) **El**, 100 min **HB**
AMS 4119D		US	Bar and rod	rem **Al**, 0.50 **Si**, 3.8–4.9 **Cu**, 0.30–0.9 **Mn**, 1.2–1.8 **Mg**, 0.10 **Cr**, 0.25 **Zn**, 0.50 **Fe**, 0.05 each, 0.15 total, others	**Solution heat treated and stress relieved**: 0.500–4.000 in. **diam**, 62 ksi (427 MPa) **TS**, 45 ksi (310 MPa) **YS**, 1% in 2 in. (50 mm) **El**, 100 min **HB**

WROUGHT ALUMINUM AND ALUMINUM ALLOYS

specification number	designation	country	product forms	chemical composition	mechanical properties and hardness values
ANSI/ASTM B 316	2024	US	Wire and rod: rivet and cold-heading	rem **Al**, 0.50 **Si**, 3.8–4.9 **Cu**, 0.30–0.9 **Mn**, 1.2–1.8 **Mg**, 0.10 **Cr**, 0.25 **Zn**, 0.15 **Ti**, 0.50 **Fe**, 0.20 **Zr+Ti**, 0.05 each, 0.15 total, others	**Annealed**: ≤1.000 in. **diam**, 35 max ksi (241 max MPa) **TS**; **H1**: ≤1.000 in. **diam**, 32 ksi (221 MPa) **TS**; **Solution heat treated and naturally aged**: 0.061–1.000 in. **diam**, 62 ksi (427 MPa) **TS**, 40 ksi (276 MPa) **YS**, 10% in 2 in. (50 mm) **El**
ANSI/ASTM B 241	2024	US	Tube: seamless	rem **Al**, 0.50 **Si**, 3.8–4.9 **Cu**, 0.30–0.9 **Mn**, 1.2–1.8 **Mg**, 0.10 **Cr**, 0.25 **Zn**, 0.15 **Ti**, 0.50 **Fe**, 0.20 **Zr+Ti**, 0.05 each, 0.15 total, others	**Annealed**: all **diam**, 35 max ksi (241 max MPa) **TS**, 19 max ksi (131 max MPa) **YS**, 12% in 2 in. (50 mm) **El**; **Solution heat treated and naturally aged**: 0.750–1.499 in. **diam**, 65 ksi (448 MPa) **TS**, 46 ksi (317 MPa) **YS**, 10% in 2 in. (50 mm) **El**; **T81**: 0.050–0.249 in. **diam**, 64 ksi (441 MPa) **TS**, 56 ksi (386 MPa) **YS**, 4% in 2 in. (50 mm) **El**
ANSI/ASTM B 221	2024	US	Bar, tube, rod, wire, and shapes	rem **Al**, 0.50 **Si**, 3.8–4.9 **Cu**, 0.30–0.9 **Mn**, 1.2–1.8 **Mg**, 0.10 **Cr**, 0.25 **Zn**, 0.15 **Ti**, 0.50 **Fe**, 0.20 **Ti+Zr**, 0.05 each, 0.15 total, others	**Annealed**: all **diam**, 35 max ksi (241 max MPa) **TS**, 19 max ksi (131 max MPa) **YS**, 12% in 2 in. (50 mm) **El**; **Solution heat treated and naturally aged (except tube)**: ≥1.500 in. **diam**, 68 ksi (469 MPa) **TS**, 48 ksi (331 MPa) **YS**, 8% in 2 in. (50 mm) **El**; **T81**: ≥1.500 in. **diam**, 66 ksi (455 MPa) **TS**, 58 ksi (400 MPa) **YS**, 5% in 2 in. (50 mm) **El**
ANSI/ASTM B 211	2024	US	Bar, rod, and wire	rem **Al**, 0.50 **Si**, 3.8–4.9 **Cu**, 0.30–0.9 **Mn**, 1.2–1.8 **Mg**, 0.10 **Cr**, 0.25 **Zn**, 0.15 **Ti**, 0.50 **Fe**, 0.20 **Ti+Zr**, 0.05 each, 0.15 total, others	**Annealed**: 0.125–8.000 in. **diam**, 35 max ksi (241 max MPa) **TS**, 16% in 2 in. (50 mm) **El**; **Solution heat treated and naturally aged**: 0.500–4.500 in. **diam**, 62 ksi (427 MPa) **TS**, 42 ksi (290 MPa) **YS**, 10% in 2 in. (50 mm) **El**; **Solution heat treated and artificially aged**: 0.125–4.000 in. **diam**, 62 ksi (427 MPa) **TS**, 50 ksi (345 MPa) **YS**, 5% in 2 in. (50 mm) **El**

WROUGHT ALUMINUM AND ALUMINUM ALLOYS

specification number	designation	country	product forms	chemical composition	mechanical properties and hardness values
ANSI/ASTM B 209	Alclad 2024	US	Sheet and plate	rem **Al**, 0.50 **Si**, 3.8–4.9 **Cu**, 0.30–0.9 **Mn**, 1.2–1.8 **Mg**, 0.10 **Cr**, 0.25 **Zn**, 0.15 **Ti**, 0.50 **Fe**, 0.20 **Ti+Zr**, 1230 cladding material, 0.05 each, 0.15 total, others	**Annealed**: 0.063–0.499 in. **diam**, 32 max ksi (221 MPa) **TS**, 14 max ksi (97 MPa) **YS**, 12% in 2 in. (50 mm) **El**; **Solution heat treated and naturally aged (sheet)**: 0.063–0.128 in. **diam**, 61 ksi (421 MPa) **TS**, 38 ksi (262 MPa) **YS**, 15% in 2 in. (50 mm) **El**; **T81**: 0.063–0.249 in. **diam**, 65 ksi (448 MPa) **TS**, 56 ksi (386 MPa) **YS**, 5% in 2 in. (50 mm) **El**
ANSI/ASTM B 210	2024	US	Tube: seamless	rem **Al**, 0.50 **Si**, 3.8–4.9 **Cu**, 0.30–0.9 **Mn**, 1.2–1.8 **Mg**, 0.10 **Cr**, 0.25 **Zn**, 0.15 **Ti**, 0.50 **Fe**; 0.20 **Ti+Zr**, 0.05 each, 0.15 total, others	**Annealed**: 0.018–0.500 in. **diam**, 32 max ksi (220 max MPa) **TS**, 15 max ksi (103 max MPa) **YS**; **T3**: 0.260–0.500 in. **diam**, 64 ksi (441 MPa) **TS**, 42 ksi (290 MPa) **YS**, 12% in 2 in. (50 mm) **El**
ANSI/ASTM B 209	2024	US	Sheet and plate	rem **Al**, 0.50 **Si**, 3.8–4.9 **Cu**, 0.30–0.9 **Mn**, 1.2–1.8 **Mg**, 0.10 **Cr**, 0.25 **Zn**, 0.15 **Ti**, 0.50 **Fe**, 0.20 **Ti+Zr**, 0.05 each, 0.15 total, others	**Annealed**: 0.500–1.750 in. **diam**, 32 max ksi (221 max MPa) **TS**, 14 max ksi (97 max MPa) **YS**, 12% in 2 in. (50 mm) **El**; **Solution heat treated and naturally aged (sheet)**: 0.021–0.249 in. **diam**, 62 ksi (427 MPa) **TS**, 40 ksi (276 MPa) **YS**, 15% in 2 in. (50 mm) **El**; **T81**: 0.010–0.249 in. **diam**, 67 ksi (462 MPa) **TS**, 58 ksi (400 MPa) **YS**, 5% in 2 in. (50 mm) **El**
ANSI/ASTM B 209	2124	US	Sheet and plate	0.20 **Si**, 3.8–4.9 **Cu**, 0.30–0.9 **Mn**, 1.2–1.8 **Mg**, 0.10 **Cr**, 0.25 **Zn**, 0.15 **Ti**, 0.30 **Fe**, 0.20 **Ti+Zr**, 0.05 each, 0.15 total, others	**T851**: 1.500–2.000 in. **diam**, 66 ksi (455 MPa) **TS**, 57 ksi (393 MPa) **YS**, 6% in 2 in. (50 mm) **El**
ASTM F 468	A 92094	US	Bolt, screw, stud	rem **Al**, 0.50 **Si**, 3.8–4.9 **Cu**, 1.2–1.8 **Mn**, 0.30–0.9 **Mg**, 0.10 **Cr**, 0.25 **Zn**, 0.15 **Ti**, 0.50 **Fe**, 0.20 **Ti+Zr**	**As manufactured**: all **diam**, 55 ksi (380 MPa) **TS**, 36 ksi (250 MPa) **YS**, 70–85 **HRB**
ASTM F 467	A 92024	US	Nut	rem **Al**, 0.50 **Si**, 3.8–4.9 **Cu**, 0.30–0.9 **Mn**, 1.2–1.8 **Mg**, 0.10 **Cr**, 0.25 **Zn**, 0.15 **Ti**, 0.50 **Fe**	**As agreed**: all **diam**, 70 min **HRB**
AS 1734	Alclad 2024	Australia	Sheet, plate	rem **Al**, 0.50 **Si**, 3.8–4.9 **Cu**, 0.9 **Mn**, 1.2–1.8 **Mg**, 0.10 **Cr**, 0.25 **Zn**, 0.15 **Ti**, 0.50 **Fe**, 0.05 each, 0.15 total, others	**Annealed**: 0.25–1.6 mm **diam**, 221 MPa **TS**, 97 MPa **YS**, 12% in 2 in. (50 mm) **El**; **Solution heat treated and naturally aged (by user)**: 0.25–1.6 mm **diam**, 393 MPa **TS**, 234 MPa **YS**, 12% in 2 in. (50 mm) **El**; **Solution heat treated and artificially aged (by user)**: 0.25–1.6 mm **diam**, 413 MPa **TS**, 324 MPa **YS**, 5% in 2 in. (50 mm) **El**

WROUGHT ALUMINUM AND ALUMINUM ALLOYS

specification number	designation	country	product forms	chemical composition	mechanical properties and hardness values
BS L.110		UK	Sheet and strip (core material)	rem **Al**, 0.50 **Si**, 3.8–4.9 **Cu**, 0.30–0.9 **Mn**, 1.2–1.8 **Mg**, 0.10 **Cr**, 0.05 **Ni**, 0.2 **Zn**, 0.05 **Sn**, 0.50 **Fe**, 0.05 **Pb**, 0.2 **Zr + Ti**, (clad with BS L.110 cladding material)	**Solution heat treated, quenched and aged**: 0.4–0.8 mm **diam**, 390 MPa **TS**, 235 MPa **YS**, 12% in 2 in. (50 mm) **El**
BS L.109		UK	Sheet and strip (core material)	rem **Al**, 0.50 **Si**, 3.8–4.9 **Cu**, 0.30–0.9 **Mn**, 1.2–1.8 **Mg**, 0.10 **Cr**, 0.05 **Ni**, 0.2 **Zn**, 0.05 **Sn**, 0.50 **Fe**, 0.05 **Pb**, 0.2 **Zr + Ti**, (clad with BS L.109 cladding material)	**Solution heat treated, quenched and aged**: 0.4–0.8 mm **diam**, 405 MPa **TS**, 270 MPa **YS**, 12% in 2 in. (50 mm) **El**
BS 2L.97		UK	Plate	rem **Al**, 0.50 **Si**, 3.8–4.9 **Cu**, 0.30–0.9 **Mn**, 1.2–1.8 **Mg**, 0.10 **Cr**, 0.05 **Ni**, 0.2 **Zn**, 0.05 **Sn**, 0.50 **Fe**, 0.05 **Pb**, 0.20 **Zr + Ti**	**Solution heat treated, quenched and naturally aged**: 6–12.5 mm **diam**, 430 MPa **TS**, 280 MPa **YS**, 10% in 2 in. (50 mm) **El**
CSA HA.4	.2024 Alclad	Canada	Plate, sheet (core)	rem **Al**, .50 **Si**, 3.8–4.9 **Cu**, 0.30–0.9 **Mn**, 1.2–1.8 **Mg**, 0.10 **Cr**, 0.25 **Zn**, 0.15 **Ti**, .50 **Fe**, 0.05 each, 0.15 total, others	**O**: 0.010–0.032 in. **diam**, 30.0 max ksi (207 max MPa) **TS**, 14.0 max ksi (97 max MPa) **YS**, 12% in 2 in. (50 mm) **El**; **T3 (sheet)**: 0.010–0.020 in. **diam**, 59.0 ksi (407 MPa) **TS**, 39.0 ksi (269 MPa) **YS**, 12% in 2 in. (50 mm) **El**; **T42**: 0.010–0.020 in. **diam**, 57.0 ksi (393 MPa) **TS**, 34.0 ksi (234 MPa) **YS**, 12% in 2 in. (50 mm) **El**
CSA HA.7	.2024	Canada	Tube, pipe: seamless	rem **Al**, 0.50 **Si**, 3.8–4.9 **Cu**, 0.30–0.9 **Mn**, 1.2–1.8 **Mg**, 0.10 **Cr**, 0.25 **Zn**, 0.15 **Ti**, 0.50 **Fe**, 0.20 **Ti + Zr**, 0.05 each, 0.15 total, other	**Drawn and O**: all **diam**, 32.0 max ksi (221 max MPa) **TS**, 15.0 ksi (103 MPa) **YS**; **Drawn and T3**: 0.015–0.024 in. **diam**, 64.0 ksi (441 MPa) **TS**, 42.0 ksi (290 MPa) **YS**, 10% in 2 in. (50 mm) **El**; **Extruded and O**: all **diam**, 35.0 max ksi (241 max MPa) **TS**, 19.0 max MPa) **YS**, 12% in 2 in. (50 mm) **El**
CSA HA.6	.2024	Canada	Rivet, rod, wire	rem **Al**, 0.50 **Si**, 3.8–4.9 **Cu**, 0.30–0.9 **Mn**, 1.2–1.8 **Mg**, 0.10 **Cr**, 0.25 **Zn**, 0.15 **Ti**, 0.50 **Fe**, 0.20 **Ti + Zr**, 0.05 each, 0.15 total, others	**H13**: 0.062–1.000 in. **diam**, 32.0 ksi (221 MPa) **TS**; **T4**: 0.062–1.000 in. **diam**, 62.0 ksi (427 MPa) **TS**, 40.0 ksi (276 MPa) **YS**, 10% in 2 in. (50 mm) **El**
CSA HA.5	.2024	Canada	Bar, rod, wire, shapes	rem **Al**, 0.50 **Si**, 3.8–4.9 **Cu**, 0.30–0.9 **Mn**, 1.2–1.8 **Mg**, 0.10 **Cr**, 0.25 **Zn**, 0.15 **Ti**, 0.50 **Fe**, 0.05 each, 0.15 total, others	**Drawn and O**: <2.500 in. **diam**, 35.0 max ksi (241 max MPa) **TS**, 16% in 2 in. (50 mm) **El**; **Drawn and T4**: <0.499 in. **diam**, 62.0 ksi (427 MPa) **TS**, 45.0 ksi (310 MPa) **YS**, 10% in 2 in. (50 mm) **El**; **Extruded and T3510**: 0.250–0.749 in. **diam**, 62.0 ksi (414 MPa) **TS**, 40.0 ksi (303 MPa) **YS**, 12% in 2 in. (50 mm) **El**

WROUGHT ALUMINUM AND ALUMINUM ALLOYS

specification number	designation	country	product forms	chemical composition	mechanical properties and hardness values
CSA HA.4	.2024	Canada	Plate, sheet	rem **Al**, 0.50 **Si**, 3.8–4.9 **Cu**, 0.30–0.9 **Mn**, 1.2–1.8 **Mg**, 0.10 **Cr**, 0.25 **Zn**, 0.15 **Ti**, 0.50 **Fe**, 0.05 each, 0.15 total, others	**O**: 0.010–0.032 in. **diam**, 32.0 max ksi (221 max MPa) **TS**, 14.0 max ksi (97 max MPa) **YS**, 12% in 2 in. (50 mm) **El**; **T3 (flat sheet)**: 0.010–0.020 in. **diam**, 64.0 ksi (441 MPa) **TS**, 42.0 ksi (290 MPa) **YS**, 12% in 2 in. (50 mm) **El**; **T42**: 0.010–0.020 in. **diam**, 62.0 ksi (427 MPa) **TS**, 38.0 ksi (262 MPa) **YS**, 12% in 2 in. (50 mm) **El**
WW-T-700/3E	2024	US	Tube	rem **Al**, 0.50 **Si**, 3.8–4.9 **Cu**, 0.30–0.9 **Mn**, 1.2–1.8 **Mg**, 0.10 **Cr**, 0.25 **Zn**, 0.50 **Fe**, 0.05 each, 0.15 total, others	**Annealed**: 0.018–0.500 in. **diam**, 32 max ksi (221 max MPa) **TS**, 15 max ksi (103 max MPa) **YS**; **Solution heat treated and temper T**: 0.018–0.024 in. **diam**, 64 ksi (441 MPa) **TS**, 42 ksi (290 MPa) **YS**, 10% in 2 in. (50 mm) **El**
QQ-A-430B	2024	US	Rod, wire	rem **Al**, 0.50 **Si**, 3.8–4.9 **Cu**, 0.30–0.9 **Mn**, 1.2–1.8 **Mg**, 0.10 **Cr**, 0.25 **Zn**, 0.50 **Fe**, 0.05 each, 0.15 total, others	**Annealed**: 1.000 max in. **diam**, 35 max ksi (241 max MPa) **TS**; **Strain hardened and temper H13**: 1.000 max in. **diam**, 32 ksi (221 MPa) **TS**
QQ-A-250/29A	2124	US	Plate	rem **Al**, 0.20 **Si**, 3.8–4.9 **Cu**, 0.30–0.9 **Mn**, 1.2–1.8 **Mg**, 0.10 **Cr**, 0.25 **Zn**, 0.15 **Ti**, 0.30 **Fe**, 0.20 **Zr**, **Ti** in **Zr** content; 0.05 each, 0.15 total, others	**Solution treated and temper T351**: 1.000–2.000 in. **diam**, 66 ksi (455 MPa) **TS**, 57 ksi (393 MPa) **YS**, 6% in 2 in. (50 mm) **El**
QQ-A-250/4E	2024	US	Plate, sheet	rem **Al**, 0.50 **Si**, 3.8–4.9 **Cu**, 0.30–0.9 **Mn**, 1.2–1.8 **Mg**, 0.10 **Cr**, 0.25 **Zn**, 0.50 **Fe**, 0.05 each, 0.15 total, others	**Annealed**: 0.010–0.499 in. **diam**, 32 max ksi (221 max MPa) **TS**, 14 max ksi (97 max MPa) **YS**, 12% in 2 in. (50 mm) **El**; **Solution heat treated and temper T3**: 0.008–0.009 in. **diam**, 63 ksi (434 MPa) **TS**, 42 ksi (290 MPa) **YS**, 10% in 2 in. (50 mm) **El**; **Solution heat treated and artificially aged**: 0.010–0.020 in. **diam**, 62 ksi (227 MPa) **TS**, 40 ksi (276 MPa) **YS**, 12% in 2 in. (50 mm) **El**
QQ-A-250/5F	2024	US	Plate, sheet	rem **Al**, 0.50 **Si**, 3.8–4.9 **Cu**, 0.30–0.9 **Mn**, 1.2–1.8 **Mg**, 0.10 **Cr**, 0.25 **Zn**, 0.50 **Fe**, 0.05 each, 0.15 total, others	**Annealed**: 0.008–0.009 in. **diam**, 30 max ksi (207 max MPa) **TS**, 14 max ksi (97 max MPa) **YS**, 10% in 2 in. (50 mm) **El**; **Solution treated and temper T361**: 0.020–0.062 in. **diam**, 61 ksi (421 MPa) **TS**, 47 ksi (324 MPa) **YS**, 8% in 2 in. (50 mm) **El**; **Solution treated and temper T851**: 0.250–0.499 in. **diam**, 65 ksi (448 MPa) **TS**, 56 ksi (386 MPa) **YS**, 5% in 2 in. **El**

WROUGHT ALUMINUM AND ALUMINUM ALLOYS

specification number	designation	country	product forms	chemical composition	mechanical properties and hardness values
QQ-A-225/6D	2024	US	Bar, rod, wire	rem **Al**, 0.50 **Si**, 3.8–4.9 **Cu**, 0.30–0.9 **Mn**, 1.2–1.8 **Mg**, 0.10 **Cr**, 0.25 **Zn**, 0.50 **Fe**, 0.05 each, 0.15 total, others	**Annealed**: 8.000 max in. **diam**, 35 max ksi (241 MPa) **TS**, 16% in 2 in. (50 mm) **El**; **Solution heat treated and naturally aged**: 0.499 max in. **diam**, 62 ksi (427 MPa) **TS**, 45 ksi (310 MPa) **YS**, 10% in 2 in. (50 mm) **El**; **Solution heat treated and artificially aged**: 6.500 max in. **diam**, 62 ksi (427 MPa) **TS**, 50 ksi (345 MPa) **YS**, 5% in 2 in. (50 mm) **El**
QQ-A-200/3D	2024	US	Bar, rod, shapes, tube, wire	rem **Al**, 0.50 **Si**, 3.8–4.9 **Cu**, 0.30–0.9 **Mn**, 1.2–1.8 **Mg**, 0.10 **Cr**, 0.25 **Zn**, 0.50 **Fe**, 0.5 each, 0.15 total, others	**Annealed**: all **diam**, 35 max ksi (241 max MPa) **TS**, 19 max ksi (131 max MPa) **YS**, 12% in 2 in. (50 mm) **El**; **Solution heat treated**: 0.249 max in. **diam**, 57 ksi (393 MPa) **TS**, 42 ksi (290 MPa) **YS**, 10% in 2 in. (50 mm) **El**; **Solution heat treated and artificially aged**: 0.050–0.249 in. **diam**, 64 ksi (441 MPa) **TS**, 56 ksi (386 MPa) **YS**, 4% in 2 in. (50 mm) **El**
NF A 50 506	2024 (A-U4G1)	France	Shapes	rem **Al**, 0.50 **Si**, 3.8–4.9 **Cu**, 0.30–0.90 **Mn**, 1.2–1.8 **Mg**, 0.10 **Cr**, 0.25 **Zn**, 0.15 **Ti**, 0.50 **Fe**	**Cold rolled**: 0.4–0.8 mm **diam**, 390 MPa **TS**, 250 MPa **YS**, 14% in 2 in. (50 mm) **El**
AIR 9051/A	A-U4G1 (2024-F)	France	Sheet, bar, billet, and plate: for forging	rem **Al**, 0.50 **Si**, 3.8–4.9 **Cu**, 0.3–0.9 **Mn**, 1.2–1.8 **Mg**, 0.10 **Cr**, 0.25 **Zn**, 0.50 **Fe**, 0.25 **Ti+Zr**, 0.05 each, 0.15 total, others	**Rolled (sheet)**: ≤12 mm **diam**, 420 MPa **TS**, 310 MPa **YS**, 12% in 2 in. (50 mm) **El**; **Solution treated (bar)**: ≤250 mm **diam**, 450 MPa **TS**, 290 MPa **YS**, 9% in 2 in. (50 mm) **El**; **Water quenched and tempered (billet and plate)**: ≤500 mm **diam**, 410 MPa **TS**, 260 MPa **YS**, 12% in 2 in. (50 mm) **El**
NF A 50 411	2024–2024 A (A-U4G1)	France	Bar, wire, and shapes	rem **Al**, 0.50 **Si**, 3.8–4.9 **Cu**, 0.30–0.90 **Mn**, 1.2–1.8 **Mg**, 0.10 **Cr**, 0.25 **Zn**, 0.15 **Ti**, 0.50 **Fe**, 0.20 **Ti+Zr**, 0.05 each, 0.15 total, others	**Aged (extruded bar)**: 450 MPa **TS**, 310 MPa **YS**, 10% in 2 in. (50 mm) **El**; **Aged (wire)**: 2–6 mm **diam**, 430 MPa **TS**, 280 MPa **YS**, 15% in 2 in. (50 mm) **El**; **Aged (shapes)**: 430 MPa **TS**, 310 MPa **YS**, 12% in 2 in. (50 mm) **El**
NF A 50 451	2024 (A-U4G1)	France	Sheet, plate, and strip	rem **Al**, 0.50 **Si**, 3.8–4.9 **Cu**, 0.30–0.90 **Mn**, 1.2–1.8 **Mg**, 0.25 **Zn**, 0.15 **Ti**, 0.50 **Fe**, 0.20 **Ti+Zr**, 0.05 each, 0.15 total, others	**Annealed**: 0.4–1.6 mm **diam**, 140 max MPa **YS**, 13% in 2 in. (50 mm) **El**; **Aged**: 0.4–1.6 mm **diam**, 430 MPa **TS**, 270 MPa **YS**, 14% in 2 in. (50 mm) **El**

WROUGHT ALUMINUM AND ALUMINUM ALLOYS

specification number	designation	country	product forms	chemical composition	mechanical properties and hardness values
IS 5902	24530	India	Wire, bar	rem **Al**, 0.5 **Si**, 3.8–4.9 **Cu**, 0.30–0.9 **Mn**, 1.2–1.8 **Mg**, 0.10 **Cr**, 0.25 **Zn**, 0.5 **Fe**	**Annealed and cold drawn:** 427 MPa **TS**
ISO R827	Al–Cu4Mg1	ISO	Extruded products	rem **Al**, 0.5 **Si**, 3.8–4.9 **Cu**, 0.3–1.2 **Mn**, 1.0–1.8 **Mg**, 0.2 **Ni**, 0.2 **Zn**, 0.5 **Fe**, 0.3 **Ti+Zr+Cr**	**Solution treated, straightened, and naturally aged:** 0.25 max in. (6 max mm) **diam**, 392 MPa **TS**, 284 MPa **YS**, 12% in 2 in. (50 mm) **El**
ISO R209	Al–Cu4Mg1	ISO	Wrought products	rem **Al**, 0.5 **Si**, 3.8–4.9 **Cu**, 0.3–1.2 **Mn**, 1.0–1.8 **Mg**, 0.2 **Ni**, 0.2 **Zn**, 0.5 **Fe**, 0.3 **Ti+Zr+Cr**	
ISO TR2778	Al–Cu4Mg1	ISO	Tube: drawn	rem **Al**, 0.5 **Si**, 3.8–4.9 **Cu**, 0.3–1.2 **Mn**, 1.0–1.8 **Mg**, 0.2 **Ni**, 0.2 **Zn**, 0.5 **Fe**, 0.3 **Ti+Zr+Cr**	**Solution heat treated and naturally aged:** all **diam**, 430 MPa **TS**, 285 MPa **YS**, 10% in 2 in. (50 mm) **El**
ISO R829	Al–Cu4Mg1	ISO	Forgings	rem **Al**, 0.5 **Si**, 3.8–4.9 **Cu**, 0.3–1.2 **Mn**, 1.0–1.8 **Mg**, 0.2 **Ni**, 0.2 **Zn**, 0.5 **Fe**, 0.3 **Ti+Zr+Cr**	**Solution treated and naturally aged:** all **diam**, 392 MPa **TS**, 265 MPa **YS**, 11% in 2 in. (50 mm) **El**
JIS H 4120	2024	Japan	Wire and rod: rivet	rem **Al**, 0.50 **Si**, 3.8–4.9 **Cu**, 0.30–0.9 **Mn**, 1.2–1.8 **Mg**, 0.10 **Cr**, 0.25 **Zn**, 0.50 **Fe**, 0.15 total, others	**Solution heat treated and age hardened:** 2–25 mm **diam**, 265 MPa **TS**, 245 MPa **YS**, 10% in 2 in. (50 mm) **El**
JIS H 4100	2024	Japan	Shapes: extruded	rem **Al**, 0.50 **Si**, 3.8–4.9 **Cu**, 0.30–0.9 **Mn**, 1.2–1.8 **Mg**, 0.10 **Cr**, 0.25 **Zn**, 0.50 **Fe**, 0.15 total, others	**Annealed:** all **diam**, 245 MPa **TS**, 127 MPa **YS**, 12% in 2 in. (50 mm) **El**; **Solution heat treated and age hardened:** <7 mm **diam**, 392 MPa **TS**, 294 MPa **YS**, 12% in 2 in. (50 mm) **El**
JIS H 4080	2024	Japan	Tube: extruded	rem **Al**, 0.50 **Si**, 3.8–4.9 **Cu**, 0.30–0.9 **Mn**, 1.2–1.8 **Mg**, 0.10 **Cr**, 0.25 **Zn**, 0.50 **Fe**, 0.15 total, others	**Annealed:** all **diam**, 245 MPa **TS**, 127 MPa **YS**, 12% in 2 in. (50 mm) **El**; **Solution heat treated and age hardened:** <7 mm **diam**, 392 MPa **TS**, 294 MPa **YS**, 10% in 2 in. (50 mm) **El**
JIS H 4040	2024	Japan	Bar: extruded	rem **Al**, 0.50 **Si**, 3.8–4.9 **Cu**, 0.30–0.9 **Mn**, 1.2–1.8 **Mg**, 0.10 **Cr**, 0.25 **Zn**, 0.50 **Fe**, 0.15 total, others	**Annealed:** all **diam**, 245 MPa **TS**, 127 MPa **YS**, 12% in 2 in. (50 mm) **El**; **Solution heat treated and age hardened:** <7 mm **diam**, 392 MPa **TS**, 294 MPa **YS**, 12% in 2 in. (50 mm) **El**
JIS H 4000	2024	Japan	Sheet, plate, and strip, and coil	rem **Al**, 0.50 **Si**, 3.8–4.9 **Cu**, 0.30–0.9 **Mn**, 1.2–1.8 **Mg**, 0.10 **Cr**, 0.25 **Zn**, 0.50 **Fe**, 0.15 total, others	**Annealed:** <1.5 mm **diam**, 216 MPa **TS**, 98 MPa **YS**, 12% in 2 in. (50 mm) **El**; **Solution heat treated and age hardened:** <1.5 mm **diam**, 431 MPa **TS**, 275 MPa **YS**, 15% in 2 in. **El**; **Solution heat treated and cold worked:** 1.5 mm **diam**, 441 MPa **TS**, 294 MPa **YS**, 15% in 2 in. **El**

WROUGHT ALUMINUM AND ALUMINUM ALLOYS

specification number	designation	country	product forms	chemical composition	mechanical properties and hardness values
MIL-T-15089C	2024	US	Tube: seamless	rem **Al**, 0.50 **Si**, 3.8–4.9 **Cu**, 0.30–0.9 **Mn**, 1.2–1.8 **Mg**, 0.10 **Cr**, 0.25 **Zn**, 0.15 **Ti**, 0.50 **Fe**, 0.20 **Zr+Ti**, 0.05 each, 0.15 total, others	**Heat treated**: all **diam**, 455 MPa **TS**, 400 MPa **YS**, 5% in 2 in. (50 mm) **El**
SABS 712	Al-Cu4Mgl	South Africa	Wrought products	rem **Al**, 0.5 **Si**, 3.8–4.9 **Cu**, 0.3–0.9 **Mn**, 1.2–1.8 **Mg**, 0.1 **Cr**, 0.25 **Zn**, 0.5 **Fe**, 0.2 **Ti+Zr**	
AS 1734	2024	Australia	Sheet, plate	rem **Al**, 0.50 **Si**, 3.8–4.9 **Cu**, 0.9 **Mn**, 1.2–1.8 **Mg**, 0.10 **Cr**, 0.25 **Zn**, 0.15 **Ti**, 0.50 **Fe**, 0.05 each, 0.15 total, others	**Annealed**: 0.25–6.00 mm **diam**, 221 MPa **TS**, 97 MPa **YS**, 12% in 2 in. (50 mm) **El**
IS 739	HG15	India	Wire	rem **Al**, 0.5–0.90 **Si**, 3.8–4.8 **Cu**, 0.3–1.2 **Mn**, 0.2–0.8 **Mg**, 0.2 **Ni**, 0.2 **Zn**, 0.05 **Sn**, 0.7 **Fe**, 0.05 **Pb**, 0.05 **Sb**	**Solution treated and naturally aged**: all **diam**, 387 MPa **TS**; **Solution and precipitation heat treated**: all **diam**, 432 MPa **TS**
IS 3436		India	Sheet, strip, coil	rem **Al**, 0.60–0.90 **Si**, 3.8–4.8 **Cu**, 0.40–1.2 **Mn**, 0.55–0.85 **Mg**, 0.20 **Ni**, 0.20 **Zn**, 0.05 **Ti**, 1.0 **Fe**, 0.05 **Pb**, < 0.30 **Ti+Cr**	**Solution treated and aged**: all **diam**, 386 MPa **TS**, 232 MPa **YS**, 15% in 2 in. (50 mm) **El**; **Solution treated and precipitation treated**: all **diam**, 416 MPa **TS**, 325 MPa **YS**, 8% in 2 in. (50 mm) **El**
COPANT 862	2130	COPANT	Wrought products	rem **Al**, 0.1–0.8 **Si**, 3.5–5.0 **Cu**, 0.1–0.8 **Mn**, 0.2–1.0 **Mg**, 0.3 **Zn**, 0.2 **Ti**, 0.1–0.8 **Fe**, 1.0–2.0 **Bi**, 0.05 each, 0.15 total, others	
ISO 2779	Al-Cu4PbMg	ISO	Wrought products	rem **Al**, 0.8 **Si**, 3.5–5.0 **Cu**, 1.0 **Mn**, 0.3–1.8 **Mg**, 0.1 **Cr**, 0.8 **Zn**, 0.2 **Ti**, 0.8 **Fe**, 1.5 **Pb**, 0.10 each, 0.30 total, others (0.8–2 **Pb+Bi+Cd**)	**Solution heat treated and naturally aged**: 0.125 in. (3 mm) **diam**, 370 MPa **TS**, 245 MPa **YS**, 10% in 2 in. (50 mm) **El**
NS 17 110	NS 17 110–54	Norway	Tube, rod, and wire	rem **Al**, 1.0 **Si**, 3.5–5.0 **Cu**, 0.5–1.0 **Mn**, 0.4–1.8 **Mg**, 1.0 **Zn**, 1.0 **Fe**, 1.0–3.0 **Pb+Sn+B+Cd+Sb**, 0.10 each, 0.30 total, others	**Hardened and naturally aged (tube)**: <6 mm **diam**, 372 MPa **TS**, 245 MPa **YS**, 12% in 2 in. (50 mm) **El**; **Hardened and naturally aged (wire)**: all **diam**, 372 MPa **TS**, 245 MPa **YS**, 8% in 2 in. (50 mm) **El**
SABS 712	Al-Cu4MglPb	South Africa	Wrought products	rem **Al**, 1.0 **Si**, 3.5–5.0 **Cu**, 1.0 **Mn**, 0.2–1.8 **Mg**, 0.4 **Cr**, 1.0 **Ni**, 1.0 **Zn**, 1.0 **Fe**, 0.5–2.0 **Pb**, 0.5 **Bi**, 0.2 **Ti+Zr**	
ISO R827	Al-Cu4MgSi	ISO	Extruded products	rem **Al**, 0.2–0.8 **Si**, 3.5–4.7 **Cu**, 0.3–1.0 **Mn**, 0.3–1.2 **Mg**, 0.2 **Ni**, 0.5 **Zn**, 0.7 **Fe**, 0.3 **Ti+Zr+Cr**	**Solution treated and naturally aged**: 0.75 max in. (20 max mm) **diam**, 353 MPa **TS**, 226 MPa **YS**, 10% in 2 in. (50 mm) **El**
ISO R209	Al-Cu4MgSi	ISO	Wrought products	rem **Al**, 0.2–0.8 **Si**, 3.5–4.7 **Cu**, 0.3–1.0 **Mn**, 0.3–1.2 **Mg**, 0.2 **Ni**, 0.5 **Zn**, 0.7 **Fe**, 0.3 **Ti+Zr+Cr**	
ISO TR2778	Al-Cu4MgSi	ISO	Tube: drawn	rem **Al**, 0.2–0.8 **Si**, 3.5–4.7 **Cu**, 0.3–1.0 **Mn**, 0.3–1.2 **Mg**, 0.2 **Ni**, 0.5 **Zn**, 0.7 **Fe**, 0.3 **Ti+Zr+Cr**	**Solution heat treated and naturally aged**: all **diam**, 385 MPa **TS**, 245 MPa **YS**, 12% in 2 in. (50 mm) **El**

WROUGHT ALUMINUM AND ALUMINUM ALLOYS

specification number	designation	country	product forms	chemical composition	mechanical properties and hardness values
ISO R829	Al–Cu4MgSi	ISO	Forgings	rem **Al**, 0.2–0.8 **Si**, 3.5–4.7 **Cu**, 0.3–1.0 **Mn**, 0.3–1.2 **Mg**, 0.2 **Ni**, 0.5 **Zn**, 0.7 **Fe**, 0.3 **Ti+Zr+Cr**	**Solution treated and naturally aged**: all **diam**, 353 MPa **TS**, 206 MPa **YS**, 11% in 2 in. (50 mm) **El**
SABS 712	Al–Cu4MgSi	South Africa	Wrought products	rem **Al**, 0.3–0.8 **Si**, 3.5–4.7 **Cu**, 0.3–0.8 **Mn**, 0.4–1.0 **Mg**, 0.1 **Cr**, 0.25 **Zn**, 0.7 **Fe**, 0.2 **Ti+Zr**	
AMS 4142B		US	Die forgings	rem **Al**, 0.9 **Si**, 3.5–4.5 **Cu**, 0.20 **Mn**, 1.2–1.8 **Mg**, 0.10 **Cr**, 1.7–2.3 **Ni**, 0.25 **Zn**, 1.0 **Fe**, 0.05 each, 0.15 total, others	**Solution and precipitation heat treated**: all **diam**, 40 ksi (276 MPa) **TS**, 28 ksi (193 MPa) **YS**, 4% in 2 in. (50 mm) **El**, 85 min **HB**
AMS 4140D		US	Die forgings	rem **Al**, 0.9 **Si**, 3.5–4.5 **Cu**, 0.20 **Mn**, 0.45–0.9 **Mg**, 0.10 **Cr**, 1.7–2.3 **Ni**, 0.25 **Zn**, 0.05 **Ti**, 1.0 **Fe**, 0.05 each, 0.15 total, others	**Solution and precipitation heat treated**: all **diam**, 55 ksi (379 MPa) **TS**, 40 ksi (276 MPa) **YS**, 7% in 2 in. (50 mm) **El**, 100 min **HB**
AMS 4118E		US	Bar, rod, and wire	rem **Al**, 0.8 **Si**, 3.5–4.5 **Cu**, 0.40–1.0 **Mn**, 0.20–0.8 **Mg**, 0.10 **Cr**, 0.25 **Zn**, 1.0 **Fe**, 0.05 each, 0.15 total, others	**Solution heat treated**: all **diam**, 55 ksi (379 MPa) **TS**, 32 ksi (221 MPa) **YS**, 12% in 2 in. (50 mm) **El**, 90 min **HB**
ANSI/ASTM B 316	2017	US	Wire and rod: rivet and cold-heading	rem **Al**, 0.20–0.8 **Si**, 3.5–4.5 **Cu**, 0.40–1.0 **Mn**, 0.40–0.8 **Mg**, 0.10 **Cr**, 0.25 **Zn**, 0.15 **Ti**, 0.7 **Fe**, 0.20 **Zr+Ti**, 0.05 each, 0.15 total, others	**Annealed**: ≤1.000 in. **diam**, 35 max ksi (241 max MPa) **TS**; **H1**: ≤1.000 in. **diam**, 30 ksi (207 MPa) **TS**; **Solution heat treated and naturally aged**: 0.061–1.000 in. **diam**, 55 ksi (379 MPa) **TS**, 32 ksi (221 MPa) **YS**, 12% in 2 in. (50 mm) **El**
ANSI/ASTM B 247	2018	US	Die forgings	rem **Al**, 0.9 **Si**, 3.5–4.5 **Cu**, 0.20 **Mn**, 0.45–0.9 **Mg**, 0.10 **Cr**, 1.7–2.3 **Ni**, 0.25 **Zn**, 1.0 **Fe**, 0.05 each, 0.15 total, others	**T61**: ≤4.000 in. **diam**, 55 ksi (379 MPa) **TS**, 40 ksi (276 MPa) **YS**, 7% in 2 in. (50 mm) **El**, 100 min **HB**
ANSI/ASTM B 247	2218	US	Die forgings	rem **Al**, 0.9 **Si**, 3.5–4.5 **Cu**, 0.20 **Mn**, 1.2–1.8 **Mg**, 0.10 **Cr**, 1.7–2.3 **Ni**, 0.25 **Zn**, 1.0 **Fe**, 0.05 each, 0.15 total, others	**T61**: ≤4.000 in. **diam**, 55 ksi (379 MPa) **TS**, 40 ksi (276 MPa) **YS**, 7% in 2 in. (50 mm) **El**, 100 min **HB**
ANSI/ASTM B 211	2017	US	Bar, rod, and wire	rem **Al**, 0.20–0.8 **Si**, 3.5–4.5 **Cu**, 0.40–1.0 **Mn**, 0.20–0.8 **Mg**, 0.10 **Cr**, 0.25 **Zn**, 0.15 **Ti**, 0.7 **Fe**, 0.20 **Ti+Zr**, 0.05 each, 0.15 total, others	**Annealed**: 0.125–8.000 in. **diam**, 35 max ksi (241 max MPa) **TS**, 16% in 2 in. (50 mm) **El**; **Solution heat treated and naturally aged**: 0.125–4.000 in. **diam**, 55 ksi (379 MPa) **TS**, 32 ksi (221 MPa) **YS**, 12% in 2 in. (50 mm) **El**
BS 4L.35		UK	Ingot and castings	rem **Al**, 0.6 **Si**, 3.5–4.5 **Cu**, 0.6 **Mn**, 1.2–1.7 **Mg**, 1.8–2.3 **Ni**, 0.1 **Zn**, 0.25 **Ti**, 0.05 **Sn**, 0.6 **Fe**, 0.05 **Pb**, 1.0 **Si+Fe**	**Heat treated, quenched and aged**: all **diam**, 220 MPa **TS**, 210 MPa **YS**

WROUGHT ALUMINUM AND ALUMINUM ALLOYS

specification number	designation	country	product forms	chemical composition	mechanical properties and hardness values
COPANT 862	2017	COPANT	Wrought products	rem **Al**, 0.2–0.80 **Si**, 3.5–4.5 **Cu**, 0.4–1.0 **Mn**, 0.4–0.8 **Mg**, 0.10 **Cr**, 0.25 **Zn**, 0.15 **Ti**, 0.7 **Fe**, 0.20 **Ti+Zr**, 0.05 each, 0.15 total, others	
CSA HA.8	.2018	Canada	Forgings	rem **Al**, 0.9 **Si**, 3.5–4.5 **Cu**, 0.20 **Mn**, 0.45–0.9 **Mg**, 0.10 **Cr**, 1.7–2.3 **Ni**, 0.25 **Zn**, 1.0 **Fe**, 0.05 each, 0.15 total, others	**T61 (die forgings)**: <4 in. **diam**, 55.0 ksi (379 MPa) **TS**, 40.0 ksi (276 MPa) **YS**, 7% in 2 in. (50 mm) **El**
QQ–A–430B	2017	US	Rod, wire	rem **Al**, 0.8 **Si**, 3.5–4.5 **Cu**, 0.40–1.0 **Mn**, 0.20–0.8 **Mg**, 0.10 **Cr**, 0.25 **Zn**, 0.7 **Fe**, 0.05 each, 0.15 total, others	**Annealed**: 1.000 max in. **diam**, 35 max ksi (241 max MPa) **TS**; **Strain hardened and temper H13**: 1.000 max in. **diam**, 30 ksi (207 MPa) **TS**
QQ–A–225/5D	2017	US	Bar, rod, wire	rem **Al**, 0.8 **Si**, 3.5–4.5 **Cu**, 0.40–1.0 **Mn**, 0.20–0.8 **Mg**, 0.10 **Cr**, 0.25 **Zn**, 0.7 **Fe**, 0.05 each, 0.15 total, others	**Annealed**: 8.000 max in. **diam**, 35 max ksi (241 max MPa) **TS**, 16% in 2 in. (50 mm) **El**; **Solution heat treated and naturally aged**: 8.000 max in. **diam**, 55 ksi (379 MPa) **TS**, 32 ksi (221 MPa) **YS**, 12% in 2 in. (50 mm) **El**; **Solution heat treated and temper T451**: 0.500–8.000 in. **diam**, 55 ksi (379 MPa) **TS**, 32 ksi (221 MPa) **YS**, 12% in 2 in. (50 mm) **El**
NF A 50 506	2017.S (A–U4G)	France	Shapes	rem **Al**, 0.20–0.80 **Si**, 3.5–4.5 **Cu**, 0.40–1.0 **Mn**, 0.40–1.0 **Mg**, 0.10 **Cr**, 0.25 **Zn**, 0.70 **Fe**, 0.25 **Ti+Zr**, 0.05 each, 0.15 total, others	**Cold rolled**: 0.8–3.2 mm **diam**, 390 MPa **TS**, 240 MPa **YS**, 15% in 2 in. (50 mm) **El**
NF A 50 411	2017 A (A–U4G)	France	Bar, wire, tube, and shapes	rem **Al**, 0.20–0.80 **Si**, 3.5–4.5 **Cu**, 0.40–1.0 **Mn**, 0.40–1.0 **Mg**, 0.10 **Cr**, 0.25 **Zn**, 0.70 **Fe**, 0.25 **Ti+Zr**, 0.05 each, 0.15 total, others	**Aged (extruded bar)**: all **diam**, 390 MPa **TS**, 255 MPa **YS**, 10% in 2 in. (50 mm) **El**; **Aged (tube)**: ≤150 mm **diam**, 390 MPa **TS**, 240 MPa **YS**, 14% in 2 in. (50 mm) **El**
AIR 9051/A	A–U4G (2017–F)	France	Sheet, bar, billet, and plate: for forging	rem **Al**, 0.25–0.8 **Si**, 3.5–4.5 **Cu**, 0.3–0.8 **Mn**, 0.5–1.0 **Mg**, 0.25 **Zn**, 0.20 **Ti**, 0.50 **Fe**, 0.25 **Ti+Zr**, 0.05 each, 0.15 total, others	**Water quenched and tempered (sheet)**: 6–12 mm **diam**, 390 MPa **TS**, 240 MPa **YS**, 13% in 2 in. (50 mm) **El**; **Solution treated (bar)**: ≤250 mm **diam**, 380 MPa **TS**, 230 MPa **YS**, 9% in 2 in. (50 mm) **El**; **Solution treated (billet and plate)**: ≤500 mm **diam**, 400 MPa **TS**, 240 MPa **YS**, 14% in 2 in. (50 mm) **El**
NF A 50 411	2030 (A–U4Pb)	France	Bar and tube	rem **Al**, 0.40 **Si**, 3.5–4.5 **Cu**, 1.0 **Mn**, 0.5–1.3 **Mg**, 0.10 **Cr**, 0.20 **Ni**, 0.50 **Zn**, 0.20 **Ti**, 0.50 **Fe**, 0.8–1.5 **Pb**, 0.05 each, 0.15 total, others	**Aged (bar)**: 370 MPa **TS**, 235 MPa **YS**, 7% in 2 in. (50 mm) **El**; **Cold finished (tube)**: ≤150 mm **diam**, 370 MPa **TS**, 235 MPa **YS**, 7% in 2 in. (50 mm) **El**

WROUGHT ALUMINUM AND ALUMINUM ALLOYS

specification number	designation	country	product forms	chemical composition	mechanical properties and hardness values
NF A 50 451	2017 A (A–U4G)	France	Sheet, plate, and strip	rem **Al**, 0.20–0.80 **Si**, 3.5–4.5 **Cu**, 0.40–1.0 **Mn**, 0.40–1.0 **Mg**, 0.10 **Cr**, 0.25 **Zn**, 0.70 **Fe**, 0.25 **Ti+Zr**, 0.05 each, 0.15 total, others	**Annealed**: 0.4–1.6 mm **diam**, 140 max MPa **YS**, 13% in 2 in. (50 mm) **El**; **Aged**: 0.4–1.6 mm **diam**, 390 MPa **TS**, 240 MPa **YS**, 15% in 2 in. (50 mm) **El**
AIR 9150/B	A–U4G	France	Wire: rivet	rem **Al**, 0.3–0.8 **Si**, 3.5–4.5 **Cu**, 0.3–0.8 **Mn**, 0.5–1.0 **Mg**, 0.25 **Zn**, 0.5 **Fe**, 0.2 **Ti+Zr**, 0.05 each, 0.15 total, others	**Cold finished or annealed**: 1.6–9.6 mm **diam**, 390 MPa **TS**, 240 MPa **YS**, 14% in 2 in. (50 mm) **El**
JIS H 4140	2017	Japan	Die forgings	rem **Al**, 0.8 **Si**, 3.5–4.5 **Cu**, 0.40–1.0 **Mn**, 0.20–0.8 **Mg**, 0.10 **Cr**, 0.7 **Fe**, 0.15 total, others	**Solution heat treated and age hardened**: ≤100 mm **diam**, 343 MPa **TS**, 216 MPa **YS**, 14% in 2 in. (50 mm) **El**
JIS H 4140	2018	Japan	Die forgings	rem **Al**, 0.9 **Si**, 3.5–4.5 **Cu**, 0.20 **Mn**, 0.45–0.9 **Mg**, 0.10 **Cr**, 1.7–2.3 **Ni**, 0.25 **Zn**, 1.0 **Fe**, 0.15 total, others	**Solution heat treated and artificially age hardened (high temperature)**: ≤100 mm **diam**, 382 MPa **TS**, 275 MPa **YS**, 10% in 2 in. (50 mm) **El**
JIS H 4140	2218	Japan	Die forgings	rem **Al**, 0.9 **Si**, 3.5–4.5 **Cu**, 0.20 **Mn**, 1.2–1.8 **Mg**, 0.10 **Cr**, 1.7–2.3 **Ni**, 0.25 **Zn**, 1.0 **Fe**, 0.15 total, others	**Solution heat treated and artificially age hardened (high temperature)**: ≤100 mm **diam**, 382 MPa **TS**, 275 MPa **YS**, 10% in 2 in. (50 mm) **El**
JIS H 4120	2017	Japan	Wire and rod: rivet	rem **Al**, 0.8 **Si**, 3.5–4.5 **Cu**, 0.40–1.0 **Mn**, 0.20–0.8 **Mg**, 0.10 **Cr**, 0.25 **Zn**, 0.7 **Fe**, 0.15 total, others	**Annealed**: ≤25 mm **diam**, 245 MPa **TS**; **3/8 hard**: ≤25 mm **diam**, 206 MPa **TS**; **Solution heat treated and age hardened**: 2–25 mm **diam**, 373 MPa **TS**, 216 MPa **YS**, 12% in 2 in. (50 mm) **El**
JIS H 4100	2017	Japan	Shapes: extruded	rem **Al**, 0.8 **Si**, 3.5–4.5 **Cu**, 0.40–1.0 **Mn**, 0.20–0.8 **Mg**, 0.10 **Cr**, 0.25 **Zn**, 0.7 **Fe**, 0.15 total, others	**Annealed**: all **diam**, 245 MPa **TS**, 127 MPa **YS**, 16% in 2 in. (50 mm) **El**; **Solution heat treated and age hardened**: all **diam**, 343 MPa **TS**, 216 MPa **YS**, 12% in 2 in. (50 mm) **El**
JIS H 4080	2017	Japan	Tube: extruded	rem **Al**, 0.8 **Si**, 3.5–4.5 **Cu**, 0.40–1.0 **Mn**, 0.20–0.8 **Mg**, 0.10 **Cr**, 0.25 **Zn**, 0.7 **Fe**, 0.15 total, others	**Annealed**: all **diam**, 245 MPa **TS**, 127 MPa **YS**, 16% in 2 in. (50 mm) **El**; **Solution heat treated and age hardened**: all **diam**, 343 MPa **TS**, 216 MPa **YS**, 12% in 2 in. (50 mm) **El**
JIS H 4040	2017	Japan	Bar: extruded	rem **Al**, 0.8 **Si**, 3.5–4.5 **Cu**, 0.40–1.0 **Mn**, 0.20–0.8 **Mg**, 0.10 **Cr**, 0.25 **Zn**, 0.7 **Fe**, 0.15 total, others	**Annealed**: all **diam**, 245 MPa **TS**, 127 MPa **YS**, 16% in 2 in. (50 mm) **El**; **Solution heat treated and age hardened**: all **diam**, 343 MPa **TS**, 216 MPa **YS**, 12% in 2 in. (50 mm) **El**

WROUGHT ALUMINUM AND ALUMINUM ALLOYS

specification number	designation	country	product forms	chemical composition	mechanical properties and hardness values
JIS H 4000	2017	Japan	Sheet, plate, and strip, and coil	rem **Al**, 0.8 **Si**, 3.5–4.5 **Cu**, 0.40–1.0 **Mn**, 0.20–0.8 **Mg**, 0.10 **Cr**, 0.25 **Zn**, 0.7 **Fe**, 0.15 total, others	**Annealed**: <1.5 mm **diam**, 216 MPa **TS**, 108 MPa **YS**, 12% in 2 in. (50 mm) **El**; **Solution heat treated and age hardened**: <1.5 mm **diam**, 353 MPa **TS**, 108 MPa **YS**, 12% in 2 in. (50 mm) **El**; **Solution heat treated and cold worked**: <1.5 mm **diam**, 373 MPa **TS**, 216 MPa **YS**, 15% in 2 in. (50 mm) **El**
IS 7793	2285/24850	India	Forgings	rem **Al**, 0.6 **Si**, 3.5–4.5 **Cu**, 0.2 **Mn**, 1.2–1.8 **Mg**, 1.7–2.3 **Ni**, 0.2 **Zn**, 0.2 **Ti**, 0.05 **Sn**, 0.7 **Fe**, 0.05 **Pb**	**As forged**: all **diam**, 345 MPa **TS**
ANSI/ASTM B 316	2117	US	Wire and rod: rivet and cold-heading	rem **Al**, 0.8 **Si**, 2.2–3.0 **Cu**, 0.20 **Mn**, 0.20–0.50 **Mg**, 0.10 **Cr**, 0.25 **Zn**, 0.7 **Fe**, 0.05 each, 0.15 total, others	**Annealed**: ≤1.000 in. **diam**, 25 max ksi (172 max MPa) **TS**; **H13**: 0.616–1.000 in. **diam**, 25 ksi (172 MPa) **TS**; **Solution heat treated and naturally aged**: 0.061–1.000 in. **diam**, 38 ksi (262 MPa) **TS**, 18 ksi (124 MPa) **YS**, 18% in 2 in. (50 mm) **El**
CSA HA.6	.2117	Canada	Rivet, rod, wire	rem **Al**, 0.8 **Si**, 2.2–3.0 **Cu**, 0.20 **Mn**, 0.20–0.50 **Mg**, 0.10 **Cr**, 0.25 **Zn**, 0.7 **Fe**, 0.05 each, 0.15 total, others	**H15**: 0.062–1.000 in. **diam**, 28.0 ksi (193 MPa) **TS**; **T4**: 0.062–1.000 in. **diam**, 38.0 ksi (262 MPa) **TS**, 18.0 ksi (124 MPa) **YS**, 18% in 2 in. (50 mm) **El**
QQ-A-430B	2117	US	Rod, wire	rem **Al**, 0.8 **Si**, 2.2–3.0 **Cu**, 0.20 **Mn**, 0.20–0.50 **Mg**, 0.10 **Cr**, 0.25 **Zn**, 0.7 **Fe**, 0.05 each, 0.15 total, others	**Annealed**: 1.000 max in. **diam**, 25 max ksi (172 max MPa) **TS**; **Strain hardened and temper H13**: 0.616–1.000 in. **diam**, 25 ksi (172 MPa) **TS**; **Strain hardened and temper H15**: 0.615 max in. **diam**, 28 ksi (193 MPa) **TS**
NF A 50 451	2117 (A-U2G)	France	Sheet, plate, and strip	rem **Al**, 0.80 **Si**, 2.2–3.0 **Cu**, 0.20 **Mn**, 0.20–0.50 **Mg**, 0.10 **Cr**, 0.25 **Zn**, 0.70 **Fe**, 0.05 each, 0.15 total, others	**Aged**: 0.4–3.2 mm **diam**, 250 MPa **TS**, 150 MPa **YS**, 22% in 2 in. (50 mm) **El**
AIR 9150/B	A-UZG	France	Wire: rivet	rem **Al**, 0.2–0.7 **Si**, 2.2–3.0 **Cu**, 0.2 **Mn**, 0.2–0.5 **Mg**, 0.1 **Cr**, 0.2 **Zn**, 0.4 **Fe**, 0.2 **Ti+Zr**, 0.05 each, 0.15 total, others	**Cold finished or annealed**: 1.6–9.6 mm **diam**, 290 MPa **TS**, 170 MPa **YS**, 25% in 2 in. (50 mm) **El**
JIS H 4120	2117	Japan	Wire and rod: rivet	rem **Al**, 0.8 **Si**, 2.2–3.0 **Cu**, 0.20 **Mn**, 0.20–0.50 **Mg**, 0.10 **Cr**, 0.25 **Zn**, 0.7 **Fe**, 0.15 total, others	**Annealed**: ≤25 mm **diam**, 177 MPa **TS**; **5/8 hard**: ≤15 mm **diam**, 196 MPa **TS**; **Solution heated treated and age hardened**: 2–25 mm **diam**, 255 MPa **TS**, 118 MPa **YS**, 18% in 2 in. (50 mm) **El**
BS 3L.86		UK	Wire	rem **Al**, 0.7 **Si**, 2.0–3.0 **Cu**, 0.2 **Mn**, 0.2–0.5 **Mg**, 0.1 **Cr**, 0.05 **Ni**, 0.2 **Zn**, 0.05 **Sn**, 0.5 **Fe**, 0.05 **Pb**, 0.20 **Zr+Ti**	**Solution heat treated, quenched and naturally aged**: ≤6 mm **diam**, 290 MPa **TS**

WROUGHT ALUMINUM AND ALUMINUM ALLOYS

specification number	designation	country	product forms	chemical composition	mechanical properties and hardness values
IS 5902	22500	India	Wire, bar	rem **Al**, 0.7 **Si**, 2.0–3.0 **Cu**, 0.2 **Mn**, 0.2–0.5 **Mg**, 0.1 **Cr**, 0.05 **Ni**, 0.2 **Zn**, 0.05 **Sn**, 0.5 **Fe**, 0.05 **Pb**, 0.2 **Ti+Zr**	**Annealed and cold drawn:** 294 MPa **TS**
ISO R209	Al–Cu2Mg	ISO	Wrought products	rem **Al**, 0.8 **Si**, 2.0–3.0 **Cu**, 0.2 **Mn**, 0.2–0.5 **Mg**, 0.1 **Cr**, 0.2 **Zn**, 0.7 **Fe**, 0.2 **Ti+Zr**	
ANSI/ASTM B 247	2618	US	Die and hand forgings	rem **Al**, 0.10–0.25 **Si**, 1.9–2.7 **Cu**, 1.3–1.8 **Mg**, 0.9–1.2 **Ni**, 0.10 **Zn**, 0.04–0.10 **Ti**, 0.9–1.3 **Fe**, 0.05 each, 0.15 total, others	**T61 (die forgings):** ≤4.000 in. **diam**, 58 ksi (400 MPa) **TS**, 45 ksi (310 MPa) **YS**, 4% in 2 in. (50 mm) **El**, 115 min **HB**; **T61 (hand forgings):** ≤2.000 in. **diam**, 58 ksi (400 MPa) **TS**, 47 ksi (324 MPa) **YS**, 7% in 2 in. (50 mm) **El**
BS 1472	HF12	UK	Forging stock and forgings	rem **Al**, 0.5–1.3 **Si**, 1.8–2.8 **Cu**, 0.5 **Mn**, 0.6–1.2 **Mg**, 0.6–1.4 **Ni**, 0.2 **Zn**, 0.2 **Ti**, 0.6–1.2 **Fe**	**Solution heat treated and naturally aged:** ≤150 mm **diam**, 310 MPa **TS**, 160 MPa **YS**, 13% in 2 in. (50 mm) **El**; **Solution heat treated and precipitation treated:** ≤150 mm **diam**, 385 MPa **TS**, 300 MPa **YS**, 6% in 2 in. (50 mm) **El**
MIL–A–22771C	2618	US	Forgings	rem **Al**, 0.25 **Si**, 1.9–2.7 **Cu**, 1.3–1.8 **Mg**, 0.9–1.2 **Ni**, 0.04–0.10 **Ti**, 0.9–1.3 **Fe**, 0.05 total, others	**T61:** ≤4 in. **diam**, 400 MPa **TS**, 310 MPa **YS**, 4% in 2 in. (50 mm) **El**, 115 **HB**
BS 1472	HF16	UK	Forging stock and forgings	rem **Al**, 0.25 **Si**, 1.8–2.7 **Cu**, 0.2 **Mn**, 1.2–1.8 **Mg**, 0.8–1.4 **Ni**, 0.2 **Zn**, 0.2 **Ti**, 0.9–1.4 **Fe**	**Solution heat treated and naturally aged:** ≤200 mm **diam**, 430 MPa **TS**, 340 MPa **YS**, 5% in 2 in. (50 mm) **El**
AIR 9051/A	A–UZGN	France	Sheet, bar, billet, and plate: for forging	rem **Al**, 0.25 **Si**, 1.8–2.7 **Cu**, 0.20 **Mn**, 1.2–1.8 **Mg**, 0.8–1.4 **Ni**, 0.15 **Zn**, 0.20 **Ti**, 0.9–1.4 **Fe**, 0.25 **Ti+Zr**, 0.05 each, 0.15 total, others	**Rolled (sheet):** ≤40 mm **diam**, 420 MPa **TS**, 360 MPa **YS**, 5% in 2 in. (50 mm) **El**; **Solution treated (bar):** ≤250 mm **diam**, 410 MPa **TS**, 340 MPa **YS**, 6% in 2 in. (50 mm) **El**; **Water quenched and tempered (billet and plate):** ≤500 mm **diam**, 410 MPa **TS**, 340 MPa **YS**, 6% in 2 in. (50 mm) **El**
JIS H 4140	2N01	Japan	Die forgings	rem **Al**, 0.50–1.3 **Si**, 1.5–2.5 **Cu**, 0.20 **Mn**, 1.2–1.8 **Mg**, 0.6–1.4 **Ni**, 0.20 **Zn**, 0.20 **Ti**, 0.6–1.5 **Fe**, 0.15 total, others	**Solution heat treated and artificially age hardened:** ≤100 mm **diam**, 373 MPa **TS**, 294 MPa **YS**, 6% in 2 in. (50 mm) **El**
COPANT 862	2062	COPANT	Wrought products	rem **Al**, 0.8–1.3 **Si**, 1.0–2.0 **Cu**, 0.6–1.0 **Mn**, 0.5–1.4 **Mg**, 0.20 **Zn**, 0.3 **Ti**, 0.7 **Fe**, 0.20 **Ni**, 0.05 each, 0.15 total, others	

WROUGHT ALUMINUM AND ALUMINUM ALLOYS

specification number	designation	country	product forms	chemical composition	mechanical properties and hardness values
MIL-A-46063E		US	Plate	rem **Al**, 0.30 **Si**, 0.10 **Cu**, 0.10–0.40 **Mn**, 2.3–3.3 **Mg**, 0.15–025 **Cr**, 3.5–4.5 **Zn**, 0.10 **Ti**, 0.40 **Fe**, 0.05 each, 0.15 total, others	**Heat treated as agreed**: ≤1.500 in. **diam**, 414 MPa **TS**, 352 MPa **YS**, 9% in 2 in. (50 mm) **El**
ISO TR2778	Al-Mn1	ISO	Tube: drawn	rem **Al**, 0.6 **Si**, 0.1 **Cu**, 1.5 **Mn**, 0.3 **Mg**, 0.2 **Zn**, 0.7 **Fe**, 0.2 **Ti+Zr+Cr**	**Annealed**: all **diam**, 90 MPa **TS**, 20% in 2 in. (50 mm) **El**; **Strain hardened**: all **diam**, 130 MPa **TS**, 3% in 2 in. (50 mm) **El**
DGN-W-30	3S-H16	Mexico	Sheet	97.7 min **Al**, 1.3 **Mn**	**Rolled**: 0.25–0.52 mm **diam**, 172 MPa **TS**, 159 MPa **YS**, 1% in 2 in. (50 mm) **El**
AMS 4067E		US	Tube: seamless	rem **Al**, 0.6 **Si**, 0.05–0.20 **Cu**, 1.0–1.5 **Mn**, 0.10 **Zn**, 0.7 **Fe**, 0.05 each, 0.15 total, others	**Strain hardened to 1/2 hard**: all **diam**, 20 ksi (138 MPa) **TS**
AMS 4063		US	Sheet: clad one side	rem **Al**, 0.6 **Si**, 0.05–0.20 **Cu**, 1.0–1.5 **Mn**, 0.10 **Zn**, 0.7 **Fe**, 0.05 each, 0.15 total, others	**Annealed**: 0.006–0.007 in. **diam**, 20 ksi (138 MPa) **TS**, 12% in 2 in. (50 mm) **El**
AMS 4064		US	Sheet: clad two sides	rem **Al**, 0.6 **Si**, 0.05–0.20 **Cu**, 1.0–1.5 **Mn**, 0.10 **Zn**, 0.7 **Fe**, 0.05 each, 0.15 total, others	**Annealed**: 0.006–0.007 in. **diam**, 20 ksi (138 MPa) **TS**, 12% in 2 in. (50 mm) **El**
AMS 4065E		US	Tube: seamless	rem **Al**, 0.6 **Si**, 0.05–0.20 **Cu**, 1.0–1.5 **Mn**, 0.10 **Zn**, 0.7 **Fe**, 0.05 each, 0.15 total, others	**Annealed**: all **diam**, 14 ksi (97 MPa) **TS**
AMS 4006E		US	Sheet and plate	rem **Al**, 0.6 **Si**, 0.05–0.20 **Cu**, 1.0–1.5 **Mn**, 0.10 **Zn**, 0.7 **Fe**, 0.05 each, 0.15 total, others	**Annealed**: 0.006–0.007 in. (0.15–0.18 mm) **diam**, 14 ksi (97 MPa) **TS**, 14% in 2 in. (50 mm) **El**
AMS 4008E		US	Sheet and plate	rem **Al**, 0.6 **Si**, 0.05–0.20 **Cu**, 1.0–1.5 **Mn**, 0.10 **Zn**, 0.7 **Fe**, 0.05 each, 0.15 total, others	**Strain hardened**: 0.009–0.12 in. (0.23–0.30 mm) **diam**, 20 ksi (138 MPa) **TS**, 1% in 2 in. (50 mm) **El**
AMS 4010A		US	Foil	rem **Al**, 0.6 **Si**, 0.05–0.20 **Cu**, 1.0–1.5 **Mn**, 0.10 **Zn**, 0.7 **Fe**, 0.05 each, 0.15 total, others	**Hard rolled, mill finish**: 6–34 in. (152–864 mm) **diam**, 25 ksi (172 MPa) **TS**
ANSI/ASTM B 547	3003	US	Tube: formed and arc-welded	rem **Al**, 0.6 **Si**, 0.05–0.20 **Cu**, 1.0–1.5 **Mn**, 0.10 **Zn**, 0.7 **Fe**, 0.05 each, 0.15 total, others	**Annealed**: 0.250–0.499 in. **diam**, 14 ksi (97 MPa) **TS**, 5 ksi (34 MPa) **YS**, 23% in 2 in. (50 mm) **El**; **H12**: 0.250–0.499 in. **diam**, 17 ksi (117 MPa) **TS**, 12 ksi (83 MPa) **YS**, 9% in 2 in. (50 mm) **El**; **H14**: 0.250–0.499 in. **diam**, 20 ksi (138 MPa) **TS**, 17 ksi (117 MPa) **YS**, 8% in 2 in. (50 mm) **El**

WROUGHT ALUMINUM AND ALUMINUM ALLOYS

specification number	designation	country	product forms	chemical composition	mechanical properties and hardness values
ANSI/ASTM B 547	Alclad 3003	US	Tube: formed and arc–welded	rem **Al**, 0.6 **Si**, 0.05–0.20 **Cu**, 1.0–1.5 **Mn**, 0.10 **Zn**, 0.7 **Fe**, 7072 cladding material, 0.05 each, 0.15 total, others	**Annealed**: 0.250–0.499 in. **diam**, 13 ksi (90 MPa) **TS**, 5 ksi (31 MPa) **YS**, 23% in 2 in. (50 mm) **El**; **H12**: 0.250–0.499 in. **diam**, 16 ksi (110 MPa) **TS**, 11 ksi (76 MPa) **YS**, 9% in 2 in. (50 mm) **El**; **H14**: 0.250–0.499 in. **diam**, 19 ksi (131 MPa) **TS**, 16 ksi (110 MPa) **YS**, 8% in 2 in. (50 mm) **El**
ANSI/ASTM B 547	3004	US	Tube: formed and arc–welded	rem **Al**, 0.30 **Si**, 0.25 **Cu**, 1.0–1.5 **Mn**, 0.8–1.3 **Mg**, 0.25 **Zn**, 0.7 **Fe**, 0.05 each, 0.15 total, others	**Annealed**: 0.250–0.500 in. **diam**, 22 ksi (152 MPa) **TS**, 9 ksi (59 MPa) **YS**, 16% in 2 in. (50 mm) **El**; **H32**: 0.125–0.500 in. **diam**, 28 ksi (193 MPa) **TS**, 21 ksi (145 MPa) **YS**, 6% in 2 in. (50 mm) **El**; **H34**: 0.125–0.500 in. **diam**, 32 ksi (221 MPa) **TS**, 25 ksi (172 MPa) **YS**, 5% in 2 in. (50 mm) **El**
ANSI/ASTM B 547	Alclad 3004	US	Tube: formed and arc–welded	rem **Al**, 0.30 **Si**, 0.25 **Cu**, 1.0–1.5 **Mn**, 0.8–1.3 **Mg**, 0.25 **Zn**, 0.7 **Fe**, 7072 cladding material, 0.05 each, 0.15 total, others	**Annealed**: 0.250–0.499 in. **diam**, 21 ksi (145 MPa) **TS**, 8 ksi (55 MPa) **YS**, 16% in 2 in. (50 mm) **El**; **H32**: 0.250–0.499 in. **diam**, 27 ksi (186 MPa) **TS**, 20 ksi (138 MPa) **YS**, 6% in 2 in. (50 mm) **El**; **H34**: 0.250–0.499 in. **diam**, 31 ksi (214 MPa) **TS**, 24 ksi (165 MPa) **YS**, 5% in 2 in. (50 mm) **El**
ANSI/ASTM B 491	3003	US	Tube: coiled, for general purposes	rem **Al**, 0.6 **Si**, 0.05–0.20 **Cu**, 1.0–1.5 **Mn**, 0.10 **Zn**, 0.7 **Fe**, 0.05 each, 0.15 total, others	**H112**: 0.032–0.050 in. **diam**, 14 ksi (97 MPa) **TS**, 5 ksi (34 MPa) **YS**, 25% in 2 in. (50 mm) **El**
ANSI/ASTM B 483	3003	US	Tube	rem **Al**, 0.6 **Si**, 0.05–0.20 **Cu**, 1.0–1.5 **Mn**, 0.10 **Zn**, 0.7 **Fe**, 0.05 each, 0.15 total, others	**Annealed**: 0.018–0.500 in. **diam**, 14 ksi (97 MPa) **TS**, 5 ksi (34 MPa) **YS**; **H12 (round tube)**: 0.018–0.500 in. **diam**, 17 ksi (117 MPa) **TS**, 12 ksi (83 MPa) **YS**; **H14 (round tube)**: 0.018–0.500 in. **diam**, 20 ksi (137 MPa) **TS**, 17 ksi (117 MPa) **YS**
ANSI/ASTM B 404	3003	US	Tube: seamless, for condensers and heat exchangers	rem **Al**, 0.6 **Si**, 0.05–0.20 **Cu**, 1.0–1.5 **Mn**, 0.10 **Zn**, 0.7 **Fe**, 0.05 each, 0.15 total, others	**Annealed**: 0.260–0.500 in. **diam**, 14 ksi (97 MPa) **TS**, 5 ksi (34 MPa) **YS**, 30% in 2 in. (50 mm) **El**; **H14**: 0.260–0.500 in. **diam**, 20 ksi (138 MPa) **TS**, 17 ksi (117 MPa) **YS**; **H25**: 0.010–0.500 in. **diam**, 22 ksi (152 MPa) **TS**, 19 ksi (131 MPa) **YS**

WROUGHT ALUMINUM AND ALUMINUM ALLOYS

specification number	designation	country	product forms	chemical composition	mechanical properties and hardness values
ANSI/ASTM B 404	Alclad 3003	US	Tube: seamless, for condensers and heat exchangers	rem **Al**, 0.6 **Si**, 0.05–0.20 **Cu**, 1.0–1.5 **Mn**, 0.10 **Zn**, 0.7 **Fe**, 7072 cladding material, 0.05 each, 0.15 total, others	**Annealed**: 0.010–0.500 in. **diam**, 13 ksi (90 MPa) **TS**, 5 ksi (31 MPa) **YS**; **H14**: 0.050–0.259 in. **diam**, 19 ksi (131 MPa) **TS**, 16 ksi (110 MPa) **YS**, 4% in 2 in. (50 mm) **El**; **H25**: 0.010–0.500 in. **diam**, 21 ksi (145 MPa) **TS**, 18 ksi (124 MPa) **YS**
ANSI/ASTM B 345	3003	US	Pipe and tube: seamless	rem **Al**, 0.6 **Si**, 0.05–0.20 **Cu**, 1.0–1.5 **Mn**, 0.10 **Zn**, 0.7 **Fe**, 0.05 each, 0.15 total, others	**H18 (pipe)**: <1 in. **diam**, 27 ksi (186 MPa) **TS**, 24 ksi (165 MPa) **YS**, 4% in 2 in. (50 mm) **El**; **Annealed (tube)**: all **diam**, 14 ksi (97 MPa) **TS**, 5 ksi (34 MPa) **YS**, 25% in 2 in. (50 mm) **El**; **H122 (tube)**: all **diam**, 14 ksi (97 MPa) **TS**, 5 ksi (34 MPa) **YS**
ANSI/ASTM B 345	Alclad 3003	US	Tube: seamless	rem **Al**, 0.6 **Si**, 0.05–0.20 **Cu**, 1.0–1.5 **Mn**, 0.10 **Zn**, 0.7 **Fe**, 7072 cladding material, 0.05 each, 0.15 total, others	**Annealed and H112**: all **diam**, 13 ksi (90 MPa) **TS**, 5 ksi (34 MPa) **YS**
ANSI/ASTM B 316	3003	US	Wire and rod: rivet and cold-heading	rem **Al**, 0.6 **Si**, 0.05–0.20 **Cu**, 1.0–1.5 **Mn**, 0.10 **Zn**, 0.7 **Fe**, 0.05 each, 0.15 total, others	**Annealed**: ≤1.000 in. **diam**, 19 max ksi (131 max MPa) **TS**; **H14**: ≤1.000 in. **diam**, 20 ksi (138 MPa) **TS**
ANSI/ASTM B 313	3003	US	Tube: welded	rem **Al**, 0.6 **Si**, 0.05–0.20 **Cu**, 1.0–1.5 **Mn**, 0.10 **Zn**, 0.7 **Fe**, 0.05 each, 0.15 total, others	**Annealed**: 0.051–0.125 in. **diam**, 14 ksi (97 MPa) **TS**, 5 ksi (34 MPa) **YS**, 25% in 2 in. (50 mm) **El**; **H12**: 0.051–0.113 in. **diam**, 17 ksi (117 MPa) **TS**, 12 ksi (83 MPa) **YS**, 6% in 2 in. (50 mm) **El**; **H14**: 0.051–0.113 in. **diam**, 20 ksi (138 MPa) **TS**, 17 ksi (117 MPa) **YS**, 5% in 2 in. (50 mm) **El**
ANSI/ASTM B 313	3004	US	Tube: welded	rem **Al**, 0.30 **Si**, 0.25 **Cu**, 1.0–1.5 **Mn**, 0.8–1.3 **Mg**, 0.25 **Zn**, 0.7 **Fe**, 0.05 each, 0.15 total, others	**Annealed**: 0.051–0.125 in. **diam**, 22 ksi **TS**, 9 ksi **YS**, 18% in 2 in. (50 mm) **El**; **H32**: 0.051–0.113 in. **diam**, 27 ksi **TS**, 20 ksi **YS**, 5% in 2 in. (50 mm) **El**; **H34**: 0.051–0.113 in. **diam**, 32 ksi **TS**, 25 ksi **YS**, 4% in 2 in. (50 mm) **El**
ANSI/ASTM B 313	Alclad 3004	US	Tube: welded	rem **Al**, 0.30 **Si**, 0.25 **Cu**, 1.0–1.5 **Mn**, 0.8–1.3 **Mg**, 0.25 **Zn**, 0.7 **Fe**, 7072 cladding material, 0.05 each, 0.15 total, others	**Annealed**: 0.051–0.125 in. **diam**, 21 ksi (145 MPa) **TS**, 8 ksi (55 MPa) **YS**, 16% in 2 in. (50 mm) **El**; **H32**: 0.051–0.113 in. **diam**, 27 ksi (186 MPa) **TS**, 20 ksi (138 MPa) **YS**, 5% in 2 in. (50 mm) **El**; **H34**: 0.051–0.113 in. **diam**, 31 ksi (214 MPa) **TS**, 24 ksi (165 MPa) **YS**, 4% in 2 in. (50 mm) **El**

WROUGHT ALUMINUM AND ALUMINUM ALLOYS

specification number	designation	country	product forms	chemical composition	mechanical properties and hardness values
ANSI/ASTM B 247	3003	US	Die forgings	rem **Al**, 0.6 **Si**, 0.05–0.20 **Cu**, 1.0–1.5 **Mn**, 0.10 **Zn**, 0.7 **Fe**, 0.05 each, 0.15 total, others	**H112**: ≤4.000 in. **diam**, 14 ksi (97 MPa) **TS**, 5 ksi (34 MPa) **YS**, 18% in 2 in. (50 mm) **El**, 25 min **HB**
ANSI/ASTM B 241	3003	US	Pipe and tube: seamless	rem **Al**, 0.6 **Si**, 0.05–0.20 **Cu**, 1.0–1.5 **Mn**, 0.10 **Zn**, 0.7 **Fe**, 0.05 each, 0.15 total, others	**Annealed (tube)**: all **diam**, 14 ksi (97 MPa) **TS**, 5 ksi (34 MPa) **YS**, 25% in 2 in. (50 mm) **El**; **H112 (tube)**: >0.062 in. **diam**, 13 ksi (90 MPa) **TS**, 5 ksi (31 MPa) **YS**, 25% in 2 in. (50 mm) **El**; **H18 (pipe)**: <1 in. **diam**, 27 ksi (186 MPa) **TS**, 24 ksi (165 MPa) **YS**
ANSI/ASTM B 241	Alclad 3003	US	Tube: seamless	rem **Al**, 0.6 **Si**, 0.05–0.20 **Cu**, 1.0–1.5 **Mn**, 0.10 **Zn**, 0.7 **Fe**, 7072 cladding material, 0.05 each, 0.15 total, others	**Annealed**: all **diam**, 13 ksi (90 MPa) **TS**, 5 ksi (31 MPa) **YS**; **H112**: >0.062 in. **diam**, 13 ksi (90 MPa) **TS**, 5 ksi (31 MPa) **YS**, 25% in 2 in. (50 mm) **El**
ANSI/ASTM B 234	3003	US	Tube: seamless, for condensers and heat exchangers	rem **Al**, 0.6 **Si**, 0.05–0.20 **Cu**, 1.0–1.5 **Mn**, 0.10 **Zn**, 0.7 **Fe**, 0.05 each, 0.15 total, others	**H14**: 0.050–0.200 in. **diam**, 20 ksi (138 MPa) **TS**, 17 ksi (117 MPa) **YS**, 4% in 2 in. (50 mm) **El**; **H25**: 0.010–0.200 in. **diam**, 22 ksi (152 MPa) **TS**, 19 ksi (131 MPa) **YS**
ANSI/ASTM B 234	Alclad 3003	US	Tube: seamless, for condensers and heat exchangers	rem **Al**, 0.6 **Si**, 0.05–0.20 **Cu**, 1.0–1.5 **Mn**, 0.10 **Zn**, 0.7 **Fe**, 7072 cladding material, 0.05 each, 0.15 total, others	**H14**: 0.050–0.200 in. **diam**, 19 ksi (131 MPa) **TS**, 16 ksi (110 MPa) **YS**, 4% in 2 in. (50 mm) **El**; **H25**: 0.010–0.200 in. **diam**, 21 ksi (145 MPa) **TS**, 18 ksi (124 MPa) **YS**
ANSI/ASTM B 221	3003	US	Bar, tube, rod, wire, and shapes	rem **Al**, 0.6 **Si**, 0.05–0.20 **Cu**, 1.0–1.5 **Mn**, 0.10 **Zn**, 0.7 **Fe**, 0.05 each, 0.15 total, others	**Annealed**: all **diam**, 14 ksi (97 MPa) **TS**, 5 ksi (34 MPa) **YS**, 25% in 2 in. (50 mm) **El**; **H112**: all **diam**, 14 ksi (97 MPa) **TS**, 5 ksi (34 MPa) **YS**
ANSI/ASTM B 221	Alclad 3003	US	Bar, tube, rod, wire, and shapes	rem **Al**, 0.6 **Si**, 0.05–0.20 **Cu**, 1.0–1.5 **Mn**, 0.10 **Zn**, 0.7 **Fe**, 7072 cladding material, 0.05 each, 0.15 total, others	**Annealed**: all **diam**, 13 ksi (90 MPa) **TS**, 5 ksi (31 MPa) **YS**; **H112**: >0.062 in. **diam**, 13 ksi (90 MPa) **TS**
ANSI/ASTM B 221	3004	US	Bar, tube, rod, wire, and shapes	rem **Al**, 0.30 **Si**, 0.25 **Cu**, 1.0–1.5 **Mn**, 0.8–1.3 **Mg**, 0.25 **Zn**, 0.7 **Fe**, 0.05 each, 0.15 total, others	**Annealed**: all **diam**, 23 ksi (159 MPa) **TS**, 9 ksi (59 MPa) **YS**

WROUGHT ALUMINUM AND ALUMINUM ALLOYS

specification number	designation	country	product forms	chemical composition	mechanical properties and hardness values
ANSI/ASTM B 211	3003	US	Bar, rod, and wire	rem **Al**, 0.6 **Si**, 0.05–0.20 **Cu**, 1.0–1.5 **Mn**, 0.7 **Fe**, 0.05 each, 0.15 total, others	**Annealed**: all **diam**, 14 ksi (97 MPa) **TS**, 5 ksi (34 MPa) **YS**, 25% in 2 in. (50 mm) **El**; **H12**: $\leq$0.374 in. **diam**, 17 ksi (117 MPa) **TS**; **H112**: all **diam**, 14 ksi (97 MPa) **TS**, 5 ksi (34 MPa) **YS**
ANSI/ASTM B 210	Alclad 3003	US	Tube: seamless	rem **Al**, 0.6 **Si**, 0.05–0.20 **Cu**, 1.0–1.5 **Mn**, 0.10 **Zn**, 0.7 **Fe**, 0.05 each, 0.15 total, others	**Annealed**: 0.050–0.259 in. **diam**, 13 ksi (90 MPa) **TS**, 5 ksi (31 MPa) **YS**, 25% in 2 in. (50 mm) **El**; **H14 (round tube)**: 0.050–0.259 in. **diam**, 19 ksi (131 MPa) **TS**, 16 ksi (110 MPa) **YS**, 4% in 2 in. (50 mm) **El**; **H18 (round tube)**: 0.010–0.500 in. **diam**, 26 ksi (179 MPa) **TS**, 23 ksi (159 MPa) **YS**
ANSI/ASTM B 209	Alclad 3003	US	Sheet and plate	rem **Al**, 0.6 **Si**, 0.05–0.20 **Cu**, 1.0–1.5 **Mn**, 0.10 **Zn**, 0.7 **Fe**, 7072 cladding material, 0.05 each, 0.15 total, others	**Annealed**: 0.250–0.499 in. **diam**, 13 ksi (90 MPa) **TS**, 5 ksi (34 MPa) **YS**, 23% in 2 in. (50 mm) **El**; **H12**: 0.250–0.499 in. **diam**, 16 ksi (90 MPa) **TS**, 11 ksi (76 MPa) **YS**, 9% in 2 in. (50 mm) **El**; **H14**: 0.250–0.499 in. **diam**, 19 ksi (131 MPa) **TS**, 16 ksi (110 MPa) **YS**, 8% in 2 in. (50 mm) **El**
ANSI/ASTM B 209	Alclad 3004	US	Sheet and plate	rem **Al**, 0.30 **Si**, 0.25 **Cu**, 1.0–1.5 **Mn**, 0.8–1.3 **Mg**, 0.25 **Zn**, 0.7 **Fe**, 0.05 each, 0.15 total, others	**Annealed**: 0.250–0.499 in. **diam**, 21 ksi (145 MPa) **TS**, 8 ksi (55 MPa) **YS**, 16% in 2 in. (50 mm) **El**; **H32**: 0.250–0.499 in. **diam**, 27 ksi (186 MPa) **TS**, 20 ksi (138 MPa) **YS**, 6% in 2 in. (50 mm) **El**; **H34**: 0.250–0.499 in. **diam**, 31 ksi (214 MPa) **TS**, 24 ksi (165 MPa) **YS**, 5% in 2 in. (50 mm) **El**
ANSI/ASTM B 210	3003	US	Tube: seamless	rem **Al**, 0.6 **Si**, 0.05–0.20 **Cu**, 1.0–1.5 **Mn**, 0.10 **Zn**, 0.7 **Fe**, 0.05 each, 0.15 total, others	**Annealed**: 0.050–0.259 in. **diam**, 14 ksi (97 MPa) **TS**, 5 ksi (34 MPa) **YS**, 25% in 2 in. (50 mm) **El**; **H14**: 0.050–0.259 in. **diam**, 20 ksi (138 MPa) **TS**, 17 ksi (117 MPa) **YS**, 4% in 2 in. (50 mm) **El**; **H16**: 0.050–0.259 in. **diam**, 24 ksi (165 MPa) **TS**, 21 ksi (145 MPa) **YS**, 4% in 2 in. (50 mm) **El**
ANSI/ASTM B 210	3303	US	Tube: seamless	rem **Al**, 0.6 **Si**, 0.5–0.20 **Cu**, 1.0–1.5 **Mn**, 0.30 **Zn**, 0.7 **Fe**, 0.05 each, 0.15 total, others	**Annealed**: 0.025–0.049 in. **diam**, 14 ksi (97 MPa) **TS**, 5 ksi (34 MPa) **YS**, 20% in 2 in. (50 mm) **El**

WROUGHT ALUMINUM AND ALUMINUM ALLOYS

specification number	designation	country	product forms	chemical composition	mechanical properties and hardness values
ANSI/ASTM B 209	3003	US	Sheet and plate	rem **Al**, 0.6 **Si**, 0.05–0.20 **Cu**, 1.0–1.5 **Mn**, 0.10 **Zn**, 0.7 **Fe**, 0.05 each, 0.15 total, others	**Annealed**: 0.500–3.000 in. **diam**, 14 ksi (97 MPa) **TS**, 5 ksi (34 MPa) **YS**, 23% in 2 in. (50 mm) **El**; **H12**: 0.500–2.000 in. **diam**, 17 ksi (117 MPa) **TS**, 12 ksi (83 MPa) **YS**, 10% in 2 in. (50 mm) **El**; **H14**: 0.500–1.000 in. **diam**, 20 ksi (138 MPa) **TS**, 17 ksi (117 MPa) **YS**, 10% in 2 in. (50 mm) **El**
ANSI/ASTM B 209	3004	US	Sheet and plate	rem **Al**, 0.30 **Si**, 0.25 **Cu**, 1.0–1.5 **Mn**, 0.8–1.3 **Mg**, 0.25 **Zn**, 0.7 **Fe**, 0.05 each, 0.15 total, others	**Annealed**: 0.250–3.000 in. **diam**, 22 ksi (152 MPa) **TS**, 9 ksi (62 MPa) **YS**, 16% in 2 in. (50 mm) **El**; **H32**: 0.114–2.000 in. **diam**, 28 ksi (193 MPa) **TS**, 21 ksi (145 MPa) **YS**, 6% in 2 in. (50 mm) **El**; **H34**: 0.114–1.000 in. **diam**, 32 ksi (221 MPa) **TS**, 25 ksi (172 MPa) **YS**, 5% in 2 in. (50 mm) **El**
ANSI/ASTM B 209	3005	US	Sheet and plate	rem **Al**, 0.6 **Si**, 0.30 **Cu**, 1.0–1.5 **Mn**, 0.20–0.6 **Mg**, 0.10 **Cr**, 0.25 **Zn**, 0.10 **Ti**, 0.7 **Fe**, 0.05 each, 0.15 total, others	**Annealed**: 0.051–0.249 in. **diam**, 17 ksi (117 MPa) **TS**, 7 ksi (48 MPa) **YS**, 20% in 2 in. (50 mm) **El**; **H12**: 0.162–0.249 in. **diam**, 20 ksi (138 MPa) **TS**, 17 ksi (117 MPa) **YS**, 5% in 2 in. (50 mm) **El**; **H14**: 0.114–0.249 in. **diam**, 24 ksi (165 MPa) **TS**, 21 ksi (145 MPa) **YS**, 4% in 2 in. (50 mm) **El**
AS 1867	3203	Australia	Tube: drawn	rem **Al**, 0.6 **Si**, 0.05 **Cu**, 1.0–1.5 **Mn**, 0.10 **Zn**, 0.7 **Fe**, 0.05 each, 0.15 total, others	**Annealed**: all **diam**, 132 max MPa **TS**, 25% in 2 in. (50 mm) **El**; **Strain hardened to 1/4 hard**: all **diam**, 117 MPa **TS**; **Strain hardened to full hard**: all **diam**, 186 MPa **TS**
AS 1865	3203	Australia	Wire, rod, bar, strip	rem **Al**, 0.6 **Si**, 0.05 **Cu**, 1.0–1.5 **Mn**, 0.10 **Zn**, 0.7 **Fe**, 0.05 each, 0.15 total, others	**Annealed**: all **diam**, 125 max MPa **TS**, 23% in 2 in. (50 mm) **El**; **Strain hardened to 1/4 hard**: 10 mm **diam**, 117 MPa **TS**; **Strain hardened to full hard**: 10 mm **diam**, 186 MPa **TS**
AS 1734	3005	Australia	Sheet, plate	rem **Al**, 0.6 **Si**, 0.30 **Cu**, 1.0–1.5 **Mn**, 0.20–0.60 **Mg**, 0.10 **Cr**, 0.25 **Zn**, 0.10 **Ti**, 0.7 **Fe**, 0.05 each, 0.15 total, others	**Annealed**: 0.15–0.20 mm **diam**, 103 MPa **TS**, 14% in 2 in. (50 mm) **El**; **Strain–hardened to 1/2 hard**: 0.15–0.25 mm **diam**, 151 MPa **TS**, 1% in 2 in. (50 mm) **El**; **Strain–hardened to full hard**: 0.30–0.80 mm **diam**, 206 MPa **TS**, 1% in 2 in. (50 mm) **El**

WROUGHT ALUMINUM AND ALUMINUM ALLOYS

specification number	designation	country	product forms	chemical composition	mechanical properties and hardness values
AS 1734	Alclad 3004	Australia	Sheet, plate	rem **Al**, 0.30 **Si**, 0.25 **Cu**, 1.0–1.5 **Mn**, 0.8–1.3 **Mg**, 0.25 **Zn**, 0.7 **Fe**, 0.05 each, 0.15 total, other, clad with alloy 7072	**Annealed**: 0.20–0.50 mm **diam**, 144 MPa **TS**, 55 MPa **YS**, 10% in 2 in. (50 mm) **El**; **Strain–hardened and stabilized to 1/4 hard**: 0.25–0.50 mm **diam**, 213 MPa **TS**, 165 MPa **YS**, 1% in 2 in. (50 mm) **El**; **Strain–hardened and stabilized to full hard**: 0.20–0.50 mm **diam**, 103 MPa **TS**, 14% in 2 in. (50 mm) **El**
AS 1734	3003	Australia	Sheet, plate	rem **Al**, 0.6 **Si**, 0.05–0.20 **Cu**, 1.0–1.5 **Mn**, 0.10 **Zn**, 0.7 **Fe**, 0.05 each, 0.15 total, others	**Annealed**: 0.30–0.80 mm **diam**, 96 MPa **TS**, 20% in 2 in. (50 mm) **El**; **Strain–hardened to 1/2 hard**: 0.25–0.30 mm **diam**, 137 MPa **TS**, 1% in 2 in. (50 mm) **El**; **Strain–hardened to full hard**: 0.15–0.50 mm **diam**, 186 MPa **TS**, 1% in 2 in. (50 mm) **El**
AS 1734	3203	Australia	Sheet, plate	rem **Al**, 0.6 **Si**, 0.05 **Cu**, 1.0–1.5 **Mn**, 0.10 **Zn**, 0.7 **Fe**, 0.05 each, 0.15 total, others	**Annealed**: 0.15–0.20 mm **diam**, 116 MPa **TS**, 14% in 2 in. (50 mm) **El**; **Strain–hardened to 1/2 hard**: 0.25–0.30 mm **diam**, 139 MPa **TS**, 1% in 2 in. (50 mm) **El**; **Strain–hardened to full hard**: 0.15–0.50 mm **diam**, 117 MPa **TS**, 1% in 2 in. (50 mm) **El**
AS 1734	3004	Australia	Sheet, plate	rem **Al**, 0.30 **Si**, 0.25 **Cu**, 1.0–1.5 **Mn**, 0.8–1.3 **Mg**, 0.25 **Zn**, 0.7 **Fe**, 0.05 each, 0.15 total, others	**Annealed**: 0.15–0.20 mm **diam**, 151 MPa **TS**, 58 MPa **YS**; **Strain–hardened and stabilized to 1/2 hard**: 0.25–0.50 mm **diam**, 220 MPa **TS**, 172 MPa **YS**, 1% in 2 in. (50 mm) **El**; **Strain–hardened and stabilized to full hard**: 0.15–0.20 mm **diam**, 262 MPa **TS**, 213 MPa **YS**
CSA HA.7.1	.3003	Canada	Tube: drawn and extruded, pipe: drawn	rem **Al**, 0.6 **Si**, 0.05–0.20 **Cu**, 1.0–1.5 **Mn**, 0.10 **Zn**, 0.7 **Fe**, 0.05 each, 0.15 total, others	**Drawn and O**: 0.025–0.049 in. **diam**, 14.0 ksi (97 MPa) **TS**, 5.0 ksi (34 MPa) **YS**, 30% in 2 in. (50 mm) **El**; **Drawn and H18**: <0.024 in. **diam**, 27.0 ksi (186 MPa) **TS**, 24.0 ksi (165 MPa) **YS**, 2% in 2 in. (50 mm) **El**; **Extruded and H112**: all **diam**, 14.0 ksi (97 MPa) **TS**, 5.0 ksi (34 MPa) **YS**

WROUGHT ALUMINUM AND ALUMINUM ALLOYS

specification number	designation	country	product forms	chemical composition	mechanical properties and hardness values
CSA HA.7	.3003	Canada	Tube, pipe: seamless	rem **Al**, 0.6 **Si**, 0.05–0.20 **Cu**, 1.0–1.5 **Mn**, 0.10 **Zn**, 0.7 **Fe**, 0.05 each, 0.15 total, others	**Drawn and O**: 0.025–0.049 in. **diam**, 14.0 ksi (97 MPa) **TS**, 5.0 ksi (34 MPa) **YS**, 30% in 2 in. (50 mm) **El**; **Drawn and H18**: <0.024 in. **diam**, 27.0 ksi (186 MPa) **TS**, 24.0 ksi (165 MPa) **YS**, 2% in 2 in. (50 mm) **El**; **Extruded and O**: all **diam**, 14.0 ksi (97 MPa) **TS**, 5.0 ksi (34 MPa) **YS**, 25% in 2 in. (50 mm) **El**
CSA HA.4	.3003	Canada	Plate, sheet	rem **Al**, 0.6 **Si**, 0.05–0.20 **Cu**, 1.0–1.5 **Mn**, 0.10 **Zn**, 0.7 **Fe**, 0.05 each, 0.15 total, others	**O**: all **diam**, 14.0 ksi (97 MPa) **TS**, 5.0 ksi (34 MPa) **YS**, 18% in 2 in. (50 mm) **El**; **H14**: all **diam**, 20.0 ksi (138 MPa) **TS**, 17.0 ksi (117 MPa) **YS**, 1% in 2 in. (50 mm) **El**; **H18**: all **diam**27.0 ksi (186 MPa) **TS**, 24.0 ksi (165 MPa) **YS**, 1% in 2 in. (50 mm) **El**
DS 3012	3003	Denmark	Rolled and drawn products	rem **Al**, 0.6 **Si**, 0.5–0.20 **Cu**, 1.0–1.5 **Mn**, 0.10 **Zn**, 0.7 **Fe**, 0.05 each, 0.15 total, others	**1/2 hard**: 0.5–6 mm **diam**, 120 MPa **TS**, 100 MPa **YS**, 5% in 2 in. (50 mm) **El**, 45 **HV**; **3/4 hard**: 0.5–6 mm **diam**, 130 MPa **TS**, 120 MPa **YS**, 4% in 2 in. (50 mm) **El**, 50 **HV**; **Hard**: ≥1.5 mm **diam**, 155 MPa **TS**, 140 MPa **YS**, 3% in 2 in. (50 mm) **El**, 60 **HV**
COPANT 862	3003	COPANT	Wrought products	rem **Al**, 0.6 **Si**, 0.05–0.20 **Cu**, 1.0–1.5 **Mn**, 0.10 **Zn**, 0.7 **Fe**, 0.05 each, 0.15 total, others	
COPANT 862	3004	COPANT	Wrought products	rem **Al**, 0.3 **Si**, 0.25 **Cu**, 1.0–1.5 **Mn**, 0.8–1.3 **Mg**, 0.25 **Zn**, 0.70 **Fe**, 0.05 each, 0.15 total, others	
COPANT 862	3005	COPANT	Wrought products	rem **Al**, 0.60 **Si**, 0.30 **Cu**, 1.0–1.5 **Mn**, 0.20–0.60 **Mg**, 0.10 **Cr**, 0.25 **Zn**, 0.10 **Ti**, 0.70 **Fe**, 0.05 each, 0.15 total, others	
WW-T-700/2d	3003	US	Tube	rem **Al**, 0.6 **Si**, 0.20 **Cu**, 1.0–1.5 **Mn**, 0.10 **Zn**, 0.7 **Fe**, 0.05 each, 0.15 total, others	**Annealed**: all **diam**, 19 max ksi (131 max MPa) **TS**; **Strain hardened, one quarter hard temper**: all **diam**, 17 ksi (117 MPa) **TS**; **Strain hardened, half hard temper**: all **diam**, 20 ksi (138 MPa) **TS**
QQ-A-430B	3003	US	Rod, wire	rem **Al**, 0.6 **Si**, 0.05–0.20 **Cu**, 1.0–1.5 **Mn**, 0.10 **Zn**, 0.7 **Fe**, 0.05 each, 0.15 total, others	**Annealed**: 1.000 max in. **diam**, 19 max ksi (131 max MPa) **TS**; **Strain hardened half hard temper**: 1.000 max in. **diam**, 20 ksi (138 MPa) **TS**

WROUGHT ALUMINUM AND ALUMINUM ALLOYS

specification number	designation	country	product forms	chemical composition	mechanical properties and hardness values
QQ-A-250/2D	3003	US	Plate, sheet	rem **Al**, 0.6 **Si**, 0.05–0.20 **Cu**, 1.0–1.5 **Mn**, 0.10 **Zn**, 0.7 **Fe**, 0.05 each, 0.15 total, others	**Annealed**: 0.006–.007 in. **diam**, 14 ksi (97 MPa) **TS**, 5 ksi (34 MPa) **YS**, 14% in 2 in. (50 mm) **El**; **Strain hardened one quarter hard temper**: 0.017–0.019 in. **diam**, 17 ksi (117 MPa) **TS**, 12 ksi (83 MPa) **YS**, 3% in 2 in. (50 mm) **El**; **Strain hardened half hard temper**: 0.009–0.012 in. **diam**, 20 ksi (138 MPa) **TS**, 17 ksi (117 MPa) **YS**, 1% in 2 in. (50 mm) **El**
QQ-A-225/2D	3003	US	Bar, rod, wire	rem **Al**, 0.6 **Si**, 0.05–0.20 **Cu**, 1.0–1.5 **Mn**, 0.10 **Zn**, 0.7 **Fe**, 0.05 each, 0.15 total, others	**Annealed**: all **diam**, 19 max ksi (131 max MPa) **TS**, 25% in 2 in. (50 mm) **El**; **Strain hardened one quarter hard temper**: 0.374 max in. **diam**, 17 ksi (117 MPa) **TS**,; Strain hardened half hard temper: 0.374 max in. **diam**, 20 ksi (138 MPa) **TS**
QQ-A-200/1C	3003	US	Bar, rod, shapes, tube, wire	rem **Al**, 0.6 **Si**, 0.05–0.20 **Cu**, 1.0–1.5 **Mn**, 0.10 **Zn**, 0.7 **Fe**, 0.5 each, 0.15 total, others	**Annealed**: all **diam**, 14 ksi (97 MPa) **TS**, 25% in **El**; **As extruded**: all **diam**, 14 ksi (97 MPa) **TS**
NF A 50 501	3005 (A-MG0.5)	France	Tube	rem **Al**, 0.6 **Si**, 0.30 **Cu**, 1.0–1.5 **Mn**, 0.20–0.60 **Mg**, 0.10 **Cr**, 0.25 **Zn**, 0.10 **Ti**, 0.7 **Fe**, 0.05 each, 0.15 total, others	**As manufactured**: all **diam**, 170 MPa **TS**, 130 MPa **YS**, 6% in 2 in. (50 mm) **El**
NF A 50 501	3004 (A-M1G)	France	Tube	rem **Al**, 0.30 **Si**, 0.25 **Cu**, 1.0–1.5 **Mn**, 0.8–1.30 **Mg**, 0.25 **Zn**, 0.70 **Fe**, 0.05 each, 0.15 total, others	**As manufactured**: all **diam**, 260 MPa **TS**, 210 MPa **YS**, 3% in 2 in. (50 mm) **El**
NF A 50 411	3003 (A-M1)	France	Bar, wire, tube, and shapes	rem **Al**, 0.6 **Si**, 0.05–0.20 **Cu**, 1.0–1.5 **Mn**, 0.10 **Zn**, 0.7 **Fe**, 0.05 each, 0.15 total, others	**Hard (drawn bar)**: ≤10 mm **diam**, 185 MPa **TS**, 145 MPa **YS**, 2% in 2 in. (50 mm) **El**; **Hard (wire)**: ≤6 mm **diam**, 180 MPa **TS**, 140 MPa **YS**, 2% in 2 in. (50 mm) **El**; **Drawn or extruded (tube)**: ≤150 mm **diam**, 95 MPa **TS**, 12% in 2 in. (50 mm) **El**
NF A 50 451	3004 (A-M 1 G)	France	Sheet, plate, and strip	rem **Al**, 0.30 **Si**, 0.25 **Cu**, 1.0–1.5 **Mn**, 0.8–1.30 **Mg**, 0.25 **Zn**, 0.70 **Fe**, 0.05 each, 0.15 total, others	**Annealed**: 0.4–3.2 mm **diam**, 155 MPa **TS**, 18% in 2 in. (50 mm) **El**; **3/4 hard**: 0.4–1.6 mm **diam**, 240 MPa **TS**, 200 MPa **YS**, 3% in 2 in. (50 mm) **El**; **Hard**: 0.4–1.6 mm **diam**, 260 MPa **TS**, 210 MPa **YS**, 3% in 2 in. (50 mm) **El**

WROUGHT ALUMINUM AND ALUMINUM ALLOYS

specification number	designation	country	product forms	chemical composition	mechanical properties and hardness values
NF A 50 451	3005 (A–MG 0.5)	France	Sheet, plate, and strip	rem **Al**, 0.6 **Si**, 0.30 **Cu**, 1.0–1.5 **Mn**, 0.20–0.60 **Mg**, 0.10 **Cr**, 0.25 **Zn**, 0.10 **Ti**, 0.7 **Fe**, 0.05 each, 0.15 total, others	**Annealed**: 0.4–1.6 mm **diam**, 115 MPa **TS**, 22% in 2 in. (50 mm) **El**; **3/4 hard**: 0.4–1.6 mm **diam**, 200 MPa **TS**, 180 MPa **YS**, 3% in 2 in. (50 mm) **El**; **hard**: 0.4–1.6 mm **diam**, 230 MPa **TS**, 200 MPa **YS**, 3% in 2 in. (50 mm) **El**
NF A 50 451	3003 (A–M1)	France	Sheet, plate, and strip	rem **Al**, 0.6 **Si**, 0.05–0.20 **Cu**, 1.0–1.5 **Mn**, 0.10 **Zn**, 0.70 **Fe**, 0.05 each, 0.15 total, others	**Annealed**: 0.4–3.2 mm **diam**, 100 MPa **TS**, 25% in 2 in. (50 mm) **El**; **3/4 hard**: 0.4–1.6 mm **diam**, 170 MPa **TS**, 150 MPa **YS**, 3% in 2 in. (50 mm) **El**; **Hard**: 0.4–1.6 mm **diam**, 190 MPa **TS**, 170 MPa **YS**, 3% in 2 in. (50 mm) **El**
IS 739	NG3	India	Wire	rem **Al**, 0.6 **Si**, 0.1 **Cu**, 1.0–1.5 **Mn**, 0.2 **Zn**, 0.7 **Fe**	**Annealed**: all **diam**, 123 max MPa **TS**; **Annealed and cold worked to hard**: all **diam**, 172 MPa **TS**
IS 5909	31000	India	Sheet, strip	rem **Al**, 0.6 **Si**, 0.10 **Cu**, 1.0–1.5 **Mn**, 0.1 **Mg**, 0.05 **Cr**, 0.2 **Ni**, 0.1 **Zn**, 0.15 **Ti**, 0.05 **Sn**, 0.7 **Fe**, 0.05 **Pb**	**Annealed**: 30 mm **diam**, 93 MPa **TS**, 30% in 2 in. (50 mm) **El**; **Half hard**: 7 mm **diam**, 137 MPa **TS**, 7% in 2 in. (50 mm) **El**; **Hard**: 3 mm **diam**, 177 MPa **TS**, 3% in 2 in. (50 mm) **El**
ISO R209	Al–Mn1Cu	ISO	Wrought products	rem **Al**, 0.6 **Si**, 0.05–0.20 **Cu**, 1.0–1.5 **Mn**, 0.2 **Zn**, 0.7 **Fe**, 0.2 **Ti+Zr+Cr**	
ISO TR2778	Al–Mn1Cu	ISO	Tube: drawn	rem **Al**, 0.6 **Si**, 0.05–0.20 **Cu**, 1.0–1.5 **Mn**, 0.2 **Zn**, 0.7 **Fe**, 0.2 **Ti+Zr+Cr**	**Annealed**: all **diam**, 95 MPa **TS**, 20% in 2 in. (50 mm) **El**; **Strain hardened**: 50 mm **diam**, 140 MPa **TS**, 3% in 2 in. (50 mm) **El**
ISO TR2136	Al–Mn1Cu	ISO	Sheet, plate	rem **Al**, 0.6 **Si**, 0.05–0.20 **Cu**, 1.0–1.5 **Mn**, 0.2 **Zn**, 0.7 **Fe**, 0.2 **Ti+Zr+Cr**, total	**Annealed**: all **diam**, 95 MPa **TS**, 20% in 2 in. (50 mm) **El**; **HB–Strain hardened**: all **diam**, 115 MPa **TS**, 4% in 2 in. (50 mm) **El**; **HD–strain hardened**: all **diam**, 140 MPa **TS**, 3% in 2 in. (50 mm) **El**
JIS H 4160	3003	Japan	Foil	rem **Al**, 0.6 **Si**, 0.05–0.20 **Cu**, 1.0–1.5 **Mn**, 0.10 **Zn**, 0.7 **Fe**, 0.15 total, others	
JIS H 4100	3003	Japan	Shapes: extruded	rem **Al**, 0.6 **Si**, 0.05–0.20 **Cu**, 1.0–1.5 **Mn**, 0.10 **Zn**, 0.7 **Fe**, 0.15 total, others	**As manufactured**: all **diam**, 98 MPa **TS**, 39 MPa **YS**
JIS H 4100	3203	Japan	Shapes: extruded	rem **Al**, 0.6 **Si**, 0.05 **Cu**, 1.0–1.5 **Mn**, 0.10 **Zn**, 0.7 **Fe**, 0.15 total, others	**As manufactured**: all **diam**, 98 MPa **TS**, 39 MPa **YS**

WROUGHT ALUMINUM AND ALUMINUM ALLOYS

specification number	designation	country	product forms	chemical composition	mechanical properties and hardness values
JIS H 4090	3003	Japan	Pipe and tube: welded	rem **Al**, 0.6 **Si**, 0.05–0.20 **Cu**, 1.0–1.5 **Mn**, 0.10 **Zn**, 0.7 **Fe**, 0.15 total, others	**Annealed**: <1.3 mm **diam**, 98 MPa **TS**, 39 MPa **YS**, 23% in 2 in. (50 mm) **El**; **1/2 hard**: <1.3 mm **diam**, 137 MPa **TS**, 118 MPa **YS**, 4% in 2 in. (50 mm) **El**; **Hard**: <1.3 mm **diam**, 186 MPa **TS**, 167 MPa **YS**, 3% in 2 in. (50 mm) **El**
JIS H 4090	3203	Japan	Pipe and tube: welded	rem **Al**, 0.6 **Si**, 0.05 **Cu**, 1.0–1.5 **Mn**, 0.10 **Zn**, 0.7 **Fe**, 0.15 total, others	**Annealed**: <1.3 mm **diam**, 98 MPa **TS**, 39 MPa **YS**, 23% in 2 in. (50 mm) **El**; **1/2 hard**: <1.3 mm **diam**, 137 MPa **TS**, 118 MPa **YS**, 4% in 2 in. (50 mm) **El**; **Hard**: <1.3 mm **diam**, 186 MPa **TS**, 167 MPa **YS**, 3% in 2 in. (50 mm) **El**
JIS Z 3263	BA11PC	Japan	Sheet: brazing	rem **Al**, 0.6 **Si**, 0.05–0.20 **Cu**, 1.0–1.5 **Mn**, 0.10 **Zn**, 0.7 **Fe**, 0.05 each, 0.15 total, others	**Annealed**: <0.3 mm **diam**, 137 MPa **TS**, 15% in 2 in. (50 mm) **El**; **1/4 hard**: ≤0.3 mm **diam**, 118 MPa **TS**, 2% in 2 in. (50 mm) **El**; **1/2 hard**: <0.3 mm **diam**, 137 MPa **TS**, 1% in 2 in. (50 mm) **El**
JIS H 4080	3003	Japan	Tube: extruded	rem **Al**, 0.6 **Si**, 0.05–0.20 **Cu**, 1.0–1.5 **Mn**, 0.10 **Zn**, 0.7 **Fe**, 0.15 total, others	**As manufactured**: all **diam**, 98 MPa **TS**, 39 MPa **YS**
JIS H 4080	3203	Japan	Tube: extruded	rem **Al**, 0.6 **Si**, 0.05 **Cu**, 1.0–1.5 **Mn**, 0.10 **Zn**, 0.7 **Fe**, 0.15 total, others	**As manufactured**: all **diam**, 98 MPa **TS**, 39 MPa **YS**
JIS H 4040	3003	Japan	Bar and wire: extruded	rem **Al**, 0.6 **Si**, 0.05–0.20 **Cu**, 1.0–1.5 **Mn**, 0.10 **Zn**, 0.7 **Fe**, 0.15 total, others	**As manufactured (bar)**: all **diam**, 98 MPa **TS**, 39 MPa **YS**; **1/2 hard**: ≤10 mm **diam**, 137 MPa **TS**; **Hard**: ≤10 mm **diam**, 186 MPa **TS**
JIS H 4040	3203	Japan	Bar and wire: extruded	rem **Al**, 0.6 **Si**, 0.05 **Cu**, 1.0–1.5 **Mn**, 0.10 **Zn**, 0.7 **Fe**, 0.15 total, others	**As manufactured (bar)**: all **diam**, 98 MPa **TS**, 39 MPa **YS**; **1/2 hard**: ≤10 mm **diam**, 137 MPa **TS**; **Hard**: ≤10 mm **diam**, 186 MPa **TS**
JIS H 4000	3003	Japan	Sheet, plate, and strip	rem **Al**, 0.6 **Si**, 0.05–0.20 **Cu**, 1.0–1.5 **Mn**, 0.10 **Zn**, 0.7 **Fe**, 0.15 total, others	**Annealed**: <1.3 mm **diam**, 98 MPa **TS**, 39 MPa **YS**, 23% in 2 in. (50 mm) **El**; **Hard**: <1.3 mm **diam**, 186 MPa **TS**, 167 MPa **YS**, 3% in 2 in. (50 mm) **El**; **Hot rolled**: <12 mm **diam**, 118 MPa **TS**, 67 MPa **YS**, 8% in 2 in. (50 mm) **El**

WROUGHT ALUMINUM AND ALUMINUM ALLOYS

specification number	designation	country	product forms	chemical composition	mechanical properties and hardness values
JIS H 4000	3203	Japan	Sheet, plate, and strip	rem **Al**, 0.6 **Si**, 0.05 **Cu**, 1.0–1.5 **Mn**, 0.10 **Zn**, 0.7 **Fe**, 0.15 total, others	**Annealed**: <1.3 mm **diam**, 98 MPa **TS**, 39 MPa **YS**, 23% in 2 in. (50 mm) **El**; **Hard**: <1.3 mm **diam**, 186 MPa **TS**, 167 MPa **YS**, 3% in 2 in. (50 mm) **El**; **Hot rolled**: <12 mm **diam**, 118 MPa **TS**, 67 MPa **YS**, 8% in 2 in. (50 mm) **El**
JIS H 4001	3004	Japan	Sheet and strip	rem **Al**, 0.30 **Si**, 0.25 **Cu**, 1.0–1.5 **Mn**, 0.8–1.3 **Mg**, 0.25 **Zn**, 0.7 **Fe**, 0.15 total, others	**Stabilized at 1/4 hard**: 0.8–1.2 mm **diam**, 196 MPa **TS**, 147 MPa **YS**, 4% in 2 in. (50 mm) **El**; **Stabilized at 1/2 hard**: 0.8–1.2 mm **diam**, 226 MPa **TS**, 177 MPa **YS**, 3% in 2 in. (50 mm) **El**; **Stabilized at 3/4 hard**: 0.8–1.2 mm **diam**, 245 MPa **TS**, 196 MPa **YS**, 3% in 2 in. (50 mm) **El**
JIS H 4001	3005	Japan	Sheet and strip	rem **Al**, 0.6 **Si**, 0.30 **Cu**, 1.0–1.5 **Mn**, 0.20–0.6 **Mg**, 0.10 **Cr**, 0.25 **Zn**, 0.10 **Ti**, 0.7 **Fe**, 0.15 total, others	**Annealed at 1/4 hard**: 0.8–1.2 mm **diam**, 137 MPa **TS**, 118 MPa **YS**, 2% in 2 in. (50 mm) **El**; **Annealed at 1/2 hard**: 0.8–1.2 mm **diam**, 167 MPa **TS**, 147 MPa **YS**, 2% in 2 in. (50 mm) **El**; **Annealed at 3/4 hard**: 0.8–1.2 mm **diam**, 196 MPa **TS**, 167 MPa **YS**, 2% in 2 in. (50 mm) **El**
MIL-B-20148C	Class 2	US	Sheet (core material)	rem **Al**, 0.6 **Si**, 0.05–0.20 **Cu**, 1.0–1.5 **Mn**, 0.10 **Zn**, 0.7 **Fe**, 0.05 each, 0.15 total, others, clad both sides with alloy 4343	**O**: 0.006–0.007 in. **diam**, 137 max MPa **TS**, 12% in 2 in. (50 mm) **El**; **H12**: 0.019–0.050 in. **diam**, 117 MPa **TS**, 4% in 2 in. (50 mm) **El**; **H14**: 0.019–0.050 in. **diam**, 137 MPa **TS**, 3% in 2 in. (50 mm) **El**
MIL-B-20148C	Class 1	US	Sheet (core material)	rem **Al**, 0.6 **Si**, 0.05–0.20 **Cu**, 1.0–1.5 **Mn**, 0.10 **Zn**, 0.7 **Fe**, 0.05 each, 0.15 total, others, clad one side with alloy 4343	**O**: 0.006–0.007 in. **diam**, 137 max MPa **TS**, 12% in 2 in. (50 mm) **El**; **H12**: 0.019–0.050 in. **diam**, 117 MPa **TS**, 4% in 2 in. (50 mm) **El**; **H14**: 0.019–0.050 in. **diam**, 137 MPa **TS**, 3% in 2 in. (50 mm) **El**
COPANT 862	3103	COPANT	Wrought products	rem **Al**, 0.50 **Si**, 0.10 **Cu**, 0.9–1.5 **Mn**, 0.30 **Mg**, 0.10 **Cr**, 0.20 **Zn**, 0.70 **Fe**, 0.10 **Ti+Zr**, 0.05 each, 0.15 total, others	
DGN-W-30	3S-H14	Mexico	Sheet	97.8 min **Al**, 1.2 **Mn**	**Rolled**: 0.53–0.80 mm **diam**, 145 MPa **TS**, 131 MPa **YS**, 3% in 2 in. (50 mm) **El**
DGN-W-30	3S-0	Mexico	3S-0	97.8 min **Al**, 1.2 **Mn**	**Rolled**: 0.25–0.52 mm **diam**, 117 MPa **TS**, 51 MPa **YS**, 18% in 2 in. (50 mm) **El**

WROUGHT ALUMINUM AND ALUMINUM ALLOYS

specification number	designation	country	product forms	chemical composition	mechanical properties and hardness values
DGN–W–30	3S–H12	Mexico	Sheet	97.8 **Al**, 1.2 **Mn**	**Rolled**: 0.53–0.80 mm **diam**, 124 MPa **TS**, 103 MPa **YS**, 3% in 2 in. (50 mm) **El**
DGN–W–30	3S–H18	Mexico	Sheet	97.8 min **Al**, 1.2 **Mn**	**Rolled**: 0.25–0.52 mm **diam**, 200 MPa **TS**, 186 MPa **YS**, 1% in 2 in. (50 mm) **El**
SIS 14 40 54	SIS Al 40 54–02	Sweden	Sheet, section, strip	rem **Al**, 0.50 **Si**, 0.10 **Cu**, 0.9–1.5 **Mn**, 0.30 **Mg**, 0.10 **Cr**, 0.20 **Zn**, 0.7 **Fe**, 0.10 **Ti+Zr**; 0.05 each, 0.15 total, others	**Annealed**: 0.35–6 mm **diam**, 95 MPa **TS**, 35 MPa **YS**, 30% in 2 in. (50 mm) **El**, 25 **HV**; **Strain hardened**: 0.35–5 in. **diam**, 120 MPa **TS**, 90 MPa **YS**, 10% in 2 in. (50 mm) **El**, 35–45 **HV**
SIS 14 40 54	SIS Al 40 54–12	Sweden	Strip	rem **Al**, 0.50 **Si**, 0.10 **Cu**, 0.9–1.5 **Mn**, 0.30 **Mg**, 0.10 **Cr**, 0.20 **Zn**, 0.7 **Fe**, 0.10 **Ti+Zr**; 0.05 each, 0.15 total, others	**Strain hardened**: 0.35–2 mm **diam**, 120 MPa **TS**, 90 MPa **YS**, 10% in 2 in. (50 mm) **El**, 35–45 **HV**
SIS 14 40 54	SIS Al 40 54–14	Sweden	Sheet, strip	rem **Al**, 0.50 **Si**, 0.10 **Cu**, 0.9–1.5 **Mn**, 0.30 **Mg**, 0.10 **Cr**, 0.20 **Zn**, 0.7 **Fe**, 0.10 **Ti+Zr**; 0.05 each, 0.15 total, others	**Strain hardened**: 0.35–4 mm **diam**, 140 MPa **TS**, 115 MPa **YS**, 6% in 2 in. (50 mm) **El**, 40–50 **HV**
SIS 14 40 54	SIS Al 40 54–16	Sweden	Sheet, strip	rem **Al**, 0.50 **Si**, 0.10 **Cu**, 0.9–1.5 **Mn**, 0.30 **Mg**, 0.10 **Cr**, 0.20 **Zn**, 0.7 **Fe**, 0.10 **Ti+Zr**; 0.05 each, 0.15 total, others	**Strain hardened**: 0.35–3 mm **diam**, 165 MPa **TS**, 140 MPa **YS**, 5% in 2 in. (50 mm) **El**, 50–60 **HV**
SIS 14 40 54	SIS Al 40 54–18	Sweden	Sheet, strip	rem **Al**, 0.50 **Si**, 0.10 **Cu**, 0.9–1.5 **Mn**, 0.30 **Mg**, 0.10 **Cr**, 0.20 **Zn**, 0.7 **Fe**, 0.10 **Ti+Zr**; 0.05 each, 0.15 total, others	**Strain hardened**: 0.35–1.6 mm **diam**, 205 MPa **TS**, 185 MPa **YS**, 60–75 **HV**
SIS 14 40 54	SIS Al 40 54–24	Sweden	Sheet, strip	rem **Al**, 0.50 **Si**, 0.10 **Cu**, 0.9–1.5 **Mn**, 0.30 **Mg**, 0.10 **Cr**, 0.20 **Zn**, 0.7 **Fe**, 0.10 **Ti+Zr**; 0.05 each, 0.15 total, others	**Strain hardened, annealed**: 0.35–3 mm **diam**, 140 MPa **TS**, 110 MPa **YS**, 12% in 2 in. (50 mm) **El**, 40–50 **HV**
SIS 14 40 54	SIS Al 40 54–26	Sweden	Sheet, strip	rem **Al**, 0.50 **Si**, 0.10 **Cu**, 0.9–1.5 **Mn**, 0.30 **Mg**, 0.10 **Cr**, 0.20 **Zn**, 0.7 **Fe**, 0.10 **Ti+Zr**; 0.05 each, 0.15 total, others	**Strain hardened, annealed**: 0.35–3 mm **diam**, 165 MPa **TS**, 135 MPa **YS**, 10% in 2 in. (50 mm) **El**, 50–60 **HV**
SIS 14 40 54	SIS Al 40 54–40	Sweden	Sheet, strip	rem **Al**, 0.50 **Si**, 0.10 **Cu**, 0.9–1.5 **Mn**, 0.30 **Mg**, 0.10 **Cr**, 0.20 **Zn**, 0.7 **Fe**, 0.10 **Ti+Zr**; 0.05 each, 0.15 total, others	**Annealed**: 0.35–6 mm **diam**, 95 MPa **TS**, 35 MPa **YS**, 30% in 2 in. (50 mm) **El**, 25–35 **HV**
SIS 14 40 54	SIS Al 40 54–42	Sweden	Sheet, strip	rem **Al**, 0.50 **Si**, 0.10 **Cu**, 0.9–1.5 **Mn**, 0.30 **Mg**, 0.10 **Cr**, 0.20 **Zn**, 0.7 **Fe**, 0.10 **Ti+Zr**; 0.05 each, 0.15 total, others	**Strain hardened**: 0.35–3 mm **diam**, 120 MPa **TS**, 90 MPa **YS**, 10% in 2 in. (50 mm) **El**, 35–45 **HV**
SIS 14 40 54	SIS Al 40 54–44	Sweden	Sheet, strip	rem **Al**, 0.50 **Si**, 0.10 **Cu**, 0.9–1.5 **Mn**, 0.30 **Mg**, 0.10 **Cr**, 0.20 **Zn**, 0.7 **Fe**, 0.10 **Ti+Zr**; 0.05 each, 0.15 total, others	**Strain hardened**: 0.35–3 mm **diam**, 140 MPa **TS**, 115 MPa **YS**, 6% in 2 in. (50 mm) **El**, 40–50 **HV**

WROUGHT ALUMINUM AND ALUMINUM ALLOYS

specification number	designation	country	product forms	chemical composition	mechanical properties and hardness values
BS 1475	NG3	UK	Wire	rem **Al**, 0.6 **Si**, 0.1 **Cu**, 0.8–1.5 **Mn**, 0.1 **Mg**, 0.2 **Cr**, 0.2 **Zn**, 0.2 **Ti**, 0.7 **Fe**, 0.2 **Ti**+**Cr**	**Annealed**: ≤10 mm **diam**, 130 max MPa **TS**; **Strain-hardened to hard**: ≤10 mm **diam**, 175 MPa **TS**
BS 1470	NS3	UK	Plate, sheet, strip	rem **Al**, 0.6 **Si**, 0.1 **Cu**, 0.8–1.5 **Mn**, 0.1 **Mg**, 0.2 **Zn**, 0.7 **Fe**, 0.2 **Cr**+**Ti**	**Annealed**: 0.2–6.0 mm **diam**, 90 MPa **TS**, 20% in 2 in. (50 mm) **El**; **Strain hardened to 1/2 hard**: 0.2–12.5 mm **diam**, 140 MPa **TS**, 3% in 2 in. (50 mm) **El**; **Strain hardened to hard**: 0.2–3.0 mm **diam**, 175 MPa **TS**, 2% in 2 in. (50 mm) **El**
BS 3L.61		UK	Sheet and strip	rem **Al**, 0.6 **Si**, 0.1 **Cu**, 0.8–1.5 **Mn**, 0.1 **Mg**, 0.05 **Cr**, 0.2 **Ni**, 0.2 **Zn**, 0.05 **Sn**, 0.7 **Fe**, 0.05 **Pb**, 0.2 **Zr**+**Ti**	0.4–0.8 mm **diam**, 130 MPa **TS**, 90 MPa **YS**, 20% in 2 in. (50 mm) **El**
BS 3L.60		UK		rem **Al**, 0.6 **Si**, 0.1 **Cu**, 0.8–1.5 **Mn**, 0.1 **Mg**, 0.05 **Cr**, 0.2 **Ni**, 0.2 **Zn**, 0.05 **Sn**, 0.7 **Fe**, 0.05 **Pb**, 0.2 **Zr**+**Ti**	**Cold rolled or cold rolled and partially annealed**: 0.4–0.8 mm **diam**, 145 MPa **TS**, 120 MPa **YS**, 5% in 2 in. (50 mm) **El**
BS 3L.59		UK	Sheet and strip	rem **Al**, 0.6 **Si**, 0.1 **Cu**, 0.8–1.5 **Mn**, 0.1 **Mg**, 0.05 **Cr**, 0.2 **Ni**, 0.2 **Zn**, 0.05 **Sn**, 0.7 **Fe**, 0.05 **Pb**, 0.2 **Zr**+**Ti**	**Cold rolled or cold rolled and partially annealed**: 0.4–0.8 mm **diam**, 195 MPa **TS**, 160 MPa **YS**, 2% in 2 in. (50 mm) **El**
SFS 2585	AlMn1	Finland	Sheet	rem **Al**, 0.6 **Si**, 0.10 **Cu**, 0.8–1.5 **Mn**, 0.3 **Mg**, 0.2 **Zn**, 0.7 **Fe**, 0.2 **Ti**+**Zr**+**Cr**, 0.05 each, 0.15 total, others	**Annealed**: all **diam**, 130 max MPa **TS**, 40 MPa **YS**, 24% in 2 in. (50 mm) **El**; **3/4 hard**: ≤6 mm **diam**, 160 MPa **TS**, 130 MPa **YS**, 4% in 2 in. (50 mm) **El**; **Hard**: ≤3 mm **diam**, 190 MPa **TS**, 150 MPa **YS**, 3% in 2 in. (50 mm) **El**
IS 7883		India	Sheet, strip	rem **Al**, 0.60 **Si**, 0.10 **Cu**, 0.8–1.5 **Mn**, 0.10 **Mg**, 0.05 **Cr**, 0.20 **Ni**, 0.10 **Zn**, 0.15 **Ti**, 0.05 **Sn**, 0.70 **Fe**, 0.05 **Pb**	**Annealed**: 20 mm **diam**, 90–130 MPa **TS**; **Half hard**: 3 mm **diam**, 135–180 MPa **TS**; **Hard**: 1 mm **diam**, 185 MPa **TS**
IS 736	31000	India	Plate	rem **Al**, 0.6 **Si**, 0.1 **Cu**, 0.8–1.5 **Mn**, 0.1 **Mg**, 0.2 **Cr**, 0.2 **Zn**, 0.2 **Ti**, 0.7 **Fe**	**As manufactured**: 23 mm **diam**, 95 MPa **TS**, 23% in 2 in. (50 mm) **El**; **Annealed**: 22 mm **diam**, 90 MPa **TS**, 22% in 2 in. (50 mm) **El**; **Half hard**: 5 mm **diam**, 130 MPa **TS**, 5% in 2 in. (50 mm) **El**
ISO R209	Al–Mn1	ISO	Wrought products	rem **Al**, 0.6 **Si**, 0.1 **Cu**, 0.8–1.5 **Mn**, 0.3 **Mg**, 0.2 **Zn**, 0.7 **Fe**, 0.2 **Ti**+**Zr**+**Cr**	

WROUGHT ALUMINUM AND ALUMINUM ALLOYS

specification number	designation	country	product forms	chemical composition	mechanical properties and hardness values
ISO TR2136	Al–Mn1	ISO	Sheet, plate	rem **Al**, 0.6 **Si**, 0.1 **Cu**, 0.8–1.5 **Mn**, 0.3 **Mg**, 0.2 **Zn**, 0.7 **Fe**, 0.2 **Ti+Zr+Cr**, total	**Annealed**: all **diam**, 90 MPa **TS**, 20% in 2 in. (50 mm) **El**; **HB–strain hardened**: all **diam**, 115 MPa **TS**, 5% in 2 in. (50 mm) **El**; **HD–strain hardened**: all **diam**, 130 MPa **TS**, 3% in 2 in. (50 mm) **El**
NS 17 405	NS 17 405–02	Norway	Sheet, strip, and plate	rem **Al**, 0.6 **Si**, 0.1 **Cu**, 0.9–1.4 **Mn**, 0.3 **Mg**, 0.2 **Zn**, 0.6 **Fe**	**Soft annealed**: ≥0.5 mm **diam**, 88 MPa **TS**, 24% in 2 in. (50 mm) **El**
NS 17 405	NS 17 405–14	Norway	Sheet, strip, and plate	rem **Al**, 0.6 **Si**, 0.1 **Cu**, 0.9–1.4 **Mn**, 0.3 **Mg**, 0.2 **Zn**, 0.6 **Fe**	**As manufactured**: 0.5–6 mm **diam**, 137 MPa **TS**, 108 MPa **YS**, 5% in 2 in. (50 mm) **El**
NS 17 405	NS 17 405–18	Norway	Sheet, strip, and plate	rem **Al**, 0.6 **Si**, 0.1 **Cu**, 0.9–1.4 **Mn**, 0.3 **Mg**, 0.2 **Zn**, 0.6 **Fe**	**Hard**: 0.5–1 mm **diam**, 186 MPa **TS**, 147 MPa **YS**, 2% in 2 in. (50 mm) **El**
SIS 14 40 55	SIS Al 40 55–00	Sweden	Wire	rem **Al**, 0.6 **Si**, 0.1 **Cu**, 0.9–1.4 **Mn**, 0.3 **Mg**, 0.2 **Zn**, 0.6 **Fe**, 0.2 grain refiners	
SIS 14 40 55	SIS Al 40 55–18	Sweden	Wire	rem **Al**, 0.6 **Si**, 0.1 **Cu**, 0.9–1.4 **Mn**, 0.3 **Mg**, 0.2 **Zn**, 0.6 **Fe**, 0.2 grain refiners	**Strain hardened**: all **diam**, 176 MPa **TS**
SABS 712	Al–Mn1	South Africa	Wrought products	rem **Al**, 0.6 **Si**, 0.1 **Cu**, 0.8–1.5 **Mn**, 0.1 **Mg**, 0.2 **Zn**, 0.7 **Fe**, 0.2 **Cr+Ti+Zr**	
NOM–W–28	Types A, C–R, C–L, and CEL	Mexico	Tube: for liquids (A) and electrical cables (C–R, C–L, and CEL)	95.05 min **Al**, 0.80 **Si**, 1.00 **Mn**, 1.00 **Mg**, 0.50 **Cr**, 0.10 **Zn**, 1.00 **Fe**	**Extruded or drawn (A)**: 3.175 mm **diam**, 97 MPa **TS**
ANSI/ASTM B 209	3105	US	Sheet and plate	rem **Al**, 0.6 **Si**, 0.30 **Cu**, 0.30–0.8 **Mn**, 0.20–0.8 **Mg**, 0.20 **Cr**, 0.40 **Zn**, 0.10 **Ti**, 0.7 **Fe**, 0.05 each, 0.15 total, others	**Annealed**: 0.032–0.080 in. **diam**, 14 ksi (97 MPa) **TS**, 5 ksi (34 MPa) **YS**, 20% in 2 in. (50 mm) **El**; **H12**: 0.051–0.080 in. **diam**, 19 ksi (131 MPa) **TS**, 15 ksi (103 MPa) **YS**, 3% in 2 in. (50 mm) **El**; **H14**: 0.051–0.080 in. **diam**, 22 ksi (152 MPa) **TS**, 18 ksi (124 MPa) **YS**, 2% in 2 in. (50 mm) **El**
COPANT 862	3105	COPANT	Wrought products	rem **Al**, 0.60 **Si**, 0.30 **Cu**, 0.3–0.80 **Mn**, 0.2–0.80 **Mg**, 0.20 **Cr**, 0.40 **Zn**, 0.10 **Ti**, 0.70 **Fe**, 0.05 each, 0.15 total, others	
ANSI/ASTM B 210	3102	US	Tube: seamless	rem **Al**, 0.40 **Si**, 0.10 **Cu**, 0.05–0.40 **Mn**, 0.30 **Zn**, 0.10 **Ti**, 0.7 **Fe**, 0.05 each, 0.15 total, others	**Annealed**: 0.018–0.049 in. **diam**, 11 ksi (76 MPa) **TS**, 4 ksi (24 MPa) **YS**, 20% in 2 in. (50 mm) **El**
ANSI/ASTM B 210	Alclad 3303	US	Tube: seamless	rem **Al**, 0.6 **Si**, 0.20 **Cu**, 0.10 **Mn**, 1.1–1.8 **Mg**, 0.10 **Cr**, 0.25 **Zn**, 0.7 **Fe**, 0.05 each, 0.15 total, others	**Annealed**: 0.025–0.049 in. **diam**, 13 ksi (90 MPa) **TS**, 5 ksi (31 MPa) **YS**, 20% in 2 in. (50 mm) **El**

WROUGHT ALUMINUM AND ALUMINUM ALLOYS

specification number	designation	country	product forms	chemical composition	mechanical properties and hardness values
COPANT 862	3205	COPANT	Wrought products	rem **Al**, 0.6 **Si**, 0.3 **Cu**, 0.2 **Cr**, 0.4 **Zn**, 0.1 **Ti**, 0.7 **Fe**, 0.05 each, 0.15 total, others	
IS 7793	4928–A/42285	India	Forgings	rem **Al**, 17.0–19.0 **Si**, 0.8–1.5 **Cu**, 0.2 **Mn**, 0.8–1.3 **Mg**, 0.8–1.3 **Ni**, 0.2 **Zn**, 0.2 **Ti**, 0.05 **Sn**, 0.7 **Fe**, 0.05 **Pb**	
AMS 4145E		US	Die forgings	rem **Al**, 11.0–13.5 **Si**, 0.5–1.3 **Cu**, 0.8–1.3 **Mg**, 0.10 **Cr**, 0.5–1.3 **Ni**, 0.25 **Zn**, 0.05 **Ti**, 1.0 **Fe**, 0.05 each, 0.15 total, others	**Solution and precipitation heat treated**: 52 ksi (359 MPa) **TS**, 42 ksi (290 MPa) **YS**, 4% in 2 in. (50 mm) **El**, 115 min **HB**
ANSI/ASTM B 247	4032	US	Die forgings	rem **Al**, 11.0–13.5 **Si**, 0.50–1.3 **Cu**, 0.8–1.3 **Mg**, 0.10 **Cr**, 0.50–1.3 **Ni**, 0.25 **Zn**, 1.0 **Fe**, 0.05 each, 0.15 total, others	**Solution heat treated and artificially aged**: ≤4.000 in. **diam**, 52 ksi (359 MPa) **TS**, 42 ksi (290 MPa) **YS**, 3% in 2 in. (50 mm) **El**, 115 min **HB**
CSA HA.8	4032	Canada	Forgings	rem **Al**, 11.0–13.5 **Si**, 0.50–1.3 **Cu**, 0.8–1.3 **Mg**, 0.10 **Cr**, 0.50–1.3 **Ni**, 0.25 **Zn**, 1.0 **Fe**, 0.05 each, 0.15 total, others	**T6 (die forgings)**: <4 in. **diam**, 52.0 ksi (359 MPa) **TS**, 42.0 ksi (290 MPa) **YS**, 3% in 2 in. (50 mm) **El**
DIN 8512	L–AlSi12 /3.2285	Germany	Hard solder	rem **Al**, 11–13.5 **Si**, 0.03 **Cu**, 0.1 **Mn**, 0.07 **Zn**, 0.03 **Ti**, 0.4 **Fe**, 0.03 each, 0.5 total, others	
JIS H 4140	4032	Japan	Die forgings	rem **Al**, 11.0–13.5 **Si**, 0.50–1.3 **Cu**, 0.8–1.3 **Mg**, 0.10 **Cr**, 0.50–1.3 **Ni**, 0.20 **Zn**, 1.0 **Fe**, 0.15 total, others	**Solution heat treated and artificially age hardened**: ≤100 mm **diam**, 363 MPa **TS**, 294 MPa **YS**, 5% in 2 in. (50 mm) **El**
AS 1734	4047	Australia	Sheet, plate	rem **Al**, 11.0–13.0 **Si**, 0.30 **Cu**, 0.15 **Mn**, 0.10 **Mg**, 0.20 **Zn**, 0.8 **Fe**, 0.05 each, 0.15 total, others	
CSA HA.6	.4047	Canada	Rod, wire: for welding and brazing	rem **Al**, 11.0–13.0 **Si**, 0.30 **Cu**, 0.15 **Mn**, 0.10 **Mg**, 0.20 **Zn**, 0.8 **Fe**, 0.0008 **Be**, 0.05 each, 0.15 total, others	
COPANT 862	4047	COPANT	Wrought products	rem **Al**, 11–13 **Si**, 0.3 **Cu**, 0.15 **Mn**, 0.10 **Mg**, 0.2 **Zn**, 0.8 **Fe**, 0.05 each, 0.15 total, others	
JIS Z 3263	BA4047	Japan	Wire: brazing filler	rem **Al**, 11.0–13.0 **Si**, 0.30 **Cu**, 0.15 **Mn**, 0.10 **Mg**, 0.20 **Zn**, 0.8 **Fe**, 0.05 each, 0.15 total, others	
MIL–B–20148C	Class 8 (4047)	US	Sheet (core material)	rem **Al**, 11.0–13.0 **Si**, 0.30 **Cu**, 0.15 **Mn**, 0.10 **Mg**, 0.20 **Zn**, 0.8 **Fe**, 0.05 each, 0.15 total, others	
IS 7793	4658/49582	India	Forgings	rem **Al**, 11.0–13.0 **Si**, 0.8–1.5 **Cu**, 0.2 **Mn**, 0.8–1.3 **Mg**, 1.5 **Ni**, 0.35 **Zn**, 0.2 **Ti**, 0.05 **Sn**, 0.8 **Fe**, 0.05 **Pb**	**As forged**: all **diam**, 295 MPa **TS**

WROUGHT ALUMINUM AND ALUMINUM ALLOYS

specification number	designation	country	product forms	chemical composition	mechanical properties and hardness values
BS 1475	NG2	UK	Wire	rem **Al**, 10.0–13.0 **Si**, 0.10 **Cu**, 0.5 **Mn**, 0.2 **Mg**, 0.2 **Zn**, 0.6 **Fe**	
IS 739	NG2	India	Wire	rem **Al**, 10.0–13.0 **Si**, 0.10 **Cu**, 0.5 **Mn**, 0.25 **Mg**, 0.2 **Ni**, 0.2 **Zn**, 0.05 **Sn**, 0.6 **Fe**, 0.05 **Pb**	
IS 733	46000	India	Bar, rod, section	rem **Al**, 10.0–13.0 **Si**, 0.1 **Cu**, 0.5 **Mn**, 0.2 **Mg**, 0.2 **Zn**, 0.6 **Fe**	**As manufactured**: ≤10 mm **diam**, 150 MPa **TS**, 10% in 2 in. (50 mm) **El**
SIS 14 42 62	SIS Al 42 62–18	Sweden	Wire	rem **Al**, 10.0–13.0 **Si**, 0.10 **Cu**, 0.5 **Mn**, 0.25 **Mg**, 0.2 **Zn**, 0.6 **Fe**	**Hard**: all **diam**, 196 MPa **TS**
SIS 14 42 62	SIS Al 42 62–00	Sweden	Wire	rem **Al**, 10.0–13.0 **Si**, 0.10 **Cu**, 0.5 **Mn**, 0.25 **Mg**, 0.2 **Zn**, 0.6 **Fe**	
DIN 8512	L–AlSiSn/3.2685	Germany	Hard solder	72 min **Al**, 10–12 **Si**, 8–12 **Sn+Cd**, 2–4 **Cu+Ni**, 0.5 total, others	
JIS Z 3263	BA24PC	Japan	Sheet: brazing	rem **Al**, 9.0–11.0 **Si**, 0.30 **Cu**, 0.05 **Mn**, 0.05 **Mg**, 0.10 **Zn**, 0.20 **Ti**, 0.8 **Fe**, 0.05 each, 0.15 total, others	**Annealed**: <0.8 mm **diam**, 147 MPa **TS**, 18% in 2 in. (50 mm) **El**
JIS Z 3263	BA4045	Japan	Wire: brazing filler	rem **Al**, 9.0–11.0 **Si**, 0.30 **Cu**, 0.05 **Mn**, 0.05 **Mg**, 0.10 **Zn**, 0.20 **Ti**, 0.8 **Fe**, 0.05 each, 0.15 total, others	
JIS Z 3263	BA4145	Japan	Wire: brazing filler	rem **Al**, 9.3–10.7 **Si**, 3.3–4.7 **Cu**, 0.15 **Mn**, 0.15 **Mg**, 0.15 **Cr**, 0.20 **Zn**, 0.8 **Fe**, 0.05 each, 0.15 total, others	
MIL–B–20148C	4045	US	Sheet (cladding material)	rem **Al**, 9.0–11.0 **Si**, 0.30 **Cu**, 0.05 **Mn**, 0.05 **Mg**, 0.10 **Zn**, 0.8 **Fe**, 0.05 each, 0.15 total, others	
AS 1734	4343	Australia	Sheet, plate	rem **Al**, 6.8–8.2 **Si**, 0.25 **Cu**, 0.10 **Mn**, 0.20 **Zn**, 0.8 **Fe**, 0.05 each, 0.15 total, others	
JIS Z 3263	BA12PC	Japan	Sheet: brazing	rem **Al**, 6.8–8.2 **Si**, 0.25 **Cu**, 0.10 **Mn**, 0.20 **Zn**, 0.8 **Fe**, 0.05 each, 0.15 total, others	**Annealed**: <0.3 mm **diam**, 137 MPa **TS**, 15% in 2 in. (50 mm) **El**; **1/4 hard**: <0.3 mm **diam**, 118 MPa **TS**2% in 2 in. (50 mm) **El**; **1/2 hard**: <0.3 mm **diam**, 137 MPa **TS**, 1% in 2 in. (50 mm) **El**
JIS Z 3263	BA22PC	Japan	Sheet: brazing	rem **Al**, 6.8–8.2 **Si**, 0.25 **Cu**, 0.10 **Mn**, 0.20 **Zn**, 0.8 **Fe**, 0.05 each, 0.15 total, others	**Annealed**: <0.8 mm **diam**, 137 MPa **TS**, 18% in 2 in. (50 mm) **El**
JIS Z 3263	BA4343	Japan	Wire: brazing filler	rem **Al**, 6.8–8.2 **Si**, 0.25 **Cu**, 0.10 **Mn**, 0.20 **Zn**, 0.8 **Fe**, 0.05 each, 0.15 total, others	

WROUGHT ALUMINUM AND ALUMINUM ALLOYS

specification number	designation	country	product forms	chemical composition	mechanical properties and hardness values
MIL-B-20148C	4343	US	Sheet (cladding material)	rem **Al**, 6.8–8.2 **Si**, 0.25 **Cu**, 0.10 **Mn**, 0.20 **Zn**, 0.8 **Fe**, 0.05 each, 0.15 total, others	**O**: 0.006–0.007 in. **diam**, 137 max MPa **TS**, 12% in 2 in. (50 mm) **El**; **H12**: 0.019–0.050 in. **diam**, 117 MPa **TS**, 4% in 2 in. (50 mm) **El**; **H14**: 0.019–0.050 in. **diam**, 137 MPa **TS**, 3% in 2 in. (50 mm) **El**
MIL-A-47193	4343	US	Sheet (cladding material)	rem **Al**, 6.8–8.2 **Si**, 0.25 **Cu**, 0.20 **Zn**, 0.80 **Fe**, 0.05 each, 0.15 total, others	
ISO 208	Al-Si7Mg	ISO	Castings	rem **Al**, 6.5–7.5 **Si**, 0.20 **Cu**, 0.6 **Mn**, 0.2–0.4 **Mg**, 0.05 **Ni**, 0.3 **Zn**, 0.2 **Ti**, 0.05 **Sn**, 0.5 **Fe**, 0.05 **Pb**	
AS 1866	4543	Australia	Rod, bar, shape	rem **Al**, 5.0–7.0 **Si**, 0.10 **Cu**, 0.05 **Mn**, 0.10–0.40 **Mg**, 0.05 **Cr**, 0.10 **Zn**, 0.10 **Ti**, 0.50 **Fe**, 0.05 each, 0.15 total, others	**Cooled and naturally aged**: all **diam**, 110 MPa **TS**, 55 MPa **YS**, 10% in 2 in. (50mm) **El**; **Cooled and artificially aged**: 6.0 mm **diam**, 151 MPa **TS**, 110 MPa **YS**, 8%in 2 in. (50 mm) **El**
BS 1475	NG21	UK	Wire	rem **Al**, 4.5–6.0 **Si**, 0.10 **Cu**, 0.5 **Mn**, 0.2 **Mg**, 0.2 **Zn**, 0.6 **Fe**	
CSA HA.6	.4043	Canada	Rod, wire: for welding and brazing	rem **Al**, 4.5–6.0 **Si**, 0.30 **Cu**, 0.05 **Mn**, 0.05 **Mg**, 0.10 **Zn**, 0.20 **Ti**, 0.8 **Fe**, 0.0008 **Be**, 0.05 each, 0.15 total, others	
COPANT 862	4043	COPANT	Wrought products	rem **Al**, 4.5–6.0 **Si**, 0.3 **Cu**, 0.05 **Mn**, 0.05 **Mg**, 0.10 **Zn**, 0.20 **Ti**, 0.8 **Fe**, 0.0008 **Be**, 0.05 each, 0.15 total, others	
IS 739	NG21	India	Wire	rem **Al**, 4.5–6.0 **Si**, 0.10 **Cu**, 0.5 **Mn**, 0.25 **Mg**, 0.2 **Ni**, 0.2 **Zn**, 0.05 **Sn**, 0.6 **Fe**, 0.05 **Pb**	
IS 733	43000	India	Bar, rod, section	rem **Al**, 4.5–6.0 **Si**, 0.1 **Cu**, 0.5 **Mn**, 0.2 **Mg**, 0.2 **Zn**, 0.6 **Fe**	**As manufactured**: ≤10 mm **diam**, 110 MPa **TS**, 18% in 2 in. (50 mm) **El**
JIS Z 3232	4043	Japan	Rod and wire: welding and electrode	rem **Al**, 4.5–6.0 **Si**, 0.30 **Cu**, 0.05 **Mn**, 0.05 **Mg**, 0.10 **Zn**, 0.20 **Ti**, 0.8 **Fe**, 0.0008 **Be**	**Quenched and tempered**: all **diam**, 167 MPa **TS**
SIS 14 42 25	SIS Al 42 25–00	Sweden	Wire	rem **Al**, 4.5–6.0 **Si**, 0.10 **Cu**, 0.5 **Mn**, 0.25 **Mg**, 0.2 **Zn**, 0.6 **Fe**	
SIS 14 42 25	SIS Al 42 25–18	Sweden	Wire	rem **Al**, 4.5–6.0 **Si**, 0.10 **Cu**, 0.5 **Mn**, 0.25 **Mg**, 0.2 **Zn**, 0.6 **Fe**	**Hard**: all **diam**, 176 MPa **TS**
SFS 2590	AlSi5	Finland	Shapes	rem **Al**, 3.5–5.5 **Si**, 0.10 **Cu**, 0.3 **Mn**, 0.7 **Mg**, 0.2 **Zn**, 0.5 **Fe**, 0.2 **Ti+Zr**, 0.05, each, 0.15 total, others	**Hot finished**: all **diam**, 130 MPa **TS**, 70 MPa **YS**, 15% in 2 in. (50 mm) **El**

WROUGHT ALUMINUM AND ALUMINUM ALLOYS

specification number	designation	country	product forms	chemical composition	mechanical properties and hardness values
IS 736	40800	India	Plate	98.0 min **Al**, 0.6–0.95 **Si**, 0.2 **Cu**, 0.1 **Mn**, 0.1 **Mg**, 0.2 **Zn**, 0.2 **Ti**, 0.6–0.95 **Fe**	**As manufactured**: 28 mm **diam**, 90 MPa **YS**, 28%in 2 in. (50 mm) **El**; **Annealed**: 30 mm **diam**, 120 MPa **TS**, 85 MPa **YS**, 30% in 2 in. (50 mm) **El**; **Half hard**: 7 mm **diam**, 160 MPa **TS**, 120 MPa **YS**, 7% in 2 in. (50 mm) **El**
BS 1475	NG61	UK	Wire	rem **Al**, 0.10 **Cu**, 0.6–1.0 **Mn**, 5.0–5.5 **Mg**, 0.05–0.20 **Cr**, 0.2 **Zn**, 0.05–0.20 **Ti**, 0.40 **Si**+ **Fe**	
ANSI/ASTM B 241	5456	US	Tube: seamless	rem **Al**, 0.10 **Cu**, 0.50–1.0 **Mn**, 4.7–5.5 **Mg**, 0.05–0.20 **Cr**, 0.25 **Zn**, 0.20 **Ti**, 0.45 **Si**+**Fe**, 0.05 each, 0.15 total, others	**Annealed**: ≤5.000 in. **diam**, 41 ksi (283 MPa) **TS**, 19 ksi (131 MPa) **YS**, 14% in 2 in. (50 mm) **El**; **H111**: ≤5.000 in. **diam**, 42 ksi (290 MPa) **TS**, 26 ksi (179 MPa) **YS**, 12% in 2 in. (50 mm) **El**; **H112**: ≤5.000 in. **diam**, 41 ksi (283 MPa) **TS**, 19 ksi (131 MPa) **YS**, 12% in 2 in. (50 mm) **El**
ANSI/ASTM B 221	5456	US	Bar, tube, rod, wire, and shapes	rem **Al**, 0.10 **Cu**, 0.50–1.0 **Mn**, 4.7–5.5 **Mg**, 0.05–0.20 **Cr**, 0.25 **Zn**, 0.20 **Ti**, 0.40 **Si**+**Fe**, 0.05 each, 0.15 total, others	**Annealed**: ≤5.000 in. **diam**, 41 ksi (283 MPa) **TS**, 19 ksi (131 MPa) **YS**, 14% in 2 in. (50 mm) **El**; **H111**: ≤5.000 in. **diam**, 42 ksi (290 MPa) **TS**, 26 ksi (179 MPa) **YS**, 12% in 2 in. (50 mm) **El**; **H112**: ≤5.000 in. **diam**, 41 ksi (283 MPa) **TS**, 19 ksi (131 MPa) **YS**, 12% in 2 in. (50 mm) **El**
ANSI/ASTM B 210	5456	US	Tube: seamless	rem **Al**, 0.25 **Si**, 0.10 **Cu**, 0.50–1.0 **Mn**, 4.7–5.5 **Mg**, 0.05–0.20 **Cr**, 0.25 **Zn**, 0.20 **Ti**, 0.40 **Fe**, 0.05 each, 0.15 total, others	**Annealed**: 0.018–0.450 in. **diam**, 41 ksi (283 MPa) **TS**, 19 ksi (131 MPa) **YS**, 14% in 2 in. (50 mm) **El**
ANSI/ASTM B 209	5456	US	Sheet and plate	rem **Al**, 0.25 **Si**, 0.10 **Cu**, 0.50–1.0 **Mn**, 4.7–5.5 **Mg**, 0.05–0.20 **Cr**, 0.25 **Zn**, 0.20 **Ti**, 0.40 **Fe**, 0.05 each, 0.15 total, others	**Annealed**: 3.001–5.000 in. **diam**, 40 ksi (276 MPa) **TS**, 17 ksi (117 MPa) **YS**, 14% in 2 in. (50 mm) **El**; **H321**: 1.501–3.000 in. **diam**, 41 ksi (283 MPa) **TS**, 29 ksi (200 MPa) **YS**, 12% in 2 in. (50 mm) **El**; **H323**: 0.126–0.249 in. **diam**, 48 ksi (331 MPa) **TS**, 36 ksi (248 MPa) **YS**, 8% in 2 in. (50 mm) **El**
CSA HA.6	.5556	Canada	Rod, wire: for welding and brazing	rem **Al**, 0.10 **Cu**, 0.50–1.0 **Mn**, 4.7–5.5 **Mg**, 0.05–0.20 **Cr**, 0.25 **Zn**, 0.05–0.20 **Ti**, 0.008 **Be**, 0.40 **Si**+**Fe** 0.05 each, 0.15 total, others	

WROUGHT ALUMINUM AND ALUMINUM ALLOYS

specification number	designation	country	product forms	chemical composition	mechanical properties and hardness values
QQ-A-250/20	5456	US	Plate, sheet	rem **Al**, 0.10 **Cu**, 0.50–1.0 **Mn**, 4.7–5.5 **Mg**, 0.05–0.20 **Cr**, 0.25 **Zn**, 0.20 **Ti**, 0.40 **Fe**, **Si** in **Fe** content; 0.05 each, 0.15 total, others	**Annealed**: 0.051–1.500 in. **diam**, 42 ksi (290 MPa) **TS**, 19 ksi (131 MPa) **YS**, 16% in 2 in. (50 mm) **El**; **As rolled**: 0.250–1.500 in. **diam**, 42 ksi (290 MPa) **TS**, 19 ksi (131 MPa) **YS**, 12% in 2 in. (50 mm) **El**; **Specially treated**: 0.188–0.624 in. **diam**, 46 ksi (317 MPa) **TS**, 33 ksi (228 MPa) **YS**, 10% in 2 in. (50 mm) **El**
QQ-A-250/9F	5456	US	Plate, sheet	rem **Al**, 0.10 **Cu**, 0.50–1.0 **Mn**, 4.7–5.5 **Mg**, 0.05–0.20 **Cr**, 0.25 **Zn**, 0.20 **Ti**, 0.40 **Fe**, **Si** in **Fe** content; 0.05 each, 0.15 total, others	**Annealed**: 0.051–1.500 in. **diam**, 42 ksi (290 MPa) **TS**, 19 ksi (131 MPa) **YS**, 16% in 2 in. (50 mm) **El**; **As rolled**: 0.250–1.500 in. **diam**, 42 ksi (290 MPa) **TS**, 19 ksi (131 MPa) **YS**, 12% in 2 in. (50 mm) **El**; **Strain hardened 1/4 hard (hot rolled) temper and stabilized**: 0.188–0.624 in. **diam**, 46 ksi (317 MPa) **TS**, 33 ksi (228 MPa) **YS**, 12% in 2 in. (50 mm) **El**
QQ-A-200/7D	5456	US	Bar, rod, shapes, tube, wire	rem **Al**, 0.10 **Cu**, 0.50–1.0 **Mn**, 4.7–5.5 **Mg**, 0.05–0.20 **Cr**, 0.25 **Zn**, 0.20 **Ti**, 0.40 **Fe**, **Si** in **Fe** content; 0.05 each, 0.15 total, others	**Annealed**: 5.000 max in. **diam**, 41 ksi (283 MPa) **TS**, 19 ksi (131 MPa) **YS**, 14% in 2 in. (50 mm) **El**; **Strain hardened**: 5.000 max in. **diam**, 42 ksi (290 MPa) **TS**, 26 ksi (179 MPa) **YS**, 12% in 2 in. (50 mm) **El**; **As extruded**: 5.000 max in. **diam**, 41 ksi (283 MPa) **TS**, 19 ksi (131 MPa) **YS**, 12% in 2 in. (50 mm) **El**
JIS Z 3232	5556	Japan	Rod and wire: welding and electrode	rem **Al**, 0.10 **Cu**, 0.50–1.0 **Mn**, 4.7–5.5 **Mg**, 0.05–0.20 **Cr**, 0.25 **Zn**, 0.05–0.20 **Ti**, 0.40 **Si + Fe**	**Annealed**: all **diam**, 275 MPa **TS**
MIL-A-46027F	5456	US	Plate	rem **Al**, 0.10 **Cu**, 0.50–1.0 **Mn**, 4.7–5.5 **Mg**, 0.05–0.20 **Cr**, 0.25 **Zn**, 0.20 **Ti**, 0.40 **Si + Fe**, 0.05 each, 0.15 total, others	
MIL-A-46083B	5456	US	Shapes	rem **Al**, 0.25 **Si**, 0.10 **Cu**, 0.50–1.0 **Mn**, 4.7–5.5 **Mg**, 0.05–0.20 **Cr**, 0.25 **Zn**, 0.20 **Ti**, 0.40 **Fe**, 0.05 each, 0.15 total, others	**Class I**: ≤2.000 in. **diam**, 310 MPa **TS**, 241 MPa **YS**, 9% in 2 in. (50 mm) **El**
AMS 4005		US	Foil	rem **Al**, 0.30 **Si**, 0.10 **Cu**, 0.05–0.20 **Mn**, 4.5–5.6 **Mg**, 0.05–0.20 **Cr**, 0.10 **Zn**, 0.40 **Fe**, 0.05 each, 0.15 total, others	**Strain hardened, mill finish**: <0.006 in. **diam**, 58 ksi (400 MPa) **TS**
AMS 4182B		US	Wire	rem **Al**, 0.30 **Si**, 0.10 **Cu**, 0.05–0.20 **Mn**, 4.5–5.6 **Mg**, 0.05–0.20 **Cr**, 0.10 **Zn**, 0.40 **Fe**, 0.05 each, 0.15 total, others	**Annealed**: all **diam**, 46 ksi (317 MPa) **TS**, 20% in 2 in. (50 mm) **El**

WROUGHT ALUMINUM AND ALUMINUM ALLOYS

specification number	designation	country	product forms	chemical composition	mechanical properties and hardness values
ANSI/ASTM B 316	5056	US	Wire and rod: rivet and cold–heading	rem **Al**, 0.30 **Si**, 0.10 **Cu**, 0.05–0.20 **Mn**, 4.5–5.6 **Mg**, 0.05–0.20 **Cr**, 0.10 **Zn**, 0.40 **Fe**, 0.05 each, 0.15 total, others	**Annealed**: ≤1.000 in. **diam**, 46 max ksi (317 max MPa) **TS**; **H32**: ≤1.000 in. **diam**, 44 ksi (303 MPa) **TS**
ANSI/ASTM B 211	5056	US	Bar, rod, and wire	rem **Al**, 0.30 **Si**, 0.10 **Cu**, 0.05–0.20 **Mn**, 4.5–5.6 **Mg**, 0.05–0.20 **Cr**, 0.10 **Zn**, 0.40 **Fe**, 0.05 each, 0.15 total, others	**Annealed**: ≥0.125 in. **diam**, 46 max ksi (317 MPa) **TS**, 20% in 2 in. (50 mm) **El**; **H111**: ≤0.374 in. **diam**, 44 ksi (303 MPa) **TS**; **H12**: ≤0.374 in. **diam**, 46 ksi (317 MPa) **TS**
ANSI/ASTM B 211	Alclad 5056	US	Bar, rod, and wire	rem **Al**, 0.30 **Si**, 0.10 **Cu**, 0.05–0.20 **Mn**, 4.5–5.6 **Mg**, 0.05–0.20 **Cr**, 0.10 **Zn**, 0.40 **Fe**, 0.05 each, 0.15 total, others, 6253 cladding material	**H192**: ≤0.374 in. **diam**, 52 ksi (359 MPa) **TS**; **H392**: ≤0.374 in. **diam**, 50 ksi (345 MPa) **TS**; **H393**: 0.120–0.192 in. **diam**, 54 ksi (372 MPa) **TS**, 47 ksi (324 MPa) **YS**
AS 1865	Alclad 5056	Australia	Wire, rod, bar, strip	rem **Al**, 0.30 **Si**, 0.10 **Cu**, 0.05–0.20 **Mn**, 4.5–5.6 **Mg**, 0.05–020 **Cr**, 0.10 **Zn**, 0.40 **Fe**, 0.05 each, 0.15 total, others, clad with Alloy 6253	
AS 1865	5056	Australia	Wire, rod, bar, strip	rem **Al**, 0.30 **Si**, 0.10 **Cu**, 0.05–0.20 **Mn**, 4.5–5.6 **Mg**, 0.05–0.20 **Cr**, 0.10 **Zn**, 0.40 **Fe**, 0.05 each, 0.15 total, others	**Annealed**: 10 mm **diam**, 246 MPa **TS**, 20% in 2 in. (50 mm) **El**; **Strain–hardened and stabilized to 1/4 hard**: 10 mm **diam**, 303 MPa **TS**; **Strain–hardened and stabilized to full hard**: 10 mm **diam**, 386 MPa **TS**
CSA HA.6	.5056	Canada	Rivet, rod, wire	rem **Al**, 0.30 **Si**, 0.10 **Cu**, 0.05–0.20 **Mn**, 4.5–5.6 **Mg**, 0.05–0.20 **Cr**, 0.10 **Zn**, 0.40 **Fe**, 0.05 each, 0.15 total, others	**O**: 0.062–1.000 in. **diam**, 38.0 ksi (262 MPa) **TS**; **H32**: 0.062–1.000 in. **diam**, 44.0 ksi (303 MPa) **TS**
COPANT 862	5056	COPANT	Wrought products	rem **Al**, 0.30 **Si**, 0.10 **Cu**, 0.05–0.20 **Mn**, 4.5–5.6 **Mg**, 0.05–0.20 **Cr**, 0.10 **Zn**, 0.40 **Fe**, 0.05 each, 0.15 total, others	
QQ–A–430B	5056	US	Rod, wire	rem **Al**, 0.30 **Si**, 0.10 **Cu**, 0.05–0.20 **Mn**, 4.5–5.6 **Mg**, 0.05–0.20 **Cr**, 0.10 **Zn**, 0.40 **Fe**, 0.05 each, 0.15 total, others	**Annealed**: 1.000 max in. **diam**, 46 max ksi (317 max MPa) **TS**; **Strain hardened and temper H32**: 1.000 max in. **diam**, 44 ksi (303 MPa) **TS**
ISO R209	Al–Mg5	ISO	Wrought products	rem **Al**, 0.5 **Si**, 0.10 **Cu**, 0.5 **Mn**, 4.5–5.6 **Mg**, 0.35 **Cr**, 0.2 **Zn**, 0.5 **Fe**, 0.2 **Ti+Zr**, 0.1–0.5 **Mn+Cr**	
JIS H 4040	5056	Japan	Bar and wire: extruded	rem **Al**, 0.30 **Si**, 0.10 **Cu**, 0.05–0.20 **Mn**, 4.5–5.6 **Mg**, 0.05–0.20 **Cr**, 0.10 **Zn**, 0.40 **Fe**, 0.15 total, others	**As manufactured (bar)**: all **diam**, 245 MPa **TS**, 118 MPa **YS**; **Annealed**: ≥6 mm **diam**, 314 MPa **TS**, 98 MPa **YS**, 20% in (50 mm) **El**; **1/4 hard**: ≥10 mm **diam**, 304 MPa **TS**

WROUGHT ALUMINUM AND ALUMINUM ALLOYS

specification number	designation	country	product forms	chemical composition	mechanical properties and hardness values
JIS H 4080	5056	Japan	Tube: extruded	rem **Al**, 0.30 **Si**, 0.10 **Cu**, 0.05–0.20 **Mn**, 4.5–5.6 **Mg**, 0.05–0.20 **Cr**, 0.10 **Zn**, 0.40 **Fe**, 0.15 total, others	**As manufactured**: all **diam**, 245 MPa **TS**, 118 MPa **YS**
JIS H 4120	5056	Japan	Wire and rod: rivet	rem **Al**, 0.30 **Si**, 0.10 **Cu**, 0.05–0.20 **Mn**, 4.5–5.6 **Mg**, 0.50 **Cr**, 0.10 **Zn**, 0.20 **Ti**, 0.40 **Fe**, 0.15 total, others	**Annealed**: ≤25 mm **diam**, 314 MPa **TS**; **Stabilized at 1/4 hard**: ≤25 mm **diam**, 304 MPa **TS**
JIS H 4140	5056	Japan	Die forgings	rem **Al**, 0.30 **Si**, 0.10 **Cu**, 0.05–0.20 **Mn**, 4.5–5.6 **Mg**, 0.05–0.20 **Cr**, 0.10 **Zn**, 0.40 **Fe**, 0.15 total, others	**As manufactured**: ≤100 mm **diam**, 245 MPa **TS**, 118 MPa **YS**, 18% in 2 in. (50 mm) **El**
BS 1475	NG6	UK	Wire	rem **Al**, 0.3 **Si**, 0.10 **Cu**, 0.5 **Mn**, 4.5–5.5 **Mg**, 0.25 **Cr**, 0.2 **Zn**, 0.2 **Ti**, 0.5 **Fe**, 0.1–0.5 **Mn+Cr**	**Annealed**: ≤10 mm **diam**, 250 MPa **TS**; **Strain–hardened to hard**: ≤10 mm **diam**, 310 MPa **TS**
BS 1473	NB6	UK	Bolt, screw	rem **Al**, 0.3 **Si**, 0.10 **Cu**, 0.5 **Mn**, 4.5–5.5 **Mg**, 0.25 **Cr**, 0.2 **Zn**, 0.2 **Ti**, 0.5 **Fe**, 0.1–0.5 **Mn+Cr**	**Strain hardened to 1/2 hard**: ≤12 mm **diam**, 310 MPa **TS**, 240 MPa **YS**
BS 1473	NR6	UK	Rivet	rem **Al**, 0.3 **Si**, 0.10 **Cu**, 0.5 **Mn**, 4.5–5.5 **Mg**, 0.25 **Cr**, 0.2 **Zn**, 0.2 **Ti**, 0.5 **Fe**, 0.1–0.5 **Mn+Cr**	**Annealed or as manufactured**: ≤25 mm **diam**, 255 MPa **TS**
BS 3L.58		UK	Wire	rem **Al**, 0.3 **Si**, 0.10 **Cu**, 0.5 **Mn**, 4.5–5.5 **Mg**, 0.25 **Cr**, 0.2 **Ni**, 0.2 **Zn**, 0.05 **Sn**, 0.5 **Fe**, 0.05 **Pb**, 0.20 **Zr+Ti**, 0.1–0.5 **Mn+Cr**	
CSA HA.6	.5356	Canada	Rod, wire: for welding and brazing	rem **Al**, 0.10 **Cu**, 0.05–0.20 **Mn**, 4.5–5.5 **Mg**, 0.05–0.20 **Cr**, 0.10 **Zn**, 0.06–0.20 **Ti**, 0.008 **Be**, 0.50 **Si+Fe**, 0.05 each, 0.15 total, others	
SFS 2589	AMg5	Finland	Shapes and sheet	rem **Al**, 0.5 **Si**, 0.10 **Cu**, 1.0 **Mn**, 4.5–5.5 **Mg**, 0.35 **Cr**, 0.2 **Zn**, 0.5 **Fe**, 1.0 **Mn+Cr**, 0.2 **Ti+Zr**, 0.05 each, 0.15 total, others	**Hot finished (shapes)**: ≤125 mm **diam**, 270 MPa **TS**, 130 MPa **YS**, 12% in 2 in. (50 mm) **El**; **Annealed (sheet)**: ≤6 mm **diam**, 300 max MPa **TS**, 120 MPa **YS**, 16% in 2 in. (50 mm) **El**; **1/2 hard (sheet)**: ≤6 mm **diam**, 300 MPa **TS**, 220 MPa **YS**, 8% in 2 in. (50 mm) **El**
IS 739	NG6	India	Wire	rem **Al**, 0.6 **Si**, 0.10 **Cu**, 0.4 **Mn**, 4.5–5.5 **Mg**, 0.25 **Cr**, 0.2 **Zn**, 0.7 **Fe**	**Annealed**: all **diam**, 245 MPa **TS**; **Annealed and cold worked to hard**: all **diam**, 387 MPa **TS**
IS 738	NT6	India	Tube: drawn	rem **Al**, 0.6 **Si**, 0.10 **Cu**, 1.0 **Mn**, 4.5–5.5 **Mg**, 0.25 **Cr**, 0.7 **Fe**	**Annealed**: all **diam**, 265 MPa **TS**, 18% in 2 in. (50 mm) **El**; **Annealed and cold worked to 1/2 hard**: all **diam**, 279 MPa **TS**, 216 MPa **YS**, 5% in 2 in. (50 mm) **El**

WROUGHT ALUMINUM AND ALUMINUM ALLOYS

specification number	designation	country	product forms	chemical composition	mechanical properties and hardness values
IS 736	55000	India	Plate	rem **Al**, 0.6 **Si**, 0.1 **Cu**, 0.5 **Mn**, 4.5–5.5 **Mg**, 0.25 **Cr**, 0.2 **Zn**, 0.2 **Ti**, 0.7 **Fe**, 0.5 **Cr+Mn**	**As manufactured**: 275 MPa **TS**, 125 MPa **YS**; **Annealed**: 16 mm **diam**, 265 MPa **TS**, 100 MPa **YS**, 16% in 2 in. (50 mm) **El**; **Half hard**: 5 mm **diam**, 355 MPa **TS**, 275 MPa **YS**, 5% in 2 in. (50 mm) **El**
IS 5902	55000	India	Wire, bar	rem **Al**, 0.6 **Si**, 0.10 **Cu**, 0.5 **Mn**, 4.5–5.5 **Mg**, 0.25 **Cr**, 0.2 **Ni**, 0.1 **Zn**, 0.2 **Ti**, 0.05 **Sn**, 0.7 **Fe**, 0.05 **Pb**	**Annealed**: all **diam**, 265 MPa **TS**
JIS Z 3232	5356	Japan	Rod and wire: welding and electrode	rem **Al**, 0.10 **Cu**, 0.05–0.20 **Mn**, 4.5–5.5 **Mg**, 0.05–0.20 **Cr**, 0.10 **Zn**, 0.06–0.20 **Ti**, 0.0008 **Be**	**Annealed**: all **diam**, 265 MPa **TS**
SIS 14 41 46	SIS Al 41 46–00	Sweden	Wire	rem **Al**, 0.5 **Si**, 0.10 **Cu**, 1.0 **Mn**, 4.5–5.5 **Mg**, 0.35 **Cr**, 0.2 **Zn**, 0.5 **Fe**, 0.2 grain refiners	
SIS 14 41 46	SIS Al 41 46–18	Sweden	Wire	rem **Al**, 0.5 **Si**, 0.10 **Cu**, 1.0 **Mn**, 4.5–5.5 **Mg**, 0.35 **Cr**, 0.2 **Zn**, 0.5 **Fe**, 0.2 grain refiners	**Hard worked**: all **diam**, 274 MPa **TS**
SIS 14 41 63	SIS Al 41 63–00	Sweden	Ingot	rem **Al**, 0.5–1.5 **Si**, 0.10 **Cu**, 0.5 **Mn**, 4.0–6.0 **Mg**, 0.05 **Ni**, 0.2 **Zn**, 0.20 **Ti**, 0.05 **Sn**, 0.50 **Fe**, 0.05 **Pb**	
SIS 14 41 63	SIS Al 41 63–03	Sweden	Sand castings	rem **Al**, 0.5–1.5 **Si**, 0.10 **Cu**, 0.5 **Mn**, 4.0–6.0 **Mg**, 0.05 **Ni**, 0.2 **Zn**, 0.20 **Ti**, 0.05 **Sn**, 0.50 **Fe**, 0.05 **Pb**	**As cast**: all **diam**, 160 MPa **TS**, 90 MPa **YS**, 2% in 2 in. (50 mm) **El**, 55–70 **HB**
SIS 14 41 63	SIS Al 41 63–06	Sweden	Chill castings	rem **Al**, 0.5–1.5 **Si**, 0.10 **Cu**, 0.5 **Mn**, 4.0–6.0 **Mg**, 0.05 **Ni**, 0.2 **Zn**, 0.20 **Ti**, 0.05 **Sn**, 0.50 **Fe**, 0.05 **Pb**	**As cast**: all **diam**, 170 MPa **TS**, 100 MPa **YS**, 2% in 2 in. (50 mm) **El**, 55–75 **HB**
NS 17 220	NS 17 220–10	Norway	Sheet, strip, plate, tube, rod, and wire	rem **Al**, 0.4 **Si**, 0.05 **Cu**, 0.55 **Mn**, 4.3–5.5 **Mg**, 0.2 **Zn**, 0.1 **Ti**, 0.40 **Fe**, 0.3 **Cr**, 0.05 each, 0.15 total, others	
NS 17 220	NS 17 220–08	Norway	Tube, rod, and profiles	rem **Al**, 0.4 **Si**, 0.05 **Cu**, 0.55 **Mn**, 4.3–5.5 **Mg**, 0.2 **Zn**, 0.1 **Ti**, 0.40 **Fe**, 0.3 **Cr**, 0.05 each, 0.15 total, others	**Extruded (tube)**: ≥3.5 mm **diam**, 235 MPa **TS**, 108 MPa **YS**, 14% in 2 in. (50 mm) **El**; **Extruded (rod)**: all **diam**, 250 MPa **TS**, 118 MPa **YS**, 13% in 2 in. (50 mm) **El**; **Extruded (profiles)**: ≥10 mm **diam**, 235 MPa **TS**

WROUGHT ALUMINUM AND ALUMINUM ALLOYS

specification number	designation	country	product forms	chemical composition	mechanical properties and hardness values
NS 17 220	NS 17 220–02	Norway	Sheet, strip, tube, rod, and wire	rem **Al**, 0.4 **Si**, 0.05 **Cu**, 0.55 **Mn**, 4.3–5.5 **Mg**, 0.2 **Zn**, 0.1 **Ti**, 0.40 **Fe**, 0.3 **Cr**, 0.05 each, 0.15 total, others	**Soft annealed (sheet and strip)**: 0.2–6 mm **diam**, 235 MPa **TS**, 108 MPa **YS**, 17% in 2 in. (50 mm) **El**; **Soft annealed (rod and wire)**: all **diam**, 235 MPa **TS**, 108 MPa **YS**, 16% in 2 in. (50 mm) **El**; **Soft annealed (tube)**: ≥10 mm **diam**, 235 MPa **TS**, 108 MPa **YS**, 14% in 2 in. (50 mm) **El**
NS 17 220	NS 17 220–34	Norway	Sheet, strip, tube, rod, and wire	rem **Al**, 0.4 **Si**, 0.05 **Cu**, 0.55 **Mn**, 4.3–5.5 **Mg**, 0.2 **Zn**, 0.1 **Ti**, 0.40 **Fe**, 0.3 **Cr**, 0.05 each, 0.15 total, others	**1/2 hard (sheet and strip)**: 0.2–6 mm **diam**, 274 MPa **TS**, 176 MPa **YS**, 9% in 2 in. (50 mm) **El**; **1/2 hard (tube)**: <10 mm **diam**, 274 MPa **TS**, 176 MPa **YS**, 9% in 2 in. (50 mm) **El**; **1/2 hard (rod and wire)**: <35 mm **diam**, 255 MPa **TS**, 147 MPa **YS**, 8% in 2 in. (50 mm) **El**
NS 17 220	NS 17 220–38	Norway	Sheet, strip, tube, rod, and wire	rem **Al**, 0.4 **Si**, 0.05 **Cu**, 0.55 **Mn**, 4.3–5.5 **Mg**, 0.2 **Zn**, 0.1 **Ti**, 0.40 **Fe**, 0.3 **Cr**, 0.05 each, 0.15 total, others	**Hard (sheet and strip)**: 0.2–6 mm **diam**, 314 MPa **TS**, 235 MPa **YS**, 4% in 2 in. (50 mm) **El**; **Hard (tube)**: <6 mm **diam**, 314 MPa **TS**, 235 MPa **YS**, 4% in 2 in. (50 mm) **El**; **Hard (rod and wire)**: <15 mm **diam**, 294 MPa **TS**, 196 MPa **YS**, 4% in 2 in. (50 mm) **El**
NS 17 220	NS 17 220–32	Norway	Sheet, strip, and plate	rem **Al**, 0.40 **Si**, 0.05 **Cu**, 4.3–5.5 **Mg**, 0.30 **Cr**, 0.2 **Zn**, 0.1 **Ti**, 0.40 **Fe**, 0.05 each, 0.15 total, others	**1/4 hard (sheet and strip)**: 4–10 mm **diam**, 265 MPa **TS**, 157 MPa **YS**, 12% in 2 in. (50 mm) **El**; **1/4 hard (strip and plate)**: 10–20 mm **diam**, 265 MPa **TS**, 147 MPa **YS**, 12% in 2 in. (50 mm) **El**
CSA HA.6	.5183	Canada	Rod, wire: for welding and brazing	rem **Al**, 0.40 **Si**, 0.10 **Cu**, 0.50–1.0 **Mn**, 4.3–5.2 **Mg**, 0.05–0.25 **Cr**, 0.25 **Zn**, 0.15 **Ti**, 0.40 **Fe**, 0.008 **Be**, 0.05 each, 0.15 total, others	
JIS Z 3232	5183	Japan	Rod and wire: welding and electrode	rem **Al**, 0.40 **Si**, 0.10 **Cu**, 0.50–1.0 **Mn**, 4.3–5.2 **Mg**, 0.05–0.25 **Cr**, 0.25 **Zn**, 0.15 **Ti**, 0.40 **Fe**, 0.0008 **Be**	**Annealed**: all **diam**, 275 MPa **TS**
AMS 4057D		US	Sheet	rem **Al**, 0.40 **Si**, 0.10 **Cu**, 0.40–1.0 **Mn**, 4.0–4.9 **Mg**, 0.05–0.25 **Cr**, 0.25 **Zn**, 0.15 **Ti**, 0.40 **Fe**, 0.05 each, 0.15 total, others	**Strain hardened to 1/4 hard and stabilized**: 0.051–0.125 in. (1.30–3.18 mm) **diam**, 45 ksi (310 MPa) **TS**, 34 ksi (234 MPa) **YS**, 8% in 2 in. (50 mm) **El**

WROUGHT ALUMINUM AND ALUMINUM ALLOYS

specification number	designation	country	product forms	chemical composition	mechanical properties and hardness values
AMS 4058D		US	Sheet	rem **Al**, 0.40 **Si**, 0.10 **Cu**, 0.40–1.0 **Mn**, 4.0–4.9 **Mg**, 0.05–0.25 **Cr**, 0.25 **Zn**, 0.15 **Ti**, 0.40 **Fe**, 0.05 each, 0.15 total, others	**Strain hardened to 1/2 hard and stabilized**: 0.051–0.125 in. (1.30–3.18 mm) **diam**, 50 ksi (345 MPa) **TS**, 39 ksi (269 MPa) **YS**, 6% in 2 in. (50 mm) **El**
AMS 4059E		US	Sheet and plate	rem **Al**, 0.40 **Si**, 0.10 **Cu**, 0.40–1.0 **Mn**, 4.0–4.9 **Mg**, 0.05–0.25 **Cr**, 0.25 **Zn**, 0.15 **Ti**, 0.40 **Fe**, 0.05 each, 0.15 total, others	**Strain hardened to 1/4 hard and stabilized**: 0.188–1.500 in. (4.78–38.10 mm) **diam**, 44 ksi (303 MPa) **TS**, 31 ksi (214 MPa) **YS**, 12% in 2 in. (50 mm) **El**
AMS 4056D		US	Sheet and plate	rem **Al**, 0.40 **Si**, 0.10 **Cu**, 0.40–1.0 **Mn**, 4.0–4.9 **Mg**, 0.05–0.25 **Cr**, 0.25 **Zn**, 0.15 **Ti**, 0.40 **Fe**, 0.05 each, 0.15 total, others	**Annealed and mill finish**: 0.051–1.500 in. (1.30–38.10 mm) **diam**, 40 ksi (276 MPa) **TS**, 18 ksi (124 MPa) **YS**, 16% in 2 in. (50 mm) **El**
ANSI/ASTM B 547	5083	US	Tube: formed and arc-welded	rem **Al**, 0.40 **Si**, 0.10 **Cu**, 0.40–1.0 **Mn**, 4.0–4.9 **Mg**, 0.05–0.25 **Cr**, 0.25 **Zn**, 0.15 **Ti**, 0.40 **Fe**, 0.05 each, 0.15 total, others	**Annealed**: 0.125–0.500 in. **diam**, 40 ksi (276 MPa) **TS**, 18 ksi (124 MPa) **YS**, 16% in 2 in. (50 mm) **El**; **H321**: 0.188–0.500 in. **diam**, 44 ksi (303 MPa) **TS**, 31 ksi (214 MPa) **YS**, 12% in 2 in. (50 mm) **El**
ANSI/ASTM B 345	5083	US	Tube: seamless	rem **Al**, 0.40 **Si**, 0.10 **Cu**, 0.30–1.0 **Mn**, 4.0–4.9 **Mg**, 0.05–0.25 **Cr**, 0.25 **Zn**, 0.15 **Ti**, 0.40 **Fe**, 0.05 each, 0.15 total, others	**Annealed**: $\leq$5.000 in. **diam**, 39 ksi (269 MPa) **TS**, 16 ksi (110 MPa) **YS**, 14% in 2 in. (50 mm) **El**; **H111**: $\leq$5.000 in. **diam**, 40 ksi (276 MPa) **TS**, 24 ksi (165 MPa) **YS**, 12% in 2 in. (50 mm) **El**; **H112**: 39 ksi (269 MPa) **TS**, 16 ksi (110 MPa) **YS**, 12% in 2 in. (50 mm) **El**
ANSI/ASTM B 247	5083	US	Die and hand forgings	rem **Al**, 0.40 **Si**, 0.10 **Cu**, 0.40–1.0 **Mn**, 4.0–4.9 **Mg**, 0.05–0.20 **Cr**, 0.25 **Zn**, 0.15 **Ti**, 0.40 **Fe**, 0.05 each, 0.15 total, others	**Annealed**: $\leq$3.000 in. **diam**, 39 ksi (269 MPa) **TS**, 16 ksi (110 MPa) **YS**, 16% in 2 in. (50 mm) **El**; **H111**: $\leq$4.000 in. **diam**, 42 ksi (290 MPa) **TS**, 22 ksi (152 MPa) **YS**, 14% in 2 in. (50 mm) **El**; **H112**: $\leq$4.000 in. **diam**, 40 ksi (276 MPa) **TS**, 18 ksi (124 MPa) **YS**, 16% in 2 in. (50 mm) **El**
ANSI/ASTM B 241	5083	US	Tube: seamless	rem **Al**, 0.40 **Si**, 0.10 **Cu**, 0.40–1.0 **Mn**, 4.0–4.9 **Mg**, 0.05–0.25 **Cr**, 0.25 **Zn**, 0.15 **Ti**, 0.40 **Fe**, 0.05 each, 0.15 total, others	**Annealed**: $\leq$5.000 in. **diam**, 39 ksi (269 MPa) **TS**, 16 ksi (110 MPa) **YS**, 14% in 2 in. (50 mm) **El**; **H111**: $\leq$5.000 in. **diam**, 40 ksi (276 MPa) **TS**, 24 ksi (165 MPa) **YS**, 12% in 2 in. (50 mm) **El**; **H112**: $\leq$5.000 in. **diam**, 39 ksi (269 MPa) **TS**, 16 ksi (110 MPa) **YS**, 12% in 2 in. (50 mm) **El**

WROUGHT ALUMINUM AND ALUMINUM ALLOYS

specification number	designation	country	product forms	chemical composition	mechanical properties and hardness values
ANSI/ASTM B 221	5083	US	Bar, tube, rod, wire, and shapes	rem **Al**, 0.40 **Si**, 0.10 **Cu**, 0.40–1.0 **Mn**, 4.0–4.9 **Mg**, 0.05–0.25 **Cr**, 0.25 **Zn**, 0.15 **Ti**, 0.40 **Fe**, 0.05 each, 0.15 total, others	**Annealed**: ≤5.000 in. **diam**, 39 ksi (269 MPa) **TS**, 16 ksi (110 MPa) **YS**, 14% in 2 in. (50 mm) **El**; **H111**: ≤5.000 in. **diam**, 40 ksi (276 MPa) **TS**, 24 ksi (165 MPa) **YS**, 12% in 2 in. (50 mm) **El**; **H112**: ≤5.000 in. **diam**, 39 ksi (269 MPa) **TS**, 16 ksi (110 MPa) **YS**, 12% in 2 in. (50 mm) **El**
ANSI/ASTM B 210	5083	US	Tube: seamless	rem **Al**, 0.40 **Si**, 0.10 **Cu**, 0.40–1.0 **Mn**, 4.0–4.9 **Mg**, 0.05–0.25 **Cr**, 0.25 **Zn**, 0.15 **Ti**, 0.40 **Fe**, 0.05 each, 0.15 total, others	**Annealed**: 0.018–0.450 in. **diam**, 39 ksi (269 MPa) **TS**, 16 ksi (110 MPa) **YS**, 14% in 2 in. (50 mm) **El**
ANSI/ASTM B 209	5083	US	Sheet and plate	rem **Al**, 0.40 **Si**, 0.10 **Cu**, 0.40–1.0 **Mn**, 4.0–4.9 **Mg**, 0.05–0.25 **Cr**, 0.25 **Zn**, 0.15 **Ti**, 0.40 **Fe**, 0.05 each, 0.15 total, others	**Annealed**: 0.051–1.500 in. **diam**, 40 ksi (276 MPa) **TS**, 18 ksi (124 MPa) **YS**, 16% in 2 in. (50 mm) **El**; **H321**: 1.501–3.000 in. **diam**, 41 ksi (283 MPa) **TS**, 29 ksi (200 MPa) **YS**, 12% in 2 in. (50 mm) **El**; **H323**: 0.126–0.249 in. **diam**, 45 ksi (310 MPa) **TS**, 34 ksi (234 MPa) **YS**, 10% in 2 in. (50 mm) **El**
AS 1866	5083	Australia	Rod, bar, shape	rem **Al**, 0.40 **Si**, 0.10 **Cu**, 0.40–1.0 **Mn**, 4.0–4.9 **Mg**, 0.05–0.25 **Cr**, 0.15 **Ti**, 0.40 **Fe**, 0.05 each, 0.15 total, others	**Annealed**: 125.0 mm **diam**, 275 MPa **TS**, 165 MPa **YS**, 10% in 2 in. (50 mm) **El**; **Strain–hardened**: 125.0 mm **diam**, 268 MPa **TS**, 110 MPa **YS**, 10% in 2 in. (50 mm) **El**
AS 1734	5083	Australia	Sheet, plate	rem **Al**, 0.40 **Si**, 0.10 **Cu**, 0.40–1.0 **Mn**, 4.0–4.9 **Mg**, 0.05–0.25 **Cr**, 0.25 **Zn**, 0.15 **Ti**, 0.40 **Fe**, 0.05 each, 0.15 total, others	**Annealed**: 1.30–40.00 mm **diam**, 275 MPa **TS**, 124 MPa **YS**, 14% in 2 in. (50 mm) **El**; **Strain–hardened less than 1/8 hard**: 6–40 mm **diam**, 289 MPa **TS**, 172 MPa **YS**, 12% in 2 in. (50 mm) **El**; **Strain–hardened less than 1/4 hard**: 6–50 mm **diam**, 303 MPa **TS**, 213 MPa **TS**, 10% in 2 in. (50 mm) **El**
BS 1471	NT8	UK	Tube: drawn	rem **Al**, 0.40 **Si**, 0.10 **Cu**, 0.5–1.0 **Mn**, 4.0–4.9 **Mg**, 0.25 **Cr**, 0.2 **Zn**, 0.2 **Ti**, 0.40 **Fe**	**Annealed**: ≤10.0 mm **diam**, 275 MPa **TS**, 125 MPa **YS**, 12% in 2 in. (50 mm) **El**; **Strain–hardened to 1/4 hard**: ≤10.0 mm **diam**, 310 MPa **TS**, 235 MPa **YS**, 5% in 2 in. (50 mm) **El**
BS 1470	NS8	UK	Plate, sheet, strip	rem **Al**, 0.40 **Si**, 0.10 **Cu**, 0.5–1.0 **Mn**, 4.0–4.9 **Mg**, 0.25 **Cr**, 0.2 **Zn**, 0.2 **Ti**, 0.40 **Fe**	**Annealed**: 0.2–25.0 mm **diam**, 275 MPa **TS**, 125 MPa **YS**, 12% in 2 in. (50 mm) **El**; **Strain hardened to 1/2 hard**: 0.2–6.0 mm **diam**, 310 MPa **TS**, 235 MPa **YS**, 5% in 2 in. (50 mm) **El**

WROUGHT ALUMINUM AND ALUMINUM ALLOYS

specification number	designation	country	product forms	chemical composition	mechanical properties and hardness values
BS 1472	NF8	UK	Forging stock and forgings	rem **Al**, 0.40 **Si**, 0.10 **Cu**, 0.5–1.0 **Mn**, 4.0–4.9 **Mg**, 0.25 **Cr**, 0.2 **Zn**, 0.2 **Ti**, 0.40 **Fe**	**As manufactured**: ≤150 mm **diam**, 280 MPa **TS**, 130 MPa **YS**, 12% in 2 in. (50 mm) **El**
BS 1474	NE8	UK	Bar, tube, section	rem **Al**, 0.40 **Si**, 0.10 **Cu**, 0.5–1.0 **Mn**, 4.0–4.9 **Mg**, 0.25 **Cr**, 0.2 **Zn**, 0.2 **Ti**, 0.40 **Fe**	**Annealed**: ≤150 mm **diam**, 275 MPa **TS**, 125 MPa **YS**, 13% in 2 in. (50 mm) **El**; **As manufactured**: ≤150 mm **diam**, 280 MPa **TS**, 130 MPa **TS**, 11% in 2 in. (50 mm) **El**
CSA HA.7	.5083	Canada	Tube, pipe: seamless	rem **Al**, 0.40 **Si**, 0.10 **Cu**, 0.40–1.0 **Mn**, 4.0–4.9 **Mg**, 0.05–0.25 **Cr**, 0.25 **Zn**, 0.15 **Ti**, 0.40 **Fe**, 0.05 each, 0.15 total, others	**Extruded and H111**: all **diam**, 40.0 ksi (276 MPa) **TS**, 24.0 ksi (165 MPa) **YS**, 12% in 2 in. (50 mm) **El**
CSA HA.5	.5083	Canada	Bar, rod, wire, shapes	rem **Al**, 0.40 **Si**, 0.10 **Cu**, 0.40–1.0 **Mn**, 4.0–4.9 **Mg**, 0.05–0.25 **Cr**, 0.25 **Zn**, 0.15 **Ti**, 0.40 **Fe**, 0.05 each, 0.15 total, others	**Extruded and O**: all **diam**, 39.0 ksi (269 MPa) **TS**, 16.0 ksi (110 MPa) **YS**, 16% in 2 in. (50 mm) **El**; **Extruded and H111**: all **diam**, 40.0 ksi (276 MPa) **TS**, 24.0 ksi (165 MPa) **YS**, 12% in 2 in. (50 mm) **El**
CSA HA.4	.5083	Canada	Plate, sheet	rem **Al**, 0.40 **Si**, 0.10 **Cu**, 0.40–1.0 **Mn**, 4.0–4.9 **Mg**, 0.05–0.25 **Cr**, 0.25 **Zn**, 0.15 **Ti**, 0.40 **Fe**, 0.05 each, 0.15 total, others	**O**: 0.051–1.500 in. **diam**, 40.0 ksi (124 MPa) **TS**, 18.0 ksi (276 MPa) **YS**, 16% in 2 in. (50 mm) **El**; **H112**: 0.051–1.500 in. **diam**, 40.0 ksi (124 MPa) **TS**, 18.0 ksi (276 MPa) **YS**, 12% in 2 in. (50 mm) **El**; **H321**: 0.188–1.500 in. **diam**, 44.0 ksi (303 MPa) **TS**, 31.0 ksi (214 MPa) **YS**, 12% in 2 in. (50 mm) **El**
DS 3012	5083	Denmark	Rolled, drawn, and extruded products	rem **Al**, 0.40 **Si**, 0.10 **Cu**, 0.40–1.0 **Mn**, 4.0–4.9 **Mg**, 0.05–0.25 **Cr**, 0.25 **Zn**, 0.15 **Ti**, 0.40 **Fe**, 0.05 each, 0.15 total, others	**Soft (rolled and drawn products)**: 1.3–25 mm **diam**, 275 MPa **TS**, 125 MPa **YS**, 16% in 2 in. (50 mm) **El**, 70 **HV**; **1/2 hard (rolled and drawn products)**: 1.3–6 mm **diam**, 345 MPa **TS**, 270 MPa **YS**, 6% in 2 in. (50 mm) **El**, 90 **HV**; **As manufactured (extruded products)**: 1.3–6 mm **diam**, 345 MPa **TS**, 270 MPa **YS**, 6% in 2 in. (50 mm) **El**

WROUGHT ALUMINUM AND ALUMINUM ALLOYS

specification number	designation	country	product forms	chemical composition	mechanical properties and hardness values
QQ-A-250/6F	5083	US	Plate, sheet	rem **Al**, 0.40 **Si**, 0.10 **Cu**, 0.30–1.0 **Mn**, 4.0–4.9 **Mg**, 0.05–0.25 **Cr**, 0.25 **Zn**, 0.15 **Ti**, 0.40 **Fe**, 0.05 each, 0.15 total, others	**Annealed**: 0.051–1.500 in. **diam**, 40 ksi (276 MPa) **TS**, 18 ksi (124 MPa) **YS**, 16% in 2 in. (50 mm) **El**; **Strain hardened 1/4 hard (hot rolled) temper and stabilized**: 0.188–1.500 in. **diam**, 44 ksi (303 MPa) **TS**, 31 ksi (214 MPa) **YS**, 12% in 2 in. (50 mm) **El**; **Strain hardened 1/4 hard (cold rolled) temper and stabilized**: 0.051–0.125 in. **diam**, 56 ksi (310 MPa) **TS**, 34 ksi (234 MPa) **YS**, 8% in 2 in. (50 mm) **El**
QQ-A-200/4C	5083	US	Bar, rod, shapes, tube, wire	rem **Al**, 0.40 **Si**, 0.10 **Cu**, 0.30–1.0 **Mn**, 4.0–4.9 **Mg**, 0.05–0.25 **Cr**, 0.25 **Zn**, 0.15 **Ti**, 0.40 **Fe**, 0.05 each, 0.15 total, others	**Annealed**: 5.000 max in. **diam**, 39 ksi (269 MPa) **TS**, 16 ksi (110 MPa) **YS**, 14% in 2 in. (50 mm) **El**; **Strain hardened**: 5.000 max in. **diam**, 40 ksi (276 MPa) **TS**, 24 ksi (165 MPa) **YS**, 12% in 2 in. (50 mm) **El**; **As extruded**: 5.000 max in. **diam**, 39 ksi (269 MPa) **TS**, 16 ksi (110 MPa) **YS**, 12% in 2 in. (50 mm) **El**
NF A 50 411	5083	France	Bar, tube, and shapes	rem **Al**, 0.40 **Si**, 0.10 **Cu**, 0.40–1.0 **Mn**, 4.0–4.9 **Mg**, 0.05–0.25 **Cr**, 0.25 **Zn**, 0.15 **Ti**, 0.40 **Fe**, 0.05 each, 0.15 total, others	**1/4 hard (drawn bar)**: ≤25 mm **diam**, 300 MPa **TS**, 200 MPa **YS**, 4% in 2 in. (50 mm) **El**; **Annealed (tube)**: ≤150 mm **diam**, 270 MPa **TS**, 110 MPa **YS**, 17% in 2 in. (50 mm) **El**; **Annealed (shapes)**: ≤110 mm **diam**, 270 MPa **TS**, 110 MPa **YS**, 18% in 2 in. (50 mm) **El**
NF A 50 451	5083	France	Sheet, plate, and strip	rem **Al**, 0.40 **Si**, 0.10 **Cu**, 0.40–1.0 **Mn**, 4.0–4.9 **Mg**, 0.05–0.25 **Cr**, 0.25 **Zn**, 0.15 **Ti**, 0.40 **Fe**, 0.05 each, 0.15 total, others	**Annealed**: 12–150 mm **diam**, 270 MPa **TS**, 120 MPa **YS**, 10% in 2 in. (50 mm) **El**; **1/4 hard**: 0.4–25 mm **diam**, 300 MPa **TS**, 210 MPa **YS**, 11% in 2 in. (50 mm) **El**
IS 736	54300	India	Plate	rem **Al**, 0.4 **Si**, 0.1 **Cu**, 0.5–1.0 **Mn**, 4.0–4.9 **Mg**, 0.25 **Cr**, 0.2 **Zn**, 0.2 **Ti**, 0.7 **Fe**	**As manufactured**: 12 mm **diam**, 280 MPa **TS**, 125 MPa **YS**, 12% in 2 in. (50 mm) **El**; **Annealed**: 16 mm **diam**, 270 MPa **TS**, 115 MPa **YS**, 16% in 2 in. (50 mm) **El**; **Half hard**: 5 mm **diam**, 355 MPa **TS**, 275 MPa **YS**, 5% in 2 in. (50 mm) **El**

WROUGHT ALUMINUM AND ALUMINUM ALLOYS

specification number	designation	country	product forms	chemical composition	mechanical properties and hardness values
IS 733	54300	India	Bar, rod, section	rem Al, 0.4 Si, 0.1 Cu, 0.5–1.0 Mn, 4.0–4.9 Mg, 0.25 Cr, 0.2 Zn, 0.2 Ti, 0.7 Fe	Annealed: ≤150 mm diam, 265 MPa TS, 125 MPa YS, 13% in 2 in. (50 mm) El; As manufactured: ≤150 mm diam, 275 MPa TS, 130 MPa YS, 11% in 2 in. (50 mm) El
ISO R827	Al–Mg4.5Mn	ISO	Extruded products	rem Al, 0.5 Si, 0.10 Cu, 0.3–1.0 Mn, 4.0–4.9 Mg, 0.25 Cr, 0.2 Zn, 0.5 Fe, 0.2 Ti+Zr	As manufactured: 5.0 max in. (127 max mm) diam, 265 MPa TS, 108 MPa YS, 12% in 2 in. (50 mm) El
ISO R209	Al–Mg4.5Mn	ISO	Wrought products	rem Al, 0.5 Si, 0.10 Cu, 0.3–1.0 Mn, 4.0–4.9 Mg, 0.25 Cr, 0.2 Zn, 0.5 Fe, 0.2 Ti+Zr, 0.15–0.9 Mn+Cr	
ISO TR2778	Al–Mg4.5Mn	ISO	Tube: drawn	rem Al, 0.5 Si, 0.10 Cu, 0.3–1.0 Mn, 4.0–4.9 Mg, 0.25 Cr, 0.2 Zn, 0.5 Fe, 0.2 Ti+Zr	Annealed: all diam, 270 MPa TS, 110 MPa YS, 12% in 2 in. (50 mm) El
ISO TR2136	Al–Mg4.5Mn	ISO	Sheet, plate	rem Al, 0.5 Si, 0.10 Cu, 0.3–1.0 Mn, 4.0–4.9 Mg, 0.25 Cr, 0.2 Zn, 0.5 Fe, 0.2 Ti+Zr	Annealed: all diam, 275 MPa TS, 125 MPa YS, 12% in 2 in. (50 mm) El; HB–Strain hardened: all diam, 305 MPa TS, 215 MPa YS, 5% in 2 in. (50 mm) El; HD– Strain hardened: all diam, 345 MPa TS, 270 MPa YS, 4% in 2 in. (50 mm) El
JIS H 4000	5083	Japan	Sheet, plate, and strip	rem Al, 0.40 Si, 0.10 Cu, 0.30–1.0 Mn, 4.0–4.9 Mg, 0.05–0.25 Cr, 0.25 Zn, 0.15 Ti, 0.40 Fe, 0.15 total, others	Annealed: <1.3 mm diam, 275 MPa TS, 196 MPa YS, 16% in 2 in. (50 mm) El; Stabilized at 1/4 hard: <1.3 mm diam, 314 MPa TS, 304 MPa YS, 8% in 2 in. (50 mm) El; Hot rolled: <12 mm diam, 275 MPa TS, 118 MPa YS, 12% in 2 in. (50 mm) El
JIS H 4040	5083	Japan	Bar: extruded	rem Al, 0.40 Si, 0.10 Cu, 0.30–1.0 Mn, 4.0–4.9 Mg, 0.05–0.25 Cr, 0.25 Zn, 0.15 Ti, 0.40 Fe, 0.15 total, others	As fabricated: ≤130 mm diam, 275 MPa TS, 108 MPa YS, 12% in 2 in. (50 mm) El; Annealed: ≤130 mm diam, 353 MPa TS, 108 MPa YS, 14% in 2 in. (50 mm) El
JIS H 4080	5083	Japan	Tube: extruded	rem Al, 0.40 Si, 0.10 Cu, 0.30–1.0 Mn, 4.0–4.9 Mg, 0.05–0.25 Cr, 0.25 Zn, 0.15 Ti, 0.40 Fe, 0.15 total, others	As manufactured: all diam, 275 MPa TS, 108 MPa YS, 12% in 2 in. (50 mm) El; Annealed: all diam, 353 MPa TS, 108 MPa YS, 14% in 2 in. (50 mm) El
JIS H 4100	5083	Japan	Shapes: extruded	rem Al, 0.40 Si, 0.10 Cu, 0.30–1.0 Mn, 4.0–4.9 Mg, 0.05–0.25 Cr, 0.25 Zn, 0.15 Ti, 0.40 Fe, 0.15 total, others	As manufactured: ≤130 mm diam, 275 MPa TS, 108 MPa YS, 12% in 2 in. (50 mm) El; Annealed: ≤130 mm diam, 353 MPa TS, 108 MPa YS, 14% in 2 in. (50 mm) El

WROUGHT ALUMINUM AND ALUMINUM ALLOYS

specification number	designation	country	product forms	chemical composition	mechanical properties and hardness values
JIS H 4140	5083	Japan	Die forgings	rem **Al**, 0.40 **Si**, 0.10 **Cu**, 0.30–1.0 **Mn**, 4.0–4.9 **Mg**, 0.05–0.25 **Cr**, 0.15 **Ti**, 0.40 **Fe**, 0.15 total, others	**As manufactured**: ≤100 mm **diam**, 275 MPa **TS**, 127 MPa **YS**, 16% in 2 in. (50 mm) **El**
MIL-A-46027F	5083	US	Plate	rem **Al**, 0.40 **Si**, 0.10 **Cu**, 0.40–1.0 **Mn**, 4.0–4.9 **Mg**, 0.05–0.25 **Cr**, 0.25 **Zn**, 0.15 **Ti**, 0.40 **Fe**, 0.05 each, 0.15 total, others	
MIL-A-46083B	5083	US	Shapes	rem **Al**, 0.40 **Si**, 0.10 **Cu**, 0.40–1.0 **Mn**, 4.0–4.9 **Mg**, 0.05–0.25 **Cr**, 0.25 **Zn**, 0.15 **Ti**, 0.40 **Fe**, 0.05 each, 0.15 total, others	**Class I**: ≤2.000 in. **diam**, 310 MPa **TS**, 241 MPa **YS**, 9% in 2 in. (50 mm) **El**
NS 17 215	NS 17 215	Norway	Sheet, strip, and plate	rem **Al**, 0.40 **Si**, 0.10 **Cu**, 0.60–1.0 **Mn**, 4.0–4.9 **Mg**, 0.2 **Zn**, 0.1 **Ti**, 0.40 **Fe**, 0.35 **Cr**, 0.05 each, 0.15 total, others	**Soft annealed (sheet and strip)**: 0.4–30 mm **diam**, 274 MPa **TS**, 123 MPa **YS**, 17% in 2 in. (50 mm) **El**; **Soft annealed (strip and plate)**: 30–50 mm **diam**, 265 MPa **TS**, 118 MPa **YS**, 17% in 2 in. (50 mm) **El**
NS 17 215	NS 17 215–32	Norway	Sheet, strip, and plate	rem **Al**, 0.40 **Si**, 0.10 **Cu**, 0.60–1.0 **Mn**, 4.0–4.9 **Mg**, 0.2 **Zn**, 0.1 **Ti**, 0.40 **Fe**, 0.05–0.25 **Cr**, 0.05 each, 0.15 total, others	**1/4 hard (sheet and strip)**: 0.4–3 mm **diam**, 304 MPa **TS**, 235 MPa **YS**, 8% in 2 in. (50 mm) **El**; **1/4 hard (strip and plate)**: 5–40 mm **diam**, 294 MPa **TS**, 206 MPa **YS**, 12% in 2 in. (50 mm) **El**
NS 17 215	NS 17 215–08	Norway	Profiles	rem **Al**, 0.40 **Si**, 0.10 **Cu**, 0.60–1.0 **Mn**, 4.0–4.9 **Mg**, 0.2 **Zn**, 0.1 **Ti**, 0.40 **Fe**, 0.05–0.25 **Cr**, 0.05 each, 0.15 total, others	**Extruded**: all **diam**, 274 MPa **TS**, 157 MPa **YS**, 12% in 2 in. (50 mm) **El**
SIS 14 41 40	SIS Al 41 40–00	Sweden	Plate, section	94.65 **Al**, 0.40 **Si**, 0.10 **Cu**, 0.40–1.0 **Mn**, 4.0–4.9 **Mg**, 0.05–0.25 **Cr**, 0.25 **Zn**, 0.15 **Ti**, 0.40 **Fe**, 0.05 each, 0.15 total, others	**Hard worked**: 6–30 mm **diam**, 275 MPa **TS**, 125 MPa **YS**, 12% in 2 in. (50 mm) **El**, 75 **HV**
SIS 14 41 40	SIS Al 41 40–02	Sweden	Sheet, plate, strip	94.65 **Al**, 0.40 **Si**, 0.10 **Cu**, 0.40–1.0 **Mn**, 4.0–4.9 **Mg**, 0.15 **Cr**, 0.25 **Zn**, 0.15 **Ti**, 0.40 **Fe**, 0.05 each, 0.15 total, others	**Annealed**: 0.5–5 mm **diam**, 270 MPa **TS**, 120 MPa **YS**, 17% in 2 in. (50 mm) **El**, 75 **HV**
SIS 14 41 40	SIS Al 41 40–12	Sweden	Plate	94.65 **Al**, 0.40 **Si**, 0.10 **Cu**, 0.40–1.0 **Mn**, 4.0–4.9 **Mg**, 0.15 **Cr**, 0.25 **Zn**, 0.15 **Ti**, 0.40 **Fe**, 0.05 each, 0.15 total, others	**Strain hardened**: all **diam**, 310 MPa **TS**, 205 MPa **YS**, 12% in 2 in. (50 mm) **El**, 95 **HV**
SIS 14 41 40	SIS Al 41 40–22	Sweden	Strip, sheet	94.65 **Al**, 0.40 **Si**, 0.10 **Cu**, 0.40–1.0 **Mn**, 4.0–4.9 **Mg**, 0.15 **Cr**, 0.25 **Zn**, 0.15 **Ti**, 0.40 **Fe**, 0.05 each, 0.15 total, others	**Strain hardened**: 0.5–2 mm **diam**, 310 MPa **TS**, 205 MPa **YS**, 13% in 2 in. (50 mm) **El**, 95 **HV**

WROUGHT ALUMINUM AND ALUMINUM ALLOYS

specification number	designation	country	product forms	chemical composition	mechanical properties and hardness values
SIS 14 41 40	SIS Al 41 40–24	Sweden	Sheet, strip	94.65 **Al**, 0.40 **Si**, 0.10 **Cu**, 0.40–1.0 **Mn**, 4.0–4.9 **Mg**, 0.15 **Cr**, 0.25 **Zn**, 0.15 **Ti**, 0.40 **Fe**, 0.05 each, 0.15 total, others	**Strain hardened**: 0.5–6 mm **diam**, 345 MPa **TS**, 270 MPa **YS**, 6% in 2 in. (50 mm) **El**, 110 **HV**
ISO R827	Al–Mg4	ISO	Extruded products	rem **Al**, 0.5 **Si**, 0.10 **Cu**, 0.8 **Mn**, 3.5–4.6 **Mg**, 0.35 **Cr**, 0.2 **Zn**, 0.5 **Fe**, 0.2 **Ti+Zr**, 0.15–0.9 **Mn+Cr**	**As manufactured**: 5.0 max in. (127 max mm) **diam**, 241 MPa **TS**, 98 MPa **YS**, 12% in 2 in. (50 mm) **El**
ISO R827	Al–Mg4	ISO	Wrought products	rem **Al**, 0.5 **Si**, 0.10 **Cu**, 0.8 **Mn**, 3.5–4.6 **Mg**, 0.35 **Cr**, 0.2 **Zn**, 0.5 **Fe**, 0.2 **Ti+Zr**, 0.15–0.9 **Mn+Cr**	
ISO TR2778	Al–Mg4	ISO	Tube: drawn	rem **Al**, 0.5 **Si**, 0.10 **Cu**, 0.8 **Mn**, 3.5–4.6 **Mg**, 0.35 **Cr**, 0.2 **Zn**, 0.5 **Fe**, 0.2 **Ti+Zr**, 0.15–0.9 **Mn+Cr**	**Annealed**: all **diam**, 240 MPa **TS**, 95 MPa **YS**, 15% in 2 in. (50 mm) **El**; **Strain hardened**: 50 mm **diam**, 260 MPa **TS**, 185 MPa **YS**, 5% in 2 in. (50 mm) **El**
ISO TR2136	Al–Mg4	ISO	Sheet, plate	rem **Al**, 0.5 **Si**, 0.10 **Cu**, 0.8 **Mn**, 3.5–4.6 **Mg**, 0.35 **Cr**, 0.2 **Zn**, 0.5 **Fe**, 0.2 **Ti+Zr**, 0.15–0.9 **Mn+Cr**	**Annealed**: all **diam**, 240 MPa **TS**, 95 MPa **YS**, 12% in 2 in. (50 mm) **El**; **HB–Strain hardened**: all **diam**, 270 MPa **TS**, 185 MPa **YS**, 5% in 2 in. (50 mm) **El**; **HD– Strain hardened**: all **diam**, 305 MPa **TS**, 230 MPa **YS**, 4% in 2 in. (50 mm) **El**
ANSI/ASTM B 547	5086	US	Tube: formed and arc-welded	rem **Al**, 0.40 **Si**, 0.10 **Cu**, 0.20–0.7 **Mn**, 3.5–4.5 **Mg**, 0.05–0.25 **Cr**, 0.25 **Zn**, 0.15 **Ti**, 0.50 **Fe**, 0.05 each, 0.15 total, others	**Annealed**: 0.250–0.500 in. **diam**, 35 ksi (241 MPa) **TS**, 14 ksi (97 MPa) **YS**, 16% in 2 in. (50 mm) **El**; **H32**: 0.505–0.500 in. **diam**, 40 ksi (276 MPa) **TS**, 28 ksi (193 MPa) **YS**, 12% in 2 in. (50 mm) **El**; **H34**: 0.250–0.500 in. **diam**, 44 ksi (303 MPa) **TS**, 34 ksi (234 MPa) **YS**, 10% in 2 in. (50 mm) **El**
ANSI/ASTM B 345	5086	US	Tube: seamless	rem **Al**, 0.40 **Si**, 0.10 **Cu**, 0.20–0.7 **Mn**, 3.5–4.5 **Mg**, 0.05–0.25 **Cr**, 0.25 **Zn**, 0.15 **Ti**, 0.50 **Fe**, 0.05 each, 0.15 total, others	**Annealed**: ≤5.000 in. **diam**, 35 ksi (241 MPa) **TS**, 14 ksi (97 MPa) **YS**, 14% in 2 in. (50 mm) **El**; **H111**: ≤5.000 in. **diam**, 36 ksi (248 MPa) **TS**, 21 ksi (145 MPa) **YS**, 12% in 2 in. (50 mm) **El**; **H112**: ≤5.000 in. **diam**, 35 ksi (241 MPa) **TS**, 14 ksi (97 MPa) **YS**, 12% in 2 in. (50 mm) **El**

WROUGHT ALUMINUM AND ALUMINUM ALLOYS

specification number	designation	country	product forms	chemical composition	mechanical properties and hardness values
ANSI/ASTM B 313	5086	US	Tube: welded	rem **Al**, 0.40 **Si**, 0.20–0.7 **Mn**, 3.5–4.5 **Mg**, 0.25 **Cr**, 0.25 **Zn**, 0.50 **Fe**, 0.05 each, 0.15 total, others	**Annealed**: 0.051–0.125 in. **diam**, 35 ksi (241 MPa) **TS**, 14 ksi (97 MPa) **YS**, 18% in 2 in. (50 mm) **El**; **H32**: 0.051–0.125 in. **diam**, 40 ksi (276 MPa) **TS**, 28 ksi (193 MPa) **YS**, 8% in 2 in. (50 mm) **El**; **H34**: 0.051–0.125 in. **diam**, 44 ksi (303 MPa) **TS**, 34 ksi (234 MPa) **YS**, 6% in 2 in. (50 mm) **El**
ANSI/ASTM B 241	5086	US	Tube: seamless	rem **Al**, 0.40 **Si**, 0.10 **Cu**, 0.20–0.7 **Mn**, 3.5–4.5 **Mg**, 0.05–0.25 **Cr**, 0.25 **Zn**, 0.15 **Ti**, 0.50 **Fe**, 0.05 each, 0.15 total, others	**Annealed**: $\leq$5.000 in. **diam**, 35 ksi (241 MPa) **TS**, 14 ksi (97 MPa) **YS**, 14% in 2 in. (50 mm) **El**; **H111**: $\leq$5.000 in. **diam**, 36 ksi (248 MPa) **TS**, 21 ksi (145 MPa) **YS**, 12% in 2 in. (50 mm) **El**; **H112**: $\leq$5.000 in. **diam**, 35 ksi (241 MPa) **TS**, 14 ksi (97 MPa) **YS**, 12% in 2 in. (50 mm) **El**
ANSI/ASTM B 221	5086	US	Bar, tube, rod, wire, and shapes	rem **Al**, 0.40 **Si**, 0.10 **Cu**, 0.20–0.7 **Mn**, 3.5–4.5 **Mg**, 0.05–0.25 **Cr**, 0.25 **Zn**, 0.15 **Ti**, 0.50 **Fe**, 0.05 each, 0.15 total, others	**Annealed**: $\leq$5.000 in. **diam**, 35 ksi (241 MPa) **TS**, 14 ksi (97 MPa) **YS**, 14% in 2 in. (50 mm) **El**; **H111**: $\leq$5.000 in. **diam**, 36 ksi (248 MPa) **TS**, 21 ksi (145 MPa) **YS**, 12% in 2 in. (50 mm) **El**; **H112**: $\leq$5.000 in. **diam**, 35 ksi (241 MPa) **TS**, 14 ksi (97 MPa) **YS**, 12% in 2 in. (50 mm) **El**
ANSI/ASTM B 210	5086	US	Tube: seamless	rem **Al**, 0.40 **Si**, 0.10 **Cu**, 0.20–0.7 **Mn**, 3.5–4.5 **Mg**, 0.05–0.25 **Cr**, 0.25 **Zn**, 0.15 **Fe**, 0.50 **Fe**, 0.05 each, 0.15 total, others	**Annealed**: 0.018–0.450 in. **diam**, 35 ksi (241 MPa) **TS**, 14 ksi (97 MPa) **YS**, 14% in 2 in. (50 mm) **El**; **H32 (round tube)**: 0.018–0.450 in. **diam**, 40 ksi (276 MPa) **TS**, 28 ksi (193 MPa) **YS**; **H34 (round tube)**: 0.018–0.450 in. **diam**, 44 ksi (303 MPa) **TS**, 34 ksi (234 MPa) **YS**
ANSI/ASTM B 209	5086	US	Sheet and plate	rem **Al**, 0.40 **Si**, 0.10 **Cu**, 0.20–0.7 **Mn**, 3.5–4.5 **Mg**, 0.05–0.25 **Cr**, 0.25 **Zn**, 0.15 **Ti**, 0.50 **Fe**, 0.05 each, 0.15 total, others	**Annealed**: 0.250–2.000 in. **diam**, 35 ksi (241 MPa) **TS**, 14 ksi (97 MPa) **YS**, 16% in 2 in. (50 mm) **El**; **H32**: 0.250–2.000 in. **diam**, 40 ksi (276 MPa) **TS**, 28 ksi (193 MPa) **YS**, 12% in 2 in. (50 mm) **El**; **H34**: 0.250–1.000 in. **diam**, 44 ksi (303 MPa) **TS**, 34 ksi (234 MPa) **YS**, 10% in 2 in. (50 mm) **El**

WROUGHT ALUMINUM AND ALUMINUM ALLOYS

specification number	designation	country	product forms	chemical composition	mechanical properties and hardness values
AS 1734	5086	Australia	Sheet, plate	rem **Al**, 0.40 **Si**, 0.10 **Cu**, 0.20–0.7 **Mn**, 3.5–4.5 **Mg**, 0.05–0.25 **Cr**, 0.25 **Zn**, 0.15 **Ti**, 0.50 **Fe**, 0.05 each, 0.15 total, others	**Annealed**: 0.50–1.30 mm **diam**, 241 MPa **TS**, 96 MPa **YS**, 15% in 2 in. (50 mm) **El**; **Strain–hardened and stabilized to 1/2 hard**: 0.25–0.50 mm **diam**, 303 MPa **TS**, 234 MPa **YS**, 4% in 2 in. (50 mm) **El**; **Strain–hardened and stabilized to full hard**: 0.15–0.50 mm **diam**, 344 MPa **TS**, 282 MPa **TS**, 3% in 2 in. (50 mm) **El**
CSA HA.4	.5086	Canada	Plate, sheet	rem **Al**, 0.40 **Si**, 0.10 **Cu**, 0.20–0.7 **Mn**, 3.5–4.5 **Mg**, 0.05–0.25 **Cr**, 0.25 **Zn**, 0.15 **Ti**, 0.50 **Fe**, 0.05 each, 0.15 total, others	**O**: 0.020–0.050 in. **diam**, 35.0 ksi (241 MPa) **TS**, 14.0 ksi (97 MPa) **YS**, 15% in 2 in. (50 mm) **El**; **H32**: 0.020–0.050 in. **diam**, 40.0 ksi (276 MPa) **TS**, 28.0 ksi (193 MPa) **YS**, 6% in 2 in. (50 mm) **El**; **H112**: 0.188–0.499 in. **diam**, 36.0 ksi (248 MPa) **TS**, 18.0 ksi (124 MPa) **YS**, 8% in 2 in. (50 mm) **El**
COPANT 862	5086	COPANT	Wrought products	rem **Al**, 0.40 **Si**, 0.10 **Cu**, 0.20–0.70 **Mn**, 3.5–4.5 **Mg**, 0.05–0.25 **Cr**, 0.25 **Zn**, 0.15 **Ti**, 0.50 **Fe**, 0.05 each, 0.15 total, others	
WW–T–700/5d	5086	US	Tube	rem **Al**, 0.40 **Si**, 0.10 **Cu**, 0.20–0.7 **Mn**, 3.5–4.5 **Mg**, 0.05–0.25 **Cr**, 0.25 **Zn**, 0.15 **Ti**, 0.50 **Fe**, 0.05 each, 0.15 total, others	**Annealed**: 0.010–0.450 in. **diam**, 35 ksi (241 MPa) **TS**, 14 ksi (97 MPa) **YS**, 14% in 2 in. (50 mm) **El**; **Strain hardened and stabilized 1/4 hard**: 0.010–0.050 in. **diam**, 40 ksi (276 MPa) **TS**, 28 ksi (193 MPa) **YS**, 6% in 2 in. (50 mm) **El**; **Strain hardened and stabilized 1/2 hard**: 0.010–0.050 in. **diam**, 44 ksi (303 MPa) **TS**, 34 ksi (234 MPa) **YS**, 5% in 2 in. (50 mm) **El**
QQ–A–250/7E	5086	US	Plate, sheet	rem **Al**, 0.40 **Si**, 0.10 **Cu**, 0.20–0.7 **Mn**, 3.5–4.5 **Mg**, 0.05–0.25 **Cr**, 0.25 **Zn**, 0.15 **Ti**, 0.50 **Fe**, 0.05 each, 0.15 total, others	**Annealed**: 0.020–0.050 in. **diam**, 35 ksi (2441MPa) **TS**, 14 ksi (97 MPa) **YS**, 15% in 2 in. (50 mm) **El**; **As rolled**: 0.188–0.499 in. **diam**, 36 ksi (248 MPa) **TS**, 18 ksi (124 MPa) **YS**, 8% in 2 in. (50 mm) **El**; **Strain hardened 1/4 hard temper and stabilized**: 0.020–0.050 in. **diam**, 40 ksi (276 MPa) **TS**, 28 ksi (193 MPa) **YS**, 6% in 2 in. (50 mm) **El**

WROUGHT ALUMINUM AND ALUMINUM ALLOYS

specification number	designation	country	product forms	chemical composition	mechanical properties and hardness values
QQ-A-200/5C	5086	US	Bar, rod, shapes, tube, wire	rem **Al**, 0.40 **Si**, 0.10 **Cu**, 0.20–0.7 **Mn**, 3.5–4.5 **Mg**, 0.05–0.25 **Cr**, 0.25 **Zn**, 0.15 **Ti**, 0.50 **Fe**, 0.05 each, 0.15 total, others	**Annealed**: 5.000 max in. **diam**, 35 ksi (241 MPa) **TS**, 14 ksi (97 MPa) **YS**, 14% in 2 in. (50 mm) **El**; **Strain hardened**: 5.000 max in. **diam**, 36 ksi (248 MPa) **TS**, 21 ksi (145 MPa) **YS**, 12% in 2 in. (50 mm) **El**; **As extruded**: 5.000 max in. **diam**, 35 ksi (241 MPa) **TS**, 14 ksi (97 MPa) **YS**, 12% in 2 in. (50 mm) **El**
NF A 50 411	5086 (A–G4MC)	France	Bar, wire, tube, and shapes	rem **Al**, 0.40 **Si**, 0.10 **Cu**, 0.20–0.70 **Mn**, 3.5–4.5 **Mg**, 0.05–0.25 **Cr**, 0.25 **Zn**, 0.15 **Ti**, 0.50 **Fe**, 0.05 each, 0.15 total, others	**1/4 hard (drawn bar)**: ≤25 mm **diam**, 270 MPa **TS**, 190 MPa **YS**, 4% in 2 in. (50 mm) **El**; **Hard (wire)**: ≤25 mm **diam**, 240 MPa **TS**, 95 MPa **YS**, 18% in 2 in. (50 mm) **El**; **Annealed (tube)**: ≤150 mm **diam**, 240 MPa **TS**, 95 MPa **YS**, 18% in 2 in. (50 mm) **El**
NF A 50 451	5086 (A–G4 MC)	France	Sheet, plate, and strip	rem **Al**, 0.40 **Si**, 0.10 **Cu**, 0.20–0.70 **Mn**, 3.5–4.5 **Mg**, 0.05–0.25 **Cr**, 0.25 **Zn**, 0.15 **Ti**, 0.50 **Fe**, 0.05 each, 0.15 total, others	**Annealed**: 0.4–6 mm **diam**, 240 MPa **TS**, 100 MPa **YS**, 10% in 2 in. (50 mm) **El**; **1/4 hard**: 0.4–1.6 mm **diam**, 280 MPa **TS**, 190 MPa **YS**, 8% in 2 in. (50 mm) **El**; **1/2 hard**: 0.4–1.6 mm **diam**, 310 MPa **TS**, 230 MPa **YS**, 7% in 2 in. (50 mm) **El**
JIS Z 3232	7N11	Japan	Rod and wire: welding and electrode	rem **Al**, 0.25 **Si**, 0.10 **Cu**, 0.20–0.7 **Mn**, 3.0–4.6 **Mg**, 0.30 **Cr**, 1.0–3.0 **Zn**, 0.20 **Ti**, 0.30 **Fe**, 0.0008 **Be**, 0.6 of one or more of **B**, **Ag**, **Zr**, and **V**	**Solution heat treated and age hardened**: all **diam**, 294 MPa **TS**
AMS 4018B		US	Sheet and plate	rem **Al**, 0.10 **Cu**, 0.10 **Mn**, 3.1–3.9 **Mg**, 0.15–0.35 **Cr**, 0.20 **Zn**, 0.20 **Ti**, 0.45 **Si+Fe**, 0.05 each, 0.15 total, others	**Annealed**: 0.020–0.031 in. **diam**, 30 ksi (207 MPa) **TS**, 11 ksi (76 MPa) **YS**, 12% in 2 in. (50 mm) **El**
AMS 4019		US	Sheet and plate	rem **Al**, 0.10 **Cu**, 0.10 **Mn**, 3.1–3.9 **Mg**, 0.15–0.35 **Cr**, 0.20 **Zn**, 0.20 **Ti**, 0.05 each, 0.15 total, others	**Strain hardened hard and stabilized**: <0.050 in. **diam**, 36 ksi (248 MPa) **TS**, 26 ksi (179 MPa) **YS**, 5% in 2 in. (50 mm) **El**
ANSI/ASTM B 547	5154	US	Tube: formed and arc-welded	rem **Al**, 0.10 **Cu**, 0.10 **Mn**, 3.1–3.9 **Mg**, 0.15–0.35 **Cr**, 0.20 **Zn**, 0.20 **Ti**, 0.45 **Si+Fe**, 0.05 each, 0.15 total, others	**Annealed**: 0.125–0.500 in. **diam**, 30 ksi (207 MPa) **TS**, 11 ksi (76 MPa) **YS**, 18% in 2 in. (50 mm) **El**; **H32**: 0.250–0.500 in. **diam**, 36 ksi (248 MPa) **TS**, 26 ksi (179 MPa) **YS**, 12% in 2 in. (50 mm) **El**; **H34**: 0.250–0.500 in. **diam**, 39 ksi (269 MPa) **TS**, 29 ksi (200 MPa) **YS**, 10% in 2 in. (50 mm) **El**

WROUGHT ALUMINUM AND ALUMINUM ALLOYS

specification number	designation	country	product forms	chemical composition	mechanical properties and hardness values
ANSI/ASTM B 313	5154	US	Tube: welded	rem **Al**, 0.10 **Cu**, 0.10 **Mn**, 3.1–3.9 **Mg**, 0.15–0.35 **Cr**, 0.20 **Zn**, 0.20 **Ti**, 0.45 **Fe+Si**, 0.05 each, 0.15 total, others	**Annealed**: 0.051–0.113 in. **diam**, 30 ksi (207 MPa) **TS**, 11 ksi (76 MPa) **YS**, 16% in 2 in. (50 mm) **El**; **H32**: 0.051–0.125 in. **diam**, 36 ksi (248 MPa) **TS**, 26 ksi (179 MPa) **YS**, 8% in 2 in. (50 mm) **El**; **H34**: 0.051–0.125 in. **diam**, 39 ksi (269 MPa) **TS**, 29 ksi (200 MPa) **YS**, 6% in 2 in. (50 mm) **El**
ANSI/ASTM B 241	5254	US	Tube: seamless	rem **Al**, 0.05 **Cu**, 0.01 **Mn**, 3.1–3.9 **Mg**, 0.15–0.35 **Cr**, 0.20 **Zn**, 0.05 **Ti**, 0.45 **Si+Fe**, 0.05 each, 0.15 total, others	**Annealed**: all **diam**, 30 ksi (207 MPa) **TS**, 11 ksi (76 MPa) **YS**; **H112**: all **diam**, 30 ksi (207 MPa) **TS**, 11 ksi (76 MPa) **YS**
ANSI/ASTM B 221	5154	US	Bar, tube, rod, wire, and shapes	rem **Al**, 0.10 **Cu**, 0.10 **Mn**, 3.1–3.9 **Mg**, 0.15–0.35 **Cr**, 0.20 **Zn**, 0.20 **Ti**, 0.40 **Si+Fe**, 0.05 each, 0.15 total, others	**Annealed**: all **diam**, 30 ksi (207 MPa) **TS**, 11 ksi (76 MPa) **YS**; **H112**: all **diam**, 30 ksi (207 MPa) **TS**, 11 ksi (76 MPa) **YS**
ANSI/ASTM B 211	5154	US	Bar, rod, and wire	rem **Al**, 0.10 **Cu**, 0.10 **Mn**, 3.1–3.9 **Mg**, 0.15–0.35 **Cr**, 0.20 **Zn**, 0.20 **Ti**, 0.45 **Si+Fe**, 0.05 each, 0.15 total, others	**Annealed**: all **diam**, 30 ksi (207 MPa) **TS**, 11 ksi (76 MPa) **YS**, 25% in 2 in. (50 mm) **El**; **H32**: ≤0.374 in. **diam**, 36 ksi (248 MPa) **TS**; **H34**: ≤0.374 in. **diam**, 39 ksi (269 MPa) **TS**
ANSI/ASTM B 210	5154	US	Tube: seamless	rem **Al**, 0.25 **Si**, 0.10 **Cu**, 0.10 **Mn**, 3.1–3.9 **Mg**, 0.15–0.35 **Cr**, 0.20 **Zn**, 0.20 **Ti**, 0.40 **Fe**, 0.05 each, 0.15 total, others	**Annealed**: 0.010–0.450 in. **diam**, 30 ksi (207 MPa) **TS**, 11 ksi (76 MPa) **YS**, 10% in 2 in. (50 mm) **El**; **H34 (round tube)**: 0.010–0.450 in. **diam**, 39 ksi (269 MPa) **TS**, 29 ksi (200 MPa) **YS**, 5% in 2 in. (50 mm) **El**; **H38 (round tube)**: 0.010–0.450 in. **diam**, 45 ksi (310 MPa) **TS**, 34 ksi (234 MPa) **YS**
ANSI/ASTM B 209	5154	US	Sheet and plate	rem **Al**, 0.25 **Si**, 0.10 **Cu**, 0.10 **Mn**, 3.1–3.9 **Mg**, 0.15–0.35 **Cr**, 0.20 **Zn**, 0.20 **Ti**, 0.40 **Fe**, 0.05 each, 0.15 total, others	**Annealed**: 0.114–3.000 in. **diam**, 30 ksi (207 MPa) **TS**, 11 ksi (76 MPa) **YS**, 18% in 2 in. (50 mm) **El**; **H32**: 0.250–2.000 in. **diam**, 36 ksi (248 MPa) **TS**, 26 ksi (179 MPa) **YS**, 12% in 2 in. (50 mm) **El**; **H34**: 0.250–1.000 in. **diam**, 39 ksi (269 MPa) **TS**, 29 ksi (200 MPa) **YS**, 10% in 2 in. (50 mm) **El**

WROUGHT ALUMINUM AND ALUMINUM ALLOYS

specification number	designation	country	product forms	chemical composition	mechanical properties and hardness values
ANSI/ASTM B 209	5254	US	Sheet and plate	rem **Al**, 0.05 **Cu**, 0.01 **Mn**, 3.1–3.9 **Mg**, 0.15–0.35 **Cr**, 0.20 **Zn**, 0.45 **Si+Fe**, 0.05 each, 0.15 total, others	**Annealed**: 0.114–3.000 in. **diam**, 30 ksi (207 MPa) **TS**, 11 ksi (76 MPa) **YS**, 18% in 2 in. (50 mm) **El**; **H32**: 0.250–2.000 in. **diam**, 36 ksi (248 MPa) **TS**, 26 ksi (179 MPa) **YS**, 12% in 2 in. (50 mm) **El**; **H34**: 0.250–1.000 in. **diam**, 39 ksi (269 MPa) **TS**, 29 ksi (200 MPa) **YS**, 10% in 2 in. (50 mm) **El**
AS 1866	5154A	Australia	Rod, bar, shape	rem **Al**, 0.50 **Si**, 0.10 **Cu**, 0.10–0.50 **Mn**, 3.1–3.9 **Mg**, 0.25 **Cr**, 0.20 **Zn**, 0.20 **Ti**, 0.50 **Fe**, 0.05 each, 0.15 total, others	**Strain–hardened**: 50.0 mm **diam**, 217 MPa **TS**, 100 MPa **YS**, 16% in 2 in. (50 mm) **El**
AS 1734	5154A	Australia	Sheet, plate	rem **Al**, 0.50 **Si**, 0.10 **Cu**, 0.10–0.50 **Mn**, 3.1–3.9 **Mg**, 0.25 **Cr**, 0.20 **Zn**, 0.20 **Ti**, 0.50 **Fe**, 0.05 each, 0.15 total, others	**Annealed**: 0.50–0.80 mm **diam**, 216 MPa **TS**, 75 MPa **YS**, 12% in 2 in. (50 mm) **El**; **Strain–hardened and stabilized to 1/4 hard**: 0.50–1.30 mm **diam**, 248 MPa **TS**, 179 MPa **YS**, 5% in 2 in. (50 mm) **El**; **Strain–hardened and stabilized to 1/2 hard**: 0.25–1.30 mm **diam**, 279 MPa **TS**, 224 MPa **TS**, 4% in 2 in. (50 mm) **El**
BS 1471	NT5	UK	Tube: drawn	rem **Al**, 0.5 **Si**, 0.10 **Cu**, 0.5 **Mn**, 3.1–3.9 **Mg**, 0.25 **Cr**, 0.2 **Zn**, 0.2 **Ti**, 0.5 **Fe**, 0.5 **Mn+Cr**	**Annealed**: ≤10.0 mm **diam**, 215 MPa **TS**, 85 MPa **YS**, 16% in 2 in. (50 mm) **El**; **Strain–hardened to 1/2 hard**: ≤10.0 mm **diam**, 245 MPa **TS**, 200 MPa **YS**, 4% in 2 in. (50 mm) **El**
BS 1475	NG5	UK	Wire	rem **Al**, 0.5 **Si**, 0.10 **Cu**, 0.5 **Mn**, 3.1–3.9 **Mg**, 0.25 **Cr**, 0.2 **Zn**, 0.2 **Ti**, 0.5 **Fe**, 0.5 **Mn+Cr**	
BS 1473	NR5	UK	Rivet	rem **Al**, 0.5 **Si**, 0.10 **Cu**, 0.5 **Mn**, 3.1–3.9 **Mg**, 0.25 **Cr**, 0.2 **Zn**, 0.2 **Ti**, 0.5 **Fe**, 0.5 **Mn+Cr**	**Annealed or as manufactured**: ≤25 mm **diam**, 215 MPa **TS**
BS 1470	NS5	UK	Plate, sheet, strip	rem **Al**, 0.5 **Si**, 0.10 **Cu**, 0.5 **Mn**, 3.1–3.9 **Mg**, 0.25 **Cr**, 0.2 **Zn**, 0.2 **Ti**, 0.5 **Fe**, 0.5 **Mn+Cr**	**Annealed**: 0.2–6.0 mm **diam**, 215 MPa **TS**, 85 MPa **YS**, 12% in 2 in. (50 mm) **El**; **Strain hardened to 1/4 hard**: 0.2–6.0 mm **diam**, 245 MPa **TS**, 165 MPa **YS**, 5% in 2 in. (50 mm) **El**; **Strain hardened to 1/2 hard**: 0.2–6.0 mm **diam**, 275 MPa **TS**, 225 MPa **YS**, 4% in 2 in. (50 mm) **El**
BS 1472	NF5	UK	Forging stock and forgings	rem **Al**, 0.5 **Si**, 0.10 **Cu**, 0.5 **Mn**, 3.1–3.9 **Mg**, 0.25 **Cr**, 0.2 **Zn**, 0.2 **Ti**, 0.5 **Fe**, 0.5 **Mn+Cr**	**As manufactured**: ≤150 mm **diam**, 215 MPa **TS**, 100 MPa **YS**, 16% in 2 in. (50 mm) **El**

WROUGHT ALUMINUM AND ALUMINUM ALLOYS

specification number	designation	country	product forms	chemical composition	mechanical properties and hardness values
BS 1474	NE5	UK	Bar, tube, section	rem **Al**, 0.5 **Si**, 0.10 **Cu**, 0.5 **Mn**, 3.1–3.9 **Mg**, 0.25 **Cr**, 0.2 **Zn**, 0.2 **Ti**, 0.5 **Fe**, 0.5 **Mn+Cr**	**Annealed**: ≤150 mm **diam**, 215 MPa **TS**, 85 MPa **YS**, 16% in 2 in. (50 mm) **El**; **As manufactured**: ≤150 mm **diam**, 215 MPa **TS**, 100 MPa **TS**, 14% in 2 in. (50 mm) **El**
CSA HA.6	.5654	Canada	Rod, wire: for welding and brazing	rem **Al**, 0.05 **Cu**, 0.01 **Mn**, 3.1–3.9 **Mg**, 0.15–0.35 **Cr**, 0.20 **Zn**, 0.05–0.15 **Ti**, 0.008 **Be**, 0.45 **Si+Fe**, 0.05 each, 0.15 total, others	
ISO R827	Al–Mg3.5	ISO	Extruded products	rem **Al**, 0.5 **Si**, 0.10 **Cu**, 0.6 **Mn**, 3.1–3.9 **Mg**, 0.35 **Cr**, 0.2 **Zn**, 0.5 **Fe**, 0.2 **Ti+Zr**	**As manufactured**: 5.0 max in. (127 max mm) **diam**, 206 MPa **TS**, 79 MPa **YS**, 12% in 2 in. (50 mm) **El**
ISO R209	Al–Mg3.5	ISO	Wrought products	rem **Al**, 0.5 **Si**, 0.10 **Cu**, 0.6 **Mn**, 3.1–3.9 **Mg**, 0.35 **Cr**, 0.2 **Zn**, 0.5 **Fe**, 0.2 **Ti+Zr**	
ISO TR2136	Al–Mg3.5	ISO	Sheet, plate	rem **Al**, 0.5 **Si**, 0.10 **Cu**, 0.6 **Mn**, 3.1–3.9 **Mg**, 0.35 **Cr**, 0.2 **Zn**, 0.5 **Fe**, 0.2 **Ti+Zr**	**Annealed**: all **diam**, 210 MPa **TS**, 75 MPa **YS**, 12% in 2 in. (50 mm) **El**; **HB–Strain hardened**: all **diam**, 235 MPa **TS**, 165 MPa **YS**, 5% in 2 in. (50 mm) **El**; **HD– Strain hardened**: all **diam**, 260 MPa **TS**, 195 MPa **YS**, 4% in 2 in. (50 mm) **El**
JIS H 4000	5154	Japan	Sheet, plate, strip, and coil	rem **Al**, 0.10 **Cu**, 0.10 **Mn**, 3.1–3.9 **Mg**, 0.15–0.35 **Cr**, 0.20 **Zn**, 0.20 **Ti**, 0.45 **Si+Fe**, 0.15 total, others	**Annealed**: <1.3 mm **diam**, 206 MPa **TS**, 78 MPa **YS**, 14% in 2 in. (50 mm) **El**; **Hard**: <1.3 mm **diam**, 314 MPa **TS**, 245 MPa **YS**, 3% in 2 in. (50 mm) **El**; **Hot rolled**: <12 mm **diam**, 226 MPa **TS**, 127 MPa **YS**, 8% in 2 in. (50 mm) **El**
JIS H 4080	5154	Japan	Tube: extruded	rem **Al**, 0.10 **Cu**, 0.10 **Mn**, 3.1–3.9 **Mg**, 0.15–0.35 **Cr**, 0.20 **Zn**, 0.20 **Ti**, 0.45 **Si+Fe**, 0.15 total, others	**As manufactured**: all **diam**, 206 MPa **TS**, 78 MPa **YS**; **Annealed**: all **diam**, 284 MPa **TS**, 78 MPa **YS**
JIS H 4120	5N02	Japan	Wire and rod: rivet	rem **Al**, 0.40 **Si**, 0.10 **Cu**, 0.30–1.0 **Mn**, 3.0–4.0 **Mg**, 0.50 **Cr**, 0.10 **Zn**, 0.20 **Ti**, 0.40 **Fe**, 0.15 total, others	**As manufactured**: ≤25 mm **diam**, 266 MPa **TS**, 20% in 2 in. (50 mm) **El**
JIS Z 3232	5154	Japan	Rod and wire: welding and electrode	rem **Al**, 0.10 **Cu**, 0.10 **Mn**, 3.1–3.9 **Mg**, 0.15–0.35 **Cr**, 0.20 **Zn**, 0.20 **Ti**, 0.0008 **Be**, 0.45 **Si+Fe**	**Annealed**: all **diam**, 206 MPa **TS**
SIS 14 41 33	SIS Al 41 33–00	Sweden	Wire	rem **Al**, 0.5 **Si**, 0.10 **Cu**, 0.6 **Mn**, 3.1–3.9 **Mg**, 0.35 **Cr**, 0.2 **Zn**, 0.5 **Fe**, 0.2 grain refiners	
SIS 14 41 33	SIS Al 41 33–18	Sweden	Wire	rem **Al**, 0.5 **Si**, 0.10 **Cu**, 0.6 **Mn**, 3.1–3.9 **Mg**, 0.35 **Cr**, 0.2 **Zn**, 0.5 **Fe**, 0.2 grain refiners	**Hard worked**: all **diam**, 255 MPa **TS**

WROUGHT ALUMINUM AND ALUMINUM ALLOYS

specification number	designation	country	product forms	chemical composition	mechanical properties and hardness values
SIS 14 41 34	SIS Al 41 34–00	Sweden	Wire	rem **Al**, 0.4 **Si**, 0.10 **Cu**, 0.05 **Mn**, 3.1–3.9 **Mg**, 0.05 **Cr**, 0.2 **Zn**, 0.05 **Ti**, 0.4 **Fe**, **Cr** in **Ti** content	
SIS 14 41 34	SIS Al 41 34–18	Sweden	Wire	rem **Al**, 0.4 **Si**, 0.10 **Cu**, 0.05 **Mn**, 3.1–3.9 **Mg**, 0.05 **Cr**, 0.2 **Zn**, 0.05 **Ti**, 0.4 **Fe**, **Cr** in **Ti** content	**Hard worked**: all **diam**, 255 MPa **TS**
SABS 712	Al–Mg3.5	South Africa	Wrought products	rem **Al**, 0.5 **Si**, 0.10 **Cu**, 0.5 **Mn**, 3.1–3.9 **Mg**, 0.25 **Cr**, 0.2 **Zn**, 0.5 **Fe**, 0.2 **Ti** + total, others	
IS 739	NG5	India	Wire	rem **Al**, 0.6 **Si**, 0.10 **Cu**, 0.6 **Mn**, 2.8–4.0 **Mg**, 0.25 **Cr**, 0.2 **Zn**, 0.7 **Fe**	
IS 733	53000	India	Bar, rod, section	rem **Al**, 0.6 **Si**, 0.1 **Cu**, 0.5 **Mn**, 2.8–4.0 **Mg**, 0.25 **Cr**, 0.2 **Zn**, 0.2 **Ti**, 0.5 **Fe**, 0.5 **Cr+Mn**	**As manufactured**: $\leq$50 mm **diam**, 215 MPa **TS**, 100 MPa **YS**, 14% in 2 in. (50 mm) **El**
AMS 4171B		US	Bar, tube, rod, wire, and shapes: extruded	rem **Al**, 0.30 **Si**, 0.40–0.8 **Cu**, 0.10–0.30 **Mn**, 2.9–3.7 **Mg**, 0.10–0.25 **Cr**, 3.8–4.8 **Zn**, 0.10 **Ti**, 0.40 **Fe**, 0.05 each, 0.15 total, others	**Solution and precipitation heat treated**: $<$0.250 in. **diam**, 75 ksi (517 MPa) **TS**, 67 ksi (462 MPa) **YS**, 7% in 2 in. (50 mm) **El**, 130 min **HB**
COPANT 862	X 5854	COPANT	Wrought products	rem **Al**, 0.30 **Si**, 0.10 **Cu**, 0.30 **Mn**, 2.5–3.7 **Mg**, 0.10 **Zn**, 0.40 **Fe**, 0.05 each, 0.15 total, others	
COPANT 862	X 5754	COPANT	Wrought products	rem **Al**, 0.40 **Si**, 0.10 **Cu**, 0.10–0.50 **Mn**, 2.6–3.6 **Mg**, 0.10 **Cr**, 0.20 **Zn**, 0.15 **Ti**, 0.05 each, 0.15 total, others	
NF A 50 411	5754 (A–G3M)	France	Bar, wire, tube, and shapes	rem **Al**, 0.40 **Si**, 0.10 **Cu**, 0.50 **Mn**, 2.6–3.6 **Mg**, 0.30 **Cr**, 0.20 **Zn**, 0.15 **Ti**, 0.40 **Fe**, 0.10–0.60 **Cr+Mn**, 0.05 each, 0.15 total, others	**Hard (drawn bar)**: $\leq$10 mm **diam**, 280 MPa **TS**, 240 MPa **YS**, 2% in 2 in. (50 mm) **El**; **Hard (wire)**: $\leq$6 mm **diam**, 280 MPa **TS**, 240 MPa **YS**, 2% in 2 in. (50 mm) **El**; **Annealed (tube)**: $\leq$150 mm **diam**, 200 MPa **TS**, 80 MPa **YS**, 17% in 2 in. (50 mm) **El**
NF A 50 451	5754 X (A–G 3M)	France	Sheet, plate, and strip	rem **Al**, 0.40 **Si**, 0.10 **Cu**, 0.10–0.50 **Mn**, 2.6–3.6 **Mg**, 0.10 **Cr**, 0.20 **Zn**, 0.15 **Ti**, 0.50 **Fe**, 0.05 each, 0.15 total, others	**Annealed**: 0.4–1.6 mm **diam**, 190 MPa **TS**, 80 MPa **YS**, 20% in 2 in. (50 mm) **El**; **1/2 hard**: 0.4–3.2 mm **diam**, 240 MPa **TS**, 160 MPa **YS**, 8% in 2 in. (50 mm) **El**; **3/4 hard**: 0.4–3.2 mm **diam**, 260 MPa **TS**, 190 MPa **YS**, 7% in 2 in. (50 mm) **El**

WROUGHT ALUMINUM AND ALUMINUM ALLOYS

specification number	designation	country	product forms	chemical composition	mechanical properties and hardness values
SFS 2588	AlMg3	Finland	Shapes, sheet, rod, tube, and wire	rem **Al**, 0.5 **Si**, 0.10 **Cu**, 0.5 **Mn**, 2.6–3.4 **Mg**, 0.35 **Cr**, 0.2 **Zn**, 0.5 **Fe**, 0.2 **Ti+Zr**, 0.05 each, 0.15 total, others	**Hot finished (shapes)**: ≤50 mm **diam**, 220 MPa **TS**, 80 MPa **YS**, 14% in 2 in. (50 mm) **El**; **Annealed (sheet)**: all **diam**, 240 max MPa **TS**, 70 MPa **YS**, 18% in 2 in. (50 mm) **El**; **Hard (rod, tube, and wire)**: ≤6 mm **diam**, 220 MPa **TS**, 180 MPa **YS**
SABS 903	SABS 903	South Africa	Sheet: corrugated and troughed	rem **Al**, 0.6 **Si**, 0.2 **Cu**, 1.5 **Mn**, 3.0 **Mg**, 0.25 **Zn**, 0.7 **Fe**, 0.20 **Ti** + others	**As manufactured**: 1.12–1.28 mm **diam**, 175 MPa **TS**
SABS 712	Al–Mg3Mn	South Africa	Wrought products	rem **Al**, 0.10 **Cu**, 1.0 **Mn**, 3.0 **Mg**, 0.20 **Cr**, 0.25 **Zn**, 0.40 **Fe+Si**, 0.20 **Ti** + total, others	
ISO R827	Al–Mg3Mn	ISO	Extruded products	rem **Al**, 0.5 **Si**, 0.10 **Cu**, 0.3–1.0 **Mn**, 2.4–3.4 **Mg**, 0.25 **Cr**, 0.2 **Zn**, 0.5 **Fe**, 0.2 **Ti+Zr**	**As manufactured**: 5.0 max in. (127 max mm) **diam**, 206 MPa **TS**, 79 MPa **YS**, 12% in 2 in. (50 mm) **El**
ISO R209	Al–Mg3Mn	ISO	Wrought products	rem **Al**, 0.5 **Si**, 0.10 **Cu**, 0.3–1.0 **Mn**, 2.4–3.4 **Mg**, 0.25 **Cr**, 0.2 **Zn**, 0.5 **Fe**, 0.2 **Ti+Zr**	
ISO TR2136	Al–Mg3Mn	ISO	Sheet, plate	rem **Al**, 0.5 **Si**, 0.10 **Cu**, 0.3–1.0 **Mn**, 2.4–3.4 **Mg**, 0.25 **Cr**, 0.2 **Zn**, 0.5 **Fe**, 0.2 **Ti+Zr**	**Annealed**: all **diam**, 200 MPa **TS**, 75 MPa **YS**, 12% in 2 in. (50 mm) **El**; **HB–Strain hardened**: all **diam**, 235 MPa **TS**, 165 MPa **YS**, 4% in 2 in. (50 mm) **El**; **HD– Strain hardened**: all **diam**, 260 MPa **TS**, 195 MPa **YS**, 2% in 2 in. (50 mm) **El**
DS 3012	5454	Denmark	Rolled and drawn products	rem **Al**, 0.10 **Cu**, 0.50–1.0 **Mn**, 2.4–3.1 **Mg**, 0.05–0.20 **Cr**, 0.25 **Zn**, 0.20 **Ti**, 0.40 **Fe+Si**, 0.05 each, 0.15 total, others	**Soft**: 0.5–25 mm **diam**, 175 MPa **TS**, 70 MPa **YS**, 18% in 2 in. (50 mm) **El**, 50 **HV**; **1/4 hard**: 0.5–25 mm **diam**, 235 MPa **TS**, 175 MPa **YS**, 7% in 2 in. (50 mm) **El**, 75 **HV**; **1/2 hard**: 0.5–6 mm **diam**, 265 MPa **TS**, 215 MPa **YS**, 4% in 2 in. (50 mm) **El**, 85 **HV**
ISO R209	Al–Mg3	ISO	Wrought products	rem **Al**, 0.5 **Si**, 0.10 **Cu**, 0.4 **Mn**, 2.4–3.1 **Mg**, 0.35 **Cr**, 0.2 **Zn**, 0.5 **Fe**, 0.2 **Ti+Zr**	
ISO TR2778	Al–Mg3	ISO	Tube: drawn	rem **Al**, 0.5 **Si**, 0.10 **Cu**, 0.4 **Mn**, 2.4–3.1 **Mg**, 0.35 **Cr**, 0.2 **Zn**, 0.5 **Fe**, 0.2 **Ti+Zr**	**Annealed**: all **diam**, 180 MPa **TS**, 70 MPa **YS**, 15% in 2 in. (50 mm) **El**; **Strain hardened**: 80 mm **diam**, 215 MPa **TS**, 140 MPa **YS**, 3% in 2 in. (50 mm) **El**

WROUGHT ALUMINUM AND ALUMINUM ALLOYS

specification number	designation	country	product forms	chemical composition	mechanical properties and hardness values
ISO TR2136	Al–Mg3	ISO	Sheet, plate	rem **Al**, 0.5 **Si**, 0.10 **Cu**, 0.4 **Mn**, 2.4–3.1 **Mg**, 0.35 **Cr**, 0.2 **Zn**, 0.5 **Fe**, 0.2 **Ti+Zr**	**Annealed**: all **diam**, 180 MPa **TS**, 75 MPa **YS**, 17% in 2 in. (50 mm) **El**; **HB–Strain hardened**: all **diam**, 215 MPa **TS**, 140 MPa **YS**, 3% in 2 in. (50 mm) **El**; **HD– Strain hardened**: all **diam**, 245 MPa **TS**, 180 MPa **YS**, 2% in 2 in. (50 mm) **El**
ANSI/ASTM B 547	5454	US	Tube: formed and arc–welded	rem **Al**, 0.10 **Cu**, 0.50–1.0 **Mn**, 2.4–3.0 **Mg**, 0.05–0.20 **Cr**, 0.25 **Zn**, 0.20 **Ti**, 0.40 **Si+Fe**, 0.05 each, 0.15 total, others	**Annealed**: 0.125–0.500 in. **diam**, 31 ksi (241 MPa) **TS**, 12 ksi (83 MPa) **YS**, 18% in 2 in. (50 mm) **El**; **H32**: 0.250–0.500 in. **diam**, 36 ksi (248 MPa) **TS**, 26 ksi (179 MPa) **YS**, 12% in 2 in. (50 mm) **El**; **H34**: 0.250–0.500 in. **diam**, 39 ksi (269 MPa) **TS**, 29 ksi (200 MPa) **YS**, 10% in 2 in. (50 mm) **El**
ANSI/ASTM B 404	5454	US	Tube: seamless, for condensers and heat exchangers	rem **Al**, 0.10 **Cu**, 0.50–1.0 **Mn**, 2.4–3.0 **Mg**, 0.05–0.20 **Cr**, 0.25 **Zn**, 0.20 **Ti**, 0.40 **Si+Fe**, 0.05 each, 0.15 total, others	**Annealed**: 0.010–0.200 in. **diam**, 31 ksi (214 MPa) **TS**, 12 ksi (83 MPa) **YS**; **H32 (round tube)**: 36 ksi (248 MPa) **TS**, 26 ksi (179 MPa) **YS**; **H34 (round tube)**: 39 ksi (269 MPa) **TS**, 29 ksi (200 MPa) **YS**
ANSI/ASTM B 241	5454	US	Tube: seamless	rem **Al**, 0.10 **Cu**, 0.50–1.0 **Mn**, 2.4–3.0 **Mg**, 0.05–0.20 **Cr**, 0.25 **Zn**, 0.20 **Ti**, 0.40 **Si+Fe**, 0.05 each, 0.15 total, others	**Annealed**: ≤5.000 in. **diam**, 31 ksi (214 MPa) **TS**, 12 ksi (83 MPa) **YS**, 14% in 2 in. (50 mm) **El**; **H111**: ≤5.000 in. **diam**, 33 ksi (228 MPa) **TS**, 19 ksi (131 MPa) **YS**, 12% in 2 in. (50 mm) **El**; **H112**: ≤5.000 in. **diam**, 31 ksi (214 MPa) **TS**, 12 ksi (83 MPa) **YS**, 12% in 2 in. (50 mm) **El**
ANSI/ASTM B 234	5454	US	Tube: seamless, for condensers and heat exchangers	rem **Al**, 0.25 **Si**, 0.10 **Cu**, 0.50–1.0 **Mn**, 2.4–3.0 **Mg**, 0.05–0.20 **Cr**, 0.25 **Zn**, 0.20 **Ti**, 0.40 **Fe**, 0.05 each, 0.15 total, others	**H32**: 0.051–0.200 in. **diam**, 36 ksi (248 MPa) **TS**, 26 ksi (179 MPa) **YS**, 8% in 2 in. (50 mm) **El**; **H34**: 0.051–0.200 in. **diam**, 39 ksi (269 MPa) **TS**, 29 ksi (200 MPa) **YS**, 6% in 2 in. (50 mm) **El**
ANSI/ASTM B 221	5454	US	Bar, tube, rod, wire, and shapes	rem **Al**, 0.10 **Cu**, 0.50–1.0 **Mn**, 2.4–3.0 **Mg**, 0.05–0.20 **Cr**, 0.25 **Zn**, 0.20 **Ti**, 0.40 **Si+Fe**, 0.05 each, 0.15 total, others	**Annealed**: ≤5.000 in. **diam**, 31 ksi (214 MPa) **TS**, 12 ksi (83 MPa) **YS**, 14% in 2 in. (50 mm) **El**; **H111**: ≤5.000 in. **diam**, 33 ksi (228 MPa) **TS**, 19 ksi (131 MPa) **YS**, 12% in 2 in. (50 mm) **El**; **H112**: ≤5.000 in. **diam**, 31 ksi (214 MPa) **TS**, 12 ksi (83 MPa) **YS**, 12% in 2 in. (50 mm) **El**

WROUGHT ALUMINUM AND ALUMINUM ALLOYS

specification number	designation	country	product forms	chemical composition	mechanical properties and hardness values
ANSI/ASTM B 209	5454	US	Sheet and plate	rem **Al**, 0.25 **Si**, 0.10 **Cu**, 0.50–1.0 **Mn**, 2.4–3.0 **Mg**, 0.05–0.20 **Cr**, 0.25 **Zn**, 0.20 **Ti**, 0.40 **Fe**, 0.05 each, 0.15 total, others	**Annealed**: 0.114–3.000 in. **diam**, 31 ksi (214 MPa) **TS**, 12 ksi (83 MPa) **YS**, 12% in 2 in. (50 mm) **El**; **H32**: 0.250–2.000 in. **diam**, 36 ksi (248 MPa) **TS**, 26 ksi (179 MPa) **YS**, 12% in 2 in. (50 mm) **El**; **H34**: 0.250–1.000 in. **diam**, 39 ksi (269 MPa) **TS**, 29 ksi (200 MPa) **YS**, 10% in 2 in. (50 mm) **El**
AS 1734	5454	Australia	Sheet, plate	rem **Al**, 0.10 **Cu**, 0.50–1.0 **Mn**, 2.4–3.0 **Mg**, 0.05–0.20 **Cr**, 0.25 **Zn**, 0.20 **Ti**, 0.05 each, 0.15 total, others	**Annealed**: 0.15–0.80 mm **diam**, 213 MPa **TS**, 82 MPa **YS**, 12% in 2 in. (50 mm) **El**; **Strain–hardened and stabilized to 1/4 hard**: 0.05–1.30 mm **diam**, 248 MPa **TS**, 179 MPa **YS**, 5% in 2 in. (50 mm) **El**; **Strain–hardened and stabilized to 1/2 hard**: 0.05–1.30 mm **diam**, 268 MPa **TS**, 199 MPa **TS**, 4% in 2 in. (50 mm) **El**
CSA HA.6	.5554	Canada	Rod, wire: for welding and brazing	rem **Al**, 0.10 **Cu**, 0.50–1.0 **Mn**, 2.4–3.0 **Mg**, 0.05–0.20 **Cr**, 0.25 **Zn**, 0.05–0.20 **Ti**, 0.008 **Be**, 0.40 **Si+Fe**, 0.05 each, 0.15 total, others	
CSA HA.4	.5454	Canada	Plate, sheet	rem **Al**, 0.10 **Cu**, 0.50–1.0 **Mn**, 2.4–3.0 **Mg**, 0.05–0.20 **Cr**, 0.25 **Zn**, 0.20 **Ti**, 0.40 **Fe+Si**, 0.05 each, 0.15 total, others	**O**: 0.020–0.031 in. **diam**, 31.0 ksi **TS**, 12.0 ksi **YS**, 12% in 2 in. (50 mm) **El**; **H32**: 0.020–0.050 in. **diam**, 36.0 ksi **TS**, 26.0 ksi **YS**, 5% in 2 in. (50 mm) **El**; **H112**: 0.250–0.499 in. **diam**, 32.0 ksi **TS**, 18.0 ksi **YS**, 8% in 2 in. (50 mm) **El**
COPANT 862	5454	COPANT	Wrought products	rem **Al**, 0.25 **Si**, 0.10 **Cu**, 0.05–1.0 **Mn**, 2.4–3.0 **Mg**, 0.05–0.20 **Cr**, 0.25 **Zn**, 0.20 **Ti**, 0.40 **Fe**, 0.05 each, 0.15 total, others	
QQ-A-250/10D	5454	US	Plate, sheet	rem **Al**, 0.10 **Cu**, 0.50–1.0 **Mn**, 2.4–3.0 **Mg**, 0.05–0.20 **Cr**, 0.25 **Zn**, 0.20 **Ti**, 0.40 **Fe**, **Si** in **Fe** content; 0.05 each, 0.15 total, others	**Annealed**: 0.020–0.031 in. **diam**, 31 ksi (214 MPa) **TS**, 12 ksi (83 MPa) **YS**, 12% in 2 in. (50 mm) **El**; **Strain hardened 1/4 hard temper and stabilized**: 0.020–0.050 in. **diam**, 36 ksi (248 MPa) **TS**, 26 ksi (179 MPa) **YS**, 5% in 2 in. (50 mm) **El**; **Strain hardened 1/2 hard temper and stabilized**: 0.020–0.050 in. **diam**, 39 ksi (269 MPa) **TS**, 29 ksi (200 MPa) **YS**, 4% in 2 in. (50 mm) **El**

WROUGHT ALUMINUM AND ALUMINUM ALLOYS

specification number	designation	country	product forms	chemical composition	mechanical properties and hardness values
QQ-A-200/6D	5454	US	Bar, rod, shapes, tube, wire	rem **Al**, 0.10 **Cu**, 0.50–1.0 **Mn**, 2.4–3.0 **Mg**, 0.05–0.20 **Cr**, 0.25 **Zn**, 0.20 **Ti**, 0.40 **Fe**, **Si** in **Fe** content; 0.05 each, 0.15 total, others	**Annealed**: 5.000 max in. **diam**, 31 ksi (214 MPa) **TS**, 12 ksi (83 MPa) **YS**, 14% in 2 in. (50 mm) **El**; **Strain hardened**: 5.000 max in. **diam**, 33 ksi (228 MPa) **TS**, 19 ksi (131 MPa) **YS**, 12% in 2 in. (50 mm) **El**; **As extruded**: 5.000 max in. **diam**, 31 ksi (214 MPa) **TS**, 12 ksi (83 MPa) **YS**, 12% in 2 in. (50 mm) **El**
NF A 50 411	5454	France	Bar, wire, and tube	rem **Al**, 0.25 **Si**, 0.10 **Cu**, 0.50–1.0 **Mn**, 2.4–3.0 **Mg**, 0.05–0.20 **Cr**, 0.25 **Zn**, 0.20 **Ti**, 0.40 **Fe**, 0.05 each, 0.15 total, others	**Annealed (extruded bar)**: 215 MPa **TS**, 80 MPa **YS**, 16% in 2 in. (50 mm) **El**; **Annealed (wire)**: ≤25 mm **diam**, 215 MPa **TS**, 80 MPa **YS**, 16% in 2 in. (50 mm) **El**; **Annealed (tube)**: ≤150 mm **diam**, 215 MPa **TS**, 80 MPa **YS**, 16% in 2 in. (50 mm) **El**
NF A 50 451	5454	France	Sheet, plate, and strip	rem **Al**, 0.10 **Cu**, 0.50–1.0 **Mn**, 2.4–3.0 **Mg**, 0.05–0.20 **Cr**, 0.25 **Zn**, 0.20 **Ti**, 0.40 **Fe+Si**, 0.05 each, 0.15 total, others	**Annealed**: 0.4–6 mm **diam**, 210 MPa **TS**, 80 MPa **YS**, 19% in 2 in. (50 mm) **El**; **1/4 hard**: 0.4–25 mm **diam**, 250 MPa **TS**, 180 MPa **YS**, 9% in 2 in. (50 mm) **El**; **1/2 hard**: 0.4–25 mm **diam**, 270 MPa **TS**, 200 MPa **YS**, 8% in 2 in. (50 mm) **El**
JIS Z 3232	5554	Japan	Rod and wire: welding and electrode	rem **Al**, 0.10 **Cu**, 0.50–1.0 **Mn**, 2.4–3.0 **Mg**, 0.05–0.20 **Cr**, 0.25 **Zn**, 0.05–0.20 **Ti**, 0.0008 **Be**, 0.40 **Si+Fe**	**Annealed**: all **diam**, 177 MPa **TS**
AMS 4069		US	Tube: seamless	rem **Al**, 0.10 **Cu**, 0.10 **Mn**, 2.2–2.8 **Mg**, 0.15–0.35 **Cr**, 0.10 **Zn**, 0.5 **Fe+Si**, 0.05 each, 0.15 total, others	**Annealed**: all **diam**, 35 ksi (241 MPa) **TS**, 20 ksi (137 MPa) **YS**, 12% in 2 in. (50 mm) **El**
AMS 4070G		US	Tube: seamless	rem **Al**, 0.10 **Cu**, 0.10 **Mn**, 2.2–2.8 **Mg**, 0.15–0.35 **Cr**, 0.10 **Zn**, 0.45 **Fe+Si**, 0.05 each, 0.15 total, others	**Annealed**: 0.010–0.450 in. **diam**, 25 ksi (172 MPa) **TS**, 10 ksi (69 MPa) **YS**
AMS 4071G		US	Tube: hydraulic, seamless	rem **Al**, 0.10 **Cu**, 0.10 **Mn**, 2.2–2.8 **Mg**, 0.15–0.35 **Cr**, 0.10 **Zn**, 0.45 **Fe+Si**, 0.05 each, 0.15 total, others	**Annealed**: 0.010–0.450 in. **diam**, 25 ksi (172 MPa) **TS**, 10 ksi (69 MPa) **YS**
AMS 4015G		US	Sheet and plate	rem **Al**, 0.25 **Si**, 0.10 **Cu**, 0.10 **Mn**, 2.2–2.8 **Mg**, 0.15–0.35 **Cr**, 0.10 **Zn**, 0.40 **Fe**, 0.05 each, 0.15 total, others	**Annealed**: 0.007–0.012 in. (0.18–0.30 mm) **diam**, 25 ksi (172 MPa) **TS**, 14% in 2 in. (50 mm) **El**
AMS 4016F		US	Sheet and plate	rem **Al**, 0.25 **Si**, 0.10 **Cu**, 0.10 **Mn**, 2.2–2.8 **Mg**, 0.15–0.35 **Cr**, 0.10 **Ti**, 0.40 **Fe**, 0.05 each, 0.15 total, others	**Strain hardened to 1/4 hard and stabilized**: 0.017–0.019 in. (0.43–0.48 mm) **diam**, 31 ksi (214 MPa) **TS**, 4% in 2 in. (50 mm) **El**

WROUGHT ALUMINUM AND ALUMINUM ALLOYS

specification number	designation	country	product forms	chemical composition	mechanical properties and hardness values
AMS 4017G		US	Sheet and plate	rem **Al**, 0.25 **Si**, 0.10 **Cu**, 0.10 **Mn**, 2.2–2.8 **Mg**, 0.15–0.35 **Cr**, 0.10 **Zn**, 0.40 **Fe**, 0.05 each, 0.15 total, others	**Strain hardened to 1/2 hard and stabilized**: 0.009–0.019 in. (0.23–0.48 mm) **diam**, 34 ksi (234 MPa) **TS**, 3% in 2 in. (50 mm) **El**
AMS 4168C		US	Bar, tube, rod, and shapes: extruded	rem **Al**, 0.40 **Si**, 1.2–2.0 **Cu**, 0.30 **Mn**, 2.1–2.9 **Mg**, 0.18–0.35 **Cr**, 5.1–6.1 **Zn**, 0.20 **Ti**, 0.50 **Fe**, 0.05 each, 0.15 total, others	**Solution and precipitation heat treated**: ≤0.250 in. **diam**, 78 ksi (537 MPa) **TS**, 70 ksi (483 MPa) **YS**, 7% in 2 in. (50 mm) **El**, 135 min **HB**
AMS 4166B		US	Bar, tube, rod, wire, and shapes: extruded	rem **Al**, 0.40 **Si**, 1.2–2.0 **Cu**, 0.30 **Mn**, 2.1–2.9 **Mg**, 0.18–0.35 **Cr**, 5.1–6.1 **Zn**, 0.20 **Ti**, 0.50 **Fe**, 0.25 **Zr+Ti**, 0.05 each, 0.15 total, others	**Solution and precipitation heat treated**: 1.499–2.999 in. (38.07–76.17 mm) **diam**, 69 ksi (476 MPa) **TS**, 59 ksi (407 MPa) **YS**, 8% in 2 in. (50 mm) **El**, 125 min **HB**
AMS 4167C		US	Bar, tube, rod, and shapes: extruded	rem **Al**, 0.40 **Si**, 1.2–2.0 **Cu**, 0.30 **Mn**, 2.1–2.9 **Mg**, 0.18–0.35 **Cr**, 5.1–6.1 **Zn**, 0.20 **Ti**, 0.50 **Fe**, 0.05 each, 0.15 total, others	**Solution and precipitation heat treated**: 0.062–0.249 in. **diam**, 68 ksi (469 MPa) **TS**, 58 ksi (400 MPa) **YS**, 7% in 2 in. (50 mm) **El**, 125 min **HB**
AMS 4147A		US	Die and hand forgings	rem **Al**, 0.40 **Si**, 1.2–2.0 **Cu**, 0.30 **Mn**, 2.1–2.9 **Mg**, 0.18–0.28 **Cr**, 5.1–6.1 **Zn**, 0.20 **Ti**, 0.50 **Fe**, 0.25 **Zr+Ti**, 0.05 each, 0.15 total, others	**Solution heat treated and aged (die forgings)**: ≤3 in. (≤76 mm) **diam**, 66 ksi (455 MPa) **TS**, 56 ksi (386 MPa) **YS**, 7% in 2 in. (50 mm) **El**, 125 min **HB**; **Solution heat treated and aged (hand forgings)**: ≤3 in. (≤76 mm) **diam**, 66 ksi (455 MPa) **TS**, 54 ksi (372 MPa) **YS**, 7% in 2 in. (50 mm) **El**, 125 min **HB**
AMS 4148		US	Die forgings: high strength	rem **Al**, 0.15 **Si**, 1.2–2.0 **Cu**, 0.10 **Mn**, 2.1–2.9 **Mg**, 0.18–0.30 **Cr**, 5.1–6.1 **Zn**, 0.10 **Ti**, 0.20 **Fe**, 0.05 each, 0.15 total, others	**Solution and precipitation heat treated**: ≤3 in. (≤76.2 mm) **diam**, 86 ksi (593 MPa) **TS**, 76 ksi (524 MPa) **YS**, 7% in 2 in. (50 mm) **El**, 140 min **HB**
AMS 4149		US	Die forgings: high strength	rem **Al**, 0.15 **Si**, 1.2–2.0 **Cu**, 0.10 **Mn**, 2.1–2.9 **Mg**, 0.18–0.30 **Cr**, 5.1–6.1 **Zn**, 0.10 **Ti**, 0.20 **Fe**, 0.05 each, 0.15 total, others	**Solution and precipitation heat treated**: ≤3 in. (≤76.2 mm) **diam**, 76 ksi (524 MPa) **TS**, 66 ksi (455 MPa) **YS**, 7% in 2 in. (50 mm) **El**, 135 min **HB**
AMS 4114E		US	Bar and rod: rolled, drawn, or cold finished	rem **Al**, 0.10 **Cu**, 0.10 **Mn**, 2.2–2.8 **Mg**, 0.15–0.35 **Cr**, 0.10 **Zn**, 0.45 **Fe+Si**, 0.05 each, 0.15 total, others	

WROUGHT ALUMINUM AND ALUMINUM ALLOYS

specification number	designation	country	product forms	chemical composition	mechanical properties and hardness values
ANSI/ASTM B 547	5052	US	Tube: formed and arc-welded	rem **Al**, 0.10 **Cu**, 0.10 **Mn**, 2.2–2.8 **Mg**, 0.15–0.35 **Cr**, 0.10 **Zn**, 0.45 **Si + Fe**, 0.05 each, 0.15 total, others	**Annealed**: 0.250–0.500 in. **diam**, 25 ksi (172 MPa) **TS**, 10 ksi (66 MPa) **YS**, 18% in 2 in. (50 mm) **El**; **H32**: 0.250–0.499 in. **diam**, 31 ksi (214 MPa) **TS**, 23 ksi (159 MPa) **YS**, 11% in 2 in. (50 mm) **El**; **H34**: 0.250–0.500 in. **diam**, 34 ksi (234 MPa) **TS**, 26 ksi (179 MPa) **YS**, 10% in 2 in. (50 mm) **El**
ANSI/ASTM B 483	5052	US	Tube: for general purposes	rem **Al**, 0.10 **Cu**, 0.10 **Mn**, 2.2–2.8 **Mg**, 0.15–0.35 **Cr**, 0.10 **Zn**, 0.45 **Si + Fe**, 0.05 each, 0.15 total, others	**Annealed**: 0.018–0.500 in. **diam**, 25 ksi (172 MPa) **TS**, 10 ksi (69 MPa) **YS**; **H32 (round tube)**: 0.018–0.500 in. **diam**, 31 ksi (214 MPa) **TS**, 23 ksi (159 MPa) **YS**; **H34 (round tube)**: 0.018–0.500 in. **diam**, 34 ksi (234 MPa) **TS**, 26 ksi (179 MPa) **YS**
ANSI/ASTM B 404	5052	US	Tube: seamless, for condensers and heat exchangers	rem **Al**, 0.10 **Cu**, 0.10 **Mn**, 2.2–2.8 **Mg**, 0.15–0.35 **Cr**, 0.10 **Zn**, 0.45 **Si + Fe**, 0.05 each, 0.15 total, others	**Annealed**: 0.018–0.450 in. **diam**, 25 ksi (175 MPa) **TS**, 10 ksi (69 MPa) **YS**; **H32 (round tube)**: 0.018–0.450 in. **diam**, 31 ksi (214 MPa) **TS**, 23 ksi (159 MPa) **YS**; **H34 (round tube)**: 0.010–0.200 in. **diam**, 39 ksi (269 MPa) **TS**, 29 ksi (200 MPa) **YS**
ANSI/ASTM B 316	5052	US	Wire and rod: rivet and cold-heading	rem **Al**, 0.10 **Cu**, 0.10 **Mn**, 2.2–2.8 **Mg**, 0.15–0.35 **Cr**, 0.10 **Zn**, 0.45 **Si + Fe**, 0.05 each, 0.15 total, others	**Annealed**: $\leq$1.000 in. **diam**, 32 max ksi (221 max MPa) **TS**; **H32**: $\leq$1.000 in. **diam**, 31 ksi (214 MPa) **TS**
ANSI/ASTM B 313	5052	US	Tube: welded	rem **Al**, 0.10 **Cu**, 0.10 **Mn**, 2.2–2.8 **Mg**, 0.15–0.35 **Cr**, 0.10 **Zn**, 0.05 each, 0.15 total, others	**Annealed**: 0.051–0.113 in. **diam**, 25 ksi (172 MPa) **TS**, 10 ksi (69 MPa) **YS**, 19% in 2 in. (50 mm) **El**; **H32**: 0.051–0.113 in. **diam**, 31 ksi (214 MPa) **TS**, 23 ksi (159 MPa) **YS**, 7% in 2 in. (50 mm) **El**; **H34**: 0.051–0.113 in. **diam**, 34 ksi (234 MPa) **TS**, 26 ksi (179 MPa) **YS**, 6% in 2 in. (50 mm) **El**
ANSI/ASTM B 241	5052	US	Tube: seamless	rem **Al**, 0.10 **Cu**, 0.10 **Mn**, 2.2–2.8 **Mg**, 0.15–0.35 **Cr**, 0.10 **Zn**, 0.45 **Si + Fe**, 0.05 each, 0.15 total, others	**Annealed**: all **diam**, 25 ksi (172 MPa) **TS**, 10 ksi (69 MPa) **YS**
ANSI/ASTM B 241	5652	US	Tube: seamless	rem **Al**, 0.04 **Cu**, 0.01 **Mn**, 2.2–2.8 **Mg**, 0.15–0.35 **Cr**, 0.10 **Zn**, 0.40 **Si + Fe**, 0.05 each, 0.15 total, others	**Annealed**: all **diam**, 25 ksi (172 MPa) **TS**, 10 ksi (69 MPa) **YS**; **H112**: all **diam**, 25 ksi (172 MPa) **TS**, 10 ksi (69 MPa) **YS**

WROUGHT ALUMINUM AND ALUMINUM ALLOYS

specification number	designation	country	product forms	chemical composition	mechanical properties and hardness values
ANSI/ASTM B 234	5052	US	Tube: seamless, for condensers and heat exchangers	rem **Al**, 0.25 **Si**, 0.10 **Cu**, 0.10 **Mn**, 2.2–2.8 **Mg**, 0.15–0.35 **Cr**, 0.10 **Zn**, 0.40 **Fe**, 0.05 each, 0.15 total, others	**H32**: 0.010–0.200 in. **diam**, 31 ksi (214 MPa) **TS**, 23 ksi (159 MPa) **YS**; **H34**: 0.010–0.200 in. **diam**, 34 ksi (234 MPa) **TS**, 26 ksi (179 MPa) **YS**
ANSI/ASTM B 221	5052	US	Bar, tube, rod, wire, and shapes	rem **Al**, 0.10 **Cu**, 0.10 **Mn**, 2.2–2.8 **Mg**, 0.15–0.35 **Cr**, 0.10 **Zn**, 0.45 **Si+Fe**, 0.05 each, 0.15 total, others	**Annealed**: all **diam**, 25 ksi (172 MPa) **TS**, 10 ksi (69 MPa) **YS**
ANSI/ASTM B 211	5052	US	Bar, rod, and wire	rem **Al**, 0.10 **Cu**, 0.10 **Mn**, 2.2–2.8 **Mg**, 0.15–0.35 **Cr**, 0.10 **Zn**, 0.45 **Si+Fe**, 0.05 each, 0.15 total, others	**Annealed**: ≥0.125 in. **diam**, 25 ksi (172 MPa) **TS**, 10 ksi (66 MPa) **YS**, 25% in 2 in. (50 mm) **El**; **H32**: 0.125–0.374 in. **diam**, 31 ksi (214 MPa) **TS**, 23 ksi (159 MPa) **YS**; **H34**: ≤0.374 in. **diam**, 34 ksi (234 MPa) **TS**, 26 ksi (179 MPa) **YS**
ANSI/ASTM B 210	5052	US	Tube: seamless	rem **Al**, 0.25 **Si**, 0.10 **Cu**, 0.10 **Mn**, 2.2–2.8 **Mg**, 0.15–0.35 **Cr**, 0.10 **Zn**, 0.40 **Fe**, 0.05 each, 0.15 total, others	**Annealed**: 0.018–0.450 in. **diam**, 25 ksi (172 MPa) **TS**, 10 ksi (69 MPa) **YS**; **H32 (round tube)**: 0.018–0.450 in. **diam**, 31 ksi (214 MPa) **TS**, 23 ksi (159 MPa) **YS**; **H34 (round tube)**: 0.018–0.450 in. **diam**, 34 ksi (234 MPa) **TS**, 26 ksi (179 MPa) **YS**
ANSI/ASTM B 209	5052	US	Sheet and plate	rem **Al**, 0.25 **Si**, 0.10 **Cu**, 0.10 **Mn**, 2.2–2.8 **Mg**, 0.15–0.35 **Cr**, 0.10 **Zn**, 0.40 **Fe**, 0.05 each, 0.15 total, others	**Annealed**: 0.250–3.000 in. **diam**, 25 ksi (172 MPa) **TS**, 10 ksi (69 MPa) **YS**, 18% in 2 in. (50 mm) **El**; **H32 or H22**: 0.500–2.000 in. **diam**, 31 ksi (207 MPa) **TS**, 23 ksi (159 MPa) **YS**, 12% in 2 in. (50 mm) **El**; **H34 or H24**: 0.250–1.000 in. **diam**, 34 ksi (234 MPa) **TS**, 26 ksi (179 MPa) **YS**, 10% in 2 in. (50 mm) **El**
ANSI/ASTM B 209	5252	US	Sheet and plate	rem **Al**, 0.08 **Si**, 0.10 **Cu**, 0.10 **Mn**, 2.2–2.8 **Mg**, 0.05 **Zn**, 0.10 **Fe**, 0.03 each, 0.10 total, others	**H24**: 0.030–0.090 in. **diam**, 30 ksi (207 MPa) **TS**, 10% in 2 in. (50 mm) **El**; **H25**: 0.030–0.090 in. **diam**, 31 ksi (214 MPa) **TS**, 9% in 2 in. (50 mm) **El**; **H28**: 0.030–0.090 in. **diam**, 38 ksi (262 MPa) **TS**, 3% in 2 in. (50 mm) **El**

WROUGHT ALUMINUM AND ALUMINUM ALLOYS

specification number	designation	country	product forms	chemical composition	mechanical properties and hardness values
ANSI/ASTM B 209	5652	US	Sheet and plate	rem **Al**, 0.04 **Cu**, 0.01 **Mn**, 2.2–2.8 **Mg**, 0.15–0.35 **Cr**, 0.10 **Zn**, 0.40 **Si+Fe**, 0.05 each, 0.15 total, others	**Annealed**: 0.250–3.000 in. **diam**, 25 ksi (172 MPa) **TS**, 10 ksi (69 MPa) **YS**, 18% in 2 in. (50 mm) **El**; **H32**: 0.500–2.000 in. **diam**, 31 ksi (214 MPa) **TS**, 23 ksi (159 MPa) **YS**, 12% in 2 in. (50 mm) **El**; **H34**: 0.250–1.000 in. **diam**, 34 ksi (234 MPa) **TS**, 26 ksi (179 MPa) **YS**, 10% in 2 in. (50 mm) **El**
AS 1734	5052	Australia	Sheet, plate	rem **Al**, 0.10 **Cu**, 0.10 **Mn**, 2.2–2.8 **Mg**, 0.15–0.35 **Cr**, 0.10 **Zn**, 0.45 **Si+Fe**, 0.05 each, 0.15 total, others	**Annealed**: 0.20–0.30 mm **diam**, 172 MPa **TS**, 65 MPa **YS**, 14% in 2 in. (50 mm) **El**; **Strain-hardened and stabilized to 1/2 hard**: 0.25–0.50 mm **diam**, 234 MPa **TS**, 179 MPa **YS**, 3% in 2 in. (50 mm) **El**; **Strain-hardened and stabilized to full hard**: 0.15–0.20 mm **diam**, 268 MPa **TS**, 220 MPa **YS**, 2% in 2 in. (50 mm) **El**
AS 1734	5252	Australia	Sheet, plate	rem **Al**, 0.08 **Si**, 0.10 **Cu**, 0.10 **Mn**, 2.2–2.8 **Mg**, 0.10 **Fe**, 0.03 each, 0.10 total, others	**Strain-hardened and partially annealed to 5/8 hard**: 0.50–0.80 mm **diam**, 213 MPa **TS**, 7% in 2 in. (50 mm) **El**; **Strain-hardened and partially annealed to 7/8 hard**: 0.50–0.80 mm **diam**, 241 MPa **TS**, 4% in 2 in. (50 mm) **El**
CSA HA.7.1	.5052	Canada	Tube: drawn and extruded, pipe: drawn	rem **Al**, 0.10 **Cu**, 0.10 **Mn**, 2.2–2.8 **Mg**, 0.15–0.35 **Cr**, 0.10 **Zn**, 0.45 **Si+Fe**, 0.05 each, 0.15 total, others	**Drawn and O**: 0.010–0.450 in. **diam**, 25.0 ksi (172 MPa) **TS**, 10.0 ksi (69 MPa) **YS**; **Drawn and H38**: 0.010–0.450 in. **diam**, 39.0 ksi (269 MPa) **TS**, 31.0 ksi (214 MPa) **YS**; **Drawn and H34**: 0.010–0.450 in. **diam**, 34.0 ksi (234 MPa) **TS**, 26.0 ksi (179 MPa) **YS**
CSA HA.7	.5052	Canada	Tube, pipe: seamless	rem **Al**, 0.10 **Cu**, 0.10 **Mn**, 2.2–2.8 **Mg**, 0.15–0.35 **Cr**, 0.10 **Zn**, 0.45 **Si+Fe**, 0.05 each, 0.15 total, others	**Drawn and O**: 0.010–0.450 in. **diam**, 25.0 ksi (172 MPa) **TS**, 10.0 ksi (69 MPa) **YS**; **Drawn and H38**: 0.010–0.450 in. **diam**, 39.0 ksi (269 MPa) **TS**, 31.0 ksi (214 MPa) **YS**; **Extruded and H111**: all **diam**, 25.0 ksi (172 MPa) **TS**, 10.0 ksi (69 MPa) **YS**, 18% in 2 in. (50 mm) **El**
CSA HA.6	.5052	Canada	Rivet, rod, wire	rem **Al**, 0.10 **Cu**, 0.10 **Mn**, 2.2–2.8 **Mg**, 0.15–0.35 **Cr**, 0.10 **Zn**, 0.45 **Si+Fe**, 0.05 each, 0.15 total, others	**O**: 0.062–1.000 in. **diam**, 32.0 max ksi (221 max MPa) **TS**; **H32**: 0.062–1.000 in. **diam**, 31.0 ksi (214 MPa) **TS**

WROUGHT ALUMINUM AND ALUMINUM ALLOYS

specification number	designation	country	product forms	chemical composition	mechanical properties and hardness values
CSA HA.5	.5052	Canada	Bar, rod, wire, shapes	rem **Al**, 0.10 **Cu**, 0.10 **Mn**, 2.2–2.8 **Mg**, 0.15–0.35 **Cr**, 0.10 **Zn**, 0.45 **Si+Fe**, 0.05 each, 0.15 total, others	**Drawn and O**: all **diam**, 25.0 ksi (172 MPa) **TS**, 9.5 ksi (66 MPa) **YS**, 25% in 2 in. (50 mm) **El**; **Drawn and H34**: <0.374 in. **diam**, 34.0 ksi (234 MPa) **TS**, 26.0 ksi (179 MPa) **YS**; **Drawn and H38**: <0.374 in. **diam**, 39.0 ksi (269 MPa) **TS**
CSA HA.4	.5052	Canada	Plate, sheet	rem **Al**, 0.10 **Cu**, 0.10 **Mn**, 2.2–2.8 **Mg**, 0.15–0.35 **Cr**, 0.10 **Zn**, 0.45 **Fe+Si**, 0.05 each, 0.15 total, others	**O**: 0.013–0.019 in. **diam**, 25.0 ksi (172 MPa) **TS**, 9.5 ksi (66 MPa) **YS**, 15% in 2 in. (50 mm) **El**; **H32**: 0.016–0.019 in. **diam**, 31.0 ksi (214 MPa) **TS**, 23.0 ksi (159 MPa) **YS**, 4% in 2 in. (50 mm) **El**; **H112**: 0.250–0.499 in. **diam**, 28.0 ksi (193 MPa) **TS**, 16.0 ksi (110 MPa) **YS**, 7% in 2 in. (50 mm) **El**
COPANT 862	5252	COPANT	Wrought products	rem **Al**, 0.08 **Si**, 0.10 **Cu**, 0.10 **Mn**, 2.2–2.8 **Mg**, 0.05 **Zn**, 0.03 each, 0.10 total, others	
COPANT 862	5052	COPANT	Wrought products	rem **Al**, 0.25 **Si**, 0.10 **Cu**, 0.10 **Mn**, 2.2–2.8 **Mg**, 0.15–0.35 **Cr**, 0.10 **Zn**, 0.40 **Fe**, 0.05 each, 0.15 total, others	
WW-T-700/4E	5052	US	Tube	rem **Al**, 0.10 **Cu**, 0.10 **Mn**, 2.2–2.8 **Mg**, 0.15–0.35 **Cr**, 0.10 **Zn**, 0.45 **Fe**, **Si** in **Fe** content; 0.05 each, 0.15 total, others	**Annealed**: 0.010–0.450 in. **diam**, 25 ksi (172 MPa) **TS**, 10 ksi (69 MPa) **YS**; **Strain hardened and stabilized 1/4 hard**: 0.10–0.450 in. **diam**, 31 ksi (214 MPa) **TS**, 23 ksi (159 MPa) **YS**; **Strain hardened and stabilized 1/2 hard**: 0.10–0.450 in. **diam**, 34 ksi (234 MPa) **TS**, 26 ksi (179 MPa) **YS**
QQ-A-250/8E	5052	US	Plate, sheet	rem **Al**, 0.10 **Cu**, 0.10 **Mn**, 2.2–2.8 **Mg**, 0.15–0.35 **Cr**, 0.10 **Zn**, 0.45 **Fe**, **Si** in **Fe** content; 0.05 each, 0.15 total, others	**Annealed**: 0.006–0.007 in. **diam**, 25 ksi (172 MPa) **TS**; **Strain hardened 1/4 hard temper and stabilized**: 0.017–0.019 in. **diam**, 31 ksi (214 MPa) **TS**, 4% in 2 in. (50 mm) **El**; **Strain hardened 1/2 hard temper and stabilized**: 0.009–0.019 in. **diam**, 34 ksi (234 MPa) **TS**, 3% in 2 in. (50 mm) **El**
QQ-A-225/7C	5052	US	Bar, rod, wire	rem **Al**, 0.10 **Cu**, 0.10 **Mn**, 2.2–2.8 **Mg**, 0.15–0.35 **Cr**, 0.10 **Zn**, 0.45 **Fe**, **Si** in **Fe** content; 0.05 each, 0.15 total, others	**Annealed**: all **diam**, 32 max ksi (221 max MPa) **TS**, 25% in 2 in. (50 mm) **El**; **Strain hardened 1/4 hard temper and stabilized**: 0.374 max in. **diam**, 31 ksi (214 MPa) **TS**; **Strain hardened 1/2 hard temper and stabilized**: 0.374 max in. **diam**, 34 ksi (234 MPa) **TS**

WROUGHT ALUMINUM AND ALUMINUM ALLOYS

specification number	designation	country	product forms	chemical composition	mechanical properties and hardness values
QQ-A-430B	5052	US	Rod, wire	rem **Al**, 0.45 **Si**, 0.10 **Cu**, 0.10 **Mn**, 2.2–2.8 **Mg**, 0.15–0.35 **Cr**, 0.10 **Zn**, **Fe** in **Si** content; 0.05 each, 0.15 total, others	**Annealed**: 1.000 max in. **diam**, 32 max ksi (221 max MPa) **TS**; **Strain hardened and temper H32**: 1.000 max in. **diam**, 31 ksi (214 MPa) **TS**
SFS 2587	AlMg2,5	Finland	Shapes and sheet	rem **Al**, 0.5 **Si**, 0.10 **Cu**, 0.5 **Mn**, 2.2–2.8 **Mg**, 0.35 **Cr**, 0.2 **Zn**, 0.5 **Fe**, 0.5 **Mn+Cr**, 0.2 **Ti+Zr**, 0.05 each, 0.15 total, others	**Hot finished (shapes)**: all **diam**, 180 MPa **TS**, 80 MPa **YS**, 14% in 2 in. (50 mm) **El**; **Annealed**: all **diam**, 220 max MPa **TS**, 70 MPa **YS**, 18% in 2 in. (50 mm) **El**; **1/2 hard**: ≤3 mm **diam**, 240 MPa **TS**, 180 MPa **YS**, 14% in 2 in. (50 mm) **El**
NF A 50 411	5052	France	Bar and tube	rem **Al**, 0.25 **Si**, 0.10 **Cu**, 0.10 **Mn**, 2.2–2.8 **Mg**, 0.15–0.35 **Cr**, 0.10 **Zn**, 0.40 **Fe**, 0.05 each, 0.15 total, others	**1/2 hard (drawn bar)**: ≤50 mm **diam**, 235 MPa **TS**, 180 MPa **YS**, 4% in 2 in. (50 mm) **El**; **Hard (tube)**: ≤50 mm **diam**, 270 MPa **TS**, 210 MPa **YS**, 2% in 2 in. (50 mm) **El**
NF A 50 451	5052	France	Sheet, plate, and strip	rem **Al**, 0.10 **Cu**, 0.10 **Mn**, 2.2–2.8 **Mg**, 0.15–0.35 **Cr**, 0.10 **Zn**, 0.45 **Fe+Si**, 0.05 each, 0.15 total, others	**Annealed**: 0.4–6 mm **diam**, 170 MPa **TS**, 60 MPa **YS**, 10% in 2 in. (50 mm) **El**; **3/4 hard**: 0.4–3.2 mm **diam**, 250 MPa **TS**, 200 MPa **YS**, 5% in 2 in. (50 mm) **El**; **Hard**: 0.4–3.2 mm **diam**, 270 MPa **TS**, 220 MPa **YS**, 4% in 2 in. (50 mm) **El**
ISO R827	Al–Mg2.5	ISO	Extruded products	rem **Al**, 0.5 **Si**, 0.10 **Cu**, 0.5 **Mn**, 2.2–2.8 **Mg**, 0.35 **Cr**, 0.2 **Zn**, 0.5 **Fe**, 0.2 **Ti+Zr**, 0.5 **Mn+Cr**	**As manufactured**: all **diam**, 172 MPa **TS**, 59 MPa **YS**, 12% in 2 in. (50 mm) **El**
ISO R209	Al–Mg2.5	ISO	Wrought products	rem **Al**, 0.5 **Si**, 0.10 **Cu**, 0.5 **Mn**, 2.2–2.8 **Mg**, 0.35 **Cr**, 0.2 **Zn**, 0.5 **Fe**, 0.2 **Ti+Zr**, 0.5 **Mn+Cr**	
ISO TR2778	Al–Mg2.5	ISO	Tube: drawn	rem **Al**, 0.5 **Si**, 0.10 **Cu**, 0.5 **Mn**, 2.2–2.8 **Mg**, 0.35 **Cr**, 0.2 **Zn**, 0.5 **Fe**, 0.2 **Ti+Zr**, 0.5 **Mn+Cr**	**Annealed**: all **diam**, 170 MPa **TS**, 65 MPa **YS**, 15% in 2 in. (50 mm) **El**; **Strain hardened**: 80 mm **diam**, 235 MPa **TS**, 180 MPa **YS**, 3% in 2 in. (50 mm) **El**
ISO TR2136	Al–Mg2.5	ISO	Sheet, plate	rem **Al**, 0.5 **Si**, 0.10 **Cu**, 0.5 **Mn**, 2.2–2.8 **Mg**, 0.35 **Cr**, 0.2 **Zn**, 0.5 **Fe**, 0.2 **Ti+Zr**, 0.5 **Mn+Cr**	**Annealed**: all **diam**, 170 MPa **TS**, 65 MPa **YS**, 16% in 2 in. (50 mm) **El**; **HB–Strain hardened**: all **diam**, 205 MPa **TS**, 145 MPa **YS**, 3% in 2 in. (50 mm) **El**; **HD– Strain hardened**: all **diam**, 235 MPa **TS**, 180 MPa **YS**, 2% in 2 in. (50 mm) **El**

WROUGHT ALUMINUM AND ALUMINUM ALLOYS

specification number	designation	country	product forms	chemical composition	mechanical properties and hardness values
JIS H 4000	5052	Japan	Sheet, plate, and strip	rem **Al**, 0.10 **Cu**, 0.10 **Mn**, 2.2–2.8 **Mg**, 0.15–0.35 **Cr**, 0.10 **Zn**, 0.45 **Si + Fe**, 0.15 total, others	**Annealed**: <1.3 mm **diam**, 177 MPa **TS**, 69 MPa **YS**, 18% in 2 in. (50 mm) **El**; **Hard**: <1.3 mm **diam**, 275 MPa **TS**, 226 MPa **YS**, 4% in 2 in. (50 mm) **El**; **Hot rolled**: <12 mm **diam**, 196 MPa **TS**, 108 MPa **YS**, 7% in 2 in. (50 mm) **El**
JIS H 4040	5052	Japan	Bar and wire: extruded	rem **Al**, 0.10 **Cu**, 0.10 **Mn**, 2.2–2.8 **Mg**, 0.15–0.35 **Cr**, 0.10 **Zn**, 0.45 **Si + Fe**, 0.15 total, others	**As manufactured (bar)**: all **diam**, 177 MPa **TS**, 69 MPa **YS**; **Annealed (bar)**: all **diam**, 245 MPa **TS**, 69 MPa **YS**; **Hard**: ≤10 mm **diam**, 275 MPa **TS**
JIS H 4080	5052	Japan	Tube: extruded	rem **Al**, 0.10 **Cu**, 0.10 **Mn**, 2.2–2.8 **Mg**, 0.15–0.35 **Cr**, 0.10 **Zn**, 0.45 **Si + Fe**, 0.15 total, others	**As manufactured**: all **diam**, 177 MPa **TS**, 69 MPa **YS**; **Annealed**: all **diam**, 245 MPa **TS**, 69 MPa **YS**, 20% in 2 in. (50 mm) **El**
JIS H 4090	5052	Japan	Pipe and tube: welded	rem **Al**, 0.10 **Cu**, 0.10 **Mn**, 2.2–2.8 **Mg**, 0.15–0.35 **Cr**, 0.45 **Si + Fe**, 0.15 total, others	**Annealed**: <1.3 mm **diam**, 177 MPa **TS**, 69 MPa **YS**, 18% in 2 in. (50 mm) **El**; **1/2 hard**: <1.3 mm **diam**, 235 MPa **TS**, 177 MPa **YS**, 4% in 2 in. (50 mm) **El**; **Hard**: <1.3 mm **diam**, 275 MPa **TS**, 226 MPa **YS**, 4% in 2 in. (50 mm) **El**
JIS H 4100	5052	Japan	Shapes: extruded	rem **Al**, 0.10 **Cu**, 0.10 **Mn**, 2.2–2.8 **Mg**, 0.15–0.35 **Cr**, 0.45 **Si + Fe**, 0.15 total, others	**As manufactured**: all **diam**, 177 MPa **TS**, 69 MPa **YS**; **Annealed**: all **diam**, 245 MPa **TS**, 69 MPa **YS**, 20% in 2 in. (50 mm) **El**
JIS H 4120	5052	Japan	Wire and rod: rivet	rem **Al**, 0.10 **Cu**, 0.10 **Mn**, 2.2–2.8 **Mg**, 0.15–0.35 **Cr**, 0.10 **Zn**, 0.45 **Si + Fe**, 0.15 total, others	**Annealed and stabilized at 1/4 hard**: ≤25 mm **diam**, 216 MPa **TS**
NS 17 210	NS 17 210–02	Norway	Sheet and strip	rem **Al**, 0.5 **Si**, 0.1 **Cu**, 0.5 **Mn**, 2.2–2.8 **Mg**, 0.2 **Zn**, 0.5 **Fe**, 0.35 **Cr**, 0.2 **Ti + Zr**	**Soft annealed**: <0.5 mm **diam**, 176 MPa **TS**, 69 MPa **YS**, 18% in 2 in. (50 mm) **El**
NS 17 210	NS 17 210–14	Norway	Sheet, strip, rod, and wire	rem **Al**, 0.5 **Si**, 0.1 **Cu**, 0.5 **Mn**, 2.2–2.8 **Mg**, 0.2 **Zn**, 0.5 **Fe**, 0.35 **Cr**, 0.2 **Ti + Zr**	**1/2 hard (sheet and strip)**: 0.5–8 mm **diam**, 235 MPa **TS**, 196 MPa **YS**, 5% in 2 in. (50 mm) **El**; **1/2 hard (rod and wire)**: <6 mm **diam**, 235 MPa **TS**, 176 MPa **YS**
NS 17 210	NS 17 210–34	Norway	Sheet and strip	rem **Al**, 0.5 **Si**, 0.1 **Cu**, 0.5 **Mn**, 2.2–2.8 **Mg**, 0.2 **Zn**, 0.5 **Fe**, 0.35 **Cr**, 0.2 **Ti + Zr**	**1/2 hard**: 0.5–3 mm **diam**, 235 MPa **TS**, 176 MPa **YS**, 6% in 2 in. (50 mm) **El**
NS 17 210	NS 17 210–18	Norway	Sheet, strip, rod, and wire	rem **Al**, 0.5 **Si**, 0.1 **Cu**, 0.5 **Mn**, 2.2–2.8 **Mg**, 0.2 **Zn**, 0.5 **Fe**, 0.35 **Cr**, 0.2 **Ti + Zr**	**Hard (sheet and strip)**: 0.5–4 mm **diam**, 275 MPa **TS**, 245 MPa **YS**, 3% in 2 in. (50 mm) **El**; **Hard (rod and wire)**: <6 mm **diam**, 275 MPa **TS**, 245 MPa **YS**

WROUGHT ALUMINUM AND ALUMINUM ALLOYS

specification number	designation	country	product forms	chemical composition	mechanical properties and hardness values
SIS 14 41 20	SIS Al 41 20–02	Sweden	Plate, sheet, strip	95.34 min **Al**, 0.25 **Si**, 0.10 **Cu**, 0.50 **Mn**, 2.2–2.8 **Mg**, 0.35 **Cr**, 0.10 **Zn**, 0.40 **Fe**, 0.05 each, 0.15 total, others	**Annealed**: 5–20 mm **diam**, 170 MPa **TS**, 65 MPa **YS**, 20% in 2 in. (50 mm) **El**, 45–60 **HV**
SIS 14 41 20	SIS Al 41 20–14	Sweden	Sheet, plate	95.34 min **Al**, 0.25 **Si**, 0.10 **Cu**, 0.50 **Mn**, 2.2–2.8 **Mg**, 0.35 **Cr**, 0.10 **Zn**, 0.40 **Fe**, 0.05 each, 0.15 total, others	**Strain hardened**: 3–5 mm **diam**, 230 MPa **TS**, 180 MPa **YS**, 8% in 2 in. (50 mm) **El**, 70–90 **HV**
SIS 14 41 20	SIS Al 41 20–24	Sweden	Sheet, strip	95.34 min **Al**, 0.25 **Si**, 0.10 **Cu**, 0.50 **Mn**, 2.2–2.8 **Mg**, 0.35 **Cr**, 0.10 **Zn**, 0.40 **Fe**, 0.05 each, 0.15 total, others	**Strain hardened, annealed**: 0.35–3 mm **diam**, 220 MPa **TS**, 170 MPa **YS**, 14% in 2 in. (50 mm) **El**, 65–85 **HV**
SIS 14 41 20	SIS Al 41 20–18	Sweden	Sheet, strip	95.34 min **Al**, 0.25 **Si**, 0.10 **Cu**, 0.50 **Mn**, 2.2–2.8 **Mg**, 0.35 **Cr**, 0.10 **Zn**, 0.40 **Fe**, 0.05 each, 0.15 total, others	**Strain hardened**: 0.35–4 mm **diam**, 280 MPa **TS**, 240 MPa **YS**, 3% in 2 in. (50 mm) **El**, 80–100 **HV**
SIS 14 41 20	SIS Al 41 20–40	Sweden	Sheet, strip	95.34 min **Al**, 0.25 **Si**, 0.10 **Cu**, 0.50 **Mn**, 2.2–2.8 **Mg**, 0.35 **Cr**, 0.10 **Zn**, 0.40 **Fe**, 0.05 each, 0.15 total, others	**Annealed**: 0.35–6 mm **diam**, 170 MPa **TS**, 65 MPa **YS**, 20% in 2 in. (50 mm) **El**, 45–60 **HV**
SIS 14 41 20	SIS Al 41 20–44	Sweden	Sheet, strip	95.34 min **Al**, 0.25 **Si**, 0.10 **Cu**, 0.50 **Mn**, 2.2–2.8 **Mg**, 0.35 **Cr**, 0.10 **Zn**, 0.40 **Fe**, 0.05 each, 0.15 total, others	**Strain hardened**: 0.35–4 mm **diam**, 230 MPa **TS**, 180 MPa **YS**, 5% in 2 in. (50 mm) **El**, 70–90 **HV**
SIS 14 41 20	SIS Al 41 20–00	Sweden	Plate, sheet	95.34 min **Al**, 0.25 **Si**, 0.10 **Cu**, 0.50 **Mn**, 2.2–2.8 **Mg**, 0.35 **Cr**, 0.10 **Zn**, 0.40 **Fe**, 0.05 each, 0.15 total, others	**Hot worked**: 5–25 mm **diam**, 190 MPa **TS**, 75 MPa **YS**, 10% in 2 in. (50 mm) **El**, 50–70 **HV**
AMS 4004		US	Foil	rem **Al**, 0.10 **Cu**, 0.10 **Mn**, 2.1–2.8 **Mg**, 0.15–0.35 **Cr**, 0.10 **Zn**, 0.45 **Fe+Si**, 0.05 each, 0.15 total, others	**Strain hardened, mill finish**: <0.006 in. **diam**, 42 ksi (290 MPa) **TS**
NS 17 205	NS 17 205–02	Norway	Sheet, strip, tube, rod, and wire	rem **Al**, 0.5 **Si**, 0.10 **Cu**, 0.5 **Mn**, 2.4 **Mg**, 0.2 **Zn**, 0.5 **Fe**, 0.25 **Cr**	**Soft annealed (sheet and strip)**: 0.2–6 mm **diam**, 157 MPa **TS**, 59 MPa **YS**; **Soft annealed (rod and wire)**: <10 mm **diam**, 167 MPa **TS**; **Soft annealed (tube)**: <10 mm **diam**, 157 MPa **TS**, 59 MPa **YS**, 18% in 2 in. (50 mm) **El**
NS 17 205	NS 17 205–34	Norway	Sheet, strip, and tube	rem **Al**, 0.5 **Si**, 0.10 **Cu**, 0.5 **Mn**, 2.4 **Mg**, 0.2 **Zn**, 0.5 **Fe**, 0.25 **Cr**	**1/2 hard (sheet and strip)**: 0.2–6 mm **diam**, 196 MPa **TS**, 127 MPa **YS**; **1/2 hard (tube)**: <10 mm **diam**, 221 MPa **TS**, 172 MPa **YS**, 5% in 2 in. (50 mm) **El**
NS 17 205	NS 17 205–08	Norway	Rod, wire, and profiles	rem **Al**, 0.5 **Si**, 0.10 **Cu**, 0.5 **Mn**, 2.4 **Mg**, 0.2 **Zn**, 0.5 **Fe**, 0.25 **Cr**	**Extruded (rod and wire)**: <150 mm **diam**, 167 MPa **TS**, 59 MPa **YS**, 16% in 2 in. (50 mm) **El**; **Hard (profiles)**: <150 mm **diam**, 167 MPa **TS**, 59 MPa **YS**, 16% in 2 in. (50 mm) **El**

WROUGHT ALUMINUM AND ALUMINUM ALLOYS

specification number	designation	country	product forms	chemical composition	mechanical properties and hardness values
NS 17 205	NS 17 205–38	Norway	Sheet, strip, plate, rod, and wire	rem **Al**, 0.5 **Si**, 0.10 **Cu**, 0.5 **Mn**, 2.4 **Mg**, 0.2 **Zn**, 0.5 **Fe**, 0.25 **Cr**	**Hard (sheet, strip, and plate)**: 0.2–12.5 mm **diam**, 221 MPa **TS**, 172 MPa **YS**; **Hard (rod and wire)**: <10 mm **diam**, 255 MPa **TS**
NS 17 205	NS 17 205–10	Norway	Sheet, strip, and plate	rem **Al**, 0.5 **Si**, 0.10 **Cu**, 0.5 **Mn**, 2.4 **Mg**, 0.2 **Zn**, 0.5 **Fe**, 0.25 **Cr**	
IS 736	52000	India	Plate	rem **Al**, 0.6 **Si**, 0.1 **Cu**, 0.5 **Mn**, 1.7–2.6 **Mg**, 0.25 **Cr**, 0.2 **Zn**, 0.2 **Ti**, 0.7 **Fe**, 0.5 **Cr+Mn**	**As manufactured**: 12 mm **diam**, 190 MPa **TS**, 12% in 2 in. (50 mm) **El**; **Annealed**: 18 mm **diam**, 175 MPa **TS**, 60 MPa **YS**, 18% in 2 in. (50 mm) **El**; **Half hard**: 5 mm **diam**, 200 MPa **TS**, 160 MPa **YS**, 5% in 2 in. (50 mm) **El**
IS 733	52000	India	Bar, rod, section	rem **Al**, 0.6 **Si**, 0.1 **Cu**, 0.5 **Mn**, 1.7–2.6 **Mg**, 0.25 **Cr**, 0.2 **Zn**, 0.2 **Ti**, 0.5 **Fe**, 0.5 **Cr+Mn**	**As manufactured**: ≤150 mm **diam**, 170 MPa **TS**, 14% in 2 in. (50 mm) **El**
AS 1867	5251	Australia	Tube: drawn	rem **Al**, 0.40 **Si**, 0.15 **Cu**, 0.10–0.50 **Mn**, 1.7–2.4 **Mg**, 0.15 **Cr**, 0.15 **Zn**, 0.15 **Ti**, 0.50 **Fe**, 0.05 each, 0.15 total, others	**Annealed**: all **diam**, 172 MPa **TS**, 68 MPa **YS**, 18% in 2 in. (50 mm) **El**; **Strain–hardened and stabilized to 1/4 hard**: all **diam**, 213 MPa **TS**, 158 MPa **YS**; **Strain–hardened and stabilized to full hard**: all **diam**, 268 MPa **TS**, 213 MPa **YS**
AS 1865	5251	Australia	Wire, rod, bar, strip	rem **Al**, 0.40 **Si**, 0.15 **Cu**, 0.10–0.50 **Mn**, 1.7–2.4 **Mg**, 0.15 **Cr**, 0.15 **Zn**, 0.15 **Ti**, 0.50 **Fe**, 0.05 each, 0.15 total, others	**Annealed**: 10 mm **diam**, 172 MPa **TS**, 65 MPa **YS**, 25% in 2 in. (50 mm) **El**; **Strain–hardened and stabilized to 1/4 hard**: 10 mm **diam**, 213 MPa **TS**, 158 MPa **TS**; **Strain–hardened and stabilized to full hard**: 10 mm **diam**, 268 MPa **TS**
AS 1734	5251	Australia	Sheet, plate	rem **Al**, 0.40 **Si**, 0.15 **Cu**, 0.10–0.50 **Mn**, 1.7–2.4 **Mg**, 0.15 **Cr**, 0.15 **Zn**, 0.15 **Ti**, 0.50 **Fe**, 0.05 each, 0.15 total, others	**Annealed**: 0.20–0.50 mm **diam**, 170 MPa **TS**15% in 2 in. (50 mm) **El**; **Strain–hardened and stabilized to 1/2 hard**: 0.25–0.50 mm **diam**, 231 MPa **TS**, 179 MPa **YS**, 3% in 2 in. (50 mm) **El**; **Strain–hardened and stabilized to full hard**: 0.20–0.80 mm **diam**, 262 MPa **TS**, 224 MPa **TS**, 3% in 2 in. (50 mm) **El**
BS 1471	NT4	UK	Tube: drawn	rem **Al**, 0.5 **Si**, 0.10 **Cu**, 0.5 **Mn**, 1.7–2.4 **Mg**, 0.25 **Cr**, 0.2 **Zn**, 0.2 **Ti**, 0.5 **Fe**, 0.5 **Mn+Cr**	**Annealed**: ≤10.0 mm **diam**, 160 MPa **TS**, 60 MPa **YS**, 18% in 2 in. (50 mm) **El**; **Strain–hardened to 1/2 hard**: ≤10.0 mm **diam**, 225 MPa **TS**, 175 MPa **YS**, 5% in 2 in. (50 mm) **El**
BS 1475	NG4	UK	Wire	rem **Al**, 0.5 **Si**, 0.10 **Cu**, 0.5 **Mn**, 1.7–2.4 **Mg**, 0.25 **Cr**, 0.2 **Zn**, 0.2 **Ti**, 0.5 **Fe**, 0.5 **Mn+Cr**	**Annealed**: ≤10 mm **diam**, 170 MPa **TS**; **Strain–hardened to hard**: ≤10 mm **diam**, 260 MPa **TS**

WROUGHT ALUMINUM AND ALUMINUM ALLOYS

specification number	designation	country	product forms	chemical composition	mechanical properties and hardness values
BS 1470	NS4	UK	Plate, sheet, strip	rem **Al**, 0.5 **Si**, 0.10 **Cu**, 0.5 **Mn**, 1.7–2.4 **Mg**, 0.25 **Cr**, 0.2 **Zn**, 0.2 **Ti**, 0.5 **Fe**, 0.5 **Mg+Cr**	**Annealed**: 0.2–6.0 mm **diam**, 160 MPa **TS**, 60 MPa **YS**, 18% in 2 in. (50 mm) **El**; **Strain hardened to 3/8 hard**: 0.2–6.0 mm **diam**, 200 MPa **TS**, 130 MPa **YS**, 4% in 2 in. (50 mm) **El**
BS 1472	NF4	UK	Forging stock and forgings	rem **Al**, 0.5 **Si**, 0.10 **Cu**, 0.5 **Mn**, 1.7–2.4 **Mg**, 0.25 **Cr**, 0.2 **Zn**, 0.2 **Ti**, 0.5 **Fe**, 0.5 **Mn+Cr**	**As manufactured**: ≤150 mm **diam**, 170 MPa **TS**, 60 MPa **YS**, 16% in 2 in. (50 mm) **El**
BS 2L.81		UK	Sheet and strip	rem **Al**, 0.5 **Si**, 0.10 **Cu**, 0.5 **Mn**, 1.7–2.4 **Mg**, 0.25 **Cr**, 0.2 **Zn**, 0.2 **Ti**, 0.05 **Sn**, 0.5 **Fe**, 0.05 **Pb**, 0.2 **Zr+Ti**, 0.5 **Mg+Cr**	**Cold rolled, or cold rolled and partially annealed**: 0.4–0.8 mm **diam**, 225 MPa **TS**, 175 MPa **YS**, 3% in 2 in. (50 mm) **El**
BS 2L.80		UK	Sheet and strip	rem **Al**, 0.5 **Si**, 0.10 **Cu**, 0.5 **Mn**, 1.7–2.4 **Mg**, 0.25 **Cr**, 0.2 **Ni**, 0.2 **Zn**, 0.05 **Sn**, 0.5 **Fe**, 0.05 **Pb**, 0.2 **Zr+Ti**, 0.5 **Mg+Cr**	**Annealed**: 0.4–2.6 mm **diam**, 160 MPa **TS**, 60 MPa **YS**, 18% in 2 in. (50 mm) **El**
BS 1474	NE4	UK	Bar, tube, section	rem **Al**, 0.5 **Si**, 0.10 **Cu**, 0.5 **Mn**, 1.7–2.4 **Mg**, 0.25 **Cr**, 0.2 **Zn**, 0.2 **Ti**, 0.05 **Fe**, 0.5 **Mn+Cr**	**As manufactued**: ≤150 mm **diam**, 170 MPa **TS**, 60 MPa **YS**, 14% in 2 in. (50 mm) **El**
BS 4L.44		UK	Forging stock, bar, section, forgings	rem **Al**, 0.5 **Si**, 0.10 **Cu**, 0.5 **Mn**, 1.7–2.4 **Mg**, 0.25 **Cr**, 0.2 **Ni**, 0.2 **Zn**, 0.05 **Sn**, 0.5 **Fe**, 0.05 **Pb**, 0.2 **Zr+Ti**, 0.5 **Mg+Cr**	
BS 3L.56		UK	Tube	rem **Al**, 0.5 **Si**, 0.10 **Cu**, 0.5 **Mn**, 1.7–2.4 **Mg**, 0.25 **Cr**, 0.2 **Ni**, 0.2 **Zn**, 0.05 **Sn**, 0.5 **Fe**, 0.05 **Pb**, 0.20 **Zr+Ti**, 0.5 **Mg+Cr**	**Softened**: all **diam**, 160 MPa **TS**, 60 MPa **YS**, 18% in 2 in. (50 mm) **El**
COPANT 862	X 5951	COPANT	Wrought products	rem **Al**, 0.40 **Si**, 0.15 **Cu**, 0.10–0.50 **Mn**, 1.7–2.4 **Mg**, 0.15 **Cr**, 0.15 **Zn**, 0.15 **Ti**, 0.50 **Fe**, 0.05 each, 0.15 total, others	
NF A 50 411	5251 (A–G2M)	France	Bar, wire, tube, and shapes	rem **Al**, 0.40 **Si**, 0.15 **Cu**, 0.10–0.50 **Mn**, 1.7–2.4 **Mg**, 0.15 **Cr**, 0.15 **Zn**, 0.15 **Ti**, 0.50 **Fe**, 0.05 each, 0.15 total, others	**Hard (drawn bar)**: ≤10 mm **diam**, 240 MPa **TS**, 200 MPa **YS**, 2% in 2 in. (50 mm) **El**; **Hard (wire)**: ≤6 mm **diam**, 240 MPa **TS**, 200 MPa **YS**, 2% in 2 in. (50 mm) **El**; **Hard (tube)**: ≤25 mm **diam**, 240 MPa **TS**, 200 MPa **YS**, 2% in 2 in. (50 mm) **El**
ISO R827	Al–Mg2	ISO	Extruded products	rem **Al**, 0.5 **Si**, 0.10 **Cu**, 0.5 **Mn**, 1.7–2.4 **Mg**, 0.35 **Cr**, 0.2 **Zn**, 0.5 **Fe**, 0.2 **Ti+Zr**, 0.5 **Mn+Cr**	**As manufactured**: all **diam**, 148 MPa **TS**, 59 MPa **YS**, 12% in 2 in. (50 mm) **El**
ISO R209	Al–Mg2	ISO	Wrought products	rem **Al**, 0.5 **Si**, 0.10 **Cu**, 0.5 **Mn**, 1.7–2.4 **Mg**, 0.35 **Cr**, 0.2 **Zn**, 0.5 **Fe**, 0.2 **Ti+Zr**	

WROUGHT ALUMINUM AND ALUMINUM ALLOYS

specification number	designation	country	product forms	chemical composition	mechanical properties and hardness values
ISO TR2778	Al–Mg2	ISO	Tube: drawn	rem **Al**, 0.5 **Si**, 0.10 **Cu**, 0.5 **Mn**, 1.7–2.4 **Mg**, 0.35 **Cr**, 0.2 **Zn**, 0.5 **Fe**, 0.2 **Ti+Zr**, 0.5 **Mn+Cr**	**Annealed**: all **diam**, 145 MPa **TS**, 60 MPa **YS**, 15% in 2 in. (50 mm) **El**; **Strain hardened**: 80 mm **diam**, 180 MPa **TS**, 110 MPa **YS**, 4% in 2 in. (50 mm) **El**
ISO TR2136	Al–Mg2	ISO	Sheet, plate	rem **Al**, 0.5 **Si**, 0.10 **Cu**, 0.5 **Mn**, 1.7–2.4 **Mg**, 0.35 **Cr**, 0.2 **Zn**, 0.5 **Fe**, 0.2 **Ti+Zr**, 0.5 **Mn+Cr**	**Annealed**: all **diam**, 150 MPa **TS**, 60 MPa **YS**, 16% in 2 in. (50 mm) **El**; **HB–Strain hardened**: all **diam**, 180 MPa **TS**, 110 MPa **YS**, 3% in 2 in. (50 mm) **El**; **HD– Strain hardened**: all **diam**, 200 MPa **TS**, 160 MPa **YS**, 2% in 2 in. (50 mm) **El**
SABS 712	Al–Mg2	South Africa	Wrought products	rem **Al**, 0.5 **Si**, 0.10 **Cu**, 0.5 **Mn**, 1.7–2.4 **Mg**, 0.25 **Cr**, 0.2 **Zn**, 0.5 **Fe**, 0.2 **Cr+Ti**, 0.5 **Cr+Mn**	
DS 3012	5051	Denmark	Rolled, drawn, and extruded products	rem **Al**, 0.40 **Si**, 0.25 **Cu**, 0.20 **Mn**, 1.7–2.2 **Mg**, 0.10 **Cr**, 0.25 **Zn**, 0.10 **Ti**, 0.7 **Fe**, 0.05 each, 0.15 total, others	**3/4 hard (rolled and drawn products)**: 0.5–6 mm **diam**, 245 MPa **TS**, 215 MPa **YS**, 4% in 2 in. (50 mm) **El**, 80 **HV**; **Hard (rolled and drawn products)**: 0.5–6 mm **diam**, 265 MPa **TS**, 235 MPa **YS**, 3% in 2 in. (50 mm) **El**, 85 **HV**; **As manufactured (extruded products)**: all **diam**, 145 MPa **TS**, 60 MPa **YS**, 12% in 2 in. (50 mm) **El**
AMS 4132B		US	Die and hand forgings and rings	rem **Al**, 0.10–0.25 **Si**, 1.9–2.7 **Cu**, 1.3–1.8 **Mg**, 0.9–1.2 **Ni**, 0.10 **Zn**, 0.04–0.10 **Ti**, 0.9–1.3 **Fe**, 0.05 each, 0.15 total, others	**Solution and precipitation heat treated (die forgings)**: ≤4 in. (102 mm) **diam**, 58 ksi (400 MPa) **TS**, 48 ksi (331 MPa) **YS**, 4% in 2 in. (50 mm) **El**, 115 min **HB**; **Solution and precipitation heat treated (rings)**: ≤4 in. (102 mm) **diam**, 55 ksi (379 MPa) **TS**, 41 ksi (283 MPa) **YS**, 6% in 2 in. (50 mm) **El**, 115 min **HB**; **Solution and precipitation heat treated (hand forgings)**: ≤2 in. (≤51 mm) **diam**, 58 ksi (400 MPa) **TS**, 47 ksi (324 MPa) **YS**, 7% in 2 in. (50 mm) **El**, 115 min **HB**
AMS 4112B		US	Bar, rod, and wire	rem **Al**, 0.50 **Si**, 3.8–4.9 **Cu**, 0.30–0.9 **Mn**, 1.2–1.8 **Mg**, 0.10 **Cr**, 0.25 **Zn**, 0.15 **Ti**, 0.50 **Fe**, 0.20 **Ti+Zr**, 0.05 each, 0.15 total, others	**Solution and precipitation heat treated**: 4.000 in. (101.60 mm) **diam**, 62 ksi (427 MPa) **TS**, 50 ksi (345 MPa) **YS**, 5% in 2 in. (50 mm) **El**, 125 min **HB**

WROUGHT ALUMINUM AND ALUMINUM ALLOYS

specification number	designation	country	product forms	chemical composition	mechanical properties and hardness values
AMS 4106		US	Sheet and plate	rem **Al**, 0.50 **Si**, 3.8–4.9 **Cu**, 0.30–0.9 **Mn**, 1.2–1.8 **Mg**, 0.10 **Cr**, 0.25 **Zn**, 0.50 **Fe**, 0.05 each, 0.15 total, others	**Solution and precipitation heat treated**: 0.063–0.249 in. **diam**, 71 ksi (490 MPa) **TS**, 66 ksi (455 MPa) **YS**, 4% in 2 in. (50 mm) **El**
AMS 4105		US	Sheet and plate	rem **Al**, 0.50 **Si**, 3.8–4.9 **Cu**, 0.30–0.9 **Mn**, 1.2–1.8 **Mg**, 0.10 **Cr**, 0.25 **Zn**, 0.50 **Fe**, 0.05 each, 0.15 total, others	**Solution and precipitation heat treated**: 0.020–0.062 in. **diam**, 70 ksi (483 MPa) **TS**, 62 ksi (427 MPa) **YS**, 3% in 2 in. (50 mm) **El**
AMS 4104		US	Sheet and plate	rem **Al**, 0.50 **Si**, 3.8–4.9 **Cu**, 0.30–0.9 **Mn**, 1.2–1.8 **Mg**, 0.10 **Cr**, 0.25 **Zn**, 0.50 **Fe**, 0.05 each, 0.15 total, others	**Solution and precipitation heat treated**: 0.020–0.062 in. **diam**, 72 ksi (496 MPa) **TS**, 66 ksi (455 MPa) **YS**, 3% in 2 in. (50 mm) **El**
AMS 4103		US	Sheet and plate	rem **Al**, 0.50 **Si**, 3.8–4.9 **Cu**, 0.30–0.9 **Mn**, 1.2–1.8 **Mg**, 0.10 **Cr**, 0.25 **Zn**, 0.50 **Fe**, 0.05 each, 0.15 total, others	**Solution and precipitation heat treated**: 0.020–0.062 in. **diam**, 72 ksi (496 MPa) **TS**, 66 ksi (455 MPa) **YS**, 3% in 2 in. (50 mm) **El**
NF A 50 451	5150 (A–85 GT)	France	Sheet, plate, and strip	rem **Al**, 0.08 **Si**, 0.10 **Cu**, 0.03 **Mn**, 1.3–1.7 **Mg**, 0.10 **Zn**, 0.6 **Ti**, 0.10 **Fe**, 0.03 each, 0.10 total, others	**Annealed**: 0.4–1.6 mm **diam**, 130 MPa **TS**, 25% in 2 in. (50 mm) **El**; **3/4 hard**: 0.4–3.2 mm **diam**, 190 MPa **TS**, 170 MPa **YS**, 3% in 2 in. (50 mm) **El**; **Hard**: 0.4–3.2 mm **diam**, 210 MPa **TS**, 190 MPa **YS**, 2% in 2 in. (50 mm) **El**
ANSI/ASTM B 547	5050	US	Tube: formed and arc–welded	rem **Al**, 0.40 **Si**, 0.20 **Cu**, 0.10 **Mn**, 1.1–1.8 **Mg**, 0.10 **Cr**, 0.25 **Zn**, 0.7 **Fe**, 0.05 each, 0.15 total, others	**Annealed**: 0.250–0.500 in. **diam**, 18 ksi (124 MPa) **TS**, 6 ksi (41 MPa) **YS**, 20% in 2 in. (50 mm) **El**; **H32**: 0.125–0.249 in. **diam**, 22 ksi (152 MPa) **TS**, 16 ksi (110 MPa) **YS**, 6% in 2 in. (50 mm) **El**; **H34**: 0.125–0.249 in. **diam**, 25 ksi (172 MPa) **TS**, 20 ksi (138 MPa) **YS**, 5% in 2 in. (50 mm) **El**
ANSI/ASTM B 483	5050	US	Tube: for general purposes	rem **Al**, 0.40 **Si**, 0.20 **Cu**, 0.10 **Mn**, 1.1–1.8 **Mg**, 0.10 **Cr**, 0.25 **Zn**, 0.7 **Fe**, 0.05 each, 0.15 total, others	**Annealed**: 0.018–0.500 in. **diam**, 18 ksi (124 MPa) **TS**, 6 ksi (41 MPa) **YS**; **H32 (round tube)**: 0.018–0.500 in. **diam**, 22 ksi (152 MPa) **TS**, 16 ksi (110 MPa) **YS**; **H34 (round tube)**: 0.018–0.500 in. **diam**, 25 ksi (172 MPa) **TS**, 20 ksi (138 MPa) **YS**

WROUGHT ALUMINUM AND ALUMINUM ALLOYS

specification number	designation	country	product forms	chemical composition	mechanical properties and hardness values
ANSI/ASTM B 313	5050	US	Tube: welded	rem **Al**, 0.40 **Si**, 0.20 **Cu**, 0.10 **Mn**, 1.1–1.8 **Mg**, 0.10 **Cr**, 0.25 **Zn**, 0.7 **Fe**, 0.45 **Fe+Si**, 0.05 each, 0.15 total, others	**Annealed**: 0.032–0.113 in. **diam**, 18 ksi (124 MPa) **TS**, 6 ksi (41 MPa) **YS**, 20% in 2 in. (50 mm) **El**; **H32**: 0.051–0.125 in. **diam**, 22 ksi (152 MPa) **TS**, 16 ksi (110 MPa) **YS**, 6% in 2 in. (50 mm) **El**; **H34**: 0.051–0.125 in. **diam**, 25 ksi (172 MPa) **TS**, 20 ksi (138 MPa) **YS**, 5% in 2 in. (50 mm) **El**
ANSI/ASTM B 210	5050	US	Tube: seamless	rem **Al**, 0.40 **Si**, 0.20 **Cu**, 0.10 **Mn**, 1.1–1.8 **Mg**, 0.10 **Cr**, 0.25 **Zn**, 0.7 **Fe**, 0.05 each, 0.15 total, others	**Annealed**: 0.018–0.500 in. **diam**, 18 ksi (124 MPa) **TS**, 6 ksi (41 MPa) **YS**; **H32 (round tube)**: 0.018–0.500 in. **diam**, 22 ksi (152 MPa) **TS**, 16 ksi (110 MPa) **YS**; **H34 (round tube)**: 0.018–0.500 in. **diam**, 25 ksi (172 MPa) **TS**, 20 ksi (138 MPa) **YS**
ANSI/ASTM B 209	5050	US	Sheet and plate	rem **Al**, 0.40 **Si**, 0.20 **Cu**, 0.10 **Mn**, 1.1–1.8 **Mg**, 0.10 **Cr**, 0.25 **Zn**, 0.7 **Fe**, 0.05 each, 0.15 total, others	**Annealed**: 0.250–3.000 in. **diam**, 18 ksi (124 MPa) **TS**, 6 ksi (41 MPa) **YS**, 20% in 2 in. (50 mm) **El**; **H32**: 0.051–0.249 in. **diam**, 22 ksi (152 MPa) **TS**, 16 ksi (110 MPa) **YS**, 6% in 2 in. (50 mm) **El**; **H34**: 0.051–0.249 in. **diam**, 25 ksi (172 MPa) **TS**, 20 ksi (138 MPa) **YS**, 5% in 2 in. (50 mm) **El**
AS 1867	5050A	Australia	Tube: drawn	rem **Al**, 0.40 **Si**, 0.20 **Cu**, 0.30 **Mn**, 1.1–1.8 **Mg**, 0.10 **Cr**, 0.25 **Zn**, 0.7 **Fe**, 0.05 each, 0.15 total, others	**Annealed**: all **diam**, 124 MPa **TS**, 41 MPa **YS**, 20% in 2 in. (50 mm) **El**; **Strain–hardened and stabilized to 1/4 hard**: all **diam**, 151 MPa **TS**, 110 MPa **YS**; **Strain–hardened and stabilized to full hard**: all **diam**, 199 MPa **TS**, 165 MPa **YS**
AS 1734	5050A	Australia	Sheet, plate	rem **Al**, 0.40 **Si**, 0.20 **Cu**, 0.30 **Mn**, 1.1–1.8 **Mg**, 0.10 **Cr**, 0.25 **Zn**, 0.7 **Fe**, 0.05 each, 0.15 total, others	**Annealed**: 0.20–0.50 mm **diam**, 124 MPa **TS**, 41 MPa **YS**, 16% in 2 in. (50 mm) **El**; **Strain–hardened to 1/2 hard**: 0.25–0.80 mm **diam**, 172 MPa **TS**, 137 MPa **YS**, 3% in 2 in. (50 mm) **El**; **Strain–hardened and stabilized to full hard**: 0.20–0.80 mm **diam**, 199 MPa **TS**, 2% in 2 in. (50 mm) **El**
COPANT 862	5050	COPANT	Wrought products	rem **Al**, 0.40 **Si**, 0.20 **Cu**, 0.10 **Mn**, 1.1–1.8 **Mg**, 0.10 **Cr**, 0.25 **Zn**, 0.70 **Fe**, 0.05 each, 0.15 total, others	

WROUGHT ALUMINUM AND ALUMINUM ALLOYS

specification number	designation	country	product forms	chemical composition	mechanical properties and hardness values
IS 736	51000–B	India	Plate	rem **Al**, 0.6 **Si**, 0.2 **Cu**, 0.7 **Mn**, 1.1–1.8 **Mg**, 0.1 **Cr**, 0.25 **Zn**, 0.7 **Fe**	**As manufactured**: 17 mm **diam**, 135 MPa **TS**, 17% in 2 in. (50 mm) **El**; **Annealed**: 19 mm **diam**, 125 MPa **TS**, 19% in 2 in. (50 mm) **El**; **Half hard**: 4 mm **diam**, 170 MPa **TS**, 4% in 2 in. (50 mm) **El**
ISO R827	Al–Mg1.5	ISO	Extruded products	rem **Al**, 0.4 **Si**, 0.20 **Cu**, 0.3 **Mn**, 1.1–1.8 **Mg**, 0.1 **Cr**, 0.2 **Zn**, 0.7 **Fe**, 0.2 **Ti+Zr**	**As manufactured**: all **diam**, 123 MPa **TS**, 12% in 2 in. (50 mm) **El**
ISO R209	Al–Mg1.5	ISO	Wrought products	rem **Al**, 0.4 **Si**, 0.20 **Cu**, 0.3 **Mn**, 1.1–1.8 **Mg**, 0.1 **Cr**, 0.2 **Zn**, 0.7 **Fe**, 0.2 **Ti+Zr**	
ISO TR2778	Al–Mg1.5	ISO	Tube: drawn	rem **Al**, 0.4 **Si**, 0.20 **Cu**, 0.3 **Mn**, 1.1–1.8 **Mg**, 0.1 **Cr**, 0.2 **Zn**, 0.7 **Fe**, 0.2 **Ti+Zr**	**Annealed**: all **diam**, 125 MPa **TS**, 18% in 2 in. (50 mm) **El**; **Strain hardened**: 50 mm **diam**, 165 MPa **TS**, 140 MPa **YS**, 3% in 2 in. (50 mm) **El**
ISO TR2136	Al–Mg1.5	ISO	Sheet, plate	rem **Al**, 0.4 **Si**, 0.20 **Cu**, 0.3 **Mn**, 1.1–1.8 **Mg**, 0.1 **Cr**, 0.2 **Zn**, 0.7 **Fe**, 0.2 **Ti+Zr**	**Annealed**: all **diam**, 125 MPa **TS**, 18% in 2 in. (50 mm) **El**; **HB–Strain hardened**: all **diam**, 145 MPa **TS**, 110 MPa **YS**, 3% in 2 in. (50 mm) **El**; **HD– Strain hardened**: all **diam**, 165 MPa **TS**, 140 MPa **YS**, 2% in 2 in. (50 mm) **El**
ANSI/ASTM B 209	5457	US	Sheet and plate	rem **Al**, 0.08 **Si**, 0.20 **Cu**, 0.15–0.45 **Mn**, 0.8–1.2 **Mg**, 0.05 **Zn**, 0.10 **Fe**, 0.05 **V**, 0.03 each, 0.10 total, others	**Annealed**: 0.030–0.090 in. **diam**, 16 ksi (110 MPa) **TS**, 20% in 2 in. (50 mm) **El**
AS 1734	5457	Australia	Sheet, plate	rem **Al**, 0.08 **Si**, 0.20 **Cu**, 0.15–0.45 **Mn**, 0.8–1.2 **Mg**, 0.03 **Zn**, 0.10 **Fe**, 0.03 each, 0.10 total, others	**Annealed**: 0.50–0.80 mm **diam**, 110 MPa **TS**, 18% in 2 in. (50 mm) **El**; **Strain–hardened and partially annealed to 5/8 hard**: 0.50–0.80 mm **diam**, 158 MPa **TS**, 4% in 2 in. (50 mm) **El**
CSA HA.4	.6061	Canada	Plate, sheet	rem **Al**, 0.40–0.8 **Si**, 0.15–0.40 **Cu**, 0.15 **Mn**, 0.8–1.2 **Mg**, 0.04–0.35 **Cr**, 0.25 **Zn**, 0.15 **Ti**, 0.7 **Fe**, 0.05 each, 0.15 total, others	**O**: 0.016–0.020 in. **diam**, 22 max ksi (152 max MPa) **TS**, 12 max ksi (83 max MPa) **YS**, 14% in 2 in. (50 mm) **El**; **T4 (sheet)**: 0.016–0.020 in. **diam**, 30.0 ksi (207 MPa) **TS**, 16.0 ksi (110 MPa) **YS**, 14% in 2 in. (50 mm) **El**; **T651 (plate)**: 0.250–0.499 in. **diam**, 42.0 ksi (290 MPa) **TS**, 35.0 ksi (241 MPa) **YS**, 10% in 2 in. (50 mm) **El**
COPANT 862	5457	COPANT	Wrought products	rem **Al**, 0.08 **Si**, 0.20 **Cu**, 0.15–0.45 **Mn**, 0.8–1.2 **Mg**, 0.05 **Zn**, 0.10 **Fe**, 0.05 **V**, 0.03 each, 0.10 total, others	

WROUGHT ALUMINUM AND ALUMINUM ALLOYS

specification number	designation	country	product forms	chemical composition	mechanical properties and hardness values
ANSI/ASTM B 531	5005	US	Rod: redraw, for electrical purposes	rem **Al**, 0.40 **Si**, 0.20 **Cu**, 0.20 **Mn**, 0.50–1.1 **Mg**, 0.10 **Cr**, 0.25 **Zn**, 0.7 **Fe**, 0.05 each, 0.15 total, others	**Annealed**: 0.375 in. (9.52 mm) **diam**, 14 ksi (97 MPa) **TS**; **H12 and H22**: 0.375 in. (9.52 mm) **diam**, 17 ksi (117 MPa) **TS**; **H14 and H24**: 20 ksi (138 MPa) **TS**
ANSI/ASTM B 483	5005	US	Tube: for general purposes	rem **Al**, 0.40 **Si**, 0.20 **Cu**, 0.20 **Mn**, 0.50–1.1 **Mg**, 0.10 **Cr**, 0.25 **Zn**, 0.7 **Fe**, 0.05 each, 0.15 total, others	**Annealed**: 0.018–0.500 in. **diam**, 15 ksi (103 MPa) **TS**, 5 ksi (34 MPa) **YS**
ANSI/ASTM B 316	5005	US	Wire and rod: rivet and cold-heading	rem **Al**, 0.30 **Si**, 0.20 **Cu**, 0.20 **Mn**, 0.50–1.1 **Mg**, 0.10 **Cr**, 0.25 **Zn**, 0.7 **Fe**, 0.05 each, 0.15 total, others	**Annealed**: $\leq$1.000 in. **diam**, 20 max ksi (138 max MPa) **TS**; **H32**: $\leq$1.000 in. **diam**, 17 ksi (117 MPa) **TS**
ANSI/ASTM B 210	5005	US	Tube: seamless	rem **Al**, 0.30 **Si**, 0.20 **Cu**, 0.20 **Mn**, 0.50–1.1 **Mg**, 0.10 **Cr**, 0.25 **Zn**, 0.7 **Fe**, 0.05 each, 0.15 total, others	**Annealed**: 0.018–0.500 in. **diam**, 15 ksi (103 MPa) **TS**, 5 ksi (34 MPa) **YS**
ANSI/ASTM B 209	5005	US	Sheet and plate	rem **Al**, 0.30 **Si**, 0.20 **Cu**, 0.20 **Mn**, 0.50–1.1 **Mg**, 0.10 **Cr**, 0.25 **Zn**, 0.7 **Fe**, 0.05 each, 0.15 total, others	**Annealed**: 0.250–3.000 in. **diam**, 15 ksi (103 MPa) **TS**, 5 ksi (34 MPa) **YS**, 22% in 2 in. (50 mm) **El**; **H12**: 0.500–2.000 in. **diam**, 18 ksi (124 MPa) **TS**, 14 ksi (97 MPa) **YS**, 10% in 2 in. (50 mm) **El**; **H14**: 0.500–1.000 in. **diam**, 21 ksi (145 MPa) **TS**, 17 ksi (117 MPa) **YS**, 10% in 2 in. (50 mm) **El**
ANSI/ASTM B 209	5657	US	Sheet and plate	rem **Al**, 0.08 **Si**, 0.10 **Cu**, 0.03 **Mn**, 0.6–1.0 **Mg**, 0.05 **Zn**, 0.10 **Fe**, 0.05 **V**, 0.03 **Ga**, 0.02 each, 0.05 total, others	**H25**: 0.030–0.090 in. **diam**, 20 ksi (138 MPa) **TS**, 8% in 2 in. (50 mm) **El**; **H26**: 0.030–0.090 in. **diam**, 22 ksi (152 MPa) **TS**, 7% in 2 in. (50 mm) **El**; **H28**: 0.030–0.090 in. **diam**, 25 ksi (172 MPa) **TS**, 5% in 2 in. (50 mm) **El**
AS 1734	5005	Australia	Sheet, plate	rem **Al**, 0.30 **Si**, 0.20 **Cu**, 0.20 **Mn**, 0.50–1.1 **Mg**, 0.10 **Cr**, 0.25 **Zn**, 0.7 **Fe**, 0.05 each, 0.15 total, others	**Annealed**: 0.15–0.20 mm **diam**, 103 MPa **TS**, 12% in 2 in. (50 mm) **El**; **Strain-hardened to 1/2 hard**: 0.25–0.80 mm **diam**, 144 MPa **TS**, 1% in 2 in. (50 mm) **El**; **Strain-hardened to full hard**: 0.15–0.80 mm **diam**, 186 MPa **TS**, 1% in 2 in. (50 mm) **El**
COPANT 862	5657	COPANT	Wrought products	rem **Al**, 0.08 **Si**, 0.10 **Cu**, 0.03 **Mn**, 0.6–1.0 **Mg**, 0.05 **Zn**, 0.10 **Fe**, 0.05 **V**, 0.03 **Ga**, 0.02 each, 0.05 total, others	
COPANT 862	5005	COPANT	Wrought products	rem **Al**, 0.30 **Si**, 0.20 **Cu**, 0.20 **Mn**, 0.5–1.10 **Mg**, 0.10 **Cr**, 0.25 **Zn**, 0.70 **Fe**, 0.05 each, 0.15 total, others	

WROUGHT ALUMINUM AND ALUMINUM ALLOYS

specification number	designation	country	product forms	chemical composition	mechanical properties and hardness values
DS 3012	5005	Denmark	Rolled and drawn products	rem **Al**, 0.40 **Si**, 0.20 **Cu**, 0.20 **Mn**, 0.50–1.1 **Mg**, 0.10 **Cr**, 0.25 **Zn**, 0.7 **Fe**, 0.05 each, 0.15 total, others	**Soft**: 0.5–25 mm **diam**, 80 MPa **TS**, 24% in 2 in. (50 mm) **El**, 30 **HV**; **1/2 hard**: 0.5–6 mm **diam**, 120 MPa **TS**, 100 MPa **YS**, 5% in 2 in. (50 mm) **El**, 45 **HV**; **3/4 hard**: 0.5–6 mm **diam**, 130 MPa **TS**, 120 MPa **YS**, 4% in 2 in. (50 mm) **El**, 55 **HV**
QQ–A–430B	5005	US	Rod, wire	rem **Al**, 0.4 **Si**, 0.20 **Cu**, 0.20 **Mn**, 0.50–1.1 **Mg**, 0.10 **Cr**, 0.25 **Zn**, 0.7 **Fe**, 0.05 each, 0.15 total, others	**Annealed**: 1.000 max **diam**, 20 max ksi (138 max MPa) **TS**; **Strain hardened and temper H32**: 1.000 max **diam**, 17 ksi (117 MPa) **TS**
SFS 2586	AlMg1	Finland	Shapes and sheet	rem **Al**, 0.4 **Si**, 0.20 **Cu**, 0.2 **Mn**, 0.5–1.1 **Mg**, 0.10 **Cr**, 0.2 **Zn**, 0.7 **Fe**, 0.2 **Ti+Zr**, 0.05 each, 0.15 total, others	**Hot finished (shapes)**: all **diam**, 100 MPa **TS**, 40 MPa **YS**, 17% in 2 in. (50 mm) **El**; **Annealed (sheet)**: all **diam**, 130 max MPa **TS**, 40 MPa **YS**, 24% in 2 in. (50 mm) **El**; **Hard (sheet)**: 3–6 mm **diam**, 160 MPa **TS**, 140 MPa **YS**, 4% in 2 in. (50 mm) **El**
NF A 50 411	5005 (A–G0.6)	France	Bar, wire, tube, and shapes	rem **Al**, 0.30 **Si**, 0.20 **Cu**, 0.20 **Mn**, 0.50–1.1 **Mg**, 0.10 **Cr**, 0.25 **Zn**, 0.7 **Fe**, 0.05 each, 0.15 total, others	**Hard (drawn bar)**: ≤10 mm **diam**, 185 MPa **TS**, 145 MPa **YS**, 2% in 2 in. (50 mm) **El**; **Hard (wire)**: ≤4 mm **diam**, 240 MPa **TS**, 200 MPa **YS**, 2% in 2 in. (50 mm) **El**; **1/2 hard (tube)**: ≤50 mm **diam**, 145 MPa **TS**, 105 MPa **YS**, 5% in 2 in. (50 mm) **El**
NF A 50 451	5005A ((A–G 0.6)	France	Sheet, plate, and strip	rem **Al**, 0.3 **Si**, 0.20 **Cu**, 0.20 **Mn**, 0.50–1.10 **Mg**, 0.10 **Cr**, 0.25 **Zn**, 0.70 **Fe**, 0.05 each, 0.15 total, others	**Annealed**: 0.4–3.2 mm **diam**, 100 MPa **TS**, 12% in 2 in. (50 mm) **El**; **3/4 hard**: 0.4–1.6 mm **diam**, 160 MPa **TS**, 140 MPa **YS**, 3% in 2 in. (50 mm) **El**; **Hard**: 0.4–1.6 mm **diam**, 180 MPa **TS**, 160 MPa **YS**, 3% in 2 in. (50 mm) **El**
IS 736	51000–A	India	Plate	rem **Al**, 0.6 **Si**, 0.2 **Cu**, 0.2 **Mn**, 0.5–1.1 **Mg**, 0.1 **Cr**, 0.25 **Zn**, 0.7 **Fe**	**As manufactured**: 20 mm **diam**, 105 MPa **TS**, 20% in 2 in. (50 mm) **El**; **Annealed**: 22 mm **diam**, 95 MPa **TS**, 22% in 2 in. (50 mm) **El**; **Half hard**: 4 mm **diam**, 140 MPa **TS**, 4% in 2 in. (50 mm) **El**
ISO R827	Al–Mg1	ISO	Extruded products	rem **Al**, 0.4 **Si**, 0.20 **Cu**, 0.2 **Mn**, 0.5–1.1 **Mg**, 0.1 **Cr**, 0.2 **Zn**, 0.7 **Fe**, 0.2 **Ti+Zr**	**As manufactured**: all **diam**, 98 MPa **TS**, 22% in 2 in. (50 mm) **El**
ISO R209	Al–Mg1	ISO	Wrought products	rem **Al**, 0.4 **Si**, 0.20 **Cu**, 0.2 **Mn**, 0.5–1.1 **Mg**, 0.1 **Cr**, 0.2 **Zn**, 0.7 **Fe**, 0.2 **Ti+Zr**	

WROUGHT ALUMINUM AND ALUMINUM ALLOYS

specification number	designation	country	product forms	chemical composition	mechanical properties and hardness values
ISO TR2778	Al–Mg1	ISO	Tube: drawn	rem **Al**, 0.4 **Si**, 0.20 **Cu**, 0.2 **Mn**, 0.5–1.1 **Mg**, 0.1 **Cr**, 0.2 **Zn**, 0.7 **Fe**, 0.2 **Ti+Zr**	**Annealed**: all **diam**, 95 MPa **TS**, 18% in 2 in. (50 mm) **El**; **Strain hardened**: all **diam**, 115 MPa **TS**, 4% in 2 in. (50 mm) **El**
ISO TR2136	Al–Mg1	ISO	Sheet, plate	rem **Al**, 0.4 **Si**, 0.20 **Cu**, 0.2 **Mn**, 0.5–1.1 **Mg**, 0.1 **Cr**, 0.2 **Zn**, 0.7 **Fe**, 0.2 **Ti+Zr**	**Annealed**: all **diam**, 95 MPa **TS**, 18% in 2 in. (50 mm) **El**; **HB–Strain hardened**: all **diam**, 115 MPa **TS**, 3% in 2 in. (50 mm) **El**; **HD– Strain hardened**: all **diam**, 140 MPa **TS**, 95 MPa **YS**, 2% in 2 in. (50 mm) **El**
JIS H 4000	5005	Japan	Sheet, plate, and strip	rem **Al**, 0.40 **Si**, 0.20 **Cu**, 0.20 **Mn**, 0.50–1.1 **Mg**, 0.10 **Cr**, 0.25 **Zn**, 0.7 **Fe**, 0.5 total, others	**Annealed**: <1.3 mm **diam**, 108 MPa **TS**, 39 MPa **YS**, 20% in 2 in. (50 mm) **El**; **Hard**: <0.8 mm **diam**, 177 MPa **TS**, 1% in 2 in. (50 mm) **El**; **Hot rolled**: <12 mm **diam**, 118 MPa **TS**, 8% in 2 in. (50 mm) **El**
SIS 14 41 06	SIS Al 41 06–14	Sweden	Strip	96.9 **Al**, 0.30 **Si**, 0.20 **Cu**, 0.20 **Mn**, 0.50–1.1 **Mg**, 0.10 **Cr**, 0.25 **Zn**, 0.7 **Fe**, 0.05 each, 0.15 total, others	**Strain hardened**: 0.35–2 mm **diam**, 145 MPa **TS**, 120 MPa **YS**, 6% in 2 in. (50 mm) **El**, 40–50 **HV**
SIS 14 41 06	SIS Al 41 06–24	Sweden	Sheet, strip	96.9 **Al**, 0.30 **Si**, 0.20 **Cu**, 0.20 **Mn**, 0.50–1.1 **Mg**, 0.10 **Cr**, 0.25 **Zn**, 0.7 **Fe**, 0.05 each, 0.15 total, others	**Strain hardened, annealed**: 0.35–3 mm **diam**, 145 MPa **TS**, 115 MPa **YS**, 10% in 2 in. (50 mm) **El**, 40–50 **HV**
SIS 14 41 06	SIS Al 41 06–26	Sweden	Sheet, strip	96.9 min **Al**, 0.30 **Si**, 0.20 **Cu**, 0.20 **Mn**, 0.50–1.1 **Mg**, 0.10 **Cr**, 0.25 **Zn**, 0.7 **Fe**, 0.05 each, 0.15 total, others	**Strain hardened, annealed**: 0.35–3 mm **diam**, 165 MPa **TS**, 135 MPa **YS**, 8% in 2 in. (50 mm) **El**, 50–60 **HV**
SIS 14 41 06	SIS Al 41 06–40	Sweden	Sheet, strip	96.9 min **Al**, 0.30 **Si**, 0.20 **Cu**, 0.20 **Mn**, 0.50–1.1 **Mg**, 0.10 **Cr**, 0.25 **Zn**, 0.7 **Fe**, 0.05 each, 0.15 total, others	**Annealed**: 0.35–6 mm **diam**, 95 MPa **TS**, 35 MPa **YS**, 25% in 2 in. (50 mm) **El**, 25–35 **HV**
SIS 14 41 06	SIS Al 41 06–44	Sweden	Sheet, strip	96.9 min **Al**, 0.30 **Si**, 0.20 **Cu**, 0.20 **Mn**, 0.50–1.1 **Mg**, 0.10 **Cr**, 0.25 **Zn**, 0.7 **Fe**, 0.05 each, 0.15 total, others	**Strain hardened**: 0.35–3 mm **diam**, 145 MPa **TS**, 120 MPa **YS**, 6% in 2 in. (50 mm) **El**, 40–50 **HV**
SIS 14 41 06	SIS Al 41 06–02	Sweden	Sheet, strip, section	96.9 min **Al**, 0.30 **Si**, 0.20 **Cu**, 0.20 **Mn**, 0.50–1.1 **Mg**, 0.10 **Cr**, 0.25 **Zn**, 0.7 **Fe**, 0.05 each, 0.15 total, others	**Annealed**: 0.35–6 mm **diam**, 95 MPa **TS**, 35 MPa **YS**, 25% in 2 in. (50 mm) **El**, 25 **HV**
SIS 14 41 06	SIS Al 41 06–18	Sweden	Strip, sheet	96.9 min **Al**, 0.30 **Si**, 0.20 **Cu**, 0.20 **Mn**, 0.50–1.1 **Mg**, 0.10 **Cr**, 0.25 **Zn**, 0.7 **Fe**, 0.05 each, 0.15 total, others	**Strain hardened**: 0.35–1.6 mm **diam**, 205 MPa **TS**, 185 MPa **YS**, 2% in 4 in. (100 mm) **El**, 60–75 **HV**

WROUGHT ALUMINUM AND ALUMINUM ALLOYS

specification number	designation	country	product forms	chemical composition	mechanical properties and hardness values
SIS 14 41 06	SIS Al 41 06–14	Sweden	Sheet, strip	96.9 min **Al**, 0.30 **Si**, 0.20 **Cu**, 0.20 **Mn**, 0.50–1.1 **Mg**, 0.10 **Cr**, 0.25 **Zn**, 0.7 **Fe**, 0.05 each, 0.15 total, others	**Strain hardened (sheet)**: 0.35–4 mm **diam**, 145 MPa **TS**, 120 MPa **YS**, 6% in 2 in. (50 mm) **El**, 40–50 **HV**; **Strain hardened (strip)**: 0.35–2 mm **diam**, 145 MPa **TS**, 120 MPa **YS**, 6% in 2 in. (50 mm) **El**, 40–50 **HV**
SIS 14 41 06	SIS Al 41 06–18	Sweden	Sheet	96.9 min **Al**, 0.30 **Si**, 0.20 **Cu**, 0.20 **Mn**, 0.50–1.1 **Mg**, 0.10 **Cr**, 0.25 **Zn**, 0.7 **Fe**, 0.05 each, 0.15 total, others	**Strain hardened, annealed**: 0.35–1.6 mm **diam**, 205 MPa **TS**, 185 MPa **YS**, 2% in 4 in. (100 mm) **El**, 60–75 **HV**
SABS 712	Al–Mg1	South Africa	Wrought products	rem **Al**, 0.10 **Si**, 0.05–0.15 **Cu**, 0.10 **Mn**, 0.5–1.1 **Mg**, 0.10 **Zn**, 0.10 **Fe**, 0.02 each, 0.05 total, others	
AIR 9051/A	A–G4MC (5086)	France	Sheet, bar, billet, and plate: for forging	rem **Al**, 0.25–0.8 **Si**, 3.5–4.5 **Cu**, 0.3–0.8 **Mn**, 0.5–1.0 **Mg**, 0.25 **Zn**, 0.20 **Ti**, 0.50 **Fe**, 0.25 **Ti+Zr**, 0.05 each, 0.15 total, others	**Rolled or annealed**: ≤120 mm **diam**, 240 MPa **TS**, 100 MPa **YS**, 16% in 2 in. (50 mm) **El**; **Annealed (bar)**: ≤250 mm **diam**, 240 MPa **TS**, 95 MPa **YS**, 16% in 2 in. (50 mm) **El**; **As cast or heat treated (billet and plate)**: ≤500 mm **diam**, 240 MPa **TS**, 95 MPa **YS**, 16% in 2 in. (50 mm) **El**
BS 1474	HE9	UK	Bar, tube, section	rem **Al**, 0.3–0.7 **Si**, 0.10 **Cu**, 0.1 **Mn**, 0.4–0.9 **Mg**, 0.10 **Cr**, 0.2 **Zn**, 0.2 **Ti**, 0.4 **Fe**	**Annealed**: ≤200 mm **diam**, 140 max MPa **TS**, 13% in 2 in. (50 mm) **El**; **As manufactured**: ≤200 mm **diam**, 100 MPa **TS**, 12% in 2 in. (50 mm) **El**; **Solution heat–treated and naturally aged**: 150–200 mm **diam**, 120 MPa **TS**, 70 MPa **YS**, 16% in 2 in. (50 mm) **El**
AMS 4110C		US	Bar and rod	rem **Al**, 0.20–0.8 **Si**, 3.5–4.5 **Cu**, 0.40–1.0 **Mn**, 0.40–0.8 **Mg**, 0.10 **Cr**, 0.25 **Zn**, 0.15 **Ti**, 0.7 **Fe**, 0.20 **Ti+Zr**, 0.05 each, 0.15 total, others	**Solution heat treated**: 0.500–8.000 in. (12.70–203.20 mm) **diam**, 55 ksi (379 MPa) **TS**, 32 ksi (221 MPa) **YS**, 12% in 2 in. (50 mm) **El**, 90 min **HB**
AS 1734	5557	Australia	Sheet, plate	rem **Al**, 0.10 **Si**, 0.15 **Cu**, 0.10–0.40 **Mn**, 0.40–0.8 **Mg**, 0.12 **Fe**, 0.03 each, 0.10 total, others	**Annealed**: 0.50–0.80 mm **diam**, 89 MPa **TS**, 17% in 2 in. (50 mm) **El**; **Strain–hardened and partially annealed to 5/8 hard**: 0.50–0.80 mm **diam**, 137 MPa **TS**, 4% in 2 in. (50 mm) **El**
COPANT 862	X 5257	COPANT	Wrought products	rem **Al**, 0.08 **Si**, 0.10 **Cu**, 0.03 **Mn**, 0.2–0.6 **Mg**, 0.03 **Zn**, 0.10 **Fe**, 0.02 each, 0.10 total, others	

WROUGHT ALUMINUM AND ALUMINUM ALLOYS

specification number	designation	country	product forms	chemical composition	mechanical properties and hardness values
JIS H 4000	5N01	Japan	Sheet, plate, and strip	rem **Al**, 0.20 **Si**, 0.20 **Cu**, 0.30 **Mn**, 0.20–0.6 **Mg**, 0.03 **Zn**, 0.30 **Fe**, 0.10 total, others	**Annealed**: <0.3 mm **diam**, 88 MPa **TS**, 10% in 2 in. (50 mm) **El**; **3/4 hard**: <0.3 mm **diam**, 147 MPa **TS**, 1% in 2 in. (50 mm) **El**; **Hard**: <0.3 mm **diam**, 167 MPa **TS**, 1% in 2 in. (50 mm) **El**
IS 738	NT5	India	Tube: drawn	rem **Al**, 0.6 **Si**, 0.10 **Cu**, 0.6 **Mn**, 2.8–4.0 **Mg**, 0.25 **Cr**, 0.7 **Fe**	**Annealed**: all **diam**, 216 MPa **TS**, 18% in 2 in. (50 mm) **El**; **Annealed and cold worked to 1/2 hard**: all **diam**, 245 MPa **TS**, 5% in 2 in. **El**
ANSI/ASTM B 234	6061	US	Tube: seamless, for condensers and heat exchangers	rem **Al**, 0.40–0.8 **Si**, 0.10 **Cu**, 0.50–1.0 **Mn**, 2.4–3.0 **Mg**, 0.05–0.20 **Cr**, 0.25 **Zn**, 0.20 **Ti**, 0.40 **Fe**, 0.05 each, 0.15 total, others	**Solution heat treated and naturally aged**: 0.050–0.200 in. **diam**, 30 ksi (207 MPa) **TS**, 16 ksi (110 MPa) **YS**, 16% in 2 in. (50 mm) **El**; **Solution heat treated and artificially aged**: 0.050–0.200 in. **diam**, 42 ksi (290 MPa) **TS**, 35 ksi (241 MPa) **YS**, 10% in 2 in. (50 mm) **El**
IS 739	NG4	India	Wire	rem **Al**, 0.6 **Si**, 0.10 **Cu**, 0.5 **Mn**, 1.7–2.8 **Mg**, 0.25 **Cr**, 0.2 **Zn**, 0.7 **Fe**	**Annealed**: all **diam**, 172 MPa **TS**; **Annealed and cold worked to hard**: all **diam**, 265 MPa **TS**
IS 738	NT4	India	Tube: drawn	rem **Al**, 0.6 **Si**, 0.10 **Cu**, 0.5 **Mn**, 1.7–2.8 **Mg**, 0.25 **Cr**, 0.7 **Fe**	**Annealed**: all **diam**, 172 MPa **TS**, 18% in 2 in. (50 mm) **El**; **Annealed and cold worked to 1/2 hard**: all **diam**, 230 MPa **TS**, 5% in 2 in. (50 mm) **El**
ANSI/ASTM B 247	6066	US	Die forgings	rem **Al**, 0.9–1.8 **Si**, 0.7–1.2 **Cu**, 0.6–1.1 **Mn**, 0.8–1.4 **Mg**, 0.40 **Cr**, 0.25 **Zn**, 0.20 **Ti**, 0.50 **Fe**, 0.05 each, 0.15 total, others	**Solution heat treated and artificially aged**: ≤4.000 in. **diam**, 50 ksi (345 MPa) **TS**, 45 ksi (310 MPa) **YS**, 8% in 2 in. (50 mm) **El**, 100 min **HB**
ANSI/ASTM B 221	6066	US	Bar, tube, rod, wire, and shapes	rem **Al**, 0.9–1.8 **Si**, 0.7–1.2 **Cu**, 0.6–1.1 **Mn**, 0.8–1.4 **Mg**, 0.40 **Cr**, 0.25 **Zn**, 0.20 **Ti**, 0.50 **Fe**, 0.05 each, 0.15 total, others	**Annealed**: all **diam**, 29 max ksi (200 max MPa) **TS**, 18 max ksi (124 max MPa) **YS**, 16% in 2 in. (50 mm) **El**; **Solution heat treated and naturally aged**: all **diam**, 40 ksi (276 MPa) **TS**, 25 ksi (172 MPa) **YS**, 14% in 2 in. (50 mm) **El**; **Solution heat treated and artificially aged**: all **diam**, 50 ksi (345 MPa) **TS**, 45 ksi (310 MPa) **YS**, 8% in 2 in. (50 mm) **El**

WROUGHT ALUMINUM AND ALUMINUM ALLOYS

specification number	designation	country	product forms	chemical composition	mechanical properties and hardness values
QQ-A-200/10D	6066	US	Bar, rod, shapes, tube, wire	rem **Al**, 0.9–1.8 **Si**, 0.7–1.2 **Cu**, 0.6–1.1 **Mn**, 0.8–1.4 **Mg**, 0.40 **Cr**, 0.25 **Zn**, 0.20 **Ti**, 0.50 **Fe**, 0.05 each, 0.15 total, others	**Annealed**: all **diam**, 29 max ksi (200 max MPa) **TS**, 18 max ksi (124 max MPa) **YS**, 16% in 2 in. (50 mm) **El**; **Solution heat treated and naturally aged**: all **diam**, 40 ksi (276 MPa) **TS**, 25 ksi (172 MPa) **YS**, 14% in 2 in. (50 mm) **El**; **Solution heat treated and artificially aged**: all **diam**, 50 ksi (345 MPa) **TS**, 45 ksi (310 MPa) **YS**, 8% in 2 in. (50 mm) **El**
COPANT 862	X 6080	COPANT	Wrought products	rem **Al**, 1.0–1.5 **Si**, 0.10 **Cu**, 0.5–1.5 **Mg**, 0.10 **Zn**, 0.10 **Ti**, 0.50 **Fe**, 0.50 **Mn + Cr**, 0.05 each, 0.15 total, others	
ANSI/ASTM B 345	6070	US	Tube: seamless	rem **Al**, 1.0–1.7 **Si**, 0.15–0.40 **Cu**, 0.40–1.0 **Mn**, 0.50–1.2 **Mg**, 0.10 **Cr**, 0.25 **Zn**, 0.15 **Ti**, 0.50 **Fe**, 0.05 each, 0.15 total, others	**Solution heat treated and artificially aged**: ≥2.999 in. **diam**, 48 ksi (331 MPa) **TS**, 45 ksi (310 MPa) **YS**, 6% in 2 in. (50 mm) **El**
MIL-A-46104		US	Bar, rod, shape, tube	rem **Al**, 1.0–1.7 **Si**, 0.15–0.40 **Cu**, 0.40–1.0 **Mn**, 0.50–1.2 **Mg**, 0.10 **Cr**, 0.25 **Zn**, 0.15 **Ti**, 0.50 **Fe**, 0.05 each, 0.15 total, others	**T6**: ≤2.999 in. **diam**, 331 MPa **TS**, 310 MPa **YS**, 6% in 2 in. (50 mm) **El**
COPANT 862	X 6082	COPANT	Wrought products	rem **Al**, 0.8–1.50 **Si**, 0.10 **Cu**, 0.20–1.0 **Mn**, 0.70–1.30 **Mg**, 0.10 **Zn**, 0.10 **Ti**, 0.50 **Fe**, 0.05 each, 0.15 total, others	
SFS 2593	AlSi1Mg	Finland	Shapes, sheet, rod, tube, and wire	rem **Al**, 0.6–1.6 **Si**, 0.10 **Cu**, 0.2–1.0 **Mn**, 0.4–1.4 **Mg**, 0.35 **Cr**, 0.2 **Zn**, 0.5 **Fe**, 0.2 **Ti + Zr**, 0.05 each, 0.15 total, others	**Hot finished (shapes)**: all **diam**, 150 MPa **TS**, 90 MPa **YS**, 12% in 2 in. (50 mm) **El**; **Annealed (except shapes)**: all **diam**, 160 max MPa **TS**, 40 MPa **YS**, 20% in 2 in. (50 mm) **El**; **Solution treated and aged (rod, tube, and wire)**: ≤6 mm **diam**, 300 MPa **TS**, 240 MPa **YS**, 8% in 2 in. (50 mm) **El**
ISO R829	Al-Si1Mg	ISO	Forgings	rem **Al**, 0.6–1.6 **Si**, 0.10 **Cu**, 0.4–1.0 **Mn**, 0.4–1.4 **Mg**, 0.35 **Cr**, 0.2 **Zn**, 0.5 **Fe**, 0.2 **Ti + Zr**	**Solution treated and naturally aged**: all **diam**, 186 MPa **TS**, 118 MPa **YS**, 15% in 2 in. (50 mm) **El**; **Solution and precipitation treated**: all **diam**, 294 MPa **TS**, 245 MPa **YS**, 8% in 2 in. (50 mm) **El**

WROUGHT ALUMINUM AND ALUMINUM ALLOYS

specification number	designation	country	product forms	chemical composition	mechanical properties and hardness values
ISO R827	Al–Si1Mg	ISO	Extruded products	rem **Al**, 0.6–1.6 **Si**, 0.10 **Cu**, 0.4–1.0 **Mn**, 0.4–1.4 **Mg**, 0.35 **Cr**, 0.2 **Zn**, 0.5 **Fe**, 0.2 **Ti+Zr**	**Solution treated and naturally aged**: 4.0 max in. (100 max mm) **diam**, 192 MPa **TS**, 118 MPa **YS**, 15% in 2 in. (50 mm) **El**; **Solution and precipitation treated**: 0.75 max in. (50 max mm) **diam**, 285 MPa **TS**, 245 MPa **YS**, 8% in 2 in. (50 mm) **El**
ISO R209	Al–Si1Mg	ISO	Wrought products	rem **Al**, 0.6–1.6 **Si**, 0.10 **Cu**, 0.4–1.0 **Mn**, 0.4–1.4 **Mg**, 0.35 **Cr**, 0.2 **Zn**, 0.5 **Fe**, 0.2 **Ti+Zr**	
ISO TR2778	Al–Si1Mg	ISO	Tube: drawn	rem **Al**, 0.6–1.6 **Si**, 0.10 **Cu**, 0.4–1.0 **Mn**, 0.4–1.4 **Mg**, 0.35 **Cr**, 0.2 **Zn**, 0.5 **Fe**, 0.2 **Ti+Zr**	**Annealed**: all **diam**, 160 max MPa **TS**, 15% in 2 in. (50 mm) **El**; **Solution heat treated and naturally aged**: all **diam**, 200 MPa **TS**, 110 MPa **YS**, 12% in 2 in. (50 mm) **El**; **Solution and precipitation treated**: all **diam**, 290 MPa **TS**, 240 MPa **YS**, 8% in 2 in. (50 mm) **El**
ISO TR2136	Al–Si1Mg	ISO	Sheet, plate	rem **Al**, 0.6–1.6 **Si**, 0.10 **Cu**, 0.4–1.0 **Mn**, 0.4–1.4 **Mg**, 0.35 **Cr**, 0.2 **Zn**, 0.5 **Fe**, 0.2 **Ti+Zr**	**Annealed**: all **diam**, 160 max MPa **TS**, 16% in 2 in. (50 mm) **El**; **Solution heat treated and naturally aged**: all **diam**, 200 MPa **TS**, 110 MPa **YS**, 15% in 2 in. (50 mm) **El**; **Solution heat treated and artificially aged**: all **diam**, 285 MPa **TS**, 235 MPa **YS**, 8% in 2 in. (50 mm) **El**
NS 17 305	NS 17 305–02	Norway	Sheet, strip, rod, and wire	rem **Al**, 0.6–1.6 **Si**, 0.10 **Cu**, 0.4–1.0 **Mn**, 0.4–1.4 **Mg**, 0.2 **Zn**, 0.5 **Fe**, 0.35 **Cr**, 0.2 **Ti+Zr**	**Soft annealed (sheet and strip)**: 0.5–25 mm **diam**, 16% in 2 in. (50 mm) **El**; **Soft annealed (rod and wire)**: <200 mm **diam**, 14% in 2 in. (50 mm) **El**
NS 17 305	NS 17 305–54	Norway	Sheet, strip, tube, rod, wire, and profiles	rem **Al**, 0.6–1.6 **Si**, 0.10 **Cu**, 0.4–1.0 **Mn**, 0.4–1.4 **Mg**, 0.2 **Zn**, 0.5 **Fe**, 0.35 **Cr**, 0.2 **Ti+Zr**	**Hardened and naturally aged (sheet and strip)**: 0.5–25 mm **diam**, 200 MPa **TS**, 108 MPa **YS**, 15% in 2 in. (50 mm) **El**; **Hardened and naturally aged (tube)**: <6 mm **diam**, 211 MPa **TS**, 113 MPa **YS**, 12% in 2 in. (50 mm) **El**; **Hardened and naturally aged (rod and wire)**: <150 mm **diam**, 186 MPa **TS**, 118 MPa **YS**, 14% in 2 in. (50 mm) **El**

WROUGHT ALUMINUM AND ALUMINUM ALLOYS

specification number	designation	country	product forms	chemical composition	mechanical properties and hardness values
NS 17 305	NS 17 305–56	Norway	Sheet, strip, tube, rod, wire, and profiles	rem **Al**, 0.6–1.6 **Si**, 0.10 **Cu**, 0.4–1.0 **Mn**, 0.4–1.4 **Mg**, 0.2 **Zn**, 0.5 **Fe**, 0.35 **Cr**, 0.2 **Ti+Zr**	**Hardened and artificially aged (sheet and strip)**: 0.5–25 mm **diam**, 284 MPa **TS**, 235 MPa **YS**, 8% in 2 in. (50 mm) **El**; **Hardened and artificially aged (tube)**: <6 mm **diam**, 304 MPa **TS**, 250 MPa **YS**, 7% in 2 in. (50 mm) **El**; **Hardened and artificially aged**: <20 mm **diam**, 289 MPa **TS**, 250 MPa **YS**, 7% in 2 in. (50 mm) **El**
NS 17 305	NS 17 305–07	Norway	Rod and wire	rem **Al**, 0.6–1.6 **Si**, 0.10 **Cu**, 0.4–1.0 **Mn**, 0.4–1.4 **Mg**, 0.2 **Zn**, 0.5 **Fe**, 0.35 **Cr**, 0.2 **Ti+Zr**	**Hot rolled**: <200 mm **diam**, 108 MPa **TS**, 12% in 2 in. (50 mm) **El**
CSA HA.6	.6053	Canada	Rivet, rod, wire	rem **Al**, 0.50–0.91 **Si**, 0.10 **Cu**, 1.1–1.4 **Mg**, 0.15–0.35 **Cr**, 0.10 **Zn**, 0.35 **Fe**, 0.05 each, 0.15 total, others	**H13**: 0.062–1.000 in. **diam**, 19.0 ksi (131 MPa) **TS**; **T61**: 0.062–1.000 in. **diam**, 30.0 ksi (207 MPa) **TS**, 20.0 ksi (137 MPa) **YS**, 14% in 2 in. (50 mm) **El**
QQ–A–430B	6053	US	Rod, wire	rem **Al**, 0.10 **Cu**, 1.1–1.4 **Mg**, 0.15–0.35 **Cr**, 0.10 **Zn**, 0.35 **Fe**, 0.05 each, 0.15 total, others; **Si** is 0.45–0.65 of **Mg** content	**Annealed**: 1.000 max in. **diam**, 19 max ksi (131 max MPa) **TS**; **Strain hardened and temper H13**: 1.000 max in. **diam**, 19 ksi (131 MPa) **TS**; **Solution heat treated and temper T61**: 0.063–1.000 in. **diam**, 30 ksi (207 MPa) **TS**, 20 ksi (138 MPa) **YS**, 14% in 2 in. (50 mm) **El**
BS 2L.85		UK	Forging stock, bar, section and forgings	rem **Al**, 0.8–1.3 **Si**, 1.0–2.0 **Cu**, 1.0 **Mn**, 0.5–1.2 **Mg**, 0.2 **Ni**, 0.2 **Zn**, 0.05 **Sn**, 0.7 **Fe**, 0.05 **Pb**, 0.20 **Zr+Ti**	**Solution heat and precipitation treated and quenched**: all **diam**, 390 MPa **TS**, 320 MPa **YS**, 8% in 2 in. (50 mm) **El**
NF A 50 411	6082 (A–SGM0.7)	France	Bar, tube, and shapes	rem **Al**, 0.7–1.30 **Si**, 0.10 **Cu**, 0.40–1.0 **Mn**, 0.6–1.20 **Mg**, 0.25 **Cr**, 0.20 **Zn**, 0.10 **Ti**, 0.50 **Fe**, 0.05 each, 0.015 total, others	**Aged (extruded bar)**: 200 MPa **TS**, 120 MPa **YS**, 15% in 2 in. (50 mm) **El**; **Aged (tube)**: ≤150 mm **diam**, 310 MPa **TS**, 282 MPa **YS**, 15% in 2 in. (50 mm) **El**; **Aged (shapes)**: 200 MPa **TS**, 120 MPa **YS**, 15% in 2 in. (50 mm) **El**
NF A 50 451	6181 (A–SG)	France	Sheet, plate, and strip	rem **Al**, 0.80–1.20 **Si**, 0.10 **Cu**, 0.15 **Mn**, 0.7–1.10 **Mg**, 0.30 **Cr**, 0.20 **Zn**, 0.10 **Ti**, 0.45 **Fe**, 0.05 each, 0.15 total, others	**Aged**: 0.4–6 mm **diam**, 210 MPa **TS**, 110 MPa **YS**, 18% in 2 in. (50 mm) **El**; **Quenched and tempered**: 0.4–3.2 mm **diam**, 250 MPa **TS**, 150 MPa **YS**, 18% in 2 in. (50 mm) **El**
SFS 2594	AlSi1MgPb	Finland	Shapes	rem **Al**, 0.6–1.4 **Si**, 0.10 **Cu**, 0.4–1.0 **Mn**, 0.6–1.2 **Mg**, 0.35 **Cr**, 0.5 **Zn**, 0.5 **Fe**, 1.0–3.0 **Pb+Sn+Cd+Bi+Sb**, 0.5 **Ti+Zr**, 0.05 each, 0.15 total, others	**Solution treated and aged**: all **diam**, 280 MPa **TS**, 180 MPa **YS**, 8% in 2 in. (50 mm) **El**

WROUGHT ALUMINUM AND ALUMINUM ALLOYS

specification number	designation	country	product forms	chemical composition	mechanical properties and hardness values
IS 738	HT19	India	Tube: drawn	rem **Al**, 0.60–1.3 **Si**, 0.10 **Cu**, 0.2 **Mn**, 0.4–1.5 **Mg**, 0.10 **Cr**, 0.6 **Fe**	**Solution treated and naturally aged:** ≤1.60 mm **diam**, 216 MPa **TS**, 108 MPa **YS**, 12% in 2 in. (50 mm) **El**; **Solution and precipitation heat treated:** ≤1.60 mm **diam**, 294 MPa **TS**, 230 MPa **YS**, 7% in 2 in. (50 mm) **El**
IS 733	64423	India	Bar, rod, section	rem **Al**, 0.7–1.3 **Si**, 0.5–1.0 **Cu**, 1.0 **Mn**, 0.5–1.3 **Mg**, 0.8 **Fe**,	**As manufactured or annealed:** all **diam**, 120 MPa **TS**, 10% in 2 in. (50 mm) **El**; **Solution treated and naturally aged:** all **diam**, 265 MPa **TS**, 155 MPa **YS**, 13% in 2 in. (50 mm) **El**; **Solution and precipitation heat treated:** >6.3 mm **diam**, 330 MPa **TS**, 265 MPa **YS**, 7% in 2 in. (50 mm) **El**
SIS 14 42 12	SIS Al 42 12–00	Sweden	Bar, rod	97.4 nom **Al**, 0.7–1.3 **Si**, 0.10 **Cu**, 0.40–1.0 **Mn**, 0.6–1.2 **Mg**, 0.25 **Cr**, 0.20 **Zn**, 0.10 **Ti**, 0.50 **Fe**, 0.05 each, 0.15 total, others	**Hot worked:** 10–200 mm **diam**, 150 MPa **TS**, 90 MPa **YS**, 12% in 2 in. (50 mm) **El**, 35–45 **HV**
SIS 14 42 12	SIS Al 42 12–02	Sweden	Plate, sheet, strip, bar, tube	97.4 nom **Al**, 0.7–1.3 **Si**, 0.10 **Cu**, 0.40–1.0 **Mn**, 0.6–1.2 **Mg**, 0.25 **Cr**, 0.20 **Zn**, 0.10 **Ti**, 0.50 **Fe**, 0.05 each, 0.15 total, others	**Annealed:** 5–30 mm **diam**, 22% in 2 in. (50 mm) **El**, 30–40 **HV**
SIS 14 42 12	SIS Al 42 12–04	Sweden	Plate, sheet, strip, bar, tube, section, forgings	97.4 nom **Al**, 0.7–1.3 **Si**, 0.10 **Cu**, 0.40–1.0 **Mn**, 0.6–1.2 **Mg**, 0.25 **Cr**, 0.20 **Zn**, 0.10 **Ti**, 0.50 **Fe**, 0.05 each, 0.15 total, others	**Naturally aged:** 5–30 mm **diam**, 205 MPa **TS**, 115 MPa **YS**, 18% in 2 in. (50 mm) **El**, 65–95 **HV**
SIS 14 42 12	SIS Al 42 12–06	Sweden	Plate, sheet, strip, bar, tube, section, forgings	97.4 **Al**, 0.7–1.3 **Si**, 0.10 **Cu**, 0.40–1.0 **Mn**, 0.6–1.2 **Mg**, 0.25 **Cr**, 0.20 **Zn**, 0.10 **Ti**, 0.05 each, 0.15 total, others	**Artificially aged:** 0.5–5 mm **diam**, 290 MPa **TS**, 245 MPa **YS**, 8% in 2 in. (50 mm) **El**, 85–120 **HV**
SIS 14 42 12	SIS Al 42 12–10	Sweden	Sheet, plate, strip	97.4 **Al**, 0.7–1.3 **Si**, 0.10 **Cu**, 0.40–1.0 **Mn**, 0.6–1.2 **Mg**, 0.25 **Cr**, 0.20 **Zn**, 0.10 **Ti**, 0.50 **Fe**, 0.05 each, 0.15 total, others	**Strain hardened:** 0.5–5 mm **diam**, 170 MPa **TS**, 140 MPa **YS**, 10% in 2 in. (50 mm) **El**, 45–70 **HV**
BS 1471	HT30	UK	Tube: drawn	rem **Al**, 0.7–1.3 **Si**, 0.10 **Cu**, 0.40–1.0 **Mn**, 0.5–1.2 **Mg**, 0.25 **Cr**, 0.2 **Zn**, 0.2 **Ti**, 0.5 **Fe**	**Solution heat treated and naturally aged:** ≤6.0 mm **diam**, 215 MPa **TS**, 115 MPa **YS**, 12% in 2 in. (50 mm) **El**; **Solution heat treated and precipitation treated:** ≤6.0 mm **diam**, 310 MPa **TS**, 255 MPa **YS**, 7% in 2 in. (50 mm) **El**
BS 1473	HB30	UK	Bolt, screw	rem **Al**, 0.7–1.3 **Si**, 0.10 **Cu**, 0.4–1.0 **Mn**, 0.5–1.2 **Mg**, 0.25 **Cr**, 0.2 **Zn**, 0.2 **Ti**, 0.5 **Fe**	**Solution heat treated and precipitation treated:** ≤6 mm **diam**, 295 MPa **TS**, 255 MPa **YS**

WROUGHT ALUMINUM AND ALUMINUM ALLOYS

specification number	designation	country	product forms	chemical composition	mechanical properties and hardness values
BS 1473	HR30	UK	Rivet	rem **Al**, 0.7–1.3 **Si**, 0.10 **Cu**, 0.4–1.0 **Mn**, 0.5–1.2 **Mg**, 0.25 **Cr**, 0.2 **Zn**, 0.2 **Ti**, 0.5 **Fe**	**Solution heat treated and naturally aged:** ≤25 mm **diam**, 200 MPa **TS**
BS 1470	HS30	UK	Plate, sheet, strip	rem **Al**, 0.7–1.3 **Si**, 0.10 **Cu**, 0.4–1.0 **Mn**, 0.5–1.2 **Mg**, 0.25 **Cr**, 0.2 **Zn**, 0.2 **Ti**, 0.5 **Fe**	**Annealed**: 0.2–3.0 mm **diam**, 155 max MPa **TS**, 16% in 2 in. (50 mm) **El**; **Solution heat treated and naturally aged:** 0.2–3.0 mm **diam**, 200 MPa **TS**, 120 MPa **YS**, 15% in 2 in. (50 mm) **El**; **Solution heat treated and precipitation treated:** 0.2–3.0 mm **diam**, 255 MPa **YS**, 8% in 2 in. (50 mm) **El**
BS 1472	HF30	UK	Forging stock and forgings	rem **Al**, 0.7–1.3 **Si**, 0.10 **Cu**, 0.40–1.0 **Mn**, 0.5–1.2 **Mg**, 0.25 **Cr**, 0.2 **Zn**, 0.2 **Ti**, 0.5 **Fe**	**Solution heat treated and naturally aged:** ≤150 mm **diam**, 185 MPa **TS**, 120 MPa **YS**, 16% in 2 in. (50 mm) **El**; **Solution heat treated and precipitation treated:** ≤150 mm **diam**, 295 MPa **TS**, 255 MPa **YS**, 8% in 2 in. (50 mm) **El**
BS 1474	HE30	UK	Bar, tube, section	rem **Al**, 0.7–1.3 **Si**, 0.10 **Cu**, 0.40–1.0 **Mn**, 0.5–1.2 **Mg**, 0.25 **Cr**, 0.2 **Zn**, 0.2 **Ti**, 0.5 **Fe**	**Annealed**: ≤200 mm **diam**, 170 max MPa **TS**, 14% in 2 in. (50 mm) **El**; **As manufactured:** ≤200 mm **diam**, 110 MPa **TS**, 12% in 2 in. (50 mm) **El**; **Solution heat treated ad naturally aged**: 150–200 mm **diam**, 170 MPa **TS**, 100 MPa **YS**
BS L.115		UK	Plate	rem **Al**, 0.7–1.3 **Si**, 0.10 **Cu**, 0.4–1.0 **Mn**, 0.5–1.2 **Mg**, 0.25 **Cr**, 0.10 **Ni**, 0.2 **Zn**, 0.2 **Ti**, 0.05 **Sn**, 0.5 **Fe**, 0.05 **Pb**	**Solution heat and precipitation treated and quenched:** ≤25 mm **diam**, 295 MPa **TS**, 240 MPa **YS**, 8% in 2 in. (50 mm) **El**
BS L.114		UK	Tube	rem **Al**, 0.7–1.3 **Si**, 0.10 **Cu**, 0.4–1.0 **Mn**, 0.5–1.2 **Mg**, 0.25 **Cr**, 0.10 **Ni**, 0.2 **Zn**, 0.2 **Ti**, 0.05 **Sn**, 0.5 **Fe**, 0.05 **Pb**	**Solution heat and precipitation treated and quenched:** ≤6.0 mm **diam**, 310 MPa **TS**, 255 MPa **YS**, 7% in 2 in. (50 mm) **El**
BS L.113		UK	Sheet, strip	rem **Al**, 0.7–1.3 **Si**, 0.10 **Cu**, 0.4–1.0 **Mn**, 0.5–1.2 **Mg**, 0.25 **Cr**, 0.10 **Ni**, 0.2 **Zn**, 0.2 **Ti**, 0.05 **Sn**, 0.5 **Fe**, 0.05 **Pb**	**Solution heat and precipitation treated and quenched:** 0.2–3.0 mm **diam**, 295 MPa **TS**, 255 MPa **YS**, 8% in 2 in. (50 mm) **El**
BS L.112		UK	Forging stock and forgings	rem **Al**, 0.7–1.3 **Si**, 0.10 **Cu**, 0.4–1.0 **Mn**, 0.5–1.2 **Mg**, 0.25 **Cr**, 0.10 **Ni**, 0.2 **Zn**, 0.2 **Ti**, 0.05 **Sn**, 0.5 **Fe**, 0.05 **Pb**	**Solution heat and precipitation treated and quenched:** all **diam**, 295 MPa **TS**, 255 MPa **YS**, 8% in 2 in. (50 mm) **El**
BS L.111		UK	Bar, section	rem **Al**, 0.7–1.3 **Si**, 0.10 **Cu**, 0.4–1.0 **Mn**, 0.5–1.2 **Mg**, 0.25 **Cr**, 0.10 **Ni**, 0.2 **Zn**, 0.2 **Ti**, 0.05 **Sn**, 0.5 **Fe**, 0.05 **Pb**	**Solution heat and precipitation treated and quenched:** 2–150 mm **diam**, 310 MPa **TS**, 270 MPa **YS**, 8% in 2 in. (50 mm) **El**

WROUGHT ALUMINUM AND ALUMINUM ALLOYS

specification number	designation	country	product forms	chemical composition	mechanical properties and hardness values
IS 739	HG30	India	Wire	rem **Al**, 0.60–1.3 **Si**, 0.10 **Cu**, 0.40–1.0 **Mn**, 0.4–1.4 **Mg**, 0.3 **Cr**, 0.1 **Zn**, 0.6 **Fe**	**Annealed**: all **diam**, 172 max MPa **TS**; **Solution treated and naturally aged**: all **diam**, 201 MPa **TS**
IS 738	HT30	India	Tube: drawn	rem **Al**, 0.60–1.3 **Si**, 0.10 **Cu**, 0.40–1.0 **Mn**, 0.4–1.4 **Mg**, 0.3 **Cr**, 0.6 **Fe**	**Solution treated and naturally aged**: ≤1.60 mm **diam**, 216 MPa **TS**, 108 MPa **YS**, 12% in 2 in. (50 mm) **El**; **Solution and precipitation heat treated**: ≤1.60 mm **diam**, 309 MPa **TS**, 246 MPa **YS**, 7% in 2 in. (50 mm) **El**
ANSI/ASTM B 209	6003	US	Sheet and plate	rem **Al**, 0.35–1.0 **Si**, 0.10 **Cu**, 0.8 **Mn**, 0.8–1.5 **Mg**, 0.35 **Cr**, 0.20 **Zn**, 0.10 **Ti**, 0.6 **Fe**, 0.05 each, 0.15 total, others	
CSA HA.4	.6003	Canada	Plate, sheet (cladding for .2014 alclad)	rem **Al**, 0.35–1.0 **Si**, 0.10 **Cu**, 0.8 **Mn**, 0.8–1.5 **Mg**, 0.35 **Cr**, 0.20 **Zn**, 0.10 **Ti**, 0.6 **Fe**, 0.05 each, 0.15 total, others	**T3 (flat sheet)**: 0.029–0.039 in. **diam**, 55.0 ksi (379 MPa) **TS**, 34.0 ksi (234 MPa) **YS**, 14% in 2 in. (50 mm) **El**; **T451 (plate)**: 0.250–0.499 in. **diam**, 57.0 ksi (393 MPa) **TS**, 15% in 2 in. (50 mm) **El**; **T6 (sheet)**: 0.020–0.039 in. **diam**, 63.0 ksi (434 MPa) **TS**, 55.0 ksi (379 MPa) **YS**, 7% in 2 in. (50 mm) **El**
NF A 50 411	6181 (A-SG)	France	Bar and shapes	rem **Al**, 0.8–1.20 **Si**, 0.10 **Cu**, 0.15 **Mn**, 0.6–1.0 **Mg**, 0.10 **Cr**, 0.20 **Zn**, 0.10 **Ti**, 0.45 **Fe**	**Aged (bar)**: 200 MPa **TS**, 100 MPa **YS**, 15% in 2 in. (50 mm) **El**; **Quenched and tempered (bar)**: 280 MPa **TS**, 240 MPa **YS**, 8% in 2 in. (50 mm) **El**; **Quenched and tempered (shapes)**: 275 MPa **TS**, 200 MPa **YS**, 8% in 2 in. (50 mm) **El**
IS 736	64430	India	Plate	rem **Al**, 0.6–1.3 **Si**, 0.1 **Cu**, 0.4–1.0 **Mn**, 0.4–1.2 **Mg**, 0.25 **Cr**, 0.1 **Zn**, 0.2 **Ti**, 0.6 **Fe**	**Solution heat treated and naturally aged**: 15 mm **diam**, 200 MPa **TS**, 115 MPa **YS**, 15% in 2 in. (50 mm) **El**; **Solution and precipitation heat treated**: 8 mm **diam**, 285 MPa **TS**, 240 MPa **YS**, 8% in 2 in. (50 mm) **El**
IS 733	64430	India	Bar, rod, section	rem **Al**, 0.6–1.3 **Si**, 0.1 **Cu**, 0.4–1.0 **Mn**, 0.4–1.2 **Mg**, 0.25 **Cr**, 0.1 **Zn**, 0.2 **Ti**, 0.6 **Fe**	**As manufactured or annealed**: all **diam**, 110 MPa **TS**, 12% in 2 in. (50 mm) **El**; **Solution treated and naturally aged**: ≤150 mm **diam**, 185 MPa **TS**, 120 MPa **YS**, 14% in 2 in. (50 mm) **El**; **Solution and precipitation heat treated**: ≤5 mm **diam**, 295 MPa **TS**, 255 MPa **YS**, 7% in 2 in. (50 mm) **El**

WROUGHT ALUMINUM AND ALUMINUM ALLOYS

specification number	designation	country	product forms	chemical composition	mechanical properties and hardness values
NF A 50 451	6081 (A–SGM 0.3)	France	Sheet, plate, and strip	rem **Al**, 0.7–1.10 **Si**, 0.10 **Cu**, 0.10–0.45 **Mn**, 0.6–1.0 **Mg**, 0.10 **Cr**, 0.20 **Zn**, 0.15 **Ti**, 0.50 **Fe**, 0.05 each, 0.15 total, others	**Aged**: 0.4–6 mm **diam**, 210 MPa **TS**, 110 MPa **YS**, 18% in 2 in. (50 mm) **El**; **Quenched and tempered**: 0.4–3.2 mm **diam**, 250 MPa **TS**, 150 MPa **YS**, 18% in 2 in. (50 mm) **El**
AMS 4150F		US	Bar, tube, rod, wire, rings, and shapes	rem **Al**, 0.40–0.8 **Si**, 0.15–0.40 **Cu**, 0.15 **Mn**, 0.8–1.2 **Mg**, 0.04–0.35 **Cr**, 0.25 **Zn**, 0.15 **Ti**, 0.7 **Fe**, 0.05 each, 0.15 total, others	**Extruded, solution and precipitation heat treated (except rings)**: ≥0.250 in. (≥6.35 mm) **diam**, 38 ksi (262 MPa) **TS**, 35 ksi (241 MPa) **YS**, 10% in 2 in. (50 mm) **El**, 80 min **HB**
AMS 4310		US	Rings: rolled or forged	rem **Al**, 0.40–0.8 **Si**, 0.15–0.40 **Cu**, 0.15 **Mn**, 0.8–1.2 **Mg**, 0.04–0.35 **Cr**, 0.25 **Zn**, 0.15 **Ti**, 0.7 **Fe**, 0.05 each, 0.15 total, others	**Solution and precipitation heat treated**: ≥4 in. (≥102 mm) **diam**, 38 ksi (262 MPa) **TS**, 35 ksi (241 MPa) **YS**, 10% in 2 in. (50 mm) **El**, 80 min **HB**
AMS 4079B		US	Tube: seamless	rem **Al**, 0.40–0.8 **Si**, 0.15–0.40 **Cu**, 0.15 **Mn**, 0.8–1.2 **Mg**, 0.04–0.35 **Cr**, 0.25 **Zn**, 0.15 **Ti**, 0.7 **Fe**, 0.05 each, 0.15 total, others	**Annealed**: 0.018–0.500 in. **diam**, 22 ksi (152 MPa) **TS**, 14 ksi (97 MPa) **YS**, 15% in 2 in. (50 mm) **El**; **Solution and precipitation heat treated**: 0.025–0.049 in. **diam**, 42 ksi (290 MPa) **TS**, 35 ksi (241 MPa) **YS**, 10% in 2 in. (50 mm) **El**, 50 min **HRB**
AMS 4080J		US	Tube: seamless	rem **Al**, 0.40–0.8 **Si**, 0.15–0.40 **Cu**, 0.15 **Mn**, 0.8–1.2 **Mg**, 0.04–0.35 **Cr**, 0.25 **Zn**, 0.15 **Ti**, 0.7 **Fe**, 0.05 each, 0.15 total, others	**Annealed**: 0.018–0.500 in. **diam**, 22 ksi (152 MPa) **TS**, 14 ksi (97 MPa) **YS**, 15% in 2 in. (50 mm) **El**; **Solution and precipitation heat treated**: 0.025–0.049 in. **diam**, 42 ksi (290 MPa) **TS**, 35 ksi (241 MPa) **YS**, 10% in 2 in. (50 mm) **El**
AMS 4081C		US	Tube: seamless	rem **Al**, 0.40–0.8 **Si**, 0.15–0.40 **Cu**, 0.15 **Mn**, 0.8–1.2 **Mg**, 0.04–0.35 **Cr**, 0.25 **Zn**, 0.15 **Ti**, 0.7 **Fe**, 0.05 each, 0.15 total, others	**Solution heat treated**: 0.025–0.049 in. **diam**, 30 ksi (207 MPa) **TS**, 16 ksi (110 MPa) **YS**, 16% in 2 in. (50 mm) **El**; **Precipitation heat treated**: 0.025–0.049 in. **diam**, 42 ksi (290 MPa) **TS**, 35 ksi (241 MPa) **YS**, 10% in 2 in. (50 mm) **El**
AMS 4082J		US	Tube: seamless	rem **Al**, 0.40–0.8 **Si**, 0.15–0.40 **Cu**, 0.15 **Mn**, 0.8–1.2 **Mg**, 0.04–0.35 **Cr**, 0.25 **Zn**, 0.15 **Ti**, 0.7 **Fe**, 0.05 each, 0.15 total, others	**Solution and precipitation heat treated**: 0.025–0.049 in. (0.64–1.24 mm) **diam**, 42 ksi (290 MPa) **TS**, 35 ksi 241 MPa **YS**, 10% in 2 in. (50 mm) **El**

WROUGHT ALUMINUM AND ALUMINUM ALLOYS

specification number	designation	country	product forms	chemical composition	mechanical properties and hardness values
AMS 4083G		US	Tube: seamless	rem **Al**, 0.40–0.8 **Si**, 0.15–0.40 **Cu**, 0.15 **Mn**, 0.8–1.2 **Mg**, 0.04–0.35 **Cr**, 0.25 **Zn**, 0.15 **Ti**, 0.7 **Fe**, 0.05 each, 0.15 total, others	**Solution and precipitation heat treated**: 0.025–0.049 in. (0.64–1.24 mm) **diam**, 42 ksi (290 MPa) **TS**, 35 ksi (241 MPa) **YS**, 10% in 2 in. (50 mm) **El**
AMS 4025F		US	Sheet and plate	rem **Al**, 0.40–0.8 **Si**, 0.15–0.40 **Cu**, 0.15 **Mn**, 0.8–1.2 **Mg**, 0.04–0.35 **Cr**, 0.25 **Zn**, 0.15 **Ti**, 0.7 **Fe**	**Annealed**: 0.006–0.012 in. (0.15–0.18 mm) **diam**, 22 ksi (152 MPa) **TS**, 10% in 2 in. (50 mm) **El**; **Solution and precipitation heat treated**: 0.006–0.007 in. (0.15–0.18 mm) **diam**, 42 ksi (290 MPa) **TS**, 35 ksi (241 MPa) **YS**, 4% in 2 in. (50 mm) **El**
AMS 4026G		US	Sheet and plate	rem **Al**, 0.40–0.8 **Si**, 0.15–0.40 **Cu**, 0.15 **Mn**, 0.8–1.2 **Mg**, 0.04–0.35 **Cr**, 0.25 **Zn**, 0.15 **Ti**, 0.7 **Fe**, 0.05 each, 0.15 total, others	**Solution heat treated**: 0.006–0.007 in. (0.15–0.18 mm) **diam**, 30 ksi (207 MPa) **TS**, 16 ksi (110 MPa) **YS**, 10% in 2 in. (50 mm) **El**; **Precipitation heat treated**: 0.006–0.007 in. (0.15–0.18 mm) **diam**, 42 ksi (290 MPa) **TS**, 35 ksi (241 MPa) **YS**, 4% in 2 in. (50 mm) **El**
AMS 4027H		US	Sheet and plate	rem **Al**, 0.40–0.8 **Si**, 0.15–0.40 **Cu**, 0.15 **Mn**, 0.8–1.2 **Mg**, 0.04–0.35 **Cr**, 0.25 **Zn**, 0.15 **Ti**, 0.7 **Fe**, 0.05 each, 0.15 total, others	**Solution and precipitation heat treated**: 0.006–0.007 in. (0.15–0.18 mm) **diam**, 42 ksi (290 MPa) **TS**, 35 ksi (241 MPa) **YS**, 4% in 2 in. (50 mm) **El**
AMS 4009		US	Foil	rem **Al**, 0.40–0.8 **Si**, 0.15–0.40 **Cu**, 0.15 **Mn**, 0.8–1.2 **Mg**, 0.04–0.35 **Cr**, 0.25 **Zn**, 0.15 **Ti**, 0.7 **Fe**, 0.05 each, 0.15 total, others	**Annealed**: all **diam**, 22 max ksi (152 max MPa) **TS**; **Solution heat treated**: all **diam**, 42 ksi (290 max MPa) **TS**
AMS 4021D		US	Sheet and plate: alclad	rem **Al**, 0.40–0.8 **Si**, 0.15–0.40 **Cu**, 0.15 **Mn**, 0.8–1.2 **Mg**, 0.04–0.35 **Cr**, 0.25 **Zn**, 0.15 **Ti**, 0.7 **Fe**, 0.05 each, 0.15 total, others	**Annealed**: 0.010–0.020 in. (0.25–0.51 mm) **diam**, 20 ksi (138 MPa) **TS**, 12 ksi (83 MPa) **YS**, 14% in 2 in. (50 mm) **El**; **Solution and precipitation heat treated**: 0.010–0.020 in. (0.25–0.51 mm) **diam**, 38 ksi (262 MPa) **TS**, 32 ksi (221 MPa) **YS**, 8% in 2 in. (50 mm) **El**
AMS 4022E		US	Sheet and plate: alclad	rem **Al**, 0.40–0.8 **Si**, 0.15–0.40 **Cu**, 0.15 **Mn**, 0.8–1.2 **Mg**, 0.04–0.35 **Cr**, 0.25 **Zn**, 0.15 **Ti**, 0.7 **Fe**, 0.05 each, 0.15 total, others	**Solution heat treated**: 0.010–0.020 in. (0.25–0.51 mm) **diam**, 27 ksi (186 MPa) **TS**, 14 ksi (97 MPa) **YS**, 14% in 2 in. (50 mm) **El**; **Precipitation heat treated**: 0.010–0.020 in. (0.25–0.51 mm) **diam**, 38 ksi (262 MPa) **TS**, 32 ksi (221 MPa) **YS**, 8% in 2 in. (50 mm) **El**

WROUGHT ALUMINUM AND ALUMINUM ALLOYS

specification number	designation	country	product forms	chemical composition	mechanical properties and hardness values
AMS 4023E		US	Sheet and plate: alclad	rem **Al**, 0.40–0.8 **Si**, 0.15–0.40 **Cu**, 0.15 **Mn**, 0.8–1.2 **Mg**, 0.04–0.35 **Cr**, 0.25 **Zn**, 0.15 **Ti**, 0.7 **Fe**, 0.05 each, 0.15 total, others	**Solution and precipitation heat treated**: 0.010–0.020 in. (0.25–0.51 mm) **diam**, 38 ksi (262 MPa) **TS**, 32 ksi (221 MPa) **YS**, 8% in 2 in. (50 mm) **El**
AMS 4146B		US	Die and hand forgings and rings	rem **Al**, 0.40–0.8 **Si**, 0.15–0.40 **Cu**, 0.15 **Mn**, 0.8–1.2 **Mg**, 0.04–0.35 **Cr**, 0.25 **Zn**, 0.15 **Ti**, 0.7 **Fe**, 0.05 each, 0.15 total, others	**Precipitation heat treated (die forgings)**: $\leq$4 in. ($\leq$102 mm) **diam**, 38 ksi (262 MPa) **TS**, 35 ksi (241 MPa) **YS**, 7% in 2 in. (50 mm) **El**, 80 min **HB**; **Precipitation heat treated (rings and hand forgings)**: $\leq$2.5 in. ($\leq$64 mm) **diam**, 38 ksi (262 MPa) **TS**, 35 ksi (241 MPa) **YS**, 10% in 2 in. (50 mm) **El**, 80 min **HB**
AMS 4127D		US	Die forgings and rings	rem **Al**, 0.40–0.8 **Si**, 0.15–0.40 **Cu**, 0.15 **Mn**, 0.80–1.2 **Mg**, 0.04–0.35 **Cr**, 0.25 **Zn**, 0.15 **Ti**, 0.7 **Fe**, 0.05 each, 0.15 total, others	**Solution and precipitation heat treated**: $\leq$4 in. ($\leq$102 mm) **diam**, 38 ksi (262 MPa) **TS**, 35 ksi (241 MPa) **YS**, 10% in 2 in. (50 mm) **El**, 80 min **HB**
AMS 4128		US	Bar and rod	rem **Al**, 0.40–0.8 **Si**, 0.15–0.40 **Cu**, 0.15 **Mn**, 0.8–1.2 **Mg**, 0.04–0.35 **Cr**, 0.25 **Zn**, 0.15 **Ti**, 0.7 **Fe**, 0.05 each, 0.15 total, others	**Solution heat treated**: $\leq$0.500 in. **diam**, 30 ksi (207 MPa) **TS**, 16 ksi (110 MPa) **YS**, 18% in 2 in. (50 mm) **El**; **Precipitation heat treated**: $\leq$0.500 in. **diam**, 42 ksi (290 MPa) **TS**, 35 ksi (241 MPa) **YS**, 10% in 2 in. (50 mm) **El**, 80 min **HB**
AMS 4129		US	Bar and rod	rem **Al**, 0.40–0.8 **Si**, 0.15–0.40 **Cu**, 0.15 **Mn**, 0.8–1.2 **Mg**, 0.04–0.35 **Cr**, 0.25 **Zn**, 0.15 **Ti**, 0.7 **Fe**, 0.05 each, 0.15 total, others	**Solution and precipitation heat treated**: $\leq$0.500 in. **diam**, 42 ksi (290 MPa) **TS**, 35 ksi (241 MPa) **YS**, 10% in 2 in. (50 mm) **El**, 80 min **HB**
AMS 4116D		US	Bar, rod, and wire	rem **Al**, 0.40–0.8 **Si**, 0.15–0.40 **Cu**, 0.15 **Mn**, 0.8–1.2 **Mg**, 0.04–0.35 **Cr**, 0.25 **Zn**, 0.15 **Ti**, 0.7 **Fe**, 0.05 each, 0.15 total, others	**Solution heat treated**: all **diam**, 30 ksi (207 MPa) **TS**, 16 ksi (110 MPa) **YS**, 18% in 2 in. (50 mm) **El**, 50–80 **HB**; **Precipitation heat treated**: all **diam**, 42 ksi (290 MPa) **TS**, 35 ksi (241 MPa) **YS**, 10% in 2 in. (50 mm) **El**
AMS 4117D		US	Bar, rod, wire, and rings	rem **Al**, 0.40–0.8 **Si**, 0.15–0.40 **Cu**, 0.15 **Mn**, 0.8–1.2 **Mg**, 0.04–0.35 **Cr**, 0.25 **Zn**, 0.15 **Ti**, 0.7 **Fe**, 0.05 each, 0.15 total, others	**Solution and precipitation heat treated**: all **diam**, 42 ksi (290 MPa) **TS**, 35 ksi (241 MPa) **YS**, 10% in 2 in. (50 mm) **El**, 80 min **HB**

WROUGHT ALUMINUM AND ALUMINUM ALLOYS

specification number	designation	country	product forms	chemical composition	mechanical properties and hardness values
AMS 4115C		US	Bar, rod, wire, and rings	rem **Al**, 0.40–0.8 **Si**, 0.15–0.40 **Cu**, 0.15 **Mn**, 0.8–1.2 **Mg**, 0.04–0.35 **Cr**, 0.25 **Zn**, 0.15 **Ti**, 0.7 **Fe**, 0.05 each, 0.15 total, others	**Annealed**: all **diam**, 22 ksi (152 MPa) **TS**, 2% in 2 in. (50 mm) **El**, 40 min **HB**; **Solution and precipitation heat treated**: all **diam**, 42 ksi (290 MPa) **TS**, 35 ksi (241 MPa) **YS**, 10% in 2 in. (50 mm) **El**, 80 min **HB**
AMS 4113		US	Shapes: structural	rem **Al**, 0.40–0.8 **Si**, 0.15–0.40 **Cu**, 0.15 **Mn**, 0.8–1.2 **Mg**, 0.04–0.35 **Cr**, 0.25 **Zn**, 0.15 **Ti**, 0.7 **Fe**, 0.05 each, 0.15 total, others	**Solution and precipitation heat treated**: $\leq$0.250 in. **diam**, 38 ksi (262 MPa) **TS**, 35 ksi (241 MPa) **YS**, 8% in 2 in. (50 mm) **El**, 80 min **HB**
ANSI/ASTM B 483	6061	US	Tube	rem **Al**, 0.40–0.8 **Si**, 0.15–0.40 **Cu**, 0.15 **Mn**, 0.8–1.2 **Mg**, 0.04–0.35 **Cr**, 0.25 **Zn**, 0.15 **Ti**, 0.7 **Fe**, 0.05 each, 0.15 total, others	**Annealed**: 0.018–0.500 in. **diam**, 22 max ksi (152 max MPa) **TS**, 14 max ksi (97 Max MPa) **YS**, 15% in 2 in. (50 mm) **El**; **Solution heat treated and naturally aged (round tube)**: 0.260–0.500 in. **diam**, 30 ksi (207 MPa) **TS**, 16 ksi (110 MPa) **YS**, 18% in 2 in. (50 mm) **El**; **Solution heat treated and artificially aged (round tube)**: 0.025–0.049 in. **diam**, 42 ksi (290 MPa) **TS**, 35 ksi (241 MPa) **YS**, 8% in 2 in. (50 mm) **El**
ANSI/ASTM B 483	6262	US	Tube: for general purposes	rem **Al**, 0.40–0.8 **Si**, 0.15–0.40 **Cu**, 0.15 **Mn**, 0.8–1.2 **Mg**, 0.04–0.14 **Cr**, 0.25 **Zn**, 0.15 **Ti**, 0.7 **Fe**, 0.05 each, 0.15 total, others	**Solution heat treated and artificially aged**: 0.260–0.500 in. **diam**, 42 ksi (290 MPa) **TS**, 35 ksi (241 MPa) **YS**, 12% in 2 in. (50 mm) **El**; **Solution heat treated, artificially aged**: 0.025–0.375 in. **diam**, 48 ksi (331 MPa) **TS**, 44 ksi (303 MPa) **YS**, 4% in 2 in. (50 mm) **El**
ANSI/ASTM B 429	6061	US	Pipe and tube: structural	rem **Al**, 0.40–0.8 **Si**, 0.15–0.40 **Cu**, 0.15 **Mn**, 0.8–1.2 **Mg**, 0.04–0.35 **Cr**, 0.25 **Zn**, 0.15 **Ti**, 0.7 **Fe**, 0.05 each, 0.15 total, others	**Solution heat treated and artificially aged**: $\leq$0.250 in. **diam**, 38 ksi (262 MPa) **TS**, 35 ksi (241 MPa) **YS**, 10% in 2 in. (50 mm) **El**

WROUGHT ALUMINUM AND ALUMINUM ALLOYS

specification number	designation	country	product forms	chemical composition	mechanical properties and hardness values
ANSI/ASTM B 404	6061	US	Tube: seamless, for condensers and heat exchangers	rem **Al**, 0.40–0.8 **Si**, 0.15–0.40 **Cu**, 0.15 **Mn**, 0.8–1.2 **Mg**, 0.04–0.35 **Cr**, 0.15 **Ti**, 0.7 **Fe**, 0.05 each, 0.15 total, others	**Annealed**: 0.018–0.500 in. **diam**, 22 max ksi (152 max MPa) **TS**, 14 max ksi (97 max MPa) **YS**, 15% in 2 in. (50 mm) **El**; **Solution heat treated and naturally aged**: 0.260–0.500 in. **diam**, 30 ksi (207 MPa) **TS**, 16 ksi (110 MPa) **YS**, 18% in 2 in. (50 mm) **El**; **Solution heat treated and artificially aged**: 0.260–0.500 in. **diam**, 42 ksi (290 MPa) **TS**, 35 ksi (241 MPa) **YS**, 12% in 2 in. (50 mm) **El**
ANSI/ASTM B 345	6061	US	Pipe and tube: seamless	rem **Al**, 0.40–0.8 **Si**, 0.15–0.40 **Cu**, 0.15 **Mn**, 0.8–1.2 **Mg**, 0.04–0.35 **Cr**, 0.25 **Zn**, 0.15 **Ti**, 0.7 **Fe**, 0.05 each, 0.15 total, others	**Solution heat treated and artificially aged (pipe)**: ≥1 **diam**, 38 ksi (262 MPa) **TS**, 35 ksi (241 MPa) **YS**; **Annealed (tube)**: all **diam**, 22 max ksi (152 max MPa) **TS**, 16 ksi (110 MPa) **YS**, 16% in 2 in. (50 mm) **El**; **Solution heat treated and naturally aged (tube)**: all **diam**, 26 ksi (179 MPa) **TS**, 16 ksi (110 MPa) **YS**, 16% in 2 in. (50 mm) **El**
ANSI/ASTM B 316	6061	US	Wire and rod: rivet and cold-heading	rem **Al**, 0.40–0.8 **Si**, 0.15–0.40 **Cu**, 0.15 **Mn**, 0.8–1.2 **Mg**, 0.04–0.35 **Cr**, 0.25 **Zn**, 0.15 **Ti**, 0.7 **Fe**, 0.05 each, 0.15 total, others	**Annealed**: ≤1.000 in. **diam**, 22 max ksi (152 max MPa) **TS**; **H13**: ≤1.000 in. **diam**, 22 ksi (152 MPa) **TS**; **Solution heat treated and artificially aged**: 0.061–1.000 in. **diam**, 42 ksi (290 MPa) **TS**, 35 ksi (241 MPa) **YS**, 10% in 2 in. (50 mm) **El**
ANSI/ASTM B 313	6061	US	Tube: welded	rem **Al**, 0.40–0.8 **Si**, 0.15–0.40 **Cu**, 0.15 **Mn**, 0.8–1.2 **Mg**, 0.04–0.35 **Cr**, 0.25 **Zn**, 0.15 **Ti**, 0.7 **Fe**, 0.05 each, 0.15 total, others	**Annealed**: 0.032–0.125 in. **diam**, 22 max ksi (152 max MPa) **TS**, 12 max ksi (83 max MPa) **YS**, 16% in 2 in. (50 mm) **El**; **Solution heat treated and naturally aged**: 0.032–0.125 in. **diam**, 30 ksi (207 MPa) **TS**, 16 ksi (110 MPa) **YS**, 16% in 2 in. (50 mm) **El**; **Solution heat treated and artificially agd**: 0.032–0.125 in. **diam**, 42 ksi (290 MPa) **TS**, 35 ksi (241 MPa) **YS**, 10% in 2 in. (50 mm) **El**

specification number	designation	country	product forms	chemical composition	mechanical properties and hardness values
ANSI/ASTM B 247	6061	US	Die and hand forgings	rem **Al**, 0.40–0.8 **Si**, 0.15–0.40 **Cu**, 0.15 **Mn**, 0.8–1.2 **Mg**, 0.04–0.35 **Cr**, 0.25 **Zn**, 0.15 **Ti**, 0.7 **Fe**, 0.05 each, 0.15 total, others	**Solution heat treated and artificially aged (die forgings)**: ≤4.000 in. **diam**, 38 ksi (262 MPa) **TS**, 35 ksi (241 MPa) **YS**, 7% in 2 in. (50 mm) **El**, 80 min **HB**; **Solution heat treated and artificially aged (hand forgings)**: ≤4.000 in. **diam**, 38 ksi (262 MPa) **TS**, 35 ksi (241 MPa) **YS**, 10% in 2 in. (50 mm) **El**
ANSI/ASTM B 241	6061	US	Tube: seamless	rem **Al**, 0.40–0.8 **Si**, 0.15–0.40 **Cu**, 0.15 **Mn**, 0.8–1.2 **Mg**, 0.04–0.35 **Cr**, 0.25 **Zn**, 0.15 **Ti**, 0.7 **Fe**, 0.05 each, 0.15 total, others	**Annealed**: all **diam**, 22 max ksi (152 max MPa) **TS**, 16 max ksi (110 max MPa) **YS**, 16% in 2 in. (50 mm) **El**; **Solution heat treated and naturally aged**: all **diam**, 26 ksi (179 MPa) **TS**, 16 ksi (110 MPa) **YS**, 16% in 2 in. (50 mm) **El**; **Solution heat treated and artificially aged**: ≥0.250 in. **diam**, 38 ksi (262 MPa) **TS**, 35 ksi (241 MPa) **YS**, 10% in 2 in. (50 mm) **El**
ANSI/ASTM B 221	6351	US	Bar, tube, rod, wire, and shapes	rem **Al**, 0.7–1.3 **Si**, 0.10 **Cu**, 0.40–0.8 **Mn**, 0.40–0.8 **Mg**, 0.20 **Zn**, 0.20 **Ti**, 0.50 **Fe**, 0.05 each, 0.15 total, others	**Cooled and naturally aged**: ≤0.499 in. **diam**, 26 ksi (179 MPa) **TS**, 13 ksi (90 MPa) **YS**, 15% in 2 in. (50 mm) **El**; **Solution heat treated and naturally aged**: ≤0.749 in. **diam**, 32 ksi (221 MPa) **TS**, 19 ksi (131 MPa) **YS**, 16% in 2 in. (50 mm) **El**; **Cooled and artificially aged**: 0.250–1.000 in. **diam**, 38 ksi (262 MPa) **TS**, 35 ksi (241 MPa) **YS**, 10% in 2 in. (50 mm) **El**
ANSI/ASTM B 221	6061	US	Bar, tube, rod, wire, and shapes	rem **Al**, 0.40–0.8 **Si**, 0:15–0.40 **Cu**, 0.15 **Mn**, 0.8–1.2 **Mg**, 0.04–0.35 **Cr**, 0.25 **Zn**, 0.15 **Ti**, 0.7 **Fe**, 0.05 each, 0.15 total, others	**Annealed**: all **diam**, 22 max ksi (152 max MPa) **TS**, 16 max ksi (110 max MPa) **YS**, 16% in 2 in. (50 mm) **El**; **Cooled and naturally aged**: ≤0.625 in. **diam**, 26 ksi (179 MPa) **TS**, 14 ksi (97 MPa) **YS**, 16% in 2 in. (50 mm) **El**; **Solution heat treted and naturally aged**: all **diam**, 26 ksi (179 MPa) **TS**, 16 ksi (110 MPa) **YS**, 16% in 2 in. (50 mm) **El**

WROUGHT ALUMINUM AND ALUMINUM ALLOYS

specification number	designation	country	product forms	chemical composition	mechanical properties and hardness values
ANSI/ASTM B 221	6262	US	Bar, tube, rod, wire, and shapes	rem **Al**, 0.40–0.8 **Si**, 0.15–0.40 **Cu**, 0.15 **Mn**, 0.8–1.2 **Mg**, 0.04–0.14 **Cr**, 0.25 **Zn**, 0.15 **Ti**, 0.7 **Fe**, 0.40–0.7 **Pb**, 0.40–0.7 **Bi**, 0.05 each, 0.15 total, others	**Solution heat treated and artificially aged**: all **diam**, 38 ksi (262 MPa) **TS**, 35 ksi (241 MPa) **YS**, 10% in 2 in. (50 mm) **El**
ANSI/ASTM B 211	6262	US	Bar, rod, and wire	rem **Al**, 0.40–0.8 **Si**, 0.15–0.40 **Cu**, 0.15 **Mn**, 0.8–1.2 **Mg**, 0.04–0.14 **Cr**, 0.25 **Zn**, 0.15 **Ti**, 0.7 **Fe**, 0.40–0.7 **Bi**, 0.05 each, 0.15 total, others	**Solution heat treated and artificially aged**: 0.125–4.000 in. **diam**, 42 ksi (290 MPa) **TS**, 35 ksi (241 MPa) **YS**, 10% in 2 in. (50 mm) **El**; **Solution heat treated, artificially aged, and cold worked**: 0.125–2.000 in. **diam**, 52 ksi (359 MPa) **TS**, 48 ksi (331 MPa) **YS**, 5% in 2 in. (50 mm) **El**
ANSI/ASTM B 211	6061	US	Bar, rod, and wire	rem **Al**, 0.40–0.8 **Si**, 0.15–0.40 **Cu**, 0.15 **Mn**, 0.8–1.2 **Mg**, 0.04–0.35 **Cr**, 0.25 **Zn**, 0.15 **Ti**, 0.7 **Fe**, 0.05 each, 0.15 total, others	**Annealed**: 0.125–8.000 in. **diam**, 22 max ksi (152 max MPa) **TS**, 18% in 2 in. (50 mm) **El**; **Solution heat treated and naturally aged**: 0.125–8.000 in. **diam**, 30 ksi (207 MPa) **TS**, 16 ksi (110 MPa) **YS**, 18% in 2 in. (50 mm) **El**; **Solution heat treated and artificially aged**: 0.125–8.000 in. **diam**, 42 ksi (290 MPa) **TS**, 35 ksi (241 MPa) **YS**, 10% in 2 in. (50 mm) **El**
ANSI/ASTM B 209	Alclad 6061	US	Sheet and plate	rem **Al**, 0.40–0.8 **Si**, 0.15–0.40 **Cu**, 0.15 **Mn**, 0.8–1.2 **Mg**, 0.04–0.35 **Cr**, 0.25 **Zn**, 0.15 **Ti**, 0.7 **Fe**, 0.05 each, 0.15 total, others	**Annealed**: 0.129–0.499 in. **diam**, 20 max ksi (138 max MPa) **TS**, 12 max ksi (83 max MPa) **YS**, 18% in 2 in. (50 mm) **El**; **Solution heat treated and naturally aged**: 0.021–0.249 in. **diam**, 27 ksi (186 MPa) **TS**, 14 ksi (97 MPa) **YS**, 16% in 2 in. (50 mm) **El**; **Solution heat treated and artificially aged**: 0.021–0.249 in. **diam**, 38 ksi (262 MPa) **TS**, 32 ksi (221 MPa) **YS**, 10% in 2 in. (50 mm) **El**

WROUGHT ALUMINUM AND ALUMINUM ALLOYS

specification number	designation	country	product forms	chemical composition	mechanical properties and hardness values
ANSI/ASTM B 210	6061	US	Tube: seamless	rem **Al**, 0.40–0.8 **Si**, 0.15–0.40 **Cu**, 0.15 **Mn**, 0.8–1.2 **Mg**, 0.04–0.35 **Cr**, 0.25 **Zn**, 0.15 **Ti**, 0.7 **Fe**, 0.05 each, 0.15 total, others	**Annealed**: 0.018–0.500 in. **diam**, 22 max ksi (152 max MPa) **TS**, 14 ksi (97 MPa) **YS**, 15% in 2 in. (50 mm) **El**; **Solution heat treated and naturally aged (round tube)**: 0.260–0.500 in. **diam**, 30 ksi (207 MPa) **TS**, 16 ksi (110 MPa) **YS**, 18% in 2 in. (50 mm) **El**; **Solution heat treated and artificially aged (round tube)**: 0.260–0.500 in. **diam**, 42 ksi (290 MPa) **TS**, 35 ksi (241 MPa) **YS**, 12% in 2 in. (50 mm) **El**
ANSI/ASTM B 210	6262	US	Tube: seamless	rem **Al**, 0.40–0.8 **Si**, 0.15–0.40 **Cu**, 0.15 **Mn**, 0.8–1.2 **Mg**, 0.04–0.14 **Cr**, 0.25 **Zn**, 0.15 **Ti**, 0.7 **Fe**, 0.40–0.7 **Pb**, 0.40–0.7 **Bi**, 0.05 each, 0.15 total, others	**Solution heat treated and artificially aged**: 0.260–0.500 in. **diam**, 42 ksi (290 MPa) **TS**, 35 ksi (241 MPa) **YS**, 12% in 2 in. (50 mm) **El**; T9: 0.025–0.375 in. **diam**, 48 ksi (331 MPa) **TS**, 44 ksi (303 MPa) **YS**, 4% in 2 in. (50 mm) **El**
ANSI/ASTM B 209	6061	US	Sheet and plate	rem **Al**, 0.40–0.8 **Si**, 0.15–0.40 **Cu**, 0.15 **Mn**, 0.8–1.2 **Mg**, 0.04–0.35 **Cr**, 0.25 **Zn**, 0.15 **Ti**, 0.7 **Fe**, 0.05 each, 0.15 total, others	**Annealed**: 1.001–3.000 in. **diam**, 22 max ksi (152 max MPa) **TS**, 16% in 2 in. (50 mm) **El**; **Solution heat treated and naturally aged**: 0.021–0.249 in. **diam**, 30 ksi (207 MPa) **TS**, 16 ksi (110 MPa) **YS**, 16% in 2 in. (50 mm) **El**; **Solution heat treated and artificially aged**: 0.021–0.249 in. **diam**, 42 ksi (290 MPa) **TS**, 35 ksi (241 MPa) **YS**, 10% in 2 in. (50 mm) **El**
ASTM F 467	A 96061	US	Nut	rem **Al**, 0.40–0.8 **Si**, 0.15–0.40 **Cu**, 0.15 **Mn**, 0.8–1.2 **Mg**, 0.04–0.35 **Cr**, 0.25 **Zn**, 0.15 **Ti**, 0.7 **Fe**	**As agreed**: all **diam**, 40 min **HRB**
ASTM F 467	A 96262	US	Nut	rem **Al**, 0.40–0.8 **Si**, 0.15–0.40 **Cu**, 0.15 **Mn**, 0.8–1.2 **Mg**, 0.04–0.14 **Cr**, 0.25 **Zn**, 0.15 **Ti**, 0.7 **Fe**	**As agreed**: all **diam**, 60 min **HRB**
AS 1867	6061	Australia	Tube: drawn	rem **Al**, 0.40–0.8 **Si**, 0.15–0.40 **Cu**, 0.15 **Mn**, 0.8–1.2 **Mg**, 0.04–0.35 **Cr**, 0.25 **Zn**, 0.15 **Ti**, 0.7 **Fe**, 0.05 each, 0.15 total, others	**Annealed**: all **diam**, 152 max MPa **TS**, 96 MPa **YS**, 15% in 2 in. (50 mm) **El**; **Solution heat treated and naturally aged**: 0.6–1.2 mm **diam**, 216 MPa **TS**, 110 MPa **YS**, 12% in 2 in. (50 mm) **El**; **Solution heat treated, cold worked, and artificially aged**: all **diam**, 342 MPa **TS**, 268 MPa **YS**, 8% in 2 in. (50 mm) **El**

WROUGHT ALUMINUM AND ALUMINUM ALLOYS

specification number	designation	country	product forms	chemical composition	mechanical properties and hardness values
AS 1866	6351	Australia	Rod, bar, shape	rem **Al**, 0.7–1.3 **Si**, 0.10 **Cu**, 0.40–0.8 **Mn**, 0.40–0.8 **Mg**, 0.20 **Zn**, 0.20 **Ti**, 0.50 **Fe**, 0.05 each, 0.15 total, others	**Solution heat–treated and naturally aged**: 150.0 mm **diam**, 185 MPa **TS**, 115 MPa **YS**, 16% in 2 in. (50 mm) **El**; **Cooled and artificially aged**: all **diam**, 262 MPa **TS**, 241 MPa **YS**, 8% in 2 in. (50 mm) **El**; **Solution heat treated and artificially aged**: 150.0 mm **diam**, 293 MPa **TS**, 255 MPa **YS**, 8% in 2 in. (50 mm) **El**
AS 1866	6262	Australia	Rod, bar, shape	rem **Al**, 0.40–0.8 **Si**, 0.15–0.40 **Cu**, 0.15 **Mn**, 0.8–1.2 **Mg**, 0.04–0.14 **Cr**, 0.25 **Zn**, 0.15 **Ti**, 0.7 **Fe**, 0.05 each, 0.15 total, others	**Solution heat treated and artificially aged**: all **diam**, 262 MPa **TS**, 241 MPa **YS**, 8% in 2 in. (50 mm) **El**
AS 1866	6061	Australia	Rod, bar, shape	rem **Al**, 0.40–0.8 **Si**, 0.15–0.40 **Cu**, 0.15 **Mn**, 0.8–1.2 **Mg**, 0.04–0.35 **Cr**, 0.25 **Zn**, 0.15 **Ti**, 0.7 **Fe**, 0.05 each, 0.15 total, others	**Annealed**: all **diam**, 152 max MPa **TS**, 100 max MPa **YS**, 14% in 2 in. (50 mm) **El**; **Solution heat treated and naturally aged**: all **diam**, 179 MPa **TS**, 110 MPa **YS**, 14% in 2 in. (50 mm) **El**; **Solution heat treated and artificially aged**: all **diam**, 262 MPa **TS**, 241 MPa **YS**, 8% in 2 in. (50 mm) **El**
AS 1865	6262	Australia	Wire, rod, bar, strip	rem **Al**, 0.40–0.8 **Si**, 0.15–0.40 **Cu**, 0.15 **Mn**, 0.8–1.2 **Mg**, 0.04–0.14 **Cr**, 0.25 **Zn**, 0.15 **Ti**, 0.7 **Fe**, 0.05 each, 0.15 total, others	**Solution heat treated and artificially aged**: 10–150 mm **diam**, 262 MPa **TS**, 241 MPa **YS**, 8% in 2 in. (50 mm) **El**; **Solution heat treated, artificially aged, and cold worked**: 3–40 mm **diam**, 358 MPa **TS**, 330 MPa **YS**, 3% in 2 in. (50 mm) **El**
AS 1865	6061	Australia	Wire, rod, bar, strip	rem **Al**, 0.40–0.8 **Si**, 0.15–0.40 **Cu**, 0.15 **Mn**, 0.8–1.2 **Mg**, 0.04–0.35 **Cr**, 0.25 **Zn**, 0.15 **Ti**, 0.7 **Fe**, 0.05 each, 0.15 total, others	**Annealed**: 10 mm **diam**, 152 max MPa **TS**; **(H13)**: 12 mm **diam**, 151 MPa **TS**; **Solution heat treated, cold worked and artificially aged**: 6 mm **diam**, 372 MPa **TS**, 324 MPa **YS**
AS 1734	6061	Australia	Sheet, plate	rem **Al**, 0.40–0.8 **Si**, 0.15–0.40 **Cu**, 0.15 **Mn**, 0.8–1.2 **Mg**, 0.04–0.35 **Cr**, 0.25 **Zn**, 0.15 **Ti**, 0.7 **Fe**, 0.05 each, 0.15 total, others	**Annealed**: 0.25–0.50 mm **diam**, 152 MPa **TS**, 83 MPa **YS**, 14% in 2 in. (50 mm) **El**; **Solution heat treated and naturally aged**: 0.25–0.50 mm **diam**, 206 MPa **TS**, 115 MPa **YS**, 14% in 2 in. (50 mm) **El**; **Solution heat treated and artificially aged**: 0.25–0.50 mm **diam**, 289 MPa **TS**, 241 MPa **YS**, 8% in 2 in. (50 mm) **El**

WROUGHT ALUMINUM AND ALUMINUM ALLOYS

specification number	designation	country	product forms	chemical composition	mechanical properties and hardness values
BS 1471	HT20	UK	Tube: drawn	rem **Al**, 0.4–0.8 **Si**, 0.15–0.40 **Cu**, 0.2–0.8 **Mn**, 0.8–1.2 **Mg**, 0.04–0.35 **Cr**, 0.2 **Zn**, 0.2 **Ti**, 0.7 **Fe**	**Strain–hardened to 1/2 hard:** ≤6.0 mm **diam**, 185 MPa **TS**, 160 MPa **YS**, 5% in 2 in. (50 mm) **El**; **Solution heat treated and naturally aged:** ≤6.0 mm **diam**, 215 MPa **TS**, 115 MPa **YS**, 12% in 2 in. (50 mm) **El**; **Solution heat treated and precipitation treated:** ≤6.0 mm **diam**, 295 MPa **TS**, 240 MPa **YS**, 7% in 2 in. (50 mm) **El**
BS 1475	HG20	UK	Wire	rem **Al**, 0.4–0.8 **Si**, 0.15–0.40 **Cu**, 0.2–0.8 **Mn**, 0.8–1.2 **Mg**, 0.04–0.35 **Cr**, 0.2 **Zn**, 0.2 **Ti**, 0.7 **Fe**	**Solution heat treated, cold worked and precipitation treated:** 6–10 mm **diam**, 355 MPa **TS**
BS 1473	HB20	UK	Bolt, screw	rem **Al**, 0.4–0.8 **Si**, 0.15–0.40 **Cu**, 0.2–0.8 **Mn**, 0.8–1.2 **Mg**, 0.04–0.35 **Cr**, 0.2 **Zn**, 0.2 **Ti**, 0.7 **Fe**	**Solution heat treated, cold worked, and precipitation treated:** ≤12 mm **diam**, 310 MPa **TS**, 245 MPa **YS**
BS 1474	HE20	UK	Bar, tube, section	rem **Al**, 0.4–0.8 **Si**, 0.15–0.40 **Cu**, 0.2–0.8 **Mn**, 0.8–1.2 **Mg**, 0.04–0.35 **Cr**, 0.2 **Zn**, 0.2 **Ti**, 0.7 **Fe**	**Solution heat treated and naturally aged:** 150 mm **diam**, 190 MPa **TS**, 115 MPa **YS**, 4% in 2 in. (50 mm) **El**; **Solution heat treated and precipitation treated:** 150 mm **diam**, 280 MPa **TS**, 240 MPa **YS**, 7% in 2 in. (50 mm) **El**
BS L.118		UK	Tube	rem **Al**, 0.40–0.8 **Si**, 0.15–0.40 **Cu**, 0.15 **Mn**, 0.8–1.2 **Mg**, 0.04–0.35 **Cr**, 0.25 **Zn**, 0.15 **Ti**, 0.7 **Fe**, 0.05 each, 0.15 total, others	**Solution heat treated, quenched and artificially aged:** ≤1.0 mm **diam**, 290 MPa **TS**, 240 MPa **YS**, 10% in 2 in. (50 mm) **El**
BS L.117		UK	Tube	rem **Al**, 0.40–0.8 **Si**, 0.15–0.40 **Cu**, 0.15 **Mn**, 0.8–1.2 **Mg**, 0.04–0.35 **Cr**, 0.25 **Zn**, 0.15 **Ti**, 0.7 **Fe**, 0.05 each, 0.15 total, others	**Solution heat treated, quenched, and artificially aged:** ≤1.0 mm **diam**, 290 MPa **TS**, 240 MPa **YS**, 10% in 2 in. (50 mm) **El**
AMS 4173		US	Bar, tube, rod, and shapes: extruded	rem **Al**, 0.40–0.8 **Si**, 0.15–0.40 **Cu**, 0.15 **Mn**, 0.8–1.2 **Mg**, 0.04–0.35 **Cr**, 0.25 **Zn**, 0.15 **Ti**, 0.7 **Fe**, 0.05 each, 0.15 total, others	**Solution and precipitation heat treated:** <0.250 in. **diam**, 38 ksi (262 MPa) **TS**,35 ksi (241 **mPa) YS**, 8% in 2 in. (50 mm) **El**, 80 min **HB**
AMS 4172		US	Bar, tube, rod, and shapes: extruded	rem **Al**, 0.40–0.8 **Si**, 0.15–0.40 **Cu**, 0.15 **Mn**, 0.8–1.2 **Mg**, 0.04–0.35 **Cr**, 0.25 **Zn**, 0.15 **Ti**, 0.7 **Fe**, 0.05 each, 0.15 total, others	**Solution heat treated:** all **diam**, 26 ksi (179 MPa) **TS**, 16 ksi (110 MPa) **YS**, 16% in 2 in. (50 mm) **El**, 55 min **HB**; **Precipitation heat treated:** <0.250 in. **diam**, 38 ksi (262 MPa) **TS**, 35 ksi (241 MPa) **YS**, 8% in 2 in. (50 mm) **El**

WROUGHT ALUMINUM AND ALUMINUM ALLOYS

specification number	designation	country	product forms	chemical composition	mechanical properties and hardness values
AMS 4161B		US	Bar, tube, rod, wire, and shapes: extruded	rem **Al**, 0.40–0.8 **Si**, 0.15–0.40 **Cu**, 0.15 **Mn**, 0.8–1.2 **Mg**, 0.04–0.35 **Cr**, 0.25 **Zn**, 0.15 **Ti**, 0.7 **Fe**, 0.05 each, 0.15 total, others	**Solution heat treated**: all **diam**, 26 ksi (179 MPa) **TS**, 16 ksi (110 MPa) **YS**, 16% in 2 in. (50 mm) **El**, 55 min **HB**; **Precipitation heat treated:** ≤0.250 in. **diam**, 38 ksi (262 MPa) **TS**, 35 ksi (241 MPa) **YS**, 10% in 2 in. (50 mm) **El**
AMS 4160C		US	Bar, tube, rod, wire, and shapes: extruded	rem **Al**, 0.40–0.8 **Si**, 0.15–0.40 **Cu**, 0.15 **Mn**, 0.8–1.2 **Mg**, 0.04–0.35 **Cr**, 0.25 **Zn**, 0.15 **Ti**, 0.7 **Fe**, 0.05 each, 0.15 total, others	**Annealed**: all **diam**, 22 ksi (152 MPa) **TS**, 16 ksi (110 MPa) **YS**, 16% in 2 in. (50 mm) **El**; **Solution and precipitation heat treated**: ≥0.250 in.(≥6.35 mm) **diam**, 38 ksi (262 MPa) **TS**, 35 ksi (241 MPa) **YS**, 10% in 2 in. (50 mm) **El**, 80 min **HB**
NF A 50 411	6061	France	Bar, wire, tube, and shapes	rem **Al**, 0.40–0.80 **Si**, 0.15–0.40 **Cu**, 0.15 **Mn**, 0.80–1.20 **Mg**, 0.04–0.35 **Cr**, 0.25 **Zn**, 0.15 **Ti**, 0.70 **Fe**, 0.05 each, 0.15 total, others	**Aged (extruded bar)**: 180 MPa **TS**, 110 MPa **YS**, 15% in 2 in. (50 mm) **El**; **Aged (wire)**: 2–12 mm **diam**, 140 MPa **TS**, 80 MPa **YS**, 16% in 2 in. (50 mm) **El**; **Quenched and tempered (tube)**: ≤150 mm **diam**, 290 MPa **TS**, 240 MPa **YS**, 8% in 2 in. (50 mm) **El**
NF A 50 451	6061	France	Sheet, plate, and strip	rem **Al**, 0.40–0.80 **Si**, 0.15–0.40 **Cu**, 0.15 **Mn**, 0.8–1.2 **Mg**, 0.04–0.35 **Cr**, 0.25 **Zn**, 0.15 **Ti**, 0.7 **Fe**, 0.05 each, 0.15 total, others	**Annealed**: 0.4–3.2 mm **diam**, 80 max MPa **YS**, 20% in 2 in. (50 mm) **El**; **Aged**: 0.4–6 mm **diam**, 210 MPa **TS**, 110 MPa **YS**; **Quenched and tempered**: 0.4–6 mm **diam**, 290 MPa **TS**, 240 MPa **YS**, 10% in 2 in. (50 mm) **El**
CSA HA.8	.6061	Canada	Forgings	rem **Al**, 0.40–0.8 **Si**, 0.15–0.40 **Cu**, 0.15 **Mn**, 0.8–1.2 **Mg**, 0.04–0.35 **Cr**, 0.25 **Zn**, 0.15 **Ti**, 0.7 **Fe**, 0.05 each, 0.15 total, others	**T6 (die forgings)**: <4 in. **diam**, 38.0 ksi (262 m pa) **TS**, 35.0 ksi (241 MPa) **YS**, 7% in 2 in. (50 mm) **El**
CSA HA 7.1	.6351	Canada	Tube: drawn and extruded, pipe: drawn	rem **Al**, 0.7–1.3 **Si**, 0.10 **Cu**, 0.40–0.8 **Mn**, 0.40–0.8 **Mg**, 0.20 **Zn**, 0.20 **Ti**, 0.50 **Fe**, 0.05 each, 0.15 total, others	**Drawn and O**: all **diam**, 22.0 max ksi (152 max MPa) **TS**, 97 MPa **YS**, 15% in 2 in. (50 mm) **El**; **Drawn and T6**: 0.015–0.049 in. **diam**, 42.0 ksi (290 MPa) **TS**, 8% in 2 in. (50 mm) **El**; **Extruded and T4**: all **diam**, 32.0 ksi (221 MPa) **TS**, 19.0 ksi (131 MPa) **YS**, 16% in 2 in. (50 mm) **El**

WROUGHT ALUMINUM AND ALUMINUM ALLOYS

specification number	designation	country	product forms	chemical composition	mechanical properties and hardness values
CSA HA.7.1	.6061	Canada	Tube: drawn and extruded, and pipe: drawn	rem **Al**, 0.40–0.8 **Si**, 0.15–0.40 **Cu**, 0.15 **Mn**, 0.8–1.2 **Mg**, 0.04–0.35 **Cr**, 0.25 **Zn**, 0.15 **Ti**, 0.7 **Fe**, 0.05 each, 0.15 total, others	**Drawn and O**: all **diam**, 22.0 max ksi (152 MPa) **TS**, 14.0 ksi (97 MPa) **YS**, 15% in 2 in. (50 mm) **El**; **Drawn and T4**: 0.015–0.049 in. **diam**, 30.0 ksi (207 MPa) **TS**, 16% in 2 in. (50 mm) **El**; **Extruded and T6510**: <0.250 in. **diam**, 38.0 ksi (262 MPa) **TS**, 35.0 ksi (241 MPa) **YS**, 10% in 2 in. (50 mm) **El**
CSA HA.7	.6061	Canada	Tube, pipe: seamless	rem **Al**, 0.40–0.8 **Si**, 0.15–0.40 **Cu**, 0.15 **Mn**, 0.8–1.2 **Mg**, 0.04–0.35 **Cr**, 0.25 **Zn**, 0.15 **Ti**, 0.7 **Fe**, 0.05 each, 0.15 total, others	**Drawn and O**: all **diam**, 22.0 max ksi (152 max MPa) **TS**, 14.0 ksi (97 MPa) **YS**, 15% in 2 in. (50 mm) **El**; **Drawn and T4**: 0.015–0.049 in. **diam**, 30.0 ksi (207 MPa) **TS**, 16.0 ksi (110 MPa) **YS**, 16% in 2 in. (50 mm) **El**; **Extruded and O**: all **diam**, 22.0 max ksi (152 max MPa) **TS**, 16.0 ksi (110 max MPa) **YS**, 16% in 2 in. (50 mm) **El**
CSA HA.7	.6351	Canada	Tube, pipe: seamless	rem **Al**, 0.7–1.3 **Si**, 0.10 **Cu**, 0.40–0.8 **Mn**, 0.40–0.8 **Mg**, 0.20 **Zn**, 0.20 **Ti**, 0.50 **Fe**, 0.05 each, 0.15 total, others	**Drawn and O**: all **diam**, 22.0 max ksi (152 max MPa) **TS**, 14.0 ksi (97 MPa) **YS**, 15% in 2 in. (50 mm) **El**; **Drawn and T6**: 0.015–0.049 in. **diam**, 42.0 ksi (290 MPa) **TS**, 37.0 ksi (255 MPa) **YS**, 8% in 2 in. (50 mm) **El**; **Extruded and T4**: all **diam**, 32.0 ksi (221 MPa) **TS**, 19.0 ksi (131 MPa) **YS**, 16% in 2 in. (50 mm) **El**
CSA HA.6	.6061	Canada	Rivet, rod, wire	rem **Al**, 0.40–0.8 **Si**, 0.15–0.40 **Cu**, 0.15 **Mn**, 0.8–1.2 **Mg**, 0.04–0.35 **Cr**, 0.25 **Zn**, 0.15 **Ti**, 0.7 **Fe**, 0.05 each, 0.15 total, others	**H13**: 0.062–1.000 in. **diam**, 22.0 ksi (152 MPa) **TS**, T6: 0.062–1.000 in. **diam**, 42.0 ksi (290 MPa) **TS**, 35.0 ksi (241 MPa) **YS**, 10% in 2 in. (50 mm) **El**
CSA HA.5	.6351	Canada	Bar, rod, wire, shapes	rem **Al**, 0.7–1.3 **Si**, 0.10 **Cu**, 0.40–0.8 **Mn**, 0.40–0.8 **Mg**, 0.20 **Zn**, 0.20 **Ti**, 0.50 **Fe**, 0.05 each, 0.15 total, others	**Extruded and T4**: all **diam**, 32.0 ksi (221 MPa) **TS**, 19.0 ksi (131 MPa) **YS**, 16% in 2 in. (50 mm) **El**; **Extruded and T6**: <0.124 in. **diam**, 42.0 ksi (290 MPa) **TS**, 37.0 ksi (255 MPa) **YS**, 8% in 2 in. (50 mm) **El**

WROUGHT ALUMINUM AND ALUMINUM ALLOYS

specification number	designation	country	product forms	chemical composition	mechanical properties and hardness values
CSA HA.5	.6061	Canada	Bar, rod, wire, shapes	rem **Al**, 0.40–0.8 **Si**, 0.15–0.40 **Cu**, 0.15 **Mn**, 0.8–1.2 **Mg**, 0.04–0.35 **Cr**, 0.25 **Zn**, 0.15 **Ti**, 0.7 **Fe**, 0.05 each, 0.15 total, others	**Drawn and O**: <2.500 in. **diam**, 22.0 max ksi (152 max MPa) **TS**, 18% in 2 in. (50 mm) **El**; **Drawn and T42**: <2.500 in. **diam**, 30.0 ksi (207 MPa) **TS**, 14.0 ksi (97 MPa) **YS**, 18% in 2 in. (50 mm) **El**; **Extruded and T6**: <0.249 in. **diam**, 38.0 ksi (262 MPa) **TS**, 35.0 ksi (241 MPa) **YS**, 8% in 2 in. (50 mm) **El**
COPANT 862	6061	COPANT	Wrought products	rem **Al**, 0.4–0.80 **Si**, 0.15–0.40 **Cu**, 0.15 **Mn**, 0.8–1.2 **Mg**, 0.04–0.35 **Cr**, 0.25 **Zn**, 0.15 **Ti**, 0.7 **Fe**, 0.05 each, 0.15 total, others	
COPANT 862	6262	COPANT	Wrought products	rem **Al**, 0.4–0.8 **Si**, 0.15–0.40 **Cu**, 0.15 **Mn**, 0.8–1.2 **Mg**, 0.04–0.14 **Cr**, 0.25 **Zn**, 0.15 **Ti**, 0.7 **Fe**, 0.4–0.7 **Pb**, 0.4–0.7 **Bi**, 0.05 each, 0.15 total, others	
COPANT 862	6351	COPANT	Wrought products	rem **Al**, 0.7–1.3 **Si**, 0.1 **Cu**, 0.4–0.8 **Mn**, 0.4–0.8 **Mg**, 0.2 **Zn**, 0.2 **Ti**, 0.5 **Fe**, 0.05 each, 0.15 total, others	
DS 3012	6351	Denmark	Rolled and extruded products	rem **Al**, 0.7–1.3 **Si**, 0.10 **Cu**, 0.40–0.8 **Mn**, 0.40–0.8 **Mg**, 0.20 **Zn**, 0.20 **Ti**, 0.50 **Fe**, 0.05 each, 0.15 total, others	**Soft (rolled products)**: 0.5–25 mm **diam**, 90 MPa **TS**, 20% in 2 in. (50 mm) **El**, 35 **HB**; **Solution annealed (rolled products)**: 0.5–25 mm **diam**, 205 MPa **TS**, 120 MPa **YS**, 15% in 2 in. (50 mm) **El**, 70 **HB**; **Artificially aged (extruded products)**: 20–75 mm **diam**, 295 MPa **TS**, 255 MPa **YS**, 8% in 2 in. (50 mm) **El**, 100 **HV**
WW-T-700/6E	6061	US	Tube	rem **Al**, 0.40–0.8 **Si**, 0.15–0.40 **Cu**, 0.15 **Mn**, 0.8–1.2 **Mg**, 0.04–0.35 **Cr**, 0.25 **Zn**, 0.15 **Ti**, 0.7 **Fe**, 0.05 each, 0.15 total, others	**Annealed**: 0.018–0.500 in. **diam**, 22 max ksi (152 max MPa) **TS**, 14 max ksi (97 max MPa) **YS**, 15% in 2 in. (50 mm) **El**; **Solution heat treated and artificially aged**: 0.025–0.049 in. **diam**, 30 ksi (207 MPa) **TS**, 16 ksi (110 MPa) **YS**, 16% in 2 in. (50 mm) **El**; **Solution heat treated and naturally aged**: 0.025–0.049 in. **diam**, 42 ksi (290 MPa) **TS**, 35 ksi (241 MPa) **YS**, 10% in 2 in. (50 mm) **El**

WROUGHT ALUMINUM AND ALUMINUM ALLOYS

specification number	designation	country	product forms	chemical composition	mechanical properties and hardness values
QQ-A-250/11E	6061	US	Plate, sheet	rem **Al**, 0.40–0.8 **Si**, 0.15–0.40 **Cu**, 0.15 **Mn**, 0.8–1.2 **Mg**, 0.15–0.35 **Cr**, 0.25 **Zn**, 0.15 **Ti**, 0.7 **Fe**, 0.05 each, 0.15 total, others	**Annealed**: 0.006–0.007 in. **diam**, 22 max ksi (152 max MPa) **TS**, 12 max ksi (83 max MPa) **YS**, 10% in 2 in. (50 mm) **El**; **Solution heat treated and naturally aged**: 0.006–0.007 in. **diam**, 30 ksi (207 MPa) **TS**, 16 ksi (110 MPa) **YS**, 10% in 2 in. (50 mm) **El**; **Solution heat treated and artificially aged**: 0.006–0.007 in. **diam**, 42 ksi (290 MPa) **TS**, 35 ksi (241 MPa) **YS**, 4% in 2 in. (50 mm) **El**
QQ-A-225/10A	6262	US	Bar, rod, wire	rem **Al**, 0.40–0.8 **Si**, 0.15–0.40 **Cu**, 0.15 **Mn**, 0.8–1.2 **Mg**, 0.04–0.14 **Cr**, 0.25 **Zn**, 0.15 **Ti**, 0.7 **Fe**, 0.40–0.7 **Pb**, 0.40–0.7 **Bi**, 0.05 each, 0.15 total, others	**Solution heat treated and artificially aged**: 8.000 in. **diam**, 42 ksi (290 MPa) **TS**, 35 ksi (241 MPa) **YS**, 10% in 2 in. (50 mm) **El**; **Solution treated, stress relieved and aged**: 0.500–8.000 in. **diam**, 42 ksi (290 MPa) **TS**, 35 ksi (241 MPa) **YS**, 10% in 2 in. (50 mm) **El**; **Solution heat treated and temper T9**: 0.125–2.000 in. **diam**, 52 ksi (359 MPa) **TS**, 48 ksi (331 MPa) **YS**, 5% in 2 in. (50 mm) **El**
QQ-A-225/8D	6061	US	Bar, rod, wire	rem **Al**, 0.40–0.8 **Si**, 0.15–0.40 **Cu**, 0.15 **Mn**, 0.8–1.2 **Mg**, 0.04–0.35 **Cr**, 0.25 **Zn**, 0.15 **Ti**, 0.7 **Fe**, 0.05 each, 0.15 total, others	**Annealed**: 8.000 max in. **diam**, 22 max ksi (152 max MPa) **TS**, 18% in 2 in. (50 mm) **El**; **Solution heat treated and naturally aged**: 8.000 max in. **diam**, 30 ksi (207 MPa) **TS**, 16 ksi (110 MPa) **YS**, 18% in 2 in. (50 mm) **El**; **Solution heat treated and artificially aged**: 8.000 max in. **diam**, 42 ksi (290 MPa) **TS**, 35 ksi (241 MPa) **YS**, 10% in 2 in. (50 mm) **El**
QQ-A-200/16A	6061	US	Shapes	rem **Al**, 0.40–0.8 **Si**, 0.15–0.40 **Cu**, 0.15 **Mn**, 0.8–1.2 **Mg**, 0.04–0.35 **Cr**, 0.25 **Zn**, 0.15 **Ti**, 0.7 **Fe**, 0.05 each, 0.15 total, others	**Solution heat treated and artificially aged**: 0.249 max in. **diam**, 38 ksi (262 MPa) **TS**, 35 ksi (241 MPa) **YS**, 8% in 2 in. (50 mm) **El**, 89–90 **HRE**

WROUGHT ALUMINUM AND ALUMINUM ALLOYS

specification number	designation	country	product forms	chemical composition	mechanical properties and hardness values
QQ-A-200/8D	6061	US	Bar, rod, shapes, tube, wire	rem **Al**, 0.40–0.8 **Si**, 0.15–0.40 **Cu**, 0.15 **Mn**, 0.8–1.2 **Mg**, 0.04–0.35 **Cr**, 0.25 **Zn**, 0.15 **Ti**, 0.7 **Fe**, 0.05 each, 0.15 total, others	**Annealed**: all **diam**, 22 max ksi (152 MPa) **TS**, 16 max ksi (110 MPa) **YS**, 16% in 2 in. (50 mm) **El**, 89–90 **HRE**; **Solution heat treated**: all **diam**, 26 ksi (179 MPa) **TS**, 16 ksi (110 MPa) **YS**, 16% in 2 in. (50 mm) **El**, 89–90 **HRE**; **Solution heat treated and artificially aged**: 0.249 max in. **diam**, 38 ksi (262 MPa) **TS**, 35 ksi (241 MPa) **YS**, 8% in 2 in. (50 mm) **El**, 89–90 **HRE**
QQ-A-430B	6061	US	Rod, wire	rem **Al**, 0.40–0.8 **Si**, 0.15–0.40 **Cu**, 0.15 **Mn**, 0.8–1.2 **Mg**, 0.04–0.35 **Cr**, 0.25 **Zn**, 0.15 **Ti**, 0.7 **Fe**, 0.05 each, 0.15 total, others	**Annealed**: 1.000 max in. **diam**, 22 max ksi (152 max MPa) **TS**; **Strain hardened and temper H13**: 1.000 max in. **diam**, 22 ksi (152 MPa) **TS**
IS 739	HG20	India	Wire	rem **Al**, 0.4–0.8 **Si**, 0.15–0.40 **Cu**, 0.2–0.8 **Mn**, 0.8–1.2 **Mg**, 0.15–0.35 **Cr**, 0.2 **Zn**, 0.7 **Fe**	**Solution and precipitation treated and drawn**: ≤6.30 mm **diam**, 372 MPa **TS**
IS 738	HT20	India	Tube: drawn	rem **Al**, 0.4–0.8 **Si**, 0.15–0.40 **Cu**, 0.2–0.8 **Mn**, 0.8–1.2 **Mg**, 0.15–0.35 **Cr**, 0.7 **Fe**	**Solution treated and naturally aged**: ≤1.60 mm **diam**, 216 MPa **TS**, 108 MPa **YS**, 12% in 2 in. (50 mm) **El**; **Solution and precipitation heat treated**: ≤1.60 mm **diam**, 294 MPa **TS**, 230 MPa **YS**, 7% in 2 in. (50 mm) **El**
ISO R827	Al–Mg1SiCu	ISO	Extruded products	rem **Al**, 0.4–0.8 **Si**, 0.15–0.40 **Cu**, 0.15 **Mn**, 0.8–1.2 **Mg**, 0.04–0.35 **Cr**, 0.25 **Zn**, 0.7 **Fe**, 0.2 **Ti+Zr**	**Solution and precipitation treated**: 6.0 max in. (150 max mm) **diam**, 265 MPa **TS**, 236 MPa **YS**, 9% in 2 in. (50 mm) **El**
ISO R209	Al–Mg1SiCu	ISO	Wrought products	rem **Al**, 0.4–0.8 **Si**, 0.15–0.40 **Cu**, 0.15 **Mn**, 0.8–1.2 **Mg**, 0.04–0.35 **Cr**, 0.25 **Zn**, 0.7 **Fe**, 0.2 **Ti+Zr**	
ISO TR2778	Al–Mg1SiCu	ISO	Tube: drawn	rem **Al**, 0.4–0.8 **Si**, 0.15–0.40 **Cu**, 0.15 **Mn**, 0.8–1.2 **Mg**, 0.04–0.35 **Cr**, 0.25 **Zn**, 0.7 **Fe**, 0.2 **Ti+Zr**	**Solution heat treated and naturally aged**: all **diam**, 200 MPa **TS**, 110 MPa **YS**, 14% in 2 in. (50 mm) **El**; **Solution and precipitation treated**: all **diam**, 285 MPa **TS**, 230 MPa **YS**, 8% in 2 in. (50 mm) **El**
ISO R829	Al–Mg1SiCu	ISO	Forgings	rem **Al**, 0.4–0.8 **Si**, 0.15–0.40 **Cu**, 0.15 **Mn**, 0.8–1.2 **Mg**, 0.04–0.35 **Cr**, 0.25 **Zn**, 0.7 **Fe**, 0.2 **Ti+Zr**	**Solution and precipitation treated**: all **diam**, 265 MPa **TS**, 236 MPa **YS**, 8% in 2 in. (50 mm) **El**

WROUGHT ALUMINUM AND ALUMINUM ALLOYS

specification number	designation	country	product forms	chemical composition	mechanical properties and hardness values
ISO TR2136	Al-Mg1SiCu	ISO	Sheet, plate	rem **Al**, 0.4–0.8 **Si**, 0.15–0.40 **Cu**, 0.15 **Mn**, 0.8–1.2 **Mg**, 0.04–0.35 **Cr**, 0.25 **Zn**, 0.2 **Ti+Zr**	**Annealed**: all **diam**, 160 max MPa **TS**, 90 max MPa **YS**, 16% in 2 in. (50 mm) **El**; **Solution heat treated and naturally aged**: all **diam**, 200 MPa **TS**, 110 MPa **YS**, 15% in 2 in. (50 mm) **El**; **Solution heat treated and artificially aged**: all **diam**, 285 MPa **TS**, 235 MPa **YS**, 8% in 2 in. (50 mm) **El**
JIS H 4180	6061	Japan	Plate	rem **Al**, 0.40–0.8 **Si**, 0.15–0.40 **Cu**, 0.15 **Mn**, 0.8–1.2 **Mg**, 0.04–0.35 **Cr**, 0.25 **Zn**, 0.15 **Ti**, 0.7 **Fe**, 0.15 total, others	**Solution heat treated and artificially age hardened**: <7 mm **diam**, 265 MPa **TS**, 245 MPa **YS**, 8% in 2 in. (50 mm) **El**
JIS H 4140	6061	Japan	Die forgings	rem **Al**, 0.40–0.8 **Si**, 0.15–0.40 **Cu**, 0.15 **Mn**, 0.8–1.2 **Mg**, 0.04–0.35 **Cr**, 0.25 **Zn**, 0.15 **Ti**, 0.7 **Fe**, 0.15 total, others	**Solution heat treated and artificially age hardened**: ≤100 mm **diam**, 265 MPa **TS**, 245 MPa **YS**, 10% in 2 in. (50 mm) **El**
JIS H 4120	6061	Japan	Wire and rod: rivet	rem **Al**, 0.40–0.8 **Si**, 0.15–0.40 **Cu**, 0.15 **Mn**, 0.8–1.2 **Mg**, 0.04–0.35 **Cr**, 0.25 **Zn**, 0.15 **Ti**, 0.7 **Fe**, 0.15 total, others	**Annealed**: ≤25 mm **diam**, 147 MPa **TS**, 3/8 hard: ≤25 mm **diam**, 157 MPa **TS**; **Solution heat treated and artificially age hardened**: 2–25 mm **diam**, 265 MPa **TS**, 245 MPa **YS**, 10% in 2 in. (50 mm) **El**
JIS H 4100	6061	Japan	Shapes: extruded	rem **Al**, 0.40–0.80 **Si**, 0.15–0.40 **Cu**, 0.15 **Mn**, 0.8–1.2 **Mg**, 0.04–0.35 **Cr**, 0.25 **Zn**, 0.15 **Ti**, 0.7 **Fe**, 0.15 total, others	**Annealed**: all **diam**, 147 MPa **TS**, 108 MPa **YS**, 16% in 2 in. (50 mm) **El**; **Solution heat treated and age hardened**: all **diam**, 177 MPa **TS**, 108 MPa **YS**, 16% in 2 in. (50 mm) **El**; **Solution heat treated and artificially aged**: <7 mm **diam**, 265 MPa **TS**, 245 MPa **YS**, 8% in 2 in. (50 mm) **El**
JIS H 4080	6061	Japan	Tube: extruded	rem **Al**, 0.40–0.8 **Si**, 0.15–0.40 **Cu**, 0.15 **Mn**, 0.8–1.2 **Mg**, 0.04–0.35 **Cr**, 0.25 **Zn**, 0.15 **Ti**, 0.7 **Fe**, 0.15 total, others	**Annealed**: all **diam**, 147 MPa **TS**, 108 MPa **YS**, 16% in 2 in. (50 mm) **El**; **Solution heat treated and age hardened**: all **diam**, 177 MPa **TS**, 108 MPa **YS**, Solution heat treated and artificially age hardened: <7 mm) **diam**, 265 MPa **TS**, 245 MPa **YS**, 8% in 2 in. (50 mm) **El**

WROUGHT ALUMINUM AND ALUMINUM ALLOYS

specification number	designation	country	product forms	chemical composition	mechanical properties and hardness values
JIS H 4040	6061	Japan	Bar: extruded	rem **Al**, 0.40–0.8 **Si**, 0.15–0.40 **Cu**, 0.15 **Mn**, 0.8–1.2 **Mg**, 0.04–0.35 **Cr**, 0.25 **Zn**, 0.15 **Ti**, 0.7 **Fe**, 0.15 total, others	**Annealed**: all **diam**, 147 MPa **TS**, 108 MPa **YS**, 16% in 2 in. (50 mm) **El**; **Solution heat treated and age hardened**: all **diam**, 177 MPa **TS**, 108 MPa **YS**, 16% in 2 in. (50 mm) **El**; **Solution heat treated and artificially age hardened**: <7 mm **diam**, 265 MPa **TS**, 245 MPa **YS**, 10% in 2 in. (50 mm) **El**
JIS H 4000	6061	Japan	Sheet, plate, strip	rem **Al**, 0.40–0.8 **Si**, 0.15–0.40 **Cu**, 0.15 **Mn**, 0.8–1.2 **Mg**, 0.04–0.35 **Cr**, 0.25 **Zn**, 0.15 **Ti**, 0.7 **Fe**, 0.15 total, others	**Annealed**: <1.5 mm **diam**, 147 MPa **TS**, 78 MPa **YS**, 16% in 2 in. (50 mm) **El**; **Solution heat treated and age hardened**: <1.5 mm **diam**, 206 MPa **TS**, 108 MPa **YS**, 16% in 2 in. (50 mm) **El**; **Solution heat treated and artificially age hardened**: <1.5 mm **diam**, 294 MPa **TS**, 245 MPa **YS**, 10% in 2 in. (50 mm) **El**
MIL-A-22771C	6061	US	Forgings	rem **Al**, 0.40–0.8 **Si**, 0.15–0.40 **Cu**, 0.15 **Mn**, 0.8–1.2 **Mg**, 0.04–0.35 **Cr**, 0.25 **Zn**, 0.15 **Ti**, 0.7 **Fe**, 0.05 total, others	**T6**: ≤4 in. **diam**, 262 MPa **TS**, 241 MPa **YS**, 7% in 2 in. (50 mm) **El**, 80 **HB**
MIL-T-7081D		US	Tube	rem **Al**, 0.40–0.8 **Si**, 0.15–0.40 **Cu**, 0.15 **Mn**, 0.8–1.2 **Mg**, 0.04–0.35 **Cr**, 0.25 **Zn**, 0.15 **Ti**, 0.7 **Fe**, 0.05 each, 0.15 total, others	**Solution heat treated**: 0.025–0.049 in. **diam**, 110 MPa **YS**, 16% in 2 in. (50 mm) **El**; **Solution heat treated and artificially aged**: 0.025–0.049 in. **diam**, 241 MPa **YS**, 10% in 2 in. (50 mm) **El**
SABS 712	Al-SiMgMn	South Africa	Wrought products	rem **Al**, 0.7–1.3 **Si**, 0.10 **Cu**, 0.40–0.8 **Mn**, 0.40–0.8 **Mg**, 0.20 **Zn**, 0.20 **Ti**, 0.50 **Fe**, 0.5 each, 0.15 total, others	
IS 736	65032	India	Plate	rem **Al**, 0.4–0.8 **Si**, 0.15–0.4 **Cu**, 0.2–0.8 **Mn**, 0.7–1.2 **Mg**, 0.15–0.35 **Cr**, 0.2 **Zn**, 0.2 **Ti**, 0.7 **Fe**	**Solution heat treated and naturally aged**: 15 mm **diam**, 200 MPa **TS**, 110 MPa **YS**, 15% in 2 in. (50 mm) **El**; **Solution and precipitation heat treated**: 8 mm **diam**, 280 MPa **TS**, 235 MPa **YS**, 8% in 2 in. (50 mm) **El**
IS 733	65032	India	Bar, rod, section	rem **Al**, 0.4–0.8 **Si**, 0.15–0.4 **Cu**, 0.2–0.8 **Mn**, 0.7–1.2 **Mg**, 0.15–0.35 **Cr**, 0.2 **Zn**, 0.2 **Ti**, 0.7 **Fe**	**As manufactured or annealed**: all **diam**, 110 MPa **TS**, 12% in 2 in. (50 mm) **El**; **Solution treated and naturally aged**: ≤150 mm **diam**, 185 MPa **TS**, 115 MPa **YS**, 14% in 2 in. (50 mm) **El**; **Solution and precipitation heat treated**: ≤150 mm **diam**, 280 MPa **TS**, 235 MPa **YS**, 7% in 2 in. (50 mm) **El**

WROUGHT ALUMINUM AND ALUMINUM ALLOYS

specification number	designation	country	product forms	chemical composition	mechanical properties and hardness values
AMS 4125F		US	Die forgings and rings	rem **Al**, 0.6–1.2 **Si**, 0.35 **Cu**, 0.20 **Mn**, 0.45–0.8 **Mg**, 0.15–0.35 **Cr**, 0.25 **Zn**, 0.15 **Ti**, 1.0 **Fe**, 0.05 each, 0.15 total others	**Solution and precipitation heat treated (die forgings):** ≤4 in. (≤102 mm) **diam**, 44 ksi (303 MPa) **TS**, 37 ksi (255 MPa) **YS**, 1% in 2 in. (50 mm) **El**, 90 min **HB**; **Solution and precipitation heat treated:** ≤2.5 in. (≤64 mm) **diam**, 44 ksi (303 MPa) **TS**, 37 ksi (255 MPa) **YS**, 5% in 2 in. (50 mm) **El**, 90 min **HB**
ANSI/ASTM B 247	6151	US	Die forgings	rem **Al**, 0.6–1.2 **Si**, 0.35 **Cu**, 0.20 **Mn**, 0.45–0.8 **Mg**, 0.15–0.35 **Cr**, 0.25 **Zn**, 0.15 **Ti**, 1.0 **Fe**, 0.05 each, 0.15 total, others	**Solution heat treated and artificially aged:** ≤4.000 in. **diam**, 44 ksi (303 MPa) **TS**, 37 ksi (255 MPa) **YS**, 10% in 2 in. (50 mm) **El**, 90 min **HB**
CSA HA.8	6151	Canada	Forgings	rem **Al**, 0.6–1.2 **Si**, 0.35 **Cu**, 0.20 **Mn**, 0.45–0.8 **Mg**, 0.15–0.35 **Cr**, 0.25 **Zn**, 0.15 **Ti**, 1.0 **Fe**, 0.05 each, 0.15 total, others	**T6 (die forgings):** <4 in. **diam**, 44.0 ksi (303 MPa) **TS**, 37.0 ksi (255 MPa) **YS**, 10% in 2 in. (50 mm) **El**
JIS H 4140	6151	Japan	Die forgings	rem **Al**, 0.6–1.2 **Si**, 0.35 **Cu**, 0.20 **Mn**, 0.45–0.8 **Mg**, 0.15–0.35 **Cr**, 0.25 **Zn**, 0.15 **Ti**, 1.0 **Fe**, 0.15 total, others	**Solution heat treated and artificially age hardened:** ≤100 mm **diam**, 304 MPa **TS**, 255 MPa **YS**, 14% in 2 in. (50 mm) **El**
MIL–A–22771C	6151	US	Forgings	rem **Al**, 0.6–1.2 **Si**, 0.35 **Cu**, 0.20 **Mn**, 0.45–0.8 **Mg**, 0.15–0.35 **Cr**, 0.25 **Zn**, 0.15 **Ti**, 1.0 **Fe**, 0.05 total, others	**T6:** ≤4 in. **diam**, 303 MPa **TS**, 255 MPa **YS**, 10% in 2 in. (50 mm) **El**, 90 **HB**
QQ–A–200/17	6162	US	Bar, rod, shapes, tube, wire	rem **Al**, 0.40–0.8 **Si**, 0.20 **Cu**, 0.10 **Mn**, 0.7–1.1 **Mg**, 0.10 **Cr**, 0.25 **Zn**, 0.10 **Ti**, 0.50 **Fe**, 0.05 each, 0.15 total, others	**Solution heat treated and artificially aged:** 1.000 max in. **diam**, 37 ksi (255 MPa) **TS**, 34 ksi (234 MPa) **YS**, 7% in 2 in. (50 mm) **El**, 89–90 **HRE**
ANSI/ASTM B 398	6201	US	Wire: for electrical purposes	rem **Al**, 0.50–0.9 **Si**, 0.10 **Cu**, 0.03 **Mn**, 0.6–0.9 **Mg**, 0.03 **Cr**, 0.10 **Zn**, 0.50 **Fe**, 0.06 **B**, 0.03 each, 0.10 total, others	**T81:** 0.0612–0.1327 in. (1.554–3.355 mm) **diam**, 44 ksi (303 MPa) **TS**, 3% in 10 in. (250 mm) **El**
ANSI/ASTM B 399	6201	US	Conductors: concentric lay–stranded	rem **Al**, 0.50–0.9 **Si**, 0.10 **Cu**, 0.03 **Mn**, 0.6–0.9 **Mg**, 0.03 **Cr**, 0.10 **Zn**, 0.50 **Fe**, 0.06 **B**, 0.03 each, 0.10 total, others	**T81:** 0.0612–0.1327 in. (1.554–3.355 mm) **diam**, 44 ksi (303 MPa) **TS**, 3% in 10 in. (250 mm) **El**
COPANT 862	6201	COPANT	Wrought products	rem **Al**, 0.5–0.9 **Si**, 0.10 **Cu**, 0.03 **Mn**, 0.6–0.9 **Mg**, 0.03 **Cr**, 0.10 **Zn**, 0.03 **Ti**, 0.50 **Fe**, 0.06 **V**, 0.03 each, 0.10 total, others	
SABS 712	Al–MgSi–EC (62 014)	South Africa	Wrought products	rem **Al**, 0.50–0.9 **Si**, 0.05 **Cu**, 0.6–0.9 **Mg**, 0.40 **Fe**, 0.06 **B**, 0.020 **Cr+Mn+Ti+V**	

WROUGHT ALUMINUM AND ALUMINUM ALLOYS

specification number	designation	country	product forms	chemical composition	mechanical properties and hardness values
COPANT 862	6261	COPANT	Wrought products	rem **Al**, 0.4–0.70 **Si**, 0.15–0.40 **Cu**, 0.20–0.35 **Mn**, 0.7–1.0 **Mg**, 0.10 **Cr**, 0.20 **Zn**, 0.10 **Ti**, 0.40 **Fe**, 0.05 each, 0.15 total, others	
ISO TR2136	Al–Cu4SiMg	ISO	Sheet, plate	rem **Al**, 0.5–1.2 **Si**, 3.8–5.0 **Cu**, 0.3–1.2 **Mn**, 0.2–0.8 **Mg**, 0.2 **Ni**, 0.2 **Zn**, 0.7 **Fe**, 0.3 **Ti+Zr+Cr**, total	**Solution heat treated and naturally aged**: all **diam**, 385 MPa **TS**, 240 MPa **YS**, 14% in 2 in. (50 mm) **El**; **Solution heat treated and artificially aged**: all **diam**, 430 MPa **TS**, 380 MPa **YS**, 6% in 2 in. (50 mm) **El**
ANSI/ASTM B 316	6053	US	Wire and rod: rivet and cold-heading	rem **Al**, 0.10 **Cu**, 1.1–1.4 **Mg**, 0.15–0.35 **Cr**, 0.10 **Zn**, 0.35 **Fe**, 0.05 each, 0.15 total, others	**Annealed**: ≤1.000 in. **diam**, 19 max ksi (131 max MPa) **TS**; **H13**: ≤1.000 in. **diam**, 19 ksi (131 MPa) **TS**; **T61**: 0.061–1.000 in. **diam**, 30 ksi (207 MPa) **TS**, 20 ksi (138 MPa) **YS**, 14% in 2 in. (50 mm) **El**
ANSI/ASTM B 221	6005	US	Bar, tube, rod, wire, and shapes	rem **Al**, 0.6–0.9 **Si**, 0.10 **Cu**, 0.10 **Mn**, 0.40–0.6 **Mg**, 0.10 **Cr**, 0.10 **Zn**, 0.10 **Ti**, 0.35 **Fe**, 0.05 each, 0.15 total, others	**Cooled and naturally aged**: ≤500 in. **diam**, 25 ksi (172 MPa) **TS**, 15 ksi (103 MPa) **YS**, 16% in 2 in. (50 mm) **El**; **Cooled and artificially aged**: 0.125–1.000 in. **diam**, 38 ksi (262 MPa) **TS**, 35 ksi (241 MPa) **YS**, 10% in 2 in. (50 mm) **El**
ANSI/ASTM B 211	6253 cladding	US	Bar, rod, and wire	rem **Al**, 0.10 **Cu**, 1.0–1.5 **Mg**, 0.15–0.35 **Cr**, 1.6–2.4 **Zn**, 0.50 **Fe**, 0.05 each, 0.15 total, others	
NF A 50 411	6005 A (A–SG0.5)	France	Bar, wire, tube, and shapes	rem **Al**, 0.50–0.90 **Si**, 0.30 **Cu**, 0.50 **Mn**, 0.40–0.70 **Mg**, 0.30 **Cr**, 0.20 **Zn**, 0.10 **Ti**, 0.35 **Fe**, 0.12–0.50 **Cr+Mn** each, 0.15 total, others	**Quenched and tempered (drawn bar)**: 265 MPa **TS**, 235 MPa **YS**, 8% in 2 in. (50 mm) **El**; **Quenched and tempered (wire)**: 2–25 mm **diam**, 265 MPa **TS**, 235 MPa **YS**, 8% in 2 in. (50 mm) **El**; **Quenched and tempered (tube)**: ≤150 mm **diam**, 265 MPa **TS**, 235 MPa **YS**, 8% in 2 in. (50 mm) **El**
COPANT 862	6005	COPANT	Wrought products	rem **Al**, 0.6–0.9 **Si**, 0.10 **Cu**, 0.10 **Mn**, 0.4–0.6 **Mg**, 0.10 **Cr**, 0.10 **Zn**, 0.10 **Ti**, 0.35 **Fe**, 0.05 each, 0.15 total, others	
IS 738	HT14	India	Tube: drawn	rem **Al**, 0.2–0.7 **Si**, 3.5–4.7 **Cu**, 0.4–1.0 **Mn**, 0.4–1.2 **Mg**, 0.7 **Fe**	**Solution treated and naturally aged**: ≤1.60 mm **diam**, 402 MPa **TS**, 279 MPa **YS**, 8% in 2 in. (50 mm) **El**
ISO TR2136	Al–Cu4MgSi	ISO	Sheet, plate	rem **Al**, 0.2–0.8 **Si**, 3.5–4.7 **Cu**, 0.3–1.0 **Mn**, 0.3–1.2 **Mg**, 0.2 **Ni**, 0.5 **Zn**, 0.7 **Fe**, 0.3 **Ti+Zr+Cr**, total	**Solution heat treated and naturally aged**: all **diam**, 385 MPa **TS**, 240 MPa **YS**, 14% in 2 in. (50 mm) **El**

WROUGHT ALUMINUM AND ALUMINUM ALLOYS

specification number	designation	country	product forms	chemical composition	mechanical properties and hardness values
BS 1471	HT15	UK	Tube: drawn	rem **Al**, 0.5–0.9 **Si**, 3.9–5.0 **Cu**, 0.4–1.2 **Mn**, 0.2–0.8 **Mg**, 0.10 **Cr**, 0.2 **Zn**, 0.2 **Ti**, 0.7 **Fe**	**Solution heat treated and naturally aged:** ≤10.0 mm **diam**, 400 MPa **TS**, 290 MPa **YS**, 8% in 2 in. (50 mm) **El**; **Solution heat treated and precipitation treated:** ≤10.0 mm **diam**, 450 MPa **TS**, 370 MPa **YS**, 6% in 2 in. (50 mm) **El**
BS 1475	HG15	UK	Wire	rem **Al**, 0.5–0.9 **Si**, 3.9–5.0 **Cu**, 0.4–1.2 **Mn**, 0.2–0.8 **Mg**, 0.10 **Cr**, 0.2 **Zn**, 0.2 **Ti**, 0.7 **Fe**	**Solution heat treated and naturally aged:** ≤10 mm **diam**, 385 MPa **TS**; **Solution heat treated and precipitation treated:** ≤10 mm **diam**, 430 MPa **TS**
BS 1473	HB15	UK	Bolt, screw	rem **Al**, 0.5–0.9 **Si**, 3.9–5.0 **Cu**, 0.4–1.2 **Mn**, 0.2–0.8 **Mg**, 0.10 **Cr**, 0.2 **Zn**, 0.2 **Ti**, 0.7 **Fe**	**Solution heat treated and precipitation treated:** ≤12 mm **diam**, 430 MPa **TS**, 390 MPa **YS**
IS 738	HT15	India	Tube: drawn	rem **Al**, 0.5–0.90 **Si**, 3.8–4.8 **Cu**, 0.3–1.2 **Mn**, 0.2–0.8 **Mg**, 0.7 **Fe**	**Solution treated and naturally aged:** ≤1.60 mm **diam**, 402 MPa **TS**, 279 MPa **YS**, 8% in 2 in. (50 mm) **El**; **Solution and precipitation heat treated:** ≤1.60 mm **diam**, 446 MPa **TS**, 353 MPa **YS**, 6% in 2 in. (50 mm) **El**
COPANT 862	8112	COPANT	Wrought products	rem **Al**, 0.4–1.0 **Si**, 0.15–0.4 **Cu**, 0.2–0.6 **Mn**, 0.3–0.7 **Mg**, 0.20 **Cr**, 1.0 **Zn**, 0.20 **Ti**, 0.5–1.0 **Fe**, 0.05 each, 0.15 total, others	
BS 1471	HT9	UK	Tube: drawn	rem **Al**, 0.3–0.7 **Si**, 0.10 **Cu**, 0.1 **Mn**, 0.4–0.9 **Mg**, 0.10 **Cr**, 0.2 **Zn**, 0.2 **Ti**, 0.40 **Fe**	**Annealed:** ≤10 mm **diam**; **Solution heat treated and naturally aged:** ≤10 mm **diam**, 155 MPa **TS**, 100 MPa **YS**, 15% in 2 in. (50 mm) **El**; **Solution heat treated and precipitation treated:** ≤10 mm **diam**, 200 MPa **TS**, 180 MPa **YS**, 8% in 2 in. (50 mm) **El**
BS 2898	E91E	UK	Bar, tube, section	rem **Al**, 0.3–0.7 **Si**, 0.05 **Cu**, 0.4–0.9 **Mg**, 0.40 **Fe**	**Solution heat treated and precipitation treated:** all **diam**, 200 MPa **TS**, 170 MPa **YS**, 8% in 2 in. (50 mm) **El**
BS 1475	HG9	UK	Wire	rem **Al**, 0.3–0.7 **Si**, 0.10 **Cu**, 0.1 **Mn**, 0.4–0.9 **Mg**, 0.10 **Cr**, 0.2 **Zn**, 0.2 **Ti**, 0.40 **Fe**	**As manufactured:** ≤10 mm **diam**; **Solution heat treated and naturally aged:** ≤10 mm **diam**, (140 MPa) **TS**; **Solution heat treated and precipitation treated:** ≤10 mm **diam**, 185 MPa **TS**

WROUGHT ALUMINUM AND ALUMINUM ALLOYS

specification number	designation	country	product forms	chemical composition	mechanical properties and hardness values
BS 1472	HF9	UK	Forging stock and forgings	rem **Al**, 0.3–0.7 **Si**, 0.10 **Cu**, 0.1 **Mn**, 0.4–0.9 **Mg**, 0.10 **Cr**, 0.2 **Zn**, 0.2 **Ti**, 0.40 **Fe**	**Solution heat treated and naturally aged**: 150–200 mm **diam**, 125 MPa **TS**, 85 MPa **YS**, 13% in 2 in. (50 mm) **El**; **Solution heat treated and precipitation treated**: 150–200 mm **diam**, 150 MPa **TS**, 130 MPa **YS**, 6% in 2 in. (50 mm) **El**
SFS 2591	AlMgSi	Finland	Shapes	rem **Al**, 0.3–0.7 **Si**, 0.10 **Cu**, 0.3 **Mn**, 0.4–0.9 **Mg**, 0.10 **Cr**, 0.2 **Zn**, 0.5 **Fe**, 0.2 **Ti+Zr**, 0.5 each, 0.15 total, others	**Hot finished**: ≤25 mm **diam**, 250 MPa **TS**, 80 MPa **YS**, 15% in 2 in. (50 mm) **El**; **Aged**: ≤25 mm **diam**, 160 MPa **TS**, 110 MPa **YS**, 10% in 2 in. (50 mm) **El**; **Solution treated and aged**: ≤25 mm **diam**, 220 MPa **TS**, 180 MPa **YS**, 10% in 2 in. (50 mm) **El**
IS 739	HG9	India	Wire	rem **Al**, 0.3–0.7 **Si**, 0.10 **Cu**, 0.30 **Mn**, 0.4–0.9 **Mg**, 0.10 **Cr**, 0.1 **Zn**, 0.6 **Fe**	**Solution treated and naturally aged**: all **diam**, 137 MPa **TS**; **Solution and precipitation heat treated**: all **diam**, 186 MPa **TS**
IS 738	HT9	India	Tube: drawn	rem **Al**, 0.3–0.7 **Si**, 0.10 **Cu**, 0.30 **Mn**, 0.4–0.9 **Mg**, 0.10 **Cr**, 0.7 **Fe**	**As manufactured**: all **diam**, 157 MPa **TS**; **Solution treated and naturally aged**: all **diam**, 157 MPa **TS**, 93 MPa **YS**, 15% in 2 in. (50 mm) **El**; **Solution and precipitation heat treated**: all **diam**, 201 MPa **TS**, 172 MPa **YS**, 8% in 2 in. (50 mm) **El**
IS 733	63400	India	Bar, rod, section	rem **Al**, 0.3–0.7 **Si**, 0.1 **Cu**, 0.3 **Mn**, 0.4–0.9 **Mg**, 0.1 **Cr**, 0.2 **Zn**, 0.2 **Ti**, 0.6 **Fe**	**Solution treated and naturally aged**: ≤150 mm **diam**, 140 MPa **TS**, 80 MPa **YS**, 14% in 2 in. (50 mm) **El**; **Precipitation heat treated**: ≤3 mm **diam**, 170 MPa **TS**, 140 MPa **YS**, 7% in 2 in. (50 mm) **El**; **Solution and precipitation heat treated**: ≤150 mm **diam**, 185 MPa **TS**, 150 MPa **YS**, 7% in 2 in. (50 mm) **El**
ISO R827	Al–MgSi	ISO	Extruded products	rem **Al**, 0.3–0.7 **Si**, 0.10 **Cu**, 0.30 **Mn**, 0.4–0.9 **Mg**, 0.10 **Cr**, 0.2 **Zn**, 0.5 **Fe**, 0.2 **Ti+Zr**	**Precipitation treated**: 1.0 max in. (25 max mm) **diam**, 147 MPa **TS**, 103 MPa **YS**, 8% in 2 in. (50 mm) **El**; **Solution and precipitation treated**: 0.5 max in. (12 max mm) **diam**, 201 MPa **TS**, 147 MPa **YS**, 8% in 2 in. (50 mm) **El**
ISO R209	Al–MgSi	ISO	Wrought products	rem **Al**, 0.3–0.7 **Si**, 0.10 **Cu**, 0.30 **Mn**, 0.4–0.9 **Mg**, 0.10 **Cr**, 0.2 **Zn**, 0.5 **Fe**, 0.2 **Ti+Zr**	

WROUGHT ALUMINUM AND ALUMINUM ALLOYS

specification number	designation	country	product forms	chemical composition	mechanical properties and hardness values
ISO TR2778	Al–MgSi	ISO	Tube: drawn	rem **Al**, 0.3–0.7 **Si**, 0.10 **Cu**, 0.30 **Mn**, 0.4–0.9 **Mg**, 0.10 **Cr**, 0.2 **Zn**, 0.5 **Fe**, 0.2 **Ti+Zr**	**Solution heat treated and naturally aged**: all **diam**, 140 MPa **TS**, 70 MPa **YS**, 14% in 2 in. (50 mm) **El**; **Solution and precipitation treated**: all **diam**, 215 MPa **TS**, 180 MPa **YS**, 8% in 2 in. (50 mm) **El**
SABS 712	Al–MgSi–EC (61 014)	South Africa	Wrought products	rem **Al**, 0.3–0.7 **Si**, 0.05 **Cu**, 0.4–0.9 **Mg**, 0.40 **Fe**, 0.020 **Cr+Mn+Ti+V**	
NS 17 310	NS 17 310–54	Norway	Tube, rod, wire, and profiles	rem **Al**, 0.35–0.7 **Si**, 0.05 **Cu**, 0.1 **Mn**, 0.4–0.8 **Mg**, 0.2 **Zn**, 0.1 **Ti**, 0.3 **Fe**, 0.05 **Cr**, 0.05 each, 0.15 total, others	**Hardened and naturally aged**: all **diam**, 127 MPa **TS**, 69 MPa **YS**, 15% in 2 in. (50 mm) **El**
NS 17 310	NS 17 310–56	Norway	Tube, rod, wire, and profiles	rem **Al**, 0.35–0.7 **Si**, 0.05 **Cu**, 0.1 **Mn**, 0.4–0.8 **Mg**, 0.2 **Zn**, 0.1 **Ti**, 0.3 **Fe**, 0.05 **Cr**, 0.05 each, 0.15 total, others	**Hardened and artificially aged**: all **diam**, 216 MPa **TS**, 157 MPa **YS**, 12% in 2 in. (50 mm) **El**
SFS 2592	E–AlMgSi	Finland	Shapes, rod, tube, and wire	rem **Al**, 0.3–0.7 **Si**, 0.5 **Cu**, 0.1 **Mn**, 0.4–0.8 **Mg**, 0.05 **Cr**, 0.2 **Zn**, 0.1–0.3 **Fe**, 0.1 **Ti+Zr**, 0.03 each, 0.15 total, others	**Solution treated and aged (shapes)**: all **diam**, 220 MPa **TS**, 180 MPa **YS**, 10% in 2 in. (50 mm) **El**; **Solution treated, cold finished, and aged (rod, tube, and wire)**: ≤6 mm **diam**, 220 MPa **TS**, 180 MPa **YS**, 8% in 2 in. (50 mm) **El**
ANSI/ASTM B 491	6063	US	Tube: coiled	rem **Al**, 0.20–0.6 **Si**, 0.10 **Cu**, 0.10 **Mn**, 0.45–0.9 **Mg**, 0.10 **Cr**, 0.10 **Zn**, 0.10 **Ti**, 0.35 **Fe**, 0.05 each, 0.15 total, others	**Cooled and naturally aged**: 0.032–0.050 in. **diam**, 17 ksi (117 MPa) **TS**, 9 ksi (83 MPa) **YS**, 12% in 2 in. (50 mm) **El**
ANSI/ASTM B 483	6063	US	Tube: for general purposes	rem **Al**, 0.20–0.6 **Si**, 0.10 **Cu**, 0.10 **Mn**, 0.45–0.9 **Mg**, 0.10 **Cr**, 0.10 **Zn**, 0.10 **Ti**, 0.35 **Fe**, 0.05 each, 0.15 total, others	**Annealed**: 0.018–0.500 in. **diam**, 19 max ksi (131 max MPa) **TS**; **Solution heat treated and naturally aged (round tube)**: 0.025–0.049 in. **diam**, 22 ksi (152 MPa) **TS**, 10 ksi (69 MPa) **YS**, 14% in 2 in. (50 mm) **El**; **Solution heat treated and artificially aged (round tube)**: 0.025–0.049 in. **diam**, 33 ksi (228 MPa) **TS**, 28 ksi (193 MPa) **YS**, 8% in 2 in. (50 mm) **El**
ANSI/ASTM B 404	6063	US	Pipe and tube: structural	rem **Al**, 0.20–0.6 **Si**, 0.10 **Cu**, 0.10 **Mn**, 0.45–0.9 **Mg**, 0.10 **Cr**, 0.10 **Ti**, 0.35 **Fe**, 0.05 each, 0.15 total, others	**Solution heat treated and artificially aged**: all **diam**, 30 ksi (207 MPa) **TS**, 25 ksi (172 MPa) **YS**, 8% in 2 in. (50 mm) **El**

WROUGHT ALUMINUM AND ALUMINUM ALLOYS

specification number	designation	country	product forms	chemical composition	mechanical properties and hardness values
ANSI/ASTM B 345	6063	US	Pipe and tube: seamless	rem **Al**, 0.20–0.6 **Si**, 0.10 **Cu**, 0.10 **Mn**, 0.45–0.9 **Mg**, 0.10 **Cr**, 0.10 **Zn**, 0.10 **Ti**, 0.35 **Fe**, 0.05 each, 0.15 total, others	**Solution heat treated and artificially aged**: ≥0.124 in. **diam**, 30 ksi (207 MPa) **TS**, 25 ksi (172 MPa) **YS**, 8% in 2 in. (50 mm) **El**; **Annealed (tube)**: all **diam**, 19 max ksi (131 max MPa) **TS**, 18% in 2 in. (50 mm) **El**; **Solution heat treated and naturally aged (tube)**: ≥0.500 in. **diam**, 19 ksi (131 MPa) **TS**, 10 ksi (69 MPa) **YS**, 14% in 2 in. (50 mm) **El**
ANSI/ASTM B 317	6101	US	Bar, rod, pipe, and shapes: for electrical purposes	rem **Al**, 0.30–0.7 **Si**, 0.10 **Cu**, 0.03 **Mn**, 0.35–0.8 **Mg**, 0.03 **Cr**, 0.10 **Zn**, 0.50 **Fe**, 0.06 **B**, 0.03 each, 0.10 total, others	**Solution heat treated and artificially aged**: 0.125–0.500 in. **diam**, 29 ksi (200 MPa) **TS**, 25 ksi (172 MPa) **YS**; **T61**: 0.750–1.499 in. **diam**, 18 ksi (124 MPa) **TS**, 11 ksi (76 MPa) **YS**; **T64**: 0.125–1.000 in. **diam**, 15 ksi (103 MPa) **TS**, 8 ksi (55 MPa) **YS**
ANSI/ASTM B 241	6063	US	Pipe and tube: seamless	rem **Al**, 0.20–0.6 **Si**, 0.10 **Cu**, 0.10 **Mn**, 0.45–0.9 **Mg**, 0.10 **Cr**, 0.10 **Zn**, 0.10 **Ti**, 0.35 **Fe**, 0.05 each, 0.15 total, others	**Annealed**: all **diam**, 19 max ksi (131 max MPa) **TS**, 18% in 2 in. (50 mm) **El**; **Cooled and naturally aged**: ≤0.500 in. **diam**, 17 ksi (117 MPa) **TS**, 9 ksi (62 MPa) **YS**, 12% in 2 in. (50 mm) **El**; **Solution heat treated and naturally aged**: ≤0.500 in. **diam**, 19 ksi (131 MPa) **TS**, 10 ksi (69 MPa) **YS**, 14% in 2 in. (50 mm) **El**
ANSI/ASTM B 221	6063	US	Bar, tube, rod, wire, and shapes	rem **Al**, 0.20–0.6 **Si**, 0.10 **Cu**, 0.10 **Mn**, 0.45–0.9 **Mg**, 0.10 **Cr**, 0.10 **Zn**, 0.10 **Ti**, 0.35 **Fe**, 0.05 each, 0.15 total, others	**Annealed**: all **diam**, 19 max ksi (131 max MPa) **TS**, 18% in 2 in. (50 mm) **El**; **Cooled and naturally aged**: 0.501–1.000 in. **diam**, 16 ksi (110 MPa) **TS**, 8 ksi (53 MPa) **YS**, 12% in 2 in. (50 mm) **El**; **Solution heat treated and naturally aged**: ≤0.500 in. **diam**, 19 ksi (131 MPa) **TS**, 10 ksi (69 MPa) **YS**, 14% in 2 in. (50 mm) **El**

WROUGHT ALUMINUM AND ALUMINUM ALLOYS

specification number	designation	country	product forms	chemical composition	mechanical properties and hardness values
ANSI/ASTM B 221	6463	US	Bar, tube, rod, wire, and shapes	rem **Al**, 0.20–0.6 **Si**, 0.20 **Cu**, 0.05 **Mn**, 0.45–0.9 **Mg**, 0.15 **Fe**, 0.05 each, 0.15 total, others	**Cooled and naturally aged**: ≤0.500 in. **diam**, 17 ksi (117 MPa) **TS**, 9 ksi (62 MPa) **YS**, 12% in 2 in. (50 mm) **El**; **Cooled and artificially aged**: ≤0.500 in. **diam**, 22 ksi (152 MPa) **TS**, 16 ksi (110 MPa) **YS**, 8% in 2 in. (50 mm) **El**; **Solution heat treated and artificially aged**: 0.125–0.500 in. **diam**, 30 ksi (207 MPa) **TS**, 25 ksi (172 MPa) **YS**, 10% in 2 in. (50 mm) **El**
ANSI/ASTM B 210	6063	US	Tube: seamless	rem **Al**, 0.20–0.6 **Si**, 0.10 **Cu**, 0.10 **Mn**, 0.45–0.9 **Mg**, 0.10 **Cr**, 0.10 **Zn**, 0.10 **Ti**, 0.35 **Fe**, 0.05 each, 0.15 total, others	**Solution heat treated and naturally aged**: 0.260–0.500 in. **diam**, 22 ksi (152 MPa) **TS**, 10 ksi (69 MPa) **YS**, 18% in 2 in. (50 mm) **El**; **Solution heat treated and artificially aged**: 0.260–0.500 in. **diam**, 33 ksi (228 MPa) **TS**, 28 ksi (193 MPa) **YS**, 12% in 2 in. (50 mm) **El**; **T832**: 0.050–0.259 in. **diam**, 40 ksi (276 MPa) **TS**, 35 ksi (241 MPa) **YS**, 5% in 2 in. (50 mm) **El**
AS 1867	6063	Australia	Tube: drawn	rem **Al**, 0.20–0.6 **Si**, 0.10 **Cu**, 0.10 **Mn**, 0.45–0.9 **Mg**, 0.10 **Cr**, 0.10 **Zn**, 0.10 **Ti**, 0.35 **Fe**, 0.05 each, 0.15 total, others	**Annealed**: all **diam**, 132 max MPa **TS**, 20% in 2 in. (50 mm) **El**; **Strain hardened to full hard**: all **diam**, 179 MPa **TS**; **Solution heat treated and artificially aged**: all **diam**, 200 MPa **TS**, 177 MPa **YS**, 8% in 2 in. (50 mm) **El**
AS 1866	6063	Australia	Rod, bar, shape	rem **Al**, 0.20–0.6 **Si**, 0.10 **Cu**, 0.10 **Mn**, 0.45–0.9 **Mg**, 0.10 **Cr**, 0.10 **Zn**, 0.10 **Ti**, 0.35 **Fe**, 0.05 each, 0.15 total, others	**Annealed**: all **diam**, 131 max MPa **TS**, 16% in 2 in. (50 mm) **El**; **Solution heat treated and naturally aged**: 150.0 mm **diam**, 131 MPa **TS**, 68 MPa **YS**, 12% in 2 in. (50 mm) **El**; **Solution heat treated and artificially aged**: 25.0 mm **diam**, 206 MPa **TS**, 172 MPa **YS**, 8% in 2 in. (50 mm) **El**
AS 1865	6063	Australia	Wire, rod, bar, strip	rem **Al**, 0.20–0.6 **Si**, 0.10 **Cu**, 0.10 **Mn**, 0.45–0.9 **Mg**, 0.10 **Cr**, 0.10 **Zn**, 0.10 **Ti**, 0.35 **Fe**, 0.05 each, 0.15 total, others	**Annealed**: all **diam**, 132 max MPa **TS**; **Strain hardened to full hard**: 10 mm **diam**, 186 MPa **TS**; **Solution heat treated and naturally aged**: all **diam**, 141 MPa **TS**, 12% in 2 in. (50 mm) **El**

WROUGHT ALUMINUM AND ALUMINUM ALLOYS

specification number	designation	country	product forms	chemical composition	mechanical properties and hardness values
AMS 4156E		US	Bar, tube, rod, wire, and shapes: extruded	rem **Al**, 0.20–0.6 **Si**, 0.10 **Cu**, 0.10 **Mn**, 0.45–0.9 **Mg**, 0.10 **Cr**, 0.10 **Zn**, 0.10 **Ti**, 0.35 **Fe**, 0.05 each, 0.15 total, others	**Solution and precipitation heat treated**: 0.125–1.000 in. **diam**, 30 ksi (207 MPa) **TS**, 25 ksi (172 MPa) **YS**, 8% in 2 in. (50 mm) **El**, 60 min **HB**
CSA HA.7.1	.6063	Canada	Tube: drawn and extruded, and pipe: drawn	rem **Al**, 0.20–0.6 **Si**, 0.10 **Cu**, 0.10 **Mn**, 0.45–0.9 **Mg**, 0.10 **Cr**, 0.10 **Zn**, 0.10 **Ti**, 0.35 **Fe**, 0.05 each, 0.15 total, others	**Drawn and O**: all **diam**, 19.0 max ksi (131 max MPa) **TS**; **Drawn and T4**: 0.015–0.049 in. **diam**, 22.0 ksi (152 MPa) **TS**, 10.0 ksi (69 MPa) **YS**, 16% in 2 in. (50 mm) **El**; **Extruded and O**: all **diam**, 19.0 max ksi (131 max MPa) **TS**, 18% in 2 in. (50 mm) **El**
CSA HA.7	.6063	Canada	Tube, pipe: seamless	rem **Al**, 0.20–0.6 **Si**, 0.10 **Cu**, 0.10 **Mn**, 0.45–0.9 **Mg**, 0.10 **Cr**, 0.10 **Zn**, 0.10 **Ti**, 0.35 **Fe**, 0.05 each, 0.15 total, others	**Drawn and O**: all **diam**, 19.0 max ksi (131 max MPa) **TS**; **Drawn and T4**: 0.015–0.049 in. **diam**, 22.0 ksi (152 MPa) **TS**, 10.0 ksi (69 MPa) **YS**, 16% in 2 in. (50 mm) **El**; **Extruded and O**: all **diam**, 19.0 max ksi (131 max MPa) **TS**, 18% in 2 in. (50 mm) **El**
CSA HA.5	.6063	Canada	Bar, rod, wire, shapes	rem **Al**, 0.20–0.6 **Si**, 0.10 **Cu**, 0.10 **Mn**, 0.45–0.9 **Mg**, 0.10 **Cr**, 0.10 **Zn**, 0.10 **Ti**, 0.35 **Fe**, 0.05 each, 0.15 total, others	**Drawn and O**: all **diam**, 17.0 max ksi (117 max MPa) **TS**; **Drawn and T4**: <1.500 in. **diam**, 18.0 ksi (124 MPa) **TS**, 8.0 ksi (55 MPa) **YS**, 18% in 2 in. (50 mm) **El**; **Extruded and T62**: <0.125–1.000 in. **diam**, 30.0 ksi (207 MPa) **TS**, 25.0 ksi (173 MPa) **YS**, 10% in 2 in. (50 mm) **El**
COPANT 862	6101	COPANT	Wrought products	rem **Al**, 0.3–0.70 **Si**, 0.10 **Cu**, 0.03 **Mn**, 0.35–0.80 **Mg**, 0.03 **Cr**, 0.10 **Zn**, 0.50 **Fe**, 0.06 **B**, 0.03 each, 0.10 total, others	
COPANT 862	6463	COPANT	Wrought products	rem **Al**, 0.2–0.6 **Si**, 0.20 **Cu**, 0.05 **Mn**, 0.45–0.90 **Mg**, 0.05 **Zn**, 0.15 **Fe**, 0.05 each, 0.15 total, others	
COPANT 862	6063	COPANT	Wrought products	rem **Al**, 0.2–0.60 **Si**, 0.10 **Cu**, 0.10 **Mn**, 0.45–0.90 **Mg**, 0.10 **Cr**, 0.10 **Zn**, 0.10 **Ti**, 0.35 **Fe**, 0.05 each, 0.15 total, others	
COPANT 862	6763	COPANT	Wrought products	rem **Al**, 0.2–0.6 **Si**, 0.04–0.16 **Cu**, 0.03 **Mn**, 0.45–0.9 **Mg**, 0.03 **Zn**, 0.08 **Fe**, 0.03 each, 0.10 total, others	

WROUGHT ALUMINUM AND ALUMINUM ALLOYS

specification number	designation	country	product forms	chemical composition	mechanical properties and hardness values
DS 3012	6063	Denmark	Extruded products	rem **Al**, 0.20–0.6 **Si**, 0.10 **Cu**, 0.10 **Mn**, 0.45–0.9 **Mg**, 0.10 **Cr**, 0.10 **Zn**, 0.10 **Ti**, 0.35 **Fe**, 0.05 each, 0.15 total, others	**Solution annealed**: ≤25 mm **diam**, 145 MPa **TS**, 105 MPa **YS**, 8% in 2 in. (50 mm) **El**, 40 **HB**; **Artificially aged**: ≤12 mm **diam**, 185 MPa **TS**, 145 MPa **YS**, 8% in 2 in. (50 mm) **El**; **75 HB**
QQ-A-200/9C	6063	US	Bar, rod, shapes, tube, wire	rem **Al**, 0.20–0.6 **Si**, 0.10 **Cu**, 0.10 **Mn**, 0.45–0.9 **Mg**, 0.10 **Cr**, 0.10 **Zn**, 0.10 **Ti**, 0.35 **Fe**, 0.05 each, 0.15 total, others	**Annealed**: all **diam**, 19 max ksi (131 max MPa) **TS**, 18% in 2 in. (50 mm) **El**; **Solution heat treated and quenched**: 0.500 max in. **diam**, 17 ksi (117 MPa) **TS**, 9 ksi (62 MPa) **YS**, 12% in 2 in. (50 mm) **El**; **Solution heat treated and naturally aged**: 0.500 max in. **diam**, 19 ksi (131 MPa) **TS**, 10 ksi (69 MPa) **YS**, 14% in 2 in. (50 mm) **El**
JIS H 4180	6101	Japan	Plate	rem **Al**, 0.30–0.7 **Si**, 0.10 **Cu**, 0.03 **Mn**, 0.35–0.8 **Mg**, 0.03 **Cr**, 0.10 **Zn**, 0.50 **Fe**, 0.06 **B**, 0.10 total, others	**Solution heat treated and artificially age hardened**: <7 mm **diam**, 196 MPa **TS**, 167 MPa **YS**, 10% in 2 in. (50 mm) **El**
JIS H 4180	6063	Japan	Plate	rem **Al**, 0.20–0.6 **Si**, 0.10 **Cu**, 0.10 **Mn**, 0.45–0.9 **Mg**, 0.10 **Cr**, 0.10 **Zn**, 0.10 **Ti**, 0.35 **Fe**, 0.15 total, others	**Solution heat treated and artificially age hardened**: 3–16 mm **diam**, 206 MPa **TS**, 167 MPa **YS**, 8% in 2 in. (50 mm) **El**
JIS H 4100	6063	Japan	Shapes: extruded	rem **Al**, 0.20–0.6 **Si**, 0.10 **Cu**, 0.10 **Mn**, 0.45–0.9 **Mg**, 0.10 **Cr**, 0.10 **Zn**, 0.10 **Ti**, 0.35 **Fe**, 0.15 total, others	**As manufactured**: all **diam**, 118 MPa **TS**, 49 MPa **YS**, 12% in 2 in. (50 mm) **El**; **Quick cooled and artificially aged**: all **diam**, 147 MPa **TS**, 108 MPa **YS**, 8% in 2 in. (50 mm) **El**; **Solution heat treated and artificially aged**: all **diam**, 206 MPa **TS**, 167 MPa **YS**, 8% in 2 in. (50 mm) **El**
JIS H 4080	6063	Japan	Tube: extruded	rem **Al**, 0.20–0.6 **Si**, 0.10 **Cu**, 0.10 **Mn**, 0.45–0.9 **Mg**, 0.10 **Cr**, 0.10 **Zn**, 0.10 **Ti**, 0.35 **Fe**, 0.15 total, others	**As manufactured**: all **diam**, 118 MPa **TS**, 49 MPa **YS**, 12% in 2 in. (50 mm) **El**; **Quick cooled and artificially aged**: all **diam**, 147 MPa **TS**, 108 MPa **YS**, 8% in 2 in. (50 mm) **El**; **Solution heat treated and artificially age hardened**: all **diam**, 206 MPa **TS**, 167 MPa **YS**, 8% in 2 in. (50 mm) **El**

WROUGHT ALUMINUM AND ALUMINUM ALLOYS

specification number	designation	country	product forms	chemical composition	mechanical properties and hardness values
JIS H 4040	6063	Japan	Bar: extruded	rem **Al**, 0.20–0.6 **Si**, 0.10 **Cu**, 0.10 **Mn**, 0.45–0.9 **Mg**, 0.10 **Cr**, 0.10 **Zn**, 0.10 **Ti**, 0.35 **Fe**, 0.15 total, others	**As manufactured**: all **diam**, 118 MPa **TS**, 49 MPa **YS**, 12% in 2 in. (50 mm) **El**; **Quick cooled and artificially aged**: all **diam**, 118 MPa **TS**, 49 MPa **YS**, 8% in 2 in. (50 mm) **El**, Solution heat treated and artificially age hardened: all **diam**, 206 MPa **TS**, 167 MPa **YS**, 8% in 2 in. (50 mm) **El**
SIS 14 41 04	SIS Al 41 04–00	Sweden	Bar, tube	98.9 **Al**, 0.20–0.6 **Si**, 0.10 **Cu**, 0.10 **Mn**, 0.45–0.9 **Mg**, 0.10 **Cr**, 0.10 **Zn**, 0.10 **Ti**, 0.35 **Fe**, 0.05 each, 0.15 total, others	**Hot worked**: 10–200 mm **diam**, 120 MPa **TS**, 60 MPa **YS**, 18% in 2 in. (50 mm) **El**, 40–70 **HV**
SIS 14 41 04	SIS Al 41 04–04	Sweden	Bar, tube, section	98.9 **Al**, 0.20–0.6 **Si**, 0.10 **Cu**, 0.10 **Mn**, 0.45–0.9 **Mg**, 0.10 **Cr**, 0.10 **Zn**, 0.10 **Ti**, 0.35 **Fe**, 0.05 each, 0.15 total, others	**Naturally aged**: 10–200 mm **diam**, 130 MPa **TS**, 70 MPa **YS**, 15% in 2 in. (50 mm) **El**, 45–75 **HV**
SIS 14 41 04	SIS Al 41 04–06	Sweden	Bar, tube, section	98.9 **Al**, 0.20–0.6 **Si**, 0.10 **Cu**, 0.10 **Mn**, 0.45–0.9 **Mg**, 0.10 **Cr**, 0.10 **Zn**, 0.10 **Ti**, 0.35 **Fe**, 0.05 each, 0.15 total, others	**Artificially aged**: 10–200 mm **diam**, 210 MPa **TS**, 170 MPa **YS**, 12% in 2 in. (50 mm) **El**, 75–105 **HV**
SABS 712	Al–MgSi (64 631)	South Africa	Wrought products	rem **Al**, 0.20–0.6 **Si**, 0.05–0.20 **Cu**, 0.05 **Mn**, 0.45–0.9 **Mg**, 0.2 **Zn**, 0.15 **Fe**, 0.05 each, 0.15 total, others	
SABS 712	Al–MgSi (60 630)	South Africa	Wrought products	rem **Al**, 0.20–0.6 **Si**, 0.10 **Cu**, 0.10 **Mn**, 0.45–0.9 **Mg**, 0.10 **Cr**, 0.10 **Zn**, 0.10 **Ti**, 0.35 **Fe**, 0.05 each, 0.15 total, others	
ANSI/ASTM B 547	6061	US	Tube: formed and arc-welded	rem **Al**, 0.15–0.40 **Cu**, 0.15 **Mn**, 0.8–1.2 **Mg**, 0.04–0.35 **Cr**, 0.25 **Zn**, 0.15 **Ti**, 0.7 **Fe**, 0.05 each, 0.15 total, others	**Solution heat treated and naturally aged**: 0.125–0.249 in. **diam**, 30 ksi (207 MPa) **TS**, 16 ksi (110 MPa) **YS**, 16% in 2 in. (50 mm) **El**; **T451**: 0.250–0.500 in. **diam**, 30 ksi (207 MPa) **TS**, 16 ksi (110 MPa) **YS**, 18% in 2 in. (50 mm) **El**; **Solution heat treated and artificially aged**: 0.250–0.499 in. **diam**, 42 ksi (290 MPa) **TS**, 35 ksi (241 MPa) **YS**, 10% in 2 in. (50 mm) **El**
AS 1866	6463A	Australia	Rod, bar, shape	rem **Al**, 0.20–0.6 **Si**, 0.25 **Cu**, 0.05 **Mn**, 0.30–0.9 **Mg**, 0.15 **Fe**, 0.05 each, 0.15 total, others	**Cooled and naturally aged**: 12.0 mm **diam**, 117 MPa **TS**, 62 MPa **YS**, 12% in 2 in. (50 mm) **El**; **Cooled and artificially aged**: 12.0 mm **diam**, 151 MPa **TS**, 110 MPa **YS**, 8% in 2 in. (50 mm) **El**; **Solution heat treated and artificially aged**: 3.0 mm **diam**, 206 MPa **TS**, 172 MPa **YS**, 8% in 2 in. (50 mm) **El**

WROUGHT ALUMINUM AND ALUMINUM ALLOYS

specification number	designation	country	product forms	chemical composition	mechanical properties and hardness values
SFS 2430	Al Mg Si	Finland	Wire	98 **Al**, 0.5 **Si**, 0.5 **Mg**	**Cold finished and heat treated:** 1.5–3.5 mm **diam**, 315 MPa **TS**, 3% in (50 mm) **El**
AMS 4054B		US	Sheet: clad one side	rem **Al**, 0.20–0.50 **Si**, 0.15–0.40 **Cu**, 0.10 **Mn**, 0.40–0.8 **Mg**, 0.20 **Zn**, 0.8 **Fe**, 0.05 each, 0.15 total, others	**Annealed**: 0.010–0.020 in. (0.25–0.51 mm) **diam**, 20 ksi (138 MPa) **TS**, 14% in 2 in. (50 mm) **El**; **Solution and precipitation heat treated**: 0.010–0.020 in. (0.25–0.51 mm) **diam**, 35 ksi (241 MPa) **TS**, 30 ksi (207 MPa) **YS**, 6% in 2 in. (50 mm) **El**
AMS 4055B		US	Sheet: clad two sides	rem **Al**, 0.20–0.50 **Si**, 0.15–0.40 **Cu**, 0.10 **Mn**, 0.40–0.8 **Mg**, 0.20 **Zn**, 0.8 **Fe**, 0.05 each, 0.15 total, others	**Annealed**: <0.020 in. (0.25–0.51 mm) **diam**, 20 ksi (138 MPa) **TS**, 14% in 2 in. (50 mm) **El**; **Solution and precipitation heat treated**: 0.010–0.20 in. (0.25–0.51 mm) **diam**, 35 ksi (241 MPa) **TS**, 30 ksi (207 MPa) **YS**, 6% in 2 in. (50 mm) **El**
JIS Z 3263	BA23PC	Japan	Sheet: brazing	rem **Al**, 0.20–0.50 **Si**, 0.15–0.40 **Cu**, 0.10 **Mn**, 0.40–0.8 **Mg**, 0.8 **Fe**, 0.05 each, 0.15 total, others	**Annealed**: <0.8 mm **diam**, 147 MPa **TS**, 18% in 2 in. (50 mm) **El**
JIS Z 3263	BA21PC	Japan	Sheet: brazing	rem **Al**, 0.20–0.50 **Si**, 0.15–0.40 **Cu**, 0.10 **Mn**, 0.40–0.8 **Mg**, 0.20 **Zn**, 0.8 **Fe**, 0.05 each, 0.15 total, others	**Annealed**: <0.8 mm **diam**, 137 MPa **TS**, 18% in 2 in. (50 mm) **El**
MIL-A-47193	6951	US	Sheet (core material)	rem **Al**, 0.20–0.50 **Si**, 0.40 **Cu**, 0.10 **Mn**, 0.40–0.80 **Mg**, 0.20 **Zn**, 0.80 **Fe**, clad with alloy 4343	**T6**: 0.125 in. **diam**, 241 MPa **TS**, 207 MPa **YS**, 8% in 2 in. (50 mm) **El**
MIL-B-20148C	Class 6	US	Sheet (core material)	rem **Al**, 0.20–0.50 **Si**, 0.15–0.40 **Cu**, 0.10 **Mn**, 0.40–0.8 **Mg**, 0.20 **Zn**, 0.8 **Fe**, 0.05 each, 0.15 total, others, clad both sides with alloy 4045	**O**: 0.020–0.031 in. **diam**, 145 max MPa **TS**, 18% in 2 in. (50 mm) **El**
MIL-B-20148C	Class 5	US	Sheet (core material)	rem **Al**, 0.20–0.50 **Si**, 0.15–0.40 **Cu**, 0.10 **Mn**, 0.40–0.8 **Mg**, 0.20 **Zn**, 0.8 **Fe**, 0.05 each, 0.15 total, others, clad one side with alloy 4045	**O**: 0.020–0.031 in. **diam**, 145 max MPa **TS**, 18% in 2 in. (50 mm) **El**
MIL-B-20148C	Class 4	US	Sheet (core material)	rem **Al**, 0.20–0.50 **Si**, 0.15–0.40 **Cu**, 0.10 **Mn**, 0.40–0.8 **Mg**, 0.20 **Zn**, 0.8 **Fe**, 0.05 each, 0.15 total, others, clad both sides with alloy 4343	**O**: 0.020–0.031 in. **diam**, 137 max MPa **TS**, 18% in 2 in. (50 mm) **El**
MIL-B-20148C	Class 3	US	Sheet (core material)	rem **Al**, 0.20–0.50 **Si**, 0.15–0.40 **Cu**, 0.10 **Mn**, 0.40–0.8 **Mg**, 0.20 **Zn**, 0.8 **Fe**, 0.05 each, 0.15 total, others, clad one side with alloy 4343	**O**: 0.020–0.031 in. **diam**, 137 max MPa **TS**, 18% in 2 in. (50 mm) **El**

WROUGHT ALUMINUM AND ALUMINUM ALLOYS

specification number	designation	country	product forms	chemical composition	mechanical properties and hardness values
NF A 50 411	6060 (A–GS)	France	Bar, wire, tube, and shapes	rem **Al**, 0.30–0.60 **Si**, 0.10 **Cu**, 0.10 **Mn**, 0.35–0.60 **Mg**, 0.05 **Cr**, 0.15 **Zn**, 0.15 **Ti**, 0.10–0.30 **Fe**, 0.05 each, 0.15 total, others	**Annealed (drawn bar):** ≤25 mm **diam**, 180 MPa **TS**, 130 MPa **YS**, 10% in 2 in. (50 mm) **El**; **Aged (wire)**: 2–12 mm **diam**, 140 MPa **TS**, 80 MPa **YS**, 16% in 2 in. (50 mm) **El**; **Quenched and tempered (tube)**: ≤150 mm **diam**, 220 MPa **TS**, 170 MPa **YS**, 10% in 2 in. (50 mm) **El**
SIS 14 41 03	SIS Al 41 03–00	Sweden	Bar, tube	98.9 **Al**, 0.30–0.6 **Si**, 0.10 **Cu**, 0.10 **Mn**, 0.35–0.6 **Mg**, 0.05 **Cr**, 0.15 **Zn**, 0.10 **Ti**, 0.10–0.30 **Fe**, 0.05 each, 0.15 total, others	**Hot worked**: 20–200 mm **diam**, 110 MPa **TS**, 50 MPa **YS**, 15% in 2 in. (50 mm) **El**
SIS 14 41 03	SIS Al 41 03–04	Sweden	Bar, tube, section	98.9 **Al**, 0.30–0.6 **Si**, 0.10 **Cu**, 0.10 **Mn**, 0.35–0.6 **Mg**, 0.05 **Cr**, 0.15 **Zn**, 0.10 **Ti**, 0.10–0.30 **Fe**, 0.05 each, 0.15 total, others	**Naturally aged**: 10–200 mm **diam**, 120 MPa **TS**, 60 MPa **YS**, 12% in 2 in. (50 mm) **El**
SIS 14 41 03	SIS Al 41 03–06	Sweden	Bar, tube, section	98.9 **Al**, 0.30–0.6 **Si**, 0.10 **Cu**, 0.10 **Mn**, 0.35–0.6 **Mg**, 0.05 **Cr**, 0.15 **Zn**, 0.10 **Ti**, 0.10–0.30 **Fe**, 0.05 each, 0.15 total, others	**Artificially aged**: 10–200 mm **diam**, 190 MPa **TS**, 150 MPa **YS**, 10% in 2 in. (50 mm) **El**, 60–95 **HV**
ANSI/ASTM B 345	6351	US	Tube: seamless	rem **Al**, 0.10 **Cu**, 0.10 **Mn**, 0.10 **Mg**, 0.8–1.3 **Zn**, 0.7 **Si**+ **Fe**, 0.05 each, 0.15 total, others	**Solution heat treated and naturally aged**: all **diam**, 32 ksi (221 MPa) **TS**, 19 ksi (131 MPa) **YS**, 16% in 2 in. (50 mm) **El**; **Solution heat treated and artificially aged**: 0.125–0.749 in. **diam**, 42 ksi (290 MPa) **TS**, 37 ksi (255 MPa) **YS**, 10% in 2 in. (50 mm) **El**
NF A 50 411	7049 A (A–Z8GU)	France	Bar	rem **Al**, 0.40 **Si**, 1.2–2.9 **Cu**, 0.50 **Mn**, 2.1–3.1 **Mg**, 0.05–0.25 **Cr**, 7.2–8.4 **Zn**, 0.50 **Fe**, 0.05 each, 0.15 total, others	**Quenched and tempered**: 610 MPa **TS**, 530 MPa **YS**, 5% in 2 in. (50 mm) **El**
AMS 4200		US	Plate	rem **Al**, 0.25 **Si**, 1.2–1.9 **Cu**, 0.20 **Mn**, 2.0–2.9 **Mg**, 0.10–0.22 **Cr**, 7.2–8.2 **Zn**, 0.10 **Ti**, 0.35 **Fe**, 0.05 each, 0.15 total, others	**Solution and precipitation heat treated**: 0.750–1.000 in. (19–25 mm) **diam**, 74 ksi (510 MPa) **TS**, 65 ksi (448 MPa) **YS**, 8% in 2 in. (50 mm) **El**
AMS 4111		US	Die and hand forgings	rem **Al**, 0.25 **Si**, 1.2–1.9 **Cu**, 0.20 **Mn**, 2.0–2.9 **Mg**, 0.10–0.22 **Cr**, 7.2–8.2 **Zn**, 0.10 **Ti**, 0.35 **Fe**, 0.05 each, 0.15 total, others	**Solution and precipitation heat treated (die forgings)**: ≤2 in. (≤51 mm) **diam**, 72 ksi (496 MPa) **TS**, 62 ksi (427 MPa) **YS**, 7% in 2 in. (50 mm) **El**, 135 min **HB**; **Solution and precipitation heat treated (hand forgings)**: 2.00–3.00 in. **diam**, 71 ksi **TS**, 61 ksi **YS**, 9% in 2 in. (50 mm) **El**, 135 min **HB**

WROUGHT ALUMINUM AND ALUMINUM ALLOYS

specification number	designation	country	product forms	chemical composition	mechanical properties and hardness values
ANSI/ASTM B 247	7049	US	Die and hand forgings	rem **Al**, 0.25 **Si**, 1.2–1.9 **Cu**, 0.20 **Mn**, 2.0–2.9 **Mg**, 0.10–0.22 **Cr**, 7.2–8.2 **Zn**, 0.10 **Ti**, 0.35 **Fe**, 0.05 each, 0.15 total, others	**T73 (die forgings)**: ≤1.000 in. **diam**, 72 ksi (496 MPa) **TS**, 62 ksi (427 MPa) **YS**, 7% in 2 in. (50 mm) **El**, 135 min **HB**; **T73 (hand forgings)**: 2.001–3.000 in. **diam**, 71 ksi (490 MPa) **TS**, 61 ksi (421 MPa) **YS**, 9% in 2 in. (50 mm) **El**; **T7352**: 1.001–3.000 in. **diam**, 71 ksi (490 MPa) **TS**, 59 ksi (407 MPa) **YS**, 9% in 2 in. (50 mm) **El**
AMS 4159		US	Bar, tube, rod, wire, and shapes: extruded	rem **Al**, 0.25 **Si**, 1.2–1.9 **Cu**, 0.20 **Mn**, 2.0–2.9 **Mg**, 0.10–0.22 **Cr**, 7.2–8.2 **Zn**, 0.10 **Ti**, 0.35 **Fe**, 0.05 each, 0.15 total, others	**Solution and precipitation heat treated**: ≤2.999 in. (≤76.17 mm) **diam**, 78 ksi (538 MPa) **TS**, 70 ksi (483 MPa) **YS**, 7% in 2 in. (50 mm) **El**
AMS 4157		US	Bar, tube, rod, wire, and shapes: extruded	rem **Al**, 0.25 **Si**, 1.2–1.9 **Cu**, 0.20 **Mn**, 2.0–2.9 **Mg**, 0.10–0.22 **Cr**, 7.2–8.2 **Zn**, 0.10 **Ti**, 0.35 **Fe**, 0.05 each, 0.15 total, others	**Solution and precipitation heat treated**: ≤2.999 in. (≤76.17 mm) **diam**, 74 ksi (510 MPa) **TS**, 64 ksi (441 MPa) **YS**, 7% in 2 in. (50 mm) **El**
AMS 4137A		US	Die and hand forgings	rem **Al**, 0.40 **Si**, 0.30–1.0 **Cu**, 0.30–0.8 **Mn**, 1.2–2.0 **Mg**, 7.0–8.0 **Zn**, 0.20 **Ti**, 0.6 **Fe**, 0.05 each, 0.15 total, others	**Solution and precipitation heat treated**: ≤3 in. **diam**, 70 ksi (483 MPa) **TS**, 60 ksi (414 MPa) **YS**, 10% in 2 in. (50 mm) **El**, 140 min **HB**
ANSI/ASTM B 247	7076	US	Die forgings	rem **Al**, 0.40 **Si**, 0.30–1.0 **Cu**, 0.30–0.8 **Mn**, 1.2–2.0 **Mg**, 7.0–8.0 **Zn**, 0.20 **Ti**, 0.6 **Fe**, 0.05 each, 0.15 total, others	**T61**: ≤4.000 in. **diam**, 70 ksi (483 MPa) **TS**, 60 ksi (414 MPa) **YS**, 10% in 2 in. (50 mm) **El**, 140 min **HB**
AMS 4051D		US	Sheet and plate: alclad	rem **Al**, 0.40 **Si**, 1.6–2.4 **Cu**, 0.30 **Mn**, 2.4–3.1 **Mg**, 0.18–0.35 **Cr**, 6.3–7.3 **Zn**, 0.20 **Ti**, 0.50 **Fe**, 0.05 each, 0.15 total, others	**Annealed**: 0.015–0.062 in. (0.38–1.57 mm) **diam**, 36 ksi (248 MPa) **TS**, 20 ksi (138 MPa) **YS**, 10% in 2 in. (50 mm) **El**; **Solution and precipitation heat treated**: 0.015–0.044 in. (0.38–1.12 mm) **diam**, 76 ksi (524 MPa) **TS**, 66 ksi (455 MPa) **YS**, 7% in 2 in. (50 mm) **El**
AMS 4052A		US	Sheet and plate: alclad	rem **Al**, 0.50 **Si**, 0.18–0.40 **Cu**, 0.30 **Mn**, 2.4–3.1 **Mg**, 0.18–0.40 **Cr**, 6.3–7.3 **Zn**, 0.20 **Ti**, 0.7 **Fe**, 0.05 each, 0.15 total, others	**Solution and precipitation heat treated plate**: 0.015–0.044 in. **diam**, 76 ksi (524 MPa) **TS**, 66 ksi (455 MPa) **YS**, 7% in 2 in. (50 mm) **El**

WROUGHT ALUMINUM AND ALUMINUM ALLOYS

specification number	designation	country	product forms	chemical composition	mechanical properties and hardness values
ANSI/ASTM B 316	7178	US	Wire and rod: rivet and cold-heading	rem **Al**, 0.40 **Si**, 1.6–2.4 **Cu**, 0.30 **Mn**, 2.4–3.1 **Mg**, 0.18–0.35 **Cr**, 6.3–7.3 **Zn**, 0.20 **Ti**, 0.50 **Fe**, 0.05 each, 0.15 total, others	**Annealed**: ≤1.000 in. **diam**, 40 max ksi (276 max MPa) **TS**; **H13**: ≤1.000 in. **diam**, 36 ksi (248 MPa) **TS**; **Solution heat treated and artificially aged**: 0.061–1.000 in. **diam**, 84 ksi (579 MPa) **TS**, 73 ksi (503 MPa) **YS**, 5% in 2 in. (50 mm) **El**
ANSI/ASTM B 241	7178	US	Tube: seamless	rem **Al**, 0.40 **Si**, 1.6–2.4 **Cu**, 0.30 **Mn**, 2.4–3.1 **Mg**, 0.18–0.35 **Cr**, 6.3–7.3 **Zn**, 0.20 **Ti**, 0.50 **Fe**, 0.05 each, 0.15 total, others	**Annealed**: all **diam**, 40 max ksi (276 max MPa) **TS**, 24 max ksi (165 max MPa) **YS**, 10% in 2 in. (50 mm) **El**; **Solution heat treated and artificially aged**: 1.500–2.499 in. **diam**, 86 ksi (593 MPa) **TS**, 77 ksi (531 MPa) **YS**, 5% in 2 in. (50 mm) **El**
ANSI/ASTM B 221	7178	US	Bar, tube, rod, wire, and shapes	rem **Al**, 0.40 **Si**, 1.6–2.4 **Cu**, 0.30 **Mn**, 2.4–3.1 **Mg**, 0.18–0.35 **Cr**, 6.3–7.3 **Zn**, 0.20 **Ti**, 0.50 **Fe**, 0.05 each, 0.15 total, others	**Annealed**: all **diam**, 40 max ksi (276 max MPa) **TS**, 24 max ksi (165 max MPa) **YS**, 10% in 2 in. (50 mm) **El**; **Solution heat treated and artificially aged**: 0.250–1.499 in. **diam**, 87 ksi (600 MPa) **TS**, 78 ksi (538 MPa) **YS**, 5% in 2 in. (50 mm) **El**; **T76**: 0.125–0.249 in. **diam**, 76 ksi (524 MPa) **TS**, 66 ksi (455 MPa) **YS**, 7% in 2 in. (50 mm) **El**
ANSI/ASTM B 209	Alclad 7178	US	Sheet and plate	rem **Al**, 0.40 **Si**, 1.6–2.4 **Cu**, 0.30 **Mn**, 2.4–3.1 **Mg**, 0.18–0.35 **Cr**, 6.3–7.3 **Zn**, 0.20 **Ti**, 0.50 **Fe**, 7072 cladding material, 0.05 each, 0.15 total, others	**Annealed**: 0.188–0.499 in. **diam**, 40 max ksi (276 MPa) **TS**, 21 max ksi (145 max MPa) **YS**, 10% in 2 in. (50 mm) **El**; **Solution heat treated and artificially aged**: 0.063–0.187 in. **diam**, 80 ksi (552 MPa) **TS**, 70 ksi (483 MPa) **YS**, 8% in 2 in. (50 mm) **El**; **T76**: 0.063–0.187 in. **diam**, 71 ksi (490 MPa) **TS**, 60 ksi (414 MPa) **YS**, 8% in 2 in. (50 mm) **El**
AMS 4158A		US	Bar, rod, wire, and shapes: extruded	rem **Al**, 0.50 **Si**, 1.6–2.4 **Cu**, 0.30 **Mn**, 2.4–3.1 **Mg**, 0.18–0.40 **Cr**, 6.3–7.3 **Zn**, 0.20 **Ti**, 0.7 **Fe**, 0.05 each, 0.15 total, others	**Solution and precipitation heat treated**: 0.249–1.499 in. **diam**, 88 ksi (607 MPa) **TS**, 79 ksi (545 MPa) **YS**, 5% in 2 in. (50 mm) **El**, 145 min **HB**

specification number	designation	country	product forms	chemical composition	mechanical properties and hardness values
ANSI/ASTM B 209	7178	US	Sheet and plate	rem **Al**, 0.40 **Si**, 1.6–2.4 **Cu**, 0.30 **Mn**, 2.4–3.1 **Mg**, 0.18–0.35 **Cr**, 6.3–7.3 **Zn**, 0.20 **Ti**, 0.50 **Fe**, 0.05 each, 0.15 total, others	**Annealed**: 0.015–0.499 in. **diam**, 40 max ksi (276 max MPa) **TS**, 21 max ksi (145 max MPa) **YS**, 10% in 2 in. (50 mm) **El**; **Solution heat treated and artificially aged**: 0.188–0.249 in. **diam**, 82 ksi (565 MPa) **TS**, 71 ksi (490 MPa) **YS**, 8% in 2 in. (50 mm) **El**; **T76**: 0.188–0.249 in. **diam**, 73 ksi (503 MPa) **TS**, 61 ksi (421 MPa) **YS**, 8% in 2 in. (50 mm) **El**
QQ-A-250/28A	7178	US	Plate, sheet	rem **Al**, 0.40 **Si**, 1.6–2.4 **Cu**, 0.30 **Mn**, 2.4–3.1 **Mg**, 0.18–0.35 **Cr**, 6.3–7.3 **Zn**, 0.20 **Ti**, 0.50 **Fe**, 0.05 each, 0.15 total, others	**Annealed**: 0.015–0.499 in. **diam**, 40 max ksi (276 max MPa) **TS**, 21 max ksi (145 max MPa) **YS**, 10% in 2 in. (50 mm) **El**; **Solution heat treated and artificially aged**: 0.015–0.044 in. **diam**, 79 ksi (545 MPa) **TS**, 69 ksi (476 MPa) **YS**, 7% in 2 in. (50 mm) **El**; **Solution heat treated and 2 step aged**: 0.045–0.062 in. **diam**, 73 ksi (503 MPa) **TS**, 62 ksi (427 MPa) **YS**, 8% in 2 in. (50 mm) **El**
QQ-A-250/21A	7178-T76	US	Plate, sheet	rem **Al**, 0.40 **Si**, 1.6–2.4 **Cu**, 0.30 **Mn**, 2.4–3.1 **Mg**, 0.18–0.35 **Cr**, 6.3–7.3 **Zn**, 0.20 **Ti**, 0.50 **Fe**, 0.05 each, 0.15 total, others	**Solution heat treated and artificially aged (sheet)**: 0.045–0.249 in. **diam**, 75 ksi (517 MPa) **TS**, 64 ksi (441 MPa) **YS**, 8% in 2 in. (50 mm) **El**; **Solution heat treated, stress relieved and aged (plate)**: 0.250–0.499 in. **diam**, 74 ksi (510 MPa) **TS**, 63 ksi (434 MPa) **YS**, 8% in 2 in. (50 mm) **El**
QQ-A-250/14E	7178	US	Plate, sheet	rem **Al**, 0.40 **Si**, 1.6–2.4 **Cu**, 0.30 **Mn**, 2.4–3.1 **Mg**, 0.18–0.35 **Cr**, 6.3–7.3 **Zn**, 0.20 **Ti**, 0.50 **Fe**, 0.05 each, 0.15 total, others	**Annealed**: 0.015–0.499 in. **diam**, 40 max ksi (276 max MPa) **TS**, 21 max ksi (145 max MPa) **YS**, 10% in 2 in. (50 mm) **El**; **Solution heat treated and artificially aged**: 0.015–0.044 in. **diam**, 83 ksi (572 MPa) **TS**, 72 ksi (496 MPa) **YS**, 7% in 2 in. (50 mm) **El**; **Solution treated, stress relieved and aged**: 0.250–0.499 in. **diam**, 84 ksi (579 MPa) **TS**, 73 ksi (503 MPa) **YS**, 8% in 2 in. (50 mm) **El**

WROUGHT ALUMINUM AND ALUMINUM ALLOYS

specification number	designation	country	product forms	chemical composition	mechanical properties and hardness values
QQ-A-250/15E	7178	US	Plate, sheet	rem **Al**, 0.40 **Si**, 1.6–2.4 **Cu**, 0.30 **Mn**, 2.4–3.1 **Mg**, 0.18–0.35 **Cr**, 6.3–7.3 **Zn**, 0.20 **Ti**, 0.50 **Fe**, 0.05 each, 0.15 total, others	**Annealed**: 0.015–0.062 in. **diam**, 36 max ksi (248 max MPa) **TS**, 20 max ksi (138 max MPa) **YS**, 10% in 2 in. (50 mm) **El**; **Solution heat treated and artificially aged**: 0.015–0.044 in. **diam**, 76 ksi (524 MPa) **TS**, 66 ksi (455 MPa) **YS**, 7% in 2 in. (50 mm) **El**; **Solution treated, stress relieved and aged**: 0.250–0.499 in. **diam**, 82 ksi (565 MPa) **TS**, 71 ksi (490 MPa) **YS**, 8% in 2 in. (50 mm) **El**
QQ-A-200/13B	7178	US	Bar, rod, shapes, tube, wire	rem **Al**, 0.40 **Si**, 1.6–2.4 **Cu**, 0.30 **Mn**, 2.4–3.1 **Mg**, 0.18–0.35 **Cr**, 6.3–7.3 **Zn**, 0.20 **Ti**, 0.50 **Fe**, 0.05 each, 0.15 total, others	**Annealed**: all **diam**, 40 max ksi (276 max MPa) **TS**, 24 max ksi (165 max MPa) **YS**, 10% in 2 in. (50 mm) **El**; **Solution heat treated and artificially aged**: 0.061 max in. **diam**, 82 ksi (565 MPa) **TS**, 76 ksi (524 MPa) **YS**, 5% in 2 in. (50 mm) **El**
QQ-A-200/14A	7178-T76	US	Bar, rod, shapes, wire	rem **Al**, 0.40 **Si**, 1.6–2.4 **Cu**, 0.30 **Mn**, 2.4–3.1 **Mg**, 0.18–0.35 **Cr**, 6.3–7.3 **Zn**, 0.20 **Ti**, 0.50 **Fe**, 0.05 each, 0.15 total, others	**Solution heat treated and artificially aged**: 0.249 max in. **diam**, 76 ksi (524 MPa) **TS**, 66 ksi (455 MPa) **YS**, 7% in 2 in. (50 mm) **El**; **Solution heat treated, aged and stretched**: 0.250–0.499 in. **diam**, 77 ksi (531 MPa) **TS**, 67 ksi (462 MPa) **YS**, 7% in 2 in. (50 mm) **El**; **Solution heat treated, aged, stretched and straightened**: 0.500–1.000 in. **diam**, 77 ksi (531 MPa) **TS**, 67 ksi (462 MPa) **YS**, 7% in 2 in. (50 mm) **El**
AMS 4050A		US	Plate	rem **Al**, 0.12 **Si**, 2.0–2.6 **Cu**, 0.10 **Mn**, 1.9–2.6 **Mg**, 0.04 **Cr**, 5.7–6.7 **Zn**, 0.06 **Ti**, 0.15 **Fe**, 0.08–0.15 **Zr**, 0.05 each, 0.15 total, others	**Solution and precipitation heat treated**: $\leq$2 in. ($\leq$51 mm) **diam**, 71 ksi (490 MPa) **TS**, 62 ksi (427 MPa) **YS**, 9% in 2 in. (50 mm) **El**
AMS 4108A		US	Hand forgings	rem **Al**, 0.12 **Si**, 2.0–2.6 **Cu**, 0.10 **Mn**, 1.9–2.6 **Mg**, 0.04 **Cr**, 5.7–6.7 **Zn**, 0.06 **Ti**, 0.15 **Fe**, 0.08–0.15 **Zr**, 0.05 each, 0.15 total, others	**Solution and precipitation heat treated**: $\leq$2 in. ($\leq$51 mm) **diam**, 72 ksi (496 MPa) **TS**, 63 ksi (434 MPa) **YS**, 9% in 2 in. (50 mm) **El**, 135 min **HB**
AMS 4107A		US	Die forgings	rem **Al**, 0.12 **Si**, 2.0–2.6 **Cu**, 0.10 **Mn**, 1.9–2.6 **Mg**, 0.04 **Cr**, 5.7–6.7 **Zn**, 0.06 **Ti**, 0.15 **Fe**, 0.08–0.15 **Zr**, 0.05 each, 0.15 total, others	**Solution and precipitation heat treated**: $\leq$2 in. ($\leq$51 mm) **diam**, 72 ksi (496 MPa) **TS**, 62 ksi (427 MPa) **YS**, 7% in 2 in. (50 mm) **El**, 135 min **HB**

WROUGHT ALUMINUM AND ALUMINUM ALLOYS

specification number	designation	country	product forms	chemical composition	mechanical properties and hardness values
ANSI/ASTM B 247	7050	US	Die and hand forgings	rem **Al**, 0.12 **Si**, 2.0–2.6 **Cu**, 0.10 **Mn**, 1.9–2.6 **Mg**, 0.04 **Cr**, 5.7–6.7 **Zn**, 0.06 **Ti**, 0.15 **Fe**, 0.08–0.15 **Zr**, 0.05 each, 0.15 total, others	**T736 (die forgings)**: ≤2.000 in. **diam**, 72 ksi (496 MPa) **TS**, 62 ksi (427 MPa) **YS**, 7% in 2 in. (50 mm) **El**, 135 min **HB**; **T73652 (hand forgings)**: ≤2.000 in. **diam**, 72 ksi (496 MPa) **TS**, 63 ksi (434 MPa) **YS**, 9% in 2 in. (50 mm) **El**
AMS 4342		US	Bar, tube, rod, wire, and shapes: extruded	rem **Al**, 0.12 **Si**, 2.0–2.6 **Cu**, 0.10 **Mn**, 1.9–2.6 **Mg**, 0.04 **Cr**, 5.7–6.7 **Zn**, 0.06 **Ti**, 0.15 **Fe**, 0.08–0.15 **Zr**, 0.05 each, 0.15 total, others	**Solution and precipitation heat treated**: <3.000 in. (<76.20 mm) **diam**, 73 ksi (503 MPa) **TS**, 63 ksi (434 MPa) **YS**, 7% in 2 in. (50 mm) **El**
AMS 4341		US	Bar, tube, rod, wire, and shapes: extruded	rem **Al**, 0.12 **Si**, 2.0–2.6 **Cu**, 0.10 **Mn**, 1.9–2.6 **Mg**, 0.04 **Cr**, 5.7–6.7 **Zn**, 0.06 **Ti**, 0.15 **Fe**, 0.08–0.15 **Zr**, 0.05 each, 0.15 total, others	**Solution and precipitation heat treated**: <3.000 in. (<76.20 mm) **diam**, 70 ksi (483 MPa) **TS**, 60 ksi (414 MPa) **YS**, 8% in 2 in. (50 mm) **El**
AMS 4340		US	Bar, tube, rod, wire, and shapes: extruded	rem **Al**, 0.12 **Si**, 2.0–2.6 **Cu**, 0.10 **Mn**, 1.9–2.6 **Mg**, 0.04 **Cr**, 5.7–6.7 **Zn**, 0.06 **Ti**, 0.15 **Fe**, 0.08–0.15 **Zr**, 0.05 each, 0.15 total, others	**Solution and precipitation heat treated**: ≤5.000 in. (≤127.00 mm) **diam**, 79 ksi (545 MPa) **TS**, 69 ksi (476 MPa) **YS**, 7% in 2 in. (50 mm) **El**
BS 2L.95		UK	Plate	rem **Al**, 0.40 **Si**, 1.2–2.0 **Cu**, 0.30 **Mn**, 2.1–2.9 **Mg**, 0.10–0.25 **Cr**, 0.05 **Ni**, 5.1–6.4 **Zn**, 0.05 **Sn**, 0.50 **Fe**, 0.05 **Pb**, 0.20 **Zr+Ti**	**Solution heat and precipitation treated and quenched**: 6–12.5 mm **diam**, 530 MPa **TS**, 450 MPa **YS**, 8% in 2 in. (50 mm) **El**
BS 2L.88		UK	Sheet and strip	rem **Al**, 0.40 **Si**, 1.2–2.0 **Cu**, 0.30 **Mn**, 2.1–2.9 **Mg**, 0.10–0.25 **Cr**, 0.05 **Ni**, 5.1–6.4 **Zn**, 0.05 **Sn**, 0.50 **Fe**, 0.05 **Pb**, 0.20 **Zr+Ti**	**Solution heat and precipitation treated and quenched**: 0.4–0.8 mm **diam**, 460 MPa **TS**, 420 MPa **YS**, 8% in 2 in. (50 mm) **El**
ISO R827	Al–Zn6MgCu	ISO	Extruded products	rem **Al**, 0.40 **Si**, 1.2–2.0 **Cu**, 0.30 **Mn**, 2.1–2.9 **Mg**, 0.10–0.35 **Cr**, 0.10 **Ni**, 5.1–6.4 **Zn**, 0.50 **Fe**, 0.30 **Ti+Zr**, 0.50 **Mn+Cr**	**Solution and precipitation treated**: 0.25 max in. (6 max mm) **diam**, 525 MPa **TS**, 471 MPa **YS**, 7% in 2 in. (50 mm) **El**
ISO R209	Al–Zn6MgCu	ISO	Wrought products	rem **Al**, 0.40 **Si**, 1.2–2.0 **Cu**, 0.30 **Mn**, 2.1–2.9 **Mg**, 0.10–0.35 **Cr**, 0.10 **Ni**, 5.1–6.4 **Zn**, 0.50 **Fe**, 0.30 **Ti+Zr**, 0.50 **Mn+Cr**	
ISO TR2778	Al–Zn6MgCu	ISO	Tube: drawn	rem **Al**, 0.40 **Si**, 1.2–2.0 **Cu**, 0.30 **Mn**, 2.1–2.9 **Mg**, 0.10–0.35 **Cr**, 0.10 **Ni**, 5.1–6.4 **Zn**, 0.50 **Fe**, 0.30 **Ti+Zr**, 0.50 **Mn+Cr**	**Solution and precipitation treated**: all **diam**, 515 MPa **TS**, 440 MPa **YS**, 7% in 2 in. (50 mm) **El**
ISO TR2136	Al–Zn6MgCu	ISO	Sheet, plate	rem **Al**, 0.40 **Si**, 1.2–2.0 **Cu**, 0.30 **Mn**, 2.1–2.9 **Mg**, 0.10–0.35 **Cr**, 0.10 **Ni**, 5.1–6.4 **Zn**, 0.50 **Fe**, 0.30 **Ti+Zr**, 0.50 **Mn+Cr**	**Solution heat treated and artificially aged**: all **diam**, 525 MPa **TS**, 450 MPa **YS**, 7% in 2 in. (50 mm) **El**

WROUGHT ALUMINUM AND ALUMINUM ALLOYS

specification number	designation	country	product forms	chemical composition	mechanical properties and hardness values
AMS 4084		US	Sheet	rem **Al**, 0.10 **Si**, 1.2–1.9 **Cu**, 0.06 **Mn**, 1.9–2.6 **Mg**, 0.18–0.25 **Cr**, 5.2–6.2 **Zn**, 0.06 **Ti**, 0.12 **Fe**, 0.05 each, 0.15 total, others	**Solution and precipitation heat treated:** ≥0.040 in. (≥1.02 mm) **diam**, 75 ksi (517 MPa) **TS**, 64 ksi (441 MPa) **YS**, 9% in 2 in. (50 mm) **El**
AMS 4085		US	Sheet	rem **Al**, 0.10 **Si**, 1.2–1.9 **Cu**, 0.06 **Mn**, 1.9–2.6 **Mg**, 0.18–0.25 **Cr**, 5.2–6.2 **Zn**, 0.06 **Ti**, 0.12 **Fe**, 0.05 each, 0.15 total, others	**Solution and precipitation heat treated:** ≥0.040 in. (1.02 mm) **diam**, 71 ksi (490 MPa) **TS**, 60 ksi (414 MPa) **YS**, 9% in 2 in. (50 mm) **El**
AMS 4089		US	Plate	rem **Al**, 0.10 **Si**, 1.2–1.9 **Cu**, 0.06 **Mn**, 1.9–2.6 **Mg**, 0.18–0.25 **Cr**, 5.2–6.2 **Zn**, 0.06 **Ti**, 0.12 **Fe**, 0.05 each, 0.15 total, others	**Solution and precipitation heat treated:** 0.250–0.499 in. (6.35–12.67 mm) **diam**, 70 ksi (483 MPa) **TS**, 60 ksi (414 MPa) **YS**, 9% in 2 in. (50 mm) **El**
AMS 4090		US	Plate	rem **Al**, 0.10 **Si**, 1.2–1.9 **Cu**, 0.06 **Mn**, 1.9–2.6 **Mg**, 0.18–0.25 **Cr**, 5.2–6.2 **Zn**, 0.06 **Ti**, 0.12 **Fe**, 0.05 each, 0.15 total, others	**Solution and precipitation heat treated:** 0.250–0.499 in. (6.35–12.67 mm) **diam**, 77 ksi (531 MPa) **TS**, 67 ksi (462 MPa) **YS**, 9% in 2 in. (50 mm) **El**
AECMA prEN2092	7075	AECMA	Sheet and strip	rem **Al**, 0.40 **Si**, 1.2–2.0 **Cu**, 0.30 **Mn**, 2.1–2.9 **Mg**, 0.18–0.28 **Cr**, 5.1–6.1 **Zn**, 0.20 **Ti**, 0.25 **Ti+Zr**	**Solution treated, quenched and artificially aged:** 0.4–0.8 mm **diam**, 485 MPa **TS**, 420 MPa **YS**, 7% in 2 in. (50 mm) **El**
AECMA prEN2126	7075–T651	AECMA	Plate	rem **Al**, 0.40 **Si**, 1.2–2.0 **Cu**, 0.30 **Mn**, 2.1–2.9 **Mg**, 0.18–0.28 **Cr**, 5.1–6.1 **Zn**, 0.20 **Ti**, 0.25 **Ti+Zr**, 0.05 each, 0.15 total others	**Solution treated, quenched and artificially aged:** 6–10 mm **diam**, 530 MPa **TS**, 470 MPa **YS**, 8% in 2 in. (50 mm) **El**
AMS 4078B		US	Plate	rem **Al**, 0.40 **Si**, 1.2–2.0 **Cu**, 0.30 **Mn**, 2.1–2.9 **Mg**, 0.18–0.35 **Cr**, 5.1–6.1 **Zn**, 0.20 **Ti**, 0.50 **Fe**, 0.25 **Zr+Ti**, 0.05 each, 0.15 total, others	**Solution and precipitation heat treated:** 0.250–1.000 in. (6.35–25.40 mm) **diam**, 69 ksi (476 MPa) **TS**, 57 ksi (393 MPa) **YS**, 7% in 2 in. (50 mm) **El**
AMS 4044E		US	Sheet and plate	rem **Al**, 0.40 **Si**, 1.2–2.0 **Cu**, 0.30 **Mn**, 2.1–2.9 **Mg**, 0.18–0.35 **Cr**, 5.1–6.1 **Zn**, 0.20 **Ti**, 0.50 **Fe**, 0.25 **Zr+Ti**, 0.05 each, 0.15 total, others	**Annealed:** 0.008–0.014 in. (0.20–0.36 mm) **diam**, 40 ksi (276 MPa) **TS**, 21 ksi (145 MPa) **YS**, 9% in 2 in. (50 mm) **El**; **Solution and precipitation heat treated:** 0.008–0.011 in. (0.20–0.28 mm) **diam**, 74 ksi (510 MPa) **TS**, 63 ksi (434 MPa) **YS**, 5% in 2 in. (50 mm) **El**
AMS 4045E		US	Sheet and plate	rem **Al**, 0.40 **Si**, 1.2–2.0 **Cu**, 0.30 **Mn**, 2.1–2.9 **Mg**, 0.18–0.35 **Cr**, 5.1–6.1 **Zn**, 0.20 **Ti**, 0.50 **Fe**, 0.05 each, 0.15 total, others	**Solution and precipitation heat treated:** 0.008–0.011 in. **diam**, 74 ksi (510 MPa) **TS**, 63 ksi (434 MPa) **YS**, 5% in 2 in. (50 mm) **El**

specification number	designation	country	product forms	chemical composition	mechanical properties and hardness values
AMS 4046C		US	Sheet and plate	rem **Al**, 0.40 **Si**, 1.2–2.0 **Cu**, 0.30 **Mn**, 2.1–2.9 **Mg**, 0.18–0.35 **Cr**, 5.1–6.1 **Zn**, 0.20 **Ti**, 0.50 **Fe**, 0.05 each, 0.15 total, others	**Solution and precipitation heat treated**: 0.012–0.039 in. **diam**, 73 ksi (503 MPa) **TS**, 63 ksi (434 MPa) **YS**, 7% in 2 in. (50 mm) **El**
AMS 4047B		US	Sheet and plate: alclad, roll tapered	rem **Al**, 0.50 **Si**, 1.2–2.0 **Cu**, 0.30 **Mn**, 2.1–2.9 **Mg**, 0.18–0.40 **Cr**, 5.1–6.1 **Zn**, 0.20 **Ti**, 0.7 **Fe**, 0.05 each, 0.15 total, others	**Solution and precipitation heat treated**: ≥3/4 in. **diam**, 72 ksi (496 MPa) **TS**, 62 ksi (427 MPa) **YS**, 8% in 2 in. (50 mm) **El**
AMS 4048F		US	Sheet and plate: alclad	0.40 **Si**, 1.2–2.0 **Cu**, 0.30 **Mn**, 2.1–2.9 **Mg**, 0.18–0.35 **Cr**, 5.1–6.1 **Zn**, 0.20 **Ti**, 0.50 **Fe**, 0.25 **Zr+Ti**, 0.05 each, 0.15 total, others	**Annealed**: 0.008–0.014 in. (0.20–0.36 mm) **diam**, 36 ksi (248 MPa) **TS**, 20 ksi (138 MPa) **YS**, 9% in 2 in. (50 mm) **El**; **Solution and precipitation heat treated**: 0.008–0.011 in. (0.20–0.28 mm) **diam**, 68 ksi (469 MPa) **TS**, 58 ksi (400 MPa) **YS**, 5% in 2 in. (50 mm) **El**
AMS 4049F		US	Sheet and plate, alclad	rem **Al**, 0.40 **Si**, 1.2–2.0 **Cu**, 0.30 **Mn**, 2.1–2.9 **Mg**, 0.18–0.35 **Cr**, 5.1–6.1 **Zn**, 0.20 **Ti**, 0.50 **Fe**, 0.05 each, 0.15 total, others	**Solution and precipitation heat treated**: 0.008–0.011 in. **diam**, 68 ksi (469 MPa) **TS**, 58 ksi (400 MPa) **YS**, 5% in 2 in. (50 mm) **El**
AMS 4141B		US	Die forgings	rem **Al**, 0.40 **Si**, 1.2–2.0 **Cu**, 0.30 **Mn**, 2.1–2.9 **Mg**, 0.18–0.35 **Cr**, 5.1–6.1 **Zn**, 0.20 **Ti**, 0.50 **Fe**, 0.25 **Ti+Zr**, 0.05 each, 0.15 total, others	**Solution and precipitation heat treated**: ≤3 in. (≤76 mm) **diam**, 66 ksi (455 MPa) **TS**, 36 ksi (386 MPa) **YS**, 7% in 2 in. (50 mm) **El**, 125 min **HB**
AMS 4139G		US	Die and hand forgings	rem **Al**, 0.40 **Si**, 1.2–2.0 **Cu**, 0.30 **Mn**, 2.1–2.9 **Mg**, 0.18–0.35 **Cr**, 5.1–6.1 **Zn**, 0.20 **Ti**, 0.5 **Fe**, 0.05 each, 0.15 total, others	**Solution and precipitation heat treated (die forgings)**: ≤3 in. (≤76 mm) **diam**, 75 ksi (517 MPa) **TS**, 65 ksi (448 MPa) **YS**, 7% in 2 in. (50 mm) **El**, 135 min **HB**; **Solution and precipitation heat treated (hand forgings)**: ≤3 in. (76 mm) **diam**, 75 ksi (517 MPa) **TS**, 63 ksi (434 MPa) **YS**, 9% in 2 in. (50 mm) **El**, 135 min **HB**
AMS 4131		US	Die and hand forgings	rem **Al**, 0.40 **Si**, 1.2–2.0 **Cu**, 0.30 **Mn**, 2.1–2.9 **Mg**, 0.18–0.35 **Cr**, 5.1–6.1 **Zn**, 0.20 **Ti**, 0.50 **Fe**, 0.05 each, 0.15 total, others	**Solution heat treated and aged (die forgings)**: ≤3 in. (≤76 mm) **diam**, 76 ksi (524 MPa) **TS**, 66 ksi (455 MPa) **YS**, 7% in 2 in. (50 mm) **El**, 135 min **HB**; **Solution heat treated and aged (hand forgings)**: ≤2 in. (≤51 mm) **diam**, 73 ksi (503 MPa) **TS**, 63 ksi (434 MPa) **YS**, 9% in 2 in. (50 mm) **El**, 135 min **HB**

WROUGHT ALUMINUM AND ALUMINUM ALLOYS

specification number	designation	country	product forms	chemical composition	mechanical properties and hardness values
AMS 4126		US	Die and hand forgings and rings	rem **Al**, 0.40 **Si**, 1.2–2.0 **Cu**, 0.30 **Mn**, 2.1–2.9 **Mg**, 0.18–0.35 **Cr**, 5.1–6.1 **Zn**, 0.20 **Ti**, 0.50 **Fe**, 0.05 each, 0.15 total, others	**Solution and precipitation heat treated (forgings)**: ≤1 in. (≤25 mm) **diam**, 75 ksi (517 MPa) **TS**, 64 ksi (441 MPa) **YS**, 10% in 2 in. (50 mm) **El**, 135 min **HB**; **Solution and precipitation heat treated (rings)**: ≤2 in. (≤51 mm) **diam**, 73 ksi (503 MPa) **TS**, 62 ksi (427 MPa) **YS**, 7% in 2 in. (50 mm) **El**, 135 min **HB**; **Solution and precipitation heat treated (hand forgings)**: ≤2 in. (≤51 mm) **diam**, 74 ksi (510 MPa) **TS**, 63 ksi (434 MPa) **YS**, 9% in 2 in. (50 mm) **El**, 135 min **HB**
AMS 4122F		US	Bar, rod, wire, and rings	rem **Al**, 0.40 **Si**, 1.2–2.0 **Cu**, 0.30 **Mn**, 2.1–2.9 **Mg**, 0.18–0.35 **Cr**, 5.1–6.1 **Zn**, 0.20 **Ti**, 0.50 **Fe**, 0.05 each, 0.15 total, others	**Solution and precipitation heat treated**: ≤3 in. (76 mm) **diam**, 77 ksi (531 MPa) **TS**, 66 ksi (455 MPa) **YS**, 7% in 2 in. (50 mm) **El**, 135 min **HB**
AMS 4123D		US	Bar and rod	rem **Al**, 0.50 **Si**, 1.2–2.0 **Cu**, 0.30 **Mn**, 2.1–2.9 **Mg**, 0.18–0.40 **Cr**, 5.1–6.1 **Zn**, 0.20 **Ti**, 0.7 **Fe**, 0.05 each, 0.15 total, others	**Solution and precipitation heat treated**: ≥0.005 in. **diam**, 77 ksi (530 MPa) **TS**, 66 ksi (455 MPa) **YS**, 7% in 2 in. (50 mm) **El**, 135 min **HB**
AMS 4124B		US	Bar, rod, and wire	rem **Al**, 0.40 **Si**, 1.2–2.0 **Cu**, 0.30 **Mn**, 2.1–2.9 **Mg**, 0.18–0.28 **Cr**, 5.1–6.1 **Zn**, 0.20 **Ti**, 0.50 **Fe**, 0.25 **Ti+Zr**, 0.05 each, 0.15 total, others	**Solution and precipitation heat treated**: ≤3.000 in. (76.20 mm) **diam**, 68 ksi (469 MPa) **TS**, 56 ksi (386 MPa) **YS**, 10% in 2 in. (50 mm) **El**, 125 min **HB**
ANSI/ASTM B 316	7075	US	Wire and rod: rivet and cold-heading	rem **Al**, 0.40 **Si**, 1.2–2.0 **Cu**, 0.30 **Mn**, 2.1–2.9 **Mg**, 0.18–0.35 **Cr**, 5.1–6.1 **Zn**, 0.20 **Ti**, 0.50 **Fe**, 0.25 **Zr+Ti**, 0.05 each, 0.15 total, others	**Annealed**: ≤1.000 in. **diam**, 40 max ksi (276 max MPa) **TS**; **H13**: ≤1.000 in. **diam**, 36 ksi (248 MPa) **TS**; **Solution heat treated and artificially aged**: 0.061–1.000 in. **diam**, 77 ksi (531 MPa) **TS**, 66 ksi (455 MPa) **YS**, 7% in 2 in. (50 mm) **El**

WROUGHT ALUMINUM AND ALUMINUM ALLOYS

specification number	designation	country	product forms	chemical composition	mechanical properties and hardness values
ANSI/ASTM B 247	7075	US	Die and hand forgings	rem **Al**, 0.40 **Si**, 1.2–2.0 **Cu**, 0.30 **Mn**, 2.1–2.9 **Mg**, 0.18–0.28 **Cr**, 5.1–6.1 **Zn**, 0.20 **Ti**, 0.50 **Fe**, 0.25 **Zr+Ti**, 0.05 each, 0.15 total, others	**Solution heat treated and artificially aged (die forgings):** ≤1.000 in. **diam**, 75 ksi (517 MPa) **TS**, 64 ksi (441 MPa) **YS**, 7% in 2 in. (50 mm) **El**, 135 min **HB**; **T73 (die forgings):** ≤3.000 in. **diam**, 66 ksi (455 MPa) **TS**, 56 ksi (386 MPa) **YS**, 7% in 2 in. (50 mm) **El**, 125 min **HB**; **Solution heat treated and artificially aged (hand forgings):** ≤2.000 in. **diam**, 74 ksi (510 MPa) **TS**, 63 ksi (434 MPa) **YS**, 9% in 2 in. (50 mm) **El**
ANSI/ASTM B 247	7175	US	Die and hand forgings	rem **Al**, 0.15 **Si**, 1.2–2.0 **Cu**, 0.10 **Mn**, 2.1–2.9 **Mg**, 0.18–0.30 **Cr**, 5.1–6.1 **Zn**, 0.10 **Ti**, 0.20 **Fe**, 0.05 each, 0.15 total, others	**T736 (die forgings):** ≤3.000 in. **diam**, 76 ksi (524 MPa) **TS**, 66 ksi (455 MPa) **YS**, 7% in 2 in. (50 mm) **El**; **T73652 (die forgings):** ≤3.000 in. **diam**, 73 ksi (503 MPa) **TS**, 63 ksi (434 MPa) **YS**, 7% in 2 in. (50 mm) **El**; **T736 (hand forgings):** ≤3.000 in. **diam**, 73 ksi (503 MPa) **TS**, 63 ksi (434 MPa) **YS**, 9% in 2 in. (50 mm) **El**
ANSI/ASTM B 241	7075	US	Tube: seamless	rem **Al**, 0.40 **Si**, 1.2–2.0 **Cu**, 0.30 **Mn**, 2.1–2.9 **Mg**, 0.18–0.35 **Cr**, 5.1–6.1 **Zn**, 0.20 **Ti**, 0.50 **Fe**, 0.25 **Zr+Ti**, 0.05 each, 0.15 total, others	**Annealed**: all **diam**, 40 max ksi (276 max MPa) **TS**, 24 max ksi (165 max MPa) **YS**, 10% in 2 in. (50 mm) **El**; **Solution heat treated and artificially aged**: 3.000–4.999 in. **diam**, 81 ksi (558 MPa) **TS**, 71 ksi (490 MPa) **YS**, 7% in 2 in. (50 mm) **El**; **T73**: 3.000–4.999 in. **diam**, 68 ksi (469 MPa) **TS**, 57 ksi (393 MPa) **YS**, 7% in 2 in. (50 mm) **El**
ANSI/ASTM B 221	7075	US	Bar, tube, rod, wire, and shapes	rem **Al**, 0.40 **Si**, 1.2–2.0 **Cu**, 0.30 **Mn**, 2.1–2.9 **Mg**, 0.18–0.35 **Cr**, 5.1–6.1 **Zn**, 0.20 **Ti**, 0.50 **Fe**, 0.05 each, 0.15 total, others	**Annealed**: all **diam**, 40 max ksi (276 max MPa) **TS**, 24 max ksi (165 max MPa) **YS**, 10% in 2 in. (50 mm) **El**; **Solution heat treated and artificially aged**: 4.500–5.000 in. **diam**, 78 ksi (538 MPa) **TS**, 68 ksi (469 MPa) **YS**; **T73**: 3.000–4.499 in. **diam**, 65 ksi (448 MPa) **TS**, 55 ksi (379 MPa) **YS**, 7% in 2 in. (50 mm) **El**

WROUGHT ALUMINUM AND ALUMINUM ALLOYS

specification number	designation	country	product forms	chemical composition	mechanical properties and hardness values
ANSI/ASTM B 211	7075	US	Bar, rod, and wire	rem **Al**, 0.40 **Si**, 1.2–2.0 **Cu**, 0.30 **Mn**, 2.1–2.9 **Mg**, 0.18–0.35 **Cr**, 5.1–6.1 **Zn**, 0.20 **Ti**, 0.50 **Fe**, 0.25 **Ti+Zr**, 0.05 each, 0.15 total, others	**Annealed**: 0.125–8.000 in. **diam**, 40 max ksi (276 max MPa) **TS**, 10% in 2 in. (50 mm) **El**; **Solution heat treated and artificially aged**: 0.125–3.000 in. **diam**, 77 ksi (531 MPa) **TS**, 66 ksi (455 MPa) **YS**, 7% in 2 in. (50 mm) **El**; **T73**: 0.125–3.000 in. **diam**, 68 ksi (469 MPa) **TS**, 56 ksi (386 MPa) **YS**, 10% in 2 in. (50 mm) **El**
ANSI/ASTM B 209	Alclad 7075	US	Sheet and plate	rem **Al**, 0.40 **Si**, 1.2–2.0 **Cu**, 0.30 **Mn**, 2.1–2.9 **Mg**, 0.18–0.28 **Cr**, 5.1–6.1 **Zn**, 0.50 **Fe**, 0.25 **Ti+Zr**, 7072 cladding material, 0.05 each, 0.15 total, others	**Annealed**: 0.188–0.499 in. **diam**, 39 max ksi (269 max MPa) **TS**, 21 max ksi (145 max MPa) **YS**, 10% in 2 in. (50 mm) **El**; **Solution heat treated and artificially aged**: 0.188–0.249 in. **diam**, 75 ksi (517 MPa) **TS**, 64 ksi (441 MPa) **YS**, 8% in 2 in. (50 mm) **El**; **T76 (sheet)**: 0.188–0.249 in. **diam**, 70 ksi (483 MPa) **TS**, 59 ksi (407 MPa) **YS**, 8% in 2 in. (50 mm) **El**
ANSI/ASTM B 209	7008 Alclad 7075	US	Sheet and plate	rem **Al**, 0.40 **Si**, 1.2–2.0 **Cu**, 0.30 **Mn**, 2.1–2.9 **Mg**, 0.18–0.28 **Cr**, 5.1–6.1 **Zn**, 0.50 **Fe**, 0.25 **Ti+Zr**, 7008 cladding material, 0.05 each, 0.15 total, others	**Annealed**: 0.015–0.499 in. **diam**, 40 max ksi (276 max MPa) **TS**, 21 max ksi (145 max MPa) **YS**, 10% in 2 in. (50 mm) **El**; **Solution heat treated and artificially aged**: 0.188–0.249 in. **diam**, 76 ksi (524 MPa) **TS**, 66 ksi (455 MPa) **YS**, 8% in 2 in. (50 mm) **El**; **T76 (sheet)**: 0.188–0.249 in. **diam**, 72 ksi (496 MPa) **TS**, 61 ksi (421 MPa) **YS**, 8% in 2 in. (50 mm) **El**
ANSI/ASTM B 209	7011 Alclad 7075	US	Sheet and plate	rem **Al**, 0.40 **Si**, 1.2–2.0 **Cu**, 0.30 **Mn**, 2.1–2.9 **Mg**, 0.18–0.28 **Cr**, 5.1–6.1 **Zn**, 0.50 **Fe**, 0.25 **Ti+Zr**, 7011 cladding material, 0.05 each, 0.15 total, others	**Annealed**: 0.015–0.499 in. **diam**, 40 max ksi (276 max MPa) **TS**, 21 max ksi (145 max MPa) **YS**, 10% in 2 in. (50 mm) **El**; **Solution heat treated and artificially aged**: 0.188–0.249 in. **diam**, 76 ksi (524 MPa) **TS**, 66 ksi (455 MPa) **YS**, 8% in 2 in. (50 mm) **El**; **T76 (sheet)**: 0.188–0.249 in. **diam**, 72 ksi (496 MPa) **TS**, 61 ksi (421 MPa) **YS**, 8% in 2 in. (50 mm) **El**
ASTM F 468	A 97075	US	Bolt, screw, stud	rem **Al**, 0.40 **Si**, 1.2–2.0 **Cu**, 2.1–2.9 **Mn**, 0.30 **Mg**, 0.18–0.35 **Cr**, 5.1–6.1 **Zn**, 0.20 **Ti**, 0.50 **Fe**, 0.25 **Ti+Zr**	**As manufactured**: all **diam**, 61 ksi (420 MPa) **TS**, 50 ksi (345 MPa) **YS**, 80–90 **HRB**

WROUGHT ALUMINUM AND ALUMINUM ALLOYS

specification number	designation	country	product forms	chemical composition	mechanical properties and hardness values
AS 1865	7075	Australia	Wire, rod, bar, strip	rem **Al**, 0.40 **Si**, 1.2–2.0 **Cu**, 0.30 **Mn**, 2.1–2.9 **Mg**, 0.18–0.35 **Cr**, 5.1–6.1 **Zn**, 0.20 **Ti**, 0.50 **Fe**, 0.05 each, 0.15 total, others	**Annealed**: 10 mm **diam**, 276 max MPa **TS**; **Strain hardened to 3/8 hard**: 10 mm **diam**, 248 MPa **TS**; **Solution heat treated and artificially aged**: 10 mm **diam**, 530 MPa **TS**, 455 MPa **YS**
AMS 4196		US	Sheet and plate: alclad	rem **Al**, 0.40 **Si**, 1.2–2.0 **Cu**, 0.30 **Mn**, 2.1–2.9 **Mg**, 0.18–0.35 **Cr**, 5.1–6.1 **Zn**, 0.20 **Ti**, 0.50 **Fe**, 0.05 each, 0.15 total, others	**Annealed**: <0.500 in. **diam**, 40 ksi (276 MPa) **TS**, 21 ksi (145 MPa) **YS**, 10% in 2 in. (50 mm) **El**; **Solution and precipitation heat treated**: 0.500–1.000 in. **diam**, 78 ksi (537 MPa) **TS**, 68 ksi (469 MPa) **YS**, 7% in 2 in. (50 mm) **El**
AMS 4197		US	Sheet and plate: alclad	rem **Al**, 0.40 **Si**, 1.2–2.0 **Cu**, 0.30 **Mn**, 2.1–2.9 **Mg**, 0.18–0.35 **Cr**, 5.1–6.1 **Zn**, 0.20 **Ti**, 0.50 **Fe**, 0.05 each, 0.15 total, others	**Solution and precipitation heat treated**: 0.500–1.000 in. **diam**, 78 ksi (537 MPa) **TS**, 68 ksi (469 MPa) **YS**, 7% in 2 in. (50 mm) **El**
AMS 4187		US	Bar, rod, and wire	rem **Al**, 0.40 **Si**, 1.2–2.0 **Cu**, 0.30 **Mn**, 2.1–2.9 **Mg**, 0.18–0.35 **Cr**, 5.1–6.1 **Zn**, 0.20 **Ti**, 0.50 **Fe**, 0.05 each, 0.15 total, others	**Annealed**: ≤8 in. (≤203 mm) **diam**, 40 ksi (276 MPa) **TS**, 10% in 2 in. (50 mm) **El**; **Solution and precipitation heat treated (rod)**: ≤4 in. (≤102 mm) **diam**, 77 ksi (531 MPa) **TS**, 66 ksi (455 MPa) **YS**, 7% in 2 in. (50 mm) **El**, 135 min **HB**
AMS 4186		US	Bar, rod, and wire	rem **Al**, 0.40 **Si**, 1.2–2.0 **Cu**, 0.30 **Mn**, 2.1–2.9 **Mg**, 0.18–0.35 **Cr**, 5.1–6.1 **Zn**, 0.20 **Ti**, 0.50 **Fe**, 0.05 each, 0.15 total, others	**Solution and precipitation heat treated (rod)**: ≤4 in. (≤102 mm) **diam**, 77 ksi (531 MPa) **TS**, 66 ksi (455 MPa) **YS**, 7% in 2 in. (50 mm) **El**, 135 min **HB**
AMS 4169D		US	Bar, tube, rod, and shapes: extruded	rem **Al**, 0.40 **Si**, 1.2–2.0 **Cu**, 0.30 **Mn**, 2.1–2.9 **Mg**, 0.18–0.35 **Cr**, 5.1–6.1 **Zn**, 0.20 **Ti**, 0.50 **Fe**, 0.05 each, 0.15 total, others	**Solution and precipitation heat treated**: <0.250 in. **diam**, 78 ksi (537 MPa) **TS**, 70 ksi (483 MPa) **YS**, 7% in 2 in. (50 mm) **El**, 135 min **HB**
AMS 4310		US	Rings: rolled or forged	rem **Al**, 0.40 **Si**, 1.2–2.0 **Cu**, 0.30 **Mn**, 2.1–2.9 **Mg**, 0.18–0.35 **Cr**, 5.1–6.1 **Zn**, 0.20 **Ti**, 0.50 **Fe**, 0.25 **Ti+Zr**, 0.05 each, 0.15 total, others	**Solution and precipitation heat treated**: ≤2 in. (≤51 mm) **diam**, 74 ksi (510 MPa) **TS**, 63 ksi (434 MPa) **YS**, 9% in 2 in. (50 mm) **El**, 135 min **HB**
AMS 4311		US	Rings: rolled or forged	rem **Al**, 0.40 **Si**, 1.2–2.0 **Cu**, 0.30 **Mn**, 2.1–2.9 **Mg**, 0.18–0.35 **Cr**, 5.1–6.1 **Zn**, 0.20 **Ti**, 0.50 **Fe**, 0.25 **Ti+Zr**, 0.05 each, 0.15 total, others	**Solution and precipitation heat treated**: ≤3 in. (≤76 mm) **diam**, 66 ksi (455 MPa) **TS**, 54 ksi (372 MPa) **YS**, 7% in 2 in. (50 mm) **El**, 130 min **HB**

WROUGHT ALUMINUM AND ALUMINUM ALLOYS

specification number	designation	country	product forms	chemical composition	mechanical properties and hardness values
ANSI/ASTM B 210	7075	US	Tube: seamless	rem **Al**, 0.40 **Si**, 1.2–2.0 **Cu**, 0.30 **Mn**, 2.1–2.9 **Mg**, 0.18–0.28 **Cr**, 5.1–6.1 **Zn**, 0.20 **Ti**, 0.50 **Fe**, 0.25 **Ti+Zr**, 0.05 each, 0.15 total, others	**Annealed**: 0.050–0.500 in. **diam**, 40 ksi (276 MPa) **TS**, 21 ksi (145 MPa) **YS**; **Solution heat treated and artificially aged**: 0.260–0.500 in. **diam**, 77 ksi (531 MPa) **TS**, 66 ksi (455 MPa) **YS**; **T73**: 0.260–0.500 in. **diam**, 66 ksi (455 MPa) **TS**, 56 ksi (386 MPa) **YS**
ANSI/ASTM B 209	7075	US	Sheet and plate	rem **Al**, 0.40 **Si**, 1.2–2.0 **Cu**, 0.30 **Mn**, 2.1–2.9 **Mg**, 0.18–0.28 **Cr**, 5.1–6.1 **Zn**, 0.20 **Ti**, 0.50 **Fe**, 0.25 **Ti+Zr**, 0.05 each, 0.15 total, others	**Annealed**: 0.015–0.499 in. **diam**, 40 max ksi (276 max MPa) **TS**, 21 max ksi (145 max MPa) **YS**, 10% in 2 in. (50 mm) **El**; **Solution heat treated and artificially aged**: 0.126–0.249 in. **diam**, 78 ksi (738 MPa) **TS**, 69 ksi (476 MPa) **YS**, 8% in 2 in. (50 mm) **El**; **T73 (sheet)**: 0.040–0.249 in. **diam**, 67 ksi (462 MPa) **TS**, 56 ksi (386 MPa) **YS**, 8% in 2 in. (50 mm) **El**
CSA HA.8	.7075	Canada	Forgings	rem **Al**, 0.40 **Si**, 1.2–2.0 **Cu**, 0.30 **Mn**, 2.1–2.9 **Mg**, 0.18–0.35 **Cr**, 5.1–6.1 **Zn**, 0.20 **Ti**, 0.50 **Fe**, 0.25 **Zr+Ti**, 0.05 each, 0.15 total, others	**T6 (die forgings)**: <1 in. **diam**, 75.0 ksi (517 MPa) **TS**, 64.0 ksi (441 MPa) **YS**, 7% in 2 in. (50 mm) **El**; **T6 (hand forgings)**: <2.000 in. **diam**, 74.0 ksi (510 MPa) **TS**, 63.0 ksi (434 MPa) **YS**, 9% in 2 in. (50 mm) **El**
CSA HA.7	.7075	Canada	Tube, pipe: seamless	rem **Al**, 0.40 **Si**, 1.2–2.0 **Cu**, 0.30 **Mn**, 2.1–2.9 **Mg**, 0.18–0.35 **Cr**, 5.1–6.1 **Zn**, 0.20 **Ti**, 0.50 **Fe**, 0.25 **Ti+Zr**, 0.05 each, 0.15 total, others	**Drawn and O**: 0.025–0.259 in. **diam**, 40.0 max ksi (276 max MPa) **TS**, 21.0 ksi (145 MPa) **YS**, 8% in 2 in. (50 mm) **El**; **Drawn and T6**: <0.259 in. **diam**, 77.0 ksi (531 MPa) **TS**, 66.0 ksi (455 MPa) **YS**, 8% in 2 in. (50 mm) **El**; **Drawn and T62**: 0.260–0.500 in. **diam**, 77.0 ksi (531 MPa) **TS**, 66.0 ksi (455 MPa) **YS**, 9% in 2 in. (50 mm) **El**
CSA HA.5	.7075	Canada	Bar, rod, wire, shapes	rem **Al**, 0.40 **Si**, 1.2–2.0 **Cu**, 0.30 **Mn**, 2.1–2.9 **Mg**, 0.18–0.35 **Cr**, 5.1–6.1 **Zn**, 0.20 **Ti**, 0.50 **Fe**, 0.05 each, 0.15 total, others	**Drawn and O**: <0.250–1.500 in. **diam**, 40.0 max ksi (276 max MPa) **TS**, 10% in 2 in. (50 mm) **El**; **Extruded and T6**: <0.249 in. **diam**, 78.0 ksi (518 MPa) **TS**, 70.0 ksi (483 MPa) **YS**, 7% in 2 in. (50 mm) **El**; **Extruded and T6510**: 0.500–2.999 in. **diam**, 81.0 ksi (558 MPa) **TS**, 72.0 ksi (496 MPa) **YS**, 7% in 2 in. (50 mm) **El**

WROUGHT ALUMINUM AND ALUMINUM ALLOYS

specification number	designation	country	product forms	chemical composition	mechanical properties and hardness values
CSA HA.4	.7075	Canada	Plate, sheet	rem **Al**, 0.40 **Si**, 1.2–2.0 **Cu**, 0.30 **Mn**, 2.1–2.9 **Mg**, 0.18–0.35 **Cr**, 5.1–6.1 **Zn**, 0.20 **Ti**, 0.50 **Fe**, 0.05 each, 0.15 total, others	**O**: 0.015–0.020 in. **diam**, 40.0 max ksi (276 max MPa) **TS**, 21.0 max ksi (145 max MPa) **YS**, 10% in 2 in. (50 mm) **El**; **T6**: 0.015–0.020 in. **diam**, 76.0 ksi (524 MPa) **TS**, 67.0 ksi (462 MPa) **YS**, 7% in 2 in. (50 mm) **El**; **T651**: 0.250–0.499 in. **diam**, 78.0 ksi (538 MPa) **TS**, 67.0 ksi (462 MPa) **YS**, 9% in 2 in. (50 mm) **El**
CSA HA.4	.7075 Alclad	Canada	Plate, sheet (core)	rem **Al**, 0.40 **Si**, 1.2–2.0 **Cu**, 0.30 **Mn**, 2.1–2.9 **Mg**, 0.18–0.35 **Cr**, 5.1–6.1 **Zn**, 0.20 **Ti**, 0.50 **Fe**, 0.05 each, 0.15 total, others	**O**: 0.015–0.032 in. **diam**, 36.0 max ksi (248 MPa) **TS**, 20.0 max ksi (138 MPa) **YS**, 10% in 2 in. (50 mm) **El**; **T6**: 0.012–0.020 in. **diam**, 70.0 ksi (483 MPa) **TS**, 60.0 ksi (414 MPa) **YS**, 7% in 2 in. (50 mm) **El**; **T62 (sheet)**: 0.021–0.039 in. **diam**, 70.0 ksi (483 MPa) **TS**, 60.0 ksi (414 MPa) **YS**, 7% in 2 in. (50 mm) **El**
COPANT 862	7075	COPANT	Wrought products	rem **Al**, 0.4 **Si**, 1.2–2.0 **Cu**, 0.3 **Mn**, 2.1–2.9 **Mg**, 0.18–0.35 **Cr**, 5.1–6.1 **Zn**, 0.20 **Ti**, 0.5 **Fe**, 0.25 **Ti+Zr**, 0.05 each, 0.15 total, others	
WW-T-700/7A	7075 (Type I–round, II–rectangular and square, III–streamline, IV–oval, V–odd shapes)	US	Tube	rem **Al**, 0.40 **Si**, 1.2–2.0 **Cu**, 0.30 **Mn**, 2.1–2.9 **Mg**, 0.18–0.35 **Cr**, 5.1–6.1 **Zn**, 0.20 **Ti**, 0.50 **Fe**, 0.05 each, 0.15 total, others	**Annealed**: 0.025–0.049 in. **diam**, 40 max ksi (276 max MPa) **TS**, 21 max ksi (145 max MPa) **YS**, 10% in 2 in. (50 mm) **El**; **Solution heat treated and artificially aged**: 0.025–0.259 in. **diam**, 77 ksi (531 MPa) **TS**, 66 ksi (455 MPa) **YS**, 8% in 2 in. (50 mm) **El**
QQ-A-430B	7075	US	Rod, wire	rem **Al**, 0.40 **Si**, 1.2–2.0 **Cu**, 0.30 **Mn**, 2.1–2.9 **Mg**, 0.18–0.35 **Cr**, 5.1–6.1 **Zn**, 0.20 **Ti**, 0.50 **Fe**, 0.05 each, 0.15 total, others	**Annealed**: 1.000 max in. **diam**, 40 max ksi (276 max MPa) **TS**; **Strain hardened and temper H13**: 1.000 max in. **diam**, 36 ksi (248 MPa) **TS**
QQ-A-250/24B	7075	US	Plate, sheet	0.40 **Si**, 1.2–2.0 **Cu**, 0.30 **Mn**, 2.1–2.9 **Mg**, 0.18–0.35 **Cr**, 5.1–6.1 **Zn**, 0.50 **Fe**, 0.05 each, 0.15 total, others; 0.25 **Ti+Zr**	**Solution treated, stress relieved and 2 step aged (plate)**: 0.250–0.499 in. **diam**, 72 ksi (496 MPa) **TS**, 61 ksi (421 MPa) **YS**, 8% in 2 in. (50 mm) **El**; **Solution treated and 2 step aged (sheet)**: 0.063–0.249 in. **diam**, 73 ksi (503 MPa) **TS**, 62 ksi (427 MPa) **YS**, 8% in 2 in. (50 mm) **El**

WROUGHT ALUMINUM AND ALUMINUM ALLOYS

specification number	designation	country	product forms	chemical composition	mechanical properties and hardness values
QQ-A-250/25A	7075	US	Plate, sheet	rem **Al**, 0.40 **Si**, 1.2–2.0 **Cu**, 0.30 **Mn**, 2.1–2.9 **Mg**, 0.18–0.35 **Cr**, 5.1–6.1 **Zn**, 0.25 **Ti**, 0.50 **Fe**, **Zr** in **Ti** content; 0.05 each, 0.15 total, others	**Solution heat treated and artificially aged (sheet):** 0.040–0.062 in. **diam**, 67 ksi (462 MPa) **TS**, 56 ksi (386 MPa) **YS**, 8% in 2 in. (50 mm) **El**; **Solution heat treated, stress relieved and aged (plate):** 0.250–0.499 in. **diam**, 69 ksi (476 MPa) **TS**, 58 ksi (400 MPa) **YS**, 8% in 2 in. (50 mm) **El**
QQ-A-250/26A	7075	US	Plate, sheet	rem **Al**, 0.40 **Si**, 1.2–2.0 **Cu**, 0.30 **Mn**, 2.1–2.9 **Mg**, 0.18–0.35 **Cr**, 5.1–6.1 **Zn**, 0.20 **Ti**, 0.50 **Fe**, 0.05 each, 0.15 total, others	**Annealed**: 0.015–0.499 in. **diam**, 40 max ksi (276 max MPa) **TS**, 21 max ksi (145 max MPa) **YS**, 10% in 2 in. (50 mm) **El**; **Solution heat treated and artificially aged**: 0.015–0.039 in. **diam**, 73 ksi (503 MPa) **TS**, 63 ksi (343 MPa) **YS**, 7% in 2 in. (50 mm) **El**; **Solution heat treated, stress relieved and artificially aged**: 0.250–0.499 in. **diam**, 76 ksi (524 MPa) **TS**, 66 ksi (455 MPa) **YS**, 9% in 2 in. (50 mm) **El**
QQ-A-250/12E	7075	US	Plate, sheet	rem **Al**, 0.40 **Si**, 1.2–2.0 **Cu**, 0.30 **Mn**, 2.1–2.9 **Mg**, 0.18–0.35 **Cr**, 5.1–6.1 **Zn**, .20 **Ti**, 0.50 **Fe**, .05 each, 0.15 total, others	**Annealed**: 0.015–0.499 in. **diam**, 40 max ksi (276 max MPa) **TS**, 21 max ksi (145 max MPa) **YS**, 10% in 2 in. (50 mm) **El**; **Solution heat treated and artificially aged**: 0.008–0.011 in. **diam**, 74 ksi (510 MPa) **TS**, 63 ksi (434 MPa) **YS**, 5% in 2 in. **El**
QQ-A-250/18E	7075	US	Plate, sheet	0.40 **Si**, 1.2–2.0 **Cu**, 0.30 **Mn**, 2.1–2.9 **Mg**, 0.18–0.35 **Cr**, 5.1–6.1 **Zn**, 0.20 **Ti**, 0.50 **Fe**, 0.05 each, 0.15 total, others	**Annealed**: 0.015–0.062 in. **diam**, 38 max ksi (262 max MPa) **TS**, 21 max ksi (145 max MPa) **YS**, 10% in 2 in. (50 mm) **El**; **Solution treated and artificially aged**: 0.012–0.039 in. **diam**, 73 ksi (503 MPa) **TS**, 63 ksi (434 MPa) **YS**, 7% in 2 in. (50 mm) **El**; **Solution treated, stress relieved and aged**: 0.250–0.499 in. **diam**, 76 ksi (524 MPa) **TS**, 66 ksi (455 MPa) **YS**, 9% in 2 in. (50 mm) **El**

WROUGHT ALUMINUM AND ALUMINUM ALLOYS

specification number	designation	country	product forms	chemical composition	mechanical properties and hardness values
QQ-A-225/9D	7075	US	Bar, rod, wire, shapes	rem **Al**, 0.40 **Si**, 1.2–2.0 **Cu**, 0.30 **Mn**, 2.1–2.9 **Mg**, 0.18–0.35 **Cr**, 5.1–6.1 **Zn**, 0.20 **Ti**, 0.50 **Fe**, 0.05 each, 0.15 total, others	**Annealed**: 8.000 max in. **diam**, 40 max ksi (276 max MPa) **TS**, 10% in 2 in. (50 mm) **El**; **Solution heat treated and artificially aged**: 4.000 max in. **diam**, 77 ksi (531 MPa) **TS**, 66 ksi (455 MPa) **YS**, 7% in 2 in. (50 mm) **El**; **Solution heat treated, stress relieved and aged**: 0.500–4.000 in. **diam**, 77 ksi (531 MPa) **TS**, 66 ksi (455 MPa) **YS**, 7% in 2 in. (50 mm) **El**
QQ-A-200/11D	7075	US	Bar, rod, shapes, tube, wire	rem **Al**, 0.40 **Si**, 1.2–2.0 **Cu**, 0.30 **Mn**, 2.1–2.9 **Mg**, 0.18–0.35 **Cr**, 5.1–6.1 **Zn**, 0.20 **Ti**, 0.50 **Fe**, 0.05 each, 0.15 total, others	**Annealed**: all **diam**, 40 max ksi (276 max MPa) **TS**, 24 max ksi (165 max MPa) **YS**, 10% in 2 in. (50 mm) **El**; **Solution heat treated and artificially aged**: 0.249 max in. **diam**, 78 ksi (538 MPa) **TS**, 70 ksi (483 MPa) **YS**, 7% in 2 in. (50 mm) **El**; **Solution heat treated, aged and temper T73**: 0.062–0.249 in. **diam**, 68 ksi (469 MPa) **TS**, 58 ksi (400 MPa) **YS**, 7% in 2 in. (50 mm) **El**
QQ-A-200/15A	7075–T76	US	Bar, rod, shapes	rem **Al**, 0.40 **Si**, 1.2–2.0 **Cu**, 0.30 **Mn**, 2.1–2.9 **Mg**, 0.18–0.35 **Cr**, 5.1–6.1 **Zn**, 0.25 **Ti**, 0.50 **Fe**, **Zr+In+Ti** content; 0.05 each, 0.15 total, others	**Solution treated and aged**: 0.125–0.249 in. **diam**, 74 ksi (510 MPa) **TS**, 64 ksi (441 MPa) **YS**, 7% in 2 in. (50 mm) **El**; **Solution treated, stress relieved and aged**: 0.250–1.000 in. **diam**, 75 ksi (517 MPa) **TS**, 65 ksi (448 MPa) **YS**, 7% in 2 in. (50 mm) **El**
NF A 50 506	7075 (A–Z5GU)	France	Shapes	rem **Al**, 0.40 **Si**, 1.2–2.0 **Cu**, 0.30 **Mn**, 2.1–2.9 **Mg**, 0.18–0.35 **Cr**, 5.1–6.1 **Zn**, 0.20 **Ti**, 0.25 **Ti+Zr**, 0.05 each, 0.15 total, others	**Cold rolled**: 0.4–0.8 mm **diam**, 480 MPa **TS**, 410 MPa **YS**, 7% in 2 in. (50 mm) **El**
AIR 9051/A	A–Z5GU (7075)	France	Sheet, bar, billet, and plate: for forgings	rem **Al**, 0.40 **Si**, 1.2–2.0 **Cu**, 0.30 **Mn**, 2.1–2.9 **Mg**, 0.18–0.35 **Cr**, 5.1–6.1 **Zn**, 0.50 **Fe**, 0.05 each, 0.15 total, others	**Rolled (sheet)**: ≤25 mm **diam**, 530 MPa **TS**, 450 MPa **YS**, 7% in 2 in. (50 mm) **El**; **Solution treated (bar)**: ≤160 mm **diam**, 540 MPa **TS**, 480 MPa **YS**, 6% in 2 in. (50 mm) **El**
NF A 50 411	7075 (A–Z5GU	France	Bar and tube	rem **Al**, 0.40 **Si**, 1.2–2.0 **Cu**, 0.30 **Mn**, 2.1–2.9 **Mg**, 0.18–0.28 **Cr**, 5.1–6.1 **Zn**, 0.20 **Ti**, 0.50 **Fe**, 0.25 **Ti+Zr**, 0.05 each, 0.15 total, others	**Quenched and tempered (extruded bar)**: 2 in. (50 mm) **El**; **Quenched and tempered (tube)**: ≤150 mm **diam**, 530 MPa **TS**, 450 MPa **YS**, 8% in 2 in. (50 mm) **El**

WROUGHT ALUMINUM AND ALUMINUM ALLOYS

specification number	designation	country	product forms	chemical composition	mechanical properties and hardness values
NF A 50 451	7075 (A–Z5GU)	France	Sheet, plate, and strip	rem **Al**, 0.40 **Si**, 1.2–2.0 **Cu**, 0.30 **Mn**, 2.1–2.9 **Mg**, 0.18–0.35 **Cr**, 5.1–6.1 **Zn**, 0.20 **Ti**, 0.5 **Fe**, 0.25 **Ti+Zr**, 0.05 each, 0.15 total, others	**Annealed**: 0.4–0.8 mm **diam**, 150 max MPa **YS**; **Quenched and tempered**: 0.8–3.2 mm **diam**, 530 MPa **TS**, 450 MPa **YS**, 9% in 2 in. (50 mm) **El**
JIS H 4140	7075	Japan	Die forgings	rem **Al**, 0.40 **Si**, 1.2–2.0 **Cu**, 0.30 **Mn**, 2.1–2.9 **Mg**, 0.18–0.35 **Cr**, 5.1–6.1 **Zn**, 0.21 **Ti**, 0.50 **Fe**, 0.15 total, others	**Solution heat treated and artificially aged hardened**: ≤75 mm **diam**, 520 MPa **TS**, 451 MPa **YS**, 10% in 2 in. (50 mm) **El**
JIS H 4100	7075	Japan	Shapes: extruded	rem **Al**, 0.40 **Si**, 1.2–2.0 **Cu**, 0.30 **Mn**, 2.1–2.9 **Mg**, 0.18–0.35 **Cr**, 5.1–6.1 **Zn**, 0.20 **Ti**, 0.50 **Fe**, 0.15 total, others	**Annealed**: all **diam**, 275 MPa **TS**, 167 MPa **YS**, 10% in 2 in. (50 mm) **El**; **Solution heat treated and artificially aged**: ≤7 mm **diam**, 539 MPa **TS**, 481 MPa **YS**, 7% in 2 in. (50 mm) **El**
JIS H 4080	7075	Japan	Tube: extruded	rem **Al**, 0.40 **Si**, 1.2–2.0 **Cu**, 0.30 **Mn**, 2.1–2.9 **Mg**, 0.18–0.35 **Cr**, 5.1–6.1 **Zn**, 0.20 **Ti**, 0.50 **Fe**, 0.15 total, others	**Annealed**: all **diam**, 275 MPa **TS**, 167 MPa **YS**, 10% in 2 in. (50 mm) **El**; **Solution heat treated and artificially aged hardened**: <7 mm **diam**, 539 MPa **TS**, 481 MPa **YS**, 7% in 2 in. (50 mm) **El**
JIS H 4040	7075	Japan	Bar: extruded	rem **Al**, 0.40 **Si**, 1.2–2.0 **Cu**, 0.30 **Mn**, 2.1–2.9 **Mg**, 0.18–0.35 **Cr**, 5.1–6.1 **Zn**, 0.20 **Ti**, 0.50 **Fe**, 0.15 total, others	**Annealed**: all **diam**, 275 MPa **TS**, 167 MPa **YS**, 10% in 2 in. (50 mm) **El**; **Solution heat treated and artificially aged hardened**: <7 mm **diam**, 539 MPa **TS**, 481 MPa **YS**, 7% in 2 in. (50 mm) **El**
JIS H 4000	7075	Japan	Sheet, plate, strip	rem **Al**, 0.40 **Si**, 1.2–2.0 **Cu**, 0.30 **Mn**, 2.1–2.9 **Mg**, 0.18–0.35 **Cr**, 5.1–6.1 **Zn**, 0.20 **Ti**, 0.50 **Fe**, 0.15 total, others	**Annealed**: <1.5 mm **diam**, 275 MPa **TS**, 147 MPa **YS**, 10% in 2 in. (50 mm) **El**; **Solution heat treated and artificially aged hardened**: <1 mm **diam**, 530 MPa **TS**, 461 MPa **YS**, 7% in 2 in. (50 mm) **El**
MIL–A–22771C	7075	US	Forgings	rem **Al**, 0.40 **Si**, 1.2–2.0 **Cu**, 0.30 **Mn**, 2.1–2.9 **Mg**, 0.18–0.35 **Cr**, 5.1–6.1 **Zn**, 0.20 **Ti**, 0.50 **Fe**, 0.05 total, others	**T6**: ≤1 **diam**, 517 MPa **TS**, 441 MPa **YS**, 7% in 2 in. (50 mm) **El**, 135 **HB**; **T73**: ≤3 **diam**, 455 MPa **TS**, 386 MPa **YS**, 7% in 2 in. (50 mm) **El**, 130 **HB**; **T7352**: ≤3 **diam**, 455 MPa **TS**, 386 MPa **YS**, 7% in 2 in. (50 mm) **El**

WROUGHT ALUMINUM AND ALUMINUM ALLOYS

specification number	designation	country	product forms	chemical composition	mechanical properties and hardness values
ANSI/ASTM B 209	7008	US	Sheet and plate	rem **Al**, 0.10 **Si**, 0.05 **Cu**, 0.05 **Mn**, 0.7–1.4 **Mg**, 0.12–0.25 **Cr**, 4.5–5.5 **Zn**, 0.05 **Ti**, 0.10 **Fe**, 0.05 each, 0.10 total, others	**Annealed**: 0.015–0.499 in. **diam**, 40 max ksi (276 max MPa) **TS**, 21 max ksi (145 max MPa) **YS**, 10% in 2 in. (50 mm) **El**; **Solution heat treated and artificially aged**: 0.188–0.249 in. **diam**, 76 ksi (524 MPa) **TS**, 66 ksi (455 MPa) **YS**, 8% in 2 in. (50 mm) **El**; **T76 (sheet)**: 0.188–0.249 in. **diam**, 72 ksi (496 MPa) **TS**, 61 ksi (421 MPa) **YS**, 8% in 2 in. (50 mm) **El**
COPANT 862	X 7073	COPANT	Wrought products	rem **Al**, 0.5 **Si**, 0.5–1.0 **Cu**, 0.1–0.6 **Mn**, 2–3 **Mg**, 0.05–0.35 **Cr**, 4–6 **Zn**, 0.10 **Ti**, 0.5 **Fe**, 0.05 each, 0.15 total, others	
ANSI/ASTM B 209	7011	US	Sheet and plate	rem **Al**, 0.15 **Si**, 0.05 **Cu**, 0.10–0.30 **Mn**, 1.0–1.6 **Mg**, 0.05–0.20 **Cr**, 4.0–5.5 **Zn**, 0.05 **Ti**, 0.20 **Fe**, 0.05 each, 0.15 total, others	**Annealed**: 0.015–0.499 in. **diam**, 40 max ksi (276 max MPa) **TS**, 21 max ksi (145 max MPa) **YS**, 10% in 2 in. (50 mm) **El**; **Solution heat treated and artificially aged**: 0.188–0.249 in. **diam**, 76 ksi (524 MPa) **TS**, 66 ksi (455 MPa) **YS**, 8% in 2 in. (50 mm) **El**; **T76 (sheet)**: 0.188–0.249 in. **diam**, 72 ksi (496 MPa) **TS**, 61 ksi (421 MPa) **YS**, 8% in 2 in. (50 mm) **El**
ANSI/ASTM B 221	7005	US	Bar, tube, rod, wire, and shapes	rem **Al**, 0.35 **Si**, 0.10 **Cu**, 0.20–0.7 **Mn**, 1.0–1.8 **Mg**, 0.06–0.20 **Cr**, 4.0–5.0 **Zn**, 0.01–0.06 **Ti**, 0.40 **Fe**, 0.08–0.20 **Zr**, 0.05 each, 0.15 total, others	**T53**: 0.125–1.000 in. **diam**, 50 ksi (345 MPa) **TS**, 44 ksi (303 MPa) **YS**, 10% in 2 in. (50 mm) **El**
COPANT 862	7005	COPANT	Wrought products	rem **Al**, 0.35 **Si**, 0.10 **Cu**, 0.2–0.7 **Mn**, 1.0–1.8 **Mg**, 0.06–0.2 **Cr**, 4–5 **Zn**, 0.01–0.06 **Ti**, 0.40 **Fe**, 0.05 each, 0.15 total, others	
DS 3012	7005	Denmark	Rolled and extruded products	rem **Al**, 0.35 **Si**, 0.10 **Cu**, 0.20–0.7 **Mn**, 1.0–1.8 **Mg**, 0.06–0.20 **Cr**, 4.0–5.0 **Zn**, 0.01–0.06 **Ti**, 0.40 **Fe**, 0.08–0.20 **Zr**, 0.05 each, 0.15 total, others	**Solution annealed**: all **diam**, 275 MPa **TS**, 145 MPa **YS**, 12% in 2 in. (50 mm) **El**, 85 **HV**; **Artificially aged**: all **diam**, 335 MPa **TS**, 275 MPa **YS**, 10% in 2 in. (50 mm) **El**, 110 **HV**
SFS 2596	AlZn5Mgl	Finland	Shapes and plate	rem **Al**, 0.5 **Si**, 0.10 **Cu**, 0.5 **Mn**, 1.0–1.4 **Mg**, 0.10–0.25 **Cr**, 4.0–5.0 **Zn**, 0.5 **Fe**, 0.1–0.2 **Ti+Zr**, 0.05 each, 0.15 total, others	**Hot finished (shapes)**: all **diam**, 240 MPa **TS**, 140 MPa **YS**, 10% in 2 in. (50 mm) **El**; **Annealed (plate)**: all **diam**, 150 max MPa **TS**, 70 MPa **YS**, 15% in 2 in. (50 mm) **El**; **Solution treated and aged**: ≤12 mm **diam**, 330 MPa **TS**, 280 MPa **YS**, 10% in 2 in. (50 mm) **El**

WROUGHT ALUMINUM AND ALUMINUM ALLOYS

specification number	designation	country	product forms	chemical composition	mechanical properties and hardness values
NF A 50 411	7020 (A-Z5G)	France	Bar, wire, tube, and shapes	rem **Al**, 0.35 **Si**, 0.20 **Cu**, 0.05–0.50 **Mn**, 1.0–1.4 **Mg**, 0.10–0.35 **Cr**, 4.0–5.0 **Zn**, 0.40 **Fe**, 0.08–0.20 **Zr**, 0.15 **Cr**+ **Mn**, 0.08–0.25 **Ti**+**Zr**, 0.05 each, 0.15 total, others	**Quenched and tempered (extruded bar)**: 340 MPa **TS**, 275 MPa **YS**, 10% in 2 in. (50 mm) **El**; **Quenched and tempered (wire)**: 2–12 mm **diam**, 340 MPa **TS**, 275 MPa **YS**, 10% in 2 in. (50 mm) **El**; **Quenched and tempered (tube)**: ≤150 mm **diam**, 295 MPa **TS**, 235 MPa **YS**, 11% in 2 in. (50 mm) **El**
IS 736	74530	India	Plate	rem **Al**, 0.4 **Si**, 0.2 **Cu**, 0.2–0.7 **Mn**, 1.0–1.5 **Mg**, 0.2 **Cr**, 4.0–5.0 **Zn**, 0.2 **Ti**, 0.7 **Fe**	**Solution heat treated and naturally aged**: 8 mm **diam**, 265 MPa **TS**, 160 MPa **YS**, 8% in 2 in. (50 mm) **El**; **Solution and precipitation heat treated**: 7 mm **diam**, 305 MPa **TS**, 255 MPa **YS**, 7% in 2 in. (50 mm) **El**
IS 733	74530	India	Bar, rod, section	rem **Al**, 0.4 **Si**, 0.2 **Cu**, 0.2–0.7 **Mn**, 1.0–1.5 **Mg**, 0.2 **Cr**, 4.0–5.0 **Zn**, 0.2 **Ti**, 0.7 **Fe**	**Solution treated and naturally aged for 30 days**: ≤6.3 mm **diam**, 255 MPa **TS**, 220 MPa **YS**, 9% in 2 in. (50 mm) **El**; **Solution and precipitation heat treated**: ≤6.3 mm **diam**, 285 MPa **TS**, 245 MPa **YS**, 7% in 2 in. (50 mm) **El**
NS 17 410	NS 17 410–02	Norway	Sheet and strip	rem **Al**, 0.5 **Si**, 0.1 **Cu**, 0.1–0.5 **Mn**, 1.0–1.4 **Mg**, 4.0–5.0 **Zn**, 0.01–0.02 **Ti**, 0.5 **Fe**, 0.1–0.25 **Cr**, 0.05 each, 0.15 total, others	**Soft annealed (sheet)**: 0.02–12 mm **diam**, 15% in 2 in. (50 mm) **El**
NS 17 410	NS 17410–54	Norway	Sheet and strip	rem **Al**, 0.5 **Si**, 0.1 **Cu**, 0.1–0.5 **Mn**, 1.0–1.4 **Mg**, 4.0–5.0 **Zn**, 0.01–0.2 **Ti**, 0.5 **Fe**, 0.1–0.25 **Cr**, 0.05 each, 0.15 total, others	**Hardened and naturally aged**: 0.2–12 mm **diam**, 314 MPa **TS**, 216 MPa **YS**, 12% in 2 in. (50 mm) **El**
NS 17 410	NS 17 410–56	Norway	Sheet, strip, rod, wire, and profiles	rem **Al**, 0.5 **Si**, 0.1 **Cu**, 0.1–0.5 **Mn**, 1.0–1.4 **Mg**, 4.0–5.0 **Zn**, 0.01–0.2 **Ti**, 0.5 **Fe**, 0.1–0.25 **Cr**, 0.05 each, 0.15 total, others	**Hardened and artificially aged (sheet and strip)**: 0.2–12 mm **diam**, 353 MPa **TS**, 274 MPa **YS**, 10% in 2 in. (50 mm) **El**; **Hardened and artificially aged (rod and wire)**: all **diam**, 353 MPa **TS**, 274 MPa **YS**, 10% in 2 in. (50 mm) **El**; **Hardened and artificially aged**: <15 mm **diam**, 353 MPa **TS**, 274 MPa **YS**, 10% in 2 in. (50 mm) **El**
SIS 14 44 25	SIS Al 44 25–00	Sweden	Bar, tube	93.6 **Al**, 0.3 **Mn**, 1.2 **Mg**, 0.2 **Cr**, 4.5 **Zn**, 0.15 **Zr**, 0.2 **Zr**+**Ti**; 0.05 each, 0.15 total, others	**Hot worked**: 10–200 mm **diam**, 240 MPa **TS**, 140 MPa **YS**, 10% in 2 in. (50 mm) **El**, 85 **HV**
SIS 14 44 25	SIS Al 44 25–02	Sweden	Plate, sheet, strip	93.6 **Al**, 0.3 **Mn**, 12 **Mg**, 0.2 **Cr**, 4.5 **Zn**, 0.15 **Zr**, 0.2 **Zr**+**Ti**; 0.05 each, 0.15 total, others	**Annealed**: 5–30 mm **diam**, 220 MPa **TS**, 150 MPa **YS**, 15% in 2 in. (50 mm) **El**, 40–65 **HV**

WROUGHT ALUMINUM AND ALUMINUM ALLOYS

specification number	designation	country	product forms	chemical composition	mechanical properties and hardness values
SIS 14 44 25	SIS Al 44 25–04	Sweden	Bar, tube, section	93.6 **Al**, 0.3 **Mn**, 1.2 **Mg**, 0.2 **Cr**, 4.5 **Zn**, 0.15 **Zr**, 0.2 **Zr+Ti**; 0.05 each, 0.15 total, others	**Naturally aged:** 10–200 mm **diam**, 275 MPa **TS**, 145 MPa **YS**, 12% in 2 in. (50 mm) **El**, 85 **HV**
SIS 14 44 25	SIS Al 44 25–06	Sweden	Plate, sheet, strip, bar, tube, section	93.6 **Al**, 0.3 **Mn**, 1.2 **Mg**, 0.2 **Cr**, 4.5 **Zn**, 0.15 **Zr**, 0.2 **Zr+Ti**; 0.05 each, 0.15 total, others	**Artificially aged:** 5–15 mm **diam**, 350 MPa **TS**, 275 MPa **YS**, 10% in 2 in. (50 mm) **El**, 100–135 **HV**
SIS 14 44 25	SIS Al 44 25–07	Sweden	Bar, tube, section	93.6 **Al**, 0.3 **Mn**, 1.2 **Mg**, 0.2 **Cr**, 4.5 **Zn**, 0.15 **Zr**, 0.2 **Zr+Ti**, 0.05 each, 0.15 total, others	**Artificially aged:** 1–25 mm **diam**, 350 MPa **TS**, 290 MPa **YS**, 10% in 2 in. (50 mm) **El**, 90–125 **HV**
JIS H 4100	7N01	Japan	Shapes: extruded	rem **Al**, 0.30 **Si**, 0.25 **Cu**, 0.20–0.9 **Mn**, 1.0–2.2 **Mg**, 0.30 **Cr**, 3.8–5.0 **Zn**, 0.20 **Ti**, 0.40 **Fe**, 0.30 **Zr**, 0.15 total, others	**Annealed:** all **diam**, 245 MPa **TS**, 147 MPa **YS**, 12% in 2 in. (50 mm) **El**; **Solution heat treated and aged hardened:** all **diam**, 314 MPa **TS**, 196 MPa **YS**, 11% in 2 in. (50 mm) **El**; **Quicked cooled and artificially aged:** all **diam**, 324 MPa **TS**, 245 MPa **YS**, 10% in 2 in. (50 mm) **El**
JIS H 4080	7N01	Japan	Tube: extruded	rem **Al**, 0.30 **Si**, 0.25 **Cu**, 0.20–0.9 **Mn**, 1.0–2.2 **Mg**, 0.30 **Cr**, 3.8–5.0 **Zn**, 0.20 **Ti**, 0.40 **Fe**, 0.30 **Zr**, 0.15 total, others	**Annealed:** 1.6–1.2 mm **diam**, 245 MPa **TS**, 147 MPa **YS**, 12% in 2 in. (50 mm) **El**; **Solution heat treated and age hardened:** 1.6–12 mm **diam**, 314 MPa **TS**, 196 MPa **YS**, 11% in 2 in. (50 mm) **El**; **Solution heat treated and artificially aged hardened:** <6 mm **diam**, 324 MPa **TS**, 235 MPa **YS**, 10% in 2 in. (50 mm) **El**
JIS H 4040	7N01	Japan	Bar: extruded	rem **Al**, 0.30 **Si**, 0.25 **Cu**, 0.20–0.9 **Mn**, 1.0–2.2 **Mg**, 0.30 **Cr**, 3.8–5.0 **Zn**, 0.20 **Ti**, 0.40 **Fe**, 0.30 **Zr**, 0.15 total, others	**Annealed:** all **diam**, 245 MPa **TS**, 147 MPa **YS**, 12% in 2 in. (50 mm) **El**; **Solution heat treated and age hardened:** all **diam**, 314 MPa **TS**, 196 MPa **YS**, 11% in 2 in. (50 mm) **El**; **Solution heat treated and artificially aged hardened:** 333 MPa **TS**, 275 MPa **YS**, 10% in 2 in. (50 mm) **El**
JIS H 4000	7N01	Japan	Sheet, plate, strip	rem **Al**, 0.30 **Si**, 0.25 **Cu**, 0.20–0.9 **Mn**, 1.0–2.2 **Mg**, 0.30 **Cr**, 3.8–5.0 **Zn**, 0.20 **Ti**, 0.40 **Fe**, 0.30 **Zr**, 0.15 total, others	**Annealed:** <3 mm **diam**, 245 MPa **TS**, 147 MPa **YS**, 12% in 2 in. (50 mm) **El**; **Solution heat treated and age hardened:** <3 mm **diam**, 314 MPa **TS**, 196 MPa **YS**, 1% in 2 in. (50 mm) **El**; **Solution heat treated and artificially aged hardened:** <3 mm **diam**, 333 MPa **TS**, 275 MPa **YS**, 10% in 2 in. (50 mm) **El**

WROUGHT ALUMINUM AND ALUMINUM ALLOYS

specification number	designation	country	product forms	chemical composition	mechanical properties and hardness values
NF A 50 501	7020 (A–Z5G)	France	Tube	rem **Al**, 0.35 **Si**, 0.20 **Cu**, 0.05–0.5 **Mn**, 0.9–1.5 **Mg**, 0.35 **Cr**, 3.7–5.0 **Zn**, 0.40 **Fe**, 0.08–0.20 **Zr**, 0.08–0.25 **Ti+Zr**, 0.05 each, 0.15 total, others	**As manufactured**: all **diam**, 340 MPa **TS**, 275 MPa **YS**, 10% in 2 in. (50 mm) **El**
NF A 50 451	7020 (A–Z5G)	France	Sheet, plate, and strip	rem **Al**, 0.35 **Si**, 0.20 **Cu**, 0.05–0.50 **Mn**, 0.9–1.5 **Mg**, 0.35 **Cr**, 3.7–5.0 **Zn**, 0.40 **Fe**, 0.08–0.20 **Zr**, 0.15 **Cr+Mn**, 0.08–0.25 **Ti+Zr**, 0.05 each, 0.15 total, others	**Aged**: 0.4–12 mm **diam**, 320 MPa **TS**, 210 MPa **YS**, 14% in 2 in. (50 mm) **El**; **Quenched and tempered**: 0.4–25 mm **diam**, 350 MPa **TS**, 280 MPa **YS**, 10% in 2 in. (50 mm) **El**
AMS 4024C		US	Sheet and plate	rem **Al**, 0.30 **Si**, 0.40–0.8 **Cu**, 0.10–0.30 **Mn**, 2.9–3.7 **Mg**, 0.10–0.25 **Cr**, 3.8–4.8 **Zn**, 0.10 **Ti**, 0.40 **Fe**, 0.05 each, 0.15 total, others	**Solution and precipitation heat treated**: 0.015–0.039 in. **diam**, 73 ksi (503 MPa) **TS**, 64 ksi (441 MPa) **YS**, 7% in 2 in. (50 mm) **El**
AMS 4136		US	Die forgings and rings	rem **Al**, 0.30 **Si**, 0.40–0.8 **Cu**, 0.10–0.30 **Mn**, 2.9–3.7 **Mg**, 0.10–0.25 **Cr**, 3.8–4.8 **Zn**, 0.10 **Ti**, 0.40 **Fe**, 0.05 each, 0.15 total, others	**Solution and precipitation heat treated**: ≤6 in. **diam**, 69 ksi (476 MPa) **TS**, 56 ksi (386 MPa) **YS**, 7% in **El**, 125 min **HB**
AMS 4138A		US	Die and hand forgings	rem **Al**, 0.30 **Si**, 0.40–0.8 **Cu**, 0.10–0.30 **Mn**, 2.9–3.7 **Mg**, 0.10–0.35 **Cr**, 3.8–4.8 **Zn**, 0.10 **Ti**, 0.40 **Fe**, 0.05 each, 0.15 total, others	**Solution and precipitation heat treated (die forgings)**: ≤7 in. (≤178 mm) **diam**, 74 ksi (510MPa) **TS**, 64 ksi (441 MPa) **YS**, 7% in 2 in. (50 mm) **El**, 135 min **HB**; **Solution and precipitation heat treated (hand forgings)**: ≤7 in. (≤178 mm) **diam**, 73 ksi (503 MPa) **TS**, 62 ksi (427 MPa) **YS**, 9% in 2 in. (50 mm) **El**, 135 min **HB**
ANSI/ASTM B 241	7079	US	Tube: seamless	rem **Al**, 0.30 **Si**, 0.40–0.8 **Cu**, 0.10–0.30 **Mn**, 2.9–3.7 **Mg**, 0.10–0.25 **Cr**, 3.8–4.8 **Zn**, 0.10 **Ti**, 0.40 **Fe**, 0.05 each, 0.15 total, others	**Annealed**: all **diam**, 40 max ksi (276 max MPa) **TS**, 24 max ksi (165 max MPa) **YS**, 10% in 2 in. (50 mm) **El**; **Solution heat treated and naturally aged**: 1.500–2.999 in. **diam**, 79 ksi (545 MPa) **TS**, 70 ksi (483 MPa) **YS**, 7% in 2 in. (50 mm) **El**
ANSI/ASTM B 221	7079	US	Bar, tube, rod, wire, and shapes	rem **Al**, 0.30 **Si**, 0.40–0.8 **Cu**, 0.10–0.30 **Mn**, 2.9–3.7 **Mg**, 0.10–0.25 **Cr**, 3.8–4.8 **Zn**, 0.10 **Ti**, 0.40 **Fe**, 0.25 **Ti+Zr**, 0.05 each, 0.15 total, others	**Annealed**: all **diam**, 42 max ksi (290 max MPa) **TS**, 24 max ksi (165 max MPa) **YS**, 10% in 2 in. (50 mm) **El**; **Solution heat treated and artificially aged**: 1.500–2.999 in. **diam**, 79 ksi (545 MPa) **TS**, 70 ksi (483 MPa) **YS**, 7% in 2 in. (50 mm) **El**

WROUGHT ALUMINUM AND ALUMINUM ALLOYS

specification number	designation	country	product forms	chemical composition	mechanical properties and hardness values
AMS 4198		US	Sheet and plate: alclad	rem **Al**, 0.30 **Si**, 0.40–0.8 **Cu**, 0.10–0.30 **Mn**, 2.9–3.7 **Mg**, 0.10–0.25 **Cr**, 3.8–4.8 **Zn**, 0.10 **Ti**, 0.40 **Fe**, 0.05 each, 0.15 total, others	**Annealed**: <0.500 in. **diam**, 40 ksi (276 MPa) **TS**, 21 ksi (145 MPa) **YS**, 10% in 2 in. (50 mm) **El**; **Solution and precipitation heat treated**: 0.500–1.500 in. **diam**, 73 ksi **TS**, 63 ksi **YS**, 8% in 2 in. (50 mm) **El**
AMS 4199		US	Sheet and plate: alclad	rem **Al**, 0.30 **Si**, 0.40–0.8 **Cu**, 0.10–0.30 **Mn**, 2.9–3.7 **Mg**, 0.10–0.25 **Cr**, 3.8–4.8 **Zn**, 0.10 **Ti**, 0.40 **Fe**, 0.05 each, 0.15 total, others	**Solution and precipitation heat treated**: 0.500–1.000 in. **diam**, 74 ksi (510 MPa) **TS**, 65 ksi (448 MPa) **YS**, 8% in 2 in. (50 mm) **El**
QQ-A-250/23A	7079	US	Sheet	rem **Al**, 0.30 **Si**, 0.40–0.8 **Cu**, 0.10–0.30 **Mn**, 2.9–3.7 **Mg**, 0.10–0.25 **Cr**, 3.8–4.8 **Zn**, 0.10 **Ti**, 0.40 **Fe**, 0.05 each, 0.15 total, others	**Annealed**: 0.020–0.062 in. **diam**, 38 max ksi (262 max MPa) **TS**, 21 max ksi (145 max MPa) **YS**, 10% in 2 in. (50 mm) **El**; **Solution heat treated and artificially aged**: 0.020–.039 in. **diam**, 69 ksi (476 MPa) **TS**, 60 ksi (414 MPa) **YS**, 7% in 2 in. (50 mm) **El**
MIL-A-22771C	7079	US	Forgings	rem **Al**, 0.30 **Si**, 0.40–0.8 **Cu**, 0.10–0.30 **Mn**, 2.9–3.7 **Mg**, 0.10–0.25 **Cr**, 3.8–4.8 **Zn**, 0.10 **Ti**, 0.40 **Fe**, 0.05 total, others	**T6**: ≤1 **diam**, 496 MPa **TS**, 427 MPa **YS**, 7% in 2 in. (50 mm) **El**, 135 **HB**
COPANT 862	7004	COPANT	Wrought products	rem **Al**, 0.25 **Si**, 0.05 **Cu**, 0.2–0.7 **Mn**, 1–2 **Mg**, 0.05 **Cr**, 3.8–4.6 **Zn**, 0.05 **Ti**, 0.35 **Fe**, 0.10–0.20 **Zr**, 0.05 each, 0.15 total, others	
COPANT 862	X 7007	COPANT	Wrought products	rem **Al**, 0.3 **Si**, 0.10 **Cu**, 0.1–0.6 **Mn**, 0.5–1.0 **Mg**, 0.05–0.35 **Cr**, 3–5 **Zn**, 0.1 **Ti**, 0.4 **Fe**, 0.05 each, 0.10 total, others	
COPANT 862	X 7010	COPANT	Wrought products	rem **Al**, 0.30 **Si**, 0.10 **Cu**, 0.1–0.6 **Mn**, 0.8–1.5 **Mg**, 0.05–0.35 **Cr**, 3–5 **Zn**, 0.10 **Ti**, 0.40 **Fe**, 0.05 each, 0.15 total, others	
COPANT 862	X 7052	COPANT	Wrought products	rem **Al**, 0.3 **Si**, 0.1 **Cu**, 0.1–0.6 **Mn**, 1.5–2.5 **Mg**, 0.05–0.35 **Cr**, 3–5 **Zn**, 0.10 **Ti**, 0.4 **Fe**, 0.05 each, 0.15 total, others	
COPANT 862	7104	COPANT	Wrought products	rem **Al**, 0.25 **Si**, 0.03 **Cu**, 0.5–0.90 **Mg**, 3.6–4.4 **Zn**, 0.10 **Ti**, 0.40 **Fe**, 0.05 each, 0.15 total, others	
MIL-A-22771C	7039	US	Forgings	rem **Al**, 0.30 **Si**, 0.10 **Cu**, 0.10–0.40 **Mn**, 2.3–3.3 **Mg**, 0.15–0.25 **Cr**, 3.5–4.5 **Zn**, 0.40 **Fe**	**T64**: ≤4 **diam**, 393 MPa **TS**, 331 MPa **YS**, 8% in 2 in. (50 mm) **El**, 125 **HB**

WROUGHT ALUMINUM AND ALUMINUM ALLOYS

specification number	designation	country	product forms	chemical composition	mechanical properties and hardness values
MIL-A-46083B	7039	US	Shapes	rem **Al**, 0.30 **Si**, 0.10 **Cu**, 0.10–0.40 **Mn**, 2.3–3.3 **Mg**, 0.15–0.25 **Cr**, 3.5–4.5 **Zn**, 0.10 **Ti**, 0.40 **Fe**, 0.05 each, 0.15 total, others	**Class II**: ≤2.000 in. **diam**, 393 MPa **TS**, 331 MPa **YS**, 8% in 2 in. (50 mm) **El**
NF A 50 411	7051 (A-Z3G2)	France	Wire and shapes	rem **Al**, 0.35 **Si**, 0.15 **Cu**, 0.10–0.45 **Mn**, 1.7–2.5 **Mg**, 0.05–0.25 **Cr**, 3.0–4.0 **Zn**, 0.15 **Ti**, 0.45 **Fe**, 0.05 each, 0.15 total, others	**Aged (wire)**: 2–12 mm **diam**, 340 MPa **TS**, 220 MPa **YS**, 15% in 2 in. (50 mm) **El**; **Aged (shapes)**: 320 MPa **TS**, 200 MPa **YS**, 12% in 2 in. (50 mm) **El**
ANSI/ASTM B 547	7072	US	Tube: formed and arc-welded	rem **Al**, 0.10 **Cu**, 0.10 **Mn**, 0.10 **Mg**, 0.8–1.3 **Zn**, 0.05 each, 0.15 total, others	
ANSI/ASTM B 404	7072	US	Cladding material	rem **Al**, 0.10 **Cu**, 0.10 **Mn**, 0.10 **Mg**, 0.8–1.3 **Zn**, 0.7 **Si**+**Fe**, 0.05 each, 0.15 total, others	
ANSI/ASTM B 313	7072	US	Cladding material	rem **Al**, 0.10 **Cu**, 0.1 **Mn**, 0.10 **Mg**, 0.8–1.3 **Zn**, 0.7 **Fe**+**Si**, 0.05 each, 0.15 total, others	
ANSI/ASTM B 241	7072	US	Cladding material	rem **Al**, 0.10 **Cu**, 0.10 **Mn**, 0.10 **Mg**, 0.8–1.3 **Zn**, 0.7 **Si**+**Fe**, 0.05 each, 0.15 total, others	
ANSI/ASTM B 234	7072	US	Cladding material	rem **Al**, 0.10 **Cu**, 0.10 **Mn**, 0.10 **Mg**, 0.8–1.3 **Zn**, 0.7 **Si**+**Fe**, 0.05 each, 0.15 total, others	
ANSI/ASTM B 221	7072	US	Bar, tube, rod, wire, and shapes	rem **Al**, 0.10 **Cu**, 0.10 **Mn**, 0.10 **Mg**, 0.8–1.3 **Zn**, 0.7 **Si**+**Fe**, 0.05 each, 0.15 total, others	
AS 1734	7072	Australia	Sheet, plate	rem **Al**, 0.10 **Cu**, 0.10 **Mn**, 0.10 **Mg**, 0.8–1.3 **Zn**, 0.7 **Si**+**Fe**, 0.05 each, 0.15 total, others	
ANSI/ASTM B 210	7072	US	Cladding material	rem **Al**, 0.10 **Cu**, 0.10 **Mn**, 0.10 **Mg**, 0.8–1.3 **Zn**, 0.7 **Si**+**Fe**, 0.05 each, 0.15 total, others	
ANSI/ASTM B 209	7072	US	Sheet and plate	rem **Al**, 0.10 **Cu**, 0.10 **Mn**, 0.10 **Mg**, 0.8–1.3 **Zn**, 0.7 **Si**+**Fe**, 0.05 each, 0.15 total, others	
CSA HA.4	.7075 Alclad	Canada	Plate, sheet (cladding)	rem **Al**, 0.10 **Cu**, 0.10 **Mn**, 0.10 **Mg**, 0.8–1.3 **Zn**, 0.7 **Si**+**Fe**, 0.05 each, 0.15 total, others	**O**: 0.015–0.032 in. **diam**, 36.0 max ksi (248 MPa) **TS**, 20.0 max ksi (138 MPa) **YS**, 10% in 2 in. (50 mm) **El**; **T6**: 0.012–0.020 in. **diam**, 70.0 ksi (483 MPa) **TS**, 60.0 ksi (414 MPa) **YS**, 7% in 2 in. (50 mm) **El**; **T62 (sheet)**: 0.021–0.039 in. **diam**, 70.0 ksi (483 MPa) **TS**, 60.0 ksi (414 MPa) **YS**, 7% in 2 in. (50 mm) **El**

WROUGHT ALUMINUM AND ALUMINUM ALLOYS

specification number	designation	country	product forms	chemical composition	mechanical properties and hardness values
COPANT 862	7072	COPANT	Wrought products	rem **Al**, 0.1 **Cu**, 0.1 **Mn**, 0.1 **Mg**, 0.8–1.3 **Zn**, 0.7 **Fe+Si**, 0.05 each, 0.15 total, others	
QQ-A-250/13E	7075	US	Plate, sheet	rem **Al**, 0.10 **Cu**, 0.10 **Mn**, 0.10 **Mg**, 0.8–1.3 **Zn**, 0.7 **Fe**, **Si+In+Fe** content; 0.05 each, 0.15 total, others	**Annealed**: 0.008–0.014 in. **diam**, 36 max ksi (248 max MPa) **TS**, 20 max ksi (138 max MPa) **YS**, 9% in 2 in. (50 mm) **El**; **Solution heat treated and artificially aged**: 0.008–0.011 in. **diam**, 68 ksi (469 MPa) **TS**, 58 ksi (400 MPa) **YS**, 5% in 2 in. (50 mm) **El**; **Solution treated, stress relieved and aged**: 0.250–0.499 in. **diam**, 75 ksi (517 MPa) **TS**, 65 ksi (448 MPa) **YS**, 9% in 2 in. (50 mm) **El**
ANSI/ASTM B 308	6061	US	Shapes: structural	rem **Al**, 0.40–0.8 **Si**, 0.15–0.40 **Cu**, 0.15 **Mn**, 0.8–1.2 **Mg**, 0.04–0.35 **Cr**, 0.25 **Zn**, 0.15 **Ti**, 0.7 **Fe**, 0.05 each, 0.15 total, others	**Solution heat treated and artificially aged**: 0.050–0.075 in. **diam**, 38 ksi (262 MPa) **TS**, 35 ksi (241 MPa) **YS**, 10% in **El**, 89 min **HRE**
ASTM F 468	A 96061	US	Bolt, screw, stud	rem **Al**, 0.40–0.8 **Si**, 0.15–0.40 **Cu**, 0.8–1.2 **Mn**, 0.15 **Mg**, 0.04–0.35 **Cr**, 0.25 **Zn**, 0.15 **Ti**, 0.7 **Fe**	**As manufactured**: all **diam**, 37 ksi (260 MPa) **TS**, 31 ksi (215 MPa) **YS**, 40–50 **HRB**
QQ-A-250/19	5086	US	Plate, sheet	rem **Al**, 0.40 **Si**, 0.10 **Cu**, 0.20–0.7 **Mn**, 3.5–4.5 **Mg**, 0.05–0.25 **Cr**, 0.25 **Zn**, 0.15 **Ti**, 0.50 **Fe**, 0.05 each, 0.15 total, others	**Annealed**: 0.020–0.050 in. **diam**, 35 ksi (241 MPa) **TS**, 14 ksi (97 MPa) **YS**, 15% in 2 in. (50 mm) **El**; **As rolled**: 0.188–0.499 in. **diam**, 36 ksi (248 MPa) **TS**, 18 ksi (124 MPa) **YS**, 8% in 2 in. (50 mm) **El**; **Specially treated**: 0.188–0.249 in. **diam**, 40 ksi (276 MPa) **TS**, 28 ksi (193 MPa) **YS**, 8% in 2 in. (50 mm) **El**

CAST ALUMINUM
AND
ALUMINUM ALLOYS

CAST ALUMINUM AND ALUMINUM ALLOYS

specification number	designation	country	product forms	chemical composition	mechanical properties and hardness values
CSA HA.2	.9999	Canada	Ingot	99.990 min **Al**	
CSA HA.2	.9995	Canada	Ingot	99.950 min **Al**	
CSA HA.2	.9990	Canada	Ingot	99.90 min **Al**, 0.07 **Si**, 0.07 **Fe**, 0.02 each, others	
SABS 711	AL 99.9	South Africa	Ingot	99.90 min **Al**, 0.07 **Si**, 0.01 **Cu**, 0.01 **Mn**, 0.01 **Mg**, 0.07 **Fe**, 0.02 each, 0.10 total, others	
AS 1874	AP185	Australia	Ingot, sand castings and permanent mould castings	99.85 min **Al**, 0.10 **Si**, 0.02 **Cu**, 0.10 **Fe**, 0.03 each, 0.10 total, others	
CSA HA.2	.9985	Canada	Ingot	99.85 min **Al**, 0.10 **Si**, 0.02 **Cu**, 0.10 **Fe**, 0.03 each, 0.10 total, others	
SABS 711	AL 99.85	South Africa	Ingot	99.85 min **Al**, 0.10 **Si**, 0.02 **Cu**, 0.02 **Mn**, 0.02 **Mg**, 0.02 **Cr**, 0.05 **Zn**, 0.10 **Fe**, 0.02 each, 0.10 total, others	
AS 1874	AP180	Australia	Ingot, sand castings and permanent mould castings	99.80 min **Al**, 0.15 **Si**, 0.02 **Cu**, 0.15 **Fe**, 0.03 each, 0.10 total, others	
CSA HA.2	.9980	Canada	Ingot	99.80 min **Al**, 0.15 **Si**, 0.02 **Cu**, 0.15 **Fe**, 0.03 each, 0.10 total, others	
SFS 2560	G-AL99.8	Finland	Ingot	99.80 **Al**, 0.15 **Si**, 0.02 **Cu**, 0.06 **Zn**, 0.15 **Fe**, 0.03 each, 0.15 total, others	
ISO R115	AL99.8	ISO	Ingot	99.80 min **Al**, 0.15 **Si**, 0.02 **Cu**, 0.06 **Zn**, 0.15 **Fe**, 0.03 each, 0.20 total, others	
SIS 14 40 20	SIS 40 20-00	Sweden	Ingot, pigs	99.80 min **Al**, 0.15 **Si**, 0.02 **Cu**, 0.06 **Zn**, 0.15 **Fe**, 0.03 each, 0.20 total, others	
SABS 711	AL 99.8	South Africa	Ingot	99.80 min **Al**, 0.15 **Si**, 0.02 **Cu**, 0.02 **Mn**, 0.02 **Mg**, 0.02 **Cr**, 0.06 **Zn**, 0.15 **Fe**, 0.02 each, 0.10 total, others	
AS 1874	AP175	Australia	Ingot, sand castings and permanent mould castings	99.75 min **Al**, 0.20 **Si**, 0.02 **Cu**, 0.20 **Fe**, 0.03 each, 0.10 total, others	
CSA HA.2	.9975	Canada	Ingot	99.75 min **Al**, 0.20 **Si**, 0.02 **Cu**, 0.20 **Fe**, 0.03 each, 0.10 total, others	
AS 1874	BP170	Australia	Ingot, sand castings and permanent mould castings	99.70 min **Al**, 0.025 **Mn+Ti+Cr+V**, 1.5 x Si min **Fe**, 0.03 each, 0.10 total, others	
AS 1874	AP170	Australia	Ingot, sand castings and permanent mould castings	99.70 min **Al**, 0.20 **Si**, 0.02 **Cu**, 0.25 **Fe**, 0.03 each, 0.10 total, others	

CAST ALUMINUM AND ALUMINUM ALLOYS

specification number	designation	country	product forms	chemical composition	mechanical properties and hardness values
CSA HA.2	.9970	Canada	Ingot	99.70 min **Al**, 0.20 **Si**, 0.02 **Cu**, 0.25 **Fe**, 0.03 each, 0.10 total, others	
SFS 2561	G–AL99.7	Finland	Ingot	99.70 **Al**, 0.20 **Si**, 0.02 **Cu**, 0.06 **Zn**, 0.25 **Fe**, 0.03 each, 0.15 total, others	
IS 23	99.7al	India	Bar, ingot	99.7 **Al**, 0.20 **Si**, 0.02 **Cu**, 0.03 **Mn**, 0.03 **Zn**, 0.25 **Fe**	
ISO R115	AL99.7	ISO	Ingot	99.70 min **Al**, 0.20 **Si**, 0.02 **Cu**, 0.06 **Zn**, 0.25 **Fe**, 0.03 each, 0.30 total, others	
SABS 992	AL–99,7A	South Africa	Pressure die castings	99.7 min **Al**, 0.20 **Si**, 0.02 **Cu**, 0.06 **Zn**, 0.25 **Fe**, 0.03 each, others	
SABS 991	AL–99,7A	South Africa	Gravity die castings	99.7 min **Al**, 0.20 **Si**, 0.02 **Cu**, 0.06 **Zn**, 0.25 **Fe**, 0.03 each, others	
SABS 990	AL–99,7A	South Africa	Sand castings	99.7 min **Al**, 0.20 **Si**, 0.02 **Cu**, 0.06 **Zn**, 0.25 **Fe**, 0.03 each, others	
SABS 989	AL–99,7A	South Africa	Ingot: for pressure die castings	99.7 min **Al**, 0.20 **Si**, 0.02 **Cu**, 0.06 **Zn**, 0.25 **Fe**, 0.03 each, others	
SABS 989	AL–99,7A	South Africa	Ingot: for sand and gravity die castings	99.7 min **Al**, 0.20 **Si**, 0.02 **Cu**, 0.06 **Zn**, 0.25 **Fe**, 0.03 each, others	
SIS 14 40 21	SIS AL40 21–00	Sweden	Ingot, pigs	99.70 min **Al**, 0.20 **Si**, 0.02 **Cu**, 0.06 **Zn**, 0.25 **Fe**, 0.03 each, 0.30 total, others	
SABS 711	AL 99.7	South Africa	Ingot	99.70 min **Al**, 0.20 **Si**, 0.02 **Cu**, 0.03 **Mn**, 0.03 **Mg**, 0.03 **Cr**, 0.06 **Zn**, 0.25 **Fe**, 0.02 each, 0.10 total, others	
CSA HA.2	.9965	Canada	Ingot	99.65 min **Al**, 0.25 **Si**, 0.30 **Fe**, 0.03 each, 0.10 total, others	
SABS 711	AL 99.65	South Africa	Ingot	99.65 min **Al**, 0.20 **Si**, 0.02 **Cu**, 0.03 **Mn**, 0.03 **Mg**, 0.03 **Cr**, 0.06 **Zn**, 0.25 **Fe**, 0.02 each, 0.10 total, others	
AS 1874	AP160	Australia	Ingot, sand castings and permanent mould castings	99.60 min **Al**, 0.10 **Si**, 0.30 **Fe**, 0.01 **Mn+Ti+Cr+V**, 0.02 each, 0.10 total, others	
IS 4026	Grade 1	India	Ingot	99.6 min **Al**, 0.13 **Si**, 0.04 **Cu**, 0.30 **Fe**, 0.02 **Ti+V**, 0.01 **Mn+Zn+Cr**	
SABS 711	AL–EC 99.6	South Africa	Ingot	99.60 min **Al**, 0.15 **Si**, 0.020 **Cu**, 0.07 **Zn**, 0.30 **Fe**, 0.02 each, 0.15 total, others	
AS 1874	BP150	Australia	Ingot, sand castings and permanent mould castings	99.50 min **Al**, 1.5 x Si min **Fe**, 0.05 each, others	

CAST ALUMINUM AND ALUMINUM ALLOYS

specification number	designation	country	product forms	chemical composition	mechanical properties and hardness values
AS 1874	AP150	Australia	Ingot, sand castings and permanent mould castings	99.50 min **Al**, 0.30 **Si**, 0.40 **Fe**, 0.03 each, 0.15 total, others	
CSA HA.2	.9950	Canada	Ingot	99.50 min **Al**, 0.30 **Si**, 0.40 **Fe**, 0.03 each, 0.15 total, others	
SFS 2562	G–AL99.5	Finland	Ingot	99.50 **Al**, 0.30 **Si**, 0.03 **Cu**, 0.07 **Zn**, 0.40 **Fe**, 0.03 each, 0.15 total, others	
IS 4026	Grade 2	India	Ingot	99.5 min **Al**, 0.15 **Si**, 0.04 **Cu**, 0.35 **Fe**, 0.02 **Ti+V**, 0.02 **Mn+Zn+Cr**	
IS 23	99.5al	India	Bar, ingot	99.5 **Al**, 0.30 **Si**, 0.03 **Cu**, 0.03 **Mn**, 0.05 **Zn**, 0.40 **Fe**	
ISO R115	AL99.5	ISO	Ingot	99.50 min **Al**, 0.30 **Si**, 0.03 **Cu**, 0.07 **Zn**, 0.40 **Fe**, 0.03 each, 0.50 total, others	
SABS 992	AL–99.5A	South Africa	Pressure die castings	99.5 min **Al**, 0.30 **Si**, 0.03 **Cu**, 0.07 **Zn**, 0.40 **Fe**, 0.03 each, others	
SABS 991	AL–99,5A	South Africa	Gravity die castings	99.5 min **Al**, 0.30 **Si**, 0.03 **Cu**, 0.07 **Zn**, 0.40 **Fe**, 0.03 each, others	
SABS 990	AL–99,5A	South Africa	Sand castings	99.5 min **Al**, 0.30 **Si**, 0.03 **Cu**, 0.07 **Zn**, 0.40 **Fe**, 0.03 each, others	
SABS 989	AL–99,5A	South Africa	Ingot: for pressure die castings	99.5 min **Al**, 0.30 **Si**, 0.03 **Cu**, 0.07 **Zn**, 0.40 **Fe**, 0.03 each, others	
SABS 989	AL–99,5A	South Africa	Ingot: for sand and gravity die castings	99.5 min **Al**, 0.30 **Si**, 0.03 **Cu**, 0.07 **Zn**, 0.40 **Fe**, 0.03 each, others	
SIS 14 40 22	SIS AL 40 22–00	Sweden	Ingot, pigs	99.50 min **Al**, 0.30 **Si**, 0.03 **Cu**, 0.07 **Zn**, 0.40 **Fe**, 0.03 each, 0.50 total, others	
SABS 711	AL 99.5	South Africa	Ingot	99.50 min **Al**, 0.30 **Si**, 0.03 **Cu**, 0.03 **Mn**, 0.03 **Mg**, 0.07 **Zn**, 0.40 **Fe**, 0.03 each, 0.15 total, others	
NF A 57 703	A5–Y4	France	Pressure die castings	99.45 min **Al**, 0.05 **Cu**, 0.05 **Mn**, 0.05 **Mg**, 0.05 **Ni**, 0.1 **Zn**, 0.1 **Ti**, 0.05 **Sn**, 0.55 **Cu+Fe+Si**	
ANSI/ASTM B 179	100.1	US	Ingot: for die castings	99.00 **Al**, 0.15 **Si**, 0.10 **Cu**, 0.05 **Zn**, 0.6–0.8 **Fe**, 0.025 **Mn+Cr+Ti**, 0.03 each, 0.10 total, others	
AS 1874	AP100	Australia	Ingot, sand castings and permanent mould castings	99.00 min **Al**, 0.25 **Si**, 0.10 **Cu**, 0.40–0.8 **Fe**, 0.05 each, others	
CSA HA.2	.9900	Canada	Ingot	99.00 min **Al**, 0.50 **Si**, 0.60 **Fe**, 0.03 each, 0.15 total, others	

CAST ALUMINUM AND ALUMINUM ALLOYS

specification number	designation	country	product forms	chemical composition	mechanical properties and hardness values
SFS 2563	G–AL99.0	Finland	Ingot	99.00 **Al**, 0.50 **Si**, 0.03 **Cu**, 0.08 **Zn**, 0.80 **Fe**, 0.03 each, 0.15 total, others	
IS 23	99al	India	Bar, ingot	99.0 **Al**, 0.50 **Si**, 0.03 **Cu**, 0.05 **Mn**, 0.06 **Zn**, 0.60 **Fe**	
ISO R115	AL99.0	ISO	Ingot	99.00 min **Al**, 0.50 **Si**, 0.03 **Cu**, 0.08 **Zn**, 0.80 **Fe**, 0.03 each, 1.00 total, others	
SIS 14 40 24	SIS AL 40 24–00	Sweden	Ingot, pigs	99.00 min **Al**, 0.50 **Si**, 0.03 **Cu**, 0.08 **Zn**, 0.80 **Fe**, 0.03 each, 1.00 total, others	
SABS 711	AL 99.0	South Africa	Ingot	99.00 min **Al**, 0.50 **Si**, 0.03 **Cu**, 0.08 **Zn**, 0.60 **Fe**, 0.03 each, 0.15 total, others	
ANSI/ASTM B 37	980A	US	Ingot, rod, and shot	98.0 min **Al**, 0.2 **Cu**, 0.5 **Mg**, 0.2 **Zn**, 2.0 total, others	
SABS 711	AL Cr2	South Africa	Ingot	96.27 min **Al**, 0.30 **Si**, 0.10 **Cu**, 2.5 **Cr**, 0.07 **Zn**, 0.60 **Fe**, 0.05 each, 0.15 total, others	
SABS 711	AlCu30	South Africa	Ingot	rem **Al**, 0.30 **Si**, 33.0 **Cu**, 0.07 **Zn**, 0.40 **Fe**, 0.05 each, 0.15 total, others	
ANSI/ASTM B 327	CG71A	US	Ingot, bar, and shot	77.0 min **Al**, 0.6 **Si**, 17.0–19.0 **Cu**, 0.50 **Mn**, 0.65–0.95 **Mg**, 0.20 **Ni**, 1.0 **Zn**, 0.2 **Sn**, 0.7 **Fe**, 0.010	
ANSI/ASTM B 179	238.1	US	Ingot: for permanent mold castings	rem **Al**, 3.5–4.5 **Si**, 9.0–11.0 **Cu**, 0.6 **Mn**, 0.20–0.35 **Mg**, 1.0 **Ni**, 1.5 **Zn**, 0.25 **Ti**, 1.2 **Fe**, 0.50 total, others	
ANSI/ASTM B 179	238.2	US	Ingot: for permanent mold castings	rem **Al**, 3.5–4.5 **Si**, 9.5–10.5 **Cu**, 0.50 **Mn**, 0.20–0.35 **Mg**, 0.50 **Ni**, 0.50 **Zn**, 0.20 **Ti**, 1.2 **Fe**, 0.50 total, others	
ANSI/ASTM B 108	238.0	US	Permanent mold castings	rem **Al**, 3.5–4.5 **Si**, 9.0–11.0 **Cu**, 0.6 **Mn**, 0.15–0.35 **Mg**, 1.0 **Ni**, 1.5 **Zn**, 0.25 **Ti**, 1.5 **Fe**, 0.50 total, others	**As fabricated**: all **diam**, 100 **HB**
QQ-A-371F	238.1 (formerly 138(2))	US	Ingot	rem **Al**, 3.5–4.5 **Si**, 9.0–11.0 **Cu**, 0.6 **Mn**, 0.20–0.35 **Mg**, 1.0 **Ni**, 1.5 **Zn**, 0.25 **Ti**, 1.2 **Fe**, 0.50 total, others	
QQ-A-371F	238.2 (formerly 138(2))	US	Ingot	rem **Al**, 3.5–4.5 **Si**, 9.5–10.5 **Cu**, 0.50 **Mn**, 0.20–0.35 **Mg**, 0.50 **Ni**, 0.50 **Zn**, 0.20 **Ti**, 1.2 **Fe**, 0.50 total, others	
QQ-A-596d	122	US	Castings	rem **Al**, 2.0 **Si**, 9.2–10.8 **Cu**, 0.50 **Mn**, 0.15–0.35 **Mg**, 0.50 **Ni**, 0.8 **Zn**, 0.25 **Ti**, 1.5 **Fe**, 0.35 total, others	**Artificially aged only and temper T551**: all **diam**, 30 ksi (207 MPa) **TS**; **Solution heat treated and artificially aged**: all **diam**, 40 ksi (276 MPa) **TS**

CAST ALUMINUM AND ALUMINUM ALLOYS

specification number	designation	country	product forms	chemical composition	mechanical properties and hardness values
ANSI/ASTM B 179	222.1	US	Ingot: for permanent mold castings	rem **Al**, 2.0 **Si**, 9.2–10.7 **Cu**, 0.50 **Mn**, 0.20–0.35 **Mg**, 0.50 **Ni**, 0.8 **Zn**, 0.25 **Ti**, 1.2 **Fe**, 0.35 total, others	
ANSI/ASTM B 618	222.0	US	Investment castings	rem **Al**, 2.0 **Si**, 9.2–10.7 **Cu**, 0.50 **Mn**, 0.15–0.35 **Mg**, 0.50 **Ni**, 0.8 **Zn**, 0.25 **Ti**, 1.5 **Fe**, 0.35 total, others	**Annealed**: ≥0.250 in. **diam**, 23 ksi (159 MPa) **TS**; **Solution heat treated and artificially aged**: ≥0.250 in. **diam**, 30 ksi (207 MPa) **TS**
ANSI/ASTM B 108	222.0	US	Permanent mold castings	rem **Al**, 2.0 **Si**, 9.2–10.7 **Cu**, 0.50 **Mn**, 0.15–0.35 **Mg**, 0.50 **Ni**, 0.8 **Zn**, 0.25 **Ti**, 1.5 **Fe**, 0.35 total, others	**T551**: all **diam**, 30 ksi (207 MPa) **TS**, 115 **HB**; **T65**: all **diam**, 40 ksi (276 MPa) **TS**, 140 **HB**
ANSI/ASTM B 26	222.0	US	Sand castings	rem **Al**, 2.0 **Si**, 9.2–10.7 **Cu**, 0.50 **Mn**, 0.15–0.35 **Mg**, 0.50 **Ni**, 0.8 **Zn**, 0.25 **Ti**, 1.5 **Fe**, 0.35 total, others	**Annealed**: ≥0.250 in. **diam**, 23 ksi (159 MPa) **TS**, 80 **HB**; **T61**: ≥0.250 in. **diam**, 30 ksi (200 MPa) **TS**, 55 **HB**
QQ-A-371F	222.1 (formerly 122)	US	Ingot	rem **Al**, 2.0 **Si**, 9.2–10.7 **Cu**, 0.50 **Mn**, 0.20–0.35 **Mg**, 0.50 **Ni**, 0.8 **Zn**, 0.25 **Ti**, 1.2 **Fe**, 0.35 total, others	
QQ-A-601E	222.0	US	Castings	rem **Al**, 2.0 **Si**, 9.2–10.7 **Cu**, 0.50 **Mn**, 0.15–0.035 **Mg**, 0.50 **Ni**, 0.8 **Zn**, 0.25 **Ti**, 1.5 **Fe**, 0.35 total, others	**Annealed and temper T2**: all **diam**, 23 ksi (159 MPa) **TS**, 20 ksi (138 MPa) **YS**, 80 **HB**; **Solution heat treated, aged and temper T61**: all **diam**, 30 ksi (207 MPa) **TS**, 40 ksi (276 MPa) **TS**, 115 **HB**
QQ-A-596d	113	US	Castings	rem **Al**, 1.0–4.0 **Si**, 6.0–8.0 **Cu**, 0.6 **Mn**, 0.10 **Mg**, 0.35 **Ni**, 2.5 **Zn**, 0.25 **Ti**, 1.4 **Fe**, 0.50 total, others	**As cast**: all **diam**, 23 ksi (159 MPa) **TS**
CSA HA.3	.CS72	Canada	Ingot	rem **Al**, 1.0–3.0 **Si**, 6.5–8.0 **Cu**, 0.50 **Mn**, 0.07 **Mg**, 0.20 **Ni**, 0.10 **Zn**, 0.20 **Ti**, 0.95–1.4 **Fe**, 0.05 each, 0.20 total, others	
CSA HA.9	.CS72	Canada	Sand castings	rem **Al**, 1.0–3.0 **Si**, 6.5–8.0 **Cu**, 0.50 **Mn**, 0.30 **Mg**, 0.20 **Ni**, 0.10 **Zn**, 0.20 **Ti**, 0.95–1.5 **Fe**, 0.05 each, 0.20 total, others	
QQ-A-371F	213.1 (formerly 113)	US	Ingot	rem **Al**, 1.0–3.0 **Si**, 6.0–8.0 **Cu**, 0.6 **Mn**, 0.10 **Mg**, 0.35 **Ni**, 2.5 **Zn**, 0.25 **Ti**, 0.9 **Fe**, 0.50 total, others	
AMS 4227A		US	Sand castings	rem **Al**, 0.50 **Si**, 7.0–9.0 **Cu**, 0.30–0.7 **Mn**, 5.5–6.5 **Mg**, 0.30–0.7 **Ni**, 0.10 **Zn**, 0.20 **Ti**, 0.50 **Fe**, 0.05 each, 0.15 total, others	**As cast**: all **diam**, 22 ksi (152 MPa) **TS**, 80 min **HB**
QQ-A-371F	A240.1 (formerly 140(2))	US	Ingot	rem **Al**, 0.50 **Si**, 7.0–9.0 **Cu**, 0.30–0.7 **Mn**, 5.6–6.5 **Mg**, 0.30–0.7 **Ni**, 0.10 **Zn**, 0.20 **Ti**, 0.40 **Fe**, 0.05 each, 0.15 total, others	

CAST ALUMINUM AND ALUMINUM ALLOYS

specification number	designation	country	product forms	chemical composition	mechanical properties and hardness values
AMS 4283D		US	Permanent mold castings	rem **Al**, 2.0–3.0 **Si**, 4.0–5.0 **Cu**, 0.30 **Mn**, 0.05 **Mg**, 0.30 **Zn**, 0.20 **Ti**, 1.2 **Fe**, 0.30 total, others	**Solution heat treated**: all **diam**, 25 ksi (172 MPa) **TS**, 1% in 2 in. (50 mm) **El**, 65–90 **HB**
AMS 4282E		US	Permanent mold castings	rem **Al**, 2.0–3.0 **Si**, 4.0–5.0 **Cu**, 0.30 **Mn**, 0.05 **Mg**, 0.30 **Zn**, 0.20 **Ti**, 1.2 **Fe**, 0.30 total, others	**Solution and precipitation heat treated**: all **diam**, 27 ksi (186 MPa) **TS**, 17 ksi (117 MPa) **YS**, 1% in 2 in. (50 mm) **El**, 80–110 **HB**
ANSI/ASTM B 179	208.1	US	Ingot: sand and permanent mold castings	rem **Al**, 2.5–3.5 **Si**, 3.5–4.5 **Cu**, 0.50 **Mn**, 0.10 **Mg**, 0.35 **Ni**, 1.0 **Zn**, 0.25 **Ti**, 0.9 **Fe**, 0.50 total, others	
ANSI/ASTM B 179	208.2	US	Ingot: sand and permanent mold castings	rem **Al**, 2.5–3.5 **Si**, 3.5–4.5 **Cu**, 0.30 **Mn**, 0.03 **Mg**, 0.20 **Zn**, 0.20 **Ti**, 0.8 **Fe**, 0.30 total, others	
ANSI/ASTM B 618	208.0	US	Investment castings	rem **Al**, 2.5–3.5 **Si**, 3.5–4.5 **Cu**, 0.50 **Mn**, 0.10 **Mg**, 1.0 **Zn**, 0.25 **Ti**, 1.2 **Fe**, 0.50 total, others	**As fabricated**: ≥0.250 in. **diam**, 19 ksi (131 MPa) **TS**, 12 ksi (83 MPa) **YS**, 2% in 2 in. (50 mm) **El**
ANSI/ASTM B 108	208.0	US	Permanent mold castings	rem **Al**, 2.5–3.5 **Si**, 3.5–4.5 **Cu**, 0.50 **Mn**, 0.10 **Mg**, 0.35 **Ni**, 1.0 **Zn**, 0.25 **Ti**, 1.2 **Fe**, 0.50 total, others	**Solution heat treated and naturally aged**: all **diam**, 33 ksi (228 MPa) **TS**, 15 ksi (103 MPa) **YS**, 5% in 2 in. (50 mm) **El**, 75 **HB**; **Solution heat treated and artificially aged**: all **diam**, 35 ksi (241 MPa) **TS**, 22 ksi (152 MPa) **YS**, 2% in 2 in. (50 mm) **El**, 90 **HB**; **Solution heat treated and stabilized**: all **diam**, 33 ksi (228 MPa) **TS**, 16 ksi (110 MPa) **YS**, 3% in 2 in. (50 mm) **El**, 80 **HB**
ANSI/ASTM B 26	208.0	US	Sand castings	rem **Al**, 2.5–3.5 **Si**, 3.5–4.5 **Cu**, 0.50 **Mn**, 0.10 **Mg**, 0.35 **Ni**, 1.0 **Zn**, 0.25 **Ti**, 1.2 **Fe**, 0.50 total, others	**As fabricated**: ≥0.250 in. **diam**, 19 ksi (131 MPa) **TS**, 12 ksi (83 MPa) **YS**, 2% in 2 in. (50 mm) **El**, 55 **HB**
CSA HA.3	.CS42	Canada	Ingot	rem **Al**, 2.0–3.0 **Si**, 4.0–5.0 **Cu**, 0.10 **Mn**, 0.03 **Mg**, 0.20 **Ti**, 0.12 **Fe**, 0.05 each, 0.15 total, others	
CSA HA.10	.CS42	Canada	Mold castings	rem **Al**, 2.0–3.0 **Si**, 4.0–5.0 **Cu**, 0.10 **Mn**, 0.03 **Mg**, 0.20 **Ti**, 0.20 **Fe**, 0.05 each, 0.15 total, others	**T4**: all **diam**, 43.0 ksi (296 MPa) **TS**, 18.0 ksi (124 MPa) **YS**, 15% in 2 in. (50 mm) **El**; **T6**: all **diam**, 49.0 ksi (338 MPa) **TS**, 28.0 ksi (193 MPa) **YS**, 8% in 2 in. (50 mm) **El**

CAST ALUMINUM AND ALUMINUM ALLOYS

specification number	designation	country	product forms	chemical composition	mechanical properties and hardness values
CSA HA.9	.CS42	Canada	Sand castings	rem **Al**, 2.0–3.0 **Si**, 4.0–5.0 **Cu**, 0.10 **Mn**, 0.03 **Mg**, 0.20 **Ti**, 0.20 **Fe**, 0.05 each, 0.15 total, others	**T4**: all **diam**, 36.0 ksi (248 MPa) **TS**, 20.0 ksi (137 MPa) **YS**, 4% in 2 in. (50 mm) **El**; **T6**: all **diam**, 41.0 ksi (283 MPa) **TS**, 30.0 ksi (207 MPa) **YS**, 2% in 2 in. (50 mm) **El**
QQ–A–371F	208.1 (formerly 108)	US	Ingot	rem **Al**, 2.5–3.5 **Si**, 3.5–4.5 **Cu**, 0.50 **Mn**, 0.10 **Mg**, 0.35 **Ni**, 1.0 **Zn**, 0.25 **Ti**, 0.9 **Fe**, 0.50 total, others	
QQ–A–371F	208.2 (formerly 108)	US	Ingot	rem **Al**, 2.5–3.5 **Si**, 3.5–4.5 **Cu**, 0.30 **Mn**, 0.03 **Mg**, 0.20 **Zn**, 0.25 **Ti**, 0.8 **Fe**, 0.30 total, others	
QQ–A–371F	B295.1 (formerly B195)	US	Ingot	rem **Al**, 2.0–3.0 **Si**, 4.0–5.0 **Cu**, 0.35 **Mn**, 0.05 **Mg**, 0.35 **Ni**, 0.50 **Zn**, 0.25 **Ti**, 0.9 **Fe**, 0.35 total, others	
QQ–A–371F	B295.2 (formerly B195)	US	Ingot	rem **Al**, 2.0–3.0 **Si**, 4.0–5.0 **Cu**, 0.30 **Mn**, 0.03 **Mg**, 0.30 **Zn**, 0.20 **Ti**, 0.8 **Fe**, 0.05 each, 0.15 total, others	
QQ–A–601E	208.0	US	Castings	rem **Al**, 2.5–3.5 **Si**, 3.5–4.5 **Cu**, 0.50 **Mn**, 0.10 **Mg**, 0.35 **Ni**, 1.0 **Zn**, 0.25 **Ti**, 1.2 **Fe**, 0.50 total, others	**As cast**: all **diam**, 19 ksi (131 MPa) **TS**, 14 ksi (97 MPa) **YS**, 1.5% in 2 in. (50 mm) **El**, 55 **HB**; **Artificially aged and temper T55**: all **diam**, 21 ksi (145 MPa) **TS**, 75 **HB**
QQ–A–596d	B195	US	Castings	rem **Al**, 2.0–3.0 **Si**, 4.0–5.0 **Cu**, 0.35 **Mn**, 0.05 **Mg**, 0.35 **Ni**, 0.50 **Zn**, 0.25 **Ti**, 1.2 **Fe**, 0.35 total, others	**Solution heat treated and temper T4**: all **diam**, 33 ksi (228 MPa) **TS**, 4.5% in 2 in. (50 mm) **El**; **Solution heat treated and artificially aged T6**: all **diam**, 35 ksi (241 MPa) **TS**, 2.0% in 2 in. (50 mm) **El**; **Solution heat treated and stabilized T7**: all **diam**, 33 ksi (228 MPa) **TS**, 3.0% in 2 in. (50 mm) **El**
IS 202	A–26	India	Sand castings	rem **Al**, 2.0–3.0 **Si**, 4.0–5.0 **Cu**, 0.30 **Mn**, 0.05 **Mg**, 0.10 **Zn**, 1.0 **Fe**, 0.20 **Ti + Nb**	
IS 202	A–26	India	Sand castings	rem **Al**, 2.0–3.0 **Si**, 4.0–5.0 **Cu**, 0.30 **Mn**, 0.05 **Mg**, 0.10 **Zn**, 1.0 **Fe**, 0.20 **Ti + Nb**	
DIN 1725 sheet 2	GK–AlCu4TiMg/3.1371.6 2	Germany	Gravity die castings	rem **Al**, 0.18 **Si**, 4.2–9.2 **Cu**, 0.05 **Mn**, 0.15–0.30 **Mg**, 0.07 **Zn**, 0.15–0.30 **Ti**, 0.20 **Fe**, 0.03 each, 0.10 total, others	**Quenched and age hardened**: all **diam**, 350 MPa **TS**, 260 MPa **YS**, 3% in 2 in. (50 mm) **El**, 90–100 **HB**; **Precipitation hardened**: all **diam**, 320 MPa **TS**, 220 MPa **YS**,8 in 2 in. (50 mm) el, 95–115 **HB**

CAST ALUMINUM AND ALUMINUM ALLOYS

specification number	designation	country	product forms	chemical composition	mechanical properties and hardness values
AMS 4230C		US	Sand castings	rem **Al**, 1.2 **Si**, 4.0–5.0 **Cu**, 0.30 **Mn**, 0.03 **Mg**, 0.30 **Zn**, 0.20 **Ti**, 1.0 **Fe**, 0.05 each, 0.15 total, others	**Solution heat treated**: all **diam**, 22 ksi (152 MPa) **TS**, 2% in 2 in. (50 mm) **El**, 50–80 **HB**
IS 202	A–25	India	Sand castings	rem **Al**, 1.2 **Si**, 4.0–5.0 **Cu**, 0.30 **Mn**, 0.03 **Mg**, 0.10 **Zn**, 1.0 **Fe**, 0.20 **Ti+Nb**	**Solution and precipitation treated**: all **diam**, 221 MPa **TS**, 128 MPa **YS**, 3% in 2 in. (50 mm) **El**
IS 202	A–25	India	Sand castings	rem **Al**, 1.2 **Si**, 4.0–5.0 **Cu**, 0.30 **Mn**, 0.03 **Mg**, 0.10 **Zn**, 1.0 **Fe**, 0.20 **Ti+Nb**	**Solution and precipitation treated**: all **diam**, 221 MPa **TS**, 128 MPa **YS**, 3% in 2 in. (50 mm) **El**
JIS H 5202	Class 1 A	Japan	Castings	rem **Al**, 1.2 **Si**, 4.0–5.0 **Cu**, 0.3 **Mn**, 0.3 **Mg**, 0.3 **Zn**, 0.25 **Ti**, 0.5 **Fe**	**As cast**: all **diam**, 157 MPa **TS**, 5% in 2 in. (50 mm) **El**; **Quenched and tempered**: all **diam**, 275 MPa **TS**, 3% in 2 in. (50 mm) **El**, 80 **HB**
JIS H 2211	Class 1A	Japan	Ingot: for castings	rem **Al**, 1.2 **Si**, 4.0–5.0 **Cu**, 0.03 **Mn**, 0.3 **Mg**, 0.03 **Ni**, 0.03 **Zn**, 0.25 **Ti**, 0.3 **Fe**	
JIS H 2117	Class 1A	Japan	Ingot: for castings	rem **Al**, 1.2 **Si**, 4.0–5.0 **Cu**, 0.3 **Mn**, 0.3 **Mg**, 0.1 **Ni**, 0.3 **Zn**, 0.25 **Ti**, 0.4 **Fe**	
ISO R208	Al–Cu4S:	ISO	Castings	rem **Al**, 1.2 **Si**, 4.0–5.0 **Cu**, 0.3 **Mn**, 0.03 **Mg**, 0.05 **Ni**, 0.3 **Zn**, 0.2 **Ti**, 0.05 **Sn**, 1.0 **Fe**, 0.05 **Pb**	
ANSI/ASTM B 179	295.1	US	Ingot: for sand castings	rem **Al**, 0.7–1.5 **Si**, 4.0–5.0 **Cu**, 0.35 **Mn**, 0.03 **Mg**, 0.35 **Zn**, 0.25 **Ti**, 0.8 **Fe**, 0.05 each, 0.15 total, others	
ANSI/ASTM B 618	295.0	US	Investment castings	rem **Al**, 0.7–1.5 **Si**, 4.0–5.0 **Cu**, 0.35 **Mn**, 0.03 **Mg**, 0.35 **Zn**, 0.25 **Ti**, 1.0 **Fe**, 0.05 each, 0.15 total, others	**Solution heat treated and naturally aged**: ≥0.250 in. **diam**, 29 ksi (200 MPa) **TS**, 13 ksi (90 MPa) **YS**, 6% in 2 in. (50 mm) **El**; **Solution heat treated and artificially aged**: ≥0.250 in. **diam**, 32 ksi (221 MPa) **TS**, 20 ksi (138 MPa) **YS**, 3% in 2 in. (50 mm) **El**; **Solution heat treated and stabilized**: ≥0.250 in. **diam**, 29 ksi (200 MPa) **TS**, 16 ksi (110 MPa) **YS**, 3% in 2 in. (50 mm) **El**

CAST ALUMINUM AND ALUMINUM ALLOYS

specification number	designation	country	product forms	chemical composition	mechanical properties and hardness values
ANSI/ASTM B 26	295.0	US	Sand castings	rem **Al**, 0.7–1.5 **Si**, 4.0–5.0 **Cu**, 0.35 **Mn**, 0.03 **Mg**, 0.35 **Zn**, 0.25 **Ti**, 1.0 **Fe**, 0.05 each, 0.15 total, others	**Solution heat treated and naturally aged**: ≥0.250 in. **diam**, 29 ksi (200 MPa) **TS**, 13 ksi (90 MPa) **YS**, 6% in 2 in. (50 mm) **El**, 60 **HB**; **Solution heat treated and artificially aged**: ≥0.250 in. **diam**, 32 ksi (221 MPa) **TS**, 20 ksi (138 MPa) **YS**, 3% in 2 in. (50 mm) **El**, 75 **HB**; **Solution heat treated and stabilized**: ≥0.250 in. **diam**, 29 ksi (200 MPa) **TS**, 16 ksi (110 MPa) **YS**, 3% in 2 in. (50 mm) **El**, 70 **HB**
QQ-A-371F	295.1 (formerly 195)	US	Ingot	rem **Al**, 0.7–1.5 **Si**, 4.0–5.0 **Cu**, 0.35 **Mn**, 0.03 **Mg**, 0.35 **Ni**, 0.25 **Ti**, 0.8 **Fe**, 0.05 each, 0.15 total, others	
QQ-A-601E	295.0	US	Castings	rem **Al**, 0.7–1.5 **Si**, 4.0–5.0 **Cu**, 0.35 **Mn**, 0.03 **Mg**, 0.35 **Zn**, 0.25 **Ti**, 1.0 **Fe**, 0.05 each, 0.15 total, others	**Solution heat treated and temper T4**: all **diam**, 29 ksi (200 MPa) **TS**, 16 ksi (110 MPa) **YS**, 6.0% in 2 in. (50 mm) **El**, 60 **HB**; **Solution heat treated, artificially aged and temper T6**: all **diam**, 32 ksi (221 MPa) **TS**, 20 ksi (138 MPa) **TS**, 3.0% in 2 in. (50 mm) **El**, 75 **HB**; **Solution heat treated, aged and temper T62**: all **diam**, 36 ksi (248 MPa) **TS**, 34 ksi (234 MPa) **YS**, 60 **HB**
ANSI/ASTM B 179	295.2	US	Ingot: for sand castings	rem **Al**, 0.7–1.2 **Si**, 4.0–5.0 **Cu**, 0.30 **Mn**, 0.03 **Mg**, 0.30 **Zn**, 0.8 **Fe**, 0.05 each, 0.15 total, others	
QQ-A-371F	295.2 (formerly 195)	US	Ingot	rem **Al**, 0.7–1.2 **Si**, 4.0–5.0 **Cu**, 0.30 **Mn**, 0.03 **Mg**, 0.30 **Ni**, 0.20 **Ti**, 0.8 **Fe**, 0.05 each, 0.15 total, others	
BS L.155		UK	Ingot and castings	rem **Al**, 1.0–1.5 **Si**, 3.8–4.5 **Cu**, 0.1 **Mn**, 0.10 **Mg**, 0.1 **Ni**, 0.1 **Zn**, 0.05–0.25 **Ti**, 0.05 **Sn**, 0.25 **Fe**, 0.05 **Pb**	**Solution heat treated, quenched and artificially aged**: all **diam**, 280 MPa **TS**, 200 MPa **YS**, 4% in 2 in. (50 mm) **El**
BS L.154		UK	Ingot and castings	rem **Al**, 1.0–1.5 **Si**, 3.8–4.5 **Cu**, 0.1 **Mn**, 0.10 **Mg**, 0.1 **Ni**, 0.1 **Zn**, 0.05–0.25 **Ti**, 0.05 **Sn**, 0.25 **Fe**, 0.05 **Pb**	**Solution heat treated and quenched**: all **diam**, 215 MPa **TS**, 160 MPa **YS**, 7% in 2 in. (50 mm) **El**
SABS 991	Al-Cu5MgTiA	South Africa	Gravity die castings	rem **Al**, 0.30 **Si**, 5.0 **Cu**, 0.10 **Mn**, 0.15–0.35 **Mg**, 0.05 **Ni**, 0.10 **Zn**, 0.05–0.30 **Ti**, 0.05 **Sn**, 0.35 **Fe**, 0.05 each, 0.15 total, others	**Solution and precipitation heat treated**: all **diam**, 325 MPa **TS**, 3% in 2 in. (50 mm) **El**, 90–120 **HB**

CAST ALUMINUM AND ALUMINUM ALLOYS

specification number	designation	country	product forms	chemical composition	mechanical properties and hardness values
AMS 4225		US	Sand castings: moderate heat resistant	rem **Al**, 0.20 **Si**, 4.5–5.5 **Cu**, 0.20–0.30 **Mn**, 1.3–1.8 **Ni**, 0.15–0.25 **Ti**, 0.30 **Fe**, 0.10–0.30 **Zr**, 0.10–0.40 **Co**, 0.10–0.40 **Sb**, 0.6 **Sb+Co**, 0.50 **Ti+Zr**, 0.05 each, 0.30 total, others	**Solution and precipitation heat treated**: all **diam**, 32 ksi (221 MPa) **TS**, 24 ksi (165 MPa) **YS**, 2% in 2 in. (50 mm) **El**, 80 min **HB**
BS 2L.77		UK	Forging stock and forgings	rem **Al**, 0.50–0.90 **Si**, 3.9–5.0 **Cu**, 0.40–1.2 **Mn**, 0.20–0.8 **Mg**, 0.10 **Cr**, 0.2 **Ni**, 0.2 **Zn**, 0.05 **Sn**, 0.5 **Fe**, 0.05 **Pb**, 0.2 **Zr+Ti**	**Solution heat and precipitation treated and quenched**: all **diam**, 450 MPa **TS**, 395 MPa **YS**, 6% in 2 in. (50 mm) **El**
AMS 4226A		US	Castings: high strength sand, permanent and composite mold	rem **Al**, 0.06 **Si**, 4.5–5.5 **Cu**, 0.20–0.50 **Mn**, 0.35 **Ti**, 0.10 **Fe**, 0.05–0.15 **V**, 0.10–0.25 **Zr**, 0.03 each, 0.10 total, others	**Solution and precipitation heat treated**: all **diam**, 55 ksi (379 MPa) **TS**, 37 ksi (255 MPa) **YS**, 5% in 2 in. (50 mm) **El**
DIN 1725 sheet 2	G–AlCu4Ti wa/3.1841.61	Germany	Sand castings	rem **Al**, 0.18 **Si**, 4.5–5.2 **Cu**, 0.05 **Mn**, 0 07 **Zn**, 0.15–0.30 **Ti**, 0.18 **Fe**, 0.03 each, 0.10 total, others	**Quenched and age hardened**: all **diam**, 300 MPa **TS**, 200 MPa **YS**, 3% in 2 in. (50 mm) **El**, 95–110 **HB**
DIN 1725 sheet 2	G–AlCu4Ti ta/3.1841.63	Germany	Sand castings	rem **Al**, 0.18 **Si**, 4.5–5.2 **Cu**, 0.05 **Mn**, 0.07 **Zn**, 0.15–0.30 **Ti**, 0.18 **Fe**, 0.03 each, 0.10 total, others	**Selectively hardened**: all **diam**, 280 MPa **TS**, 180 MPa **YS**, 5% in 2 in. (50 mm) **El**, 85–105 **HB**
DIN 1725 sheet 2	GK–AlCu4Ti wa/3.1841.62	Germany	Gravity die castings	rem **Al**, 0.18 **Si**, 4.5–5.2 **Cu**, 0.05 **Mn**, 0.07 **Zn**, 0.15–0.30 **Ti**, 0.18 **Fe**, 0.03 each, 0.10 total, others	**Quenched and age hardened**: all **diam**, 330 MPa **TS**, 220 MPa **YS**, 7% in 2 in. (50 mm) **El**, 95–110 **HB**
DIN 1725 sheet 2	GK–AlCu4Ti ta/3.1841.64	Germany	Gravity die castings	rem **Al**, 0.18 **Si**, 4.5–5.2 **Cu**, 0.05 **Mn**, 0.07 **Zn**, 0.15–0.30 **Ti**, 0.18 **Fe**, 0.03 each, 0.10 total, others	**Selectively hardened**: all **diam**, 320 MPa **TS**, 180 MPa **YS**, 8% in 2 in. (50 mm) **El**, 90–105 **HB**
CSA HA.9	.C4	Canada	Sand castings	rem **Al**, 0.25 **Si**, 4.5–5.0 **Cu**, 0.10 **Mn**, 0.05 **Mg**, 0.10 **Ni**, 0.10 **Zn**, 0.05–0.20 **Ti**, 0.25 **Fe**, 0.05 each, 0.15 total, others	**T4**: all **diam**, 34.0 ksi (234 MPa) **TS**, 8% in 2 in. (50 mm) **El**; **T6**: all **diam**, 40.0 ksi (276 MPa) **TS**, 3.5% in 2 in. (50 mm) **El**
CSA HA.3	.CG50	Canada	Ingot	rem **Al**, 0.30 **Si**, 4.3–5.0 **Cu**, 0.10 **Mn**, 0.05 **Mg**, 0.05 **Ni**, 0.10 **Zn**, 0.15–0.30 **Ti**, 0.35 **Fe**, 0.05 each, 0.15 total, others	
CSA HA.9	.CG50	Canada	Sand castings	rem **Al**, 0.30 **Si**, 4.3–5.0 **Cu**, 0.10 **Mn**, 0.15–0.35 **Mg**, 0.05 **Ni**, 0.10 **Zn**, 0.15–0.30 **Ti**, 0.40 **Fe**, 0.05 each, 0.15 total, others	**T4**: all **diam**, 45.0 ksi (310 MPa) **TS**, 28.0 ksi (193 MPa) **YS**, 7% in 2 in. (50 mm) **El**;
SABS 990	Al–Cu5MgTiA	South Africa	Sand castings	rem **Al**, 0.30 **Si**, 4.2–5.0 **Cu**, 0.10 **Mn**, 0.15–0.35 **Mg**, 0.05 **Ni**, 0.10 **Zn**, 0.05–0.30 **Ti**, 0.05 **Sn**, 0.35 **Fe**, 0.05 each, 0.15 total, others	**Solution and precipitation heat treated**: all **diam**, 295 MPa **TS**, 2% in 2 in. (50 mm) **El**

CAST ALUMINUM AND ALUMINUM ALLOYS

specification number	designation	country	product forms	chemical composition	mechanical properties and hardness values
SABS 989	Al–Cu5MgTiA	South Africa	Ingot: for sand and gravity die castings	rem **Al**, 0.25 **Si**, 4.2–5.0 **Cu**, 0.10 **Mn**, 0.20–0.35 **Mg**, 0.05 **Ni**, 0.10 **Zn**, 0.05–0.30 **Ti**, 0.05 **Sn**, 0.30 **Fe**, 0.05 each, 0.15 total, others	**Sand cast, solution and precipitation heat treated**: all **diam**, 295 MPa **TS**, 2% in 2 in. (50 mm) **El**; **Chill cast, solution and precipitation heat treataed**: all **diam**, 325 MPa **TS**, 3% in 2 in. (50 mm) **El**
SFS 2564	G–AlCu4Ti	Finland	Die castings	rem **Al**, 0.35 **Si**, 4.0–5.0 **Cu**, 0.10 **Mn**, 0.05 **Mg**, 0.10 **Ni**, 0.2 **Zn**, 0.05–0.35 **Ti**, 0.05 **Sn**, 0.35 **Fe**, 0.03 each, 0.10 total, others	**Solution heat treated and aged**: $\leq$20 mm **diam**, 320 MPa **TS**, 220 MPa **TS**, 4% in 2 in. (50 mm) **El**
ISO R2147	Al–Cu4MgTi	ISO	Sand castings: test pieces	rem **Al**, 0.35 **Si**, 4.0–5.0 **Cu**, 0.10 **Mn**, 0.15–0.35 **Mg**, 0.05 **Ni**, 0.20 **Zn**, 0.35 **Ti**, 0.05 **Sn**, 0.40 **Fe**, 0.05 **Pb**	**Solution heat treated and naturally aged**: all **diam** 290 MPa **TS**, 4% in 2 in. (50 mm) **El**
ISO R164	Al–Cu4Mg Ti	ISO	Castings	rem **Al**, 0.35 **Si**, 4.0–5.0 **Cu**, 0.10 **Mn**, 0.15–0.35 **Mg**, 0.5 **Ni**, 0.20 **Zn**, 0.35 **Ti**, 0.05 **Sn**, 0.40 **Fe**, 0.05 **Pb**	
ISO R164	Al–Cu4 Ti	ISO	Castings	rem **Al**, 0.35 **Si**, 4.0–5.0 **Cu**, 0.10 **Mn**, 0.05 **Mg**, 0.10 **Ni**, 0.2 **Zn**, 0.05–0.35 **Ti**, 0.05 **Sn**, 0.40 **Fe**, 0.05 **Pb**	
ANSI/ASTM B 618	204.0	US	Investment castings	rem **Al**, 0.20 **Si**, 4.2–5.0 **Cu**, 0.10 **Mn**, 0.15–0.35 **Mg**, 0.05 **Ni**, 0.10 **Zn**, 0.15–0.30 **Ti**, 0.05 **Sn**, 0.35 **Fe**, 0.05 each, 0.15 total, others	**Solution heat treated and naturally aged**: $\geq$0.250 in. **diam**, 45 ksi (310 MPa) **TS**, 28 ksi (193 MPa) **YS**, 6% in 2 in. (50 mm) **El**
ANSI/ASTM B 108	204.0	US	Permanent mold castings	rem **Al**, 0.20 **Si**, 4.2–5.0 **Cu**, 0.10 **Mn**, 0.15–0.35 **Mg**, 0.05 **Ni**, 0.10 **Zn**, 0.15–0.30 **Ti**, 0.05 **Sn**, 0.35 **Fe**, 0.05 each, 0.30 total, others	**Solution heat treated and naturally aged**: all **diam**, 48 ksi (331 MPa) **TS**, 29 ksi (200 MPa) **YS**, 8% in 2 in. (50 mm) **El**
ANSI/ASTM B 26	204.0	US	Sand castings	rem **Al**, 0.20 **Si**, 4.2–5.0 **Cu**, 0.10 **Mn**, 0.15–0.35 **Mg**, 0.05 **Ni**, 0.10 **Zn**, 0.15–0.30 **Ti**, 0.05 **Sn**, 0.35 **Fe**, 0.05 each, 0.15 total, others	**Solution heat treated and naturally aged**: $\geq$0.250 in. **diam**, 45 ksi (310 MPa) **TS**, 28 ksi (193 MPa) **YS**, 6% in 2 in. (50 mm) **El**
BS 2L.92		UK	Ingot and castings	rem **Al**, 0.25 **Si**, 4.0–5.0 **Cu**, 0.10 **Mn**, 0.10 **Mg**, 0.10 **Ni**, 0.10 **Zn**, 0.25 **Ti**, 0.05 **Sn**, 0.25 **Fe**, 0.05 **Pb**	**Heat treated and quenched**: all **diam**, 280 MPa **TS**, 200 MPa **YS**, 4% in 2 in. (50 mm) **El**
BS 2L.91		UK	Ingot and castings	rem **Al**, 0.25 **Si**, 4.0–5.0 **Cu**, 0.10 **Mn**, 0.10 **Mg**, 0.10 **Ni**, 0.10 **Zn**, 0.25 **Ti**, 0.05 **Sn**, 0.25 **Fe**, 0.05 **Pb**	**Heat treated and quenched**: all **diam**, 220 MPa **TS**, 165 MPa **YS**, 7% in 2 in. (50 mm) **El**
CSA HA.3	.C4	Canada	Ingot	rem **Al**, 0.25 **Si**, 4.0–5.0 **Cu**, 0.10 **Mn**, 0.05 **Mg**, 0.10 **Ni**, 0.10 **Zn**, 0.05–0.20 **Ti**, 0.20 **Fe**, 0.05 each, 0.15 total, others	
IS 202	A–11	India	Sand castings	rem **Al**, 0.25 **Si**, 4.0–5.0 **Cu**, 0.10 **Mn**, 0.10 **Mg**, 0.10 **Ni**, 0.10 **Zn**, 0.05 **Sn**, 0.25 **Fe**, 0.05 **Pb**, 0.05–0.30 **Ti+Nb**	**Solution and precipitation treated**: all **diam**, 278 MPa **TS**, 186 MPa **YS**, 4% in 2 in. (50 mm) **El**

CAST ALUMINUM AND ALUMINUM ALLOYS

specification number	designation	country	product forms	chemical composition	mechanical properties and hardness values
IS 202	A–11	India	Sand castings	rem **Al**, 0.25 **Si**, 4.0–5.0 **Cu**, 0.10 **Mn**, 0.10 **Mg**, 0.10 **Ni**, 0.10 **Zn**, 0.05 **Sn**, 0.25 **Fe**, 0.05 **Pb**, 0.05–0.30 **Ti+Nb**	**Solution and precipitation treated**: all **diam**, 278 MPa **TS**, 186 MPa **YS**, 4% in 2 in. (50 mm) **El**
DIN 1725 sheet 2	G–AlCu4TiMg/3.137 1.42	Germany	Sand castings	rem **Al**, 0.18 **Si**, 4.2–4.9 **Cu**, 0.05 **Mn**, 0.15–0.30 **Mg**, 0.07 **Zn**, 0.15–0.30 **Ti**, 0.20 **Fe**	**Quenched and age hardened**: all **diam**, 350 MPa **TS**, 240 MPa **YS**, 3% in 2 in. (50 mm) **El**, 95–125 **HB**; **Precipitation hardened**: all **diam**, 300 MPa **TS**, 220 MPa **YS**, 5% in 2 in. (50 mm) **El**, 90–115 **HB**
AMS 4220D		US	Sand castings	rem **Al**, 0.6 **Si**, 3.7–4.5 **Cu**, 0.10 **Mn**, 1.2–1.7 **Mg**, 0.15–0.25 **Cr**, 1.8–2.3 **Ni**, 0.10 **Zn**, 0.07–0.18 **Ti**, 0.8 **Fe**, 0.05 each, 0.15 total, others	**Solution heat treated and overaged**: all **diam**, 20 ksi (138 MPa) **TS**, 70–85 **HB**
ANSI/ASTM B 179	201.2	US	Ingot: for sand castings	rem **Al**, 0.10 **Si**, 4.0–5.2 **Cu**, 0.20–0.50 **Mn**, 0.15–0.35 **Ti**, 0.10 **Fe**, 0.40–1.2 **Ag**, 0.05 each, 0.10 total, others	
ANSI/ASTM B 179	204.2	US	Ingot: for sand and permanent mold castings	rem **Al**, 0.15 **Si**, 4.2–4.9 **Cu**, 0.05 **Mn**, 0.20–0.35 **Mg**, 0.03 **Ni**, 0.05 **Zn**, 0.15–0.25 **Ti**, 0.05 **Sn**, 0.10–0.20 **Fe**, 0.05 each, 0.15 total, others	
ANSI/ASTM B 179	242.1	US	Ingot: for sand and permanent mold castings	rem **Al**, 0.7 **Si**, 3.5–4.5 **Cu**, 0.35 **Mn**, 1.3–1.8 **Mg**, 0.25 **Cr**, 1.7–2.3 **Ni**, 0.35 **Zn**, 0.25 **Ti**, 0.8 **Fe**, 0.05 each, 0.15 total, others	
ANSI/ASTM B 618	201.0	US	Investment castings	rem **Al**, 0.10 **Si**, 4.0–5.2 **Cu**, 0.20–0.50 **Mn**, 0.15–0.55 **Mg**, 0.15–0.35 **Ti**, 0.15 **Fe**, 0.40–1.0 **Ag**, 0.05 each, 0.10 total, others	**Solution heat treated and artificially aged**: ≥0.250 in. **diam**, 60 ksi (414 MPa) **TS**, 50 ksi (345 MPa) **YS**, 5% in 2 in. (50 mm) **El**; **Solution heat treated and stabilized**: ≥0.250 in. **diam**, 60 ksi (414 MPa) **TS**, 50 ksi (345 MPa) **YS**, 3% in 2 in. (50 mm) **El**
ANSI/ASTM B 618	242.0	US	Investment castings	rem **Al**, 0.7 **Si**, 3.5–4.5 **Cu**, 0.35 **Mn**, 1.2–1.8 **Mg**, 0.25 **Cr**, 1.7–2.3 **Ni**, 0.35 **Zn**, 0.25 **Ti**, 1.0 **Fe**, 0.05 each, 0.15 total, others	**Annealed**: ≥0.250 in. **diam**, 23 ksi (159 MPa) **TS**; **T61**: ≥0.250 in. **diam**, 32 ksi (221 MPa) **TS**, 20 ksi (138 MPa) **YS**
ANSI/ASTM B 108	242.0	US	Permanent mold castings	rem **Al**, 0.7 **Si**, 3.5–4.5 **Cu**, 0.35 **Mn**, 1.2–1.8 **Mg**, 0.25 **Cr**, 1.7–2.3 **Ni**, 0.35 **Zn**, 0.25 **Ti**, 1.0 **Fe**, 0.05 each, 0.15 total, others	**T571**: all **diam**, 34 ksi (234 MPa) **TS**, 105 **HB**; **T61**: all **diam**, 40 ksi (276 MPa) **TS**, 110 **HB**
ANSI/ASTM B 26	242.0	US	Sand castings	rem **Al**, 0.7 **Si**, 3.5–4.5 **Cu**, 0.35 **Mn**, 1.2–1.8 **Mg**, 0.25 **Cr**, 1.7–2.3 **Ni**, 0.35 **Zn**, 0.25 **Ti**, 1.0 **Fe**, 0.05 each, 0.15 total, others	**Annealed**: ≥0.250 in. **diam**, 23 ksi (159 MPa) **TS**, 70 **HB**; **T61**: ≥0.250 in. **diam**, 32 ksi (221 MPa) **TS**, 20 ksi (138 MPa) **YS**, 105 **HB**

CAST ALUMINUM AND ALUMINUM ALLOYS

specification number	designation	country	product forms	chemical composition	mechanical properties and hardness values
ANSI/ASTM B 26	201.0	US	Sand castings	rem **Al**, 0.10 **Si**, 4.0–5.2 **Cu**, 0.20–0.50 **Mn**, 0.15–0.55 **Mg**, 0.15–0.35 **Ti**, 0.15 **Fe**, 0.05 each, 0.10 total, others	**Solution heat treated and artificially aged:** ≥0.250 in. **diam**, 60 ksi (414 MPa) **TS**, 50 ksi (345 MPa) **YS**, 5% in 2 in. (50 mm) **El**; **Solution heat treated and stabilized:** ≥0.250 in. **diam**, 60 ksi (414 MPa) **TS**, 50 ksi (345 MPa) **YS**, 3% in 2 in. (50 mm) **El**
AS 1874	AP201	Australia	Ingot, sand castings and permanent mold castings	rem **Al**, 0.20 **Si**, 4.0–5.0 **Cu**, 0.05 **Mn**, 0.05 **Mg**, 0.05 **Ni**, 0.10 **Zn**, 0.20 **Ti**, 0.25 **Fe**, 0.05 each, 0.20 total, others	**Solution heat treated and artificially aged**: all **diam**, 275 MPa **TS**, 4% in 2 in. (50 mm) **El**
QQ-A-371F	242.1 (formerly 142)	US	Ingot	rem **Al**, 0.7 **Si**, 3.5–4.5 **Cu**, 0.35 **Mn**, 1.3–1.8 **Mg**, 0.25 **Cr**, 1.7–2.3 **Ni**, 0.35 **Zn**, 0.25 **Ti**, 0.8 **Fe**, 0.05 each, 0.15 total, others	
QQ-A-371F	A242.1 (formerly A142(2))	US	Ingot	rem **Al**, 0.6 **Si**, 3.7–4.5 **Cu**, 0.10 **Mn**, 1.3–1.7 **Mg**, 0.15–0.25 **Cr**, 1.8–2.3 **Ni**, 0.10 **Zn**, 0.07–0.20 **Ti**, 0.6 **Fe**, 0.05 each, 0.15 total, others	
QQ-A-601E	242.0	US	Castings	rem **Al**, 0.7 **Si**, 3.5–4.5 **Cu**, 0.35 **Mn**, 1.2–1.8 **Mg**, 0.25 **Cr**, 1.7–2.3 **Ni**, 0.35 **Zn**, 0.25 **Ti**, 1.0 **Fe**, 0.05 each 0.15 total, others	**Annealed and temper T21**: all **diam**, 23 ksi (159 MPa) **TS**, 18 ksi (124 MPa) **YS**, 70 **HB**; **Solution heat treated, aged and temper T571**: all **diam**, 29 ksi (200 MPa) **TS**, 30 ksi (207 MPa) **TS**, 85 **HB**; **Solution heat treated, stabilized and temper T77**: all **diam**, 24 ksi (165 MPa) **TS**, 13 ksi (90 MPa) **TS**, 1.0% in 2 in. (50 mm) **El**, 75 **HB**
QQ-A-596d	142	US	Castings	rem **Al**, 0.7 **Si**, 3.5–4.5 **Cu**, 0.35 **Mn**, 1.2–1.8 **Mg**, 0.25 **Cr**, 1.7–2.3 **Ni**, 0.35 **Zn**, 0.25 **Ti**, 1.0 **Fe**, 0.05 each, 0.15 total, others	**Artificially aged only and temper T571**: all **diam**, 34 ksi (234 MPa) **TS**; **Solution heat treated and artificially aged T61**: all **diam**, 40 ksi (276 MPa) **TS**
ISO R164	Al–Cu4Ni2Mg2	ISO	Castings	rem **Al**, 0.7 **Si**, 3.5–4.5 **Cu**, 0.6 **Mn**, 1.2–1.8 **Mg**, 0.2 **Cr**, 1.7–2.3 **Ni**, 0.1 **Zn**, 0.2 **Ti**, 0.05 **Sn**, 0.7 **Fe**, 0.05 **Pb**	
AMS 4222E		US	Sand castings	rem **Al**, 0.6 **Si**, 3.5–4.5 **Cu**, 0.10 **Mn**, 1.2–1.8 **Mg**, 1.7–2.3 **Ni**, 0.10 **Zn**, 0.07–0.18 **Ti**, 0.8 **Fe**, 0.05 each, 0.15 total, others	**Solution heat treated and overaged**: all **diam**, 17 ksi (117 MPa) **TS**, 60–85 **HB**
QQ-A-371F	242.2 (formerly 142)	US	Ingot	rem **Al**, 0.6 **Si**, 3.5–4.5 **Cu**, 0.10 **Mn**, 1.3–1.8 **Mg**, 1.7–2.3 **Ni**, 0.10 **Zn**, 0.20 **Ti**, 0.6 **Fe**, 0.05 each, 0.15 total, others	

CAST ALUMINUM AND ALUMINUM ALLOYS

specification number	designation	country	product forms	chemical composition	mechanical properties and hardness values
IS 7793	2285/24850	India	Castings	rem **Al**, 0.6 **Si**, 3.5–4.5 **Cu**, 0.2 **Mn**, 1.2–1.8 **Mg**, 1.7–2.3 **Ni**, 0.2 **Zn**, 0.2 **Ti**, 0.05 **Sn**, 0.7 **Fe**	**Chill cast**: all **diam**, 225 MPa **TS**, 90–130 **HB**
JIS H 5202	Class 5 A	Japan	Castings	rem **Al**, 0.6 **Si**, 3.5–4.5 **Cu**, 0.3 **Mn**, 1.2–1.8 **Mg**, 1.7–2.3 **Ni**, 0.1 **Zn**, 0.2 **Ti**, 0.8 **Fe**	**As cast**: all **diam**, 216 MPa **TS**; **Tempered**: all **diam**, 196 MPa **TS**, 65 **HB**; **Quenched and tempered**: all **diam**, 294 MPa **TS**, 110 **HB**
JIS H 2117	Class 5A	Japan	Ingot: for castings	rem **Al**, 0.6 **Si**, 3.5–4.5 **Cu**, 0.3 **Mn**, 1.3–1.8 **Mg**, 1.7–2.3 **Ni**, 0.1 **Zn**, 0.2 **Ti**, 0.7 **Fe**	
AMS 4223		US	Castings: sand, permanent and composite mold	rem **Al**, 0.05 **Si**, 4.0–5.0 **Cu**, 0.20–0.40 **Mn**, 0.15–0.35 **Mg**, 0.15–0.35 **Ti**, 0.10 **Fe**, 0.40–1.0 **Ag**, 0.03 each, 0.10 total, others	**Solution heat treated and naturally aged**: all **diam**, 50 ksi (345 MPa) **TS**, 30 ksi (207 MPa) **YS**, 12% in 2 in. (50 mm) **El**, 80 min **HB**
AMS 4228A		US	Castings: high strength sand, permanent and composite mold	rem **Al**, 0.05 **Si**, 4.0–5.0 **Cu**, 0.20–0.40 **Mn**, 0.15–0.35 **Mg**, 0.15–0.35 **Ti**, 0.10 **Fe**, 0.40–1.0 **Ag**, 0.03 each, 0.10 total, others	**Solution and precipitation heat treated**: all **diam**, 60 ksi (414 MPa) **TS**, 50 ksi (345 MPa) **YS**, 5% in 2 in. (50 mm) **El**, 110 min **HB**
AMS 4229A		US	Castings: high strength sand, permanent and composite mold	rem **Al**, 0.05 **Si**, 4.0–5.0 **Cu**, 0.20–0.40 **Mn**, 0.15–0.35 **Mg**, 0.15–0.35 **Ti**, 0.10 **Fe**, 0.40–1.0 **Ag**, 0.03 each, 0.10 total, others	**Solution heat treated and overaged**: all **diam**, 60 ksi (414 MPa) **TS**, 50 ksi (345 MPa) **YS**, 3% in 2 in. (50 mm) **El**, 110 min **HB**
MIL–A–21180C	A201.0	US	Castings	rem **Al**, 0.05 **Si**, 4.0–5.0 **Cu**, 0.20–0.40 **Mn**, 0.15–0.35 **Mg**, 0.15–0.35 **Ti**, 0.10 **Fe**, 0.03 each, 0.10 total, others	**Class 1**: all **diam**, 414 MPa **TS**, 345 MPa **TS**, 3% in 2 in. (50 mm) **El**; **Class 2**: all **diam**, 414 MPa **TS**, 345 MPa **TS**, 5% in 2 in. (50 mm) **El**
QQ–A–371F	A242.2 (formerly A142(2))	US	Ingot	rem **Al**, 0.35 **Si**, 3.7–4.5 **Cu**, 0.10 **Mn**, 1.3–1.7 **Mg**, 0.15–0.25 **Cr**, 1.8–2.3 **Ni**, 0.10 **Zn**, 0.07–0.20 **Ti**, 0.6 **Fe**, 0.05 each, 0.15 total, others	
JIS H 2211	Class 5A	Japan	Ingot: for castings	rem **Al**, 0.4 **Si**, 3.5–4.5 **Cu**, 0.03 **Mn**, 1.3–1.8 **Mg**, 1.7–2.3 **Ni**, 0.03 **Zn**, 0.2 **Ti**, 0.4 **Fe**	
AMS 4224		US	Sand castings	rem **Al**, 0.35 **Si**, 3.5–4.5 **Cu**, 0.15–0.45 **Mn**, 1.8–2.3 **Mg**, 0.20–0.40 **Cr**, 1.9–2.3 **Ni**, 0.05 **Zn**, 0.06–0.20 **Ti**, 0.40 **Fe**, 0.06–0.20 **V**, 0.05 each, 0.15 total, others	**Stabilized**: all **diam**, 18 ksi (124 MPa) **TS**, 1% in 2 in. (50 mm) **El**, 70 min **HB**
ANSI/ASTM B 179	242.2	US	Ingot: for sand and permanent mold castings	rem **Al**, 0.06 **Si**, 3.5–4.5 **Cu**, 0.10 **Mn**, 1.3–1.8 **Mg**, 1.7–2.3 **Ni**, 0.10 **Zn**, 0.20 **Ti**, 0.6 **Fe**, 0.05 each, 0.15 total, others	

CAST ALUMINUM AND ALUMINUM ALLOYS

specification number	designation	country	product forms	chemical composition	mechanical properties and hardness values
SABS 989	Al–Si24CuMgNiCrA	South Africa	Ingot: for pistons and gravity die castings	rem **Al**, 23.0–26.0 **Si**, 0.8–1.4 **Cu**, 0.2 **Mn**, 1.0–1.3 **Mg**, 0.8–1.3 **Ni**, 0.2 **Zn**, 0.2 **Ti**, 0.6 **Fe**, 0.05 each, 0.15 total, others	**Chill cast, solution and precipitation heat treated**: all **diam**, 175 MPa **TS**
IS 7793	4928–B	India	Castings	rem **Al**, 23.0–26.0 **Si**, 0.8–1.5 **Cu**, 0.2 **Mn**, 0.8–1.3 **Mg**, 0.3–0.6 **Cr**, 0.8–1.3 **Ni**, 0.2 **Zn**, 0.2 **Ti**, 0.05 **Sn**, 0.7 **Fe**, 0.05 **Pb**	**Chill cast**: all **diam**, 165 MPa **TS**, 90–125 **HB**
SABS 991	Al–Si24CuMgNiCr	South Africa	Gravity die castings	rem **Al**, 23.0–26.0 **Si**, 0.8–1.5 **Cu**, 0.2 **Mn**, 0.8–1.3 **Mg**, 1.3 **Ni**, 0.2 **Zn**, 0.2 **Ti**, 0.7 **Fe**, 0.05 each, 0.15 total, others	**Solution and precipitation heat treated**: all **diam**, 175 MPa **TS**, 90–125 **HB**
QQ–A–371F	390.2 (formerly 390)	US	Ingot	rem **Al**, 16.0–18.0 **Si**, 4.0–5.0 **Cu**, 0.10 **Mn**, 0.50–0.65 **Mg**, 0.10 **Zn**, 0.20 **Ti**, 0.6–1.0 **Fe**, 0.10 each, 0.20 total, others	
QQ–A–371F	A390.1 (formerly A390)	US	Ingot	rem **Al**, 16.0–18.0 **Si**, 4.0–5.0 **Cu**, 0.10 **Mn**, 0.50–0.65 **Mg**, 0.10 **Zn**, 0.20 **Ti**, 0.40 **Fe**, 0.10 each, 0.20 total, others	
SABS 989	Al–Si18CuMgNiA	South Africa	Ingot: for pistons and gravity die castings	rem **Al**, 17.0–19.0 **Si**, 0.8–1.4 **Cu**, 0.2 **Mn**, 1.0–1.3 **Mg**, 0.8–1.3 **Ni**, 0.2 **Zn**, 0.2 **Ti**, 0.6 **Fe**, 0.05 each, 0.15 total, others	**Chill cast, solution and precipitation heat treated**: all **diam**, 225 MPa **TS**
IS 7793	4928–A/49285	India	Castings	rem **Al**, 17.0–19.0 **Si**, 0.8–1.5 **Cu**, 0.2 **Mn**, 0.8–1.3 **Mg**, 0.8–1.3 **Ni**, 0.2 **Zn**, 0.2 **Ti**, 0.05 **Sn**, 0.7 **Fe**, 0.05 **Pb**	**Chill cast**: all **diam**, 175 MPa **TS**, 90–125 **HB**
SABS 991	Al–Si18CuMgNi	South Africa	Gravity die castings	rem **Al**, 17.0–19.0 **Si**, 0.8–1.5 **Cu**, 0.2 **Mn**, 0.8–1.3 **Ni**, 0.2 **Zn**, 0.2 **Ti**, 0.7 **Fe**, 0.05 each, 0.15 total, others	**Solution and precipitation heat treated**: all **diam**, 225 MPa **TS**, 90–125 **HB**
ANSI/ASTM B 179	384.1/SC114A	US	Ingot: for die castings	rem **Al**, 10.5–12.0 **Si**, 3.0–4.5 **Cu**, 0.50 **Mn**, 0.10 **Mg**, 0.50 **Ni**, 2.9 **Zn**, 0.35 **Sn**, 1.0 **Fe**, 0.50 total, others	
ANSI/ASTM B 179	384.2	US	Ingot: for die castings	rem **Al**, 10.5–12.0 **Si**, 3.0–4.5 **Cu**, 0.10 **Mn**, 0.10 **Mg**, 0.10 **Ni**, 0.10 **Zn**, 0.10 **Sn**, 0.6–1.0 **Fe**, 0.20 total, others	
AS 1874	AP315	Australia	Ingot, sand castings and permanent mould castings	rem **Al**, 10.5–12.0 **Si**, 3.0–4.5 **Cu**, 0.10 **Mn**, 0.10 **Mg**, 0.10 **Ni**, 0.10 **Zn**, 0.10 **Sn**, 0.6–1.0 **Fe**, 0.20 total, others	
AS 1874	AS315	Australia	Ingot, sand castings and permanent mould castings	rem **Al**, 10.5–12.0 **Si**, 3.0–4.5 **Cu**, 0.5 **Mn**, 0.10 **Mg**, 0.5 **Ni**, 1.0 **Zn**, 1.3 **Fe**, 0.50 total, others	
QQ–A–371F	384.1 (formerly 384)	US	Ingot	rem **Al**, 10.5–12.0 **Si**, 3.0–4.5 **Cu**, 0.50 **Mn**, 0.10 **Mg**, 0.50 **Ni**, 2.9 **Zn**, 0.35 **Ti**, 1.0 **Fe**, 0.50 total, others	

CAST ALUMINUM AND ALUMINUM ALLOYS

specification number	designation	country	product forms	chemical composition	mechanical properties and hardness values
QQ-A-371F	384.2 (formerly 384)	US	Ingot	rem **Al**, 10.5–12.0 **Si**, 3.0–4.5 **Cu**, 0.10 **Mn**, 0.10 **Mg**, 0.10 **Ni**, 0.10 **Zn**, 0.10 **Sn**, 0.6–1.0 **Fe**, 0.20 total, others	
QQ-A-591E	384.0	US	Castings	rem **Al**, 10.5–12.0 **Si**, 3.0–4.5 **Cu**, 0.50 **Mn**, 0.10 **Mg**, 0.50 **Ni**, 3.0 **Zn**, 0.35 **Sn**, 1.3 **Fe**, 0.50 total, others	**Die cast**: all **diam**, 48 ksi (331 MPa) **TS**, 24 ksi (165 MPa) **YS**, 2.5% in 2 in. (50 mm) **El**
IS 7793	4658/49582	India	Castings	rem **Al**, 11.0–13.0 **Si**, 0.8–1.5 **Cu**, 0.2 **Mn**, 0.8–1.3 **Mg**, 1.5 **Ni**, 0.35 **Zn**, 0.2 **Ti**, 0.05 **Sn**, 0.8 **Fe**, 0.05 **Pb**	**Chill cast**: all **diam**, 195 MPa **TS**, 90–140 **HB**
UNI 7363	GD–AlSi12Cu2FeZn	Italy	Die castings	rem **Al**, 11.0–12.5 **Si**, 1.75–2.5 **Cu**, 0.50 **Mn**, 0.30 **Mg**, 0.30 **Ni**, 1.40 **Zn**, 0.20 **Ti**, 0.10 **Sn**, 0.7–1.0 **Fe**, 0.15 **Pb**	**As cast**: all **diam**, 265 MPa **TS**, 155 MPa **YS**, 1% in 2 in. (50 mm) **El**, 85–100 **HB**
UNI 5076	GD–AlSi12Cu2Fe	Italy	Pressure die castings	rem **Al**, 11.0–12.5 **Si**, 1.75–2.5 **Cu**, 0.50 **Mn**, 0.30 **Mg**, 0.30 **Ni**, 0.8 **Zn**, 0.15 **Ti**, 0.10 **Sn**, 0.70–1.0 **Fe**, 0.15 **Pb**	**As cast**: all **diam**, 265 MPa **TS**, 155 MPa **YS**, 1% in 2 in. (50 mm) **El**, 85–100 **HB**
SABS 989	Al–Si12Ni2MgCuA	South Africa	Ingot: for pistons and gravity die castings	rem **Al**, 11.0–13.0 **Si**, 0.5–1.3 **Cu**, 0.5 **Mn**, 1.0–1.5 **Mg**, 0.7–2.5 **Ni**, 0.1 **Zn**, 0.2 **Ti**, 0.1 **Sn**, 0.7 **Fe**, 0.1 **Pb**, 0.05 each, 0.15 total, others	**Chill cast, solution and precipitation heat treated**: all **diam**, 275 MPa **TS**
ANSI/ASTM B 179	A332.2	US	Ingot: for permanent mold castings	rem **Al**, 11.0–13.0 **Si**, 0.50–1.5 **Cu**, 0.10 **Mn**, 0.9–1.3 **Mg**, 2.0–3.0 **Ni**, 0.10 **Zn**, 0.20 **Ti**, 0.9 **Fe**, 0.05 each, 0.15 total, others	
CSA HA.3	.SN122	Canada	Ingot	rem **Al**, 11.0–13.0 **Si**, 0.50–1.5 **Cu**, 0.10 **Mn**, 0.9–1.3 **Mg**, 2.0–3.0 **Ni**, 0.10 **Zn**, 0.20 **Ti**, 0.9 **Fe**, 0.05 each, 0.15 total, others	
QQ-A-371F	A332.2 (formerly A132)	US	Ingot	rem **Al**, 11.0–13.0 **Si**, 0.50–1.5 **Cu**, 0.10 **Mn**, 0.9–1.3 **Mg**, 2.0–3.0 **Ni**, 0.10 **Zn**, 0.20 **Ti**, 0.9 **Fe**, 0.05 each, 0.15 total, others	
JIS H 2211	Class 8A	Japan	Ingot: for castings	rem **Al**, 11.0–13.0 **Si**, 0.8–1.3 **Cu**, 0.03 **Mn**, 0.8–1.3 **Mg**, 1.0–2.5 **Ni**, 0.03 **Zn**, 0.2 **Ti**, 0.4 **Fe**	
JIS H 2117	Class 8A	Japan	Ingot: for castings	rem **Al**, 11.0–13.0 **Si**, 0.8–1.3 **Cu**, 0.1 **Mn**, 0.8–1.3 **Mg**, 1.0–2.5 **Ni**, 0.1 **Zn**, 0.2 **Ti**, 0.7 **Fe**	
ANSI/ASTM B 179	A332.1	US	Ingot: for permanent mold castings	rem **Al**, 11.0–13.0 **Si**, 0.50–1.5 **Cu**, 0.35 **Mn**, 0.8–1.3 **Mg**, 2.0–3.0 **Ni**, 0.35 **Zn**, 0.25 **Ti**, 0.9 **Fe**, 0.05 each, others	

CAST ALUMINUM AND ALUMINUM ALLOYS

specification number	designation	country	product forms	chemical composition	mechanical properties and hardness values
QQ–A–371F	A332.1 (formerly A132)	US	Ingot	rem **Al**, 11.0–13.0 **Si**, 0.50–1.5 **Cu**, 0.35 **Mn**, 0.8–1.3 **Mg**, 2.0–3.0 **Ni**, 0.35 **Zn**, 0.25 **Ti**, 0.9 **Fe**, 0.05 each, others	
JIS H 5202	Class 8 A	Japan	Castings	rem **Al**, 11.0–13.0 **Si**, 0.8–1.3 **Cu**, 0.1 **Mn**, 0.7–1.3 **Mg**, 1.0–2.5 **Ni**, 0.1 **Zn**, 0.2 **Ti**, 0.8 **Fe**	**As cast**: all **diam**, 177 MPa **TS**; **Tempered**: all **diam**, 216 MPa **TS**, 90 **HB**; **Quenched and tempered**: all **diam**, 275 MPa **TS**, 110 **HB**
ANSI/ASTM B 108	A322.0	US	Permanent mold castings	rem **Al**, 11.0–13.0 **Si**, 0.50–1.5 **Cu**, 0.35 **Mn**, 0.7–1.3 **Mg**, 2.0–3.0 **Ni**, 0.35 **Zn**, 0.25 **Ti**, 1.2 **Fe**, 0.05 each, others	**T551**: all **diam**, 31 ksi (214 MPa) **TS**, 105 **HB**; **T65**: all **diam**, 40 ksi (276 MPa) **TS**, 125 **HB**
CSA HA.10	.SN122	Canada	Mold castings	rem **Al**, 11.0–13.0 **Si**, 0.50–1.5 **Cu**, 0.35 **Mn**, 0.7–1.3 **Mg**, 2.0–3.0 **Ni**, 0.35 **Zn**, 0.25 **Ti**, 1.2 **Fe**, 0.05 each, others	**T5A**: all **diam**, 31.0 ksi (214 MPa) **TS**; **T6A**: all **diam**, 40.0 ksi (276 MPa) **TS**
SABS 991	Al–Si12Ni2MgCuA	South Africa	Gravity die castings	rem **Al**, 11.0–13.0 **Si**, 0.5–1.3 **Cu**, 0.5 **Mn**, 0.7–1.5 **Mg**, 0.7–2.5 **Ni**, 0.1 **Zn**, 0.2 **Ti**, 0.1 **Sn**, 0.8 **Fe**, 0.05 each, 0.15 total, others	**Solution and precipitation heat treated**: all **diam**, 275 MPa **TS**, 100/150 **HB**; **Full heat treatment and stabilized**: all **diam**, 200 MPa **TS**, 65–85 **HB**
SABS 989	Al–Si12Cu2MgMnA	South Africa	Ingot: for gravity die castings	rem **Al**, 11.0–12.5 **Si**, 1.0–2.0 **Cu**, 0.5–0.9 **Mn**, 0.45–1.0 **Mg**, 0.05 **Ni**, 1.0 **Zn**, 0.25 **Ti**, 0.7 **Fe**	**Chill cast, solution and precipitation heat treated**: all **diam**, 290 MPa **TS**
SABS 991	Al–Si12Cu2MgMnA	South Africa	Gravity die castings	rem **Al**, 11.0–12.5 **Si**, 1.0–2.0 **Cu**, 0.5–0.9 **Mn**, 0.40–1.0 **Mg**, 0.05 **Ni**, 1.0 **Zn**, 0.25 **Ti**, 0.9 **Fe**, 0.50 total, others	**Precipitation heat treated**: all **diam**, 220 MPa **TS**; **Solution and precipitation heat treated**: all **diam**, 290 MPa **TS**
SFS 2565	G–AlSi12Cu	Finland	Die castings	rem **Al**, 11.0–13.5 **Si**, 1.2 **Cu**, 0.5 **Mn**, 0.3 **Mg**, 0.30 **Ni**, 0.5 **Zn**, 0.2 **Ti**, 0.1 **Sn**, 0.90 **Fe**, 0.20 **Pb**, 0.05 each, 0.15 total, others	**As cast**: $\leq$20 mm **diam**, 180 MPa **TS**, 90 MPa **YS**, 2% in 2 in. (50 mm) **El**
JIS H 5302	Class 12	Japan	Die castings	rem **Al**, 9.6–12.0 **Si**, 1.5–3.5 **Cu**, 0.5 **Mn**, 0.3 **Mg**, 0.5 **Ni**, 1.0 **Zn**, 0.3 **Sn**, 1.3 **Fe**	
JIS H 2118	Class 12	Japan	Ingot: for die castings	rem **Al**, 9.6–12.0 **Si**, 1.5–3.5 **Cu**, 0.5 **Mn**, 0.3 **Mg**, 0.5 **Ni**, 1.0 **Zn**, 0.3 **Sn**, 0.9 **Fe**	
SABS 989	Al–Si10Cu3MgNiA	South Africa	Ingot: for pistons	rem **Al**, 8.5–10.5 **Si**, 2.0–4.0 **Cu**, 0.5 **Mn**, 0.7–1.5 **Mg**, 0.5–1.5 **Ni**, 0.5 **Zn**, 0.2 **Sn**, 1.0 **Fe**, 0.05 each, 0.5 total, others	**Chill cast and precipitation heat treated**: all **diam**, 215 MPa **TS**
ANSI/ASTM B 179	F332.1	US	Ingot: for permanent mold castings	rem **Al**, 8.5–10.5 **Si**, 2.0–4.0 **Cu**, 0.50 **Mn**, 0.6–1.5 **Mg**, 0.50 **Ni**, 1.0 **Zn**, 0.25 **Ti**, 0.9 **Fe**, 0.50 total, others	

CAST ALUMINUM AND ALUMINUM ALLOYS

specification number	designation	country	product forms	chemical composition	mechanical properties and hardness values
AS 1874	AS305	Australia	Ingot, sand castings and permanent mould castings	rem **Al**, 8.5–10.5 **Si**, 2.0–4.0 **Cu**, 0.5 **Mn**, 0.6–1.5 **Mg**, 0.10 **Cr**, 0.5 **Ni**, 1.0 **Zn**, 0.25 **Ti**, 0.15 **Sn**, 0.9 **Fe**, 0.25 **Pb**, 0.05 each, 0.20 total, others	**Cooled and artificially aged**: all **diam**, 215 MPa **TS**
QQ–A–371F	F332.1 (formerly F132)	US	Ingot	rem **Al**, 8.5–10.5 **Si**, 2.0–4.0 **Cu**, 0.50 **Mn**, 0.6–1.5 **Mg**, 0.50 **Ni**, 1.0 **Zn**, 0.25 **Ti**, 0.9 **Fe**, 0.50 total, others	
DIN 1725 sheet 2	GD–AlSi12 (Cu)/3.2982.05	Germany	Pressure die casting	rem **Al**, 11.0–13.5 **Si**, 1.00 **Cu**, 0.20–0.50 **Mn**, 0.30 **Mg**, 0.20 **Ni**, 0.50 **Zn**, 0.15 **Ti**, 0.10 **Sn**, 1.30 **Fe**, 0.05 each, 0.15 total, others	**As cast**: all **diam**, 220 MPa **TS**, 140 MPa **YS**, 1% in 4 in. (100 mm) **El**, 60–80 **HB**
DIN 1725 sheet 2	GK–AlSi12 (Cu)/3.2583.02	Germany	Gravity die castings	rem **Al**, 11.0–13.5 **Si**, 1.0 **Cu**, 0.2–0.5 **Mn**, 0.30 **Mg**, 0.20 **Ni**, 0.50 **Zn**, 0.15 **Ti**, 0.10 **Sn**, 0.8 **Fe**, 0.05 each, 0.15 total, others	**As cast**: all **diam**, 180 MPa **TS**, 90 MPa **YS**, 2% in 2 in. (50 mm) **El**, 55–75 **HB**
DIN 1725 sheet 2	G–AlSi12 (Cu)/3.2583.01	Germany	Sand castings	rem **Al**, 11.0–13.5 **Si**, 1.0 **Cu**, 0.2–0.5 **Mn**, 0.30 **Mg**, 0.20 **Ni**, 0.50 **Zn**, 0.15 **Ti**, 0.10 **Sn**, 0.80 **Fe**, 0.20 **Pb**, 0.05 each, 0.15 total, others	**As cast**: all **diam**, 150 MPa **TS**, 80 MPa **YS**, 1% in 2 in. (50 mm) **El**, 50–65 **HB**
JIS H 2211	Class 8B	Japan	Ingot: for castings	rem **Al**, 8.5–10.5 **Si**, 2.0–4.0 **Cu**, 0.03 **Mn**, 0.6–1.5 **Mg**, 0.5–1.5 **Ni**, 0.03 **Zn**, 0.2 **Ti**, 0.4 **Fe**	
JIS H 2211	Class 8C	Japan	Ingot: for castings	rem **Al**, 8.5–10.5 **Si**, 2.0–4.0 **Cu**, 0.03 **Mn**, 0.6–1.5 **Mg**, 0.03 **Ni**, 0.03 **Zn**, 0.2 **Ti**, 0.4 **Fe**	
JIS H 2117	Class 8B	Japan	Ingot: for castings	rem **Al**, 8.5–10.5 **Si**, 2.0–4.0 **Cu**, 0.5 **Mn**, 0.6–1.5 **Mg**, 0.5–1.5 **Ni**, 0.5 **Zn**, 0.2 **Ti**, 0.8 **Fe**	
JIS H 2117	Class 8C	Japan	Ingot: for castings	rem **Al**, 8.5–10.5 **Si**, 2.0–4.0 **Cu**, 0.5 **Mn**, 0.6–1.5 **Mg**, 0.5 **Ni**, 0.5 **Zn**, 0.2 **Ti**, 0.8 **Fe**	
ANSI/ASTM B 108	F332.0	US	Permanent mold castings	rem **Al**, 8.5–10.5 **Si**, 2.0–4.0 **Cu**, 0.50 **Mn**, 0.50–1.5 **Mg**, 0.50 **Ni**, 1.0 **Zn**, 0.25 **Ti**, 1.2 **Fe**, 0.50 total, others	**Cooled and artificially aged**: all **diam**, 31 ksi (214 MPa) **TS**, 105 **HB**
QQ–A–596d	F132	US	Castings	rem **Al**, 8.5–10.5 **Si**, 2.0–4.0 **Cu**, 0.50 **Mn**, 0.50–1.5 **Mg**, 0.50 **Ni**, 1.0 **Zn**, 0.25 **Ti**, 1.2 **Fe**, 0.50 total, others	**Artificially aged only and temper T5**: all **diam**, 31 ksi (214 MPa) **TS**
JIS H 5202	Class 8 B	Japan	Castings	rem **Al**, 8.5–10.5 **Si**, 2.0–4.0 **Cu**, 0.5 **Mn**, 0.5–1.5 **Mg**, 0.5–1.5 **Ni**, 0.5 **Zn**, 0.2 **Ti**, 1.0 **Fe**	**As cast**: all **diam**, 177 MPa **TS**; **Tempered**: all **diam**, 206 MPa **TS**, 90 **HB**; **Quenched and tempered**: all **diam**, 275 MPa **TS**, 110 **HB**
JIS H 5202	Class 8 C	Japan	Castings	rem **Al**, 8.5–10.5 **Si**, 2.0–4.0 **Cu**, 0.5 **Mn**, 0.5–1.5 **Mg**, 0.5 **Zn**, 0.2 **Ti**, 1.0 **Fe**	**As cast**: 177 MPa **TS**; **Tempered**: 206 MPa **TS**, 90 **HB**; **Quenched and tempered**: 275 MPa **TS**, 110 **HB**

CAST ALUMINUM AND ALUMINUM ALLOYS

specification number	designation	country	product forms	chemical composition	mechanical properties and hardness values
SABS 991	Al–Si10Cu3MgNiA	South Africa	Gravity die castings	rem **Al**, 8.5–10.5 **Si**, 2.0–4.0 **Cu**, 0.5 **Mn**, 0.5–1.5 **Mg**, 0.5–1.5 **Ni**, 0.5 **Zn**, 0.2 **Sn**, 1.2 **Fe**, 0.05 each, 0.5 total, others	**Precipitation heat treated**: all **diam**, 215 MPa **TS**, 95–125 **HB**
QQ–A–371F	F332.2 (formerly F132)	US	Ingot	rem **Al**, 8.5–10.0 **Si**, 2.0–4.0 **Cu**, 0.10 **Mn**, 0.9–1.3 **Mg**, 0.10 **Ni**, 0.10 **Zn**, 0.20 **Ti**, 0.6 **Fe**, 0.30 total, others	
QQ–A–596d	336.0	US	Castings	rem **Al**, 11.0–13.0 **Si**, 0.50–1.5 **Cu**, 0.7–1.8 **Mn**, 0.35 **Mg**, 2.0–3.0 **Ni**, 0.35 **Zn**, 0.25 **Ti**, 1.3 **Fe**, 0.05 each, others	**Artificially aged only and temper T551**: all **diam**, 31 ksi (214 MPa) **TS**; **Solution heat treated and artificially aged**: all **diam**, 40 ksi (276 MPa) **TS**
JIS H 2212	Class 12	Japan	Ingot: for die castings	rem **Al**, 9.6–12.0 **Si**, 1.5–3.5 **Cu**, 0.03 **Mn**, 0.03 **Mg**, 0.03 **Ni**, 0.03 **Zn**, 0.03 **Sn**, 0.3–0.6 **Fe**	
JIS H 5302	Class 1	Japan	Die castings	rem **Al**, 11.0–13.0 **Si**, 1.0 **Cu**, 0.3 **Mn**, 0.3 **Mg**, 0.5 **Ni**, 0.5 **Zn**, 0.1 **Sn**, 1.3 **Fe**	
ANSI/ASTM B 179	383.1/SC102A	US	Ingot: for die castings	rem **Al**, 9.5–11.5 **Si**, 2.0–3.0 **Cu**, 0.50 **Mn**, 0.10 **Mg**, 0.30 **Ni**, 2.9 **Zn**, 0.15 **Sn**, 0.6–1.0 **Fe**, 0.50 total, others	
ANSI/ASTM B 179	383.2	US	Ingot: for die castings	rem **Al**, 9.5–11.5 **Si**, 2.0–3.0 **Cu**, 0.10 **Mn**, 0.10 **Mg**, 0.10 **Ni**, 0.10 **Zn**, 0.10 **Sn**, 0.6–1.0 **Fe**, 0.20 total, others	
QQ–A–591E	383.0	US	Castings	rem **Al**, 9.5–11.5 **Si**, 2.0–3.0 **Cu**, 0.50 **Mn**, 0.10 **Mg**, 0.30 **Ni**, 3.0 **Zn**, 0.15 **Sn**, 1.3 **Fe**, 0.50 total, others	**Die cast**: all **diam**, 45 ksi (310 MPa) **TS**, 22 ksi (152 MPa) **YS**, 3.5% in 2 in. (50 mm) **El**
ANSI/ASTM B 85	383.0/SC102A	US	Die castings	rem **Al**, 9.5–11.5 **Si**, 2.0–3.0 **Cu**, 0.50 **Mn**, 0.10 **Mg**, 0.30 **Ni**, 3.0 **Zn**, 0.15 **Sn**, 1.3 **Fe**, 0.50 total, others	**As cast**: all **diam**, 45 ksi (310 MPa) **TS**, 22 ksi (150 MPa) **YS**, 4% in 2 in. (50 mm) **El**
ANSI/ASTM B 179	331.1	US	Ingot: for permanent mold castings	rem **Al**, 8.0–10.0 **Si**, 3.0–4.0 **Cu**, 0.50 **Mn**, 0.10–0.50 **Mg**, 0.50 **Ni**, 1.0 **Zn**, 0.25 **Ti**, 0.8 **Fe**, 0.50 total, others	
QQ–A–371F	333.1 (formerly 333)	US	Ingot	rem **Al**, 8.0–10.0 **Si**, 3.0–4.0 **Cu**, 0.50 **Mn**, 0.10–0.50 **Mg**, 0.50 **Ni**, 1.0 **Zn**, 0.25 **Ti**, 0.8 **Fe**, 0.50 total, others	
ANSI/ASTM B 108	333.0	US	Permanent mold castings	rem **Al**, 8.0–10.0 **Si**, 3.0–4.0 **Cu**, 0.50 **Mn**, 0.05–0.50 **Mg**, 0.50 **Ni**, 1.0 **Zn**, 0.25 **Ti**, 1.0 **Fe**, 0.50 total, others	**As fabricated**: all **diam**, 28 ksi (193 MPa) **TS**, 90 **HB**; **Cooled and artificially aged**: all **diam**, 30 ksi (207 MPa) **TS**, 100 **HB**; **Solution heat treated and artificially aged**: all **diam**, 35 ksi (241 MPa) **TS**, 105 **HB**

CAST ALUMINUM AND ALUMINUM ALLOYS

specification number	designation	country	product forms	chemical composition	mechanical properties and hardness values
QQ A 596d	333	US	Castings	rem **Al**, 8.0–10.0 **Si**, 3.0–4.0 **Cu**, 0.50 **Mn**, 0.05–0.50 **Mg**, 0.50 **Ni**, 1.0 **Zn**, 0.25 **Ti**, 1.0 **Fe**, 0.50 total, others	**As cast**: all **diam**, 28 ksi (193 MPa) **TS**; **Aged and temper T5**: all **diam**, 30 ksi (207 MPa) **TS**; **Solution heat treated and artificially aged–T6**: all **diam**, 35 ksi (242 MPa) **TS**
UNI 5075	GD AlSi8.5Cu3.5Fe	Italy	Pressure die casting	rem **Al**, 8.0–9.5 **Si**, 3.0–4.0 **Cu**, 0.30 **Mn**, 0.30 **Mg**, 0.30 **Ni**, 0.90 **Zn**, 0.15 **Ti**, 0.10 **Sn**, 0.7–1.0 **Fe**, 0.15 **Pb**	**As cast**: all **diam**, 215 MPa **TS**, 145 MPa **YS**, 1% in 2 in. (50 mm) **El**, 85–105 **HB**
AS 1874	AS301	Australia	Ingot, sand castings and permanent mould castings	rem **Al**, 5.5–6.5 **Si**, 5.5–6.5 **Cu**, 0.6 **Mn**, 0.25–0.6 **Mg**, 0.5 **Ni**, 0.8 **Zn**, 0.20 **Ti**, 0.05 **Sn**, 0.8 **Fe**, 0.05 each, 0.20 total, others	
AS 1874	AP301	Australia	Ingot, sand castings and permanent mould castings	rem **Al**, 5.5–6.5 **Si**, 5.5–6.5 **Cu**, 0.05 **Mn**, 0.25–0.50 **Mg**, 0.05 **Cr**, 0.05 **Ni**, 0.10 **Zn**, 0.05 **Ti**, 0.05 **Sn**, 0.40 **Fe**, 0.05 each, 0.20 total, others	**Cooled and artificially aged**: all **diam**, 220 MPa **TS**
AS 1874	AS313	Australia	Ingot, sand castings and permanent mould castings	rem **Al**, 7.5–9.5 **Si**, 3.0–4.0 **Cu**, 0.5 **Mn**, 0.30 **Mg**, 0.10 **Cr**, 0.5 **Ni**, 3.0 **Zn**, 0.20 **Ti**, 0.25 **Sn**, 1.3 **Fe**, 0.35 **Pb**, 0.05 each, 0.20 total, others	
NF A 57 703	A–S9 U3–Y4	France	Pressure die castings	rem **Al**, 7.5–10.0 **Si**, 2.5–4.0 **Cu**, 0.5 **Mn**, 0.3 **Mg**, 0.5 **Ni**, 1.2 **Zn**, 0.2 **Ti**, 0.2 **Sn**, 1.3 **Fe**, 0.2 **Pb**	**As cast**: 200 MPa **TS**, 1% in 2 in. (50 mm) **El**
AS 1874	AS307	Australia	Ingot, sand castings and permanent mould castings	rem **Al**, 9.0–11.5 **Si**, 0.7–2.5 **Cu**, 0.5 **Mn**, 0.30 **Mg**, 0.10 **Cr**, 0.5 **Ni**, 2.0 **Zn**, 0.20 **Ti**, 0.25 **Sn**, 1.0 **Fe**, 0.35 **Pb**, 0.05 each, 0.20 total, others	
SABS 992	Al–Si10Cu2FeA	South Africa	Pressure die castings	rem **Al**, 9.0–11.5 **Si**, 0.7–2.5 **Cu**, 0.5 **Mn**, 0.30 **Mg**, 1.0 **Ni**, 1.2 **Zn**, 0.2 **Ti**, 0.2 **Sn**, 1.0 **Fe**, 0.3 **Pb**, 0.15 total, others	**As cast**: all **diam**, 215 MPa **TS**, 70 MPa **YS**, 2% in 2 in. (50 mm) **El**, 65–90 **HB**
SABS 989	Al–Si10Cu2FeA	South Africa	Ingot: for pressure die castings	rem **Al**, 9.0–11.5 **Si**, 0.7–2.5 **Cu**, 0.5 **Mn**, 0.30 **Mg**, 1.0 **Ni**, 1.2 **Zn**, 0.2 **Ti**, 0.2 **Sn**, 0.5–0.9 **Fe**, 0.3 **Pb**, 0.15 total, others	**Chill cast**: all **diam**, 145 MPa **TS**
AMS 4291C		US	Die castings	rem **Al**, 7.5–9.5 **Si**, 3.0–4.0 **Cu**, 0.50 **Mn**, 0.10 **Mg**, 0.50 **Ni**, 3.0 **Zn**, 0.35 **Sn**, 1.3 **Fe**, 0.50 total, others	
ANSI/ASTM B 179	380.2/SC84C	US	Ingot: for die castings	rem **Al**, 7.5–9.5 **Si**, 3.0–4.0 **Cu**, 0.10 **Mn**, 0.10 **Mg**, 0.10 **Ni**, 0.10 **Zn**, 0.10 **Sn**, 0.7–1.1 **Fe**, 0.20 total, others	
ANSI/ASTM B 179	A380.1/SC84A–B	US	Ingot: for die castings	rem **Al**, 7.5–9.5 **Si**, 3.0–4.0 **Cu**, 0.50 **Mn**, 0.10 **Mg**, 0.50 **Ni**, 2.9 **Zn**, 0.35 **Sn**, 1.0 **Fe**, 0.50 total, others	

CAST ALUMINUM AND ALUMINUM ALLOYS

specification number	designation	country	product forms	chemical composition	mechanical properties and hardness values
ANSI/ASTM B 179	A380.2	US	Ingot: for die castings	rem **Al**, 7.5–9.5 **Si**, 3.0–4.0 **Cu**, 0.10 **Mn**, 0.10 **Mg**, 0.10 **Ni**, 0.10 **Zn**, 0.6 **Fe**, 0.05 each, 0.15 total, others	
AS 1874	BP313	Australia	Ingot, sand castings and permanent mould castings	rem **Al**, 7.5–9.5 **Si**, 3.0–4.0 **Cu**, 0.10 **Mn**, 0.10 **Mg**, 0.10 **Ni**, 0.10 **Zn**, 0.6 **Fe**, 0.05 each, 0.15 total, others	
QQ–A-371F	380.2 (formerly 380)	US	Ingot	rem **Al**, 7.5–9.5 **Si**, 3.0–4.0 **Cu**, 0.10 **Mn**, 0.10 **Mg**, 0.10 **Ni**, 0.10 **Zn**, 0.10 **Sn**, 0.7–1.1 **Fe**, 0.20 total, others	
QQA–371F	A380.1 (formerly A380)	US	Ingot	rem **Al**, 7.5–9.5 **Si**, 3.0–4.0 **Cu**, 0.50 **Mn**, 0.10 **Mg**, 0.50 **Ni**, 2.9 **Zn**, 0.35 **Sn**, 1.0 **Fe**, 0.50 total, others	
QQ–A-371F	A380.2 (formerly A380)	US	Ingot	rem **Al**, 7.5–9.5 **Si**, 3.0–4.0 **Cu**, 0.10 **Mn**, 0.10 **Mg**, 0.10 **Ni**, 0.10 **Zn**, 0.6 **Fe**, 0.05 each, 0.15 total, others	
QQ–A-591E	A380.0	US	Castings	rem **Al**, 7.5–9.5 **Si**, 3.0–4.0 **Cu**, 0.50 **Mn**, 0.10 **Mg**, 0.50 **Ni**, 3.0 **Zn**, 0.35 **Sn**, 1.3 **Fe**, 0.50 total, others	**Die cast**: all **diam**, 47 ksi (324 MPa) **TS**, 23 ksi (159 MPa) **YS**, 3.5% in 2 in. (50 mm) **El**
QQ–A-591E	380.0	US	Castings	rem **Al**, 7.5–9.5 **Si**, 3.0–4.0 **Cu**, 0.50 **Mn**, 0.10 **Mg**, 0.50 **Ni**, 3.0 **Zn**, 0.35 **Sn**, 2.0 **Fe**, 0.50 total, others	**Die cast**: all **diam**, 46 ksi (317 MPa) **TS**, 23 ksi (159 MPa) **YS**, 2.5% in 2 in. (50 mm) **El**
SABS 992	Al–Si8Cu4FeA	South Africa	Pressure die castings	rem **Al**, 7.5–9.5 **Si**, 3.0–4.0 **Cu**, 0.5 **Mn**, 0.1 **Mg**, 0.5 **Ni**, 3.0 **Zn**, 0.2 **Ti**, 0.2 **Sn**, 1.3 **Fe**, 0.3 **Pb**, 0.15 total, others	**As cast**: all **diam**, 230 MPa **TS**, 110 MPa **YS**, 1% in 2 in. (50 mm) **El**, 85 **HB**
SABS 989	Al–Si8Cu4FeA	South Africa	Ingot: for pressure die castings	rem **Al**, 7.5–9.5 **Si**, 3.0–4.0 **Cu**, 0.5 **Mn**, 0.1 **Mg**, 0.5 **Ni**, 3.0 **Zn**, 0.2 **Ti**, 0.2 **Sn**, 0.5–1.2 **Fe**, 0.3 **Pb**, 0.15 total, others	**Chill cast**: all **diam**, 180 MPa **TS**, 1.5% in 2 in. (50 mm) **El**
ANSI/ASTM B 85	380.0/SC84.B	US	Die castings	rem **Al**, 7.5–9.5 **Si**, 3.0–4.0 **Cu**, 0.50 **Mn**, 0.10 **Mg**, 0.50 **Ni**, 3.0 **Zn**, 0.35 **Sn**, 2.0 **Fe**, 0.50 other, total	**As cast**: all **diam**, 46 ksi (320 MPa) **TS**, 23 ksi (160 MPa) **YS**, 3% in 2 in. (50 mm) **El**
ANSI/ASTM B 85	A380.0/SC84A	US	Die castings	rem **Al**, 7.5–9.5 **Si**, 3.0–4.0 **Cu**, 0.50 **Mn**, 0.10 **Mg**, 0.50 **Ni**, 3.0 **Zn**, 0.35 **Sn**, 1.3 **Fe**, 0.50 total, others	**As cast**: all **diam**, 47 ksi (320 MPa) **TS**, 23 ksi (160 MPa) **YS**, 4% in 2 in. (50 mm) **El**
DS 3002	4254	Denmark	Ingot: for castings	rem **Al**, 7.5–10.0 **Si**, 2.0–4.0 **Cu**, 0.5 **Mn**, 0.3 **Mg**, 0.3 **Ni**, 3.0 **Zn**, 0.2 **Ti**, 0.2 **Sn**, 1.10 **Fe**, 0.3 **Pb**	**Pressure die cast**: all **diam**, 245 MPa **TS**, 176 MPa **YS**, 1% in 2 in. (50 mm) **El**, 70–100 **HB**
SFS 2568	G–AlSi9Cu3Fe	Finland	Pressure castings	rem **Al**, 7.5–10.0 **Si**, 2.0–4.0 **Cu**, 0.5 **Mn**, 0.3 **Mg**, 0.3 **Ni**, 1.2 **Zn**, 0.2 **Ti**, 0.2 **Sn**, 1.30 **Fe**, 0.3 **Pb**, 0.05 each, 0.15 total, others	**As cast**: ≤20 mm **diam**, 250 MPa **TS**, 180 MPa **YS**, 1% in 2 in. (50 mm) **El**
NS 17 530	NS 17 530–00	Norway	Ingot: for casting	88 **Al**, 7.5–10.0 **Si**, 2.0–4.0 **Cu**, 0.5 **Mn**, 0.3 **Mg**, 0.3 **Ni**, 1.2 **Zn**, 0.20 **Ti**, 0.2 **Sn**, 1.1 **Fe**, 0.3 **Pb**	

CAST ALUMINUM AND ALUMINUM ALLOYS

specification number	designation	country	product forms	chemical composition	mechanical properties and hardness values
NS 17 530	NS 17 530–05	Norway	Castings: pressure die, chill, and sand	88 Al, 7.5–10.0 Si, 2.0–4.0 Cu, 0.5 Mn, 0.3 Mg, 0.3 Ni, 1.2 Zn, 0.20 Ti, 0.2 Sn, 1.3 Fe, 0.3 Pb	**As cast (pressure die castings)**: all **diam**, 245 MPa **TS**, 176 MPa **YS**, 1% in 2 in. (50 mm) **El**
NS 17 532	NS 17 532–00	Norway	Ingot: casting	86 Al, 7.5–10.0 Si, 2.0–4.0 Cu, 0.5 Mn, 0.3 Mg, 0.3 Ni, 3.0 Zn, 0.20 Ti, 0.2 Sn, 1.1 Fe, 0.3 Pb	
NS 17 532	NS 17 532–05	Norway	Pressure die castings	86 Al, 7.5–10.0 Si, 2.0–4.0 Cu, 0.5 Mn, 0.3 Mg, 0.3 Ni, 3.0 Zn, 0.20 Ti, 0.2 Sn, 1.3 Fe, 0.3 Pb	**As cast**: all **diam**, 245 MPa **TS**, 176 MPa **YS**, 1% in 2 in. (50 mm) **El**
SIS 14 42 54	SIS Al 42 54–00	Sweden	Ingot	rem Al, 7.5–10.0 Si, 2.0–4.0 Cu, 0.5 Mn, 0.3 Mg, 0.3 Ni, 1.2–3.0 Zn, 0.20 Ti, 0.2 Sn, 1.1 Fe, 0.3 Pb	
SIS 14 42 54	SIS Al 42 54–10	Sweden	Pressure die castings	rem Al, 7.5–10.0 Si, 2.0–4.0 Cu, 0.5 Mn, 0.3 Mg, 0.3 Ni, 1.2–3.0 Zn, 0.20 Ti, 0.2 Sn, 1.3 Fe, 0.3 Pb	**As cast**: all **diam**, 250 MPa **TS**, 180 MPa **YS**, 1% in 2 in. (50 mm) **El**, 70–110 **HB**
SIS 14 42 52	SIS Al 42 52–00	Sweden	Ingot	rem Al, 7.5–10.0 Si, 2.0–4.0 Cu, 0.5 Mn, 0.3 Mg, 0.3 Ni, 1.2 Zn, 0.20 Ti, 0.2 Sn, 1.1 Fe, 0.3 Pb	
SIS 14 42 52	SIS Al 42 52–10	Sweden	Pressure die castings	rem Al, 7.5–10.0 Si, 2.0–4.0 Cu, 0.5 Mn, 0.3 Mg, 0.3 Ni, 1.2 Zn, 0.20 Ti, 0.2 Sn, 1.3 Fe, 0.3 Pb	**As cast**: all **diam**, 250 MPa **TS**, 180 MPa **YS**, 1% in 2 in. (50 mm) **El**, 70–110 **HB**
SABS 989	Al–Si12MgMnA	South Africa	Ingot: for sand and gravity die castings	rem Al, 10.0–13.0 Si, 0.08 Cu, 0.3–0.7 Mn, 0.25–0.6 Mg, 0.1 Ni, 0.1 Zn, 0.2 Ti, 0.05 Sn, 0.5 Fe, 0.1 Pb, 0.05 each, 0.15 total, others	**Sand cast, solution and precipitation heat treated**: all **diam**, 240 MPa **TS**; **Chill cast, solution and precipitation heat treated**: all **diam**, 295 MPa **TS**
IS 202	A–9	India	Sand castings	rem Al, 10.0–13.0 Si, 0.10 Cu, 0.30–0.70 Mn, 0.20–0.60 Mg, 0.10 Ni, 0.10 Zn, 0.05 Sn, 0.60 Fe, 0.10 Pb, 0.20 Ti + Nb	**Solution and precipitation treated**: all **diam**, 239 MPa **TS**, 201 MPa **YS**
JIS H 5202	Class 4 B	Japan	Castings	rem Al, 7.0–10.0 Si, 2.0–4.0 Cu, 0.5 Mn, 0.5 Mg, 0.3 Ni, 1.0 Zn, 0.2 Ti, 1.0 Fe	**As cast**: all **diam**, 177 MPa **TS**; **Quenched and tempered**: all **diam**, 245 MPa **TS**, 90 **HB**
JIS H 2117	Class 4B	Japan	Ingot: for castings	rem Al, 7.0–10.0 Si, 2.0–4.0 Cu, 0.5 Mn, 0.5 Mg, 0.3 Ni, 1.0 Zn, 0.2 Ti, 0.8 Fe	
SABS 991	Al–Si12MgMnA	South Africa	Gravity die castings	rem Al, 10.0–13.0 Si, 0.1 Cu, 0.3–0.7 Mn, 0.2–0.6 Mg, 0.1 Ni, 0.1 Zn, 0.2 Ti, 0.05 Sn, 0.6 Fe, 0.1 Pb, 0.05 each, 0.15 total, others	**Precipitation heat treated**: all **diam**, 230 MPa **TS**, 2% in 2 in. (50 mm) **El**; **Solution and precipitation heat treated**: all **diam**, 295 MPa **TS**
SABS 990	Al–Si12MgMnA	South Africa	Sand castings	rem Al, 10.0–13.0 Si, 0.1 Cu, 0.3–0.7 Mn, 0.2–0.6 Mg, 0.1 Ni, 0.1 Zn, 0.2 Ti, 0.05 Sn, 0.6 Fe, 0.1 Pb, 0.05 each, 0.15 total, others	**Precipitation heat treated**: all **diam**, 170 MPa **TS**, 1.5% in 2 in. (50 mm) **El**; **Solution and precipitation heat treated**: all **diam**, 240 MPa **TS**
ISO R164	Al–Si8Cu3Fe	ISO	Pressure die castings	rem Al, 7.0–9.5 Si, 2.5–4.5 Cu, 0.6 Mn, 0.15 Mg, 0.3 Ni, 1.2 Zn, 0.2 Ti, 0.2 Sn, 1.3 Fe, 0.3 Pb	

CAST ALUMINUM AND ALUMINUM ALLOYS

specification number	designation	country	product forms	chemical composition	mechanical properties and hardness values
JIS H 5302	Class 10	Japan	Die castings	rem **Al**, 7.5–9.5 **Si**, 2.0–4.0 **Cu**, 0.5 **Mn**, 0.3 **Mg**, 0.5 **Ni**, 1.0 **Zn**, 0.3 **Sn**, 1.3 **Fe**	
JIS H 2118	Class 10	Japan	Ingot: for die castings	rem **Al**, 7.5–9.5 **Si**, 2.0–4.0 **Cu**, 0.5 **Mn**, 0.3 **Mg**, 0.5 **Ni**, 1.0 **Zn**, 0.3 **Sn**, 0.9 **Fe**	
DIN 1725 sheet 2	GD–AlSi8Cu3/3.2162.05	Germany	Pressure die casting	rem **Al**, 7.5–9.5 **Si**, 2.0–3.5 **Cu**, 0.2–0.5 **Mn**, 0.30 **Mg**, 0.30 **Ni**, 1.20 **Zn**, 0.15 **Ti**, 0.10 **Sn**, 0.80 **Fe**, 0.20 **Pb**, 0.05 each, 0.15 total, others	**As cast**: 240 MPa **TS**, 160 MPa **YS**, 0.5% in 4 in. (100 mm) **El**, 80–110 **HB**
JIS H 2212	Class 10	Japan	Ingot: for die castings	rem **Al**, 7.5–9.5 **Si**, 2.0–4.0 **Cu**, 0.03 **Mn**, 0.03 **Mg**, 0.03 **Ni**, 0.03 **Zn**, 0.03 **Sn**, 0.3–0.6 **Fe**	
JIS H 2211	Class 4B	Japan	Ingot: for castings	rem **Al**, 7.0–10.0 **Si**, 2.0–4.0 **Cu**, 0.03 **Mn**, 0.03 **Mg**, 0.03 **Ni**, 0.03 **Zn**, 0.03 **Ti**, 0.3 **Fe**	
DIN 1725 sheet 2	GK–AlSi8Cu3/3.2161.02	Germany	Gravity die castings	rem **Al**, 7.5–9.5 **Si**, 2.0–3.5 **Cu**, 0.20–0.50 **Mn**, 0.10–0.30 **Mg**, 0.30 **Ni**, 1.20 **Zn**, 0.15 **Ti**, 0.10 **Sn**, 0.80 **Fe**, 0.20 **Pb**, 0.05 each, 0.15 total, others	**As cast**: all **diam**, 170 MPa **TS**, 110 MPa **YS**, 1% in 2 in. (50 mm) **El**, 70–100 **HB**
DIN 1725 sheet 2	G–AlSi8Cu3/3.2161.0 1	Germany	Sand castings	rem **Al**, 7.5–9.5 **Si**, 2.0–3.5 **Cu**, 0.20–0.50 **Mn**, 0.10–0.30 **Mg**, 0.30 **Ni**, 1.20 **Zn**, 0.15 **Ti**, 0.10 **Sn**, 0.80 **Fe**, 0.20 **Pb**, 0.05 each, 0.15 total, others	**As cast**: all **diam**, 160 MPa **TS**, 100 MPa **YS**, 1% in 2 in. (50 mm) **El**, 65–90 **HB**
ANSI/ASTM B 179	354.1	US	Ingot: for permanent mold castings	rem **Al**, 8.6–9.4 **Si**, 1.6–2.0 **Cu**, 0.10 **Mn**, 0.45–0.6 **Mg**, 0.10 **Zn**, 0.20 **Ti**, 0.15 **Fe**, 0.05 each, 0.15 total, others	
QQ–A–371F	354.1 (formerly 354)	US	Ingot	rem **Al**, 8.6–9.4 **Si**, 1.6–2.0 **Cu**, 0.10 **Mn**, 0.45–0.6 **Mg**, 0.10 **Zn**, 0.20 **Ti**, 0.15 **Fe**, 0.05 each, 0.15 total, others	
ANSI/ASTM B 108	354.0	US	Permanent mold castings	rem **Al**, 8.6–9.4 **Si**, 1.6–2.0 **Cu**, 0.10 **Mn**, 0.40–0.6 **Mg**, 0.10 **Zn**, 0.20 **Ti**, 0.20 **Fe**, 0.05 each, 0.15 total, others	**T61**: all **diam**, 43 ksi (297 MPa) **TS**, 33 ksi (228 MPa) **YS**, 2% in 2 in. (50 mm) **El**; **T62**: all **diam**, 43 ksi (297 MPa) **TS**, 33 ksi (228 MPa) **YS**, 2% in 2 in. (50 mm) **El**
MIL–A–21180C	354.0	US	Castings	rem **Al**, 8.6–9.4 **Si**, 1.6–2.0 **Cu**, 0.10 **Mn**, 0.40–0.6 **Mg**, 0.10 **Zn**, 0.20 **Ti**, 0.20 **Fe**, 0.05 each, 0.15 total, others	**Class 1**: all **diam**, 324 MPa **TS**, 248 MPa **YS**, 3% in 2 in. (50 mm) **El**; **Class 2**: all **diam**, 345 MPa **TS**, 290 MPa **YS**, 2% in 2 in. (50 mm) **El**
SABS 991	Al–Si9A	South Africa	Gravity die castings	rem **Al**, 7.0–11.0 **Si**, 1.6 **Cu**, 0.5 **Mn**, 0.3 **Mg**, 0.4 **Ni**, 1.50 **Zn**, 0.15 **Ti**, 0.2 **Sn**, 1.0 **Fe**, 0.2 **Pb**, 0.05 each, others	**As cast**: all **diam**, 165 MPa **TS**, 1% in 2 in. (50 mm) **El**

CAST ALUMINUM AND ALUMINUM ALLOYS

specification number	designation	country	product forms	chemical composition	mechanical properties and hardness values
SABS 990	Al–Si9A	South Africa	Sand castings	rem **Al**, 7.0–11.0 **Si**, 1.6 **Cu**, 0.5 **Mn**, 0.3 **Mg**, 0.4 **Ni**, 1.50 **Zn**, 0.15 **Ti**, 0.2 **Sn**, 1.0 **Fe**, 0.2 **Pb**, 0.05 each, 0.15 total, others	**As cast**: all **diam**, 150 MPa **TS**, 1% in 2 in. (50 mm) **El**
SABS 989	Al–Si9A	South Africa	Ingot: for sand and gravity die casting	rem **Al**, 7.0–11.0 **Si**, 1.6 **Cu**, 0.5 **Mn**, 0.3 **Mg**, 0.4 **Ni**, 1.20 **Zn**, 0.15 **Ti**, 0.2 **Sn**, 0.8 **Fe**, 0.2 **Pb**, 0.05 each, 0.15 total, others	**Sand cast**: all **diam**, 150 MPa **TS**, 1% in 2 in. (50 mm) **El**; **Chill cast**: all **diam**, 165 MPa **TS**, 1% in 2 in. (50 mm) **El**
DIN 1725 sheet 2	GD–AlSi10Mg (Cu)/3.2983.05	Germany	Pressure die castings	rem **Al**, 9.0–11.0 **Si**, 0.30 **Cu**, 0.20–0.50 **Mn**, 0.20–0.50 **Mg**, 0.10 **Ni**, 0.30 **Zn**, 0.15 **Ti**, 0.60 **Fe**, 0.05 each, 0.15 total, others	**As cast**: all **diam**, 220 MPa **TS**, 140 MPa **YS**, 1% in 4 in. (100 mm) **El**, 70–90 **HB**
DIN 1725 sheet 2	GK–AlSi10Mg (Cu)/3.2383.02	Germany	Gravity die castings	rem **Al**, 9.0–11.0 **Si**, 0.30 **Cu**, 0.20–0.50 **Mn**, 0.20–0.50 **Mg**, 0.10 **Ni**, 0.30 **Zn**, 0.15 **Ti**, 0.60 **Fe**, 0.05 each, 0.15 total, others	**As cast**: all **diam**, 200 MPa **TS**, 100 MPa **YS**, 1% in 2 in. (50 mm) **El**, 65–85 **HB**
DIN 1725 sheet 2	G–AlSi10Mg (Cu) wa/3.2383.61	Germany	Sand castings	rem **Al**, 9.0–11.0 **Si**, 0.30 **Cu**, 0.20–0.50 **Mn**, 0.20–0.50 **Mg**, 0.10 **Ni**, 0.30 **Zn**, 0.15 **Ti**, 0.60 **Fe**, 0.05 each, 0.15 total, others	**Quenched and age hardened**: all **diam**, 220 MPa **TS**, 180 MPa **YS**, 1% in 2 in. (50 mm) **El**, 80–110 **HB**
DIN 1725 sheet 2	G–AlSi10Mg (Cu)/3.2383.01	Germany	Sand castings	rem **Al**, 9.0–11.0 **Si**, 0.30 **Cu**, 0.20–0.50 **Mn**, 0.20–0.50 **Mg**, 0.10 **Ni**, 0.30 **Zn**, 0.15 **Ti**, 0.6 **Fe**, 0.05 each, 0.15 total, others	**As cast**: all **diam**, 180 MPa **TS**, 90 MPa **YS**, 1% in 2 in. (50 mm) **El**, 55–65 **HB**
DIN 1725 sheet 2	GK–AlSi10Mg wa/3.2381.62	Germany	Gravity die castings	rem **Al**, 9.0–11.0 **Si**, 0.30 **Cu**, 0.40 **Mn**, 0.20–0.50 **Mg**, 0.10 **Zn**, 0.15 **Ti**, 0.50 **Fe**, 0.05 each, 0.15 total, others	**Quench and age hardened**: all **diam**, 240 MPa **TS**, 210 MPa **YS**, 1% in 2 in. (50 mm) **El**, 85–115 **HB**
ANSI/ASTM B 179	A360.1/SG100A–B	US	Ingot: for die castings	rem **Al**, 9.0–10.0 **Si**, 0.6 **Cu**, 0.35 **Mn**, 0.45–0.6 **Mg**, 0.50 **Ni**, 0.40 **Zn**, 0.15 **Sn**, 1.0 **Fe**, 0.25 total, others	
QQ–A–371F	A360.1 (formerly A360)	US	Ingot	rem **Al**, 9.0–10.0 **Si**, 0.6 **Cu**, 0.35 **Mn**, 0.45–0.6 **Mg**, 0.50 **Ni**, 0.40 **Zn**, 1.0 **Fe**, 0.25 total, others	
QQ–A–591E	360.0	US	Castings	rem **Al**, 9.0–10.0 **Si**, 0.6 **Cu**, 0.35 **Mn**, 0.40–0.6 **Mg**, 0.50 **Ni**, 0.50 **Zn**, 0.15 **Sn**, 2.0 **Fe**, 0.25 total, others	**Die cast**: all **diam**, 44 ksi (303 MPa) **TS**, 25 ksi (172 MPa) **YS**, 2.5% in 2 in. (50 mm) **El**
QQ–A–591E	A360.0	US	Castings	rem **Al**, 9.0–10.0 **Si**, 0.6 **Cu**, 0.35 **Mn**, 0.40–0.6 **Mg**, 0.50 **Ni**, 0.50 **Zn**, 0.15 **Sn**, 1.3 **Fe**, 0.25 total, others	**Die cast**: all **diam**, 46 ksi (317 MPa) **TS**, 24 ksi (165 MPa) **YS**, 3.5% in 2 in. (50 mm) **El**
JIS H 5302	Class 3	Japan	Die castings	rem **Al**, 9.0–10.0 **Si**, 0.6 **Cu**, 0.3 **Mn**, 0.4–0.6 **Mg**, 0.5 **Ni**, 0.5 **Zn**, 0.1 **Sn**, 1.3 **Fe**	

CAST ALUMINUM AND ALUMINUM ALLOYS

specification number	designation	country	product forms	chemical composition	mechanical properties and hardness values
JIS H 2118	Class 3	Japan	Ingot: for die castings	rem **Al**, 9.0–10.0 **Si**, 0.6 **Cu**, 0.3 **Mn**, 0.40–0.6 **Mg**, 0.5 **Ni**, 0.5 **Zn**, 0.1 **Sn**, 0.9 **Fe**	
ANSI/ASTM B 85	360.0/SG100B	US	Die castings	rem **Al**, 9.0–10.0 **Si**, 0.6 **Cu**, 0.35 **Mn**, 0.40–0.6 **Mg**, 0.50 **Ni**, 0.50 **Zn**, 0.15 **Sn**, 2.0 **Fe**, 0.25 total, others	**As cast**: all **diam**, 44 ksi (300 MPa) **TS**, 25 ksi (170 MPa) **YS**, 3% in 2 in. (50 mm) **El**
ANSI/ASTM B 85	A 360.0/SG100A	US	Die castings	rem **Al**, 9.0–10.0 **Si**, 0.6 **Cu**, 0.35 **Mn**, 0.40–0.6 **Mg**, 0.50 **Ni**, 0.50 **Zn**, 0.15 **Sn**, 1.3 **Fe**, 0.25 total, others	**As cast**: all **diam**, 46 ksi (320 MPa) **TS**, 24 ksi (170 MPa) **YS**, 4% in 2 in. (50 mm) **El**
DS 3002	4253	Denmark	Ingot: for castings	rem **Al**, 9.0–11.0 **Si**, 0.20 **Cu**, 0.5 **Mn**, 0.20–0.40 **Mg**, 0.1 **Ni**, 0.3 **Zn**, 0.20 **Ti**, 0.05 **Sn**, 0.50 **Fe**, 0.05 **Pb**	**Sand cast**: all **diam**, 235 MPa **TS**, 196 MPa **YS**, 1% in 2 in. (50 mm) **El**, 75–105 **HB**; **Chill cast**: all **diam**, 255 MPa **TS**, 216 MPa **YS**, 1% in 2 in. (50 mm) **El**, 80–110 **HB**
UNI 5074	GD–AlSi9MgFe	Italy	Pressure die casting	rem **Al**, 9.0–10.0 **Si**, 0.50 **Cu**, 0.35 **Mn**, 0.40–0.60 **Mg**, 0.50 **Ni**, 0.40 **Zn**, 0.15 **Ti**, 0.15 **Sn**, 0.7–1.2 **Fe**, 0.15 **Pb**	**As cast**: all **diam**, 195 MPa **TS**, 145 MPa **YS**, 2% in 2 in. (50 mm) **El**, 70–90 **HB**; **Quenched and tempered**: all **diam**, 235 MPa **TS**, 175 MPa **YS**, 1.5% in 2 in. (50 mm) **El**, 80–95 **HB**
NS 17 520	NS 17 520–00	Norway	Ingot: for casting	90 **Al**, 9.0–11.0 **Si**, 0.20 **Cu**, 0.5 **Mn**, 0.20–0.40 **Mg**, 0.1 **Ni**, 0.3 **Zn**, 0.20 **Ti**, 0.05 **Sn**, 0.50 **Fe**, 0.05 **Pb**	
NS 17 520	NS 17 520–41	Norway	Sand castings	90 **Al**, 9.0–11.0 **Si**, 0.20 **Cu**, 0.5 **Mn**, 0.20–0.40 **Mg**, 0.1 **Ni**, 0.3 **Zn**, 0.20 **Ti**, 0.05 **Sn**, 0.50 **Fe**, 0.05 **Pb**	**Hardened and artificially aged**: all **diam**, 235 MPa **TS**, 196 MPa **YS**, 1% in 2 in. (50 mm) **El**
NS 17 520	NS 17 520–42	Norway	Chill castings	90 **Al**, 9.0–11.0 **Si**, 0.20 **Cu**, 0.5 **Mn**, 0.20–0.40 **Mg**, 0.1 **Ni**, 0.3 **Zn**, 0.20 **Ti**, 0.05 **Sn**, 0.50 **Fe**, 0.05 **Pb**	**Hardened and artificially aged**: all **diam**, 255 MPa **TS**, 216 MPa **YS**, 1% in 2 in. (50 mm) **El**
NS 17 520	NS 17 520–XX	Norway	Pressure die castings	90 **Al**, 9.0–11.0 **Si**, 0.20 **Cu**, 0.5 **Mn**, 0.20–0.40 **Mg**, 0.1 **Ni**, 0.3 **Zn**, 0.20 **Ti**, 0.05 **Sn**, 0.50 **Fe**, 0.05 **Pb**	
SIS 14 42 53	SIS Al 42 53–07	Sweden	Chill castings	rem **Al**, 9.0–11.0 **Si**, 0.20 **Cu**, 0.5 **Mn**, 0.20–0.40 **Mg**, 0.3 **Zn**, 0.20 **Ti**, 0.05 **Sn**, 0.50 **Fe**, 0.05 **Pb**	**Solution treated, artificially aged**: all **diam**, 250 MPa **TS**, 220 MPa **YS**, 1% in 2 in. (50 mm) **El**, 80–110 **HB**
SIS 14 42 53	SIS Al 42 53–00	Sweden	Ingot	rem **Al**, 9.0–11.0 **Si**, 0.20 **Cu**, 0.5 **Mn**, 0.20–0.40 **Mg**, 0.1 **Ni**, 0.3 **Zn**, 0.20 **Ti**, 0.05 **Sn**, 0.50 **Fe**, 0.05 **Pb**	
SIS 14 42 53	SIS Al 42 53–04	Sweden	Sand castings	rem **Al**, 9.0–11.0 **Si**, 0.20 **Cu**, 0.5 **Mn**, 0.20–0.40 **Mg**, 0.1 **Ni**, 0.3 **Zn**, 0.20 **Ti**, 0.05 **Sn**, 0.50 **Fe**, 0.05 **Pb**	**Solution treated, artificially aged**: all **diam**, 240 MPa **TS**, 200 MPa **YS**, 1% in 2 in. (50 mm) **El**, 75–105 **HB**

CAST ALUMINUM AND ALUMINUM ALLOYS

specification number	designation	country	product forms	chemical composition	mechanical properties and hardness values
DIN 1725 sheet 2	GD–AlSi10Mg/3.2382.05	Germany	Pressure die castings	rem **Al**, 9.0–11.0 **Si**, 0.10 **Cu**, 0.40 **Mn**, 0.20–0.50 **Mg**, 0.10 **Zn**, 0.15 **Ti**, 1.0 **Fe**	**As cast**: all **diam**, 220 MPa **TS**, 140 MPa **YS**, 1% in 4 in. (100 mm) **El**, 70–90 **HB**
DIN 1725 sheet 2	GD–AlSi6Cu4/3.2152.05	Germany	Pressure die casting	rem **Al**, 5.0–7.5 **Si**, 3.0–5.0 **Cu**, 0.3–0.6 **Mn**, 0.10–0.30 **Mg**, 0.30 **Ni**, 2.0 **Zn**, 0.15 **Ti**, 0.10 **Sn**, 1.30 **Fe**, 0.30 **Pb**, 0.05 each, 0.15 total, others	**As cast**: all **diam**, 220 MPa **TS**, 150 MPa **YS**, 0.5% in 4 in. (100 mm) **El**, 70–100 **HB**
DIN 1725 sheet 2	GK–AlSi6Cu4/3.2151.02	Germany	Gravity die casting	rem **Al**, 5.0–7.5 **Si**, 3.0–5.0 **Cu**, 0.3–0.6 **Mn**, 0.1–0.3 **Mg**, 0.30 **Ni**, 2.00 **Zn**, 0.15 **Ti**, 0.10 **Sn**, 1.00 **Fe**, 0.30 **Pb**, 0.05 each, 0.15 total, others	**As cast**: all **diam**, 180 MPa **TS**, 120 MPa **YS**, 1% in 2 in. (50 mm) **El**, 70–100 **HB**
DIN 1725 sheet 2	G–AlSi6Cu4/3.2151.0 1	Germany	Sand castings	rem **Al**, 5.0–7.5 **Si**, 3.0–5.0 **Cu**, 0.3–0.6 **Mn**, 0.1–0.3 **Mg**, 0.30 **Ni**, 2.00 **Zn**, 0.15 **Ti**, 0.10 **Sn**, 1.00 **Fe**, 0.30 **Pb**, 0.05 each, 0.15 total, others	**As cast**: all **diam**, 160 MPa **TS**, 100 MPa **YS**, 1% in 2 in. (50 mm) **El**, 60–80 **HB**
DIN 1725 sheet 2	GK–AlSi10Mg (Cu) Wa/3.2383.62	Germany	Gravity die castings	rem **Al**, 9.0–11.0 **Si**, 0.05 **Cu**, 0.20–0.50 **Mn**, 0.20–0.50 **Mg**, 0.10 **Ni**, 0.30 **Zn**, 0.15 **Ti**, 0.60 **Fe**, 0.05 each, 0.15 total, others	**Quenched and age hardened**: all **diam**, 240 MPa **TS**, 210 MPa **YS**, 1% in 2 in. (50 mm) **El**, 85–115 **HB**
DIN 1725 sheet 2	GK–AlSi10Mg/3.2381.02	Germany	Gravity die castings	rem **Al**, 9.0–11.0 **Si**, 0.05 **Cu**, 0.40 **Mn**, 0.20–0.50 **Mg**, 0.10 **Zn**, 0.15 **Ti**, 0.50 **Fe**, 0.05 each, 0.15 total, others	**As cast**: all **diam**, 180 MPa **TS**, 90 MPa **YS**, 2% in 2 in. (50 mm) **El**, 60–80 **HB**
DIN 1725 sheet 2	G–AlSi10Mg wa/3.2381.61	Germany	Sand castings	rem **Al**, 9.0–11.0 **Si**, 0.05 **Cu**, 0.40 **Mn**, 0.20–0.50 **Mg**, 0.10 **Zn**, 0.15 **Ti**, 0.50 **Fe**, 0.05 each, 0.15 total, others	**Quenched and age hardened**: all **diam**, 220 MPa **TS**, 180 MPa **YS**, 1% in 2 in. (50 mm) **El**, 80–100 **HB**
DIN 1725 sheet 2	G–AlSi10Mg/3.2381. 01	Germany	Sand castings	rem **Al**, 9.0–11.0 **Si**, 0.05 **Cu**, 0.40 **Mn**, 0.20–0.50 **Mg**, 0.10 **Zn**, 0.15 **Ti**, 0.5 **Fe**, 0.05 each, 0.15 total, other	**As cast**: all **diam**, 170 MPa **TS**, 80 MPa **YS**, 2% in 2 in. (50 mm) **El**, 50–60 **HB**
ISO R164	Al–Si10Mg	ISO	Castings	rem **Al**, 9.0–11.0 **Si**, 0.10 **Cu**, 0.6 **Mn**, 0.15–0.40 **Mg**, 0.1 **Ni**, 0.1 **Zn**, 0.15 **Ti**, 0.05 **Sn**, 0.70 **Fe**, 0.05 **Pb**	
QQ–A–596d	319	US	Castings	rem **Al**, 5.5–7.0 **Si**, 3.5–4.5 **Cu**, 0.50 **Mn**, 0.10 **Mg**, 0.35 **Ni**, 1.0 **Zn**, 0.25 **Ti**, 1.0 **Fe**, 0.50 total, others	**As cast**: all **diam**, 28 ksi (193 MPa) **TS**, 1.5% in 2 in. (50 mm) **El**; **Solution heat treated and artificially aged–T6**: all **diam**, 34 ksi (234 MPa) **TS**, 2.0% in 2 in. (50 mm) **El**
SFS 2570	AlSi6Cu4	Finland	Die castings	rem **Al**, 5.0–7.0 **Si**, 3.0–5.0 **Cu**, 0.2–0.6 **Mn**, 0.3 **Mg**, 0.3 **Ni**, 2.0 **Zn**, 0.2 **Ti**, 0.1 **Sn**, 1.30 **Fe**, 0.2 **Pb**, 0.05 each, 0.15 total, others	**As cast**: $\leq$20 mm **diam**, 180 MPa **TS**, 120 MPa **YS**, 1% in 2 in. (50 mm) **El**
ISO R164	Al–Si6Cu4	ISO	Castings	rem **Al**, 5.0–7.0 **Si**, 3.0–5.0 **Cu**, 0.2–0.6 **Mn**, 0.3 **Mg**, 0.3 **Ni**, 2.0 **Zn**, 0.2 **Ti**, 0.1 **Sn**, 1.3 **Fe**, 0.2 **Pb**	

CAST ALUMINUM AND ALUMINUM ALLOYS

specification number	designation	country	product forms	chemical composition	mechanical properties and hardness values
SABS 989	Al–Si6Cu4MnMgA	South Africa	Ingot: for sand and gravity die castings	rem **Al**, 5.0–7.0 **Si**, 3.0–5.0 **Cu**, 0.3–0.6 **Mn**, 0.15–0.3 **Mg**, 0.3 **Ni**, 2.0 **Zn**, 0.2 **Ti**, 0.1 **Sn**, 0.8 **Fe**, 0.2 **Pb**, 0.05 each, 0.15 total, others	**Sand cast**: all **diam**, 155 MPa **TS**, 1% in 2 in. (50 mm) **El**; **Chill cast**: all **diam**, 170 MPa **TS**, 1% in 2 in. (50 mm) **El**
SABS 991	Al–Si6Cu4MnMgA	South Africa	Gravity die castings	rem **Al**, 5.0–7.0 **Si**, 3.0–5.0 **Cu**, 0.3–0.6 **Mn**, 0.1–0.3 **Mg**, 0.3 **Ni**, 2.0 **Zn**, 0.2 **Ti**, 0.1 **Sn**, 1.0 **Fe**, 0.2 **Pb**, 0.05 each, 0.15 total, others	**As cast**: all **diam**, 170 MPa **TS**, 1% in 2 in. (50 mm) **El**
SABS 990	Al–Si6Cu4MnMgA	South Africa	Sand castings	rem **Al**, 5.0–7.0 **Si**, 3.0–5.0 **Cu**, 0.3–0.6 **Mn**, 0.1–0.3 **Mg**, 0.3 **Ni**, 2.0 **Zn**, 0.2 **Ti**, 0.1 **Sn**, 1.0 **Fe**, 0.2 **Pb**, 0.05 each, 0.15 total, others	**As cast**: all **diam**, 155 MPa **TS**, 1% in 2 in. (50 mm) **El**
ANSI/ASTM B 179	360.2/SG100C	US	Ingot: for die castings	rem **Al**, 9.0–10.0 **Si**, 0.10 **Cu**, 0.10 **Mn**, 0.45–0.6 **Mg**, 0.10 **Ni**, 0.10 **Zn**, 0.10 **Sn**, 0.7–1.1 **Fe**, 0.20 total, others	
ANSI/ASTM B 179	A360.2	US	Ingot: for die castings	rem **Al**, 9.0–10.0 **Si**, 0.10 **Cu**, 0.05 **Mn**, 0.45–0.6 **Mg**, 0.05 **Zn**, 0.6 **Fe**, 0.05 each, 0.15 total, others	
QQ–A–371F	360.2 (formerly 360)	US	Ingot	rem **Al**, 9.0–10.0 **Si**, 0.10 **Cu**, 0.10 **Mn**, 0.45–0.6 **Mg**, 0.10 **Ni**, 0.10 **Zn**, 0.10 **Sn**, 0.7–1.1 **Fe**, 0.20 total, others	
QQ–A–371F	308.1 (formerly A108)	US	Ingot	rem **Al**, 5.0–6.0 **Si**, 4.0–5.0 **Cu**, 0.50 **Mn**, 0.10 **Mg**, 1.0 **Zn**, 0.25 **Ti**, 0.8 **Fe**, 0.50 total, others	
QQ–A–371F	308.2 (formerly A108)	US	Ingot	rem **Al**, 5.0–6.0 **Si**, 4.0–5.0 **Cu**, 0.30 **Mn**, 0.10 **Mg**, 0.50 **Zn**, 0.20 **Ti**, 0.8 **Fe**, 0.5 total, others	
QQ–A–596d	A108	US	Castings	rem **Al**, 5.0–6.0 **Si**, 4.0–5.0 **Cu**, 0.50 **Mn**, 0.10 **Mg**, 1.0 **Zn**, 0.25 **Ti**, 1.0 **Fe**, 0.50 total, others	**As cast**: all **diam**, 24 ksi (165 MPa) **TS**
ANSI/ASTM B 179	328.1	US	Ingot: for sand and permanent mold castings	rem **Al**, 7.5–8.5 **Si**, 1.0–2.0 **Cu**, 0.20–0.6 **Mn**, 0.25–0.6 **Mg**, 0.35 **Cr**, 0.25 **Ni**, 1.5 **Zn**, 0.25 **Ti**, 0.8 **Fe**, 0.50 total, others	
QQ–A–371F	328.1 (formerly RedX–8)	US	Ingot	rem **Al**, 7.5–8.5 **Si**, 1.0–2.0 **Cu**, 0.20–0.6 **Mn**, 0.25–0.6 **Mg**, 0.35 **Cr**, 0.25 **Ni**, 1.5 **Zn**, 0.25 **Ti**, 0.8 **Fe**, 0.50 total, others	

CAST ALUMINUM AND ALUMINUM ALLOYS

specification number	designation	country	product forms	chemical composition	mechanical properties and hardness values
ANSI/ASTM B 618	328.0	US	Investment castings	rem **Al**, 7.5–8.5 **Si**, 1.0–2.0 **Cu**, 0.20–0.6 **Mn**, 0.20–0.6 **Mg**, 0.35 **Cr**, 0.25 **Ni**, 1.5 **Zn**, 0.25 **Ti**, 1.0 **Fe**, 0.50 total, others	**As fabricated:** ≥0.250 in. **diam**, 25 ksi (172 MPa) **TS**, 14 ksi (97 MPa) **YS**, 1% in 2 in. (50 mm) **El**; **Solution heat treated and artificially aged:** ≥0.250 in. **diam**, 34 ksi (234 MPa) **TS**, 21 ksi (145 MPa) **YS**, 1% in 2 in. (50 mm) **El**
ANSI/ASTM B 26	328.0	US	Sand castings	rem **Al**, 7.5–8.5 **Si**, 1.0–2.0 **Cu**, 0.20–0.6 **Mn**, 0.20–0.6 **Mg**, 0.35 **Cr**, 0.25 **Ni**, 1.5 **Zn**, 0.25 **Ti**, 1.0 **Fe**, 0.50 total, others	**As fabricated:** ≥0.250 in. **diam**, 25 ksi (172 MPa) **TS**, 14 ksi (97 MPa) **YS**, 1% in 2 in. (50 mm) **El**, 60 **HB**; **Solution heat treated and artificially aged:** ≥0.250 in. **diam**, 34 ksi (234 MPa) **TS**, 21 ksi (145 MPa) **YS**, 1% in 2 in. (50 mm) **El**, 80 **HB**
QQ-A-601E	328.0	US	Castings	rem **Al**, 7.5–8.5 **Si**, 1.0–2.0 **Cu**, 0.20–0.6 **Mn**, 0.20–0.6 **Mg**, 0.35 **Cr**, 0.25 **Ni**, 1.5 **Zn**, 0.25 **Ti**, 1.0 **Fe**, 0.50 total, others	**As cast**: all **diam**, 25 ksi (172 MPa) **TS**, 1.0% in 2 in. (50 mm) **El**, 60 **HB**; **Solution heat treated, artificially aged and T6**: all **diam**, 34 ksi (234 MPa) **TS**, 1.0% in 2 in. (50 mm) **El**, 85 **HB**
DIN 1725 sheet 2	GK-AlSi9Mg wa/3.2373.62	Germany	Gravity die castings	rem **Al**, 9.0–10.0 **Si**, 0.05 **Cu**, 0.05 **Mn**, 0.20–0.40 **Mg**, 0.07 **Zn**, 0.15 **Ti**, 0.18 **Fe**, 0.03 each, 0.10 total, others	**Quenched and age hardened**: all **diam**, 260 MPa **TS**, 200 MPa **YS**, 4% in 2 in. (50 mm) **El**, 80–115 **HB**
DIN 1725 sheet 2	G-AlSi9Mg wa/3.2373.61	Germany	Sand castings	rem **Al**, 9.0–10.0 **Si**, 0.05 **Cu**, 0.05 **Mn**, 0.20–0.40 **Mg**, 0.07 **Zn**, 0.15 **Ti**, 0.18 **Fe**, 0.03 each, 0.10 total, others	**Quenched and age hardened**: all **diam**, 250 MPa **TS**, 200 MPa **YS**, 2% in 2 in. (50 mm) **El**, 75–100 **HB**
ANSI/ASTM B 108	359.0	US	Permanent mold castings	rem **Al**, 8.5–9.5 **Si**, 0.20 **Cu**, 0.10 **Mn**, 0.50–0.7 **Mg**, 0.10 **Zn**, 0.20 **Ti**, 0.20 **Fe**, 0.05 each, 0.15 total, others	**T61**: all **diam**, 40 ksi (276 MPa) **TS**, 30 ksi (207 MPa) **YS**, 3% in 2 in. (50 mm) **El**; **T62**: all **diam**, 40 ksi (276 MPa) **TS**, 30 ksi (207 MPa) **YS**, 3% in 2 in. (50 mm) **El**
DS 3002	4251	Denmark	Ingot: for castings	rem **Al**, 6.0–8.0 **Si**, 2.0–3.0 **Cu**, 0.5 **Mn**, 0.3 **Mg**, 0.3 **Ni**, 2.0 **Zn**, 0.20 **Ti**, 0.1 **Sn**, 0.7 **Fe**, 0.2 **Pb**	**Sand cast**: all **diam**, 147 MPa **TS**, 98 MPa **YS**, 2% in 2 in. (50 mm) **El**, 60–80 **HB**; **Chill cast**: all **diam**, 167 MPa **TS**, 108 MPa **YS**, 2% in 2 in. (50 mm) **El**, 70–90 **HB**
JIS H 5202	Class 4 A	Japan	Castings	rem **Al**, 8.0–10.0 **Si**, 0.2 **Cu**, 0.3–0.8 **Mn**, 0.4–0.8 **Mg**, 0.2 **Zn**, 0.2 **Ti**, 0.5 **Fe**	**As cast**: all **diam**, 177 MPa **TS**, 3% in 2 in. (50 mm) **El**; **Quenched and tempered**: all **diam**, 245 MPa **TS**, 2% in 2 in. (50 mm) **El**, 90 **HB**

CAST ALUMINUM AND ALUMINUM ALLOYS

specification number	designation	country	product forms	chemical composition	mechanical properties and hardness values
JIS H 2117	Class 4A	Japan	Ingot: for castings	rem **Al**, 8.0–10.0 **Si**, 0.2 **Cu**, 0.3–0.8 **Mn**, 0.40–0.8 **Mg**, 0.1 **Ni**, 0.2 **Zn**, 0.2 **Ti**, 0.4 **Fe**	
MIL–A–21180C	359.0	US	Castings	rem **Al**, 8.5–9.5 **Si**, 0.20 **Cu**, 0.10 **Mn**, 0.50–0.7 **Mg**, 0.10 **Zn**, 0.20 **Ti**, 0.20 **Fe**, 0.05 each, 0.15 total, others	**Class 1**: all **diam**, 310 MPa **TS**, 241 MPa **YS**, 4% in 2 in. (50 mm) **El**; **Class 2**: all **diam**, 324 MPa **TS**, 262 MPa **YS**, 3% in 2 in. (50 mm) **El**
NS 17 535	NS 17 535–00	Norway	Ingot: for casting	90 **Al**, 6.0–8.0 **Si**, 2.0–3.0 **Cu**, 0.5 **Mn**, 0.3 **Mg**, 0.3 **Ni**, 2.0 **Zn**, 0.20 **Ti**, 0.1 **Sn**, 0.7 **Fe**, 0.2 **Pb**	
NS 17 535	NS 17 535–01	Norway	Sand castings	90 **Al**, 6.0–8.0 **Si**, 2.0–3.0 **Cu**, 0.5 **Mn**, 0.3 **Mg**, 0.3 **Ni**, 2.0 **Zn**, 0.20 **Ti**, 0.1 **Sn**, 0.7 **Fe**, 0.2 **Pb**	**As cast**: all **diam**, 147 MPa **TS**, 98 MPa **YS**, 2% in 2 in. (50 mm) **El**
NS 17 535	NS 17 535–02	Norway	Chill castings	90 **Al**, 6.0–8.0 **Si**, 2.0–3.0 **Cu**, 0.5 **Mn**, 0.3 **Mg**, 0.3 **Ni**, 2.0 **Zn**, 0.20 **Ti**, 0.1 **Sn**, 0.7 **Fe**, 0.2 **Pb**	**As cast**: all **diam**, 167 MPa **TS**, 108 MPa **YS**, 2% in 2 in. (50 mm) **El**
SIS 14 42 51	SIS Al 42 51–03	Sweden	Sand castings	rem **Al**, 6.0–8.0 **Si**, 2.0–3.0 **Cu**, 0.5 **Mn**, 0.3 **Mg**, 0.3 **Ni**, 2.0 **Zn**, 0.20 **Ti**, 0.1 **Sn**, 0.7 **Fe**, 0.2 **Pb**	**As cast**: all **diam**, 150 MPa **TS**, 100 MPa **YS**, 2% in 2 in. (50 mm) **El**, 60–80 **HB**
SIS 14 42 51	SIS Al 42 51–00	Sweden	Ingot	rem **Al**, 6.0–8.0 **Si**, 2.0–3.0 **Cu**, 0.5 **Mn**, 0.3 **Mg**, 0.3 **Ni**, 2.0 **Zn**, 0.20 **Ti**, 0.1 **Sn**, 0.7 **Fe**, 0.2 **Pb**	
SIS 14 42 51	SIS Al 42 51–06	Sweden	Chill castings	rem **Al**, 6.0–8.0 **Si**, 2.0–3.0 **Cu**, 0.5 **Mn**, 0.3 **Mg**, 0.3 **Ni**, 2.0 **Zn**, 0.20 **Ti**, 0.1 **Sn**, 0.7 **Fe**, 0.2 **Pb**	**As cast**: all **diam**, 170 MPa **TS**, 110 MPa **YS**, 2% in 2 in. (50 mm) **El**, 70–90 **HB**
ANSI/ASTM B 179	359.2	US	Ingot: for permanent mold castings	rem **Al**, 8.5–9.5 **Si**, 0.10 **Cu**, 0.10 **Mn**, 0.55–0.7 **Mg**, 0.10 **Zn**, 0.20 **Ti**, 0.12 **Fe**, 0.05 each, 0.15 total, others	
QQ–A–371F	359.2 (formerly 359)	US	Ingot	rem **Al**, 8.5–9.5 **Si**, 0.10 **Cu**, 0.10 **Mn**, 0.55–0.7 **Mg**, 0.10 **Zn**, 0.20 **Ti**, 0.12 **Fe**, 0.05 each, 0.15 total, others	
JIS H 2211	Class 4A	Japan	Ingot: for castings	rem **Al**, 8.0–10.0 **Si**, 0.05 **Cu**, 0.3–0.8 **Mn**, 0.40–0.8 **Mg**, 0.03 **Ni**, 0.03 **Zn**, 0.03 **Ti**, 0.3 **Fe**	
ANSI/ASTM B 618	319.0	US	Investment castings	rem **Al**, 5.5–6.5 **Si**, 3.0–4.0 **Cu**, 0.50 **Mn**, 0.10 **Mg**, 0.35 **Ni**, 1.0 **Zn**, 0.25 **Ti**, 1.0 **Fe**, 0.50 total, others	**As fabricated**: ≥0.250 in. **diam**, 23 ksi (159 MPa) **TS**, 13 ksi (90 MPa) **YS**, 2% in 2 in. (50 mm) **El**; **Solution heat treated and artificially aged**: ≥0.250 in. **diam**, 32 ksi (221 MPa) **TS**, 20 ksi (138 MPa) **YS**, 3% in 2 in. (50 mm) **El**
ANSI/ASTM B 179	319.1	US	Ingot: for sand and permanent mold castings	rem **Al**, 5.5–6.5 **Si**, 3.0–4.0 **Cu**, 0.50 **Mn**, 0.10 **Mg**, 0.35 **Ni**, 1.0 **Zn**, 0.25 **Ti**, 0.8 **Fe**, 0.50 total, others	

CAST ALUMINUM AND ALUMINUM ALLOYS

specification number	designation	country	product forms	chemical composition	mechanical properties and hardness values
ANSI/ASTM B 179	319.2	US	Ingot: for sand and permanent mold castings	rem **Al**, 5.5–6.5 **Si**, 3.0–4.0 **Cu**, 0.10 **Mn**, 0.10 **Mg**, 0.10 **Ni**, 0.10 **Zn**, 0.20 **Ti**, 0.6 **Fe**, 0.20 total, others	
ANSI/ASTM B 108	319.0	US	Permanent mold castings	rem **Al**, 5.5–6.5 **Si**, 3.0–4.0 **Cu**, 0.50 **Mn**, 0.10 **Mg**, 0.35 **Ni**, 1.0 **Zn**, 0.25 **Ti**, 1.0 **Fe**, 0.50 total, others	**As fabricated**: all **diam**, 27 ksi (186 MPa) **TS**, 14 ksi (97 MPa) **YS**, 3% in 2 in. (50 mm) **El**, 95 **HB**
ANSI/ASTM B 26	319.0	US	Sand castings	rem **Al**, 5.5–6.5 **Si**, 3.0–4.0 **Cu**, 0.50 **Mn**, 0.10 **Mg**, 0.35 **Ni**, 1.0 **Zn**, 0.25 **Ti**, 1.0 **Fe**, 0.50 total, others	**As fabricated**: ≥0.250 in. **diam**, 23 ksi (159 MPa) **TS**, 13 ksi (90 MPa) **YS**, 2% in 2 in. (50 mm) **El**, 70 **HB**; **Solution heat treated and artificially aged**: ≥0.250 in. **diam**, 32 ksi (221 MPa) **TS**, 20 ksi (138 MPa) **YS**, 3% in 2 in. (50 mm) **El**, 80 **HB**
AS 1874	AP303	Australia	Ingot, sand castings and permanent mould castings	rem **Al**, 5.5–6.5 **Si**, 3.0–4.0 **Cu**, 0.50 **Mn**, 0.10 **Mg**, 0.10 **Ni**, 0.10 **Zn**, 0.20 **Ti**, 0.6 **Fe**, 0.20 total, others	**Cooled and naturally aged**: all **diam**, 155 MPa **TS**; **Cooled and artificially aged**: all **diam**, 170 MPa **TS**, 1% in 2 in. (50 mm) **El**; **Solution heat treated and artificially aged**: all **diam**, 210 MPa **TS**, 1.5% in 2 in. (50 mm) **El**
QQ–A–371F	319.1 (formerly 319)	US	Ingot	rem **Al**, 5.5–6.5 **Si**, 3.0–4.0 **Cu**, 0.50 **Mn**, 0.10 **Mg**, 0.35 **Ni**, 1.0 **Zn**, 0.25 **Ti**, 0.8 **Fe**, 0.50 total, others	
QQ–A–371F	319.2 (formerly 319)	US	Ingot	rem **Al**, 5.5–6.5 **Si**, 3.0–4.0 **Cu**, 0.10 **Mn**, 0.10 **Mg**, 0.10 **Ni**, 0.10 **Zn**, 0.20 **Ti**, 0.6 **Fe**, 0.20 total, others	
QQ–A–601E	319.0	US	Castings	rem **Al**, 5.5–6.5 **Si**, 3.0–4.0 **Cu**, 0.50 **Mn**, 0.10 **Mg**, 0.35 **Ni**, 1.0 **Zn**, 0.25 **Ti**, 1.0 **Fe**, 0.50 total, others	**As cast**: all **diam**, 23 ksi (159 MPa) **TS**, 18 ksi (124 MPa) **YS**, 70 **HB**; **Artificially aged and temper T5**: all **diam**, 25 ksi (172 MPa) **TS**, 26 ksi (179 MPa) **YS**, 80 **HB**; **Solution heat treated, artificially aged and temper T6**: all **diam**, 31 ksi (214 MPa) **TS**, 24 ksi (166 MPa) **YS**, 1.5% in 2 in. (50 mm) **El**, 80 **HB**
SIS 14 42 30	SIS Al 42 30–00	Sweden	Ingot	rem **Al**, 5.0–7.0 **Si**, 2.0–4.5 **Cu**, 0.2–0.6 **Mn**, 0.3 **Mg**, 0.3 **Ni**, 2.0 **Zn**, 0.2 **Ti**, 0.1 **Sn**, 0.95 **Fe**, 0.2 **Pb**	
SIS 14 42 30	SIS Al 42 30–03	Sweden	Sand castings	rem **Al**, 5.0–7.0 **Si**, 2.0–4.5 **Cu**, 0.2–0.6 **Mn**, 0.3 **Mg**, 0.3 **Ni**, 2.0 **Zn**, 0.2 **Ti**, 0.1 **Sn**, 0.95 **Fe**, 0.2 **Pb**	**As cast**: all **diam**, 190 MPa **TS**, 110 MPa **YS**, 0.5% in 2 in. (50 mm) **El**, 50–80 **HB**
SIS 14 42 30	SIS Al 42 30–06	Sweden	Chill castings	rem **Al**, 5.0–7.0 **Si**, 2.0–4.5 **Cu**, 0.2–0.6 **Mn**, 0.3 **Mg**, 0.3 **Ni**, 2.0 **Zn**, 0.2 **Ti**, 0.1 **Sn**, 0.95 **Fe**, 0.2 **Pb**	**As cast**: all **diam**, 140 MPa **TS**, 120 ksi **YS**, 05% in 2 in. (50 mm) **El**, 60–90 **HB**

CAST ALUMINUM AND ALUMINUM ALLOYS

specification number	designation	country	product forms	chemical composition	mechanical properties and hardness values
JIS H 5202	Class 2 B	Japan	Castings	rem **Al**, 5.0–7.0 **Si**, 2.0–4.0 **Cu**, 0.5 **Mn**, 0.5 **Mg**, 0.3 **Ni**, 1.0 **Zn**, 0.2 **Ti**, 1.0 **Fe**	**As cast**: all **diam**, 157 MPa **TS**, 1% in 2 in. (50 mm) **El**; **Quenched and tempered**: all **diam**, 245 MPa **TS**, 1% in 2 in. (50 mm) **El**, 90 **HB**
JIS H 2117	Class 2B	Japan	Ingot: for castings	rem **Al**, 5.0–7.0 **Si**, 2.0–4.0 **Cu**, 0.5 **Mn**, 0.5 **Mg**, 0.3 **Ni**, 1.0 **Zn**, 0.2 **Ti**, 0.8 **Fe**	
AS 1874	AS317	Australia	Ingot, sand castings and permanent mould castings	rem **Al**, 6.0–8.0 **Si**, 1.5–2.5 **Cu**, 0.20–0.6 **Mn**, 0.35 **Mg**, 0.10 **Cr**, 0.35 **Ni**, 1.0 **Zn**, 0.20 **Ti**, 0.15 **Sn**, 0.8 **Fe**, 0.25 **Pb**, 0.05 each, 0.20 total, others	**As cast (sand castings)**: all **diam**, 140 MPa **TS**, 1% in 2 in. (50 mm) **El**
QQ–A–371F	A360.2 (formerly A360)	US	Ingot	rem **Al**, 7.5–9.5 **Si**, 0.20 **Cu**, 0.10 **Mn**, 0.25–0.40 **Mg**, 0.25–0.50 **Cr**, 0.15 **Ni**, 0.15 **Zn**, 0.15 **Sn**, 0.7–1.1 **Fe**, 0.05 each, 0.15 total, others	
QQ–A–371F	364.2 (formerly 364)	US	Ingot	rem **Al**, 7.5–9.5 **Si**, 0.20 **Cu**, 0.10 **Mn**, 0.25–0.40 **Mg**, 0.25–0.50 **Cr**, 0.15 **Ni**, 0.15 **Zn**, 0.15 **Sn**, 0.7–1.1 **Fe**, 0.02–0.04 **Be**, 0.05 each, 0.15 total, others	
MIL–C–47140		US	Castings	rem **Al**, 7.60–8.60 **Si**, 0.20 **Cu**, 0.20 **Mn**, 0.40–0.60 **Mg**, 0.20 **Cr**, 0.20 **Zn**, 0.10–0.20 **Ti**, 0.40 **Fe**, 0.10–0.30 **Be**, 0.05 each, 0.15 total, others	**Solution heat treated and quenched (sand casting)**: all **diam**, 193 MPa **TS**, 110 MPa **YS**, 6% in 2 in. (50 mm) **El**, 6 **HRE**; **Solution heat treated water quenched and aged (mold casting)**: all **diam**, 276 MPa **TS**, 221 MPa **YS**, 2 **HRE**; **Solution heat treated, oil quenched and aged (mold casting)**: all **diam**, 207 MPa **TS**, 165 MPa **YS**, 2 **HRE**
QQ–A–371F	B358.2 (formerly Tens–50(2))	US	Ingot	rem **Al**, 7.6–8.6 **Si**, 0.10 **Cu**, 0.10 **Mn**, 0.45–0.6 **Mg**, 0.05 **Cr**, 0.10 **Zn**, 0.12–0.20 **Ti**, 0.20 **Fe**, 0.15–0.30 **Be**, 0.05 each, 0.15 total, others	
JIS H 5202	Class 2 A	Japan	Castings	rem **Al**, 4.0–5.0 **Si**, 3.5–4.5 **Cu**, 0.5 **Mn**, 0.2 **Mg**, 0.5 **Zn**, 0.2 **Ti**, 0.8 **Fe**	**As cast**: all **diam**, 177 MPa **TS**, 2% in 2 in. (50 mm) **El**; **Quenched and tempered**: all **diam**, 275 MPa **TS**, 1% in 2 in. (50 mm) **El**, 90 **HB**
JIS H 2117	Class 2A	Japan	Ingot: for castings	rem **Al**, 4.0–5.0 **Si**, 3.5–4.5 **Cu**, 0.5 **Mn**, 0.2 **Mg**, 0.3 **Ni**, 0.5 **Zn**, 0.2 **Ti**, 0.7 **Fe**	
ISO R164	Al–Si5Cu3	ISO	Castings	rem **Al**, 4.0–6.5 **Si**, 2.0–4.5 **Cu**, 0.2–0.7 **Mn**, 0.15 **Mg**, 0.3 **Ni**, 0.5 **Zn**, 0.2 **Ti**, 0.05 **Sn**, 1.0 **Fe**, 0.1 **Pb**	
ISO R164	Al–Si5Cu3Fe	ISO	Pressure die castings	rem **Al**, 4.0–6.5 **Si**, 2.0–4.5 **Cu**, 0.2–0.7 **Mn**, 0.15 **Mg**, 0.3 **Ni**, 0.5 **Zn**, 0.2 **Ti**, 0.2 **Sn**, 1.3 **Fe**, 0.3 **Pb**	

CAST ALUMINUM AND ALUMINUM ALLOYS

specification number	designation	country	product forms	chemical composition	mechanical properties and hardness values
SIS 14 42 31	SIS Al 42 31–00	Sweden	Ingot	rem **Al**, 4.0–6.5 **Si**, 2.0–4.5 **Cu**, 0.2–0.7 **Mn**, 0.15 **Mg**, 0.3 **Ni**, 0.5 **Zn**, 0.2 **Ti**, 0.05 **Sn**, 0.9 **Fe**, 0.1 **Pb**	
SIS 14 42 31	SIS Al 42 31–03	Sweden	Sand castings	86.09 min **Al**, 4.0–6.5 **Si**, 2.0–4.5 **Cu**, 0.2–0.7 **Mn**, 0.15 **Mg**, 0.3 **Ni**, 0.5 **Zn**, 0.2 **Ti**, 0.05 **Sn**, 0.9 **Fe**, 0.1 **Pb**	**As cast**: all **diam**, 140 MPa **TS**, 110 MPa **YS**, 2% in 2 in. (50 mm) **El**, 50–70 **HB**
SIS 14 42 31	SIS Al 42 31–06	Sweden	Chill castings	86.09 min **Al**, 4.0–6.5 **Si**, 2.0–4.5 **Cu**, 0.2–0.7 **Mn**, 0.15 **Mg**, 0.3 **Ni**, 0.5 **Zn**, 0.2 **Ti**, 0.05 **Sn**, 0.9 **Fe**, 0.1 **Pb**	**As cast**: all **diam**, 160 MPa **TS**, 120 MPa **YS**, 2% in 2 in. (50 mm) **El**, 60–80 **HB**
QQ–A–371F	324.1 (formerly 324(2))	US	Ingot	rem **Al**, 7.0–8.0 **Si**, 0.40–0.6 **Cu**, 0.50 **Mn**, 0.45–0.7 **Mg**, 0.30 **Ni**, 1.0 **Zn**, 0.20 **Ti**, 0.9 **Fe**, 0.15 each, 0.20 total, others	
JIS H 2211	Class 2A	Japan	Ingot: for castings	rem **Al**, 4.0–5.0 **Si**, 3.5–4.5 **Cu**, 0.03 **Mn**, 0.03 **Mg**, 0.03 **Ni**, 0.03 **Zn**, 0.03 **Ti**, 0.3 **Fe**	
AS 1874	AS303	Australia	Ingot, sand castings and permanent mould castings	rem **Al**, 4.0–6.0 **Si**, 2.0–4.0 **Cu**, 0.7 **Mn**, 0.15 **Mg**, 0.10 **Cr**, 0.30 **Ni**, 0.50 **Zn**, 0.20 **Ti**, 0.15 **Sn**, 0.8 **Fe**, 0.15 **Pb**, 0.05 each, 0.20 total, others	**Cooled and naturally aged**: all **diam**, 155 MPa **TS**; **Solution heat–treated and artificially aged**: all **diam**, 210 MPa **TS**, 1.5% in 2 in. (50 mm) **El**
CSA HA.3	.SC53	Canada	Ingot	rem **Al**, 4.0–6.0 **Si**, 2.0–4.0 **Cu**, 0.30–0.6 **Mn**, 0.15 **Mg**, 0.30 **Ni**, 0.20 **Zn**, 0.20 **Ti**, 0.7 **Fe**, 0.05 each, 0.30 total, others	
CSA HA.10	.SC53	Canada	Mold castings	rem **Al**, 4.0–6.0 **Si**, 2.0–4.0 **Cu**, 0.30–0.7 **Mn**, 0.15 **Mg**, 0.30 **Ni**, 0.20 **Zn**, 0.20 **Ti**, 0.8 **Fe**, 0.05 each, 0.30 total, others	
CSA HA.9	.SC53	Canada	Sand castings	rem **Al**, 4.0–6.0 **Si**, 2.0–4.0 **Cu**, 0.30–0.6 **Mn**, 0.15 **Mg**, 0.30 **Ni**, 0.20 **Zn**, 0.20 **Ti**, 0.8 **Fe**, 0.05 each, 0.30 total, others	
IS 202	A–4	India	Sand castings	rem **Al**, 4.0–6.0 **Si**, 2.0–4.0 **Cu**, 0.30–0.70 **Mn**, 0.15 **Mg**, 0.30 **Ni**, 0.50 **Zn**, 0.05 **Sn**, 0.80 **Fe**, 0.10 **Pb**, 0.20 **Ti+ Nb**	**As cast**: all **diam**, 119 MPa **TS**, 83 MPa **YS**, 2% in 2 in. (50 mm) **El**
IS 202	A–4	India	Sand castings	rem **Al**, 4.0–6.0 **Si**, 2.0–4.0 **Cu**, 0.30–0.70 **Mn**, 0.15 **Mg**, 0.30 **Ni**, 0.50 **Zn**, 0.05 **Sn**, 0.80 **Fe**, 0.10 **Pb**, 0.20 **Ti+ Nb**	**As cast**: 119 MPa **TS**, 84 MPa **YS**, 2% in 2 in. (50 mm) **El**
SABS 991	Al–Si5Cu3MnA	South Africa	Gravity die castings	rem **Al**, 4.0–6.0 **Si**, 2.0–4.0 **Cu**, 0.3–0.7 **Mn**, 0.15 **Mg**, 0.3 **Ni**, 0.5 **Zn**, 0.2 **Ti**, 0.05 **Sn**, 0.8 **Fe**, 0.1 **Pb**, 0.05 each, 0.15 total, others	**As cast**: all **diam**, 155 MPa **TS**, 2% in 2 in. (50 mm) **El**; **Solution and precipitation heat treated**: all **diam**, 275 MPa **TS**
SABS 990	Al–Si5Cu3MnA	South Africa	Sand castings	rem **Al**, 4.0–6.0 **Si**, 2.0–4.0 **Cu**, 0.3–0.7 **Mn**, 0.15 **Mg**, 0.3 **Ni**, 0.5 **Zn**, 0.2 **Ti**, 0.05 **Sn**, 0.8 **Fe**, 0.1 **Pb**, 0.05 each, 0.15 total, others	**As cast**: all **diam**, 140 MPa **TS**, 2% in 2 in. (50 mm) **El**; **Solution and precipitation heat treated**: all **diam**, 230 MPa **TS**

CAST ALUMINUM AND ALUMINUM ALLOYS

specification number	designation	country	product forms	chemical composition	mechanical properties and hardness values
SABS 989	Al–Si5Cu3MnA	South Africa	Ingot: for sand and gravity die castings	rem **Al**, 4.0–6.0 **Si**, 2.0–4.0 **Cu**, 0.3–0.7 **Mn**, 0.15 **Mg**, 0.3 **Ni**, 0.5 **Zn**, 0.2 **Ti**, 0.05 **Sn**, 0.7 **Fe**, 0.1 **Pb**, 0.05 each, 0.15 total, others	**Sand cast**: all **diam**, 140 MPa **TS**, 2% in 2 in. (50 mm) **El**; **Chill cast**: all **diam**, 155 MPa **TS**, 2% in 2 in. (50 mm) **El**; **Chill cast, solution and precipitation heat treated**: all **diam**, 275 MPa **TS**
ANSI/ASTM B 108	A357.0	US	Permanent mold castings	rem **Al**, 6.5–7.5 **Si**, 0.20 **Cu**, 0.10 **Mn**, 0.40–0.7 **Mg**, 0.10 **Zn**, 0.10–0.20 **Ti**, 0.20 **Fe**, 0.04–0.07 **Be**, 0.05 each, 0.15 total, others	**T61**: all **diam**, 41 ksi (283 MPa) **TS**, 31 ksi (214 MPa) **YS**, 3% in 2 in. (50 mm) **El**
MIL–A–21180C	A357.0	US	Castings	rem **Al**, 6.5–7.5 **Si**, 0.20 **Cu**, 0.10 **Mn**, 0.40–0.7 **Mg**, 0.10 **Zn**, 0.10–0.20 **Ti**, 0.20 **Fe**, 0.04–0.07 **Be**, 0.05 each, 0.15 total, others	**Class 1**: all **diam**, 310 MPa **TS**, 241 MPa **YS**, 3% in 2 in. (50 mm) **El**; **Class 2**: all **diam**, 345 MPa **TS**, 276 MPa **YS**
ANSI/ASTM B 179	356.1	US	Ingot: for sand and permanent mold castings	rem **Al**, 6.7–7.5 **Si**, 0.25 **Cu**, 0.35 **Mn**, 0.25–0.40 **Mg**, 0.35 **Zn**, 0.25 **Ti**, 0.50 **Fe**, 0.05 each, 0.15 total, others	
QQ–A–371F	A357.2 (formerly A357)	US	Ingot	rem **Al**, 6.5–7.5 **Si**, 0.10 **Cu**, 0.05 **Mn**, 0.45–0.7 **Mg**, 0.05 **Zn**, 0.10–0.20 **Ti**, 0.12 **Fe**, 0.04–0.07 **Be**, 0.03 each, 0.10 total, others	
AS 1874	AP603	Australia	Ingot, sand castings and permanent mould castings	rem **Al**, 6.5–7.5 **Si**, 0.05 **Cu**, 0.03 **Mn**, 0.47–0.7 **Mg**, 0.05 **Zn**, 0.20 **Ti**, 0.15 **Fe**, 0.05 each, 0.15 total, others	**Solution heat–treated for 10h at 540°C, quenched and aged (sand castings)**: all **diam**, 270 MPa **TS**, 1% in 2 in. (50 mm) **El**; **Solution heat–treated for 8h at 540°C, quenched and aged (permanent mould castings)**: all **diam**, 290 MPa **TS**, 3% in 2 in. (50 mm) **El**
QQ–A–596d	356	US	Castings	rem **Al**, 6.5–7.5 **Si**, 0.25 **Cu**, 0.20–0.40 **Mn**, 0.35 **Mg**, 0.35 **Zn**, 0.25 **Ti**, 0.6 **Fe**, 0.05 each, 0.15 total, others	**Solution heat treated and artificially aged–T6**: all **diam**, 33 ksi (228 MPa) **TS**, 3.0% in 2 in. (50 mm) **El**; **Solution heat treated and stabilized and T7**: all **diam**, 29 ksi (200 MPa) **TS**, 4.0% in 2 in. (50 mm) **El**; **Artificially aged only and temper T51**: all **diam**, 25 ksi (172 MPa) **TS**
QQ–A–371F	357.1 (formerly 357)	US	Ingot	rem **Al**, 6.5–7.5 **Si**, 0.05 **Cu**, 0.03 **Mn**, 0.45–0.6 **Mg**, 0.05 **Zn**, 0.20 **Ti**, 0.12 **Fe**, 0.05 each, 0.15 total, others	

CAST ALUMINUM AND ALUMINUM ALLOYS

specification number	designation	country	product forms	chemical composition	mechanical properties and hardness values
ANSI/ASTM B 618	356.0	US	Investment castings	rem **Al**, 6.5–7.5 **Si**, 0.25 **Cu**, 0.35 **Mn**, 0.20–0.40 **Mg**, 0.35 **Zn**, 0.25 **Ti**, 0.6 **Fe**, 0.05 each, 0.15 total, others	**As fabricated**: ≥0.250 in. **diam**, 19 ksi (131 MPa) **TS**, 2% in 2 in. (50 mm) **El**; **Solution heat treated and artificially aged**: ≥0.250 in. **diam**, 30 ksi (207 MPa) **TS**, 20 ksi (138 MPa) **YS**, 3% in 2 in. (50 mm) **El**; **Solution heat treated and stabilized**: ≥0.250 in. **diam**, 31 ksi (214 MPa) **TS**
ANSI/ASTM B 618	A356.0	US	Investment castings	rem **Al**, 6.5–7.6 **Si**, 0.20 **Cu**, 0.10 **Mn**, 0.20–0.40 **Mg**, 0.10 **Zn**, 0.20 **Ti**, 0.20 **Fe**, 0.05 each, 0.15 total, others	**Solution heat treated and artificially aged**: 34 ksi (234 MPa) **TS**, 24 ksi (166 MPa) **YS**, 4% in 2 in. (50 mm) **El**
ANSI/ASTM B 108	356.0	US	Permanent mold castings	rem **Al**, 6.5–7.5 **Si**, 0.25 **Cu**, 0.35 **Mn**, 0.20–0.40 **Mg**, 0.35 **Zn**, 0.25 **Ti**, 0.6 **Fe**, 0.05 each, 0.15 total, others	**As fabricated**: all **diam**, 21 ksi (145 MPa) **TS**, 3% in 2 in. (50 mm) **El**; **Solution heat treated and artificially aged**: all **diam**, 33 ksi (228 MPa) **TS**, 22 ksi (152 MPa) **YS**, 3% in 2 in. (50 mm) **El**, 85 **HB**; **T71**: all **diam**, 25 ksi (172 MPa) **TS**, 3% in 2 in. (50 mm) **El**, 70 **HB**
ANSI/ASTM B 26	356.0	US	Sand castings	rem **Al**, 6.5–7.5 **Si**, 0.25 **Cu**, 0.35 **Mn**, 0.20–0.40 **Mg**, 0.35 **Zn**, 0.25 **Ti**, 0.6 **Fe**, 0.05 each, 0.15 total, others	**As fabricated**: ≥0.250 in. **diam**, 19 ksi (131 MPa) **TS**, 2% in 2 in. (50 mm) **El**, 55 **HB**; **Solution heat treated and artificially aged**: ≥0.250 in. **diam**, 30 ksi (207 MPa) **TS**, 20 ksi (138 MPa) **YS**, 3% in 2 in. (50 mm) **El**, 70 **HB**; **Solution heat treated and stabilized**: ≥0.250 in. **diam**, 31 ksi (214 MPa) **TS**, 75 **HB**
QQ-A-601E	356.0	US	Castings	rem **Al**, 6.5–7.5 **Si**, 0.25 **Cu**, 0.35 **Mn**, 0.20–0.40 **Mg**, 0.35 **Zn**, 0.25 **Ti**, 0.6 **Fe**, 0.05 each, 0.15 total, others	**Artificially aged only and T51**: all **diam**, 23 ksi (159 MPa) **TS**, 20 ksi (138 MPa) **YS**, 60 **HB**; **Solution heat treated, aged and T6**: all **diam**, 30 ksi (207 MPa) **TS**, 20 ksi (138 MPa) **YS**, 3.0% in 2 in. (50 mm) **El**, 70 **HB**; **Solution heat treated, stabilized and T7**: all **diam**, 31 ksi (214 MPa) **TS**, 29 ksi (200 MPa) **YS**, 75 **HB**
JIS H 2117	Class 4C	Japan	Ingot: for castings	rem **Al**, 6.5–7.5 **Si**, 0.2 **Cu**, 0.3 **Mn**, 0.25–0.4 **Mg**, 0.1 **Ni**, 0.3 **Zn**, 0.2 **Ti**, 0.4 **Fe**	
ANSI/ASTM B 108	A356.0	US	Permanent mold castings	rem **Al**, 6.5–7.5 **Si**, 0.20 **Cu**, 0.10 **Mn**, 0.20–0.40 **Mg**, 0.10 **Zn**, 0.20 **Ti**, 0.20 **Fe**, 0.05 each, 0.15 total, others	**T61**: all **diam**, 28 ksi (193 MPa) **TS**, 26 ksi (179 MPa) **YS**, 3% in 2 in. (50 mm) **El**

CAST ALUMINUM AND ALUMINUM ALLOYS

specification number	designation	country	product forms	chemical composition	mechanical properties and hardness values
ANSI/ASTM B 26	A356.0	US	Sand castings	rem **Al**, 6.5–7.5 **Si**, 0.20 **Cu**, 0.10 **Mn**, 0.20–0.40 **Mg**, 0.10 **Zn**, 0.20 **Ti**, 0.20 **Fe**, 0.05 each, 0.15 total, others	**Solution heat treated and artificially aged**: ≥0.250 **diam**, 34 ksi (234 MPa) **TS**, 24 ksi (166 MPa) **YS**, 4% in 2 in. (50 mm) **El**
CSA HA.10	.SG70P	Canada	Mold castings	rem **Al**, 6.5–7.5 **Si**, 0.20 **Cu**, 0.10 **Mn**, 0.20–0.40 **Mg**, 0.10 **Zn**, 0.20 **Ti**, 0.20 **Fe**, 0.05 each, 0.15 total, others	**T6A**: all **diam**, 38.0 ksi (262 MPa) **TS**, 26.0 ksi (179 MPa) **YS**, 5% in 2 in. (50 mm) **El**
CSA HA.10	.SG70N	Canada	Mold castings	rem **Al**, 6.5–7.5 **Si**, 0.20 **Cu**, 0.35 **Mn**, 0.20–0.40 **Mg**, 0.35 **Zn**, 0.25 **Ti**, 0.50 **Fe**, 0.05 each, 0.15 total, others	**T6**: all **diam**, 33.0 ksi (228 MPa) **TS**, 22.0 ksi (152 MPa) **YS**, 3% in 2 in. (50 mm) **El**
CSA HA.9	.SG70P	Canada	Sand castings	rem **Al**, 6.5–7.5 **Si**, 0.20 **Cu**, 0.10 **Mn**, 0.20–0.40 **Mg**, 0.10 **Zn**, 0.20 **Ti**, 0.20 **Fe**, 0.05 each, 0.15 total, others	**T6**: all **diam**, 34.0 ksi (234 MPa) **TS**, 24.0 ksi (165 MPa) **YS**, 3.5% in 2 in. (50 mm) **El**; **T6A**: 38.0 ksi (262 MPa) **TS**, 26.0 ksi (179 MPa) **YS**, 5% in 2 in. (50 mm) **El**
CSA HA.9	.SG70N	Canada	Sand castings	rem **Al**, 6.5–7.5 **Si**, 0.20 **Cu**, 0.35 **Mn**, 0.20–0.40 **Mg**, 0.35 **Zn**, 0.25 **Ti**, 0.50 **Fe**, 0.05 each, 0.15 total, others	**T6**: all **diam**, 30.0 ksi (207 MPa) **TS**, 20.0 ksi (138 MPa) **YS**, 3% in 2 in. (50 mm) **El**; **T5**: all **diam**, 159 MPa **TS**
DS 3002	4244	Denmark	Ingot: for castings	rem **Al**, 6.5–7.5 **Si**, 0.20 **Cu**, 0.5 **Mn**, 0.20–0.40 **Mg**, 0.1 **Ni**, 0.3 **Zn**, 0.20 **Ti**, 0.05 **Sn**, 0.50 **Fe**, 0.05 **Pb**	**Sand cast**: all **diam**, 235 MPa **TS**, 196 MPa **YS**, 1% in 2 in. (50 mm) **El**, 75–105 **HB**; **Chill cast**: all **diam**, 255 MPa **TS**, 216 MPa **YS**, 1% in 2 in. (50 mm) **El**, 80–110 **HB**
QQ-A-601E	A356.0	US	Castings	rem **Al**, 6.5–7.5 **Si**, 0.20 **Cu**, 0.10 **Mn**, 0.20–0.40 **Mg**, 0.10 **Zn**, 0.20 **Ti**, 0.20 **Fe**, 0.05 each, 0.15 total, others	**Solution heat-treated, artificially aged and T6**: all **diam**, 34 ksi (234 MPa) **TS**, 24 ksi (165 MPa) **YS**, 3.5% in 2 in. (50 mm) **El**, 75 **HB**
IS 202	A-27	India	Sand castings	rem **Al**, 6.5–7.5 **Si**, 0.20 **Cu**, 0.10 **Mn**, 0.20–0.40 **Mg**, 0.10 **Zn**, 0.5 **Fe**, 0.20 **Ti + Nb**	**Solution and precipitation treated**: all **diam**, 207 MPa **TS**, 123 MPa **YS**, 3% in 2 in. (50 mm) **El**
IS 202	A-27	India	Sand castings	rem **Al**, 6.5–7.5 **Si**, 0.20 **Cu**, 0.10 **Mn**, 0.20–0.40 **Mg**, 0.10 **Zn**, 0.5 **Fe**, 0.20 **Ti + Nb**	**Solution and precipitation treated**: 207 MPa **TS**, 123 MPa **YS**, 3% in 2 in. (50 mm) **El**
ISO R2147	Al-Si7Mg	ISO	Sand castings: test pieces	rem **Al**, 6.5–7.5 **Si**, 0.20 **Cu**, 0.6 **Mn**, 0.2–0.4 **Mg**, 0.05 **Ni**, 0.3 **Zn**, 0.2 **Ti**, 0.05 **Sn**, 0.5 **Fe**, 0.05 **Pb**	**Solution heat treated and artificially aged**: all **diam**, 230 MPa **TS**, 1.5% in 2 in. (50 mm) **El**
JIS H 5202	Class 4 C	Japan	Castings	rem **Al**, 6.5–7.5 **Si**, 0.2 **Cu**, 0.3 **Mn**, 0.20–0.4 **Mg**, 0.3 **Zn**, 0.2 **Ti**, 0.5 **Fe**	**As cast**: all **diam**, 157 MPa **TS**, 3% in 2 in. (50 mm) **El**; **Tempered**: all **diam**, 177 MPa **TS**, 3% in 2 in. (50 mm) **El**, 65 **HB**; **Quenched and tempered**: all **diam**, 226 MPa **TS**, 3% in 2 in. (50 mm) **El**, 85 **HB**

CAST ALUMINUM AND ALUMINUM ALLOYS

specification number	designation	country	product forms	chemical composition	mechanical properties and hardness values
MIL-A-21180C	A356.0	US	Castings	rem **Al**, 6.5–7.5 **Si**, 0.20 **Cu**, 0.10 **Mn**, 0.20–0.40 **Mg**, 0.10 **Zn**, 0.20 **Ti**, 0.20 **Fe**, 0.05 each, 0.15 total, others	**Class 1**: all **diam**, 262 MPa **TS**, 193 MPa **YS**, 5% in 2 in. (50 mm) **El**; **Class 2**: all **diam**, 276 MPa **TS**, 207 MPa **YS**, 3% in 2 in. (50 mm) **El**; **Class 3**: all **diam**, 310 MPa **TS**, 234 MPa **YS**, 3% in 2 in. (50 mm) **El**
NS 17 525	NS 17 525–00	Norway	Ingot: for casting	93 **Al**, 6.5–7.5 **Si**, 0.20 **Cu**, 0.5 **Mn**, 0.20–0.40 **Mg**, 0.1 **Ni**, 0.3 **Zn**, 0.20 **Ti**, 0.05 **Sn**, 0.50 **Fe**, 0.05 **Pb**	
NS 17 525	NS 17 525–41	Norway	Sand castings	93 **Al**, 6.5–7.5 **Si**, 0.20 **Cu**, 0.5 **Mn**, 0.20–0.40 **Mg**, 0.1 **Ni**, 0.3 **Zn**, 0.20 **Ti**, 0.05 **Sn**, 0.50 **Fe**, 0.05 **Pb**	
NS 17 525	NS 17 525–42	Norway	Chill castings	93 **Al**, 6.5–7.5 **Si**, 0.20 **Cu**, 0.20–0.40 **Mg**, 0.1 **Ni**, 0.3 **Zn**, 0.20 **Ti**, 0.05 **Sn**, 0.50 **Fe**, 0.05 **Pb**	**Hardened and artificially aged**: all **diam**, 255 MPa **TS**, 216 MPa **YS**, 1% in 2 in. (50 mm) **El**
SIS 14 42 44	SIS Al 42 44–00	Sweden	Ingot	rem **Al**, 6.5–7.5 **Si**, 0.20 **Cu**, 0.5 **Mn**, 0.20–0.40 **Mg**, 0.1 **Ni**, 0.3 **Zn**, 0.20 **Ti**, 0.05 **Sn**, 0.50 **Fe**, 0.05 **Pb**	
SIS 14 42 44	SIS Al 42 44–04	Sweden	Sand castings	rem **Al**, 6.5–7.5 **Si**, 0.20 **Cu**, 0.5 **Mn**, 0.20–0.40 **Mg**, 0.1 **Ni**, 0.3 **Zn**, 0.20 **Ti**, 0.05 **Sn**, 0.50 **Fe**, 0.05 **Pb**	**Solution treated and artificially aged**: all **diam**, 240 MPa **TS**, 200 MPa **YS**, 2% in 50 in. **El**, 75–105 **HB**
SIS 14 42 44	SIS Al 42 44–07	Sweden	Chill castings	rem **Al**, 6.5–7.5 **Si**, 0.20 **Cu**, 0.5 **Mn**, 0.2–0.40 **Mg**, 0.1 **Ni**, 0.32 **Zn**, 0.20 **Ti**, 0.05 **Sn**, 0.50 **Fe**, 0.05 **Pb**	**Solution treated, artificially aged**: all **diam**, 250 MPa **TS**, 220 MPa **YS**, 1% in 2 in. (50 mm) **El**, 80–110 **HB**
ANSI/ASTM B 179	A356.2	US	Ingot: for sand and permanent mold castings	rem **Al**, 6.5–7.5 **Si**, 0.10 **Cu**, 0.05 **Mn**, 0.30–0.40 **Mg**, 0.05 **Zn**, 0.20 **Ti**, 0.12 **Fe**, 0.05 each, 0.15 total, others	
ANSI/ASTM B 179	356.2	US	Ingot: for sand and permanent mold castings	rem **Al**, 6.5–7.5 **Si**, 0.10 **Cu**, 0.05 **Mn**, 0.30–0.40 **Mg**, 0.05 **Zn**, 0.20 **Ti**, 0.13–0.25 **Fe**, 0.05 each, 0.15 total, others	
CSA HA.3	.SG70P	Canada	Ingot	rem **Al**, 6.5–7.5 **Si**, 0.10 **Cu**, 0.05 **Mn**, 0.30–0.40 **Mg**, 0.05 **Zn**, 0.20 **Ti**, 0.11 **Fe**, 0.05 each, 0.15 total, others	
CSA HA.3	.SG70N	Canada	Ingot	rem **Al**, 6.5–7.5 **Si**, 0.10 **Cu**, 0.05 **Mn**, 0.30–0.40 **Mg**, 0.05 **Zn**, 0.20 **Ti**, 0.12–0.25 **Fe**, 0.05 each, 0.15 total, others	
QQ-A-371F	356.2 (formerly 356)	US	Ingot	rem **Al**, 6.5–7.5 **Si**, 0.10 **Cu**, 0.05 **Mn**, 0.30–0.40 **Mg**, 0.05 **Zn**, 0.20 **Ti**, 0.12–0.25 **Fe**, 0.05 each, 0.15 total, others	

CAST ALUMINUM AND ALUMINUM ALLOYS

specification number	designation	country	product forms	chemical composition	mechanical properties and hardness values
QQ-A-371F	A356.2 (formerly A356)	US	Ingot	rem **Al**, 6.5–7.5 **Si**, 0.10 **Cu**, 0.05 **Mn**, 0.30–0.40 **Mg**, 0.05 **Zn**, 0.20 **Ti**, 0.12 **Fe**, 0.05 each, 0.15 total, others	
SABS 989	Al–Si7MgA	South Africa	Ingot: for sand and gravity die castings	rem **Al**, 6.5–7.5 **Si**, 0.1 **Cu**, 0.3 **Mn**, 0.25–0.4 **Mg**, 0.05 **Ni**, 0.1 **Zn**, 0.2 **Ti**, 0.05 **Sn**, 0.45 **Fe**, 0.05 **Pb**, 0.05 each, 0.15 total, others	**Sand cast**: all **diam**, 125 MPa **TS**, 2% in 2 in. (50 mm) **El**; **Sand cast, solution and precipitation heat treated**: all **diam**, 230 MPa **TS**; **Chill cast**: all **diam**, 160 MPa **TS**, 3% in 2 in. (50 mm) **El**
SABS 991	Al–Si7MgA	South Africa	Gravity die castings	rem **Al**, 6.5–7.5 **Si**, 0.1 **Cu**, 0.3 **Mn**, 0.2–0.4 **Mg**, 0.05 **Ni**, 0.1 **Zn**, 0.2 **Ti**, 0.05 **Sn**, 0.5 **Fe**, 0.05 **Pb**, 0.05 each, 0.15 total, others	**As cast**: all **diam**, 160 MPa **TS**, 3% in 2 in. (50 mm) **El**; **Precipitation heat treated**: all **diam**, 185 MPa **TS**, 2% in 2 in. (50 mm) **El**, all **HV**; **Solution heat treated and stabilized**: 230 MPa **TS**, 5% in 2 in. (50 mm) **El**
SABS 990	Al–Si7MgA	South Africa	Sand castings	rem **Al**, 6.5–7.5 **Si**, 0.1 **Cu**, 0.3 **Mn**, 0.2–0.4 **Mg**, 0.05 **Ni**, 0.1 **Zn**, 0.2 **Ti**, 0.05 **Sn**, 0.5 **Fe**, 0.05 **Pb**, 0.05 each, 0.15 total, others	**As cast**: all **diam**, 125 MPa **TS**, 2% in 2 in. (50 mm) **El**; **Precipitation heat treated**: all **diam**, 145 MPa **TS**, 1% in 2 in. (50 mm) **El**; **Solution heat treated and stabilized**: all **diam**, 160 MPa **TS**, 2.5% in 2 in. (50 mm) **El**
JIS H 2211	Class 4C	Japan	Ingot: for castings	rem **Al**, 6.5–7.5 **Si**, 0.05 **Cu**, 0.03 **Mn**, 0.25–0.4 **Mg**, 0.03 **Ni**, 0.03 **Zn**, 0.03 **Ti**, 0.3 **Fe**	
DIN 1725 sheet 2	GK–AlSi7Mg wa/3.2371.62	Germany	Gravity die castings	rem **Al**, 6.5–7.5 **Si**, 0.05 **Cu**, 0.05 **Mn**, 0.20–0.40 **Mg**, 0.07 **Zn**, 0.15 **Ti**, 0.18 **Fe**, 0.03 each, 0.10 total, others	**Quenched and age hardened**: all **diam**, 250 MPa **TS**, 200 MPa **YS**, 5% in 2 in. (50 mm) **El**, 80–105 **HB**
DIN 1725 sheet 2	G–AlSi7Mg wa/3.2371.61	Germany	Sand castings	rem **Al**, 6.5–7.5 **Si**, 0.05 **Cu**, 0.05 **Mn**, 0.20–0.40 **Mg**, 0.07 **Zn**, 0.15 **Ti**, 0.18 **Fe**, 0.03 each, 0.10 total, others	**Quenched and age hardened**: all **diam**, 230 MPa **TS**, 190 MPa **YS**, 2% in 2 in. (50 mm) **El**, 75–105 **HB**
QQ–A–596d	A356	US	Castings	rem **Al**, 6.5–7.5 **Si**, 0.20 **Cu**, 0.20–0.40 **Mn**, 0.10 **Mg**, 0.10 **Zn**, 0.20 **Ti**, 0.20 **Fe**, 0.05 each, 0.15 total, others	**Solution heat treated and artificially aged–T61**: all **diam**, 37 ksi (255 MPa) **TS**, 5.0% in 2 in. (50 mm) **El**
QQ–A–596d	357	US	Castings	rem **Al**, 6.5–7.5 **Si**, 0.05 **Cu**, 0.45–0.6 **Mn**, 0.03 **Mg**, 0.05 **Zn**, 0.20 **Ti**, 0.15 **Fe**, 0.05 each, 0.15 total, others	**Solution heat treated and artificially aged–T6**: all **diam**, 45 ksi (310 MPa) **TS**, 3.0% in 2 in. (50 mm) **El**
ISO R164	Al–Si5Cul	ISO	Castings	rem **Al**, 4.5–6.0 **Si**, 1.0–1.5 **Cu**, 0.5 **Mn**, 0.3–0.6 **Mg**, 0.3 **Ni**, 0.5 **Zn**, 0.2 **Ti**, 0.1 **Sn**, 0.8 **Fe**, 0.2 **Pb**	
ANSI/ASTM B 179	C355.2	US	Ingot: for sand and permanent mold castings	rem **Al**, 4.5–5.5 **Si**, 1.0–1.5 **Cu**, 0.05 **Mn**, 0.50–0.6 **Mg**, 0.05 **Zn**, 0.20 **Ti**, 0.13 **Fe**, 0.05 each, 0.15 total, others	

CAST ALUMINUM AND ALUMINUM ALLOYS

specification number	designation	country	product forms	chemical composition	mechanical properties and hardness values
ANSI/ASTM B 179	355.2	US	Ingot: for sand and permanent mold castings	rem **Al**, 4.5–5.5 **Si**, 1.0–1.5 **Cu**, 0.05 **Mn**, 0.50–0.6 **Mg**, 0.05 **Zn**, 0.20 **Ti**, 0.14–0.25 **Fe**, 0.05 each, 0.15 total, others	
CSA HA.3	.SC51P	Canada	Ingot	rem **Al**, 4.5–5.5 **Si**, 1.0–1.5 **Cu**, 0.05 **Mn**, 0.50–0.6 **Mg**, 0.05 **Zn**, 0.20 **Ti**, 0.13 **Fe**, 0.05 each, 0.15 total, others	
CSA HA.3	.SC51N	Canada	Ingot	rem **Al**, 4.5–5.5 **Si**, 1.0–1.5 **Cu**, 0.05 **Mn**, 0.50–0.6 **Mg**, 0.05 **Zn**, 0.20 **Ti**, 0.14–0.25 **Fe**, 0.05 each, 0.15 total, others	
QQ–A–371F	355.2 (formerly 355)	US	Ingot	rem **Al**, 4.5–5.5 **Si**, 1.0–1.5 **Cu**, 0.05 **Mn**, 0.50–0.6 **Mg**, 0.05 **Zn**, 0.14–0.25 **Fe**, 0.05 each, 0.15 total, others	
ANSI/ASTM B 179	355.1	US	Ingot: for sand and permanent mold castings	rem **Al**, 4.5–5.5 **Si**, 1.0–1.5 **Cu**, 0.50 **Mn**, 0.45–0.6 **Mg**, 0.25 **Cr**, 0.35 **Zn**, 0.25 **Ti**, 0.50 **Fe**, 0.05 each, 0.15 total, others	
QQ–A–371F	355.1 (formerly 355)	US	Ingot	rem **Al**, 4.5–5.5 **Si**, 1.0–1.5 **Cu**, 0.25–0.50 **Mn**, 0.45–0.6 **Mg**, 0.25 **Cr**, 0.30 **Zn**, 0.50 **Fe**, 0.05 each, 0.15 total, others	
QQ–A–371F	C355.2 (formerly C355)	US	Ingot	rem **Al**, 4.5–5.5 **Si**, 1.0–1.5 **Cu**, 0.05 **Mn**, 0.45–0.6 **Mg**, 0.05 **Zn**, 0.20 **Ti**, 0.13 **Fe**, 0.05 each, 0.15 total others	
QQ–A–371F	356.1 (formerly 356)	US	Ingot	rem **Al**, 4.5–5.5 **Si**, 1.0–1.5 **Cu**, 0.05 **Mn**, 0.45–0.6 **Mg**, 0.05 **Zn**, 0.20 **Ti**, 0.13 **Fe**, 0.05 each, 0.15 total, others	
ANSI/ASTM B 618	355.0	US	Investment castings	rem **Al**, 4.5–5.5 **Si**, 1.0–1.5 **Cu**, 0.50 **Mn**, 0.40–0.6 **Mg**, 0.25 **Cr**, 0.35 **Zn**, 0.25 **Ti**, 0.6 **Fe**, 0.05 each, 0.15 total, others	**Solution heat treated and artificially aged**: ≥0.250 in. **diam**, 32 ksi (221 MPa) **TS**, 20 ksi (138 MPa) **YS**, 2% in 2 in. (50 mm) **El**; **T51**: ≥0.250 in. **diam**, 25 ksi (172 MPa) **TS**, 18 ksi (124 MPa) **YS**; **T71**: ≥0.250 in. **diam**, 30 ksi (207 MPa) **TS**, 22 ksi (152 MPa) **YS**
ANSI/ASTM B 618	C355.0	US	Investment castings	rem **Al**, 4.5–5.5 **Si**, 1.0–1.5 **Cu**, 0.10 **Mn**, 0.40–0.6 **Mg**, 0.10 **Zn**, 0.20 **Ti**, 0.20 **Fe**, 0.05 each, 0.15 total, others	**Solution heat treated and artificially aged**: ≥0.250 in. **diam**, 36 ksi (248 MPa) **TS**, 25 ksi (172 MPa) **YS**, 2.5% in 2 in. (50 mm) **El**
ANSI/ASTM B 108	355.0	US	Permanent mold castings	rem **Al**, 4.5–5.5 **Si**, 1.0–1.5 **Cu**, 0.50 **Mn**, 0.40–0.6 **Mg**, 0.25 **Cr**, 0.35 **Zn**, 0.25 **Ti**, 0.6 **Fe**, 0.05 each, 0.15 total, others	**T51**: all **diam**, 27 ksi (186 MPa) **TS**, 75 **HB**; **T62**: all **diam**, 42 ksi (290 MPa) **TS**, 105 **HB**; **Solution heat treated and stabilized**: all **diam**, 36 ksi (248 MPa) **TS**, 90 **HB**

CAST ALUMINUM AND ALUMINUM ALLOYS

specification number	designation	country	product forms	chemical composition	mechanical properties and hardness values
ANSI/ASTM B 108	C355.0	US	Permanent mold castings	rem **Al**, 4.5–5.5 **Si**, 1.0–1.5 **Cu**, 0.10 **Mn**, 0.40–0.6 **Mg**, 0.10 **Zn**, 0.20 **Ti**, 0.20 **Fe**, 0.05 each, 0.15 total, others	**T61**: all **diam**, 37 ksi (255 MPa) **TS**, 30 ksi (207 MPa) **YS**, 1% in 2 in. (50 mm) **El**
ANSI/ASTM B 26	C355.0	US	Sand castings	rem **Al**, 4.5–5.5 **Si**, 1.0–1.5 **Cu**, 0.10 **Mn**, 0.40–0.6 **Mg**, 0.10 **Zn**, 0.20 **Ti**, 0.20 **Fe**, 0.05 each, 0.15 total, others	**Solution heat treated and artificially aged**: ≥0.250 in. **diam**, 36 ksi (248 MPa) **TS**, 25 ksi (172 MPa) **YS**, 3% in 2 in. (50 mm) **El**
ANSI/ASTM B 26	355.0	US	Sand castings	rem **Al**, 4.5–5.5 **Si**, 1.0–1.5 **Cu**, 0.50 **Mn**, 0.40–0.6 **Mg**, 0.25 **Cr**, 0.35 **Zn**, 0.25 **Ti**, 0.6 **Fe**, 0.05 each, 0.15 total, others	**Solution heat treated and artificially aged**: ≥0.250 in. **diam**, 32 ksi (221 MPa) **TS**, 20 ksi (138 MPa) **YS**, 2% in 2 in. (50 mm) **El**, 80 **HB**; **T51**: ≥0.250 in. **diam**, 25 ksi (172 MPa) **TS**, 18 ksi (124 MPa) **YS**, 65 **HB**; **T71**: ≥0.250 in. **diam**, 30 ksi (207 MPa) **TS**, 22 ksi (152 MPa) **YS**, 75 **HB**
BS 3L.78		UK	Ingot and castings	rem **Al**, 4.5–5.5 **Si**, 1.0–1.5 **Cu**, 0.5 **Mn**, 0.4–0.6 **Mg**, 0.25 **Ni**, 0.10 **Zn**, 0.25 **Ti**, 0.05 **Sn**, 0.6 **Fe**, 0.05 **Pb**	**Heat treated and quenched**: all **diam**, 250 MPa **TS**, 220 MPa **YS**
CSA HA.10	.SC51N	Canada	Mold castings	rem **Al**, 4.5–5.5 **Si**, 1.0–1.5 **Cu**, 0.30 **Mn**, 0.40–0.6 **Mg**, 0.35 **Zn**, 0.25 **Ti**, 0.6 **Fe**, 0.05 each, 0.15 total, others	**T6**: all **diam**, 37.0 ksi (255 MPa) **TS**, 23.0 ksi (159 MPa) **YS**, 1.5% in 2 in. (50 mm) **El**; **T6B**: all **diam**, 42.0 ksi (290 MPa) **TS**; **T5**: all **diam**, 27.0 ksi (186 MPa) **TS**
CSA HA.10	.SC51P	Canada	Mold castings	rem **Al**, 4.5–5.5 **Si**, 1.0–1.5 **Cu**, 0.10 **Mn**, 0.40–0.6 **Mg**, 0.10 **Zn**, 0.20 **Ti**, 0.20 **Fe**, 0.05 each, 0.15 total, others	**T6A**: all **diam**, 40.0 ksi (276 MPa) **TS**, 30.0 ksi (207 MPa) **YS**, 3% in 2 in. (50 mm) **El**
CSA HA.9	.SG51N	Canada	Sand castings	rem **Al**, 4.5–5.5 **Si**, 1.0–1.5 **Cu**, 0.30 **Mn**, 0.40–0.6 **Mg**, 0.35 **Zn**, 0.25 **Ti**, 0.6 **Fe**, 0.05 each, 0.15 total, others	**T6**: all **diam**, 32.0 ksi (221 MPa) **TS**, 20.0 ksi (138 MPa) **YS**, 2% in 2 in. (50 mm) **El**; **T6A**: all **diam**, 37.0 ksi (255 MPa) **TS**; **T5**: all **diam**, 25.0 ksi (172 MPa) **TS**
CSA HA.9	.SC51P	Canada	Sand castings	rem **Al**, 4.5–5.5 **Si**, 1.0–1.5 **Cu**, 0.10 **Mn**, 0.40–0.6 **Mg**, 0.10 **Zn**, 0.20 **Ti**, 0.20 **Fe**, 0.05 each, 0.15 total, others	**T6**: all **diam**, 36.0 ksi (248 MPa) **TS**, 25.0 ksi (172 MPa) **YS**, 2.5% in 2 in. (50 mm) **El**; **T6A**: all **diam**, 40.0 ksi (276 MPa) **TS**, 30.0 ksi (207 MPa) **YS**, 3% in 2 in. (50 mm) **El**

CAST ALUMINUM AND ALUMINUM ALLOYS

specification number	designation	country	product forms	chemical composition	mechanical properties and hardness values
QQ–A–601E	355.0	US	Castings	rem **Al**, 4.5–5.5 **Si**, 1.0–1.5 **Cu**, 0.50 **Mn**, 0.40–0.6 **Mg**, 0.25 **Cr**, 0.35 **Zn**, 0.25 **Ti**, 0.6 **Fe**, 0.05 each, 0.15 total, others	**Artificially aged only and temper T51**: all **diam**, 25 ksi (172 MPa) **TS**, 23 ksi (159 MPa) **YS**, 65 **HB**; **Solution heat treated, artificially aged and temper T6**: all **diam**, 32 ksi (221 MPa) **TS**, 20 ksi (138 MPa) **YS**, 2.0% in 2 in. (50 mm) **El**, 80 **HB**; **Solution heat treated, stabilized and temper T71**: all **diam**, 30 ksi (207 MPa) **TS**, 29 ksi (200 MPa) **YS**, 75 **HB**
QQ–A–601E	C355.0	US	Castings	rem **Al**, 4.5–5.5 **Si**, 1.0–1.5 **Cu**, 0.10 **Mn**, 0.40–0.6 **Mg**, 0.10 **Zn**, 0.20 **Ti**, 0.20 **Fe**, 0.05 each, 0.15 total, others	**Solution heat treated, artificially aged and T6**: all **diam**, 36 ksi (248 MPa) **TS**, 25 ksi (172 MPa) **YS**, 2.5% in 2 in. (50 mm) **El**, 85 **HB**
QQ–A–596d	355	US	Castings	rem **Al**, 4.5–5.5 **Si**, 1.0–1.5 **Cu**, 0.40–0.6 **Mn**, 0.50 **Mg**, 0.25 **Cr**, 0.35 **Zn**, 0.25 **Ti**, 0.8 **Fe**, 0.05 each, 0.15 total, others	**Solution heat treated and artificially aged and T6**: all **diam**, 37 ksi (255 MPa) **TS**, 1.5% in 2 in. (50 mm) **El**; **Artificially aged only and temper T51**: all **diam**, 27 ksi (186 MPa) **TS**; **Solution heat treated and artificially aged and T62**: all **diam**, 42 ksi (290 MPa) **TS**
JIS H 2211	Class 4D	Japan	Ingot: for castings	rem **Al**, 4.5–5.5 **Si**, 1.0–1.5 **Cu**, 0.03 **Mn**, 0.40–0.6 **Mg**, 0.03 **Ni**, 0.03 **Zn**, 0.03 **Ti**, 0.3 **Fe**	
JIS H 2117	Class 4D	Japan	Ingot: for castings	rem **Al**, 4.5–5.5 **Si**, 1.0–1.5 **Cu**, 0.5 **Mn**, 0.40–0.6 **Mg**, 0.1 **Ni**, 0.3 **Zn**, 0.2 **Ti**, 0.5 **Fe**	
MIL–A–21180C	C355.0	US	Castings	rem **Al**, 4.5–5.5 **Si**, 1.0–1.5 **Cu**, 0.10 **Mn**, 0.40–0.6 **Mg**, 0.10 **Zn**, 0.20 **Ti**, 0.20 **Fe**, 0.05 each, 0.15 total, others	**Class 1**: all **diam**, 283 MPa **TS**, 214 MPa **YS**, 3% in 2 in. (50 mm) **El**; **Class 2**: all **diam**, 303 MPa **TS**, 228 MPa **YS**, 3% in 2 in. (50 mm) **El**; **Class**: all **diam**, 345 MPa **TS**, 276 MPa **YS**, 2% in 2 in. (50 mm) **El**
QQ–A–596d	C355	US	Castings	rem **Al**, 4.5–5.5 **Si**, 1.0–1.5 **Cu**, 0.40–0.6 **Mn**, 0.10 **Mg**, 0.10 **Zn**, 0.20 **Ti**, 0.20 **Fe**, 0.05 each, 0.15 total, others	**Solution heat treated and artificially aged–T61**: all **diam**, 40 ksi (276 MPa) **TS**, 3.0% in 2 in. (50 mm) **El**
DIN 1725 sheet 2	GK–AlSi5Mg wa/3.2341.62	Germany	Gravity die castings	rem **Al**, 5.0–6.0 **Si**, 0.05 **Cu**, 0.4 **Mn**, 0.4–0.8 **Mg**, 0.10 **Zn**, 0.20 **Ti**, 0.5 **Fe**, 0.05 each, 0.15 total, others	**Quenched and age hardened**: all **diam**, 260 MPa **TS**, 240 MPa **YS**, 1% in 2 in. (50 mm) **El**, 90–110 **HB**
DIN 1725 sheet 2	GK–AlSi5Mg ka/3.2341.42	Germany	Gravity die castings	rem **Al**, 5.0–6.0 **Si**, 0.05 **Cu**, 0.4 **Mn**, 0.4–0.8 **Mg**, 0.10 **Zn**, 0.20 **Ti**, 0.5 **Fe**, 0.05 each, 0.15 total, others	**As cast and precipitation hardened**: all **diam**, 210 MPa **TS**, 160 MPa **YS**, 2% in 2 in. (50 mm) **El**, 70–90 **HB**

CAST ALUMINUM AND ALUMINUM ALLOYS

specification number	designation	country	product forms	chemical composition	mechanical properties and hardness values
DIN 1725 sheet 2	GK–AlSi5Mg/3.2341.02	Germany	Gravity die castings	rem **Al**, 5.0–6.0 **Si**, 0.05 **Cu**, 0.40 **Mn**, 0.40–0.80 **Mg**, 0.10 **Zn**, 0.20 **Ti**, 0.50 **Fe**, 0.05 each, 0.15 total, others	**As cast**: all **diam**, 160 MPa **TS**, 120 MPa **YS**, 1% in 2 in. (50 mm) **El**, 60–75 **HB**
DIN 1725 sheet 2	G–AlSi5Mg wa/3.2341.61	Germany	Sand castings	rem **Al**, 5.0–6.0 **Si**, 0.05 **Cu**, 0.40 **Mn**, 0.40–0.80 **Mg**, 0.10 **Zn**, 0.20 **Ti**, 0.50 **Fe**	**Quenched and age hardened**: all **diam**, 240 MPa **TS**, 220 MPa **YS**, 0.5% in 2 in. (50 mm) **El**, 80–110 **HB**
DIN 1725 sheet 2	G–AlSi5Mg ka/3.2341.41	Germany	Sand castings	rem **Al**, 5.0–6.0 **Si**, 0.05 **Cu**, 0.40 **Mn**, 0.40–0.80 **Mg**, 0.10 **Zn**, 0.20 **Ti**, 0.50 **Fe**, 0.05 each, 0.15 total, others	**Precipitation hardened**: all **diam**, 180 MPa **TS**, 150 MPa **YS**, 2% in 2 in. (50 mm) **El**, 70–85 **HB**
DIN 1725 sheet 2	G–AlSi5Mg/3.2341.0 1	Germany	Sand castings	rem **Al**, 5.0–6.0 **Si**, 0.05 **Cu**, 0.40 **Mn**, 0.40–0.80 **Mg**, 0.10 **Zn**, 0.20 **Ti**, 0.50 **Fe**, 0.05 each, 0.15 total, others	**As cast**: all **diam**, 140 MPa **TS**, 100 MPa **YS**, 1% in 2 in. (50 mm) **El**, 55–70 **HB**
ISO R2147	Al–Si5Mg	ISO	Sand castings: test pieces	rem **Al**, 3.5–6.0 **Si**, 0.1 **Cu**, 0.6 **Mn**, 0.4–0.9 **Mg**, 0.1 **Ni**, 0.1 **Zn**, 0.2 **Ti**, 0.05 **Sn**, 0.6 **Fe**, 0.1 **Pb**	**Solution heat treated and artificially aged**: all **diam**, 230 MPa **TS**, 1% in 2 in. (50 mm) **El**
ISO R164	Al–Si5Mg	ISO	Castings	rem **Al**, 3.5–6.0 **Si**, 0.1 **Cu**, 0.6 **Mn**, 0.4–0.9 **Mg**, 0.1 **Ni**, 0.1 **Zn**, 0.2 **Ti**, 0.05 **Sn**, 0.6 **Fe**, 0.1 **Pb**	
ISO R164	Al–Si5Mg Fe	ISO	Pressure die castings	rem **Al**, 3.5–6.0 **Si**, 0.1 **Cu**, 0.6 **Mn**, 0.4–0.9 **Mg**, 0.1 **Ni**, 0.1 **Zn**, 0.2 **Ti**, 0.05 **Sn**, 1.3 **Fe**, 0.1 **Pb**	
SABS 991	Al–Si5MgA	South Africa	Gravity die castings	rem **Al**, 3.5–6.0 **Si**, 0.1 **Cu**, 0.5 **Mn**, 0.4–0.8 **Mg**, 0.1 **Ni**, 0.1 **Zn**, 0.2 **Ti**, 0.05 **Sn**, 0.6 **Fe**, 0.1 **Pb**, 0.05 each, 0.15 total, others	**Precipitation heat treated**: all **diam**, 185 MPa **TS**, 2% in 2 in. (50 mm) **El**; **Solution heat treated**: all **diam**, 230 MPa **TS**, 5% in 2 in. (50 mm) **El**; **Solution and precipitation heat treated**: all **diam**, 275 MPa **TS**, 2% in 2 in. (50 mm) **El**
SABS 990	Al–Si5MgA	South Africa	Sand castings	rem **Al**, 3.5–6.0 **Si**, 0.1 **Cu**, 0.5 **Mn**, 0.4–0.8 **Mg**, 0.1 **Ni**, 0.1 **Zn**, 0.2 **Ti**, 0.05 **Sn**, 0.6 **Fe**, 0.1 **Pb**, 0.05 each, 0.15 total, others	**Precipitation heat treated**: all **diam**, 145 MPa **TS**, 1% in 2 in. (50 mm) **El**; **Solution heat treated**: all **diam**, 160 MPa **TS**, 2.5% in 2 in. (50 mm) **El**
BS 3L.52		UK	Ingot and castings	rem **Al**, 0.6–2.0 **Si**, 1.3–3.0 **Cu**, 0.1 **Mn**, 0.5–1.7 **Mg**, 0.5–2.0 **Ni**, 0.1 **Zn**, 0.25 **Ti**, 0.05 **Sn**, 0.8–1.4 **Fe**, 0.05 **Pb**	**Heat treated and quenched**: all **diam**, 280 MPa **TS**, 245 MPa **YS**
BS 3L.51		UK	Ingot and castings	rem **Al**, 1.5–2.8 **Si**, 0.8–2.0 **Cu**, 0.1 **Mn**, 0.05–0.20 **Mg**, 0.8–1.7 **Ni**, 0.1 **Zn**, 0.25 **Ti**, 0.05 **Sn**, 0.8–1.4 **Fe**, 0.05 **Pb**	**Heat treated**: all **diam**, 160 MPa **TS**, 125 MPa **YS**, 2% in 2 in. (50 mm) **El**
SABS 711	Al Si20	South Africa	Ingot	rem **Al**, 18.0–22.0 **Si**, 0.10 **Cu**, 0.07 **Zn**, 0.70 **Fe**, 0.05 each, 0.15 total, others	

CAST ALUMINUM AND ALUMINUM ALLOYS

specification number	designation	country	product forms	chemical composition	mechanical properties and hardness values
AS 1874	CP401	Australia	Ingot, sand castings and permanent mould castings	rem **Al**, 12.0–13.0 **Si**, 0.10 **Cu**, 0.05 **Mn**, 0.05 **Mg**, 0.05 **Ni**, 0.10 **Zn**, 0.40 **Fe**, 0.05 each, 0.20 total, others	**As cast (sand castings)**: all **diam**, 160 MPa **TS**, 5% in 2 in. (50 mm) **El**
DS 3002	4260	Denmark	Ingot: for castings	rem **Al**, 11.0–13.5 **Si**, 0.6 **Cu**, 0.5 **Mn**, 0.3 **Mg**, 0.2 **Ni**, 0.5 **Zn**, 0.20 **Ti**, 0.1 **Sn**, 0.7 **Fe**, 0.1 **Pb**	**Sand cast**: all **diam**, 147 MPa **TS**, 78 MPa **YS**, 2% in 2 in. (50 mm) **El**, 50–70 **HB**; **Chill cast**: all **diam**, 157 MPa **TS**, 88 MPa **YS**, 2% in 2 in. (50 mm) **El**, 55–75 **HB**
DS 3002	4261	Denmark	Ingot: for castings	rem **Al**, 11.0–13.5 **Si**, 0.20 **Cu**, 0.5 **Mn**, 0.1 **Mg**, 0.1 **Ni**, 0.3 **Zn**, 0.20 **Ti**, 0.05 **Sn**, 0.6 **Fe**, 0.1 **Pb**	**Sand cast**: all **diam**, 167 MPa **TS**, 78 MPa **YS**, 4% in 2 in. (50 mm) **El**, 45–65 **HB**; **Chill cast**: all **diam**, 176 MPa **TS**, 88 MPa **YS**, 5% in 2 in. (50 mm) **El**, 50–70 **HB**
SFS 2566	G–AlSi12	Finland	Die castings	rem **Al**, 11.0–13.5 **Si**, 0.10 **Cu**, 0.5 **Mn**, 0.10 **Mg**, 0.1 **Ni**, 0.1 **Zn**, 0.15 **Ti**, 0.05 **Sn**, 0.60 **Fe**, 0.1 **Pb**, 0.05 each, 0.15 total, others	**As cast**: $\leq$20 mm **diam**, 180 MPa **TS**, 90 MPa **YS**, 5% in 2 in. (50 mm) **El**
NF A 57 703	A–S12–Y4	France	Pressure die castings	rem **Al**, 11.0–13.5 **Si**, 0.6 **Cu**, 0.3 **Mn**, 0.2 **Mg**, 0.5 **Ni**, 0.5 **Zn**, 0.2 **Ti**, 0.1 **Sn**, 1.3 **Fe**	**As cast**: all **diam**, 170 MPa **TS**, 2% in 2 in. (50 mm) **El**
DIN 1725 sheet 2	GD–AlSi12/3.2582.05	Germany	Pressure die castings	rem **Al**, 11.0–13.5 **Si**, 0.10 **Cu**, 0.40 **Mn**, 0.05 **Mg**, 0.10 **Zn**, 0.15 **Ti**, 1.0 **Fe**, 0.05 each, 0.15 total, other	**As cast**: all **diam**, 220 MPa **TS**, 140 MPa **YS**, 1% in 4 in. (100 mm) **El**, 60–80 **HB**
DIN 1725 sheet 2	GK–AlSi12g/3.2581.45	Germany	Gravity die castings	rem **Al**, 11.0–13.5 **Si**, 0.05 **Cu**, 0.40 **Mn**, 0.05 **Mg**, 0.10 **Zn**, 0.15 **Ti**, 0.50 **Fe**, 0.05 each, 0.15 total, others	**Annealed and quenched**: all **diam**, 180 MPa **TS**, 80 MPa **YS**, 6% in 2 in. (50 mm) **El**, 50–60 **HB**
DIN 1725 sheet 2	GK–AlSi12/3.2581.02	Germany	Gravity die castings	rem **Al**, 11.0–13.5 **Si**, 0.05 **Cu**, 0.40 **Mn**, 0.05 **Mg**, 0.10 **Zn**, 0.15 **Ti**, 0.50 **Fe**	**As cast**: all **diam**, 160 MPa **TS**, 80 MPa **YS**, 6% in 2 in. (50 mm) **El**, 50–60 **HB**
DIN 1725 sheet 2	G–AlSi12g/3.2581.44	Germany	Sand casting	rem **Al**, 11.0–13.5 **Si**, 0.05 **Cu**, 0.40 **Mn**, 0.05 **Mg**, 0.10 **Zn**, 0.15 **Ti**, 0.50 **Fe**, 0.05 each, 0.15 total, others	**Annealed and quenched**: all **diam**, 160 MPa **TS**, 80 MPa **YS**, 6% in 2 in. (50 mm) **El**, 50–60 **HB**
DIN 1725 sheet 2	G–AlSi12/3.2581.01	Germany	Sand castings	rem **Al**, 11.0–13.5 **Si**, 0.05 **Cu**, 0.4 **Mn**, 0.05 **Mg**, 0.10 **Zn**, 0.15 **Ti**, 0.5 **Fe**, 0.05 each, 0.15 total, others	**As cast**: all **diam**, 160 MPa **TS**, 70 MPa **YS**, 5% in 2 in. (50 mm) **El**, 45–60 **HB**
ISO R2147	Al–Si12	ISO	Sand castings: test pieces	rem **Al**, 11.0–13.5 **Si**, 0.10 **Cu**, 0.5 **Mn**, 0.10 **Mg**, 0.1 **Ni**, 0.1 **Zn**, 0.15 **Ti**, 0.05 **Sn**, 0.70 **Fe**, 0.1 **Pb**	**As cast**: all **diam**, 160 MPa **TS**, 4% in 2 in. (50 mm) **El**
ISO R164	Al–Si12	ISO	Castings	rem **Al**, 11.0–13.5 **Si**, 0.10 **Cu**, 0.5 **Mn**, 0.10 **Mg**, 0.1 **Ni**, 0.1 **Zn**, 0.15 **Ti**, 0.05 **Sn**, 0.70 **Fe**, 0.1 **Pb**	

CAST ALUMINUM AND ALUMINUM ALLOYS

specification number	designation	country	product forms	chemical composition	mechanical properties and hardness values
ISO R164	Al–Si12Fe	ISO	Pressure die castings	rem **Al**, 11.0–13.5 **Si**, 0.10 **Cu**, 0.5 **Mn**, 0.10 **Mg**, 0.1 **Ni**, 0.1 **Zn**, 0.15 **Ti**, 0.05 **Sn**, 1.3 **Fe**, 0.1 **Pb**	
ISO R164	Al–Si12Cu	ISO	Castings	rem **Al**, 11.0–13.5 **Si**, 1.2 **Cu**, 0.5 **Mn**, 0.3 **Mg**, 0.2 **Ni**, 0.5 **Zn**, 0.2 **Ti**, 0.1 **Sn**, 0.8 **Fe**, 0.1 **Pb**	
ISO R164	Al–Si12CuFe	ISO	Pressure die castings	rem **Al**, 11.0–13.5 **Si**, 1.2 **Cu**, 0.5 **Mn**, 0.3 **Mg**, 0.2 **Ni**, 0.5 **Zn**, 0.2 **Ti**, 0.1 **Sn**, 1.3 **Fe**, 0.1 **Pb**	
UNI 5079	GD–AlSi13Fe	Italy	Pressure die castings	rem **Al**, 11.5–13.0 **Si**, 0.80 **Cu**, 0.30 **Mn**, 0.30 **Mg**, 0.20 **Ni**, 0.50 **Zn**, 0.15 **Ti**, 0.10 **Sn**, 0.7–1.0 **Fe**, 0.15 **Pb**	**As cast**: all **diam**, 225 MPa **TS**, 130 MPa **YS**, 1.5% in 2 in. (50 mm) **El**, 75–95 **HB**
NS 17 512	NS 17 512–00	Norway	Ingot: for casting	88 **Al**, 11.0–13.5 **Si**, 0.6 **Cu**, 0.5 **Mn**, 0.3 **Mg**, 0.2 **Ni**, 0.5 **Zn**, 0.20 **Ti**, 0.1 **Sn**, 0.7 **Fe**, 0.1 **Pb**	
NS 17 512	NS 17 512–01	Norway	Sand castings	88 **Al**, 11.0–13.5 **Si**, 0.6 **Cu**, 0.5 **Mn**, 0.3 **Mg**, 0.2 **Ni**, 0.5 **Zn**, 0.20 **Ti**, 0.1 **Sn**, 0.7 **Fe**, 0.1 **Pb**	**As cast**: all **diam**, 147 MPa **TS**, 78 MPa **YS**, 2% in 2 in. (50 mm) **El**
NS 17 512	NS 17 512–02	Norway	Chill castings	88 **Al**, 11.0–13.5 **Si**, 0.6 **Cu**, 0.5 **Mn**, 0.3 **Mg**, 0.2 **Ni**, 0.5 **Zn**, 0.20 **Ti**, 0.1 **Sn**, 0.7 **Fe**, 0.1 **Pb**	**As cast**: all **diam**, 157 MPa **TS**, 88 MPa **YS**, 2% in 2 in. (50 mm) **El**
NS 17 512	NS 17 512–05	Norway	Pressure die castings	88 **Al**, 11.0–13.5 **Si**, 0.6 **Cu**, 0.5 **Mn**, 0.3 **Mg**, 0.2 **Ni**, 0.5 **Zn**, 0.20 **Ti**, 0.1 **Sn**, 1.3 **Fe**, 0.1 **Pb**	**As cast**: all **diam**, 196 MPa **TS**, 147 MPa **YS**, 2% in 2 in. (50 mm) **El**
SIS 14 42 60	SIS Al 42 60–06	Sweden	Chill castings	rem **Al**, 11.0–13.5 **Si**, 0.6 **Cu**, 0.5 **Mn**, 0.3 **Mg**, 0.5 **Zn**, 0.20 **Ti**, 0.1 **Sn**, 0.7 **Fe**, 0.1 **Pb**	**As cast**: all **diam**, 160 MPa **TS**, 90 MPa **YS**, 2% in 2 in. (50 mm) **El**, 55–75 **HB**
SIS 14 42 60	SIS Al 42 60–03	Sweden	Sand castings	rem **Al**, 11.0–13.5 **Si**, 0.6 **Cu**, 0.5 **Mn**, 0.3 **Mg**, 0.2 **Ni**, 0.5 **Zn**, 0.20 **Ti**, 0.1 **Sn**, 0.7 **Fe**, 0.1 **Pb**	**As cast**: all **diam**, 150 MPa **TS**, 80 MPa **YS**, 2% in 2 in. (50 mm) **El**, 50–70 **HB**
SIS 14 42 60	SIS Al 42 60–00	Sweden	Ingot	rem **Al**, 11.0–13.5 **Si**, 0.6 **Cu**, 0.5 **Mn**, 0.3 **Mg**, 0.2 **Ni**, 0.5 **Zn**, 0.20 **Ti**, 0.1 **Sn**, 0.7 **Fe**, 0.1 **Pb**	
SIS 14 42 61	SIS Al 42 61–06	Sweden	Chill castings	rem **Al**, 11.0–13.5 **Si**, 0.20 **Cu**, 0.5 **Mn**, 0.1 **Mg**, 0.1 **Ni**, 0.3 **Zn**, 0.20 **Ti**, 0.05 **Sn**, 0.6 **Fe**, 0.1 **Pb**	**As cast**: all **diam**, 180 MPa **TS**, 90 MPa **YS**, 5% in 2 in. (50 mm) **El**, 50–70 **HB**
SIS 14 42 61	SIS Al 42 61–03	Sweden	Sand castings	rem **Al**, 11.0–13.5 **Si**, 0.20 **Cu**, 0.5 **Mn**, 0.1 **Mg**, 0.1 **Ni**, 0.3 **Zn**, 0.20 **Ti**, 0.05 **Sn**, 0.6 **Fe**, 0.1 **Pb**	**As cast**: all **diam**, 170 MPa **TS**, 80 MPa **YS**, 4% in 2 in. (50 mm) **El**, 45–65 **HB**
SIS 14 42 61	SIS 42 61–00	Sweden	Ingot	rem **Al**, 11.0–13.5 **Si**, 0.20 **Cu**, 0.5 **Mn**, 0.1 **Mg**, 0.1 **Ni**, 0.3 **Zn**, 0.20 **Ti**, 0.05 **Sn**, 0.6 **Fe**, 0.1 **Pb**	
ANSI/ASTM B 179	413.2/S12C	US	Ingot: for die castings	rem **Al**, 11.0–13.0 **Si**, 0.10 **Cu**, 0.10 **Mn**, 0.07 **Mg**, 0.10 **Ni**, 0.10 **Zn**, 0.10 **Sn**, 0.7–1.1 **Fe**, 0.20 each, others	

CAST ALUMINUM AND ALUMINUM ALLOYS

specification number	designation	country	product forms	chemical composition	mechanical properties and hardness values
ANSI/ASTM B 179	A413.1/S12A–B	US	Ingot: for die castings	rem **Al**, 11.0–13.0 **Si**, 1.0 **Cu**, 0.35 **Mn**, 0.10 **Mg**, 0.50 **Ni**, 0.40 **Zn**, 0.15 **Sn**, 1.0 **Fe**, 0.25 total, others	
ANSI/ASTM B 179	A413.2	US	Ingot: for die castings	rem **Al**, 11.0–13.0 **Si**, 0.10 **Cu**, 0.05 **Mn**, 0.03 **Mg**, 0.05 **Ni**, 0.05 **Zn**, 0.05 **Sn**, 0.6 **Fe**, 0.10 total, others	
AS 1874	BP401	Australia	Ingot, sand castings and permanent mould castings	rem **Al**, 11.0–13.0 **Si**, 0.10 **Cu**, 0.10 **Mn**, 0.05 **Mg**, 0.05 **Ni**, 0.10 **Zn**, 0.20 **Ti**, 0.40 **Fe**, 0.05 each, 0.15 total, others	**As cast (sand castings)**: all **diam**, 160 MPa **TS**, 5% in 2 in. (50 mm) **El**
AS 1874	BS401	Australia	Ingot, sand castings and permanent mould castings	rem **Al**, 11.0–13.0 **Si**, 0.15 **Cu**, 0.5 **Mn**, 0.10 **Mg**, 0.10 **Cr**, 0.10 **Ni**, 0.15 **Zn**, 0.20 **Ti**, 0.05 **Sn**, 0.6 **Fe**, 0.15 **Pb**, 0.05 each, 0.20 total, others	**As cast (sand castings)**: all **diam**, 160 MPa **TS**, 5% in 2 in. (50 mm) **El**
CSA HA.3	.S12N	Canada	Ingot	rem **Al**, 11.0–13.0 **Si**, 0.05 **Cu**, 0.30 **Mn**, 0.05 **Zn**, 0.6 **Fe**, 0.05 each, 0.15 total, others	
CSA HA.9	.S12N	Canada	Sand castings	rem **Al**, 11.0–13.0 **Si**, 0.10 **Cu**, 0.30 **Mn**, 0.10 **Zn**, 0.6 **Fe**, 0.05 each, 0.15 total, others	
QQ-A-371F	413.2 (formerly 13)	US	Ingot	rem **Al**, 11.0–13.0 **Si**, 0.10 **Cu**, 0.10 **Mn**, 0.07 **Mg**, 0.10 **Ni**, 0.10 **Zn**, 0.10 **Sn**, 0.7–1.1 **Fe**, 0.20 total, others	
QQ-A-371F	A413.1 (formerly A13)	US	Ingot	rem **Al**, 11.0–13.0 **Si**, 0.6 **Cu**, 0.35 **Mn**, 0.10 **Mg**, 0.50 **Ni**, 0.40 **Zn**, 0.15 **Sn**, 1.0 **Fe**, 0.25 total, others	
QQ-A-371F	A413.2 (formerly A13)	US	Ingot	rem **Al**, 11.0–13.0 **Si**, 0.10 **Cu**, 0.05 **Mn**, 0.03 **Mg**, 0.05 **Ni**, 0.05 **Sn**, 0.6 **Fe**, 0.10 total, others	
QQ-A-591E	A413.0	US	Castings	rem **Al**, 11.0–13.0 **Si**, 0.6 **Cu**, 0.35 **Mn**, 0.10 **Mg**, 0.50 **Ni**, 0.50 **Zn**, 0.15 **Sn**, 1.3 **Fe**, 0.25 total, others	**Die cast**: all **diam**, 42 ksi (290 MPa) **TS**, 19 ksi (131 MPa) **YS**, 3.5% in 2 in. (50 mm) **El**
QQ-A-591E	413.0	US	Castings	rem **Al**, 11.0–13.0 **Si**, 0.6 **Cu**, 0.35 **Mn**, 0.10 **Mg**, 0.50 **Ni**, 0.50 **Zn**, 0.15 **Sn**, 2.0 **Fe**, 0.25 total, others	**Die cast**: all **diam**, 43 ksi (296 MPa) **TS**, 21 ksi (145 MPa) **YS**, 2.5% in 2 in. (50 mm) **El**
JIS H 2212	Class 1	Japan	Ingot: for die castings	rem **Al**, 11.0–13.0 **Si**, 0.05 **Cu**, 0.03 **Mn**, 0.03 **Mg**, 0.03 **Ni**, 0.03 **Zn**, 0.03 **Sn**, 0.3–0.6 **Fe**	
JIS H 2118	Class 1	Japan	Ingot: for die castings	rem **Al**, 11.0–13.0 **Si**, 1.0 **Cu**, 0.3 **Mn**, 0.3 **Mg**, 0.5 **Ni**, 0.5 **Zn**, 0.1 **Sn**, 0.9 **Fe**	
ANSI/ASTM B 85	413.0/S12B	US	Die castings	rem **Al**, 11.0–13.0 **Si**, 1.0 **Cu**, 0.35 **Mn**, 0.10 **Mg**, 0.50 **Ni**, 0.50 **Zn**, 0.15 **Sn**, 2.0 **Fe**, 0.25 total, others	**As cast**: all **diam**, 43 ksi (300 MPa) **TS**, 21 ksi (140 MPa) **YS**, 3% in 2 in. (50 mm) **El**

CAST ALUMINUM AND ALUMINUM ALLOYS

specification number	designation	country	product forms	chemical composition	mechanical properties and hardness values
ANSI/ASTM B 85	A413.0/S12A	US	Die castings	rem **Al**, 11.0–13.0 **Si**, 1.0 **Cu**, 0.35 **Mn**, 0.10 **Mg**, 0.50 **Ni**, 0.50 **Zn**, 0.15 **Sn**, 1.3 **Fe**, 0.25 total, others	**As cast**: all **diam**, 42 ksi (290 MPa) **TS**, 19 ksi (130 MPa) **YS**, 4% in 2 in. (50 mm) **El**
AS 1874	AS401	Australia	Ingot, sand castings and permanent mould castings	rem **Al**, 10.0–13.0 **Si**, 0.6 **Cu**, 0.5 **Mn**, 0.25 **Mg**, 0.10 **Cr**, 0.5 **Ni**, 0.40 **Zn**, 0.20 **Ti**, 0.15 **Sn**, 1.0 **Fe**, 0.15 **Pb**, 0.25 total, others	
AS 1874	AS607	Australia	Ingot, sand castings and permanent mould castings	rem **Al**, 10.0–13.0 **Si**, 0.15 **Cu**, 0.30–0.7 **Mn**, 0.20–0.6 **Mg**, 0.15 **Ni**, 0.15 **Zn**, 0.25 **Ti**, 0.05 **Sn**, 0.6 **Fe**, 0.05 each, 0.20 total, others	**Cooled and naturally aged**: all **diam**, 190 MPa **TS**, 3% in 2 in. (50 mm) **El**; **Cooled and artificially aged (sand castings)**: all **diam**, 170 MPa **TS**, 1.5% in 2 in. (50 mm) **El**; **Solution heat–treated and artificially aged**: all **diam**, 240 MPa **TS**
IS 202	A–6	India	Sand castings	rem **Al**, 10.0–13.0 **Si**, 0.10 **Cu**, 0.50 **Mn**, 0.10 **Mg**, 0.10 **Ni**, 0.10 **Zn**, 0.05 **Sn**, 0.60 **Fe**, 0.10 **Pb**	**As cast**: all **diam**, 162 MPa **TS**, 54 MPa **YS**, 5% in 2 in. (50 mm) **El**
IS 202	A–9	India	Sand castings	rem **Al**, 10.0–13.0 **Si**, 0.10 **Cu**, 0.30–0.70 **Mn**, 0.20–0.60 **Mg**, 0.10 **Ni**, 0.10 **Zn**, 0.05 **Sn**, 0.60 **Fe**, 0.10 **Pb**, 0.20 **Ti+Nb**	**Solution and precipitation treated**: 239 MPa **TS**, 201 MPa **YS**
IS 202	A–6	India	Sand castings	rem **Al**, 10.0–13.0 **Si**, 0.10 **Cu**, 0.50 **Mn**, 0.10 **Mg**, 0.10 **Ni**, 0.10 **Zn**, 0.05 **Sn**, 0.60 **Fe**, 0.10 **Pb**	**As cast**: 162 MPa **TS**, 54 MPa **YS**, 5% in 2 in. (50 mm) **El**
JIS H 5202	Class 3 A	Japan	Castings	rem **Al**, 10.0–13.0 **Si**, 0.2 **Cu**, 0.3 **Mn**, 0.1 **Mg**, 0.3 **Zn**, 0.8 **Fe**	**As cast**: all **diam**, 177 MPa **TS**, 5% in 2 in. (50 mm) **El**, 50 **HB**
JIS H 2211	Class 3A	Japan	Ingot: for castings	rem **Al**, 10.0–13.0 **Si**, 0.05 **Cu**, 0.03 **Mn**, 0.03 **Mg**, 0.03 **Ni**, 0.03 **Zn**, 0.03 **Ti**, 0.3 **Fe**	
JIS H 2117	Class 3A	Japan	Ingot: for castings	rem **Al**, 10.0–13.0 **Si**, 0.2 **Cu**, 0.3 **Mn**, 0.1 **Mg**, 0.1 **Ni**, 0.3 **Zn**, 0.2 **Ti**, 0.7 **Fe**	
SABS 992	Al–Si12FeA	South Africa	Pressure die castings	rem **Al**, 10.0–13.0 **Si**, 0.6 **Cu**, 0.5 **Mn**, 0.10 **Mg**, 0.50 **Ni**, 0.50 **Zn**, 0.15 **Ti**, 0.15 **Sn**, 1.2 **Fe**, 0.20 total, others	**As cast**: all **diam**, 205 MPa **TS**, 115 MPa **YS**, 1% in 2 in. (50 mm) **El**, 70–90 **HB**
SABS 991	Al–Si12B	South Africa	Gravity die castings	rem **Al**, 10.0–13.0 **Si**, 0.10 **Cu**, 0.5 **Mn**, 0.10 **Mg**, 0.1 **Ni**, 0.1 **Zn**, 0.20 **Ti**, 0.05 **Sn**, 0.7 **Fe**, 0.1 **Pb**, 0.05 each, 0.15 total, others	**As cast**: all **diam**, 185 MPa **TS**, 7% in 2 in. (50 mm) **El**
SABS 991	Al–Si12C	South Africa	Gravity die castings	rem **Al**, 10.0–13.0 **Si**, 0.4 **Cu**, 0.5 **Mn**, 0.15 **Mg**, 0.1 **Ni**, 0.2 **Zn**, 0.2 **Ti**, 0.05 **Sn**, 0.7 **Fe**, 0.1 **Pb**, 0.05 each, 0.15 total, others	**As cast**: all **diam**, 185 MPa **TS**, 5% in 2 in. (50 mm) **El**
SABS 991	Al–Si12A	South Africa	Gravity die castings	rem **Al**, 10.0–13.0 **Si**, 0.05 **Cu**, 0.5 **Mn**, 0.05 **Mg**, 0.10 **Zn**, 0.15 **Ti**, 0.5 **Fe**, 0.05 each, 0.15 total, others	**As cast**: all **diam**, 235 MPa **TS**, 7% in 2 in. (50 mm) **El**

CAST ALUMINUM AND ALUMINUM ALLOYS

specification number	designation	country	product forms	chemical composition	mechanical properties and hardness values
SABS 990	Al–Si12A	South Africa	Sand castings	rem **Al**, 10.0–13.0 **Si**, 0.05 **Cu**, 0.5 **Mn**, 0.05 **Mg**, 0.10 **Zn**, 0.15 **Ti**, 0.5 **Fe**, 0.05 each, 0.15 total, others	**As cast**: all **diam**, 175 MPa **TS**, 6% in 2 in. (50 mm) **El**
SABS 990	Al–Si12B	South Africa	Sand castings	rem **Al**, 10.0–13.0 **Si**, 0.12 **Cu**, 0.5 **Mn**, 0.10 **Mg**, 0.1 **Ni**, 0.1 **Zn**, 0.2 **Ti**, 0.05 **Sn**, 0.7 **Fe**, 0.05 each, 0.15 total, others	**As cast**: all **diam**, 160 MPa **TS**, 5% in 2 in. (50 mm) **El**
SABS 989	Al–Si12FeA	South Africa	Ingot: for pressure die castings	rem **Al**, 10.0–13.0 **Si**, 0.6 **Cu**, 0.5 **Mn**, 0.10 **Mg**, 0.50 **Ni**, 0.50 **Zn**, 0.15 **Ti**, 0.15 **Sn**, 0.5–1.0 **Fe**, 0.20 total, others	**Chill cast**: all **diam**, 195 MPa **TS**, 3.0% in 2 in. (50 mm) **El**
SABS 989	Al–Si12B	South Africa	Ingot: for sand and gravity die castings	rem **Al**, 10.0–13.0 **Si**, 0.08 **Cu**, 0.5 **Mn**, 0.10 **Mg**, 0.1 **Ni**, 0.1 **Zn**, 0.20 **Ti**, 0.05 **Sn**, 0.6 **Fe**, 0.1 **Pb**, 0.05 each, 0.15 total, others	**Sand cast**: all **diam**, 160 MPa **TS**, 5% in 2 in. (50 mm) **El**; **Chill cast**: all **diam**, 185 MPa **TS**, 7% in 2 in. (50 mm) **El**
SABS 989	Al–Si12C	South Africa	Ingot: for gravity die castings	rem **Al**, 10.0–13.0 **Si**, 0.38 **Cu**, 0.5 **Mn**, 0.15 **Mg**, 0.1 **Ni**, 0.2 **Zn**, 0.2 **Ti**, 0.05 **Sn**, 0.6 **Fe**, 0.1 **Pb**, 0.05 each, 0.15 total, others	**Chill cast**: all **diam**, 185 MPa **TS**, 5% in 2 in. (50 mm) **El**
SABS 989	Al–Si12A	South Africa	Ingot: for sand and gravity die castings	rem **Al**, 10.0–13.0 **Si**, 0.03 **Cu**, 0.5 **Mn**, 0.05 **Mg**, 0.10 **Zn**, 0.15 **Ti**, 0.4 **Fe**, 0.05 each, 0.15 total, others	**Sand cast**: all **diam**, 175 MPa **TS**, 6% in 2 in. (50 mm) **El**; **Chill cast**: all **diam**, 200 MPa **TS**, 7% in 2 in. (50 mm) **El**
SFS 2567	G–AlSi10Mg	Finland	Die castings	rem **Al**, 9.0–11.0 **Si**, 0.10 **Cu**, 0.4 **Mn**, 0.20–0.50 **Mg**, 0.1 **Ni**, 0.1 **Zn**, 0.20 **Ti**, 0.05 **Sn**, 0.50 **Fe**, 0.05 **Pb**, 0.05 each, 0.15 total, others	**Solution treated and aged**: ≤20 mm **diam**, 260 MPa **TS**, 220 MPa **YS**, 1% in 2 in. (50 mm) **El**
NF A 57 703	A–59 G–Y4	France	Pressure die castings	rem **Al**, 9–11 **Si**, 0.3 **Cu**, 0.5 **Mn**, 0.15–0.9 **Mg**, 0.5 **Ni**, 0.5 **Zn**, 0.2 **Ti**, 0.1 **Sn**, 1.3 **Fe**	**As cast**: all **diam**, 180 MPa **TS**, 1% in 2 in. (50 mm) **El**
ISO R2147	Al–Si10Mg	ISO	Sand castings: test pieces	rem **Al**, 9.0–11.0 **Si**, 0.10 **Cu**, 0.6 **Mn**, 0.15–0.40 **Mg**, 0.1 **Ni**, 0.1 **Zn**, 0.15 **Ti**, 0.05 **Sn**, 0.70 **Fe**, 0.05 **Pb**	**Solution heat treated and artificially aged**: all **diam**, 220 MPa **TS**, 1% in 2 in. (50 mm) **El**
SIS 14 42 55	SIS Al 42 55–03	Sweden	Sand castings	rem **Al**, 9.0–11.0 **Si**, 0.20 **Cu**, 0.5 **Mn**, 0.1 **Mg**, 0.1 **Ni**, 0.3 **Zn**, 0.20 **Ti**, 0.05 **Sn**, 0.70 **Fe**, 0.1 **Pb**	**As cast**: all **diam**, 170 MPa **TS**, 80 MPa **YS**, 4% in 2 in. (50 mm) **El**, 50–70 **HB**
SIS 14 42 55	SIS Al 42 55–06	Sweden	Chill castings	86.74 min **Al**, 9.0–11.0 **Si**, 0.20 **Cu**, 0.5 **Mn**, 0.1 **Mg**, 0.1 **Ni**, 0.3 **Zn**, 0.20 **Ti**, 0.05 **Sn**, 0.70 **Fe**, 0.1 **Pb**	**As cast**: all **diam**, 180 MPa **TS**, 90 MPa **YS**, 5% in 2 in. (50 mm) **El**, 55–75 **HB**
SIS 14 42 55	SIS Al 42 55–00	Sweden	Ingot	rem **Al**, 9.0–11.0 **Si**, 0.20 **Cu**, 0.5 **Mn**, 0.1 **Mg**, 0.1 **Ni**, 0.3 **Zn**, 0.20 **Ti**, 0.05 **Sn**, 0.65 **Fe**, 0.1 **Pb**	
AMS 4290F		US	Die castings	rem **Al**, 9.0–10.0 **Si**, 0.6 **Cu**, 0.30 **Mn**, 0.40–0.6 **Mg**, 0.50 **Ni**, 0.50 **Zn**, 0.10 **Sn**, 2.0 **Fe**, 0.20 total, others	

CAST ALUMINUM AND ALUMINUM ALLOYS

specification number	designation	country	product forms	chemical composition	mechanical properties and hardness values
AS 1874	BP605	Australia	Ingot, sand castings and permanent mould castings	rem **Al**, 9.0–10.0 **Si**, 0.10 **Cu**, 0.05 **Mn**, 0.45–0.6 **Mg**, 0.05 **Zn**, 0.6 **Fe**, 0.05 each, 0.15 total, others	
AS 1874	AS605	Australia	Ingot, sand castings and permanent mould castings	rem **Al**, 9.0–10.0 **Si**, 0.6 **Cu**, 0.35 **Mn**, 0.45–0.6 **Mg**, 0.5 **Ni**, 0.5 **Zn**, 0.15 **Sn**, 0.7–1.1 **Fe**, 0.05 each, 0.25 total, others	
JIS H 2212	Class 3	Japan	Ingot: for die castings	rem **Al**, 9.0–10.0 **Si**, 0.05 **Cu**, 0.03 **Mn**, 0.40–0.6 **Mg**, 0.03 **Ni**, 0.03 **Zn**, 0.03 **Sn**, 0.3–0.6 **Fe**	
SIS 14 42 47	SIS Al 42 47–00	Sweden	Ingot	rem **Al**, 8.0–10.0 **Si**, 0.20 **Cu**, 0.5 **Mn**, 0.5 **Mg**, 0.1 **Ni**, 0.3 **Zn**, 0.20 **Ti**, 0.05 **Sn**, 0.8–1.3 **Fe**, 0.1 **Pb**	
SIS 14 42 47	SIS Al 42 47–10	Sweden	Pressure die castings	rem **Al**, 8.0–10.0 **Si**, 0.20 **Cu**, 0.5 **Mn**, 0.5 **Mg**, 0.1 **Ni**, 0.3 **Zn**, 0.20 **Ti**, 0.05 **Sn**, 0.8–1.3 **Fe**, 0.1 **Pb**	**As cast**: all **diam**, 240 ksi (1655 MPa) **TS**, 110 ksi (758 MPa) **YS**, 2% in 2 in. (50 mm) **El**, 60–80 **HB**
AMS 4286B		US	Permanent mold castings	rem **Al**, 6.5–7.5 **Si**, 0.20 **Cu**, 0.30 **Mn**, 0.20–0.40 **Mg**, 0.30 **Zn**, 0.20 **Ti**, 0.6 **Fe**, 0.05 each, 0.15 total, others	**Artificially aged**: 19 ksi (131 MPa) **TS**, 50 min **HB**
AMS 4285		US	Centrifugal castings	rem **Al**, 6.5–7.5 **Si**, 0.20 **Cu**, 0.30 **Mn**, 0.20–0.40 **Mg**, 0.30 **Zn**, 0.20 **Ti**, 0.6 **Fe**, 0.05 each, 0.15 total, others	**Solution and precipitation heat treated**: 25 ksi (172 MPa) **TS**, 17 ksi (117 MPa) **YS**, 1% in 2 in. (50 mm) **El**, 65–95 **HB**
AMS 4284D		US	Permanent mold castings	rem **Al**, 6.5–7.5 **Si**, 0.20 **Cu**, 0.30 **Mn**, 0.20–0.40 **Mg**, 0.30 **Zn**, 0.20 **Ti**, 0.6 **Fe**, 0.05 each, 0.15 total, others	**Solution and precipitation heat treated**: 25 ksi (172 MPa) **TS**, 17 ksi (117 MPa) **YS**, 1% in 2 in. (50 mm) **El**, 65–95 **HB**
AMS 4261A		US	Investment castings	rem **Al**, 6.5–7.5 **Si**, 0.20 **Cu**, 0.30 **Mn**, 0.20–0.40 **Mg**, 0.30 **Zn**, 0.20 **Ti**, 0.50 **Fe**, 0.05 each, 0.15 total, others	**Precipitation heat treated**: 18 ksi (124 MPa) **TS**, 12 ksi (83 MPa) **YS**, 1% in 2 in. (50 mm) **El**, 50 min **HB**
AMS 4260B		US	Investment castings	rem **Al**, 6.5–7.5 **Si**, 0.20 **Cu**, 0.30 **Mn**, 0.20–0.40 **Mg**, 0.30 **Zn**, 0.20 **Ti**, 0.50 **Fe**, 0.05 each, 0.15 total, others	**Solution and precipitation heat treated**: 25 ksi (172 MPa) **TS**, 17 ksi (117 MPa) **YS**, 1% in 2 in. (50 mm) **El**
AMS 4219		US	Castings: premium quality, high strength	rem **Al**, 6.5–7.5 **Si**, 0.20 **Cu**, 0.10 **Mn**, 0.45–0.75 **Mg**, 0.10 **Zn**, 0.20 **Ti**, 0.20 **Fe**, 0.25 **Be**, 0.05 each, 0.15 total, others	**Solution and precipitation heat treated**: 38 ksi (262 MPa) **TS**, 30 ksi (207 MPa) **YS**, 2% in 2 in. (50 mm) **El**, 80–115 **HB**
AMS 4218C		US	Castings: premium grade	rem **Al**, 6.5–7.5 **Si**, 0.20 **Cu**, 0.10 **Mn**, 0.20–0.40 **Mg**, 0.10 **Zn**, 0.20 **Ti**, 0.20 **Fe**, 0.05 each, 0.15 total, others	**Solution and precipitation heat treated**: 32 ksi (221 MPa) **TS**, 22 ksi (152 MPa) **YS**, 2% in 2 in. (50 mm) **El**, 70–105 **HB**

CAST ALUMINUM AND ALUMINUM ALLOYS

specification number	designation	country	product forms	chemical composition	mechanical properties and hardness values
AMS 4217D		US	Sand castings	rem **Al**, 6.5–7.5 **Si**, 0.20 **Cu**, 0.30 **Mn**, 0.20–0.40 **Mg**, 0.30 **Zn**, 0.20 **Ti**, 0.6 **Fe**, 0.05 each, 0.15 total, others	**Solution and precipitation heat treated**: 23 ksi (159 MPa) **TS**, 15 ksi (103 MPa) **YS**, 1% in 2 in. (50 mm) **El**, 65–95 **HB**
ANSI/ASTM B 179	A444.2	US	Ingot: for permanent mold castings	rem **Al**, 6.5–7.5 **Si**, 0.05 **Cu**, 0.05 **Mn**, 0.05 **Mg**, 0.05 **Zn**, 0.20 **Ti**, 0.12 **Fe**, 0.05 each, 0.15 total, others	
ANSI/ASTM B 108	A444.0	US	Permanent mold castings	rem **Al**, 6.5–7.5 **Si**, 0.10 **Cu**, 0.10 **Mn**, 0.05 **Mg**, 0.10 **Zn**, 0.20 **Ti**, 0.20 **Fe**, 0.05 each, 0.15 total, others	**Solution heat treated and naturally aged**: all **diam**, 20 ksi (138 MPa) **TS**, 20% in 2 in. (50 mm) **El**
AS 1874	AP601	Australia	Ingot, sand castings and permanent mould castings	rem **Al**, 6.5–7.5 **Si**, 0.05 **Cu**, 0.05 **Mn**, 0.30–0.40 **Mg**, 0.05 **Zn**, 0.20 **Ti**, 0.12–0.20 **Fe**, 0.05 each, 0.15 total, others	**Cooled and naturally aged (sand castings)**: all **diam**, 130 MPa **TS**, 2% in 2 in. (50 mm) **El**; **Cooled and artificially aged (sand castings)**: all **diam**, 155 MPa **TS**; **Solution heat–treated and artificially aged**: all **diam**, 205 MPa **TS**, 3% in 2 in. (50 mm) **El**
AS 1874	BP601	Australia	Ingot, sand castings and permanent mould castings	rem **Al**, 6.5–7.5 **Si**, 0.05 **Cu**, 0.05 **Mn**, 0.30–0.40 **Mg**, 0.05 **Zn**, 0.20 **Ti**, 0.11 **Fe**, 0.05 each, 0.15 total, others	**Solution heat–treated and artificially aged for 4h at 155°C (sand castings)**: all **diam**, 205 MPa **TS**, 3% in 2 in. (50 mm) **El**; **Solution heat–treated and artificially aged for 8 h (sand castings)**: all **diam**, 255 MPa **TS**, 5% in 2 in. (50 mm) **El**
AS 1874	CP601	Australia	Ingot, sand castings and permanent mould castings	rem **Al**, 6.5–7.5 **Si**, 0.05 **Cu**, 0.05 **Mn**, 0.30–0.40 **Mg**, 0.05 **Zn**, 0.20 **Ti**, 0.12–0.20 **Fe**, 0.05 each, 0.15 total, others	**Cooled and naturally aged (sand castings)**: all **diam**, 130 MPa **TS**, 2% in 2 in. (50 mm) **El**; **Cooled and artificially aged (sand castings)**: all **diam**, 155 MPa **TS**; **Solution heat–treated and artificially aged (sand castings)**: all **diam**, 205 MPa **TS**, 3% in 2 in. (50 mm) **El**
AS 1874	AS601	Australia	Ingot, sand castings and permanent mould castings	rem **Al**, 6.5–7.5 **Si**, 0.25 **Cu**, 0.35 **Mn**, 0.30–0.50 **Mg**, 0.35 **Zn**, 0.25 **Ti**, 0.50 **Fe**, 0.05 each, 0.15 total, others	**Cooled and naturally aged**: all **diam**, 130 MPa **TS**, 2% in 2 in. (50 mm) **El**; **Solution heat–treated and artificially aged**: all **diam**, 205 MPa **TS**, 3% in 2 in. (50 mm) **El**
BS 2L.99		UK	Ingot and castings	rem **Al**, 6.5–7.5 **Si**, 0.10 **Cu**, 0.10 **Mn**, 0.20–0.45 **Mg**, 0.10 **Ni**, 0.10 **Zn**, 0.20 **Ti**, 0.05 **Sn**, 0.20 **Fe**, 0.05 **Pb**	**Heat treated and quenched**: all **diam**, 230 MPa **TS**, 185 MPa **YS**, 2% in 2 in. (50 mm) **El**
QQ–A–371F	A444.2 (formerly A344(2))	US	Ingot	rem **Al**, 6.5–7.5 **Si**, 0.05 **Cu**, 0.05 **Mn**, 0.05 **Mg**, 0.05 **Zn**, 0.20 **Ti**, 0.12 **Fe**, 0.05 each, 0.15 total, others	

CAST ALUMINUM AND ALUMINUM ALLOYS

specification number	designation	country	product forms	chemical composition	mechanical properties and hardness values
SFS 2569	G-AlSi7Mg	Finland	Die castings	rem **Al**, 6.5–7.5 **Si**, 0.10 **Cu**, 0.5 **Mn**, 0.2–0.4 **Mg**, 0.05 **Ni**, 0.3 **Zn**, 0.2 **Ti**, 0.05 **Sn**, 0.50 **Fe**, 0.05 **Pb**, 0.05 each, 0.15 total, others	**Solution treated and aged:** 260 MPa **TS**, 220 MPa **YS**, 1% in 2 in. (50 mm) **El**
ANSI/ASTM B 179	C443.1/S5C	US	Ingot: for die castings	rem **Al**, 4.5–6.0 **Si**, 0.6 **Cu**, 0.35 **Mn**, 0.10 **Mg**, 0.50 **Ni**, 0.40 **Zn**, 0.15 **Sn**, 1.0 **Fe**, 0.25 total, others	
ANSI/ASTM B 179	C443.2	US	Ingot: for die castings	rem **Al**, 4.5–6.0 **Si**, 0.10 **Cu**, 0.10 **Mn**, 0.05 **Mg**, 0.10 **Zn**, 0.7–1.1 **Fe**, 0.05 each, 0.15 total, others	
ANSI/ASTM B 179	443.1	US	Ingot: for sand and permanent mold castings	rem **Al**, 4.5–6.0 **Si**, 0.6 **Cu**, 0.50 **Mn**, 0.05 **Mg**, 0.25 **Cr**, 0.50 **Zn**, 0.25 **Ti**, 0.6 **Fe**, 0.35 total, others	
ANSI/ASTM B 179	443.2	US	Ingot: for sand and permanent mold castings	rem **Al**, 4.5–6.0 **Si**, 0.10 **Cu**, 0.10 **Mn**, 0.05 **Mg**, 0.10 **Zn**, 0.20 **Ti**, 0.6 **Fe**, 0.05 each, 0.15 total, others	
ANSI/ASTM B 179	A443.1	US	Ingot: for sand and permanent mold castings	rem **Al**, 4.5–6.0 **Si**, 0.30 **Cu**, 0.50 **Mn**, 0.05 **Mg**, 0.25 **Cr**, 0.50 **Zn**, 0.25 **Ti**, 0.6 **Fe**, 0.35 total, others	
ANSI/ASTM B 179	B443.1	US	Ingot: for sand and permanent mold castings	rem **Al**, 4.5–6.0 **Si**, 0.15 **Cu**, 0.35 **Mn**, 0.05 **Mg**, 0.35 **Zn**, 0.25 **Ti**, 0.6 **Fe**, 0.05 each, 0.15 total, others	
ANSI/ASTM B 108	443.0	US	Permanent mold castings	rem **Al**, 4.5–6.0 **Si**, 0.6 **Cu**, 0.50 **Mn**, 0.05 **Mg**, 0.25 **Cr**, 0.50 **Zn**, 0.25 **Ti**, 0.8 **Fe**, 0.35 total, others	**As fabricated:** 21 ksi (145 MPa) **TS**, 7 ksi (49 MPa) **YS**, 2% in 2 in. (50 mm) **El**, 45 **HB**
ANSI/ASTM B 618	443.0	US	Investment castings	rem **Al**, 4.5–6.0 **Si**, 0.6 **Cu**, 0.50 **Mn**, 0.05 **Mg**, 0.25 **Cr**, 0.50 **Zn**, 0.25 **Ti**, 0.8 **Fe**, 0.35 total, others	**As fabricated:** ≥0.250 in. **diam**, 17 ksi (117 MPa) **TS**, 7 ksi (48 MPa) **YS**, 3% in 2 in. (50 mm) **El**
ANSI/ASTM B 618	A443.0	US	Investment castings	rem **Al**, 4.5–6.0 **Si**, 0.30 **Cu**, 0.50 **Mn**, 0.05 **Mg**, 0.25 **Cr**, 0.50 **Zn**, 0.25 **Ti**, 0.8 **Fe**, 0.35 total, others	**As fabricated:** ≥0.250 in. **diam**, 17 ksi (117 MPa) **TS**, 7 ksi (48 MPa) **YS**, 3% in 2 in. (50 mm) **El**
ANSI/ASTM B 618	B443.0	US	Investment castings	rem **Al**, 4.5–6.0 **Si**, 0.15 **Cu**, 0.35 **Mn**, 0.05 **Mg**, 0.35 **Zn**, 0.25 **Ti**, 0.8 **Fe**, 0.05 each, 0.15 total, others	**As fabricated:** ≥0.250 in. **diam**, 17 ksi (117 MPa) **TS**, 6 ksi (41 MPa) **YS**, 3% in 2 in. (50 mm) **El**
ANSI/ASTM B 108	A443.0	US	Permanent mold castings	rem **Al**, 4.5–6.0 **Si**, 0.30 **Cu**, 0.50 **Mn**, 0.05 **Mg**, 0.25 **Cr**, 0.50 **Zn**, 0.25 **Ti**, 0.8 **Fe**, 0.35 total, others	**As fabricated:** all **diam**, 21 ksi (145 MPa) **TS**, 7 ksi (49 MPa) **YS**, 2% in 2 in. (50 mm) **El**, 45 **HB**
ANSI/ASTM B 108	B443.0	US	Permanent mold castings	rem **Al**, 4.5–6.0 **Si**, 0.15 **Cu**, 0.35 **Mn**, 0.05 **Mg**, 0.35 **Zn**, 0.25 **Ti**, 0.8 **Fe**, 0.05 each, 0.15 total, others	**As fabricated:** all **diam**, 21 ksi (145 MPa) **TS**, 6 ksi (41 MPa) **YS**, 3% in 2 in. (50 mm) **El**, 45 **HB**
ANSI/ASTM B 26	443.0	US	Sand castings	rem **Al**, 4.5–6.0 **Si**, 0.6 **Cu**, 0.50 **Mn**, 0.05 **Mg**, 0.25 **Cr**, 0.50 **Zn**, 0.25 **Ti**, 0.8 **Fe**, 0.35 total, others	**As fabricated:** ≥0.250 in. **diam**, 17 ksi (117 MPa) **TS**, 7 ksi (48 MPa) **YS**, 3% in 2 in. (50 mm) **El**, 40 **HB**

CAST ALUMINUM AND ALUMINUM ALLOYS

specification number	designation	country	product forms	chemical composition	mechanical properties and hardness values
ANSI/ASTM B 26	A443.0	US	Sand castings	rem **Al**, 4.5–6.0 **Si**, 0.30 **Cu**, 0.50 **Mn**, 0.05 **Mg**, 0.25 **Cr**, 0.50 **Zn**, 0.25 **Ti**, 0.8 **Fe**, 0.35 total, others	**As fabricated**: ≥0.250 in. **diam**, 17 ksi (117 MPa) **TS**, 7 ksi (48 MPa) **YS**, 3% in 2 in. (50 mm) **El**, 40 **HB**
ANSI/ASTM B 26	B443.0	US	Sand castings	rem **Al**, 4.5–6.0 **Si**, 0.15 **Cu**, 0.35 **Mn**, 0.05 **Mg**, 0.35 **Zn**, 0.25 **Ti**, 0.8 **Fe**, 0.05 each, 0.15 total, others	**As fabricated**: ≥0.250 in. **diam**, 17 ksi (117 MPa) **TS**, 6 ksi (41 MPa) **YS**, 3% in 2 in. (50 mm) **El**, 40 **HB**
AS 1874	AP403	Australia	Ingot, sand castings and permanent mould castings	rem **Al**, 4.5–6.0 **Si**, 0.10 **Cu**, 0.10 **Mn**, 0.05 **Mg**, 0.10 **Zn**, 0.20 **Ti**, 0.6 **Fe**, 0.05 each, 0.15 total, others	**As cast (sand castings)**: all **diam**, 115 MPa **TS**, 3% in 2 in. (50 mm) **El**
CSA HA.3	.S5	Canada	Ingot	rem **Al**, 4.5–6.0 **Si**, 0.10 **Cu**, 0.10 **Mn**, 0.05 **Mg**, 0.10 **Zn**, 0.20 **Ti**, 0.6 **Fe**, 0.05 each, 0.15 total, others	
CSA HA.10	.S5	Canada	Mold castings	rem **Al**, 4.5–6.0 **Si**, 0.10 **Cu**, 0.10 **Mn**, 0.05 **Mg**, 0.10 **Zn**, 0.20 **Ti**, 0.8 **Fe**, 0.05 each, 0.15 total, others	
CSA HA.9	.S5	Canada	Sand castings	rem **Al**, 4.5–6.0 **Si**, 0.10 **Cu**, 0.10 **Mn**, 0.05 **Mg**, 0.10 **Zn**, 0.20 **Ti**, 0.8 **Fe**, 0.05 each, 0.15 total, others	
QQ–A–371F	443.2 (formerly 43)	US	Ingot	rem **Al**, 4.5–6.0 **Si**, 0.10 **Cu**, 0.10 **Mn**, 0.05 **Mg**, 0.10 **Zn**, 0.20 **Ti**, 0.6 **Fe**, 0.05 each, 0.15 total, others	
QQ–A–371F	C443.1 (formerly A43(2)	US	Ingot	rem **Al**, 4.5–6.0 **Si**, 0.6 **Cu**, 0.35 **Mn**, 0.10 **Mg**, 0.50 **Ni**, 0.40 **Zn**, 0.15 **Sn**, 1.0 **Fe**, 0.25 total, others	
QQ–A–371F	C443.2 (formerly A43(2))	US	Ingot	rem **Al**, 4.5–6.0 **Si**, 0.10 **Cu**, 0.10 **Mn**, 0.05 **Mg**, 0.10 **Zn**, 0.7–1.1 **Fe**, 0.05 each, 0.15 total, others	
QQ–A–601E	B443.0	US	Castings	rem **Al**, 4.5–6.0 **Si**, 0.15 **Cu**, 0.35 **Mn**, 0.05 **Mg**, 0.35 **Zn**, 0.25 **Ti**, 0.8 **Fe**, 0.05 each, 0.15 total, others	**As cast**: all **diam**, 17 ksi (117 MPa) **TS**, 8 ksi (55 MPa) **YS**, 3.0% in 2 in. (50 mm) **El**, 40 **HB**
QQ–A–596d	43	US	Castings	rem **Al**, 4.5–6.0 **Si**, 0.15 **Cu**, 0.35 **Mn**, 0.05 **Mg**, 0.35 **Zn**, 0.25 **Ti**, 0.8 **Fe**, 0.05 each, others	**As cast**: all **diam**, 21 ksi (145 MPa) **TS**, 5.0% in 2 in. (50 mm) **El**
QQ–A–591E	443.0	US	Castings	rem **Al**, 4.5–6.0 **Si**, 0.6 **Cu**, 0.35 **Mn**, 0.10 **Mg**, 0.50 **Ni**, 0.50 **Zn**, 0.15 **Sn**, 2.0 **Fe**, 0.25 total, others	**Die cast**: all **diam**, 33 ksi (228 MPa) **TS**, 14 ksi (97 MPa) **YS**, 9.0% in 2 in. (50 mm) **El**
UNI 5077	GD–AlSi5Fe	Italy	Pressure die castings	rem **Al**, 4.5–6.0 **Si**, 0.40 **Cu**, 0.30 **Mn**, 0.20 **Mg**, 0.30 **Ni**, 0.50 **Zn**, 0.15 **Ti**, 0.10 **Sn**, 0.7–1.0 **Fe**, 0.10 **Pb**	**As cast**: all **diam**, 175 MPa **TS**, 100 MPa **YS**, 2.5% in 2 in. (50 mm) **El**, 50–65 **HB**
ANSI/ASTM B 85	C443.0/S5C	US	Die castings	rem **Al**, 4.5–6.0 **Si**, 0.6 **Cu**, 0.35 **Mn**, 0.10 **Mg**, 0.50 **Ni**, 0.50 **Zn**, 0.15 **Sn**, 2.0 **Fe**, 0.25 total, others	**As cast**: all **diam**, 33 ksi (230 MPa) **TS**, 14 ksi (100 MPa) **YS**, 9% in 2 in. (50 mm) **El**

CAST ALUMINUM AND ALUMINUM ALLOYS

specification number	designation	country	product forms	chemical composition	mechanical properties and hardness values
AMS 4215C		US	Castings: premium grade sand, permanent and composite mold	rem **Al**, 4.5–5.5 **Si**, 1.0–1.5 **Cu**, 0.10 **Mn**, 0.40–0.6 **Mg**, 0.10 **Zn**, 0.20 **Ti**, 0.20 **Fe**, 0.05 each, 0.15 total, others	**Solution and precipitation heat treated**: 35 ksi (241 MPa) **TS**, 28 ksi (193 MPa) **YS**, 1% in 2 in. (50 mm) **El**, 75–110 **HB**
AMS 4214D		US	Sand castings	rem **Al**, 4.5–5.5 **Si**, 1.0–1.5 **Cu**, 0.50 **Mn**, 0.4–0.6 **Mg**, 0.25 **Cr**, 0.30 **Zn**, 0.25 **Ti**, 0.6 **Fe**, 0.05 each, 0.15 total, others	**Solution heat treated and overaged**: 23 ksi (159 MPa) **TS**, 1% in 2 in. (50 mm) **El**, 65–86 **HB**
AMS 4210F		US	Sand castings	rem **Al**, 4.5–5.5 **Si**, 1.0–1.5 **Cu**, 0.50 **Mn**, 0.4–0.6 **Mg**, 0.25 **Cr**, 0.30 **Zn**, 0.25 **Ti**, 0.6 **Fe**, 0.05 each, 0.15 total, others	**Solution and precipitation heat treated**: 24 ksi (165 MPa) **TS**, 15 ksi (103 MPa) **YS**, 1% in 2 in. (50 mm) **El**, 65–95 **HB**
AMS 4210F		US	Sand castings	rem **Al**, 4.5–5.5 **Si**, 1.0–1.5 **Cu**, 0.50 **Mn**, 0.4–0.6 **Mg**, 0.25 **Cr**, 0.30 **Zn**, 0.25 **Ti**, 0.6 **Fe**, 0.05 each, 0.15 total, others	**Stress relieved**: 19 ksi (131 MPa) **TS**, 55 min **HB**
AMS 4281D		US	Permanent mold castings	rem **Al**, 4.5–5.5 **Si**, 1.0–1.5 **Cu**, 0.50 **Mn**, 0.40–0.6 **Mg**, 0.25 **Cr**, 0.35 **Zn**, 0.25 **Ti**, 0.6 **Fe**, 0.05 each, 0.15 total, others	**Solution and precipitation heat treated**: 27 ksi (191 MPa) **TS**, 17 ksi (117 MPa) **YS**, 1% in 2 in. (50 mm) **El**, 80–110 **HB**
AMS 4280F		US	Permanent mold castings	rem **Al**, 4.5–5.5 **Si**, 1.0–1.5 **Cu**, 0.50 **Mn**, 0.40–0.6 **Mg**, 0.25 **Cr**, 0.35 **Zn**, 0.25 **Ti**, 0.6 **Fe**, 0.05 each, 0.15 total, others	**Solution heat treated and overaged**: 26 ksi (176 MPa) **TS**, 20 ksi (138 MPa) **YS**, 70–95 **HB**
AS 1874	AP309	Australia	Ingot, sand castings and permanent mould castings	rem **Al**, 4.5–5.5 **Si**, 1.0–1.5 **Cu**, 0.05 **Mn**, 0.50–0.6 **Mg**, 0.05 **Zn**, 0.20 **Ti**, 0.14–0.25 **Fe**, 0.05 each, 0.15 total, others	**Cooled and artificially aged (sand castings)**: all **diam**, 170 MPa **TS**; **Solution heat-treated and aged for 4h at 155°C**: all **diam**, 220 MPa **TS**, 2% in 2 in. (50 mm) **El**; **Solution heat-treated and aged for 10h at 170°C**: all **diam**, 275 MPa **TS**
AS 1874	AP311	Australia	Ingot, sand castings and permanent mould castings	rem **Al**, 4.0–6.0 **Si**, 1.0–1.5 **Cu**, 0.05 **Mn**, 0.05 **Mg**, 0.10 **Zn**, 0.20 **Ti**, 0.15 **Fe**, 0.05 each, 0.20 total, others	
QQ-A-371F	443.1 (formerly 43)	US	Ingot	rem **Al**, 4.0–6.0 **Si**, 0.50 **Mn**, 0.05 **Mg**, 0.25 **Cr**, 0.50 **Zn**, 0.25 **Ti**, 0.6 **Fe**, 0.35 total, others	
ISO R164	Al-Si5	ISO	Castings	rem **Al**, 4.0–6.0 **Si**, 0.10 **Cu**, 0.5 **Mn**, 0.1 **Mg**, 0.1 **Ni**, 0.1 **Zn**, 0.20 **Ti**, 0.1 **Sn**, 0.8 **Fe**, 0.1 **Pb**	
ISO R164	Al-Si5Fe	ISO	Pressure die castings	rem **Al**, 4.0–6.0 **Si**, 0.10 **Cu**, 0.5 **Mn**, 0.1 **Mg**, 0.1 **Ni**, 0.1 **Zn**, 0.20 **Ti**, 0.1 **Sn**, 1.3 **Fe**, 0.1 **Pb**	

CAST ALUMINUM AND ALUMINUM ALLOYS

specification number	designation	country	product forms	chemical composition	mechanical properties and hardness values
SABS 989	Al–Si5MgA	South Africa	Ingot: for sand and gravity die castings	rem **Al**, 3.5–6.0 **Si**, 0.08 **Cu**, 0.5 **Mn**, 0.5–0.8 **Mg**, 0.1 **Ni**, 0.1 **Zn**, 0.2 **Ti**, 0.05 **Sn**, 0.5 **Fe**, 0.1 **Pb**, 0.05 each, 0.15 total, others	**Sand cast**: all **diam**, 125 MPa **TS**, 2% in 2 in. (50 mm) **El**; **Chill cast**: all **diam**, 160 MPa **TS**, 3% in 2 in. (50 mm) **El**; **Chill cast, solution and precipitation heat treated**: all **diam**, 275 MPa **TS**, 2% in 2 in. (50 mm) **El**
AS 1874	AS801	Australia	Ingot, sand castings and permanent mould castings	rem **Al**, 2.0–3.0 **Si**, 0.70–1.3 **Cu**, 0.30 **Mn**, 0.10 **Mg**, 0.30–0.7 **Ni**, 0.20 **Ti**, 5.5–7.0 **Sn**, 0.50 **Fe**, 0.30 total, others	
JIS H 2211	Class 7B	Japan	Ingot: for castings	rem **Al**, 0.2 **Si**, 0.05 **Cu**, 0.03 **Mn**, 9.6–11.0 **Mg**, 0.03 **Ni**, 0.03 **Zn**, 0.2 **Ti**, 0.2 **Fe**	
SABS 989	Al–Mg10A	South Africa	Ingot: for sand and gravity die castings	rem **Al**, 0.25 **Si**, 0.08 **Cu**, 0.10 **Mn**, 9.6–11.0 **Mg**, 0.10 **Ni**, 0.10 **Zn**, 0.2 **Ti**, 0.05 **Sn**, 0.35 **Fe**, 0.05 **Pb**, 0.05 each, 0.15 total, others	**Sand cast and solution heat treated**: all **diam**, 275 MPa **TS**, 8% in 2 in. (50 mm) **El**; **Chill cast and solution heat treated**: 310 MPa **TS**, 12% in 2 in. (50 mm) **El**
AS 1874	AS505	Australia	Ingot, sand castings and permanent mould castings	rem **Al**, 0.25 **Si**, 0.15 **Cu**, 0.10 **Mn**, 9.5–11.0 **Mg**, 0.10 **Ni**, 0.10 **Zn**, 0.20 **Ti**, 0.05 **Sn**, 0.35 **Fe**, 0.05 each, 0.15 total, others	**Solution heat–treated and naturally aged**: 280 MPa **TS**, 8% in 2 in. (50 mm) **El**
IS 202	A–10	India	Sand castings	rem **Al**, 0.25 **Si**, 0.10 **Cu**, 0.10 **Mn**, 9.5–11.0 **Mg**, 0.10 **Ni**, 0.10 **Zn**, 0.05 **Sn**, 0.35 **Fe**, 0.05 **Pb**, 0.20 **Ti+Nb**	**Solution treated**: all **diam**, 278 MPa **TS**, 152 MPa **YS**, 8% in 2 in. (50 mm) **El**
IS 202	A–10	India	Sand castings	rem **Al**, 0.25 **Si**, 0.10 **Cu**, 9.5–11.0 **Mg**, 0.10 **Ni**, 0.10 **Zn**, 0.05 **Sn**, 0.35 **Fe**, 0.05 **Pb**, 0.20 **Ti+Nb**	**Solution treated**: 278 MPa **TS**, 152 MPa **YS**, 8% in 2 in. (50 mm) **El**
SABS 991	Al–Mg10A	South Africa	Gravity die castings	rem **Al**, 0.25 **Si**, 0.1 **Cu**, 0.1 **Mn**, 9.5–11.0 **Mg**, 0.1 **Ni**, 0.1 **Zn**, 0.15 **Ti**, 0.05 **Sn**, 0.4 **Fe**, 0.05 each, 0.15 total, others	**Solution heat treated**: all **diam**, 310 MPa **TS**, 12% in 2 in. (50 mm) **El**
SABS 990	Al–Mg10A	South Africa	Sand castings	rem **Al**, 0.25 **Si**, 0.1 **Cu**, 0.1 **Mn**, 9.5–11.0 **Mg**, 0.1 **Ni**, 0.1 **Zn**, 0.15 **Ti**, 0.05 **Sn**, 0.4 **Fe**, 0.05 **Pb**, 0.05 each, 0.15 total, others	**Solution heat treated**: all **diam**, 275 MPa **TS**, 8% in 2 in. (50 mm) **El**
ANSI/ASTM B 179	520.2	US	Ingot: for sand castings	rem **Al**, 0.15 **Si**, 0.20 **Cu**, 0.10 **Mn**, 9.6–10.6 **Mg**, 0.10 **Zn**, 0.20 **Ti**, 0.20 **Fe**, 0.05 each, 0.15 total, others	
CSA HA.3	.G10	Canada	Ingot	rem **Al**, 0.15 **Si**, 0.20 **Cu**, 0.10 **Mn**, 9.6–10.6 **Mg**, 0.10 **Zn**, 0.20 **Ti**, 0.20 **Fe**, 0.05 each, 0.15 total, others	
QQ–A–371F	520.2 (formerly 220)	US	Ingot	rem **Al**, 0.15 **Si**, 0.20 **Cu**, 0.10 **Mn**, 9.6–10.6 **Mg**, 0.10 **Zn**, 0.20 **Ti**, 0.20 **Fe**, 0.05 each, 0.15 total, others	

CAST ALUMINUM AND ALUMINUM ALLOYS

specification number	designation	country	product forms	chemical composition	mechanical properties and hardness values
AMS 4240D		US	Sand castings	rem **Al**, 0.25 **Si**, 0.25 **Cu**, 0.15 **Mn**, 9.5–10.6 **Mg**, 0.15 **Zn**, 0.25 **Ti**, 0.30 **Fe**, 0.05 each, 0.15 total, others	**Solution heat treated**: all **diam**, 32 ksi (217 MPa) **TS**, 17 ksi (114 MPa) **YS**, 3% in 2 in. (50 mm) **El**
ANSI/ASTM B 618	520.0	US	Investment castings	rem **Al**, 0.25 **Si**, 0.25 **Cu**, 0.15 **Mn**, 9.5–10.6 **Mg**, 0.15 **Zn**, 0.25 **Ti**, 0.30 **Fe**, 0.05 each, 0.15 total, others	**Solution heat treated and naturally aged**: ≥0.250 in. **diam**, 42 ksi (290 MPa) **TS**, 22 ksi (152 MPa) **YS**, 12% in 2 in. (50 mm) **El**
CSA HA.9	.G10	Canada	Sand castings	rem **Al**, 0.25 **Si**, 0.25 **Cu**, 0.15 **Mn**, 9.5–10.6 **Mg**, 0.15 **Zn**, 0.25 **Ti**, 0.30 **Fe**, 0.05 each, 0.15 total, others	**T4**: all **diam**, 42.0 ksi (290 MPa) **TS**, 22.0 ksi (152 MPa) **YS**, 12% in 2 in. (50 mm) **El**
QQ-A-601E	520.0	US	Castings	rem **Al**, 0.25 **Si**, 0.25 **Cu**, 0.15 **Mn**, 9.5–10.6 **Mg**, 0.15 **Zn**, 0.25 **Ti**, 0.30 **Fe**, 0.05 each, 0.15 total, others	**Solution heat treated and temper T4**: all **diam**, 42 ksi (290 MPa) **TS**, 22 ksi (152 MPa) **YS**, 12.0 in 2 in. (50 mm) **El**, 75 **HB**
ANSI/ASTM B 26	520.0	US	Sand castings	rem **Al**, 0.25 **Si**, 0.25 **Cu**, 0.15 **Mn**, 9.5–10.5 **Mg**, 0.15 **Zn**, 0.25 **Ti**, 0.30 **Fe**, 0.05 each, 0.15 total, others	**Solution heat treated and naturally aged**: ≥0.250 in. **diam**, 42 ksi (290 MPa) **TS**, 22 ksi (152 MPa) **YS**, 12% in 2 in. (50 mm) **El**, 75 **HB**
DIN 1725 sheet 2	G–AlMg10ho/3.3591. 43	Germany	Sand castings	rem **Al**, 0.30 **Si**, 0.05 **Cu**, 0.30 **Mn**, 9.0–11.0 **Mg**, 0.10 **Zn**, 0.15 **Ti**, 0.05 **Sn**, 0.3 **Fe**, 0.05 **Pb**, 0.05 each, 0.15 total, others	**As cast and homogenized**: all **diam**, 220 MPa **TS**, 140 MPa **YS**, 6% in 2 in. (50 mm) **El**, 75–90 **HB**
ISO R2147	Al–Mg10	ISO	Sand castings: test pieces	rem **Al**, 0.30 **Si**, 0.10 **Cu**, 0.3 **Mn**, 9.0–11.0 **Mg**, 0.10 **Ni**, 0.10 **Zn**, 0.15 **Ti**, 0.05 **Sn**, 0.3 **Fe**, 0.05 **Be**, 0.05 **Pb**	**Solution heat treated**: all **diam**, 270 MPa **TS**, 10% in 2 in. (50 mm) **El**
ISO R164	Al–Mg10	ISO	Castings	rem **Al**, 0.30 **Si**, 0.10 **Cu**, 0.3 **Mn**, 9.0–11.0 **Mg**, 0.10 **Ni**, 0.10 **Zn**, 0.15 **Ti**, 0.05 **Sn**, 0.3 **Fe**, 0.05 **Be**, 0.05 **Pb**	
NF A 57 703	A–G10–Y4	France	Pressure die castings	rem **Al**, 1.0 **Si**, 0.2 **Cu**, 0.6 **Mn**, 8.5–11.0 **Mg**, 0.1 **Ni**, 0.4 **Zn**, 0.2 **Ti**, 0.1 **Sn**, 1.3 **Fe**	
DIN 1725 sheet 2	GD–AlMg9/3.3292.05	Germany	Pressure die castings	rem **Al**, 2.5 **Si**, 0.05 **Cu**, 0.2–0.5 **Mn**, 7.0–10.0 **Mg**, 0.1 **Zn**, 0.15 **Ti**, 1.0 **Fe**, 0.05 each, 0.15 total, others	**As cast**: all **diam**, 200 MPa **TS**, 140 MPa **YS**, 1% in 4 in. (100 mm) **El**, 70–100 **HB**
SABS 989	Al–Mg8FeA	South Africa	Ingot: for pressure die castings	rem **Al**, 0.35 **Si**, 0.25 **Cu**, 0.35 **Mn**, 8.5 **Mg**, 0.15 **Ni**, 0.15 **Zn**, 0.15 **Sn**, 0.5–1.0 **Fe**, 0.25 total, others	
ANSI/ASTM B 179	518.1/G8A	US	Ingot: for die castings	rem **Al**, 0.35 **Si**, 0.25 **Cu**, 0.35 **Mn**, 7.6–8.5 **Mg**, 0.15 **Ni**, 0.15 **Zn**, 0.15 **Sn**, 1.0 **Fe**, 0.25 total, others	
ANSI/ASTM B 179	518.2	US	Ingot: for die castings	rem **Al**, 0.25 **Si**, 0.10 **Cu**, 0.10 **Mn**, 7.6–8.5 **Mg**, 0.05 **Ni**, 0.05 **Sn**, 0.7 **Fe**, 0.10 total, others	

CAST ALUMINUM AND ALUMINUM ALLOYS

specification number	designation	country	product forms	chemical composition	mechanical properties and hardness values
QQ-A-371F	518.1 (formerly 218)	US	Ingot	rem **Al**, 0.35 **Si**, 0.25 **Cu**, 0.35 **Mn**, 7.6–8.5 **Mg**, 0.15 **Ni**, 0.15 **Zn**, 0.15 **Sn**, 1.0 **Fe**, 0.25 total, others	
QQ-A-371F	518.2 (formerly 218)	US	Ingot	rem **Al**, 0.25 **Si**, 0.10 **Cu**, 0.10 **Mn**, 7.6–8.5 **Mg**, 0.05 **Ni**, 0.7 **Fe**, 0.05 **Bi**, 0.10 total, others	
QQ-A-591E	518.0	US	Castings	rem **Al**, 0.35 **Si**, 0.25 **Cu**, 0.35 **Mn**, 7.5–8.5 **Mg**, 0.15 **Ni**, 0.15 **Zn**, 0.15 **Sn**, 1.8 **Fe**, 0.25 total, others	**Die cast**: all **diam**, 45 ksi (310 MPa) **TS**, 28 ksi (193 MPa) **YS**, 5.0% in 2 in. (50 mm) **El**
SABS 992	Al-Mg8FeA	South Africa	Pressure die castings	rem **Al**, 0.35 **Si**, 0.25 **Cu**, 0.35 **Mn**, 7.5–8.5 **Mg**, 0.15 **Ni**, 0.15 **Zn**, 0.15 **Sn**, 1.8 **Fe**, 0.25 total, others	**As cast**: 195 MPa **TS**, 115 MPa **YS**, 1% in 2 in. (50 mm) **El**, 60–80 **HB**
UNI 5080	GD-ALMg 7.5 Fe	Italy	Pressure die castings	rem **Al**, 0.30 **Si**, 0.05 **Cu**, 0.40 **Mn**, 7.0–8.0 **Mg**, 0.05 **Ni**, 0.10 **Zn**, 0.20 **Ti**, 0.05 **Sn**, 0.7–1.0 **Fe**, 0.005 min **Be**, 0.05 **Pb**	**As cast**: all **diam**, 195 MPa **TS**, 120 MPa **YS**, 1.5% in 2 in. (50 mm) **El**, 60–80 **HB**
ANSI/ASTM B 179	535.2	US	Ingot: for sand castings	rem **Al**, 0.10 **Si**, 0.05 **Cu**, 0.10–0.25 **Mn**, 6.6–7.5 **Mg**, 0.10–0.25 **Ti**, 0.10 **Fe**, 0.002 **B**, 0.003–0.007 **Be**, 0.05 each, 0.15 total, others	
QQ-A-371F	535.2 (formerly Almag 35)	US	Ingot	rem **Al**, 0.10 **Si**, 0.05 **Cu**, 0.10–0.25 **Mn**, 6.6–7.5 **Mg**, 0.10–0.25 **Ti**, 0.10 **Fe**, 0.002 **B**, 0.003–0.007 **Be**, 0.05 each, including **Be+Ti+H**; 0.15 total, others	
QQ-A-371F	A535.1 (formerly A218)	US	Ingot	rem **Al**, 0.20 **Si**, 0.10 **Cu**, 0.10–0.25 **Mn**, 6.6–7.5 **Mg**, 0.25 **Ti**, 0.15 **Fe**, 0.05 each, 0.15 total, others	
QQ-A-371F	B535.2 (formerly B218)	US	Ingot	rem **Al**, 0.10 **Si**, 0.05 **Cu**, 0.05 **Mn**, 6.6–7.5 **Mg**, 0.10–0.25 **Ti**, 0.12 **Fe**, 0.05 each, 0.15 total, others	
AMS 4238A		US	Sand castings	rem **Al**, 0.20 **Si**, 0.10 **Cu**, 0.10–0.25 **Mn**, 6.2–7.5 **Mg**, 0.10–0.25 **Ti**, 0.25 **Fe**, 0.40 **Fe+Si**, 0.5 each, 0.15 total, others	**As cast**: all **diam**, 27 ksi (186 MPa) **TS**, 14 ksi (97 MPa) **YS**, 3% in 2 in. (50 mm) **El**, 70 min **HB**
AMS 4239		US	Sand castings	rem **Al**, 0.20 **Si**, 0.10 **Cu**, 0.10–0.25 **Mn**, 6.2–7.5 **Mg**, 0.10–0.25 **Ti**, 0.25 **Fe**, 0.40 **Fe+Si**, 0.05 each, 0.15 others, total	**Stabilized**: all **diam**, 27 ksi (186 MPa) **TS**, 14 ksi (97 MPa) **YS**, 3% in 2 in. (50 mm) **El**, 70 min **HB**
ANSI/ASTM B 618	535.0	US	Investment castings	rem **Al**, 0.15 **Si**, 0.05 **Cu**, 0.10–0.25 **Mn**, 6.2–7.5 **Mg**, 0.10–0.25 **Ti**, 0.15 **Fe**, 0.002 **B**, 0.003–0.007 **Be**, 0.05 each, 0.15 total, others	**As fabricated**: $\geq$0.250 in. **diam**, 35 ksi (241 MPa) **TS**, 18 ksi (124 MPa) **YS**, 9% in 2 in. (50 mm) **El**

CAST ALUMINUM AND ALUMINUM ALLOYS

specification number	designation	country	product forms	chemical composition	mechanical properties and hardness values
ANSI/ASTM B 108	535.0	US	Permanent mold castings	rem **Al**, 0.15 **Si**, 0.05 **Cu**, 0.10–0.25 **Mn**, 6.2–7.5 **Mg**, 0.10–0.25 **Ti**, 0.15 **Fe**, 0.002 **B**, 0.003–0.007 **Be**	**As fabricated**: all **diam**, 35 ksi (241 MPa) **TS**, 18 ksi (124 MPa) **YS**, 8% in 2 in. (50 mm) **El**
ANSI/ASTM B 26	535.0	US	Sand castings	rem **Al**, 0.15 **Si**, 0.05 **Cu**, 0.10–0.25 **Mn**, 6.2–7.5 **Mg**, 0.10–0.25 **Ti**, 0.15 **Fe**, 0.05 each, 0.15 total, others	**As fabricated**: $\geq$0.250 in. **diam**, 35 ksi (241 MPa) **TS**, 18 ksi (124 MPa) **YS**, 9% in 2 in. (50 mm) **El**, 70 **HB**
QQ-A-601E	535.0	US	Castings	rem **Al**, 0.15 **Si**, 0.05 **Cu**, 0.10–0.25 **Mn**, 6.2–7.5 **Mg**, 0.10–0.25 **Ti**, 0.15 **Fe**, 0.05 each, 0.15 total, others	**As cast**: all **diam**, 35 ksi (242 MPa) **TS**, 18 ksi (124 MPa) **YS**, 9.0% in 2 in. (50 mm) **El**, 70 **HB**; Annealed and temper T4: all **diam**, 35 ksi (242 MPa) **TS**, 18 ksi (124 MPa) **YS**, 9.0% in 2 in. (50 mm) **El**
NF A 57 703	A-G6-Y4	France	Pressure die castings	rem **Al**, 1.0 **Si**, 0.2 **Cu**, 0.6 **Mn**, 5.0–8.5 **Mg**, 0.1 **Ni**, 0.4 **Zn**, 0.2 **Ti**, 0.1 **Sn**, 1.3 **Fe**	
JIS H 2212	Class 5	Japan	Ingot: for die castings	rem **Al**, 0.3 **Si**, 0.05 **Cu**, 0.03 **Mn**, 4.1–8.5 **Mg**, 0.03 **Ni**, 0.03 **Zn**, 0.03 **Sn**, 0.3–0.6 **Fe**	
JIS H 2118	Class 5	Japan	Ingot: for die castings	rem **Al**, 0.3 **Si**, 0.2 **Cu**, 0.3 **Mn**, 4.1–8.5 **Mg**, 0.1 **Ni**, 0.1 **Zn**, 0.1 **Sn**, 1.1 **Fe**	
JIS H 5302	Class 5	Japan	Die castings	rem **Al**, 0.3 **Si**, 0.2 **Cu**, 0.3 **Mn**, 4.0–8.5 **Mg**, 0.1 **Ni**, 0.1 **Zn**, 0.1 **Sn**, 1.8 **Fe**	
DS 3002	4163	Denmark	Ingot: for castings	rem **Al**, 0.5–1.5 **Si**, 0.10 **Cu**, 0.5 **Mn**, 4.0–6.0 **Mg**, 0.05 **Ni**, 0.2 **Zn**, 0.20 **Ti**, 0.05 **Sn**, 0.50 **Fe**, 0.05 **Pb**	**Sand cast**: all **diam**, 157 MPa **TS**, 88 MPa **YS**, 2% in 2 in. (50 mm) **El**, 55–70 **HB**; **Chill cast**: all **diam**, 167 MPa **TS**, 98 MPa **YS**, 2% in 2 in. (50 mm) **El**, 55–75 **HB**
DS 3002	4162	Denmark	Ingot: for castings	rem **Al**, 0.40 **Si**, 0.10 **Cu**, 0.40 **Mn**, 4.0–6.0 **Mg**, 0.05 **Ni**, 0.2 **Zn**, 0.20 **Ti**, 0.05 **Sn**, 0.50 **Fe**, 0.05 **Pb**	**Sand cast**: all **diam**, 157 MPa **TS**, 88 MPa **YS**, 2% in 2 in. (50 mm) **El**, 55–70 **HB**; **Chill cast**: all **diam**, 167 MPa **TS**, 98 MPa **YS**, 2% in 2 in. (50 mm) **El**, 55–75 **HB**
SFS 2572	G-AlMg5Si1	Finland	Die castings	rem **Al**, 0.5–1.5 **Si**, 0.10 **Cu**, 0.5 **Mn**, 4.0–6.0 **Mg**, 0.05 **Ni**, 0.2 **Zn**, 0.2 **Ti**, 0.05 **Sn**, 0.50 **Fe**, 0.05 **Pb**, 0.05 each, 0.15 total, others	**As cast**: $\leq$20 mm **diam**, 180 MPa **TS**, 110 MPa **YS**, 2% in 2 in. (50 mm) **El**
AIR 9150/B	A-G5MC	France	Wire: rivet	rem **Al**, 0.3 **Si**, 0.1 **Cu**, 0.05–0.7 **Mn**, 4.5–5.5 **Mg**, 0.1 **Zn**, 0.4 **Fe**, 0.2 **Ti**+**Zr**, 0.05 each, 0.15 total, others	**Cold finished or annealed**: 1.6–9.6 mm **diam**, 290 MPa **TS**, 140 MPa **YS**, 18% in 2 in. (50 mm) **El**
DIN 1725 sheet 2	GK-AlMg5Si/3.3261.02	Germany	Gravity die castings	rem **Al**, 0.9–1.5 **Si**, 0.05 **Cu**, 0.4 **Mn**, 4.5–5.5 **Mg**, 0.10 **Zn**, 0.20 **Ti**, 0.5 **Fe**, 0.05 each, 0.15 total, others	**As cast**: all **diam**, 180 MPa **TS**, 110 MPa **YS**, 2% in 2 in. (50 mm) **El**, 65–85 **HB**

CAST ALUMINUM AND ALUMINUM ALLOYS

specification number	designation	country	product forms	chemical composition	mechanical properties and hardness values
DIN 1725 sheet 2	G–AlMg5Si/3.3261.0 1	Germany	Sand castings	rem **Al**, 0.9–1.5 **Si**, 0.05 **Cu**, 0.4 **Mn**, 4.5–5.5 **Mg**, 0.10 **Zn**, 0.20 **Ti**, 0.5 **Fe**, 0.05 each, 0.15 total, others	**As cast**: all **diam**, 160 MPa **TS**, 110 MPa **YS**, 2% in 2 in. (50 mm) **El**, 60–75 **HB**
DIN 1725 sheet 2	GK–AlMg5/3.3561.02	Germany	Gravity die castings	rem **Al**, 0.5 **Si**, 0.05 **Cu**, 0.4 **Mn**, 4.5–5.5 **Mg**, 0.10 **Zn**, 0.20 **Ti**, 0.5 **Fe**, 0.05 each, 0.15 total, others	**As cast**: all **diam**, 180 MPa **TS**, 100 MPa **YS**, 4% in 2 in. (50 mm) **El**, 60–75 **HB**
DIN 1725 sheet 2	G–AlMg5/3.3561.01	Germany	Sand castings	rem **Al**, 0.5 **Si**, 0.05 **Cu**, 0.4 **Mn**, 4.5–5.5 **Mg**, 0.10 **Zn**, 0.20 **Ti**, 0.5 **Fe**, 0.05 each, 0.15 total, others	**As cast**: all **diam**, 160 MPa **TS**, 100 MPa **YS**, 3% in 2 in. (50 mm) **El**, 55–70 **HB**
NS 17 550	NS 17550–00	Norway	Ingot: for casting	94 **Al**, 0.5–1.5 **Si**, 0.10 **Cu**, 0.5 **Mn**, 4.0–6.0 **Mg**, 0.05 **Ni**, 0.2 **Zn**, 0.20 **Ti**, 0.05 **Sn**, 0.50 **Fe**, 0.05 **Pb**	
NS 17 550	NS 17 550–02	Norway	Sand castings	94 **Al**, 0.10 **Cu**, 0.5 **Mn**, 4.0–6.0 **Mg**, 0.2 **Zn**, 0.2 **Ti**, 0.05 **Sn**, 0.50 **Fe**, 0.05 **Pb**	**As cast**: all **diam**, 167 MPa **TS**, 98 MPa **YS**, 2% in 2 in. (50 mm) **El**
NS 17 552	NS 17552–00	Norway	Ingot: for casting	95 **Al**, 0.40 **Si**, 0.10 **Cu**, 0.40 **Mn**, 4.0–6.0 **Mg**, 0.05 **Ni**, 0.2 **Zn**, 0.20 **Ti**, 0.05 **Sn**, 0.50 **Fe**, 0.05 **Pb**	
NS 17 552	NS 17552–01	Norway	Sand castings	95 **Al**, 0.40 **Si**, 0.10 **Cu**, 0.40 **Mn**, 4.0–6.0 **Mg**, 0.05 **Ni**, 0.2 **Zn**, 0.20 **Ti**, 0.05 **Sn**, 0.50 **Fe**, 0.05 **Pb**	**As cast**: all **diam**, 157 MPa **TS**, 88 MPa **YS**, 2% in 2 in. (50 mm) **El**
NS 17 552	NS 17552–02	Norway	Chill castings	95 **Al**, 0.40 **Si**, 0.10 **Cu**, 0.40 **Mn**, 4.0–6.0 **Mg**, 0.05 **Ni**, 0.2 **Zn**, 0.20 **Ti**, 0.05 **Sn**, 0.50 **Fe**, 0.05 **Pb**	**As cast**: all **diam**, 167 MPa **TS**, 98 MPa **YS**, 2% in 2 in. (50 mm) **El**
JIS H 2211	Class 7A	Japan	Ingot: for castings	rem **Al**, 0.2 **Si**, 0.05 **Cu**, 0.6 **Mn**, 3.6–5.5 **Mg**, 0.03 **Ni**, 0.03 **Zn**, 0.2 **Ti**, 0.2 **Fe**	
JIS H 2117	Class 7A	Japan	Ingot: for castings	rem **Al**, 0.30 **Si**, 0.1 **Cu**, 0.6 **Mn**, 3.6–5.5 **Mg**, 0.1 **Ni**, 0.1 **Zn**, 0.2 **Ti**, 0.30 **Fe**, 0.50 **Fe+Si**	
JIS H 5202	Class 7A	Japan	Castings	rem **Al**, 0.3 **Si**, 0.1 **Cu**, 0.6 **Mn**, 3.5–5.5 **Mg**, 0.1 **Zn**, 0.2 **Ti**, 0.4 **Fe**	**As cast**: all **diam**, 216 MPa **TS**, 12% in 2 in. (50 mm) **El**, 60 **HB**
SABS 991	Al–Mg4MnA	South Africa	Gravity die castings	rem **Al**, 0.3 **Si**, 0.1 **Cu**, 0.3–0.7 **Mn**, 3.0–6.0 **Mg**, 0.1 **Ni**, 0.1 **Zn**, 0.2 **Ti**, 0.05 **Sn**, 0.6 **Fe**, 0.05 **Pb**, 0.05 each, 0.15 total, others	**As cast**: all **diam**, 170 MPa **TS**, 5% in 2 in. (50 mm) **El**
SABS 990	Al–Mg4MnA	South Africa	Sand castings	rem **Al**, 0.3 **Si**, 0.1 **Cu**, 0.7 **Mn**, 3.0–6.0 **Mg**, 0.1 **Ni**, 0.1 **Zn**, 0.2 **Ti**, 0.05 **Sn**, 0.6 **Fe**, 0.05 **Pb**, 0.05 each, 0.15 total, others	**As cast**: all **diam**, 140 MPa **TS**, 3% in 2 in. (50 mm) **El**
ANSI/ASTM B 179	514.1	US	Ingot: for sand castings	rem **Al**, 0.35 **Si**, 0.15 **Cu**, 0.35 **Mn**, 3.6–4.5 **Mg**, 0.15 **Zn**, 0.25 **Ti**, 0.40 **Fe**, 0.05 each, 0.15 total, others	

CAST ALUMINUM AND ALUMINUM ALLOYS

specification number	designation	country	product forms	chemical composition	mechanical properties and hardness values
ANSI/ASTM B 179	514.2	US	Ingot: for sand castings	0.30 **Si**, 0.10 **Cu**, 0.10 **Mn**, 3.6–4.5 **Mg**, 0.10 **Zn**, 0.20 **Ti**, 0.30 **Fe**, 0.05 each, 0.15 total, others	
ANSI/ASTM B 179	A514.2	US	Ingot: for permanent mold castings	rem **Al**, 0.30 **Si**, 0.10 **Cu**, 0.10 **Mn**, 3.6–4.5 **Mg**, 1.4–2.2 **Zn**, 0.20 **Ti**, 0.30 **Fe**, 0.05 each, 0.15 total, others	
ANSI/ASTM B 179	B514.2	US	Ingot: for sand and permanent mold castings	rem **Al**, 1.4–2.2 **Si**, 0.10 **Cu**, 0.10 **Mn**, 3.6–4.5 **Mg**, 0.10 **Zn**, 0.20 **Ti**, 0.30 **Fe**, 0.05 each, 0.15 total, others	
AS 1874	AP501	Australia	Ingot, sand castings and permanent mould castings	rem **Al**, 0.30 **Si**, 0.10 **Cu**, 0.50 **Mn**, 3.6–4.5 **Mg**, 0.10 **Zn**, 0.20 **Ti**, 0.30 **Fe**, 0.05 each, 0.15 total, others	**As cast (sand castings)**: all **diam**, 150 MPa **TS**, 6% in 2 in. (50 mm) **El**
CSA HA.3	.GS40	Canada	Ingot	rem **Al**, 0.30–0.7 **Si**, 0.10 **Cu**, 0.10 **Mn**, 3.6–4.5 **Mg**, 0.10 **Zn**, 0.20 **Ti**, 0.30 **Fe**, 0.05 each, 0.15 total, others	
QQ-A-371F	514.1 (formerly 214)	US	Ingot	rem **Al**, 0.35 **Si**, 0.15 **Cu**, 0.35 **Mn**, 3.6–4.5 **Mg**, 0.15 **Zn**, 0.25 **Ti**, 0.40 **Fe**, 0.05 each, 0.15 total, others	
QQ-A-371F	514.2 (formerly 214)	US	Ingot	rem **Al**, 0.30 **Si**, 0.10 **Cu**, 0.10 **Mn**, 3.6–4.5 **Mg**, 0.10 **Zn**, 0.20 **Ti**, 0.30 **Fe**, 0.05 each, 0.15 total, others	
QQ-A-371F	A514.2 (formerly A214)	US	Ingot	rem **Al**, 0.30 **Si**, 0.10 **Cu**, 0.10 **Mn**, 3.6–4.5 **Mg**, 1.4–2.2 **Zn**, 0.20 **Ti**, 0.30 **Fe**, 0.05 each, 0.15 total, others	
QQ-A-371F	B514.2 (formerly B214)	US	Ingot	rem **Al**, 1.4–2.2 **Si**, 0.10 **Cu**, 0.10 **Mn**, 3.6–4.5 **Mg**, 0.10 **Zn**, 0.20 **Ti**, 0.30 **Fe**, 0.05 each, 0.15 total, others	
QQ-A-371F	F514.1 (formerly F214)	US	Ingot	rem **Al**, 0.30–0.7 **Si**, 0.15 **Cu**, 0.35 **Mn**, 3.6–4.5 **Mg**, 0.15 **Zn**, 0.25 **Ti**, 0.40 **Fe**, 0.05 each, 0.15 total, others	
QQ-A-371F	F514.2 (formerly F214)	US	Ingot	rem **Al**, 0.30–0.7 **Si**, 0.10 **Cu**, 0.10 **Mn**, 3.6–4.5 **Mg**, 0.10 **Zn**, 0.20 **Ti**, 0.30 **Fe**, 0.05 each, 0.15 total, others	
ANSI/ASTM B 108	A514.0	US	Permanent mold castings	rem **Al**, 0.30 **Si**, 0.10 **Cu**, 0.30 **Mn**, 3.5–4.5 **Mg**, 1.4–2.2 **Zn**, 0.20 **Ti**, 0.40 **Fe**, 0.05 each, 0.15 total, others	**As fabricated**: all **diam**, 22 ksi (152 MPa) **TS**, 12 ksi (83 MPa) **YS**, 3% in 2 in. (50 mm) **El**, 60 **HB**
ANSI/ASTM B 618	514.0	US	Investment castings	rem **Al**, 0.35 **Si**, 0.15 **Cu**, 0.35 **Mn**, 3.5–4.5 **Mg**, 0.15 **Zn**, 0.25 **Ti**, 0.50 **Fe**, 0.05 each, 0.15 total, others	**As fabricated**: $\geq$0.250 in. **diam**, 22 ksi (152 MPa) **TS**, 9 ksi (62 MPa) **YS**, 6% in 2 in. (50 mm) **El**
ANSI/ASTM B 618	B514.0	US	Investment castings	rem **Al**, 1.4–2.2 **Si**, 0.35 **Cu**, 0.8 **Mn**, 3.5–4.5 **Mg**, 0.25 **Cr**, 0.35 **Zn**, 0.25 **Ti**, 0.6 **Fe**, 0.05 each, 0.15 total, others	**As fabricated**: $\geq$0.250 in. **diam**, 17 ksi (117 MPa) **TS**, 10 ksi (69 MPa) **YS**

CAST ALUMINUM AND ALUMINUM ALLOYS

specification number	designation	country	product forms	chemical composition	mechanical properties and hardness values
ANSI/ASTM B 108	B514.0	US	Permanent mold castings	rem **Al**, 1.4–2.2 **Si**, 0.35 **Cu**, 0.8 **Mn**, 3.5–4.5 **Mg**, 0.25 **Cr**, 0.35 **Zn**, 0.25 **Ti**, 0.6 **Fe**, 0.05 each, 0.15 total, others	**As fabricated**: all **diam**, 19 ksi (131 MPa) **TS**, 2% in 2 in. (50 mm) **El**
ANSI/ASTM B 26	514.0	US	Sand castings	rem **Al**, 0.35 **Si**, 0.15 **Cu**, 0.35 **Mn**, 3.5–4.5 **Mg**, 0.15 **Zn**, 0.25 **Ti**, 0.50 **Fe**, 0.05 each, 0.15 total, others	**As fabricated**: ≥0.250 in. **diam**, 22 ksi (152 MPa) **TS**, 9 ksi (62 MPa) **YS**, 6% in 2 in. (50 mm) **El**, 50 **HB**
ANSI/ASTM B 26	B514.0	US	Sand castings	rem **Al**, 1.4–2.2 **Si**, 0.35 **Cu**, 0.8 **Mn**, 3.5–4.5 **Mg**, 0.25 **Cr**, 0.35 **Zn**, 0.25 **Ti**, 0.6 **Fe**, 0.05 each, 0.15 total, others	**As fabricated**: ≥0.250 in. **diam**, 17 ksi (117 MPa) **TS**, 10 ksi (69 MPa) **YS**, 50 **HB**
CSA HA.9	.GS40	Canada	Sand castings	rem **Al**, 0.30–0.7 **Si**, 0.10 **Cu**, 0.35 **Mn**, 3.5–4.5 **Mg**, 0.15 **Zn**, 0.25 **Ti**, 0.50 **Fe**, 0.05 each, 0.15 total, others	
QQ-A-601E	514.0	US	Castings	rem **Al**, 0.35 **Si**, 0.15 **Cu**, 0.35 **Mn**, 3.5–4.5 **Mg**, 0.15 **Zn**, 0.25 **Ti**, 0.50 **Fe**, 0.05 each, 0.15 total, others	**As cast**: all **diam**, 22 ksi (152 MPa) **TS**, 12 ksi (83 MPa) **YS**, 6.0% in 2 in. (50 mm) **El**, 50 **HB**
QQ-A-601E	B514.0	US	Castings	rem **Al**, 1.4–2.2 **Si**, 0.35 **Cu**, 0.8 **Mn**, 3.5–4.5 **Mg**, 0.25 **Cr**, 0.35 **Zn**, 0.25 **Ti**, 0.6 **Fe**, 0.05 each, 0.15 total, others	**As cast**: all **diam**, 17 ksi (117 MPa) **TS**, 10 ksi (69 MPa) **YS**, 50 **HB**
QQ-A-596d	A214	US	Castings	rem **Al**, 0.30 **Si**, 0.10 **Cu**, 0.30 **Mn**, 3.5–4.5 **Mg**, 1.4–2.2 **Zn**, 0.20 **Ti**, 0.40 **Fe**, 0.05 each, 0.15 total, others	**As cast**: all **diam**, 22 ksi (152 MPa) **TS**, 2.5% in 2 in. (50 mm) **El**
QQ-A-371F	L514.2 (formerly L214)	US	Ingot	rem **Al**, 0.50–1.0 **Si**, 0.10 **Cu**, 0.40–0.6 **Mn**, 2.7–4.0 **Mg**, 0.05 **Zn**, 0.6–1.0 **Fe**, 0.05 each, 0.15 total, others	
JIS H 2212	Class 6	Japan	Ingot: for die castings	rem **Al**, 1.0 **Si**, 0.05 **Cu**, 0.4–0.6 **Mn**, 2.6–4.0 **Mg**, 0.03 **Ni**, 0.03 **Zn**, 0.03 **Sn**, 0.3–0.6 **Fe**	
JIS H 2118	Class 6	Japan	Ingot: for die castings	rem **Al**, 1.0 **Si**, 0.1 **Cu**, 0.4–0.6 **Mn**, 2.6–4.0 **Mg**, 0.1 **Ni**, 0.4 **Zn**, 0.1 **Sn**, 0.6 **Fe**	
SFS 2571	G–AlMg3	Finland	Die castings	rem **Al**, 0.5 **Si**, 0.10 **Cu**, 0.6 **Mn**, 2.5–4.0 **Mg**, 0.1 **Cr**, 0.05 **Ni**, 0.2 **Zn**, 0.2 **Ti**, 0.05 **Sn**, 0.50 **Fe**, 0.05 **Pb**, 0.05 each, 0.15 total, others	**As cast**: ≤20 mm **diam**, 150 MPa **TS**, 70 MPa **YS**, 5% in 2 in. (50 mm) **El**
ISO R2147	Al–Mg3	ISO	Sand castings: test pieces	rem **Al**, 0.5 **Si**, 0.10 **Cu**, 0.6 **Mn**, 2.0–4.5 **Mg**, 0.1 **Cr**, 0.05 **Ni**, 0.2 **Zn**, 0.2 **Ti**, 0.05 **Sn**, 0.5 **Fe**, 0.05 **Pb**	**As cast**: all **diam**, 150 MPa **TS**, 5% in 2 in. (50 mm) **El**
ISO R164	Al–Mg3	ISO	Castings	rem **Al**, 0.5 **Si**, 0.10 **Cu**, 0.6 **Mn**, 2.0–4.5 **Mg**, 0.1 **Cr**, 0.05 **Ni**, 0.2 **Zn**, 0.2 **Ti**, 0.05 **Sn**, 0.5 **Fe**, 0.05 **Pb**	
ISO R164	Al–Mg3Si	ISO	Castings	rem **Al**, 1.3 **Si**, 0.10 **Cu**, 0.6 **Mn**, 2.0–4.5 **Mg**, 0.4 **Cr**, 0.05 **Ni**, 0.2 **Zn**, 0.2 **Ti**, 0.05 **Sn**, 0.5 **Fe**, 0.05 **Pb**	

CAST ALUMINUM AND ALUMINUM ALLOYS

specification number	designation	country	product forms	chemical composition	mechanical properties and hardness values
JIS H 5302	Class 6	Japan	Die castings	rem **Al**, 1.0 **Si**, 0.1 **Cu**, 0.4–0.6 **Mn**, 2.5–4.0 **Mg**, 0.1 **Ni**, 0.4 **Zn**, 0.1 **Sn**, 0.8 **Fe**	
SABS 989	Al–Mg3A	South Africa	Ingot: for sand castings	rem **Al**, 1.3 **Si**, 0.03 **Cu**, 0.5 **Mn**, 2.8–3.5 **Mg**, 0.05 **Ni**, 0.10 **Zn**, 0.2 **Ti**, 0.05 **Sn**, 0.4 **Fe**, 0.05 **Pb**, 0.05 each, 0.15 total, others	**Sand cast**: all **diam**, 140 MPa **TS**, 3% in 2 in. (50 mm) **El**; **Sand cast, solution and precipitation heat treated**: all **diam**, 250 MPa **TS**, 2% in 2 in.(50 mm) **El**; **Chill cast**: all **diam**, 145 MPa **TS**, 3% in 2 in.(50 mm) **El**
DIN 1725 sheet 2	GK–AlMg3(Cu)/3.3543.02	Germany	Gravity die castings	rem **Al**, 1.3 **Si**, 0.3 **Cu**, 0.6 **Mn**, 2.0–4.0 **Mg**, 0.3 **Zn**, 0.20 **Ti**, 0.6 **Fe**, 0.05 each, 0.15 total, others	**As cast**: all **diam**, 90 MPa **TS**, 140 MPa **YS**, 3% in 2 in. (50 mm) **El**, 50–65 **HB**
DIN 1725 sheet 2	G–AlMg3(Cu)/3.3543.01	Germany	Sand castings	rem **Al**, 1.3 **Si**, 0.3 **Cu**, 0.6 **Mn**, 2.0–4.0 **Mg**, 0.3 **Zn**, 0.20 **Ti**, 0.6 **Fe**, 0.05 each, 0.15 total, others	**As cast**: all **diam**, 140 MPa **TS**, 80 MPa **YS**, 2% in 2 in. (50 mm) **El**, 50–65 **HB**
DIN 1725 sheet 2	GK–AlMg3Siwa/3.3241.62	Germany	Gravity die castings	rem **Al**, 0.9–1.3 **Si**, 0.05 **Cu**, 0.4 **Mn**, 2.5–3.5 **Mg**, 0.10 **Zn**, 0.20 **Ti**, 0.5 **Fe**, 0.05 each, 0.15 total, others	**Quenched and age hardened**: all **diam**, 220 MPa **TS**, 120 MPa **YS**, 3% in 2 in. (50 mm) **El**, 65–90 **HB**
DIN 1725 sheet 2	GK–AlMg3Si/3.3241.02	Germany	Gravity die castings	rem **Al**, 0.9–1.3 **Si**, 0.05 **Cu**, 0.4 **Mn**, 2.5–3.5 **Mg**, 0.10 **Zn**, 0.20 **Ti**, 0.5 **Fe**, 0.05 each, 0.15 total, others	**As cast**: all **diam**, 150 MPa **TS**, 80 MPa **YS**, 4% in 2 in. (50 mm) **El**, 50–65 **HB**
DIN 1725 sheet 2	G–AlMg3Siwa/3.3241.61	Germany	Sand castings	rem **Al**, 0.9–1.3 **Si**, 0.05 **Cu**, 0.4 **Mn**, 2.5–3.5 **Mg**, 0.10 **Zn**, 0.20 **Ti**, 0.5 **Fe**, 0.05 each, 0.15 total, others	**Quenched and age hardened**: all **diam**, 200 MPa **TS**, 120 MPa **YS**, 2% in 2 in. (50 mm) **El**, 65–90 **HB**
DIN 1725 sheet 2	G–AlMg3Si/3.3241.01	Germany	Sand castings	rem **Al**, 0.9–1.3 **Si**, 0.05 **Cu**, 0.4 **Mn**, 2.5–3.5 **Mg**, 0.10 **Zn**, 0.20 **Ti**, 0.5 **Fe**, 0.05 each, 0.15 total, others	**As cast**: all **diam**, 140 MPa **TS**, 80 MPa **YS**, 3% in 2 in. (50 mm) **El**, 50–60 **HB**
DIN 1725 sheet 2	GK–AlMg3/3.3541.02	Germany	Gravity die castings	rem **Al**, 0.5 **Si**, 0.05 **Cu**, 0.4 **Mn**, 2.5–3.5 **Mg**, 0.10 **Zn**, 0.20 **Ti**, 0.5 **Fe**, 0.05 each, 0.15 total, others	**As cast**: all **diam**, 150 MPa **TS**, 70 MPa **YS**, 5% in 2 in. (50 mm) **El**, 50–60 **HB**
DIN 1725 sheet 2	G–AlMg3/3.3541.01	Germany	Sand castings	rem **Al**, 0.5 **Si**, 0.05 **Cu**, 0.4 **Mn**, 2.5–3.5 **Mg**, 0.10 **Zn**, 0.20 **Ti**, 0.5 **Fe**, 0.05 each, 0.15 total, others	**As cast**: all **diam**, 140 MPa **TS**, 70 MPa **YS**, 3% in 2 in. (50 mm) **El**, 50–60 **HB**
SABS 990	Al–Mg3A	South Africa	Sand castings	rem **Al**, 1.3 **Si**, 0.05 **Cu**, 0.5 **Mn**, 2.5–3.5 **Mg**, 0.05 **Ni**, 0.10 **Zn**, 0.2 **Ti**, 0.5 **Sn**, 0.5 **Fe**, 0.05 **Pb**, 0.05 each, 0.15 total, others	**As cast**: all mm **diam**, 140 MPa **TS**, 3% in 2 in. (50 mm) **El**; Solution and precipitation heat treated: all **diam**, 205 MPa **TS**, 2% in 2 in (50 mm) **El**
ANSI/ASTM B 327	G1C	US	Ingot, bar, and shot	95.0 min **Al**, 0.7 **Si**, 2.0 **Cu**, 0.50 **Mn**, 0.75–1.10 **Mg**, 0.20 **Cr**, 0.20 **Ni**, 1.0 **Zn**, 0.02 **Sn**, 0.8 **Fe**, 0.020 **Pb**, 0.010 **Cd**	

CAST ALUMINUM AND ALUMINUM ALLOYS

specification number	designation	country	product forms	chemical composition	mechanical properties and hardness values
ANSI/ASTM B 618	713.0	US	Investment castings	rem **Al**, 0.25 **Si**, 0.40–1.0 **Cu**, 0.6 **Mn**, 0.20–0.50 **Mg**, 0.35 **Cr**, 0.15 **Ni**, 7.0–8.0 **Zn**, 0.25 **Ti**, 1.1 **Fe**, 0.10 each, 0.15 total, others	**Cooled and naturally and artificially aged:** ≤0.250 in. **diam**, 30 ksi (207 MPa) **TS**, 22 ksi (152 MPa) **YS**, 3% in 2 in. (50 mm) **El**
ANSI/ASTM B 179	713.1	US	Ingot: for sand castings	rem **Al**, 0.25 **Si**, 0.40–1.0 **Cu**, 0.6 **Mn**, 0.25–0.50 **Mg**, 0.35 **Cr**, 0.15 **Ni**, 7.0–8.0 **Zn**, 0.25 **Ti**, 0.10 each, 0.25 total, others	
ANSI/ASTM B 108	713.0	US	Permanent mold castings	rem **Al**, 0.25 **Si**, 0.40–1.0 **Cu**, 0.6 **Mn**, 0.20–0.50 **Mg**, 0.35 **Cr**, 0.15 **Ni**, 7.0–8.0 **Zn**, 0.25 **Ti**, 0.10 each, 0.25 total, others	**Cooled and naturally or artificially aged**: all **diam**, 32 ksi (221 MPa) **TS**, 22 ksi (152 MPa) **YS**, 4% in 2 in. (50 mm) **El**
ANSI/ASTM B 26	713.0	US	Sand castings	rem **Al**, 0.25 **Si**, 0.40–1.0 **Cu**, 0.6 **Mn**, 0.20–0.50 **Mg**, 0.35 **Cr**, 0.15 **Ni**, 7.0–8.0 **Zn**, 0.25 **Ti**, 1.1 **Fe**, 0.10 each, 0.25 total, others	**Cooled and artificially aged:** ≤0.250 in. **diam**, 30 ksi (207 MPa) **TS**, 22 ksi (152 MPa) **YS**, 3% in 2 in. (50 mm) **El**, 75 **HB**
QQ-A-371F	713.1 (formerly Tenzaloy)	US	Ingot	rem **Al**, 0.25 **Si**, 0.40–1.0 **Cu**, 0.6 **Mn**, 0.25–0.50 **Mg**, 0.35 **Cr**, 0.15 **Ni**, 7.0–8.0 **Zn**, 0.25 **Ti**, 0.8 **Fe**, 0.10 each, 0.25 total, others	
QQ-A-601E	713.0	US	Castings	rem **Al**, 0.25 **Si**, 0.40–1.0 **Cu**, 0.6 **Mn**, 0.20–0.50 **Mg**, 0.35 **Cr**, 0.15 **Ni**, 7.0–8.0 **Zn**, 0.25 **Ti**, 1.1 **Fe**, 0.10 each, 0.25 total, others	**As cast or artificially aged and T5**: all **diam**, 32 ksi (221 MPa) **TS**, 22 ksi (152 MPa) **YS**, 3.0% in 2 in. (50 mm) **El**, 75 **HB**
QQ-A-596d	Tenzaloy (613)	US	Castings	rem **Al**, 0.25 **Si**, 0.40–1.0 **Cu**, 0.6 **Mn**, 0.20–0.50 **Mg**, 0.35 **Cr**, 0.15 **Ni**, 7.0–8.0 **Zn**, 0.25 **Ti**, 1.8 **Fe**, 0.10 each, 0.25 total, others	**Aged and temper T5**: all **diam**, 32 ksi (221 MPa) **TS**, 4.0% in 2 in. (50 mm) **El**
ANSI/ASTM B 618	771.0	US	Investment castings	rem **Al**, 0.15 **Si**, 0.10 **Cu**, 0.10 **Mn**, 0.8–1.0 **Mg**, 0.06–0.20 **Cr**, 6.5–7.5 **Zn**, 0.10–0.20 **Ti**, 0.15 **Fe**, 0.05 each, 0.15 total, others	**Cooled and artificially aged:** ≤0.250 in. **diam**, 42 ksi (290 MPa) **TS**, 38 ksi (262 MPa) **YS**, 2% in 2 in. (50 mm) **El**; **Solution heat treated and artificially aged:** ≤0.250 in. **diam**, 42 ksi (290 MPa) **TS**, 35 ksi (241 MPa) **YS**, 5% in 2 in. (50 mm) **El**; **T71:** ≤0.250 in. **diam**, 48 ksi (331 MPa) **TS**, 45 ksi (310 MPa) **YS**, 2% in 2 in. (50 mm) **El**
ANSI/ASTM B 327	ZG71A	US	Ingot, bar, and shot	87.0 min **Al**, 0.7 **Si**, 1.7 **Cu**, 0.50 **Mn**, 0.65–1.05 **Mg**, 0.20 **Cr**, 0.20 **Ni**, 6.5–7.5 **Zn**, 0.02 **Sn**, 0.8 **Fe**, 0.020 **Pb**, 0.010 **Cd**	

CAST ALUMINUM AND ALUMINUM ALLOYS

specification number	designation	country	product forms	chemical composition	mechanical properties and hardness values
ANSI/ASTM B 179	771.2	US	Ingot: for sand castings	rem **Al**, 0.10 **Si**, 0.10 **Cu**, 0.10 **Mn**, 0.85–1.0 **Mg**, 0.06–0.20 **Cr**, 6.5–7.5 **Zn**, 0.10–0.20 **Ti**, 0.10 **Fe**, 0.05 each, 0.15 total, others	
ANSI/ASTM B 26	771.0	US	Sand castings	rem **Al**, 0.15 **Si**, 0.10 **Cu**, 0.10 **Mn**, 0.8–1.0 **Mg**, 0.06–0.20 **Cr**, 6.5–7.5 **Zn**, 0.10–0.20 **Ti**, 0.15 **Fe**, 0.05 each, 0.15 total, others	**Cooled and artificially aged**: ≤0.250 in. **diam**, 42 ksi (290 MPa) **TS**, 38 ksi (262 MPa) **YS**, 2% in 2 in. (50 mm) **El**, 100 **HB**; **T51**: ≤0.250 in. **diam**, 32 ksi (221 MPa) **TS**, 27 ksi (186 MPa) **YS**, 3% in 2 in. (50 mm) **El**, 85 **HB**; **Solution heat treated and artificially aged**: ≤0.250 in. **diam**, 42 ksi (290 MPa) **TS**, 35 ksi (241 MPa) **YS**, 5% in 2 in. (50 mm) **El**, 90 **HB**
AS 1874	AP703	Australia	Ingot, sand castings and permanent mould castings	rem **Al**, 0.15 **Si**, 0.10 **Cu**, 0.10 **Mn**, 0.65–0.85 **Mg**, 0.06–0.15 **Cr**, 6.5–7.5 **Zn**, 0.10–0.25 **Ti**, 0.15 **Fe**, 0.05 each, 0.20 total, others	
QQ-A-371F	771.2 (formerly Precedent 71A)	US	Ingot	rem **Al**, 0.10 **Si**, 0.10 **Cu**, 0.10 **Mn**, 0.85–1.0 **Mg**, 0.06–0.20 **Cr**, 6.5–7.5 **Zn**, 0.10–0.20 **Ti**, 0.10 **Fe**, 0.05 each, 0.15 total, others	
QQ-A-601E	771.0	US	Castings	rem **Al**, 0.15 **Si**, 0.10 **Cu**, 0.10 **Mn**, 0.8–1.0 **Mg**, 0.06–0.20 **Cr**, 6.5–7.5 **Zn**, 0.10–0.20 **Ti**, 0.15 **Fe**, 0.05 each, 0.15 total, others	**Annealed and T2**: all **diam**, 36 ksi (248 MPa) **TS**, 27 ksi (186 MPa) **YS**, 1.5% in 2 in. (50 mm) **El**, 87 **HB**; **Artificially aged and T5**: all **diam**, 42 ksi (290 MPa) **TS**, 38 ksi (262 MPa) **YS**, 2.0% in 2 in (50 mm) **El**, 100 **HB**; **Solution heat treated, artificially aged and T6**: all **diam**, 42 ksi (290 MPa) **TS**, 35 ksi (242 MPa) **YS**, 5.0% in 2 in (50 mm) **El**, 98 **HB**
ANSI/ASTM B 179	A712.1	US	Ingot: for sand castings	rem **Al**, 0.15 **Si**, 0.35–0.65 **Cu**, 0.05 **Mn**, 0.65–0.8 **Mg**, 6.0–7.0 **Zn**, 0.25 **Ti**, 0.40 **Fe**, 0.05 each, 0.15 total, others	
ANSI/ASTM B 618	A712.0	US	Investment castings	rem **Al**, 0.15 **Si**, 0.35–0.65 **Cu**, 0.05 **Mn**, 0.6–0.8 **Mg**, 6.0–7.0 **Zn**, 0.25 **Ti**, 0.50 **Fe**, 0.05 each, 0.15 total, others	**Cooled and naturally aged**: ≤0.250 in. **diam**, 32 ksi (221 MPa) **TS**, 20 ksi (138 MPa) **YS**, 2% in 2 in. (50 mm) **El**
ANSI/ASTM B 179	C712.1	US	Ingot: for permanent mold castings	rem **Al**, 0.30 **Si**, 0.35–0.65 **Cu**, 0.05 **Mn**, 0.30–0.45 **Mg**, 6.0–7.0 **Zn**, 0.20 **Ti**, 0.7–1.1 **Fe**, 0.05 each, 0.15 total, others	
ANSI/ASTM B 108	C712.0	US	Permanent mold castings	rem **Al**, 0.30 **Si**, 0.35–0.65 **Cu**, 0.05 **Mn**, 0.25–0.45 **Mg**, 6.0–7.0 **Zn**, 0.20 **Ti**, 0.7 **Fe**, 0.05 each, 0.15 total, others	**Cooled and naturally aged**: all **diam**, 28 ksi (193 MPa) **TS**, 18 ksi (124 MPa) **YS**, 7% in 2 in. (50 mm) **El**, 70 **HB**

CAST ALUMINUM AND ALUMINUM ALLOYS

specification number	designation	country	product forms	chemical composition	mechanical properties and hardness values
ANSI/ASTM B 26	A712.0	US	Sand castings	rem **Al**, 0.15 **Si**, 0.35–0.65 **Cu**, 0.05 **Mn**, 0.6–0.8 **Mg**, 6.0–7.0 **Zn**, 0.25 **Ti**, 0.50 **Fe**, 0.05 each, 0.15 total, others	**Cooled and artificially aged**: ≤0.250 in. **diam**, 32 ksi (221 MPa) **TS**, 20 ksi (138 MPa) **YS**, 2% in 2 in. (50 mm) **El**, 75 **HB**
CSA HA.3	.ZG61P	Canada	Ingot	rem **Al**, 0.15 **Si**, 0.35–0.65 **Cu**, 0.05 **Mn**, 0.65–0.8 **Mg**, 6.0–7.0 **Zn**, 0.20 **Ti**, 0.40 **Fe**, 0.05 each, 0.15 total, others	
CSA HA.9	.ZG61P	Canada	Sand castings	rem **Al**, 0.15 **Si**, 0.35–0.65 **Cu**, 0.05 **Mn**, 0.6–0.8 **Mg**, 6.0–7.0 **Zn**, 0.25 **Ti**, 0.50 **Fe**, 0.05 each, 0.15 total, others	**T5**: all **diam**, 32.0 ksi (221 MPa) **TS**, 20.0 ksi (138 MPa) **YS**, 2% in 2 in. (50 mm) **El**
QQ-A-371F	A712.1 (formerly A612)	US	Ingot	rem **Al**, 0.15 **Si**, 0.35–0.65 **Cu**, 0.05 **Mn**, 0.65–0.8 **Mg**, 6.0–7.0 **Zn**, 0.25 **Ti**, 0.40 **Fe**, 0.05 each, 0.15 total, others	
QQ-A-371F	C712.1 (formerly C612)	US	Ingot	rem **Al**, 0.30 **Si**, 0.35–0.65 **Cu**, 0.05 **Mn**, 0.30–0.45 **Mg**, 6.0–7.0 **Zn**, 0.20 **Ti**, 0.7–1.1 **Fe**, 0.05 each, 0.15 total, others	
QQ-A-601E	A712.0	US	Castings	rem **Al**, 0.15 **Si**, 0.35–0.65 **Cu**, 0.05 **Mn**, 0.6–0.8 **Mg**, 6.0–7.0 **Zn**, 0.25 **Ti**, 0.50 **Fe**, 0.05 each, 0.15 total, others	**As cast**: all **diam**, 32 ksi (221 MPa) **TS**, 20 ksi (138 MPa) **YS**, 2.0% in 2 in. (50 mm) **El**, 75 **HB**
NS 17 570	NS 17 570–00	Norway	Ingot: for casting	93 **Al**, 0.3 **Si**, 0.20–0.50 **Cu**, 0.40 **Mn**, 0.60–0.80 **Mg**, 0.05 **Ni**, 6.0 **Zn**, 0.25 **Ti**, 0.05 **Sn**, 0.7 **Fe**, 0.05 **Pb**, 0.3–0.6 **Cr**	
NS 17 570	NS 17 570–31	Norway	Castings: sand and chill	93 **Al**, 0.3 **Si**, 0.20–0.50 **Cu**, 0.40 **Mn**, 0.60–0.80 **Mg**, 0.05 **Ni**, 6.0 **Zn**, 0.25 **Ti**, 0.05 **Sn**, 0.7 **Fe**, 0.05 **Pb**, 0.3–0.6 **Cr**	**Naturally aged (sand castings)**: all **diam**, 216 MPa **TS**, 167 MPa **YS**, 4% in 2 in. (50 mm) **El**
ANSI/ASTM B 618	D712.0	US	Investment castings	rem **Al**, 0.30 **Si**, 0.25 **Cu**, 0.10 **Mn**, 0.50–0.65 **Mg**, 0.40–0.6 **Cr**, 5.0–6.5 **Zn**, 0.15–0.25 **Ti**, 0.50 **Fe**, 0.05 each, 0.20 total, others	**Cooled and naturally or artificially aged**: ≤0.250 in. **diam**, 34 ksi (234 MPa) **TS**, 25 ksi (172 MPa) **YS**, 4% in 2 in. (50 mm) **El**
ANSI/ASTM B 179	D712.2	US	Ingot: for sand castings	rem **Al**, 0.15 **Si**, 0.25 **Cu**, 0.10 **Mn**, 0.50–0.65 **Mg**, 0.40–0.6 **Cr**, 5.0–6.5 **Zn**, 0.15–0.25 **Ti**, 0.40 **Fe**, 0.05 each, 0.15 total, others	
ANSI/ASTM B 26	D172.0	US	Sand castings	rem **Al**, 0.30 **Si**, 0.25 **Cu**, 0.10 **Mn**, 0.50–0.65 **Mg**, 0.40–0.6 **Cr**, 5.0–6.5 **Zn**, 0.15–0.25 **Ti**, 0.50 **Fe**, 0.05 each, 0.20 total, others	**Cooled and artificially aged**: ≤0.250 in. **diam**, 34 ksi (224 MPa) **TS**, 25 ksi ksi (172 MPa) **YS**, 4% in 2 in. (50 mm) **El**, 75 **HB**
CSA HA.3	.ZG61N	Canada	Ingot	rem **Al**, 0.15 **Si**, 0.25 **Cu**, 0.10 **Mn**, 0.50–0.65 **Mg**, 0.40–0.6 **Cr**, 5.0–6.5 **Zn**, 0.15–0.25 **Ti**, 0.40 **Fe**, 0.05 each, 0.20 total, others	

CAST ALUMINUM AND ALUMINUM ALLOYS

specification number	designation	country	product forms	chemical composition	mechanical properties and hardness values
CSA HA.9	.ZG61N	Canada	Sand castings	rem **Al**, 0.30 **Si**, 0.25 **Cu**, 0.10 **Mn**, 0.50–0.65 **Mg**, 0.40–0.6 **Cr**, 5.0–6.5 **Zn**, 0.15–0.25 **Ti**, 0.50 **Fe**, 0.05 each, 0.20 total, others	**T5**: all **diam**, 34.0 ksi (234 MPa) **TS**, 25.0 ksi (172 MPa) **YS**, 4% in 2 in. (50 mm) **El**
QQ-A-601E	D712.0	US	Castings	rem **Al**, 0.30 **Si**, 0.25 **Cu**, 0.10 **Mn**, 0.50–0.65 **Mg**, 0.40–0.6 **Cr**, 5.0–6.5 **Zn**, 0.15–0.25 **Ti**, 0.50 **Fe**, 0.05 each, 0.20 total, others	**As cast or artificially aged and T5**: all **diam**, 34 ksi (234 MPa) **TS**, 25 ksi (173 MPa) **YS**, 4.0% in 2 in. (50 mm) **El**, 75 **HB**
BS L.162		UK	Forgings	rem **Al**, 0.40 **Si**, 1.2–2.0 **Cu**, 0.30 **Mn**, 2.1–2.9 **Mg**, 0.18–0.28 **Cr**, 5.1–6.1 **Zn**, 0.20 **Ti**, 0.50 **Fe**, 0.25 **Zr** + **Ti**, 0.05 each, 0.15 total, others	**Solution heat treated, quenched and artificially aged**: ≤75 mm **diam**, 455 MPa **TS**, 370 MPa **YS**, 7% in 2 in. (50 mm) **El**
BS L.161		UK	Forgings	rem **Al**, 0.40 **Si**, 1.2–2.0 **Cu**, 0.30 **Mn**, 2.1–2.9 **Mg**, 0.18–0.28 **Cr**, 5.1–6.1 **Zn**, 0.20 **Ti**, 0.50 **Fe**, 0.25 **Zr** + **Ti**, 0.05 each, 0.15 total, others	**Solution heat treated, quenched and artificially aged**: ≤75 mm **diam**, 455 MPa **TS**, 385 MPa **YS**, 7% in 2 in. (50 mm) **El**
BS L.160		UK	Bar, section	rem **Al**, 0.40 **Si**, 1.2–2.0 **Cu**, 0.30 **Mn**, 2.1–2.9 **Mg**, 0.18–0.28 **Cr**, 5.1–6.1 **Zn**, 0.20 **Ti**, 0.50 **Fe**, 0.25 **Zr** + **Ti**, 0.05 each, 0.15 total, others	**Solution heat treated, quenched and artificially aged**: ≤10 mm **diam**, 470 MPa **TS**, 400 MPa **YS**, 7% in 2 in. (50 mm) **El**
DS 3002	4438	Denmark	Ingot: for castings	rem **Al**, 0.3 **Si**, 0.20–0.50 **Cu**, 0.40 **Mn**, 0.60–0.80 **Mg**, 0.3–0.6 **Cr**, 0.05 **Ni**, 5.0–6.0 **Zn**, 0.15–0.25 **Ti**, 0.05 **Sn**, 0.7 **Fe**, 0.05 **Pb**	**Sand cast**: all **diam**, 216 MPa **TS**, 167 MPa **YS**, 4% in 2 in. (50 mm) **El**, 70–90 **HB**
SFS 2573	G–AlZn5Mg	Finland	Castings	rem **Al**, 0.3 **Si**, 0.20–0.50 **Cu**, 0.40 **Mn**, 0.60–0.80 **Mg**,0.3–0.6 **Cr**, 0.05 **Ni**, 5.0–6.0 **Zn**, 0.15–0.25 **Ti**, 0.05 **Sn**, 0.70 **Fe**, 0.05 **Pb**, 0.05 each, 0.15 total, others	**Aged**: ≤20 mm **diam**, 220 MPa **TS**, 170 MPa **YS**, 4% in 2 in. (50 mm) **El**
SIS 14 44 38	SIS Al 44 38–00	Sweden	Ingot	rem **Al**, 0.3 **Si**, 0.20–0.50 **Cu**, 0.40 **Mn**, 0.60–0.80 **Mg**, 0.3–0.6 **Cr**, 0.05 **Ni**, 5.0–6.0 **Zn**, 0.15–0.25 **Ti**, 0.05 **Sn**, 0.7 **Fe**, 0.05 **Pb**	
SIS 14 44 38	SIS Al 44 38–04	Sweden	Sand castings	rem **Al**, 0.3 **Si**, 0.20–0.50 **Cu**, 0.40 **Mn**, 0.60–0.80 **Mg**, 0.3–0.6 **Cr**, 0.05 **Ni**, 5.0–6.0 **Zn**, 0.15–0.25 **Ti**, 0.05 **Sn**, 0.7 **Fe**, 0.05 **Pb**	**Cast and naturally aged**: all **diam**, 220 MPa **TS**, 170 MPa **YS**, 4% in 2 in. (50 mm) **El**, 70–90 **HB**
SIS 14 44 38	SIS Al 44 38–07	Sweden	Chill castings	rem **Al**, 0.3 **Si**, 0.20–0.50 **Cu**, 0.40 **Mn**, 0.60–0.80 **Mg**, 0.3–0.6 **Cr**, 0.05 **Ni**, 5.0–6.0 **Zn**, 0.15–0.25 **Ti**, 0.05 **Sn**, 0.7 **Fe**, 0.05 **Pb**	**Cast, naturally aged**: all **diam**, 230 MPa **TS**, 180 MPa **YS**, 4% in 2 in. (50 mm) **El**, 70–100 **HB**
AS 1874	AP701	Australia	Ingot, sand castings and permanent mould castings	rem **Al**, 0.15 **Si**, 0.10 **Cu**, 0.10 **Mn**, 0.50–0.75 **Mg**, 0.40–0.6 **Cr**, 0.10 **Ni**, 4.8–5.7 **Zn**, 0.15–0.25 **Ti**, 0.05 **Sn**, 0.40 **Fe**, 0.05 **Pb**, 0.05 each, 0.20 total, others	**Cooled and naturally aged**: all **diam**, 215 MPa **TS**, 4% in 2 in. (50 mm) **El**; **Cooled and artificially aged**: all **diam**, 215 MPa **TS**, 4% in 2 in. (50 mm) **El**

CAST ALUMINUM AND ALUMINUM ALLOYS

specification number	designation	country	product forms	chemical composition	mechanical properties and hardness values
ISO R164	Al–Zn5Mg	ISO	Castings	rem **Al**, 0.30 **Si**, 0.35 **Cu**, 0.4 **Mn**, 0.20–0.70 **Mg**, 0.15–0.60 **Cr**, 0.05 **Ni**, 4.5–6.0 **Zn**, 0.10–0.30 **Ti**, 0.05 **Sn**, 1.0 **Fe**, 0.05 **Pb**	
ANSI/ASTM B 179	707.1	US	Ingot: for sand and permanent mold castings	rem **Al**, 0.20 **Si**, 0.20 **Cu**, 0.40–0.6 **Mn**, 1.9–2.4 **Mg**, 0.20–0.40 **Cr**, 4.0–4.5 **Zn**, 0.25 **Ti**, 0.6 **Fe**, 0.05 each, 0.15 total, others	
ANSI/ASTM B 618	707.0	US	Investment castings	rem **Al**, 0.20 **Si**, 0.20 **Cu**, 0.40–0.6 **Mn**, 1.8–2.4 **Mg**, 0.20–0.40 **Cr**, 4.0–4.5 **Zn**, 0.25 **Ti**, 0.8 **Fe**, 0.05 each, 0.15 total, others	**Cooled and naturally and artificially aged:** ≤0.250 in. **diam**, 33 ksi (228 MPa) **TS**, 22 ksi (152 MPa) **YS**, 2% in 2 in. (50 mm) **El**; **Solution heat treated and stabilized:** ≤0.250 in. **diam**, 37 ksi (255 MPa) **TS**, 30 ksi (207 MPa) **YS**, 1% in 2 in. (50 mm) **El**
ANSI/ASTM B 26	707.0	US	Sand castings	rem **Al**, 0.20 **Si**, 0.20 **Cu**, 0.40–0.6 **Mn**, 1.8–2.4 **Mg**, 0.20–0.40 **Cr**, 4.0–4.5 **Zn**, 0.25 **Ti**, 0.8 **Fe**, 0.05 each, 0.15 total, others	**Cooled and artificially aged:** ≤0.250 in. **diam**, 33 ksi (228 MPa) **TS**, 22 ksi (152 MPa) **YS**, 2% in 2 in. (50 mm) **El**; **Solution heat treated and stabilized:** ≤0.250 in. **diam**, 37 ksi (255 MPa) **TS**, 30 ksi (207 MPa) **YS**, 1% in 2 in. (50 mm) **El**, 80 **HB**,
QQ-A-371F	707.1 (formerly Ternalloy7)	US	Ingot	rem **Al**, 0.20 **Si**, 0.20 **Cu**, 0.40–0.6 **Mn**, 1.9–2.4 **Mg**, 0.20–0.40 **Cr**, 4.0–4.5 **Zn**, 0.25 **Ti**, 0.6 **Fe**, 0.05 each, 0.15 total, others	
QQ-A-601E	707.0	US	Castings	rem **Al**, 0.20 **Si**, 0.20 **Cu**, 0.40–0.6 **Mn**, 1.8–2.4 **Mg**, 0.20–0.40 **Cr**, 4.0–4.5 **Zn**, 0.25 **Ti**, 0.8 **Fe**, 0.05 each, 0.15 total others	**As cast or artificially aged and T5**: all **diam**, 33 ksi (228 MPa) **TS**, 33 ksi (228 MPa) **TS**, 22 ksi (152 MPa) **YS**, 2.0% in 2 in. (50 mm) **El**, 85 **HB**; **Solution heat treated, stabilized and T7**: all **diam**, 37 ksi (255 MPa) **TS**, 30 ksi (207 MPa) **YS**, 1.0% in 2 in. (50 mm) **El**, 80 **HB**
QQ-A-596d	Ternalloy 7 (607)	US	Castings	rem **Al**, 0.20 **Si**, 0.20 **Cu**, 0.40–0.6 **Mn**, 1.8–2.4 **Mg**, 0.20–0.40 **Cr**, 4.0–4.5 **Zn**, 0.25 **Ti**, 0.8 **Fe**, 0.35 each, others	**Aged and temper T5**: all **diam**, 42 ksi (290 MPa) **TS**, 4.0% in 2 in. (50 mm) **El**; **Solution heat treated and stabilized T7**: all **diam**, 45 ksi (310 MPa) **TS**, 3.0% in 2 in (50 mm) **El**
QQ-A-596d	Ternalloy 5 (603)	US	Castings	rem **Al**, 0.20 **Si**, 0.20 **Cu**, 0.40–0.6 **Mn**, 1.4–1.8 **Mg**, 0.20–0.40 **Cr**, 2.7–3.8 **Zn**, 0.25 **Ti**, 0.8 **Fe**, 0.05 each, 0.15 total, others	**Aged and temper T5**: all **diam**, 37 ksi (255 MPa) **TS**, 10% in 2 in. (50 mm) **El**

CAST ALUMINUM AND ALUMINUM ALLOYS

specification number	designation	country	product forms	chemical composition	mechanical properties and hardness values
ANSI/ASTM B 179	705.1	US	Ingot: for sand and permanent mold castings	rem **Al**, 0.20 **Si**, 0.20 **Cu**, 0.40–0.6 **Mn**, 1.5–1.8 **Mg**, 0.20–0.40 **Cr**, 2.7–3.3 **Zn**, 0.25 **Ti**, 0.6 **Fe**, 0.05 each, 0.15 total, others	
ANSI/ASTM B 618	705.0	US	Investment castings	rem **Al**, 0.20 **Si**, 0.20 **Cu**, 0.40–0.6 **Mn**, 1.4–1.8 **Mg**, 0.20–0.40 **Cr**, 2.7–3.3 **Zn**, 0.25 **Ti**, 0.8 **Fe**, 0.05 each, 0.15 total, others	**Cooled and naturally or artificially aged**: ≤0.250 in. **diam**, 30 ksi (207 MPa) **TS**, 17 ksi (117 MPa) **YS**, 5% in 2 in. (50 mm) **El**
ANSI/ASTM B 108	705.0	US	Permanent mold castings	rem **Al**, 0.20 **Si**, 0.20 **Cu**, 0.40–0.6 **Mn**, 1.4–1.8 **Mg**, 0.20–0.40 **Cr**, 2.7–3.3 **Zn**, 0.25 **Ti**, 0.8 **Fe**, 0.05 each, 0.15 total, others	**Cooled and naturally or artificially aged**: all **diam**, 37 ksi (255 MPa) **TS**, 17 ksi (117 MPa) **YS**, 10% in 2 in. (50 mm) **El**
ANSI/ASTM B 26	705.0	US	Sand castings	rem **Al**, 0.20 **Si**, 0.20 **Cu**, 0.40–0.6 **Mn**, 1.4–1.8 **Mg**, 0.20–0.40 **Cr**, 2.7–3.3 **Zn**, 0.25 **Ti**, 0.8 **Fe**, 0.05 each, 0.15 total, others	**Cooled and artificially aged**: ≤0.250 in. **diam**, 30 ksi (207 MPa) **TS**, 17 ksi (117 MPa) **YS**, 5% in 2 in. (50 mm) **El**, 65 **HB**
QQ-A-371F	705.1 (formerly Ternalloy5)	US	Ingot	rem **Al**, 0.20 **Si**, 0.20 **Cu**, 0.40–0.6 **Mn**, 1.5–1.8 **Mg**, 0.20–0.40 **Cr**, 2.7–3.3 **Zn**, 0.25 **Ti**, 0.6 **Fe**, 0.05 each, 0.15 total, others	
QQ-A-371F	D712.2 (formerly 40E)	US	Ingot	rem **Al**, 0.20 **Si**, 0.20 **Cu**, 0.40–0.6 **Mn**, 1.5–1.8 **Mg**, 0.20–0.40 **Cr**, 2.7–3.3 **Zn**, 0.25 **Ti**, 0.6 **Fe**, 0.05 each, 0.15 total, others	
QQ-A-601E	705.0	US	Castings	rem **Al**, 0.20 **Si**, 0.20 **Cu**, 0.40–0.6 **Mn**, 1.4–1.8 **Mg**, 0.20–0.40 **Cr**, 2.7–3.3 **Zn**, 0.25 **Ti**, 0.8 **Fe**, 0.05 each, 0.15 total, others	**As cast or artificially aged and T5**: all **diam**, 30 ksi (207 MPa) **TS**, 17 ksi (117 MPa) **YS**, 5.0% in 2 in. (50 mm) **El**, 65 **HB**
ANSI/ASTM B 108	707.0	US	Permanent mold castings	rem **Al**, 0.20 **Si**, 0.20 **Cu**, 0.40–0.6 **Mn**, 1.8–2.4 **Mg**, 0.20–0.40 **Cr**, 4.0–4.5 **Zn**, 0.25 **Ti**, 0.8 **Fe**, 0.05 each, 0.15 total, others	**Cooled and naturally or artificially aged**: all **diam**, 42 ksi (290 MPa) **TS**, 25 ksi (173 MPa) **YS**, 4% in 2 in. (50 mm) **El**; **Solution heat treated and stabilized**: all **diam**, 45 ksi (310 MPa) **TS**, 35 ksi (241 MPa) **YS**, 3% in 2 in. (50 mm) **El**
IS 6754	89 200	India	Ingot for wrought product	rem **Al**, 0.7 **Si**, 0.7–1.3 **Cu**, 0.7 **Mn**, 17.5–22.5 **Sn**, 0.7 **Fe**, 1.0 **Fe+Si+Mn**, 0.5 total, others	
JIS H 5402	Class 1	Japan	Castings: for bearings	rem **Al**, 0.5–1.0 **Cu**, 0.5 **Mg**, 1.0 **Ni**, 10.0–13.0 **Sn**, 2.0 other elements, total	**Annealed**: all **diam**, 30–40 **HV**
JIS H 5402	Class 2	Japan	Castings: for bearings	rem **Al**, 2.0–3.0 **Cu**, 1.0 **Mg**, 1.5 **Ni**, 6.0–9.0 **Sn**, 2.0 other elements, total	**Annealed**: all **diam**, 45–55 **HV**

CAST ALUMINUM AND ALUMINUM ALLOYS

specification number	designation	country	product forms	chemical composition	mechanical properties and hardness values
IS 6754	8482	India	Ingot for cast product	rem **Al**, 0.35–0.85 **Si**, 0.7–1.3 **Cu**, 0.75–1.25 **Mg**, 1.5–1.8 **Ni**, 6.5–7.5 **Sn**, 0.60 **Fe**	
AMS 4275C		US	Permanent mold castings	rem **Al**, 0.7 **Si**, 0.7–1.3 **Cu**, 0.10 **Mn**, 0.10 **Mg**, 0.7–1.3 **Ni**, 0.20 **Ti**, 5.5–7.0 **Sn**, 0.7 **Fe**, 0.30 others, total	**Stress relieved**: 35–50 **HB**
ANSI/ASTM B 179	850.1	US	Ingot: for sand and permanent mold castings	rem **Al**, 0.7 **Si**, 0.7–1.3 **Cu**, 0.10 **Mn**, 0.10 **Mg**, 0.7–1.3 **Ni**, 0.20 **Ti**, 5.5–7.0 **Sn**, 0.50 **Fe**, 0.30 others, total	
ANSI/ASTM B 179	A850.1	US	Ingot: for sand and permanent mold castings	rem **Al**, 2.0–3.0 **Si**, 0.7–1.3 **Cu**, 0.10 **Mn**, 0.10 **Mg**, 0.30–0.7 **Ni**, 0.20 **Ti**, 5.5–7.0 **Sn**, 0.50 **Fe**, 0.30 others, total	
ANSI/ASTM B 618	850.0	US	Investment castings	rem **Al**, 0.7 **Si**, 0.7–1.3 **Cu**, 0.10 **Mn**, 0.10 **Mg**, 0.7–1.3 **Ni**, 0.20 **Ti**, 5.5–7.0 **Sn**, 0.7 **Fe**, 0.30 others, total	**Cooled and artificially aged**: $\geq$0.250 in. **diam**, 16 ksi (110 MPa) **TS**, 5% in 2 in. (50 mm) **El**
ANSI/ASTM B 618	B850.0	US	Investment castings	rem **Al**, 0.40 **Si**, 1.7–2.3 **Cu**, 0.10 **Mn**, 0.6–0.9 **Mg**, 0.9–1.5 **Ni**, 0.20 **Ti**, 5.5–7.0 **Sn**, 0.7 **Fe**, 0.30 others, each	**Cooled and artificially aged**: $\geq$0.250 in. **diam**, 24 ksi (166 MPa) **TS**, 18 ksi (124 MPa) **YS**
ANSI/ASTM B 618	A850.0	US	Investment castings	rem **Al**, 2.0–3.0 **Si**, 0.7–1.3 **Cu**, 0.10 **Mn**, 0.10 **Mg**, 0.3–0.7 **Ni**, 0.20 **Ti**, 5.5–7.0 **Sn**, 0.7 **Fe**, 0.30 others, total	**Cooled and artificially aged**: $\geq$0.250 in. **diam**, 17 ksi (117 MPa) **TS**, 3% in 2 in. (50 mm) **El**
ANSI/ASTM B 179	B850.1	US	Ingot: for sand and permanent mold castings	rem **Al**, 0.40 **Si**, 1.7–2.3 **Cu**, 0.10 **Mn**, 0.7–0.9 **Mg**, 0.9–1.5 **Ni**, 0.20 **Ti**, 5.5–7.0 **Sn**, 0.50 **Fe**, 0.30 others, total	
ANSI/ASTM B 108	850.0	US	Permanent mold castings	rem **Al**, 0.7 **Si**, 0.7–1.3 **Cu**, 0.10 **Mn**, 0.10 **Mg**, 0.7–1.3 **Ni**, 0.20 **Ti**, 5.5–7.0 **Sn**, 0.7 **Fe**, 0.30 others, total	**Cooled and artificially aged**: 18 ksi (124 MPa) **TS**, 8% in 2 in. (50 mm) **El**
ANSI/ASTM B 108	A850.0	US	Permanent mold castings	rem **Al**, 2.0–3.0 **Si**, 0.7–1.3 **Cu**, 0.10 **Mn**, 0.10 **Mg**, 0.3–0.7 **Ni**, 0.20 **Ti**, 5.5–7.0 **Sn**, 0.7 **Fe**, 0.30 others, total	**Cooled and artificially aged**: all **diam**, 17 ksi (117 MPa) **TS**, 3% in 2 in. (50 mm) **El**; **Solution heat treated and artificially aged**: all **diam**, 18 ksi (124 MPa) **TS**, 8% in 2 in. (50 mm) **El**
ANSI/ASTM B 108	B850.0	US	Permanent mold castings	rem **Al**, 0.40 **Si**, 1.7–2.3 **Cu**, 0.10 **Mn**, 0.6–0.9 **Mg**, 0.9–1.5 **Ni**, 0.20 **Ti**, 5.5–7.0 **Sn**, 0.7 **Fe**, 0.30 others, total	**Cooled and artificially aged**: all **diam**, 27 ksi (186 MPa) **TS**, 3% in 2 in. (50 mm) **El**
ANSI/ASTM B 26	850.0	US	Sand castings	rem **Al**, 0.7 **Si**, 0.7–1.3 **Cu**, 0.10 **Mn**, 0.10 **Mg**, 0.7–1.3 **Ni**, 0.20 **Ti**, 5.5–7.0 **Sn**, 0.7 **Fe**, 0.30 others, total	**Cooled and artificially aged**: $\geq$0.250 in. **diam**, 16 ksi (110 MPa) **TS**, 5% in 2 in. (50 mm) **El**, 45 **HB**
ANSI/ASTM B 26	A850.0	US	Sand castings	rem **Al**, 2.0–3.0 **Si**, 0.7–1.3 **Cu**, 0.10 **Mn**, 0.10 **Mg**, 0.3–0.7 **Ni**, 0.20 **Ti**, 5.5–7.0 **Sn**, 0.7 **Fe**, 0.30 others, total	**Cooled and artificially aged**: $\geq$0.250 in. **diam**, 17 ksi (117 MPa) **TS**, 3% in 2 in. (50 mm) **El**, 45 **HB**

CAST ALUMINUM AND ALUMINUM ALLOYS

specification number	designation	country	product forms	chemical composition	mechanical properties and hardness values
ANSI/ASTM B 26	B850.0	US	Sand castings	rem **Al**, 0.40 **Si**, 0.7–1.3 **Cu**, 0.10 **Mn**, 0.6–0.9 **Mg**, 0.9–1.5 **Ni**, 0.20 **Ti**, 5.5–7.0 **Sn**, 0.7 **Fe**, 0.30 others, total	**Cooled and artificially aged:** ≥0.250 in. **diam**, 24 ksi (166 MPa) **TS**, 18 ksi (124 MPa) **YS**, 60 **HB**
QQ-A-371F	850.1 (formerly 750)	US	Ingot	rem **Al**, 0.7 **Si**, 0.7–1.3 **Cu**, 0.10 **Mn**, 0.10 **Mg**, 0.7–1.3 **Ni**, 0.20 **Ti**, 5.5–7.0 **Sn**, 0.50 **Fe**, 0.30 total, others	
QQ-A-371F	A850.1 (formerly A750)	US	Ingot	rem **Al**, 2.0–3.0 **Si**, 0.7–1.3 **Cu**, 0.10 **Mn**, 0.10 **Mg**, 0.30–0.7 **Ni**, 0.20 **Ti**, 5.5–7.0 **Sn**, 0.50 **Fe**, 0.30 total, others	
QQ-A-601F	850.0	US	Castings	rem **Al**, 0.7 **Si**, 0.7–1.3 **Cu**, 0.10 **Mn**, 0.10 **Mg**, 0.7–1.3 **Ni**, 0.20 **Ti**, 5.5–7.0 **Sn**, 0.7 **Fe**, 0.30 total, others	**Artificially aged and T5:** all in. **diam**, 16 ksi (110 MPa) **TS**, 11 ksi (76 MPa) **YS**, 5.0% in 2 in. (50 mm) **El**, 45 **HB**
QQ-A-601E	A850.0	US	Castings	rem **Al**, 2.0–3.0 **Si**, 0.7–1.3 **Cu**, 0.10 **Mn**, 0.30–0.7 **Ni**, 0.20 **Ti**, 5.5–7.0 **Sn**, 0.7 **Fe**, 0.30 total, others	**Artificially aged and T5:** all in. **diam**, 17 ksi (117 MPa) **TS**, 11 ksi (76 MPa) **YS**,5.0% in 2 in. (50 mm) **El**, 45 **HB**
QQ-A-601E	B850.0	US	Castings	rem **Al**, 0.40 **Si**, 1.7–2.3 **Cu**, 0.10 **Mn**, 0.6–0.9 **Mg**, 0.9–1.5 **Ni**, 0.20 **Ti**, 5.5–7.0 **Sn**, 0.7 **Fe**, 0.30 total, others	**Artificially aged and T5:** all in. **diam**, 24 ksi (165 MPa) **TS**, 18 ksi (124 MPa) **YS**, 65 **HB**
QQ-A-596d	850.0	US	Castings	rem **Al**, 0.7 **Si**, 0.7–1.3 **Cu**, 0.10 **Mn**, 0.7–1.3 **Ni**, 0.20 **Ti**, 5.5–7.0 **Sn**, 0.7 **Fe**, 0.30 total, others	**Aged and temper T5:** all in. **diam**, 18 ksi (124 MPa) **TS**, 8.0% in 2 in. (50 mm) **El**
QQ-A-596d	A750	US	Castings	rem **Al**, 2.0–3.0 **Si**, 0.7–1.3 **Cu**, 0.10 **Mn**, 0.30–0.7 **Ni**, 0.20 **Ti**, 5.5–7.0 **Sn**, 0.7 **Fe**, 0.30 total, others	**Aged and temper T5:** all in. **diam**, 17 ksi (117 MPa) **TS**, 3.0% in 2 in. (50 mm) **El**
QQ-a-596d	B750	US	Castings	rem **Al**, 0.40 **Si**, 1.7–2.3 **Cu**, 0.10 **Mn**, 0.6–0.9 **Mg**, 0.9–1.5 **Ni**, 0.20 **Ti**, 5.5–7.0 **Sn**, 0.7 **Fe**, 0.30 total, others	**Aged and temper T5:** all in. **diam**, 27 ksi (186 MPa) **TS**, 3.0% in 2 in. (50 mm) **El**
IS 6754	8328	India	Ingot for cast product	rem **Al**, 0.7 **Si**, 0.7–1.3 **Cu**, 0.10 **Mn**, 0.3 **Mg**, 0.7–1.3 **Ni**, 0.20 **Ti**, 5.5–7.0 **Sn**, 0.7 **Fe**, 0.30 total, others	
IS 6754	83 428	India	Ingot for wrought product	rem **Al**, 1.0–2.0 **Si**, 0.7–1.3 **Cu**, 0.10 **Mn**, 0.3 **Mg**, 0.2–0.7 **Ni**, 0.10 **Ti**, 5.5–7.0 **Sn**, 0.7 **Fe**, 0.15 total, others	
QQ-A-371F	B850.1 (formerly B750)	US	Ingot	rem **Al**, 0.40 **Si**, 1.7–2.3 **Cu**, 0.10 **Mn**, 0.7–0.9 **Mg**, 0.9–1.5 **Ni**, 0.20 **Ti**, 5.5–7.0 **Sn**, 0.50 **Fe**, 0.30 total, others	

MISCELLANEOUS MASTER AND OTHER CAST ALUMINUM ALLOYS

specification number	designation	country	product forms	chemical composition	mechanical properties and hardness values
DIN 1725 sheet 3	V–AlBe5/3.0841	Germany	Ingot	rem **Al**, 0.20 **Si**, 0.05 **Cu**, 0.03 **Mn**, 0.05 **Mg**, 0.03 **Cr**, 0.10 **Zn**, 0.02 **Ti**, 0.40 **Fe**, 4.5–6.0 **B**, 0.05 each, 0.2 total, others	
DIN 1725 sheet 3	V–AlB4/3.0831	Germany	Ingot	rem **Al**, 0.20 **Si**, 0.02 **Cu**, 0.02 **Mn**, 0.02 **Mg**, 0.02 **Cr**, 0.02 **Ni**, 0.03 **Zn**, 0.02 **Ti**, 0.30 **Fe**, 3.5–4.5 **B**, 0.02 **V**, 0.02 **Zr**, 0.05 each, 0.10 total, others	
DIN 1725 sheet 3	V–AlB3/3.0821	Germany	Ingot	rem **Al**, 0.2 **Si**, 0.02 **Cu**, 0.02 **Mn**, 0.02 **Mg**, 0.02 **Cr**, 0.02 **Ni**, 0.03 **Zn**, 0.02 **Ti**, 0.30 **Fe**, 2.5–3.4 **B**, 0.05 each, 0.10 total, others	
DIN 1725 sheet 3	V–AlCr5/3.0551	Germany	Ingot	rem **Al**, 0.40 **Si**, 0.15 **Cu**, 0.35 **Mn**, 0.50 **Mg**, 4.0–6.0 **Cr**, 0.10 **Ni**, 0.15 **Zn**, 0.10 **Ti**, 0.10 **Sn**, 0.45 **Fe**, 0.10 **Pb**, 0.05 each, 0.15 total, others	
DIN 1725 sheet 3	V–AlCu50/3.1191	Germany	Ingot	rem **Al**, 0.40 **Si**, 48.0–52.0 **Cu**, 0.35 **Mn**, 0.30 **Mg**, 0.10 **Cr**, 0.20 **Ni**, 0.30 **Zn**, 0.10 **Ti**, 0.10 **Sn**, 0.45 **Fe**, 0.20 **Pb**, 0.05 each, 0.15 total, others	
DIN 1725 sheet 3	V–AlFe5/3.0941	Germany	Ingot	rem **Al**, 0.40 **Si**, 0.15 **Cu**, 0.35 **Mn**, 0.40 **Mg**, 0.10 **Cr**, 0.10 **Ni**, 0.2 **Zn**, 0.1 **Ti**, 0.10 **Sn**, 4.0–6.0 **Fe**, 0.15 **Pb**, 0.05 each, 0.15 total, others	
SABS 711	Al Mn10	South Africa	Ingot	rem **Al**, 0.45 **Si**, 0.10 **Cu**, 11.0 **Mn**, 0.07 **Zn**, 0.70 **Fe**, 0.05 each, 0.15 total, others	
DIN 1725 sheet 3	V–AlMo10/3.0571	Germany	Ingot	rem **Al**, 0.40 **Si**, 0.2 **Cu**, 9.0–11.0 **Mn**, 0.50 **Mg**, 0.10 **Cr**, 0.20 **Ni**, 0.2 **Zn**, 0.1 **Ti**, 0.10 **Sn**, 0.45 **Fe**, 0.10 **Pb**, 0.05 each, 0.15 total, others	
SABS 711	Al Mn5	South Africa	Ingot	rem **Al**, 0.45 **Si**, 0.10 **Cu**, 6.0 **Mn**, 0.07 **Zn**, 0.70 **Fe**, 0.05 each, 0.15 total, others	
DIN 1725 sheet 3	VR–AlSi20/3.2292	Germany	Ingot	rem **Al**, 18.0–21.0 **Si**, 0.05 **Cu**, 0.10 **Mn**, 0.05 **Mg**, 0.05 **Cr**, 0.05 **Ni**, 0.10 **Zn**, 0.05 **Ti**, 0.03 **Fe**, 0.05 each, 0.15 total, others	
DIN 1725 sheet 3	V–AlSi20/3.2291	Germany	Ingot	rem **Al**, 18.0–21.0 **Si**, 0.20 **Cu**, 0.35 **Mn**, 0.40 **Mg**, 0.10 **Cr**, 0.20 **Ni**, 0.2 **Zn**, 0.1 **Ti**, 0.10 **Sn**, 0.45 **Fe**, 0.10 **Pb**, 0.05 each, 0.15 total, others	
DIN 1725 sheet 3	V–AlSi12/3.2581	Germany	Ingot	rem **Al**, 11.0–13.5 **Si**, 0.05 **Cu**, 0.4 **Mn**, 0.05 **Mg**, 0.10 **Zn**, 0.15 **Ti**, 0.5 **Fe**, 0.05 each, 0.15 total, others	

MISCELLANEOUS MASTER AND OTHER CAST ALUMINUM ALLOYS

specification number	designation	country	product forms	chemical composition	mechanical properties and hardness values
DIN 1725 sheet 3	V–AlTi10/3.0881	Germany	Ingot	rem **Al**, 0.20 **Si**, 0.02 **Cu**, 0.02 **Mn**, 0.02 **Mg**, 0.02 **Cr**, 0.04 **Ni**, 9.0–11.0 **Ti**, 0.30 **Fe**, 0.30 **V**, 0.05 each, 0.15 total, others	
DIN 1725 sheet 3	V–AlTi5B1/3.0861	Germany	Ingot	rem **Al**, 0.20 **Si**, 0.02 **Cu**, 0.02 **Mn**, 0.02 **Mg**, 0.02 **Cr**, 0.04 **Ni**, 0.03 **Zn**, 5.0–6.2 **Ti**, 0.30 **Fe**, 0.9–1.4 **B**, 0.20 **V**, 0.02 **Zr** 0.05 each, 0.10 total, others	
DIN 1725 sheet 3	V–AlTi5/3.0851	Germany	Ingot	rem **Al**, 0.50 **Si**, 0.15 **Cu**, 0.35 **Mn**, 0.50 **Mg**, 0.10 **Cr**, 0.10 **Ni**, 0.15 **Zn**, 4.5–6.0 **Ti**, 0.10 **Sn**, 0.45 **Fe**, 0.25 **V**, 0.05 each, 0.15 total, others	
DIN 1725 sheet 3	V–AlZr6/3.0862	Germany	Ingot	rem **Al**, 0.20 **Si**, 0.02 **Cu**, 0.02 **Mn**, 0.02 **Mg**, 0.02 **Cr**, 0.04 **Ni**, 0.03 **Zn**, 0.02 **Ti**, 0.30 **Fe**, 0.02 **V**, 5.0–6.5 **Zr**, 0.05 each, 0.10 total, others	

WROUGHT COPPER AND COPPER ALLOYS

WROUGHT COPPER AND COPPER ALLOYS

specification number	designation	country	product forms	chemical composition	mechanical properties and hardness values
ANSI/ASTM F 68		US	Wrought products	99.99 min **Cu**, 0.0010 **Pb**, 0.0001 **Zn**, 0.0003 **P**, 0.0015 **S**, 0.0010 **Bi**, 0.0001 **Ca**, 0.0001 **Hg**, 0.0005 **O**, 0.0010 **Se**, 0.0010 **Te**, 0.0040 max **Sb + As + Bi + Mn + Se + Fe + Sn**	
ANSI/ASTM B 111	C 10100	US	Tube: seamless	99.99 min **Cu**, 0.0010 **Pb**, 0.0001 **Zn**, 0.0003 **P**, 0.0018 **S**, 0.0001 **Hg**, 0.0001 **Cd**, 0.0010 **Se**, 0.0010 **Bi**, 0.0010 **O**, 0.0040 **Se + Te + Bi + As + Sb + Sn + Mn**	**Light drawn**: all **diam**, 36 ksi (250 MPa) **TS**, 30 ksi (205 MPa) **YS**; **Hard drawn**: all **diam**, 45 ksi (310 MPa) **TS**, 40 ksi (275 MPa) **YS**
ANSI/ASTM B 170	Grade 1	US	Bar, billet, cake	99.99 min **Cu**, 0.0001 **Zn**, 0.001 **Fe**, 0.0003 **P**, 0.0018 **S**, 0.0001 **Ca**, 0.0001 **Hg**, 0.001 **Se**, 0.001 **Te**, 0.001 **Bi**, 0.001 **O**	
ANSI/ASTM B 280	C 10100	US	Tube: seamless	99.99 min **Cu**, 0.0010 **Pb**, 0.0001 **Zn**, 0.0003 **P**, 0.0018 **S**, 0.0001 **Hg**, 0.0001 **Cd**, 0.0010 **Se**, 0.0010 **Te**, 0.0010 **Bi**, 0.0010 **O**, 0.0040 **Se + Te + Bi + S + Sb + Sn + Mg**	**Annealed (coiled lengths)**: all **diam**, 30 ksi (205 MPa) **TS**, 40% in 2 in. (50.8 mm) **El**; **Cold drawn (straight lengths)**: all **diam**, 36 ksi (250 MPa) **TS**
ANSI/ASTM B 359	C 10100	US	Tube: seamless	99.99 min **Cu**, 0.0010 **Pb**, 0.0001 **Zn**, 0.0003 **P**, 0.0018 **S**, 0.0001 **Hg**, 0.0001 **Cd**, 0.0010 **Se**, 0.0010 **Bi**, 0.0010 **O**	
ANSI/ASTM B 75	C 10100	US	Tube: seamless	99.99 min **Cu**, 0.0010 **Pb**, 0.0001 **Zn**, 0.0003 **P**, 0.0018 **S**, 0.0001 **Hg**, 0.0001 **Cd**, 0.0010 **Se**, 0.0010 **Te**, 0.0010 **Bi**, 0.0010 **O**, 0.0040 **Se + Te + Bi + As + Sb + Sn + Mn**	**Drawn**: all **diam**, 36 ksi (250 MPa) **TS**, 30 **HR30T**; **Soft anneal**: all **diam**, 60 **HR15T**; **Light anneal**: all **diam**, 65 **HR15T**
ANSI/ASTM B 447	C 10100	US	Tube	99.99 min **Cu**, 0.0010 **Pb**, 0.0001 **Zn**, 0.0003 **P**, 0.0018 **S**, 0.0001 **Hg**, 0.0001 **Cd**, 0.0010 **Se**, 0.0010 **Bi**, 0.0010 **O**, 0.0040 **Se + Te + Bi + As + Sb + Sn + Mg**	**Annealed (soft)**: all **diam**, 60 max **HR15T**; **As welded (from hard strip)**: all **diam**, 45 ksi (310 MPa) **TS**, 54–62 **HR30T**; **Welded and cold drawn (drawn)**: all **diam**, 36 ksi (250 MPa) **TS**, 30 min **HR30T**
ANSI/ASTM B 152	C 10100	US	Sheet, strip, plate, bar	99.99 min **Cu**, 0.0010 **Pb**, 0.0001 **Zn**, 0.003 **P**, 0.0018 **S**, 0.0001 **Hg**, 0.0001 **Cd**, 0.0010 **Se**, 0.0010 **Te**, 0.0010 **Bi**, 0.0010 **O**, 0.0040 **Se + Te + Bi + As + Sb + Sn + Mn**	**Hard**: >0.020 in. (>0.508 mm) **diam**, 43 ksi (295 MPa) **TS**, 86–93 **HRF**; **Spring**: >0.020 in. (>0.508 mm) **diam**, 50 ksi (345 MPa) **TS**, 91–97 **HRF**; **Hot rolled**: >0.020 in. (>0.508 mm) **diam**, 30 ksi (205 MPa) **TS**, 75 max **HRF**

WROUGHT COPPER AND COPPER ALLOYS

specification number	designation	country	product forms	chemical composition	mechanical properties and hardness values
AS 1567	101A	Australia	Rod, bar, section	99.99 min **Cu**, 0.0003 **P**, 0.0040 total, others	**Soft annealed**: 12–50 mm **diam**, 230 max MPa **TS**, 45% in 2 in. (50 mm) **El**; **1/2 hard**: 25–50 mm **diam**, 230 MPa **TS**, 22% in 2 in. (50 mm) **El**; **Hard**: 10–12 mm **diam**, 280 MPa **TS**, 8% in 2 in. (50 mm) **El**
BS 2873	C101	UK	Wire	99.99 min **Cu**, 0.005 **Pb**, 0.0010 **Bi**, 0.03 total, others	**Annealed**: ≤0.125 mm **diam**, 10% in 2 in. (50 mm) **El**
CDA C 101		US	Plate, sheet, strip, pipe, rod, shapes, tube, wire	99.99 min **Cu**, 0.0010 **Pb**, 0.0001 **Zn**, 0.0003 **P**, 0.0018 **S**, 0.0010 **Te**, 0.0001 **Hg**, 0.0001 **Cd**, 0.0010 **Se**, 0.0010 **Bi**, 0.0010 **O**, 0.0040 **Se+Te+Bi+As+Sb+Sn+Mn**	**Hard (plate, sheet, strip)**: 0.040 in. (1.016 mm) **diam**, 50 ksi (345 MPa) **TS**, 45 ksi (310 MPa) **YS**, 6% in 2 in. (50 mm) **El**, 90 **HRF**; **Hard (rod)**: 1.0 in. (25.4 mm) **diam**, 55 ksi (379 MPa) **TS**, 50 ksi (345 MPa) **YS**, 10% in 2 in. (50 mm) **El**, 94 **HRF**; **Hard (shapes)**: 0.500 in. (12.7 mm) **diam**, 40 ksi (276 MPa) **TS**, 32 ksi (221 MPa) **YS**, 30% in 2 in. (50 mm) **El**, 35 **HRB**
CDA C109		US	Plate, sheet, strip, rod, wire, tube, shapes	99.99 min **Cu**, 0.0010 **Pb**, 0.0005 **Zn**, 0.0005 **P**, 0.0015 **S**, 0.005–0.015 **B**, 0.0003 **Cd**, 0.0001 **Hg**, **Ag+B** included in **Cu** content, 0.0040 **Se+Te+Bi+As+Sb+Sn+Mn**	**Hard (plate, sheet, strip)**: 0.040 in. (1.016 mm) **diam**, 50 ksi (345 MPa) **TS**, 45 ksi (310 MPa) **YS**, 6% in 2 in. (50 mm) **El**, 90 **HRF**; **Hard (rod)**: 1.0 in. (25.4 mm) **diam**, 48 ksi (331 MPa) **TS**, 44 ksi (303 MPa) **YS**, 16% in 2 in. (50 mm) **El**, 87 **HRF**; **Hard (shapes)**: 0.500 in. (12.7 mm) **diam**, 40 ksi (276 MPa) **TS**, 32 ksi (221 MPa) **YS**, 30% in 2 in. (50 mm) **El**, 35 **HRB**
COPANT 521	Cu OF Gr 1	COPANT	Tube	99.99 min **Cu**, 0.001 **Pb**, 0.0003 **P**, 0.0018 **S**, 0.001 O_2, 0.001 **Bi**, 0.001 **Hg**, 0.001 **Se**, 0.001 **Te**, 0.0040 **As+Mn+Sb+Se+Sn+Te**, silver included in copper content	**Annealed**: 10.3–324 mm od **diam**, 255 MPa **TS**, 25% in 2 in. (50 mm) **El**
COPANT 536	Cu OFE (Gr 1)	COPANT	Sheet	99.99 min **Cu**, 0.001 **Sn**, 0.001 **Pb**, 0.0001 **Zn**, 0.001 **Sb**, 0.001 **Mn**, 0.0003 **P**, 0.0018 **S**, 0.001 **As**, 0.001 **Bi**, 0.001 **Se**, 0.001 O_2, 0.0001 **Cd**, 0.0001 **Hg**, 0.001 **Te**, 0.004 total, impurities, 0.010 total, others	**Annealed**: all **diam**, 205 MPa **TS**; **Tempered**: all **diam**, 225 MPa **TS**
COPANT 162	C 101 00 (=OFE)	COPANT	Wrought products	99.99 min **Cu**, 0.001 **Pb**, 0.0001 **Zn**, 0.0003 **P**, 0.0018 **S**, 0.0010 **Te**, 0.001 **Bi**, 0.001 **Se**, 0.001 O_2, 0.0001 **Hg**, 0.0001 **Cd**	**Rolled**: 50 mm **diam**, 345 MPa **TS**, 140 MPa **YS**, 35% in 2 in. (50 mm) **El**

WROUGHT COPPER AND COPPER ALLOYS

specification number	designation	country	product forms	chemical composition	mechanical properties and hardness values
JIS H 3510	C 1011	Japan	Sheet, plate, strip, pipe, tube, rod, bar, and wire: for electron devices	99.98 min **Cu**, 0.001 **Pb**, 0.0001 **Zn**, 0.0003 **P**, 0.0018 **S**, 0.001 **Bi**, 0.0001 **Cd**, 0.0001 **Hg**, 0.001 O_2, 0.001 **Se**, 0.001 **Te**	**Annealed**: 0.3–12 mm **diam**, 196 MPa **TS**, 40% in 2 in. (50 mm) **El**; **1/2 hard**: 0.3–12 mm **diam**, 245 MPa **TS**, 15% in 2 in. (50 mm) **El**; **Hard**: 0.3–10 mm **diam**, 275 MPa **TS**
COPANT 536	Cu BDHC	COPANT	Sheet	99.97 min **Cu**, 0.001 **Sn**, 0.001 **Pb**, 0.0005 **Zn**, 0.001 **Fe**, 0.001 **Sb**, 0.001 **Mn**, 0.0005 **P**, 0.0015 **S**, 0.002 **Ag**, 0.005–0.015 **B**, 0.001 **As**, 0.001 **Bi**, 0.003 **Cd**, 0.004 total, others	**Annealed**: all **diam**, 205 MPa **TS**; **Tempered**: all **diam**, 225 MPa **TS**
COPANT 528	Cu BDHC	COPANT	Tube	99.97 min **Cu**, 0.001 **Pb**, 0.0005 **Zn**, 0.001 **Sb**, 0.001 **Mn**, 0.0005 **P**, 0.0015 **S**, 0.002 **Ag**, 0.005–0.015 **B**, 0.001 **As**, 0.001 **Bi**, 0.0003 **Cd**, 0.0001 **Hg**, 0.001 **Se**, 0.001 **Te**, ≤0.004 **As + Sb + Bi + Te + Sn + Mn**	**Annealed (regular)**: 0.45–9.7 mm **diam**, 205 MPa **TS**, 60 MPa **YS**, 40% in 2 in. (50 mm) **El**; **Tempered to 1/2 hard**: all in. (0.45–9.7 mm) **diam**, 245 MPa **TS**; **Tempered to full hard**: all in. (0.45–9.7 mm) **diam**, 310 MPa **TS**, 275 MPa **YS**
JIS H 3300	C 1020	Japan	Pipe and tube: seamless	99.96 min **Cu**	**Annealed**: 4–100 mm **diam**, 206 MPa **TS**, 40% in 2 in. (50 mm) **El**, 60 max **HR15T**; **1/2 hard**: 4–100 mm **diam**, 245 MPa **TS**, 30–60 **HR30T**; **Hard**: ≤25 mm **diam**, 314 MPa **TS**, 55 min **HR30T**
JIS H 3250	C 1020	Japan	Rod and bar	99.96 min **Cu**	**As manufactured**: ≥6 mm **diam**, 196 MPa **TS**, 25% in 2 in. (50 mm) **El**; **Annealed**: 6–25 mm **diam**, 196 MPa **TS**, 30% in 2 in. (50 mm) **El**; **Hard**: 6–25 mm **diam**, 275 MPa **TS**
JIS H 3140	C 1020	Japan	Bar: bus	99.96 min **Cu**	**Annealed**: 2–30 mm **diam**, 196 MPa **TS**, 35% in 2 in. (50 mm) **El**; **1/2 hard**: 2–20 mm **diam**, 245 MPa **TS**, 15% in 2 in. (50 mm) **El**; **Hard**: 2–20 mm **diam**, 275 MPa **TS**
JIS H 3100	C1020	Japan	Sheet, plate, and strip	99.96 min **Cu**	**Annealed**: 0.3–30 mm **diam**, 196 MPa **TS**, 35% in 2 in. (50 mm) **El**; **1/2 hard**: 0.3–20 mm **diam**, 245 MPa **TS**, 15% in 2 in. (50 mm) **El**; **Hard**: 0.3–10 mm **diam**, 275 MPa **TS**
JIS H 2121		Japan	Cathode: electrolytic	99.96 min **Cu**, 0.005 **Pb**, 0.01 **Fe**, 0.005 **Sb**, 0.010 **S**, silver included as copper	

WROUGHT COPPER AND COPPER ALLOYS

specification number	designation	country	product forms	chemical composition	mechanical properties and hardness values
JIS H 2125		Japan	Billet and cake: oxygen-free electrolytic	99.96 min **Cu**, silver included as copper	
AMS 4501A		US	Sheet, strip, and plate	99.95 min **Cu**, 0.05 total others, silver included in copper content	**Cold rolled to 1/8 hard**: all **diam**, 32 ksi (221 MPa) **TS**
ANSI/ASTM B 75	C 10200	US	Tube: seamless	99.95 min **Cu**	**Drawn**: all **diam**, 36 ksi (250 MPa) **TS**, 30 **HR30T**; **Soft anneal**: all **diam**, 60 **HR15T**; **Light anneal**: all **diam**, 65 **HR15T**
ANSI/ASTM B 68	C 102	US	Tube: seamless	99.95 min **Cu**	
ANSI/ASTM B 88	C 10200	US	Tube: seamless	99.95 min **Cu**	**Annealed coil**: all **diam**, 30 ksi (205 MPa) **TS**, 50 **HRF**; **Annealed straight length**: all **diam**, 30 ksi (205 MPa) **TS**, 55 **HRF**; **Drawn**: all **diam**, 36 ksi (250 MPa) **TS**, (all MPa) **YS**, 30 **HR30T**
ANSI/ASTM B 447	C 10300	US	Tube	99.95 min **Cu**, 0.001–0.005 **P**, silver and phosphorus included in copper content	**Annealed (soft)**: all **diam**, 60 max **HR15T**; **As welded (from hard strip)**: all **diam**, 45 ksi (310 MPa) **TS**, 54–62 **HR30T**; **Welded and cold drawn (drawn)**: all **diam**, 36 ksi (250 MPa) **TS**, 30 min **HR30T**
ANSI/ASTM B 447	C 10800	US	Tube	99.95 min **Cu**, 0.005–0.012 **P**, silver and phosphorus included in copper content	**Annealed (soft)**: all **diam**, 60 max **HR15T**; **As welded (from hard strip)**: all **diam**, 45 ksi (310 MPa) **TS**, 54–62 **HR30T**; **Welded and cold drawn (drawn)**: all **diam**, 36 ksi (250 MPa) **TS**, 30 min **HR30T**
ANSI/ASTM B 447	C 10200	US	Tube	99.95 min **Cu**, silver included in copper content	**Annealed (soft)**: all **diam**, 60 max **HR15T**; **As welded (from hard strip)**: all **diam**, 45 ksi (310 MPa) **TS**, 54–62 **HR30T**; **Welded and cold drawn (drawn)**: all **diam**, 36 ksi (250 MPa) **TS**, 30 min **HR30T**
ANSI/ASTM B 543	C 10800	US	Tube: welded	99.95 min **Cu**, 0.05–0.012 **P**, silver and phosphorus included in copper content	**As welded from annealed strip**: all **diam**, 32 ksi (220 MPa) **TS**, 15 ksi (105 MPa) **YS**; **Welded and annealed**: all **diam**, 30 ksi (205 MPa) **TS**, 9 ksi (62 MPa) **YS**; **Fully finished–hard drawn**: all **diam**, 45 ksi (310 MPa) **TS**, 40 ksi (275 MPa) **YS**
ANSI/ASTM B 111	C 10300	US	Tube: seamless	99.95 min **Cu**, 0.001–0.005 **P**, phosphorous included in copper content	**Light drawn**: all **diam**, 36 ksi (250 MPa) **TS**, 30 ksi (205 MPa) **YS**; **Hard drawn**: all **diam**, 45 ksi (310 MPa) **TS**, 40 ksi (275 MPa) **YS**

WROUGHT COPPER AND COPPER ALLOYS

specification number	designation	country	product forms	chemical composition	mechanical properties and hardness values
ANSI/ASTM B 111	C 10800	US	Tube: seamless	99.95 min **Cu**, 0.005–0.012 **P**, phosphorous included in copper content	**Light drawn**: all **diam**, 36 ksi (250 MPa) **TS**, 30 ksi (205 MPa) **YS**; **Hard drawn**: all **diam**, 45 ksi (310 MPa) **TS**, 40 ksi (275 MPa) **YS**
ANSI/ASTM B 170	Grade 2	US	Bar, billet, cake	99.95 min **Cu**, silver included in copper content	
ANSI/ASTM B 280	C 10300	US	Tube: seamless	99.95 min **Cu**, 0.001–0.005 **P**, silver and phosphorous included in copper content	**Annealed (coiled lengths)**: all **diam**, 30 ksi (205 MPa) **TS**, 40% in 2 in. (50.8 mm) **El**; **Cold drawn (straight lengths)**: all **diam**, 36 ksi (250 MPa) **TS**
ANSI/ASTM B 280	C 10800	US	Tube: seamless	99.95 min **Cu**, 0.005–0.012 **P**, silver and phosphorous included in copper content	**Annealed (coiled lengths)**: all **diam**, 30 ksi (205 MPa) **TS**, 40% in 2 in. (50.8 mm) **El**; **Cold drawn (straight lengths)**: all **diam**, 36 ksi (250 MPa) **TS**
ANSI/ASTM B 280	C 10200	US	Tube: seamless	99.95 min **Cu**, silver included in copper content	**Annealed (coiled lengths)**: all **diam**, 30 ksi (205 MPa) **TS**, 40% in 2 in. (50.8 mm) **El**; **Cold drawn (straight lengths)**: all **diam**, 36 ksi (250 MPa) **TS**
ANSI/ASTM B 302	108	US	Pipe: threadless	99.95 min **Cu**, 0.005–0.012 **P**, silver and phosphorous included in copper content	
ANSI/ASTM B 302	103	US	Pipe: threadless	99.95 min **Cu**, 0.001–0.005 **P**, silver and phosphorous included in copper content	
ANSI/ASTM B 359	C 10800	US	Tube: seamless	99.95 min **Cu**, 0.005–0.012 **P**, phosphorus included in copper	**Annealed**: all **diam**, 30 ksi (205 MPa) **TS**, 9 ksi (62 MPa) **YS**; **Light drawn**: all **diam**, 36 ksi (250 MPa) **TS**, 30 ksi (205 MPa) **YS**
ANSI/ASTM B 359	C 10300	US	Tube: seamless	99.95 min **Cu**, 0.001–0.005 **P**, phosphorus included in copper	**Annealed**: all **diam**, 30 ksi (205 MPa) **TS**, 9 ksi (62 MPa) **YS**; **Light drawn**: all **diam**, 36 ksi (250 MPa) **TS**, 30 ksi (205 MPa) **YS**
ANSI/ASTM B 359	C 10200	US	Tube: seamless	99.95 min **Cu**	**Annealed**: all **diam**, 30 ksi (205 MPa) **TS**, 9 ksi (62 MPa) **YS**; **Light drawn**: all **diam**, 36 ksi (250 MPa) **TS**, 30 ksi (205 MPa) **YS**
ANSI/ASTM B 306	C 10800	US	Tube	99.95 min **Cu**, 0.005–0.012 **P**, silver and phosphorus included in copper content	
ANSI/ASTM B 306	C 10300	US	Tube	99.95 min **Cu**, 0.001–0.015 **P**, silver and phosphorous included in copper content	
ANSI/ASTM B 372	CA 102	US	Tube: seamless	99.95 min **Cu**, silver included in copper content	**As agreed**: 30 **HR30T**

WROUGHT COPPER AND COPPER ALLOYS

specification number	designation	country	product forms	chemical composition	mechanical properties and hardness values
ANSI/ASTM B 395	C 10800	US	Tube: seamless	99.95 min **Cu**, 0.005–0.012 **P**, silver and phosphorus included in copper content.	**Light drawn**: <0.048 in. (<1.21 mm) **diam**, 36 ksi (250 MPa) **TS**, 30 ksi (205 MPa) **YS**, 12% in 2 in. (50.8 mm) **El**
ANSI/ASTM B 395	C 10300	US	Tube: seamless	99.95 min **Cu**, 0.001–0.005 **P**, silver and phosphorus included in copper content.	**Light drawn**: <0.048 in. (<1.21 mm) **diam**, 36 ksi (250 MPa) **TS**, 30 ksi (205 MPa) **YS**, 12% in 2 in. (50.8 mm) **El**
ANSI/ASTM B 395	C 10200	US	Tube: seamless	99.95 min **Cu**, silver included in copper content.	**Light drawn**: ≤0.048 in. (<1.21 mm) **diam**, 36 ksi (250 MPa) **TS**, 30 ksi (205 MPa) **YS**, 12% in 2 in. (50.8 mm) **El**
ANSI/ASTM B 379	108	US	Shapes	99.95 min **Cu**, 0.005–0.012 **P**, silver and phosphorus included in copper content.	
ANSI/ASTM B 379	103	US	Shapes	99.95 min **Cu**, 0.001–0.005 **P**, silver and phosphorus included in copper content.	
ANSI/ASTM B 12	C 102	US	Rod	99.95 min **Cu**	**Hot and/or cold worked**: all **diam**, 30 ksi (205 MPa) **TS**, 35% in 1 in. (25.4 mm) **El**
ANSI/ASTM B 12	C 108	US	Rod	99.95 min **Cu**, silver and phosphorous included in copper content	
ANSI/ASTM B 12	C 103	US	Rod	99.95 min **Cu**, silver and phosphorous included in copper content	
ANSI/ASTM B 42	C 10200	US	Pipe: seamless	99.95 min **Cu**, silver included in copper content	
ANSI/ASTM B 42	C 10300	US	Pipe: seamless	99.95 min **Cu**, 0.001–0.005 **P**, silver included in copper content	
ANSI/ASTM B 42	C 10800	US	Pipe: seamless	99.95 min **Cu**, 0.005–0.012 **P**, silver included in copper content	
ANSI/ASTM B 75	C 10300	US	Tube: seamless	99.95 **Cu**, 0.001–0.005 **P**, phosphorus included in copper content	**Drawn**: all **diam**, 36 ksi (250 MPa) **TS**, 30 **HR30T**; **Soft anneal**: all **diam**, 60 **HR15T**; **Light anneal**: all **diam**, 65 **HR15T**
ANSI/ASTM B 88	C 10800	US	Tube: seamless	99.95 min **Cu**, 0.005–0.012 **P**, silver and phosphorous included in copper content	**Annealed coils**: all **diam**, 30 ksi (205 MPa) **TS**, 50 **HRF**; **Annealed straight lengths**: all **diam**, 30 ksi (205 MPa) **TS**, 55 **HRF**; **Drawn**: all **diam**, 36 ksi (250 MPa) **TS**, 30 **HR30T**
ANSI/ASTM B 88	C 10300	US	Tube: seamless	99.95 min **Cu**, 0.001–0.005 **P**, silver and phosphorous included in copper content	**Annealed coils**: all **diam**, 30 ksi (205 MPa) **TS**, 50 **HRF**; **Annealed straight length**: all **diam**, 30 ksi (205 MPa) **TS**, 55 **HRF**; **Drawn**: all **diam**, 36 ksi (250 MPa) **TS**, 30 **HR30T**

WROUGHT COPPER AND COPPER ALLOYS

specification number	designation	country	product forms	chemical composition	mechanical properties and hardness values
ANSI/ASTM B 75	C 10800	US	Tube: seamless	99.95 **Cu**, 0.005–0.012 **P**, phosphorus included in copper content	**Drawn**: all **diam**, 36 ksi (250 MPa) **TS**, 30 **HR30T**; **Soft anneal**: all **diam**, 60 **HR15T**; **Light anneal**: all **diam**, 65 **HR15T**
AMS 4701C		US	Wire	99.95 min **Cu**, 0.001 **O**, silver included in copper content	**Cold drawn or cold rolled, and annealed**: >0.2893 in. (>7.348 mm) **diam**, 35% in 10 in. (250 mm) **El**
ANSI/ASTM B 379	CN 103	US	Shapes	99.95 min **Cu**, 0.001–0.005 **P**, silver and phosphorus included in copper content	
ANSI/ASTM B 379	CN 108	US	Shapes	99.95 min **Cu**, 0.005–0.012 **P**, silver and phosphorus included in copper content	
ANSI/ASTM B 111	C 10200	US	Tube: seamless, stock: ferrule	99.95 min **Cu**	**Light drawn**: all **diam**, 36 ksi (250 MPa) **TS**, 30 ksi (205 MPa) **YS**; **Hard drawn**: all **diam**, 45 ksi (310 MPa) **TS**, 40 ksi (275 MPa) **YS**
ANSI/ASTM B 152	C 10800	US	Sheet, strip, plate, bar	99.95 min **Cu**, 0.005–0.012 **P**, silver and phosphorous included in copper content	**Hard**: >0.020 in. (>0.508 mm) **diam**, 43 ksi (295 MPa) **TS**, 86–93 **HRF**; **Spring**: >0.020 in. (>0.508 mm) **diam**, 50 ksi (345 MPa) **TS**, 91–97 **HRF**; **Hot rolled**: >0.020 in. (>0.508 mm) **diam**, 30 ksi (205 MPa) **TS**, 75 max **HRF**
ANSI/ASTM B 152	C 10200	US	Sheet, strip, plate, bar	99.95 min **Cu**, silver included in copper content	**Hard**: >0.020 in. (>0.508 mm) **diam**, 43 ksi (295 MPa) **TS**, 86–93 **HRF**; **Spring**: >0.020 in. (>0.508 mm) **diam**, 50 ksi (345 MPa) **TS**, 91–97 **HRF**; **Hot rolled**: >0.020 in. (>0.508 mm) **diam**, 30 ksi (205 MPa) **TS**, 75 max **HRF**
ANSI/ASTM B 152	C 10300	US	Sheet, strip, plate, bar	99.95 min **Cu**, 0.001–0.005 **P**, silver and phosphorous included in copper content	**Hard**: >0.020 in. (>0.508 mm) **diam**, 43 ksi (295 MPa) **TS**, 86–93 **HRF**; **Spring**: >0.020 in. (>0.508 mm) **diam**, 50 ksi (345 MPa) **TS**, 91–97 **HRF**; **Hot rolled**: >0.020 in. (>0.508 mm) **diam**, 30 ksi (205 MPa) **TS**, 75 max **HRF**
AS 1567	102A	Australia	Rod, bar, section	99.95 min **Cu**	**Soft annealed**: 12–50 mm **diam**, 230 max MPa **TS**, 45% in 2 in. (50 mm) **El**; **1/2 hard**: 25–50 mm **diam**, 230 MPa **TS**, 22% in 2 in. (50 mm) **El**; **Hard**: 10–12 mm **diam**, 280 MPa **TS**, 8% in 2 in. (50 mm) **El**

WROUGHT COPPER AND COPPER ALLOYS

specification number	designation	country	product forms	chemical composition	mechanical properties and hardness values
BS 2871	C 103	UK	Tube	99.95 min **Cu**, 0.005 **Pb**, 0.0010 **Bi**, 0.03 total, others	**Annealed**: all **diam**, 200 MPa **TS**, 40% in 2 in. (50 mm) **El**, 60 **HV**; **As drawn**: all **diam**, 270 MPa **TS**, 100 **HV**
BS 2870	C 103	UK	Sheet, strip, foil	99.95 min **Cu**, 0.005 **Pb**, 0.0010 **Bi**, 0.03 total, others	**Annealed**: 0.6–10.0 mm **diam**, 148 MPa **TS**, 35% in 2 in. (50 mm) **El**; **1/2 hard**: 0.6–1.3 mm **diam**, 172 MPa **TS**, 10% in 2 in. (50 mm) **El**; **Hard**: 0.6–2.7 mm **diam**, 217 MPa **TS**
BS 2873	C103	UK	Wire	99.95 min **Cu**, 0.005 **Pb**, 0.0010 **Bi**, 0.03 total, others	
BS 2875	C103	UK	Plate	99.95 min **Cu**, 0.005 **Pb**, 0.0010 **Bi**, 0.03 total, others	**As manufactured or annealed**: ≥10 mm **diam**, 211 MPa **TS**, 35% in 2 in. (50 mm) **El**; **Hard**: 10–16 mm **diam**, 279 MPa **TS**, 15% in 2 in. (50 mm) **El**
BS 2874	C103	UK	Rod, section	99.95 min **Cu**, 0.005 **Pb**, 0.0010 **Bi**, 0.03 total, others	**Annealed**: 3–6 mm **diam**, 260 max MPa **TS**, 32% in 2 in. (50 mm) **El**; **1/2 hard**: 3–6 mm **diam**, 290 MPa **TS**, 4% in 2 in. (50 mm) **El**; **Hard**: 3–6 mm **diam**, 348 MPa **TS**
CDA C 102		US	Plate, sheet, strip, rod, wire, tube, shapes	99.95 min **Cu**, silver included in copper content	**Hard (plate, sheet, strips)**: 1.0 in. (25.4 mm) **diam**, 45 ksi (310 MPa) **TS**, 40 ksi (276 MPa) **YS**, 20% in 2 in. (50 mm) **El**, 85 **HRF**; **Hard (rod)**: 1.0 in. (25.4 mm) **diam**, 48 ksi (331 MPa) **TS**, 44 ksi (303 MPa) **YS**, 16% in 2 in. (50 mm) **El**, 87 **HRF**; **Hard (shapes)**: 0.500 in. (12.7 mm) **diam**, 40 ksi (276 MPa) **TS**, 32 ksi (221 MPa) **YS**, 30% in 2 in. (50 mm) **El**, 35 **HRB**
COPANT 521	Cu OF Gr 2	COPANT	Tube	99.95 min **Cu**, 0.005 **Pb**, 0.001 **Bi**, 0.03 total, others, silver included in copper content	**Annealed**: 10.3–324 mm od **diam**, 255 MPa **TS**, 25% in 2 in. (50 mm) **El**; **Hard tempered**: ≤102 mm od **diam**, 275 MPa **TS**, 3% in 2 in. (50 mm) **El**
COPANT 521	Cu Ag OF	COPANT	Tube	99.95 min **Cu**, 0.002 **Ag**, 0.015–0.04 **O_2**, silver included in copper content	**Annealed**: 10.3–324 mm od **diam**, 255 MPa **TS**, 25% in 2 in. (50 mm) **El**; **Hard tempered**: ≤102 mm od **diam**, 275 MPa **TS**, 3% in 2 in. (50 mm) **El**

WROUGHT COPPER AND COPPER ALLOYS

specification number	designation	country	product forms	chemical composition	mechanical properties and hardness values
COPANT 672	Cu OF	COPANT	Bar	99.95 min **Cu**, 0.001 **P**, 0.03 total, others, silver included in copper content	**Cold finished**: 6.35 mm **diam**, 345 MPa **TS**; **Annealed**: all **diam**, 255 MPa **TS**, 25% in 2 in. (50 mm) **El**; **Drawn temper**: 9.53–12 mm **diam**, 275 MPa **TS**, 12% in 2 in. (50 mm) **El**
COPANT 536	Cu OF (Gr 2)	COPANT	Sheet	99.95 **Cu**, silver included with copper	
COPANT 162	C 108 00 (=OFLP)	COPANT	Wrought products	99.95 min **Cu**, 0.005–0.012 **P**, silver and phosphorus included in copper content	
COPANT 162	C 103 00 (=OFXLP)	COPANT	Wrought products	99.95 min **Cu**, silver and phosphorus included in copper content	
COPANT 162	C 102 00 (=OF)	COPANT	Wrought products	99.95 min **Cu**, silver included in copper content	
COPANT 522	Cu OF Gr 2	COPANT	Tube	99.95 min **Cu**, 0.005 **Pb**, 0.001 **Bi**, 0.03 total, others, silver included in copper content	**Annealed (fluids piping)**: 0.25–12 mm **diam**, 205 MPa **TS**; **Tempered (fluids piping)**: all **diam**, 245 MPa **TS**
ISO/R1337	Cu–OF	ISO	Plate, sheet, strip, rod, bar, extruded sections, tube, wire, forgings	99.95 min **Cu**, silver included in copper content	
MIL–T–24107A	108	US	Tube: seamless	99.95 min **Cu**, 0.005–0.012 **P**, silver and phosphorus included in copper content	**Annealed**: all **diam**, 207 MPa **TS**, 62 MPa **YS**
MIL–T–24107A	102	US	Tube: seamless	99.95 min **Cu**, silver included in copper content	**Annealed**: all **diam**, 207 MPa **TS**, 62 MPa **YS**
NS 16 011	Cu–OF	Norway	Sheet, strip, plate, bar, tube, rod, wire, profiles, and forgings	99.95 min **Cu**, silver included in copper content	
SIS 14 50 11	SIS Cu 50 11–10	Sweden	Tube	99.95 min **Cu**, **Ag** in **Cu** content	**Strain hardened**: all **diam**, 250 MPa **TS**, 200 MPa **YS**, 10% in 2 in. (50 mm) **El**, 85 min **HV**
SIS 14 50 11	SIS Cu 50 11–04	Sweden	Strip, bar, wire, rod, tube	99.95 min **Cu**, **Ag** in **Cu** content	**Strain hardened**: 0.5–2.5 mm **diam**, 250 MPa **TS**, 180 MPa **YS**, 10% in 2 in. (50 mm) **El**, 75–100 **HV**
SIS 14 50 11	SIS Cu 50 11–03	Sweden	Tube	99.95 min **Cu**, **Ag** in **Cu** content	**Strain hardened**: 5 max mm **diam**, 250 MPa **TS**, 180 MPa **YS**, 20% in 2 in. (50 mm) **El**, 75–95 **HV**
SIS 14 50 11	SIS Cu 50 11–02	Sweden	Strip, bar, wire, rod, tube	99.95 min **Cu**, **Ag** in **Cu** content	**Annealed**: 0.5–1.5 mm **diam**, 220 MPa **TS**, 40 MPa **YS**, 30% in 2 in. (50 mm) **El**, 40–60 **HV**

WROUGHT COPPER AND COPPER ALLOYS

specification number	designation	country	product forms	chemical composition	mechanical properties and hardness values
ANSI/ASTM B 68	C 103	US	Tube: seamless	99.94 min **Cu**, 0.001–0.005 **P**, silver included in copper content	
ANSI/ASTM B 372	CA 103	US	Tube: seamless	99.94 min **Cu**, 0.001–0.005 **P**, 99.95 min **Cu + P + Ag**	**As agreed**: 30 **HR30T**
COPANT 522	Cu Ag OF	COPANT	Tube	99.94 min **Cu**, 0.002 **Ag**	**Tempered (fluids piping)**: all **diam**, 245 MPa **TS**
MIL-R-19631B	MIL-RCu-1	US	Rod	99.94 min **Cu**, 0.003 **Fe**, 0.002 **Sb**, 0.001 **P**, 0.001 **As**, 0.050 total, others, silver included in copper content	**As milled**: all **diam**, 172 MPa **TS**
MIL-T-24107A	103	US	Tube: seamless	99.94 min **Cu**, 0.001–0.005 **P**, silver included in copper content	**Annealed**: all **diam**, 207 MPa **TS**, 62 MPa **YS**
ANSI/ASTM B 68	C 108	US	Tube: seamless	99.93 min **Cu**, 0.005–0.012 **P**, silver included in copper content	
AMS 4700		US	Wire: bare, high purity	99.93 min **Cu**, 0.0010 **Pb**, 0.0005 **Zn**, 0.0005 **P**, 0.0018 **S**, 0.0520 **Ag**, 0.0150 **B**, 0.0010 **Bi**, 0.0003 **Cd**, 0.0001 **Hg**, 0.0010 **Se**, 0.0010 **Te**, 0.0040 **As + Bi + Mn + Se + Sb + Sn + Te**, 99.90 min **Cu + Ag + B**	
CDA C119		US	Plate, sheet, strip, rod, pipe, tube, shapes	99.93 min **Cu**, 0.002–0.010 **P**, silver included in copper content	**Hard (plate, sheet, strip)**: 0.040 in. (1.016 mm) **diam**, 50 ksi (345 MPa) **TS**, 45 ksi (310 MPa) **YS**, 6% in 2 in. (50 mm) **El**, 90 **HRF**; **Hard (rod)**: 1.0 in. (25.4 mm) **diam**, 48 ksi (331 MPa) **TS**, 44 ksi (303 MPa) **YS**, 16% in 2 in. (50 mm) **El**, 87 **HRF**; **Hard (shapes)**: 0.500 in. (12.7 mm) **diam**, 40 ksi (276 MPa) **TS**, 32 ksi (221 MPa) **YS**, 30% in 2 in. (50 mm) **El**, 35 **HRB**
COPANT 522	Cu DLP	COPANT	Tube	99.93 min **Cu**, 0.005–0.014 **P**, silver included in copper content	**Annealed (fluids piping)**: 0.25–12 mm **diam**, 205 MPa **TS**; **Tempered (fluids piping)**: all **diam**, 245 MPa **TS**
COPANT 521	Cu DLP	COPANT	Tube	99.93 min **Cu**, 0.005–0.014 **P**, silver included in copper content	**Annealed**: 10.3–324 mm od **diam**, 255 MPa **TS**, 25% in 2 in. (50 mm) **El**; **Hard tempered**: ≤102 mm od **diam**, 275 MPa **TS**, 3% in 2 in. (50 mm) **El**
COPANT 524	Cu DLP	COPANT	Tube	99.93 min **Cu**, 0.005–0.014 **P**, silver included in copper content	**Annealed**: all **diam**, 255 MPa **TS**, 25% in 2 in. (50 mm) **El**; **Hard tempered**: 0.25–12 mm **diam**, 275 MPa **TS**, 3% in 2 in. (50 mm) **El**

WROUGHT COPPER AND COPPER ALLOYS

specification number	designation	country	product forms	chemical composition	mechanical properties and hardness values
COPANT 536	Cu BLP	COPANT	Sheet	99.93 min **Cu**, 0.005–0.014 **P**, silver included in copper content	**Annealed**: all **diam**, 205 MPa **TS**; **Tempered**: all **diam**, 225 MPa **TS**
COPANT 536	Cu DPHC	COPANT	Sheet	99.93 min **Cu**, 0.015 **Sb**, 0.10 **Si**, 0.05 **As**, 0.05 O_2, silver included in copper content	**Annealed**: all **diam**, 205 MPa **TS**; **Tempered**: all **diam**, 225 MPa **TS**
COPANT 523	Cu DLP	COPANT	Tube	99.93 min **Cu**, 0.005–0.014 **P**, silver included in copper content	
COPANT 162	C 105 00 (=OFS)	COPANT	Wrought products	99.92 min **Cu**, 0.034 min **Ag**	
DGN-W-23	Type OF	Mexico	Tube: seamless, for refrigeration service	99.92 min **Cu**	**Cold finished**: 3.18–15.92 mm **diam**, 30 ksi (207 MPa) **TS**, 40% in 2 in. (50 mm) **El**
AMS 4500E		US	Sheet, strip, and plate	99.90 min **Cu**, 0.10 total others, silver included in copper content	**Soft annealed**: ≥0.030 in. (≥0.76 mm) **diam**, 65 max **HRF**
ANSI/ASTM B 68	C 120	US	Tube: seamless	99.90 min **Cu**, 0.004–0.012 **P**	
ANSI/ASTM B 68	C 122	US	Tube: seamless	99.9 min **Cu**, 0.015–0.040 **P**	
ANSI/ASTM B 75	C 12200	US	Tube: seamless	99.9 min **Cu**, 0.015–0.040 **P**	**Drawn**: all **diam**, 36 ksi (250 MPa) **TS**, 30 **HR30T**; **Soft anneal**: all **diam**, 60 **HR15T**; **Light anneal**: all **diam**, 65 **HR15T**
ANSI/ASTM B 75	C 12000	US	Tube: seamless	99.90 min **Cu**, 0.004–0.012 **P**	**Drawn**: all **diam**, 36 ksi (250 MPa) **TS**, 30 **HR30T**; **Soft anneal**: all **diam**, 60 **HR15T**; **Light anneal**: all **diam**, 65 **HR15T**
ANSI/ASTM B 101		US	Sheet: lead coated	99.9 **Cu**, silver included in copper	**Cold rolled (high yield)**: all **diam**, 34 ksi (240 MPa) **TS**, 28 ksi (195 MPa) **YS**, 30 **HR30T**; **Cold rolled**: all **diam**, 32 ksi (225 MPa) **TS**, 20 ksi (140 MPa) **YS**, 60 **HRF**; **Soft**: all **diam**, 30 ksi (210 MPa) **TS**, 65 max **HRF**
ANSI/ASTM B 5		US	Bar, cake, slab, billet, ingot	99.90 min **Cu**, silver included in copper	
ANSI/ASTM B 4		US	Bar, cake, slab, billet, ingot	99.90 min **Cu**, silver included in copper	
ANSI/ASTM B 88	C 12200	US	Tube: seamless	99.9 min **Cu**, 0.015–0.040 **P**	**Annealed coils**: all **diam**, 30 ksi (205 MPa) **TS**, 50 **HRF**; **Annealed straight lengths**: all **diam**, 30 ksi (205 MPa) **TS**, 55 **HRF**; **Drawn**: all **diam**, 36 ksi (250 MPa) **TS**, 30 **HR30T**

WROUGHT COPPER AND COPPER ALLOYS

specification number	designation	country	product forms	chemical composition	mechanical properties and hardness values
ANSI/ASTM B 88	C 12000	US	Tube: seamless	99.90 min **Cu**, 0.004–0.012 **P**	**Annealed coils**: all **diam**, 30 ksi (205 MPa) **TS**, 50 **HRF**; **Annealed straight lengths**: all **diam**, 30 ksi (205 MPa) **TS**, 55 **HRF**; **Drawn**: all **diam**, 36 ksi (250 MPa) **TS**, 30 **HR30T**
ANSI/ASTM B 447	C 11000	US	Tube	99.90 min **Cu**, silver included in copper content	**Annealed (soft)**: all **diam**, 60 max **HR15T**; **As welded (from hard strip)**: all **diam**, 45 ksi (310 MPa) **TS**, 54–62 **HR30T**; **Welded and cold drawn (drawn)**: all **diam**, 36 ksi (250 MPa) **TS**, 30 min **HR30T**
ANSI/ASTM B 447	C 12000	US	Tube	99.90 min **Cu**, 0.004–0.012 **P**, silver included in copper content	**Annealed (soft)**: all **diam**, 60 max **HR15T**; **As welded (from hard strip)**: all **diam**, 45 ksi (310 MPa) **TS**, 54–62 **HR30T**; **Welded and cold drawn (drawn)**: all **diam**, 30 min **HR30T**
ANSI/ASTM B 447	C 12200	US	Tube	99.9 min **Cu**, 0.015–0.040 **P**, silver included in copper content	**Annealed (soft)**: all **diam**, 60 max **HR15T**; **As welded (from hard strip)**: all **diam**, 45 ksi (310 MPa) **TS**, 54–62 **HR30T**; **Welded and cold drawn (drawn)**: all **diam**, 30 min **HR30T**
ASTM F 468	C 11000	US	Bolt, screw, stud	99.9 min **Cu**	**As manufactured**: all **diam**, 30 ksi (210 MPa) **TS**, 10 ksi (70 MPa) **YS**, 15% in 2 in. (50 mm) **El**, 65–90 **HRF**
ASTM F 467	C 11000	US	Nut	99.9 min **Cu**	**As agreed**: all **diam**, 65 min **HRF**
ANSI/ASTM B 543	C 12200	US	Tube: welded	99.9 min **Cu**, 0.015–0.040 **P**, silver included in copper content	**As welded from annealed strip**: all **diam**, 32 ksi (220 MPa) **TS**, 15 ksi (105 MPa) **YS**; **Welded and annealed**: all **diam**, 30 ksi (205 MPa) **TS**, 9 ksi (62 MPa) **YS**; **Fully finished–hard drawn**: all **diam**, 45 ksi (310 MPa) **TS**, 40 ksi (275 MPa) **YS**
ANSI/ASTM B 283	C 11000	US	Die forgings	99.90 min **Cu**, silver included in copper content	
ANSI/ASTM B 280	C 12000	US	Tube: seamless	99.90 min **Cu**, 0.004–0.012 **P**, silver included in copper content	**Annealed (coiled lengths)**: all **diam**, 30 ksi (205 MPa) **TS**, 40% in 2 in. (50.8 mm) **El**; **Cold drawn (straight lengths)**: all **diam**, 36 ksi (250 MPa) **TS**
ANSI/ASTM B 280	C 12200	US	Tube: seamless	99.9 min **Cu**, 0.015–0.040 **P**, silver included in copper content	**Annealed (coiled lengths)**: all **diam**, 30 ksi (205 MPa) **TS**, 40% in 2 in. (50.8 mm) **El**; **Cold drawn (straight lengths)**: all **diam**, 36 ksi (250 MPa) **TS**

WROUGHT COPPER AND COPPER ALLOYS

specification number	designation	country	product forms	chemical composition	mechanical properties and hardness values
ANSI/ASTM B 302	122	US	Pipe: threadless	99.9 min **Cu**, 0.015–0.040 **P**, silver included in copper content	
ANSI/ASTM B 302	120	US	Pipe: threadless	99.90 min **Cu**, 0.004–0.012 **P**, silver included in copper content	
ANSI/ASTM B 359	C 12200	US	Tube: seamless	99.9 min **Cu**, 0.015–0.040 **P**, 99.95 min **Cu+P**	**Annealed**: all **diam**, 30 ksi (205 MPa) **TS**, 9 ksi (62 MPa) **YS**; **Light drawn**: all **diam**, 36 ksi (250 MPa) **TS**, 30 ksi (205 MPa) **YS**
ANSI/ASTM B 359	C 12000	US	Tube: seamless	99.90 min **Cu**, 0.004–0.012 **P**, 99.95 min **Cu+P**	**Annealed**: all **diam**, 30 ksi (205 MPa) **TS**, 9 ksi (62 MPa) **YS**; **Light drawn**: all **diam**, 36 ksi (250 MPa) **TS**, 30 ksi (205 MPa) **YS**
ANSI/ASTM B 306	C 12200	US	Tube	99.9 min **Cu**, 0.015–0.040 **P**, silver included in copper content	
ANSI/ASTM B 306	C 12000	US	Tube	99.90 min **Cu**, 0.004–0.012 **P**, silver included in copper content	
ANSI/ASTM B 372	CA 120	US	Tube: seamless	99.90 min **Cu**, 0.004–0.012 **P**, silver included in copper content	**As agreed**: 30 **HR30T**
ANSI/ASTM B 370		US	Sheet, strip	99.9 min **Cu**, silver included in copper content.	**Cold rolled**: >0.020 in. (>0.508 mm) **diam**, 32 ksi (220 MPa) **TS**, 20 ksi (135 MPa) **YS**, 60–82 **HRF**; **Cold rolled, high yield**: >0.020 in. (>0.508 mm) **diam**, 34 ksi (235 MPa) **TS**, 28 ksi (190 MPa) **YS**; **Soft**: >0.020 in. (>0.508 mm) **diam**, 30 ksi (205 MPa) **TS**, 65 max **HRF**
ANSI/ASTM B 379	122	US	Shapes	99.9 min **Cu**, 0.015–0.040 **P**, silver included in copper content.	
ANSI/ASTM B 379	120	US	Shapes	99.90 min **Cu**, 0.004–0.012 **P**, silver included in copper content	
ANSI/ASTM B 395	C 12200	US	Tube: seamless	99.9 min **Cu**, 0.015–0.040 **P**, silver included in copper content	**Light drawn**: <0.048 in. (<1.21 mm) **diam**, 36 ksi (250 MPa) **TS**, 30 ksi (205 MPa) **YS**, 12% in 2 in. (50.8 mm) **El**
ANSI/ASTM B 395	C 12000	US	Tube: seamless	99.90 min **Cu**, 0.004–0.012 **P**, silver included in copper content	**Light drawn**: <0.048 in. (<1.21 mm) **diam**, 36 ksi (250 MPa) **TS**, 30 ksi (205 MPa) **YS**, 12% in 2 in. (50.8 mm) **El**

WROUGHT COPPER AND COPPER ALLOYS

specification number	designation	country	product forms	chemical composition	mechanical properties and hardness values
ANSI/ASTM B 12	C 125	US	Rod	99.90 min **Cu**	**Hot and/or cold worked**: all **diam**, 30 ksi (205 MPa) **TS**, 35% in 1 in. (25.4 mm) **El**
ANSI/ASTM B 12	C 122	US	Rod	99.9 min **Cu**	**Hot and/or cold worked**: all **diam**, 30 ksi (205 MPa) **TS**, 35% in 1 in. (25.4 mm) **El**
ANSI/ASTM B 12	C 120	US	Rod	99.90 min **Cu**	**Hot and/or cold worked**: all **diam**, 30 ksi (205 MPa) **TS**, 35% in 1 in. (25.4 mm) **El**
ANSI/ASTM B 12	C 110	US	Rod	99.90 min **Cu**	**Hot and/or cold worked**: all **diam**, 30 ksi (205 MPa) **TS**, 35% in 1 in. (25.4 mm) **El**
ANSI/ASTM B 11	C 12200	US	Plate	99.9 min **Cu**, 0.015–0.040 **P**, silver included with copper content	**Hot rolled**: all **diam**, 30 ksi (205 MPa) **TS**, 30% in 8 in. (203 mm) **El**
ANSI/ASTM B 11	C 11000	US	Plate	99.90 min **Cu**, silver included with copper content.	**Hot rolled**: all **diam**, 30 ksi (205 MPa) **TS**, 30% in 8 in. (203 mm) **El**
ANSI/ASTM B 42	C 12000	US	Pipe: seamless	99.90 min **Cu**, 0.004–0.012 **P**, silver included in copper content	
ANSI/ASTM B 42	C 12200	US	Pipe: seamless	99.9 min **Cu**, 0.015–0.040 **P**, silver included in copper content	
AMS 4602		US	Bar and rod: oxygen free	99.90 min **Cu**	**Cold drawn or cold rolled to hard**: 2.000–3.000 in. **diam**, 33 ksi (228 MPa) **TS**, 15% in 2 in. (50 mm) **El**
ANSI/ASTM B 379	CN 120	US	Shapes	99.90 min **Cu**, 0.004–0.012 **P**, silver included in copper content	
ANSI/ASTM B 379	CN 122	US	Shapes	99.9 min **Cu**, 0.015–0.040 **P**, silver included in copper content	
ANSI/ASTM B 111	C 12200	US	Tube: seamless, stock: ferrule	99.9 min **Cu**, 0.015–0.040 **P**	**Light drawn**: all **diam**, 36 ksi (250 MPa) **TS**, 30 ksi (205 MPa) **YS**; **Hard drawn**: all **diam**, 45 ksi (310 MPa) **TS**, 40 ksi (275 MPa) **YS**
ANSI/ASTM B 111	C 12000	US	Tube: seamless, stock: ferrule	99.90 min **Cu**, 0.004–0.012 **P**	**Light drawn**: all **diam**, 36 ksi (250 MPa) **TS**, 30 ksi (205 MPa) **YS**; **Hard drawn**: all **diam**, 45 ksi (310 MPa) **TS**, 40 ksi (275 MPa) **YS**
ANSI/ASTM B 124	C 11000	US	Rod, bar, shapes	99.90 min **Cu**	

WROUGHT COPPER AND COPPER ALLOYS

specification number	designation	country	product forms	chemical composition	mechanical properties and hardness values
ANSI/ASTM B 152	C 11100	US	Sheet, strip, plate, bar	99.90 min **Cu**, silver included in copper content	**Hard**: >0.020 in. (>0.508 mm) **diam**, 43 ksi (295 MPa) **TS**, 86–93 **HRF**; **Spring**: >0.020 in. (>0.508 mm) **diam**, 50 ksi (345 MPa) **TS**, 91–97 **HRF**; **Hot rolled**: >0.020 in. (>0.508 mm) **diam**, 30 ksi (205 MPa) **TS**, 75 max **HRF**
ANSI/ASTM B 152	C 12300	US	Sheet, strip, plate, bar	99.90 min **Cu**, 0.015–0.040 **P**, 0.0136 min **Ag**, silver included in copper content	**Hard**: >0.020 in. (>0.508 mm) **diam**, 43 ksi (295 MPa) **TS**, 86–93 **HRF**; **Spring**: >0.020 in. (>0.508 mm) **diam**, 50 ksi (345 MPa) **TS**, 91–97 **HRF**; **Hot rolled**: >0.020 in. (>0.508 mm) **diam**, 30 ksi (205 MPa) **TS**, 75 max **HRF**
ANSI/ASTM B 152	C 11600	US	Sheet, strip, plate, bar	99.90 min **Cu**, 0.085 min **Ag**, silver included in copper content	**Hard**: >0.020 in. (>0.508 mm) **diam**, 43 ksi (295 MPa) **TS**, 86–93 **HRF**; **Spring**: >0.020 in. (>0.508 mm) **diam**, 50 ksi (345 MPa) **TS**, 91–97 **HRF**; **Hot rolled**: >0.020 in. (>0.508 mm) **diam**, 30 ksi (205 MPa) **TS**, 75 max **HRF**
ANSI/ASTM B 152	C 12000	US	Sheet, strip, plate, bar	99.90 min **Cu**, 0.004–0.012 **P**, silver included in copper content	**Hard**: >0.020 in. (>0.508 mm) **diam**, 43 ksi (295 MPa) **TS**, 86–93 **HRF**; **Spring**: >0.020 in. (>0.508 mm) **diam**, 50 ksi (345 MPa) **TS**, 91–97 **HRF**; **Hot rolled**: >0.020 in. (>0.508 mm) **diam**, 30 ksi (205 MPa) **TS**, 75 max **HRF**
ANSI/ASTM B 152	C 12200	US	Sheet, strip, plate, bar	99.9 min **Cu**, 0.015–0.040 **P**, silver included in copper content	**Hard**: >0.020 in. (>0.508 mm) **diam**, 43 ksi (295 MPa) **TS**, 86–93 **HRF**; **Spring**: >0.020 in. (>0.508 mm) **diam**, 50 ksi (345 MPa) **TS**, 91–97 **HRF**; **Hot rolled**: >0.020 in. (>0.508 mm) **diam**, 30 ksi (205 MPa) **TS**, 75 max **HRF**
AS 1567	110A	Australia	Rod, bar, section	99.90 min **Cu**, 0.05, others	**Soft annealed**: 12–50 mm **diam**, 230 max MPa **TS**, 45% in 2 in. (50 mm) **El**; **1/2 hard**: 25–50 mm **diam**, 230 MPa **TS**, 22% in 2 in. (50 mm) **El**; **Hard**: 10–12 mm **diam**, 280 MPa **TS**, 8% in 2 in. (50 mm) **El**

WROUGHT COPPER AND COPPER ALLOYS

specification number	designation	country	product forms	chemical composition	mechanical properties and hardness values
AS 1567	120C	Australia	Rod, bar, section	99.90 min **Cu**, 0.003–0.012 **P**	**As manufactured**: >6 mm **diam**, 230 MPa **TS**, 13% in 2 in. (50 mm) **El**; **Soft annealed**: >6 mm **diam**, 210 MPa **TS**, 33% in 2 in. (50 mm) **El**
AS 1572	122A	Australia	Tube	99.90 min **Cu**, 0.015–0.040 **P**	**Annealed**: 55 max **HV**; **Half hard**: 75–100 **HV**; **Hard**: 100 **HV**
AS 1567	122A	Australia	Rod, bar, section	99.90 min **Cu**, 0.015–0.040 **P**	**Soft annealed**: >6 mm **diam**, 210 MPa **TS**, 33% in 2 in. (50 mm) **El**; **As manufactured**: >6 mm **diam**, 230 MPa **TS**, 13% in 2 in. (50 mm) **El**
BS 2871	C 101	UK	Tube	99.90 min **Cu**, 0.005 **Pb**, 0.0010 **Bi**, 0.03 total, others	**Annealed**: all **diam**, 200 MPa **TS**, 40% in 2 in. (50 mm) **El**, 60 **HV**; **As drawn**: all **diam**, 270 MPa **TS**, 100 **HV**
BS 2871	C 102	UK	Tube	99.90 min **Cu**, 0.005 **Pb**, 0.0025 **Bi**, 0.04 total, others	**Annealed**: all **diam**, 200 MPa **TS**, 40% in 2 in. (50 mm) **El**, 60 **HV**; **As drawn**: all **diam**, 270 MPa **TS**, 100 **HV**
BS 2870	C 101	UK	Sheet, strip, foil	99.90 min **Cu**, 0.005 **Pb**, 0.0010 **Bi**, 0.03 total, others	**Annealed**: 0.6–10.0 mm **diam**, 148 MPa **TS**, 35% in 2 in. (50 mm) **El**; **1/2 hard**: 0.6–1.3 mm **diam**, 172 MPa **TS**, 10% in 2 in. (50 mm) **El**; **Hard**: 0.6–2.7 mm **diam**, 217 MPa **TS**
BS 2870	C 102	UK	Sheet, strip, foil	99.90 min **Cu**, 0.005 **Pb**, 0.0025 **Bi**, 0.04 total, others	**Annealed**: 0.6–10.0 mm **diam**, 148 MPa **TS**, 35% in 2 in. (50 mm) **El**; **1/2 hard**: 0.6–1.3 mm **diam**, 172 MPa **TS**, 10% in 2 in. (50 mm) **El**; **Hard**: 0.6–2.7 mm **diam**, 217 MPa **TS**
BS 2873	C102	UK	Wire	99.90 min **Cu**, 0.005 **Pb**, 0.0025 **Bi**, 0.04 total, others	**Hard**: ≤1.60 mm **diam**, 456 MPa **TS**
BS 2875	C101	UK	Plate	99.90 min **Cu**, 0.005 **Pb**, 0.0010 **Bi**, 0.03 total, others	**As manufactured or annealed**: ≥10 mm **diam**, 211 MPa **TS**, 35% in 2 in. (50 mm) **El**; **Hard**: 10–16 mm **diam**, 279 MPa **TS**, 15% in 2 in. (50 mm) **El**
BS 2875	C102	UK	Plate	99.90 min **Cu**, 0.005 **Pb**, 0.0025 **Bi**, 0.04 total, others	**As manufactured or annealed**: ≥10 mm **diam**, 211 MPa **TS**, 35% in 2 in. (50 mm) **El**; **Hard**: 10–16 mm **diam**, 279 MPa **TS**, 15% in 2 in. (50 mm) **El**

WROUGHT COPPER AND COPPER ALLOYS

specification number	designation	country	product forms	chemical composition	mechanical properties and hardness values
BS 2874	C101	UK	Rod, section	99.90 min **Cu**, 0.005 **Pb**, 0.0010 **Bi**, 0.03 total, others	**Annealed**: 3–6 mm **diam**, 260 max MPa **TS**, 32% in 2 in. (50 mm) **El**; **1/2 hard**: 3–6 mm **diam**, 290 MPa **TS**, 4% in 2 in. (50 mm) **El**; **Hard**: 3–6 mm **diam**, 348 MPa **TS**
BS 2874	C102	UK	Rod, section	99.90 min **Cu**, 0.005 **Pb**, 0.0025 **Bi**, 0.04 total, others	**Annealed**: 3–6 mm **diam**, 260 max MPa **TS**, 32% in 2 in. (50 mm) **El**; **1/2 hard**: 3–6 mm **diam**, 290 MPa **TS**, 4% in 2 in. (50 mm) **El**; **Hard**: 3–6 mm **diam**, 348 MPa **TS**
CSA HC 4.1	Cu–DHP	Canada	Bar	99.90 min **Cu**, 0.015–0.040 **P**	**Cold rolled**: all **diam**, 32 ksi (220 MPa) **TS**; **Full hard**: all **diam**, 43 ksi (296 MPa) **TS**; **Hot rolled and annealed**: all **diam**, 30 ksi (206 MPa) **TS**
CSA HC 4.1	Cu–DHP	Canada	Strip	99.90 min **Cu**, 0.015–0.040 **P**	**Cold rolled**: all **diam**, 32 ksi (220 MPa) **TS**; **Spring**: all **diam**, 50 ksi (344 MPa) **TS**; **Hot rolled and annealed**: all **diam**, 30 ksi (206 MPa) **TS**
CSA HC 4.1	Cu–DHP	Canada	Sheet	99.90 min **Cu**, 0.015–0.040 **P**	**Cold rolled**: all **diam**, 32 ksi (220 MPa) **TS**; **Full hard**: all **diam**, 43 ksi (296 MPa) **TS**; **Hot rolled and annealed**: all **diam**, 30 ksi (206 MPa) **TS**
CSA HC 4.1	Cu–DHP	Canada	Sheet	99.90 min **Cu**, 0.015–0.040 **P**	**Cold rolled**: all **diam**, 32 ksi (220 MPa) **TS**; **Full hard**: all **diam**, 43 ksi (296 MPa) **TS**; **Hot rolled and annealed**: all **diam**, 30 ksi **TS**
CSA HC 4.1	Cu–ETP	Canada	Sheet	99.90 min **Cu**	**Cold rolled**: all **diam**, 32 ksi (220 MPa) **TS**; **Full hard**: all **diam**, 43 ksi (296 MPa) **TS**; **Hot rolled and annealed**: all **diam**, 30 ksi (206 MPa) **TS**
CSA H C 4 1	Cu–ETP	Canada	Strip	99.9 min **Cu**	**Cold rolled**: all **diam**, 32 ksi (220 MPa) **TS**; **Spring**: all **diam**, 50 ksi (344 MPa) **TS**; **Hot rolled and annealed**: all **diam**, 30 ksi (206 MPa) **TS**
CSA HC 4.1	Cu–ETP	Canada	Plate	99.90 min **Cu**	**Cold rolled**: all **diam**, 32 ksi (220 MPa) **TS**; **Full hard**: all **diam**, 43 ksi (296 MPa) **TS**; **Hot rolled and annealed**: all **diam**, 30 ksi (206 MPa) **TS**

WROUGHT COPPER AND COPPER ALLOYS

specification number	designation	country	product forms	chemical composition	mechanical properties and hardness values
CSA HC 4.1	Cu–ETP	Canada	Bar	99.90 min **Cu**	**Cold rolled**: all **diam**, 32 ksi (220 MPa) **TS**; **Full hard**: all **diam**, 43 ksi (296 MPa) **TS**; **Hot rolled and annealed**: all **diam**, 30 ksi (206 MPa) **TS**
CSA HC 4.1	Cu–STP	Canada	Sheet	99.90 min **Cu**	**Cold rolled**: all **diam**, 32 ksi (220 MPa) **TS**; **Full hard**: all **diam**, 43 ksi (296 MPa) **TS**; **Hot rolled and annealed**: all **diam**, 30 ksi (206 MPa) **TS**
CSA HC 4 1	Cu–STP	Canada	Plate	99.90 min **Cu**	**Cold rolled**: all **diam**, 32 ksi (220 MPa) **TS**; **Full hard**: all **diam**, 43 ksi (296 MPa) **TS**; **Hot rolled and annealed**: all **diam**, 30 ksi (206 MPa) **TS**
CSA HC 4.1	Cu–STP	Canada	Strip	99.90 min **Cu**	**Cold rolled**: all **diam**, 32 ksi (220 MPa) **TS**; **Spring**: all **diam**, 50 ksi (344 MPa) **TS**; **Hot rolled and annealed**: all **diam**, 30 ksi (206 MPa) **TS**
CSA HC 4.1	Cu–STP	Canada	Bar	99.90 min **Cu**	**Cold rolled**: all **diam**, 32 ksi (220 MPa) **TS**; **Full hard**: all **diam**, 43 ksi (296 MPa) **TS**; **Hot rolled and annealed**: all **diam**, 30 ksi (206 MPa) **TS**
CDA C 110		US	Plate, sheet, strip, rod, wire, tube, shapes	99.90 min **Cu**, 0.04 **O**, silver included in copper content	**Hard (plate, sheet, strip)**: 0.040 in. (1.016 mm) **diam**, 50 ksi (345 MPa) **TS**, 45 ksi (310 MPa) **YS**, 6% in 2 in. (50 mm) **El**, 90 **HRF**; **Hard (rod)**: 1.0 in. (25.4 mm) **diam**, 48 ksi (331 MPa) **TS**, 44 ksi (303 MPa) **YS**, 16% in 2 in. (50 mm) **El**, 87 **HRF**; **Hard (shapes)**: 0.500 in. (12.7 mm) **diam**, 40 ksi (276 MPa) **TS**, 32 ksi (221 MPa) **YS**, 30% in 2 in. (50 mm) **El**, 35 **HRB**
CDA C111		US	Plate, sheet, strip, wire	99.90 min **Cu**, 0.04 **O**, silver included in copper content. 0.01 **Cd**, other elements	**Hard (wire)**: 0.080 in. (2.032 mm) **diam**, 66 ksi (455 MPa) **TS**, 1.5% in 60 in. (1524 mm) **El**

WROUGHT COPPER AND COPPER ALLOYS

specification number	designation	country	product forms	chemical composition	mechanical properties and hardness values
CDA C 120		US	Plate, sheet, strip, rod, pipe, tube, shapes	99.90 min **Cu**, 0.004–0.012 **P**, silver included in copper content	**Hard (plate, sheet, strip):** 0.040 in. (1.016 mm) **diam**, 50 ksi (345 MPa) **TS**, 45 ksi (310 MPa) **YS**, 6% in 2 in. (50 mm) **El**, 90 **HRF**; **Hard (rod):** 1.0 in. (25.4 mm) **diam**, 48 ksi (331 MPa) **TS**, 44 ksi (303 MPa) **YS**, 16% in 2 in. (50 mm) **El**, 87 **HRF**; **Hard (shapes):** 0.500 in. (12.7 mm) **diam**, 40 ksi (276 MPa) **TS**, 32 ksi (221 MPa) **YS**, 30% in 2 in. (50 mm) **El**, 35 **HRB**
CDA C121		US	Plate, sheet, strip, rod, pipe, tube, shapes	99.90 min **Cu**, 0.005–0.012 **P**, 0.014 **Ag**, silver included in copper content.	**Hard (plate, sheet, strip):** 0.040 in. (1.016 mm) **diam**, 50 ksi (345 MPa) **TS**, 45 ksi (310 MPa) **YS**, 6% in 2 in. (50 mm) **El**, 90 **HRF**; **Hard (rod):** 1.0 in. (25.4 mm) **diam**, 48 ksi (331 MPa) **TS**, 44 ksi (303 MPa) **YS**, 16% in 2 in. (50 mm) **El**, 87 **HRF**; **Hard (shapes):** 0.500 in. (12.7 mm) **diam**, 40 ksi (276 MPa) **TS**, 32 ksi (221 MPa) **YS**, 30% in 2 in. (50 mm) **El**, 35 **HRB**
CDA C122		US	Plate, sheet, strip, tube, pipe	99.90 min **Cu**, 0.015–0.040 **P**, silver included in copper content	**Hard (plate, sheet, strip):** 0.040 in. (1.016 mm) **diam**, 50 ksi (345 MPa) **TS**, 45 ksi (310 MPa) **YS**, 6% in 2 in. (50 mm) **El**, 90 **HRF**; **Hard drawn (tube):** 1.0 in. (25.4 mm) **diam**, 55 ksi (379 MPa) **TS**, 50 ksi (345 MPa) **YS**, 8% in 2 in. (50 mm) **El**, 95 **HRF**; **Hard (pipe):** 0.75 in. (19.05 mm) **diam**, 50 ksi (345 MPa) **TS**, 45 ksi (310 MPa) **YS**, 10% in 2 in. (50 mm) **El**, 90 **HRF**
COPANT 521	Cu DHP	COPANT	Tube	99.90 min **Cu**, 0.015–0.05 **P**, silver included in copper content	**Hard tempered**: all **diam**, 315 MPa **TS**, 275 MPa **YS**
COPANT 521	Cu ETP	COPANT	Tube	99.90 min **Cu**, 0.004 **Pb**, 0.003 **Sb**, 0.002 **Ag**, 10.015–0.04 O_2, 0.012 **As**, 0.003 **Bi**, 0.025 **Te**, 0.03 total, others, silver included in copper content	**Annealed:** 10.3–324 mm od **diam**, 255 MPa **TS**, 25% in 2 in. (50 mm) **El**; **Hard tempered:** ≤102 mm od **diam**, 275 MPa **TS**, 3% in 2 in. (50 mm) **El**

WROUGHT COPPER AND COPPER ALLOYS

specification number	designation	country	product forms	chemical composition	mechanical properties and hardness values
COPANT 672	Cu ETP	COPANT	Bar	99.90 min **Cu**, 0.004 **Pb**, 0.025 **Fe**, 0.003 **Sb**, 0.002 **Ag**, 0.015–0.04 O_2, 0.012 **As**, 0.003 **Bi**, 0.04 total, impurities, silver included in copper content	**Cold finished**: 6.35 mm **diam**, 345 MPa **TS**; **Annealed**: all **diam**, 255 MPa **TS**, 25% in 2 in. (50 mm) **El**; **Drawn temper**: 9.53–12 mm **diam**, 275 MPa **TS**, 12% in 2 in. (50 mm) **El**
COPANT 521	Cu Ag TBHC	COPANT	Tube	99.90 min **Cu**, 0.002 **Ag**, 0.015–0.04 O_2, silver included in copper content	**Annealed**: 10.3–324 mm od **diam**, 255 MPa **TS**, 25% in 2 in. (50 mm) **El**; **Hard tempered**: ≤102 mm od **diam**, 275 MPa **TS**, 3% in 2 in. (50 mm) **El**
COPANT 672	Cu RFTPHC	COPANT	Bar	99.90 min **Cu**, silver included in copper content	**Cold finished**: 6.35 mm **diam**, 345 MPa **TS**; **Annealed**: all **diam**, 255 MPa **TS**, 25% in 2 in. (50 mm) **El**; **Drawn temper**: 9.53–12 mm **diam**, 12% in 2 in. (50 mm) **El**
COPANT 522	Cu DHP	COPANT	Tube	99.90 min **Cu**, 0.015–0.05 **P**	**Annealed (fluids piping)**: 0.25–12 mm **diam**, 205 MPa **TS**, Tempered (fluids piping): all **diam**, 245 MPa **TS**
COPANT 536	Cu ETPHC	COPANT	Sheet	99.90 min **Cu**, 0.005 **Pb**, 0.001 **Bi**, 0.03 total, others, silver included in copper content	**Annealed**: all **diam**, 205 MPa **TS**; **Tempered**: all **diam**, 225 MPa **TS**
COPANT 162	C 122 00 (=DHP)	COPANT	Wrought products	99.9 min **Cu**, 0.015–0.040 **P**, silver included in copper content	
COPANT 162	C 120 00 (=DLP)	COPANT	Wrought products	99.90 min **Cu**, 0.004–0.012 **P**, silver included in copper content	
COPANT 162	C 117 00	COPANT	Wrought products	99.9 min **Cu**, 0.04 **P**, 0.004–0.02 **B**, silver and boron included in copper content	
COPANT 162	C 111 00	COPANT	Wrought products	99.90 min **Cu**, silver included in copper content	
COPANT 162	C 110 00 (=ETP, FRHC, CRTP)	COPANT	Wrought products	99.90 min **Cu**, silver included in copper content	
COPANT 672	Cu QTP	COPANT	Bar	99.90 min **Cu**, 0.0002 **Pb**, 0.01 **Fe**, 0.005 **Sb**, 0.0013 **Al**, 0.001 **S**, 0.002 **Ag**, 0.015–0.04 O_2, 0.001 **As**, tr **Bi**, tr **Se**+**Zn** included in copper content	**Cold finished**: 6.35 mm **diam**, 345 MPa **TS**; **Annealed**: all **diam**, 255 MPa **TS**, 25% in 2 in. (50 mm) **El**; **Drawn temper**: 9.53–12 mm **diam**, 275 MPa **TS**, 12% in 2 in. (50 mm) **El**

WROUGHT COPPER AND COPPER ALLOYS

specification number	designation	country	product forms	chemical composition	mechanical properties and hardness values
COPANT 672	Cu DHP	COPANT	Bar	99.90 min **Cu**, 0.015–0.05 **P**, silver included in copper content	**Cold finished**: 6.35 mm **diam**, 345 MPa **TS**; **Annealed**: all **diam**, 255 MPa **TS**, 25% in 2 in. (50 mm) **El**; **Drawn temper**: 9.53–12 mm **diam**, 275 MPa **TS**, 12% in 2 in. (50 mm) **El**
COPANT 526	Cu DHP	COPANT	Tube	99.90 min **Cu**, 0.015–0.050 **P**, silver included in copper content	**Straight**: 0.76–2.79 mm **diam**, 245 MPa **TS**; **In coil**: all **diam**, 205 MPa **TS**, 40% in 2 in. (50 mm) **El**
COPANT 525	Cu DHP	COPANT	Tube	99.90 min **Cu**, 0.015–0.05 **P**, silver included in copper content	**As drawn**: 0.584–0.826 mm **diam**, 390 MPa **TS**
COPANT 536	Cu DHP	COPANT	Sheet	99.90 min **Cu**, 0.01 **Fe**, 0.005 **Sb**, 0.10 **Ni**, 0.015–0.05 **P**, 0.05 **As**, 0.003 **Bi**, 0.010 **Se + Te**, 0.010 **Te**, silver included in copper content	**Annealed**: all **diam**, 205 MPa **TS**; **Tempered**: all **diam**, 225 MPa **TS**
COPANT 536	Cu QTP	COPANT	Sheet	99.90 min **Cu**, 0.015–0.04 O_2, silver included in copper content	**Annealed**: all **diam**, 205 MPa **TS**; **Tempered**: all **diam**, 225 MPa **TS**
COPANT 536	Cu FRTPHC	COPANT	Sheet	99.90 min **Cu**, 0.005 **Pb**, 0.0025 **As**, 0.04 total, others (excl **Ag**, O_2), silver included in copper content	**Annealed**: all **diam**, 205 MPa **TS**; **Tempered**: all **diam**, 225 MPa **TS**
COPANT 536	Cu ETP	COPANT	Sheet	99.90 min **Cu**, 0.004 **Pb**, 0.003 **Sb**, 0.015–0.04 O_2, 0.012 **As**, 0.003 **Bi**, 0.025 **Te**, 0.04 total, others (excl **Ag** and O_2), silver included in copper content	**Annealed**: all **diam**, 205 MPa **TS**; **Tempered**: all **diam**, 225 MPa **TS**, 135 MPa **YS**
COPANT 536	Cu Catb	COPANT	Sheet	99.90 min **Cu**, 0.005 **Pb**, 0.001 **Bi**, 0.03 total, others, silver included in copper content	**Annealed**: all **diam**, 205 MPa **TS**; **Tempered**: all **diam**, 225 MPa **TS**, 135 MPa **YS**
COPANT 564		COPANT	Wire	99.90 min **Cu**, silver included in copper content	**1/2 hard temper**: 1.0–1.06 mm **diam**, 365 MPa **TS**; **Hard temper**: 1.0–1.29 mm **diam**, 450 MPa **TS**
COPANT 430	Cu FRTPHC	COPANT	Wire	99.90 min **Cu**, silver included in copper content	**Annealed (regular)**: $\leq$0.26 mm **diam**, 20% in 2 in. (50 mm) **El**; **Roll temper**: 0.51–3.18 mm **diam**, 295 MPa **TS**, 10% in 2 in. (50 mm) **El**
DS 3003	5010	Denmark	Rolled, drawn, extruded, and forged products	99.90 min **Cu**, silver included in copper content	
WW-P-377d		US	Pipe: seamless	99.90 **Cu**, 0.040 **P**, **Ag** in **Cu** content	
WW-T-799E	Types K,L,M	US	Tube	99.9 min **Cu**, 0.04 **P**, **Ag** in **Cu** content	

WROUGHT COPPER AND COPPER ALLOYS

specification number	designation	country	product forms	chemical composition	mechanical properties and hardness values
ISO/R1337	Cu–DLP	ISO	Plate, sheet, strip, rod, bar, extruded sections, tube	99.90 min **Cu**, 0.005–0.012 **P**, silver included in copper content	
ISO/R1337	Cu–FRHC	ISO	Plate, sheet, strip, rod, bar, extruded sections, tube, wire, forgings	99.90 min **Cu**, silver included in copper content	
ISO/R1337	CuETP	ISO	Plate, sheet, strip, rod, bar, extruded sections, tube, wire, forgings	99.90 min **Cu**, silver included in copper content	
JIS C 2801	Class 1	Japan	Bar: commutator	99.9 min **Cu**	**As manufactured**: <5 mm **diam**, 108 min **HV**
JIS H 3300	C 1220	Japan	Pipe and tube: seamless	99.90 min **Cu**, 0.015–0.040 **P**	**Annealed**: 4–250 mm **diam**, 206 MPa **TS**, 40% in 2 in. (50 mm) **El**, 60 max **HR15T**; **1/2 hard**: 4–250 mm **diam**, 245 MPa **TS**, 30–60 **HR30T**; **Hard**: ≤25 mm **diam**, 314 MPa **TS**, 55 min **HR30T**
JIS H 3300	C 1201	Japan	Pipe and tube: seamless	99.90 min **Cu**, 0.004–0.015 **P**	**Annealed**: 4–250 mm **diam**, 206 MPa **TS**, 40% in (50 mm) **El**, 60 max **HR15T**; **1/2 hard**: 4–250 mm **diam**, 245 MPa **TS**, 30–60 **HR30T**; **Hard**: ≤25 mm **diam**, 314 MPa **TS**, 55 min **HR30T**
JIS H 3300	C 1100	Japan	Pipe and tube: seamless	99.90 min **Cu**	**Annealed**: 5–250 mm **diam**, 206 MPa **TS**, 40% in 2 in. (50 mm) **El**; **1/2 hard**: 5–250 mm **diam**, 245 MPa **TS**, 30–60 **HR30T**; **Hard**: 5–100 mm **diam**, 275 MPa **TS**, 80 min **HRF**
JIS H 3260	C 1220	Japan	Rod and bar	99.90 min **Cu**, 0.015–0.040 **P**	**Annealed**: 0.5–2 mm **diam**, 196 MPa **TS**, 15% in 2 in. (50 mm) **El**; **1/2 hard**: 0.5–12 mm **diam**, 255 MPa **TS**; **Hard**: 0.5–10 mm **diam**, 343 MPa **TS**
JIS H 3260	C 1201	Japan	Rod and bar	99.90 min **Cu**, 0.004–0.015 **P**	**Annealed**: 0.5–2 mm **diam**, 196 MPa **TS**, 15% in 2 in. (50 mm) **El**; **1/2 hard**: 0.5–12 mm **diam**, 255 MPa **TS**; **Hard**: 0.5–10 mm **diam**, 343 MPa **TS**

WROUGHT COPPER AND COPPER ALLOYS

specification number	designation	country	product forms	chemical composition	mechanical properties and hardness values
JIS H 3260	C 1100	Japan	Rod and bar	99.90 min **Cu**	**Annealed**: 0.5–2 mm **diam**, 196 MPa **TS**, 15% in 2 in. (50 mm) **El**; **1/2 hard**: 0.5–12 mm **diam**, 255 MPa **TS**; **Hard**: 0.5–10 mm **diam**, 343 MPa **TS**
JIS H 3250	C 1220	Japan	Rod and bar	99.90 min **Cu**, 0.015–0.040 **P**	**As manufactured**: ≥6 mm **diam**, 196 MPa **TS**, 25% in 2 in. (50 mm) **El**; **Annealed**: 6–25 mm **diam**, 196 MPa **TS**, 30% in 2 in. (50 mm) **El**; **Hard**: 6–25 mm **diam**, 275 MPa **TS**
JIS H 3250	C 1201	Japan	Rod and bar	99.90 min **Cu**, <0.015 **P**	**As manufactured**: ≥6 mm **diam**, 196 MPa **TS**, 25% in 2 in. (50 mm) **El**; **Annealed**: 6–25 mm **diam**, 196 MPa **TS**, 30% in 2 in. (50 mm) **El**; **Hard**: 6–25 mm **diam**, 275 MPa **TS**
JIS H 3250	C 1100	Japan	Rod and bar	99.90 min **Cu**	**As manufactured**: ≥6 mm **diam**, 196 MPa **TS**, 25% in 2 in. (50 mm) **El**; **Annealed**: 6–25 mm **diam**, 196 MPa **TS**, 30% in 2 in. (50 mm) **El**; **Hard**: 6–25 mm **diam**, 275 MPa **TS**
JIS H 3140	C 1100	Japan	Bar: bus	99.90 min **Cu**	**Annealed**: 2–30 mm **diam**, 196 MPa **TS**, 35% in 2 in. (50 mm) **El**; **1/2 hard**: 2–20 mm **diam**, 245 MPa **TS**, 15% in 2 in. (50 mm) **El**; **Hard**: 2–10 mm **diam**, 275 MPa **TS**
JIS H 3100	C1100	Japan	Sheet, plate, and strip	99.90 min **Cu**	**Annealed**: 0.5–30 mm **diam**, 196 MPa **TS**, 35% in 2 in. (50 mm) **El**; **1/2 hard**: 0.5–20 mm **diam**, 245 MPa **TS**, 15% in 2 in. (50 mm) **El**; **Hard**: 0.5–10 mm **diam**, 275 MPa **TS**, 90 min **HV**
JIS H 3100	C1201	Japan	Sheet, plate, and strip	99.90 min **Cu**, 0.015 **P**	**Annealed**: 0.3–30 mm **diam**, 196 MPa **TS**, 35% in 2 in. (50 mm) **El**; **1/2 hard**: 0.3–20 mm **diam**, 245 MPa **TS**, 15% in 2 in. (50 mm) **El**; **Hard**: 0.3–10 mm **diam**, 275 MPa **TS**

WROUGHT COPPER AND COPPER ALLOYS

specification number	designation	country	product forms	chemical composition	mechanical properties and hardness values
JIS H 3100	C1220	Japan	Sheet, plate, and strip	99.90 min **Cu**, 0.015–0.040 **P**	**Annealed**: 0.3–30 mm **diam**, 196 MPa **TS**, 35% in 2 in. (50 mm) **El**; **1/2 hard**: 0.3–20 mm **diam**, 245 MPa **TS**, 15% in 2 in. (50 mm) **El**; **Hard**: 0.3–10 mm **diam**, 275 MPa **TS**
JIS H 2122		Japan	Wire: electrical	99.90 min **Cu**, silver included as copper	
JIS H 2123		Japan	Billet and cake: tough–pitch	99.90 min **Cu**, silver included as copper	
JIS H 2124		Japan	Billet and cake: phosphorus–deoxidized	99.90 min **Cu**, 0.004–0.040 **P**, silver included as copper	
NOM–W–17	Type L	Mexico	Tube: seamless	99.9 min **Cu**, 0.015–0.04 **P**, **Ag** included in **Cu**	**Tempered**: 6.35–304.8 mm **diam**, 30 ksi (207 MPa) **TS**, 50 max **HRF**; **Cold drawn**: 6.35–304.8 mm **diam**, 36 ksi (248 MPa) **TS**, 20 min **HRB**
NOM–W–17	Type K	Mexico	Tube: seamless	99.9 min **Cu**, 0.015–0.04 **P**, **Ag** included in **Cu**	**Tempered**: 6.35–304.8 mm **diam**, 30 ksi (207 MPa) **TS**, 50 max **HRF**; **Cold drawn**: 6.35–304.8 mm **diam**, 36 ksi (248 MPa) **TS**, 20 min **HRB**
NOM–W–17	Type M	Mexico	Tube: seamless	99.9 min **Cu**, 0.015–0.04 **P**, **Ag** included in **Cu**	**Tempered**: 6.35–304.8 mm **diam**, 30 ksi (207 MPa) **TS**, 50 max **HRF**; **Cold drawn**: 6.35–304.8 mm **diam**, 36 ksi (248 MPa) **TS**, 20 min **HRB**
NOM–W–17	Type R	Mexico	Tube: seamless	99.9 min **Cu**, 0.015–0.04 **P**, **Ag** included in **Cu**	**Tempered**: 6.35–304.8 mm **diam**, 30 ksi (207 MPa) **TS**, 50 max **HRF**; **Cold drawn**: 6.35–304.8 mm **diam**, 36 ksi (248 MPa) **TS**, 20 min **HRB**
NOM–W–17	Type RR	Mexico	Tube: seamless	99.9 min **Cu**, 0.015–0.04 **P**, **Ag** included in **Cu**	**Tempered**: 6.35–304.8 mm **diam**, 30 ksi (207 MPa) **TS**, 50 max **HRF**; **Cold drawn**: 6.35–304.8 mm **diam**, 36 ksi (207 MPa) **TS**, 20 min **HRB**
NOM–W–17	Type RRR	Mexico	Tube: seamless	99.9 min **Cu**, 0.015–0.04 **P**, **Ag** included in **Cu**	**Tempered**: 12.7–203.2 mm **diam**, 30 ksi (207 MPa) **TS**, 50 max **HRF**; **Cold drawn**: 12.7–203.2 mm **diam**, 36 ksi (207 MPa) **TS**, 20 min **HRB**

WROUGHT COPPER AND COPPER ALLOYS

specification number	designation	country	product forms	chemical composition	mechanical properties and hardness values
NOM-W-17	Special: for residential water lines	Mexico	Tube: seamless	99.9 min **Cu**, 0.015–0.04 **P**, **Ag** included in **Cu**	**Tempered**: 12.7–203.2 mm **diam**, 30 ksi (207 MPa) **TS**, 50 max **HRF**; **Cold drawn**: 12.7–203.2 mm **diam**, 36 ksi (207 MPa) **TS**, 20 min **HRB**
NOM-W-18		Mexico	Tube: seamless	99.9 min **Cu**, 0.015–0.040 **P**	**Cold finished and annealed**: 9.525–155.575 mm **diam**, 250 MPa **TS**, 40% in 2 in. (50 mm) **El**
DGN-W-23	Type DLP	Mexico	Tube: seamless, for refrigeration service	99.9 min **Cu**, 0.004–0.012 **P**	**Cold finished**: 3.18–15.92 mm **diam**, 30 ksi (207 MPa) **TS**, 40% in 2 in. (50 mm) **El**
DGN-W-23	Type DHP	Mexico	Tube: seamless, for refrigeration service	99.9 min **Cu**, 0.015–0.040 **P**	**Cold finished**: 3.18–15.92 mm **diam**, 30 ksi (207 MPa) **TS**, 40% in 2 in. (50 mm) **El**
DGN-W-349		Mexico	Ingot: electrolytic	99.9 min **Cu**	
MIL-T-22214A	Alloy no. 122	US	Tube	99.9 min **Cu**, 0.015–0.040 **P**	
MIL-T-24107A	122	US	Tube: seamless	99.90 min **Cu**, 0.015–0.040 **P**, silver included in copper content	**Annealed**: all **diam**, 207 MPa **TS**, 62 MPa **YS**
MIL-T-24107A	120	US	Tube: seamless	99.90 **Cu**, 0.004–0.012 **P**, silver included in copper content	**Annealed**: all **diam**, 207 MPa **TS**, 62 MPa **YS**
MIL-C-12166		US	Rod	99.90 min **Cu**	**Hot rolled or extruded**: 0.252 in. **diam**, 310 MPa **TS**, 1.75% in 2 in. (50 mm) **El**
NS 16010	Cu-ETP and Cu-FRHC	Norway	Sheet, strip, plate, bar, rod, wire, profiles, and forgings	99.9 min **Cu**, silver included in copper content	
SABS 460	Cu-ETP	South Africa	Tube	99.90 min **Cu**, 0.005 **Pb**, 0.001 **Bi**, 0.03 total, others, **Ag** included in **Cu**	**Annealed**: all **diam**, 200–250 MPa **TS**, 40% in 2 in. (50 mm) **El**, 60 max **HV**
SABS 460	Cu-FRHC	South Africa	Tube	99.90 min **Cu**, 0.005 **Pb**, 0.002 **Bi**, 0.04 total, others, **Ag** included in **Cu**	**As drawn**: all **diam**, 270 MPa **TS**, 100 **HV**
SIS 14 50 10	SIS Cu 50 10-25	Sweden	Wire	99.90 min **Cu**, **Ag** in **Cu** content, 0.02–0.06 **O**	**Strain hardened**: 80 min mm **diam**, 363 MPa **TS**, 340 MPa **YS**, 3% in 8 in. (200 mm) **El**
SIS 14 50 10	SIS Cu 50 10-24	Sweden	Section	99.90 min **Cu**, **Ag** in **Cu** content, 0.02–0.06 **O**	**Strain hardened**: all **diam**, 270 MPa **TS**, 220 MPa **YS**, 10% in 2 in. (50 mm) **El**, 85–110 **HV**
SIS 14 50 10	SIS Cu 50 10-12	Sweden	Sheet, strip	99.90 min **Cu**, **Ag** in **Cu** content, 0.02–0.06 **O**	**Annealed**: 0.2–0.5 mm **diam**, 220 MPa **TS**, 40 MPa **YS**, 20% in 4 in. (100 mm) **El**, 40–60 **HV**

WROUGHT COPPER AND COPPER ALLOYS

specification number	designation	country	product forms	chemical composition	mechanical properties and hardness values
SIS 14 50 10	SIS Cu 50 10–10	Sweden	Section, wire	99.90 min **Cu**, **Ag** in **Cu** content, 0.02–0.06 **O**	**Strain hardened**: all **diam**, 230 MPa **TS**, 120 MPa **YS**, 10% in 2 in. (50 mm) **El**, 65 min **HV**
SIS 14 50 10	SIS Cu 50 10–05	Sweden	Plate, sheet, strip	99.90 min **Cu**, **Ag** in **Cu** content, 0.02–0.06 **O**	**Strain hardened**: 5 mm **diam**, 280 MPa **TS**, 250 MPa **YS**, 10% in 2 in. (50 mm) **El**, 90–105 **HV**
SIS 14 50 10	SIS Cu 50 10–04	Sweden	Plate, sheet, strip, bar, wire, rod	99.90 min **Cu**, **Ag** in **Cu** content, 0.02–0.06 **O**	**Strain hardened**: 5 mm **diam**, 250 MPa **TS**, 180 MPa **YS**, 20% in 2 in. (50 mm) **El**, 75–95 **HV**
SIS 14 50 10	SIS Cu 50 10–02	Sweden	Plate, sheet, strip, bar, wire	99.90 min **Cu**, **Ag** in **Cu** content, 0.02–0.06 **O**	**Annealed**: 5 mm **diam**, 220 MPa **TS**, 40 MPa **YS**, 40% in 2 in. (50 mm) **El**, 40–65 **HV**
SIS 14 50 10	SIS Cu 50 10–00	Sweden	Bar, section, wire, forgings	99.90 min **Cu**, **Ag** in **Cu** content, 0.02–0.06 **O**	**Hot worked**: all **diam**, 230 MPa **TS**, 60 MPa **YS**, 45% in 2 in. (50 mm) **El**, 60 **HV**
SABS 804/807	SABS 804	South Africa	Bar	99.90 min **Cu**, 0.005 **Pb**, 0.0010 **Bi**, 0.03 total, others	
SABS 804/807	SABS 805	South Africa	Bar	99.90 min **Cu**, 0.005 **Pb**, 0.005 **Fe**, 0.0010 **Bi**, 0.0010 **Se**, 0.0005 **Te**, 0.03 total, others	
SABS 804/807	SABS 806	South Africa	Bar	99.90 min **Cu**, 0.01 **Sn**, 0.005 **Pb**, 0.01 **Fe**, 0.001 **Sb**, 0.015 **Ni**, 0.0020 **Bi**, 0.007 **Se**, 0.003 **Te**, 0.007 **As**, 0.007 each, 0.04 total, others	
SFS 2910	CuAg0.1 (OF)	Finland	Sheet, strip, bar, rod, wire, and shapes	0.08–0.12 **Ag**, 99.90 min **Cu + Ag**	**Cold finished (sheet and strip)**: 0.2–0.5 mm **diam**, 250 MPa **TS**, 180 MPa **YS**, 5% in 2 in. (50 mm) **El**; **Annealed (rod and wire)**: 0.2–1 mm **diam**, 210 MPa **TS**, 40 MPa **YS**, 20% in 2 in. (50 mm) **El**; **Drawn (bar)**: 2.5–5 mm **diam**, 340 MPa **TS**, 320 MPa **YS**, 5% in 2 in. (50 mm) **El**
NF A 53–301	Cu/a 1	France	Bar and shapes	99.90 min **Cu + Ag**	**1/4 hard**: all **diam**, 230 MPa **TS**, 20% in 2 in. (50 mm) **El**; **1/2 hard**: $\leq$70 mm **diam**, 250 MPa **TS**, 15% in 2 in. (50 mm) **El**; **3/4 hard**: $\leq$60 mm **diam**, 270 MPa **TS**, 10% in 2 in. (50 mm) **El**
NF A 53–301	Cu/a 2	France	Bar and shapes	99.90 **Cu + Ag**	**1/4 hard**: all **diam**, 230 MPa **TS**, 20% in 2 in. (50 mm) **El**; **1/2 hard**: $\leq$70 mm **diam**, 250 MPa **TS**, 15% in 2 in. (50 mm) **El**; **3/4 hard**: $\leq$60 mm **diam**, 270 MPa **TS**, 10% in 2 in. (50 mm) **El**

WROUGHT COPPER AND COPPER ALLOYS

specification number	designation	country	product forms	chemical composition	mechanical properties and hardness values
SFS 2906	Cu–DLP	Finland	Sheet and strip	0.005–0.012 **P**, 99.90 min **Cu + Ag**	**Annealed**: 0.2–0.5 mm **diam**, 220 MPa **TS**, 140 MPa **YS**, 20% in 2 in. (50 mm) **El**; **Cold finished**: 0.2–0.5 mm **diam**, 250 MPa **TS**, 180 MPa **YS**, 5% in 2 in. (50 mm) **El**
SFS 2908	Cu–ETP and Cu–FRHC	Finland	Sheet, strip, rod, bar, wire, shapes, and forgings	0.02–0.06 **O**, 99.90 min **Cu + Ag**	**Annealed (sheet, strip, and bar)**: 1.5–2.5 mm **diam**, 220 MPa **TS**, 40 MPa **YS**, 40% in 2 in. (50 mm) **El**; **Hot finished (rod, wire, and bar)**: ≥5 mm **diam**, 230 MPa **TS**, 60 MPa **YS**, 45% in 2 in. (50 mm) **El**; **Drawn (shapes)**: all **diam**, 230 MPa **TS**, 120 MPa **YS**, 10% in 2 in. (50 mm) **El**
DS 3003	5030	Denmark	Rolled, drawn, extruded, and forged products	0.08–0.12 **Ag**, 0.06 **O**, 99.90 min **Cu + Ag**	
COPANT 162	C 123 00	COPANT	Wrought products	99.89 min **Cu**, 0.015–0.025 **P**, 0.014 min **Ag**	
COPANT 162	C 121 00	COPANT	Wrought products	99.89 min **Cu**, 0.005–0.012 **P**, 0.014 min **Ag**	
COPANT 522	Cu Ag TP	COPANT	Tube	99.89 min **Cu**, 0.002 **Ag**	**Tempered (fluids piping)**: all **diam**, 245 MPa **TS**
COPANT 522	Cu ETP	COPANT	Tube	99.89 min **Cu**, 0.004 **Pb**, 0.003 **Sb**, 0.002 **Ag**, 0.012 **As**, 0.003 **Bi**, 0.025 **Te**	**Annealed (fluids piping)**: 0.25–12 mm **diam**, 205 MPa **TS**; **Tempered (fluids piping)**: all **diam**, 245 MPa **TS**
COPANT 430	Cu QTP	COPANT	Wire	99.89 min **Cu**, 0.0002 **Pb**, 0.010 **Fe**, 0.0013 **Ni**, 0.001 **S**, 0.002 **Ag**, 0.015–0.04 O_2, 0.001 **As**	**Annealed (regular)**: ≤0.26 mm **diam**, 20% in 2 in. (50 mm) **El**; **Roll temper**: 0.51–3.18 mm **diam**, 295 MPa **TS**, 10% in 2 in. (50 mm) **El**
COPANT 430	Cu ETP	COPANT	Wire	99.89 min **Cu**, 0.004 **Pb**, 0.003 **Sb**, 0.002 **Ag**, 0.015–0.04 O_2, 0.012 **As**, 0.003 **Bi**, 0.025 **Te**, 0.04 total, others	**Annealed (regular)**: ≤0.26 mm **diam**, 20% in 2 in. (50 mm) **El**; **Roll temper**: 0.51–3.18 mm **diam**, 295 MPa **TS**, 10% in 2 in. (50 mm) **El**
COPANT 530	Cu QTP	COPANT	Tube	99.89 min **Cu**, 0.0002 **Pb**, 0.01 **Fe**, 0.0005 **Sb**, 0.0013 **Ni**, 0.001 **S**, 0.002 **Ag**, 0.015–0.04 O_2, 0.001 **As**, 0.04 total, others (excl **Ag** and O_2)	**Annealed**: 0.01–5/8 mm **diam**, 205 MPa **TS**; **Tempered to 1/2 hard**: 0.01–5/8 mm **diam**, 245 MPa **TS**, 30 min **HR30T**; **Tempered to full hard**: 0.51–3.04 mm **diam**, 315 MPa **TS**
ANSI/ASTM B 216		US	Wrought products: fire-refined	99.88 min **Cu**, 0.004 **Pb**, 0.003 **Sb**, 0.05 **Ni**, 0.012 **As**, 0.025 **Se + Te**, 0.003 **Bi**, silver included in copper content	
ANSI/ASTM B 11	C 12500	US	Plate	99.88 min **Cu**, silver included with copper content	**Hot rolled**: all **diam**, 30 ksi (205 MPa) **TS**, 30% in 8 in. (203 mm) **El**

WROUGHT COPPER AND COPPER ALLOYS

specification number	designation	country	product forms	chemical composition	mechanical properties and hardness values
ANSI/ASTM B 216	C 1300	US	Bar, billet, cake	99.88 min **Cu**, 0.004 **Pb**, 0.003 **Sb**, 0.05 **Ni**, 0.012 **As**, 0.003 **Bi**, silver included in copper content, 0.025 **Se**+**Te**	
ANSI/ASTM B 152	C 12500	US	Sheet, strip, plate, bar	99.88 min **Cu**, 0.004 **Pb**, 0.003 **Sb**, 0.05 **Ni**, 0.012 **As**, 0.003 **Bi**, 0.025 **Se**+**Te**, silver included in copper content	**Hard**: >0.020 in. (>0.508 mm) **diam**, 43 ksi (295 MPa) **TS**, 86–93 **HRF**; **Spring**: >0.020 in. (>0.508 mm) **diam**, 50 ksi (345 MPa) **TS**, 91–97 **HRF**; **Hot rolled**: >0.020 in. (>0.508 mm) **diam**, 75 max **HRF**
COPANT 672	Cu RFTP	COPANT	Bar	99.88 **Cu**, 0.004 **Pb**, 0.003 **Sb**, 0.05 **Al**, 0.015–0.04 O_2, 0.012 **As**, 0.003 **Bi**, 0.025 **Se**, silver included in copper content	**Cold finished**: 6.35 mm **diam**, 345 MPa **TS**; **Annealed**: all **diam**, 255 MPa **TS**, 25% in 2 in. (50 mm) **El**; **Drawn temper**: 9.53–12 mm **diam**, 275 MPa **TS**, 12% in 2 in. (50 mm) **El**
COPANT 536	Cu FRTP	COPANT	Sheet	99.88 min **Cu**, 0.01 **Sn**, 0.004 **Pb**, 0.01 **Fe**, 0.003 **Sb**, 0.05 **Ni**, 0.012 **As**, 0.003 **Bi**, 0.025 **Se**+**Te**, 0.05 total, others (excl **Ni**, **Ag**, O_2), silver included in copper content	**Hot rolled**: 0.2–1.5 mm **diam**; **Annealed**: all **diam**, 205 MPa **TS**; **Tempered**: all **diam**, 225 MPa **TS**
COPANT 536	Cu FRTP	COPANT	Sheet	99.88 min **Cu**, 0.01 **Sn**, 0.004 **Pb**, 0.01 **Fe**, 0.003 **Sb**, 0.05 **Ni**, 0.012 **As**, 0.003 **Bi**, 0.025 **Se**+**Te**, silver included in copper content	**Annealed**: all **diam**, 205 MPa **TS**; **Tempered**: all **diam**, 225 MPa **TS**
COPANT 162	C 125 00 (=FRTP)	COPANT	Wrought products	99.88 min **Cu**, 0.004 **Pb**, 0.003 **Sb**, 0.050 **Ni**, 0.012 **As**, 0.025 **Te**, 0.003 **Bi**, silver included in copper content	
COPANT 672	Cu Ag TP	COPANT	Bar	99.88 min **Cu**, 0.02 **Ag**, 0.015–0.04 O_2	**Cold finished**: 6.35 mm **diam**, 345 MPa **TS**; **Annealed**: all **diam**, 255 MPa **TS**, 25% in 2 in. (50 mm) **El**; **Drawn temper**: 9.53–12 mm **diam**, 275 MPa **TS**, 12% in 2 in. (50 mm) **El**
QQ-C-576b		US	Plate, bar, sheet, strip	99.88 min **Cu**, **Ag** in **Cu** content	**Cold rolled–half hard**: all **diam**, 37 ksi (255 MPa) **TS**, 77–89 **HRF**; **Hot rolled**: all **diam**, 30 ksi (207 MPa) **TS**, 75 **HRF**; **Hot rolled and annealed**: all **diam**, 30 ksi (207 MPa) **TS**, 65 **HRF**

WROUGHT COPPER AND COPPER ALLOYS

specification number	designation	country	product forms	chemical composition	mechanical properties and hardness values
QQ-C-502c		US	Rod, shapes, flat wire, strip, bar	99.88 min **Cu**, **Ag** in **Cu** content	**Soft temper (shapes)**: all **diam**, 35 ksi (242 MPa) **TS**, 25% in 2 in. (50 mm) **El**; **Hard temper (shapes)**: all **diam**, 32 ksi (221 MPa) **TS**, 15% in 2 in. (50 mm) **El**; **Hard temper (flat wire and strip)**: 0.020–0.125 in. **diam**, 43 ksi (296 MPa) **TS**, 10% in 2 in. (50 mm) **El**, 85–97 **HRF**
BS 2871	C 106	UK	Tube	99.85 min **Cu**, 0.01 **Sn**, 0.010 **Pb**, 0.030 **Fe**, 0.005 **Sb**, 0.10 **Ni**, 0.013–0.050 **P**, 0.05 **As**, 0.0030 **Bi**, 0.010 **Te**, 0.020 **Se+Te**, 0.06 total, others	**As drawn**: 105 **HV**; **Half hard**: 80–100 **HV**; **Annealed**: 60 max **HV**
BS 2871	C 106	UK	Tube	99.85 min **Cu**, 0.01 **Sn**, 0.010 **Pb**, 0.030 **Fe**, 0.10 **Ni**, 0.013–0.050 **P**, 0.05 **As**, 0.005 **Sb**, 0.0030 **Bi**, 0.010 **Te**, 0.020 **Se+Te**, 0.060 total, others	**Annealed**: 200 MPa **TS**, 40% in 2 in. (50 mm) **El**, 60 max **HV**; **1/2 hard**: 250 MPa **TS**, 30% in 2 in. (50 mm) **El**, 75–100 **HV**; **As drawn**: 280 MPa **TS**, 100 **HV**
BS 2870	C 104	UK	Sheet, strip, foil	99.85 min **Cu**, 0.01 **Sn**, 0.010 **Pb**, 0.01 **Fe**, 0.005 **Sb**, 0.05 **Ni**, 0.02 **As**, 0.0030 **Bi**, 0.10 **O**, 0.020 **Se**, 0.010 **Te**, 0.05 total, others	**Manufactured or annealed**: 0.6–10.0 mm **diam**, 148 MPa **TS**, 35% in 2 in. (50 mm) **El**; **1/2 hard**: 0.6–1.3 mm **diam**, 172 MPa **TS**, 10% in 2 in. (50 mm) **El**; **Hard**: 0.6–2.7 mm **diam**, 217 MPa **TS**
BS 2873	C106	UK	Wire	99.85 min **Cu**, 0.01 **Sn**, 0.010 **Pb**, 0.030 **Fe**, 0.005 **Sb**, 0.10 **Ni**, 0.013–0.050 **P**, 0.05 **As**, 0.0030 **Bi**, 0.020 **Se+Te**, 0.010 **Te**, 0.06 total, others	
BS 2870	C 106	UK	Sheet, strip, foil	99.85 min **Cu**, 0.01 **Sn**, 0.10 **Pb**, 0.030 **Fe**, 0.005 **Sb**, 0.10 **Ni**, 0.013–0.050 **P**, 0.05 **As**, 0.0030 **Bi**, 0.020 **Se+Te**, 0.010 **Te**, 0.06 total, others	**As manufactured or annealed**: 0.6–10.0 mm **diam**, 148 MPa **TS**, 35% in 2 in. (50 mm) **El**; **1/2 hard**: 0.6–1.3 mm **diam**, 172 MPa **TS**, 10% in 2 in. (50 mm) **El**; **Hard**: 0.6–2.7 mm **diam**, 217 MPa **TS**
BS 2875	C104	UK	Plate	99.85 min **Cu**, 0.01 **Sn**, 0.010 **Pb**, 0.01 **Fe**, 0.005 **Sb**, 0.05 **Ni**, 0.02 **As**, 0.0030 **Bi**, 0.10 **O**, 0.20 **Se**, 0.010 **Te**, 0.05 total, others	**As manufactured or annealed**: ≥10 mm **diam**, 211 MPa **TS**, 35% in 2 in. (50 mm) **El**; **Hard**: 10–16 mm **diam**, 279 MPa **TS**, 15% in 2 in. (50 mm) **El**
BS 2875	C106	UK	Plate	99.85 min **Cu**, 0.01 **Sn**, 0.010 **Pb**, 0.030 **Fe**, 0.005 **Sb**, 0.10 **Ni**, 0.013–0.050 **P**, 0.05 **As**, 0.0030 **Bi**, 0.020 **Se+Te**, 0.010 **Te**, 0.06 total, others	**As manufactured or annealed**: ≥10 mm **diam**, 211 MPa **TS**, 35% in 2 in. (50 mm) **El**; **Hard**: 10–16 mm **diam**, 279 MPa **TS**, 15% in 2 in. (50 mm) **El**

WROUGHT COPPER AND COPPER ALLOYS

specification number	designation	country	product forms	chemical composition	mechanical properties and hardness values
BS 2874	C106	UK	Rod, section	99.85 min **Cu**, 0.01 **Sn**, 0.010 **Pb**, 0.030 **Fe**, 0.005 **Sb**, 0.10 **Ni**, 0.013–0.050 **P**, 0.05 **As**, 0.0030 **Bi**, 0.020 **Se+Te Te**, 0.06 total, others	**Annealed**: ≥6 mm **diam**, 216 MPa **TS**, 33% in 2 in. (50 mm) **El**; **As manufactured; ≥6 mm diam, 230 MPa TS, 13% in 2 in. (50 mm) El**
ISO/R1337	CuDHP	ISO	Plate, sheet, strip, rod, bar, extruded sections, tube	99.85 min **Cu**, 0.013–0.050 **P**, silver included in copper content	
ISO/R1337	Cu–FRTP	ISO	Plate, sheet, strip, rod, bar, extruded sections	99.85 min **Cu**, silver included in copper content	
ISO/R1336	CuAg0.05	ISO	Strip, rod, bar, wire	99.85 min **Cu**, 0.02–0.08 **Ag**, 0.06 **O**, 0.1 total, others	
ISO/R1336	CuAg0.1	ISO	Strip, rod, bar, wire	99.85 min **Cu**, 0.08–0.12 **Ag**, 0.06 **O**, 0.1 total, others	
NS 16 013	Cu–FRTP	Norway	Sheet, strip, plate, bar, and profiles	99.85 min **Cu**, silver included in copper content	
NS 16 015	Cu–DHP	Norway	Sheet, strip, plate, bar, tube, and profiles	99.85 min **Cu**, 0.015–0.050 **P**	
SABS 460	Cu–DHP	South Africa	Tube	99.85 min **Cu**, 0.01 **Sn**, 0.010 **Pb**, 0.030 **Fe**, 0.005 **Sb**, 0.10 **Ni**, 0.013–0.050 **P**, 0.05 **As**, 0.003 **Bi**, 0.020 **Se+Te**, 0.060 total, others, **Ag** included in **Cu**	**Annealed**: all **diam**, 200 MPa **TS**, 40% in 2 in. (50 mm) **El**, 60 max **HV**; **Half–hard**: all **diam**, 250 MPa **TS**, 30% in 2 in. (50 mm) **El**, 75–100 **HV**; **As drawn**: all **diam**, 450 MPa **TS**, 130 **HV**
VSM 11557 (Part 1)	Cu–DHP	Switzerland	Tube	99.85 min **Cu**, 0.013–0.050 **P**	**1/2 hard**: all **diam**, 245 MPa **TS**, 147 MPa **YS**, 25% in 2 in. (50 mm) **El**, 70 **HB**
SIS 14 50 15	SIS Cu 50 15–04	Sweden	Tube	99.85 min **Cu**, 0.015–0.050 **P**, **Ag** in **Cu** content	**Strain hardened**: 0.7 mm **diam**, 380 MPa **TS**, 360 MPa **YS**, 3% in 2 in. (50 mm) **El**, 110–130 **HV**
SIS 14 50 15	SIS Cu 50 15–03	Sweden	Plate, sheet, strip, tube	99.85 min **Cu**, 0.015–0.050 **P**, **Ag** in **Cu** content	**Strain hardened**: 5–20 mm **diam**, 240 MPa **TS**, 100 MPa **YS**, 30% in 2 in. (50 mm) **El**, 60–85 **HV**
SIS 14 50 15	SIS Cu 50 15–10	Sweden	Tube	99.85 min **Cu**, 0.015–0.050 **P**, **Ag** in **Cu** content	**Strain hardened**: 5 mm **diam**, 250 MPa **TS**, 200 MPa **YS**, 10% in 2 in. (50 mm) **El**, 85 min **HV**
SIS 14 50 15	SIS Cu 50 15–23	Sweden	Tube	99.85 min **Cu**, 0.015–0.050 **P**, **Ag** in **Cu** content	**Strain hardened**: 3 mm **diam**, 240 MPa **TS**, 180 MPa **YS**, 30% in 2 in.(50 mm) **El**, 75–95 **HV**
SIS 14 50 15	SIS Cu 50 15–02	Sweden	Plate, sheet, strip, tube	99.85 min **Cu**, 0.015–0.050 **P**, **Ag** in **Cu** content	**Annealed**: 5–20 mm **diam**, 210 MPa **TS**, 44 MPa **YS**, 45% in 2 in. (50 mm) **El**, 40–65 **HV**

WROUGHT COPPER AND COPPER ALLOYS

specification number	designation	country	product forms	chemical composition	mechanical properties and hardness values
SIS 14 50 13	SIS Cu 50 13–12	Sweden	Sheet, strip	99.85 min **Cu, Ag** in **Cu** content, 0.02–0.06 **O**	**Annealed**: 0.2–0.5 mm **diam**, 220 MPa **TS**, 40 MPa **YS**, 20% in 4 in. (100 mm) **El**, 40–60 **HV**
SIS 14 50 13	SIS Cu 50 13–10	Sweden	Strip	99.85 min **Cu, Ag** in **Cu** content, 0.02–0.06 **O**	
SIS 14 50 13	SIS Cu 50 13–05	Sweden	Plate, sheet, strip	99.85 min **Cu, Ag** in **Cu** content, 0.02–0.06 **O**	**Strain hardened**: 5 min mm **diam**, 280 MPa **TS**, 250 MPa **YS**, 10% in 2 in. (50 mm) **El**, 90–105 **HV**
SIS 14 50 13	SIS Cu 50 13–04	Sweden	Sheet, plate, strip	99.85 min **Cu, Ag** in **Cu** content, 0.02–0.06 **O**	**Strain hardened**: 0.2–0.5 mm **diam**, 250 MPa **TS**, 180 MPa **YS**, 5% in 4 in. (100 mm) **El**, 75–95 **HV**
SIS 14 50 13	SIS Cu 50 13–02	Sweden	Plate, sheet, strip	99.85 min **Cu, Ag** in **Cu** content, 0.02–0.06 **O**	**Annealed**: 2.5 min mm **diam**, 220 MPa **TS**, 40 MPa **YS**, 40% in 2 in. (50 mm) **El**, 40–65 **HV**
SFS 2907	Cu–DHP	Finland	Tube and shapes	0.015–0.050 **P**, 99.85 min **Cu+ Ag**	**Annealed**: all **diam**, 210 MPa **TS**, 40 MPa **YS**, 40% in 2 in. (50 mm) **El**; **Cold finished (tube)**: ≤5 mm **diam**, 250 MPa **TS**, 180 MPa **YS**, 20% in 2 in. (50 mm) **El**
DS 3003	5013	Denmark	Rolled, drawn, extruded, and forged products	99.85 **Cu+Ag**	
DS 3003	5015	Denmark	Rolled, drawn, extruded, and forged products	0.015–0.050 **P**, 99.85 min **Cu+ Ag**	
NS 16 032	CuAg0.1	Norway	Profiles: extruded	99.83 min **Cu**, 0.08–0.12 **Ag**	
ISO 1637	CuAg0.05(P)	ISO	Solid products: straight lengths	99.805 min **Cu**, 0.001–0.005 **P**, 0.02–0.08 **Ag**, 0.1 **P**, + total, others	**Annealed**: 5 mm **diam**, 35% in 2 in. (50 mm) **El**, 60 max **HV**; **HA**: 5–40 mm **diam**, 250 MPa **TS**, 15% in 2 in. (50 mm) **El**, 90 **HV**; **HB–strain hardened**: 5–20 mm **diam**, 280 MPa **TS**, 5% in 2 in. (50 mm) **El**, 105 **HV**
ISO 1638	CuAg0.05(P)	ISO	Solid product: drawn, on coils or reels	99.805 min **Cu**, 0.001–0.005 **P**, 0.02–0.08 **Ag**, 0.1 **P**, + total, others	**Annealed**: 1–1.5 mm **diam**, 210 MPa **TS**, 25% in 2 in. (50 mm) **El**; **HC**: 1–5 mm **diam**, 390 MPa **TS**; **HD–strain hardened**: 1–3 mm **diam**, 420 MPa **TS**
ISO/R1336	CuAg0.05(P)	ISO	Strip, rod, bar, wire	99.805 min **Cu**, 0.001–0.005 **P**, 0.02–0.08 **Ag**, 0.1 total, others	
ISO 1634	CuAg0.05(P)	ISO	Plate, sheet, strip	99.805 min **Cu**, 0.001–0.005 **P**, 0.02–0.08 **Ag**, 0.1 **P**, + total, others	**Strain hardened (HB)**: all **diam**, 280 MPa **TS**, 8% in 2 in. (50 mm) **El**, 100 **HV**

WROUGHT COPPER AND COPPER ALLOYS

specification number	designation	country	product forms	chemical composition	mechanical properties and hardness values
ISO 1637	CuAg0.1	ISO	Solid products: straight lengths	99.8 min **Cu**, 0.08–0.12 **Ag**, 0.06 **O**, 0.1 **O**, + total, others	**Annealed**: 5 mm **diam**, 35% in 2 in. (50 mm) **El**, 60 max **HV**; **HA**: 5–40 mm **diam**, 250 MPa **TS**, 15% in 2 in. (50 mm) **El**, 90 **HV**; **HB–strain hardened**: 5–20 mm **diam**, 280 MPa **TS**, 5% in 2 in. (50 mm) **El**, 105 **HV**
ISO 1637	CuAg0.1(P)	ISO	Solid products: straight lengths	99.765 min **Cu**, 0.001–0.005 **P**, 0.08–0.12 **Ag**, 0.1 **P**, + total, others	**Annealed**: 5 mm **diam**, 35% in 2 in. (50 mm) **El**, 60 max **HV**; **HA**: 5–40 mm **diam**, 250 MPa **TS**, 15% in 2 in. (50 mm) **El**, 90 **HV**; **HB–strain hardened**: 5–20 mm **diam**, 280 MPa **TS**, 5% in 2 in. (50 mm) **El**, 105 **HV**
ISO 1638	CuAg0.1(P)	ISO	Solid product: drawn, on coils or reels	99.765 min **Cu**, 0.001–0.005 **P**, 0.08–0.12 **Ag**, 0.1 **P**, + total, others	**Annealed**: 1–1.5 mm **diam**, 210 MPa **TS**, 25% in 2 in. (50 mm) **El**; **HC**: 1–5 mm **diam**, 390 MPa **TS**; **HD–strain hardened**: 1–3 mm) **diam**, 420 MPa **TS**
ISO/R1336	CuAg0.1(P)	ISO	Strip, rod, bar, wire	99.765 min **Cu**, 0.001–0.005 **P**, 0.08–0.12 **Ag**, 0.1 total, others	
ISO 1634	CuAg0.1 (P)	ISO	Plate, sheet, strip	99.765 min **Cu**, 0.001–0.005 **P**, 0.08–0.12 **Ag**, 0.1 **P**, + total, others	**HC**: all **diam**, 340 MPa **TS**, 4% in 2 in. (50 mm) **El**, 110 **HV**
COPANT 536	Cu CAST 1	COPANT	Sheet	99.75 **Cu**, 0.025 **Sn**, 0.10 **Pb**, 0.01 **Fe**, 0.012 **Sb**, 0.01 **S**, 0.0075 **As**, 0.003 **Bi**, 0.04 **Se**, 0.10 O_2, silver included in copper content	**Annealed**: all **diam**, 205 MPa **TS**; **Tempered**: all **diam**, 225 MPa **TS**
ISO 1637	CuAg0.05	ISO	Solid products: straight lengths	99.75 min **Cu**, 0.02–0.08 **Ag**, 0.06 **O**, 0.1 **O**, + total, others	**Annealed**: 5 mm **diam**, 35% in 2 in. (50 mm) **El**, 60 max **HV**; **HA**: 5–40 mm **diam**, 250 MPa **TS**, 15% in 2 in. (50 mm) **El**, 90 **HV**; **HB–strain hardened**: 5–20 mm **diam**, 280 MPa **TS**, 5% in 2 in. (50 mm) **El**, 105 **HV**
ISO 1638	CuAg0.05	ISO	Solid product: drawn, on coils or reels	99.75 min **Cu**, 0.02–0.08 **Ag**, 0.06 **O**, 0.1 **O**, + total, others	**Annealed**: 1–1.5 mm **diam**, 210 MPa **TS**, 25% in 2 in. (50 mm) **El**; **HC**: 1–5 mm **diam**, 390 MPa **TS**; **HD–strain hardened**: 1–3 mm **diam**, 420 MPa **TS**
ISO 1634	CuAg0.05	ISO	Plate, sheet, strip	99.75 min **Cu**, 0.02–0.08 **Ag**, 0.06 **O**, 0.1 **O** and total, others	**Annealed**: all **diam**, 30% in 2 in. (50 mm) **El**, 60 max **HV**

WROUGHT COPPER AND COPPER ALLOYS

specification number	designation	country	product forms	chemical composition	mechanical properties and hardness values
JIS H 3300	C 1221	Japan	Pipe and tube: seamless	99.75 min **Cu**, 0.004–0.040 **P**	**Annealed**: 4–250 mm **diam**, 206 MPa **TS**, 40% in 2 in. (50 mm) **El**, 60 max **HR15T**; **1/2 hard**: 4–250 mm **diam**, 245 MPa **TS**, 30–60 **HR30T**; **Hard**: ≤25 mm **diam**, 314 MPa **TS**, 55 min **HR30T**
JIS H 3260	C 1221	Japan	Rod and bar	99.75 min **Cu**, 0.004–0.040 **P**	**Annealed**: 0.5–2 mm **diam**, 196 MPa **TS**, 15% in 2 in. (50 mm) **El**; **1/2 hard**: 0.5–12 mm **diam**, 255 MPa **TS**; **Hard**: 0.5–10 mm **diam**, 343 MPa **TS**
JIS H 3250	C 1221	Japan	Rod and bar	99.75 min **Cu**, 0.004–0.040 **P**	**As manufactured**: ≥6 mm **diam**, 196 MPa **TS**, 25% in 2 in. (50 mm) **El**; **Annealed**: 6–25 mm **diam**, 196 MPa **TS**, 30% in 2 in. (50 mm) **El**; **Hard**: 6–25 mm **diam**, 275 MPa **TS**
JIS H 3100	C1221	Japan	Sheet, plate, and strip	99.75 min **Cu**, 0.004–0.040 **P**	**Annealed**: 0.3–30 mm **diam**, 196 MPa **TS**, 35% in 2 in. (50 mm) **El**; **1/2 hard**: 0.3–20 mm **diam**, 245 MPa **TS**, 15% in 2 in. (50 mm) **El**; **Hard**: ≥0.5 mm **diam**, 275 MPa **TS**, 90 min **HV**
SABS 804/807	SABS 807	South Africa	Bar	99.75 min **Cu**, 0.03 **Sn**, 0.02 **Pb**, 0.04 **Fe**, 0.007 **Sb**, 0.10 **Ni**, 0.005 **Bi**, 0.014 **As**, 0.015 **Se+Te**, 0.010 each, 0.10 total, others	
NF A 53-301	Cu/a 3	France	Bar and shapes	99.75 min **Cu+Ag**	**1/4 hard**: all **diam**, 230 MPa **TS**, 20% in 2 in. (50 mm) **El**; **1/2 hard**: ≤70 mm **diam**, 250 MPa **TS**, 15% in 2 in. (50 mm) **El**; **3/4 hard**: ≤60 mm **diam**, 270 MPa **TS**, 10% in 2 in. (50 mm) **El**
ISO 1638	CuAg0.1	ISO	Solid product: drawn, on coils or reels	99.71 min **Cu**, 0.08–0.12 **Ag**, 0.06 **O**, 0.1 **O**, + total, others	**Annealed**: 1–1.5 mm **diam**, 210 MPa **TS**, 25% in 2 in. (50 mm) **El**; **HC**: 1–5 mm **diam**, 390 MPa **TS**; **HD–strain hardened**: 1–3 mm **diam**, 420 MPa **TS**
ANSI/ASTM B 451		US	Foil, strip, sheet	99.50 min **Cu**, silver included in copper content	**Rolled and annealed**: all **diam**, 15 ksi (105 MPa) **TS**, 5% in 2 in. (50 mm) **El**; **Light cold rolled**: all **diam**, 32 ksi (220 MPa) **TS**, 5% in 2 in. (50 mm) **El**; **As rolled**: all **diam**, 50 ksi (345 MPa) **TS**

WROUGHT COPPER AND COPPER ALLOYS

specification number	designation	country	product forms	chemical composition	mechanical properties and hardness values
COPANT 536	Cu CAST 2	COPANT	Sheet	99.50 min **Cu**, 0.05 **Sn**, 0.30 **Pb**, 0.01 **Fe**, 0.012 **Sb**, 0.10 **Ni**, 0.01 **S**, 0.10 **As**, 0.003 **Bi**, 0.04 **Se**, silver included in copper content	**Annealed**: all **diam**, 205 MPa **TS**; **Tempered**: all **diam**, 225 MPa **TS**
JIS Z 3202	YCu Copper	Japan	Rod: bare welding	99.5 min **Cu**, 0.03 **Pb**, 0.1 **P**	
BS 2871	C 107	UK	Tube	99.20 min **Cu**, 0.01 **Sn**, 0.010 **Pb**, 0.030 **Fe**, 0.15 **Ni**, 0.013–0.050 **P**, 0.30–0.50 **As**, 0.01 **Sb**, 0.0030 **Bi**, 0.010 **Te**, 0.020 **Se+Te**, 0.070 total, others	**Annealed**: 200 MPa **TS**, 40% in 2 in. (50 mm) **El**, 60 max **HV**; **1/2 hard**: 250 MPa **TS**, 30% in 2 in. (50 mm) **El**, 75–100 **HV**; **As drawn**: 280 MPa **TS**, 100 **HV**
COPANT 523	Cu DHP	COPANT	Tube	99.00 min **Cu**, 0.015–0.05 **P**, silver included in copper content	
ISO 1634	CuAg0.1	ISO	Plate, sheet, strip	98.81 min **Cu**, 0.08–0.12 **Ag**, 0.06 **O**, 1.0 **O** and total, others	**Strain hardened (HA)**: all **diam**, 250 MPa **TS**, 16% in 2 in. (50 mm) **El**, 80 **HV**
DS 3003	5011	Denmark	Rolled, drawn, extruded, and forged products	99.5 min **Cu+Ag**	
DS 3003	5031	Denmark	Rolled, drawn, extruded, and forged products	0.08–0.12 **Ag**, 99.5 min **Cu+Ag**	
SFS 2905	Cu–OF	Finland	Sheet, strip, bar, rod, wire, tube, shapes, and forgings: for electrical purposes	99.5 min **Cu+Ag**	**Annealed (sheet and strip)**: 0.2–0.5 mm **diam**, 220 MPa **TS**, 40 MPa **YS**, 20% in 2 in. (50 mm) **El**; **Drawn (rod and wire)**: 0.2–2.5 mm **diam**, 420 MPa **TS**, 400 MPa **YS**, 1% in 2 in. (50 mm) **El**; **Hot finished (forgings)**: all **diam**, 230 MPa **TS**, 60 MPa **YS**, 45% in 2 in. (50 mm) **El**
NF A 53–301	Cu/b	France	Bar and shapes	99.90 min **Cu+Ag**	**1/4 hard**: all **diam**, 230 MPa **TS**, 20% in 2 in. (50 mm) **El**; **1/2 hard**: ≤70 mm **diam**, 250 MPa **TS**, 15% in 2 in. (50 mm) **El**; **3/4 hard**: ≤60 mm **diam**, 270 MPa **TS**, 10% in 2 in. (50 mm) **El**
NF A 53–301	Cu/c 1	France	Bar and shapes	99.92 min **Cu+Ag**	**1/4 hard**: all **diam**, 230 MPa **TS**, 20% in 2 in. (50 mm) **El**; **1/2 hard**: ≤70 mm **diam**, 250 MPa **TS**, 15% in 2 in. (50 mm) **El**; **3/4 hard**: ≤60 mm **diam**, 270 MPa **TS**, 10% in 2 in. (50 mm) **El**

WROUGHT COPPER AND COPPER ALLOYS

specification number	designation	country	product forms	chemical composition	mechanical properties and hardness values
NF A 53-301	Cu/c 2	France	Bar and shapes	99.96 min **Cu+Ag**	**1/4 hard**: all **diam**, 230 MPa **TS**, 20% in 2 in. (50 mm) **El**; **1/2 hard**: ≤70 mm **diam**, 250 MPa **TS**, 15% in 2 in. (50 mm) **El**; **3/4 hard**: ≤60 mm **diam**, 10% in 2 in. (50 mm) **El**
IS 6912	High conductivity copper	India	Forgings	0.005 **Pb**, 0.0025 **Bi**, 99.90 min **Cu+Ag**, 0.04 total, others	**Annealed**: ≥6 mm **diam**, 210 MPa **TS**, 45% in 2 in. (50 mm) **El**
IS 6912	Electrolytic tough pitch copper	India	Forgings	0.005 **Pb**, 0.001 **Bi**, 99.90 min **Cu+Ag** total, others	**Annealed**: ≥6 mm **diam**, 210 MPa **TS**, 45% in 2 in. (50 mm) **El**
ISO 1637	Cu–ETP	ISO	Solid products: straight lengths	99.90 min **Cu+Ag**	**Annealed**: 5 mm **diam**, 35% in 2 in. (50 mm) **El**, 60 max **HV**; **HA**: 5–40 mm **diam**, 250 MPa **TS**, 15% in 2 in. (50 mm) **El**, 90 **HV**; **HB–strain hardened**: 5–20 mm **diam**, 280 MPa **TS**, 5% in 2 in. (50 mm) **El**, 105 **HV**
ISO 1637	Cu–FRHC	ISO	Solid products: straight lengths	99.90 min **Cu+Ag**	**Annealed**: 5 mm **diam**, 35% in 2 in. (50 mm) **El**, 60 max **HV**; **HA**: 5–40 mm **diam**, 250 MPa **TS**, 15% in 2 in. (50 mm) **El**, 90 **HV**; **HB–strain hardened**: 5–20 mm **diam**, 280 MPa **TS**, 5% in 2 in. (50 mm) **El**, 105 **HV**
ISO 1637	Cu–FRTP	ISO	Solid products: straight lengths	99.85 min **Cu+Ag**	**Annealed**: 5 mm **diam**, 35% in 2 in. (50 mm) **El**, 60 max **HV**; **HA**: 5–40 mm **diam**, 250 MPa **TS**, 15% in 2 in. (50 mm) **El**, 90 **HV**; **HB–strain hardened**: 5–20 mm **diam**, 280 MPa **TS**, 5% in 2 in. (50 mm) **El**, 105 **HV**
ISO 1637	Cu–OF	ISO	Solid products: straight lengths	99.95 min **Cu+Ag**	**Annealed**: 5 mm **diam**, 35% in 2 in. (50 mm) **El**, 60 max **HV**; **HA**: 5–40 mm **diam**, 250 MPa **TS**, 15% in 2 in. (50 mm) **El**, 90 **HV**; **HB–strain hardened**: 5–20 mm **diam**, 280 MPa **TS**, 5% in 2 in. (50 mm) **El**, 105 **HV**
ISO 1637	Cu–DLP	ISO	Solid products: straight lengths	0.005–0.012 **P**, 99.90 min **Cu+Ag**	**Annealed**: 5 mm **diam**, 35% in 2 in. (50 mm) **El**, 60 max **HV**; **HA**: 5–40 mm **diam**, 250 MPa **TS**, 15% in 2 in. (50 mm) **El**, 90 **HV**; **HB–strain hardened**: 5–20 mm **diam**, 280 MPa **TS**, 5% in 2 in. (50 mm) **El**, 105 **HV**

WROUGHT COPPER AND COPPER ALLOYS

specification number	designation	country	product forms	chemical composition	mechanical properties and hardness values
ISO 1637	Cu–DHP	ISO	Solid products: straight lengths	0.013–0.050 **P**, 99.85 min **Cu+Ag**	**Annealed**: 5 mm **diam**, 35% in 2 in. (50 mm) **El**, 60 max **HV**; **HA**: 5–40 mm **diam**, 250 MPa **TS**, 15% in 2 in. (50 mm) **El**, 90 **HV**; **HB–strain hardened:** 5–20 mm **diam**, 280 MPa **TS**, 5% in 2 in. (50 mm) **El**, 105 **HV**
ISO 1635	Cu–ETP	ISO	Tube	99.90 min **Cu+Ag**, total	**Annealed**: all **diam**, 35% in 2 in. (50 mm) **El**, 65 max **HV**; **HA**: 200 max mm **diam**, 250 MPa **TS**, 20% in 2 in. (50 mm) **El**, 80 **HV**; **Strain hardened (HB)**: 100 max mm **diam**, 290 MPa **TS**, 10% in 2 in. (50 mm) **El**, 100 **HV**
ISO 1638	Cu–ETP	ISO	Solid product: drawn, on coils or reels	99.90 min **Cu+Ag**	**Annealed**: 1–1.5 mm **diam**, 210 MPa **TS**, 25% in 2 in. (50 mm) **El**; **HC**: 1–5 mm **diam**, 390 MPa **TS**; **HD–strain hardened**: 1–3 mm **diam**, 420 MPa **TS**
ISO 1638	Cu–FRHC	ISO	Solid product: drawn, on coils or reels	99.85 min **Cu+Ag**	**HC**: 1–5 mm **diam**, 390 MPa **TS**; **HD–strain hardened**: 1–3 mm **diam**, 420 MPa **TS**; **Annealed**: 1–1.5 mm **diam**, 210 MPa **TS**, 25% in 2 in. (50 mm) **El**
ISO 1635	Cu–FRHC	ISO	Tube	99.90 min **Cu+Ag**, total	**Annealed**: all **diam**, 35% in 2 in. (50 mm) **El**, 65 max **HV**; **HA**: 200 max mm **diam**, 250 MPa **TS**, 20% in 2 in. (50 mm) **El**, 80 **HV**; **Strain hardened (HB)**: 100 max mm **diam**, 290 MPa **TS**, 10% in 2 in. (50 mm) **El**, 100 **HV**
ISO 1635	Cu–DHP	ISO	Tube	0.013–0.050 **P**, 99.85 min **Cu+Ag**, total	**Annealed**: all **diam**, 35% in 2 in. (50 mm) **El**, 65 max **HV**; **HA**: 200 max mm **diam**, 250 MPa **TS**, 20% in 2 in. (50 mm) **El**, 80 **HV**; **Strain hardened (HB)**: 100 max mm **diam**, 290 MPa **TS**, 10% in 2 in. (50 mm) **El**, 100 **HV**
ISO 1634	Cu–DHP	ISO	Plate, sheet, strip	0.005–0.012 **P**, 99.90 min **Cu+Ag**, total	**HC**: all **diam**, 340 MPa **TS**, 4% in 2 in. (50 mm) **El**, 110 **HV**
ISO 1634	Cu–OF	ISO	Plate, sheet, strip	99.95 min **Cu+Ag**, total	**Strain hardened (HB)**: all **diam**, 280 MPa **TS**, 8% in 2 in. (50 mm) **El**, 100 **HV**
ISO 1634	Cu–FRTP	ISO	Plate, sheet, strip	99.85 **Cu+Ag**, total	**HA**: all **diam**, 250 MPa **TS**, 16% in 2 in. (50 mm) **El**, 80 **HV**

WROUGHT COPPER AND COPPER ALLOYS

specification number	designation	country	product forms	chemical composition	mechanical properties and hardness values
ISO 1634	Cu–FRHC	ISO	Plate, sheet, strip	99.90 min **Cu + Ag**, total	**M**: all **diam**, 220 MPa **TS**, 35% in 2 in. (50 mm) **El**, 75 max **HV**
ISO 1634	Cu ETP	ISO	Plate, sheet, strip	99.90 min **Cu + Ag**, total	**Annealed**: all **diam**, 30% in 2 in. (50 mm) **El**, 60 max **HV**
JIS C 2801	Class 2	Japan	Bar: commutator	0.15–0.25 **Ag**, 99.9 min **Cu + Ag**	**As manufactured**: <5 mm **diam**, 110 min **HV**
ANSI/ASTM B 152	C 10400	US	Sheet, strip, plate, bar	99.95 min **Cu**, 0.0272 min **Ag**, silver included in copper content	**Hard**: >0.020 in. (>0.508 mm) **diam**, 43 ksi (295 MPa) **TS**, 86–93 **HRF**; **Spring**: >0.020 in. (>0.508 mm) **diam**, 50 ksi (345 MPa) **TS**, 91–97 **HRF**; **Hot rolled**: >0.020 in. (>0.508 mm) **diam**, 30 ksi (205 MPa) **TS**, 75 max **HRF**
ANSI/ASTM B 152	C 10500	US	Sheet, strip, plate, bar	99.95 min **Cu**, 0.034 min **Ag**, silver included in copper content	**Hard**: >0.020 in. (>0.508 mm) **diam**, 43 ksi (295 MPa) **TS**, 86–93 **HRF**; **Spring**: >0.020 in. (>0.508 mm) **diam**, 50 ksi (345 MPa) **TS**, 91–97 **HRF**; **Hot rolled**: >0.020 in. (>0.508 mm) **diam**, 30 ksi (205 MPa) **TS**, 75 max **HRF**
ANSI/ASTM B 152	C 10700	US	Sheet, strip, plate, bar	99.95 min **Cu**, 0.085 min **Ag**, silver included in copper content	**Hard**: >0.020 in. (>0.508 mm) **diam**, 43 ksi (295 MPa) **TS**, 86–93 **HRF**; **Spring**: >0.020 in. (>0.508 mm) **diam**, 50 ksi (345 MPa) **TS**, 91–97 **HRF**; **Hot rolled**: >0.020 in. (>0.508 mm) **diam**, 30 ksi (205 MPa) **TS**, 75 max **HRF**
CDA C 107		US	Plate, sheet, strip, rod, wire, shapes, tube	99.95 min **Cu**, 0.085 **Ag**, silver included in copper content	**Hard (plate, sheet, strip)**: 0.040 in. (1.016 mm) **diam**, 50 ksi (345 MPa) **TS**, 45 ksi (310 MPa) **YS**, 6% in 2 in. (50 mm) **El**, 90 **HRF**; **Hard (rod)**: 1.0 in. (25.4 mm) **diam**, 48 ksi (331 MPa) **TS**, 44 ksi (303 MPa) **YS**, 16% in 2 in. (50 mm) **El**, 87 **HRF**; **Hard (shapes)**: 0.500 in. (12.7 mm) **diam**, 40 ksi (276 MPa) **TS**, 32 ksi (221 MPa) **YS**, 30% in 2 in. (50 mm) **El**, 35 **HRB**

WROUGHT COPPER AND COPPER ALLOYS

specification number	designation	country	product forms	chemical composition	mechanical properties and hardness values
CDA C 105		US	Plate, sheet, strip, rod, wire, shapes, tube	99.95 min **Cu**, 0.034 **Ag**, silver included in copper content	**Hard (plate, sheet, strip)**: 0.040 in. (1.016 mm) **diam**, 50 ksi (345 MPa) **TS**, 45 ksi (310 MPa) **YS**, 6% in 2 in. (50 mm) **El**, 90 **HRF**; **Hard (rod)**: 1.0 in. (25.4 mm) **diam**, 48 ksi (331 MPa) **TS**, 44 ksi (303 MPa) **YS**, 16% in 2 in. (50 mm) **El**, 87 **HRF**; **Hard (shapes)**: 0.500 in. (12.7 mm) **diam**, 40 ksi (276 MPa) **TS**, 32 ksi (221 MPa) **YS**, 30% in 2 in. (50 mm) **El**, 35 **HRB**
CDA C 104		US	Plate, sheet, strip, rod, wire, shapes, tube	99.95 min **Cu**, 0.027 **Ag**, silver included in copper content	**Hard (plate, sheet, strip)**: 0.040 in. (1.016 mm) **diam**, 50 ksi (345 MPa) **TS**, 45 ksi (310 MPa) **YS**, 6% in 2 in. (50 mm) **El**, 90 **HRF**; **Hard (rod)**: 1.0 in. (25.4 mm) **diam**, 48 ksi (331 MPa) **TS**, 44 ksi (303 MPa) **YS**, 16% in 2 in. (50 mm) **El**, 87 **HRF**; **Hard (shapes)**: 0.500 in. (12.7 mm) **diam**, 40 ksi (276 MPa) **TS**, 32 ksi (221 MPa) **YS**, 30% in 2 in. (50 mm) **El**, 35 **HRB**
COPANT 162	C 104 00 (=OFS)	COPANT	Wrought products	99.95 min **Cu**, 0.027 **Ag**, silver included in copper content	
ANSI/ASTM B 152	C 11400	US	Sheet, strip, plate, bar	99.90 min **Cu**, 0.034 min **Ag**, silver included in copper content	**Hard**: >0.020 in. (>0.508 mm) **diam**, 43 ksi (295 MPa) **TS**, 86–93 **HRF**; **Spring**: >0.020 in. (>0.508 mm) **diam**, 50 ksi (345 MPa) **TS**, 91–97 **HRF**; **Hot rolled**: >0.020 in. (>0.508 mm) **diam**, 30 ksi (205 MPa) **TS**, 75 max **HRF**
ANSI/ASTM B 152	C 11300	US	Sheet, strip, plate, bar	99.90 min **Cu**, 0.0272 min **Ag**, silver included in copper content	**Hard**: >0.020 in. (>0.508 mm) **diam**, 43 ksi (295 MPa) **TS**, 86–93 **HRF**; **Spring**: >0.020 in. (>0.508 mm) **diam**, 50 ksi (345 MPa) **TS**, 91–97 **HRF**; **Hot rolled**: >0.020 in. (>0.508 mm) **diam**, 30 ksi (205 MPa) **TS**, 75 max **HRF**
AS 1567	116A	Australia	Rod, bar, section	99.90 min **Cu**, 0.085 **Ag** min, 0.05 **O**, others	**Soft annealed**: 12–50 mm **diam**, 230 max MPa **TS**, 45% in 2 in. (50 mm) **El**; **1/2 hard**: 25–50 mm **diam**, 230 MPa **TS**, 22% in 2 in. (50 mm) **El**; **Hard**: 10–12 mm **diam**, 280 MPa **TS**, 8% in 2 in. (50 mm) **El**

WROUGHT COPPER AND COPPER ALLOYS

specification number	designation	country	product forms	chemical composition	mechanical properties and hardness values
CDA C 116		US	Plate, sheet, strip, rod, wire, shapes	99.90 min **Cu**, 0.085 **Ag**, 0.04 **O**, silver included in copper content	**Hard (plate, sheet, strip)**: 0.040 in. (1.016 mm) **diam**, 50 ksi (345 MPa) **TS**, 45 ksi (310 MPa) **YS**, 6% in 2 in. (50 mm) **El**, 90 **HRF**; **Hard (rod)**: 1.0 in. (25.4 mm) **diam**, 48 ksi (331 MPa) **TS**, 44 ksi (303 MPa) **YS**, 16% in 2 in. (50 mm) **El**, 87 **HRF**; **Hard (shapes)**: 0.500 in. (12.7 mm) **diam**, 40 ksi (276 MPa) **TS**, 32 ksi (221 MPa) **YS**, 30% in 2 in. (50 mm) **El**, 35 **HRB**
CDA C 115		US	Plate, sheet, strip, rod, wire, shapes	99.90 min **Cu**, 0.054 **Ag**, 0.04 **O**, silver included in copper content	**Hard (plate, sheet, strip)**: 0.040 in. (1.016 mm) **diam**, 50 ksi (345 MPa) **TS**, 45 ksi (310 MPa) **YS**, 6% in 2 in. (50 mm) **El**, 90 **HRF**; **Hard (rod)**: 1.0 in. (25.4 mm) **diam**, 48 ksi (331 MPa) **TS**, 44 ksi (303 MPa) **YS**, 16% in 2 in. (50 mm) **El**, 87 **HRF**; **Hard (shapes)**: 0.500 in. (12.7 mm) **diam**, 40 ksi (276 MPa) **TS**, 32 ksi (221 MPa) **YS**, 30% in 2 in. (50 mm) **El**, 35 **HRB**
CDA C 114		US	Plate, sheet, strip, rod, wire, shapes	99.90 min **Cu**, 0.034 **Ag**, 0.04 **O**, silver included in copper content	**Hard (plate, sheet, strip)**: 0.040 in. (1.016 mm) **diam**, 50 ksi (345 MPa) **TS**, 45 ksi (310 MPa) **YS**, 6% in 2 in. (50 mm) **El**, 90 **HRF**; **Hard (rod)**: 1.0 in. (25.4 mm) **diam**, 48 ksi (331 MPa) **TS**, 44 ksi (303 MPa) **YS**, 16% in 2 in. (50 mm) **El**, 87 **HRF**; **Hard (shapes)**: 0.500 in. (12.7 mm) **diam**, 40 ksi (276 MPa) **TS**, 32 ksi (221 MPa) **YS**, 30% in 2 in. (50 mm) **El**, 35 **HRB**
CDA C 113		US	Plate, sheet, strip, rod, wire, shapes	99.90 min **Cu**, 0.027 **Ag**, 0.04 **O**, silver included in copper content	**Hard (plate, sheet, strip)**: 0.040 in. (1.016 mm) **diam**, 50 ksi (345 MPa) **TS**, 45 ksi (310 MPa) **YS**, 6% in 2 in. (50 mm) **El**, 90 **HRF**; **Hard (rod)**: 1.0 in. (25.4 mm) **diam**, 48 ksi (331 MPa) **TS**, 44 ksi (303 MPa) **YS**, 16% in 2 in. (50 mm) **El**, 87 **HRF**; **Hard (shapes)**: 0.500 in. (12.7 mm) **diam**, 40 ksi (276 MPa) **TS**, 32 ksi (221 MPa) **YS**, 30% in 2 in. (50 mm) **El**, 35 **HRB**
COPANT 162	C 113 00 (=STP)	COPANT	Wrought products	99.90 min **Cu**, 0.027 min **Ag**, silver included in copper content	

WROUGHT COPPER AND COPPER ALLOYS

specification number	designation	country	product forms	chemical composition	mechanical properties and hardness values
SIS 14 50 30	SIS Cu 50 30–24	Sweden	Rod	99.90 min **Cu**, 0.08–0.12 **Ag**, **Ag** in **Cu** content	**Strain hardened**: all **diam**, 270 MPa **TS**, 220 MPa **YS**, 10% in 2 in. (50 mm) **El**, 85–110 **HV**
CDA C 130		US	Plate, sheet, strip, rod, wire, shapes	99.88 min **Cu**, 0.004 **Pb**, 0.003 **Sb**, 0.050 **Ni**, 0.085 **Ag**, 0.012 **As**, 0.003 **Bi**, silver included in copper content 0.025 **Te + Se**	**Hard (plate, sheet, strip)**: 0.040 in. (1.016 mm) **diam**, 50 ksi (345 MPa) **TS**, 45 ksi (310 MPa) **YS**, 6% in 2 in. (50 mm) **El**, 90 **HRF**; **Hard (rod)**: 1.0 in. (25.4 mm) **diam**, 48 ksi (331 MPa) **TS**, 44 ksi (303 MPa) **YS**, 16% in 2 in. (50 mm) **El**, 87 **HRF**; **Hard (shapes)**: 0.500 in. (12.7 mm) **diam**, 40 ksi (276 MPa) **TS**, 32 ksi (221 MPa) **YS**, 30% in 2 in. (50 mm) **El**, 35 **HRB**
CDA C 129		US	Plate, sheet, strip, rod, wire, shapes	99.88 min **Cu**, 0.004 **Pb**, 0.003 **Sb**, 0.050 **Ni**, 0.054 **Ag**, 0.012 **As**, 0.003 **Bi**, silver included in copper content 0.025 **Te + Se**	**Hard (plate, sheet, strip)**: 0.040 in. (1.016 mm) **diam**, 50 ksi (345 MPa) **TS**, 45 ksi (310 MPa) **YS**, 6% in 2 in. (50 mm) **El**, 90 **HRF**; **Hard (rod)**: 1.0 in. (25.4 mm) **diam**, 48 ksi (331 MPa) **TS**, 44 ksi (303 MPa) **YS**, 16% in 2 in. (50 mm) **El**, 87 **HRF**; **Hard (shapes)**: 0.500 in. (12.7 mm) **diam**, 40 ksi (276 MPa) **TS**, 32 ksi (221 MPa) **YS**, 30% in 2 in. (50 mm) **El**, 35 **HRB**
CDA C 128		US	Plate, sheet, strip, rod, wire, shapes	99.88 min **Cu**, 0.004 **Pb**, 0.003 **Sb**, 0.050 **Ni**, 0.034 min **Ag**, 0.012 **As**, 0.003 **Bi**, silver included in copper content 0.025 **Te + Se**	**Hard (plate, sheet, strip)**: 0.040 in. (1.016 mm) **diam**, 50 ksi (345 MPa) **TS**, 45 ksi (310 MPa) **YS**, 6% in 2 in. (50 mm) **El**, 90 **HRF**; **Hard (rod)**: 1.0 in. (25.4 mm) **diam**, 48 ksi (331 MPa) **TS**, 44 ksi (303 MPa) **YS**, 16% in 2 in. (50 mm) **El**, 87 **HRF**; **Hard (shapes)**: 0.500 in. (12.7 mm) **diam**, 40 ksi (276 MPa) **TS**, 32 ksi (221 MPa) **YS**, 30% in 2 in. (50 mm) **El**, 35 **HRB**

WROUGHT COPPER AND COPPER ALLOYS

specification number	designation	country	product forms	chemical composition	mechanical properties and hardness values
CDA C 127		US	Plate, sheet, strip, rod, wire, shapes	99.88 min **Cu**, 0.004 **Pb**, 0.003 **Sb**, 0.050 **Ni**, 0.027 min **Ag**, 0.012 **As**, 0.003 **Bi**, silver included in copper content 0.025 **Te+Se**	**Hard (plate, sheet, strip)**: 0.040 in. (1.016 mm) **diam**, 50 ksi (345 MPa) **TS**, 45 ksi (310 MPa) **YS**, 6% in 2 in. (50 mm) **El**, 90 **HRF**; **Hard (rod)**: 1.0 in. (25.4 mm) **diam**, 48 ksi (331 MPa) **TS**, 44 ksi (303 MPa) **YS**, 16% in 2 in. (50 mm) **El**, 87 **HRF**; **Hard (shapes)**: 0.500 in. (12.7 mm) **diam**, 40 ksi (276 MPa) **TS**, 32 ksi (221 MPa) **YS**, 30% in 2 in. (50 mm) **El**, 35 **HRB**
COPANT 162	C 114 00 (=STP)	COPANT	Wrought products	99.87 min **Cu**, 0.034 min **Ag**	
COPANT 162	C 107 00 (=OFS)	COPANT	Wrought products	99.87 min **Cu**, 0.085 min **Ag**	
COPANT 162	C 127 00 (=FRSTP)	COPANT	Wrought products	99.85 min **Cu**, 0.004 **Pb**, 0.003 **Sb**, 0.050 **Ni**, 0.027 min **Ag**, 0.012 **As**, 0.025 **Te**, 0.003 **Bi**	
COPANT 162	C 115 00 (=STP)	COPANT	Wrought products	99.85 min **Cu**, 0.054 min **Ag**	
COPANT 162	C 129 00 (=FRSTP)	COPANT	Wrought products	99.83 min **Cu**, 0.004 **Pb**, 0.003 **Sb**, 0.050 **Ni**, 0.054 min **Ag**, 0.012 **As**, 0.025 **Te**, 0.003 **Bi**	
COPANT 162	C 116 00 (=STP)	COPANT	Wrought products	99.82 min **Cu**, 0.085 min **Ag**	
COPANT 162	C 130 00 (=FRSTP)	COPANT	Wrought products	99.80 min **Cu**, 0.80 min **Sn**, 0.004 **Pb**, 0.003 **Sb**, 0.050 **Ni**, 0.085 min **Ag**, 0.012 **As**, 0.025 **Te**, 0.003 **Bi**	
COPANT 162	C 128 00 (=FRSTP)	COPANT	Wrought products	99.76 min **Cu**, 0.004 **Pb**, 0.003 **Sb**, 0.050 **Ni**, 0.034 **Ag**, 0.012 **As**, 0.025 **Te**, 0.003 **Bi**	
CDA C 155		US	Plate, sheet, strip	99.75 min **Cu**, 0.10–0.12 **Mn**, 0.040–0.080 **P**, 0.027–0.10 **Ag**, silver included in copper content	**Light anneal**: 0.040 in. (1.016 mm) **diam**, 40 ksi (276 MPa) **TS**, 18 ksi (124 MPa) **YS**, 34% in 2 in. (50 mm) **El**, 70 **HRF**; **Hard**: 0.040 in. (1.016 mm) **diam**, 62 ksi (427 MPa) **TS**, 57 ksi (393 MPa) **YS**, 5% in 2 in. (50 mm) **El**, 97 **HRF**; **Quarter hard**: 0.200 in. (5.08 mm) **diam**, 45 ksi (310 MPa) **TS**, 36 ksi (248 MPa) **YS**, 28% in 2 in. (50 mm) **El**, 89 **HRF**
COPANT 162	C 155 00	COPANT	Wrought products	99.65 min **Cu**, 0.04–0.08 **P**, 0.027–0.10 **Ag**, 0.08–0.13 **Mg**	

WROUGHT COPPER AND COPPER ALLOYS

specification number	designation	country	product forms	chemical composition	mechanical properties and hardness values
BS 2870	C 105	UK	Sheet, strip, foil	99.20 min **Cu**, 0.03 **Sn**, 0.02 **Pb**, 0.02 **Fe**, 0.01 **Sb**, 0.15 **Ni**, 0.30–0.50 **Ag**, 0.0050 **Bi**, 0.10 **O**, 0.030 **Si**+**Te**	**As manufactured or annealed**: 0.6–10.0 mm **diam**, 148 MPa **TS**, 35% in 2 in. (50 mm) **El**; **1/2 hard**: 0.6–1.3 mm **diam**, 172 MPa **TS**, 10% in 2 in. (50 mm) **El**; **Hard**: 0.6–2.7 mm **diam**, 217 MPa **TS**
COPANT 527	Cu As DMP	COPANT	Tube	98.9 min **Cu**, 0.015–0.050 **P**, 0.15–0.50 **Ag**, 99.5 min, total elements	**Tempered**: 0.8–3.4 mm **diam**, 245 MPa **TS**, 205 MPa **YS**; **1/2 tempered**: all in. (0.8–3.4 mm) **diam**, 310 MPa **TS**, 275 MPa **YS**
MIL–C–19311B	185	US	Forgings, rod, bar, strip	98.63 min **Cu**, 0.015 **Pb**, 0.40–1.0 **Cr**, 0.04 **P**, 0.08–0.12 **Ag**	**Solution heat treated and artificially aged**: ≤1.50 in. **diam**, 434 MPa **TS**, 386 MPa **YS**, 13% in 2 in. (50 mm) **El**, 72 **HRB**
JIS Z 3264	BCuP–5	Japan	Ribbon, wire, and bar: brazing filler	78.99 min **Cu**, 4.8–5.3 **P**, 14.5–15.5 **Ag**, 0.2 total, others	
CDA C 125		US	Plate, sheet, strip, rod, wire, shapes	99.88 min **Cu**, 0.004 **Pb**, 0.003 **Sb**, 0.050 **Ni**, 0.012 **As**, 0.003 **Bi**, silver included in copper content, 0.025 **Te**+**Se**	**Hard (plate, sheet, strip)**: 0.040 in. (1.016 mm) **diam**, 50 ksi (345 MPa) **TS**, 45 ksi (310 MPa) **YS**, 6% in 2 in. (50 mm) **El**, 90 **HRF**; **Hard (rod)**: 1.0 in. (25.4 mm) **diam**, 48 ksi (331 MPa) **TS**, 44 ksi (303 MPa) **YS**, 16% in 2 in. (50 mm) **El**, 87 **HRF**; **Hard (shapes)**: 0.500 in. (12.7 mm) **diam**, 40 ksi (276 MPa) **TS**, 32 ksi (221 MPa) **YS**, 30% in 2 in. (50 mm) **El**, 35 **HRB**
ANSI/ASTM B 75	C 14200	US	Tube: seamless	99.40 min **Cu**, 0.015–0.40 **P**, 0.15–0.50 **As**	**Drawn**: all **diam**, 36 ksi (250 MPa) **TS**, 30 **HR30T**; **Soft anneal**: all **diam**, 60 **HR15T**; **Light anneal**: all **diam**, 65 **HR15T**
ANSI/ASTM B 447	C 14200	US	Tube	99.40 min **Cu**, 0.015–0.040 **P**, 0.15–0.50 **As**, silver included in copper content	**Annealed (soft)**: all **diam**, 60 max **HR15T**; **As welded (from hard strip)**: all **diam**, 45 ksi (310 MPa) **TS**, 54–65 **HR30T**; **Welded and cold drawn (drawn)**: all **diam**, 36 ksi (250 MPa) **TS**, 30 min **HR30T**
ANSI/ASTM B 359	C 14200	US	Tube: seamless	99.40 min **Cu**, 0.015–0.040 **P**, 0.15–0.50 **As**, 99.95 min **Cu**+**P**+**As**	**Annealed**: all **diam**, 30 ksi (205 MPa) **TS**, 9 ksi (62 MPa) **YS**; **Light drawn**: all **diam**, 36 ksi (250 MPa) **TS**, 30 ksi (205 MPa) **YS**
ANSI/ASTM B 379	142	US	Shapes	99.4 min **Cu**, 0.015–0.040 **P**, 0.15–0.5 **As**, 99.9 min **Cu**+**P**+**As**	

WROUGHT COPPER AND COPPER ALLOYS

specification number	designation	country	product forms	chemical composition	mechanical properties and hardness values
ANSI/ASTM B 395	C 14200	US	Tube: seamless	99.40 min **Cu**, 0.015–0.040 **P**, 0.15–0.50 **As**, silver included in copper content	**Light drawn**: <0.048 in. (<1.21 mm) **diam**, 36 ksi (250 MPa) **TS**, 30 ksi (205 MPa) **YS**, 12% in 2 in. (50.8 mm) **El**
ANSI/ASTM B 11	C 14200	US	Plate	99.40 min **Cu**, 0.015–0.040 **P**, 0.15–0.50 **As**, silver included with copper content	**Hot rolled**: all **diam**, 31 ksi (215 MPa) **TS**, 35% in 8 in. (203 mm) **El**
ANSI/ASTM B 379	CN 142	US	Shapes	99.4 min **Cu**, 0.015–0.040 **P**, 0.15–0.5 **As**, 99.9 min **Cu + P + As**	
ANSI/ASTM B 111	C 14200	US	Tube: seamless, and stock: ferrule	99.40 min **Cu**, 0.015–0.040 **P**, 0.15–0.50 **As**	**Light drawn**: all **diam**, 36 ksi (250 MPa) **TS**, 30 ksi (205 MPa) **YS**; **Hard drawn**: 45 ksi (310 MPa) **TS**, 40 ksi (275 MPa) **YS**
CDA C 142		US	Plate, sheet, strip, tube, pipe	99.40 min **Cu**, 0.015–0.040 **P**, 0.15–0.50 **As**, silver included in copper content	**0.050 mm (plate, sheet, strip)**: 0.040 in. (1.016 mm) **diam**, 32 ksi (221 MPa) **TS**, 10 ksi (69 MPa) **YS**, 45% in 2 in. (50 mm) **El**, 40 **HRF**; **Hard drawn (tube)**: 1.0 in. (25.4 mm) **diam**, 55 ksi (379 MPa) **TS**, 50 ksi (345 MPa) **YS**, 8% in 2 in. (50 mm) **El**, 95 **HRF**; **Hard (pipe)**: 0.75 in. **diam**, 50 ksi (345 MPa) **TS**, 45 ksi (310 MPa) **YS**, 10% in 2 in. (60 mm) **El**, 90 **HRF**
COPANT 162	C 142 00 (= DPA)	COPANT	Wrought products	99.40 min **Cu**, 0.015–0.040 **P**, 0.15–0.50 **As**, silver included in copper content	
COPANT 162	C 141 00 (= ATP)	COPANT	Wrought products	99.40 min **Cu**, 0.15–0.50 **As**, silver included in copper content	
COPANT 528	Cu As TPDHP	COPANT	Tube	99.4 min **Cu**, 0.015–0.05 **P**, 0.15–0.50 **As**, silver included in copper content	**Annealed (regular)**: 0.45–9.7 mm **diam**, 60 MPa **YS**; **Tempered to 1/2 hard**: all in. (0.45–9.7 mm) **diam**, 245 MPa **TS**; **Tempered to full hard**: all in. (0.45–9.7 mm) **diam**, 275 MPa **YS**
MIL–T–24107A	142	US	Tube: seamless	99.40 min **Cu**, 0.015–0.040 **P**, 0.15–0.50 **As**, silver included in copper content	**Annealed**: all **diam**, 207 MPa **TS**, 62 MPa **YS**
ANSI/ASTM B 12	C 142	US	Rod	99.38 min **Cu**, 0.15–0.50 **As**, 99.88 min **Cu + Ag + As**	**Hot and/or cold worked**: all **diam**, 31 ksi (215 MPa) **TS**, 40% in 1 in. (25.4 mm) **El**
BS 2870	C 107	UK	Sheet, strip, foil	99.20 min **Cu**, 0.01 **Sn**, 0.010 **Pb**, 0.030 **Fe**, 0.01 **Sb**, 0.15 **Ni**, 0.013–0.050 **P**, 0.30–0.50 **As**, 0.0030 **Bi**, 0.020 **Se + Te**, 0.010 **Te**, 0.07 total, others	**As manufactured or annealed**: 0.6–10.0 mm **diam**, 148 MPa **TS**, 35% in 2 in. (50 mm) **El**; **1/2 hard**: 0.6–1.3 mm **diam**, 172 MPa **TS**, 10% in 2 in. (50 mm) **El**; **Hard**: 0.6–2.7 mm **diam**, 217 MPa **TS**

WROUGHT COPPER AND COPPER ALLOYS

specification number	designation	country	product forms	chemical composition	mechanical properties and hardness values
BS 2875	C 107	UK	Plate	99.20 min **Cu**, 0.01 **Sn**, 0.010 **Pb**, 0.030 **Fe**, 0.01 **Sb**, 0.15 **Ni**, 0.013–0.050 **P**, 0.30–0.50 **As**, 0.0030 **Bi**, 0.020 **Se+Te**, 0.010 **Te**, 0.07 total, others	**As manufactured or annealed**: ≥10 mm **diam**, 211 MPa **TS**, 35% in 2 in. (50 mm) **El**; **Hard**: 10–16 mm **diam**, 279 MPa **TS**, 15% in 2 in. (50 mm) **El**
BS 2875	C 105	UK	Plate	99.20 min **Cu**, 0.03 **Sn**, 0.02 **Pb**, 0.02 **Fe**, 0.01 **Sb**, 0.15 **Ni**, 0.30–0.50 **As**, 0.0050 **Bi**, 0.010 **O**, 0.030 **Se+Te**	**As manufactured or annealed**: ≥10 mm **diam**, 221 MPa **TS**, 35% in 2 in. (50 mm) **El**; : 10–16 mm **diam**, 279 MPa **TS**, 15% in 2 in. (50 mm) **El**
VSM 11557 (Part 1)	Cu–DPA	Switzerland	Tube	99.2 min **Cu**, 0.013–0.050 **P**, 0.15–0.50 **As**	**1/2 hard**: all **diam**, 245 MPa **TS**, 147 MPa **YS**, 25% in 2 in. (50 mm) **El**, 70 **HB**
ISO 1637	CuAs(P)	ISO	Solid products: straight lengths	99.14 min **Cu**, 0.013–0.050 **P**, 0.15–0.50 **As**, 0.3 **P** + others, total	**Annealed**: 5 mm **diam**; 35% in 2 in. (50 mm) **El**, 60 max **HV**; **HA**: 5–40 mm **diam**, 250 MPa **TS**, 15% in 2 in. (50 mm) **El**, 90 **HV**; **HB – strain hardened**: 5–20 mm **diam**, 280 MPa **TS**, 5% in 2 in. (50 mm) **El**, 105 **HV**
ISO 1640	CuAs(P)	ISO	Forgings	99.14 min **Cu**, 0.013–0.050 **P**, 0.15–0.50 **As**, 0.3 **P** + others, total	**As cast**: all **diam**, 220 MPa **TS**, 40% in 2 in. (50 mm) **El**, 80 **HV**
ISO/R1336	CuAs(P) [Cu–DPA]	ISO	Plate, sheet, rod, bar, tube, forgings	99.14 min **Cu**, 0.013–0.050 **P**, 0.15–0.50 **As**, 0.3 total, others	
ISO 1634	CuAs(P)	ISO	Plate, sheet, strip	99.14 min **Cu**, 0.013–0.050 **P**, 0.15–0.50 **As**, 0.3 **P** + others, total	**Annealed**: all **diam**, 30% in 2 in. (50 mm) **El**, 60 max **HV**; **M**: all **diam**, 220 MPa **TS**, 35% in 2 in. (50 mm) **El**, 75 max **HV**; **Strain hardened**: all **diam**, 280 MPa **TS**, 8% in 2 in. (50 mm) **El**, 100 **HV**
ISO 1635	CuAs(P)	ISO	Tube	99.14 min **Cu**, 0.013–0.050 **P**, 0.15–0.50 **As**, 0.3 **P** + others, total	**Annealed**: all **diam**, 35% in 2 in. (50 mm) **El**, 65 max **HV**; **HA**: all **diam**, 250 MPa **TS**, 20% in 2 in. (50 mm) **El**, 80 **HV**; **Strain hardened (HB)**: all **diam**, 290 MPa **TS**, 10% in 2 in. (50 mm) **El**, 100 **HV**
IS 5743	Cu As 10	India	Bar, ingot, shot	88.5 **Cu**, 9.0–11.0 **As**	
IS 6912	Phosphorus deoxidized arsenical copper	India	Forgings	0.01 **Sn**, 0.010 **Pb**, 0.03 **Fe**, 0.01 **Sb**, 0.15 **Ni**, 0.02–0.10 **P**, 0.003 **Bi**, 0.01 **Te**, 0.20–050 **As**, 0.02 **Te+Se**, 99.20 min **Cu+Ag**, 0.07 total, others	**Annealed**: ≥6 mm **diam**, 215 MPa **TS**, 55% in 2 in. (50 mm) **El**

WROUGHT COPPER AND COPPER ALLOYS

specification number	designation	country	product forms	chemical composition	mechanical properties and hardness values
CDA C143		US	Plate, sheet, strip	99.90 min **Cu**, 0.05–0.15 **Cd**	**Quarter hard:** 0.040 in. (1.016 mm) **diam**, 40 ksi (276 MPa) **TS**, 32 ksi (221 MPa) **YS**, 25% in 2 in. (50 mm) **El**, 70 **HRF**; **Hard**: 0.040 in. (1.016 mm) **diam**, 52 ksi (358 MPa) **TS**, 47 ksi (324 MPa) **YS**, 6% in 2 in. (50 mm) **El**, 91 **HRF**; **Spring**: 0.040 in. (1.016 mm) **diam**, 58 ksi (400 MPa) **TS**, 54 ksi (372 MPa) **YS**, 4% in 2 in. (50 mm) **El**, 95 **HRF**
COPANT 162	C 143 00	COPANT	Wrought products	99.90 min **Cu**, 0.05–0.15 **Cd**, silver included in copper content	
COPANT 162	C 162 00	COPANT	Wrought products	99.8 min **Cu**, 0.02 **Fe**, 0.7–1.2 **Cd**, silver included in copper content	
COPANT 162	C 164 00	COPANT	Wrought products	99.8 min **Cu**, 0.20–0.40 **Sn**, 0.02 **Fe**, 0.6–0.9 **Cd**	
COPANT 162	C 165 00	COPANT	Wrought products	99.8 min **Cu**, 0.5–0.7 **Sn**, 0.02 **Fe**, 0.6–1.0 **Cd**	
CDA C162		US	Plate, sheet, strip, rod, wire	99.0 min **Cu**, 0.02 **Fe**, 0.7–1.2 **Cd**, 99.8 min **Cu + Ag + Cd + Fe**	**Hard (plate, sheet, strip)**: 0.040 in. (1.016 mm) **diam**, 60 ksi (414 MPa) **TS**, 45 ksi (310 MPa) **YS**, 5% in 2 in. (50 mm) **El**, 64 **HRB**; **Hard (rod)**: 0.500 in. (12.7 mm) **diam**, 73 ksi (503 MPa) **TS**, 70 ksi (483 MPa) **YS**, 9% in 2 in. (50 mm) **El**, 73 **HRB**; **Hard (wire)**: 0.080 in. **diam**, 70 ksi (483 MPa) **TS**, 55 ksi (379 MPa) **YS**, 6% in 2 in. (50 mm) **El**
AS 1567	162B	Australia	Rod, bar, section	98.75 min **Cu**, 0.5–1.2 **Cd**, others	**Soft annealed**: >6 mm **diam**; **1/2 hard**: 12–25 mm **diam**, 290 MPa **TS**, 15% in 2 in. (50 mm) **El**
BS 2873	C108	UK	Wire	98.75 min **Cu**, 0.5–1.2 **Cd**, 0.05 total, others	
BS 2875	C108	UK	Plate	98.75 min **Cu**, 0.5–1.2 **Cd**, 0.05 total, others	**Hard**: 10–16 mm **diam**, 309 MPa **TS**, 13% in 2 in. (50 mm) **El**

WROUGHT COPPER AND COPPER ALLOYS

specification number	designation	country	product forms	chemical composition	mechanical properties and hardness values
CDA C165		US	Plate, sheet, strip, rod, wire, bar	98.6 min **Cu**, 0.50–0.7 **Sn**, 0.02 **Fe**, 0.6–1.0 **Cd**, 99.8 min **Cu + Ag + Cd + Sn + Fe**	**Hard (plate, sheet, strip)**: 0.040 in. (1.016 mm) **diam**, 66 ksi (455 MPa) **TS**, 5% in 2 in. (50 mm) **El**, 79 **HRB**; **Hard (rod)**: 0.500 in. (12.7 mm) **diam**, 65 ksi (441 MPa) **TS**, 55 ksi (379 MPa) **YS**, 15% in 2 in. (50 mm) **El**, 75 **HRB**; **Hard (wire)**: 0.080 in. (2.03 mm) **diam**, 95 ksi (655 MPa) **TS**
SIS 14 50 55	SIS Cu 50 55–25	Sweden	Sections: for overhead lines	98.39 min **Cu**, 0.7–1.3 **Cd**, 0.3 total, others	**Strain annealed**: 80 mm **diam**, 431 MPa **TS**, 370 MPa **YS**, 2.5% in 8 in. (200 mm) **El**
ISO 1637	CuCd1	ISO	Solid products: straight lenghts	98.39 min **Cu**, 0.7–1.3 **Cd**, 0.3 total, others	**HA**: 18–30 mm **diam**, 350 MPa **TS**, 10% in 2 in. (50 mm) **El**, 110 **HV**; **HB – strain hardened**: 5–18 mm **diam**, 410 MPa **TS**, 8% in 2 in. (50 mm) **El**, 125 **HV**
ISO 1638	CuCd1	ISO	Solid products: drawn, on coils or reels	98.39 min **Cu**, 0.7–1.3 **Cd**, 0.3 total, others	**HC**: 1–5 mm **diam**, 490 MPa **TS**; **HD – strain hardened**: 1–3 mm **diam**, 590 MPa **TS**
ISO/R1336	CuCd1	ISO	Rod, bar, wire	98.39 min **Cu**, 0.7–1.3 **Cd**, 0.3 total, others	
ANSI/ASTM B 124	C 14700	US	Rod, bar, shapes	99.90 min **Cu**, 0.20–0.50 **S**, silver and sulfur included in copper content	
ANSI/ASTM B 283	C 14700	US	Did forgings	99.90 min **Cu**, 0.20–0.50 **S**, silver and sulfur included in copper content	
ANSI/ASTM B 301	C 14700	US	Rod, bar, shapes	99.4 min **Cu**, 0.20–0.50 **S**, silver included in copper	**Half hard (rod)**: 0.063–0.25 in. (1.59–6.35 mm) **diam**, 38 ksi (260 MPa) **TS**, 30 ksi (205 MPa) **YS**, 8% in 1 in. (25.4 mm) **El**; **Hard (rod)**: 0.063–0.25 in. (1.59–6.35 mm) **diam**, 48 ksi (330 MPa) **TS**, 40 ksi (275 MPa) **YS**, 4% in 1 in. (25.4 mm) **El**; **Hard (bar)**: 0.188–0.375 in. (4.78–9.52 mm) **diam**, 42 ksi (290 MPa) **TS**, 35 ksi (240 MPa) **YS**, 10% in 1 in. (25.4 mm) **El**

WROUGHT COPPER AND COPPER ALLOYS

specification number	designation	country	product forms	chemical composition	mechanical properties and hardness values
ANSI/ASTM B 301	C 14710	US	Rod, bar	99.3 min **Cu**, 0.05 **Pb**, 0.010–0.030 **P**, 0.05–0.15 **S**, silver included with copper	**Half hard (rod)**: 0.063–0.25 in. (1.59–6.35 mm) **diam**, 38 ksi (260 MPa) **TS**, 30 ksi (205 MPa) **YS**, 8% in 1 in. (25.4 mm) **El**; **Hard (rod)**: 0.063–0.25 in. (1.59–6.35 mm) **diam**, 48 ksi (330 MPa) **TS**, 40 ksi (275 MPa) **YS**, 4% in 1 in. (25.4 mm) **El**; **Hard (bar)**: 0.188–0.375 in. (4.78–9.52 mm) **diam**, 42 ksi (290 MPa) **TS**, 35 ksi (240 MPa) **YS**, 10% in 1 in. (25.4 mm) **El**
ANSI/ASTM B 301	C 14720	US	Rod, bar	98.9 min **Cu**, 0.10 **Pb**, 0.010–0.030 **P**, 0.20–0.50 **S**, 99.50 min **Cu+Ag+Pb+P+S**	**Half hard (rod)**: 0.063–0.25 in. (1.59–6.35 mm) **diam**, 38 ksi (260 MPa) **TS**, 30 ksi (205 MPa) **YS**, 8% in 1 in. (25.4 mm) **El**; **Hard (rod)**: 0.063–0.25 in. (1.59–6.35 mm) **diam**, 48 ksi (330 MPa) **TS**, 40 ksi (275 MPa) **YS**, 4% in 1 in. (25.4 mm) **El**; **Hard (bar)**: 0.188–0.375 in. (4.78–9.52 mm) **diam**, 42 ksi (290 MPa) **TS**, 35 ksi (240 MPa) **YS**, 10% in 1 in. (25.4 mm) **El**
BS 2874	C111	UK	Rod, section	99.19 min **Cu**, 0.3–0.6 **S**, 0.20 total, others	**Annealed**: ≤6 mm **diam**, 216 MPa **TS**, 28% in 2 in. (50 mm) **El**; **As manufactured**: 6 mm **diam**, 260 MPa **TS**, 8% in 2 in. (50 mm) **El**
AS 1567–1974	147B	Australia	Rod, bar, section	99.80 min **Cu**, 0.30–0.60 **S**, others	**Soft annealed**: >6 mm **diam**, 210 MPa **TS**, 28% in 2 in. (50 mm) **El**; **As manufactured**: 6–50 mm **diam**, 260 MPa **TS**, 8% in 2 in. (50 mm) **El**
CDA C 147		US	Plate, sheet, strip, rod	99.6 min **Cu**, 0.20–0.50 **S**, 99.90 min **Cu+Ag+S**	**Half hard (rod)**: 0.500 in. (1.016 mm) **diam**, 43 ksi (296 MPa) **TS**, 40 ksi (276 MPa) **YS**, 20% in 2 in. (50 mm) **El**, 43 **HRB**; **Hard (rod)**: 1.000 in. (25.4 mm) **diam**, 46 ksi (317 MPa) **TS**, 43 ksi (296 MPa) **YS**, 11% in 2 in. (50 mm) **El**, 46 **HRB**; **Extra hard (rod)**: 0.375 in. (9.525 mm) **diam**, 57 ksi (393 MPa) **TS**, 55 ksi (379 MPa) **YS**, 8% in 2 in. (50 mm) **El**

WROUGHT COPPER AND COPPER ALLOYS

specification number	designation	country	product forms	chemical composition	mechanical properties and hardness values
COPANT 162	C 14700	COPANT	Wrought products	99.4 min **Cu**, 0.20–0.50 **S**, silver included in copper content	
ISO/R1336	CuS(P0.01)	ISO	Rod, bar	99.378 min **Cu**, 0.004–0.012 **P**, 0.20–0.50 **S**, 0.1 total, others (incl. **P**)	
ISO 1637	CuS(P0.1)	ISO	Solid products: straight lenghts	99.378 min **Cu**, 0.004–0.012 **P**, 0.20–0.50 **S**, 0.1 **P** + others, total	**Annealed**: 5 mm **diam**, 28% in 2 in. (50 mm) **El**, 70 max **HV**; **HA**: 5–40 mm **diam**, 250 MPa **TS**, 10% in 2 in. (50 mm) **El**, 90 **HV**
ISO 1637	CuS(P0.03)	ISO	Solid products: straight lenghts	99.34 min **Cu**, 0.013–0.050 **P**, 0.20–0.50 **S**, 0.1 **P** + others, total	**Annealed**: 5 mm **diam**, 28% in 2 in. (50 mm) **El**, 70 max **HV**; **HA**: 5–40 mm **diam**, 250 MPa **TS**, 10% in 2 in. (50 mm) **El**, 90 **HV**
ISO/R1336	CuS(P0.03)	ISO	Rod, bar	99.34 min **Cu**, 0.013–0.050 **P**, 0.20–0.50 **S**, 0.1 total, others	
CDA C 150		US	Plate, sheet, strip, rod, wire	99.80 min **Cu**, 0.10–0.20 **Zr**	**Solution heat treated and aged (rod)**: 1.000 in. (25.4 mm) **diam**, 62 ksi (427 MPa) **TS**, 60 ksi (414 MPa) **YS**, 15% in 2 in. (50 mm) **El**; **Solution heat treated and aged (wire)**: 0.090 in. (2.286 mm) **diam**, 30 ksi (207 MPa) **TS**, 13 ksi (90 MPa) **YS**, 49% in 2 in. (50 mm) **El**; **Mill annealed (wire)**: 0.250 in. (6.35 mm) **diam**, 37 ksi (255 MPa) **TS**, 11 ksi (76 MPa) **YS**, 50% in 2 in. (50 mm) **El**, 40 **HRF**
COPANT 162	C 150 00	COPANT	Wrought products	99.8 min **Cu**, 0.10–0.20 **Zr**, silver included in copper content	
COPANT 533	Cu Be 2 (Gr 1)	COPANT	Strip, sheet, plate, bar	99.5 min **Cu**, 0.20 min **Ni+Co**, 1.8–2.0 **Be**, 0.60 **Ni+Co+Fe**	**Hot or cold finished and annealed**: all **diam**, 410 MPa **TS**, 35% in 2 in. (50 mm) **El**, 45–78 **HRB**; **Hot or cold finished to full hard**: all **diam**, 690 MPa **TS**, 2% in 2 in. (50 mm) **El**, 96–102 **HRB**; **Precipitation hardened to full hard**: all **diam**, 1315 MPa **TS**, 39 **HRC**

WROUGHT COPPER AND COPPER ALLOYS

specification number	designation	country	product forms	chemical composition	mechanical properties and hardness values
COPANT 533	Cu Be 2 (Gr 2)	COPANT	Strip, sheet, plate, bar	99.5 min **Cu**, 0.2 min **Ni + Co**, 1.6–1.79 **Be**, 0.60 **Ni + Co + Fe**	**Hot or cold finished and annealed**: all **diam**, 410 MPa **TS**, 35% in 2 in. (50 mm) **El**, 45–78 **HRB**; **Hot or cold finished to full hard**: all **diam**, 690 MPa **TS**, 2% in 2 in. (50 mm) **El**, 96–102 **HRB**; **Precipitation hardened to full hard**: all **diam**, 1315 MPa **TS**, 40 **HRC**
COPANT 162	C 170 00	COPANT	Wrought products	99.5 **Cu**, 0.20 min **Ni + Co**, 1.6–1.79 **Be**, 0.60 **Ni + Fe + Co**	
COPANT 162	C 172 00	COPANT	Wrought products	99.5 **Cu**, 0.20 min **Ni + Co**, 1.8–2.0 **Be**, 0.60 **Ni + Fe + Co**	
COPANT 162	C 173 00	COPANT	Wrought products	99.5 **Cu**, 0.2–0.6 **Pb**, 0.20 min **Ni + Co**, 1.8–2.0 **Be**, 0.60 **Ni + Fe + Co**	
COPANT 162	C 175 00	COPANT	Wrought products	99.5 **Cu**, 0.10 **Fe**, 2.4–2.7 **Co**, 0.4–0.7 **Be**	
COPANT 162	C 176 00	COPANT	Wrought products	99.5 **Cu**, 0.10 **Fe**, 1.4–1.7 **Co**, 0.25–0.50 **Be**, 0.9–1.1 **Ag**	
COPANT 162	C 177 00	COPANT	Wrought products	99.5 **Cu**, 0.10 **Fe**, 2.4–2.7 **Co**, 0.4–0.7 **Be**, 0.4–0.6 **Te**	
ANSI/ASTM B 534		US	Plate, sheet, strip, bar	99.5 **Cu**, 0.10 **Fe**, 2.4–2.7 **Co**, 0.40–0.7 **Be**, beryllium and cobalt included in copper content	**Hard**: >0.045 in. (>1.14 mm) **diam**, 70 ksi (486 MPa) **TS**, 20% in 2 in. (50 mm) **El**, 45 max **HRB**; **Solution and precipitation heat treated**: >0.045 in. (>1.14 mm) **diam**, 100 ksi (689 MPa) **TS**, 8% in 2 in. (50 mm) **El**, 92–100 max **HRB**; **Hard, then precipitation heat treated**: >0.045 in. (>1.14 mm) **diam**, 110 ksi (758 MPa) **TS**, 5% in 2 in. (50 mm) **El**, 95–102 mmax **HRB**
NF A 51-109	Cu Be 1.7	France	Sheet, bar, and strip	99.05 min **Cu**, 0.25 **Ni + Co**, 1.70 **Be**	**Soft**: ≤10 mm **diam**, 410 MPa **TS**, 35% in 2 in. (50 mm) **El**, 90–130 **HV**; **1/2 hard**: ≤10 mm **diam**, 580 MPa **TS**, 6% in 2 in. (50 mm) **El**, 180–220 **HV**; **Hard**: ≤10 mm **diam**, 690 MPa **TS**, 2% in 2 in. (50 mm) **El**, 215–255 **HV**
ISO 1637	CuCo2Be	ISO	Solid products: straight lenghts	98.49 min **Cu**, 2.0–2.8 **Co**, 0.4–0.7 **Be**, 0.5 **Ni + Fe**, 0.5 total, others	**Solution heat treated and artificially aged**: 60 max mm **diam**, 700 MPa **TS**, 500 MPa **YS**, 8% in 2 in. (50 mm) **El**, 195 **HV**

WROUGHT COPPER AND COPPER ALLOYS

specification number	designation	country	product forms	chemical composition	mechanical properties and hardness values
CDA C 170		US	Plate, sheet, strip, rod	98.3 min **Cu**, 0.20 min **Ni+Co**, 1.60–1.79 **Be**, 0.6 **Ni+Fe+Co**	**Hard (plate, sheet, strip)**: <0.188 in. (<4.775 mm) **diam**, 110 ksi (758 MPa) **TS**, 104 ksi (717 MPa) **YS**, 5% in 2 in. (50 mm) **El**, 99 **HRB**; **Precipitation heat treated (plate, sheet, strip)**: <0.188 in. (<4.775 mm) **diam**, 190 ksi (1310 MPa) **TS**, 170 ksi (1172 MPa) **YS**, 3% in 2 in. (50 mm) **El**, 40 **HRC**; **Solution heat treated (rod)**: all **diam**, 68 ksi (469 MPa) **TS**, 62 **HRB**
CDA C 172		US	Plate, sheet, strip, rod, wire	98.1 min **Cu**, 0.20 min **Ni+Co**, 1.80–2.00 **Be**, 0.6 **Ni+Fe+Co**	**Hard (plate, sheet, strip)**: <0.188 in. (<4.775 mm) **diam**, 110 ksi (758 MPa) **TS**, 104 ksi (717 MPa) **YS**, 5% in 2 in. (50 mm) **El**, 99 **HRB**; **Solution heat treated (rod)**: all **diam**, 68 ksi (469 MPa) **TS**, 25 ksi (173 MPa) **YS**, 48% in 2 in. (50 mm) **El**, 62 **HRB**; **Solution heat treated (wire)**: all **diam**, 68 ksi (469 MPa) **TS**, 28 ksi (193 MPa) **YS**, 35% in 2 in. (50 mm) **El**
NF A 51-109	Cu Be 1.9	France	Sheet, bar, and strip	97.85 min **Cu**, 0.25 **Ni+Co**, 1.90 **Be**	**Soft**: $\leq$10 mm **diam**, 410 MPa **TS**, 35% in 2 in. (50 mm) **El**, 90–130 **HV**; **1/2 hard**: $\leq$10 mm **diam**, 580 MPa **TS**, 6% in 2 in. (50 mm) **El**, 180–220 **HV**; **Hard**: $\leq$10 mm **diam**, 690 MPa **TS**, 2% in 2 in. (50 mm) **El**, 215–255 **HV**
CDA C 173		US	Plate, sheet, strip, rod, wire	97.7 min **Cu**, 0.20–0.6 **Pb**, 0.20 min **Ni+Co**, 1.80–2.00 **Be**, 0.6 **Ni+Fe+Co**.	**Hard (plate, sheet, strip)**: <0.188 in. (<4.775 mm) **diam**, 110 ksi (758 MPa) **TS**, 104 ksi (717 MPa) **YS**, 5% in 2 in. (50 mm) **El**, 99 **HRB**; **Solution heat treated (rod)**: all **diam**, 68 ksi (469 MPa) **TS**, 25 ksi (173 MPa) **YS**, 48% in 2 in. (50 mm) **El**, 62 **HRB**; **Solution heat treated (wire)**: all **diam**, 68 ksi (469 MPa) **TS**, 28 ksi (193 MPa) **YS**, 35% in 2 in. (50 mm) **El**

WROUGHT COPPER AND COPPER ALLOYS

specification number	designation	country	product forms	chemical composition	mechanical properties and hardness values
BS 2870	CB 101	UK	Sheet, strip, foil	97.19 min **Cu**, 0.05–0.40 **Ni+Co**, 1.7–1.9 **Be**, 0.50 total, others	**Solution heat treated**: ≤10.0 mm **diam**, 293 MPa **TS**, 40% in 2 in. (50 mm) **El**, 85–120 **HV**; **Solution heat and precipitation treated**: ≤10.0 mm **diam**, 350 **HV**; **Solution heat treated, cold worked and precipitation treated to 1/2 hard**: ≤10.0 mm **diam**, 350 **HV**
BS 2873	CB101	UK	Wire	97.19 min **Cu**, 0.05–0.40 **Ni+Co**, 1.7–1.9 **Be**, 0.50 total, others	**Solution heat treated**: 0.5–10.0 mm **diam**, 382 MPa **TS**, 30% in 2 in. (50 mm) **El**; **Solution heat treated and cold worked**: ≤3.0 mm **diam**, 755 MPa **TS**; **Solution heat treated and precipitatioin treated**: 0.5–10.0 mm **diam**, 1029 MPa **TS**
QQ-C-533B	Copper alloy No. 170	US	Strip	97.11 min **Cu**, 0.20 **Co**, 0.20 min **Ni**, 1.60–1.79 **Be**, 0.6 **Ni+Co+Fe**, total	**Cold rolled and solution heat treated**: all **diam**, 60 ksi (414 MPa) **TS**, 35% in 2 in. (50 mm) **El**, 45–78 **HRB**; **Hot or cold rolled, solution treated and cold rolled 1/4 hard**: all **diam**, 75 ksi (517 MPa) **TS**, 10% in 2 in. (50 mm) **El**, 68–90 **HRB**; **One half hard**: all **diam**, 85 ksi (586 MPa) **TS**, 5% in 2 in. (50 mm) **El**, 88–96 **HRB**
ANSI/ASTM B 196	CA 170	US	Rod, bar	97.1 min **Cu**, 0.20 min **Ni+Co**, 1.60–1.79 **Be**, 0.6 **Ni+Co+Fe**	**Hard**: <0.375 in.(<9.53 mm) **diam**, 95 ksi (660 MPa) **TS**, 92–103 **HRB**; **Solution and precipitation heat treated**: all **diam**, 150 ksi (1030 MPa) **TS**, 32–39 **HRC**; **Hard, then precipitation heat treated**: <0.375 in.(<9.53) mm) **diam**, 175 ksi (1210 MPa) **TS**, 36–41 **HRC**

WROUGHT COPPER AND COPPER ALLOYS

specification number	designation	country	product forms	chemical composition	mechanical properties and hardness values
ANSI/ASTM B 194	CA 17000	US	Plate, sheet, strip, bar	97.1 min **Cu**, 0.20 min **Ni + Co**, 1.60–1.79 **Be**, 0.6 **Ni + Co + Fe**	**Hard**: >0.032 in.(>0.813 mm) **diam**, 100 ksi (690 MPa) **TS**, 2% in 2 in. (50 mm) **El**, 96–102 **HRB**; **Solution and precipitation heat treated**: >0.032 in. (>0.813 mm) **diam**, 150 ksi (1030 MPa) **TS**, 130 ksi (890 MPa) **YS**, 33 min **HRC**; **Hard, then precipitation heat treated**: >0.032 in.(>0.813) mm) **diam**, 180 ksi (1240 MPa) **TS**, 155 ksi (1070 MPa) **YS**, 39 min **HRC**
ASTM B 570	CA 170	US	Forgings	97.1 min **Cu**, 0.20 min **Ni + Co**, 1.60–1.79 **Be**, 0.6 **Ni + Co + Fe**	**As manufactured**: all **diam**, 60 ksi (410 MPa) **TS**, 45–85 **HRB**; **Precipitation heat treated**: all **diam**, 150 ksi (1030 MPa) **TS**, 32–39 **HRB**
JIS H 3130	C1700	Japan	Sheet, plate, and strip: for spring	97.1 min **Cu**, 0.20 **Ni + Co**, 1.6–1.8 **Be**	**Annealed**: 0.16–1.6 mm **diam**, 412 MPa **TS**, 35% in 2 in. (50 mm) **El**, 90–160 **HV**; **1/2 hard**: 0.16–1.6 mm **diam**, 588 MPa **TS**, 5% in 2 in. (50 mm) **El**, 180–240 **HV**; **Hard**: 0.16–1.6 mm **diam**, 686 MPa **TS**, 2% in 2 in. (50 mm) **El**, 210–270 **HV**
ISO 1638	CuBe1.7	ISO	Solid products: drawn, on coils or reels	96.99 min **Cu**, 0.20–0.60 **Ni + Co**, 1.6–1.8 **Be**, 0.20–0.60 **Co + Ni + Fe**, total	**Solution heat treated and naturally aged**: 1–5 mm **diam**, 390 MPa **TS**, 30% in 2 in. (50 mm) **El**; **TD**: 1–3 mm **diam**, 780 MPa **TS**; **TH**: 1–3 mm **diam**, 1230 MPa **TS**
ANSI/ASTM B 197		US	Wire	96.9 min **Cu**, 0.20 min **Ni + Co**, 1.80–2.00 **Be**, 0.6 **Ni + Co + Fe**	**Annealed**: all **diam**, 58 ksi (400 MPa) **TS**; **Solution and precipitation heat treated**: all **diam**, 160 ksi (1103 MPa) **TS**; **Half hard**: all **diam**, 185 ksi (1276 MPa) **TS**

WROUGHT COPPER AND COPPER ALLOYS

specification number	designation	country	product forms	chemical composition	mechanical properties and hardness values
ANSI/ASTM B 196	CA 173	US	Rod, bar	96.9 min **Cu**, 0.20–0.6 **Pb**, 0.20 min **Ni+Co**, 1.80–2.00 **Be**, 0.6 **Ni+Co+Fe**	**Hard**: <0.375 in.(<9.53 mm) **diam**, 95 ksi (660 MPa) **TS**, 92–103 **HRB**; **Solution and precipitation heat treated**: all **diam**, 165 ksi (1140 MPa) **TS**, 36–40 **HRC**; **Hard, then precipitation heat treated**: <0.375 in.(<9.53) mm) **diam**, 185 ksi (1280 MPa) **TS**, 39–45 **HRC**
ANSI/ASTM B 196	CA 172	US	Rod, bar	96.9 min **Cu**, 0.20 min **Ni+Co**, 1.80–2.00 **Be**, 0.6 **Ni+Co+Fe**	**Hard**: <0.375 in.(<9.53 mm) **diam**, 95 ksi (660 MPa) **TS**, 92–103 **HRB**; **Solution and precipitation heat treated**: all **diam**, 165 ksi (1140 MPa) **TS**, 36–40 **HRC**; **Hard, then precipitation heat treated**: <0.375 in.(<9.53) mm) **diam**, 185 ksi (1280 MPa) **TS**, 39–45 **HRC**
ANSI/ASTM B 194	CA 17200	US	Plate, sheet, strip, bar	96.9 min **Cu**, 0.20 min **Ni+Co**, 1.80–2.00 **Be**, 0.6 **Ni+Co+Fe**	**Hard**: >0.032 in.(>0.813 mm) **diam**, 100 ksi (690 MPa) **TS**, 2% in 2 in. (50 mm) **El**, 96–102 **HRB**; **Solution and precipitation heat treated**: >0.032 in. (>0.813 mm) **diam**, 165 ksi (1140 MPa) **TS**, 140 ksi (960 MPa) **YS**, 36 min **HRC**; **Hard, then precipitation heat treated**: >0.032 in.(>0.813) mm) **diam**, 190 ksi (1310 MPa) **TS**, 165 ksi (1140 MPa) **YS**, 40 min **HRC**
ASTM B 570	CA 172	US	Forgings	96.9 min **Cu**, 0.20 min **Ni+Co**, 1.80–2.00 **Be**, 0.6 **Ni+Co+Fe**	**As manufactured**: all **diam**, 60 ksi (410 MPa) **TS**, 45–85 **HRB**; **Precipitation heat treated**: all **diam**, 165 ksi (1140 MPa) **TS**, 32–39 **HRB**
AMS 4530D		US	Sheet, strip, and plate	96.9 min **Cu**, 0.20 min **Co**, 1.8–2.0 **Be**, 99.5 **Cu** + total named elements, 0.60 **Ni+Co+Fe**	**Solution heat treated**: >0.075 in. **diam**, 60 ksi (414 MPa) **TS**, 35% in 2 in. (50 mm) **El**, 45–78 **HRB**; **Precipitation heat treated**: 0.020–0.065 in. **diam**, 165 ksi (1138 MPa) **TS**, 140 ksi (965 MPa) **YS**, 3% in 2 in. (50 mm) **El**, 56–61 **HR30N**

WROUGHT COPPER AND COPPER ALLOYS

specification number	designation	country	product forms	chemical composition	mechanical properties and hardness values
AMS 4532C		US	Sheet and strip	96.9 min **Cu**, 0.20 min **Co**, 1.8–2.0 **Be**, 99.5 **Cu** + total named elements, 0.60 **Ni**+ **Co**+**Fe**	**Solution heat treated and cold rolled to 1/2 hard:** 0.045–0.188 in. **diam**, 85 ksi (586 MPa) **TS**, 5% in 2 in. (50 mm) **El**, 88–96 **HR15C**; **Precipitation heat treated:** 0.060–0.188 in. **diam**, 185 ksi (1276 MPa) **TS**, 160 ksi (1103 MPa) **YS**, 1% in 2 in. (50 mm) **El**, 38 min **HRC**
AMS 4650F		US	Bar, rod, and forgings	96.9 min **Cu**, 0.20 min **Co**, 1.8–2.0 **Be**, 99.5 **Cu** + total named elements, 0.60 **Ni**+ **Co**+**Fe**	**Hot or cold worked and solution heat treated (bar and rod):** >0.311 in. **diam**, 60 ksi (414 MPa) **TS**, 35% in 2 in. (50 mm) **El**, 45–85 **HRB**; **Precipitation heat treated (bar and rod):** 0.188 in. **diam**, 165 ksi (1138 MPa) **TS**, 145 ksi (100 MPa) **YS**, 3% in 2 in. (50 mm) **El**, 36–42 **HRC**
AMS 4651A		US	Bar and rod	96.9 min **Cu**, 0.20 min **Co**, 1.8–2.0 **Be**, 99.5 **Cu** + total named elements, 0.60 **Ni**+ **Co**+**Fe**, 0.50 others, total	**Solution heat treated and cold rolled to hard:** ≤0.375 in. (≤9.52 mm) **diam**, 95 ksi (655 MPa) **TS**, 10% in 2 in. (50 mm) **El**, 92–103 **HRB**; **Precipitation heat treated:** ≤0.375 in. (≤9.52 mm) **diam**, 185 ksi (1276 MPa) **TS**, 145 ksi (1000 MPa) **YS**, 1% in 2 in. (50 mm) **El**, 39–45 **HRC**
AMS 4725C		US	Wire	96.9 min **Cu**, 0.20 min **Co**, 1.8–2.0 **Be**, 99.5 min **Cu** + total named elements, 0.60 **Ni**+**Co**+**Fe**	**Cold drawn or rolled and solution heat treated:** all **diam**, 58 ksi (400 MPa) **TS**, 35% in 2 in. (50 mm) **El**; **Precipitation heat treated:** all **diam**, 165 ksi (1138 MPa) **TS**, 3% in 2 in. (50 mm) **El**
CDA C 175		US	Plate, sheet, strip, rod	96.9 min **Cu**, 0.10 **Fe**, 2.4–2.7 **Co**, 0.40–0.7 **Be**	**Solution heat treated (plate, sheet, strip):** <0.188 in. (<4.775 mm) **diam**, 45 ksi (310 MPa) **TS**, 25 ksi (172 MPa) **YS**, 28% in 2 in. (50 mm) **El**, 32 **HRB**; **Hard (plate, sheet, strip):** <0.188 in. (<4.775 mm) **diam**, 115 ksi (793 MPa) **TS**, 110 ksi (758 MPa) **YS**, 8% in 2 in. (50 mm) **El**, 98 **HRB**; **Solution heat treated (rod):** all **diam**, 45 ksi (310 MPa) **TS**, 25 ksi (172 MPa) **YS**, 28% in 2 in. (50 mm) **El**, 35 **HRB**

WROUGHT COPPER AND COPPER ALLOYS

specification number	designation	country	product forms	chemical composition	mechanical properties and hardness values
QQ–C–533B	Copper alloy No. 172	US	Strip	96.9 min **Cu**, 0.20 min **Co**, 0.20 min **Ni**, 1.80–2.00 **Be**, **Cu** includes **Be** plus others; 0.6 **Ni + Co + Fe**, total	**Cold rolled and solution heat treated**: all **diam**, 60 ksi (414 MPa) **TS**, 35% in 2 in. (50 mm) **El**, 45–78 **HRB**; **Hot or cold rolled, solution treated and cold rolled 1/4 hard**: all **diam**, 75 ksi (517 MPa) **TS**, 10% in 2 in. (50 mm) **El**, 68–90 **HRB**; **One half hard**: all **diam**, 85 ksi (586 MPa) **TS**, 5% in 2 in. (50 mm) **El**, 88–96 **HRB**
QQ–C–530C	Copper alloy No. 172	US	Bar, rod, wire	96.9 min **Cu**, 0.20 min **Co**, 0.20 min **Ni**, 1.80–2.00 **Be**, 0.6 **Ni + Co + Fe**	**Hot or cold worked and solution heat treated**: 0.020–0.311 in. **diam**, 60 ksi (414 MPa) **TS**; **Hot worked, solution treated and hard worked**: 0.020–0.249 in. **diam**, 95 ksi (655 MPa) **TS**; **Temper AT**: 0.020–0.249 in. **diam**, 165 ksi (1138 MPa) **TS**, 130 ksi (896 MPa) **YS**, 3% in 2 in. (50 mm) **El**
JIS H 3270	C1720	Japan	Rod, bar, and wire	96.9 min **Cu**, 0.20 **Ni + Co**, 1.8–2.0 **Be**, 0.6 **Ni + Co + Fe**	**Annealed**: <6 mm **diam**, 412 MPa **TS**, 90–190 **HV**; **Hard**: <6 mm **diam**, 647 MPa **TS**, 180–300 **HV**
JIS 3130	C1720	Japan	Sheet, plate, and strip: for spring	96.9 min **Cu**, 0.20 **Ni + Co**, 1.8–2.0 **Be**, 99.5 min **Cu + Be + Ni + Co + Fe**, 0.20 min **Ni + Co**, 0.6 **Ni + Co + Fe**	**Annealed**: 0.16–1.6 mm **diam**, 412 MPa **TS**, 35% in 2 in. (50 mm) **El**, 90–160 **HV**; **1/2 hard**: 0.16–1.6 mm **diam**, 588 MPa **TS**, 5% in 2 in. (50 mm) **El**, 180–240 **HV**; **Hard**: 0.16–1.6 mm **diam**, 686 MPa **TS**, 2% in 2 in. (50 mm) **El**, 210–270 **HV**
ISO 1638	CuCo2Be	ISO	Solid products: drawn, on coils or reels	96.49 min **Cu**, 2.0–2.8 **Co**, 0.4–0.7 **Be**	**Solution heat treated and naturally aged**: 1–3 mm **diam**, 290 MPa **TS**, 25% in 2 in. (50 mm) **El**; **TD**: 1–3 mm **diam**, 490 MPa **TS**, 3% in 2 in. (50 mm) **El**; **Solution heat treated and artificially aged**: 1–3 mm **diam**, 640 MPa **TS**, 8% in 2 in. (50 mm) **El**
ISO/R1187	CuBe1.7CoNi	ISO	Plate, sheet, strip, rod, bar, tube, wire	96.49 min **Cu**, 0.20–0.60 **Ni + Co**, 1.6–1.8 **Be**, 0.20–0.60 **Co + Ni + Fe**, 0.5 total, others	

WROUGHT COPPER AND COPPER ALLOYS

specification number	designation	country	product forms	chemical composition	mechanical properties and hardness values
ISO 1634	CuBe1.7	ISO	Plate, sheet, strip	96.49 min **Cu**, 0.20–0.60 **Ni + Co**, 1.6–1.8 **Be**, 0.20–0.60 **Co + Ni + Fe**, 0.5 total, others	**Solution heat treated and naturally aged**: all **diam**, 440 MPa **TS**, 200 MPa **YS**, 35% in 2 in. (50 mm) **El**, 85 **HV**; **Solution and precipitation treated**: all **diam**, 1180 MPa **TS**, 980 MPa **YS**, 2% in 2 in. (50 mm) **El**, 340 **HV**; **TH**: all **diam**, 1280 MPa **TS**, 1080 MPa **YS**, 360 **HV**
QQ–C–530C	Copper alloy No. 173	US	Bar, rod, wire	96.3 min **Cu**, 0.20–0.6 **Pb**, 0.20 min **Co**, 0.20 min **Ni**, 1.80–2.00 **Be**, 0.6 **Ni + Co + Fe**	**Hot or cold worked and solution heat treated**: 0.020–0.311 in. **diam**, 60 ksi (414 MPa) **TS**; **Hot worked, solution treated and hard worked**: 0.020–0.249 in. **diam**, 95 ksi (655 MPa) **TS**; **Temper AT**: 0.020–0.249 in. **diam**, 165 ksi (1138 MPa) **TS**, 130 ksi (896 MPa) **YS**, 3% in 2 in. (50 mm) **El**
ISO/R1187	CuBe2CoNi	ISO	Plate, sheet, strip, rod, bar, tube, wire, forgings	96.19 min **Cu**, 0.20–0.60 **Ni + Co**, 1.8–2.1 **Be**, 0.20–0.60 **Co + Ni + Fe**, 0.5 total, others	
ISO 1634	CuBe2	ISO	Plate, sheet, strip	96.19 min **Cu**, 0.20–0.60 **Ni + Co**, 1.8–2.1 **Be**, 0.20–0.60 **Co + Ni + Fe**, 0.5 total, others	**Solution heat treated and naturally aged**: all **diam**, 490 MPa **TS**, 250 MPa **YS**, 35% in 2 in. (50 mm) **El**, 95 **HV**; **Solution and precipitation treated**: all **diam**, 1280 MPa **TS**, 1080 MPa **YS**, 360 **HV**; **TH**: all **diam**, 1370 MPa **TS**, 1180 MPa **YS**, 380 **HV**
ANSI/ASTM B 441		US	Rod, bar	96.1 min **Cu**, 2.4–2.7 **Co**, 0.40–0.7 **Be**, 99.5 min **Cu + Be + Co**	**Annealed**: all **diam**, 35 ksi (241 MPa) **TS**, 20–50 **HRB**; **Solution & precipitation heat treated**: all **diam**, 100 ksi (689 MPa) **TS**, 92–100 **HRB**; **Hard, then precipitation heat treated**: all **diam**, 110 ksi (758 MPa) **TS**, 95–102 **HRB**
MIL–C–46087A		US	Bar and rod	96.1 min **Cu**, 2.4–2.7 **Co**, 0.40–0.7 **Be**	**Cold rolled or drawn and solution heat treated**: all **diam**, 379 max MPa **TS**, 20% in 2 in. (50 mm) **El**; **Cold rolled or drawn, solution heat treated and rolled or drawn hard**: all **diam**, 448 MPa **TS**, 10% in 2 in. (50 mm) **El**; **Cold rolled or drawn, solution heat treated and precipitation hardened**: all **diam**, 689 MPa **TS**, 517 MPa **YS**, 10% in 2 in. (50 mm) **El**
IS 5743	Cu Be 4	India	Bar, ingot, shot	94.5–96.5 **Cu**, 3.0–5.0 **Be**	

WROUGHT COPPER AND COPPER ALLOYS

specification number	designation	country	product forms	chemical composition	mechanical properties and hardness values
ISO/R1187	CuCo2Be	ISO	Strip, rod, bar, wire	95.49 min **Cu**, 2.0–2.8 **Co**, 0.4–0.7 **Be**, 0.5 **Ni + Fe**, 0.5 total, others	
ISO 1634	CuCo2Be	ISO	Plate, sheet, strip	95.49 min **Cu**, 2.0–2.8 **Co**, 0.4–0.7 **Be**, 0.5 **Ni + Fe**, 0.5 total, others	**Solution heat treated and naturally aged**: all **diam**, 340 MPa **TS**, 150 MPa **YS**, 25% in 2 in. (50 mm) **El**, 70 **HV**; **Solution and precipitation treated**: all **diam**, 690 MPa **TS**, 490 MPa **YS**, 8% in 2 in. (50 mm) **El**, 195 **HV**; **TH**: all **diam**, 780 MPa **TS**, 640 MPa **YS**, 5% in 2 in. (50 mm) **El**, 215 **HV**
MIL–W–81822		US	Wire	99.26 **Cu**, 0.001 **Pb**, 0.03 **Zn**, 0.03 **Fe**, 0.40 **Cd**, 0.28 **Cr**	
MIL–W–82598		US	Wire	98.78–99.30 **Cu**, 0.02 **Fe**, 0.7–1.2 **Cd**	**Hard**: 0.0226 **diam**, 586 MPa **TS**
COPANT 162	C 182 00	COPANT	Wrought products	99.1 min **Cu**, 0.05 **Pb**, 0.10 **Fe**, 0.10 **Si**, 0.6–1.2 **Cr**	
CDA C 182		US	Plate, sheet, strip, rod, tube	99.1 **Cu**, 0.05 **Pb**, 0.10 **Si**, 0.6–1.2 **Cr**	**Solution heat treated (plate, sheet, strip)**: 0.040 in. (1.016 mm) **diam**, 34 ksi (234 MPa) **TS**, 19 ksi (131 MPa) **YS**, 40% in 2 in. (50 mm) **El**, 16 **HRB**; **Solution heat treated and aged (plate)**: 2.0 in. (50.8 mm) **diam**, 58 ksi (400 MPa) **TS**, 42 ksi (290 MPa) **YS**, 25% in 2 in. (50 mm) **El**, 70 **HRB**; **Solution heat treated (rod)**: 0.500 in. (12.7 mm) **diam**, 45 ksi (310 MPa) **TS**, 14 ksi (97 MPa) **YS**, 40% in 2 in. (50 mm) **El**
COPANT 162	C 185 00	COPANT	Wrought products	98.8 **Cu**, 0.015 **Pb**, 0.4–1.0 **Cr**, 0.04 **P**	
CDA C 185		US	Plate, sheet, strip, rod, tube	98.6 **Cu**, 0.015 **Pb**, 0.40–1.0 **Cr**, 0.04 **P**, 0.08–0.12 **Ag**	**Solution heat treated (plate, sheet, strip)**: 0.040 in. (1.016 mm) **diam**, 34 ksi (234 MPa) **TS**, 19 ksi (131 MPa) **YS**, 40% in 2 in. (50 mm) **El**, 16 **HRB**; **Solution heat treated and aged (plate)**: 2.0 in. (50.8 mm) **diam**, 58 ksi (400 MPa) **TS**, 42 ksi (290 MPa) **YS**, 25% in 2 in. (50 mm) **El**, 70 **HRB**; **Solution heat treated (rod)**: 0.500 in. (12.7 mm) **diam**, 45 ksi (310 MPa) **TS**, 14 ksi (97 MPa) **YS**, 40% in 2 in. (50 mm) **El**
COPANT 162	C 184 00	COPANT	Wrought products	98.6 **Cu**, 0.15 **Fe**, 0.10 **Si**, 0.4–1.2 **Cr**, 0.005 **As**	

WROUGHT COPPER AND COPPER ALLOYS

specification number	designation	country	product forms	chemical composition	mechanical properties and hardness values
ISO 1637	CuCr1	ISO	Solid products: straight lengths	98.49 min **Cu**, 0.3–1.2 **Cr**, 0.3 total, others	**Solution and precipitation treated**: 5–80 mm **diam**, 370 MPa **TS**, 270 MPa **YS**, 18% in 2 in. (50 mm) **El**, 100 **HV**; **TH**: 5–25 mm **diam**, 440 MPa **TS**, 350 MPa **YS**, 10% in 2 in. (50 mm) **El**, 125 **HV**; **TL**: 5–25 mm **diam**, 500 MPa **TS**, 440 MPa **YS**, 5% in 2 in. (50 mm) **El**, 130 **HV**
ISO 1640	CuCr1	ISO	Forgings	98.49 **Cu**, 0.03–1.2 **Cr**, 0.3 total, others	**Solution treated and artificially aged**: all **diam**, 340 MPa **TS**, 14% in 2 in. (50 mm) **El**, 100 **HV**
ISO R1336	CuCr1	ISO	Plate, sheet, rod, bar, extruded sections, tube, wire, forgings	98.49 **Cu**, 0.03–1.2 **Cr**, 0.3 total, others	
MIL–C–19311B	182	US	Forgings, rod, bar, strip	98.05 min **Cu**, 0.05 **Pb**, 0.10 **Fe**, 0.10 **Si**, 0.6–1.2 **Cr**, silver included in copper content	
MIL–C–19310A		US	Castings	98.00 **Cu**, 0.14–1.5 **Cr**, 0.50 total, others	**Heat treated and artificially aged**: all **diam**, 276 MPa **TS**, 207 MPa **YS**, 15% in 2 in. (50 mm) **El**, 89 **HB**, 60 **HRB**
MIL–C–19311B	184	US	Forgings, rod, bar, strip	97.64 **Cu**, 0.7 **Zn**, 0.15 **Fe**, 0.10 **Si**, 0.40–1.2 **Cr**, 0.05 **P**, 0.005 **As**, 0.005 **Ca**, 0.05 **Li**, silver included in copper content	
CDA C 184		US	Plate, sheet, strip, rod, tube	97.54 **Cu**, 0.7 **Zn**, 0.15 **Fe**, 0.10 **Si**, 0.40–1.2 **Cr**, 0.05 **P**, 0.005 **As**, 0.005 **Ca**, 0.05 **Li**	**Solution heat treated (plate, sheet, strip)**: 0.040 in. (1.016 mm) **diam**, 34 ksi (234 MPa) **TS**, 19 ksi (131 MPa) **YS**, 40% in 2 in. (50 mm) **El**, 16 **HRB**; **Solution heat treated and aged (plate)**: 2.0 in. (50.8 mm) **diam**, 58 ksi (400 MPa) **TS**, 42 ksi (290 MPa) **YS**, 25% in 2 in. (50 mm) **El**, 70 **HRB**; **Solution heat treated (rod)**: 0.500 in. (12.7 mm) **diam**, 45 ksi (310 MPa) **TS**, 14 ksi (97 MPa) **YS**, 40% in 2 in. (50 mm) **El**
IS 5743	Cu Cr 10	India	Bar, ingot, shot	88.5–89.5 **Cu**, 0.05 **Fe**, 9.0–11.0 **Cr**	
NF A 51 122		France	Tube	0.13–0.15 **Pb**, 99.90 min **Cu+Ag**	**Annealed**: 0.40–1 mm **diam**, 200 MPa **TS**, 80 MPa **YS**, 40% in 2 in. (50 mm) **El**, 40–50 **HV**; **1/8 hard**: 0.40–1 mm **diam**, 220 MPa **TS**, 80 MPa **YS**, 40% in 2 in. (50 mm) **El**, 50–60 **HV**

WROUGHT COPPER AND COPPER ALLOYS

specification number	designation	country	product forms	chemical composition	mechanical properties and hardness values
COPANT 162	C 187 00	COPANT	Wrought products	99.9 **Cu**, 0.8–1.5 **Pb**	
CDA C 187		US	Rod	99.0 **Cu**, 0.8–1.5 **Pb**	**Half hard**: 1.0 in. (25.4 mm) **diam**, 42 ksi (290 MPa) **TS**, 38 ksi (262 MPa) **YS**, 25% in 2 in. (50 mm) **El**, 42 **HRB**; **Hard**: 1.0 in. (25.4 mm) **diam**, 48 ksi (331 MPa) **TS**, 42 ksi (290 MPa) **YS**, 15% in 2 in. (50 mm) **El**, 48 **HRB**
ANSI/ASTM B 301	C 18700	US	Rod, bar	98.4 min **Cu**, 0.8–1.5 **Pb**	**Half hard (rod)**: 0.063–0.25 in. (1.59–6.35 mm) **diam**, 38 ksi (260 MPa) **TS**, 30 ksi (205 MPa) **YS**, 8% in 1 in. (25.4 mm) **El**; **Hard (rod)**: 0.063–0.25 in. (1.59–6.35 mm) **diam**, 48 ksi (330 MPa) **TS**, 40 ksi (275 MPa) **YS**, 4% in 1 in. (25.4 mm) **El**; **Hard (bar)**: 0.188–0.375 in. (4.78–9.52 mm) **diam**, 42 ksi (290 MPa) **TS**, 35 ksi (240 MPa) **YS**, 10% in 1 in. (25.4 mm) **El**
CDA C 189		US	Rod, wire	98.7 **Cu**, 0.6–0.9 **Sn**, 0.02 **Pb**, 0.10 **Zn**, 0.15–0.40 **Si**, 0.01 **Al**, 0.10–0.30 **Mn**, 0.05 **P**	**Soft (rod)**: 0.125 in. (3.175 mm) **diam**, 38 ksi (262 MPa) **TS**, 9 ksi (62 MPa) **YS**, 48% in 2 in. (50 mm) **El**; **Hard (rod)**: 0.125 in. (3.175 mm) **diam**, 56 ksi (386 MPa) **TS**, 52 ksi (359 MPa) **YS**, 14% in 2 in. (50 mm) **El**; **Hard (wire)**: 0.045 in. (1.143 mm) **diam**, 95 ksi (655 MPa) **TS**
COPANT 162	C 189 00	COPANT	Wrought products	99.9 **Cu**, 0.6–0.9 **Sn**, 0.02 **Pb**, 0.10 **Zn**, 0.15–0.40 **Si**, 0.01 **Al**, 0.10–0.30 **Mn**, 0.05 **P**	
JIS H 3100	C1401	Japan	Sheet, plate, and strip	99.30 min **Cu**, 0.10–0.20 **Ni**	**Hard**: $\geq$0.5 mm **diam**, 90 min **HV**
CDA 190		US	Rod, wire, flat wire	98.6 **Cu**, 0.05 **Pb**, 0.8 **Zn**, 0.10 **Fe**, 0.9–1.3 **Ni**, 0.15–0.35 **P**, 99.5 min **Cu** + others	**Solution treated (rod)**: 1.0 in. **diam**, 38 ksi (262 MPa) **TS**, 20 ksi (138 MPa) **YS**, 50% in 2 in. (50 mm) **El**, 45 **HRB**; **Age hardened (wire)**: 0.080 in. **diam**, 65 ksi (448 MPa) **TS**, 40 ksi (276 MPa) **YS**, 25% in 2 in. (50 mm) **El**; **Heat treated–half hard (flat wire)**: 0.50 in. **diam**, 76 ksi (524 MPa) **TS**, 70 ksi (483 MPa) **YS**, 4% in 2 in. (50 mm) **El**, 83 **HRB**

WROUGHT COPPER AND COPPER ALLOYS

specification number	designation	country	product forms	chemical composition	mechanical properties and hardness values
COPANT 162	C 190 00	COPANT	Wrought products	98.2 **Cu**, 0.05 **Pb**, 0.8 **Zn**, 0.10 **Fe**, 0.9–1.3 **Ni**, 0.15–0.35 **P**	
CDA 191		US	Rod, forgings	98.2 **Cu**, 0.10 **Pb**, 0.50 **Zn**, 0.20 **Fe**, 0.9–1.3 **Ni**, 0.15–0.35 **P**, 0.35–0.6 **Te**, 99.5 min **Cu** + others	**Heat treated–hard 35% (rod)**: 0.250 in. **diam**, 84 ksi (579 MPa) **TS**, 77 ksi (531 MPa) **YS**, 10% in 2 in. (50 mm) **El**, 85 **HRB**; **As forged and quenched (forgings)**: all **diam**, 36 ksi (248 MPa) **TS**, 35 **HRF**; **As age hardened (forgings)**: all **diam**, 58 ksi (400 MPa) **TS**, 97 **HRF**
COPANT 162	C 191 00	COPANT	Wrought products	98.2 **Cu**, 0.10 **Pb**, 0.20 **Fe**, 0.9–1.3 **Ni**, 0.15–0.35 **P**, 0.35–0.6 **Te**	
ISO R1187	CuNi1Si	ISO	Strip, rod, bar, wire	97.19 min **Cu**, 0.4–0.7 **Si**, 1.0–1.6 **Ni**, 0.5 total, others	
ISO 1637	CuNi1Si	ISO	Solid products: straight lengths	97.19 min **Cu**, 0.4–0.7 **Si**, 1.0–1.6 **Ni**, 0.5 total, others	**Solution heat treated and naturally aged**: 30 max mm **diam**, 450MPa **TS**, 290MPa **YS**, 9% in 2 in. (50 mm) **El**, 110 **HV**; **Solution heat treated and artificially aged**: 30 max mm **diam**, 630MPa **TS**, 540MPa **YS**, 12% in 2 in. (50 mm) **El**, 160 **HV**
ISO 1634	CuNi1Si	ISO	Plate, sheet, strip	97.19 min **Cu**, 0.4–0.7 **Si**, 1.0–1.6 **Ni**, 0.5 total, others	**TD**: 3 mm **diam**, 450MPa **TS**, 340MPa **YS**, 8% in 2 in. (50 mm) **El**, 120 **HV**; **TH**: 3 mm **diam**, 640MPa **TS**, 540MPa **YS**, 10% in 2 in. (50 mm) **El**, 180 **HV**
ISO 1634	CuNi2Si	ISO	Plate, sheet, strip	94.59 min **Cu**, 0.5–0.8 **Si**, 1.6–2.5 **Ni**, 0.5 total, others	**TD**: 200 max mm **diam**, 460MPa **TS**, 360MPa **YS**, 6% in 2 in. (50 mm) **El**, 135 **HV**; **TH**: 200 max mm **diam**, 680MPa **TS**, 590MPa **YS**, 8% in 2 in. (50 mm) **El**, 190 **HV**
ANSI/ASTM B 469	C 19200	US	Tube: seamless	98.7 min **Cu**, 0.8–1.2 **Fe**, 0.01–0.04 **P**, silver included in copper content	**As manufactured**: <0.25 in. (<6.350 mm) **diam**, 40 ksi (275 MPa) **TS**, 170 ksi (275 MPa) **YS**, 18% in 2 in. (50.8 mm) **El**
ASTM B 585		US	Tube: seamless	98.7 min **Cu**, 0.8–1.2 **Fe**, 0.01–0.04 **P**	**Annealed**: all **diam**, 38 ksi (260 MPa) **TS**, 30% in 2 in. (50 mm) **El**, 80 max **HRF**; **Cold worked**: all **diam**, 45 ksi (310 MPa) **TS**, 45 min **HR30T**
ANSI/ASTM B 359	C 19200	US	Tube: seamless	98.7 min **Cu**, 0.8–1.2 **Fe**, 0.01–0.04 **P**	**Annealed**: all **diam**, 38 ksi (260 MPa) **TS**, 12 ksi (85 MPa) **YS**

WROUGHT COPPER AND COPPER ALLOYS

specification number	designation	country	product forms	chemical composition	mechanical properties and hardness values
ANSI/ASTM B 395	C 19200	US	Tube: seamless	98.7 min **Cu**, 0.8–1.2 **Fe**, 0.01–0.04 **P**, silver included in copper content	**Light drawn**: 0.048 in. (1.21 mm) **diam**, 40 ksi (275 MPa) **TS**, 35 ksi (240 MPa) **YS**, 12% in 2 in. (50.8 mm) **El**; **Annealed**: 0.048 in. (1.21 mm) **diam**, 38 ksi (260 MPa) **TS**, 12 ksi (85 MPa) **YS**, 12% in 2 in. (50.8 mm) **El**
ANSI/ASTM B 111	C 19200	US	Tube: seamless, and stock; ferrule	98.7 min **Cu**, 0.8–1.2 **Fe**, 0.01–0.04 **P**	**Light drawn**: all **diam**, 40 ksi (275 MPa) **TS**, 35 ksi (240 MPa) **YS**; **Hard drawn**: all **diam**, 48 ksi (330 MPa) **TS**, 43 ksi (295 MPa) **YS**
CDA 192		US	Tube	98.7 min **Cu**, 0.8–1.2 **Fe**, 0.01–0.04 **P**	**Light annealed**: 1.875 in. **diam**, 42 ksi (290 MPa) **TS**, 22 ksi (152 MPa) **YS**, 30% in 2 in. (50 mm) **El**; **Hard drawn (40%)**: 1.875 in. **diam**, 56 ksi (386 MPa) **TS**, 52 ksi (359 MPa) **YS**, 7% in 2 in. (50 mm) **El**; **Light drawn**: 0.1875 in. **diam**, 42 ksi (290 MPa) **TS**, 30 ksi (207 MPa) **YS**, 35% in 2 in. (50 mm) **El**
COPANT 162	C 192 00	COPANT	Wrought products	98.7 min **Cu**, 0.8–1.2 **Fe**, 0.01–0.04 **P**	
ANSI/ASTM B 543	C 19400	US	Tube: welded	97.0–97.8 **Cu**, 0.03 **Pb**, 0.05–0.20 **Zn**, 2.1–2.6 **Fe**, 0.015–0.15 **P**, silver included in copper content	**As welded from annealed strip**: all **diam**, 45 ksi (310 MPa) **TS**, 22 ksi (150 MPa) **YS**; **Welded and annealed**: all **diam**, 45 ksi (310 MPa) **TS**, 15 ksi (105 MPa) **YS**; **Fully finished–hard drawn**: all **diam**, 63 ksi (435 MPa) **TS**, 56 ksi (385 MPa) **YS**
ASTM B 586		US	Tube: welded	97.0–97.8 **Cu**, 0.03 **Pb**, 0.05–0.20 **Zn**, 2.1–2.6 **Fe**, 0.015–0.15 **P**	**Annealed**: 0.016–0.028 in. (0.406–0.711 mm) **diam**, 45 ksi (310 MPa) **TS**, 64 max **HR30T**; **Cold worked**: 0.016–0.028 in. (0.406–0.711 mm) **diam**, 60 ksi (415 MPa) **TS**, 66 min **HR30T**
ANSI/ASTM B 465		US	Plate, sheet, strip, bar	97.0 min **Cu**, 0.03 **Pb**, 0.05–0.20 **Zn**, 2.1–2.6 **Fe**, 0.015–0.15 **P**	**Annealed**: all **diam**, 45 ksi (310 MPa) **TS**; **Hard**: 0.020–0.036 in. (0.508–0.914 mm) **diam**, 60 ksi (414 MPa) **TS**, 67–73 **HRB**; **Spring**: 0.020–0.036 in. (0.508–0.914 mm) **diam**, 70 ksi (483 MPa) **TS**, 73–78 **HRB**
COPANT 162	C 194 00	COPANT	Wrought products	97.0 min **Cu**, 0.03 **Pb**, 0.05–0.20 **Zn**, 2.1–2.6 **Fe**, 0.015–0.15 **P**	

WROUGHT COPPER AND COPPER ALLOYS

specification number	designation	country	product forms	chemical composition	mechanical properties and hardness values
CDA 194		US	Tube, strip	97.0 min **Cu**, 0.03 **Sn**, 0.03 **Pb**, 0.05–0.20 **Zn**, 2.1–2.6 **Fe**, 0.015–0.15 **P**, 0.15 total, others	**Soft annealed (tube)**: 1 in. **diam**, 45 ksi (310 MPa) **TS**, 24 ksi (165 MPa) **YS**, 28% in 2 in. (50 mm) **El**, 38 **HRB**; **Half hard (strip)**: 0.040 in. **diam**, 60 ksi (414 MPa) **TS**, 53 ksi (365 MPa) **YS**, 9% in 2 in. (50 mm) **El**, 68 **HRB**; **Hard temper (strip)**: 0.0405 in. **diam**, 67 ksi (462 MPa) **TS**, 63 ksi (434 MPa) **YS**, 4% in 2 in. (50 mm) **El**, 73 **HRB**
COPANT 162	C 195 00	COPANT	Wrought products	96.16 min **Cu**, 0.4–0.7 **Sn**, 0.02 **Pb**, 0.20 **Zn**, 1.3–1.7 **Fe**, 0.10 **Al**, 0.6–1.0 **Co**, 0.08–0.12 **P**	
CDA 195		US	Strip: rolled	96.0 min **Cu**, 0.40–0.7 **Sn**, 0.02 **Pb**, 0.20 **Zn**, 1.3–1.7 **Fe**, 0.02 **Al**, 0.6–1.0 **Co**, 0.08–0.12 **P**, 0.05 each, 0.10 total, others	**Precipitation heat treated**: 0.040 in. **diam**, 80 ksi (552 MPa) **TS**, 65 ksi (448 MPa) **YS**, 15% in 2 in. (50 mm) **El**; **Precipitation heat treated and cold rolled half hard**: 0.040 in. **diam**, 86 ksi (593 MPa) **TS**, 80 ksi (552 MPa) **YS**, 10% in 2 in. (50 mm) **El**; **Precipitation heat treated and cold rolled spring**: 0.040 in. **diam**, 93 ksi (641 MPa) **TS**, 90 ksi (621 MPa) **YS**, 4% in 2 in. (50 mm) **El**
IS 5743	Cu Fe 15	India	Bar, ingot, shot	83.5–85.5 **Cu**, 14.0–16.0 **Fe**	
IS 5743	Cu Fe 20	India	Bar, ingot, shot	78.5–80.5 **Cu**, 19.0–21.0 **Fe**	
JIS Z 3264	BCuP–1	Japan	Ribbon, wire, and bar: brazing filler	94.49 min **Cu**, 4.8–5.3 **P**, 0.2 other elements, total	
JIS Z 3264	BCuP–2	Japan	Ribbon, wire, and bar: brazing filler	92.29 min **Cu**, 6.8–7.5 **P**, 0.2 other elements, total	
IS 5743	Cu P10	India	Bar, ingot, shot	88.5–90.5 **Cu**, 0.25 **Fe**, 9.0–11.0 **P**	
JIS Z 3264	BCuP–3	Japan	Ribbon, wire, and bar: brazing filler	86.79 min **Cu**, 5.8–6.7 **P**, 4.7–6.3 **Ag**, 0.2 other elements, total	
JIS Z 3264	BCuP–4	Japan	Ribbon, wire, and bar: brazing filler	85.79 min **Cu**, 6.8–7.7 **P**, 4.7–6.3 **Ag**, 0.2 other elements, total	
IS 5743	Cu P14	India	Bar, ingot, shot	84.5–86.5 **Cu**, 0.20 **Fe**, 13.0–15.0 **P**	
COPANT 162	C 145•00 (=DPTE)	COPANT	Wrought products	99.90 min **Cu**, 0.004–0.012 **P**, 0.40–0.60 **Te**, silver included in copper content	

WROUGHT COPPER AND COPPER ALLOYS

specification number	designation	country	product forms	chemical composition	mechanical properties and hardness values
CDA C 145		US	Plate, sheet, strip, rod, wire, tube	99.50 **Cu**, 0.004–0.012 **P**, 0.40–0.60 **Te**, 99.90 min **Cu**+**Ag**+**Te**+**P**	**Hard (rod)**: 0.500 in. (12.7 mm) **diam**, 48 ksi (331 MPa) **TS**, 44 ksi (303 MPa) **YS**, 15% in 2 in. (50 mm) **El**, 48 **HRB**; **Hard (wire)**: 0.080 in. (2.032 mm) **diam**, 56 ksi (386 MPa) **TS**, 51 ksi (352 MPa) **YS**, 3% in 2 in. (50 mm) **El**; **Light drawn (tube)**: 1.0 in. (25.4 mm) **diam**, 40 ksi (276 MPa) **TS**, 32 ksi (221 MPa) **YS**, 20% in 2 in. (50 mm) **El**, 35 **HRB**
ANSI/ASTM B 124	C 14500	US	Rod, bar, shapes	99.30 min **Cu**, 0.004–0.012 **P**, 0.40–0.6 **Te**, silver included in copper content	
ANSI/ASTM B 283	C 14500	US	Die forgings	99.30 min **Cu**, 0.004–0.012 **P**, 0.40–0.60 **Te**, silver and included in copper content	
ANSI/ASTM B 301	C 14500	US	Rod, bar	99.3 **Cu**, 0.004–0.012 **P**, 0.40–0.6 **Te**, silver included with copper	**Half hard (rod)**: 0.063–0.25 in. (1.59–6.35 mm) **diam**, 38 ksi (260 MPa) **TS**, 30 ksi (205 MPa) **YS**, 8% in 1 in. (25.4 mm) **El**; **Hard (rod)**: 0.063–0.25 in. (1.59–6.35 mm) **diam**, 48 ksi (330 MPa) **TS**, 40 ksi (275 MPa) **YS**, 4% in 1 in. (25.4 mm) **El**; **Hard (bar)**: 0.188–0.375 in. (4.78–9.52 mm) **diam**, 42 ksi (290 MPa) **TS**, 35 ksi (240 MPa) **YS**, 10% in 1 in. (25.4 mm) **El**
AS 1567	145C	Australia	Rod, bar, section	99.20 min **Cu**, 0.004–0.015 **P**, 0.30–0.70 **Te**, others	**As manufactured**: 6–50 mm **diam**, 260 MPa **TS**, 8% in 2 in. (50 mm) **El**; **Soft annealed**: >6 mm **diam**, 210 MPa **TS**, 28% in 2 in. (50 mm) **El**
BS 2874	C109	UK	Rod, section	99.20 min **Cu**, 0.30–0.70 **Te**, 0.20 total, others	**Annealed**: ≤6 mm **diam**, 216 MPa **TS**, 28% in 2 in. (50 mm) **El**; **As manufactured**: 6–50 mm **diam**, 260 MPa **TS**, 8% in 2 in. (50 mm) **El**
ISO 1637	CuTe	ISO	Solid products: straight lengths	98.99 min **Cu**, 0.3–0.8 **Te**, 0.2 total, others	**Annealed**: 5 mm **diam**, 28% in 2 in. (50 mm) **El**, 70 max **HV**; **HA**: 5–40 mm **diam**, 250 MPa **TS**, 10% in 2 in. (50 mm) **El**, 90 **HV**
ISO R1336	CuTe	ISO	Rod, bar	98.99 min **Cu**, 0.3–0.8 **Te**, 0.00 **O**, 0.2 total, others	

WROUGHT COPPER AND COPPER ALLOYS

specification number	designation	country	product forms	chemical composition	mechanical properties and hardness values
ISO 1637	CuTe(P)	ISO	Solid products: straight lengths	98.978 min **Cu**, 0.004–0.012 **P**, 0.3–0.8 **Te**, 0.2 **P** + others, total	**Annealed**: 5 mm **diam**, 28% in 2 in. (50 mm) **El**, 70 max **HV**; **HA**: 5–40 mm **diam**, 250 MPa **TS**, 10% in 2 in. (50 mm) **El**, 90 **HV**
ISO R1336	CuTe(P)	ISO	Rod, bar	98.978 min **Cu**, 0.004–0.012 **P**, 0.3–0.8 **Te**, 0.2 total, others	
JIS H 3100	C 2051	Japan	Sheet, plate, and strip	98.0–99.0 **Cu**, 0.05 **Pb**, rem **Zn**, 0.05 **Fe**	**As manufactured**: <0.35 mm **diam**, 216 MPa **TS**, 38% in 2 in. (50 mm) **El**
COPANT 162	C 205 00	COPANT	Wrought products	97.0–98.0 **Cu**, 0.02 **Pb**, rem **Zn**, 0.05 **Fe**	
COPANT 803	C 260 00	COPANT	Bar	97.0–98.0 **Cu**, 0.02 **Pb**, rem **Zn**, 0.05 **Fe**	**1/4 hard temper**: 380 MPa **TS**, 25% in 2 in. (50 mm) **El**
BS 2870	CZ 125	UK	Sheet, strip, foil	95.0–98.0 **Cu**, 0.02 **Pb**, rem **Zn**, 0.05 **Fe**, 0.25 total, others	**Annealed**: ≤10.0 mm **diam**, 75 max **HV**
IS 3167		India	Strip	95–98 **Cu**, 0.01 **Sn**, 0.02 **Pb**, rem **Zn**, 0.05 **Fe**, 0.01 **Sb**, 0.10 **Ni**, 0.005 total, others	
ASTM F 587	CA 210	US	Tube: welded	94.0–96.0 **Cu**, 0.05 **Pb**, rem **Zn**, 0.05 **Fe**	**Welded from annealed strip**: all **diam**, 34 ksi (235 MPa) **TS**, 7 min **HR30T**; **Cold reduced or light drawn**: all **diam**, 37 ksi (255 MPa) **TS**, 34 min **HR30T**; **Soft anneal**: <0.045 in. (<1.14 mm) **diam**, 17 **HR30T**
ANSI/ASTM B 36	C 21000	US	Plate, sheet, strip, bar	94.0–96.0 **Cu**, 0.03 **Pb**, rem **Zn**, 0.05 **Fe**	**Quarter hard**: 0.020 in. (0.508 mm) **diam**, 37 ksi (255 MPa) **TS**, 20 **HRB**; **Hard**: 0.020 in. (0.508 mm) **diam**, 50 ksi (345 MPa) **TS**, 57 **HRB**; **Spring**: 0.020 in. (0.508 mm) **diam**, 60 ksi (415 MPa) **TS**, 68 **HRB**
ANSI/ASTM B 134	CA 210	US	Wire	94.0–96.0 **Cu**, 0.05 **Pb**, rem **Zn**, 0.05 **Fe**	**Quarter hard**: >0.020 in. (>0.508 mm) **diam**, 35 ksi (240 MPa) **TS**; **Hard**: >0.020 in. (>0.508 mm) **diam**, 61 ksi (420 MPa) **TS**; **Spring**: >0.020 in. (>0.508 mm) **diam**, 72 ksi (495 MPa) **TS**
CSA HC.4.2	HC 4.Z5	Canada	Sheet	94.0–96.0 **Cu**, 0.03 **Pb**, rem **Zn**, 0.05 **Fe**, 0.10 other elements	**Half hard**: all **diam**, 37 ksi (255 MPa) **TS**; **Full hard**: all **diam**, 50 ksi (344 MPa) **TS**; **Spring**: all **diam**, 60 ksi (413 MPa) **TS**

WROUGHT COPPER AND COPPER ALLOYS

specification number	designation	country	product forms	chemical composition	mechanical properties and hardness values
CDA 210		US	Strip, wire	94.0–96.0 **Cu**, 0.05 **Pb**, rem **Zn**, 0.05 **Fe**, 0.10 total, others	**Quarter hard temper (rolled strip)**: 0.040 in. **diam**, 42 ksi (290 MPa) **TS**, 25% in 2 in. (50 mm) **El**, 38 **HRB**; **Half hard temper (rolled strip)**: 0.040 in. **diam**, 48 ksi (331 MPa) **TS**, 12% in 2 in. (50 mm) **El**, 52 **HRB**; **Hard temper (rolled strip)**: 0.040 in. **diam**, 56 ksi (386 MPa) **TS**, 5% in 2 in. (50 mm) **El**, 64 **HRB**
CSA HC.4.2	HC 4.Z5	Canada	Plate	94.0–96.0 **Cu**, 0.03 **Pb**, rem **Zn**, 0.05 **Fe**, 0.01 other elements	**Half hard**: all **diam**, 42 ksi (289 MPa) **TS**; **Full hard**: all **diam**, 50 ksi (344 MPa) **TS**; **Spring**: all **diam**, 60 ksi (413 MPa) **TS**
CSA HC.4.2	HC 4.Z5	Canada	Bar	94.0–96.0 **Cu**, 0.03 **Pb**, rem **Zn**, 0.05 **Fe**, 0.10 other elements	**Half hard**: all **diam**, 42 ksi (289 MPa) **TS**; **Full hard**: all **diam**, 50 ksi (344 MPa) **TS**; **Spring**: all **diam**, 60 ksi (413 MPa) **TS**
COPANT 162	C 210 00	COPANT	Wrought products	94.0–96.0 **Cu**, 0.05 **Pb**, rem **Zn**, 0.05 **Fe**	
COPANT 612	Cu Zn 5	COPANT	Strip, sheet, plate, foil, bar	94–96 **Cu**, 0.03 **Pb**, rem **Zn**, 0.05 **Fe**, 0.10 total, others	**1/4 hard temper**: all **diam**, 255 MPa **TS**; **1/2 hard temper**: all **diam**, 295 MPa **TS**; **Hard temper**: all **diam**, 345 MPa **TS**
QQ-W-321d	210	US	Wire	94.0–96.0 **Cu**, 0.05 **Pb**, rem **Zn**, 0.05 **Fe**, 0.10 total, others; **Co** in **Ni**	**Eighth hard temper**: all **diam**, 35 ksi (242 MPa) **TS**; **Quarter hard temper**: all **diam**, 41 ksi (283 MPa) **TS**; **Half hard temper**: all **diam**, 49 ksi (338 MPa) **TS**
NF A 51 101	CuZn 5	France	Sheet and strip	95.0 **Cu**, 5.0 **Zn**	**Stress hardened, Temper 1**: all **diam**, 250 MPa **TS**
ISO 1638	CuZn5	ISO	Solid product: drawn, on coils or reels	94.0–96.0 **Cu**, 0.05 **Pb**, rem **Zn**, 0.1 **Fe**, 0.3 **Ni** included in **Cu**; 0.3 **Fe+Pb** + total, others	**Annealed**: 1–5 mm **diam**, 220 MPa **TS**, 30% in 2 in. (50 mm) **El**; **HC**: 1–5 mm **diam**, 320 MPa **TS**, 5% in 2 in. (50 mm) **El**
ISO 1634	CuZn5	ISO	Plate, sheet, strip	94.0–96.0 **Cu**, 0.05 **Pb**, rem **Zn**, 0.1 **Fe**, **Cu** total includes 0.3 **Ni**; 0.3 **Fe+Pb** + total, others	**Annealed**: all **diam**, 33% in 2 in. (50 mm) **El**, 75 max **HV**; **HA**: all **diam**, 260 MPa **TS**, 19% in 2 in. (50 mm) **El**, 85 **HV**; **HB – strain hardened**: 0.2–5 mm **diam**, 310 MPa **TS**, 8% in 2 in. (50 mm) **El**, 110 **HV**
ISO 426 1	CuZn5	ISO	Plate, sheet, strip, tube, wire	94.0–96.0 **Cu**, 0.05 **Pb**, rem **Zn**, 0.1 **Fe**, 0.3 **Pb+Fe**	

WROUGHT COPPER AND COPPER ALLOYS

specification number	designation	country	product forms	chemical composition	mechanical properties and hardness values
JIS H 3260	C 2100	Japan	Rod and bar	94.0–96.0 **Cu**, 0.05 **Pb**, rem **Zn**, 0.05 **Fe**	**Annealed**: ≥0.5 mm **diam**, 206 MPa **TS**, 20% in 2 in. (50 mm) **El**; **1/2 hard**: 0.5–10 mm **diam**, 324 MPa **TS**; **Hard**: 0.5–10 mm **diam**, 412 MPa **TS**
JIS H 3100	C 2100	Japan	Sheet, plate, and strip	94.0–96.0 **Cu**, 0.05 **Pb**, rem **Zn**, 0.05 **Fe**	**Annealed**: 0.3–30 mm **diam**, 206 MPa **TS**, 33% in 2 in. (50 mm) **El**; **1/2 hard**: 0.3–20 mm **diam**, 255 MPa **TS**, 18% in 2 in. (50 mm) **El**; **Hard**: 0.3–10 mm **diam**, 284 MPa **TS**
DGN–W–24	Type No. 1	Mexico	Wire	94–96 **Cu**, 0.05 **Pb**, rem **Zn**, 0.05 **Fe**	**3/4 hard**: <19.05 mm **diam**, 57 ksi (393 MPa) **TS**; **Hard**: <12.7 mm **diam**, 61 ksi (421 MPa) **TS**; **Spring hard**: <6.35mm **diam**, 72 ksi (496 MPa) **TS**
DGN–W–27	TU–95	Mexico	Sheet and strip	94–96 **Cu**, 0.03 **Pb**, rem **Zn**, 0.2 **Ni**	**Hard**: all **diam**, 50 max ksi (343 max MPa) **TS**, 80–100 **HB**, 48–64 **HRB**; **1/2 hard**: all **diam**, 43 max ksi (294 max MPa) **TS**, 71–80 **HB**, 36–48 **HRB**; **1/4 hard**: all **diam**, 40 max ksi (274 max MPa) **TS**, 65–71 **HB**, 25–36 **HRB**
DGN–W–27	TU–90	Mexico	Sheet and strip	89–92 **Cu**, 0.08 **Pb**, rem **Zn**, 0.2 **Ni**	**Hard**: all **diam**, 50 max ksi (343 max MPa) **TS**, 80–100 **HB**, 48–64 **HRB**; **1/2 hard**: all **diam**, 43 max ksi (294 max MPa) **TS**, 71–80 **HB**, 36–48 **HRB**; **1/4 hard**: all **diam**, 40 max ksi (274 max MPa) **TS**, 65–71 **HB**, 25–36 **HRB**
ANSI/ASTM B 372	CA 220	US	Tube: seamless	89.0–91.0 **Cu**, 0.05 **Pb**, rem **Zn**, 0.05 **Fe**, silver included in copper content	**Cold drawn and annealed**: 43–66 **HR30T**
ASTM F 587	CA 220	US	Tube: welded	89.0–91.0 **Cu**, 0.05 **Pb**, rem **Zn**, 0.05 **Fe**	**Welded from annealed strip**: all **diam**, 37 ksi (255 MPa) **TS**, 10 min **HR30T**; **Cold reduced or light drawn**: all **diam**, 40 ksi (275 MPa) **TS**, 38 min **HR30T**; **Soft anneal**: <0.045 in. (<1.14 mm) **diam**, 30 **HR30T**
ANSI/ASTM B 36	C 22000	US	Plate, sheet, strip, bar	89.0–91.0 **Cu**, 0.05 **Pb**, rem **Zn**, 0.05 **Fe**	**Quarter hard**: 0.020 in. (0.508 mm) **diam**, 40 ksi (275 MPa) **TS**, 27 **HRB**; **Hard**: 0.020 in. (0.508 mm) **diam**, 57 ksi (395 MPa) **TS**, 65 **HRB**; **Spring**: 0.020 in. (0.508 mm) **diam**, 69 ksi (475 MPa) **TS**, 76 **HRB**

WROUGHT COPPER AND COPPER ALLOYS

specification number	designation	country	product forms	chemical composition	mechanical properties and hardness values
ANSI/ASTM B 134	CA 220	US	Wire	89.0–91.0 **Cu**, 0.05 **Pb**, rem **Zn**, 0.05 **Fe**	**Quarter hard**: >0.020 in. (>0.508 mm) **diam**, 45 ksi (310 MPa) **TS**; **Hard**: >0.020 in. (>0.508 mm) **diam**, 70 ksi (485 MPa) **TS**; **Spring**: >0.020 in. (>0.508 mm) **diam**, 84 ksi (580 MPa) **TS**
ANSI/ASTM B 131	C 22000	US	Cups: bullet jacket	89.0–91.0 **Cu**, 0.05 **Pb**, rem **Zn**, 0.05 **Fe**	
ANSI/ASTM B 130	C 22000	US	Strip	89.0–91.0 **Cu**, 0.05 **Pb**, rem **Zn**, 0.05 **Fe**	**Hard**: >0.012 in. (>0.305 mm) **diam**, 57 ksi (395 MPa) **TS**, 60 **HR30T**; **Spring**: >0.012 in. (>0.305 mm) **diam**, 69 ksi (475 MPa) **TS**, 68 **HR30T**; **Intermediate anneal**: >0.100 in. (>2.54 mm) **diam**, 36 ksi (250 MPa) **TS**, 35% in 2 in. (50.8 mm) **El**
ANSI/ASTM B 135	CA 220	US	Tube: seamless	89.0–91.0 **Cu**, 0.05 **Pb**, rem **Zn**, 0.05 **Fe**	**Drawn**: all **diam**, 40 ksi (275 MPa) **TS**, 38 **HR30T**; **Hard drawn**: >0.020 in. (0.508 mm) **diam**, 52 ksi (360 MPa) **TS**, 55 **HR30T**; **Spring anneal**: >0.045 in. (>1.14 mm) **diam**, 30 **HR30T**
AS 1567	220A	Australia	Rod, bar, section	89.0–91.0 **Cu**, 0.05 **Pb**, rem **Zn**, 0.05 **Fe**, 0.10 total, others	**As manufactured**: >6 mm **diam**, 280 MPa **TS**, 24% in 2 in. (50 mm) **El**
AS 1566	220A	Australia	Plate, bar, sheet, strip, foil	89.0–91.0 **Cu**, 0.05 **Pb**, rem **Zn**, 0.05 **Fe**, 0.10 total, others	**Annealed (sheet, strip)**: all **diam**, 245 MPa **TS**, 40% in 2 in. (50 mm) **El**, 70 **HV**; **Strain hardened to 1/2 hard (sheet, strip)**: all **diam**, 310 MPa **TS**, 95–115 **HV**; **Strain hardened to hard (sheet, strip)**: all **diam**, 360 MPa **TS**, 110–135 **HV**
BS 2870	CZ 101	UK	Sheet, strip, foil	89.0–91.0 **Cu**, 0.10 **Pb**, rem **Zn**, 0.10 **Fe**, 0.40 total, others	**Annealed**: ≤10.0 mm **diam**, 172 MPa **TS**, 35% in 2 in. (50 mm) **El**, 75 max **HV**; **1/2 hard**: ≤3.5 mm **diam**, 217 MPa **TS**, 7% in 2 in. (50 mm) **El**, 95 **HV**; **Hard**: ≤10.0 mm **diam**, 245 MPa **TS**, 3% in 2 in. (50 mm) **El**, 110 **HV**
BS 2873	CZ101	UK	Wire	89.0–91.0 **Cu**, 0.10 **Pb**, rem **Zn**, 0.10 **Fe**, 0.40 total, others	

WROUGHT COPPER AND COPPER ALLOYS

specification number	designation	country	product forms	chemical composition	mechanical properties and hardness values
CDA 220		US	Rod, tube, wire, plate, sheet, strip	89.0–91.0 **Cu**, 0.05 **Pb**, rem **Zn**, 0.05 **Fe**, 0.10 total, others	**Eighth hard temper (rod)**: 0.500 in. **diam**, 45 ksi (310 MPa) **TS**, 25% in 2 in. (50 mm) **El**, 42 **HRB**; **Hard drawn 35% (tube)**: 1.0 in. **diam**, 60 ksi (414 MPa) **TS**, 6% in 2 in. (50 mm) **El**, 69 **HRB**; **As hot rolled (flat products: plate, sheet, strip)**: 0.040 in. **diam**, 39 ksi (269 MPa) **TS**, 44% in 2 in. (50 mm) **El**, 60 **HRB**
CSA HC.4.2	HC 4.Z10	Canada	Plate	89.0–91.0 **Cu**, 0.05 **Pb**, rem **Zn**, 0.05 **Fe**, 0.10 other elements	**Half hard**: all **diam**, 47 ksi (324 MPa) **TS**; **Full hard**: all **diam**, 57 ksi (393 MPa) **TS**; **Spring**: all **diam**, 69 ksi (475 MPa) **TS**
CSA HC.4.2	HC 4.Z10	Canada	Sheet	89.0–91.0 **Cu**, 0.05 **Pb**, rem **Zn**, 0.05 **Fe**, 0.10 other elements	**Half hard**: all **diam**, 47 ksi (324 MPa) **TS**; **Full hard**: all **diam**, 57 ksi (393 MPa) **TS**; **Spring**: all **diam**, 69 ksi (475 MPa) **TS**
CSA HC.4.2	HC 4.Z10	Canada	Bar	89.0–91.0 **Cu**, 0.05 **Pb**, rem **Zn**, 0.05 **Fe**, 0.10 other elements	**Half hard**: all **diam**, 47 ksi (324 MPa) **TS**; **Full hard**: all **diam**, 57 ksi (393 MPa) **TS**; **Spring**: all **diam**, 69 ksi (475 MPa) **TS**
CSA HC.4.2	HC 4.Z10	Canada	Strip	89.0–91.0 **Cu**, 0.05 **Pb**, rem **Zn**, 0.05 **Fe**, 0.10 other elements	**Half hard**: all **diam**, 47 ksi (324 MPa) **TS**; **Full hard**: all **diam**, 57 ksi (393 MPa) **TS**; **Spring**: all **diam**, 69 ksi (475 MPa) **TS**
COPANT 162	C 220 00	COPANT	Wrought products	89.0–91.0 **Cu**, 0.05 **Pb**, rem **Zn**, 0.05 **Fe**	
COPANT 612	Cu Zn 10	COPANT	Strip, sheet, plate, foil, bar	89–91 **Cu**, 0.05 **Pb**, rem **Zn**, 0.05 **Fe**, 0.10 total, others	**1/4 hard temper**: all **diam**, 275 MPa **TS**; **1/2 hard temper**: all **diam**, 325 MPa **TS**; **Hard temper**: all **diam**, 390 MPa **TS**
QQ-W-321d	220	US	Wire	89.0–91.0 **Cu**, 0.05 **Pb**, rem **Zn**, 0.05 **Fe**, 0.10 total, others; **Co** in **Ni**	**Eighth hard temper**: all **diam**, 38 ksi (262 MPa) **TS**; **Quarter hard temper**: all **diam**, 310 MPa **TS**, 45% in 2 in. (50 mm) **El**; **Half hard temper**: all **diam**, 56 ksi (386 MPa) **TS**
SFS 2915	CuZn10	Finland	Sheet and strip	89.0–91.0 **Cu**, 0.05 **Pb**, 10 **Zn**, 0.05 **Fe**	**Annealed**: 0.2–0.5 mm **diam**, 260 MPa **TS**, 60 MPa **YS**, 25% in 2 in. (50 mm) **El**; **Cold finished**: 0.2–0.5 mm **diam**, 290 MPa **TS**, 190 MPa **YS**, 2% in 2 in. (50 mm) **El**

WROUGHT COPPER AND COPPER ALLOYS

specification number	designation	country	product forms	chemical composition	mechanical properties and hardness values
NF A 51 104	CuZn 10	France	Bar, wire, and shapes	89.0–91.0 **Cu**, 0.05 **Pb**, rem **Zn**, 0.1 **Fe**, 0.4 total, others	**Mill condition**: ≤50 mm **diam**, 320 MPa **TS**, 20% in 2 in. (50 mm) **El**
NF A 51 103	CuZn 10	France	Tube	89.0–91.0 **Cu**, rem **Zn**, 0.05 each, 0.40 total, others	**1/2 hard**: ≤80 mm **diam**, 280 MPa **TS**, 35% in 2 in. (50 mm) **El**, 85–115 **HV**; **3/4 hard**: ≤80 mm **diam**, 300 MPa **TS**, 25% in 2 in. (50 mm) **El**, 95–125 **HV**; **Hard**: ≤80 mm **diam**, 350 MPa **TS**, 7% in 2 in. (50 mm) **El**, 110–140 **HV**
NF A 51 101	CuZn 10	France	Sheet and strip	90.0 **Cu**, 10.0 **Zn**	**Stress hardened, Temper 1**: all **diam**, 270 MPa **TS**
ISO 1638	CuZn10	ISO	Solid product: drawn, on coils or reels	89.0–91.0 **Cu**, 0.05 **Pb**, rem **Zn**, 0.1 **Fe**, 0.3 **Ni** included in **Cu**; 0.4 **Fe+Pb** + total, others	**Annealed**: 1–5 mm **diam**, 240 MPa **TS**, 30% in 2 in. (50 mm) **El**; **HC**: 1–5 mm **diam**, 350 MPa **TS**, 5% in 2 in. (50 mm) **El**
ISO 1634	CuZn10	ISO	Plate, sheet, strip	89.0–91.0 **Cu**, 0.05 **Pb**, rem **Zn**, 0.1 **Fe**, **Cu** includes 0.3 **Ni**; 0.4 **Fe+Pb** + total, others	**Annealed**: all **diam**, 35% in 2 in. (50 mm) **El**, 75 max **HV**; **Specially annealed (OS35)**: 0.2–5 mm **diam**, 35% in 2 in. (50 mm) **El**, 75 max **HV**; **Strain hardened (HB)**: 5–10 mm **diam**, 340 MPa **TS**, 10% in 2 in. (50 mm) **El**, 115 **HV**
ISO 426 1	CuZn10	ISO	Plate, sheet, strip, rod, bar, tube, wire	89.0–91.0 **Cu**, 0.05 **Pb**, rem **Zn**, 0.1 **Fe**, 0.4 **Pb+Fe**	
JIS H 3300	C 2200	Japan	Pipe and tube: seamless	89.0–91.0 **Cu**, 0.05 **Pb**, rem **Zn**, 0.05 **Fe**	**Annealed**: 10–150 mm OD **diam**, 226 MPa **TS**, 35% in 2 in. (50 mm) **El**, 70 max **HRF**; **1/2 hard**: 10–150 mm OD **diam**, 275 MPa **TS**, 15% in 2 in. (50 mm) **El**, 38 min **HR30T**; **Hard**: 10–100 mm OD **diam**, 363 MPa **TS**, 55 min **HR30T**
JIS H 3260	C 2200	Japan	Rod and bar	89.0–91.0 **Cu**, 0.05 **Pb**, rem **Zn**, 0.05 **Fe**	**Annealed**: ≥0.5 mm **diam**, 226 MPa **TS**, 20% in 2 in. (50 mm) **El**; **1/2 hard**: 0.5–12 mm **diam**, 343 MPa **TS**; **Hard**: 0.5–10 mm **diam**, 471 MPa **TS**

WROUGHT COPPER AND COPPER ALLOYS

specification number	designation	country	product forms	chemical composition	mechanical properties and hardness values
JIS H 3100	C 2200	Japan	Sheet, plate, and strip	89.0–91.0 **Cu**, 0.05 **Pb**, rem **Zn**, 0.05 **Fe**	**Annealed**: 0.3–30 mm **diam**, 226 MPa **TS**, 35% in 2 in. (50 mm) **El**; **1/2 hard**: 0.3–20 mm **diam**, 275 MPa **TS**, 20% in 2 in. (50 mm) **El**; **Hard**: 0.3–10 mm **diam**, 324 MPa **TS**
DGN-W-24	Type No. 2	Mexico	Wire	89–91 **Cu**, 0.05 **Pb**, rem **Zn**, 0.05 **Fe**	**3/4 hard**: <19.05 mm **diam**, 64 ksi (441 MPa) **TS**; **Hard**: <12.7 mm **diam**, 70 ksi (483 MPa) **TS**; **Spring hard**: <6.35mm **diam**, 84 ksi (579 MPa) **TS**
NS 16 106	CuZn10	Norway	Sheet, strip, plate, bar, tube, and profiles	90 **Cu**, 10 **Zn**	
CDA 226		US	Wire, flat wire, strip	86.0–89.0 **Cu**, 0.05 **Pb**, rem **Zn**, 0.05 **Fe**, 0.15 total, others	**Hard temper (wire)**: 0.080 in. **diam**, 83 ksi (572 MPa) **TS**, 5% in 2 in. (50 mm) **El**; **Half hard temper (flat wire, strip)**: 0.040 in. **diam**, 54 ksi (372 MPa) **TS**, 12% in 2 in. (50 mm) **El**, 61 **HRB**; **Quarter hard temper (flat wire, strip)**: 0.040 in. **diam**, 47 ksi (324 MPa) **TS**, 25% in 2 in. (50 mm) **El**, 47 **HRB**
COPANT 162	C 226 00	COPANT	Wrought products	86.0–89.0 **Cu**, 0.05 **Pb**, rem **Zn**, 0.05 **Fe**	
DGN-W-27	TU-87	Mexico	Sheet and strip	86–89 **Cu**, 0.08 **Pb**, rem **Zn**, 0.2 **Ni**	**Hard**: all **diam**, 60 max ksi (412 MPa) **TS**, 70–90 **HB**, 45–57 **HRB**; **1/2 hard**: all **diam**, 47 max ksi (323 MPa) **TS**, 65–70 **HB**, 29–35 **HRB**; **1/4 hard**: all **diam**, 44 max ksi (304 max MPa) **TS**, 60–65 **HB**, 17–29 **HRB**
ANSI/ASTM B 395	C 23000	US	Tube: seamless	84.0–86.0 **Cu**, 0.06 **Pb**, rem **Zn**, 0.05 **Fe**, silver included in copper content	**Annealed**: <0.048 in. (<1.21 mm) **diam**, 40 ksi (275 MPa) **TS**, 12 ksi (85 MPa) **YS**, 12% in 2 in. (50 mm) **El**
ANSI/ASTM B 359	C 23000	US	Tube: seamless	84.0–86.0 **Cu**, 0.06 **Pb**, rem **Zn**, 0.05 **Fe**	**Annealed**: all **diam**, 40 ksi (275 MPa) **TS**, 12 ksi (85 MPa) **YS**
ASTM F 587	CA 230	US	Tube: welded	84.0–86.0 **Cu**, 0.05 **Pb**, rem **Zn**, 0.05 **Fe**	**Welded from annealed strip**: all **diam**, 40 ksi (275 MPa) **TS**, 24 min **HR30T**; **Cold reduced or light drawn**: all **diam**, 44 ksi (305 MPa) **TS**, 43 min **HR30T**; **Soft anneal**: <0.045 in. (<1.14 mm) **diam**, 36 **HR30T**

WROUGHT COPPER AND COPPER ALLOYS

specification number	designation	country	product forms	chemical composition	mechanical properties and hardness values
ANSI/ASTM B 543	C 23000	US	Tube: welded	84.0–86.0 **Cu**, 0.06 **Pb**, rem **Zn**, 0.05 **Fe**, silver included in copper content	**As welded from annealed strip**: all **diam**, 42 ksi (290 MPa) **TS**, 20 ksi (140 MPa) **YS**; **Welded and annealed**: all **diam**, 40 ksi (275 MPa) **TS**, 12 ksi (85 MPa) **YS**; **Fully finished–hard drawn**: all **diam**, 70 ksi (485 MPa) **TS**, 58 ksi (400 MPa) **YS**
ANSI/ASTM B 36	C 23000	US	Plate, sheet, strip, bar	84.0–86.0 **Cu**, 0.05 **Pb**, rem **Zn**, 0.05 **Fe**	**Quarter hard**: 0.020 in. (0.508 mm) **diam**, 44 ksi (305 MPa) **TS**, 33 **HRB**; **Hard**: 0.020 in. (0.508 mm) **diam**, 63 ksi (435 MPa) **TS**, 72 **HRB**; **Spring**: 0.020 in. (0.508 mm) **diam**, 78 ksi (540 MPa) **TS**, 82 **HRB**
ANSI/ASTM B 111	C 23000	US	Tube: seamless, and stock: ferrule	84.0–86.0 **Cu**, 0.06 **Pb**, rem **Zn**, 0.05 **Fe**	**Annealed**: all **diam**, 40 ksi (275 MPa) **TS**, 12 ksi (85 MPa) **YS**
ANSI/ASTM B 134	CA 230	US	Wire	84.0–86.0 **Cu**, 0.05 **Pb**, rem **Zn**, 0.05 **Fe**	**Quarter hard**: >0.020 in. (>0.508 mm) **diam**, 53 ksi (365 MPa) **TS**; **Hard**: >0.020 in. (>0.508 mm) **diam**, 83 ksi (570 MPa) **TS**; **Spring**: >0.020 in. (>0.508 mm) **diam**, 100 ksi (690 MPa) **TS**
ANSI/ASTM B 135	CA 230	US	Tube: seamless	84.0–86.0 **Cu**, 0.06 **Pb**, rem **Zn**, 0.05 **Fe**	**Drawn**: all **diam**, 44 ksi (305 MPa) **TS**, 43 **HR30T**; **Hard drawn**: >0.020 in. (0.508 mm) **diam**, 57 ksi (395 MPa) **TS**, 65 **HR30T**; **Soft anneal**: >0.045 in. (>1.14 mm) **diam**, 36 **HR30T**
AS 1566	230A	Australia	Plate, bar, sheet, strip, foil	84.0–86.0 **Cu**, 0.05 **Pb**, rem **Zn**, 0.05 **Fe**, 0.15 total, others	**Annealed (sheet, strip)**: all **diam**, 255 MPa **TS**, 45% in 2 in. (50 mm) **El**, 70 **HV**; **Strain hardened to 1/2 hard (sheet, strip)**: all **diam**, 320 MPa **TS**, 95–115 **HV**; **Strain hardened to full hard (sheet, strip)**: all **diam**, 370 MPa **TS**, 110–135 **HV**
BS 2870	CZ 102	UK	Sheet, strip, foil	84.0–86.0 **Cu**, 0.10 **Pb**, rem **Zn**, 0.10 **Fe**, 0.40 total, others	**Annealed**: ≤10.0 mm **diam**, 172 MPa **TS**, 35% in 2 in. (50 mm) **El**, 75 max **HV**; **1/2 hard**: ≤3.5 mm **diam**, 228 MPa **TS**, 7% in 2 in. (50 mm) **El**, 95 **HV**; **Hard**: ≤10.0 mm **diam**, 259 MPa **TS**, 3% in 2 in. (50 mm) **El**, 110 **HV**

WROUGHT COPPER AND COPPER ALLOYS

specification number	designation	country	product forms	chemical composition	mechanical properties and hardness values
BS 2873	CZ102	UK	Wire	84.0–86.0 **Cu**, 0.10 **Pb**, rem **Zn**, 0.10 **Fe**, 0.40 total, others	**Annealed**: 0.5–10.0 mm **diam**, 286 MPa **TS**, 25% in 2 in. (50 mm) **El**; **1/2 hard**: 0.5–10.0 mm **diam**, 432 MPa **TS**; **Hard**: 0.5–10.0 mm **diam**, 588 MPa **TS**
CDA 230		US	Wire, tube, pipe, sheet, strip	84.0–86.0 **Cu**, 0.05 **Pb**, rem **Zn**, 0.05 **Fe**, 0.15 total, others	**Quarter hard temper (wire)**: 0.080 in. **diam**, 59 ksi (407 MPa) **TS**, 11% in 2 in. (50 mm) **El**; **Light drawn 15% (tube)**: 1.0 in. **diam**, 50 ksi (345 MPa) **TS**, 30% in 2 in. (50 mm) **El**, 55 **HRB**; **Hard temper (sheet, strip)**: 0.040 in. **diam**, 70 ksi (483 MPa) **TS**, 5% in 2 in. (50 mm) **El**, 77 **HRB**
CSA HC.4.2	HC 4.Z15	Canada	Strip	84.0–86.0 **Cu**, 0.05 **Pb**, rem **Zn**, 0.05 **Fe**, 0.15 other elements	**Half hard**: all **diam**, 51 ksi (351 MPa) **TS**; **Full hard**: all **diam**, 63 ksi (434 MPa) **TS**; **Spring**: all **diam**, 78 ksi (537 MPa) **TS**
CSA HC.4.2	HC 4.Z15	Canada	Sheet	84.0–86.0 **Cu**, 0.05 **Pb**, rem **Zn**, 0.05 **Fe**, 0.15 other elements	**Half hard**: all **diam**, 51 ksi (351 MPa) **TS**; **Full hard**: all **diam**, 63 ksi (434 MPa) **TS**; **Spring**: all **diam**, 78 ksi (537 MPa) **TS**
CSA HC.4.2	HC 4.Z15	Canada	Plate	84.0–86.0 **Cu**, 0.05 **Pb**, rem **Zn**, 0.05 **Fe**, 0.15 other elements	**Half hard**: all **diam**, 51 ksi (351 MPa) **TS**; **Full hard**: all **diam**, 63 ksi (434 MPa) **TS**; **Spring**: all **diam**, 78 ksi (537 MPa) **TS**
CSA HC.4.2	HC 4.Z15	Canada	Bar	84.0–86.0 **Cu**, 0.05 **Pb**, rem **Zn**, 0.05 **Fe**, 0.15 other elements	**Half hard**: all **diam**, 51 ksi (351 MPa) **TS**; **Full hard**: all **diam**, 63 ksi (434 MPa) **TS**; **Spring**: all **diam**, 78 ksi (537 MPa) **TS**
COPANT 162	C 230 00	COPANT	Wrought products	84.0–86.0 **Cu**, 0.05 **Pb**, rem **Zn**, 0.05 **Fe**	
COPANT 521	Cu Zn 15	COPANT	Tube	84–86 **Cu**, 0.05 **Pb**, rem **Zn**, 0.05 **Fe**, 0.15 total, others	**Annealed**: 10.3–324 mm OD **diam**, 275 MPa **TS**, 80 MPa **YS**, 0.5% in 2 in. (50 mm) **El**
COPANT 612	Cu Zn 15	COPANT	Strip, sheet, plate, foil, bar	84–86 **Cu**, 0.05 **Pb**, rem **Zn**, 0.05 **Fe**, 0.15 total, others	**1/4 hard temper**: all **diam**, 305 MPa **TS**; **1/2 hard temper**: all **diam**, 350 MPa **TS**; **Hard temper**: all **diam**, 430 MPa **TS**
DS 3003	5112	Denmark	Rolled, drawn, extruded, and forged products	84.0–86.0 **Cu**, 0.05 **Pb**, 15 **Zn**, 0.1 **Fe**, 0.04 total, others	

WROUGHT COPPER AND COPPER ALLOYS

specification number	designation	country	product forms	chemical composition	mechanical properties and hardness values
QQ-W-321d	230	US	Wire	84.0–86.0 **Cu**, 0.05 **Pb**, rem **Zn**, 0.05 **Fe**, 0.15 total, others; **Co** in **Ni**	**Eighth hard temper**: all **diam**, 43 ksi (296 MPa) **TS**; **Quarter hard temper**: all **diam**, 53 ksi (365 MPa) **TS**; **Half hard temper**: all **diam**, 66 ksi (455 MPa) **TS**
QQ-B-626D	230	US	Rod, shapes, forgings, bar, strip	84.0–86.0 **Cu**, 0.05 **Pb**, rem **Zn**, 0.05 **Fe**, 0.15 total, others	**Quarter hard temper**: all **diam**, 44 ksi (303 MPa) **TS**; **Half hard temper**: all **diam**, 51 ksi (352 MPa) **TS**; **Hard temper**: all **diam**, 63 ksi (434 MPa) **TS**
QQ-B-613D	230	US	Plate, bar, sheet, strip	84.0–86.0 **Cu**, 0.05 **Pb**, rem **Zn**, 0.05 **Fe**, 0.15 total, others	**Quarter hard temper**: all **diam**, 44 ksi (303 MPa) **TS**; **Half hard temper**: all **diam**, 51 ksi (352 MPa) **TS**; **Hard temper**: all **diam**, 63 ksi (434 MPa) **TS**
SFS 2916	CuZn15	Finland	Sheet, strip, bar, rod, wire, tube, and shapes	84.0–86.0 **Cu**, 0.05 **Pb**, 15 **Zn**, 0.1 **Fe**	**Annealed (sheet and strip)**: 0.2–0.5 mm **diam**, 270 MPa **TS**, 70 MPa **YS**, 20% in 2 in. (50 mm) **El**; **Drawn (rod and wire)**: 0.2–1 mm **diam**, 370 MPa **TS**, 290 MPa **YS**, 3% in 2 in. (50 mm) **El**; **Cold finished (tube)**: ≤5 mm **diam**, 370 MPa **TS**, 290 MPa **YS**, 10% in 50 mm **El**
NF A 51 104	CuZn 15	France	Bar, wire, and shapes	84.0–86.0 **Cu**, 0.05 **Pb**, rem **Zn**, 0.1 **Fe**, 0.4 total, others	**Mill condition**: ≤50 mm **diam**, 330 MPa **TS**, 25% in 2 in. (50 mm) **El**
NF A 51 101	CuZn 15	France	Sheet and strip	85.0 **Cu**, 15.0 **Zn**	**Stress hardened, Temper 1**: all **diam**, 300 MPa **TS**
ISO 1638	CuZn15	ISO	Solid product: drawn, on coils or reels	84.0–86.0 **Cu**, 0.05 **Pb**, rem **Zn**, 0.1 **Fe**, 0.3 **Ni** included in **Cu**; 0.4 **Fe+Pb** + total, others	**Annealed**: 1–5 mm **diam**, 260 MPa **TS**, 30% in 2 in. (50 mm) **El**; **HC**: 1–5 mm **diam**, 270 MPa **TS**, 5% in 2 in. (50 mm) **El**
ISO 1637	CuZn15	ISO	Solid product: straight lengths	84.0–86.0 **Cu**, 0.05 **Pb**, rem **Zn**, 0.1 **Fe**, 0.3 **Ni** included in **Cu**; 0.4 **Fe+Pb** + total, others	**Annealed**: 5 mm **diam**, 40% in 2 in. (50 mm) **El**, 85 max **HV**; **HA**: all **diam**, 310 MPa **TS**, 25% in 2 in. (50 mm) **El**, 100 **HV**
ISO 1635	CuZn15	ISO	Tube	84.0–86.0 **Cu**, 0.05 **Pb**, rem **Zn**, 0.1 **Fe**, **Cu** includes 0.3 **Ni**; 0.4 **Fe+Pb** + total, others	**Annealed**: all **diam**, 40% in 2 in. (50 mm) **El**, 80 max **HV**; **Strain hardened (HB)**: all **diam**, 390 MPa **TS**, 17% in 2 in. (50 mm) **El**, 90–140 **HV**

WROUGHT COPPER AND COPPER ALLOYS

specification number	designation	country	product forms	chemical composition	mechanical properties and hardness values
ISO 1634	CuZn15	ISO	Plate, sheet, strip	84.0–86.0 **Cu**, 0.05 **Pb**, rem **Zn**, 0.1 **Fe**, **Cu** includes 0.3 **Ni**; 0.4 **Fe**+**Pb** + total, others	**Annealed**: all **diam**, 35% in 2 in. (50 mm) **El**, 85 max **HV**; **Specially annealed (OS35)**: 0.2–5 mm **diam**, 35% in 2 in. (50 mm) **El**, 80 max **HV**; **Strain hardened (HB)**: 5–10 mm **diam**, 360 MPa **TS**, 12% in 2 in. (50 mm) **El**, 120 **HV**
ISO 426 1	CuZn15	ISO	Plate, sheet, strip, rod, bar, tube, wire	84.0–86.0 **Cu**, 0.05 **Pb**, rem **Zn**, 0.1 **Fe**, 0.4 **Pb**+**Fe**	
JIS H 3300	C 2300	Japan	Pipe and tube: seamless	84.0–86.0 **Cu**, 0.05 **Pb**, rem **Zn**, 0.05 **Fe**	**Annealed**: 10–150 mm OD **diam**, 275 MPa **TS**, 35% in 2 in. (50 mm) **El**, 36 max **HR30T**; **1/2 hard**: 1–150 mm OD **diam**, 304 MPa **TS**, 20% in 2 in. (50 mm) **El**, 43 min **HR30T**; **Hard**: 10–100 mm OD **diam**, 392 MPa **TS**, 65 min **HR30T**
JIS H 3260	C 2300	Japan	Rod and bar	84.0–86.0 **Cu**, 0.05 **Pb**, rem **Zn**, 0.05 **Fe**	**Annealed**: $\geq$0.5 mm **diam**, 245 MPa **TS**, 20% in 2 in. (50 mm) **El**; **1/2 hard**: 0.5–12 mm **diam**, 373 MPa **TS**; **Hard**: 0.5–10 mm **diam**, 520 MPa **TS**
JIS H 3100	C 2300	Japan	Sheet, plate, and strip	84.0–86.0 **Cu**, 0.05 **Pb**, rem **Zn**, 0.05 **Fe**	**Annealed**: 0.3–30 mm **diam**, 245 MPa **TS**, 40% in 2 in. (50 mm) **El**; **1/2 hard**: 0.3–20 mm **diam**, 294 MPa **TS**, 23% in 2 in. (50 mm) **El**; **Hard**: 0.3–10 mm **diam**, 343 MPa **TS**
DGN–W–19	Type No. 1	Mexico	Tube	84–86 **Cu**, 0.06 **Pb**, rem **Zn**, 0.05 **Fe**	**1/4 hard**: all **diam**, 44 ksi (303 MPa) **TS**, 45–82 **HRB**; **1/2 hard**: all **diam**, 44 ksi (303 MPa) **TS**, 45 min **HRB**; **Hard**: $\leq$25.4 mm **diam**, 57 ksi (394 MPa) **TS**, 70 min **HRB**
DGN–W–24	Type No. 3	Mexico	Wire	84–86 **Cu**, 0.05 **Pb**, rem **Zn**, 0.05 **Fe**	**3/4 hard**: $<$19.05 mm **diam**, 76 ksi (524 MPa) **TS**; **Hard**: $<$12.7 mm **diam**, 83 ksi (572 MPa) **TS**; **Spring hard**: $<$6.35mm **diam**, 100 ksi (689 MPa) **TS**
MIL–T–20168B		US	Tube: seamless	84–86 **Cu**, 0.06 **Pb**, rem **Zn**, 0.05 **Fe**	
NS 16 108	CuZn15	Norway	Sheet, strip, plate, bar, tube, rod, wire, and profiles	85 **Cu**, 15 **Zn**	

WROUGHT COPPER AND COPPER ALLOYS

specification number	designation	country	product forms	chemical composition	mechanical properties and hardness values
SIS 14 51 12	SIS Brass 51 12–05	Sweden	Plate, sheet, strip	84.0–86.0 **Cu**, 0.05 **Pb**, 15 **Zn**, 0.1 **Fe**, 0.4 total, others	**Strain hardened**: 5 mm **diam**, 360 MPa **TS**, 280 MPa **YS**, 10% in 2 in. (50 mm) **El**
SIS 14 51 12	SIS Brass 51 12–04	Sweden	Sheet, strip, wire, rod, plate	84.0–86.0 **Cu**, 0.05 **Pb**, 15 **Zn**, 0.1 **Fe**, 0.4 total, others	**Strain hardened**: 0.2–0.5 mm **diam**, 310 MPa **TS**, 200 MPa **YS**, 8% in 4 in. (100 mm) **El**, 90–120 **HV**
SIS 14 51 12	SIS Brass 51 12–02	Sweden	Rod, wire, strip, sheet, plate	84.0–86.0 **Cu**, 0.05 **Pb**, 15 **Zn**, 0.1 **Fe**, 0.4 total, others	**Strain hardened**: 2.5–5.0 mm **diam**, 250 MPa **TS**, 70 MPa **YS**, 35% in 4 in. (100 mm) **El**
WW–P–351a	230	US	Pipe: red brass	83.0–86.0 **Cu**, 0.06 **Pb**, rem **Zn**, 0.05 **Fe**	**Annealed**: all **diam**, 40 ksi (276 MPa) **TS**, 12 ksi (83 MPa) **YS**
WW–T–791A	Grade A	US	Tube	83.0–86.0 **Cu**, 0.06 **Pb**, rem **Zn**, 0.05 **Fe**	
DGN–W–27	TU–85	Mexico	Sheet and strip	83–86 **Cu**, 0.08 **Pb**, rem **Zn**, 0.2 **Ni**	**Hard**: all **diam**, 60 max ksi (412 max MPa) **TS**, 70–90 **HB**, 45–57 **HRB**; **1/2 hard**: all **diam**, 47 max ksi (323 max MPa) **TS**, 65–70 **HB**, 29–35 **HRB**; **1/4 hard**: all **diam**, 44 max ksi (304 max MPa) **TS**, 60–65 **HB**, 17–29 **HRB**
COPANT 162	C 234000	COPANT	Wrought products	81.0–84.0 **Cu**, 0.05 **Pb**, rem **Zn**, 0.05 **Fe**	
ANSI/ASTM B 36	C 24000	US	Plate, sheet, strip, bar	78.5–81.5 **Cu**, 0.05 **Pb**, rem **Zn**, 0.05 **Fe**	**Quarter hard**: 0.020 in. (0.508 mm) **diam**, 48 ksi (330 MPa) **TS**, 38 **HRB**; **Hard**: 0.020 in. (0.508 mm) **diam**, 63 ksi (435 MPa) **TS**, 72 **HRB**; **Spring**: 0.020 in. (0.508 mm) **diam**, 85 ksi (585 MPa) **TS**, 87 **HRB**
ANSI/ASTM B 134	CA 240	US	Wire	78.5–81.5 **Cu**, 0.05 **Sn**, 0.05 **Pb**, rem **Zn**	**Quarter hard**: >0.020 in. (>0.508 mm) **diam**, 62 ksi (425 MPa) **TS**; **Hard**: >0.020 in. (>0.508 mm) **diam**, 100 ksi (690 MPa) **TS**; **Spring**: >0.020 in. (>0.508 mm) **diam**, 116 ksi (800 MPa) **TS**
BS 2870	CZ 103	UK	Sheet, strip, foil	79.0–81.0 **Cu**, 0.10 **Pb**, rem **Zn**, 0.10 **Fe**, 0.40 total, others	**Annealed**: ≤10.0 mm **diam**, 186 MPa **TS**, 40% in 2 in. (50 mm) **El**, 80 max **HV**; **1/2 hard**: ≤3.5 mm **diam**, 238 MPa **TS**, 10% in 2 in. (50 mm) **El**, 95 **HV**; **Hard**: ≤10.0 mm **diam**, 288 MPa **TS**, 5% in 2 in. (50 mm) **El**, 120 **HV**

WROUGHT COPPER AND COPPER ALLOYS

specification number	designation	country	product forms	chemical composition	mechanical properties and hardness values
BS 2873	CZ103	UK	Wire	79.0–81.0 **Cu**, 0.10 **Pb**, rem **Zn**, 0.10 **Fe**, 0.40 total, others	**Annealed**: 0.5–10.0 mm **diam**, 309 MPa **TS**, 30% in 2 in. (50 mm) **El**; **1/2 hard**: 0.5–10.0 mm **diam**, 461 MPa **TS**; **Hard**: 0.5–10.0 mm **diam**, 608 MPa **TS**
BS 2874	CZ103	UK	Rod, section	79.0–81.0 **Cu**, 0.10 **Pb**, rem **Zn**, 0.10 **Fe**, 0.40 total, others	**As manufactured**: >6.0 mm **diam**, 309 MPa **TS**, 24% in 2 in. (50 mm) **El**
CDA 240		US	Wire, flatwire, strip	78.5–81.5 **Cu**, 0.05 **Pb**, rem **Zn**, 0.05 **Fe**, 0.15 total, others	**Half hard temper (wire)**: 0.080 in. **diam**, 82 ksi (565 MPa) **TS**, 8% in 2 in. (50 mm) **El**; **Hard temper (flat wire, strip)**: 0.040 in. **diam**, 74 ksi (510 MPa) **TS**, 7% in 2 in. (50 mm) **El**, 82 **HRB**; **Spring temper (flat wire, strip)**: 0.040 in. **diam**, 91 ksi (627 MPa) **TS**, 3% in 2 in. (50 mm) **El**, 91 **HRB**
CSA HC.4.2	HC 4.Z20	Canada	Sheet	78.5–81.5 **Cu**, 0.05 **Pb**, rem **Zn**, 0.05 **Fe**, 0.15 other elements	**Half hard**: all **diam**, 55 ksi (379 MPa) **TS**; **Full hard**: all **diam**, 68 ksi (468 MPa) **TS**; **Spring**: all **diam**, 85 ksi (586 MPa) **TS**
CSA HC.4.2	HC 4.Z20	Canada	Strip	78.5–81.5 **Cu**, 0.05 **Pb**, rem **Zn**, 0.05 **Fe**, 0.15 other elements	**Half hard**: all **diam**, 55 ksi (379 MPa) **TS**; **Full hard**: all **diam**, 68 ksi (468 MPa) **TS**; **Spring**: all **diam**, 85 ksi (586 MPa) **TS**
CSA HC.4.2	HC 4.Z 20	Canada	Plate	78.5–81.5 **Cu**, 0.05 **Pb**, rem **Zn**, 0.05 **Fe**, 0.15 other elements	**Half hard**: all **diam**, 55 ksi (379 MPa) **TS**; **Full hard**: all **diam**, 68 ksi (468 MPa) **TS**; **Spring**: all **diam**, 85 ksi (586 MPa) **TS**
CSA HC.4.2	HC 4.Z 20	Canada	Bar	78.5–81.5 **Cu**, 0.05 **Pb**, rem **Zn**, 0.05 **Fe**, 0.15 other elements	**Half hard**: all **diam**, 55 ksi (379 MPa) **TS**; **Full hard**: all **diam**, 68 ksi (468 MPa) **TS**; **Spring**: all **diam**, 85 ksi (586 MPa) **TS**
COPANT 612	Cu Zn 20	COPANT	Strip, sheet, plate, foil, bar	78.5–81.5 **Cu**, 0.05 **Pb**, rem **Zn**, 0.05 **Fe**, 0.15 total, others	**1/4 hard temper**: all **diam**, 335 MPa **TS**; **1/2 hard temper**: all **diam**, 380 MPa **TS**; **Hard temper**: all **diam**, 470 MPa **TS**
DS 3003	5114	Denmark	Rolled, drawn, extruded, and forged products	78.5–81.5 **Cu**, 0.05 **Pb**, 20 **Zn**, 0.1 **Fe**, 0.04 total, others	

WROUGHT COPPER AND COPPER ALLOYS

specification number	designation	country	product forms	chemical composition	mechanical properties and hardness values
QQ-W-321d	240	US	Wire	78.5–81.5 **Cu**, 0.05 **Pb**, rem **Zn**, 0.05 **Fe**, 0.15 total, others; **Co** in **Ni**	**Eighth hard temper**: all **diam**, 50 ksi (345 MPa) **TS**; **Quarter hard temper**: all **diam**, 62 ksi (427 MPa) **TS**; **Half hard temper**: all **diam**, 78 ksi (538 MPa) **TS**
QQ-B-626D	240	US	Rod, shapes, forgings, bar, strip	78.5–81.5 **Cu**, 0.05 **Pb**, rem **Zn**, 0.05 **Fe**, 0.15 total, others	**Quarter hard temper**: all **diam**, 48 ksi (331 MPa) **TS**; **Half hard temper**: all **diam**, 55 ksi (379 MPa) **TS**; **Hard temper**: all **diam**, 68 ksi (469 MPa) **TS**
QQ-B-613D	240	US	Plate, bar, sheet, strip	78.5–81.5 **Cu**, 0.05 **Pb**, rem **Zn**, 0.05 **Fe**, 0.15 total, others	**Quarter hard temper**: all **diam**, 48 ksi (331 MPa) **TS**; **Half hard temper**: all **diam**, 55 ksi (379 MPa) **TS**; **Hard temper**: all **diam**, 68 ksi (469 MPa) **TS**
SFS 2917	CuZn20	Finland	Sheet, strip, and wire	78.5–81.5 **Cu**, 0.05 **Pb**, 20 **Zn**, 0.1 **Fe**	**Annealed (sheet and strip)**: 0.2–0.5 mm **diam**, 280 MPa **TS**, 80 MPa **YS**, 20% in 2 in. (50 mm) **El**; **Cold finished (sheet and strip)**: 0.2–0.5 mm **diam**, 320 MPa **TS**, 200 MPa **YS**, 10% in 2 in. (50 mm) **El**
NF A 51 104	CuZn 20	France	Bar, wire, and shapes	78.5–81.5 **Cu**, 0.05 **Pb**, rem **Zn**, 0.1 **Fe**, 0.4 total, others	**Mill condition**: $\leq$50 mm **diam**, 330 MPa **TS**, 25% in 2 in. (50 mm) **El**
NF A 51 101	CuZn 20	France	Sheet and strip	80.0 **Cu**, 20.0 **Zn**	**Stress hardened, Temper 1**: all **diam**, 330 MPa **TS**
IS 4170	CuZn20	India	Rod	79–81 **Cu**, 0.01 **Pb**, rem **Zn**, 0.05 **Fe**, 0.3 total, others	**As manufactured**: $\geq$5 mm **diam**, 314 MPa **TS**, 25% in 2 in. (50 mm) **El**; **Annealed**: 245 MPa **TS**, 50% in 2 in. (50 mm) **El**, 90 **HV**
ISO 1638	CuZn20	ISO	Solid product: drawn, on coils or reels	78.5–81.5 **Cu**, 0.05 **Pb**, rem **Zn**, 0.1 **Fe**, 0.3 **Ni** included in **Cu**; 0.4 **Fe+Pb** + total, others	**Annealed**: 1–5 mm **diam**, 260 MPa **TS**, 35% in 2 in. (50 mm) **El**; **HC**: 1–5 mm **diam**, 390 MPa **TS**, 5% in 2 in. (50 mm) **El**
ISO 1634	CuZn20	ISO	Plate	78.5–81.5 **Cu**, 0.05 **Pb**, rem **Zn**, 0.1 **Fe**, **Cu** includes 0.3 **Ni**; 0.4 **Fe+Pb** + total, others	**Annealed**: all **diam**, 40% in 2 in. (50 mm) **El**, 80 max **HV**; **Specially annealed**: 0.2–5 mm **diam**, 40% in 2 in. (50 mm) **El**, 85 max **HV**; **Strain hardened (HB)**: 5–10 mm **diam**, 380 MPa **TS**, 12% in 2 in. (50 mm) **El**, 125 **HV**
ISO 426 1	CuZn20	ISO	Plate, sheet, strip, rod, bar, tube, wire	78.5–81.5 **Cu**, 0.05 **Pb**, rem **Zn**, 0.1 **Fe**, 0.4 **Pb+Fe**	

WROUGHT COPPER AND COPPER ALLOYS

specification number	designation	country	product forms	chemical composition	mechanical properties and hardness values
JIS H 3260	C 2400	Japan	Rod and bar	78.5–81.5 **Cu**, 0.05 **Pb**, rem **Zn**, 0.05 **Fe**	**Annealed**: ≥0.5 mm **diam**, 255 MPa **TS**, 20% in 2 in. (50 mm) **El**; **1/2 hard**: 0.5–12 mm **diam**, 373 MPa **TS**; **Hard**: 0.5–10 mm **diam**, 588 MPa **TS**
JIS H 3100	C 2400	Japan	Sheet, plate, and strip	78.5–81.5 **Cu**, 0.05 **Pb**, rem **Zn**, 0.05 **Fe**	**Annealed**: 0.3–30 mm **diam**, 255 MPa **TS**, 44% in 2 in. (50 mm) **El**; **1/2 hard**: 0.3–20 mm **diam**, 314 MPa **TS**, 25% in 2 in. (50 mm) **El**; **Hard**: 0.3–10 mm **diam**, 363 MPa **TS**
DGN-W-24	Type No. 4	Mexico	Wire	78.5–81.5 **Cu**, 0.05 **Pb**, rem **Zn**, 0.05 **Fe**	**3/4 hard**: <19.05 mm **diam**, 90 ksi (621 MPa) **TS**; **Hard**: <12.7 mm **diam**, 100 ksi (689 MPa) **TS**; **Spring hard**: <6.35mm **diam**, 116 ksi (800 MPa) **TS**
NS 16 110	CuZn20	Norway	Sheet, strip, plate, and bar	80 **Cu**, 20 **Zn**	
SIS 14 51 14	SIS Brass 51 14-05	Sweden	Plate, sheet, strip	78.5–81.5 **Cu**, 0.05 **Pb**, 20 **Zn**, 0.1 **Fe**, 0.4 total, others	**Strain hardened**: 5 mm **diam**, 380 MPa **TS**, 310 MPa **YS**, 2% in 2 in. (50 mm) **El**, 125–145 **HV**
SIS 14 51 14	SIS Brass 51 14-04	Sweden	Plate, sheet, strip	78.5–81.5 **Cu**, 0.05 **Pb**, 20 **Zn**, 0.1 **Fe**, 0.4 total, others	**Strain hardened**: 5 mm **diam**, 320 MPa **TS**, 200 MPa **YS**, 30% in 2 in. (50 mm) **El**, 90–120 **HV**
SIS 14 51 14	SIS Brass 51 14-02	Sweden	Plate, sheet, strip	78.5–81.5 **Cu**, 0.05 **Pb**, 20 **Zn**, 0.1 **Fe**, 0.4 total, others	**Annealed**: 5 mm **diam**, 280 MPa **TS**, 80 MPa **YS**, 45% in 2 in. (50 mm) **El**, 60–85 **HV**
COPANT 162	C 240 00	COPANT	Wrought products	78.0–81.5 **Cu**, 0.05 **Pb**, rem **Zn**, 0.05 **Fe**	
SFS 2928	CuZn20A12	Finland	Tube	76.0–79.0 **Cu**, 0.07 **Pb**, 20 **Zn**, 0.07 **Fe**, 1.8–2.3 **Al**, 0.010 **P**, 0.020–0.035 **As**, 0.035 **As+Pb**	**Annealed**: ≤10 mm **diam**, 330 MPa **TS**, 80 MPa **YS**, 50% in 2 in. (50 mm) **El**
JIS H 3300	C 6872	Japan	Pipe and tube: seamless	76.0–79.0 **Cu**, 0.07 **Pb**, rem **Zn**, 0.06 **Fe**, 1.8–2.5 **Al**, 0.20 **Mn**, 0.20–1.0 **Ni**, 0.10 **Cr**, 0.02–0.06 **As**	**Annealed**: 5–250 mm OD **diam**, 373 MPa **TS**, 40% in 2 in. (50 mm) **El**
JIS H 3300	C 6871	Japan	Pipe and tube: seamless	76.0–79.0 **Cu**, 0.07 **Pb**, rem **Zn**, 0.06 **Fe**, 0.20–0.50 **Si**, 1.8–2.5 **Al**, 0.02–0.06 **As**	**Annealed**: 5–250 mm OD **diam**, 373 MPa **TS**, 40% in 2 in. (50 mm) **El**
JIS H 3300	C 6870	Japan	Pipe and tube: seamless	76.0–79.0 **Cu**, 0.07 **Pb**, rem **Zn**, 0.06 **Fe**, 1.8–2.5 **Al**, 0.02–0.06 **As**	**Annealed**: 5–250 mm OD **diam**, 373 MPa **TS**, 40% in 2 in. (50 mm) **El**
COPANT 162	C 250 00	COPANT	Wrought products	74.0–76.0 **Cu**, 0.05 **Pb**, rem **Zn**, 0.05 **Fe**	

WROUGHT COPPER AND COPPER ALLOYS

specification number	designation	country	product forms	chemical composition	mechanical properties and hardness values
IS 4413	CuZn30	India	Wire	68–79 **Cu**, 0.03 **Pb**, rem **Zn**, 0.05 **Fe**, 0.30 total, others	**Annealed**: all **diam**, 309 MPa **TS**, 45% in 2 in. (50 mm) **El**; **1/2 hard**: all **diam**, 461 MPa **TS**; **Hard**: all **diam**, 618 MPa **TS**
BS 2875	CZ105	UK	Plate	70.0–73.0 **Cu**, 0.075 **Pb**, rem **Zn**, 0.06 **Fe**, 0.02–0.06 **As**, 0.30 total, others	**As manufactured or annealed**: >10.0 mm **diam**, 279 MPa **TS**, 40% in 2 in. (50 mm) **El**; **Hard**: 10–16 mm **diam**, 358 MPa **TS**, 18% in 2 in. (50 mm) **El**
IS 1545	CuZn30As	India	Tube	70.0–73.0 **Cu**, 0.075 **Pb**, rem **Zn**, 0.06 **Fe**, 0.30 total, others	**Annealed**: all **diam**, 372 max MPa **TS**; **As drawn**: all **diam**, 387 MPa **TS**
JIS H 3300	C 4430	Japan	Pipe and tube: seamless	70.0–73.0 **Cu**, 0.9–1.2 **Sn**, 0.07 **Pb**, rem **Zn**, 0.06 **Fe**, 0.02–0.06 **As**	**Annealed**: 5–250 mm OD **diam**, 314 MPa **TS**, 30% in 2 in. (50 mm) **El**
DGN-W-25	LA-70	Mexico	Sheet and strip	70–71 **Cu**, rem **Zn**, 0.03 **Fe**, 0.02 **Ni**	**Spring hard**: all **diam**, 80 max ksi (552 max MPa) **TS**; **Hard**: all **diam**, 64 max ksi (441 max MPa) **TS**; **1/2 hard**: all **diam**, 55 max ksi (380 max MPa) **TS**
AMS 4505E		US	Sheet, strip, and plate	68.50–71.50 **Cu**, 0.7 **Pb**, rem **Zn**, 0.05 **Fe**, 0.05 each, 0.15 total, others	
AMS 4507D		US	Sheet, strip, and plate	68.50–71.50 **Cu**, 0.07 **Pb**, rem **Zn**, 0.05 **Fe**, 0.05 each, 0.15 total, others	**Cold rolled to 1/2 hard**: all **diam**, 57 ksi (393 MPa) **TS**, 56–68 **HR30T**
ANSI/ASTM B 19	C 26000	US	Sheet, strip, plate, bar, disk	68.5–71.5 **Cu**, 0.07 **Pb**, rem **Zn**, 0.05 **Fe**	**Hard**: all **diam**, 71 ksi (490 MPa) **TS**, 79 **HRB**; **Spring**: all **diam**, 91 ksi (625 MPa) **TS**, 89 **HRB**; **As hot rolled**: >0.500 in. (>12.7 mm) **diam**, 40 ksi (275 MPa) **TS**, 60% in 2 in. (50 mm) **El**
ASTM F 587	CA 260	US	Tube: welded	68.5–71.5 **Cu**, 0.07 **Pb**, rem **Zn**, 0.05 **Fe**	**Welded from annealed strip**: all **diam**, 48 ksi (330 MPa) **TS**, 25 min **HR30T**; **Cold reduced or drawn**: all **diam**, 54 ksi (370 MPa) **TS**, 53 min **HR30T**; **Soft anneal**: <0.030 in. (<0.762 mm) **diam**, 40 **HR30T**

WROUGHT COPPER AND COPPER ALLOYS

specification number	designation	country	product forms	chemical composition	mechanical properties and hardness values
ANSI/ASTM B 569		US	Strip	68.5–71.5 **Cu**, 0.07 **Pb**, rem **Zn**, 0.05 **Fe**	**Quarter hard**: all **diam**, 49 ksi (340 MPa) **TS**, 35 ksi (240 MPa) **YS**, 12% in 2 in. (50 mm) **El**; **Half hard**: all **diam**, 57 ksi (395 MPa) **TS**, 45 ksi (310 MPa) **YS**, 10% in 2 in. (50 mm) **El**; **Annealed**: all **diam**, 58 ksi (400 MPa) **TS**, 25 ksi (170 MPa) **YS**, 25% in 2 in. (50 mm) **El**
ANSI/ASTM B 36	C 26000	US	Plate, sheet, strip, bar	68.5–71.5 **Cu**, 0.07 **Pb**, rem **Zn**, 0.05 **Fe**	**Quarter hard**: 0.020 in. (0.508 mm) **diam**, 49 ksi (340 MPa) **TS**, 40 **HRB**; **Hard**: 0.020 in. (0.508 mm) **diam**, 71 ksi (490 MPa) **TS**, 79 **HRB**; **Spring**: 0.020 in. (0.508 mm) **diam**, 91 ksi (625 MPa) **TS**, 89 **HRB**
ANSI/ASTM B 134	CA 260	US	Wire	68.5–71.5 **Cu**, 0.07 **Pb**, rem **Zn**, 0.05 **Fe**	**Quarter hard**: >0.020 in. (>0.508 mm) **diam**, 62 ksi (425 MPa) **TS**; **Hard**: >0.020 in. (>0.508 mm) **diam**, 102 ksi (700 MPa) **TS**; **Spring**: >0.020 in. (>0.508 mm) **diam**, 120 ksi (830 MPa) **TS**
ANSI/ASTM B 129	CA 260	US	Cups: cartridge case	68.5–71.5 **Cu**, 0.07 **Pb**, rem **Zn**, 0.05 **Fe**	
ANSI/ASTM B 135	CA 260	US	Tube: seamless	68.5–71.5 **Cu**, 0.07 **Pb**, rem **Zn**, 0.05 **Fe**	**Drawn**: all **diam**, 54 ksi (370 MPa) **TS**, 53 **HR30T**; **Hard drawn**: >0.020 in. (0.508 mm) **diam**, 66 ksi (455 MPa) **TS**, 70 **HR30T**; **Soft anneal**: >0.030 in. (>0.762 mm) **diam**, 40 **HR30T**
AS 1567	259D	Australia	Rod, bar, section	69.0–71.0 **Cu**, 0.07 **Pb**, rem **Zn**, 0.06 **Fe**, 0.02–0.06 **As**, 0.30 total, others	**As manufactured**: >6 mm **diam**, 340 MPa **TS**, 28% in 2 in. (50 mm) **El**; **Soft annealed**: >6 mm **diam**, 280 MPa **TS**, 45% in 2 in. (50 mm) **El**
AS 1567	260A	Australia	Rod, bar, section	68.5–71.5 **Cu**, 0.07 **Pb**, rem **Zn**, 0.05 **Fe**, 0.15 total, others	**Soft annealed**: >6 mm **diam**, 280 MPa **TS**, 45% in 2 in. (50 mm) **El**; **As manufactured**: >6 mm **diam**, 340 MPa **TS**, 28% in 2 in. (50 mm) **El**
AS 1566	260A	Australia	Plate, bar, sheet, strip, foil	68.5–71.5 **Cu**, 0.07 **Pb**, rem **Zn**, 0.05 **Fe**, 0.15 total, others	**Annealed (sheet, strip)**: all **diam**, 290 MPa **TS**, 55% in 2 in. (50 mm) **El**, 70 **HV**; **Strain hardened to 1/2 hard (sheet, strip)**: all **diam**, 360 MPa **TS**, 110–135 **HV**: Strain hardened to full hard (sheet, strip): all **diam**, 420 MPa **TS**, 135–165 **HV**

WROUGHT COPPER AND COPPER ALLOYS

specification number	designation	country	product forms	chemical composition	mechanical properties and hardness values
AS 1568	259D	Australia	Forgings	69.0–71.0 **Cu**, 0.07 **Pb**, rem **Zn**, 0.06 **Fe**, 0.02–0.06 **As**, 0.30 total, others	**As manufactured**: ≥6 mm **diam**, 280 MPa **TS**, 45% in 2 in. (50 mm) **El**
AS 1572	259D	Australia	Tube	69.0–71.0 **Cu**, 0.07 **Pb**, rem **Zn**, 0.06 **Fe**, 0.02–0.06 **As**, 0.30 total, others	**Annealed**: all **diam**, 80 **HV**; **Half–hard**: all **diam**, 95–130 **HV**; **Hard**: all **diam**, 150 **HV**
BS 2871	CZ 126	UK	Tube	69.0–71.0 **Cu**, 0.07 **Pb**, rem **Zn**, 0.06 **Fe**, 0.02–0.06 **As**, 0.30 total, others	**As drawn**: 150 **HV**; **Temper annealed**: 80–105 **HV**; **Annealed**: 75 max **HV**
BS 2871	CZ 126	UK	Tube	69.0–71.0 **Cu**, 0.07 **Pb**, rem **Zn**, 0.06 **Fe**, 0.02–0.06 **As**, 0.30 total, others	**Annealed**: 280 MPa **TS**, 40% in 2 in. (50 mm) **El**, 75 max **HV**; **Temper annealed**: 320 MPa **TS**, 35% in 2 in. (50 mm) **El**, 80–105 **HV**; **As drawn**: 400 MPa **TS**, 130 **HV**
BS 2870	CZ 106	UK	Sheet, strip, foil	68.5–71.5 **Cu**, 0.05 **Pb**, rem **Zn**, 0.05 **Fe**, 0.30 total, others	**Annealed**: ≤10.0 mm **diam**, 197 MPa **TS**, 50% in 2 in. (50 mm) **El**, 80 max **HV**; **1/2 hard**: ≤3.5 mm **diam**, 245 MPa **TS**, 20% in 2 in. (50 mm) **El**, 100 **HV**; **Hard**: ≤10.0 mm **diam**, 5% in 2 in. (50 mm) **El**, 125 **HV**
BS 2873	CZ106	UK	Wire	68.5–71.5 **Cu**, 0.05 **Pb**, rem **Zn**, 0.05 **Fe**, 0.30 total, others	**Annealed**: 0.5–10.0 mm **diam**, 309 MPa **TS**, 45% in 2 in. (50 mm) **El**; **1/2 hard**: 0.5–10.0 mm **diam**, 461 MPa **TS**; **Hard**: 0.5–10.0 mm **diam**, 608 MPa **TS**
BS 2875	CZ106	UK	Plate	68.5–71.5 **Cu**, 0.05 **Pb**, rem **Zn**, 0.05 **Fe**, 0.30 total, others	**As manufactured or annealed**: >10.0 mm **diam**, 279 MPa **TS**, 40% in 2 in. (50 mm) **El**; **Hard**: 10–16 mm **diam**, 358 MPa **TS**, 18% in 2 in. (50 mm) **El**
BS 2874	CZ106	UK	Rod, section	68.5–71.5 **Cu**, 0.05 **Pb**, rem **Zn**, 0.05 **Fe**, 0.30 total, others	**Annealed**: >6.0 mm **diam**, 279 MPa **TS**, 45% in 2 in. (50 mm) **El**; **As manufactured**: >6 mm **diam**, 339 MPa **TS**, 28% in 2 in. (50 mm) **El**
CDA 260		US	Rod, tube, wire, sheet, bar, flat wire, strip	68.5–71.5 **Cu**, 0.07 **Pb**, rem **Zn**, 0.05 **Fe**, 0.15 total, others	**Half hard 20% (rod)**: 1.0 in. **diam**, 70 ksi (483 MPa) **TS**, 30% in 2 in. (50 mm) **El**, 80 **HRB**; **Hard drawn 35% (tube)**: 1.0 in. **diam**, 78 ksi (538 MPa) **TS**, 8% in 2 in. (50 mm) **El**, 82 **HRB**; **Spring temper (wire)**: 0.080 in. **diam**, 130 ksi (896 MPa) **TS**, 3% in 2 in. (50 mm) **El**

WROUGHT COPPER AND COPPER ALLOYS

specification number	designation	country	product forms	chemical composition	mechanical properties and hardness values
CSA HC.4.2	HC 4.Z 30	Canada	Plate	68.5–71.5 **Cu**, 0.07 **Pb**, rem **Zn**, 0.05 **Fe**, 0.15 other elements	**Half hard**: all **diam**, 57 ksi (393 MPa) **TS**; **Full hard**: all **diam**, 71 ksi (489 MPa) **TS**; **Spring**: all **diam**, 91 ksi (627 MPa) **TS**
CSA HC.4.2	HC 4.Z 30	Canada	Sheet	68.5–71.5 **Cu**, 0.07 **Pb**, rem **Zn**, 0.05 **Fe**, 0.15 other elements	**Half hard**: all **diam**, 57 ksi (393 MPa) **TS**; **Full hard**: all **diam**, 71 ksi (489 MPa) **TS**; **Spring**: all **diam**, 91 ksi (627 MPa) **TS**
CSA HC.4.2	HC 4.Z 30	Canada	Strip	68.5–71.5 **Cu**, 0.07 **Pb**, rem **Zn**, 0.05 **Fe**, 0.15 other elements	**Half hard**: all **diam**, 57 ksi (393 MPa) **TS**; **Full hard**: all **diam**, 71 ksi (489 MPa) **TS**; **Spring**: all **diam**, 91 ksi (627 MPa) **TS**
CSA HC.4.2	HC 4.Z 30	Canada	Bar	68.5–71.5 **Cu**, 0.07 **Pb**, rem **Zn**, 0.05 **Fe**, 0.15 other elements	**Half hard**: all **diam**, 57 ksi (393 MPa) **TS**; **Full hard**: all **diam**, 71 ksi (489 MPa) **TS**; **Spring**: all **diam**, 91 ksi (627 MPa) **TS**
COPANT 162	C 260 00	COPANT	Wrought products	68.5–71.5 **Cu**, 0.07 **Pb**, rem **Zn**, 0.05 **Fe**	
COPANT 162	C 261 00	COPANT	Wrought products	68.5–71.5 **Cu**, 0.05 **Pb**, rem **Zn**, 0.05 **Fe**, 0.02–0.05 **P**	
COPANT 612	Cu Zn 30	COPANT	Strip, sheet, plate, foil, bar	68.5–71.5 **Cu**, 0.07 **Pb**, rem **Zn**, 0.05 **Fe**, 0.15 total, others	**1/4 hard temper**: all **diam**, 345 MPa **TS**; **1/2 hard temper**: all **diam**, 390 MPa **TS**; **Hard temper**: all **diam**, 570 MPa **TS**
DS 3003	5122	Denmark	Rolled, drawn, extruded, and forged products	68.5–71.5 **Cu**, 0.05 **Pb**, 30 **Zn**, 0.1 **Fe**, 0.04 total, others	
QQ–B–626D	260	US	Rod, shapes, forgings, bar, strip	68.5–71.5 **Cu**, 0.07 **Pb**, rem **Zn**, 0.05 **Fe**, 0.15 total, others	**Soft temper**: all **diam**, 42 ksi (290 MPa) **TS**, 30% in 4 in. (100 mm) **El**; **Half hard temper**: 0.5 max in. **diam**, 57 ksi (393 MPa) **TS**, 15% in 4 in. (100 mm) **El**; **Hard temper**: 0.25 max **diam**, 80 ksi (552 MPa) **TS**, 8% in 4 in. (100 mm) **El**
QQ–W–321d	260	US	Wire	68.5–71.5 **Cu**, 0.07 **Pb**, rem **Zn**, 0.05 **Fe**, 0.15 total, others; **Co** in **Ni**	**Eighth hard temper**: all **diam**, 50 ksi (345 MPa) **TS**; **Quarter hard temper**: all **diam**, 62 ksi (427 MPa) **TS**; **Half hard temper**: all **diam**, 79 ksi (545 MPa) **TS**
QQ–B–613D	260	US	Plate, bar, sheet, strip	68.5–71.5 **Cu**, 0.07 **Pb**, rem **Zn**, 0.05 **Fe**, 0.15 total, others	**Quarter hard temper**: all **diam**, 49 ksi (338 MPa) **TS**; **Half hard temper**: all **diam**, 57 ksi (393 MPa) **TS**; **Hard temper**: all **diam**, 71 ksi (490 MPa) **TS**

WROUGHT COPPER AND COPPER ALLOYS

specification number	designation	country	product forms	chemical composition	mechanical properties and hardness values
SFS 2918	CuZn30	Finland	Sheet and strip	68.5–71.5 **Cu**, 0.05 **Pb**, 30 **Zn**, 0.1 **Fe**	**Annealed**: 0.2–0.5 mm **diam**, 300 MPa **TS**, 90 MPa **YS**, 25% in 2 in. (50 mm) **El**; **Cold finished**: 0.2–0.5 mm **diam**, 330 MPa **TS**, 150 MPa **YS**, 15% in 2 in. (50 mm) **El**
NF A 51 104	CuZn 30	France	Bar, wire, and shapes	68.5–71.5 **Cu**, 0.05 **Pb**, rem **Zn**, 0.1 **Fe**, 0.4 total, others	**Mill condition**: ≤50 mm **diam**, 340 MPa **TS**, 30% in 2 in. (50 mm) **El**
NF A 51 103	CuZn 30	France	Tube	68.5–71.5 **Cu**, rem **Zn**, 0.05 each, 0.40 total, others	**1/2 hard**: ≤80 mm **diam**, 370 MPa **TS**, 40% in 2 in. (50 mm) **El**, 100–135 **HV**; **3/4 hard**: ≤80 mm **diam**, 410 MPa **TS**, 30% in 2 in. (50 mm) **El**, 125–160 **HV**; **Hard**: ≤80 mm **diam**, 480 MPa **TS**, 14% in 2 in. (50 mm) **El**, 145–185 **HV**
NF A 51 101	CuZn 30	France	Sheet and strip	70.0 **Cu**, 30.0 **Zn**	**Stress hardened, Temper 1**: all **diam**, 330 MPa **TS**
IS 3168	CuZn30	India	Strip, foil	68.5–71.5 **Cu**, 0.05 **Pb**, rem **Zn**, 0.05 **Fe**, 0.20 total, others; nickel included in copper	**As rolled**: 0.035 mm **diam**, 274 MPa **TS**, 45% in 2 in. (50 mm) **El**, 60–80 **HV**
IS 407	Alloy No. 1/CuZn30As	India	Tube	68.5–71.5 **Cu**, 0.07 **Pb**, rem **Zn**, 0.05 **Fe**, 0.30 total, others; nickel included in copper	**Annealed**: all **diam**, 284 MPa **TS**; **Half hard**: all **diam**, 372 MPa **TS**, 99 **HV**; **Hard**: all **diam**, 451 MPa **TS**, 135 **HV**
IS 4170	CuZn30	India	Rod	68–72 **Cu**, 0.03 **Pb**, rem **Zn**, 0.03 **Fe**, 0.30 total, others	**As manufactured**: ≥5 mm **diam**, 343 MPa **TS**, 25% in 2 in. (50 mm) **El**, 100 **HV**; **Annealed**: 274 MPa **TS**, 50% in 2 in. (50 mm) **El**, 90 **HV**
IS 410	CuZn30	India	Plate, sheet, strip, foil	68.5–71.5 **Cu**, 0.05 **Pb**, rem **Zn**, 0.05 **Fe**, 0.20 total, others	**Annealed**: all **diam**, 274 MPa **TS**, 45% in 2 in. (50 mm) **El**, 85 max **HV**; **Half hard**: all **diam**, 392 MPa **TS**, 20% in 2 in. (50 mm) **El**, 100 **HV**; **Hard**: all **diam**, 461 MPa **TS**, 3% in 2 in. (50 mm) **El**, 135 **HV**
ISO 1638	CuZn30	ISO	Solid product: drawn, on coils or reels	68.5–71.5 **Cu**, 0.01 **Pb**, rem **Zn**, 0.1 **Fe**, 0.3 **Ni** included in **Cu**; 0.4 **Fe+Pb** + total, others	**Annealed**: 1–5 mm **diam**, 280 MPa **TS**, 35% in 2 in. (50 mm) **El**; **HC**: 1–5 mm **diam**, 420 MPa **TS**, 7% in 2 in. (50 mm) **El**
ISO 1635	CuZn30	ISO	Tube	68.5–71.5 **Cu**, 0.05 **Pb**, rem **Zn**, 0.1 **Fe**, **Cu** includes 0.3 **Ni**; 0.4 **Fe+Pb** + total, others	**Annealed**: all **diam**, 45% in 2 in. (50 mm) **El**, 90 max **HV**; **Strain hardened (HB)**: all **diam**, 440 MPa **TS**, 20% in 2 in. (50 mm) **El**, 100 **HV**

WROUGHT COPPER AND COPPER ALLOYS

specification number	designation	country	product forms	chemical composition	mechanical properties and hardness values
ISO 1634	CuZn30	ISO	Plate, sheet, strip	68.5–71.5 **Cu**, 0.05 **Pb**, rem **Zn**, 0.1 **Fe**, **Cu** includes 0.3 **Ni**; 0.4 **Fe**+**Pb** + total, others	**Annealed**: all **diam**, 45% in 2 in. (50 mm) **El**, 80 max **HV**; **Specially annealed (OS25)**: 0.2–2 mm **diam**, 35% in 2 in. (50 mm) **El**, 100 max **HV**; **Strain hardened**: 5–10 mm **diam**, 410 MPa **TS**, 15% in 2 in. (50 mm) **El**, 135 **HV**
ISO 426 1	CuZn30	ISO	Plate, sheet, strip, rod, bar, tube, wire	68.5–71.5 **Cu**, 0.05 **Pb**, rem **Zn**, 0.1 **Fe**, 0.4 **Pb**+**Fe**	
JIS H 3320	C 2600	Japan	Pipe and tube: welded	68.5–71.5 **Cu**, 0.07 **Pb**, rem **Zn**, 0.05 **Fe**	**Annealed**: 4–76.2 mm OD **diam**, 275 MPa **TS**, 45% in 2 in. (50 mm) **El**, 80 max **HV**; **1/2 hard**: 4–76.2 mm OD **diam**, 373 MPa **TS**, 20% in 2 in. (50 mm) **El**, 110 max **HV**; **Hard**: 4–76.2 mm OD **diam**, 451 MPa **TS**, 150 **HV**
JIS H 3300	C 2600	Japan	Pipe and tube: seamless	68.5–71.5 **Cu**, 0.07 **Pb**, rem **Zn**, 0.05 **Fe**	**Annealed**: 4–250 mm OD **diam**, 275 MPa **TS**, 45% in 2 in. (50 mm) **El**; **1/2 hard**: 4–250 mm OD **diam**, 373 MPa **TS**, 20% in 2 in. (50 mm) **El**, 53 min **HR30T**; **Hard**: 4–250 mm OD **diam**, 451 MPa **TS**
JIS H 3260	C 2600	Japan	Rod and bar	68.5–71.5 **Cu**, 0.07 **Pb**, rem **Zn**, 0.05 **Fe**	**Annealed**: ≥0.5 mm **diam**, 275 MPa **TS**, 20% in 2 in. (50 mm) **El**; **Hard**: 0.5–10 mm **diam**, 686 MPa **TS**; **Extra hard**: 0.5–10 mm **diam**, 785 MPa **TS**
JIS H 3250	C 2600	Japan	Rod and bar	68.5–71.5 **Cu**, 0.07 **Pb**, rem **Zn**, 0.05 **Fe**	**As manufactured**: ≥6 mm **diam**, 275 MPa **TS**, 35% in 2 in. (50 mm) **El**; **Annealed**: 6–75 mm **diam**, 275 MPa **TS**, 45% in 2 in. (50 mm) **El**; **Hard**: 6–20 mm **diam**, 412 MPa **TS**
JIS H 3100	C 2600	Japan	Sheet, plate, and strip	68.5–71.5 **Cu**, 0.07 **Pb**, rem **Zn**, 0.05 **Fe**	**Annealed**: <1 mm **diam**, 275 MPa **TS**, 40% in 2 in. (50 mm) **El**; **Hard**: 0.3–10 mm **diam**, 412 MPa **TS**, 105–175 **HV**; **Extra hard**: 0.3–10 mm **diam**, 520 MPa **TS**, 145 **HV**
DGN-W-24	Type No. 6	Mexico	Wire	68.5–71.5 **Cu**, 0.07 **Pb**, rem **Zn**, 0.05 **Fe**	**3/4 hard**: <19.05 mm **diam**, 92 ksi (634 MPa) **TS**; **Hard**: <12.7 mm **diam**, 102 ksi (703 MPa) **TS**; **Spring hard**: <6.35 mm **diam**, 120 ksi (827 MPa) **TS**

WROUGHT COPPER AND COPPER ALLOYS

specification number	designation	country	product forms	chemical composition	mechanical properties and hardness values
DGN–W–19	Type No. 2	Mexico	Tube	68.5–71.5 **Cu**, 0.07 **Pb**, rem **Zn**, 0.05 **Fe**	**1/2 hard**: all **diam**, 54 ksi (372 MPa) **TS**, 55 min **HRB**; **Hard**: ≤25.4 mm **diam**, 66 ksi (455 MPa) **TS**, 75 min **HRB**
MIL–T–20219C		US	Tube	68.5–71.5 **Cu**, 0.07 **Pb**, rem **Zn**, 0.05 **Fe**, 0.15 total, others	**Hard drawn**: 455 MPa **TS**, 67 **HR30T**; **Light drawn**: 345 MPa **TS**, 34–70 **HR30T**
NS 16 115	CuZn30	Norway	Sheet, strip, plate, and bar	70 **Cu**, 30 **Zn**	
SABS 460	Cu–Zn30As	South Africa	Tube	69.0–71.0 **Cu**, 0.070 **Pb**, rem **Zn**, 0.06 **Fe**, 0.02–0.06 **As**, 0.30 total, others	**Annealed**: all **diam**, 280 MPa **TS**, 40% in 2 in. (50 mm) **El**, 75 max **HV**; **Specially annealed**: all **diam**, 320 MPa **TS**, 35% in 2 in. (50 mm) **El**, 80–105 **HV**; **As drawn**: all **diam**, 400 MPa **TS**, 40% in 2 in. (50 mm) **El**, 130 **HV**
VSM 11557	Cu Zn 30	Switzerland	Tube	69.0–71.0 **Cu**, 0.05 **Pb**, rem **Zn**, 0.05 **Fe**, 0.2 **Ni**, 0.015 **P**, 0.30 total, others	**1/2 hard**: all **diam**, 343 MPa **TS**, 137 MPa **YS**, 40% in 2 in. (50 mm) **El**, 80 **HB**
VSM 11557	Cu Zn 30 As	Switzerland	Tube	69.0–71.0 **Cu**, 0.05 **Sn**, 0.05 **Pb**, rem **Zn**, 0.05 **Fe**, 0.02 **Al**, 0.2 **Ni**, 0.015 **P**, 0.20–0.045 **As**, 0.30 total, others	**1/2 hard**: all **diam**, 343 MPa **TS**, 137 MPa **YS**, 40% in 2 in. (50 mm) **El**, 80 **HB**
MIL–C–50D	260	US	Sheet, strip, plate, bar, disc	68.5–71.5 **Cu**, 0.07 **Pb**, rem **Zn**, 0.05 **Fe**, 0.15 total, others	**Annealed**: 0.020–0.050 in. **diam**, 310 MPa **TS**, 40% in 2 in. (50 mm) **El**; **Hot rolled**: 0.250–0.500 in. **diam**, 296 MPa **TS**, 55% in 2 in. (50 mm) **El**
SIS 14 51 22	SIS Brass 51 22–05	Sweden	Sheet, plate, strip	68.5–71.5 **Cu**, 0.05 **Pb**, 30 **Zn**, 0.1 **Fe**, 0.4 total, others	**Strain hardened**: 0.2–0.5mm **diam**, 410 MPa **TS**, 330 MPa **YS**, 5% in 4 in. (100 mm) **El**, 125–150 **HV**
SIS 14 51 22	SIS Brass 51 22–04	Sweden	Plate, sheet, strip	68.5–71.5 **Cu**, 0.05 **Pb**, 30 **Zn**, 0.1 **Fe**, 0.4 total, others	**Strain hardened**: 5 mm **diam**, 340 MPa **TS**, 220 MPa **YS**, 35% in 2 in. (50 mm) **El**, 95–125 **HV**
SIS 14 51 22	SIS Brass 51 22–02	Sweden	Plate, sheet, strip	68.5–71.5 **Cu**, 0.05 **Pb**, 30 **Zn**, 0.1 **Fe**, 0.4 total, others	**Annealed**: 5 mm **diam**, 300 MPa **TS**, 90 MPa **YS**, 50% in 2 in. (50 mm) **El**, 65–90 **HV**
MIL–T–6945	II	US	Tube: seamless	68–71 **Cu**, 0.075 **Pb**, rem **Zn**, 0.06 **Fe**	**Hard drawn and relief annealed**: all **diam**, 462 MPa **TS**, 310 MPa **YS**, 15% in 2 in. (50 mm) **El**
COPANT 162	C 262 00	COPANT	Wrought products	67.0–70.0 **Cu**, 0.07 **Pb**, rem **Zn**, 0.05 **Fe**	

WROUGHT COPPER AND COPPER ALLOYS

specification number	designation	country	product forms	chemical composition	mechanical properties and hardness values
SIS 14 51 22	SIS Brass 51 22–03	Sweden	Plate, strip, sheet	68.5 **Cu**, 0.05 **Pb**, 30 **Zn**, 0.1 **Fe**, 0.4 total, others	**Strain hardened**: 5 mm **diam**, 330 MPa **TS**, 150 MPa **YS**, 40% in 2 in. (50 mm) **El**, 75–105 **HV**
NF A 51 104	CuZn 33	France	Bar, wire, and shapes	65.5–68.5 **Cu**, 0.1 **Pb**, rem **Zn**, 0.1 **Fe**, 0.4 total, others	**Mill condition**: ≤50 mm **diam**, 350 MPa **TS**, 32% in 2 in. (50 mm) **El**
NF A 51 101	CuZn 33	France	Sheet and strip	67.0 **Cu**, 33.0 **Zn**	**Stress hardened, Temper 1**: all **diam**, 330 MPa **TS**
ISO 1638	CuZn33	ISO	Solid product: drawn, on coils or reels	65.5–68.5 **Cu**, 0.1 **Pb**, rem **Zn**, 0.1 **Fe**, 0.3 **Ni** included in **Cu**; 0.4 **Fe+Pb** + total, others	**Annealed**: 1–5 mm **diam**, 280 MPa **TS**, 35% in 2 in. (50 mm) **El**; **HC**: 1–5 mm **diam**, 430 MPa **TS**, 7% in 2 in. (50 mm) **El**
ISO 1634	CuZn33	ISO	Plate, sheet, strip	65.5–68.5 **Cu**, 0.01 **Pb**, rem **Zn**, 0.1 **Fe**, **Cu** includes 0.3 **Ni**; 0.4 **Fe+Pb** + total, others	**Annealed**: all **diam**, 45% in 2 in. (50 mm) **El**, 80 max **HV**; **Specially annealed**: 0.2–5 mm **diam**, 45% in 2 in. (50 mm) **El**, 85 max **HV**; **Strain hardened (HB)**: 5–10 mm **diam**, 420 MPa **TS**, 15% in 2 in. (50 mm) **El**, 140 **HV**
ISO 426 1	CuZn33	ISO	Plate, sheet, strip, rod, bar, tube, wire	65.5–68.5 **Cu**, 0.1 **Pb**, rem **Zn**, 0.1 **Fe**, 0.4 **Pb+Fe**	
IS 3168	CuZn33	India	Strip, foil	64.5–68.5 **Cu**, 0.10 **Pb**, rem **Zn**, 0.05 **Fe**, 0.20 total, others; nickel included in copper	**As rolled**: 0.035 mm **diam**, 274 MPa **TS**, 45% in 2 in. (50 mm) **El**, 60–80 **HV**
ASTM F 587	CA 268	US	Tube: welded	64.0–68.5 **Cu**, 0.15 **Pb**, rem **Zn**, 0.05 **Fe**	**Welded from annealed strip**: all **diam**, 48 ksi (330 MPa) **TS**, 25 min **HR30T**; **Cold reduced or drawn**: all **diam**, 54 ksi (370 MPa) **TS**, 53 min **HR30T**; **Soft anneal**: <0.030 in. (<0.762 mm) **diam**, 40 **HR30T**
ANSI/ASTM B 36	C 26800	US	Plate, sheet, strip, bar	64.0–68.5 **Cu**, 0.15 **Pb**, rem **Zn**, 0.05 **Fe**	**Quarter hard**: 0.020 in. (0.508 mm) **diam**, 49 ksi (340 MPa) **TS**, 40 **HRB**; **Hard**: 0.020 in. (0.508 mm) **diam**, 68 ksi (470 MPa) **TS**, 76 **HRB**; **Spring**: 0.020 in. (0.508 mm) **diam**, 86 ksi (595 MPa) **TS**, 87 **HRB**
CDA 268		US	Sheet	64.0–68.5 **Cu**, 0.15 **Pb**, rem **Zn**, 0.05 **Fe**, 0.15 total, others	**Hard temper**: 0.040 in. **diam**, 74 ksi (510 MPa) **TS**, 8% in 2 in. (50 mm) **El**, 80 **HRB**; **Extra hard temper**: 0.040 in. **diam**, 85 ksi (586 MPa) **TS**, 5% in 2 in. (50 mm) **El**, 87 **HRB**; **Spring temper**: 0.040 in. **diam**, 91 ksi (627 MPa) **TS**, 3% in 2 in. (50 mm) **El**, 90 **HRB**

WROUGHT COPPER AND COPPER ALLOYS

specification number	designation	country	product forms	chemical composition	mechanical properties and hardness values
CSA HC.4.2	HC 4.Z 34	Canada	Bar	64.0–68.5 **Cu**, 0.15 **Pb**, rem **Zn**, 0.05 **Fe**, 0.15 other elements	**Half hard**: all **diam**, 55 ksi (379 MPa) **TS**; **Full hard**: all **diam**, 68 ksi (468 MPa) **TS**; **Spring**: all **diam**, 86 ksi (592 MPa) **TS**
CSA HC.4.2	HC 4.Z 34	Canada	Sheet	64.0–68.5 **Cu**, 0.15 **Pb**, rem **Zn**, 0.05 **Fe**, 0.15 other elements	**Half hard**: all **diam**, 55 ksi (379 MPa) **TS**; **Full hard**: all **diam**, 68 ksi (468 MPa) **TS**; **Spring**: all **diam**, 86 ksi (592 MPa) **TS**
CSA HC.4.2	HC 4.Z 34	Canada	Strip	64.0–68.5 **Cu**, 0.15 **Pb**, rem **Zn**, 0.05 **Fe**, 0.15 other elements	**Half hard**: all **diam**, 55 ksi (379 MPa) **TS**; **Full hard**: all **diam**, 68 ksi (468 MPa) **TS**; **Spring**: all **diam**, 86 ksi (592 MPa) **TS**
CSA HC.4.2	HC 4.Z 34	Canada	Plate	64.0–68.5 **Cu**, 0.15 **Pb**, rem **Zn**, 0.05 **Fe**, 0.15 other elements	**Half hard**: all **diam**, 55 ksi (379 MPa) **TS**; **Full hard**: all **diam**, 68 ksi (468 MPa) **TS**; **Spring**: all **diam**, 86 ksi (592 MPa) **TS**
COPANT 162	C 268 00	COPANT	Wrought products	64.0–68.5 **Cu**, 0.15 **Pb**, rem **Zn**, 0.05 **Fe**	
COPANT 612	Cu Zn 34	COPANT	Strip, sheet, plate, foil, bar	64–68.5 **Cu**, 0.15 **Pb**, rem **Zn**, 0.05 **Fe**, 0.15 total, others	**1/4 hard temper**: all **diam**, 345 MPa **TS**; **1/2 hard temper**: all **diam**, 380 MPa **TS**; **Hard temper**: all **diam**, 470 MPa **TS**
QQ-B-626D	268	US	Rod, shapes, forgings, bar, strip	64.0–68.5 **Cu**, 0.15 **Pb**, rem **Zn**, 0.05 **Fe**, 0.15 total, others	**Soft temper**: all **diam**, 42 ksi (290 MPa) **TS**, 30% in 4 in. (100 mm) **El**; **Half hard temper**: 0.5 max in. **diam**, 57 ksi (393 MPa) **TS**, 15% in 4 in. (100 mm) **El**; **Hard temper**: 0.25 max **diam**, 80 ksi (552 MPa) **TS**, 10% in 4 in. (100 mm) **El**
QQ-B-613D	268	US	Plate, bar, sheet, strip	64.0–68.5 **Cu**, 0.15 **Pb**, rem **Zn**, 0.05 **Fe**, 0.15 total, others	**Quarter hard temper**: all **diam**, 49 ksi (338 MPa) **TS**; **Half hard temper**: all **diam**, 55 ksi (379 MPa) **TS**; **Hard temper**: all **diam**, 68 ksi (469 MPa) **TS**
JIS H 3320	C 2680	Japan	Pipe and tube: welded	64.0–68.0 **Cu**, 0.07 **Pb**, rem **Zn**, 0.05 **Fe**	**Annealed**: 0.3–3 mm **diam**, 294 MPa **TS**, 40% in 2 in. (50 mm) **El**, 80 max **HV**; **1/2 hard**: 4–76.2 mm OD **diam**, 373 MPa **TS**, 20% in 2 in. (50 mm) **El**, 110 max **HV**; **Hard**: 4–76.2 mm OD **diam**, 451 MPa **TS**, 115 **HV**

WROUGHT COPPER AND COPPER ALLOYS

specification number	designation	country	product forms	chemical composition	mechanical properties and hardness values
AMS 4712A		US	Wire	63.0–68.5 **Cu**, 0.1 **Pb**, rem **Zn**, 0.05 **Fe**	
AMS 4713A		US	Wire	63.0–68.5 **Cu**, 0.1 **Pb**, rem **Zn**, 0.05 **Fe**	**Cold drawn or rolled to 1/8 hard**: all **diam**, 50 ksi (345 MPa) **TS**
AMS 4710B		US	Wire: tinned	63.0–68.5 **Cu**, 0.10 **Pb**, rem **Zn**, 0.05 **Fe**	
ASTM F 468	C 27000	US	Bolt, screw, stud	63.0–68.5 **Cu**, 0.10 **Pb**, rem **Zn**, 0.07 **Fe**	**As manufactured (test specimen)**: all **diam**, 55 ksi (380 MPa) **TS**, 50 ksi (345 MPa) **TS**, 35% in 2 in. (50 mm) **El**, 55–80 **HRF**
ASTM F 467	C 27000	US	Nut	63.0–68.5 **Cu**, 0.10 **Pb**, rem **Zn**, 0.07 **Fe**	**As agreed**: all **diam**, 55 min **HRF**
ASTM F 587	CA 270	US	Tube: welded	63.0–68.5 **Cu**, 0.10 **Pb**, rem **Zn**, 0.07 **Fe**	**Welded from annealed strip**: all **diam**, 48 ksi (330 MPa) **TS**, 25 min **HR30T**; **Cold reduced or drawn**: all **diam**, 54 ksi (370 MPa) **TS**, 53 min **HR30T**; **Soft anneal**: <0.030 in. (<0.762 mm) **diam**, 40 **HR30T**
ANSI/ASTM B 134	CA 270	US	Wire	63.0–68.5 **Cu**, 0.10 **Pb**, rem **Zn**, 0.07 **Fe**	**Quarter hard**: >0.020 in. (>0.508 mm) **diam**, 62 ksi (425 MPa) **TS**; **Hard**: >0.020 in. (>0.508 mm) **diam**, 102 ksi (700 MPa) **TS**; **Spring**: >0.020 in. (>0.508 mm) **diam**, 120 ksi (830 MPa) **TS**
ANSI/ASTM B 135	CA 270	US	Tube: seamless	63.0–68.5 **Cu**, 0.10 **Pb**, rem **Zn**, 0.07 **Fe**	**Drawn**: all **diam**, 54 ksi (370 MPa) **TS**, 53 **HR30T**; **Hard drawn**: >0.020 in. (0.508 mm) **diam**, 66 ksi (455 MPa) **TS**, 70 **HR30T**; **Soft anneal**: >0.030 in. (>0.762 mm) **diam**, 40 **HR30T**
CDA 270		US	Rod, wire, plate, sheet, flat wire, strip	63.0–68.5 **Cu**, 0.10 **Pb**, rem **Zn**, 0.07 **Fe**, 0.20 total, others	**Eighth hard temper (rod)**: 1.0 in. **diam**, 55 ksi (379 MPa) **TS**, 48% in 2 in. (50 mm) **El**, 55 **HRB**; **Quarter hard temper (wire)**: 0.080 in. **diam**, 70 ksi (483 MPa) **TS**, 20% in 2 in. (50 mm) **El**; **Half hard temper (flat products)**: 0.040 in. **diam**, 61 ksi (421 MPa) **TS**, 23% in 2 in. (50 mm) **El**, 70 **HRB**
COPANT 162	C 270 00	COPANT	Wrought products	63.0–68.5 **Cu**, 0.10 **Pb**, rem **Zn**, 0.07 **Fe**	

WROUGHT COPPER AND COPPER ALLOYS

specification number	designation	country	product forms	chemical composition	mechanical properties and hardness values
QQ-W-321d	270	US	Wire	63.0–68.5 **Cu**, 0.10 **Pb**, rem **Zn**, 0.05 **Fe**, 0.15 total, others; **Co** in **Ni**	**Eighth hard temper**: all **diam**, 50 ksi (345 MPa) **TS**; **Quarter hard temper**: all **diam**, 62 ksi (427 MPa) **TS**; **Half hard temper**: all **diam**, 79 ksi (545 MPa) **TS**
DGN-W-24	Type No. 7	Mexico	Wire	63–68.5 **Cu**, 0.10 **Pb**, rem **Zn**, 0.05 **Fe**	**3/4 hard**: <19.05 mm **diam**, 92 ksi (634 MPa) **TS**; **Hard**: <12.7 mm **diam**, 102 ksi (703 MPa) **TS**; **Spring hard**: <6.35 mm **diam**, 120 ksi (827 MPa) **TS**
DGN-W-24	Type No. 7	Mexico	Wire	63–68.5 **Cu**, 0.10 **Pb**, rem **Zn**, 0.05 **Fe**	**3/4 hard**: <19.05 mm **diam**, 92 ksi (634 MPa) **TS**; **Hard**: <12.7 mm **diam**, 102 ksi (703 MPa) **TS**; **Spring hard**: <6.35 mm **diam**, 120 ksi (827 MPa) **TS**
BS 2870	CZ 107	UK	Sheet, strip, foil	64.0–67.0 **Cu**, 0.10 **Pb**, rem **Zn**, 0.10 **Fe**, 0.40 total, others	**Annealed**: ≤10.0 mm **diam**, 197 MPa **TS**, 45% in 2 in. (50 mm) **El**, 80 max **HV**; **1/2 hard**: ≤3.5–10.0 mm **diam**, 272 MPa **TS**, 20% in 2 in. (50 mm) **El**, 110 **HV**; **Hard**: ≤10.0 mm **diam**, 324 MPa **TS**, 5% in 2 in. (50 mm) **El**, 135 **HV**
BS 2873	CZ107	UK	Wire	64.0–67.0 **Cu**, 0.10 **Pb**, rem **Zn**, 0.10 **Fe**, 0.40 total, others	**Annealed**: 0.5–10.0 mm **diam**, 319 MPa **TS**, 35% in 2 in. (50 mm) **El**; **1/2 hard**: 0.5–10.0 mm **diam**, 461 MPa **TS**; **Hard**: 0.5–10.0 mm **diam**, 608 MPa **TS**
JIS H 3300	C 2700	Japan	Pipe and tube: seamless	63.0–67.0 **Cu**, 0.07 **Pb**, rem **Zn**, 0.05 **Fe**	**Annealed**: 4–250 mm OD **diam**, 294 MPa **TS**, 40% in 2 in. (50 mm) **El**; **1/2 hard**: 4–250 mm OD **diam**, 373 MPa **TS**, 20% in 2 in. (50 mm) **El**, 53 min **HR30T**; **Hard**: 4–250 mm OD **diam**, 451 MPa **TS**
JIS H 3260	C 2700	Japan	Rod and bar	63.0–67.0 **Cu**, 0.07 **Pb**, rem **Zn**, 0.05 **Fe**	**Annealed**: ≥0.5 mm **diam**, 294 MPa **TS**, 20% in 2 in. (50 mm) **El**; **Hard**: 0.5–10 mm **diam**, 686 MPa **TS**; **Extra hard**: 0.5–10 mm **diam**, 785 MPa **TS**
JIS H 3250	C 2700	Japan	Rod and bar	63.0–67.0 **Cu**, 0.07 **Pb**, rem **Zn**, 0.05 **Fe**	**As manufactured**: ≥6 mm **diam**, 294 MPa **TS**, 30% in 2 in. (50 mm) **El**; **Annealed**: 6–75 mm **diam**, 294 MPa **TS**, 40% in 2 in. (50 mm) **El**; **Hard**: 6–20 mm **diam**, 412 MPa **TS**

WROUGHT COPPER AND COPPER ALLOYS

specification number	designation	country	product forms	chemical composition	mechanical properties and hardness values
SFS 2936	CuNi12Zn24	Finland	Sheet, strip, and wire	62.0–66.0 **Cu**, 0.05 **Pb**, 24 **Zn**, 0.3 **Fe**	**Annealed (sheet and strip)**: 0.2–0.5 mm **diam**, 340 MPa **TS**, 120 MPa **YS**, 20% in 2 in. (50 mm) **El**; **Cold finished (wire)**: 0.1–2.5 mm **diam**, 740 MPa **TS**, 720 MPa **YS**, 1% in 2 in. (50 mm) **El**
NF A 51 104	CuZn 37	France	Bar, wire, and shapes	62.5–65.5 **Cu**, 0.3 **Pb**, rem **Zn**, 0.2 **Fe**, 0.5 total, others	**Mill condition**: ≤50 mm **diam**, 370 MPa **TS**, 30% in 2 in. (50 mm) **El**
NF A 51 101	CuZn 36	France	Sheet and strip	64.0 **Cu**, 36.0 **Zn**	**Stress hardened, Temper 1**: all **diam**, 330 MPa **TS**
DS 3003	5150	Denmark	Rolled, drawn, extruded, and forged products	62.0–65.5 **Cu**, 0.3 **Pb**, 37 **Zn**, 0.2 **Fe**, 0.05 total, others	
SFS 2919	CuZn37	Finland	Sheet, strip, bar, rod, wire, tube, and shapes	62.0–65.5 **Cu**, 0.3 **Pb**, 37 **Zn**, 0.2 **Fe**	**Annealed (sheet and strip)**: 0.2–0.5 mm **diam**, 310 MPa **TS**, 100 MPa **YS**, 20% in 2 in. (50 mm) **El**; **Cold finished (rod and wire)**: 1–2.5 mm **diam**, 360 MPa **TS**, 240 MPa **YS**, 12% in 2 in. (50 mm) **El**; **Drawn (shapes)**: ≤5 mm **diam**, 370 MPa **TS**, 240 MPa **YS**, 5% in 2 in. (50 mm) **El**
NF A 51 103	CuZn 36	France	Tube	62.0–65.5 **Cu**, rem **Zn**, 0.05 each, 0.40 total, others	**1/2 hard**: ≤80 mm **diam**, 380 MPa **TS**, 35% in 2 in. (50 mm) **El**, 105–140 **HV**; **3/4 hard**: ≤50 mm **diam**, 400 MPa **TS**, 25% in 2 in. (50 mm) **El**, 125–160 **HV**; **Hard**: ≤80 mm **diam**, 480 MPa **TS**, 12% in 2 in. (50 mm) **El**, 145–180 **HV**
ISO 1638	CuZn37	ISO	Solid product: drawn, on coils or reels	62.0–65.5 **Cu**, 0.3 **Pb**, rem **Zn**, 0.2 **Fe**, 0.3 **Ni** included in **Cu**; 0.5 **Fe+Pb** + total, others	**Annealed**: 1–5 mm **diam**, 290 MPa **TS**, 30% in 2 in. (50 mm) **El**; **HB – strain hardened**: 1–3 mm **diam**, 440 MPa **TS**, 4% in 2 in. (50 mm) **El** HC: 1–3 mm **diam**, 540 MPa **TS**, 2% in 2 in. (50 mm) **El**
ISO 1637	CuZn37	ISO	Solid product: straight lengths	62.0–65.5 **Cu**, 0.03 **Pb**, rem **Zn**, 0.2 **Fe**, 0.3 **Ni** included in **Cu**; 0.5 **Fe+Pb** + total, others	**Annealed**: 5 mm **diam**, 40% in 2 in. (50 mm) **El**, 85 max **HV**; **HA**: all **diam**, 360 MPa **TS**, 35% in 2 in. (50 mm) **El**, 110 **HV**; **HB – strain hardened**: 5–12 mm **diam**, 430 MPa **TS**, 15% in 2 in. (50 mm) **El**, 140 **HV**

WROUGHT COPPER AND COPPER ALLOYS

specification number	designation	country	product forms	chemical composition	mechanical properties and hardness values
ISO 1635	CuZn37	ISO	Tube	62.0–65.5 **Cu**, 0.3 **Pb**, rem **Zn**, 0.2 **Fe**, **Cu** includes 0.3 **Ni**; 0.5 **Fe**+**Pb** + total, others	**Annealed**: all **diam**, 40% in 2 in. (50 mm) **El**, 90 max **HV**; **Strain hardened (HB)**: all **diam**, 470 MPa **TS**, 13% in 2 in. (50 mm) **El**, 120 **HV**
ISO 1634	CuZn37	ISO	Plate, sheet, strip	62.0–65.5 **Cu**, 0.3 **Pb**, rem **Zn**, 0.2 **Fe**, **Cu** includes 0.3 **Ni**; 0.5 **Fe**+**Pb** + total, others	**Annealed**: all **diam**, 40% in 2 in. (50 mm) **El**, 80 max **HV**; **Specially annealed (OS25)**: 0.2–2 mm **diam**, 30% in 2 in. (50 mm) **El**, 100 max **HV**; **Strain hardened (HB)**: 5–10 mm **diam**, 430 MPa **TS**, 15% in 2 in. (50 mm) **El**, 140 **HV**
ISO 426 1	CuZn37	ISO	Plate, sheet, strip, rod, bar, tube, wire, extruded sections	62.0–65.5 **Cu**, 0.3 **Pb**, rem **Zn**, 0.2 **Fe**, 0.5 **Pb**+**Fe**	
SABS 460	Cu–Zn37	South Africa	Tube	62.0–65.5 **Cu**, 0.02 **Pb**, rem **Zn**, 0.1 total, others	**Annealed**: all **diam**, 300 MPa **TS**, 40% in 2 in. (50 mm) **El**, 75 max **HV**; **Specially annealed**: all **diam**, 325–390 MPa **TS**, 80–105 **HV**; **As drawn**: all **diam**, 480 MPa **TS**, 150 **HV**
SIS 14 51 50	SIS Brass 51 50–02	Sweden	Plate, sheet, strip, bar, wire, rod, tube	62.0–65.5 **Cu**, 0.3 **Pb**, 37 **Zn**, 0.2 **Fe**, 0.5 total, others	**Annealed**: 5 min mm **diam**, 310 MPa **TS**, 100 MPa **YS**, 45% in 2 in. (50 mm) **El**, 65–90 **HV**
SIS 14 51 50	SIS Brass 51 50–03	Sweden	Plate, sheet, strip, bar, wire, rod	62.0–65.5 **Cu**, 0.3 **Pb**, 37 **Zn**, 0.2 **Fe**, 0.5 total, others	**Strain hardened**: 5 min mm **diam**, 340 MPa **TS**, 180 MPa **YS**, 35% in 2 in. (50 mm) **El**, 75–100 **HV**
SIS 14 51 50	SIS Brass 51 50–04	Sweden	Plate, sheet, strip, bar, wire, rod, tube	62.0–65.5 **Cu**, 0.3 **Pb**, 37 **Zn**, 0.2 **Fe**, 0.5 total, others	**Strain hardened**: 5 min mm **diam**, 360 MPa **TS**, 240 MPa **YS**, 30% in 2 in. (50 mm) **El**, 100–130 **HV**
SIS 14 51 50	SIS Brass 51 50–05	Sweden	Plate, sheet, strip	62.0–65.5 **Cu**, 0.3 **Pb**, 37 **Zn**, 0.2 **Fe**, 0.5 total, others	**Strain hardened**: 5 min mm **diam**, 430 MPa **TS**, 350 MPa **YS**, 10% in 2 in. (50 mm) **El**, 130–155 **HV**
SIS 14 51 50	SIS Brass 51 50–07	Sweden	Sheet, strip, wire, rod	62.0–65.5 **Cu**, 0.3 **Pb**, 37 **Zn**, 0.2 **Fe**, 0.5 total, others	**Strain hardened**: 0.1–0.5 **diam**, 520 MPa **TS**, 470 MPa **YS**, 1% in 4 in. (100 mm) **El**, 160–185 **HV**
SIS 14 51 50	SIS Brass 51 50–10	Sweden	Tube, strip, wire	62.0–65.5 **Cu**, 0.3 **Pb**, 37 **Zn**, 0.2 **Fe**, 0.5 total, others	**Strain hardened**: all **diam**, 370 MPa **TS**, 250 MPa **YS**, 15% in 2 in. (50 mm) **El**, 105 min **HV**

WROUGHT COPPER AND COPPER ALLOYS

specification number	designation	country	product forms	chemical composition	mechanical properties and hardness values
SIS 14 51 50	SIS Brass 51 50–11	Sweden	Plate, sheet, strip	62.0–65.5 **Cu**, 0.3 **Pb**, 37 **Zn**, 0.2 **Fe**, 0.5 total, others	**Strain hardened**: 5 min mm **diam**, 400 MPa **TS**, 290 MPa **YS**, 15% in 2 in. (50 mm) **El**, 115–145 **HV**
ASTM F 587	CA 272	US	Tube: welded	62.0–65.0 **Cu**, 0.07 **Pb**, rem **Zn**, 0.07 **Fe**	**Welded from annealed strip**: all **diam**, 48 ksi (330 MPa) **TS**, 25 min **HR30T**; **Cold reduced or drawn**: all **diam**, 54 ksi (370 MPa) **TS**, 53 min **HR30T**; **Soft anneal**: <0.030 in. (<0.762 mm) **diam**, 40 **HR30T**
ANSI/ASTM B 36	C 27200	US	Plate, sheet, strip, bar	62.0–65.0 **Cu**, 0.07 **Pb**, rem **Zn**, 0.07 **Fe**	**Quarter hard**: 0.020 in. (0.508 mm) **diam**, 49 ksi (340 MPa) **TS**, 40 **HRB**; **Hard**: 0.020 in. (0.508 mm) **diam**, 70 ksi (485 MPa) **TS**, 76 **HRB**; **Extra hard**: 0.020 in. (0.508 mm) **diam**, 81 ksi (560 MPa) **TS**, 82 **HRB**
ANSI/ASTM B 135	CA 272	US	Tube: seamless	62.0–65.0 **Cu**, 0.07 **Pb**, rem **Zn**, 0.07 **Fe**	**Drawn**: all **diam**, 54 ksi (370 MPa) **TS**, 53 **HR30T**; **Hard drawn**: >0.020 in. (0.508 mm) **diam**, 66 ksi (455 MPa) **TS**, 70 **HR30T**; **Soft anneal**: >0.030 in. (>0.762 mm) **diam**, 40 **HR30T**
BS 2870	CZ 108	UK	Sheet, strip, foil	62.0–65.0 **Cu**, 0.30 **Pb**, rem **Zn**, 0.40 total, others	**Annealed**: ≤10.0 mm **diam**, 197 MPa **TS**, 40% in 2 in. (50 mm) **El**, 80 max **HV**; **1/4 hard**: ≤10.0 mm **diam**, 238 MPa **TS**, 30% in 2 in. (50 mm) **El**, 75 **HV**; **1/2 hard**: ≤3.5 mm **diam**, 272 MPa **TS**, 15% in 2 in. (50 mm) **El**, 110 **HV**
BS 2871	CZ 108	UK	Tube	62.0–65.0 **Cu**, 0.30 **Pb**, rem **Zn**, 0.60 total, others	**Annealed**: 300 MPa **TS**, 40% in 2 in. (50 mm) **El**, 80 max **HV**; **Temper annealed**: 350 MPa **TS**, 35% in 2 in. (50 mm) **El**, 85–110 **HV**; **As drawn**: 450 MPa **TS**, 130 **HV**
BS 2873	CZ108	UK	Wire	62.0–65.0 **Cu**, 0.30 **Pb**, rem **Zn**, 0.60 total, others	**Annealed**: 0.5–10.0 mm **diam**, 319 MPa **TS**, 35% in 2 in. (50 mm) **El**; **1/2 hard**: 0.5–10.0 mm **diam**, 461 MPa **TS**; **Hard**: 0.5–10.0 mm **diam**, 608 MPa **TS**
COPANT 612	Cu Zn 36	COPANT	Strip, sheet, plate, foil, bar	62–65 **Cu**, 0.17 **Pb**, rem **Zn**, 0.07 **Fe**, 0.20 total, others	**1/4 hard temper**: all **diam**, 345 MPa **TS**; **1/2 hard temper**: all **diam**, 380 MPa **TS**; **Hard temper**: all **diam**, 480 MPa **TS**

WROUGHT COPPER AND COPPER ALLOYS

specification number	designation	country	product forms	chemical composition	mechanical properties and hardness values
COPANT 162	C 272 00	COPANT	Wrought products	62–65 **Cu**, 0.07 **Pb**, rem **Zn**, 0.07 **Fe**	
AS 1566	272C	Australia	Plate, bar, sheet, strip, foil	61.0–65.0 **Cu**, 0.07 **Pb**, rem **Zn**, 0.05 **Fe**, 0.20 total, others	**Annealed (sheet, strip)**: all **diam**, 290 MPa **TS**, 45% in 2 in. (50 mm) **El**, 70 **HV**; **Strain hardened to 1/2 hard (sheet, strip)**: all **diam**, 390 MPa **TS**, 110–135 **HV**: Strain hardened to hard (sheet, strip): all **diam**, 470 MPa **TS**, 135–165 **HV**
AS 1572	272C	Australia	Tube	61.0–65.0 **Cu**, 0.07 **Pb**, rem **Zn**, 0.05 **Fe**, 0.20 total, others	**Annealed**: all **diam**, 80 **HV**; **Half–hard**: all **diam**, 95–130 **HV**; **Hard**: all **diam**, 150 **HV**
IS 410	CuZn37	India	Plate, sheet, strip, foil	61.5–64.5 **Cu**, 0.30 **Pb**, rem **Zn**, 0.10 **Fe**, 0.55 total, others	**Annealed**: all **diam**, 274 MPa **TS**, 45% in 2 in. (50 mm) **El**, 85 max **HV**; **Half hard**: all **diam**, 392 MPa **TS**, 20% in 2 in. (50 mm) **El**, 100 **HV**; **Hard**: all **diam**, 461 MPa **TS**, 3% in 2 in. (50 mm) **El**, 135 **HV**
NS 16 120	CuZn37	Norway	Sheet, strip, plate, bar, tube, rod, wire, and profiles	63 **Cu**, 37 **Zn**	
IS 4413	CuZn37	India	Wire	61.5–64 **Cu**, 0.30 **Pb**, rem **Zn**, 0.60 total, others	**Annealed**: all **diam**, 323 MPa **TS**, 40% in 2 in. (50 mm) **El**; **1/2 hard**: all **diam**, 461 MPa **TS**; **Hard**: all **diam**, 618 MPa **TS**
IS 3168	CuZn37	India	Strip, foil	61.5–64.0 **Cu**, 0.15 **Pb**, rem **Zn**, 0.05 **Fe**, 0.25 total, others; nickel included in copper	**As rolled**: 0.035 mm **diam**, 274 MPa **TS**, 45% in 2 in. (50 mm) **El**, 60–80 **HV**
ANSI/ASTM B 134	CA 274	US	Wire	61.0–64.0 **Cu**, 0.10 **Pb**, rem **Zn**, 0.05 **Fe**	**Quarter hard**: >0.020 in. (>0.508 mm) **diam**, 62 ksi (425 MPa) **TS**; **Hard**: >0.020 in. (>0.508 mm) **diam**, 102 ksi (700 MPa) **TS**; **Spring**: >0.020 in. (>0.508 mm) **diam**, 120 ksi (830 MPa) **TS**
COPANT 162	C 274 00	COPANT	Wrought products	61–64 **Cu**, 0.10 **Pb**, rem **Zn**, 0.05 **Fe**	
QQ-W-321d	274	US	Wire	61.0–64.0 **Cu**, 0.10 **Pb**, rem **Zn**, 0.05 **Fe**, 0.20 total, others; **Co** in **Ni**	**Eighth hard temper**: all **diam**, 50 ksi (345 MPa) **TS**; **Quarter hard temper**: all **diam**, 62 ksi (427 MPa) **TS**; **Half hard temper**: all **diam**, 79 ksi (545 MPa) **TS**

WROUGHT COPPER AND COPPER ALLOYS

specification number	designation	country	product forms	chemical composition	mechanical properties and hardness values
DGN-W-24	Type No. 8	Mexico	Wire	61-64 **Cu**, 0.10 **Pb**, rem **Zn**, 0.05 **Fe**	**3/4 hard**: <19.05 mm **diam**, 92 ksi (634 MPa) **TS**; **Hard**: <12.7 mm **diam**, 102 ksi (703 MPa) **TS**; **Spring hard**: <6.35 mm **diam**, 120 ksi (827 MPa) **TS**
ANSI/ASTM B 111	C 28000	US	Tube: seamless, and stock: ferrule	59.0-63.0 **Cu**, 0.30 **Pb**, rem **Zn**, 0.07 **Fe**	**Annealed**: all **diam**, 50 ksi (345 MPa) **TS**, 20 ksi (140 MPa) **YS**
ANSI/ASTM B 135	CA 280	US	Tube: seamless	59.0-63.0 **Cu**, 0.30 **Pb**, rem **Zn**, 0.07 **Fe**	**Drawn**: all **diam**, 54 ksi (370 MPa) **TS**, 55 **HR30T**; **Light anneal**: >0.030 in. (0.762 mm) **diam**, 60 **HR30T**
AS 1567	280A	Australia	Rod, bar, section	59.0-63.0 **Cu**, 0.30 **Pb**, rem **Zn**, 0.07 **Fe**, 0.20 total, others	**As manufactured**: 6-50 mm **diam**, 340 MPa **TS**, 160 MPa **YS**, 26% in 2 in. (50 mm) **El**
AS 1568	280A	Australia	Forgings	59.0-63.0 **Cu**, 0.30 **Pb**, rem **Zn**, 0.07 **Fe**, 0.20 total, others	**As manufactured**: ≥6 mm **diam**, 310 MPa **TS**, 25% in 2 in. (50 mm) **El**
AS 1572	280A	Australia	Tube	59.0-63.0 **Cu**, 0.30 **Pb**, rem **Zn**, 0.07 **Fe**, 0.20 total, others	**Annealed**: all **diam**, 80 **HV**; **Half-hard**: all **diam**, 95-130 **HV**; **Hard**: all **diam**, 150 **HV**
CDA 280		US	Tube, rod, plate, sheet, bar, strip	59.0-63.0 **Cu**, 0.30 **Pb**, rem **Zn**, 0.07 **Fe**, 0.20 total, others	**As hot rolled (plate, sheet, bar, strip)**: 0.040 in. **diam**, 54 ksi (372 MPa) **TS**, 45% in 2 in. (50 mm) **El**, 85 **HRF**; **Soft annealed (rod)**: 1.0 in. **diam**, 54 ksi (372 MPa) **TS**, 50% in 2 in. (50 mm) **El**, 80 **HRF**; **Hard drawn 30% (tube)**: 1.0 in. **diam**, 74 ksi (510 MPa) **TS**, 10% in 2 in. (50 mm) **El**, 80 **HRB**
COPANT 162	C 280 00	COPANT	Wrought products	59-63 **Cu**, 0.30 **Pb**, rem **Zn**, 0.07 **Fe**	
COPANT 527	Cu Zn 39	COPANT	Tube	59-63 **Cu**, 0.30 **Pb**, rem **Zn**, 0.07 **Fe**, 0.20 total, others	**Annealed**: 0.8-3.4 mm **diam**, 345 MPa **TS**, 135 MPa **YS**
WW-T-791A	Grade C	US	Tube	59.0-63.0 **Cu**, 0.30 **Pb**, rem **Zn**, 0.07 **Fe**	
IS 407	Alloy No. 2/CuZn39	India	Tube	59.0-63.0 **Cu**, 0.80 **Pb**, rem **Zn**, 0.07 **Fe**, 0.06 **As**, 0.30 total, others; nickel included in copper	**Annealed**: all **diam**, 284 MPa **TS**; **Half hard**: all **diam**, 372 MPa **TS**, 105 **HV**; **Hard**: all **diam**, 451 MPa **TS**
JIS H 3300	C 2800	Japan	Pipe and tube: seamless	59.0-63.0 **Cu**, 0.10 **Pb**, rem **Zn**, 0.07 **Fe**	**Annealed**: 10-250 mm OD **diam**, 314 MPa **TS**, 35% in 2 in. (50 mm) **El**; **1/2 hard**: 10-250 mm OD **diam**, 373 MPa **TS**, 15% in 2 in. (50 mm) **El**, 55 min **HR30T**; **Hard**: 10-250 mm OD **diam**, 451 MPa **TS**

WROUGHT COPPER AND COPPER ALLOYS

specification number	designation	country	product forms	chemical composition	mechanical properties and hardness values
JIS H 3260	C 2800	Japan	Rod and bar	59.0–63.0 **Cu**, 0.10 **Pb**, rem **Zn**, 0.07 **Fe**	**Annealed**: ≥0.5 mm **diam**, 314 MPa **TS**, 20% in 2 in. (50 mm) **El**; **3/4 hard**: 0.5–10 mm **diam**, 539 MPa **TS**; **Hard**: 0.5–10 mm **diam**, 686 MPa **TS**
JIS H 3250	C 2800	Japan	Rod and bar	59.0–63.0 **Cu**, 0.10 **Pb**, rem **Zn**, 0.07 **Fe**	**As manufactured**: ≥6 mm **diam**, 314 MPa **TS**, 25% in 2 in. (50 mm) **El**; **Annealed**: 6–75 mm **diam**, 314 MPa **TS**, 35% in 2 in. (50 mm) **El**; **Hard**: 6–20 mm **diam**, 451 MPa **TS**
DGN–W–19	Type No. 5	Mexico	Tube	59–63 **Cu**, 0.30 **Pb**, rem **Zn**, 0.07 **Fe**	**1/2 hard**: all **diam**, 54 ksi (372 MPa) **TS**, 58 min **HRB**
AMS 4611D		US	Rod and bar: naval	59.0–62.0 **Cu**, 0.50–1.0 **Sn**, 0.20 **Pb**, rem **Zn**, 0.10 **Fe**, 0.10 total, others	**Cold finished, 1/2 hard**: ≤0.500 in. (≤12.70 mm) **diam**, 60 ksi (414 MPa) **TS**, 27 ksi (186 MPa) **YS**
AMS 4612E		US	Rod and bar: naval	59.00–62.00 **Cu**, 0.50–1.0 **Sn**, 0.20 **Pb**, rem **Zn**, 0.10 **Fe**, 0.05 each, 0.10 total, others	**Cold finished, hard**: ≤1.000 in. (≤25.40 mm) **diam**, 67 ksi (462 MPa) **TS**, 45 ksi (310 MPa) **YS**, 13% in 2 in. (50 mm) **El**, 70–95 **HRB**
BS 2874	CZ109	UK	Rod, section	59.0–62.0 **Cu**, 0.10 **Pb**, rem **Zn**, 0.02 **Fe**, 0.30 total, others	**As manufactured**: 6–50 mm **diam**, 339 MPa **TS**, 26% in 2 in. (50 mm) **El**
BS 2872	CZ 109	UK	Forging stock, forgings	59.0–62.0 **Cu**, 0.10 **Pb**, rem **Zn**, 0.30 total, others	**As manufactured**: >6.0 mm **diam**, 309 MPa **TS**, 25% in 2 in. (50 mm) **El**
NF A 51 104	CuZn 40	France	Bar, wire, and shapes	59.0–62.0 **Cu**, 0.3 **Pb**, rem **Zn**, 0.2 **Fe**, 0.5 total, others	**Mill condition**: ≤50 mm **diam**, 390 MPa **TS**, 20% in 2 in. (50 mm) **El**
NF A 51 103	CuZn 40	France	Tube	59.0–62.0 **Cu**, rem **Zn**, 0.05 each, 0.40 total, others	**1/2 hard**: ≤80 mm **diam**, 430 MPa **TS**, 33% in 2 in. (50 mm) **El**, 110–150 **HV**; **3/4 hard**: ≤80 mm **diam**, 470 MPa **TS**, 23% in 2 in. (50 mm) **El**, 130–170 **HV**; **Hard**: ≤80 mm **diam**, 520 MPa **TS**, 10% in 2 in. (50 mm) **El**, 150–190 **HV**
IS 4170	CuZn40	India	Rod	59–62 **Cu**, 0.75 **Pb**, rem **Zn**, 0.1 **Fe**, 0.3 total, others	**As manufactured**: ≥5 mm **diam**, 343 MPa **TS**, 25% in 2 in. (50 mm) **El**; **Annealed**: 274 MPa **TS**, 30% in 2 in. (50 mm) **El**, 90 **HV**
IS 6912	Lead free brass	India	Forgings	59.0–62.0 **Cu**, 0.10 **Pb**, rem **Zn**, 0.02 **Fe**, 0.20 total, others	**As manufactured**: ≥6 mm **diam**, 310 MPa **TS**, 30% in 2 in. (50 mm) **El**

WROUGHT COPPER AND COPPER ALLOYS

specification number	designation	country	product forms	chemical composition	mechanical properties and hardness values
ISO 1639	CuZn40	ISO	Extruded sections	59.0–62.0 **Cu**, 0.03 **Pb**, rem **Zn**, 0.2 **Fe**, 0.3 **Ni** included in **Cu**; 0.5 **Fe**+**Pb** + total, others	**M**: all **diam**, 370 MPa **TS**, 35% in 2 in. (50 mm) **El**, 130 **HV**
ISO 1637	CuZn40	ISO	Solid product: straight lengths	59.0–62.0 **Cu**, 0.03 **Pb**, rem **Zn**, 0.2 **Fe**, 0.3 **Ni** included in **Cu**; 0.5 **Fe**+**Pb** + total, others	**Annealed**: 5 mm **diam**, 30% in 2 in. (50 mm) **El**, 95 max **HV**; **M**: 5 mm **diam**, 370 MPa **TS**, 40% in 2 in. (50 mm) **El**, 120 max **HV**
ISO 1635	CuZn40	ISO	Tube	59.0–62.0 **Cu**, 0.03 **Pb**, rem **Zn**, 0.2 **Fe**, **Cu** includes 0.3 **Ni**; 0.5 **Fe**+**Pb** + total, others	**Annealed**: all **diam**, 25% in 2 in. (50 mm) **El**, 100 max **HV**; **Strain hardened (HB)**: all **diam**, 510 MPa **TS**, 10% in 2 in. (50 mm) **El**, 130 **HV**
ISO 1634	CuZn40	ISO	Plate, sheet, strip	59.0–62.0 **Cu**, 0.3 **Pb**, rem **Zn**, 0.2 **Fe**, **Cu** includes 0.3 **Ni**; 0.5 **Fe**+**Pb** + total, others	**Annealed**: all **diam**, 35% in 2 in. (50 mm) **El**, 100 max **HV**; **M**: all **diam**, 370 MPa **TS**, 40% in 2 in. (50 mm) **El**, 105 max **HV**; **Strain hardened (HA)**: all **diam**, 390 MPa **TS**, 20% in 2 in. (50 mm) **El**, 125 **HV**
ISO 426 1	CuZn40	ISO	Plate, sheet, strip, rod, bar, tube, wire, extruded sections	59.0–62.0 **Cu**, 0.3 **Pb**, rem **Zn**, 0.2 **Fe**, 0.5 **Pb**+**Fe**	
NF A 51 101	CuZn 40	France	Sheet and strip	60.0 **Cu**, 40 **Zn**	**Stress hardened, Temper 1**: all **diam**, 360 MPa **TS**
IS 410	CuZn40	India	Plate, sheet, strip, foil	58.5–61.5 **Cu**, 0.30 **Pb**, rem **Zn**, 0.15 **Fe**, 0.75 total, others	**Annealed**: all **diam**, 274 MPa **TS**, 30% in 2 in. (50 mm) **El**, 85 max **HV**; **Half hard**: all **diam**, 421 MPa **TS**, 12% in 2 in. (50 mm) **El**, 100 **HV**; **Hard**: all **diam**, 490 MPa **TS**, 3% in 2 in. (50 mm) **El**, 125 **HV**
JIS Z 3262	BCuZn-1	Japan	Wire: filler	58–62 **Cu**, 0.05 **Pb**, rem **Zn**, 0.1 **Al**	
JIS H 3250	C 3712	Japan	Rod and bar	58.0–62.0 **Cu**, 0.10–1.0 **Pb**, rem **Zn**, 0.8 **Fe**+**Sn**	**As manufactured**: $\geq$6 mm **diam**, 314 MPa **TS**, 15% in 2 in. (50 mm) **El**
COPANT 162	C 282 00	COPANT	Wrought products	58–61 **Cu**, 0.05 **Sn**, 0.03 **Pb**, rem **Zn**, 0.05 **Fe**, 0.005 **Al**, 0.12–0.22 **P**, 0.005 **Al**+**Si**	
JIS Z 3262	BCuZn-2	Japan	Wire: filler	57–61 **Cu**, 0.5–1.5 **Sn**, 0.05 **Pb**, rem **Zn**, 0.02 **Al**	
JIS Z 3262	BCuZn-3	Japan	Wire: filler	56–60 **Cu**, 1.0 **Sn**, 0.05 **Pb**, rem **Zn**, 0.25–1.25 **Fe**, 0.25 **Si**, 0.01 **Al**, 1.0 **Mn**, 1.0 **Ni**	
COPANT 162	C 298 00	COPANT	Wrought products	49–52 **Cu**, 0.50 **Pb**, rem **Zn**, 0.10 **Fe**, 0.10 **Al**	

WROUGHT COPPER AND COPPER ALLOYS

specification number	designation	country	product forms	chemical composition	mechanical properties and hardness values
COPANT 527	Cu Zn 15	COPANT	Tube	84–86 **Cu+Ag**, 0.06 **Pb**, rem **Zn**, 0.05 **Fe**, 0.15 total, others	**Annealed**: 0.8–3.4 mm **diam**, 275 MPa **TS**, 80 MPa **YS**
COPANT 162	C 310 00	COPANT	Wrought products	89–91 **Cu**, 0.3–0.7 **Pb**, rem **Zn**, 0.10 **Fe**	
COPANT 612	Cu Zn 10 Pb	COPANT	Strip, sheet, plate, foil, bar	89–91 **Cu**, 0.3–0.7 **Pb**, rem **Zn**, 0.10 **Fe**, 0.50 total, others	**1/4 hard temper**: all **diam**, 275 MPa **TS**; **1/2 hard temper**: all **diam**, 325 MPa **TS**; **Hard temper**: all **diam**, 390 MPa **TS**
NF A 51 108	Cu Sn 5 Zn 4	France	Sheet and strip	89.3 min **Cu**, 3–5 **Sn**, 0.1 **Pb**, 3–5 **Zn**, 0.1 **Fe**, 0.2 **P**, 0.3 total, others	**Annealed**: $\leq$20 mm **diam**, 310 MPa **TS**, 150 MPa **YS**, 50% in 2 in. (50 mm) **El**, 80–110 **HV**; **1/2 hard**: $\leq$20 mm **diam**, 420 MPa **TS**, 350 MPa **YS**, 15% in 2 in. (50 mm) **El**, 140–170 **HV**; **Hard**: $\leq$20 mm **diam**, 570 MPa **TS**, 5% in 2 in. (50 mm) **El**, 180–210 **HV**
ANSI/ASTM B 140	CA 314	US	Rod, bar, shapes	87.5–90.5 **Cu**, 1.3–2.5 **Pb**, rem **Zn**, 0.10 **Fe**, 0.7 **Ni**	**Soft**: all **diam**, 35 ksi (240 MPa) **TS**, 10 ksi (70 MPa) **YS**, 25% in 2 in. (50.8 mm) **El**; **Half hard**: <0.50 in. (<12.7 mm) **diam**, 50 ksi (345 MPa) **TS**, 30 ksi (205 MPa) **YS**, 7% in 2 in. (50.8 mm) **El**
ANSI/ASTM B 140	CA 316	US	Rod, bar, shapes	87.5–90.5 **Cu**, 1.3–2.5 **Pb**, rem **Zn**, 0.10 **Fe**, 0.7–1.2 **Ni**, 0.04–0.10 **P**	**Half hard**: <0.50 in. (<12.7 mm) **diam**, 50 ksi (345 MPa) **TS**, 30 ksi (205 MPa) **YS**, 7% in 2 in. (50.8 mm) **El**; **Hard**: <2 in. (<50.8 mm) **diam**, 60 ksi (415 MPa) **TS**, 50 ksi (345 MPa) **YS**, 6% in 2 in. (50.8 mm) **El**
CDA 314		US	Rod, bar	87.5–90.5 **Cu**, 1.3–2.5 **Pb**, rem **Zn**, 0.10 **Fe**, 0.50 total, others	**Half hard drawn temper (bar)**: 0.250 in. **diam**, 55 ksi (379 MPa) **TS**, 12% in 2 in. (50 mm) **El**, 61 **HRB**; **Annealed 0.050 mm temper (rod)**: 1.0 in. **diam**, 37 ksi (255 MPa) **TS**, 45% in 2 in. (50 mm) **El**, 55 **HRF**; **Half hard temper 25% (rod)**: 0.500 in. **diam**, 55 ksi (379 MPa) **TS**, 14% in 2 in. (50 mm) **El**, 61 **HRB**

WROUGHT COPPER AND COPPER ALLOYS

specification number	designation	country	product forms	chemical composition	mechanical properties and hardness values
CDA 316		US	Rod, bar	87.5–90.5 **Cu**, 1.3–2.5 **Pb**, rem **Zn**, 0.10 **Fe**, 0.7–1.2 **Ni**, 0.10 **P**, 0.50 total, others	**Hard drawn (bar)**: 0.250 in. **diam**, 63 ksi (434 MPa) **TS**, 12% in 2 in. (50 mm) **El**, 70 **HRB**; **Hard temper 38% (rod)**: 0.500 in. **diam**, 67 ksi (462 MPa) **TS**, 13% in 2 in. (50 mm) **El**, 72 **HRB**; **Hard temper 38% (rod)**: 1.0 in. **diam**, 65 ksi (448 MPa) **TS**, 15% in 2 in. (50 mm) **El**, 70 **HRB**
COPANT 162	C 314 00	COPANT	Wrought products	87.5–90.5 **Cu**, 1.3–2.5 **Pb**, rem **Zn**, 0.10 **Fe**, 0.7 **Ni**	
COPANT 162	C 316 00	COPANT	Wrought products	87.5–90.5 **Cu**, 1.3–2.5 **Pb**, rem **Zn**, 0.10 **Fe**, 0.7–1.2 **Ni**	
NF A 51 108	Cu Sn 4 Zn 4 Pb 4	France	Sheet and strip	85.9 min **Cu**, 3.0–4.5 **Sn**, 3.0–4.5 **Pb**, 3.0–4.5 **Zn**, 0.1 **Fe**, 0.20 **P**, 0.3 total, others	**Annealed**: $\leq$20 mm **diam**, 320 MPa **TS**, 2 in. (50 mm) **El**, 80–100 **HV**; **1/2 hard**: $\leq$20 mm **diam**, 400 MPa **TS**, 25% in 2 in. (50 mm) **El**, 125–155 **HV**; **Hard**: $\leq$20 mm **diam**, 500 MPa **TS**, 3% in 2 in. (50 mm) **El**, 160–180 **HV**
ANSI/ASTM B 140	CA 320	US	Rod, bar, shapes	83.5–86.5 **Cu**, 1.5–2.2 **Pb**, rem **Zn**, 0.10 **Fe**, 0.25 **Ni**	**Soft**: all **diam**, 35 ksi (240 MPa) **TS**, 10 ksi (70 MPa) **YS**, 25% in 2 in. (50.8 mm) **El**; **Half hard**: <0.50 in. (<12.7 mm) **diam**, 50 ksi (345 MPa) **TS**, 30 ksi (205 MPa) **YS**, 7% in 2 in. (50.8 mm) **El**
COPANT 162	C 320 00	COPANT	Wrought products	83.5–86.5 **Cu**, 1.5–2.2 **Pb**, rem **Zn**, 0.10 **Fe**, 0.25 **Ni**	
BS 2874	CZ 104	UK	Rod, section	79.0–81.0 **Cu**, 0.1–1.0 **Pb**, rem **Zn**, 0.60 total, others	**As manufactured**: $\geq$6 mm **diam**, 309 MPa **TS**, 22% in 2 in. (50 mm) **El**
COPANT 162	C 325 00	COPANT	Wrought products	72–74.5 **Cu**, 2.5–3.0 **Pb**, rem **Zn**, 0.10 **Fe**	
AS 1567	351D	Australia	Rod, bar, section	69.0–72.0 **Cu**, 0.3–0.7 **Pb**, rem **Zn**, 0.02–0.06 **As**, 0.50 total, others	**Soft annealed**: >6 mm **diam**, 260 MPa **TS**, 38% in 2 in. (50 mm) **El**; **As manufactured**: >6 mm **diam**, 310 MPa **TS**, 25% in 2 in. (50 mm) **El**
AMS 4555D		US	Tube: seamless	65.00–71.50 **Cu**, 0.15 **Sn**, 0.80 **Pb**, rem **Zn**, 0.07 **Fe**, 0.05 each, 0.15 total, others	**Annealed**: all **diam**, 28–53 **HR30T**
AMS 4558C		US	Tube: seamless	65.00–68.00 **Cu**, 1.30–2.00 **Pb**, rem **Zn**, 0.07 **Fe**, 0.50 total, others	**Drawn**: >0.045 in. (>1.14 mm) **diam**, 55 min **HRB**

WROUGHT COPPER AND COPPER ALLOYS

specification number	designation	country	product forms	chemical composition	mechanical properties and hardness values
ANSI/ASTM B 135	CA 332	US	Tube: seamless	65.0–68.0 **Cu**, 1.3–2.0 **Pb**, rem **Zn**, 0.07 **Fe**	**Drawn**: all **diam**, 54 ksi (370 MPa) **TS**, 53 **HR30T**; **Hard drawn**: >0.020 in. (>0.508 mm) **diam**, 66 ksi (455 MPa) **TS**, 70 **HR30T**; **Soft anneal**: >0.030 in. (>0.762 mm) **diam**, 40 **HR30T**
ANSI/ASTM B 135	CA 330	US	Tube: seamless	65.0–68.0 **Cu**, 0.20–0.8 **Pb**, rem **Zn**, 0.07 **Fe**	**Drawn**: all **diam**, 54 ksi (370 MPa) **TS**, 53 **HR30T**; **Hard drawn**: >0.020 in. (>0.508 mm) **diam**, 66 ksi (455 MPa) **TS**, 70 **HR30T**; **Soft anneal**: >0.030 in. (>0.762 mm) **diam**, 40 **HR30T**
CDA 330		US	Tube	65.0–68.0 **Cu**, 0.20–0.8 **Pb**, rem **Zn**, 0.07 **Fe**, 0.50 total, others	**Annealed – 0.050 mm temper**: 1.0 in. **diam**, 47 ksi (324 MPa) **TS**, 60% in 2 in. (50 mm) **El**, 64 **HRF**; **Annealed – 0.025 mm temper**: 1.0 in. **diam**, 52 ksi (359 MPa) **TS**, 52% in 2 in. (50 mm) **El**, 75 **HRF**; **Hard drawn 35%**: 1.0 in. **diam**, 75 ksi (517 MPa) **TS**, 7% in 2 in. (50 mm) **El**, 80 **HRB**
CDA 332		US	Tube	65.0–68.0 **Cu**, 1.3–2.0 **Pb**, rem **Zn**, 0.07 **Fe**, 0.50 total, others	**Annealed 0.025 mm**: 1.0 in. **diam**, 52 ksi (359 MPa) **TS**, 50% in 2 in. (50 mm) **El**, 75 **HRF**; **Hard drawn 35%**: 1.0 in. **diam**, 75 ksi (517 MPa) **TS**, 7% in 2 in. (50 mm) **El**, 80 **HRB**
COPANT 162	C 330 00	COPANT	Wrought products	65–68 **Cu**, 0.2–0.8 **Pb**, rem **Zn**, 0.07 **Fe**	
COPANT 162	C 331 00	COPANT	Wrought products	65–68 **Cu**, 0.7–1.2 **Pb**, rem **Zn**, 0.06 **Fe**	
COPANT 162	C 332 00	COPANT	Wrought products	65–68 **Cu**, 1.3–2.0 **Pb**, rem **Zn**, 0.07 **Fe**	
WW-T-791A	Grade B	US	Tube	65.0–68.0 **Cu**, 0.20–0.8 **Pb**, rem **Zn**, 0.07 **Fe**	
DGN-W-19	Type No. 3	Mexico	Tube	65–68 **Cu**, 0.3–0.8 **Pb**, rem **Zn**, 0.07 **Fe**	**1/2 hard**: all **diam**, 54 ksi (372 MPa) **TS**, 55 min **HRB**; **Hard**: ≤25.4 mm **diam**, 66 ksi (455 MPa) **TS**, 75 min **HRB**
DGN-W-19	Type No. 4	Mexico	Tube	65–68 **Cu**, 1.3–2.0 **Pb**, rem **Zn**, 0.07 **Fe**	**1/2 hard**: all **diam**, 54 ksi (372 MPa) **TS**, 55 min **HRB**; **Hard**: 66 ksi (455 MPa) **TS**, 75 min **HRB**
MIL-T-46072A	330	US	Tube: seamless	65.0–68.0 **Cu**, 0.2–0.8 **Pb**, rem **Zn**, 0.07 **Fe**, 0.50 total, others	**Drawn**: all **diam**, 372 MPa **TS**, 53 **HR30T**; **Hard drawn**: ≤1 mm **diam**, 455 MPa **TS**, 70 **HR30T**

WROUGHT COPPER AND COPPER ALLOYS

specification number	designation	country	product forms	chemical composition	mechanical properties and hardness values
MIL–T–46072A	331	US	Tube: seamless	65.0–68.0 **Cu**, 0.7–1.2 **Pb**, rem **Zn**, 0.06 **Fe**, 0.50 total, others	**Drawn**: all **diam**, 372 MPa **TS**, 53 **HR30T**; **Hard drawn**: $\leq$1 mm **diam**, 455 MPa **TS**, 70 **HR30T**
MIL–T–46072A	332	US	Tube: seamless	65.0–68.0 **Cu**, 1.3–2.0 **Pb**, rem **Zn**, 0.07 **Fe**, 0.50 total, others	**Drawn**: all **diam**, 372 MPa **TS**, 53 **HR30T**; **Hard drawn**: $\leq$1 mm **diam**, 455 MPa **TS**, 70 **HR30T**
MIL–T–6945	III	US	Tube: seamless	65–68 **Cu**, 0.70 **Pb**, rem **Zn**, 0.05 **Fe**	**Hard drawn and relief annealed**: all **diam**, 462 MPa **TS**, 310 MPa **YS**, 15% in 2 in. (50 mm) **El**
CSA HC.4.3	HC 4.ZP 341	Canada	Plate	65.5–66.5 **Cu**, 0.8–1.4 **Pb**, rem **Zn**, 0.10 **Fe**, 0.50 other elements	**Quarter hard**: all **diam**, 49 ksi (337 MPa) **TS**; **Half hard**: all **diam**, 55 ksi (379 MPa) **TS**
QQ–B–626D	Composition 11	US	Rod, shapes, forgings, bar, strip	58.5–71.5 **Cu**, 3.7 **Pb**, rem **Zn**, 0.50 **Fe**, 0.50 total, others	**Half hard temper**: all **diam**, 60–90 **HRB**
QQ–B–613D	Composition 11	US	Plate, bar, sheet, strip	58.5–71.5 **Cu**, 3.7 **Pb**, rem **Zn**, 0.50 **Fe**, 0.50 total, others	**Soft annealed**: all **diam**, 54–75 **HRF**; **Light annealed**: all **diam**, 67–87 **HRF**; **Quarter hard temper**: all **diam**, 40–65 **HRB**
DGN–W–25	LA–65	Mexico	Sheet and strip	64–66 **Cu**, 0.3 **Sn**, 0.5 **Pb**, rem **Zn**, 0.2 **Fe**, 0.2 **Al**, 0.2 **Mn**, 0.5 **Ni**	**Spring hard**: all **diam**, 80 max ksi (552 max MPa) **TS**; **Hard**: all **diam**, 64 max ksi (441 max MPa) **TS**; **1/2 hard**: 55 max ksi (380 max MPa) **TS**
ANSI/ASTM B 121	C 34200	US	Plate, sheet, strip, bar	62.5–66.5 **Cu**, 1.5–2.5 **Pb**, rem **Zn**, 0.10 **Fe**	**Hard**: >0.012 in. (>0.305 mm) **diam**, 68 ksi (470 MPa) **TS**, 54 **HR30T**; **Spring**: 0.012 in. (>0.305 mm) **diam**, 86 ksi (595 MPa) **TS**, 75 **HR30T**; **Annealed**: >0.015 in. (>0.381 mm) **diam**, 12 **HR30T**
ANSI/ASTM B 121	C 34000	US	Plate, sheet, strip, bar	62.5–66.5 **Cu**, 0.8–1.4 **Pb**, rem **Zn**, 0.10 **Fe**	**Hard**: >0.012 in. (>0.305 mm) **diam**, 68 ksi (470 MPa) **TS**, 54 **HR30T**; **Spring**: >0.012 in. (>0.305 mm) **diam**, 86 ksi (595 MPa) **TS**, 75 **HR30T**; **Annealed**: >0.015 in. (>0.381 mm) **diam**, 12 **HR30T**
ANSI/ASTM B 121	C 33500	US	Plate, sheet, strip, bar	62.5–66.5 **Cu**, 0.30–0.8 **Pb**, rem **Zn**, 0.10 **Fe**	**Hard**: >0.012 in. (>0.305 mm) **diam**, 68 ksi (470 MPa) **TS**, 54 **HR30T**; **Spring**: >0.012 in. (>0.305 mm) **diam**, 86 ksi (595 MPa) **TS**, 75 **HR30T**; **Annealed**: >0.015 in. (>0.381 mm) **diam**, 12 **HR30T**

WROUGHT COPPER AND COPPER ALLOYS

specification number	designation	country	product forms	chemical composition	mechanical properties and hardness values
ANSI/ASTM B 453	CA 335	US	Rod	62.5–66.5 **Cu**, 0.30–0.8 **Pb**, rem **Zn**, 0.15 **Fe**	**Soft**: 0.50–1.0 in. (12.7–25.4 mm) **diam**, 44 ksi (305 MPa) **TS**, 15 ksi (105 MPa) **YS**, 25% in 2 in. (50.8 mm) **El**, 45 max **HRB**; **Quarter hard**: 0.50–1.0 in. (12.7–25.4 mm) **diam**, 50 ksi (345 MPa) **TS**, 20 ksi (140 MPa) **YS**, 15% in 2 in. (50.8 mm) **El**, 50–75 **HRB**; **Half hard**: 0.50–1.0 in. (12.7–25.4 mm) **diam**, 55 ksi (380 MPa) **TS**, 25 ksi (170 MPa) **YS**, 10% in 2 in. (50.8 mm) **El**, 60–80 **HRB**
ANSI/ASTM B 453	CA 340	US	Rod	62.5–66.5 **Cu**, 0.8–1.4 **Pb**, rem **Zn**, 0.15 **Fe**	**Soft**: 0.50–1.0 in. (12.7–25.4 mm) **diam**, 44 ksi (305 MPa) **TS**, 15 ksi (105 MPa) **YS**, 25% in 2 in. (50.8 mm) **El**, 45 max **HRB**; **Quarter hard**: 0.50–1.0 in. (12.7–25.4 mm) **diam**, 50 ksi (345 MPa) **TS**, 20 ksi (140 MPa) **YS**, 15% in 2 in. (50.8 mm) **El**, 50–75 **HRB**; **Half hard**: 0.50–1.0 in. (12.7–25.4 mm) **diam**, 55 ksi (380 MPa) **TS**, 25 ksi (170 MPa) **YS**, 10% in 2 in. (50.8 mm) **El**, 60–80 **HRB**
AS 1567	335C	Australia	Rod, bar, section	62.5–66.5 **Cu**, 0.3–0.8 **Pb**, rem **Zn**, 0.10 **Fe**, 0.02–0.06 **As**, 0.50 total, others	**As manufactured**: >6 mm **diam**, 310 MPa **TS**, 25% in 2 in. (50 mm) **El**; **Soft annealed**: >6 mm **diam**, 260 MPa **TS**, 38% in 2 in. (50 mm) **El**
BS 2870	CZ 118	UK	Sheet, strip, foil	63.0–66.0 **Cu**, 0.75–1.5 **Pb**, rem **Zn**, 0.30 total, others	**1/2 hard**: ≤6.0 mm **diam**, 259 MPa **TS**, 10% in 2 in. (50 mm) **El**, 110 **HV**; **Hard**: ≤6.0 mm **diam**, 303 MPa **TS**, 5% in 2 in. (50 mm) **El**, 140 **HV**; **Extra hard**: ≤6.0 mm **diam**, 359 MPa **TS**, 3% in 2 in. (50 mm) **El**, 165 **HV**

WROUGHT COPPER AND COPPER ALLOYS

specification number	designation	country	product forms	chemical composition	mechanical properties and hardness values
CDA 335		US	Rod, plate, bar, strip	62.5–66.5 **Cu**, 0.30–0.8 **Pb**, rem **Zn**, 0.10 **Fe**, 0.50 total, others	**Quarter hard temper (plate, bar, strip)**: 0.040 in. **diam**, 54 ksi (372 MPa) **TS**, 43% in 2 in. (50 mm) **El**, 55 **HRB**; **Half hard temper (plate, bar, strip)**: 0.040 in. **diam**, 61 ksi (421 MPa) **TS**, 23% in 2 in. (50 mm) **El**, 70 **HRB**; **Hard temper (plate, bar, strip)**: 0.040 in. **diam**, 74 ksi (510 MPa) **TS**, 8% in 2 in. (50 mm) **El**, 80 **HRB**
CDA 340		US	Rod, wire, shapes, sheet, bar, strip	62.5–66.5 **Cu**, 0.8–1.4 **Pb**, rem **Zn**, 0.10 **Fe**, 0.50 total, others	**Quarter hard temper – 10% (rod)**: 1.0 in. **diam**, 55 ksi (379 MPa) **TS**, 40% in 2 in. (50 mm) **El**, 60 **HRB**; **Half hard temper (wire)**: 0.080 in. **diam**, 88 ksi (607 MPa) **TS**, 15% in 2 in. (50 mm) **El**; **Hard temper (sheet, bar, strip)**: 0.040 in. **diam**, 74 ksi (510 MPa) **TS**, 7% in 2 in. (50 mm) **El**, 80 **HRB**
CDA 342		US	Rod, plate, sheet, bar, strip	62.5–66.5 **Cu**, 1.5–2.5 **Pb**, rem **Zn**, 0.1 **Fe**, 0.05 total, others	**Half hard – 20% (rod)**: 1.0 in. **diam**, 58 ksi (400 MPa) **TS**, 25% in 2 in. (50 mm) **El**, 75 **HRB**; **Hard temper (plate, sheet, bar, strip)**: 0.040 in. **diam**, 74 ksi (510 MPa) **TS**, 7% in 2 in. (50 mm) **El**, 80 **HRB**; **Extra hard temper (plate, sheet, bar, strip)**: 0.040 in. **diam**, 85 ksi (586 MPa) **TS**, 5% in 2 in. (50 mm) **El**, 87 **HRB**
CSA HC.4.3	HC 4.ZP 341	Canada	Sheet	62.5–66.5 **Cu**, 0.8–1.4 **Pb**, rem **Zn**, 0.10 **Fe**, 0.50 other elements	**Half hard**: all **diam**, 55 ksi (379 MPa) **TS**; **Full hard**: all **diam**, 68 ksi (468 MPa) **TS**; **Spring**: all **diam**, 86 ksi (592 MPa) **TS**
CSA HC.4.3	HC 4.ZP 341	Canada	Bar	62.5–66.5 **Cu**, 0.8–1.4 **Pb**, rem **Zn**, 0.10 **Fe**, 0.50 other elements	**Half hard**: all **diam**, 55 ksi (379 MPa) **TS**; **Full hard**: all **diam**, 68 ksi (468 MPa) **TS**; **Spring**: all **diam**, 86 ksi (592 MPa) **TS**
CSA HC.4.3	HC 4.ZP 341	Canada	Strip	62.5–66.5 **Cu**, 0.8–1.4 **Pb**, rem **Zn**, 0.10 **Fe**, 0.50 other elements	**Half hard**: all **diam**, 55 ksi (379 MPa) **TS**; **Full hard**: all **diam**, 68 ksi (468 MPa) **TS**; **Spring**: all **diam**, 86 ksi (592 MPa) **TS**
COPANT 162	C 335 00	COPANT	Wrought products	62.5–66.5 **Cu**, 0.3–0.8 **Pb**, rem **Zn**, 0.10 **Fe**	

WROUGHT COPPER AND COPPER ALLOYS

specification number	designation	country	product forms	chemical composition	mechanical properties and hardness values
COPANT 162	C 340 00	COPANT	Wrought products	62.5–66.5 **Cu**, 0.8–1.4 **Pb**, rem **Zn**, 0.10 **Fe**	
COPANT 162	C 342 00	COPANT	Wrought products	62.5–66.5 **Cu**, 1.5–2.5 **Pb**, rem **Zn**, 0.10 **Fe**	
COPANT 612	Cu Zn 34 Pb 2	COPANT	Strip, sheet, plate, foil, bar	62.5–66.5 **Cu**, 1.5–2.5 **Pb**, rem **Zn**, 0.10 **Fe**, 0.50 total, others	**1/4 hard temper**: all **diam**, 345 MPa **TS**; **1/2 hard temper**: all **diam**, 380 MPa **TS**; **Hard temper**: all **diam**, 470 MPa **TS**
COPANT 612	Cu Zn 35 Pb 1	COPANT	Strip, sheet, plate, foil, bar	62.5–66.5 **Cu**, 0.8–1.4 **Pb**, rem **Zn**, 0.10 **Fe**, 0.50 total, others	**1/4 hard temper**: all **diam**, 345 MPa **TS**; **1/2 hard temper**: all **diam**, 380 MPa **TS**; **Hard temper**: all **diam**, 470 MPa **TS**
COPANT 612	Cu Zn 36 Pb	COPANT	Strip, sheet, plate, foil, bar	62.5–66.5 **Cu**, 0.3–0.7 **Pb**, rem **Zn**, 0.10 **Fe**, 0.50 total, others	**1/4 hard temper**: all **diam**, 345 MPa **TS**; **1/2 hard temper**: all **diam**, 380 MPa **TS**; **Hard temper**: all **diam**, 470 MPa **TS**
QQ–B–626D	342	US	Rod, shapes, forgings, bar, strip	62.5–66.5 **Cu**, 1.5–2.5 **Pb**, rem **Zn**, 0.10 **Fe**, 0.50 total, others	**Quarter hard temper**: all **diam**, 49 ksi (338 MPa) **TS**; **Half hard temper**: all **diam**, 55 ksi (379 MPa) **TS**; **Hard temper**: all **diam**, 68 ksi (469 MPa) **TS**
QQ–B–613D	342	US	Plate, bar, sheet, strip	62.5–66.5 **Cu**, 1.5–2.5 **Pb**, rem **Zn**, 0.10 **Fe**, 0.50 total, others	**Quarter hard temper**: all **diam**, 49 ksi (338 MPa) **TS**; **Half hard temper**: all **diam**, 55 ksi (379 MPa) **TS**; **Hard temper**: all **diam**, 68 ksi (469 MPa) **TS**
COPANT 162	C 344 00	COPANT	Wrought products	62–66 **Cu**, 0.5–1.0 **Pb**, rem **Zn**, 0.10 **Fe**	
COPANT 162	C 347 00	COPANT	Wrought products	62.5–64.5 **Cu**, 1.0–1.8 **Pb**, rem **Zn**, 0.10 **Fe**	
IS 2704	CuZn35Pb1	India	Wire	62.0–65.0 **Cu**, 0.75–1.50 **Pb**, rem **Zn**, 0.50 total, others	**Quarter hard**: all **diam**, 323 MPa **TS**, 30% in 2 in. (50 mm) **El**, 110–130 **HV**; **Half hard**: all **diam**, 402 MPa **TS**, 20% in 2 in. (50 mm) **El**, 131–150 **HV**; **Hard**: all **diam**, 500 MPa **TS**, 15% in 2 in. (50 mm) **El**, 151–175 **HV**
DGN–W–25	LA–63	Mexico	Sheet and strip	63–64 **Cu**, 0.3 **Sn**, 0.5 **Pb**, rem **Zn**, 0.2 **Fe**, 0.2 **Al**, 0.2 **Mn**, 0.5 **Ni**	**Spring hard**: all **diam**, 80 max ksi (552 max MPa) **TS**; **Hard**: all **diam**, 64 max ksi (441 max MPa) **TS**; **1/2 hard**: 55 max ksi (380 max MPa) **TS**

WROUGHT COPPER AND COPPER ALLOYS

specification number	designation	country	product forms	chemical composition	mechanical properties and hardness values
ANSI/ASTM B 453	CA 345	US	Rod	62.0–64.0 **Cu**, 1.5–2.8 **Pb**, rem **Zn**, 0.15 **Fe**	**Soft**: 0.50–1.0 in. (12.7–25.4 mm) **diam**, 44 ksi (305 MPa) **TS**, 15 ksi (105 MPa) **YS**, 25% in 2 in. (50.8 mm) **El**, 45 max **HRB**; **Quarter hard**: 0.50–1.0 in. (12.7–25.4 mm) **diam**, 50 ksi (345 MPa) **TS**, 20 ksi (140 MPa) **YS**, 15% in 2 in. (50.8 mm) **El**, 50–75 **HRB**; **Half hard**: 0.50–1.0 in. (12.7–25.4 mm) **diam**, 55 ksi (380 MPa) **TS**, 25 ksi (170 MPa) **YS**, 10% in 2 in. (50.8 mm) **El**, 60–80 **HRB**
COPANT 162	C 345 00	COPANT	Wrought products	62–64 **Cu**, 1.5–2.8 **Pb**, rem **Zn**, 0.10 **Fe**	
JIS H 3100	C6711	Japan	Sheet, plate, and strip	61.0–65.0 **Cu**, 0.7–1.5 **Sn**, 0.10–1.0 **Pb**, rem **Zn**, 0.05–1.0 **Mn**, 1.0 **Fe+Al+Si**	**Hard**: 0.25–1.5 mm **diam**, 190 min **HV**
JIS H 3100	C2720	Japan	Sheet, plate, and strip	62.0–64.0 **Cu**, 0.07 **Pb**, rem **Zn**, 0.07 **Fe**	**Annealed**: <1 mm **diam**, 275 MPa **TS**, 40% in 2 in. (50 mm) **El**; **1/2 hard**: 0.3–20 mm **diam**, 353 MPa **TS**, 28% in 2 in. (50 mm) **El**, 85–145 **V**; **Hard**: 0.3–10 mm **diam**, 412 MPa **TS**, 105 min **HV**
NS 16 150	CuZn36Pb1	Norway	Strip, bar, tube, rod, and wire	63 **Cu**, 1 **Pb**, 36 **Zn**	
ANSI/ASTM B 453	CA 353	US	Rod	61.0–64.5 **Cu**, 2.0–3.0 **Pb**, rem **Zn**, 0.15 **Fe**	**Soft**: 0.50–1.0 in. (12.7–25.4 mm) **diam**, 44 ksi (305 MPa) **TS**, 15 ksi (105 MPa) **YS**, 25% in 2 in. (50.8 mm) **El**, 45 max **HRB**; **Quarter hard**: 0.50–1.0 in. (12.7–25.4 mm) **diam**, 50 ksi (345 MPa) **TS**, 20 ksi (140 MPa) **YS**, 15% in 2 in. (50.8 mm) **El**, 50–75 **HRB**; **Half hard**: 0.50–1.0 in. (12.7–25.4 mm) **diam**, 55 ksi (380 MPa) **TS**, 25 ksi (170 MPa) **YS**, 10% in 2 in. (50.8 mm) **El**, 60–80 **HRB**

WROUGHT COPPER AND COPPER ALLOYS

specification number	designation	country	product forms	chemical composition	mechanical properties and hardness values
ANSI/ASTM B 453	CA 350	US	Rod	61.0–64.0 **Cu**, 0.8–1.4 **Pb**, rem **Zn**, 0.15 **Fe**	**Soft**: 0.50–1.0 in. (12.7–25.4 mm) **diam**, 44 ksi (305 MPa) **TS**, 15 ksi (105 MPa) **YS**, 25% in 2 in. (50.8 mm) **El**, 45 max **HRB**; **Quarter hard**: 0.50–1.0 in. (12.7–25.4 mm) **diam**, 50 ksi (345 MPa) **TS**, 20 ksi (140 MPa) **YS**, 15% in 2 in. (50.8 mm) **El**, 50–75 **HRB**; **Half hard**: 0.50–1.0 in. (12.7–25.4 mm) **diam**, 55 ksi (380 MPa) **TS**, 25 ksi (170 MPa) **YS**, 10% in 2 in. (50.8 mm) **El**, 60–80 **HRB**
AS 1567	353C	Australia	Rod, bar, section	61.0–64.0 **Cu**, 1.0–2.5 **Pb**, rem **Zn**, 0.10 **Fe**, 0.50 total, others	**As manufactured**: 6–50 mm **diam**, 350 MPa **TS**, 22% in 2 in. (50 mm) **El**
BS 2871	CZ 119	UK	Tube	61.0–64.0 **Cu**, 1.0–2.5 **Pb**, rem **Zn**, 0.30 total, others	**Annealed**: 300 MPa **TS**, 75 max **HV**; **As drawn**: 380 MPa **TS**, 130 **HV**
BS 2870	CZ 119	UK	Sheet, strip, foil	61.0–64.0 **Cu**, 1.0–2.5 **Pb**, rem **Zn**, 0.30 total, others	**1/2 hard**: $\leq$6.0 mm **diam**, 259 MPa **TS**, 10% in 2 in. (50 mm) **El**, 110 **HV**; **Hard**: $\leq$6.0 mm **diam**, 303 MPa **TS**, 5% in 2 in. (50 mm) **El**, 140 **HV**; **Extra hard**: $\leq$6.0 mm **diam**, 359 MPa **TS**, 3% in 2 in. (50 mm) **El**, 165 **HV**
BS 2873	CZ 119	UK	Wire	61.0–64.0 **Cu**, 1.0–2.5 **Pb**, rem **Zn**, 0.30 total, others	
BS 2874	CZ 119	UK	Rod, section	61.0–64.0 **Cu**, 1.0–2.5 **Pb**, rem **Zn**, 0.30 total, others	**As manufactured**: 6–50 mm **diam**, 348 MPa **TS**, 22% in 2 in. (50 mm) **El**
CDA 349		US	Rod, wire	61.0–64.0 **Cu**, 0.10–0.50 **Pb**, rem **Zn**, 0.10 **Fe**, 0.50 total, others	**0.035 mm temper (rod)**: 0.250 in. **diam**, 53 ksi (366 MPa) **TS**, 50% in 2 in. (50 mm) **El**, 75 **HRF**; **0.015 mm temper (wire)**: 0.250 in. **diam**, 55 ksi (379 MPa) **TS**, 48% in 2 in. (50 mm) **El**, 70 **HRF**; **Quarter hard temper – 21% (wire)**: 0.250 in. **diam**, 68 ksi (469 MPa) **TS**, 18% in 2 in. (50 mm) **El**, 72 **HRB**
CSA HC.4.3	HC 4.ZP 342	Canada	Plate	60.5–64.5 **Cu**, 2.0–3.0 **Pb**, rem **Zn**, 0.10 **Fe**, 0.50 other elements	**Quarter hard**: all **diam**, 49 ksi (337 MPa) **TS**; **Half hard**: all **diam**, 55 ksi (379 MPa) **TS**
CSA HC.4.3	HC 4.ZP 342	Canada	Bar	60.5–64.5 **Cu**, 2.0–3.0 **Pb**, rem **Zn**, 0.10 **Fe**, 0.50 other elements	**Half hard**: all **diam**, 55 ksi (379 MPa) **TS**; **Full hard**: all **diam**, 68 ksi (468 MPa) **TS**; **Spring**: all **diam**, 86 ksi (592 MPa) **TS**

WROUGHT COPPER AND COPPER ALLOYS

specification number	designation	country	product forms	chemical composition	mechanical properties and hardness values
CSA HC.4.3	HC 4.ZP 342	Canada	Strip	60.5–64.5 **Cu**, 2.0–3.0 **Pb**, rem **Zn**, 0.10 **Fe**, 0.50 other elements	**Half hard**: all **diam**, 55 ksi (379 MPa) **TS**; **Full hard**: all **diam**, 68 ksi (468 MPa) **TS**; **Spring**: all **diam**, 86 ksi (592 MPa) **TS**
CSA HC.4.3	HC 4.ZP 342	Canada	Sheet	60.5–64.5 **Cu**, 2.0–3.0 **Pb**, rem **Zn**, 0.10 **Fe**, 0.50 other elements	**Half hard**: all **diam**, 55 ksi (379 MPa) **TS**; **Full hard**: all **diam**, 68 ksi (468 MPa) **TS**; **Spring**: all **diam**, 86 ksi (592 MPa) **TS**
COPANT 162	C 348 00	COPANT	Wrought products	61.5–63.5 **Cu**, 0.4–0.8 **Pb**, rem **Zn**, 0.10 **Fe**	
COPANT 162	C 349 00	COPANT	Wrought products	61–64 **Cu**, 0.10–0.50 **Pb**, rem **Zn**, 0.10 **Fe**	
COPANT 612	Cu Zn 35 Pb 2	COPANT	Strip, sheet, plate, foil, bar	60.5–64.5 **Cu**, 2.0–3.0 **Pb**, rem **Zn**, 0.10 **Fe**, 0.50 total, others	**1/4 hard temper**: all **diam**, 345 MPa **TS**; **1/2 hard temper**: all **diam**, 380 MPa **TS**; **Hard temper**: all **diam**, 470 MPa **TS**
DS 3003	5140	Denmark	Rolled, drawn, extruded, and forged products	61.0–64.0 **Cu**, 0.5–1.5 **Pb**, 36 **Zn**, 0.2 **Fe**, 0.3 total, others	
SFS 2923	CuZn36Pb1	Finland	Rod, wire, tube, shapes, and forgings	61.0–64.0 **Cu**, 0.5–1.5 **Pb**, 36 **Zn**, 0.2 **Fe**, 0.030–0.06 **As**	**Hot finished (except tube)**: all **diam**, 360 MPa **TS**, 150 MPa **YS**, 40% in 2 in. (50 mm) **El**; **Annealed (rod and wire)**: 1–2.5 mm **diam**, 310 MPa **TS**, 100 MPa **YS**, 25% in 2 in. (50 mm) **El**; **Cold finished (rod and wire)**: 1–2.5 mm **diam**, 360 MPa **TS**, 240 MPa **YS**, 10% in 2 in. (50 mm) **El**
ISO 426/II	CuZn36Pb1	ISO	Plate, sheet, strip, rod, bar, tube, wire	61.0–64.0 **Cu**, 0.5–1.5 **Pb**, rem **Zn**, 0.2 **Fe**, 0.3 total, others	
ISO 1634	CuZn35Pb2	ISO	Plate, sheet, strip	61.0–64.0 **Cu**, 1.5–2.5 **Pb**, rem **Zn**, 0.2 **Fe**, **Cu** includes 0.3 **Ni**; 0.3 total, others	**Annealed**: all **diam**, 35% in 2 in. (50 mm) **El**, 90 max **HV**; **Strain hardened (HB)**: 0.3–5 mm **diam**, 430 MPa **TS**, 10% in 2 in. (50 mm) **El**, 150 **HV**
ISO 1634	CuZn36Pb1	ISO	Plate, sheet, strip	61.0–64.0 **Cu**, 0.5–1.5 **Pb**, rem **Zn**, 0.2 **Fe**, **Cu** includes 0.3 **Ni**; 0.3 total, others	**Annealed**: 0.3–10 mm **diam**, 35% in 2 in. (50 mm) **El**, 90 max **HV**; **Strain hardened (HB)**: 0.3–5 mm **diam**, 430 MPa **TS**, 10% in 2 in. (50 mm) **El**, 150 **HV**
ISO 1637	CuZn35Pb2	ISO	Solid products: straight lengths	61.0–64.0 **Cu**, 1.5–2.5 **Pb**, rem **Zn**, 0.2 **Fe**, **Cu** includes 0.3 **Ni**; 0.3 total, others	**M**: 5 mm **diam**, 360 MPa **TS**, 30% in 2 in. (50 mm) **El**, 100 max **HV**; **HA**: 5–15 mm **diam**, 350 MPa **TS**, 20% in 2 in. (50 mm) **El**, 120 **HV**

WROUGHT COPPER AND COPPER ALLOYS

specification number	designation	country	product forms	chemical composition	mechanical properties and hardness values
ISO 1637	CuZn36Pb1	ISO	Solid products: straight lengths	61.0–64.0 **Cu**, 0.5–1.5 **Pb**, rem **Zn**, 0.2 **Fe**, **Cu** includes 0.3 **Ni**; 0.5 total, others	**M**: 5 mm **diam**, 360 MPa **TS**, 30% in 2 in. (50 mm) **El**, 100 max **HV**; **HA**. 5–15 mm **diam**, 450 MPa **TS**, 20% in 2 in. (50 mm) **El**, 120 **HV**
ISO 1638	CuZn35Pb2	ISO	Solid product: drawn, on coils or reels	61.0–64.0 **Cu**, 1.5–2.5 **Pb**, rem **Zn**, 0.2 **Fe**, **Cu** includes 0.3 **Ni**; 0.3 total, others	**Annealed**: 1–5 mm **diam**, 340 MPa **TS**, 25% in 2 in. (50 mm) **El**; **HB - strain hardened**: 1–3 mm **diam**, 390 MPa **TS**, 10% in 2 in. (50 mm) **El**
ISO 1638	CuZn36Pb1	ISO	Solid product: drawn, on coils or reels	61.0–64.0 **Cu**, 0.5–1.5 **Pb**, rem **Zn**, 0.2 **Fe**, **Cu** includes 0.3 **Ni**; 0.3 total, others	**Annealed**: 1–5 mm **diam**, 340 MPa **TS**, 25% in 2 in. (50 mm) **El**; **HB - strain hardened**: 1–3 mm **diam**, 390 MPa **TS**, 10% in 2 in. (50 mm) **El**
ISO 426/II	CuZn35Pb2	ISO	Plate, sheet, strip, rod, bar, tube, wire	61.0–64.0 **Cu**, 1.5–2.5 **Pb**, rem **Zn**, 0.2 **Fe**, 0.3 total, others	
JIS H 3250	C 4622	Japan	Rod and bar	61.0–64.0 **Cu**, 0.7–1.5 **Sn** rem **Zn**, 0.8 **Pb+Fe**	**As manufactured**: 6–50 mm **diam**, 343 MPa **TS**, 20% in 2 in. (50 mm) **El**
JIS H 3100	C3560	Japan	Sheet, plate, and strip	61.0–64.0 **Cu**, 2.0–3.0 **Pb**, rem **Zn**, 0.10 **Fe**	**1/4 hard**: 0.3–10 mm **diam**, 343 MPa **TS**, 18% in 2 in. (50 mm) **El**; **1/2 hard**: 0.3–10 mm **diam**, 373 MPa **TS**, 10% in 2 in. (50 mm) **El**; **Hard**: 0.3–10 mm **diam**, 422 MPa **TS**, 2 in. (50 mm) **El**
SABS 460	Cu–Zn36Pb1	South Africa	Tube	61.0–64.0 **Cu**, 0.5–1.5 **Pb**, rem **Zn**, 0.30 total, others	**Annealed**: all **diam**, 300 MPa **TS**, 75 max **HV**; **As drawn**: all **diam**, 380 MPa **TS**, 130 **HV**
ANSI/ASTM B 453	CA 356	US	Rod	60.0–64.5 **Cu**, 2.0–3.0 **Pb**, rem **Zn**, 0.15 **Fe**	**Soft**: 0.50–1.0 in. (12.7–25.4 mm) **diam**, 44 ksi (305 MPa) **TS**, 15 ksi (105 MPa) **YS**, 25% in 2 in. (50.8 mm) **El**, 45 max **HRB**; **Quarter hard**: 0.50–1.0 in. (12.7–25.4 mm) **diam**, 50 ksi (345 MPa) **TS**, 20 ksi (140 MPa) **YS**, 15% in 2 in. (50.8 mm) **El**, 50–75 **HRB**; **Half hard**: 0.50–1.0 in. (12.7–25.4 mm) **diam**, 55 ksi (380 MPa) **TS**, 25 ksi (170 MPa) **YS**, 10% in 2 in. (50.8 mm) **El**, 60–80 **HRB**
NF A 51 105	CuZn35 Pb2	France	Bar, wire, and shapes	61.0–63.0 **Cu**, 1.5–2.25 **Pb**, rem **Zn**, 0.20 **Fe**, 0.30 total, others	

WROUGHT COPPER AND COPPER ALLOYS

specification number	designation	country	product forms	chemical composition	mechanical properties and hardness values
JIS H 3260	C 3501	Japan	Rod and bar	60.0–64.0 **Cu**, 0.7–1.7 **Pb**, rem **Zn**, 0.20 **Fe**, 0.40 **Fe+Sn**	**Annealed**: ≥0.5 mm **diam**, 294 MPa **TS**, 20% in 2 in. (50 mm) **El**; **1/2 hard**: 0.5–12 mm **diam**, 343 MPa **TS**; **Hard**: 0.5–10 mm **diam**, 422 MPa **TS**
ANSI/ASTM B 121	C 35000	US	Plate, sheet, strip, bar	59.0–64.5 **Cu**, 1.3–2.3 **Pb**, rem **Zn**, 0.10 **Fe**	**Hard**: >0.012 in. (>0.305 mm) **diam**, 68 ksi (470 MPa) **TS**, 54 **HR30T**; **Spring**: 0.012 in. (>0.305 mm) **diam**, 86 ksi (595 MPa) **TS**, 75 **HR30T**; **Annealed**: >0.015 in. (>0.381 mm) **diam**, 12 **HR30T**
ANSI/ASTM B 121	C 35300	US	Plate, sheet, strip, bar	59.0–64.5 **Cu**, 1.3–2.3 **Pb**, rem **Zn**, 0.10 **Fe**	**Hard**: >0.012 in. (>0.305 mm) **diam**, 68 ksi (470 MPa) **TS**, 54 **HR30T**; **Spring**: >0.012 in. (>0.305 mm) **diam**, 86 ksi (595 MPa) **TS**, 75 **HR30T**; **Annealed**: >0.015 in. (>0.381 mm) **diam**, 12 **HR30T**
ANSI/ASTM B 121	C 35600	US	Plate, sheet, strip, bar	59.0–64.5 **Cu**, 2.0–3.0 **Pb**, rem **Zn**, 0.10 **Fe**	**Hard**: >0.012 in. (>0.305 mm) **diam**, 68 ksi (470 MPa) **TS**, 54 **HR30T**; **Spring**: >0.012 in. (>0.305 mm) **diam**, 86 ksi (595 MPa) **TS**, 75 **HR30T**; **Annealed**: >0.015 in. (>0.381 mm) **diam**, 12 **HR30T**
CDA 353		US	Rod, plate, sheet, bar, strip	59.0–64.5 **Cu**, 1.3–2.3 **Pb**, rem **Zn**, 0.1 **Fe**, 0.5 total, others	**Half hard – 20% (rod)**: 1.0 in. **diam**, 58 ksi (400 MPa) **TS**, 25% in 2 in. (50 mm) **El**, 75 **HRB**; **Hard temper (plate, sheet, bar, strip)**: 0.040 in. **diam**, 74 ksi (510 MPa) **TS**, 7% in 2 in. (50 mm) **El**, 80 **HRB**; **Extra hard temper (plate, sheet, bar, strip)**: 0.040 in. **diam**, 85 ksi (586 MPa) **TS**, 5% in 2 in. (50 mm) **El**, 87 **HRB**
CDA 356		US	Rod, bar, strip	59.0–64.5 **Cu**, 2.0–3.0 **Pb**, rem **Zn**, 0.10 **Fe**, 0.50 total, others	**Hard temper (rolled bar, rolled strip)**: 0.040 in. **diam**, 74 ksi (510 MPa) **TS**, 68 ksi (469 MPa) **YS**, 8% in 2 in. (50 mm) **El**, 81 **HRB**; **Extra hard temper (rolled bar, rolled strip)**: 0.040 in. **diam**, 88 ksi (607 MPa) **TS**, 77 ksi (531 MPa) **YS**, 5% in 2 in. (50 mm) **El**, 85 **HRB**; **Spring temper (rolled bar, rolled strip)**: 0.040 in. **diam**, 92 ksi (634 MPa) **TS**, 83 ksi (572 MPa) **YS**, 4% in 2 in. (50 mm) **El**, 87 **HRB**

WROUGHT COPPER AND COPPER ALLOYS

specification number	designation	country	product forms	chemical composition	mechanical properties and hardness values
CSA HC.4.3	HC 4.ZP 352	Canada	Bar	59.0–64.5 **Cu**, 1.3–2.3 **Pb**, rem **Zn**, 0.10 **Fe**, 0.50 other elements	**Half hard**: all **diam**, 55 ksi (379 MPa) **TS**; **Full hard**: all **diam**, 68 ksi (468 MPa) **TS**; **Spring**: all **diam**, 86 ksi (592 MPa) **TS**
CSA HC.4.3	HC 4 ZP 352	Canada	Strip	59.0–64.5 **Cu**, 1.3–2.3 **Pb**, rem **Zn**, 0.10 **Fe**, 0.50 other elements	**Half hard**: all **diam**, 55 ksi (379 MPa) **TS**; **Full hard**: all **diam**, 68 ksi (468 MPa) **TS**; **Spring**: all **diam**, 86 ksi (592 MPa) **TS**
CSA HC.4.3	HC 4.ZP 352	Canada	Plate	59.0–64.5 **Cu**, 1.3–2.3 **Pb**, rem **Zn**, 0.10 **Fe**, 0.50 other elements	**Quarter hard**: all **diam**, 49 ksi (337 MPa) **TS**; **Half hard**: all **diam**, 55 ksi (379 MPa) **TS**
CSA HC.4.3	HC 4.ZP 352	Canada	Sheet	59.0–64.5 **Cu**, 1.3–2.3 **Pb**, rem **Zn**, 0.10 **Fe**, 0.50 other elements	**Half hard**: all **diam**, 55 ksi (379 MPa) **TS**; **Full hard**: all **diam**, 68 ksi (468 MPa) **TS**; **Spring**: all **diam**, 86 ksi (592 MPa) **TS**
COPANT 162	C 353 00	COPANT	Wrought products	59–64.5 **Cu**, 1.3–2.3 **Pb**, rem **Zn**, 0.10 **Fe**	
COPANT 162	C 356 00	COPANT	Wrought products	59–64.5 **Cu**, 2.0–3.0 **Pb**, rem **Zn**, 0.10 **Fe**	
COPANT 612	Cu Zn 36 Pb 2	COPANT	Strip, sheet, plate, foil, bar	59–64.5 **Cu**, 1.3–2.3 **Pb**, rem **Zn**, 0.10 **Fe**, 0.50 total, others	**1/4 hard temper**: all **diam**, 345 MPa **TS**; **1/2 hard temper**: all **diam**, 380 MPa **TS**; **Hard temper**: all **diam**, 470 MPa **TS**
QQ-B-626D	353	US	Rod, shapes, forgings, bar, strip	59.0–64.5 **Cu**, 1.3–2.3 **Pb**, rem **Zn**, 0.10 **Fe**, 0.50 total, others	**Quarter hard temper**: all **diam**, 49 ksi (338 MPa) **TS**; **Half hard temper**: all **diam**, 55 ksi (379 MPa) **TS**; **Hard temper**: all **diam**, 68 ksi (469 MPa) **TS**
QQ-B-613D	353	US	Plate, bar, sheet, strip	59.0–64.5 **Cu**, 1.2–2.3 **Pb**, rem **Zn**, 0.10 **Fe**, 0.50 total, others	**Quarter hard temper**: all **diam**, 49 ksi (338 MPa) **TS**; **Half hard temper**: all **diam**, 55 ksi (379 MPa) **TS**; **Hard temper**: all **diam**, 68 ksi (469 MPa) **TS**
AMS 4610J		US	Rod and bar: free cutting	60.0–63.0 **Cu**, 2.5–3.7 **Pb**, rem **Zn**, 0.35 **Fe**, 0.50 total, others	**Cold finished, 1/2 hard**: $\leq$1.000 in. ($\leq$12.70 mm) **diam**, 60 ksi (414 MPa) **TS**, 28 ksi (193 MPa) **YS**, 10% in 2 in. (50 mm) **El**, 50–80 **HRB**;

WROUGHT COPPER AND COPPER ALLOYS

specification number	designation	country	product forms	chemical composition	mechanical properties and hardness values
ANSI/ASTM B 16	C 36000	US	Rod, bar, shapes	60.0–63.0 **Cu**, 2.5–3.7 **Pb**, rem **Zn**, 0.35 **Fe**	**Soft**: 1 in. (25.4 mm) **diam**, 48 ksi (330 MPa) **TS**, 20 ksi (140 MPa) **YS**, 15% in 1 in. (25.4 mm) **El**; **Half–hard**: 1/2 in. (12.7 mm) **diam**, 57 ksi (395 MPa) **TS**, 25 ksi (170 MPa) **YS**, 7% in 1 in. (25.4 mm) **El**; **Hard**: 1/16–3/16 in. (1.59–4.78 mm) **diam**, 80 ksi (550 MPa) **TS**, 45 ksi (310 MPa) **YS**
AS 1567	360A	Australia	Rod, bar, section	60.0–63.0 **Cu**, 2.5–3.7 **Pb**, rem **Zn**, 0.35 **Fe**, 0.50 total, others	
BS 2874	CZ 124	UK	Rod, section	60.0–63.0 **Cu**, 2.5–3.7 **Pb**, rem **Zn**, 0.35 **Fe**, 0.50 total, others	**As manufactured (round and hexagonal rod)**: 6.0–25 mm **diam**, 328 MPa **TS**, 128 MPa **YS**, 12% in 2 in. (50 mm) **El** 1/2 hard: 6.0–12 mm **diam**, 402 MPa **TS**, 162 MPa **YS**, 6% in 2 in. (50 mm) **El** Hard: 3.0–5.0 mm **diam**, 549 MPa **TS**, 290 MPa **YS**
CDA 350		US	Rod, strip	59.0–64.0 **Cu**, 0.8–1.4 **Pb**, rem **Zn**, 0.10 **Fe**, 0.50 total, others	**Half hard temper – 20% (rod)**: 0.500 in. **diam**, 70 ksi (483 MPa) **TS**, 22% in 2 in. (50 mm) **El**, 80 **HRB**; **Hard temper (rolled strip)**: 0.040 in. **diam**, 73 ksi (503 MPa) **TS**, 60 ksi (414 MPa) **YS**, 10% in 2 in. (50 mm) **El**, 80 **HRB**; **Extra hard temper (rolled strip)**: 0.040 in. **diam**, 84 ksi (579 MPa) **TS**, 69 ksi (476 MPa) **TS**, 5% in 2 in. (50 mm) **El**, 86 **HRB**
CDA 360		US	Rod, shapes, bar	60.0–63.0 **Cu**, 2.5–3.7 **Pb**, rem **Zn**, 0.35 **Fe**, 0.50 total, others	**Soft anneal (rod)**: 1.0 in. **diam**, 49 ksi (338 MPa) **TS**, 53% in 2 in. (50 mm) **El**, 68 **HRF**; **As extruded (shapes)**: 0.500 in. **diam**, 49 ksi (338 MPa) **TS**, 50% in 2 in. (50 mm) **El**, 68 **HRF**; **Quarter hard – 11% (drawn bar)**: 0.250 in. **diam**, 56 ksi (386 MPa) **TS**, 20% in 2 in. (50 mm) **El**, 62 **HRB**
COPANT 162	C 350 00	COPANT	Wrought products	59–64 **Cu**, 0.8–1.4 **Pb**, rem **Zn**, 0.10 **Fe**	
COPANT 162	C 360 00	COPANT	Wrought products	60–63 **Cu**, 2.5–3.7 **Pb**, rem **Zn**, 0.35 **Fe**	
COPANT 162	C 362 00	COPANT	Wrought products	60–63 **Cu**, 3.5–4.5 **Pb**, rem **Zn**, 0.15 **Fe**	

WROUGHT COPPER AND COPPER ALLOYS

specification number	designation	country	product forms	chemical composition	mechanical properties and hardness values
QQ-B-626D	360	US	Rod, shapes, forgings, bar, strip	60.0–63.0 **Cu**, 2.5–3.7 **Pb**, rem **Zn**, 0.35 **Fe**, 0.50 total, others	**Soft and as extruded**: 1 max in. **diam**, 48 ksi (331 MPa) **TS**, 20 ksi (138 MPa) **YS**, 15% in 4 in. (100 mm) **El**;
SFS 2922	CuZn36Pb3	Finland	Rod, bar, wire, and shapes	60.0–63.0 **Cu**, 2.5–3.5 **Pb**, 36 **Zn**, 0.35 **Fe**, 0.030–0.06 **As**	**Drawn (rod, bar, and wire)**: 2.5–5 mm **diam**, 400 MPa **TS**, 320 MPa **YS**, 5% in 2 in. (50 mm) **El**
IS 319	Type II	India	Bar, rod, section	60.0–63.0 **Cu**, 2.5–3.7 **Pb**, rem **Zn**, 0.35 **Fe**, 0.15 total, others	**Annealed**: 10–25 mm **diam**, 335 MPa **TS**, 15% in 2 in. (50 mm) **El**; **Half hard**: 10–12 mm **diam**, 395 MPa **TS**, 7% in 2 in. (50 mm) **El**; **Hard**: 10–12 mm **diam**, 550 MPa **TS**
ISO 1637	CuZn36Pb3	ISO	Solid products: straight lengths	60.0–63.0 **Cu**, 2.5–3.7 **Pb**, rem **Zn**, 0.35 **Fe**, **Cu** includes 0.3 **Ni**; 0.5 total, others	**M**: 5 mm **diam**, 360 MPa **TS**, 30% in 2 in. (50 mm) **El**, 110 max **HV**; **HA**: 5–75 mm **diam**, 310 MPa **TS**, 20% in 2 in. (50 mm) **El**, 105 **HV**; **HB – strain hardened**: 5–15 mm **diam**, 410 MPa **TS**, 15% in 2 in. (50 mm) **El**, 125 **HV**
COPANT 419	Cu Zn 35 Pb 3	COPANT	Bar	60.0–63.0 **Cu + Ag**, 2.5 **Pb**, rem **Zn**, 0.35 **Fe**	**Annealed**: ≤25 mm **diam**, 335 MPa **TS**, 140 MPa **YS**, 15% in 2 in. (50 mm) **El**; **1/2 hard/hard**: ≤13 mm **diam**, 390 MPa **TS**, 175 MPa **YS**, 7% in 2 in. (50 mm) **El**; **Extra hard**: 5–13 mm **diam**, 4% in 2 in. (50 mm) **El**
ISO 426/II	CuZn36Pb3	ISO	Rod, bar, extruded sections, tube, wire	60.0–63.0 **Cu**, 2.5–3.7 **Pb**, rem **Zn**, 0.35 **Fe**, 0.5 total, others	
DGN-W-20		Mexico	Bar and shapes	60–63 **Cu**, 2.5–3.7 **Pb**, rem **Zn**, 0.35 **Fe**	**Soft**: >25.4 mm **diam**, 48 ksi (331 MPa) **TS**, 20 ksi (138 MPa) **YS**, 15% in 2 in. (50 mm) **El**, 35–60 min **HRB**
DS 3003	5165	Denmark	Rolled, drawn, extruded, and forged products	60–62 **Cu**, 0.5–1.5 **Pb**, 38 **Zn**, 0.2 **Fe**, 0.3 total, others	
SFS 2924	CuZn38Pb1	Finland	Bar, rod, and wire	60.0–62.0 **Cu**, 0.5–1.5 **Pb**, 38 **Zn**, 0.2 **Fe**	**Drawn (bar)**: 2.5–10 mm **diam**, 410 MPa **TS**, 240 MPa **YS**, 20% in 2 in. (50 mm) **El**; **Cold finished (rod and wire)**: 2.5–5 mm **diam**, 370 MPa **TS**, 200 MPa **YS**, 20% in 2 in. (50 mm) **El**
NF A 51 105	CuZn36 Pb3	France	Bar, wire, and shapes	60.0–62.0 **Cu**, 2.5–3.5 **Pb**, rem **Zn**, 0.35 **Fe**, 0.5 total, others	**Rolled, extruded, or drawn**: ≤7 mm **diam**, 450 MPa **TS**, 6% in 2 in. (50 mm) **El**

WROUGHT COPPER AND COPPER ALLOYS

specification number	designation	country	product forms	chemical composition	mechanical properties and hardness values
JIS H 3250	C 3602	Japan	Rod and bar	59.0–63.0 **Cu**, 1.8–3.7 **Pb**, rem **Zn**, 0.50 **Fe**, 1.3 **Fe+Sn**	**As manufactured**: 6–75 mm **diam**, 314 MPa **TS**, 75 min **HV**
JIS H 3250	C 3601	Japan	Rod and bar	59.0–63.0 **Cu**, 1.8–3.7 **Pb**, rem **Zn**, 0.30 **Fe**, 0.50 **Fe+Sn**	**Annealed**: 6–75 mm **diam**, 294 MPa **TS**, 25% in 2 in. (50 mm) **El**; **1/2 hard**: 6–50 mm **diam**, 343 MPa **TS**, 95 min **HV**; **Hard**: 6–20 mm **diam**, 451 MPa **TS**, 130 min **HV**
NS 16 145	CuZn38Pb1	Norway	Strip, bar, tube, rod, wire, profiles, and forgings	61 **Cu**, 1 **Pb**, 38 **Zn**	
SIS 14 51 65	SIS Brass 51 65–00	Sweden	Bar, tube, sections, forgings	60.0–62.0 **Cu**, 0.5–1.5 **Pb**, 38 **Zn**, 0.2 **Fe**, 0.3 total, others	**Hot worked**: all **diam**, 370 MPa **TS**, 130 MPa **YS**, 35% in 2 in. (50 mm) **El**, 85 **HV**
SIS 14 51 65	SIS Brass 51 65–02	Sweden	Strip, bar	60.0–62.0 **Cu**, 0.5–1.5 **Pb**, 38 **Zn**, 0.2 **Fe**, 0.3 total, others	**Annealed**: 0.5–2.5 mm **diam**, 330 MPa **TS**, 110 MPa **YS**, 20% in 2 in. (50 mm) **El**, 20–40 **HV**
SIS 14 51 65	SIS Brass 51 65–04	Sweden	Strip, bar, wire, tube	60.0–62.0 **Cu**, 0.5–1.5 **Pb**, 38 **Zn**, 0.2 **Fe**, 0.3 total, others	**Strain hardened**: 0.5–2.5 mm **diam**, 410 MPa **TS**, 240 MPa **YS**, 10% in 2 in. (50 mm) **El**, 110–145 **HV**
SIS 14 51 65	SIS Brass 51 65–10	Sweden	Wire, section	60.0–62.0 **Cu**, 0.5–1.5 **Pb**, 38 **Zn**, 0.2 **Fe**, 0.3 total, others	**Strain hardened**: 5 max mm **diam**, 390 MPa **TS**, 200 MPa **YS**, 10% in 2 in. (50 mm) **El**, 105 **HV**
ANSI/ASTM B 135	CA 370	US	Tube: seamless	59.0–62.0 **Cu**, 0.9–1.4 **Pb**, rem **Zn**, 0.15 **Fe**	**Drawn**: all **diam**, 54 ksi (370 MPa) **TS**, 55 **HR30T**; **Light anneal**: <0.030 in. (<0.762 mm) **diam**, 60 **HR30T**
BS 2870	CZ 123	UK	Sheet, strip, foil	59.0–62.0 **Cu**, 0.3–0.8 **Pb**, rem **Zn**, 0.30 total, others	**As manufactured**: ≤10.0 mm **diam**, 259 MPa **TS**, 20% in 2 in. (50 mm) **El**
BS 2875	CZ 123	UK	Plate	59.0–62.0 **Cu**, 0.3–0.8 **Pb**, rem **Zn**, 0.30 total, others	**As manufactured**: 10–25 mm **diam**, 339 MPa **TS**, 18% in 2 in. (50 mm) **El**
BS 2874	CZ 123	UK	Rod, section	59.0–62.0 **Cu**, 0.30–0.80 **Pb**, rem **Zn**, 0.02 **Sb**, 0.30 total, others	**As manufactured**: 6–50 mm **diam**, 339 MPa **TS**, 24% in 2 in. (50 mm) **El**
BS 2872	CZ 123	UK	Forging stock, forgings	59.0–62.0 **Cu**, 0.3–0.8 **Pb**, rem **Zn**, 0.30 total, others	**As manufactured or annealed**: ≥6 mm **diam**, 309 MPa **TS**, 25% in 2 in. (50 mm) **El**
CDA 370		US	Tube	59.0–62.0 **Cu**, 0.9–1.4 **Pb**, rem **Zn**, 0.15 **Fe**, 0.50 total, others	**Light anneal**: 1.5 in. **diam**, 54 ksi (372 MPa) **TS**, 40% in 2 in. (50 mm) **El**, 80 **HRF**; **Hard drawn – 35%**: 1.5 in. **diam**, 80 ksi (552 MPa) **TS**, 6% in 2 in. (50 mm) **El**, 85 **HRB**; **Hard drawn – 25%**: 2.0 in. **diam**, 70 ksi (483 MPa) **TS**, 10% in 2 in. (50 mm) **El**, 75 **HRB**

WROUGHT COPPER AND COPPER ALLOYS

specification number	designation	country	product forms	chemical composition	mechanical properties and hardness values
COPANT 162	C 370 00	COPANT	Wrought products	59–62 **Cu**, 0.9–1.4 **Pb**, rem **Zn**, 0.15 **Fe**	
IS 6912	60/40 brass	India	Forgings	59.0–62.0 **Cu**, 0.30–0.80 **Pb**, rem **Zn**, 0.10 **Fe**, 0.30 total, others	**As manufactured**: ≥6 mm **diam**, 310 MPa **TS**, 30% in 2 in. (50 mm) **El**
JIS H 3250	C 4641	Japan	Rod and bar	59.0–62.0 **Cu**, 0.50–1.0 **Sn**, rem **Zn**, 1.0 **Pb+Fe**	**As manufactured**: 6–50 mm **diam**, 343 MPa **TS**, 20% in 2 in. (50 mm) **El**
JIS H 3100	C2801	Japan	Sheet, plate, and strip	59.0–62.0 **Cu**, 0.10 **Pb**, rem **Zn**, 0.07 **Fe**	**Annealed**: <1 mm **diam**, 324 MPa **TS**, 35% in 2 in. (50 mm) **El**; **1/2 hard**: 0.3–20 mm **diam**, 412 MPa **TS**, 15% in 2 in. (50 mm) **El**, 105–160 **V**; **Hard**: 0.3–10 mm **diam**, 471 MPa **TS**, 130 min **HV**
DGN-W-19	Type No. 6	Mexico	Tube	59–62 **Cu**, 0.90–1.40 **Pb**, rem **Zn**, 0.15 **Fe**	**1/2 hard**: all **diam**, 54 ksi (372 MPa) **TS**, 58 min **HRB**
MIL-T-46072A	370	US	Tube: seamless	59.0–62.0 **Cu**, 0.9–1.4 **Pb**, rem **Zn**, 0.15 **Fe**, 0.50 total, others	**Drawn**: all **diam**, 372 MPa **TS**, 55 **HR30T**
AMS 4614E		US	Forgings: free cutting	58.0–62.0 **Cu**, 1.5–2.5 **Pb**, rem **Zn**, 0.30 **Fe**, 0.05 each, 0.50 total, others	
AS 1568	365C	Australia	Forgings	58.0–62.0 **Cu**, 0.4–0.9 **Pb**, rem **Zn**, 0.50 total, others	**As manufactured**: ≥6 mm **diam**, 310 MPa **TS**, 25% in 2 in. (50 mm) **El**
AS 1567	365C	Australia	Rod, bar, section	58.0–62.0 **Cu**, 0.4–0.9 **Pb**, rem **Zn**, 0.50 total, others	**As manufactured**: 6–50 mm **diam**, 340 MPa **TS**, 24% in 2 in. (50 mm) **El**
CDA 377		US	Rod, shapes	58.0–62.0 **Cu**, 1.5–2.5 **Pb**, rem **Zn**, 0.30 **Fe**, 0.50 total, others	**As extruded**: 1.0 in. **diam**, 52 ksi (359 MPa) **TS**, 45% in 2 in. (50 mm) **El**, 78 **HRF**
COPANT 162	C 371 00	COPANT	Wrought products	58–62 **Cu**, 0.6–1.2 **Pb**, rem **Zn**, 0.15 **Fe**	
COPANT 162	C 377 00	COPANT	Wrought products	58–62 **Cu**, 1.5–2.5 **Pb**, rem **Zn**, 0.30 **Fe**	
DS 3003	5163	Denmark	Rolled, drawn, extruded, and forged products	59.0–61.0 **Cu**, 0.3–0.8 **Pb**, 40 **Zn**, 0.2 **Fe**, 0.5 total, others	
DS 3003	5167	Denmark	Rolled, drawn, extruded, and forged products	59.0–61.0 **Cu**, 1.5–2.5 **Pb**, 38 **Zn**, 0.2 **Fe**, 0.5 total, others	
QQ-B-626D	377	US	Rod, shapes, forgings, bar, strip	58.0–62.0 **Cu**, 1.5–2.5 **Pb**, rem **Zn**, 0.30 **Fe**, 0.50 total, others	
SFS 2925	CuZn40Pb	Finland	Sheet	59.0–61.0 **Cu**, 0.3–0.8 **Pb**, 40 **Zn**, 0.2 **Fe**	**Hot rolled**: ≥2.5 mm **diam**, 340 MPa **TS**, 140 MPa **YS**, 30% in 2 in. (50 mm) **El**; **Cold finished**: ≥2.5 mm **diam**, 350 MPa **TS**, 200 MPa **YS**, 20% in 2 in. (50 mm) **El**

WROUGHT COPPER AND COPPER ALLOYS

specification number	designation	country	product forms	chemical composition	mechanical properties and hardness values
NF A 51 105	CuZn38 Pb2	France	Bar, wire, and shapes	59.0–61.0 **Cu**, 1.5–2.5 **Pb**, rem **Zn**, 0.20 **Fe**, 0.50 total, others	**Rolled, extruded, or drawn**: ≤7 mm **diam**, 450 MPa **TS**, 6% in 2 in. (50 mm) **El**
ISO 1635	CuZn38Pb2	ISO	Tube	59.0–61.0 **Cu**, 1.5–2.5 **Pb**, rem **Zn**, 0.2 **Fe**, **Cu** includes 0.3 **Ni**; 0.5 total, others	**Annealed**: all **diam**, 30% in 2 in. (50 mm) **El**, 100 max **HV**; **Strain hardened (HB)**: all **diam**, 470 MPa **TS**, 13% in 2 in. (50 mm) **El**, 120 **HV**
ISO 1634	CuZn38Pb2	ISO	Plate, sheet, strip	59.0–61.0 **Cu**, 1.5–2.5 **Pb**, rem **Zn**, 0.2 **Fe**, **Cu** includes 0.3 **Ni**; 0.5 total, others	**Annealed**: all **diam**, 35% in 2 in. (50 mm) **El**, 90 max **HV**; **Strain hardened**: 0.3–5 mm **diam**, 430 MPa **TS**, 10% in 2 in. (50 mm) **El**, 150 **HV**
ISO 1634	CuZn40Pb	ISO	Plate, sheet, strip	59.0–61.0 **Cu**, 0.3–0.8 **Pb**, rem **Zn**, 0.2 **Fe**, **Cu** includes 0.3 **Ni**; 0.5 total, others	**M**: all **diam**, 370 MPa **TS**, 40% in 2 in. (50 mm) **El**, 110 max **HV**; **HA**: all **diam**, 390 MPa **TS**, 20% in 2 in. (50 mm) **El**, 125 **HV**
ISO 1637	CuZn38Pb2	ISO	Solid products: straight lengths	59.0–61.0 **Cu**, 1.5–2.5 **Pb**, rem **Zn**, 0.2 **Fe**, **Cu** includes 0.3 **Ni**; 0.5 total, others	**M**: 5 mm **diam**, 360 MPa **TS**, 35% in 2 in. (50 mm) **El**, 100 max **HV**; **HA**: 5–15 mm **diam**, 350 MPa **TS**, 20% in 2 in. (50 mm) **El**, 120 **HV**
ISO 1637	CuZn40Pb	ISO	Solid products: straight lengths	59.0–61.0 **Cu**, 0.3–0.8 **Pb**, rem **Zn**, 0.2 **Fe**, **Cu** includes 0.3 **Ni**; 0.5 total, others	**M**: 5 mm **diam**, 370 MPa **TS**, 35% in 2 in. (50 mm) **El**, 120 max **HV**; **HA**: 5–50 mm **diam**, 350 MPa **TS**, 25% in 2 in. (50 mm) **El**, 120 **HV**; **HB – strain hardened**: 5–15 mm **diam**, 440 MPa **TS**, 15% in 2 in. (50 mm) **El**, 140 **HV**
ISO 1638	CuZn38Pb2	ISO	Solid product: drawn, on coils or reels	59.0–61.0 **Cu**, 1.5–2.5 **Pb**, rem **Zn**, 0.2 **Fe**, **Cu** includes 0.3 **Ni**; 0.5 total, others	**Annealed**: 1–5 mm **diam**, 340 MPa **TS**, 25% in 2 in. (50 mm) **El**; **HB – strain hardened**: 1–3 mm **diam**, 390 MPa **TS**, 7% in 2 in. (50 mm) **El**
ISO 1638	CuZn40Pb	ISO	Solid product: drawn, on coils or reels	59.0–61.0 **Cu**, 0.3–0.8 **Pb**, rem **Zn**, 0.2 **Fe**, **Cu** includes 0.3 **Ni**; 0.5 total, others	**Annealed**: 1–5 mm **diam**, 340 MPa **TS**, 25% in 2 in. (50 mm) **El**; **HC**: 1–5 mm **diam**, 450 MPa **TS**, 5% in 2 in. (50 mm) **El**
ISO 426/II	CuZn38Pb2	ISO	Plate, sheet, strip, rod, bar, extruded sections, tubes, wire, forgings	59.0–61.0 **Cu**, 1.5–2.5 **Pb**, rem **Zn**, 0.2 **Fe**, 0.5 total, others	
ISO 426/II	CuZn40Pb	ISO	Plate, sheet, strip, rod, bar, wire	59.0–61.0 **Cu**, 0.3–0.8 **Pb**, rem **Zn**, 0.2 **Fe**, 0.5 total, others	
JIS H 3100	C6712	Japan	Sheet, plate, and strip	58.0–62.0 **Cu**, 0.10–1.0 **Pb**, rem **Zn**, 0.05–1.0 **Mn**, 1.0 **Fe** + **Al** + **Si**	**Hard**: 0.25–1.5 mm **diam**, 160 min **HV**

WROUGHT COPPER AND COPPER ALLOYS

specification number	designation	country	product forms	chemical composition	mechanical properties and hardness values
JIS H 3100	C3710	Japan	Sheet, plate, and strip	58.0–62.0 **Cu**, 0.6–1.2 **Pb**, rem **Zn**, 0.10 **Fe**	**1/4 hard**: 0.3–10 mm **diam**, 373 MPa **TS**, 20% in 2 in. (50 mm) **El**; **1/2 hard**: 0.3–10 mm **diam**, 422 MPa **TS**, 13% in 2 in. (50 mm) **El**; **Hard**: 0.3–10 mm **diam**, 471 MPa **TS**
JIS H 3100	C3713	Japan	Sheet, plate, and strip	58.0–62.0 **Cu**, 1.0–2.0 **Pb**, rem **Zn**, 0.10 **Fe**	**1/4 hard**: 0.3–10 mm **diam**, 373 MPa **TS**, 18% in 2 in. (50 mm) **El**; **1/2 hard**: 0.3–10 mm **diam**, 422 MPa **TS**, 10% in 2 in. (50 mm) **El**; **Hard**: 0.3–10 mm **diam**, 471 MPa **TS**
NS 16 140	CuZn40Pb	Norway	Sheet, strip, plate, and bar	60 **Cu**, 0.5 **Pb**, 40 **Zn**	
SIS 14 51 63	SIS Brass 51 63–04	Sweden	Plate, sheet	59.0–61.0 **Cu**, 0.3–0.8 **Pb**, 40 **Zn**, 0.2 **Fe**, 0.5 total, others	**Strain hardened**: 2.5–5 mm **diam**, 350 MPa **TS**, 200 MPa **YS**, 20% in 2 in. (50 mm) **El**, 100–140 **HV**
SIS 14 51 63	SIS Brass 51 63–00	Sweden	Plate, sheet	59.0–61.0 **Cu**, 0.3–0.8 **Pb**, 40 **Zn**, 0.2 **Fe**, 0.5 total, others	**Hot worked**: 5 min mm **diam**, 340 MPa **TS**, 140 MPa **YS**, 30% in 2 in. (50 mm) **El**, 80–110 **HV**
ANSI/ASTM B 283	C 37700	US	Die forgings	58.0–61.0 **Cu**, 1.5–2.5 **Pb**, rem **Zn**, 0.30 **Fe**	**As manufactured**: <1.5 in. (<38.1 mm) **diam**, 50 ksi (345 MPa) **TS**, 18 ksi (124 MPa) **YS**, 25% in 2 in. (50 mm) **El**
ANSI/ASTM B 124	C 37700	US	Rod, bar, shapes	58.0–61.0 **Cu**, 1.5–2.5 **Pb**, rem **Zn**, 0.30 **Fe**	
ANSI/ASTM B 171	C 36500	US	Plate	58.0–61.0 **Cu**, 0.25 **Sn**, 0.40–0.9 **Pb**, rem **Zn**, 0.15 **Fe**, silver included in copper content	**As manufactured**: <2.0 in. (<50.8 mm) **diam**, 50 ksi (345 MPa) **TS**, 20 ksi (140 MPa) **YS**, 35% in 2 in. (50.8 mm) **El**
AS 1566	370C	Australia	Plate, bar, sheet, strip, foil	58.0–61.0 **Cu**, 0.8–1.4 **Pb**, rem **Zn**, 0.15 **Fe**, 0.50 total, others	**Annealed (sheet, strip)**: all **diam**, 310 MPa **TS**, 25% in 2 in. (50 mm) **El**, 85 **HV**; **Strain hardened to 3/4 hard (sheet, strip)**: all **diam**, 510 MPa **TS**, 130–150 **HV**; **Strain hardened to full hard (sheet, strip)**: all **diam**, 570 MPa **TS**, 140–165 **HV**
CDA 365, 366, 367, 368		US	Plate	58.0–61.0 **Cu**, 0.25 **Sn**, 0.40–0.9 **Pb**, rem **Zn**, 0.15 **Fe**, 0.02–0.10 **As**, **Sb** or **P**; 0.10 total, others	**Hot rolled**: 1.0 in. **diam**, 54 ksi (372 MPa) **TS**, 45% in 2 in. (50 mm) **El**, 80 **HRF**
COPANT 162	C 365 00	COPANT	Wrought products	58–61 **Cu**, 0.25 **Sn**, 0.4–0.9 **Pb**, rem **Zn**, 0.15 **Fe**	
COPANT 162	C 362 00	COPANT	Wrought products	58–61 **Cu**, 0.25 **Sn**, 0.4–0.9 **Pb**, rem **Zn**, 0.15 **Fe**, 0.02–0.10 **As**	

WROUGHT COPPER AND COPPER ALLOYS

specification number	designation	country	product forms	chemical composition	mechanical properties and hardness values
COPANT 162	C 366 00	COPANT	Wrought products	58–61 **Cu**, 0.25 **Sn**, 0.4–0.9 **Pb**, rem **Zn**, 0.15 **Fe**, 0.02–0.10 **Sb**	
COPANT 162	C 368 00	COPANT	Wrought products	58–61 **Cu**, 0.25 **Sn**, 0.4–0.9 **Pb**, rem **Zn**, 0.15 **Fe**, 0.02–0.10 **P**	
BS 2870	CZ 120	UK	Sheet, strip, foil	58.0–60.0 **Cu**, 1.5–2.5 **Pb**, rem **Zn**, 0.30 total, others	**1/2 hard**: ≤6.0 mm **diam**, 10% in 2 in. (50 mm) **El**, 110 **HV**; **Hard**: ≤0.03 mm **diam**, 359 MPa **TS**, 5% in 2 in. (50 mm) **El**, 140 **HV**; **Extra hard**: ≤6.0 mm **diam**, 396 MPa **TS**, 3% in 2 in. (50 mm) **El**, 165 **HV**
NF A 51 105	CuZn39 Pb2	France	Bar, wire, and shapes	58.0–60.0 **Cu**, 1.5–2.5 **Pb**, rem **Zn**, 0.35 **Fe**, 0.5 total, others	**Rolled, extruded, or drawn**: ≤7 mm **diam**, 500 MPa **TS**, 4% in 2 in. (50 mm) **El**
NF A 51 101	CuZn39 Pb2	France	Sheet and strip	59.0 **Cu**, 2.0 **Pb**, 39.0 **Zn**	**Stress hardened, Temper 2**: all **diam**, 400 MPa **TS**
JIS H 3250	C 3771	Japan	Rod and bar	57.0–61.0 **Cu**, 0.50–2.5 **Pb**, rem **Zn**, 1.0 **Fe+Sn**	**As manufactured**: ≥6 mm **diam**, 314 MPa **TS**, 15% in 2 in. (50 mm) **El**
JIS H 3250	C 3604	Japan	Rod and bar	57.0–61.0 **Cu**, 1.8–3.7 **Pb**, rem **Zn**, 0.35 **Fe**, 0.6 **Fe+Sn**	**As manufactured**: 6–75 mm **diam**, 333 MPa **TS**, 80 min **HV**
JIS H 3250	C 3603	Japan	Rod and bar	57.0–61.0 **Cu**, 1.8–3.7 **Pb**, rem **Zn**, 0.35 **Fe**, 0.50 **Fe+Sn**	**Annealed**: 6–75 mm **diam**, 314 MPa **TS**, 20% in 2 in. (50 mm) **El**; **1/2 hard**: 6–50 mm **diam**, 363 MPa **TS**, 100 min **HV**; **Hard**: 6–20 mm **diam**, 451 MPa **TS**, 130 min **HV**
NS 16 135	CuZn39Pb2	Norway	Sheet, strip, plate, bar, rod, wire, profiles, and forgings	59 **Cu**, 2 **Pb**, 39 **Zn**	
AS 1567	377C	Australia	Rod, bar, section	56.5–60.5 **Cu**, 1.0–2.5 **Pb**, rem **Zn**, 0.30 **Fe**, 0.50 total, others	**As manufactured**: 6–80 mm **diam**, 380 MPa **TS**, 18% in 2 in. (50 mm) **El**; **Hard**: 6–40 mm **diam**, 460 MPa **TS**, 18% in 2 in. (50 mm) **El**
DS 3003	5168	Denmark	Rolled, drawn, extruded, and forged products	57.5–59.5 **Cu**, 1.5–2.5 **Pb**, 39 **Zn**, 0.35 **Fe**, 0.7 total, others	
SFS 2921	CuZn39Pb2	Finland	Rod, bar, wire, tube, shapes, and forgings	57.5–59.5 **Cu**, 1.5–2.5 **Pb**, 39 **Zn**, 0.35 **Fe**	**Hot finished**: ≥5 mm **diam**, 410 MPa **TS**, 150 MPa **YS**, 30% in 2 in. (50 mm) **El**; **Drawn**: all **diam**, 390 MPa **TS**, 180 MPa **YS**, 10% in 2 in. (50 mm) **El**
ISO 1634	CuZn39Pb2	ISO	Plate, sheet, strip	57.0–60.0 **Cu**, 1.5–2.5 **Pb**, rem **Zn**, 0.35 **Fe**, **Cu** includes 0.3 **Ni**; 0.7 total, others	**Strain hardened**: 0.3–5 mm **diam**, 500 MPa **TS**, 8% in 2 in. (50 mm) **El**, 160 **HV**
ISO 1637	CuZn39Pb2	ISO	Solid products: straight lengths	57.0–60.0 **Cu**, 1.5–2.5 **Pb**, rem **Zn**, 0.35 **Fe**, **Cu** includes 0.3 **Ni**; 0.7 total, others	**M**: 5 mm **diam**, 370 MPa **TS**, 25% in 2 in. (50 mm) **El**, 120 max **HV**; **HA**: all **diam**, 390 MPa **TS**, 15% in 2 in. (50 mm) **El**, 130 **HV**

WROUGHT COPPER AND COPPER ALLOYS

specification number	designation	country	product forms	chemical composition	mechanical properties and hardness values
ISO 1639	CuZn39Pb2	ISO	Solid product: drawn, on coils or reels	57.0–60.0 **Cu**, 1.5–2.5 **Pb**, rem **Zn**, 0.35 **Fe**, 0.7 total, others	**M**: all **diam**, 370 MPa **TS**, 24% in 2 in. (50 mm) **El**, 140 **HV**; **HA**: all **diam**, 390 MPa **TS**, 12% in 2 in. (50 mm) **El**, 150 **HV**
ISO 1640	CuZn39Pb2	ISO	Forgings	57.0–60.0 **Cu**, 1.5–2.5 **Pb**, rem **Zn**, 0.35 **Fe**, 0.3 **Ni** included in **Cu**; 0.7 total, others	**M**: all **diam**, 390 MPa **TS**, 25% in 2 in. (50 mm) **El**, 140 **HV**
ISO 426/II	CuZn39Pb2	ISO	Plate, sheet, strip, rod, bar, extruded sections, tube, wire, forgings	57.0–60.0 **Cu**, 1.5–2.5 **Pb**, rem **Zn**, 0.35 **Fe**, 0.7 total, others	
SIS 14 51 68	SIS Brass 51 68–00	Sweden	Bar	57.5–59.5 **Cu**, 1.5–2.5 **Pb**, 39 **Zn**, 0.35 **Fe**, 0.7 total, others	**Hot worked**: 5 mm **diam**, 410 MPa **TS**, 150 MPa **YS**, 30% in 2 in. (50 mm) **El**, 100 **HV**
SIS 14 51 68	SIS Brass 51 68–04	Sweden	Bar	57.5–59.5 **Cu**, 1.5–2.5 **Pb**, 39 **Zn**, 0.35 **Fe**, 0.7 total, others	**Strain hardened**: 5–10 mm **diam**, 430 MPa **TS**, 290 MPa **YS**, 15% in 2 in. (50 mm) **El**, 120–155 **HV**
SIS 14 51 68	SIS Brass 51 68–06	Sweden	Plate, sheet	57.5–59.5 **Cu**, 1.5–2.5 **Pb**, 39 **Zn**, 0.35 **Fe**, 0.7 total, others	**Strain hardened**: 5–10 mm **diam**, 500 MPa **TS**, 430 MPa **YS**, 10% in 2 in. (50 mm) **El**, 145–185 **HV**
AS 1568	377C	Australia	Forgings	56.5–60.0 **Cu**, 1.0–2.5 **Pb**, rem **Zn**, 0.30 **Fe**, 0.50 total, others	**As manufactured**: ≥6 mm **diam**, 310 MPa **TS**, 20% in 2 in. (50 mm) **El**
BS 2874	CZ 122	UK	Rod, section	56.5–60.0 **Cu**, 1.0–2.5 **Pb**, rem **Zn**, 0.3 **Fe**, 0.02 **Sb**, 0.7 total, others	**As manufactured**: 6–80 mm **diam**, 383 MPa **TS**, 18% in 2 in. (50 mm) **El** Hard: 6–40 mm **diam**, 466 MPa **TS**, 18% in 2 in. (50 mm) **El**
BS 2872	CZ 122	UK	Forging stock, forgings	56.5–60.0 **Cu**, 1.0–2.5 **Pb**, rem **Zn**, 0.3 **Fe**, 0.75 total, others	**As manufactured or annealed**: ≥6 mm **diam**, 309 MPa **TS**, 20% in 2 in. (50 mm) **El**
IS 6912	Leaded brass	India	Forgings	56.5–60.0 **Cu**, 1.0–2.5 **Pb**, rem **Zn**, 0.30 **Fe**, 0.02 **Sb**, 0.75 total, others	**As manufactured**: ≥6 mm **diam**, 310 MPa **TS**, 25% in 2 in. (50 mm) **El**
IS 3488	CuZn42Pb2	India	Bar, rod, section	56.5–60.0 **Cu**, 1.00–2.50 **Pb**, rem **Zn**, 0.02 **Sb**, 0.25 total, others	
JIS H 3250	C 6782	Japan	Rod and bar	56.0–60.5 **Cu**, 0.50 **Pb**, rem **Zn**, 0.10–1.0 **Fe**, 0.20–2.0 **Al**, 1.0–3.0 **Mn**	**As manufactured**: 6–50 mm **diam**, 461 MPa **TS**, 20% in 2 in. (50 mm) **El**
AS 1567	385C	Australia	Rod, bar, section	56.0–60.0 **Cu**, 2.5–4.5 **Pb**, rem **Zn**, 0.75 total, others	**As manufactured**: 6–80 mm **diam**, 380 MPa **TS**, 12% in 2 in. (50 mm) **El**
DS 3003	5170	Denmark	Rolled, drawn, extruded, and forged products	57.0–59.0 **Cu**, 2.5–3.5 **Pb**, 39 **Zn**, 0.35 **Fe**, 0.7 total, others	

WROUGHT COPPER AND COPPER ALLOYS

specification number	designation	country	product forms	chemical composition	mechanical properties and hardness values
SFS 2920	CuZn39Pb3	Finland	Bar, rod, wire, tube, shapes, and forgings	57.0–59.0 **Cu**, 2.5–3.5 **Pb**, 39 **Zn**, 0.35 **Fe**	**Hot finished:** ≥5 mm **diam**, 410 MPa **TS**, 150 MPa **YS**, 30% in 2 in. (50 mm) **El**; **Drawn (tube and shapes):** all **diam**, 390 MPa **TS**, 180 MPa **YS**, 10% in 2 in. (50 mm) **El**
NF A 51 105	CuZn40 Pb3	France	Bar, wire, and shapes	57.0–59.0 **Cu**, 2.5–3.5 **Pb**, rem **Zn**, 0.35 **Fe**, 0.7 total, others	**Rolled, extruded, or drawn:** ≤7 mm **diam**, 500 MPa **TS**, 4% in 2 in. (50 mm) **El**
NS 16 130	CuZn39Pb3	Norway	Strip, bar, tube, rod, wire, profiles, and forgings	58 **Cu**, 3 **Pb**, 39 **Zn**	
SIS 14 51 10	SIS Brass 51 70–00	Sweden	Bar, tube, section, forgings	57.0–59.0 **Cu**, 2.5–3.5 **Pb**, 39 **Zn**, 0.35 **Fe**, 0.7 total, others	**Hot worked:** all **diam**, 410 MPa **TS**, 150 MPa **YS**, 30% in 2 in. (50 mm) **El**, 100 **HV**
SIS 14 51 70	SIS Brass 51 70–04	Sweden	Wire, rod, bar	57.0–59.0 **Cu**, 2.5–3.5 **Pb**, 39 **Zn**, 0.35 **Fe**, 0.7 total, others	**Strain hardened:** 2.5–5 mm **diam**, 480 MPa **TS**, 350 MPa **YS**, 8% in 2 in. (50 mm) **El**, 125–170 **HV**
SIS 14 51 70	SIS Brass 51 70–10	Sweden	Tube, section	57.0–59.0 **Cu**, 2.5–3.5 **Pb**, 39 **Zn**, 0.35 **Fe**, 0.7 total, others	**Strain hardened:** all **diam**, 390 MPa **TS**, 180 MPa **YS**, 10% in 2 in. (50 mm) **El**, 105 **HV**
SIS 14 51 73	SIS Brass 51 73–00	Sweden	Section	57.0–59.0 **Cu**, 0.4–1.2 **Pb**, 41 **Zn**, 0.35 **Fe**, 0.7 total, others	**Hot worked:** all **diam**, 410 MPa **TS**, 150 MPa **YS**, 30% in 2 in. (50 mm) **El**, 100 **HV**
SIS 14 51 73	SIS Brass 51 73–10	Sweden	Section	57.0–59.0 **Cu**, 0.4–1.2 **Pb**, 41 **Zn**, 0.35 **Fe**, 0.7 total, others	**Strain hardened:** all **diam**, 390 MPa **TS**, 180 MPa **YS**, 10% in 2 in. (50 mm) **El**, 105 **HV**
ASTM B 455	CA 380	US	Shapes	55.0–60.0 **Cu**, 0.30 **Sn**, 1.0–2.5 **Pb**, rem **Zn**, 0.35 **Fe**	**As agreed:** all **diam**, 48 ksi (331 MPa) **TS**, 16 ksi (110 MPa) **YS**, 15% in 2 in. (50.8 mm) **El**, 42 min **HRB**
ASTM B 455	CA 385	US	Shapes	55.0–60.0 **Cu**, 2.0–3.8 **Pb**, rem **Zn**, 0.35 **Fe**	**As agreed:** all **diam**, 48 ksi (331 MPa) **TS**, 16 ksi (110 MPa) **YS**, 15% in 2 in. (50.8 mm) **El**, 42 min **HRB**
AS 1567	380C	Australia	Rod, bar, section	55.0–60.0 **Cu**, 1.5–3.0 **Pb**, rem **Zn**, 0.30 **Fe**, 0.50 total, others	**As manufactured:** 6–80 mm **diam**, 380 MPa **TS**, 12% in 2 in. (50 mm) **El**
BS 2874	CZ 121	UK	Rod, section	56.0–59.0 **Cu**, 2.0–3.5 **Pb**, rem **Zn**, 0.02 **Sb**, 0.75 total, others	**As manufactured:** 6–80 mm **diam**, 383 MPa **TS**, 12% in 2 in. (50 mm) **El**
CDA 385		US	Rod, shapes	55.0–60.0 **Cu**, 2.0–3.8 **Pb**, rem **Zn**, 0.35 **Fe**, 0.50 total, others	**As extruded (shapes):** 1.0 in. **diam**, 60 ksi (414 MPa) **TS**, 30% in 2 in. (50 mm) **El**, 65 **HRB**
COPANT 162	C 380 00	COPANT	Wrought products	55–60 **Cu**, 0.30 **Sn**, 1.5–2.5 **Pb**, rem **Zn**, 0.35 **Fe**, 0.50 **Al**	
COPANT 162	C 385 00	COPANT	Wrought products	55–60 **Cu**, 2.0–3.8 **Pb**, rem **Zn**, 0.35 **Fe**	

WROUGHT COPPER AND COPPER ALLOYS

specification number	designation	country	product forms	chemical composition	mechanical properties and hardness values
COPANT 419	Cu Zn 40 Pb 2	COPANT	Bar	55.0–60.0 **Cu + Ag**, 0.50 **Sn**, 1.5–2.5 **Pb**, rem **Zn**, 0.30 **Fe**	**Annealed**: ≤25 mm **diam**, 305 MPa **TS**, 130 MPa **YS**, 20% in 2 in. (50 mm) **El**; **1/2 hard/hard**: ≤13 mm **diam**, 345 MPa **TS**, 175 MPa **YS**, 10% in 2 in. (50 mm) **El**
COPANT 419	Cu Zn 39 Pb 3	COPANT	Bar	55.0–60.0 **Cu + Ag**, 2.0–3.8 **Pb**, rem **Zn**, 0.35 **Fe**	**1/2 hard**: all **diam**, 350 MPa **TS**, 20% in 2 in. (50 mm) **El**; **Hard**: all **diam**, 450 MPa **TS**, 8% in 2 in. (50 mm) **El**; **Extra hard**: all **diam**, 530 MPa **TS**, 4% in 2 in. (50 mm) **El**
IS 319	Type I	India	Bar, rod, section	56.0–59.0 **Cu**, 2.0–3.5 **Pb**, rem **Zn**, 0.35 **Fe**, 0.35 total, others	**Annealed**: 10–25 mm **diam**, 345 MPa **TS**, 12% in 2 in. (50 mm) **El**; **Half hard**: 10–12 mm **diam**, 405 MPa **TS**, 4% in 2 in. (50 mm) **El**; **Hard**: 10–12 mm **diam**, 550 MPa **TS**
ISO 1637	CuZn39Pb3	ISO	Solid products: straight lengths	56.0–59.0 **Cu**, 2.5–3.5 **Pb**, rem **Zn**, 0.35 **Fe**, **Cu** includes 0.3 **Ni**; 0.7 total, others	**M**: 5 mm **diam**, 380 MPa **TS**, 24% in 2 in. (50 mm) **El**, 130 max **HV**; **HA**: 5–75 mm **diam**, 360 MPa **TS**, 18% in 2 in. (50 mm) **El**, 120 **HV**; **HB – strain hardened**: 5–15 mm **diam**, 440 MPa **TS**, 12% in 2 in. (50 mm) **El**, 145 **HV**
ISO 1639	CuZn39Pb3	ISO	Solid product: drawn, on coils or reels	56.0–59.0 **Cu**, 2.5–3.5 **Pb**, rem **Zn**, 0.35 **Fe**, 0.3 **Ni** included in **Cu**; 0.7 total, others	**M**: all **diam**, 380 MPa **TS**, 20% in 2 in. (50 mm) **El**, 145 **HV**
ISO 426/II	CuZn39Pb3	ISO	Rod, bar, extruded sections, tube, wire, forgings	56.0–59.0 **Cu**, 2.5–3.5 **Pb**, rem **Zn**, 0.35 **Fe**, 0.7 total, others	
JIS H 3250	C 6783	Japan	Rod and bar	55.0–59.0 **Cu**, 0.50 **Pb**, rem **Zn**, 0.10–1.0 **Fe**, 0.20–2.0 **Al**, 0.50–2.5 **Mn**	**As manufactured**: 6–50 mm **diam**, 539 MPa **TS**, 12% in 2 in. (50 mm) **El**
DS 3003	5175	Denmark	Rolled, drawn, extruded, and forged products	54.0–58.0 **Cu**, 0.2–1.5 **Pb**, 43 **Zn**, 0.5 **Fe**, 0.5 total, others	
DS 3003	5272	Denmark	Rolled, drawn, extruded, and forged products	54.0–58.0 **Cu**, 0.2–1.5 **Pb**, 43 **Zn**, 0.5 **Fe**, 0.2–0.8 **Al**, 0.5 total, others	
NS 16 125	CuZn43Pb2	Norway	Profiles: extruded	56 **Cu**, 1 **Pb**, 43 **Zn**	
SIS 14 52 72	SIS Brass 52 72–00	Sweden	Section	54.0–58.0 **Cu**, 0.2–1.5 **Pb**, 43 **Zn**, 0.5 **Fe**, 0.2–0.8 **Al**, 0.5 total, others	**Hot worrked**: all **diam**, 490 MPa **TS**, 160 MPa **YS**, 20% in 2 in. (50 mm) **El**, 130 **HV**
ISO 1639	CuZn43Pb2	ISO	Solid product: drawn, on coils or reels	54.0–57.0 **Cu**, 1.0–2.5 **Pb**, rem **Zn**, 0.5 **Fe**, 0–1.0 **Al**, 0.3 **Ni** included in **Cu**; 0.5 total, others	**M**: all **diam**, 440 MPa **TS**, 15% in 2 in. (50 mm) **El**, 150 **HV**

WROUGHT COPPER AND COPPER ALLOYS

specification number	designation	country	product forms	chemical composition	mechanical properties and hardness values
ISO 426/II	CuZn43Pb2	ISO	Extruded sections, forgings	54.0–57.0 **Cu**, 1.0–2.5 **Pb**, rem **Zn**, 0.5 **Fe**, 0–1.0 **Al**, 0.5 total, others	
JIS Z 3262	BCuZn–4	Japan	Wire: filler	48–55 **Cu**, 0.50 **Pb**, rem **Zn**, 0.10 **Fe**	
JIS Z 3262	BCuZn–0	Japan	Wire: filler	32–36 **Cu**, 0.50 **Pb**, rem **Zn**, 0.10 **Fe**	
ASTM B 591	CA 405	US	Plate, sheet, strip, bar	94.0–96.0 **Cu**, 0.7–1.3 **Sn**, 0.05 **Pb**, rem **Zn**, 0.05 **Fe**	**Quarter hard rolled**: 0.012–0.028 in. (0.30–0.71 mm) **diam**, 41 ksi (283 MPa) **TS**, 36–56 **HR30T**; **Hard rolled**: 0.012–0.028 in. (0.30–0.71 mm) **diam**, 58 ksi (400 MPa) **TS**, 60–68 **HR30T**; **Annealed**: 0.035 in. (0.889 mm) **diam**, 2–16 **HR30T**
ASTM B 591	CA 408	US	Plate, sheet, strip, bar	94.0–96.0 **Cu**, 1.8–2.2 **Sn**, 0.05 **Pb**, rem **Zn**, 0.05 **Fe**	**Quarter hard rolled**: 0.012–0.028 in. (0.30–0.71 mm) **diam**, 44 ksi (303 MPa) **TS**, 44–59 **HR30T**; **Hard rolled**: 0.012–0.028 in. (0.30–0.71 mm) **diam**, 62 ksi (427 MPa) **TS**, 66–72 **HR30T**; **Annealed**: 0.035 in. (0.889 mm) **diam**, 18–26 **HR30T**
COPANT 162	C 405 00	COPANT	Wrought products	94–96 **Cu**, 0.7–1.3 **Sn**, 0.05 **Pb**, rem **Zn**, 0.05 **Fe**	
COPANT 162	C 408 00	COPANT	Wrought products	94–96 **Cu**, 1.8–2.2 **Sn**, 0.05 **Pb**, rem **Zn**, 0.05 **Fe**	
COPANT 162	C 409 00	COPANT	Wrought products	92–94 **Cu**, 0.5–0.8 **Sn**, 0.05 **Pb**, rem **Zn**, 0.05 **Fe**	
COPANT 162	C 410 00	COPANT	Wrought products	91–93 **Cu**, 2.0–2.8 **Sn**, 0.05 **Pb**, rem **Zn**, 0.05 **Fe**	
ASTM B 591	CA 411	US	Plate, sheet, strip, bar	89.0–93.0 **Cu**, 0.30–0.7 **Sn**, 0.10 **Pb**, rem **Zn**, 0.05 **Fe**	**Quarter hard rolled**: 0.012–0.028 in. (0.30–0.71 mm) **diam**, 42 ksi (290 MPa) **TS**, 37–57 **HR30T**; **Hard rolled**: 0.012–0.028 in. (0.30–0.71 mm) **diam**, 61 ksi (421 MPa) **TS**, 62–70 **HR30T**; **Annealed**: 0.035 in. (0.889 mm) **diam**, 15–26 **HR30T**

WROUGHT COPPER AND COPPER ALLOYS

specification number	designation	country	product forms	chemical composition	mechanical properties and hardness values
ASTM B 591	CA 413	US	Plate, sheet, strip, bar	89.0–93.0 **Cu**, 0.7–1.3 **Sn**, 0.10 **Pb**, rem **Zn**, 0.05 **Fe**	**Quarter hard rolled**: 0.012–0.028 in. (0.30–0.71 mm) **diam**, 45 ksi (310 MPa) **TS**, 40–58 **HR30T**; **Hard rolled**: 0.012–0.028 in. (0.30–0.71 mm) **diam**, 65 ksi (448 MPa) **TS**, 63–70 **HR30T**; **Annealed**: 0.035 in. (0.889 mm) **diam**, 17–27 **HR30T**
ASTM B 591	CA 415	US	Plate, sheet, strip, bar	89.0–93.0 **Cu**, 1.5–2.2 **Sn**, 0.10 **Pb**, rem **Zn**, 0.05 **Fe**	**Quarter hard rolled**: 0.012–0.028 in. (0.30–0.71 mm) **diam**, 46 ksi (317 MPa) **TS**, 48–65 **HR30T**; **Hard rolled**: 0.012–0.028 in. (0.30–0.71 mm) **diam**, 64 ksi (441 MPa) **TS**, 69–72 **HR30T**; **Annealed**: 0.035 in. (0.889 mm) **diam**, 20–28 **HR30T**
ANSI/ASTM B 508	C 41100	US	Strip	89.0–93.0 **Cu**, 0.30–0.7 **Sn**, 0.10 **Pb**, rem **Zn**, 0.05 **Fe**	
CDA 413		US	Rod, wire, strip	89.0–93.0 **Cu**, 1.3 **Sn**, 0.10 **Pb**, rem **Zn**, 0.05 **Fe**, 0.15 total, others	**0.050 mm temper (rod)**: 0.500 in. **diam**, 44 ksi (303 MPa) **TS**, 45% in 2 in. (50 mm) **El**, 60 **HRF**; **Half hard temper (wire)**: 0.080 in. **diam**, 67 ksi (462 MPa) **TS**, 10% in 2 in. (50 mm) **El**; **Hard temper (rolled strip)**: 0.040 in. **diam**, 71 ksi (490 MPa) **TS**, 4% in 2 in. (50 mm) **El**, 79 **HRB**
COPANT 162	C 413 00	COPANT	Wrought products	89–93 **Cu**, 0.7–1.3 **Sn**, 0.10 **Pb**, rem **Zn**, 0.05 **Fe**	
COPANT 162	C 415 00	COPANT	Wrought products	89–93 **Cu**, 1.5–2.2 **Sn**, 0.10 **Pb**, rem **Zn**, 0.05 **Fe**	
CDA 411		US	Wire, strip	89.0–92.0 **Cu**, 0.7 **Sn**, 0.10 **Pb**, rem **Zn**, 0.05 **Fe**, 0.15 total, others	**Hard temper (wire)**: 0.100 in. **diam**, 102 ksi (703 MPa) **TS**, 1.0% in 60 in. (1500 mm) **El**; **Extra hard temper (rolled strip)**: 0.040 in. **diam**, 72 ksi (496 MPa) **TS**, 69 ksi (476 MPa) **YS**, 3% in 2 in. (50 mm) **El**, 78 **HRB**; **Spring temper (rolled strip)**: 0.040 in. **diam**, 80 ksi (552 MPa) **TS**, 72 ksi (496 MPa) **YS**, 2% in 2 in. (50 mm) **El**, 81 **HRB**
COPANT 162	C 411 00	COPANT	Wrought products	89–92 **Cu**, 0.3–0.7 **Sn**, 0.10 **Pb**, rem **Zn**, 0.05 **Fe**	
COPANT 162	C 419 00	COPANT	Wrought products	89–92 **Cu**, 4.8–5.5 **Sn**, 0.10 **Pb**, rem **Zn**, 0.05 **Fe**	

WROUGHT COPPER AND COPPER ALLOYS

specification number	designation	country	product forms	chemical composition	mechanical properties and hardness values
COPANT 162	C 420 00	COPANT	Wrought products	89–91 **Cu**, 1.5–2.0 **Sn**, rem **Zn**, 0.25 **P**	
ISO 427	Cu Sn4Zn4	ISO	Strip, rod, bar, tube, wire	89.39 min **Cu**, 3.0–5.0 **Sn**, 0.1 **Pb**, 3.0–5.0 **Zn**, 0.1 **Fe**, 0.1 **P**, 0.3 total, others	
ISO 1637	Cu Sn4Zn4	ISO	Solid products: straight lengths	89.39 min **Cu**, 3.0–5.0 **Sn**, 0.1 **Pb**, 3.0–5.0 **Zn**, 0.1 **Fe**, 0–0.1 **P**, 0.3 total, others	**As cast**: 5 mm **diam**, 360 MPa **TS**, 50% in 2 in. (50 mm) **El**, 110 max **HV**
ASTM B 591	CA 425	US	Plate, sheet, strip, bar	87.0–90.0 **Cu**, 1.5–3.0 **Sn**, 0.05 **Pb**, rem **Zn**, 0.05 **Fe**, 0.35 **P**	**Quarter hard rolled**: 0.012–0.028 in. (0.30–0.71 mm) **diam**, 49 ksi (338 MPa) **TS**, 45–65 **HR30T**; **Hard rolled**: 0.012–0.028 in. (0.30–0.71 mm) **diam**, 70 ksi (483 MPa) **TS**, 70–74 **HR30T**; **Annealed**: 0.035 in. (0.889 mm) **diam**, 28–35 **HR30T**
CDA 425		US	Sheet, strip	87.0–90.0 **Cu**, 1.5–3.0 **Sn**, 0.05 **Pb**, rem **Zn**, 0.05 **Fe**, 0.35 **P**, 0.15 total, others	**Quarter hard temper (sheet, rolled strip)**: 0.040 in. **diam**, 57 ksi (393 MPa) **TS**, 22% in 2 in. (50 mm) **El**, 60 **HRB**; **Half hand temper (sheet, rolled strip)**: 0.040 in. **diam**, 62 ksi (427 MPa) **TS**, 18% in 2 in. (50 mm) **El**, 70 **HRB**; **Hard temper (heet, rolled strip)**: 0.040 in. **diam**, 73 ksi (503 MPa) **TS**, 8% in 2 in. (50 mm) **El**, 83 **HRB**
COPANT 162	C 425 00	COPANT	Wrought products	87–90 **Cu**, 1.5–3.0 **Sn**, 0.05 **Pb**, rem **Zn**, 0.05 **Fe**, 0.35 **P**	
COPANT 162	C 421 00	COPANT	Wrought products	87.5–89 **Cu**, 2.2–3.0 **Sn**, 0.05 **Pb**, rem **Zn**, 0.05 **Fe**, 0.15–0.35 **Mn**, 0.35 **P**	
ASTM B 591	CA 422	US	Plate, sheet, strip, bar	86.0–89.0 **Cu**, 0.8–1.4 **Sn**, 0.05 **Pb**, rem **Zn**, 0.05 **Fe**, 0.35 **P**	**Quarter hard rolled**: 0.012–0.028 in. (0.30–0.71 mm) **diam**, 47 ksi (324 MPa) **TS**, 43–62 **HR30T**; **Hard rolled**: 0.012–0.028 in. (0.30–0.71 mm) **diam**, 67 ksi (462 MPa) **TS**, 67–71 **HR30T**; **Annealed**: 0.035 in. (0.889 mm) **diam**, 24–29 **HR30T**
COPANT 162	C 422 00	COPANT	Wrought products	86–89 **Cu**, 0.8–1.4 **Sn**, 0.05 **Pb**, rem **Zn**, 0.05 **Fe**, 0.35 **P**	
COPANT 162	C 430 00	COPANT	Wrought products	85–89 **Cu**, 1.7–2.7 **Sn**, 0.10 **Pb**, rem **Zn**, 0.05 **Fe**	
COPANT 162	C 432 00	COPANT	Wrought products	85–89 **Cu**, 0.4–0.6 **Sn**, 0.05 **Pb**, rem **Zn**, 0.05 **Fe**, 0.35 **P**	

WROUGHT COPPER AND COPPER ALLOYS

specification number	designation	country	product forms	chemical composition	mechanical properties and hardness values
COPANT 162	C 476 00	COPANT	Wrought products	86–88 **Cu**, 1.8–2.2 **Sn**, 1.8–2.2 **Pb**, rem **Zn**, 0.05 **Fe**, 0.05–0.15 **Mn**, 0.03–0.07 **P**	
ASTM B 591	CA 430	US	Plate, sheet, strip, bar	84.0–87.0 **Cu**, 1.7–2.7 **Sn**, 0.10 **Pb**, rem **Zn**, 0.05 **Fe**	**Quarter hard rolled**: 0.012–0.028 in. (0.30–0.71 mm) **diam**, 47 ksi (324 MPa) **TS**, 47–64 **HR30T**; **Hard rolled**: 0.012–0.028 in. (0.30–0.71 mm) **diam**, 72 ksi (496 MPa) **TS**, 68–75 **HR30T**; **Annealed**: 0.035 in. (0.889 mm) **diam**, 20–39 **HR30T**
ASTM B 591	CA 434	US	Plate, sheet, strip, bar	84.0–87.0 **Cu**, 0.40–1.0 **Sn**, 0.05 **Pb**, rem **Zn**, 0.05 **Fe**	**Quarter hard rolled**: 0.012–0.028 in. (0.30–0.71 mm) **diam**, 45 ksi (310 MPa) **TS**, 45–61 **HR30T**; **Hard rolled**: 0.012–0.028 in. (0.30–0.71 mm) **diam**, 68 ksi (469 MPa) **TS**, 65–74 **HR30T**; **Annealed**: 0.035 in. (0.889 mm) **diam**, 19–25 **HR30T**
NF A 51 108	Cu Sn Zn 9	France	Sheet and strip	85.30 min **Cu**, 2–4 **Sn**, 0.1 **Pb**, 7.5–10 **Zn**, 0.1 **Fe**, 0.2 **P**, 0.3 total, others	**Annealead**: ≤20 mm **diam**, 310 MPa **TS**, 150 MPa **YS**, 40% in 2 in. (50 mm) **El**, 75–105 **HV**; **1/2 hard**: ≤20 mm **diam**, 460 MPa **TS**, 400 MPa **YS**, 15% in 2 in. (50 mm) **El**, 140–170 **HV**; **Hard**: ≤20 mm **diam**, 610 MPa **TS**, 580 MPa **YS**, 4% in 2 in. (50 mm) **El**, 190–210 **HV**
COPANT 162	C 434 00	COPANT	Wrought products	84–86 **Cu**, 0.5–1.0 **Sn**, 0.05 **Pb**, rem **Zn**, 0.05 **Fe**	
COPANT 162	C 436 00	COPANT	Wrought products	80–83 **Cu**, 0.2–0.5 **Sn**, 0.05 **Pb**, rem **Zn**, 0.05 **Fe**	
CDA 435		US	Tube, strip	79.0–83.0 **Cu**, 0.6–1.2 **Sn**, 0.10 **Pb**, rem **Zn**, 0.05 **Fe**, 0.15 total, others	**Hard temper – 35% (tube)**: 1 in. **diam**, 75 ksi (517 MPa) **TS**, 10% in 2 in. (50 mm) **El**; **0.025 mm temper (rolled strip)**: 0.040 in. **diam**, 49 ksi (338 MPa) **TS**, 46% in 2 in. (50 mm) **El**, 70 **HRF**; **Half hard temper (rolled strip)**: 0.040 in. **diam**, 65 ksi (448 MPa) **TS**, 16% in 2 in. (50 mm) **El**, 72 **HRB**
COPANT 162	C 435 00	COPANT	Wrought products	79–83 **Cu**, 0.6–1.2 **Sn**, 0.10 **Pb**, rem **Zn**, 0.05 **Fe**	
COPANT 162	C 438 00	COPANT	Wrought products	79–82 **Cu**, 1.0–1.5 **Sn**, 0.05 **Pb**, rem **Zn**, 0.05 **Fe**	

WROUGHT COPPER AND COPPER ALLOYS

specification number	designation	country	product forms	chemical composition	mechanical properties and hardness values
NS 16 210	CuZn20Al1	Norway	Tube	77 **Cu**, 1 **Sn**, 21–38 **Zn**, 2 **Al**, 0.04 **As**	
ANSI/ASTM B 395	C 44500	US	Tube: seamless	70.0–73.0 **Cu**, 0.9–1.2 **Sn**, 0.07 **Pb**, rem **Zn**, 0.06 **Fe**, 0.02–0.10 **P**, silver included in copper content	**Annealed**: <0.048 in. (<1.21 mm) **diam**, 45 ksi (310 MPa) **TS**, 15 ksi (105 MPa) **YS**, 12% in 2 in. (50.8 mm) **El**
ANSI/ASTM B 395	C 44400	US	Tube: seamless	70.0–73.0 **Cu**, 0.9–1.2 **Sn**, 0.07 **Pb**, rem **Zn**, 0.06 **Fe**, 0.02–0.10 **Sb**, silver included in copper content	**Annealed**: <0.048 in. (<1.21 mm) **diam**, 45 ksi (310 MPa) **TS**, 15 ksi (105 MPa) **YS**, 12% in 2 in. (50.8 mm) **El**
ANSI/ASTM B 395	C 44300	US	Tube: seamless	70.0–73.0 **Cu**, 0.9–1.2 **Sn**, 0.07 **Pb**, rem **Zn**, 0.06 **Fe**, 0.02–0.10 **As**, silver included in copper content	**Annealed**: <0.048 in. (<1.21 mm) **diam**, 45 ksi (310 MPa) **TS**, 15 ksi (105 MPa) **YS**, 12% in 2 in. (50.8 mm) **El**
ANSI/ASTM B 359	C 44400	US	Tube: seamless	70.0–73.0 **Cu**, 0.9–1.2 **Sn**, 0.07 **Pb**, rem **Zn**, 0.06 **Fe**, 0.02–0.10 **Sb**	**Annealed**: all **diam**, 45 ksi (310 MPa) **TS**, 15 ksi (105 MPa) **YS**
ANSI/ASTM B 359	C 44500	US	Tube: seamless	70.0–73.0 **Cu**, 0.9–1.2 **Sn**, 0.07 **Pb**, rem **Zn**, 0.06 **Fe**, 0.02–0.10 **P**	**Annealed**: all **diam**, 45 ksi (310 MPa) **TS**, 15 ksi (105 MPa) **YS**
ANSI/ASTM B 359	C 44300	US	Tube: seamless	70.0–73.0 **Cu**, 0.9–1.2 **Sn**, 0.07 **Pb**, rem **Zn**, 0.06 **Fe**, 0.02–0.10 **As**	**Annealed**: all **diam**, 45 ksi (310 MPa) **TS**, 15 ksi (105 MPa) **YS**
ANSI/ASTM B 171	C 44300, C 44400, C 44500	US	Plate	70.0–73.0 **Cu**, 0.9–1.2 **Sn**, 0.07 **Pb**, rem **Zn**, 0.06 **Fe**, silver included in copper content	**As manufactured**: <4.0 in. (<102.0 mm) **diam**, 45 ksi (310 MPa) **TS**, 15 ksi (105 MPa) **YS**, 35% in 2 in. (50.8 mm) **El**
ANSI/ASTM B 543	C 44300	US	Tube: welded	70.0–73.0 **Cu**, 0.8–1.2 **Sn**, 0.07 **Pb**, rem **Zn**, 0.06 **Fe**, 0.02–0.10 **As**, silver included in copper content	**As welded from annealed strip**: all **diam**, 45 ksi (310 MPa) **TS**, 15 ksi (105 MPa) **YS**; **Welded and annealed**: all **diam**, 45 ksi (310 MPa) **TS**, 15 ksi (105 MPa) **YS**; **Fully finished – annealed**: all **diam**, 45 ksi (310 MPa) **TS**, 15 ksi (105 MPa) **YS**
ANSI/ASTM B 543	C 44400	US	Tube: welded	70.0–73.0 **Cu**, 0.8–1.2 **Sn**, 0.07 **Pb**, rem **Zn**, 0.06 **Fe**, 0.02–0.10 **Sb**, silver included in copper content	**As welded from annealed strip**: all **diam**, 45 ksi (310 MPa) **TS**, 15 ksi (105 MPa) **YS**; **Welded and annealed**: all **diam**, 45 ksi (310 MPa) **TS**, 15 ksi (105 MPa) **YS**; **Fully finished – annealed**: all **diam**, 45 ksi (310 MPa) **TS**, 15 ksi (105 MPa) **YS**

WROUGHT COPPER AND COPPER ALLOYS

specification number	designation	country	product forms	chemical composition	mechanical properties and hardness values
ANSI/ASTM B 543	C 44500	US	Tube: welded	70.0–73.0 **Cu**, 0.8–1.2 **Sn**, 0.07 **Pb**, rem **Zn**, 0.06 **Fe**, 0.02–0.10 **P**, silver included in copper content	**As welded from annealed strip**: all **diam**, 45 ksi (310 MPa) **TS**, 15 ksi (105 MPa) **YS**; **Welded and annealed**: all **diam**, 45 ksi (310 MPa) **TS**, 15 ksi (105 MPa) **YS**; **Fully finished - annealed**: all **diam**, 45 ksi (310 MPa) **TS**, 15 ksi (105 MPa) **YS**
ANSI/ASTM B 111	C 44300	US	Tube: seamless, and stock: ferrule	70.0–73.0 **Cu**, 0.9–1.2 **Sn**, 0.07 **Pb**, rem **Zn**, 0.06 **Fe**, 0.02–0.10 **As**	**Annealed**: all **diam**, 45 ksi (310 MPa) **TS**, 15 ksi (105 MPa) **YS**
ANSI/ASTM B 111	C 44400	US	Tube: seamless, and stock: ferrule	70.0–73.0 **Cu**, 0.9–1.2 **Sn**, 0.07 **Pb**, rem **Zn**, 0.06 **Fe**, 0.02–0.10 **Sb**	**Annealed**: all **diam**, 45 ksi (310 MPa) **TS**, 15 ksi (105 MPa) **YS**
ANSI/ASTM B 111	C 44500	US	Tube: seamless, and stock: ferrule	70.0–73.0 **Cu**, 0.9–1.2 **Sn**, 0.07 **Pb**, rem **Zn**, 0.06 **Fe**, 0.02–0.10 **P**	**Annealed**: all **diam**, 45 ksi (310 MPa) **TS**, 15 ksi (105 MPa) **YS**
AS 1572	443C	Australia	Tube	70.0–73.0 **Cu**, 0.8–1.2 **Sn**, 0.07 **Pb**, rem **Zn**, 0.06 **Fe**, 0.02–0.06 **As**, 0.15 total, others	**Annealed**: all **diam**, 80 **HV**; **Quarter–hard**: 80–105 **HV**; **Hard**: all **diam**, 150 **HV**
BS 2871	CZ 111	UK	Tube	70.0–73.0 **Cu**, 1.0–1.5 **Sn**, 0.07 **Pb**, rem **Zn**, 0.06 **Fe**, 0.02–0.06 **As**, 0.30 total, others	**As drawn**: 150 **HV**; **Temper annealed**: 80–105 **HV**; **Annealed**: 75 max **HV**
CDA 443, 444, 445		US	Plate, wire, tube	70.0–73.0 **Cu**, 0.9–1.2 **Sn**, 0.07 **Pb**, rem **Zn**, 0.06 **Fe**, 0.02–0.10 **As**, **Sb** or **P**; 0.15 total, others	**As hot rolled (plate)**: 1.0 in. **diam**, 48 ksi (331 MPa) **TS**, 65% in 2 in. (50 mm) **El**, 70 **HRF**; **0.025 mm temper (tube)**: 1.0 in. **diam**, 53 ksi (365 MPa) **TS**, 65% in 2 in. (50 mm) **El**, 75 **HRF**; **0.015 mm temper (wire)**: 0.080 in. **diam**, 55 ksi (379 MPa) **TS**, 60% in 2 in. (50 mm) **El**
COPANT 162	C 443 00	COPANT	Wrought products	70–73 **Cu**, 0.9–1.2 **Sn**, 0.07 **Pb**, rem **Zn**, 0.06 **Fe**, 0.02–0.10 **As**, for rolled products, min 0.80 **Sn**	
COPANT 162	C 444 00	COPANT	Wrought products	70–73 **Cu**, 0.9–1.2 **Sn**, 0.07 **Pb**, rem **Zn**, 0.06 **Fe**, 0.02–0.10 **Sb**	
COPANT 162	C 455 00	COPANT	Wrought products	70–73 **Cu**, 0.9–1.2 **Sn**, 0.07 **Pb**, rem **Zn**, 0.06 **Fe**, 0.02–0.10 **P**	
COPANT 527	Cu Zn 28 Sn 1 As	COPANT	Tube	70–73 **Cu**, 0.9–1.2 **Sn**, 0.07 **Pb**, rem **Zn**, 0.06 **Fe**, 0.02–0.10 **As**, 0.15 total, others	**Annealed**: 0.8–3.4 mm **diam**, 310 MPa **TS**, 105 MPa **YS**
COPANT 527	Cu Zn 28 Sn 1 P	COPANT	Tube	70–73 **Cu**, 0.9–1.2 **Sn**, 0.07 **Pb**, rem **Zn**, 0.06 **Fe**, 0.02–0.10 **P**, 0.15 total, others	**Annealed**: 0.8–3.4 mm **diam**, 310 MPa **TS**, 105 MPa **YS**
COPANT 527	Cu Zn 28 Sn 1 Sb	COPANT	Tube	70–73 **Cu**, 0.9–1.2 **Sn**, 0.07 **Pb**, rem **Zn**, 0.06 **Fe**, 0.02–0.10 **Sb**, 0.15 total, others	**Annealed**: 0.8–3.4 mm **diam**, 310 MPa **TS**, 105 MPa **YS**
IS 1545	Cu Zn29Sn1As	India	Tube	70.0–73.0 **Cu**, 1.00–1.50 **Sn**, 0.075 **Pb**, rem **Zn**, 0.06 **Fe**, 0.02–0.06 **As**, 0.30 total, others	**Annealed**: all **diam**, 372 max MPa **TS**; **As drawn**: all **diam**, 387 max MPa **TS**

WROUGHT COPPER AND COPPER ALLOYS

specification number	designation	country	product forms	chemical composition	mechanical properties and hardness values
ISO 1635	CuZn28Sn1	ISO	Tube	70.0–73.0 **Cu**, 0.9–1.3 **Sn**, 0.07 **Pb**, rem **Zn**, 0.07 **Fe**, 0.02–0.06 **As**, 0.3 **Fe + Pb**, total	**Annealed**: all **diam**, 45% in 2 in. (50 mm) **El**, 110 max **HV**; **Strain hardened (HB)**: all **diam**, 420 MPa **TS**, 35% in 2 in. (50 mm) **El**, 100 **HV**
ISO 1634	CuZn28Sn1	ISO	Plate, sheet, strip	70.0–73.0 **Cu**, 0.9–1.3 **Sn**, 0.07 **Pb**, rem **Zn**, 0.07 **Fe**, 0.02–0.06 **As**, 0.3 **Fe + Pb + P**, total	**M**: all **diam**, 330 MPa **TS**, 40% in 2 in. (50 mm) **El**, 100 max **HV**; **HA**: all **diam**, 390 MPa **TS**, 30% in 2 in. (50 mm) **El**, 120 **HV**
ISO 426/1	CuZn28Sn1	ISO	Plate, sheet, tube	70.0–73.0 **Cu**, 0.9–1.3 **Sn**, 0.07 **Pb**, rem **Zn**, 0.07 **Fe**, 0.02–0.06 **As**, 0.3 **Fe + Pb**, total	
SABS 460	Cu–Zn28Sn1	South Africa	Tube	70.0–73.0 **Cu**, 1.0–1.5 **Sn**, 0.075 **Pb**, rem **Zn**, 0.06 **Fe**, 0.02–0.06 **As**, 0.30 total, others	**Annealed**: all **diam**, 280 MPa **TS**, 75 max **HV**; **Specially annealed**: all **diam**, 340 MPa **TS**, 80–105 **HV**; **As drawn**: all **diam**, 500 MPa **TS**, 150 **HV**
SIS 14 52 20	SIS Brass 52 20–12	Sweden	Tube	70.0–73.0 **Cu**, 0.9–1.3 **Sn**, 0.07 **Pb**, 28 **Zn**, 0.07 **Fe**, 0.010 **P**, 0.020–0.035 **As**, 0.035 **As + P**, total; 0.3 total, others	**Annealed**: all **diam**, 370 MPa **TS**, 150 MPa **YS**, 45% in 2 in. (50 mm) **El**, 85–110 **HV**
VSM 11557	Cu Zn 28 Sn 1	Switzerland	Tube	70.0–72.5 **Cu**, 1.0–1.3 **Sn**, 0.07 **Pb**, rem **Zn**, 0.06 **Fe**, 0.5 **Ni**, 0.015 **P**, 0.020–0.045 **As**, 0.30 total, others	**Quenched**: all **diam**, 323 MPa **TS**, 98 MPa **YS**, 50% in 2 in. (50 mm) **El**, 65 **HB**; **1/2 hard**: all **diam**, 372 MPa **TS**, 47 MPa **YS**, 40% in 2 in. (50 mm) **El**, 90 **HB**
ANSI/ASTM B 21	CA 462	US	Rod, bar, shapes	62.0–65.0 **Cu**, 0.50–1.0 **Sn**, 0.20 **Pb**, rem **Zn**, 0.10 **Fe**	**Soft, rod and bar**: all **diam**, 48 ksi (330 MPa) **TS**, 16 ksi (110 MPa) **YS**, 30% in 1 in. (25.4 mm) **El**; **For cold heading and cold forming, rod and bar**: all **diam**, 48 ksi (330 MPa) **TS**, 16 ksi (110 MPa) **YS**, 30% in 1 in. (25.4 mm) **El**; **Hard, rod and bar**: <0.500 in. (<12.7 mm) **diam**, 64 ksi (440 MPa) **TS**, 40 ksi (275 MPa) **YS**, 13% in 1 in. (25.4 mm) **El**
COPANT 420	Cu Zn 36 Sn	COPANT	Bar	62.0–65.0 **Cu + Ag**, 0.5–1.0 **Sn**, 0.20 **Pb**, rem **Zn**, 0.10 **Fe**, 0.10 total, others	**Hard (extruded)**: all **diam**, 345 MPa **TS**, 140 MPa **YS**, 30% in 2 in. (50 mm) **El**; **Annealed**: all **diam**, 335 MPa **TS**, 110 MPa **YS**, 30% in 2 in. (50 mm) **El**; **1/2 hard**: 13–25 in. **diam**, 380 MPa **TS**, 190 MPa **YS**, 25% in 2 in. (50 mm) **El**
ASTM F 468	C 46200	US	Bolt, screw, stud	62.0–65.0 **Cu**, 0.5–1.0 **Sn**, 0.20 **Pb**, rem **Zn**, 0.10 **Fe**	**As manufactured**: all **diam**, 50 ksi (340 MPa) **TS**, 25 ksi (170 MPa) **YS**, 20% in 2 in. (50 mm) **El**, 65–90 **HRB**

WROUGHT COPPER AND COPPER ALLOYS

specification number	designation	country	product forms	chemical composition	mechanical properties and hardness values
ASTM F-468	C 46200	US	Nut	62.0–65.0 **Cu**, 0.5–1.0 **Sn**, 0.20 **Pb**, rem **Zn**, 0.10 **Fe**	**As agreed**: all **diam**, 65 min **HRB**
COPANT 162	C 462 00	COPANT	Wrought products	62–65 **Cu**, 0.5–1.0 **Sn**, 0.20 **Pb**, rem **Zn**, 0.10 **Fe**	
QQ-B-639B	462	US	Plate, bar, sheet, strip	62.0–65.0 **Cu**, 0.50–1.0 **Sn**, 0.20 **Pb**, rem **Zn**, 0.10 **Fe**, 0.10 total, others	**Soft temper**: all **diam**, 45 ksi (310 MPa) **TS**, 16 ksi (110 MPa) **YS**, 30% in 1 in. (25 mm) **El**; **Half hard**: all **diam**, 54 ksi (372 MPa) **TS**, 25 ksi (172 MPa) **YS**, 20% in 1 in. (25 mm) **El**
QQ-B-637a	462	US	Rod, wire, shapes, forgings, bar, flat wire, strip	62.0–65.0 **Cu**, 0.5–1.0 **Sn**, 0.20 **Pb**, rem **Zn**, 0.10 **Fe**, 0.10 total, others	**As extruded**: all **diam**, 50 ksi (345 MPa) **TS**, 20 ksi (138 MPa) **YS**, 30% in 1 in. (25 mm) **El**; **Soft temper**: all **diam**, 48 ksi (331 MPa) **TS**, 16 ksi (110 MPa) **YS**, 30% in 1 in. (25 mm) **El** Cold heading and cold forming: all **diam**, 48 ksi (331 MPa) **TS**, 18 ksi (124 MPa) **YS**, 22% in 1 in. (25 mm) **El**
IS 6912	Naval brass	India	Forgings	61.0–64.0 **Cu**, 1.0–1.5 **Sn**, 0.20 **Pb**, rem **Zn**, 0.10 **Fe**, 0.75 total, others, nickel included in copper content	**As manufactured**: ≥6 mm **diam**, 340 MPa **TS**, 20% in 2 in. (50 mm) **El**
JIS H 3100	C4621	Japan	Sheet, plate, and strip	61.0–64.0 **Cu**, 0.7–1.5 **Sn**, 0.20 **Pb**, rem **Zn**, 0.10 **Fe**	**As manufactured**: <20 mm **diam**, 373 MPa **TS**, 20% in 2 in. (50 mm) **El**
AS 1567	464B	Australia	Rod, bar, section	61.0–63.5 **Cu**, 1.0–1.4 **Sn**, rem **Zn**, 0.75 total, others	**As manufactured**: 6–20 mm **diam**, 400 MPa **TS**, 18% in 2 in. (50 mm) **El**
AS 1566	464B	Australia	Plate, bar, sheet, strip, foil	61.0–63.5 **Cu**, 1.0–1.4 **Sn**, rem **Zn**, 0.75 total, others	
AS 1568	464B	Australia	Forgings	61.0–63.5 **Cu**, 1.0–1.4 **Sn**, rem **Zn**, 0.75 total, others	**As manufactured**: ≥6 mm **diam**, 340 MPa **TS**, 15% in 2 in. (50 mm) **El**
BS 2870	CZ 112	UK	Sheet, strip, foil	61.0–63.5 **Cu**, 1.0–1.4 **Sn**, rem **Zn**, 0.75 total, others	**As manufactured or annealed**: ≤10.0 mm **diam**, 238 MPa **TS**, 25% in 2 in. (50 mm) **El**; **Hard**: ≤10.0 mm **diam**, 288 MPa **TS**, 20% in 2 in. (50 mm) **El**
BS 2875	CZ 112	UK	Plate	61.0–63.5 **Cu**, 1.0–1.4 **Sn**, rem **Zn**, 0.75 total, others	**As manufactured or annealed**: 10–25 mm **diam**, 358 MPa **TS**, 18% in 2 in. (50 mm) **El**; **Hard**: 10–12.5 mm **diam**, 402 MPa **TS**, 18% in 2 in. (50 mm) **El**
BS 2874	CZ 112	UK	Rod, section	61.0–63.5 **Cu**, 1.0–1.4 **Sn**, rem **Zn**, 0.75 total, others	**As manufactured**: 6–20 mm **diam**, 402 MPa **TS**, 18% in 2 in. (50 mm) **El**
BS 2872	CZ 112	UK	Forging stock, forgings	61.0–63.5 **Cu**, 1.0–1.4 **Sn**, rem **Zn**, 0.75 total, others	**As manufactured or annealed**: ≥6 mm **diam**, 339 MPa **TS**, 15% in 2 in. (50 mm) **El**

WROUGHT COPPER AND COPPER ALLOYS

specification number	designation	country	product forms	chemical composition	mechanical properties and hardness values
DS 3003	5240	Denmark	Rolled, drawn, extruded, and forged products	59.5–63.5 **Cu**, 0.5–1.5 **Sn**, 0.2 **Pb**, 38.0 **Zn**, 0.2 **Fe**, 0.5 others, total	
ISO 1634	CuZn38Sn1	ISO	Plate, sheet, strip	59.5–63.5 **Cu**, 0.5–1.5 **Sn**, 0.2 **Pb**, rem **Zn**, 0.2 **Fe**, 0.5 **Fe**+**Pb**, total	**M**: all **diam**, 370 MPa **TS**, 30% in 2 in. (50 mm) **El**, 120 max **HV**; **HA**: all **diam**, 390 MPa **TS**, 25% in 2 in. (50 mm) **El**, 125 **HV**
ISO 426/1	CuZn38Sn1	ISO	Plate, sheet, rod, bar, extruded sections, tube, forgings	59.5–63.5 **Cu**, 0.5–1.5 **Sn**, 0.2 **Pb**, rem **Zn**, 0.2 **Fe**, 0.5 **Fe**+**Pb**, total	
ANSI/ASTM B 283	C 46400	US	Die forgings	59.0–62.0 **Cu**, 0.5–1.0 **Sn**, 0.20 **Pb**, rem **Zn**, 0.10 **Fe**	
ANSI/ASTM B 283	C 48500	US	Die forgings	59.0–62.0 **Cu**, 0.5–1.0 **Sn**, 1.3–2.2 **Pb**, rem **Zn**, 0.10 **Fe**	
ANSI/ASTM B 171	C 46400	US	Plate	59.0–62.0 **Cu**, 0.50–1.0 **Sn**, 0.20 **Pb**, rem **Zn**, 0.10 **Fe**, silver included in copper content	**As manufactured**: <3.0 in. (<76.2 mm) **diam**, 50 ksi (345 MPa) **TS**, 20 ksi (140 MPa) **YS**, 35% in 2 in. (50.8 mm) **El**
ASTM F 468	C 46400	US	Bolt, screw, stud	59.0–62.0 **Cu**, 0.5–1.0 **Sn**, 0.20 **Pb**, rem **Zn**, 0.10 **Fe**	**As manufactured**: all **diam**, 50 ksi (340 MPa) **TS**, 15 ksi (105 MPa) **YS**, 25% in 2 in. (50 mm) **El**, 55–75 **HRB**
ASTM F 468	C 46400	US	Nut	59.0–62.0 **Cu**, 0.5–1.0 **Sn**, 0.20 **Pb**, rem **Zn**, 0.10 **Fe**	**As agreed**: all **diam**, 55 min **HRB**
ANSI/ASTM B 124	C 48200	US	Rod, bar, shapes	59.0–62.0 **Cu**, 0.50–1.0 **Sn**, 0.40–1.0 **Pb**, rem **Zn**, 0.10 **Fe**	
ANSI/ASTM B 124	C 48500	US	Rod, bar, shapes	59.0–62.0 **Cu**, 0.5–1.0 **Sn**, 1.3–2.2 **Pb**, rem **Zn**, 0.10 **Fe**	
ANSI/ASTM B 124	C 46400	US	Rod, bar, shapes	59.0–62.0 **Cu**, 0.5–1.0 **Sn**, 0.20 **Pb**, rem **Zn**, 0.10 **Fe**	
AS 1567	486D	Australia	Rod, bar, section	59.0–62.0 **Cu**, 0.75–1.5 **Sn**, 1.0–2.5 **Pb**, rem **Zn**, 0.02–0.25 **As**	**As manufactured**: 6–20 mm **diam**, 380 MPa **TS**, 15% in 2 in. (50 mm) **El**
AS 1568	486D	Australia	Forgings	59.0–62.0 **Cu**, 0.75–1.5 **Sn**, rem **Zn**, 0.02–0.25 **As**, 0.5 total, others	**As manufactured**: $\geq$6 mm **diam**, 320 MPa **TS**, 20% in 2 in. (50 mm) **El**
CSA HC.4.9	HC.4.ZT381P	Canada	Bar, plate	59.0–62.0 **Cu**, 0.50–1.0 **Sn**, 0.20 **Pb**, rem **Zn**, 0.10 **Fe**, 0.10 total, others	**Soft**: 0.375 in. **diam**, 52 ksi (402 MPa) **TS**, 20 ksi (138 MPa) **YS**, 30% in 1 in. (25 mm) **El**; **Half hard**: 0.375 in. **diam**, 60 ksi (414 MPa) **TS**, 35 ksi (241 MPa) **YS**, 20% in 1 in. (25 mm) **El**

WROUGHT COPPER AND COPPER ALLOYS

specification number	designation	country	product forms	chemical composition	mechanical properties and hardness values
CDA 482		US	Rod, shapes, bar	59.0–62.0 **Cu**, 0.50–1.0 **Sn**, 0.40–1.0 **Pb**, rem **Zn**, 0.10 **Fe**, 0.10 total, others	**Soft annealed (rod)**: 1.0 in. **diam**, 57 ksi (393 MPa) **TS**, 40% in 2 in. (50 mm) **El**, 55 **HRB**; **As extruded (bar)**: 0.375 in. **diam**, 63 ksi (434 MPa) **TS**, 34% in 2 in. (50 mm) **El**, 60 **HRB**; **Quarter hard – 4% (bar)**: 1.5 in. **diam**, 66 ksi (455 MPa) **TS**, 32% in 2 in. (50 mm) **El**, 75 **HRB**
CDA 485		US	Rod, bar, shapes	59.0–62.0 **Cu**, 0.50–1.0 **Sn**, 1.3–2.2 **Pb**, rem **Zn**, 0.10 **Fe**, 0.10 total, others	**Soft anneal (rod)**: 1.0 in. **diam**, 57 ksi (393 MPa) **TS**, 40% in 2 in. (50 mm) **El**, 55 **HRB**; **Quarter hard – 8% (rod)**: 1.0 in. **diam**, 69 ksi (476 MPa) **TS**, 20% in 2 in. (50 mm) **El**, 78 **HRB**; **Half hard – 20% (rod)**: 1.0 in. **diam**, 75 ksi (517 MPa) **TS**, 15% in 2 in. (50 mm) **El**, 82 **HRB**
CDA 464, 465, 466, 467		US	Rod, tube, shapes, strip, bar, plate	59.0–62.0 **Cu**, 0.50–1.0 **Sn**, 0.20 **Pb**, rem **Zn**, 0.10 **Fe**, 0.02–0.10 **As**, **Sb** or **P**; 0.10 total, others	**Soft anneal (rod)**: 0.250 in. **diam**, 58 ksi (400 MPa) **TS**, 45% in 2 in. (50 mm) **El**, 56 **HRB**; **Hard drawn – 35% (tube)**: 0.375 in. **diam**, 88 ksi (608 MPa) **TS**, 18% in 2 in. (50 mm) **El**, 95 **HRB**; **As hot rolled (strip, bar, plate)**: 1.0 in. **diam**, 55 ksi (379 MPa) **TS**, 50% in 2 in. (50 mm) **El**, 55 **HRB**
CSA HC.4.9	HC.4.ZP372T	Canada	Bar, plate, sheet, strip	59.0–62.0 **Cu**, 0.50–1.0 **Sn**, 1.3–2.2 **Pb**, rem **Zn**, 0.10 **Fe**, 0.10 total, others	**Soft**: 0.375 in. **diam**, 52 ksi (359 MPa) **TS**, 20 ksi (138 MPa) **YS**, 30% in 1 in. (25 mm) **El**; **Half hard**: 0.375 in. **diam**, 60 ksi (414 MPa) **TS**, 35 ksi (241 MPa) **YS**, 20% in 1 in. (25 mm) **El**
CSA HC.4.9	HC.4.ZT391	Canada	Bar, plate, sheet, strip	59.0–62.0 **Cu**, 0.50–1.0 **Sn**, 0.20 **Pb**, rem **Zn**, 0.10 **Fe**, 0.10 total, others	**Soft**: 0.375 in. **diam**, 52 ksi (359 MPa) **TS**, 20 ksi (138 MPa) **YS**, 30% in 1 in. (25 mm) **El**; **Half hard**: 0.375 in. **diam**, 60 ksi (414 MPa) **TS**, 35 ksi (241 MPa) **YS**, 20% in 1 in. (25 mm) **El**
COPANT 162	C 464 00	COPANT	Wrought products	59–62 **Cu**, 0.5–1.0 **Sn**, 0.20 **Pb**, rem **Zn**, 0.10 **Fe**	
COPANT 162	C 465 00	COPANT	Wrought products	59–62 **Cu**, 0.5–1.0 **Sn**, 0.20 **Pb**, rem **Zn**, 0.10 **Fe**, 0.02–0.10 **As**	

WROUGHT COPPER AND COPPER ALLOYS

specification number	designation	country	product forms	chemical composition	mechanical properties and hardness values
COPANT 162	C 466 00	COPANT	Wrought products	59–62 **Cu**, 0.5–1.0 **Sn**, 0.20 **Pb**, rem **Zn**, 0.10 **Fe**, 0.02–0.10 **Sb**	
COPANT 162	C 467 00	COPANT	Wrought products	59–62 **Cu**, 0.5–1.0 **Sn**, 0.20 **Pb**, rem **Zn**, 0.10 **Fe**, 0.02–0.10 **P**	
COPANT 162	C 482 00	COPANT	Wrought products	59–62 **Cu**, 0.50–1.0 **Sn**, 0.4–1.0 **Pb**, rem **Zn**, 0.10 **Fe**	
COPANT 162	C 485 00	COPANT	Wrought products	59–62 **Cu**, 0.5–1.0 **Sn**, 1.3–2.2 **Pb**, rem **Zn**, 0.10 **Fe**	
QQ–B–639B	464	US	Plate, bar, sheet, strip	59.0–62.0 **Cu**, 0.50–1.0 **Sn**, 0.20 **Pb**, rem **Zn**, 0.10 **Fe**, 0.10 total, others	**Soft temper**: 0.375–30 in. **diam**, 52 ksi (359 MPa) **TS**, 20 ksi (138 MPa) **YS**, 30% in 1 in. (25 mm) **El**; **Half hard**: 0.375–30 in. **diam**, 60 ksi (414 MPa) **TS**, 35 ksi (241 MPa) **YS**, 20% in 1 in. (25 mm) **El** Hard: all **diam**, 65 ksi (448 MPa) **TS**, 50 ksi (345 MPa) **YS**, 10% in 1 in. (25 mm) **El**
QQ–B–639B	482	US	Plate, bar, sheet, strip	59.0–62.0 **Cu**, 0.50–1.0 **Sn**, 0.4–1.0 **Pb**, rem **Zn**, 0.10 **Fe**, 0.10 total, others	**Soft temper**: 0.375–30 in. **diam**, 52 ksi (359 MPa) **TS**, 20 ksi (138 MPa) **YS**, 30% in 1 in. (25 mm) **El**; **Half hard**: 0.375–30 in. **diam**, 60 ksi (414 MPa) **TS**, 35 ksi (241 MPa) **YS**, 20% in 1 in. (25 mm) **El**
QQ–B–639B	485	US	Plate, bar, sheet, strip	59.0–62.0 **Cu**, 0.50–1.0 **Sn**, 1.3–2.2 **Pb**, rem **Zn**, 0.10 **Fe**, 0.10 total, others	**Soft temper**: 0.375–30 in. **diam**, 52 ksi (359 MPa) **TS**, 20 ksi (138 MPa) **YS**, 30% in 1 in. (25 mm) **El**; **Half hard**: 0.375–30 in. **diam**, 60 ksi (414 MPa) **TS**, 35 ksi (241 MPa) **YS**, 20% in 1 in. (25 mm) **El** Hard: all **diam**, 65 ksi (448 MPa) **TS**, 50 ksi (345 MPa) **YS**, 10% in 1 in. (25 mm) **El**
QQ–B–637a	485	US	Rod, wire, shapes, forgings, bar, flat wire, strip	59.0–62.0 **Cu**, 0.5–1.0 **Sn**, 1.3–2.2 **Pb**, rem **Zn**, 0.10 **Fe**, 0.10 total, others	**As extruded**: all **diam**, 52 ksi (359 MPa) **TS**, 20 ksi (138 MPa) **YS**, 30% in 1 in. (25 mm) **El**; **Soft temper**: 1,000 max in. **diam**, 54 ksi (372 MPa) **TS**, 20 ksi (138 MPa) **YS**, 20% in 1 in. (25 mm) **El** Half hard or light annealed: 1,000 max **diam**, 60 ksi (414 MPa) **TS**, 27 ksi (186 MPa) **YS**, 12% in 1 in. (25 mm) **El**

WROUGHT COPPER AND COPPER ALLOYS

specification number	designation	country	product forms	chemical composition	mechanical properties and hardness values
QQ-B-637a	482	US	Rod, wire, shapes, forgings, bar, flat wire, strip	59.0–62.0 **Cu**, 0.5–1.0 **Sn**, 0.4–1.0 **Pb**, rem **Zn**, 0.10 **Fe**, 0.10 total, others	**As extruded**: all **diam**, 52 ksi (359 MPa) **TS**, 20 ksi (138 MPa) **YS**, 25% in 1 in. (25 mm) **El**; **Soft temper**: 1,000 max in. **diam**, 54 ksi (372 MPa) **TS**, 20 ksi (138 MPa) **YS**, 25% in 1 in. (25 mm) **El** Half hard or light annealed: 1,000 max **diam**, 60 ksi (414 MPa) **TS**, 27 ksi (186 MPa) **YS**, 18% in 1 in. (25 mm) **El**
QQ-B-637a	464	US	Rod, wire, shapes, forgings, bar, flat wire, strip	59.0–62.0 **Cu**, 0.5–1.0 **Sn**, 0.20 **Pb**, rem **Zn**, 0.10 **Fe**, 0.10 total, others	**As extruded**: all **diam**, 52 ksi (359 MPa) **TS**, 20 ksi (138 MPa) **YS**, 30% in 1 in. (25 mm) **El**; **Soft temper**: 1,000 max in. **diam**, 54 ksi (372 MPa) **TS**, 20 ksi (138 MPa) **YS**, 30% in 1 in. (25 mm) **El** Half hard or light annealed: 1,000 max **diam**, 60 ksi (414 MPa) **TS**, 27 ksi (186 MPa) **YS**, 22% in 1 in. (25 mm) **El**
JIS H 3100	C4640	Japan	Sheet, plate, and strip	59.0–62.0 **Cu**, 0.50–1.0 **Sn**, 0.20 **Pb**, rem **Zn**, 0.10 **Fe**	**As manufactured**: <20 mm **diam**, 373 MPa **TS**, 25% in 2 in. (50 mm) **El**
COPANT 420	Cu Zn 39 Sn	COPANT	Bar	59.0–62.0 **Cu + Ag**, 0.5–1.0 **Sn**, 0.20 **Pb**, rem **Zn**, 0.10 **Fe**, 0.10 total, others	**Hard**: all **diam**, 360 MPa **TS**, 140 MPa **YS**, 50% in 2 in. (50 mm) **El**; **Annealed**: ≤25 mm **diam**, 370 MPa **TS**, 140 MPa **YS**, 30% in 2 in. (50 mm) **El**; **1/2 hard**: 13–25 mm **diam**, 410 MPa **TS**, 185 MPa **YS**, 25% in 2 in. (50 mm) **El**
COPANT 420	Cu Zn 39 Sn Pb	COPANT	Bar	59.0–62.0 **Cu + Ag**, 0.5–1.0 **Sn**, 0.4–1.0 **Pb**, rem **Zn**	**Annealed**: ≤25 mm **diam**, 370 MPa **TS**, 140 MPa **YS**, 25% in 2 in. (50 mm) **El**; **1/2 hard**: 25–50 mm **diam**, 400 MPa **TS**, 175 MPa **YS**, 20% in 2 in. (50 mm) **El**; **Hard**: ≤25 mm **diam**, 460 MPa **TS**, 315 MPa **YS**, 11% in 2 in. (50 mm) **El**
COPANT 420	Cu Zn 37 Pb 2 Sn	COPANT	Bar	59.0–62.0 **Cu + Ag**, 0.5–1.0 **Sn**, 1.3–2.2 **Pb**, rem **Zn**, 0.10 **Fe**, 0.10 total, others	**Annealed**: ≤25 mm **diam**, 370 MPa **TS**, 140 MPa **YS**, 20% in 2 in. (50 mm) **El**; **1/2 hard**: 25–50 mm **diam**, 400 MPa **TS**, 175 MPa **YS**, 20% in 2 in. (50 mm) **El**; **Hard**: ≤25 mm **diam**, 460 MPa **TS**, 315 MPa **YS**, 11% in 2 in. (50 mm) **El**
MIL-T-6945	I	US	Tube: seamless	59–62 **Cu**, 0.5–1.5 **Sn**, 0.20 **Pb**, rem **Zn**, 0.10 **Fe**, 0.10 total, others	**Hard drawn and relief annealed**: all **diam**, 462 MPa **TS**, 310 MPa **YS**, 15% in 2 in. (50 mm) **El**

WROUGHT COPPER AND COPPER ALLOYS

specification number	designation	country	product forms	chemical composition	mechanical properties and hardness values
BS 2874	CZ 113	UK	Rod, section	57.50–60.50 **Cu**, 0.60–1.25 **Sn**, rem **Zn**, 0.75 total, others	**As manufactured**: 6–20 mm **diam**, 402 MPa **TS**, 16% in 2 in. (50 mm) **El**
COPANT 162	C 470 00	COPANT	Wrought products	57–61 **Cu**, 0.25–1.0 **Sn**, 0.05 **Pb**, rem **Zn**, 0.01 **Al**	
JIS Z 3202	YCuZnSn Naval brass	Japan	Rod: bare welding	57–61 **Cu**, 0.5–1.5 **Sn**, 0.05 **Pb**, rem **Zn**, 0.02 **Al**	
MIL-R-19631B	MIL-RBCuZn-A	US	Rod	57.0–61.0 **Cu**, 0.25–1.00 **Sn**, 0.05 **Pb**, rem **Zn**, 0.01 **Al**, 0.50 total, others, silver included in copper content	**As milled**: all **diam**, 345 MPa **TS**
IS 6912	High tensile brass	India	Forgings	56.0–61.0 **Cu**, 1.0 **Sn**, 0.50–1.5 **Pb**, rem **Zn**, 0.20–1.25 **Fe**, 0.02 **Sb**, 0.20–2.0 **Al**, 0.50–2.0 **Mn**, 0.50 total, others	**As manufactured**: ≥6 mm **diam**, 465 MPa **TS**, 195 MPa **YS**, 20% in 2 in. (50 mm) **El**
AS 1568	686D	Australia	Forgings	56.0–60.0 **Cu**, 0.2–1.0 **Sn**, 0.5–1.5 **Pb**, rem **Zn**, 0.5–1.2 **Fe**, 0.3–1.5 **Al**, 0.3–2.00 **Mn**, 0.50 total, others	**As manufactured**: ≥6 mm **diam**, 460 MPa **TS**, 190 MPa **YS**, 15% in 2 in. (50 mm) **El**
BS 2874	CZ 114	UK	Rod, section	56.0–60.0 **Cu**, 0.2–1.0 **Sn**, 0.5–1.5 **Pb**, rem **Zn**, 0.25–1.2 **Fe**, 0.02 **Sb**, 1.5 **Al**, 0.30–2.0 **Mn**, 0.50 total, others	**As manufactured**: 6–80 mm **diam**, 466 MPa **TS**, 240 MPa **YS**, 18% in 2 in. (50 mm) **El**; **Hard**: 6–40 mm **diam**, 539 MPa **TS**, 290 MPa **YS**, 12% in 2 in. (50 mm) **El**
BS 2874	CZ 115	UK	Rod, section	56.0–60.0 **Cu**, 0.6–1.1 **Sn**, 0.5–1.5 **Pb**, rem **Zn**, 0.25–1.2 **Fe**, 0.3–2.0 **Mn**, 0.50 total, others	**As manufactured (hot-worked)**: 466 MPa **TS**, 196 MPa **YS**, 20% in 2 in. (50 mm) **El**; **As manufactured (cold-worked and stress-relieved)**: 6–40 mm **diam**, 539 MPa **TS**, 279 MPa **YS**, 12% in 2 in. (50 mm) **El**
BS 2872	CZ 115	UK	Forging stock, forgings	56.0–60.0 **Cu**, 0.6–1.1 **Sn**, 0.5–1.5 **Pb**, rem **Zn**, 0.25–1.2 **Fe**, 0.3–2.0 **Mn**, 0.50 total, others	**As manufactured or annealed**: ≥6 mm **diam**, 466 MPa **TS**, 196 MPa **YS**, 15% in 2 in. (50 mm) **El**
BS 2872	CZ 114	UK	Forging stock, forgings	56.0–60.0 **Cu**, 0.20–1.0 **Sn**, 0.5–1.5 **Pb**, rem **Zn**, 0.25–1.2 **Fe**, 1.5 **Al**, 0.3–2.0 **Mn**, 0.50 total, others	**As manufactured or annealed**: ≥6 mm **diam**, 466 MPa **TS**, 196 MPa **YS**, 15% in 2 in. (50 mm) **El**
IS 6912	High tensile brass (soldering quality)	India	Forgings	56.0–60.0 **Cu**, 0.50–1.5 **Sn**, 0.20–1.5 **Pb**, rem **Zn**, 0.20–1.25 **Fe**, 0.02 **Sb**, 0.20 **Al**, 0.25–2.0 **Mn**, 0.50 total, others	**As manufactured**: ≥6 mm **diam**, 465 MPa **TS**, 195 MPa **YS**, 20% in 2 in. (50 mm) **El**
AS 1568	671D	Australia	Forgings	56.0–59.0 **Cu**, 0.5–1.5 **Sn**, 0.5–1.5 **Pb**, rem **Zn**, 0.4–1.0 **Al**, 0.5–1.5 **Mn**, 0.5–1.5 **Ni**, 0.5 total, others	**As manufactured**: ≥6 mm **diam**, 460 MPa **TS**, 190 MPa **YS**, 15% in 2 in. (50 mm) **El**
JIS Z 3262	BCuZn-5	Japan	Wire: filler	50–53 **Cu**, 3.0–4.5 **Sn**, 0.50 **Pb**, rem **Zn**, 0.10 **Fe**	
COPANT 162	C 472 00	COPANT	Wrought products	49–52 **Cu**, 3–4 **Sn**, 0.50 **Pb**, rem **Zn**, 0.10 **Fe**	

WROUGHT COPPER AND COPPER ALLOYS

specification number	designation	country	product forms	chemical composition	mechanical properties and hardness values
ANSI/ASTM B 159	C 52400	US	Wire	88.2–90.4 **Cu**, 9.0–11.0 **Sn**, 0.05 **Pb**, 0.20 **Zn**, 0.10 **Fe**, 0.03–0.35 **P**, 99.5 min **Cu**+**Sn**+**P**	**Soft**: all **diam**, 60 ksi (415 MPa) **TS**; **Hard**: all **diam**, 135 ksi (930 MPa) **TS**
ANSI/ASTM B 139	C 52400	US	Rod, bar, shapes	88.15 min **Cu**, 9.0–11.0 **Sn**, 0.05 **Pb**, 0.20 **Zn**, 0.10 **Fe**, 0.03–0.35 **P**	**Soft**: <0.25 in. (<6.35 mm) **diam**, 60 ksi (415 MPa) **TS**; **Hard**: <0.25 in. (<6.35 mm) **diam**, 105 ksi (720 MPa) **TS**
ANSI/ASTM B 103	C 52400	US	Plate, sheet, strip, bar	88.15 min **Cu**, 9.0–11.0 **Sn**, 0.05 **Pb**, 0.20 **Zn**, 0.10 **Fe**, 0.03–0.35 **P**	**Soft**: >0.039 in. (>0.991 mm) **diam**, 58 ksi (400 MPa) **TS**, 35 **HRB**; **Hard**: >0.039 in. (>0.991 mm) **diam**, 94 ksi (650 MPa) **TS**, 94 **HRB**; **Spring**: >0.039 in. (>0.991 mm) **diam**, 115 ksi (790 MPa) **TS**, 99 **HRB**
CDA 524		US	Rod, wire, strip	99.5 min **Cu**, 9.0–11.0 **Sn**, 0.05 **Pb**, 0.20 **Zn**, 0.10 **Fe**, 0.03–0.35 **P**, **Sn**+**P** in **Cu** content	**0.035 mm (rolled strip)**: 0.040 in. **diam**, 66 ksi (455 MPa) **TS**, 68% in 2 in. (50 mm) **El**, 55 **HRB**; **Half hard temper (rolled strip)**: 0.040 in. **diam**, 83 ksi (572 MPa) **TS**, 32% in 2 in. (50 mm) **El**, 92 **HRB**; **Hard temper (wire)**: 0.080 in. **diam**, 140 ksi (965 MPa) **TS**
COPANT 422	Cu Sn 10 P	COPANT	Bar	9–11 **Sn**, 0.05 **Pb**, 0.20 **Zn**, 0.10 **Fe**, 0.03–0.35 **P**, rem **Cu**+**Ag**, 0.5 total impurities	**Annealed (round)**: ≤6 mm **diam**, 410 MPa **TS**; **Hard (round)**: ≤6 mm **diam**, 725 MPa **TS**; **Hard (hex)**: 6–13 mm **diam**, 655 MPa **TS**
COPANT 162	C 524 00	COPANT	Wrought products	88.15 min **Cu**, 9–11 **Sn**, 0.05 **Pb**, 0.20 **Zn**, 0.10 **Fe**, 0.03–0.35 **P**, 99.5 min **Cu**+**Sn**+**P**	
COPANT 424	Cu Sn 10 P	COPANT	Wire	88.15 min **Cu**, 9.0–11.0 **Sn**, 0.05 **Pb**, 0.20 **Zn**, 0.10 **Fe**, 0.035–0.35 **P**, 99.5 min **Cu**+**Sn**+**P**	**Annealed**: 6.35 mm **diam**, 410 MPa **TS**; **1/2 hard**: 6.35 mm **diam**, 745 MPa **TS**; **Hard**: 6.35 mm **diam**, 930 MPa **TS**
QQ-B-750	Composition D	US	Bar, plate, rod, sheet, strip, wire, shapes	rem **Cu**, 9.0–11.0 **Sn**, 0.05 **Pb**, 0.20 **Zn**, 0.10 **Fe**, 0.03–0.35 **P**, 99.5 min **Cu**+**Sn**+**P**	**Soft temper**: 0.25 max in. **diam**, 60 ksi (414 MPa) **TS**, 35 **HRB**; **Hard temper**: 0.25 max in. **diam**, 105 ksi (724 MPa) **TS**, 94 **HRB**
ISO 1638	CuSn10	ISO	Solid product: drawn, on coils or reels	rem **Cu**, 9.0–11.0 **Sn**, 0.05 **Pb**, 0.3 **Zn**, 0.1 **Fe**, 0.3 **Ni**, 0.01–0.4 **P**, 0.3 total, others	**Annealed**: 1–5 mm **diam**, 410 MPa **TS**, 50% in 2 in. (50 mm) **El**; **HC**: 1–3 mm **diam**, 640 MPa **TS**, 2% in 2 in. (50 mm) **El**

WROUGHT COPPER AND COPPER ALLOYS

specification number	designation	country	product forms	chemical composition	mechanical properties and hardness values
ISO 1637	CuSn10	ISO	Solid products: straight lengths	rem **Cu**, 9.0–11.0 **Sn**, 0.05 **Pb**, 0.3 **Zn**, 0.1 **Fe**, 0.3 **Ni**, 0.01–0.4 **P**, 0.3 total, others	**HA**: 15–50 mm **diam**, 540 MPa **TS**, 360 MPa **YS**, 15% in 2 in. (50 mm) **El**; **HB - strain hardened**: 5–15 mm **diam**, 590 MPa **TS**, 490 MPa **YS**, 10% in 2 in. (50 mm) **El**
ISO 1634	CuSn10	ISO	Plate, sheet, strip	rem **Cu**, 9.0–11.0 **Sn**, 0.05 **Pb**, 0.3 **Zn**, 0.1 **Fe**, 0.3 **Ni**, 0.01–0.4 **P**, 0.3 total, others	**Strain hardened (HB)**: all **diam**, 540 MPa **TS**, 440 MPa **YS**, 15% in 2 in. (50 mm) **El**; **Strain hardened (HD)**: all **diam**, 740 MPa **TS**, 460 MPa **YS**, 1% in 2 in. (50 mm) **El**
ISO 427	CuSn10	ISO	Plate, sheet, strip, rod, bar, tube, wire	rem **Cu**, 9.0–11.0 **Sn**, 0.05 **Pb**, 0.3 **Zn**, 0.1 **Fe**, 0.3 **Ni**, 0.01–0.4 **P**, 0.3 total, others	
NF A 51 108	Cu Sn 9 P	France	Sheet and strip	rem **Cu**, 7.5–10 **Sn**, 0.1 **Pb**, 0.5 **Zn**, 0.1 **Fe**, 0.35 **P**, 0.03 total, others	**Annealed**: $\leq$20 mm **diam**, 360 MPa **TS**, 180 MPa **YS**, 50% in 2 in. (50 mm) **El**, 95–125 **HV**; **1/2 hard**: $\leq$20 mm **diam**, 530 MPa **TS**, 450 MPa **YS**, 20% in 2 in. (50 mm) **El**, 165–195 **HV**; **Hard**: $\leq$20 mm **diam**, 700 MPa **TS**, 680 MPa **YS**, 5% in 2 in. (50 mm) **El**, 210–240 **HV**
NF A 51 111	Cu Sn 8 P	France	Wire	rem **Cu**, 7.5–9.0 **Sn**, 0.05 **Pb**, 0.3 **Zn**, 0.1 **Fe**, 0.3 **Ni**, 0.1–0.4 **P**, 0.3 total, others	**Annealed**: 0.05–3.0 mm **diam**, 360 MPa **TS**, 55% in 2 in. (50 mm) **El**
ISO 427	Cu Sn8	ISO	Plate, sheet, strip, rod, bar, tube, wire	rem **Cu**, 7.5–9.0 **Sn**, 0.05 **Pb**, 0.3 **Zn**, 0.1 **Fe**, 0.3 **Ni**, 0.01–0.4 **P**, 0.3 total, others	
ISO 1638	CuSn8	ISO	Solid product: drawn, on coils or reels	rem **Cu**, 7.5–9.0 **Sn**, 0.05 **Pb**, 0.3 **Zn**, 0.1 **Fe**, 0.3 **Ni**, 0.01–0.4 **P**, 0.3 total, others	**Annealed**: 1–5 mm **diam**, 390 MPa **TS**, 40% in 2 in. (50 mm) **El**; **HC**: 1–3 mm **diam**, 590 MPa **TS**, 5% in 2 in. (50 mm) **El**
ISO 1637	CuSn8	ISO	Solid products: straight lengths	rem **Cu**, 7.5–9.0 **Sn**, 0.05 **Pb**, 0.3 **Zn**, 0.1 **Fe**, 0.3 **Ni**, 0.01–0.4 **P**, 0.3 total, others	**HA**: 5–50 mm **diam**, 490 MPa **TS**, 340 MPa **YS**, 20% in 2 in. (50 mm) **El**; **HB - strain hardened**: 5–15 mm **diam**, 550 MPa **TS**, 470 MPa **YS**, 10% in 2 in. (50 mm) **El**
ISO 1635	CuSn8	ISO	Tube	rem **Cu**, 7.5–9.0 **Sn**, 0.05 **Pb**, 0.3 **Zn**, 0.1 **Fe**, 0.3 **Ni**, 0.01–0.4 **P**, 0.3 total, others	**Annealed**: all **diam**, 50% in 2 in. (50 mm) **El**, 120 max **HV**; **Strain hardened (HB)**: all **diam**, 570 MPa **TS**, 18% in 2 in. (50 mm) **El**, 140 **HV**
ISO 1634	CuSn8	ISO	Plate, sheet, strip	rem **Cu**, 7.5–9.0 **Sn**, 0.05 **Pb**, 0.3 **Zn**, 0.1 **Fe**, 0.3 **Ni**, 0.01–0.4 **P**, 0.3 total, others	**Annealed**: all **diam**, 50% in 2 in. (50 mm) **El**, 110 max **HV**; **Strain hardened (HD)**: all **diam**, 720 MPa **TS**, 650 MPa **YS**, 1% in 2 in. (50 mm) **El**

WROUGHT COPPER AND COPPER ALLOYS

specification number	designation	country	product forms	chemical composition	mechanical properties and hardness values
ANSI/ASTM B 159	C 52100	US	Wire	90.2–92.5 **Cu**, 7.0–9.0 **Sn**, 0.05 **Pb**, 0.20 **Zn**, 0.10 **Fe**, 0.03–0.35 **P**	**Soft**: all **diam**, 53 ksi (365 MPa) **TS**; **Hard**: all **diam**, 125 ksi (860 MPa) **TS**
ANSI/ASTM B 139	C 52100	US	Rod, bar, shapes	90.15 min **Cu**, 7.0–9.0 **Sn**, 0.05 **Pb**, 0.20 **Zn**, 0.10 **Fe**, 0.03–0.35 **P**	**Soft**: <0.25 in. (<6.35 mm) **diam**, 53 ksi (365 MPa) **TS**; **Hard**: <0.25 in. (<6.35 mm) **diam**, 105 ksi (720 MPa) **TS**
ANSI/ASTM B 103	C 52100	US	Plate, sheet, strip, bar	90.15 min **Cu**, 7.0–9.0 **Sn**, 0.05 **Pb**, 0.20 **Zn**, 0.10 **Fe**, 0.03–0.35 **P**	**Soft**: >0.039 in. (>0.991 mm) **diam**, 53 ksi (365 MPa) **TS**, 29 **HRB**; **Hard**: >0.039 in. (>0.991 mm) **diam**, 85 ksi (585 MPa) **TS**, 91 **HRB**; **Spring**: >0.039 in. (>0.991 mm) **diam**, 105 ksi (720 MPa) **TS**, 97 **HRB**
CDA 521		US	Rod, wire, strip	99.5 min **Cu**, 7.0–9.0 **Sn**, 0.05 **Pb**, 0.20 **Zn**, 0.10 **Fe**, 0.03–0.35 **P**, **Sn**+**P** in **Cu** content	**Half hard temper – 20% (rod)**: 0.500 in. **diam**, 80 ksi (552 MPa) **TS**, 33% in 2 in. (50 mm) **El**, 85 **HRB**; **Hard temper (wire)**: 0.080 in. **diam**, 130 ksi (896 MPa) **TS**; **Spring temper (rolled strip)**: 0.040 in. **diam**, 112 ksi (772 MPa) **TS**, 3% in 2 in. (50 mm) **El**, 98 **HRB**
COPANT 162	C 521 00	COPANT	Wrought products	90.15 min **Cu**, 7–9 **Sn**, 0.05 **Pb**, 0.20 **Zn**, 0.10 **Fe**, 0.03–0.35 **P**	
COPANT 162	C 529 00	COPANT	Wrought products	88.15 min **Cu**, 7.0–9.0 **Sn**, 0.05 **Pb**, 0.20 **Zn**, 0.10 **Fe**, 1.0–2.0 **Mn**, 0.03–0.35 **P**	
COPANT 424	Cu Sn 8 P	COPANT	Wire	90.15 min **Cu**, 7.0–9.0 **Sn**, 0.05 **Pb**, 0.20 **Zn**, 0.10 **Fe**, 0.035–0.35 **P**	**Annealed**: 6.35 mm **diam**, 360 MPa **TS**; **1/2 hard**: 6.35 mm **diam**, 655 MPa **TS**; **Hard**: 6.35 mm **diam**, 860 MPa **TS**
COPANT 422	Cu Sn 8 P	COPANT	Bar	rem **Cu**, 7.0–9.0 **Sn**, 0.05 **Pb**, 0.20 **Zn**, 0.10 **Fe**, 0.03–0.35 **P**, 0.5 total impurities, silver included in copper content	**Annealed (round)**: <6 mm **diam**, 360 MPa **TS**; **Hard (round)**: <6 mm **diam**, 725 MPa **TS**; **Hard (hex)**: 6–13 mm **diam**, 590 MPa **TS**, 12% in 2 in. (50 mm) **El**

WROUGHT COPPER AND COPPER ALLOYS

specification number	designation	country	product forms	chemical composition	mechanical properties and hardness values
JIS H 3130	C 5210	Japan	Sheet, plate, and strip: for spring	90.0–92.32 **Cu**, 7.0–9.0 **Sn**, 0.05 **Pb**, 0.20 **Zn**, 0.10 **Fe**, 0.03–0.35 **P**	**1/2 hard**: ≥0.15 mm **diam**, 471 MPa **TS**, 21% in 2 in. (50 mm) **El**, 120–195 **HV**; **Hard**: ≥0.15 mm **diam**, 588 MPa **TS**, 20% in 2 in. (50 mm) **El**, 170–220 **HV**; **Extra hard**: 686 MPa **TS**, 11% in 2 in. (50 mm) **El**, 200–240 **HV**
JIS H 3110	C 5212	Japan	Sheet, plate, and strip	90.2–92.5 **Cu**, 7.0–9.0 **Sn**, 0.03–0.35 **P**	**Annealed**: 0.3–5 mm **diam**, 343 MPa **TS**, 45% in 2 in. (50 mm) **El**; **Hard**: 0.3–5 mm **diam**, 637 MPa **TS**, 8% in 2 in. (50 mm) **El**, 185 min **HV**; **Extra hard**: 0.3–5 mm **diam**, 686 MPa **TS**, 5% in 2 in. (50 mm) **El**, 200 min **HV**
JIS H 3270	C 5212	Japan	Rod and bar	90.15–92.47 **Cu**, 7.0–9.0 **Sn**, 0.03–0.35 **P**	**1/2 hard**: <13 mm **diam**, 490 MPa **TS**, 13% in 2 in. (50 mm) **El**, 140 min **HV**; **Hard**: <13 mm **diam**, 686 MPa **TS**, 10% in 2 in. (50 mm) **El**, 180 min **HV**
JIS Z 3231	DCuSnB	Japan	Wire: covered electrode	rem **Cu**, 7.0–9.0 **Sn**, 0.02 **Pb**, 0.30 **P**, 0.50 **Si + Mn + Pb + Al + Fe + Ni + Zn**	**As drawn**: 3.2–6 mm **diam**, 274 MPa **TS**, 12% in 2 in. (50 mm) **El**
BS 2870	PB103	UK	Sheet, strip, foil	rem **Cu**, 6.0–7.5 **Sn**, 0.02 **Pb**, 0.02–0.40 **P**, 0.20 total, others	**Annealed**: ≤10.0 mm **diam**, 238 MPa **TS**, 50% in 2 in. (50 mm) **El**, 90 max **HV**; **1/2 hard**: ≤10.0 mm **diam**, 369 MPa **TS**, 310 MPa **YS**, 12% in 2 in. (50 mm) **El**, 170 **HV**; **Hard**: ≤6.0 mm **diam**, 434 MPa **TS**, 393 MPa **YS**, 6% in 2 in. (50 mm) **El**, 200 **HV**
BS 2873	PB103	UK	Wire	rem **Cu**, 6.0–7.5 **Sn**, 0.02 **Pb**, 0.02–0.40 **P**, 0.20 total, others	**Annealed**: 0.5–10.0 mm **diam**, 367 MPa **TS**, 50% in 2 in. (50 mm) **El**; **1/2 hard**: 0.5–10.0 mm **diam**, 588 MPa **TS**; **Hard**: 0.5–10.0 mm **diam**, 741 MPa **TS**
IS 7814	Grade III	India	Sheet, strip, foil	rem **Cu**, 5.6–7.5 **Sn**, 0.02 **Pb**, 0.02–0.40 **P**, 0.2 total, others	**Annealed (sheet)**: all **diam**, 340 MPa **TS**, 50% in 2 in. (50 mm) **El**, 90 max **HV**; **1/2 hard (sheet)**: all **diam**, 465 MPa **TS**, 12% in 2 in. (50 mm) **El**, 150 **HV**; **Hard (sheet)**: all **diam**, 540 MPa **TS**, 6% in 2 in. (50 mm) **El**, 165 **HV**

WROUGHT COPPER AND COPPER ALLOYS

specification number	designation	country	product forms	chemical composition	mechanical properties and hardness values
IS 7814	Grade III	India	Sheet, strip, foil	rem **Cu**, 5.6–7.5 **Sn**, 0.02 **Pb**, 0.02–0.40 **P**, 0.2 total, others	**Annealed (strip)**: 340 MPa **TS**, 50% in 2 in. (50 mm) **El**, 90 max **HV**; **1/2 hard (strip)**: 520 MPa **TS**, 12% in 2 in. (50 mm) **El**, 165 **HV**; **Hard (strip)**: all **diam**, 620 MPa **TS**, 6% in 2 in. (50 mm) **El**, 200 **HV**
IS 7608	Grade II	India	Wire	rem **Cu**, 5.6–7.5 **Sn**, 0.05 **Pb**, 0.02–0.40 **P**, 0.20 total, others	**Annealed**: 0.45–10.0 mm **diam**, 370 MPa **TS**, 50% in 2 in. (50 mm) **El**; **1/2 hard**: 0.45–10.0 mm **diam**, 590 MPa **TS**; **Hard**: 0.45–6.3 mm **diam**, 740 MPa **TS**
DS 3003	5428	Denmark	Rolled, drawn, extruded, and forged products	94 **Cu**, 5.5–7.5 **Sn**, 0.05 **Pb**, 0.3 **Zn**, 0.1 **Fe**, 0.02–0.4 **P**	
NF A 51 111	Cu Sn 6 P	France	Wire	rem **Cu**, 5 5–7.5 **Sn**, 0.05 **Pb**, 0.3 **Zn**, 0.1 **Fe**, 0.3 **Ni**, 0.1–0.4 **P**, 0.3 total, others	**Annealed**: 0.05–3.0 mm **diam**, 335 MPa **TS**, 50% in 2 in. (50 mm) **El**
ISO 427	Cu Sn6	ISO	Plate, sheet, strip, rod, bar, tube, wire	rem **Cu**, 5.5–7.5 **Sn**, 0.05 **Pb**, 0.3 **Zn**, 0.1 **Fe**, 0.3 **Ni**, 0.01–0.4 **P**, 0.3 total, others	
ISO 1638	CuSn6	ISO	Solid product: drawn, on coils or reels	rem **Cu**, 5.5–7.5 **Sn**, 0.05 **Pb**, 0.3 **Zn**, 0.1 **Fe**, 0.3 **Ni**, 0.01–0.4 **P**, 0.3 total, others	**Annealed**: 1–5 mm **diam**, 370 MPa **TS**, 40% in 2 in. (50 mm) **El**; **HE**: 1–3 mm **diam**, 740 MPa **TS**
ISO 1637	CuSn6	ISO	Solid products: straight lengths	rem **Cu**, 5.5–7.5 **Sn**, 0.05 **Pb**, 0.3 **Zn**, 0.1 **Fe**, 0.3 **Ni**, 0.01–0.4 **P**, 0.3 total, others	**HA**: 5–50 mm **diam**, 450 MPa **TS**, 290 MPa **YS**, 20% in 2 in. (50 mm) **El**; **HB – strain hardened**: 5–50 mm **diam**, 520 MPa **TS**, 410 MPa **YS**, 10% in 2 in. (50 mm) **El**
ISO 1634	CuSn6	ISO	Plate, sheet, strip	rem **Cu**, 5.5–7.5 **Sn**, 0.05 **Pb**, 0.3 **Zn**, 0.1 **Fe**, 0.3 **Ni**, 0.1–0.4 **P**, 0.3 total, others	**Annealed**: all **diam**, 50% in 2 in. (50 mm) **El**, 95 max **HV**; **Strain hardened**: all **diam**, 690 MPa **TS**, 620 MPa **YS**, 1% in 2 in. (50 mm) **El**
SFS 2933	CuSn6	Finland	Sheet, strip, rod, and wire	94 **Cu**, 5.5–7.5 **Sn**, 0.05 **Pb**, 0.3 **Zn**, 0.1 **Fe**, 0.02–0.4 **P**	**Annealed (sheet and strip)**: 0.2–0.5 mm **diam**, 340 MPa **TS**, 130 MPa **YS**, 30% in 2 in. (50 mm) **El**; **Cold finished (sheet and strip)**: 0.1–0.5 mm **diam**, 470 MPa **TS**, 390 MPa **YS**, 15% in 2 in. (50 mm) **El**; **Drawn (rod and wire)**: 0.2–2.5 mm **diam**, 470 MPa **TS**, 390 MPa **YS**, 5% in 2 in. (50 mm) **El**
SIS 14 54 28	SIS Bronze 54 28–07	Sweden	Sheet, strip, wire	94 **Cu**, 5.5–7.5 **Sn**, 0.05 **Pb**, 0.3 **Zn**, 0.1 **Fe**, 0.02–0.4 **P**, 0.3 total, others	**Strain hardened**: 0.1–0.5 mm **diam**, 670 MPa **TS**, 630 MPa **YS**, 5% in 4 in. (100 mm) **El**, 200–225 **HV**

WROUGHT COPPER AND COPPER ALLOYS

specification number	designation	country	product forms	chemical composition	mechanical properties and hardness values
SIS 14 54 28	SIS Bronze 54 28–06	Sweden	Sheet, strip	94 **Cu**, 5.5–7.5 **Sn**, 0.05 **Pb**, 0.3 **Zn**, 0.1 **Fe**, 0.02–0.4 **P**, 0.3 total, others	**Strain hardened**: 0.1–0.5 mm **diam**, 550 MPa **TS**, 490 MPa **YS**, 5% in 2 in. (50 mm) **El**, 180–200 **HV**
SIS 14 54 28	SIS Bronze 54 28–05	Sweden	Sheet, strip	94 **Cu**, 5.5–7.5 **Sn**, 0.05 **Pb**, 0.3 **Zn**, 0.1 **Fe**, 0.02–0.4 **P**, 0.3 total, others	**Strain hardened**: 0.1–0.5 mm **diam**, 470 MPa **TS**, 390 MPa **YS**, 15% in 4 in. (100 mm) **El**, 155–180 **HV**
SIS 14 54 28	SIS Bronze 54 28–04	Sweden	Wire, rod	94 **Cu**, 5.5–7.5 **Sn**, 0.05 **Pb**, 0.3 **Zn**, 0.1 **Fe**, 0.02–0.4 **P**, 0.3 total, others	**Strain hardened**: 0.2–2.5 mm **diam**, 470 MPa **TS**, 390 MPa **YS**, 5% in 2 in. (50 mm) **El**, 155–180 **HV**
SIS 14 54 28	SIS Bronze 54 28–02	Sweden	Plate, sheet, strip, wire	94 **Cu**, 5.5–7.5 **Sn**, 0.05 **Pb**, 0.3 **Zn**, 0.1 **Fe**, 0.02–0.4 **P**, 0.3 total, others	**Annealed**: 5 mm **diam**, 340 MPa **TS**, 130 MPa **YS**, 55% in 2 in. (50 mm) **El**, 70–95 **HV**
COPANT 422	Cu Sn 6 P	COPANT	Bar	rem **Cu**, 4.2–8.5 **Sn**, 0.05 **Pb**, 0.30 **Zn**, 0.10 **Fe**, 0.03–0.35 **P**, 0.5 total impurities, silver included in copper content	**Annealed (round)**: ≤6 mm **diam**, 275 MPa **TS**; **Hard (round)**: ≤6 mm **diam**, 550 MPa **TS**; **Spring (round)**: ≤0.6 mm **diam**, 860 MPa **TS**
NF A 51 108	Cu Sn 6 P	France	Sheet and strip	rem **Cu**, 5–7.5 **Sn**, 0.1 **Pb**, 0.5 **Zn**, 0.1 **Fe**, 0.35 **P**, 0.3 total, others	**Annealed**: ≤20 mm **diam**, 330 MPa **TS**, 160 MPa **YS**, 50% in 2 in. (50 mm) **El**, 90–120 **HV**; **1/2 hard**: ≤20 mm **diam**, 460 MPa **TS**, 350 MPa **YS**, 20% in 2 in. (50 mm) **El**, 150–180 **HV**; **Hard**: ≤20 mm **diam**, 620 MPa **TS**, 560 MPa **YS**, 5% in 2 in. (50 mm) **El**, 190–240 **HV**
JIS H 3270	C 5191	Japan	Rod and bar	92.15–93.97 **Cu**, 5.5–7.0 **Sn**, 0.03–0.35 **P**	**1/2 hard**: <13 mm **diam**, 461 MPa **TS**, 13% in 2 in. (50 mm) **El**, 135 min **HV**; **Hard**: <13 mm **diam**, 588 MPa **TS**, 10% in 2 in. (50 mm) **El**, 165 min **HV**
JIS H 3110	C 5191	Japan	Sheet, plate, and strip	92.15–94 **Cu**, 5.5–7.0 **Sn**, 0.03–0.35 **P**	**Annealed**: 0.3–5 mm **diam**, 314 MPa **TS**, 42% in 2 in. (50 mm) **El**; **Hard**: 0.3–5 mm **diam**, 588 MPa **TS**, 8% in 2 in. (50 mm) **El**, 170 min **HV**; **Extra hard**: 0.3–5 mm **diam**, 637 MPa **TS**, 5% in 2 in. (50 mm) **El**, 85 min **HV**
COPANT 162	C 519 00	COPANT	Wrought products	92.15 min **Cu**, 5–7 **Sn**, 0.05 **Pb**, 0.30 **Zn**, 0.10 **Fe**, 0.03–0.35 **P**	

WROUGHT COPPER AND COPPER ALLOYS

specification number	designation	country	product forms	chemical composition	mechanical properties and hardness values
JIS Z 3231	DCuSnA	Japan	Wire: covered electrode	rem **Cu**, 5.0–7.0 **Sn**, 0.02 **Pb**, 0.30 **P**, 0.50 **Si+Mn+Pb+Al+Fe+Ni+Zn**	**As drawn**: 3.2–6 mm **diam**, 245 MPa **TS**, 15% in 2 in. (50 mm) **El**
AS 1567	518B	Australia	Rod, bar, section	93.80 **Cu**, 4.5–6.0 **Sn**, 0.02–0.40 **P**	**As manufactured**: 6–20 mm **diam**, 460 MPa **TS**, 350 MPa **YS**, 2% in 2 in. (50 mm) **El**
AS 1566	518B	Australia	Plate, bar, sheet, strip, foil	93.7 min **Cu**, 4.5–6.0 **Sn**, 0.02 **Pb**, 0.02–0.40 **P**	**Annealed (sheet, strip)**: all **diam**, 310 MPa **TS**, 45% in 2 in. (50 mm) **El**, 85 **HV**; **Strain hardened to 1/2 hard (sheet, strip)**: all **diam**, 500 MPa **TS**, 160–180 **HV**; **Strain hardened to full hard (sheet, strip)**: all **diam**, 590 MPa **TS**, 180–200 **HV**
BS 2870	PB102	UK	Sheet, strip, foil	rem **Cu**, 4.5–6.0 **Sn**, 0.02 **Pb**, 0.02–0.40 **P**, 0.20 total, others	**Annealed**: ≤10.0 mm **diam**, 217 MPa **TS**, 45% in 2 in. (50 mm) **El**, 85 max **HV**; **1/2 hard**: ≤10.0 mm **diam**, 348 MPa **TS**, 296 MPa **YS**, 10% in 2 in. (50 mm) **El**, 160 **HV**; **Hard**: ≤6.0 mm **diam**, 414 MPa **TS**, 379 MPa **YS**, 4% in 2 in. (50 mm) **El**, 180 **HV**
BS 2875	PB102	UK	Plate	rem **Cu**, 4.5–6.0 **Sn**, 0.02 **Pb**, 0.02–0.40 **P**, 0.20 total, others	**As manufactured or annealed**: ≤10.0 mm **diam**, 309 MPa **TS**, 35% in 2 in. (50 mm) **El**; **Hard**: 10–16 mm **diam**, 432 MPa **TS**, 12% in 2 in. (50 mm) **El**
BS 2873	PB102	UK	Wire	rem **Cu**, 4.5–6.0 **Sn**, 0.02 **Pb**, 0.02–0.40 **P**, 0.20 total, others	**Annealed**: 0.5–10.0 mm **diam**, 336 MPa **TS**, 40% in 2 in. (50 mm) **El**; **1/2 hard**: 0.5–10.0 mm **diam**, 541 MPa **TS**; **Hard**: 0.5–10.0 mm **diam**, 701 MPa **TS**
BS 2874	PB102	UK	Rod, section	rem **Cu**, 4.5–6.0 **Sn**, 0.02 **Pb**, 0.02–0.40 **P**, 0.20 total, others	**As manufactured**: 6–20 mm **diam**, 495 MPa **TS**, 412 MPa **YS**, 12% in 2 in. (50 mm) **El**
IS 7811		India	Rod and bar	rem **Cu**, 4.6–5.5 **Sn**, 0.02 **Pb**, 0.02–0.40 **P**, 0.2 total, others	**As manufactured**: 10–18 mm **diam**, 495 MPa **TS**, 410 MPa **YS**, 10% in 2 in. (50 mm) **El**
IS 7814	Grade II	India	Sheet, strip, foil	rem **Cu**, 4.6–5.5 **Sn**, 0.02 **Pb**, 0.02–0.40 **P**, 0.2 total, others	**Annealed (sheet)**: all **diam**, 310 MPa **TS**, 45% in 2 in. (50 mm) **El**, 85 max **HV**; **1/2 hard (sheet)**: all **diam**, 460 MPa **TS**, 10% in 2 in. (50 mm) **El**, 135 **HV**; **Hard (sheet)**: all **diam**, 525 MPa **TS**, 4% in 2 in. (50 mm) **El**, 155 **HV**

WROUGHT COPPER AND COPPER ALLOYS

specification number	designation	country	product forms	chemical composition	mechanical properties and hardness values
IS 7814	Grade II	India	Sheet, strip, foil	rem **Cu**, 4.6–5.5 **Sn**, 0.02 **Pb**, 0.02–0.40 **P**, 0.2 total, others	**Annealed (strip)**: 310 MPa **TS**, 45% in 2 in. (50 mm) **El**, 85 max **HV**; **1/2 hard (strip)**: 490 MPa **TS**, 10% in 2 in. (50 mm) **El**, 155 **HV**; **Hard (strip)**: 585 MPa **TS**, 4% in 2 in. (50 mm) **El**, 175 **HV**
IS 7608	Grade I	India	Wire	rem **Cu**, 4.6–5.5 **Sn**, 0.05 **Pb**, 0.02–0.40 **P**, 0.20 total, others	**Annealed**: 0.45–10.0 mm **diam**, 340 MPa **TS**, 40% in 2 in. (50 mm) **El**; **1/2 hard**: 0.45–10.0 mm **diam**, 540 MPa **TS**; **Hard**: 0.45–6.3 mm **diam**, 700 MPa **TS**
AMS 4510D		US	Sheet, strip, and plate	93.0 min **Cu**, 4.2–5.8 **Sn**, 0.05 **Pb**, 0.30 **Zn**, 0.10 **Fe**, 0.01 **Sb**, 0.03–0.35 **P**	**Cold rolled to spring hard**: ≥0.040 in. (≥1.02 mm) **diam**, 91 ksi (627 MPa) **TS**, 90–96 **HRB**
ASTM F 468	C 51000	US	Bolt, screw, stud	rem **Cu**, 4.2–5.8 **Sn**, 0.05 **Pb**, 0.30 **Zn**, 0.10 **Fe**, 0.03–0.35 **P**	**As manufactured (test specimen)**: all **diam**, 55 ksi (380 MPa) **TS**, 30 ksi (205 MPa) **YS**, 15% in 2 in. (50 mm) **El**, 60–95 **HRB**
ASTM F 467	C 51000	US	Nut	rem **Cu**, 4.2–5.8 **Sn**, 0.05 **Pb**, 0.30 **Zn**, 0.10 **Fe**, 0.03–0.35 **P**	**As agreed**: all **diam**, 60 min **HRB**
ANSI/ASTM B 159	C 51000	US	Wire	93.4–95.3 **Cu**, 4.2–5.8 **Sn**, 0.05 **Pb**, 0.30 **Zn**, 0.10 **Fe**, 0.03–0.35 **P**	**Soft**: all **diam**, 43 ksi (295 MPa) **TS**; **Hard**: all **diam**, 108 ksi (745 MPa) **TS**; **Spring**: 0.25–0.37 in. (6.35–9.52 mm) **diam**, 120 ksi (830 MPa) **TS**, 5% in 2 in. (50 mm) **El**
ANSI/ASTM B 139	C 51000	US	Rod, bar, shapes	93.35 min **Cu**, 4.2–5.8 **Sn**, 0.05 **Pb**, 0.30 **Zn**, 0.10 **Fe**, 0.03–0.35 **P**	**Soft**: <0.25 in. (<6.35 mm) **diam**, 40 ksi (275 MPa) **TS**; **Hard**: <0.25 in. (<6.35 mm) **diam**, 80 ksi (550 MPa) **TS**; **Spring**: <0.026 in. (<0.660 mm) **diam**, 125 ksi (860 MPa) **TS**
ANSI/ASTM B 103	C 51000	US	Plate, sheet, strip, bar	93.35 min **Cu**, 4.2–5.8 **Sn**, 0.05 **Pb**, 0.30 **Zn**, 0.10 **Fe**, 0.03–0.35 **P**	**Soft**: >0.039 in. (>0.991 mm) **diam**, 43 ksi (295 MPa) **TS**, 16 **HRB**; **Hard**: >0.039 in. (>0.991 mm) **diam**, 76 ksi (525 MPa) **TS**, 86 **HRB**; **Spring**: >0.039 in. (>0.991 mm) **diam**, 95 ksi (655 MPa) **TS**, 94 **HRB**
ANSI/ASTM B 100	CA 510	US	Plate, sheet	93.4–95.3 **Cu**, 4.2–5.8 **Sn**, 0.05 **Pb**, 0.30 **Zn**, 0.10 **Fe**, 0.03–0.35 **P**	**Rolled or annealed**: 1/4 in. (6.35 mm) **diam**, 60 ksi (415 MPa) **TS**, 25 ksi (170 MPa) **YS**, 10% in 2 in. (50.8 mm) **El**, 75 **HRB**

WROUGHT COPPER AND COPPER ALLOYS

specification number	designation	country	product forms	chemical composition	mechanical properties and hardness values
CDA 510		US	Rod, wire, flat wire, strip	99.5 min **Cu**, 4.2–5.8 **Sn**, 0.05 **Pb**, 0.30 **Zn**, 0.10 **Fe**, 0.35 **P**, **Sn+P** in **Cu** content	**Half hard temper – 20% (rod)**: 0.500 in. **diam**, 75 ksi (517 MPa) **TS**, 25% in 2 in. (50 mm) **El**, 80 **HRB**; **Hard temper (wire)**: 0.080 in. **diam**, 110 ksi (758 MPa) **TS**, 5% in 2 in. (50 mm) **El**; **Extra hard temper (flat wire, strip)**: 0.040 in. **diam**, 92 ksi (634 MPa) **TS**, 6% in 2 in. (50 mm) **El**, 93 **HRB**
COPANT 162	C 510 00	COPANT	Wrought products	93.35 min **Cu**, 4.2–5.8 **Sn**, 0.05 **Pb**, 0.30 **Zn**, 0.10 **Fe**, 0.03–0.35 **P**	
COPANT 162	C 518 00	COPANT	Wrought products	93.15 min **Cu**, 4–6 **Sn**, 0.02 **Pb**, 0.01 **Al**, 0.10–0.35 **P**, 99.5 min **Cu+Sn+P**	
COPANT 424	Cu Sn 5 P	COPANT	Wire	93.35 min **Cu**, 4.2–5.8 **Sn**, 0.05 **Pb**, 0.30 **Zn**, 0.10 **Fe**, 0.035–0.35 **P**, 99.5 min **Cu+Sn+P**	**Annealed**: 12.7 mm **diam**, 295 MPa **TS**; **1/2 hard**: 12.7 mm **diam**, 550 MPa **TS**; **Hard**: 12.7 mm **diam**, 745 MPa **TS**
MIL-R-19631B	MIL-RCuSn-A	US	Rod	93.5 min **Cu**, 4.0–6.0 **Sn**, 0.02 **Pb**, 0.01 **Al**, 0.10–0.35 **P**, 0.50 total, others, silver included in the copper content	**As milled**: all **diam**, 241 MPa **TS**
AMS 4720B		US	Wire	93.0 min **Cu**, 3.5–5.8 **Sn**, 0.05 **Pb**, 0.30 **Zn**, 0.10 **Fe**, 0.03–0.35 **P**	**Cold drawn, spring hard**: 0.250–0.375 in. **diam**, 120 ksi (827 MPa) **TS**, 5% in 2 in. (50 mm) **El**
AMS 4625D		US	Rod, bar, and tube	93.0 min **Cu**, 3.5–5.8 **Sn**, 0.05 **Pb**, 0.30 **Zn**, 0.10 **Fe**, 0.03–0.35 **P**	**Cold finished, hard (rod and bar)**: 0.25–0.50 in. **diam**, 70 ksi (483 MPa) **TS**, 13% in 2 in. (50 mm) **El**; **Cold finished, hard (tube)**: >1.0 in. **diam**, 55 ksi (379 MPa) **TS**, 12% in 2 in. (50 mm) **El**
ANSI/ASTM B 139	C 53400	US	Rod, bar, shapes	93.3 min **Cu**, 3.5–5.8 **Sn**, 0.05 **Pb**, 0.20 **Zn**, 0.10 **Fe**, 0.03–0.35 **P**	**Hard**: >0.063 in. (>1.59 mm) **diam**, 65 ksi (450 MPa) **TS**, 10% in 1 in. (25.4 mm) **El**
COPANT 422	Cu Sn 5 Pb P	COPANT	Bar	3.5–5.8 **Sn**, 0.8–1.0 **Pb**, 0.30 **Zn**, 0.10 **Fe**, 0.03–0.35 **P**, rem **Cu+Ag**, 0.5 total impurities	**Hard (round & hex)**: 1.6–7 mm **diam**, 450 MPa **TS**; **Hard (square & flat)**: 6–10 mm **diam**, 380 MPa **TS**, 10% in 2 in. (50 mm) **El**
QQ-W-321d	510	US	Wire	99.5 min **Cu**, 3.5–5.8 **Sn**, 0.05 **Pb**, 0.30 **Zn**, 0.10 **Fe**, 0.03–0.35 **P**, **Cu** content includes **Sn+P**; **Co** in **Ni**	**Spring temper**: 0.250–0.375 in. **diam**, 120 ksi (827 MPa) **TS**, 5.0% in 4 in. (100 mm) **El**

WROUGHT COPPER AND COPPER ALLOYS

specification number	designation	country	product forms	chemical composition	mechanical properties and hardness values
QQ-B-750	Composition A	US	Bar, plate, rod, sheet, strip, wire, shapes	rem **Cu**, 3.5–5.8 **Sn**, 0.05 **Pb**, 0.30 **Zn**, 0.10 **Fe**, 0.03–0.35 **P**, 99.5 min **Cu + Sn + P**	**Soft temper**: 0.25 max in. **diam**, 40 ksi (276 MPa) **TS**, 7 **HRB**; **Hard temper**: 0.25 max in. **diam**, 80 ksi (552 MPa) **TS**, 82 **HRB**; **Spring temper**: 0.26 max in. **diam**, 125 ksi (862 MPa) **TS**, 90 **HRB**
NF A 51 111	Cu Sn 4 P	France	Wire	rem **Cu**, 3.0–5.5 **Sn**, 0.05 **Pb**, 0.3 **Zn**, 0.1 **Fe**, 0.3 **Ni**, 0.1–0.4 **P**, 0.3 total, others	**Annealed**: 0.05–3.0 mm **diam**, 295 MPa **TS**, 45% in 2 in. (50 mm) **El**
ISO 427	Cu Sn4	ISO	Plate, sheet, strip, rod, bar, tube, wire	rem **Cu**, 3.0–5.5 **Sn**, 0.05 **Pb**, 0.3 **Zn**, 0.1 **Fe**, 0.3 **Ni**, 0.01–0.4 **P**, 0.3 total, others	
ISO 1638	CuSn4	ISO	Solid product: drawn, on coils or reels	rem **Cu**, 3.0–5.5 **Sn**, 0.05 **Pb**, 0.3 **Zn**, 0.1 **Fe**, 0.3 **Ni**, 0.1–0.4 **P**, 0.3 total, others	**Annealed**: 1–5 mm **diam**, 310 MPa **TS**, 40% in 2 in. (50 mm) **El**; **HC**: 1–3 mm **diam**, 490 MPa **TS**, 3% in 2 in. (50 mm) **El**; **HE**: 1–3 mm **diam**, 690 MPa **TS**
ISO 1637	CuSn4	ISO	Solid products: straight lengths	rem **Cu**, 3.0–5.5 **Sn**, 0.05 **Pb**, 0.3 **Zn**, 0.1 **Fe**, 0.3 **Ni**, 0.01–0.4 **P**, 0.3 total, others	**HA**: 5–100 mm **diam**, 380 MPa **TS**, 250 MPa **YS**, 20% in 2 in. (50 mm) **El**; **HB – strain hardened**: 5–50 mm **diam**, 490 MPa **TS**, 360 MPa **YS**, 12% in 2 in. (50 mm) **El**; **HC**: 5–15 mm **diam**, 510 MPa **TS**, 390 MPa **YS**, 10% in 2 in. (50 mm) **El**
ISO 1634	CuSn4	ISO	Plate, sheet, strip	rem **Cu**, 3.0–5.5 **Sn**, 0.05 **Pb**, 0.3 **Zn**, 0.1 **Fe**, 0.3 **Ni**, 0.1–0.4 **P**, 0.3 total, others	**Annealed**: all **diam**, 45% in 2 in. (50 mm) **El**, 90 max **HV**; **Strain hardened**: all **diam**, 640 MPa **TS**, 570 MPa **YS**, 1% in 2 in. (50 mm) **El**
JIS H 3270	C 5101	Japan	Rod and bar	93.7–96.5 **Cu**, 3.0–5.5 **Sn**, 0.03–0.35 **P**, 99.5 min **Cu + Sn + P**	**Hard**: <13 mm **diam**, 451 MPa **TS**, 10% in 2 in. (50 mm) **El**, 125 min **HV**
JIS H 3110	C 5101	Japan	Sheet, plate, and strip	93.65–96.47 **Cu**, 3.0–5.5 **Sn**, 0.03–0.35 **P**, 99.5 min **Cu + Sn + P**	**Annealed**: 0.3–5 mm **diam**, 294 MPa **TS**, 38% in 2 in. (50 mm) **El**; **Hard**: 0.3–5 mm **diam**, 490 MPa **TS**, 7% in 2 in. (50 mm) **El**, 135 min **HV**; **Extra hard**: 0.3–5 mm **diam**, 539 MPa **TS**, 4% in 2 in. (50 mm) **El**, 150 min **HV**
ANSI/ASTM B 103	C 51100	US	Plate, sheet, strip, bar	94.25 min **Cu**, 3.5–4.9 **Sn**, 0.05 **Pb**, 0.30 **Zn**, 0.10 **Fe**, 0.03–0.35 **P**	**Soft**: >0.039 in. (>0.991 mm) **diam**, 40 ksi (275 MPa) **TS**, 7 **HRB**; **Hard**: >0.039 in. (>0.991 mm) **diam**, 72 ksi (495 MPa) **TS**, 82 **HRB**; **Spring**: >0.039 in. (>0.991 mm) **diam**, 91 ksi (625 MPa) **TS**, 90 **HRB**

WROUGHT COPPER AND COPPER ALLOYS

specification number	designation	country	product forms	chemical composition	mechanical properties and hardness values
ANSI/ASTM B 100	CA 511	US	Plate, sheet	94.3 min **Cu**, 3.5–4.9 **Sn**, 0.05 **Pb**, 0.30 **Zn**, 0.10 **Fe**, 0.03–0.35 **P**, 99.5 min **Cu + Sn + P**	**Rolled or annealed**: 1/4 in. (6.35 mm) **diam**, 60 ksi (415 MPa) **TS**, 25 ksi (170 MPa) **YS**, 10% in 2 in. (50.8 mm) **El**, 75 **HRB**
CDA 511		US	Strip: rolled	99.5 min **Cu**, 3.5–4.9 **Sn**, 0.05 **Pb**, 0.30 **Zn**, 0.10 **Fe**, 0.03–0.35 **P**, **Sn + P** in **Cu** content	**0.050 mm temper**: 0.040 in. **diam**, 46 ksi (317 MPa) **TS**, 16 ksi (110 MPa) **YS**, 48% in 2 in. (50 mm) **El**, 70 **HRF**; **Three quarter hard temper**: 0.040 in. **diam**, 74 ksi (510 MPa) **TS**, 72 ksi (496 MPa) **YS**, 11% in 2 in. (50 mm) **El**, 84 **HRB**; **Extra spring**: 0.040 in. **diam**, 103 ksi (710 MPa) **TS**, 98 ksi (676 MPa) **YS**, 2% in 2 in. (50 mm) **El**, 95 **HRB**
COPANT 162	C 511 00	COPANT	Wrought products	94.25 min **Cu**, 3.5–4.9 **Sn**, 0.05 **Pb**, 0.30 **Zn**, 0.10 **Fe**, 0.03–0.35 **P**, 99.5 min **Cu + Sn + P**	
NF A 51 108	Cu Sn 4 P	France	Sheet and strip	rem **Cu**, 3–5 **Sn**, 0.1 **Pb**, 0.5 **Zn**, 0.1 **Fe**, 0.35 **P**, 0.3 total, others	**Annealed**: ≤20 mm **diam**, 300 MPa **TS**, 150 MPa **YS**, 50% in 2 in. (50 mm) **El**, 80–110 **HV**; **1/2 hard**: ≤20 mm **diam**, 420 MPa **TS**, 350 MPa **YS**, 30% in 2 in. (50 mm) **El**, 140–170 **HV**; **Hard**: ≤20 mm **diam**, 560 MPa **TS**, 500 MPa **YS**, 15% in 2 in. (50 mm) **El**, 175–205 **HV**
ISO 1634	CuSn4Zn4	ISO	Plate, sheet, strip	rem **Cu**, 3.0–5.0 **Sn**, 0.1 **Pb**, 3.0–5.0 **Zn**, 0.1 **Fe**, 0–0.1 **P**, 0.3 total, others	**Annealed**: all **diam**, 40% in 2 in. (50 mm) **El**, 100 max **HV**; **Strain hardened (HD)**: all **diam**, 640 MPa **TS**, 520 MPa **YS**, 3% in 2 in. (50 mm) **El**

WROUGHT COPPER AND COPPER ALLOYS

specification number	designation	country	product forms	chemical composition	mechanical properties and hardness values
BS 2870	PB101	UK	Sheet, strip, foil	rem **Cu**, 3.0–4.5 **Sn**, 0.02 **Pb**, 0.02–0.40 **P**, 0.20 total, others	**Annealed**: ≤10.0 mm **diam**, 207 MPa **TS**, 40% in 2 in. (50 mm) **El**, 80 max **HV**; **1/2 hard**: ≤10.0 mm **diam**, 324 MPa **TS**, 372 MPa **YS**, 8% in 2 in. (50 mm) **El**, 150 **HV**; **Hard**: ≤6.0 mm **diam**, 379 MPa **TS**, 338 MPa **YS**, 4% in 2 in. (50 mm) **El**, 180 **HV**
BS 2875	PB101	UK	Plate	rem **Cu**, 3.0–4.5 **Sn**, 0.02 **Pb**, 0.02–0.40 **P**, 0.20 total, others	**As manufactured or annealed**: ≥10 mm **diam**, 279 MPa **TS**, 30% in 2 in. (50 mm) **El**; **Hard**: 10–16 mm **diam**, 412 MPa **TS**, 12% in 2 in. (50 mm) **El**
IS 7814	Grade I	India	Sheet, strip, foil	rem **Cu**, 3.0–4.5 **Sn**, 0.02 **Pb**, 0.02–0.40 **P**, 0.2 total, others	**Annealed (sheet)**: all **diam**, 295 MPa **TS**, 40% in 2 in. (50 mm) **El**, 80 max **HV**; **1/2 hard (sheet)**: all **diam**, 430 MPa **TS**, 8% in 2 in. (50 mm) **El**, 130 **HV**; **Hard (sheet)**: all **diam**, 490 MPa **TS**, 4% in 2 in. (50 mm) **El**, 150 **HV**
IS 7814	Grade I	India	Sheet, strip, foil	rem **Cu**, 3.0–4.5 **Sn**, 0.02 **Pb**, 0.02–0.40 **P**, 0.2 total, others	**Annealed (strip)**: all **diam**, 295 MPa **TS**, 40% in 2 in. (50 mm) **El**, 80 max **HV**; **1/2 hard (strip)**: all **diam**, 460 MPa **TS**, 8% in 2 in. (50 mm) **El**, 150 **HV**; **Hard (strip)**: all **diam**, 540 MPa **TS**, 4% in 2 in. (50 mm) **El**, 180 **HV**
COPANT 162	C 509 00	COPANT	Wrought products	95.4 min **Cu**, 2.5–3.8 **Sn**, 0.05 **Pb**, 0.30 **Zn**, 0.10 **Fe**, 0.03–0.30 **P**, 99.5 min **Cu+Sn+P**	
COPANT 162	C 508 00	COPANT	Wrought products	96.03 min **Cu**, 2.6–3.4 **Sn**, 0.05 **Pb**, 0.10 **Fe**, 0.01–0.07 **P**, 99.5 min **Cu+Sn+P**	
COPANT 162	C 526 00	COPANT	Wrought products	93.85 min **Cu**, 2.2–3.3 **Sn**, 0.05 **Pb**, 0.20 **Zn**, 0.10 **Fe**, 1.0–2.0 **Mn**, 0.03–0.35 **P**, 99.5 min **Cu+Sn+P+Mn**	
COPANT 162	C 507 00	COPANT	Wrought products	97.2 min **Cu**, 1.5–2.0 **Sn**, 0.05 **Pb**, 0.10 **Fe**, 0.30 **P**, 99.5 min **Cu+Sn+P**	
ISO 427	Cu Sn2	ISO	Strip, rod, bar, tube, wire	rem **Cu**, 1.0–2.5 **Sn**, 0.05 **Pb**, 0.3 **Zn**, 0.1 **Fe**, 0.3 **Ni**, 0.01–0.3 **P**, 0.3 total, others	
ISO 1638	CuSn2	ISO	Solid product: drawn, on coils or reels	rem **Cu**, 1.0–2.5 **Sn**, 0.05 **Pb**, 0.3 **Zn**, 0.1 **Fe**, 0.3 **Ni**, 0.01–0.3 **P**, 0.3 total, others	**Annealed**: 1–5 mm **diam**, 260 MPa **TS**, 35% in 2 in. (50 mm) **El**; **HC**: 1–3 mm **diam**, 360 MPa **TS**, 10% in 2 in. (50 mm) **El**

WROUGHT COPPER AND COPPER ALLOYS

specification number	designation	country	product forms	chemical composition	mechanical properties and hardness values
ISO 1634	CuSn2	ISO	Plate, sheet, strip	rem **Cu**, 1.0–2.5 **Sn**, 0.05 **Pb**, 0.3 **Zn**, 0.1 **Fe**, 0.3 **Ni**, 0.01–0.3 **P**, 0.3 total, others	**Annealed**: all **diam**, 40% in 2 in. (50 mm) **El**, 75 max **HV**
ANSI/ASTM B 508	C 50500	US	Strip	97.65 min **Cu**, 1.0–1.7 **Sn**, 0.05 **Pb**, 0.30 **Zn**, 0.05 **Fe**, 0.03–0.15 **P**	
CDA 505		US	Wire, strip	99.5 min **Cu**, 1.0–1.7 **Sn**, 0.05 **Pb**, 0.10 **Fe**, 0.03–0.35 **P**, **Sn + P** in **Cu** content	**Half hard temper (rolled strip)**: 0.040 in. **diam**, 55 ksi (379 MPa) **TS**, 16% in 2 in. (50 mm) **El**, 64 **HRB**; **Spring temper (rolled strip)**: 0.040 in. **diam**, 75 ksi (517 MPa) **TS**, 4% in 2 in. (50 mm) **El**, 79 **HRB**; **Hard temper – 80% (wire)**: 0.080 in. **diam**, 79 ksi (545 MPa) **TS**, 1% in 10 in. (250 mm) **El**
COPANT 162	C 505 00	COPANT	Wrought products	97.45 min **Cu**, 1.0–1.7 **Sn**, 0.05 **Pb**, 0.30 **Zn**, 0.10 **Fe**, 0.03–0.35 **P**, 99.5 min **Cu + Sn + P**	
COPANT 162	C 502 00	COPANT	Wrought products	97.96 min **Cu**, 1.0–1.5 **Sn**, 0.05 **Pb**, 0.10 **Fe**, 0.04 **P**, 99.5 min **Cu + Sn + P**	
ISO 1637	CuZn38Sn1	ISO	Solid products: straight lengths	59.5–63.5 **Cu**, 0.5–1.5 **Sn**, 0.2 **Pb**, rem **Zn**, 0.2 **Fe**, 0.5 **Fe + Pb** + total, others	**As cast**: 5 mm **diam**, 390 MPa **TS**, 35% in 2 in. (50 mm) **El**, 120 max **HV**
MIL-R-19631B	MIL-RCu-2	US	Rod	98.0 min **Cu**, 1.0 **Sn**, 0.05 **Pb**, 0.5 **Si**, 0.01 **Al**, 0.5 **Mn**, 0.15 **P**, 0.50 total, others, silver included in the copper content	**As milled**: all **diam**, 172 MPa **TS**
COPANT 162	C 548 00	COPANT	Wrought products	86.85 min **Cu**, 4.0–6.0 **Sn**, 4.0–6.0 **Pb**, 0.30 **Zn**, 0.10 **Fe**, 0.03–0.35 **P**, 99.5 min **Cu + Sn + P + Pb + Zn**	
QQ-B-750	Composition B	US	Bar, plate, rod, sheet, strip, wire, shapes	rem **Cu**, 3.5–4.5 **Sn**, 3.5–4.5 **Pb**, 1.5–4.5 **Zn**, 0.10 **Fe**, 0.01–0.50 **P**, 99.5 min **Cu + Sn + P**	**Hard temper**: 0.25–0.50 in. **diam**, 60 ksi (414 MPa) **TS**, 10% in 2 in. (50 mm) **El**
JIS H 3270	C5441	Japan	Rod and bar	85.5–90.0 **Cu**, 3.5–4.5 **Sn**, 3.5–4.5 **Pb**, 1.5–4.5 **Zn**, 0.01–0.50 **P**	
AMS 4520E		US	Strip	86.7 min **Cu**, 3.5–4.5 **Sn**, 3.5–4.5 **Pb**, 1.5–4.0 **Zn**, 0.10 **Fe**	**Cold rolled**: all **diam**, 16% in 2 in. (50 mm) **El**, 72–79 **HRB**
ANSI/ASTM B 139	C 54400	US	Rod, bar, shapes	85.5 min **Cu**, 3.5–4.5 **Sn**, 3.5–4.5 **Pb**, 1.5–4.5 **Zn**, 0.10 **Fe**, 0.01–0.50 **P**, tin, phosphorous, lead, and zinc included in copper content	**Hard**: >0.063 in. (>1.59 mm) **diam**, 65 ksi (450 MPa) **TS**, 10% in 1 in. (25.4 mm) **El**

WROUGHT COPPER AND COPPER ALLOYS

specification number	designation	country	product forms	chemical composition	mechanical properties and hardness values
ANSI/ASTM B 103	C 54400	US	Plate, sheet, strip, bar	85.95 min **Cu**, 3.5–4.5 **Sn**, 3.5–4.5 **Pb**, 1.5–4.5 **Zn**, 0.10 **Fe**, 0.01–0.50 **P**	**Soft**: >0.039 in. (>0.991 mm) **diam**, 40 ksi (275 MPa) **TS**, 7 **HRB**; **Hard**: >0.039 in. (>0.991 mm) **diam**, 72 ksi (495 MPa) **TS**, 82 **HRB**; **Spring**: >0.039 in. (>0.991 mm) **diam**, 91 ksi (625 MPa) **TS**, 90 **HRB**
COPANT 162	C 544 00	COPANT	Wrought products	85.5 min **Cu**, 3.5–4.5 **Sn**, 3.5–4.5 **Pb**, 1.5–4.5 **Zn**, 0.10 **Fe**, 0.01–0.50 **P**, 99.5 min **Cu+Sn+P+Pb+Zn**	
COPANT 162	C 546 00	COPANT	Wrought products	85.5 min **Cu**, 3.5–4.5 **Sn**, 3.5–4.5 **Pb**, 1.5–4.5 **Zn**, 0.10 **Fe**, 0.50 **P**, 99.5 min **Cu+Sn+P+Pb+Zn**	
ANSI/ASTM B 103	C 53200	US	Plate, sheet, strip, bar	90.1 min **Cu**, 4.0–5.5 **Sn**, 2.5–4.0 **Pb**, 0.20 **Zn**, 0.10 **Fe**, 0.03–0.35 **P**	**Soft**: >0.039 in. (>0.991 mm) **diam**, 40 ksi (275 MPa) **TS**, 7 **HRB**; **Hard**: >0.039 in. (>0.991 mm) **diam**, 72 ksi (495 MPa) **TS**, 82 **HRB**; **Spring**: >0.039 in. (>0.991 mm) **diam**, 91 ksi (625 MPa) **TS**, 90 **HRB**
COPANT 162	C 532 00	COPANT	Wrought products	90.1 min **Cu**, 4.0–5.5 **Sn**, 2.5–4.0 **Pb**, 0.20 **Zn**, 0.10 **Fe**, 0.03–0.35 **P**, 99.5 min **Cu+Sn+P+Pb**	
JIS H 3270	C5341	Japan	Rod and bar	91.85–95.17 **Cu**, 3.5–5.8 **Sn**, 0.8–1.5 **Pb**, 0.03–0.35 **P**, 99.5 min **Cu+Sn+Pb+P**	**Hard**: <13 mm **diam**, 412 MPa **TS**, 10% in 2 in. (50 mm) **El**, 115 min **HV**
ANSI/ASTM B 103	C 53400	US	Plate, sheet, strip, bar	92.6 min **Cu**, 3.5–5.8 **Sn**, 0.8–1.2 **Pb**, 0.30 **Zn**, 0.10 **Fe**, 0.03–0.35 **P**	**Soft**: >0.039 in. (>0.991 mm) **diam**, 40 ksi (275 MPa) **TS**, 7 **HRB**; **Hard**: >0.039 in. (>0.991 mm) **diam**, 72 ksi (495 MPa) **TS**, 82 **HRB**; **Spring**: >0.039 in. (>0.991 mm) **diam**, 91 ksi (625 MPa) **TS**, 90 **HRB**
COPANT 162	C 534 00	COPANT	Wrought products	92.15 min **Cu**, 3.5–5.8 **Sn**, 0.8–1.2 **Pb**, 0.30 **Zn**, 0.10 **Fe**, 0.03–0.35 **P**, 99.5 min **Cu+Sn+P+Pb**	
IS 5743	Cu Al 50 (master alloy)	India	Bar, ingot, shot	48.5–50.5 **Cu**, 49.0–51.0 **Al**	
CDA 625		US	Rod, bar	82.7 **Cu**, 3.5–5.0 **Fe**, 12.5–13.5 **Al**, 2.0 **Mn**, 99.5 min **Cu**, + others	**As extruded**: all **diam**, 100 ksi (690 MPa) **TS**, 1% in 2 in. (50 mm) **El**, 29 **HRB**
COPANT 162	C 625 00	COPANT	Wrought products	79.0–84.0 **Cu**, 3.5–5.0 **Fe**, 12.5–13.5 **Al**, 2.0 **Mn**	
COPANT 162	C 622 00	COPANT	Wrought products	83.2–86.0 **Cu**, 0.02 **Pb**, 0.02 **Zn**, 3.0–4.2 **Fe**, 0.10 **Si**, 11–12 **Al**	

WROUGHT COPPER AND COPPER ALLOYS

specification number	designation	country	product forms	chemical composition	mechanical properties and hardness values
MIL-R-19631B	MIL-RCuAl-B	US	Rod	rem **Cu**, 0.02 **Pb**, 0.02 **Zn**, 3.0–4.25 **Fe**, 0.10 **Si**, 11.0–12.0 **Al**, 0.50 total, others, silver included in copper content	**As milled**: all **diam**, 483 MPa **TS**
ANSI/ASTM B 150	C62400	US	Rod, bar, shapes	rem **Cu**, 0.20 **Sn**, 2.0–4.5 **Fe**, 0.25 **Si**, 10.0–11.5 **Al**, 0.30 **Mn**	**As agreed (round rod and bar)**: <0.50 in. (<12.7 mm) **diam**, 95 ksi (655 MPa) **TS**, 45 ksi (310 MPa) **YS**, 8% in 2 in. (50.8 mm) **El**
CDA 624		US	Rod, bar	86 **Cu**, 0.20 **Sn**, 2.0–4.5 **Fe**, 0.25 **Si**, 10.0–11.5 **Al**, 0.30 **Mn**, 99.5 min **Cu**, + others	**Light anneal (bar)**: 1.0 in. **diam**, 100 ksi (690 MPa) **TS**, 14% in 2 in. (50 mm) **El**, 92 **HRB**; **As extruded (bar)**: 3.0 in. **diam**, 90 ksi (621 MPa) **TS**, 18% in 2 in. (50 mm) **El**, 87 **HRB**; **Half hard–10% (rod)**: 1.0 in. **diam**, 105 ksi (724 MPa) **TS**, 14% in 2 in. (50 mm) **El**, 92 **HRB**
COPANT 162	C 624 00	COPANT	Wrought products	82.8–88.0 **Cu**, 0.2 **Sn**, 2.0–4.5 **Fe**, 0.25 **Si**, 10–11.5 **Al**, 0.30 **Mn**	
AMS 4590		US	Bar, tube, and rod: extruded	rem **Cu**, 0.25 **Sn**, 0.03 **Pb**, 0.30 **Zn**, 4.0–5.5 **Fe**, 0.15 **Si**, 10.0–11.0 **Al**, 0.20 **Co**, 1.5 **Mn**, 4.2–6.0 **Ni**, 0.05 **Cr**, 0.05 each, 0.15 total, others	**Solution heat treated and tempered**: ≤1.00 in. (25.4 mm) **diam**, 135 ksi (931 MPa) **TS**, 100 ksi (690 MPa) **YS**, 6% in 2 in. (50 mm) **El**, 26 min **HRC**
JIS H 3250	C 6241	Japan	Rod and bar	80.0–87.0 **Cu**, 3.0–5.0 **Fe**, 9.0–12.0 **Al**, 0.50–2.0 **Mn**, 0.50–2.0 **Ni**	**As manufactured**: ≥6 mm **diam**, 686 MPa **TS**, 10% in 2 in. (50 mm) **El**, 210 min **HB**
AMS 4640 C		US	Bar, tube, rod, shapes, and forgings	78.0 min **Cu**, 0.20 **Sn**, 0.30 **Zn**, 2.0–3.5 **Fe**, 9.7–10.9 **Al**, 1.5 **Mn**, 4.5–5.5 **Ni**	**Quenched and tempered (forgings)**: ≤1.0 in. **diam**, 110 ksi (758 MPa) **TS**, 70 ksi (483 MPa) **YS**, 10% in 2 in. (50 mm) **El**, 201–248 **HB**; **Hot or cold finished and tempered (bar, tube, rod, and shapes)**: ≤1.0 in. **diam**, 110 ksi (758 MPa) **TS**, 70 ksi (483 MPa) **YS**, 10% in 2 in. (50 mm) **El**, 187–241 **HB**
AMS 4635B		US	Rod, bar, and forgings	84.5 min **Cu**, 0.20 **Sn**, 2.0–4.0 **Fe**, 9.0–11.0 **Al**, 0.30 **Mn**	**As rolled or extruded, and stress relieved**: all **diam**, 155–190 **HB**, 90–96 **HRB**
ANSI/ASTM B 171	C 63000	US	Plate	76.75–82.25 **Cu**, 0.20 **Sn**, 0.30 **Zn**, 2.0–4.0 **Fe**, 0.25 **Si**, 9.0–11.0 **Al**, 1.5 **Mn**, 4.0–5.5 **Ni + Co**	**As manufactured**: <2.0 in. (50.8 mm) **diam**, 90 ksi (620 MPa) **TS**, 36 ksi (250 MPa) **YS**, 10% in 2 in. (50.8 mm) **El**

WROUGHT COPPER AND COPPER ALLOYS

specification number	designation	country	product forms	chemical composition	mechanical properties and hardness values
ASTM F 468	C63000	US	Bolt, screw, stud	78.0 min **Cu**, 0.20 **Sn**, 2.0–4.0 **Fe**, 0.25 **Si**, 9.0–11.0 **Al**, 1.5 **Mn**, 4.0–4.5 **Ni**	**As manufactured**: all **diam**, 100 ksi (690 MPa) **TS**, 50 ksi (345 MPa) **YS**, 5% in 2 in. (50 mm) **El**, 85–100 **HRB**
ASTM F 467	C63000	US	Nut	78.0 min **Cu**, 0.20 **Sn**, 2.0–4.0 **Fe**, 0.25 **Si**, 9.0–11.0 **Al**, 1.5 **Mn**, 4.0–4.5 **Ni**	**As agreed**: all **diam**, 85 min **HRB**
ANSI/ASTM B 150	C63000	US	Rod, bar, shapes	rem **Cu**, 0.20 **Sn**, 0.30 **Zn**, 2.0–4.0 **Fe**, 0.25 **Si**, 9.0–11.0 **Al**, 1.5 **Mn**, 4.0–5.5 **Ni+Co**	**Standard strength (rod and bar)**: <0.50 in. (<12.7 mm) **diam**, 100 ksi (690 MPa) **TS**, 50 ksi (345 MPa) **YS**, 5% in 2 in. (50.8 mm) **El**; **High strength (round rod and bar)**: <1.0 in. (<25.4 mm) **diam**, 110 ksi (760 MPa) **TS**, 68 ksi (470 MPa) **YS**, 10% in 2 in. (50.8 mm) **El**; **Standard strength (shapes)**: all **diam**, 85 ksi (585 MPa) **TS**, 43 ksi (295 MPa) **YS**, 10% in 2 in. (50.8 mm) **El**
ANSI/ASTM B 124	C63000	US	Rod, bar, shapes	rem **Cu**, 0.20 **Sn**, 0.30 **Zn**, 2.0–4.0 **Fe**, 0.25 **Si**, 9.0–11.0 **Al**, 1.5 **Mn**, 4.0–5.5 **Ni+Co**	
ANSI/ASTM B 283	C 63000	US	Die forgings	78.0–85.0 **Cu**, 0.20 **Sn**, 0.30 **Zn**, 2.0–4.0 **Fe**, 0.25 **Si**, 9.0–11.0 **Al**, 1.5 **Mn**, 4.0–5.5 **Ni+Co**	
CDA 630		US	Rod, bar	82 **Cu**, 0.20 **Sn**, 0.30 **Zn**, 2.0–4.0 **Fe**, 0.25 **Si**, 9.0–11.0 **Al**, 1.5 **Mn**, 4.0–5.5 **Ni**, 99.5 min **Cu**, + others	**Half hard–10% (bar)**: 1.0 in. **diam**, 110 ksi (758 MPa) **TS**, 15% in 2 in. (50 mm) **El**, 97 **HRB**; **As extruded (bar)**: 3.0 in. **diam**, 90 ksi (621 MPa) **TS**, 15% in 2 in. (50 mm) **El**, 96 **HRB**; **As extruded (rod)**: 4.0 in. **diam**, 100 ksi (690 MPa) **TS**, 15% in 2 in. (50 mm) **El**, 96 **HRB**
COPANT 162	C 630 00	COPANT	Wrought products	78.0–85.0 **Cu**, 0.20 **Sn**, 0.30 **Zn**, 2.0–4.0 **Fe**, 0.25 **Si**, 9–11 **Al**, 1.5 **Mn**, 4.0–5.5 **Ni**	
QQ–C–450A	630	US	Plate, sheet, strip, bar	76.75 min **Cu**, 0.20 **Sn**, 0.30 **Zn**, 2.0–4.0 **Fe**, 0.25 **Si**, 9.0–11.0 **Al**, 1.5 **Mn**, 4.0–5.5 **Ni**	**Soft temper**: all **diam**, 90 ksi (621 MPa) **TS**, 36 ksi (248 MPa) **YS**, 10% in 2 in. (50 mm) **El**

WROUGHT COPPER AND COPPER ALLOYS

specification number	designation	country	product forms	chemical composition	mechanical properties and hardness values
QQ-C-465A	630	US	Rod, wire, strip, bar, shapes, forgings	78.0–85.0 **Cu**, 0.20 **Sn**, 0.30 **Zn**, 2.0–4.0 **Fe**, 0.25 **Si**, 9.0–11.0 **Al**, 1.5 **Mn**, 4.0–5.5 **Ni**, 99.5 min of above elements	**Annealed (rod and bar)**: 0.50–1.0 in. **diam**, 100 ksi (689 MPa) **TS**, 50 ksi (345 MPa) **YS**, 5% in 2 in. (50 mm) **El**; **As extruded (shapes)**: all **diam**, 85 ksi (586 MPa) **TS**, 42.5 ksi (293 MPa) **YS**, 10% in 2 in. (50 mm) **El**
ISO 1640	CuAl10Fe5Ni5	ISO	Forgings	rem **Cu**, 0.05 **Pb**, 0.5 **Zn**, 2.0–6.0 **Fe**, 8.5–11.5 **Al**, 0–2.0 **Mn**, 4.0–6.0 **Ni**, 0.5 **Pb**+ **Zn**, total	**As cast**: all **diam**, 740 MPa **TS**, 290 MPa **YS**, 10% in 2 in. (50 mm) **El**
ISO 1637	CuAl10Fe5Ni5	ISO	Solid products: straight lengths	rem **Cu**, 0.05 **Pb**, 0.5 **Zn**, 2.0–6.0 **Fe**, 8.5–11.5 **Al**, 0–2.0 **Mn**, 4.0–6.0 **Ni**, 0.5 **Pb**+ **Zn**, total	**As cast**: 10 mm **diam**, 690 MPa **TS**, 290 MPa **YS**, 12% in 2 in. (50 mm) **El**; **HA**: 5–50 mm **diam**, 740 MPa **TS**, 340 MPa **YS**, 10% in 2 in. (50 mm) **El**
ISO 428	CuAl10Fe5Ni5	ISO	Plate, sheet, rod, bar, extruded sections, tube, forgings	rem **Cu**, 0.05 **Pb**, 0.5 **Zn**, 2.0–6.0 **Fe**, 8.5–11.5 **Al**, 0–2.0 **Mn**, 4.0–6.0 **Ni**, 0.5 **Pb**+ **Zn**, total	
MIL-R-19631B	MIL-RCuAl-A2	US	Rod	rem **Cu**, 0.02 **Pb**, 0.02 **Zn**, 1.5 **Fe**, 0.10 **Si**, 9.0–11.0 **Al**, 0.50 total, others, silver included in copper content	**As milled**: all **diam**, 448 MPa **TS**
ANSI/ASTM B 150	C62300	US	Rod, bar, shapes	rem **Cu**, 0.6 **Sn**, 2.0–4.0 **Fe**, 0.25 **Si**, 8.5–11.0 **Al**, 0.50 **Mn**, 1.0 **Ni**+**Co**	**As agreed (round rod and bar)**: <0.50 in. (<12.7 mm) **diam**, 90 ksi (620 MPa) **TS**, 50 ksi (345 MPa) **YS**, 12% in 2 in. (50.8 mm) **El**; **As agreed (shapes)**: all **diam**, 75 ksi (515 MPa) **TS**, 30 ksi (205 MPa) **YS**, 20% in 2 in. (50.8 mm) **El**
ANSI/ASTM B 124	C62300	US	Rod, bar, shapes	rem **Cu**, 0.6 **Sn**, 2.0–4.0 **Fe**, 0.25 **Si**, 8.5–11.0 **Al**, 0.50 **Mn**, 1.0 **Ni**+**Co**	
ANSI/ASTM B 283	C 62300	US	Die forgings	82.2–89.5 **Cu**, 0.6 **Sn**, 2.0–4.0 **Fe**, 0.25 **Si**, 8.5–11.0 **Al**, 0.50 **Mn**, 1.0 **Ni**+**Co**	**As manufactured**: <1.5 in. (<38.1 mm) **diam**, 75 ksi (517 MPa) **TS**, 30 ksi (207 MPa) **YS**, 20% in 2 in. (50 mm) **El**
AS 1567	627D	Australia	Rod, bar, section	76.0 min **Cu**, 0.10 **Sn**, 0.05 **Pb**, 0.40 **Zn**, 4.0–6.0 **Fe**, 0.1 **Si**, 8.5–11.0 **Al**, 0.5 **Mn**, 4.0–6.0 **Ni**, 0.05 **Mg**	**As manufactured**: 6–10 mm **diam**, 700 MPa **TS**, 310 MPa **YS**, 10% in 2 in. (50 mm) **El**
AS 1568	627D	Australia	Forgings	75.3 **Cu**, 0.10 **Sn**, 0.05 **Pb**, 0.40 **Zn**, 4.0–6.0 **Fe**, 0.1 **Si**, 8.5–11.0 **Al**, 0.5 **Mn**, 4.0–6.0 **Ni**, 0.05 **Mg**	**As manufactured**: 6–10 mm **diam**, 700 MPa **TS**, 310 MPa **YS**, 10% in 2 in. (50 mm) **El**
BS 2874	CA 104	UK	Rod, section	rem **Cu**, 0.10 **Sn**, 0.05 **Pb**, 0.40 **Zn**, 4.0–6.0 **Fe**, 0.10 **Si**, 8.5–11.0 **Al**, 0.50 **Mn**, 4.0–6.0 **Ni**, 0.05 **Mg**, 0.50 total, others	**As manufactured**: 6.0–10 mm **diam**, 701 MPa **TS**, 402 MPa **YS**, 10% in 2 in. (50 mm) **El**

WROUGHT COPPER AND COPPER ALLOYS

specification number	designation	country	product forms	chemical composition	mechanical properties and hardness values
BS 2872	CA 104	UK	Forging stock, forgings	rem **Cu**, 0.10 **Sn**, 0.05 **Pb**, 0.40 **Zn**, 4.0–6.0 **Fe**, 0.10 **Si**, 8.5–11.0 **Al**, 0.50 **Mn**, 4.0–6.0 **Ni**, 0.05 **Mg**, 0.50 total, others	**As manufactured or annealed**: 6–10 mm **diam**, 696 MPa **TS**, 392 MPa **YS**, 10% in 2 in. (50 mm) **El**
CDA 618		US	Rod	89 **Cu**, 0.02 **Pb**, 0.02 **Zn**, 0.50–1.5 **Fe**, 0.10 **Si**, 8.5–11.0 **Al**, 99.5 min **Cu**, + others	**Half hard temper–15%**: 1.0 in. **diam**, 85 ksi (586 MPa) **TS**, 23% in 2 in. (50 mm) **El**, 89 **HRB**
CDA 623		US	Rod, bar	87 **Cu**, 0.6 **Sn**, 2.0–4.0 **Fe**, 0.25 **Si**, 8.5–11.0 **Al**, 0.50 **Mn**, 1.0 **Ni**, 99.5 min **Cu**, + others	**As extruded (rod)**: 4.0 in. **diam**, 75 ksi (517 MPa) **TS**, 35% in 2 in. (50 mm) **El**, 80 **HRB**; **As extruded (bar)**: 4.0 in. **diam**, 75 ksi (517 MPa) **TS**, 35% in 2 in. (50 mm) **El**, 80 **HRB**; **Half hard–15% (drawn bar)**: 0.5 in. **diam**, 98 ksi (676 MPa) **TS**, 22% in 2 in. (50 mm) **El**, 89 **HRB**
BS 2872	CA 104	UK	Forging stock, forgings	rem **Cu**, 0.10 **Sn**, 0.05 **Pb**, 0.40 **Zn**, 4.0–6.0 **Fe**, 0.10 **Si**, 8.5–11.0 **Al**, 0.50 **Mn**, 4.0–6.0 **Ni**, 0.05 **Mg**, 0.50 total, others	**As manufactured or annealed**: 6–10 mm **diam**, 696 MPa **TS**, 392 MPa **YS**, 10% in 2 in. (50 mm) **El**
BS 2874	CA104	UK	Rod, section	rem **Cu**, 0.10 **Sn**, 0.05 **Pb**, 0.40 **Zn**, 4.0–6.0 **Fe**, 0.10 **Si**, 8.5–11.0 **Al**, 0.50 **Mn**, 4.0–6.0 **Ni**, 0.05 **Mg**, 0.50 total, others	**As manufactured**: 6.0–10 mm **diam**, 701 MPa **TS**, 402 MPa **YS**, 10% in 2 in. (50 mm) **El**
COPANT 162	C 618 00	COPANT	Wrought products	86.9–91.0 **Cu**, 0.02 **Pb**, 0.02 **Zn**, 0.5–1.5 **Fe**, 0.10 **Si**, 8.5–11.0 **Al**	
COPANT 162	C 623 00	COPANT	Wrought products	82.2–89.5 **Cu**, 0.6 **Sn**, 2.0–4.0 **Fe**, 0.25 **Si**, 8.5–11.0 **Al**, 0.5 **Mn**, 1.0 **Ni**	
IS 6912	10% aluminum bronze	India	Forgings	rem **Cu**, 0.10 **Sn**, 0.05 **Pb**, 0.40 **Zn**, 4.0–6.0 **Fe**, 0.10 **Si**, 8.5–11.0 **Al**, 0.50 **Mn**, 4.0–6.0 **Ni**, 0.05 **Mg**	**As manufactured**: 6–12.5 mm **diam**, 695 MPa **TS**, 400 MPa **YS**, 12% in 2 in. (50 mm) **El**
ISO 1640	CuAl10Fe3	ISO	Forgings	rem **Cu**, 0.05 **Pb**, 0.5 **Zn**, 2.0–4.0 **Fe**, 8.5–11.0 **Al**, 0–2.0 **Mn**, 0–1.0 **Ni**, 0.5 **Pb**+ **Zn**, total	**As cast**: all **diam**, 590 MPa **TS**, 250 MPa **YS**, 15% in 2 in. (50 mm) **El**
ISO 1639	CuAl10Fe3	ISO	Solid product: drawn, on coils or reels	rem **Cu**, 0.05 **Pb**, 0.5 **Zn**, 2.0–4.0 **Fe**, 8.5–11.0 **Al**, 0–2.0 **Mn**, 0–1.0 **Ni**, 0.5 **Pb**+ **Zn**, total	**As cast**: all **diam**, 570 MPa **TS**, 200 MPa **YS**, 15% in 2 in. (50 mm) **El**
ISO 1637	CuAl10Fe3	ISO	Solid products: straight lengths	rem **Cu**, 0.05 **Pb**, 0.5 **Zn**, 2.0–4.0 **Fe**, 8.5–11.0 **Al**, 0–2.0 **Mn**, 0–1.0 **Ni**, 0.5 **Pb**+ **Zn**, total	**As cast**: 10 mm **diam**, 540 MPa **TS**, 200 MPa **YS**, 20% in 2 in. (50 mm) **El**; **HA**: 5–50 mm **diam**, 590 MPa **TS**, 250 MPa **YS**, 15% in 2 in. (50 mm) **El**
ISO 428	CuAl10Fe3	ISO	Rod, bar, extruded sections, forgings	rem **Cu**, 0.05 **Pb**, 0.5 **Zn**, 2.0–4.0 **Fe**, 8.5–11.0 **Al**, 0–0.20 **Mn**, 0–1.0 **Ni**, 0.5 **Pb**+ **Zn**, total	

WROUGHT COPPER AND COPPER ALLOYS

specification number	designation	country	product forms	chemical composition	mechanical properties and hardness values
BS 2875	CA105	UK	Plate	78.0–85.0 **Cu**, 0.10 **Sn**, 0.05 **Pb**, 0.40 **Zn**, 1.5–3.5 **Fe**, 0.15 **Si**, 8.5–10.5 **Al**, 0.5–2.0 **Mn**, 4.0–7.0 **Ni**, 0.05 **Mg**, 0.50 total, others	**As manufactured**: 10–85 mm **diam**, 618 MPa **TS**, 10% in 2 in. (50 mm) **El**
JIS H 3100	C6301	Japan	Sheet, plate, and strip	77.0–84.0 **Cu**, 3.5–6.0 **Fe**, 8.5–10.5 **Al**, 0.50–2.0 **Mn**, 3.5–6.0 **Ni**	**As manufactured**: <50 mm **diam**, 637 MPa **TS**, 15% in 2 in. (50 mm) **El**
JIS 3100	C6280	Japan	Sheet, plate, and strip	78.0–85.0 **Cu**, 1.5–3.5 **Fe**, 8.0–11.0 **Al**, 0.50–2.0 **Mn**, 4.0–7.0 **Ni**	**As manufactured**: <50 mm **diam**, 618 MPa **TS**, 10% in 2 in. (50 mm) **El**
JIS H 3250	C 6191	Japan	Rod and bar	81.0–88.0 **Cu**, 3.0–5.0 **Fe**, 8.0–11.0 **Al**, 0.50–2.0 **Mn**, 0.50–2.0 **Ni**	**As manufactured**: ≥6 mm **diam**, 686 MPa **TS**, 15% in 2 in. (50 mm) **El**, 170 min **HB**
AS 1567	623B	Australia	Rod, bar, section	88.3 **Cu**, 0.10 **Sn**, 0.05 **Pb**, 0.40 **Zn**, 0.1 **Si**, 8.8–10.0 **Al**, 0.5 **Mn**, 0.05 **Mg**	**As manufactured**: <6 mm **diam**, 520 MPa **TS**, 220 MPa **YS**, 22% in 2 in. (50 mm) **El**
AS 1568	623B	Australia	Forgings	84.3 min **Cu**, 0.10 **Sn**, 0.05 **Pb**, 0.40 **Zn**, 0.1 **Si**, 8.8–10.0 **Al**, 0.5 **Mn**, 0.05 **Mg**, 4.0 **Fe+Ti**	**As manufactured**: ≤6 mm **diam**, 520 MPa **TS**, 20% in 2 in. (50 mm) **El**
BS 2874	CA 103	UK	Rod, section	rem **Cu**, 0.10 **Sn**, 0.05 **Pb**, 0.40 **Zn**, 4.0 **Fe**, 0.10 **Si**, 8.8–10.0 **Al**, 0.50 **Mn**, 4.0 **Ni**, 0.05 **Mg**, 0.50 total, others	**As manufactured**: ≥6 mm **diam**, 525 MPa **TS**, 216 MPa **YS**, 22% in 2 in. (50 mm) **El**
BS 2872	CA 103	UK	Forging stock, forgings	rem **Cu**, 0.10 **Sn**, 0.05 **Pb**, 0.40 **Zn**, 0.10 **Si**, 8.8–10.0 **Al**, 0.50 **Mn**, 4.01 **Fe+Ni**, 0.05 **Mg**, 0.50 total, others	**As manufactured or annealed**: ≥6 mm **diam**, 525 MPa **TS**, 211 ksi **YS**, 20% in 2 in. (50 mm) **El**
BS 2874	CA103	UK	Rod, section	rem **Cu**, 0.10 **Sn**, 0.05 **Pb**, 0.40 **Zn**, 4.0 **Fe**, 0.10 **Si**, 8.8–10.0 **Al**, 0.50 **Mn**, 4.0 **Ni**, 0.05 **Mg**, 0.50 total, others	**As manufactured**: ≥6 mm **diam**, 525 MPa **TS**, 216 MPa **YS**, 22% in 2 in. (50 mm) **El**
IS 6912	9% aluminum bronze	India	Forgings	rem **Cu**, 0.10 **Sn**, 0.05 **Pb**, 0.40 **Zn**, 0.10 **Si**, 8.8–10.0 **Al**, 0.50 **Mn**, 4.0 **Fe+Ni**, 0.05 **Mg**	**As manufactured**: ≥6 mm **diam**, 525 MPa **TS**, 215 MPa **YS**, 25% in 2 in. (50 mm) **El**
ANSI/ASTM B 150	C61900	US	Rod, bar, shapes	rem **Cu**, 0.6 **Sn**, 0.8 **Zn**, 3.0–4.5 **Fe**, 8.5–10.0 **Al**, 0.02 **P**	**As agreed (round rod and bar)**: <0.50 in. (<12.7 mm) **diam**, 90 ksi (620 MPa) **TS**, 50 ksi (345 MPa) **YS**, 15% in 2 in. (50.8 mm) **El**; **As agreed (shapes)**: all **diam**, 75 ksi (515 MPa) **TS**, 30 ksi (205 MPa) **YS**, 20% in 2 in. (50.8 mm) **El**
ANSI/ASTM B 124	C61900	US	Rod, bar, shapes	0.6 **Sn**, 0.02 **Pb**, 0.8 **Zn**, 3.0–4.5 **Fe**, 8.5–10.0 **Al**	
ANSI/ASTM B 283	C 61900	US	Die forgings	83.6–88.5 **Cu**, 0.6 **Sn**, 0.02 **Pb**, 0.8 **Zn**, 3.0–4.5 **Fe**, 8.5–10.0 **Al**	**As manufactured**: <1.5 in. (<38.1 mm) **diam**, 75 ksi (517 MPa) **TS**, 30 ksi (207 MPa) **YS**, 20% in 2 in. (50 mm) el

WROUGHT COPPER AND COPPER ALLOYS

specification number	designation	country	product forms	chemical composition	mechanical properties and hardness values
CDA 619		US	Strip: rolled	86.5 **Cu**, 0.6 **Sn**, 0.02 **Pb**, 0.8 **Zn**, 3.0–4.5 **Fe**, 8.5–10.0 **Al**, 99.5 min **Cu**, + others	**Annealed**: 0.040 in. **diam**, 92 ksi (634 MPa) **TS**, 60 ksi (414 MPa) **YS**, 30% in 2 in. (50 mm) **El**, 72 **HR30T**; **Hard temper**: 0.040 in. **diam**, 121 ksi (834 MPa) **TS**, 102 ksi (703 MPa) **YS**, 8% in 2 in. (50 mm) **El**, 98 **HRB**; **Spring temper**: 0.040 in. **diam**, 130 ksi (896 MPa) **TS**, 108 ksi (745 MPa) **YS**, 4% in 2 in. (50 mm) **El**, 100 **HRB**
COPANT 162	C 619 00	COPANT	Wrought products	83.6–88.5 **Cu**, 0.6 **Sn**, 0.02 **Pb**, 0.8 **Zn**, 3.0–4.5 **Fe**, 8.5–10.0 **Al**	
ANSI/ASTM B 150	C63200	US	Rod, bar, shapes	rem **Cu**, 3.0–5.0 **Fe**, 0.10 **Si**, 8.5–9.5 **Al**, 3.5 **Mn**, 4.0–6.6 **Ni+Co**, 0.02 **P**	**As agreed (rod and bar)**: <1.0 in. (<25.4 mm) **diam**, 90 ksi (620 MPa) **TS**, 50 ksi (345 MPa) **YS**, 18% in 2 in. (50.8 mm) **El**
CDA 632		US	Rod, plate, forgings	82 **Cu**, 0.02 **Pb**, 3.0–5.0 **Fe**, 0.10 **Si**, 8.5–9.5 **Al**, 3.5 **Mn**, 4.0–5.0 **Ni**, 99.5 min **Cu**, + others	**Light anneal (rod)**: 1.0 in. **diam**, 105 ksi (724 MPa) **TS**, 22% in 2 in. (50 mm) **El**, 96 **HRB**; **Light anneal (plate)**: 0.250 in. **diam**, 100 ksi (690 MPa) **TS**, 20% in 2 in. (50 mm) **El**, 92 **HRB**; **Light anneal (forgings)**: 5.0 in. **diam**, 100 ksi (690 MPa) **TS**, 22% in 2 in. (50 mm) **El**, 92 **HRB**
COPANT 162	C 632 00	COPANT	Wrought products	75.9–84.5 **Cu**, 0.02 **Pb**, 3.0–5.0 **Fe**, 0.10 **Si**, 8.5–9.5 **Al**, 3.5 **Mn**, 4.0–5.5 **Ni**	
ISO 1640	CuAl9Mn2	ISO	Forgings	rem **Cu**, 0.05 **Pb**, 0.5 **Zn**, 0–1.5 **Fe**, 8.0–10.0 **Al**, 1.5–3.0 **Mn**, 0–0.8 **Ni**, 1.0 **Pb+Zn**, + total, others	**As cast**: all **diam**, 540 MPa **TS**, 200 MPa **YS**, 25% in 2 in. (50 mm) **El**
ISO 1637	CuAl9Mn2	ISO	Solid products: straight lengths	rem **Cu**, 0.05 **Pb**, 0.5 **Zn**, 0–1.5 **Fe**, 8.0–10.0 **Al**, 1.5–3.0 **Mn**, 0–0.8 **Ni**, 1.0 **Pb+Zn**, total	**As cast**: 5 mm **diam**, 490 MPa **TS**, 180 MPa **YS**, 20% in 2 in. (50 mm) **El**; **HA**: 5–50 mm **diam**, 510 MPa **TS**, 200 MPa **YS**, 20% in 2 in. (50 mm) **El**; **HB–strain hardened**: 15–50 mm **diam**, 610 MPa **TS**, 250 MPa **YS**, 15% in 2 in. (50 mm) **El**
ISO 428	CuAl9Mn2	ISO	Rod, bar, forgings	rem **Cu**, 0.05 **Pb**, 0.5 **Zn**, 0–1.5 **Fe**, 8.0–10.0 **Al**, 1.5–3.0 **Mn**, 0–0.8 **Ni**, 1.0 **Pb+Zn**, total	
MIL–B–24059		US	Rod, shapes, forgings	78 min **Cu**, 0.02 **Pb**, 3.0–5.0 **Fe**, 0.10 **Si**, 8.5–9.5 **Al**, 3.5 **Mn**, 4.0–5.5 **Ni**, 0.50 total, others	**Extruded and annealed**: <1 in. **diam**, 621 MPa **TS**, 345 MPa **YS**, 18% in 2 in. (50 mm) **El**

WROUGHT COPPER AND COPPER ALLOYS

specification number	designation	country	product forms	chemical composition	mechanical properties and hardness values
AMS 4632C		US	Rod and bar	87.0 min **Cu**, 1.5 **Fe**, 7.0–10.0 **Al**, 2.0 total, others	**Hot rolled, drawn, or extruded**: ≤0.5 in. **diam**, 80 ksi (552 MPa) **TS**, 5% in 2 in. (50 mm) **El**, 165 min **HB**, 91 min **HRB**
AMS 4630E		US	Rod, bar, tube, and forgings	87.0 min **Cu**, 1.5 **Fe**, 7.0–10.0 **Al**, 2.0 total, others	**Hot rolled and stress relieved**: ≤0.5 in. **diam**, 80 ksi (552 MPa) **TS**, 40 ksi (276 MPa) **YS**, 15% in 2 in. (50 mm) **El**, 130 min **HB**, 80 min **HRB**
JIS Z 3231	DCuAlNi	Japan	Wire: covered electrode	rem **Cu**, 0.02 **Pb**, 2.0–6.0 **Fe**, 1.0 **Si**, 7.0–10.0 **Al**, 2.0 **Mn**, 2.0 **Ni**, 0.50 **Pb+Zn**	**As drawn**: 3.2–6 mm **diam**, 490 MPa **TS**, 13% in 2 in. (50 mm) **El**, 120 **HB**
JIS Z 3231	DCuAlA	Japan	Wire: covered electrode	rem **Cu**, 0.02 **Pb**, 1.5 **Fe**, 1.0 **Si**, 7.0–10.0 **Al**, 2.0 **Mn**, 0.5 **Ni**, 0.50 **Pb+Zn**	**As drawn**: 3.2–6 mm **diam**, 392 MPa **TS**, 15% in 2 in. (50 mm) **El**, 100 **HB**
JIS H 3250	C 6161	Japan	Rod and bar	83.0–90.0 **Cu**, 2.0–4.0 **Fe**, 7.0–10.0 **Al**, 0.50–2.0 **Mn**, 0.50–2.0 **Ni**	**As manufactured**: ≥6 mm **diam**, 588 MPa **TS**, 25% in 2 in. (50 mm) **El**, 130 min **HB**
JIS H 3100	C6161	Japan	Sheet, plate, and strip	83.0–90.0 **Cu**, 2.0–4.0 **Fe**, 7.0–10.0 **Al**, 0.50–2.0 **Mn**, 0.50–2.0 **Ni**	**Annealed**: $<$50 mm **diam**, 490 MPa **TS**, 35% in 2 in. (50 mm) **El**; **1/2 hard**: 0.8–50 mm **diam**, 637 MPa **TS**, 25% in 2 in. (50 mm) **El**; **Hard**: 0.8–50 mm **diam**, 686 MPa **TS**, 10% in 2 in. (50 mm) **El**
ISO 1637	CuAl8	ISO	Solid products: straight lengths	rem **Cu**, 0.1 **Pb**, 0.5 **Zn**, 0.5 **Fe**, 7.0–9.0 **Al**, 0–0.5 **Mn**, 0–0.8 **Ni**, 0.8 **Fe+Pb+Zn**, total	**As cast**: 5 mm **diam**, 440 MPa **TS**, 150 MPa **YS**, 40% in 2 in. (50 mm) **El**; **HB–strain hardened**: 5–50 mm **diam**, 570 MPa **TS**, 390 MPa **YS**, 20% in 2 in. (50 mm) **El**; **HC**: 5–15 mm **diam**, 610 MPa **TS**, 440 MPa **YS**, 18% in 2 in. (50 mm) **El**
ISO 1634	CuAl8	ISO	Plate, sheet, strip	rem **Cu**, 0.1 **Pb**, 0.5 **Zn**, 0.5 **Fe**, 7.0–9.0 **Al**, 0.0–0.5 **Mn**, 0.0–0.8 **Ni**, 0.8 **Fe+Pb+Zn**, total	**As cast**: all **diam**, 470 MPa **TS**, 35% in 2 in. (50 mm) **El**, 125 max **HV**; **Annealed**: all **diam**, 35% in 2 in. (50 mm) **El**, 115 max **HV**
ISO 428	CuAl8	ISO	Plate, sheet, strip, rod, bar, tube, wire, forgings	rem **Cu**, 0.1 **Pb**, 0.5 **Zn**, 0.5 **Fe**, 7.0–9.0 **Al**, 0–0.5 **Mn**, 0–0.8 **Ni**, 0.8 **Fe+Pb+Zn**, total	
AMS 4631D		US	Rod, bar, and forgings	89.00 min **Cu**, 1.60–2.25 **Si**, 6.50–8.50 **Al**, 0.05 each, 0.15 total, others	**Hot rolled, drawn, or extruded, and stress relieved**: ≤0.500 in. (≤12.70 mm) **diam**, 90 ksi (621 MPa) **TS**, 45 ksi (310 MPa) **YS**, 15% in 2 in. (50 mm) **El**, 130 min **HB**

WROUGHT COPPER AND COPPER ALLOYS

specification number	designation	country	product forms	chemical composition	mechanical properties and hardness values
ISO 1637	CuAl8Fe3	ISO	Solid products: straight lengths	rem **Cu**, 0.05 **Pb**, 0.5 **Zn**, 1.5–3.5 **Fe**, 6.5–8.5 **Al**, 0–0.8 **Mn**, 0–1.0 **Ni**, 0.5 **Pb**+ **Zn**, total	**As cast**: 5 mm **diam**, 510 MPa **TS**, 200 MPa **YS**, 25% in 2 in. (50 mm) **El**; **HA**: 5–50 mm **diam**, 540 MPa **TS**, 220 MPa **YS**, 20% in 2 in. (50 mm) **El**; **HB–strain hardened**: 5–15 mm **diam**, 590 MPa **TS**, 250 MPa **YS**, 20% in 2 in. (50 mm) **El**
ISO 1634	CuAl8Fe3	ISO	Plate, sheet, strip	rem **Cu**, 0.05 **Pb**, 0.5 **Zn**, 1.5–3.5 **Fe**, 6.5–8.5 **Al**, 0–0.8 **Mn**, 0–1.0 **Ni**, 0.5 **Pb**+ **Zn**, total	**As cast**: all **diam**, 510 MPa **TS**, 195 MPa **YS**, 28% in 2 in. (50 mm) **El**
ISO 428	CuAl8Fe3	ISO	Plate, sheet, rod, bar	rem **Cu**, 0.05 **Pb**, 0.5 **Zn**, 1.5–3.5 **Fe**, 6.5–8.5 **Al**, 0–0.8 **Mn**, 0–1.0 **Ni**, 0.5 **Pb**+ **Zn**, total	
ANSI/ASTM B 169	C61000	US	Plate, sheet, strip, bar	90.0–93.0 **Cu**, 0.02 **Pb**, 0.20 **Zn**, 0.50 **Fe**, 0.10 **Si**, 6.0–8.5 **Al**	**Soft**: all **diam**, 50 ksi (345 MPa) **TS**, 20 ksi (140 MPa) **YS**, 30% in 2 in. (50.8 mm) **El**; **Hard**: >0.50 in. (>12.7 mm) **diam**, 55 ksi (380 MPa) **TS**, 22 ksi (150 MPa) **YS**, 25% in 2 in. (50.8 mm) **El**
BS 2875	CA 106	UK	Plate	rem **Cu**, 0.10 **Sn**, 0.05 **Pb**, 0.40 **Zn**, 2.0–3.5 **Fe**, 0.15 **Si**, 6.5–8.0 **Al**, 0.50 **Mn**, 0.50 **Ni**, 0.05 **Mg**, 0.50 total, others	**As manufactured**: 10–50 mm **diam**, 481 MPa **TS**, 31% in 2 in. (50 mm) **El**
BS 2874	CA 106	UK	Rod, section	rem **Cu**, 0.10 **Sn**, 0.05 **Pb**, 0.40 **Zn**, 2.0–3.5 **Fe**, 0.15 **Si**, 6.5–8.0 **Al**, 0.50 **Mn**, 0.50 **Ni**, 0.05 **Mg**, 0.50 total, others	**Annealed**: ≥6 mm **diam**, 466 MPa **TS**, 196 MPa **YS**, 30% in 2 in. (50 mm) **El**; **As manufactured**: 6–10 mm **diam**, 539 MPa **TS**, 240 MPa **YS**, 30% in 2 in. (50 mm) **El**
BS 2872	CA 106	UK	Forging stock, forgings	rem **Cu**, 0.10 **Sn**, 0.05 **Pb**, 0.40 **Zn**, 2.0–3.5 **Fe**, 0.15 **Si**, 6.5–8.0 **Al**, 0.50 **Mn**, 0.50 **Ni**, 0.05 **Mg**, 0.50 total, others	**As manufactured or annealed**: 6–10 mm **diam**, 539 MPa **TS**, 245 MPa **YS**, 30% in 2 in. (50 mm) **El**
CDA 610		US	Rod, wire	92 **Cu**, 0.02 **Pb**, 0.20 **Zn**, 0.50 **Fe**, 0.10 **Si**, 6.0–8.5 **Al**, 99.5 min **Cu**, + others	**Soft anneal (rod)**: 1.0 in. **diam**, 70 ksi (483 MPa) **TS**, 65% in 2 in. (50 mm) **El**, 60 **HRB**; **Hard–40% (rod)**: 1.0 in. **diam**, 80 ksi (552 MPa) **TS**, 25% in 2 in. (50 mm) **El**, 85 **HRB**
BS 2874	CA106	UK	Rod, section	rem **Cu**, 0.10 **Sn**, 0.05 **Pb**, 0.40 **Zn**, 2.0–3.5 **Fe**, 0.15 **Si**, 6.5–8.0 **Al**, 0.50 **Mn**, 0.50 **Ni**, 0.05 **Mg**, 0.50 total, others	**Annealed**: ≥6 mm **diam**, 466 MPa **TS**, 196 MPa **YS**, 30% in 2 in. (50 mm) **El**; **As manufactured**: 6–10 mm **diam**, 539 MPa **TS**, 240 MPa **YS**, 30% in 2 in. (50 mm) **El**

WROUGHT COPPER AND COPPER ALLOYS

specification number	designation	country	product forms	chemical composition	mechanical properties and hardness values
BS 2872	CA 106	UK	Forging stock, forgings	rem **Cu**, 0.10 **Sn**, 0.05 **Pb**, 0.40 **Zn**, 2.0–3.5 **Fe**, 0.15 **Si**, 6.5–8.0 **Al**, 0.50 **Mn**, 0.50 **Ni**, 0.05 **Mg**, 0.50 total, others	**As manufactured or annealed**: 6–10 mm **diam**, 539 MPa **TS**, 245 MPa **YS**, 30% in 2 in. (50 mm) **El**
COPANT 162	C 610 00	COPANT	Wrought products	90.0–93.0 **Cu**, 0.02 **Pb**, 0.20 **Zn**, 0.50 **Fe**, 0.10 **Si**, 6.0–8.5 **Al**	
QQ–C–450A	610	US	Plate, sheet, strip, bar	90.27 min **Cu**, 0.02 **Pb**, 0.20 **Zn**, 0.50 **Fe**, 0.10 **Si**, 6.0–8.5 **Al**	**Hard temper**: <0.0625 in. **diam**, 65 ksi (448 MPa) **TS**, 27 ksi (186 MPa) **YS**, 8% in 2 in. (50 mm) **El**; **Soft temper**: all **diam**, 50 ksi (345 MPa) **TS**, 20 ksi (138 MPa) **YS**, 30% in 2 in. (50 mm) **El**
IS 6912	7% aluminum bronze	India	Forgings	rem **Cu**, 0.10 **Sn**, 0.05 **Pb**, 0.40 **Zn**, 2.0–3.5 **Fe**, 0.15 **Si**, 6.5–8.0 **Al**, 0.50 **Mn**, 0.50 **Ni**, 0.05 **Mg**	**As manufactured**: 6–12.5 mm **diam**, 540 MPa **TS**, 240 MPa **YS**, 35% in 2 in. (50 mm) **El**
ANSI/ASTM B 171	C61400	US	Plate	88.0–92.5 **Cu**, 0.01 **Pb**, 0.20 **Zn**, 1.5–3.5 **Fe**, 6.0–8.0 **Al**, 1.0 **Mn**, 0.015 **P**, silver included in copper content.	**As manufactured**: <2.0 in. (<50.8 mm) **diam**, 70 ksi (485 MPa) **TS**, 30 ksi (205 MPa) **YS**, 35% in 2 in. (50.8 mm) **El**
ANSI/ASTM B 169	C61400	US	Plate, sheet, strip, bar	88.0–92.5 **Cu**, 0.01 **Pb**, 0.20 **Zn**, 1.5–3.5 **Fe**, 6.0–8.0 **Al**, 1.0 **Mn**, 0.015 **P**	**Soft**: 2–5 in. (50.8–127 mm) **diam**, 65 ksi (450 MPa) **TS**, 28 ksi (195 MPa) **YS**, 35% in 2 in. (50.8 mm) **El**; **Hard**: 0.50–1.0 in. (12.7–25.4 mm) **diam**, 70 ksi (485 MPa) **TS**, 40 ksi (275 MPa) **YS**, 30% in 2 in. (50.8 mm) **El**
ASTM F 468	C61400	US	Bolt, screw, stud	88.0 min **Cu**, 1.5–3.5 **Fe**, 6.0–8.0 **Al**, 1.0 **Mn**	**As manufactured**: all **diam**, 75 ksi (520 MPa) **TS**, 35 ksi (240 MPa) **YS**, 30% in 2 in. (50 mm) **El**, 70–95 **HRB**
ASTM F 467	C61400	US	Nut	88.0 min **Cu**, 1.5–3.5 **Fe**, 6.0–8.0 **Al**, 1.0 **Mn**	**As agreed**: all **diam**, 70 **HRB**
ANSI/ASTM B 608	CA 614	US	Pipe: welded	88.0–92.5 **Cu**, 0.01 **Pb**, 0.20 **Zn**, 1.5–3.5 **Fe**, 6.0–8.0 **Al**, 1.0 **Mn**, 0.015 **P**, 99.5 min **Cu+Al+Fe+Mn**, Silver included in copper content	**Annealed**: all **diam**, 70 ksi (485 MPa) **TS**
ANSI/ASTM B 150	C61400	US	Rod, bar, shapes	rem **Cu**, 0.015 **Pb**, 0.20 **Zn**, 1.5–3.5 **Fe**, 6.0–8.0 **Al**, 1.0 **Mn**, 0.01 **P**	**As agreed (rod and bar)**: <0.50 in. (<12.7 mm) **diam**, 80 ksi (550 MPa) **TS**, 40 ksi (275 MPa) **YS**, 30% in 2 in. (50.8 mm) **El**
ANSI/ASTM B 315	CA 614	US	Pipe and tube: seamless	88.0–92.5 **Cu**, 0.01 **Pb**, 0.20 **Zn**, 1.5–3.5 **Fe**, 1.0 **Si**, 6.0–8.0 **Al**, 1.0 **Mn** 0.015 **P**	**Annealed**: all **diam**, 65 ksi (447 MPa) **TS**, 28 ksi (193 MPa) **YS**, 30% in 2 in. (50 mm) **El**
ANSI/ASTM B 111	C 61400	US	Tube: seamless, and stock: ferrula	88.0–92.5 **Cu**, 0.01 **Pb**, 0.20 **Zn**, 1.5–3.5 **Fe**, 6.0–8.0 **Al**, 1.0 **Mn**, 0.015 **Pb**	**Annealed**: all **diam**, 70 ksi (480 MPa) **TS**, 30 ksi (205 MPa) **YS**

WROUGHT COPPER AND COPPER ALLOYS

specification number	designation	country	product forms	chemical composition	mechanical properties and hardness values
CDA 614		US	Rod, wire, tube, pipe, shapes, plate, sheet, bar	91 **Cu**, 0.01 **Pb**, 0.20 **Zn**, 1.5–3.5 **Fe**, 6.0–8.0 **Al**, 1.0 **Mn**, 0.015 **P**, 99.5 min **Cu**, + others	**Hard temper (rod)**: 0.500 in. **diam**, 85 ksi (586 MPa) **TS**, 35% in 2 in. (50 mm) **El**, 91 **HRB**; **Soft temper (plate, sheet, rolled bar)**: 0.125 in. **diam**, 82 ksi (565 MPa) **TS**, 40% in 2 in. (50 mm) **El**, 84 **HRB**; **Hard temper (plate, sheet, rolled bar)**: 0.125 in. **diam**, 89 ksi (614 MPa) **TS**, 32% in 2 in. (50 mm) **El**, 87 **HRB**
COPANT 162	C 613 00	COPANT	Wrought products	86.5–93.8 **Cu**, 0.2–0.5 **Sn**, 3.5 **Fe**, 6.0–8.0 **Al**, 0.50 **Mn**, 0.50 **Ni**	
COPANT 162	C 614 00	COPANT	Wrought products	88.0–92.5 **Cu**, 0.01 **Pb**, 0.20 **Zn**, 1.5–3.5 **Fe**, 6.0–8.0 **Al**, 1.0 **Mn**, 0.015 **P**	
QQ–C–450A	614	US	Plate, sheet, strip, bar	86.78 min **Cu**, 0.01 **Pb**, 0.20 **Zn**, 1.5–3.5 **Fe**, 6.0–8.0 **Al**, 1.0 **Mn**, 0.015 **P**	**Hard temper**: all **diam**, 80 ksi (552 MPa) **TS**, 45 ksi (310 MPa) **YS**, 25% in 2 in. (50 mm) **El**; **Soft temper**: all **diam**, 72 ksi (496 MPa) **TS**, 32 ksi (221 MPa) **YS**, 35% in 2 in. (50 mm) **El**
QQ–C–465A	614	US	Rod, wire, strip, bar, shapes, forgings	88.0–92.5 **Cu**, 0.01 **Pb**, 0.20 **Zn**, 1.5–3.5 **Fe**, 6.0–8.0 **Al**, 1.0 **Mn**, 0.015 **P**, 99.5 min of above elements	**Hard temper (flat wire and strip)**: 0.125 in. **diam**, 85 ksi (586 MPa) **TS**, 55 ksi (379 MPa) **YS**, 30% in 2 in. (50 mm) **El**; **Annealed (flat wire and strip)**: 0.125 in. **diam**, 75 ksi (517 MPa) **TS**, 42 ksi (290 MPa) **YS**, 30% in 2 in. (50 mm) **El**; **Stress relieved (rod)**: 0.5 in. **diam**, 80 ksi (552 MPa) **TS**, 40 ksi (276 MPa) **YS**, 30% in 2 in. (50 mm) **El**
QQ–C–450A	613	US	Plate, sheet, strip, bar	86.5 min **Cu**, 0.20–0.50 **Sn**, 3.5 **Fe**, 6.0–8.0 **Al**, 0.50 **Mn**, 0.50 **Ni**	**Hard temper**: all **diam**, 85 ksi (586 MPa) **TS**, 55 ksi (379 MPa) **YS**, 30% in 2 in. (50 mm) **El**; **Soft temper**: all **diam**, 72 ksi (496 MPa) **TS**, 32 ksi (221 MPa) **YS**, 30% in 2 in. (50 mm) **El**
ASTM F 468	C64200	US	Bolt, screw, stud	88.65 min **Cu**, 0.20 **Sn**, 0.50 **Zn**, 0.30 **Fe**, 1.5–2.2 **Si**, 6.3–7.6 **Al**, 0.10 **Mn**, 0.25 **Ni**, 0.15 **As**	**As manufactured**: all **diam**, 75 ksi (520 MPa) **TS**, 35 ksi (240 MPa) **YS**, 10% in 2 in. (50 mm) **El**, 75–95 **HRB**

WROUGHT COPPER AND COPPER ALLOYS

specification number	designation	country	product forms	chemical composition	mechanical properties and hardness values
ASTM F 467	C64200	US	Nut	88.65 min **Cu**, 0.20 **Sn**, 0.50 **Pb**, 0.50 **Zn**, 0.30 **Fe**, 1.5–2.2 **Si**, 6.3–7.6 **Al**, 0.10 **Mn**, 0.25 **Ni**, 0.15 **As**	**As agreed**: all **diam**, 75 min **HRB**
ANSI/ASTM B 150	C64200	US	Rod, bar, shapes	rem **Cu**, 0.20 **Sn**, 0.50 **Zn**, 0.30 **Fe**, 1.5–2.2 **Si**, 6.3–7.6 **Al**, 0.10 **Mn**, 0.25 **Ni+Co**, 0.05 **P**, 0.15 **As**	**As agreed (rod and bar)**: >0.50 in. (>12.7 mm) **diam**, 90 ksi (620 MPa) **TS**, 45 ksi (310 MPa) **YS**, 9% in 2 in. (50.8 mm) **El**
ANSI/ASTM B 283	C 64200	US	Die forgings	88.2–92.2 **Cu**, 0.20 **Sn**, 0.05 **Pb**, 0.50 **Zn**, 0.30 **Fe**, 1.5–2.2 **Si**, 6.3–7.6 **Al**, 0.10 **Mn**, 0.25 **Ni+Co**, 0.15 **As**	**As manufactured**: <1.5 in. (<38.1 mm) **diam**, 70 ksi (483 MPa) **TS**, 25 ksi (172 MPa) **YS**, 35% in 2 in. (50 mm) **El**
CDA 642		US	Rod, bar, forgings	91.2 nom **Cu**, 0.20 **Sn**, 0.05 **Pb**, 0.50 **Zn**, 0.30 **Fe**, 1.5–2.2 **Si**, 6.3–7.6 **Al**, 0.10 **Mn**, 0.25 **Ni**, 0.15 **As**, 99.5 min **Cu**, + others	**As extruded (rod, bar)**: 0.750 in. **diam**, 75 ksi (517 MPa) **TS**, 32% in 2 in. (50 mm) **El**, 77 **HRB**; **Light anneal (rod)**: 0.500 in. **diam**, 92 ksi (634 MPa) **TS**, 22% in 2 in. (50 mm) **El**, 98 **HRB**; **As forged (forgings)**: 2.0 in. **diam**, 79 ksi (545 MPa) **TS**, 30% in 2 in. (50 mm) **El**, 78 **HRB**
COPANT 162	C 642 00	COPANT	Wrought products	88.2–92.2 **Cu**, 0.20 **Sn**, 0.05 **Pb**, 0.50 **Zn**, 0.30 **Fe**, 1.5–2.2 **Si**, 6.3–7.6 **Al**, 0.10 **Mn**, 0.25 **Ni**	
QQ-C-465A	642	US	Rod, wire, strip, bar, shapes, forgings	80.0–93.0 **Cu**, 0.20 **Sn**, 0.05 **Pb**, 0.50 **Zn**, 0.30 **Fe**, 1.5–2.2 **Si**, 6.3–7.6 **Al**, 0.10 **Mn**, 0.25 **Ni**, 0.15 **As**, 99.5 min of above elements	**Annealed or stress relieved**: 0.50 in. **diam**, 80 ksi (552 MPa) **TS**, 40 ksi (276 MPa) **YS**, 15% in 2 in. (50 mm) **El**; **Hot rolled or forged and turned**: all **diam**, 70 ksi (483 MPa) **TS**, 30 ksi (207 MPa) **YS**, 15% in 2 in. (50 mm) **El**; **As extruded or annealed**: all **diam**, 70 ksi (483 MPa) **TS**, 30 ksi (207 MPa) **YS**, 15% in 2 in. (50 mm) **El**
QQ-C-425A	642	US	Rod	80.0–93.0 **Cu**, 0.20 **Sn**, 0.05 **Pb**, 0.50 **Zn**, 0.30 **Fe**, 1.5–2.2 **Si**, 6.3–7.6 **Al**, 0.10 **Mn**, 0.25 **Ni**, 0.15 **As**, 0.5 total, others	**Annealed or stress relieved**: 0.50 in. **diam**, 80 ksi (552 MPa) **TS**, 40 ksi (276 MPa) **YS**, 15% in 2 in. (50 mm) **El**
BS 2871	CA 102	UK	Tube	rem **Cu**, 6.0–7.5 **Al**, 1.0–2.5 **Ni+Fe+Mn**, 0.50	**As drawn**: 150 **HV**; **Annealed**: 110 max **HV**
BS 2875	CA 102	UK	Plate	rem **Cu**, 6.0–7.5 **Al**, 1.0–2.5 **Fe+Ni+Mn**, 0.50 total, others	**As manufactured**: 10–50 mm **diam**, 481 MPa **TS**, 31% in 2 in. (50 mm) **El**

WROUGHT COPPER AND COPPER ALLOYS

specification number	designation	country	product forms	chemical composition	mechanical properties and hardness values
BS 2875	CA102	UK	Plate	rem **Cu**, 6.0–7.5 **Al**, 1.0–2.5 **Fe + Ni + Mn**, 0.50 total, others	**As manufactured**: 10–50 mm **diam**, 481 MPa **TS**, 31% in 2 in. (50 mm) **El**
IS 1545	CuAl7	India	Tube	rem **Cu**, 6.0–7.5 **Al**, 1.0–2.5 **Fe + Mn + Ni**, 0.50 total, others	**Annealed**: all **diam**, 461 MPa **TS**; **As drawn**: all **diam**, 481 MPa **TS**
ANSI/ASTM B 150	C64210	US	Rod, bar, shapes	rem **Cu**, 0.20 **Sn**, 0.50 **Zn**, 0.30 **Fe**, 1.5–2.0 **Si**, 6.3–7.0 **Al**, 0.10 **Mn**, 0.25 **Ni + Co**, 0.05 **P**, 0.15 **As**	**As agreed (rod and bar)**: >1.0 in. (>25.4 mm) **diam**, 90 ksi (620 MPa) **TS**, 45 ksi (310 MPa) **YS**, 9% in 2 in. (50.8 mm) **El**
ANSI/ASTM B 124	C64210	US	Rod, bar, shapes	rem **Cu**, 0.20 **Sn**, 0.05 **Pb**, 0.50 **Zn**, 0.30 **Fe**, 1.5–2.0 **Si**, 6.3–7.0 **Al**, 0.10 **Mn**, 0.25 **Ni + Co**, 0.15 **P**	
ANSI/ASTM B 124	C64200	US	Rod, bar, shapes	rem **Cu**, 0.20 **Sn**, 0.05 **Pb**, 0.50 **Zn**, 0.30 **Fe**, 1.5–2.0 **Si**, 6.3–7.0 **Al**, 0.10 **Mn**, 0.25 **Ni + Co**, 0.15 **P**	
ANSI/ASTM B 283	C 64210	US	Die forgings	88.2–92.2 **Cu**, 0.20 **Sn**, 0.05 **Pb**, 0.50 **Zn**, 0.30 **Fe**, 1.5–2.0 **Si**, 6.3–7.0 **Al**, 0.10 **Mn**, 0.25 **Ni + Co**, 0.15 **As**	
AS 1567	643D	Australia	Rod, bar, section	87.5 min **Cu**, 1.0 **Fe**, 1.5–3.0 **Si**, 5.5–7.5 **Al**, 0.5 **Mn**	**As manufactured**: ≤50 mm **diam**, 550 MPa **TS**, 260 MPa **YS**, 20% in 2 in. (50 mm) **El**
ANSI/ASTM B 395	C60800	US	Tube: seamless	93.0 min **Cu**, 0.10 **Pb**, 0.10 **Fe**, 5.0–6.5 **Al**, 0.02–0.35 **As**, silver included in copper content.	**Annealed**: <0.048 in. (<1.21 mm) **diam**, 50 ksi (345 MPa) **TS**, 19 ksi (130 MPa) **YS**, 12% in 2 in. (50.8 mm) **El**
ANSI/ASTM B 359	C60800	US	Tube: seamless	93.0 min **Cu**, 0.10 **Pb**, 0.10 **Fe**, 5.0–6.5 **Al**, 0.02–0.35 **As**	**Annealed**: all **diam**, 50 ksi (345 MPa) **TS**, 19 ksi (130 MPa) **YS**
ANSI/ASTM B 111	C 60800	US	Tube: seamless, and stock: ferrule	93.0 min **Cu**, 0.10 **Pb**, 0.10 **Fe**, 5.0–6.5 **Al**, 0.02–0.35 **As**	**Annealed**: all **diam**, 50 ksi (345 MPa) **TS**, 19 ksi (130 MPa) **YS**
CDA 608		US	Tube	93.0 min **Cu**, 0.10 **Pb**, 0.10 **Fe**, 5.0–6.5 **Al**, 0.02–0.35 **As**	**0.025 mm temper**: 1.0 in. **diam**, 60 ksi (414 MPa) **TS**, 55% in 2 in. (50 mm) **El**, 77 **HRF**
COPANT 162	C 608 00	COPANT	Wrought products	92.5–94.8 **Cu**, 0.10 **Pb**, 0.10 **Fe**, 5.0–6.5 **Al**, 0.2–0.35 **As**	
COPANT 527	Cu Al 6	COPANT	Tube	93 min **Cu**, 0.10 **Pb**, 0.10 **Fe**, 5.0–6.5 **Al**, 0.35 **As**	**Annealed**: 0.8–3.4 mm **diam**, 345 MPa **TS**, 125 MPa **YS**

WROUGHT COPPER AND COPPER ALLOYS

specification number	designation	country	product forms	chemical composition	mechanical properties and hardness values
ANSI/ASTM B 169	C60600	US	Plate, sheet, strip, bar	92.0–96.0 **Cu**, 0.50 **Fe**, 4.0–7.0 **Al**	**Soft**: all **diam**, 45 ksi (310 MPa) **TS**, 17 ksi (115 MPa) **YS**, 40% in 2 in. (50.8 mm) **El**; **Hard**: >0.50 in. (>12.7 mm) **diam**, 50 ksi (345 MPa) **TS**, 20 ksi (140 MPa) **YS**, 30% in 2 in. (50.8 mm) **El**
COPANT 162	C 606 00	COPANT	Wrought products	92–96 **Cu**, 0.50 **Fe**, 4.0–7.0 **Al**	
QQ-C-450A	606	US	Plate, sheet, strip, bar	92.0 min **Cu**, 0.50 **Fe**, 4.0–7.0 **Al**	**Hard temper**: <0.0625 in. **diam**, 60 ksi (414 MPa) **TS**, 24 ksi (165 MPa) **YS**, 8% in 2 in. (50 mm) **El**; **Soft temper**: all **diam**, 45 ksi (310 MPa) **TS**, 17 ksi (117 MPa) **YS**, 40% in 2 in. (50 mm) **El**
QQ-C-465A	606	US	Rod, wire, strip, bar, shapes, forgings	92.0–96.0 **Cu**, 0.50 **Fe**, 4.0–7.0 **Al**, 99.5 min of above elements	**Stress relieved (rod)**: 0.5 max in. **diam**, 80 ksi (552 MPa) **TS**, 40 ksi (276 MPa) **YS**, 30% in 2 in. (50 mm) **El**; **Hard temper (flat wire and strip)**: all **diam**, 60 ksi (414 MPa) **TS**, 24 ksi (165 MPa) **YS**, 25% in 2 in. (50 mm) **El**; **Annealed (flat wire and strip)**: all **diam**, 45 ksi (310 MPa) **TS**, 17 ksi (117 MPa) **YS**, 40% in 2 in. (50 mm) **El**
ISO 428	CuAl5	ISO	Plate, sheet, strip, rod, bar, tube, wire	rem **Cu**, 0.1 **Pb**, 0.5 **Zn**, 0.5 **Fe**, 4.0–6.5 **Al**, 0–0.5 **Mn**, 0–0.8 **Ni**, 0.04 **As**, 0.8 **Fe+Pb+Zn**, total	
ISO 1635	CuAl5	ISO	Tube	rem **Cu**, 0.1 **Pb**, 0.5 **Zn**, 0.5 **Fe**, 4.0–6.5 **Al**, 0–0.5 **Mn**, 0–0.8 **Ni**, 0–0.4 **As**, 0.8 **Fe+Pb+Zn**, total	**As cast**: all **diam**, 410 MPa **TS**, 100 MPa **YS**, 40% in 2 in. (50 mm) **El**; **Annealed**: all **diam**, 45% in 2 in. (50 mm) **El**, 110 max **HV**
ISO 1634	CuAl5	ISO	Plate, sheet, strip	rem **Cu**, 0.1 **Pb**, 0.5 **Zn**, 0.5 **Fe**, 4.0–6.5 **Al**, 0–0.5 **Mn**, 0–0.8 **Ni**, 0–0.4 **As**, 0.8 **Fe+Pb+Zn**, total	**As cast**: all **diam**, 410 MPa **TS**, 45% in 2 in. (50 mm) **El**, 110 max **HV**; **Annealed**: all **diam**, 45% in 2 in. (50 mm) **El**, 100 max **HV**
BS 2870	CA 101	UK	Sheet, strip, foil	rem **Cu**, 0.02 **Pb**, 4.5–5.5 **Al**, 0.50 total, others	**As manufactured**: 2.7–10.0 mm **diam**, 238 MPa **TS**, 40% in 2 in. (50 mm) **El**
VSM 11557	Cu Al 5 As	Switzerland	Tube	93.0 min **Cu**, 0.02 **Pb**, 0.3 **Zn**, 0.1 **Fe**, 4.5–5.5 **Al**, 0.2 **Ni**, 0.2–0.35 **As**, 0.50 total, others	**Quenched**: all **diam**, 314 MPa **TS**, 108 MPa **YS**, 45% in 2 in. (50 mm) **El**, 80 **HB**; **1/2 hard**: all **diam**, 343 MPa **TS**, 137 MPa **YS**, 40% in 2 in. (50 mm) **El**, 100 **HB**
COPANT 162	C 636 00	COPANT	Wrought products	93.2–96.3 **Cu**, 0.20 **Sn**, 0.05 **Pb**, 0.35 **Zn**, 0.10 **Fe**, 0.7–1.3 **Si**, 3.0–4.0 **Al**, 0.15 **Ni**, 0.15 **As**	

WROUGHT COPPER AND COPPER ALLOYS

specification number	designation	country	product forms	chemical composition	mechanical properties and hardness values
COPANT 162	C 634 00	COPANT	Wrought products	94.9–97.1 **Cu**, 0.20 **Sn**, 0.05 **Pb**, 0.35 **Zn**, 0.10 **Fe**, 0.25–0.45 **Si**, 2.6–3.2 **Al**, 0.15 **Ni**, 0.15 **As**	
CDA 638		US	Strip: rolled	95 **Cu**, 0.05 **Pb**, 0.50 **Zn**, 0.05 **Fe**, 1.5–2.1 **Si**, 2.5–3.1 **Al**, 0.25–0.55 **Co**, 0.10 **Mn**, 0.10 **Ni**, 99.5 min **Cu**, + others	**Soft anneal**: 0.040 in. **diam**, 82 ksi (565 MPa) **TS**, 54 ksi (372 MPa) **YS**, 36% in 2 in. (50 mm) **El**, 86 **HRB**; **Half hard temper**: 0.040 in. **diam**, 106 ksi (731 MPa) **TS**, 88 ksi (607 MPa) **YS**, 15% in 2 in. (50 mm) **El**, 96 **HRB**; **Hard temper**: 0.040 in. **diam**, 120 ksi (827 MPa) **TS**, 101 ksi (696 MPa) **YS**, 7% in 2 in. (50 mm) **El**, 99 **HRB**
COPANT 162	C 638 00	COPANT	Wrought products	93.0–95.7 **Cu**, 0.05 **Pb**, 0.50 **Zn**, 0.05 **Fe**, 1.5–2.1 **Si**, 2.5–3.1 **Al**, 0.25–0.55 **Co**, 0.10 **Mn**, 0.10 **Ni**	
COPANT 162	C 607 00	COPANT	Wrought products	94.6–96.0 **Cu**, 1.7–2.0 **Sn**, 0.01 **Pb**, 2.3–2.9 **Al**, 99.5 min **Cu + Al + Fe + Ag**	
VSM 11557	Cu Zn 21 Al 2	Switzerland	Tube	76.0–78.0 **Cu**, 0.07 **Pb**, rem **Zn**, 0.06 **Fe**, 1.8–2.3 **Al**, 0.5 **Ni**, 0.015 **P**, 0.02–0.045 **As**, 0.05 **P + As**, 030 total, others	**Quenched**: all **diam**, 333 MPa **TS**, 108 MPa **YS**, 50% in 2 in. (50 mm) **El**, 70 **HB**; **1/2 hard**: all **diam**, 392 MPa **TS**, 157 MPa **YS**, 40% in 2 in. (50 mm) **El**, 95 **HB**
CDA 613		US	Rod, tube, pipe, shapes, plate, sheet, bar	92.7 **Cu**, 0.20–0.50 **Sn**, 3.5 **Fe**, 6.0–8.0 **Al**, 0.50 **Mn**, 0.50 **Ni**, 99.5 min **Cu**, + others	**Hard temper–25% (rod)**: 0.500 in. **diam**, 85 ksi (586 MPa) **TS**, 35% in 2 in. (50 mm) **El**, 91 **HRB**; **Soft anneal (plate, sheet, drawn bar)**: 0.125 in. **diam**, 80 ksi (552 MPa) **TS**, 40% in 2 in. (50 mm) **El**, 82 **HRB**
IS 5743	Cu Si 15	India	Bar, ingot, shot	83.5–85.5 **Cu**, 14.0–16.0 **Si**	
IS 5743	Cu Si 10	India	Bar, ingot, shot	88.5–90.5 **Cu**, 9.0–11.0 **Si**	
COPANT 162	C 656 00	COPANT	Wrought products	94.0 min **Cu**, 1.5 **Sn**, 0.02 **Pb**, 1.5 **Zn**, 0.50 **Fe**, 2.8–4.0 **Si**, 0.01 **Al**, 1.5 **Mn**	
MIL-R-19631B	MIL-RCuSi-A	US	Rod	94.0 min **Cu**, 1.5 **Sn**, 0.02 **Pb**, 1.5 **Zn**, 0.5 **Fe**, 2.8–4.0 **Si**, 0.01 **Al**, 1.5 **Mn**, 0.50 total, others, silver included in copper content	**As milled**: all **diam**, 345 MPa **TS**
ANSI/ASTM B 96	CA 658	US	Plate, sheet	94.8 min **Cu**, 0.05 **Pb**, 0.25 **Fe**, 2.8–3.8 **Si**, 0.50–1.3 **Mn**, 0.6 **Ni**	

WROUGHT COPPER AND COPPER ALLOYS

specification number	designation	country	product forms	chemical composition	mechanical properties and hardness values
ANSI/ASTM B 97	C 65500	US	Plate, sheet, strip, bar	94.8 min **Cu**, 0.05 **Pb**, 1.5 **Zn**, 0.8 **Fe**, 2.8–3.8 **Si**, 0.50–1.5 **Mn**, 0.6 **Ni**.	**Hard**: all **diam**, 85 ksi (586 MPa) **TS**, 88 **HRB**; **Spring**: all **diam**, 102 ksi (703 MPa) **TS**, 94 **HRB**; **Hot rolled**: all **diam**, 55 ksi (380 MPa) **TS**, 72 **HRF**
ANSI/ASTM B 98	C 65500	US	Rod, bar, shapes	94.8 min **Cu**, 0.05 **Pb**, 1.5 **Zn**, 0.8 **Fe**, 2.8–3.8 **Si**, 0.50–1.5 **Mn**, 0.6 **Ni**.	**Soft**: all **diam**, 52 ksi (360 MPa) **TS**, 15 ksi (105 MPa) **YS**, 35% in 1 in. (25.4 mm) **El**; **Half hard (rod and bar)**: 2 in. (50.8 mm) **diam**, 70 ksi (485 MPa) **TS**, 38 ksi (260 MPa) **YS**, 20% in 1 in. (25.4 mm) **El**; **Hard (rod and bar)**: 1/4 in. (6.35 mm) **diam**, 90 ksi (615 MPa) **TS**, 55 ksi (380 MPa) **YS**, 8% in 1 in. (25.4 mm) **El**
ANSI/ASTM B 98	C 65800	US	Rod, bar, shapes	94.8 min **Cu**, 0.05 **Pb**, 0.25 **Fe**, 2.8–3.8 **Si**, 0.01 **Al**, 0.50–1.3 **Mn**, 0.6 **Ni**	**Soft**: all **diam**, 52 ksi (360 MPa) **TS**, 15 ksi (105 MPa) **YS**, 35% in 1 in. (25.4 mm) **El**; **Half hard (rod and bar)**: 2 in. (50.8 mm) **diam**, 70 ksi (485 MPa) **TS**, 38 ksi (260 MPa) **YS**, 20% in 1 in. (25.4 mm) **El**; **Hard (rod and bar)**: 1/4 in. (6.35 mm) **diam**, 90 ksi (615 MPa) **TS**, 55 ksi (380 MPa) **YS**, 8% in 1 in. (25.4 mm) **El**
ANSI/ASTM B 99	C 65500	US	Wire	94.8 min **Cu**, 0.05 **Pb**, 1.5 **Zn**, 0.8 **Fe**, 2.8–3.8 **Si**, 0.50–1.5 **Mn**, 0.6 **Ni**	**Annealed**: 1/4 in. (6.35 mm) **diam**, 38 ksi (260 MPa) **TS**, 40% in 2 in. (50.8 mm) **El**; **Hard**: 1/2 in. (12.7 mm) **diam**, 90 ksi (620 MPa) **TS**, 8% in 2 in. (50.8 mm) **El**; **Spring**: 1/4 in. (6.35 mm) **diam**, 100 ksi (690 MPa) **TS**, 6% in 2 in. (50.8 mm) **El**
ANSI/ASTM B 96	CA 655	US	Plate, sheet	94.8 min **Cu**, 0.05 **Pb**, 1.5 **Zn**, 0.8 **Fe**, 2.8–3.8 **Si**, 0.50–1.5 **Mn**, 0.6 **Ni**	
ANSI/ASTM B 315	CA 658	US	Pipe and tube: seamless	94.8 min **Cu**, 0.05 **Pb**, 0.25 **Fe**, 2.8–3.8 **Si**, 0.01 **Al**, 0.50–1.5 **Mn**, 0.6 **Ni**	**Annealed**: all **diam**, 50 ksi (345 MPa) **TS**, 15 ksi (103 MPa) **YS**, 35% in 2 in. (50 mm) **El**
ANSI/ASTM B 315	CA 655	US	Pipe and tube: seamless	94.8 min **Cu**, 0.05 **Pb**, 1.5 **Zn**, 0.8 **Fe**, 2.8–3.8 **Si**, 0.50–1.5 **Mn**, 0.6 **Ni**	**Annealed**: all **diam**, 50 ksi (345 MPa) **TS**, 15 ksi (103 MPa) **YS**, 35% in 2 in. (50 mm) **El**

WROUGHT COPPER AND COPPER ALLOYS

specification number	designation	country	product forms	chemical composition	mechanical properties and hardness values
ANSI/ASTM B 283	C 65500	US	Die forgings	94.8 min **Cu**, 0.05 **Pb**, 1.5 **Zn**, 0.8 **Fe**, 2.8–3.8 **Si**, 1.5 **Mn**, 0.6 **Ni+Co**	
ANSI/ASTM B 124	C 65500	US	Rod, bar, shapes	rem **Cu**, 0.05 **Pb**, 1.5 **Zn**, 0.8 **Fe**, 2.8–3.8 **Si**, 1.5 **Mn**, 0.6 **Ni+Co**	
ASTM F 468	C 65500	US	Bolt, screw, stud	94.8 min **Cu**, 0.05 **Pb**, 1.5 **Zn**, 0.8 **Fe**, 2.8–3.8 **Si**, 1.5 **Mn**, 0.6 **Ni**	**As manufactured (test specimen)**: all **diam**, 50 ksi (340 MPa) **TS**, 15 ksi (105 MPa) **YS**, 20% in 2 in. (50 mm) **El**, 60–80 **HRB**
ASTM F 467	C 65500	US	Nut	94.8 min **Cu**, 0.05 **Pb**, 1.5 **Zn**, 0.8 **Fe**, 2.8–3.8 **Si**, 1.5 **Mn**, 0.6 **Ni**	**As agreed**: all **diam**, 60 min **HRB**
ANSI/ASTM B 100	CA 655	US	Plate, sheet	94.8 min **Cu**, 0.05 **Pb**, 1.5 **Zn**, 0.8 **Fe**, 2.8–3.8 **Si**, 1.5 **Mn**, 0.6 **Ni**	**Rolled or annealed**: 1/4 in. (6.35 mm) **diam**, 60 ksi (415 MPa) **TS**, 25 ksi (170 MPa) **YS**, 10% in 2 in. (50.8 mm) **El**, 75 **HRB**
AS 1567	655A	Australia	Rod, bar, section	91.25 min **Cu**, 0.05 **Pb**, 1.5 **Zn**, 0.8 **Fe**, 2.8–3.8 **Si**, 1.5 **Mn**, 0.6 **Ni**	**As manufactured**: 6–20 mm **diam**, 480 MPa **TS**, 15% in 2 in. (50 mm) **El**; **Soft annealed**: 6–70 mm **diam**, 370 MPa **TS**, 40% in 2 in. (50 mm) **El**
AS 1566	655A	Australia	Plate, bar, sheet, strip, foil	91.25 **Cu**, 0.05 **Pb**, 1.5 **Zn**, 0.8 **Fe**, 2.8–3.8 **Si**, 1.5 **Mn**, 0.6 **Ni**	**Annealed (sheet, strip)**: all **diam**, 360 MPa **TS**, 50% in 2 in. (50 mm) **El**, 85 **HV**; **Strain hardened to 1/4 hard (sheet, strip)**: all **diam**, 370 MPa **TS**, 110–140 **HV**
AS 1568	655A	Australia	Forgings	91.25 **Cu**, 0.05 **Pb**, 1.5 **Zn**, 0.8 **Fe**, 2.8–3.8 **Si**, 1.5 **Mn**, 0.6 **Ni**	**As manufactured**: $\geq$6 mm **diam**, 340 MPa **TS**, 20% in 2 in. (50 mm) **El**
CSA HC.4.7	HC.4.S3	Canada	Bar, sheet, strip, plate	94.8 min **Cu**, 0.05 **Pb**, 1.5 **Zn**, 1.6 **Fe**, 2.8–3.8 **Si**, 1.5 **Mn**, 0.6 **Ni**, 0.5 total, others	**Annealed**: 70 mm **diam**, 359 MPa **TS**; **Quarter hard**: all **diam**, 427 MPa **TS**; **Hot rolled**: all **diam**, 379 MPa **TS**
COPANT 162	C 655 00	COPANT	Wrought products	94.8 min **Cu**, 0.05 **Pb**, 1.5 **Zn**, 0.8 **Fe**, 2.8–3.8 **Si**, 1.5 **Mn**, 0.6 **Ni**	
COPANT 521	Cu Si 3.5	COPANT	Tube	93 min **Cu**, 0.05 **Pb**, 1.5 **Zn**, 0.8 **Fe**, 2.8–3.8 **Si**, 1.5 **Mn**, 0.6 **Ni**	**Annealed**: 10.3–324 mm OD **diam**, 345 MPa **TS**, 110 MPa **YS**, 35% in 2 in. (50 mm) **El**
COPANT 528	Cu Si 3.5	COPANT	Tube	rem **Cu**, 0.05 **Pb**, 1.5 **Zn**, 0.8 **Fe**, 2.8–3.8 **Si**, 1.5 **Mn**, 0.6 **Ni**	**Annealed**: 0.45–9.7 mm **diam**, 345 MPa **TS**, 105 MPa **YS**, 35% in 2 in. (50 mm) **El**

WROUGHT COPPER AND COPPER ALLOYS

specification number	designation	country	product forms	chemical composition	mechanical properties and hardness values
QQ-C-591E	655	US	Rod, wire, shapes, forgings, strip, sheet, bar, plate	94.8 min **Cu**, 0.05 **Pb**, 1.5 **Zn**, 0.8 **Fe**, 2.8–3.8 **Si**, 1.5 **Mn**, 0.6 **Ni**, 99.5 min **Cu** + sum of named elements	**Soft temper**: all **diam**, 52 ksi (359 MPa) **TS**, 15 ksi (103 MPa) **YS**, 35% in 2 in. (50 mm) **El**; **Half hard temper**: 2 in. **diam**, 70 ksi (483 MPa) **TS**, 38 ksi (262 MPa) **YS**, 17% in 2 in. (50 mm) **El**; **Hard temper**: 0.25 in. **diam**, 90 ksi (621 MPa) **TS**, 55 ksi (379 MPa) **YS**, 8% in 2 in. (50 mm) **El**
JIS Z 3231	DCuSiB	Japan	Wire: covered electrode	92.0 min **Cu**, 0.02 **Pb**, 2.5–4.0 **Si**, 3.0 **Mn**, 0.30 **P**, 0.50 **Pb**+**Al**+**Ni**+**Zn**	**As drawn**: 3.2–6 mm **diam**, 274 MPa **TS**, 20% in 2 in. (50 mm) **El**
AMS 4616C		US	Bar, tube, and forgings	87.9 min **Cu**, 1.50–4.00 **Zn**, 1.00–2.00 **Fe**, 2.40–4.00 **Si**, 1.00 **Mn**, 0.10 **P**	**Hot rolled and stress relieved**: all **diam**, 56 ksi (386 MPa) **TS**, 20 ksi (138 MPa) **YS**, 30% in 2 in. (50 mm) **El**, 90 min **HB**, 55 min **HRB**
COPANT 162	C 658 00	COPANT	Wrought products	94.8 min **Cu**, 0.05 **Pb**, 0.6 **Fe**, 2.8–3.6 **Si**, 0.01 **Al**, 0.5–1.3 **Mn**	
COPANT 521	Cu Si 3 Mn 1	COPANT	Tube	94 min **Cu**, 0.05 **Pb**, 1.5 **Zn**, 0.8 **Fe**, 2.8–3.6 **Si**, 0.01 **Al**, 0.5–1.35 **Mn**	**Annealed**: 10.3–324 mm OD **diam**, 345 MPa **TS**, 110 MPa **YS**, 35% in 2 in. (50 mm) **El**
COPANT 528	Cu Si 3 Mn 1	COPANT	Tube	rem **Cu**, 0.05 **Pb**, 0.8 **Fe**, 2.8–3.6 **Si**, 0.01 **Al**, 0.5–1.35 **Mn**	**Annealed**: 0.45–9.7 mm **diam**, 105 MPa **YS**, 35% in 2 in. (50 mm) **El**
AMS 4665A		US	Tube: seamless	95.7 min **Cu**, 0.05 **Pb**, 0.25 **Fe**, 2.8–3.5 **Si**, 99.5 min **Cu** + total named elements, 1.5 **Mn** or **Zn**	**Soft annealed**: all **diam**, 50 ksi (345 MPa) **TS**, 35% in 2 in. (50 mm) **El**
AMS 4615D		US	Rod and bar	94.80 min **Cu**, 0.70 **Sn**, 0.05 **Pb**, 2.80–3.50 **Si**, 1.60 **Fe**, **Mn**, or **Zn**	**Cold finished, hard**: $\leq$0.250 in. ($\leq$6.35 mm) **diam**, 85 ksi (586 MPa) **TS**, 50 ksi (345 MPa) **YS**, 8% in 2 in. (50 mm) **El**
ANSI/ASTM B 98	C 66100	US	Rod, bar, shapes	94.0 min **Cu**, 0.20–0.8 **Pb**, 1.5 **Zn**, 0.25 **Fe**, 2.8–3.5 **Si**, 1.5 **Mn**	**Soft**: all **diam**, 52 ksi (360 MPa) **TS**, 15 ksi (105 MPa) **YS**, 35% in 1 in. (25.4 mm) **El**; **Half hard (rod and bar)**: 2 in. (50.8 mm) **diam**, 70 ksi (485 MPa) **TS**, 38 ksi (260 MPa) **YS**, 20% in 1 in. (25.4 mm) **El**; **Hard (rod and bar)**: 1/4 in. (6.35 mm) **diam**, 90 ksi (615 MPa) **TS**, 55 ksi (380 MPa) **YS**, 8% in 1 in. (25.4 mm) **El**
ASTM F 468	C 66100	US		94.0 min **Cu**, 0.20–0.8 **Pb**, 1.5 **Zn**, 0.25 **Fe**, 2.8–3.5 **Si**, 1.5 **Mn**	**As manufactured**: all **diam**, 70 ksi (480 MPa) **TS**, 35 ksi (240 MPa) **YS**, 75–95 **HRB**
ASTM F 467	C 66100	US	Nut	94.0 min **Cu**, 0.20–0.8 **Pb**, 1.5 **Zn**, 0.25 **Fe**, 2.8–3.5 **Si**, 1.5 **Mn**	**As agreed**: all **diam**, 75 min **HRB**

WROUGHT COPPER AND COPPER ALLOYS

specification number	designation	country	product forms	chemical composition	mechanical properties and hardness values
COPANT 162	C 661 00	COPANT	Wrought products	94.0 min **Cu**, 0.20–0.8 **Pb**, 1.5 **Zn**, 0.25 **Fe**, 2.8–3.5 **Si**, 1.5 **Mn**	
QQ–C–591E	661	US	Rod, wire, shapes, forgings, strip, sheet, bar, plate	94.0 min **Cu**, 0.2–0.8 **Pb**, 0.25 **Fe**, 2.8–3.5 **Si**, 1.5 **Mn**, 99.5 min **Cu** + sum of named elements	**Soft temper**: all **diam**, 52 ksi (359 MPa) **TS**, 15 ksi (103 MPa) **YS**, 35% in 2 in. (50 mm) **El**; **Half hard temper**: 2 max in. **diam**, 70 ksi (483 MPa) **TS**, 38 ksi (262 MPa) **YS**, 17% in 2 in. (50 mm) **El**; **Hard temper**: 0.25 max in. **diam**, 90 ksi (621 MPa) **TS**, 55 ksi (379 MPa) **YS**, 8% in 2 in. (50 mm) **El**
MIL–T–8231A		US	Tube	rem **Cu**, 0.05 **Pb**, 0.24 **Fe**, 2.8–3.5 **Si**, 1.5 **Mn** or **Zn**, 0.5 total, others	**Annealed**: all **diam**, 345 MPa **TS**, 35% in 2 in. (50 mm) **El**
ISO 1637	CuSi3Mn1	ISO	Solid products: straight lengths	rem **Cu**, 0.03 **Pb**, 0.5 **Zn**, 0.3 **Fe**, 2.7–3.5 **Si**, 0.7–1.5 **Mn**, 0.3 **Ni**, 1.0 **Fe**+**Ni**+**Pb**+**Zn**, total	**M**: 5 mm **diam**, 410 MPa **TS**, 120 MPa **YS**, 30% in 2 in. (50 mm) **El**
ISO 1634	CuSi3Mn1	ISO	Plate, sheet, strip	rem **Cu**, 0.03 **Pb**, 0.5 **Zn**, 0.3 **Fe**, 2.7–3.5 **Si**, 0.7–1.5 **Mn**, 0.3 **Ni**, 1.0 **Fe**+**Ni**+**Pb**+**Zn**, total	**M**: all **diam**, 410 MPa **TS**, 120 MPa **YS**, 40% in 2 in. (50 mm) **El**
ISO R1187	CuSi3Mn1	ISO	Plate, sheet, strip, rod, bar, tube, wire, forgings	rem **Cu**, 0.03 **Pb**, 0.5 **Zn**, 0.3 **Fe**, 2.7–3.5 **Si**, 0.7–1.5 **Mn**, 0.3 **Ni**, 1.0 **Fe**+**Ni**+**Pb**+**Zn**, total	
CDA 655		US	Rod, wire, tube, plate, sheet, strip	97 **Cu**, 0.05 **Pb**, 1.5 **Zn**, 0.8 **Fe**, 2.3–3.8 **Si**, 1.5 **Mn**, 0.6 **Ni**, 99.5 min **Cu**, + others	**Hard temper (plate, sheet, strip)**: 0.040 in. **diam**, 94 ksi (648 MPa) **TS**, 8% in 2 in. (50 mm) **El**, 93 **HRB**; **Extra hard–50% (rod)**: 1.0 in. **diam**, 108 ksi (745 MPa) **TS**, 13% in 2 in. (50 mm) **El**, 95 **HRB**; **Spring–80 (wire)**: 0.080 in. **diam**, 145 ksi (1000 MPa) **TS**, 3% in 2 in. (50 mm) **El**
BS 2870	CS 101	UK	Sheet, strip, foil	rem **Cu**, 0.25 **Fe**, 2.75–3.25 **Si**, 0.75–1.25 **Mn**, 0.50 total, others	**As manufactured**: 2.7–10.0 mm **diam**, 259 MPa **TS**, 50% in 2 in. (50 mm) **El**
BS 2873	CS 101	UK	Wire	rem **Cu**, 0.25 **Fe**, 2.75–3.25 **Si**, 0.75–1.25 **Mn**, 0.50 total, others	
BS 2875	CS 101	UK	Plate	rem **Cu**, 0.25 **Fe**, 2.75–3.25 **Si**, 0.75–1.25 **Mn**, 0.50 total, others	**As manufactured**: $\geq$10 mm **diam**, 339 MPa **TS**, 31% in 2 in. (50 mm) **El**; **Annealed**: $\geq$10 mm **diam**, 319 MPa **TS**, 40% in 2 in. (50 mm) **El**
BS 2872	CS 101	UK	Forging stock, forgings	rem **Cu**, 0.25 **Fe**, 2.75–3.25 **Si**, 0.75–1.25 **Mn**, 0.50 total, others	**As manufactured or annealed**: $\geq$6 mm **diam**, 339 MPa **TS**, 20% in 2 in. (50 mm) **El**

WROUGHT COPPER AND COPPER ALLOYS

specification number	designation	country	product forms	chemical composition	mechanical properties and hardness values
BS 2874	CS 101	UK	Rod	rem **Cu**, 0.25 **Fe**, 2.75–3.25 **Si**, 0.75–1.25 **Mn**, 0.50 total, others	**Annealed**: 6–70 mm **diam**, 367 MPa **TS**, 40% in 2 in. (50 mm) **El**; **As manufactured**: 6–20 mm **diam**, 539 MPa **TS**, 15% in 2 in. (50 mm) **El**
IS 6912	Copper–silicon	India	Forgings	rem **Cu**, 0.25 **Fe**, 2.75–3.25 **Si**, 0.75–1.25 **Mn**, 0.25 total, others	**As manufactured**: ≥6 mm **diam**, 345 MPa **TS**, 25% in 2 in. (50 mm) **El**
JIS H 3300	C 6561	Japan	Pipe and tube: seamless	93.5–95.5 **Cu**, 0.50–1.5 **Sn**, 2.5–3.5 **Si**	**Annealed**: 10–60 mm **diam**, 343 MPa **TS**, 55% in 2 in. (50 mm) **El**; **1/2 hard**: 10–60 mm **diam**, 441 MPa **TS**, 2 in. **El**; **Hard**: 10–60 mm **diam**, 539 MPa **TS**
COPANT 521	Cu Si 1.5	COPANT	Tube	96 min **Cu**, 0.05 **Pb**, 0.8 **Fe**, 0.8–2.6 **Si**, 0.7 **Mn**	**Annealed**: 10.3–324 mm OD **diam**, 275 MPa **TS**, 70 MPa **YS**, 35% in 2 in. (50 mm) **El**; **Tempered**: <102 mm od **diam**, 345 MPa **TS**, 275 MPa **YS**, 7% in 2 in. (50 mm) **El**
JIS Z 3231	DCuSiA	Japan	Wire: covered electrode	93.0 min **Cu**, 0.02 **Pb**, 1.0–2.0 **Si**, 3.0 **Mn**, 0.30 **P**, 0.50 **Pb+Al+Ni+Zn**	**As drawn**: 3.2–6 mm **diam**, 245 MPa **TS**, 22% in 2 in. (50 mm) **El**
ANSI/ASTM B 97	C 65100	US	Plate, sheet, strip, bar	96.0 min **Cu**, 0.05 **Pb**, 1.5 **Zn**, 0.8 **Fe**, 0.8–2.0 **Si**, 0.7 **Mn**,	**Quarter hard**: all **diam**, 42 ksi (290 MPa) **TS**, 48 **HRB**; **Hard**: all **diam**, 60 ksi (415 MPa) **TS**, 74 **HRB**; **Spring**: all **diam**, 71 ksi (490 MPa) **TS**, 81 **HRB**
ANSI/ASTM B 98	C 65100	US	Rod, bar, shapes	96.0 min **Cu**, 0.05 **Pb**, 1.5 **Zn**, 0.8 **Fe**, 0.8–2.0 **Si**, 0.7 **Mn**	**Soft**: all **diam**, 40 ksi (275 MPa) **TS**, 12 ksi (85 MPa) **YS**, 30% in 1 in. (25.4 mm) **El**; **Hard (rod)**: 1/2 in. (12.7 mm) **diam**, 65 ksi (450 MPa) **TS**, 35 ksi (240 MPa) **YS**, 8% in 1 in. (25.4 mm) **El**; **Extra hard (rod)**: 1/2 in. (12.7 mm) **diam**, 85 ksi (585 MPa) **TS**, 55 ksi (380 MPa) **YS**, 6% in 1 in. (25.4 mm) **El**
ANSI/ASTM B 99	C 65100	US	Wire	96.0 min **Cu**, 0.05 **Pb**, 1.5 **Zn**, 0.8 **Fe**, 0.8–2.0 **Si**, 0.7 **Mn**,	**Annealed**: 1/4 in. (6.35 mm) **diam**, 38 ksi (260 MPa) **TS**, 40% in 2 in. (50.8 mm) **El**; **Hard**: 1/2 in. (12.7 mm) **diam**, 90 ksi (620 MPa) **TS**, 8% in 2 in. (50.8 mm) **El**; **Spring**: 1/4 in. (6.35 mm) **diam**, 100 ksi (690 MPa) **TS**, 6% in 2 in. (50.8 mm) **El**

WROUGHT COPPER AND COPPER ALLOYS

specification number	designation	country	product forms	chemical composition	mechanical properties and hardness values
ANSI/ASTM B 315	CA 651	US	Pipe and tube: seamless	96.0 min **Cu**, 0.05 **Pb**, 1.5 **Zn**, 0.8 **Fe**, 0.8–2.0 **Si**, 0.7 **Mn**	**Annealed**: all **diam**, 40 ksi (275 MPa) **TS**, 10 ksi (69 MPa) **YS**, 35% in 2 in. (50 mm) **El**; **Hard drawn**: all **diam**, 50 ksi (345 MPa) **TS**, 40 ksi (275 MPa) **YS**, 7% in 2 in. (50 mm) **El**
ASTM F 468	C 65100	US	Bolt, screw, stud	96.0 min **Cu**, 0.05 **Pb**, 1.5 **Zn**, 0.8 **Fe**, 0.8–2.0 **Si**, 0.7 **Mn**	**As manufactured**: 0.250–0.750 in. (0.635–1.905 mm) **diam**, 70 ksi (480 MPa) **TS**, 53 ksi (365 MPa) **YS**, 8% in 2 in. (50 mm) **El**, 75–95 **HRB**
ASTM F 467	C 65100	US	Nut	96.0 min **Cu**, 0.05 **Pb**, 1.5 **Zn**, 0.8 **Fe**, 0.8–2.0 **Si**, 0.7 **Mn**	**As agreed**: all **diam**, 75 min **HRB**
CDA 651		US	Rod, wire, tube	98.5 nom **Cu**, 0.05 **Pb**, 1.5 **Zn**, 0.8 **Fe**, 0.8–2.0 **Si**, 0.7 **Mn**, 99.5 min **Cu**, + others	**0.035 mm temper (rod)**: 1.0 in. **diam**, 40 ksi (276 MPa) **TS**, 50% in 2 in. (50 mm) **El**, 55 **HRF**; **Half hard temper (wire)**: 0.080 in. **diam**, 80 ksi (552 MPa) **TS**, 15% in 2 in. (50 mm) **El**; **Hard drawn–35% (tube)**: 1.0 in. **diam**, 65 ksi (448 MPa) **TS**, 20% in 2 in. (50 mm) **El**, 75 **HRB**
CSA HC.4.7	HC.4.S2 (651)	Canada	Bar, sheet, strip, plate	96.0 min **Cu**, 0.05 **Pb**, 1.5 **Zn**, 0.8 **Fe**, 0.8–2.0 **Si**, 0.7 **Mn**, 0.5 total, others	**Annealed**: 70 mm **diam**, 262 MPa **TS**; **Quarter hard**: all **diam**, 331 MPa **TS**; **Full hard**: all **diam**, 414 MPa **TS**
COPANT 162	C 651 00	COPANT	Wrought products	96.0 min **Cu**, 0.05 **Pb**, 1.5 **Zn**, 0.8 **Fe**, 0.8–2.0 **Si**, 0.7 **Mn**	
COPANT 528	Cu Si 1.5	COPANT	Tube	rem **Cu**, 0.05 **Pb**, 1.5 **Zn**, 0.8 **Fe**, 0.8–2.0 **Si**, 0.7 **Mn**	**Annealed**: 0.45–9.7 mm **diam**, 275 MPa **TS**, 70 MPa **YS**, 35% in 2 in. (50 mm) **El**; **Tempered to full hard**: all (0.45–9.7 mm) **diam**, 345 MPa **TS**, 275 MPa **YS**, 7% in 2 in. (50 mm) **El**
QQ-C-591E	651	US	Rod, wire, shapes, forgings, strip, sheet, bar, plate	96.0 min **Cu**, 0.05 **Pb**, 1.5 **Zn**, 0.8 **Fe**, 0.8–2.0 **Si**, 0.7 **Mn**, 0.6 **Ni**, 99.5 min **Cu** + sum of named elements	**Soft temper**: all **diam**, 40 ksi (276 MPa) **TS**, 12 ksi (83 MPa) **YS**, 30% in 2 in. (50 mm) **El**; **Half hard temper**: 2 in. **diam**, 55 ksi (379 MPa) **TS**, 20 ksi (138 MPa) **YS**, 12% in 2 in. (50 mm) **El**; **Hard temper**: 2 in. **diam**, 60 ksi (414 MPa) **TS**, 40 ksi (276 MPa) **YS**, 10% in 2 in. (50 mm) **El**

WROUGHT COPPER AND COPPER ALLOYS

specification number	designation	country	product forms	chemical composition	mechanical properties and hardness values
QQ-C-591E	692	US	Rod, wire, shapes, forgings, strip, sheet, bar, plate	89.0–91.0 **Cu**, 0.05 **Pb**, rem **Zn**, 0.05 **Fe**, 0.8–1.5 **Si**, 99.5 **Cu** + sum of named elements	**Soft temper**: all **diam**, 40 ksi (276 MPa) **TS**, 12 ksi (83 MPa) **YS**, 30% in 2 in. (50 mm) **El**; **Half hard temper**: 2 max in. **diam**, 55 ksi (379 MPa) **TS**, 20 ksi (138 MPa) **YS**, 12% in 2 in. (50 mm) **El**; **Hard temper**: 2 max **diam**, 60 ksi (414 MPa) **TS**, 40 ksi (276 MPa) **YS**, 10% in 2 in. (50 mm) **El**
COPANT 162	C 649 00	COPANT	Wrought products	96.2 min **Cu**, 1.2–1.6 **Sn**, 0.05 **Pb**, 0.20 **Zn**, 0.10 **Fe**, 0.8–1.2 **Si**, 0.10 **Al**, 0.10 **Ni**	
ANSI/ASTM B 411	c 64700	US	Rod, bar, wire	97.0 min **Cu**, 0.10 **Pb**, 0.50 **Zn**, 0.10 **Fe**, 0.40–0.8 **Si**, 1.6–2.2 **Ni+Co**	**Precipitation heat treated (round)**: 0.094–1.5 in. (2.38–38.1 mm) **diam**, 90 ksi (620 MPa) **TS**, 75 ksi (515 MPa) **YS**, 8% in 1 in. (25.4 mm) **El**; **Precipitation heat treated (square)**: 0.188–1.0 in. (4.78–25.4 mm) **diam**, 90 ksi (620 MPa) **TS**, 75 ksi (515 MPa) **YS**, 8% in 1 in. (25.4 mm) **El**; **Precipitation heat treated (rectangle)**: 0.188–1.5 in. (4.78–38.1 mm) **diam**, 80 ksi (550 MPa) **TS**, 70 ksi (485 MPa) **YS**, 8% in 1 in. (25.4 mm) **El**
ANSI/ASTM B 422	C 64700	US	Sheet, strip	97.0 min **Cu**, 0.10 **Pb**, 0.50 **Zn**, 0.10 **Fe**, 0.40–0.8 **Si**, 1.6–2.2 **Ni+Co**	**Precipitation heat treated**: all **diam**, 80 ksi (550 MPa) **TS**, 65 ksi (450 MPa) **YS**, 3% in 2 in. (50.8 mm) **El**
COPANT 162	C 647 00	COPANT	Wrought products	97.0 min **Cu**, 0.10 **Pb**, 0.50 **Zn**, 0.10 **Fe**, 0.4–0.8 **Si**, 1.6–2.2 **Ni**	
QQ-C-591E	647	US	Rod, wire, shapes, forgings, strip, sheet, bar, plate	97.0 min **Cu**, 0.10 **Pb**, 0.50 **Zn**, 0.10 **Fe**, 0.40–0.8 **Si**, 1.6–2.2 **Ni**, 99.5 min **Cu** + sum of named elements	**Cold forming**: 0.09375–0.25 in. **diam**, 60 ksi (414 MPa) **TS**, 50 ksi (345 MPa) **YS**, 10% in 2 in. (50 mm) **El**; **Hard temper**: 1.5 in. **diam**, 90 ksi (621 MPa) **TS**, 80 ksi (552 MPa) **YS**, 8% in 2 in. (50 mm) **El**
IS 5743	Cu Mn 30 (master alloy)	India	Bar, ingot, shot	99.5 min **Cu**, 29.0–31.0 **Mn**, manganese included in copper content	
BS 2871	CZ 127	UK	Tube	81.0–86.0 **Cu**, 0.10 **Sn**, 0.05 **Pb**, rem **Zn**, 0.25 **Fe**, 0.70–1.20 **Al**, 0.10 **Mn**, 0.80–1.40 **Ni**, 0.50 total, others	**Annealed**: 430 MPa **TS**, 40% in 2 in. (50 mm) **El**, 140 max **HV**; **As drawn**: 550 MPa **TS**, 150 **HV**

WROUGHT COPPER AND COPPER ALLOYS

specification number	designation	country	product forms	chemical composition	mechanical properties and hardness values
CDA 694		US	Rod	80.0–83.0 **Cu**, 0.30 **Pb**, rem **Zn**, 0.20 **Fe**, 3.5–4.5 **Si**, 0.50 total, others	**Soft anneal**: 0.500 in. **diam**, 90 ksi (621 MPa) **TS**, 20% in 2 in. (50 mm) **El**, 85 **HRB**; **Soft anneal**: 2.0 in. **diam**, 80 ksi (552 MPa) **TS**, 25% in 2 in. (50 mm) **El**, 85 **HRB**; **Eighth hard temper–7%**: 0.75 in. **diam**, 100 ksi (690 MPa) **TS**, 21% in 2 in. (50 mm) **El**, 95 **HRB**
ANSI/ASTM B 371	CA 694	US	Rod	80.0–83.0 **Cu**, 0.30 **Pb**, rem **Zn**, 0.20 **Fe**, 3.5–4.5 **Si**	**Annealed**: <1.0 in. (<25.4 mm) **diam**, 80 ksi (550 MPa) **TS**, 40 ksi (250 MPa) **YS**, 15% in 1 in. (25.4 mm) **El**
ANSI/ASTM B 111	C 68700	US	Tube: seamless, and stock: ferrule	76.0–79.0 **Cu**, 0.07 **Pb**, rem **Zn**, 0.06 **Fe**, 1.8–2.5 **Al**, 0.02–0.10 **As**	**Annealed**: all **diam**, 50 ksi (345 MPa) **TS**, 18 ksi (125 MPa) **YS**
DS 3003	5217	Denmark	Rolled, drawn, extruded, and forged products	76.0–79.0 **Cu**, 0.07 **Pb**, 20 **Zn**, 0.07 **Fe**, 1.8–2.3 **Al**, 0.020–0.035 **As**, 0.3 others, total	
ANSI/ASTM B 543	C 68700	US	Tube: welded	76.0–79.0 **Cu**, 0.07 **Pb**, rem **Zn**, 0.06 **Fe**, 1.8–2.5 **Al**, 0.02–0.10 **As**, Silver included in copper content	**As welded from annealed strip**: all **diam**, 50 ksi (345 MPa) **TS**, 18 ksi (125 MPa) **YS**; **Welded and annealed**: all **diam**, 50 ksi (345 MPa) **TS**, 18 ksi (125 MPa) **YS**; **Fully finished–annealed**: all **diam**, 50 ksi (345 MPa) **TS**, 18 ksi (125 MPa) **YS**
COPANT 527	Cu Zn 21 Al 2 As	COPANT	Tube	76–79 **Cu**, 0.07 **Pb**, rem **Zn**, 0.06 **Fe**, 1.8–2.5 **Al**, 0.02–0.10 **As**, 0.15 total, others	**Annealed**: 0.8–3.4 mm **diam**, 345 MPa **TS**, 125 MPa **YS**
CDA 687		US	Tube	76.0–79.0 **Cu**, 0.07 **Pb**, rem **Zn**, 0.06 **Fe**, 1.8–2.5 **Al**, 0.02–0.10 **As**, 0.15 total, others	**0.025 mm temper**: 1.0 in. **diam**, 60 ksi (414 MPa) **TS**, 55% in 2 in. (50 mm) **El**, 77 **HRF**
ISO 1635	CuZn20Al2	ISO	Tube	76.0–79.0 **Cu**, 0.07 **Pb**, rem **Zn**, 0.07 **Fe**, 1.8–2.5 **Al**, 0.015 **P**, 0.02–0.06 **As**, 0.3 **Fe + Pb + P**, total	**Annealed**: all **diam**, 45% in 2 in. (50 mm) **El**, 90 max **HV**; **Strain hardened**: all **diam**, 490 MPa **TS**, 20% in 2 in. (50 mm) **El**, 120 **HV**
ISO 1634	CuZn20Al2	ISO	Plate, sheet, strip	76.0–79.0 **Cu**, 0.07 **Pb**, rem **Zn**, 0.07 **Fe**, 1.8–2.5 **Al**, 0.015 **P**, 0.02–0.06 **As**, 0.3 **Fe + Pb + P**, total	**M**: all **diam**, 330 MPa **TS**, 42% in 2 in. (50 mm) **El**, 100 max **HV**; **HA**: all **diam**, 390 MPa **TS**, 30% in 2 in. (50 mm) **El**, 110 **HV**
ISO 426/1	CuZn20Al2	ISO	Plate, sheet, tube	76.0–79.0 **Cu**, 0.07 **Pb**, rem **Zn**, 0.07 **Fe**, 1.8–2.5 **Al**, 0.015 **P**, 0.2–0.6 **As**, 0.3 **Fe + Pb + P**, total	
ANSI/ASTM B 359	C 68700	US	Tube: seamless	76.0–79.0 **Cu**, 0.07 **Pb**, rem **Zn**, 0.06 **Fe**, 1.8–2.5 **Al**, 0.02–0.10 **As**	**Annealed**: all **diam**, 50 ksi (345 MPa) **TS**, 18 ksi (125 MPa) **YS**

WROUGHT COPPER AND COPPER ALLOYS

specification number	designation	country	product forms	chemical composition	mechanical properties and hardness values
ANSI/ASTM B 395	C 68700	US	Tube: seamless	76.0–79.0 **Cu**, 0.07 **Pb**, rem **Zn**, 0.06 **Fe**, 1.8–2.5 **Al**, 0.02–0.10 **As**, silver included in copper content	**Annealed**: <0.048 in. (<1.21 mm) **diam**, 50 ksi (345 MPa) **TS**, 18 ksi (125 MPa) **YS**, 12% in 2 in. (50.8 mm) **El**
ANSI/ASTM B 371	CA 697	US	Rod	75.0–80.0 **Cu**, 0.50–1.5 **Pb**, rem **Zn**, 0.20 **Fe**, 2.5–3.5 **Si**, 0.40 **Mn**	**<1.0 in. (<25.4 mm) diam, 65 ksi (450 MPa) TS, 32 ksi (220 MPa) YS, 20% in 1 in. (25.4 mm) El**
SIS 14 52 17	SIS Brass 52 17–02	Sweden	Tube	76.0–79.0 **Cu**, 0.07 **Pb**, 20 **Zn**, 0.07 **Fe**, 1.8–2.3 **Al**, 0.010 **P**, 0.020–0.035 **As**, 0.035 **As+P**, 0.3 total, others	**Annealed**: 10 max mm **diam**, 330 MPa **TS**, 80 MPa **YS**, 50% in 2 in. (50 mm) **El**, 65–90 **HV**
SIS 14 52 17	SIS Brass 52 17–12	Sweden	Tube	76.0–79.0 **Cu**, 0.07 **Pb**, 20 **Zn**, 0.07 **Fe**, 1.8–2.3 **Al**, 0.010 **P**, 0.020–0.035 **As**, 0.035 **As+P**, 0.3 total, others	**Annealed**: 3 max mm **diam**, 390 MPa **TS**, 160 MPa **YS**, 45% in 2 in. (50 mm) **El**, 85–110 **HV**
SABS 460	Cu–Zn21Al2	South Africa	Tube	76.0–78.0 **Cu**, 0.070 **Pb**, rem **Zn**, 0.06 **Fe**, 1.8–2.3 **Al**, 0.02–0.06 **As**, 0.30 total, others	**Annealed**: all **diam**, 300 MPa **TS**, 40% in 2 in. (50 mm) **El**, 75 max **HV**; **Specially annealed**: all **diam**, 350 MPa **TS**, 35% in 2 in. (50 mm) **El**, 85–110 **HV**; **As drawn**: all **diam**, 450 MPa **TS**, 130 **HV**
BS 2871	CZ 110	UK	Tube	76.0–78.0 **Cu**, 0.07 **Pb**, rem **Zn**, 0.06 **Fe**, 1.80–2.30 **Al**, 0.02–0.06 **As**, 0.30 total, others	**As drawn**: 150 **HV**; **Temper annealed**: 85–110 **HV**; **Annealed**: 75 max **HV**
BS 2870	CZ 110	UK	Sheet, strip, foil	76.0–78.0 **Cu**, 0.07 **Pb**, rem **Zn**, 0.06 **Fe**, 1.80–2.30 **Al**, 0.02–0.06 **As**, 0.30 total, others	**As manufactured**: ≤10.0 mm **diam**, 238 MPa **TS**, 45% in 2 in. (50 mm) **El**; **Annealed**: ≤10.0 mm **diam**, 217 MPa **TS**, 50% in 2 in. (50 mm) **El**, 80 max **HV**
BS 2871	CZ 110	UK	Tube	76.0–78.0 **Cu**, 0.07 **Pb**, rem **Zn**, 0.06 **Fe**, 1.80–2.30 **Al**, 0.02–0.06 **As**, 0.30 total, others	**Annealed**: 300 MPa **TS**, 40% in 2 in. (50 mm) **El**, 75 max **HV**; **Temper annealed**: 350 MPa **TS**, 35% in 2 in. (50 mm) **El**, 85–110 **HV**; **As drawn**: 450 MPa **TS**, 130 **HV**
BS 2875	CZ 110	UK	Plate	76.0–78.0 **Cu**, 0.07 **Pb**, rem **Zn**, 0.06 **Fe**, 1.80–2.30 **Al**, 0.02–0.06 **As**, 0.30 total, others	**As manufactured**: ≥10 mm **diam**, 279 MPa **TS**, 36% in 2 in. (50 mm) **El**; **Annealed**: ≥10 mm **diam**, 279 MPa **TS**, 40% in 2 in. (50 mm) **El**
AS 1572	687B	Australia	Tube	76.0–78.0 **Cu**, 0.07 **Pb**, rem **Zn**, 0.06 **Fe**, 1.80–2.30 **Al**, 0.02–0.06 **As**, 0.30 total, others	**Annealed**: all **diam**, 85 **HV**; **Half-hard**: all **diam**, 95–130 **HV**; **Hard**: all **diam**, 150 **HV**
IS 1545	CuZn21Al2As	India	Tube	76.0–78.0 **Cu**, 0.075 **Pb**, rem **Zn**, 0.06 **Fe**, 1.8–2.30 **Al**, 0.02–0.06 **As**, 0.30 total, others	**Annealed**: all **diam**, 402 MPa **TS**; **As drawn**: all **diam**, 416 MPa **TS**

WROUGHT COPPER AND COPPER ALLOYS

specification number	designation	country	product forms	chemical composition	mechanical properties and hardness values
CDA 688		US	Strip: rolled	72.3–74.7 **Cu**, 0.05 **Pb**, rem **Zn**, 0.05 **Fe**, 3.0–3.8 **Al**, 0.25–0.55 **Co**, 0.10 total, others	**Half hard temper**: 0.040 in. **diam**, 93 ksi (641 MPa) **TS**, 75 ksi (517 MPa) **YS**, 20% in 2 in. (50 mm) **El**, 90 **HRB**; **Hard temper**: 0.040 in. **diam**, 109 ksi (752 MPa) **TS**, 97 ksi (670 MPa) **YS**, 6% in 2 in. (50 mm) **El**, 96 **HRB**; **Spring temper**: 0.040 in. **diam**, 122 ksi (841 MPa) **TS**, 110 ksi (758 MPa) **YS**, 2% in 2 in. (50 mm) **El**, 98 **HRB**
CDA 667		US	Wire, sheet, strip	68.5–71.5 **Cu**, 0.07 **Pb**, rem **Zn**, 0.10 **Fe**, 0.8–1.5 **Mn**, 0.50 total, others	**Hard temper (sheet, rolled strip)**: 0.040 in. **diam**, 77.5 ksi (535 MPa) **TS**, 74 ksi (510 MPa) **YS**, 10% in 2 in. (50 mm) **El**, 84 **HRB**; **Spring temper (sheet, rolled strip)**: 0.040 in. **diam**, 96 ksi (662 MPa) **TS**, 90 ksi (621 MPa) **YS**, 2.5% in 2 in. (50 mm) **El**, 91 **HRB**; **Half hard temper (wire)**: 0.080 in. **diam**, 90 ksi (621 MPa) **TS**, 5% in 2 in. (50 mm) **El**
ASTM B 291		US	Sheet, strip	68.5–71.5 **Cu**, 0.07 **Pb**, rem **Zn**, 0.10 **Fe**, 0.8–1.5 **Mn**	**Rolled, hard**: >0.020 in. (>0.508 mm) **diam**, 72 ksi (495 MPa) **TS**, 80–88 **HRB**; **Rolled, spring**: >0.020 in. (>0.508 mm) **diam**, 91 ksi (625 MPa) **TS**, 89–94 **HRB**; **Annealed**: >0.020 in. (>0.508 mm) **diam**, 54–66 **HRF**
BS 2872	CZ 116	UK	Forging stock, forgings	64.0–68.0 **Cu**, rem **Zn**, 0.25–1.2 **Fe**, 4.0–5.0 **Al**, 0.3–2.0 **Mn**, 0.50 total, others	**As manufactured or annealed**: ≥6 mm **diam**, 539 MPa **TS**, 294 MPa **YS**, 12% in 2 in. (50 mm) **El**
BS 2874	CZ 116	UK	Rod, section	64.0–68.0 **Cu**, rem **Zn**, 0.25–1.2 **Fe**, 4.0–5.0 **Al**, 0.3–2.0 **Mn**, 0.50 total, others	**As manufactured**: 6–100 mm **diam**, 539 MPa **TS**, 290 MPa **YS**, 12% in 2 in. (50 mm) **El**
SIS 14 52 34	SIS Brass 52 34–00	Sweden	Bar, section, forgings, strip, wire	64.0–68.0 **Cu**, 24 **Zn**, 1.0–3.0 **Fe**, 3.5–5.5 **Al**, 2.5–4.5 **Mn**	**Hot worked**: 25 max mm **diam**, 670 MPa **TS**, 370 MPa **YS**, 10% in 2 in. (50 mm) **El**, 180–230 **HV**

WROUGHT COPPER AND COPPER ALLOYS

specification number	designation	country	product forms	chemical composition	mechanical properties and hardness values
ANSI/ASTM B 138	CA 670	US	Rod, bar, shapes	63.0–68.0 **Cu**, 0.50 **Sn**, 0.20 **Pb**, rem **Zn**, 2.0–4.0 **Fe**, 3.0–6.0 **Al**, 2.5–5.0 **Mn**	**Soft**: all **diam**, 85 ksi (585 MPa) **TS**, 45 ksi (310 MPa) **YS**, 10% in 1 in. (25.4 mm) **El**; **Half hard**: all **diam**, 105 ksi (720 MPa) **TS**, 60 ksi (415 MPa) **YS**, 7% in 1 in. (25.4 mm) **El**; **Hard**: all **diam**, 115 ksi (790 MPa) **TS**, 68 ksi (470 MPa) **YS**, 5% in 1 in. (25.4 mm) **El**
COPANT 421	Cu Zn 40Fe 1 Sn 1 Mn	COPANT	Bar and shapes	63.0–68.0 **Cu**, 0.50 **Sn**, 0.20 **Pb**, rem **Zn**, 2.0–4.0 **Fe**, 3.0–6.0 **Al**, 2.5–5.0 **Mn**, 0.10 total, others	**Annealed**: all **diam**, 380 MPa **TS**, 150 MPa **YS**, 20% in 2 in. (50 mm) **El**; **1/2 hard**: ≤25 mm **diam**, 500 MPa **TS**, 245 MPa **YS**, 73% in 2 in. (50 mm) **El**; **Hard (flat bar)**: ≤25 mm **diam**, 520 MPa **TS**, 360 MPa **YS**, 8% in 2 in. (50 mm) **El**
QQ–B–728	Class B	US	Rod, shapes, forgings, flat wire, strip, sheet, bar, plate	63–68 **Cu**, 0.5 **Sn**, 0.20 **Pb**, rem **Zn**, 2.0–4.0 **Fe**, 3.0–6.0 **Al**, 2.5–5.0 **Mn**, 0.10 total, others	**Soft temper**: all **diam**, 85 ksi (586 MPa) **TS**, 45 ksi (310 MPa) **YS**, 10% in 2 in. (50 mm) **El**; **Half hard**: all **diam**, 105 ksi (724 MPa) **TS**, 60 ksi (414 MPa) **YS**, 7% in 2 in. (50 mm) **El**; **Hard temper**: all **diam**, 115 ksi (793 MPa) **TS**, 68 ksi (469 MPa) **YS**, 5% in 2 in. (50 mm) **El**
COPANT 421	Cu Zn 23 Al 4 Mn Fe 3	COPANT	Bar	63.0–68.0 **Cu**, 0.50 **Sn**, 0.20 **Pb**, rem **Zn**, 2.0–4.0 **Fe**, 3.0–6.0 **Al**, 2.5–5.0 **Mn**, 0.10 total, others	**Annealed**: all **diam**, 590 MPa **TS**, 315 MPa **YS**, 10% in 2 in. (50 mm) **El**; **1/2 hard**: all **diam**, 725 MPa **TS**, 410 MPa **YS**, 7% in 2 in. (50 mm) **El**; **Hard**: all **diam**, 795 MPa **TS**, 470 MPa **YS**, 5% in 2 in. (50 mm) **El**
NF A 51 106	CuZn +, Class 1	France	Bar, wire, and shapes	52.0–70.0 **Cu**, 2.0 **Sn**, 3.0 **Pb**, rem **Zn**, 3.0 **Fe**, 5.0 **Al**, 4.0 **Mn**, 5.0 **Ni**, 1.0 others, total	**As manufactured**: ≤12 mm **diam**, 500 MPa **TS**, 260 MPa **YS**, 5% in 2 in. (50 mm) **El**
NF A 51 106	CuZn +, Class 2	France	Bar, wire, and shapes	52.0–70.0 **Cu**, 2.0 **Sn**, 3.0 **Pb**, rem **Zn**, 3.0 **Fe**, 5.0 **Al**, 4.0 **Mn**, 5.0 **Ni**, 1.0 others, total	**As manufactured**: ≤12 mm **diam**, 600 MPa **TS**, 300 MPa **YS**, 7% in 2 in. (50 mm) **El**
SFS 2929	CuZn35Mn2AlFe	Finland	Rod, tube, shapes, and forgings	58.0–62.0 **Cu**, 0.5–1.5 **Sn**, 0.2–1.0 **Pb**, 35 **Zn**, 0.5–1.5 **Fe**, 0.2–1.0 **Al**, 1.0–3.0 **Mn**	**Hot finished**: all **diam**, 450 MPa **TS**, 200 MPa **YS**, 35% in 2 in. (50 mm) **El**; **Drawn (rod, tube, and shapes)**: 25–50 mm **diam**, 470 MPa **TS**, 270 MPa **YS**, 15% in 2 in. (50 mm) **El**
DS 3003	5238	Denmark	Rolled, drawn, extruded, and forged products	58.0–62.0 **Cu**, 0.5–1.5 **Sn**, 0.2–1.0 **Pb**, 35 **Zn**, 0.5–1.5 **Fe**, 0.2–1.0 **Al**, 1.0–3.0 **Mn**	

WROUGHT COPPER AND COPPER ALLOYS

specification number	designation	country	product forms	chemical composition	mechanical properties and hardness values
SIS 14 52 38	SIS Brass 52 38–00	Sweden	Bar, tube, section, forgings	58.0–62.0 **Cu**, 0.5–1.5 **Sn**, 0.2–1.0 **Pb**, 35 **Zn**, 0.5–1.5 **Fe**, 0.2–1.0 **Al**, 1.0–3.0 **Mn**	**Hot worked**: all **diam**, 430 MPa **TS**, 160 MPa **YS**, 25% in 2 in. (50 mm) **El**, 110–150 **HV**
SIS 14 52 38	SIS Brass 52 38–04	Sweden	Wire, bar	58.0–62.0 **Cu**, 0.5–1.5 **Sn**, 0.2–1.0 **Pb**, 35 **Zn**, 0.5–1.5 **Fe**, 0.2–1.0 **Al**, 1.0–3.0 **Mn**	**Strain hardened**: 2.5–5 mm **diam**, 530 MPa **TS**, 390 MPa **YS**, 5% in 2 in. (50 mm) **El**, 155–190 **HV**
IS 320	Alloy 3	India	Rod, section	57.0–61.0 **Cu**, 1.0 **Sn**, 0.75–1.0 **Pb**, rem **Zn**, 0.25–1.0 **Fe**, 0.5–2.0 **Al**, 0.5–1.2 **Mn**, 0.5 total, others	**As agreed**: 520 MPa **TS**, 274 MPa **YS**, 15% in 2 in. (50 mm) **El**
CDA 674		US	Rod, bar, shapes	57.0–60.0 **Cu**, 0.30 **Sn**, 0.50 **Pb**, rem **Zn**, 0.35 **Fe**, 0.50–1.5 **Si**, 0.50–2.0 **Al**, 2.0–3.5 **Mn**, 0.25 **Ni**, 0.10 total, others	**Soft anneal (rod)**: 0.750 in. **diam**, 70 ksi (483 MPa) **TS**, 28% in 2 in. (50 mm) **El**, 78 **HRB**; **Half hard, stress relieved (rod)**: 0.750 in. **diam**, 92 ksi (634 MPa) **TS**, 20% in 2 in. (50 mm) **El**, 87 **HRB**; **As extruded (bar)**: 2 in. **diam**, 78 ksi (538 MPa) **TS**, 26% in 2 in. (50 mm) **El**, 78 **HRB**
CDA 675		US	Rod, shapes	57.0–60.0 **Cu**, 0.50–1.5 **Sn**, 0.20 **Pb**, rem **Zn**, 0.8–2.0 **Fe**, 0.25 **Al**, 0.05–0.50 **Mn**, 0.10 total, others	**Soft anneal (rod)**: 1.0 in. **diam**, 65 ksi (448 MPa) **TS**, 33% in 2 in. (50 mm) **El**, 65 **HRB**; **Half hard–20% (rod)**: 1.0 in. **diam**, 84 ksi (579 MPa) **TS**, 90% in 2 in. (50 mm) **El**, 90 **HRB**; **Quarter hard–10% (rod)**: 2.0 in. **diam**, 72 ksi (496 MPa) **TS**, 77% in 2 in. (50 mm) **El**, 77 **HRB**
QQ–B–728	Class A	US	Rod, shapes, forgings, flat wire, strip, sheet, bar, plate	57–60 **Cu**, 0.5–1.5 **Sn**, 0.20 **Pb**, rem **Zn**, 0.8–2.0 **Fe**, 0.25 **Al**, 0.05–0.50 **Mn**, 0.10 total, others	**Soft temper**: all **diam**, 55 ksi (379 MPa) **TS**, 22 ksi (152 MPa) **YS**, 20% in 2 in. (50 mm) **El**; **Half hard**: 1.000 in. **diam**, 72 ksi (496 MPa) **TS**, 36 ksi (248 MPa) **YS**, 13% in 2 in. (50 mm) **El**; **Hard temper**: 1.000 in. **diam**, 80 ksi (552 MPa) **TS**, 56 ksi (386 MPa) **YS**, 8% in 2 in. (50 mm) **El**
ANSI/ASTM B 124	C 67500	US	Rod, bar, shapes	57.0–60.0 **Cu**, 0.5–1.5 **Sn**, 0.20 **Pb**, rem **Zn**, 0.8–2.0 **Fe**, 0.25 **Al**, 0.05–0.5 **Mn**	

WROUGHT COPPER AND COPPER ALLOYS

specification number	designation	country	product forms	chemical composition	mechanical properties and hardness values
ANSI/ASTM B 138	CA 675	US	Rod, bar, shapes	57.0–60.0 **Cu**, 0.50–1.5 **Sn**, 0.20 **Pb**, rem **Zn**, 0.8–2.0 **Fe**, 0.25 **Al**, 0.05–0.50 **Mn**	**Soft**: all **diam**, 55 ksi (380 MPa) **TS**, 22 ksi (150 MPa) **YS**, 20% in 1 in. (25.4 mm) **El**; **Half hard (excluding shapes)**: <1 in. (<25.4 mm) **diam**, 72 ksi (495 MPa) **TS**, 36 ksi (250 MPa) **YS**, 13% in 1 in. (25.4 mm) **El**; **Hard (excluding shapes)**: <1 in. (<25.4 mm) **diam**, 80 ksi (550 MPa) **TS**, 56 ksi (385 MPa) **YS**, 8% in 1 in. (25.4 mm) **El**
ASTM F 467	C 67500	US	Nut	57.0–60.0 **Cu**, 0.5–1.5 **Sn**, 0.20 **Pb**, rem **Zn**, 0.8–2.0 **Fe**, 0.05–0.5 **Mn**	**As agreed**: all **diam**, 60 min **HRB**
ANSI/ASTM B 283	C 67500	US	Die forgings	57.0–60.0 **Cu**, 0.5–1.5 **Sn**, 0.20 **Pb**, rem **Zn**, 0.8–2.0 **Fe**, 0.25 **Al**, 0.05–0.5 **Mn**	
ASTM F 468	C 67500	US	Bolt, screw, stud	57.0–60.0 **Cu**, 0.5–1.5 **Sn**, 0.20 **Pb**, rem **Zn**, 0.8–2.0 **Fe**, 0.25 **Al**, 0.05–0.5 **Mn**	**As manufactured**: all **diam**, 55 ksi (380 MPa) **TS**, 25 ksi (170 MPa) **YS**, 60–90 **HRB**
ISO 1640	CuZn39AlFeMn	ISO	Forgings	56.0–61.0 **Cu**, 1.2 **Sn**, 1.5 **Pb**, rem **Zn**, 0.2–1.5 **Fe**, 0.2–1.5 **Al**, 0.2–2.0 **Mn**, 2.0 **Ni**, 0.5 total, others	**M**: all **diam**, 500 MPa **TS**, 200 MPa **YS**, 15% in 2 in. (50 mm) **El**
ISO 1639	CuZn39AlFeMn	ISO	Extruded sections	56.0–61.0 **Cu**, 1.2 **Sn**, 1.5 **Pb**, rem **Zn**, 0.2–1.5 **Fe**, 0.2–1.5 **Al**, 0.2–2.0 **Mn**, 2.0 **Ni**, 0.5 total, others	**M**: all **diam**, 490 MPa **TS**, 180 MPa **YS**, 15% in 2 in. (50 mm) **El**
ISO 1637	CuZn39AlFeMn	ISO	Solid products: straight lengths	56.0–61.0 **Cu**, 1.2 **Sn**, 1.5 **Pb**, rem **Zn**, 0.2–1.5 **Fe**, 0.2–1.5 **Al**, 0.2–2.0 **Mn**, 2.0 **Ni**, 0.5 total, others	**M**: 5 mm **diam**, 470 MPa **TS**, 180 MPa **YS**, 18% in 2 in. (50 mm) **El**; **HA**: 75–150 mm **diam**, 500 MPa **TS**, 200 MPa **YS**, 18% in 2 in. (50 mm) **El**; **HB–strain hardened**: 5–75 mm **diam**, 540 MPa **TS**, 250 MPa **YS**, 18% in 2 in. (50 mm) **El**
ISO 1635	CuZn39AlFeMn	ISO	Tube	56.0–61.0 **Cu**, 1.2 **Sn**, 1.5 **Pb**, rem **Zn**, 0.2–1.5 **Fe**, 0.2–1.5 **Al**, 0.2–2.0 **Mn**, 2.0 **Ni**, 0.5 total, others	**M**: 500 MPa **TS**, 200 MPa **YS**, 18% in 2 in. (50 mm) **El**
ISO 426/1	CuZn39AlFeMn	ISO	Rod, bar, extruded sections, tube, wire, forging	56.0–61.0 **Cu**, 1.2 **Sn**, 1.5 **Pb**, rem **Zn**, 0.2–1.5 **Fe**, 0.2–1.5 **Al**, 0.2–2.0 **Mn**, 2.0 **Ni**, 0.5 total, others	
AS 1567	686D	Australia	Rod, bar, section	56.0–60.0 **Cu**, 0.2–1.0 **Sn**, 0.5–1.5 **Pb**, rem **Zn**, 0.5–1.2 **Fe**, 0.3–1.5 **Al**, 0.3–2.0 **Mn**, 0.50 total, others	**As manufactured**: 6–80 mm **diam**, 460 MPa **TS**, 240 MPa **YS**, 18% in 2 in. (50 mm) **El**
AS 1567	678C	Australia	Rod, bar, section	56.0–60.0 **Cu**, 0.3–1.0 **Sn**, 0.1 **Pb**, rem **Zn**, 0.5–1.2 **Fe**, 0.3–1.3 **Al**, 0.3–2.0 **Mn**, 0.50 total, others	**As manufactured**: 6–80 mm **diam**, 460 MPa **TS**, 240 MPa **YS**, 18% in 2 in. (50 mm) **El**

WROUGHT COPPER AND COPPER ALLOYS

specification number	designation	country	product forms	chemical composition	mechanical properties and hardness values
MIL–R–19631B	MIL–RCuZn–B	US	Rod	56.0–60.0 **Cu**, 0.75–1.10 **Sn**, 0.05 **Pb**, rem **Zn**, 0.25–1.25 **Fe**, 0.04–0.15 **Si**, 0.01 **Al**, 0.01–0.50 **Mn**, 0.2–0.8 **Ni+Co**, 0.50 total, others, silver included in copper content	**As milled**: all **diam**, 386 MPa **TS**
MIL–R–19631B	MIL–RCuZn–C	US	Rod	56.0–60.0 **Cu**, 0.75–1.10 **Sn**, 0.05 **Pb**, rem **Zn**, 0.25–1.25 **Fe**, 0.04–0.15 **Si**, 0.01 **Al**, 0.01–0.50 **Mn**, 0.50 total, others, silver included in copper content	**As milled**: all **diam**, 386 MPa **TS**
IS 320	Alloy 1	India	Rod, section	56.0–59.0 **Cu**, 0.75–1.75 **Sn**, 0.5 **Pb**, rem **Zn**, 1.25 **Fe**, 0.2 **Al**, 2.0 **Mn**, 0.5 total, others	**Hot worked**: 461 MPa **TS**, 236 MPa **YS**, 20% in 2 in. (50 mm) **El**; **Drawn and stress relieved**: 520 MPa **TS**, 274 MPa **YS**, 15% in 2 in. (50 mm) **El**
AS 1567	671D	Australia	Rod, bar, section	56.0–59.0 **Cu**, 0.5–1.5 **Sn**, 0.5–1.5 **Pb**, rem **Zn**, 0.4–1.0 **Al**, 0.5–1.5 **Mn**, 0.5 total, others	**As manufactured**: 6–80 mm **diam**, 460 MPa **TS**, 240 MPa **YS**, 18% in 2 in. (50 mm) **El**
IS 320	Alloy 2	India	Rod, section	56.0–59.0 **Cu**, 0.5 **Sn**, 1.0 **Pb**, rem **Zn**, 0.7–1.2 **Fe**, 0.2–1.2 **Al**, 0.5–1.2 **Mn**, 0.5 total, others	**As agreed**: 461 MPa **TS**, 236 MPa **YS**, 20% in 2 in. (50 mm) **El**
AS 1567	683D	Australia	Rod, bar, section	56.0–59.0 **Cu**, 0.6–1.1 **Sn**, 1.5 **Pb**, rem **Zn**, 0.5–1.2 **Fe**, 0.2 **Al**, 0.3–2.0 **Mn**, 0.50 total, others	**Hot worked**: all **diam**, 460 MPa **TS**, 190 MPa **YS**, 20% in 2 in. (50 mm) **El**; **Cold worked**: 6–40 mm **diam**, 510 MPa **TS**, 260 MPa **YS**, 12% in 2 in. (50 mm) **El**
ASTM B 592		US	Plate, sheet, strip, bar	49.75–55.75 **Cu**, 0.05 **Pb**, 21.3–24.1 **Zn**, 0.05 **Fe**, 3.0–3.8 **Al**, 0.25–0.55 **Co**, 25.1–27.1 **Zn+Al**	**Annealed**: >0.012 in. (>0.31 mm) **diam**, 77 ksi (530 MPa) **TS**, 63–74 **HR30T**; **Hard**: >0.012 in. (>0.31 mm) **diam**, 106 ksi (730 MPa) **TS**, 82–83 **HR30T**; **Spring**: >0.012 in. (>0.31 mm) **diam**, 123 ksi (850 MPa) **TS**, 83–84 **HR30T**
IS 5743	Cu Ni 51 (master alloy)	India	Bar, ingot, shot	47.5–49.5 **Cu**, 0.40 **Fe**, 50.0–52.0 **Ni**	
ANSI/ASTM F 96	C 71590	US	Wrought products	67.0 min **Cu**, 0.001 **Pb**, 0.001 **Zn**, 0.005 **Fe**, 0.02 **Si**, 0.002 **Al**, 0.05 **Co**, 0.001 **Mn**, 29.0–33.0 **Ni**, 0.001 **P**, 0.003 **S**, 0.03 **C**	
BS 2871	CN 107	UK	Tube	rem **Cu**, 0.01 **Pb**, 0.40–1.00 **Fe**, 0.50–1.50 **Mn**, 30.0–32.0 **Ni**, 0.08 **S**, 0.06 **C**, 0.30 total, others	**As drawn**: 150 **HV**; **Annealed**: 90–120 **HV**
BS 2871	CN 107	UK	Tube	rem **Cu**, 0.01 **Pb**, 0.40–1.00 **Fe**, 0.50–1.50 **Mn**, 30.0–32.0 **Ni**, 0.08 **S**, 0.06 **C**, 0.30 total, others	**Annealed**: 370 MPa **TS**, 30% in 2 in. (50 mm) **El**, 110 max **HV**; **As drawn**: 500 MPa **TS**, 140 **HV**

WROUGHT COPPER AND COPPER ALLOYS

specification number	designation	country	product forms	chemical composition	mechanical properties and hardness values
BS 2870	CN 107	UK	Sheet, strip, foil	rem **Cu**, 0.01 **Pb**, 0.40–1.00 **Fe**, 0.50–1.50 **Mn**, 30.0–32.0 **Ni**, 0.08 **S**, 0.06 **C**, 0.30 total, others	**Annealed**: 0.6–2.0 mm **diam**, 259 MPa **TS**, 30% in 2 in. (50 mm) **El**
BS 2875	CN 107	UK	Plate	rem **Cu**, 0.01 **Pb**, 0.4–1.0 **Fe**, 0.5–1.5 **Mn**, 30.0–32.0 **Ni**, 0.08 **S**, 0.06 **C**, 0.30 total, others	**As manufactured or annealed**: ≥10 mm **diam**, 309 MPa **TS**, 27% in 2 in. (50 mm) **El**
CDA 715		US	Rod, tube, plate, sheet, strip	69.5 **Cu**, 0.05 **Pb**, 1.0 **Zn**, 0.40–0.7 **Fe**, 1.0 **Mn**, 29.0–33.0 **Ni**, 99.5 min **Cu**, + others	**As hot rolled (plate, sheet, strip)**: 1.0 in. **diam**, 55 ksi (379 MPa) **TS**, 45% in 2 in. (50 mm) **El**, 35 **HRB**; **0.025 mm temper (tube)**: 1.0 in. **diam**, 60 ksi (414 MPa) **TS**, 45% in 2 in. (50 mm) **El**, 45 **HRB**; **Half hard temper–20% (rod)**: 1.0 in. **diam**, 75 ksi (517 MPa) **TS**, 15% in 2 in. (50 mm) **El**, 80 **HRB**
COPANT 162	C 715 00	COPANT	Wrought products	65 min **Cu**, 0.05 **Pb**, 1.0 **Zn**, 0.4–0.7 **Fe**, 1.0 **Mn**, 29–33 **Ni**, silver included in copper	
COPANT 162	C 717 00	COPANT	Wrought products	64.8–69.8 **Cu**, 0.4–1.0 **Fe**, 29–33 **Ni**, 0.3–0.7 **Be**	
COPANT 531	Cu Ni 31 Fe	COPANT	Tube	0.05 **Pb**, 1.0 **Zn**, 0.4–0.7 **Fe**, 1.0 **Mn**, 29–33 **Ni**, 0.02 **P**, 0.02 **S**, 65 min **Cu+Ag**, 0.5 total, others	**Annealed**: 3.18–127 mm **diam**, 345 MPa **TS**, 125 MPa **YS**; **Full tempered**: 480 MPa **TS**, 315 MPa **YS**
DS 3003	5682	Denmark	Rolled, drawn, extruded, and forged products	67 **Cu**, 0.5 **Zn**, 0.4–1.0 **Fe**, 0.5–1.5 **Mn**, 30.0–32.0 **Ni**, 0.05 **S**, 0.06 **C**, 0.05 **Pb+Sn**, 0.2 others, total	
IS 1545	CuNi31Mn1Fe	India	Tube	rem **Cu**, 0.01 **Pb**, 0.40–1.00 **Fe**, 0.50–1.50 **Mn**, 30.0–32.0 **Ni**, 0.08 **S**, 0.06 **C**, 0.30 total, others	**Annealed**: all **diam**, 461 MPa **TS**; **As drawn**: all **diam**, 481 MPa **TS**
JIS H 3300	C 7150	Japan	Pipe and tube: seamless	63.75–68.85 **Cu**, 0.05 **Pb**, 1.0 **Zn**, 0.40–0.7 **Fe**, 0.20–1.0 **Mn**, 29.0–33.0 **Ni**, 99.5 min **Cu+Ni+Fe+Mn**	**Annealed**: 5–50 mm **diam**, 363 MPa **TS**, 30% in 2 in. (50 mm) **El**
JIS H 3100	C 7150	Japan	Sheet, plate, and strip	0.05 **Pb**, 1.0 **Zn**, 0.40–0.7 **Fe**, 0.20–1.0 **Mn**, 29.0–33.0 **Ni** 99.5 min **Cu+Ni+Fe+Mn**	**As manufactured**: 0.5–50 mm **diam**, 343 MPa **TS**, 35% in 2 in. (50 mm) **El**
ANSI/ASTM B 122	C 71500	US	Plate, sheet, strip, bar	65.0 min **Cu**, 0.05 **Pb**, 1.0 **Zn**, 0.40–1.0 **Fe**, 1.0 **Mn**, 29.0–33.0 **Ni+Co**, silver included in copper content	**Hard**: >0.012 in. (>0.305 mm) **diam**, 75 ksi (515 MPa) **TS**, 72–76 **HR30T**; **Spring**: >0.012 in. (>0.305 mm) **diam**, 84 ksi (580 MPa) **TS**, 74–77 **HR30T**; **Annealed**: >0.015 in. (>0.381 mm) **diam**, 31–46 **HR30T**

WROUGHT COPPER AND COPPER ALLOYS

specification number	designation	country	product forms	chemical composition	mechanical properties and hardness values
ASTM F 468	C 71500	US	Bolt, screw, stud	65.0 min **Cu**, 0.05 **Pb**, 1.00 **Zn**, 0.40–0.7 **Fe**, 1.00 **Mn**, 29.0–33.0 **Ni + Co**	**As manufactured**: all **diam**, 55 ksi (380 MPa) **TS**, 20 ksi (140 MPa) **YS**, 60–95 **HRB**
ANSI/ASTM B 608	CA 715	US	Pipe: welded	65.0 min **Cu**, 0.02 **Pb**, 0.50 **Zn**, 0.4–1.0 **Fe**, 1.0 **Mn**, 29.0–33.0 **Ni + Co**, 0.02 **P**, 0.02 **S**, 0.05 **C**, silver included in copper content	**Annealed**: all **diam**, 50 ksi (345 MPa) **TS**;
ASTM F 467	C 71500	US	Nut	65.0 min **Cu**, 0.05 **Pb**, 1.0 **Zn**, 0.40–0.7 **Fe**, 1.0 **Mn**, 29.0–33.0 **Ni + Co**	**As agreed**: all **diam**, 60 min **HRB**
ANSI/ASTM F 96	C 71580	US	Wrought products	65.0 min **Cu**, 0.05 **Pb**, 0.05 **Zn**, 0.5 **Fe**, 0.15 **Si**, 0.05 **Al**, 0.3 **Mn**, 29.0–33.0 **Ni + Co**, 0.03 **P**, 0.024 **S**, 0.07 **C**,	
ASTM B 552	CA 715	US	Tube: seamless, welded	65.0 min **Cu**, 0.05 **Pb**, 1.0 **Zn**, 0.40–1.0 **Fe**, 1.0 **Mn**, 29.0–33.0 **Ni + Co**	**Annealed**: all **diam**, 52 ksi (360 MPa) **TS**; **Cold worked**: all **diam**, 60 ksi (415 MPa) **TS**
ANSI/ASTM B 543	C 71500	US	Tube: welded	65.0 min **Cu**, 0.05 **Pb**, 1.0 **Zn**, 0.40–1.0 **Fe**, 1.0 **Mn**, 29.0–33.0 **Ni + Co**	**As welded from annealed strip**: all **diam**, 52 ksi (360 MPa) **TS**, 18 ksi (125 MPa) **YS**; **Welded and annealed**: all **diam**, 52 ksi (360 MPa) **TS**, 18 ksi (125 MPa) **YS**; **Fully finished–hard drawn**: all **diam**, 72 ksi (495 MPa) **TS**, 50 ksi (345 MPa) **YS**
ANSI/ASTM B 467	CA 715	US	Pipe, tube	65.0 min **Cu**, 0.05 **Pb**, 1.0 **Zn**, 0.40–1.0 **Fe**, 1.0 **Mn**, 29.0–33.0 **Ni + Co**, 0.02 **P**, 0.02 **S**	**Annealed**: <4.5 in. (<114 mm) **diam**, 50 ksi (345 MPa) **TS**, 20 ksi (140 MPa) **YS**, 30% in 2 in. (50.8 mm) **El**; **Drawn**: <2.0 in. (<50.8 mm) **diam**, 72 ksi (495 MPa) **TS**, 50 ksi (345 MPa) **YS**, 12% in 2 in. (50.8 mm) **El**
ANSI/ASTM B 466	CA 715	US	Pipe and tube: seamless	65.0 min **Cu**, 0.05 **Pb**, 1.0 **Zn**, 0.40–1.0 **Fe**, 1.0 **Mn**, 29.0–33.0 **Ni + Co**, 0.02 **S**	**Annealed**: >0.020 in. (>0.508 mm) **diam**, 50 ksi (345 MPa) **TS**, 18 ksi (125 MPa) **YS**, 51 max **HR30T**; **Hard drawn**: >0.020 in. (>0.508 mm) **diam**, 70 ksi (485 MPa) **TS**, 45 ksi (310 MPa) **YS**, 70 min **HR30T**
MIL–T–16420K	715	US	Tube: seamless and welded	65.0 min **Cu**, 0.02 **Pb**, 0.50 **Zn**, 0.40–1.0 **Fe**, 1.0 **Mn**, 29.0–33.0 **Ni + Co**, 0.02 **P**, 0.02 **S**, 0.05 **C**, silver included in copper content	**Annealed**: ≤4.5 in. **diam**, 345 MPa **TS**, 124 MPa **YS**, 30% in 2 in. (50 mm) **El**

WROUGHT COPPER AND COPPER ALLOYS

specification number	designation	country	product forms	chemical composition	mechanical properties and hardness values
MIL-R-19631B	MIL-RCuNi	US	Rod	rem **Cu**, 1.2 **Sn**, 0.02 **Pb**, 1.0 **Zn**, 0.40–0.70 **Fe**, 0.25 **Si**, 1.25 **Mn**, 29.0–33.0 **Ni**, 0.50 **Ti**, 0.50 total, others, silver included in copper content	**As milled**: all **diam**, 345 MPa **TS**
NS 16 415	CuNi30Mn1Fe	Norway	Tube	67 **Cu**, 0.7 **Fe**, 1 **Mn**, 31 **Ni**	
VSM 11557	Cu Ni 30 Fe Mn	Switzerland	Tube	rem **Cu**, 0.5 **Zn**, 0.4–1.0 **Fe**, 0.5–1.5 **Mn**, 30.0–32.0 **Ni**, 0.05 **S**, 0.06 **C**, 0.05 **Pb+Sn**, 0.1 total, others	**1/2 hard**: all **diam**, 363 MPa **TS**, 118 MPa **YS**, 30% in 2 in. (50 mm) **El**, 100 **HB**
COPANT 532	Cu Ni 31 Fe	COPANT	Bar	65 min **Cu**, 0.05 **Pb**, 1.0 **Zn**, 0.4–0.7 **Fe**, 1.0 **Mn**, 29–33 **Ni**, 99.5 min cu and all other specified elements, 0.5 total, others	**Full tempered**: >0.5 mm **diam**, 550 MPa **TS**; **1/4 hard**: 0.5–13 mm **diam**, 410 MPa **TS**
ANSI/ASTM B 402	C 71500	US	Plate, sheet	65.0 min **Cu**, 0.05 **Pb**, 1.0 **Zn**, 0.40–1.0 **Fe**, 1.0 **Mn**, 29.0–33.0 **Ni+Co**	**As manufactured**: <2.5 in. (<63.5 mm) **diam**, 50 ksi (345 MPa) **TS**, 20 ksi (140 MPa) **YS**, 30% in 2 in. (50.8 mm) **El**
ANSI/ASTM B 395	C 71500	US	Tube: seamless	65.0 min **Cu**, 0.05 **Pb**, 1.0 **Zn**, 0.40–1.0 **Fe**, 1.0 **Mn**, 29.0–33.0 **Ni+Co**	**Annealed**: <0.048 in. (<1.21 mm) **diam**, 52 ksi (360 MPa) **TS**, 18 ksi (125 MPa) **YS**, 12% in 2 in. (50.8 mm) **El**; **Drawn, stress relieved**: <0.048 in. (<1.21 mm) **diam**, 72 ksi (495 MPa) **TS**, 50 ksi (345 MPa) **YS**, 12% in 2 in. (50.8 mm) **El**
ANSI/ASTM B 151	CA 715	US	Rod, bar	65.0 min **Cu**, 1.0 **Zn**, 0.40–1.0 **Fe**, 1.0 **Mn**, 29.0–33.0 **Ni+Co**, 0.05 **P**, silver included in copper content	**Soft**: <0.50 in. (<12.7 mm) **diam**, 52 ksi (360 MPa) **TS**, 18 ksi (125 MPa) **YS**, 30% in 2 in. (50 mm) **El**; Quarter hard: <0.50 in. (<12.7 mm) **diam**, 65 ksi (450 MPa) **TS**, 50 ksi (345 MPa) **YS**, 10% in 2 in. (50 mm) **El**; **Hard**: <0.50 in. (<12.7 mm) **diam**, 80 ksi (550 MPa) **TS**, 60 ksi (415 MPa) **YS**, 8% in 2 in. (50 mm) **El**
ANSI/ASTM B 171	C 71500	US	Plate	65.0 min **Cu**, 0.05 **Pb**, 1.0 **Zn**, 0.40–1.0 **Fe**, 1.0 **Mn**, 29.0–33.0 **Ni+Co**, silver included in copper content	**As manufactured**: <2.5 in. (<63.5 mm) **diam**, 50 ksi (345 MPa) **TS**, 20 ksi (140 MPa) **YS**, 35% in 2 in. (50.8 mm) **El**
ANSI/ASTM B 359	C 71500	US	Tube: seamless	65.0 min **Cu**, 0.05 **Pb**, 1.0 **Zn**, 0.40–1.0 **Fe**, 1.0 **Mn**, 29.0–33.0 **Ni+Co**	**Annealed**: all **diam**, 52 ksi (360 MPa) **TS**, 18 ksi (125 MPa) **YS**

WROUGHT COPPER AND COPPER ALLOYS

specification number	designation	country	product forms	chemical composition	mechanical properties and hardness values
ANSI/ASTM B 111	C 71500	US	Tube: seamless	65.0 min **Cu**, 0.05 **Pb**, 1.0 **Zn**, 0.40–1.0 **Fe**, 1.0 **Mn**, 29.0–33.0 **Ni+Co**	**Annealed**: all **diam**, 52 ksi (360 MPa) **TS**, 18 ksi (125 MPa) **YS**; **Drawn, stress-relieved**: >0.048 in. (>1.21 mm) **diam**, 72 ksi (495 MPa) **TS**, 50 ksi (345 MPa) **YS**, 12% in 2 in. (50.8 mm) **El**
BS 2871	CN 108	UK	Tube	rem **Cu**, 1.7–2.3 **Fe**, 1.5–2.5 **Mn**, 29.0–32.0 **Ni**, 0.30 total, others	**As drawn**: 150 **HV**; **Annealed**: 90–120 **HV**
COPANT 162	C 719 00	COPANT	Wrought products	62.76–67.04 **Cu**, 0.08–0.20 **Zn**, 0.25 **Fe**, 0.5–1.0 **Mn**, 29–32 **Ni**, 2.6–3.2 **Cr**, 0.01 **C**, 0.02–0.08 **Ti**	
ISO 429	CuNi30Mn1Fe	ISO	Plate, sheet, strip, rod, bar, tube	rem **Cu**, 0.5 **Zn**, 0.4–1.0 **Fe**, 0.5–1.5 **Mn**, 29.0–32.0 **Ni**, 0.08 **S**, 0.06 **C**, 0.05 **Sn+Pb**, 0.2 total, others	
ISO 1637	CuNi30Mn1Fe	ISO	Solid products: straight lengths	rem **Cu**, 0.5 **Zn**, 0.4–1.0 **Fe**, 0.5–1.5 **Mn**, 29.0–32.0 **Ni**, 0.08 **S**, 0.06 **C**, 0.5 **Co** included in **Ni**; 0.05 **Sn+Pb**; 0.2 total, others	**Annealed**: 5 mm **diam**, 40% in 2 in. (50 mm) **El**, 110 **HV**; **HB-strain hardened**: 5–15 mm **diam**, 420 MPa **TS**, 20% in 2 in. (50 mm) **El**, 120 **HV**
ISO 1635	CuNi30Mn1Fe	ISO	Tube	rem **Cu**, 0.5 **Zn**, 0.4–1.0 **Fe**, 0.5 **Co**, 0.5–1.5 **Mn**, 29.0–32.0 **Ni**, 0.08 **S**, 0.06 **C**, **Co** included in **Ni**; 0.5 **Sn+Pb**; 0.2 total, others	**Annealed**: 35% in 2 in. (50 mm) **El**, 115 max **HV**
ISO 1634	CuNi30Mn1Fe	ISO	Plate, sheet, strip	rem **Cu**, 0.5 **Zn**, 0.4–1.0 **Fe**, 0.5–1.5 **Mn**, 29.0–32.0 **Ni**, 0.08 **S**, 0.06 **C**, 0.05 **Sn+Pb**, 0.2 total, others	**Annealed**: all **diam**, 30% in 2 in. (50 mm) **El**, 105 max **HV**
MIL-C-15726E	70–30	US	Rod, wire, strip, sheet, bar, plate, forgings	65.0 **Cu**, 0.05 **Pb**, 1.00 **Zn**, 0.40–0.70 **Fe**, 1.0 **Mn**, 29.0–32.0 **Ni**, 0.02 **P**, 0.02 **S**	**Soft (rod)**: ≤5 in. **diam**, 310 MPa **TS**, 124 MPa **YS**, 30% in 2 in. (50 mm) **El**; **Hard (rod)**: 1–3 in. **diam**, 310 MPa **TS**, 124 MPa **YS**, 30% in 2 in. (50 mm) **El**; **Hot forged (forgings)**: ≤6 in. **diam**, 345 MPa **TS**, 138 MPa **YS**, 30% in 2 in. (50 mm) **El**
MIL-T-15005E	70–30	US	Tube	65.0 **Cu**, 0.05 **Pb**, 1.00 **Zn**, 0.40–0.70 **Fe**, 1.0 **Mn**, 29.0–32.0 **Ni**	
MIL-T-22214A	Alloy no. 715	US	Tube	65.0 **Cu**, 0.05 **Pb**, 1.00 **Zn**, 0.40–0.70 **Fe**, 1.0 **Mn**, 29.0–32.0 **Ni**, 0.020 **P**, 0.020 **S**	
SABS 460	Cu-Ni30Mn1	South Africa	Tube	rem **Cu**, 0.01 **Pb**, 0.4–1.0 **Fe**, 0.5–1.5 **Mn**, 29.0–32 **Ni**, 0.08 **S**, 0.05 **C**, 0.30 total, others	**Annealed**: all **diam**, 370 MPa **TS**, 30% in 2 in. (50 mm) **El**, 110 max **HV**; **As drawn**: all **diam**, 500 MPa **TS**, 140 **HV**
BS 2870	CN 106	UK	Sheet, strip, foil	69.0–71.0 **Cu**, 0.01 **Pb**, 0.30 **Fe**, 0.05–0.50 **Mn**, 29.0–31.0 **Ni**, 0.03 **S**, 0.06 **C**, 0.1 total, others	**Annealed**: 0.6–2.0 mm **diam**, 259 MPa **TS**, 30% in 2 in. (50 mm) **El**

WROUGHT COPPER AND COPPER ALLOYS

specification number	designation	country	product forms	chemical composition	mechanical properties and hardness values
COPANT 527	Cu Ni 31 Fe	COPANT	Tube	65 min **Cu**, 0.05 **Pb**, 1.0 **Zn**, 0.4–0.7 **Fe**, 1.0 **Mn**, 29–30 **Ni**, 99.5 min, total elements	**Annealed**: 0.8–3.4 mm **diam**, 360 MPa **TS**, 125 MPa **YS**; **Tempered**: 0.8–3.4 mm **diam**, 500 MPa **TS**, 345 MPa **YS**, 12% in 2 in. (50 mm) **El**
BS 2870	CN 105	UK	Sheet, strip, foil	rem **Cu**, 0.20 **Zn**, 0.30 **Fe**, 0.05–0.40 **Mn**, 24.0–26.0 **Ni**, 0.02 **S**, 0.05 **C**, 0.35 total, others	**Annealed**: 0.6–2.0 mm **diam**, 238 MPa **TS**, 30% in 2 in. (50 mm) **El**
COPANT 162	C 713 00	COPANT	Wrought products	70.75–73.75 **Cu**, 0.05 **Pb**, 1.0 **Zn**, 0.20 **Fe**, 1.0 **Mn**, 23.5–26.5 **Ni**	
ISO 429	CuNi25	ISO	Plate, sheet, strip, wire	rem **Cu**, 0.5 **Zn**, 0.3 **Fe**, 0–0.5 **Mn**, 24.0–26.0 **Ni**, 0.02 **S**, 0.5 **C**, 0.05 **Sn+Pb**, 0.1 total, others	
ISO 1634	CuNi25	ISO	Plate, sheet, strip	rem **Cu**, 0.5 **Zn**, 0.3 **Fe**, 0–0.5 **Mn**, 24.0–26.0 **Ni**, 0.02 **S**, 0.05 **C**, 0.05 **Sn+Pb**, 0.1 total, others	**Annealed**: all **diam**, 35% in 2 in. (50 mm) **El**, 95 max **HV**
COPANT 162	C 711 00	COPANT	Wrought products	75–77 **Cu**, 0.05 **Pb**, 0.20 **Zn**, 0.10 **Fe**, 0.15 **Mn**, 22–24 **Ni**	
JIS C 2532	Class 30	Japan	Wire, ribbon, and sheet: low resistance	72.0–77.0 **Cu**, 0.5 **Fe**, 1.5 **Mn**, 20.0–25.0 **Ni+Co**, 99.0 min **Cu+Ni+Mn**	**Annealed**: <0.2 mm **diam**, 295 MPa **TS**, 20% in 2 in. (50 mm) **El**
AMS 4732		US	Wire and ribbon: for electronic applications	77.0–79.5 **Cu**, 0.35 **Mn**, 21.5–22.5 **Ni**, 0.008 **S**, 0.05 **Ti**	
CDA 710		US	Wire, tube, strip	79 **Cu**, 0.05 **Pb**, 1.0 **Zn**, 1.0 **Fe**, 1.0 **Mn**, 19.0–23.0 **Ni**, 99.5 min **Cu**, + others	**Quarter hard temper (rolled strip)**: 0.040 in. **diam**, 60 ksi (414 MPa) **TS**, 49 ksi (338 MPa) **YS**, 20% in 2 in. (50 mm) **El**, 58 **HRB**; **Light drawn (tube)**: 1.0 in. **diam**, 68 ksi (469 MPa) **TS**, 14% in 2 in. (50 mm) **El**, 76 **HRB**; **Extra spring temper (wire)**: 0.080 in. **diam**, 95 ksi (655 MPa) **TS**, 5% in 2 in. (50 mm) **El**
COPANT 162	C 710 00	COPANT		73.45–77.45 **Cu**, 0.05 **Pb**, 1.0 **Zn**, 1.0 **Fe**, 1.0 **Mn**, 19–23 **Ni**	
COPANT 527	Cu Ni 21 Fe	COPANT	Tube	74 min **Cu**, 0.05 **Pb**, 1.0 **Zn**, 0.5–1.0 **Fe**, 1.0 **Mn**, 19–23 **Ni**, 99.5 min, total elements	**Annealed**: 0.8–3.4 mm **diam**, 315 MPa **TS**, 105 MPa **YS**
ANSI/ASTM B 111	C 71000	US	Tube: seamless	74.0 min **Cu**, 0.05 **Pb**, 1.0 **Zn**, 0.50–1.0 **Fe**, 1.0 **Mn**, 19.0–23.0 **Ni+Co**	**Annealed**: all **diam**, 45 ksi (310 MPa) **TS**, 16 ksi (110 MPa) **YS**

WROUGHT COPPER AND COPPER ALLOYS

specification number	designation	country	product forms	chemical composition	mechanical properties and hardness values
ANSI/ASTM B 206	CA 710	US	Wire	74.0 min **Cu**, 0.05 **Pb**, 1.0 **Zn**, 1.0 **Fe**, 1.0 **Mn**, 19.0–23.0 **Ni + Co**	**Quarter hard**: 0.020–0.250 in. (0.508–6.35 mm) **diam**, 55 ksi (380 MPa) **TS**; Hard: 0.020–0.250 in. (0.508–6.35 mm) **diam**, 77 ksi (530 MPa) **TS**, Spring: 0.020–0.0253 in. (0.508–0.643 mm) **diam**, 90 ksi (620 MPa) **TS**
ANSI/ASTM B 359	C 71000	US	Tube: seamless	74.0 min **Cu**, 0.05 **Pb**, 1.0 **Zn**, 0.50–1.0 **Fe**, 1.0 **Mn**, 19.0–23.0 **Ni + Co**	**Annealed**: all **diam**, 45 ksi (310 MPa) **TS**, 16 ksi (110 MPa) **YS**
ANSI/ASTM B 395	C 71000	US	Tube: seamless	74.0 min **Cu**, 0.05 **Pb**, 1.0 **Zn**, 0.50–1.0 **Fe**, 1.0 **Mn**, 19.0–23.0 **Ni + Co**	**Annealed**: <0.048 in. (<1.21 mm) **diam**, 45 ksi (310 MPa) **TS**, 16 ksi (110 MPa) **YS**, 12% in 2 in. (50.8 mm) **El**
COPANT 531	Cu Ni 21 Fe	COPANT	Tube	0.05 **Pb**, 1.0 **Zn**, 0.5–1.0 **Fe**, 1.0 **Mn**, 19–23 **Ni**, 0.02 **P**, 0.02 **S**, 74 min **Cu + As**, 0.5 total, others	**Annealed**: 3.18–127 mm **diam**, 315 MPa **TS**, 105 MPa **YS**; **Full tempered**: all **diam**, 380 MPa **TS**, 295 MPa **YS**
JIS H 3300	C 7100	Japan	Pipe and tube: seamless	73.5–78.8 **Cu**, 0.05 **Pb**, 1.0 **Zn**, 0.50–1.0 **Fe**, 0.20–1.0 **Mn**, 19.0–23.0 **Ni**	**Annealed**: 5–50 mm **diam**, 314 MPa **TS**, 30% in 2 in. (50 mm) **El**
ANSI/ASTM B 467	CA 710	US	Pipe, tube	74.0 min **Cu**, 0.05 **Pb**, 1.0 **Zn**, 0.5–1.0 **Fe**, 1.0 **Mn**, 19.0–23.0 **Ni + Co**, 0.02 **P**, 0.02 **S**	**Annealed**: <2.0 in. (<50.8 mm) **diam**, 45 ksi (310 MPa) **TS**, 16 ksi (110 MPa) **YS**, 30% in 2 in. (50.8 mm) **El**
ANSI/ASTM B 466	CA 710	US	Pipe and tube: seamless	74.0 min **Cu**, 0.05 **Pb**, 1.0 **Zn**, 0.5–1.0 **Fe**, 1.0 **Mn**, 19.0–23.0 **Ni + Co**, 0.02 **S**	**Annealed**: >0.020 in. (>0.508 mm) **diam**, 45 ksi (310 MPa) **TS**, 16 ksi (110 MPa) **YS**, 48 max **HR30T**; **Hard drawn**: >0.020 in. (>0.508 mm) **diam**, 55 ksi (380 MPa) **TS**, 43 ksi (295 MPa) **YS**, 67 min **HR30T**
ASTM F 467	C 71000	US	Nut	74.0 min **Cu**, 0.05 **Pb**, 1.00 **Zn**, 0.60 **Fe**, 1.00 **Mn**, 19.0–23.0 **Ni + Co**	**As agreed**: all **diam**, 50 min **HRB**
ANSI/ASTM B 122	C 73200	US	Plate, sheet, strip, bar	70.0 min **Cu**, 0.05 **Pb**, 3.0–6.0 **Zn**, 0.6 **Fe**, 1.0 **Mn**, 19.0–23.0 **Ni + Co**, silver included in copper content	**Hard**: >0.012 in. (>0.305 mm) **diam**, 73 ksi (505 MPa) **TS**, 71–76 **HR30T**; **Spring**: >0.012 in. (>0.305 mm) **diam**, 82 ksi (565 MPa) **TS**, 74–78 **HR30T**; **Annealed**: >0.015 in. (>0.381mm) **diam**, 28–40 **HR30T**

WROUGHT COPPER AND COPPER ALLOYS

specification number	designation	country	product forms	chemical composition	mechanical properties and hardness values
ANSI/ASTM B 122	C 71000	US	Plate, sheet, strip, bar	74.0 min **Cu**, 0.05 **Pb**, 1.0 **Zn**, 1.0 **Fe**, 1.0 **Mn**, 19.0–23.0 **Ni+Co**, silver included in copper content	**Hard**: >0.012 in. (>0.305 mm) **diam**, 67 ksi (460 MPa) **TS**, 67–73 **HR30T**; **Spring**: >0.012 in. (>0.305 mm) **diam**, 76 ksi (525 MPa) **TS**, 71–75 **HR30T**; **Annealed**: >0.015 in. (>0.381 mm) **diam**, 28–40 **HR30T**
ASTM F 468	C 71000	US	Bolt, screw, stud	74.0 min **Cu**, 0.05 **Pb**, 1.00 **Zn**, 0.60 **Fe**, 1.00 **Mn**, 19.0–23.0 **Ni+Co**	**As manufactured**: all **diam**, 45 ksi (310 MPa) **TS**, 15 ksi (105 MPa) **YS**, 50–85 **HRB**
ISO 1634	CuNi20Mn1Fe	ISO	Plate, sheet, strip	rem **Cu**, 0.50 **Zn**, 0.4–1.0 **Fe**, 0.5–1.5 **Mn**, 19.0–22.0 **Ni+Co**, 0.05 **C**, 0.05 **Sn+Pb**, 0.1 total, other	**M**: all **diam**, 115 MPa **TS**, 350 MPa **YS**, 35% in 2 in. (50 mm) **El**
ISO 429	CuNi20Mn1Fe	ISO	Plate, sheet, strip, tube	rem **Cu**, 0.5 **Zn**, 0.4–1.0 **Fe**, 0.5 **Co**, 0.5–1.5 **Mn**, 19.0–22.0 **Ni**, 0.05 **S**, 0.05 **C**, **Co** included in **Ni**; 0.05 **Sn+Pb**; 0.1 total, others	**Annealed**: all **diam**, 35% in 2 in. (50 mm) **El**, 110 max **HV**
BS 2870	CN 104	UK	Sheet, strip, foil	79.0–81.0 **Cu**, 0.01 **Pb**, 0.30 **Fe**, 0.05–0.50 **Mn**, 19.0–21.0 **Ni**, 0.02 **S**, 0.05 **C**, 0.1 total, others	**Annealed**: 0.6–2.0 mm **diam**, 217 MPa **TS**, 35% in 2 in. (50 mm) **El**
ISO 429	CuNi20	ISO	Plate, sheet, strip, rod, bar, extruded sections	rem **Cu**, 0.2 **Zn**, 0.3 **Fe**, 0.0–0.5 **Mn**, 19.0–21.0 **Ni**, 0.05 **S**, 0.05 **C**, 0.05 **Sn+Pb**, 0.1 total, others	
ISO 1634	CuNi20	ISO	Plate, sheet, strip	rem **Cu**, 0.2 **Zn**, 0.3 **Fe**, 0–0.5 **Mn**, 19.0–21.0 **Ni**, 0.05 **S**, 0.05 **C**, 0.05 **Sn+Pb**, 0.1 total, others	**M**: 330 MPa **TS**, 95 MPa **YS**, 35% in 2 in. (50 mm) **El**
BS 2870	CN 103	UK	Sheet, strip, foil	84.0–86.0 **Cu**, 0.01 **Pb**, 0.25 **Fe**, 0.05–0.50 **Mn**, 14.0–16.0 **Ni**, 0.02 **S**, 0.05 **C**, 0.30 total, others	**Annealed**: 0.6–2.0 mm **diam**, 197 MPa **TS**, 35% in 2 in. (50 mm) **El**
COPANT 162	C 709 00	COPANT	Wrought products	80.75–83.75 **Cu**, 0.05 **Pb**, 1.0 **Zn**, 0.6 **Fe**, 0.6 **Mn**, 13.5–16.5 **Ni**	
COPANT 162	C 708 00	COPANT	Wrought products	86.5–88.5 **Cu**, 0.05 **Pb**, 0.20 **Zn**, 0.10 **Fe**, 0.15 **Mn**, 10.5–12.5 **Ni**	
AS 1566	706B	Australia	Plate, bar, sheet, strip, foil	85.69 **Cu**, 0.01 **Pb**, 1.0–2.0 **Fe**, 0.5–1.0 **Mn**, 10.0–11.0 **Ni**, sulfur and carbon included in copper content	**Annealed (sheet, strip)**: all **diam**, 290 MPa **TS**
BS 2871	CN 102	UK	Tube	rem **Cu**, 0.01 **Pb**, 1.00–2.00 **Fe**, 0.50–1.00 **Mn**, 10.0–11.0 **Ni**, 0.05 **S**, 0.05 **C**, 0.30 total, others	**As drawn**: 150 **HV**; **Annealed**: 80–110 **HV**
BS 2871	CN 102	UK	Tube	rem **Cu**, 0.01 **Pb**, 1.00–2.00 **Fe**, 0.50–1.00 **Mn**, 10.0–11.0 **Ni**, 0.05 **S**, 0.05 **C**, 0.30 total, others	**Annealed**: 300 MPa **TS**, 30% in 2 in. (50 mm) **El**, 100 max **HV**; **As drawn**: 430 MPa **TS**, 130 **HV**

WROUGHT COPPER AND COPPER ALLOYS

specification number	designation	country	product forms	chemical composition	mechanical properties and hardness values
BS 2870	CN 102	UK	Sheet, strip, foil	rem **Cu**, 0.01 **Pb**, 1.00–2.00 **Fe**, 0.50–1.00 **Mn**, 10.0–11.0 **Ni**, 0.05 **S**, 0.05 **C**, 0.30 total, others	**As manufactured**: ≤10.0 mm **diam**, 217 MPa **TS**, 30% in 2 in. (50 mm) **El**, 90 max **HV**; **Annealed**: ≤10.0 mm **diam**, 197 MPa **TS**, 40% in 2 in. (50 mm) **El**, 90 max **HV**
BS 2875	CN 102	UK	Plate	rem **Cu**, 0.01 **Pb**, 1.0–2.0 **Fe**, 0.50–1.0 **Mn**, 10.0–11.0 **Ni**, 0.05 **S**, 0.05 **C**, 0.30 total, others	**As manufactured**: ≥10 mm **diam**, 279 MPa **TS**, 27% in 2 in. (50 mm) **El**; **Annealed**: ≥10 mm **diam**, 279 MPa **TS**, 36% in 2 in. (50 mm) **El**
IS 1545	CuNi10Fe1	India	Tube	rem **Cu**, 0.01 **Pb**, 1.00–2.00 **Fe**, 0.50–1.00 **Mn**, 10.0–11.0 **Ni**, 0.05 **S**, 0.05 **C**, 0.30 total, others	**Annealed**: all **diam**, 372 max MPa **TS**; **As drawn**: all **diam**, 387 MPa **TS**
JIS C 2532	Class 15	Japan	Wire, ribbon, and sheet: low resistance	85.0–89.0 **Cu**, 0.5 **Fe**, 1.5 **Mn**, 8.0–12.0 **Ni + Co**, 99.0 min **Cu + Ni + Mn**	**Annealed**: <0.2 mm **diam**, 245 MPa **TS**, 20% in 2 in. (50 mm) **El**
ANSI/ASTM F 96	C 70690	US	Wrought products	89.0 min **Cu**, 0.001 **Pb**, 0.001 **Zn**, 0.005 **Fe**, 0.02 **Si**, 0.002 **Al**, 0.02 **Co**, 0.001 **Mn**, 9.0–11.0 **Ni**, 0.001 **P**, 0.003 **S**, 0.03 **C**	
ANSI/ASTM B 469	C 70600	US	Tube: seamless	86.5 **Cu**, 0.05 **Pb**, 1.0 **Zn**, 1.0–1.8 **Fe**, 1.0 **Mn**, 9.0–11.0 **Ni**	**As manufactured**: <0.25 in. (<6.350 mm) **diam**, 45 ksi (310 MPa) **TS**, 30 ksi (207 MPa) **YS**, 15% in 2 in. (50.8 mm) **El**
CDA 706		US	Tube, plate, sheet, strip	88.6 nom **Cu**, 0.05 **Pb**, 1.0 **Zn**, 1.0–1.8 **Fe**, 1.0 **Mn**, 9.0–11.0 **Ni**, 99.5 min **Cu**, + others	**0.025 mm temper (tube)**: 1.0 in. **diam**, 44 ksi (303 MPa) **TS**, 42% in 2 in. (50 mm) **El**, 15 **HRB**; **Light drawn (tube)**: 1.0 in. **diam**, 60 ksi (455 MPa) **TS**, 10% in 2 in. (50 mm) **El**, 72 **HRB**
COPANT 162	C 706 00	COPANT	Wrought products	86.5 min **Cu**, 0.05 **Pb**, 1.0 **Zn**, 1.0–1.8 **Fe**, 1.0 **Mn**, 9–11 **Ni**, silver included in copper	
ANSI/ASTM B 395	C 70600	US	Tube: seamless	86.5 min **Cu**, 0.05 **Pb**, 1.0 **Zn**, 1.0–1.8 **Fe**, 9.0–11.0 **Ni + Co**	**Annealed**: <0.048 in. (<1.21 mm) **diam**, 40 ksi (275 MPa) **TS**, 15 ksi (105 MPa) **YS**, 12% in 2 in. (50.8 mm) **El**; **Light drawn**: <0.048 in. (<1.21 mm) **diam**, 45 ksi (310 MPa) **TS**, 35 ksi (240 MPa) **YS**, 12% in 2 in. (50.8 mm) **El**
ANSI/ASTM B 402	C 70600	US	Plate, sheet	86.5 min **Cu**, 0.05 **Pb**, 1.0 **Zn**, 1.0–1.8 **Fe**, 1.0 **Mn**, 9.0–11.0 **Ni + Co**	**As manufactured**: <2.5 in. (<63.5 mm) **diam**, 40 ksi (275 MPa) **TS**, 15 ksi (105 MPa) **YS**, 30% in 2 in. (50.8 mm) **El**

WROUGHT COPPER AND COPPER ALLOYS

specification number	designation	country	product forms	chemical composition	mechanical properties and hardness values
ANSI/ASTM B 359	C 70600	US	Tube: seamless	86.5 min **Cu**, 0.05 **Pb**, 1.0 **Zn**, 1.0–1.8 **Fe**, 1.0 **Mn**, 9.0–11.0 **Ni + Co**	**Annealed**: all **diam**, 40 ksi (275 MPa) **TS**, 15 ksi (105 MPa) **YS**
ANSI/ASTM B 171	C 70600	US	Plate	86.5 min **Cu**, 0.05 **Pb**, 1.0 **Zn**, 1.0–1.8 **Fe**, 1.0 **Mn**, 9.0–11.0 **Ni + Co**, silver included in copper content	**As manufactured**: <2.5 in. (<63.5 mm) **diam**, 40 ksi (275 MPa) **TS**, 15 ksi (105 MPa) **YS**, 30% in 2 in. (50.8 mm) **El**
ANSI/ASTM B 151	CA 706	US	Rod, bar	86.5 min **Cu**, 0.05 **Pb**, 1.0 **Zn**, 1.0–1.8 **Fe**, 1.0 **Mn**, 9.0–11.0 **Ni + Co**, 0.02 **P**, 0.02 **S**, silver included in copper content	**Soft**: all **diam**, 38 ksi (260 MPa) **TS**, 15 ksi (105 MPa) **YS**, 30% in 2 in. (50.8 mm) **El**; **Hard**: <0.375 in. (<9.53 mm) **diam**, 60 ksi (415 MPa) **TS**, 38 ksi (260 MPa) **YS**, 10% in 2 in. (50.8 mm) **El**
ANSI/ASTM B 111	C 70600	US	Tube: seamless	86.5 min **Cu**, 0.05 **Pb**, 1.0 **Zn**, 1.0–1.8 **Fe**, 1.0 **Mn**, 9.0–11.0 **Ni + Co**	**Annealed**: all **diam**, 40 ksi (275 MPa) **TS**, 15 ksi (105 MPa) **YS**; **Light drawn**: all **diam**, 45 ksi (310 MPa) **TS**, 35 ksi (240 MPa) **YS**
COPANT 162	C 707 00	COPANT	Wrought products	88.5 min **Cu**, 0.05 **Fe**, 0.50 **Mn**, 9.5–10.5 **Ni**	
COPANT 527	Cu Ni 10 Fe 1	COPANT	Tube	0.05 **Pb**, 1.0 **Zn**, 1.0–1.8 **Fe**, 1.0 **Mn**, 9–11 **Ni**, 86.5 min **Cu + Ag**, 99.5 min, total elements	**Annealed**: 0.8–3.4 mm **diam**, 275 MPa **TS**, 105 MPa **YS**
COPANT 531	Cu Ni 10 Fe 1	COPANT	Tube	0.05 **Pb**, 1.0 **Zn**, 1.0–1.8 **Fe**, 1.0 **Mn**, 9–11 **Ni**, 0.02 **P**, 0.02 **S**, 86.5 min **Cu + Ag**, 0.5 total, others	**Annealed**: 3.18–127 mm **diam**, 265 MPa **TS**, 90 MPa **YS**; **1/4 temper**: all **diam**, 315 MPa **TS**, 245 MPa **YS**; **Full temper**: all **diam**, 345 MPa **TS**, 275 MPa **YS**
DS 3003	5667	Denmark	Rolled, drawn, extruded, and forged products	88 **Cu**, 1.0–1.8 **Fe**, 0.5–1.0 **Mn**, 9.0–11.0 **Ni**, 0.05 **S**, 0.05 **C**, 0.05 **Pb + Sn**, 0.1 others, total	
ISO 429	CuNi10Fe1Mn	ISO	Plate, sheet, strip, rod, bar, tube	rem **Cu**, 0.5 **Zn**, 1.0–2.0 **Fe**, 0.3–1.0 **Mn**, 9.0–11.0 **Ni**, 0.05 **S**, 0.05 **C**, 0.05 **Sn + Pb**, 0.1 total, others	
ISO 1635	CuNi10Fe1Mn	ISO	Tube	rem **Cu**, 0.5 **Zn**, 1.0–2.0 **Fe**, 0.5 **Co**, 0.3–1.0 **Mn**, 9.0–11.0 **Ni**, 0.05 **S**, 0.05 **C**, 0.05 **Sn + Pb**; **Co** included in **Ni**; 0.1 total, others	**Annealed**: all **diam**, 30% in 2 in. (50 mm) **El**, 110 max **HV**
ISO 1634	CuNi10Fe1Mn	ISO	Plate, sheet, strip	rem **Cu**, 0.5 **Zn**, 1.0–2.0 **Fe**, 0.3–1.0 **Mn**, 9.0–11.0 **Ni**, 0.05 **S**, 0.05 **C**, 0.05 **Sn + Pb**, 0.1 total, others	**As cast**: all **diam**, 330 MPa **TS**, 95 MPa **YS**, 30% in 2 in. (50 mm) **El**
JIS H 3300	C 7060	Japan	Pipe and tube: seamless	84.6–88.3 **Cu**, 0.05 **Pb**, 1.0 **Zn**, 1.0–1.8 **Fe**, 0.20–1.0 **Mn**, 9.0–11.0 **Ni**, 99.5 min **Cu + Ni + Fe + Mn**	**Annealed**: 5–50 mm **diam**, 275 MPa **TS**, 30% in 2 in. (50 mm) **El**

WROUGHT COPPER AND COPPER ALLOYS

specification number	designation	country	product forms	chemical composition	mechanical properties and hardness values
JIS H 3100	C 7060	Japan	Sheet, plate, and strip	86–89 **Cu**, 0.05 **Pb**, 1.0 **Zn**, 1.0–1.8 **Fe**, 0.20–1.0 **Mn**, 9.0–11.0 **Ni**, 99.5 min **Cu + Ni + Fe + Mn**	**As manufactured**: 0.5–50 mm **diam**, 275 MPa **TS**, 30% in 2 in. (50 mm) **El**
MIL–C–15726E	90–10	US	Rod, wire, strip, sheet, bar, plate, forgings	84.91 min **Cu**, 0.05 **Pb**, 1.00 **Zn**, 1.00–1.75 **Fe**, 0.75 **Mn**, 9.0–11.0 **Ni**, 0.02 **P**, 0.02 **S**	**Soft (rod)**: ≤5 in. **diam**, 262 MPa **TS**, 103 MPa **YS**, 30% in 2 in. (50 mm) **El**; **Hard (rod)**: 1–3 in. **diam**, 276 MPa **TS**, 103 MPa **YS**, 30% in 2 in. (50 mm) **El**; **Hot forged (forgings)**: ≤6 in. **diam**, 310 MPa **TS**, 124 MPa **YS**, 30% in 2 in. (50 mm) **El**
MIL–T–15005E	90–10	US	Tube	84.96 min **Cu**, 0.05 **Pb**, 1.00 **Zn**, 1.00–1.75 **Fe**, 0.75 **Mn**, 9.0–11.0 **Ni**	
MIL–T–22214A	Alloy no. 706	US	Tube	84.91 min **Cu**, 0.05 **Pb**, 1.00 **Zn**, 1.00–1.75 **Fe**, 0.75 **Mn**, 9.0–11.0 **Ni**, 0.020 **P**, 0.020 **S**	
NS 16 410	CuNi10Fe1Mn	Norway	Tube	88 **Cu**, 1.5 **Fe**, 0.7 **Mn**, 10 **Ni**	
ANSI/ASTM B 122	C 70600	US	Plate, sheet, strip, bar	86.5 min **Cu**, 0.05 **Pb**, 1.0 **Zn**, 1.0–1.8 **Fe**, 1.0 **Mn**, 9.0–11.0 **Ni + Co**, silver included in copper content	**Hard**: >0.012 in. (>0.305 mm) **diam**, 71 ksi (490 MPa) **TS**, 67–74 **HR30T**; **Spring**: 70.012 in. (>0.305 mm) **diam**, 78 ksi (540 MPa) **TS**, 72–78 **HR30T**; **Annealed**: >0.015 in. (>0.381 mm) **diam**, 15–34 **HR30T**
ANSI/ASTM B 608	CA 706	US		86.5 min **Cu**, 0.02 **Pb**, 0.50 **Zn**, 1.0–1.8 **Fe**, 1.0 **Mn**, 9.0–11.0 **Ni + Co**, 0.02 **P**, 0.02 **S**, 0.05 **C**, silver included in copper content	**Annealed**: all **diam**, 40 ksi (275 MPa) **TS**
ANSI/ASTM B 543	C 70600	US	Tube: welded	86.5 min **Cu**, 0.05 **Pb**, 1.0 **Zn**, 1.0–1.8 **Fe**, 1.0 **Mn**, 9.0–11.0 **Ni + Co**	**As welded from annealed strip**: all **diam**, 45 ksi (310 MPa) **TS**, 30 ksi (205 MPa) **YS**; **Welded and annealed**: all **diam**, 40 ksi (275 MPa) **TS**, 15 ksi (105 MPa) **YS**; **Fully finished–annealed**: all **diam**, 40 ksi (275 MPa) **TS**, 15 ksi (105 MPa) **YS**
ASTM B 552	CA 706	US	Tube: seamless, welded	86.5 min **Cu**, 0.05 **Pb**, 1.0 **Zn**, 1.0–1.8 **Fe**, 1.0 **Mn**, 9.0–11.0 **Ni + Co**	**Annealed**: all **diam**, 40 ksi (275 MPa) **TS**; **Cold worked**: all **diam**, 50 ksi (345 MPa) **TS**

WROUGHT COPPER AND COPPER ALLOYS

specification number	designation	country	product forms	chemical composition	mechanical properties and hardness values
ANSI/ASTM B 466	CA 706	US	Pipe and tube: seamless	86.5 min **Cu**, 0.05 **Pb**, 1.0 **Zn**, 1.0–1.8 **Fe**, 1.0 **Mn**, 9.0–11.0 **Ni+Co**, 0.02 **S**	**Annealed**: >0.020 in. (>0.508 mm) **diam**, 38 ksi (260 MPa) **TS**, 13 ksi (90 MPa) **YS**, 45 max **HR30T**; **Light drawn**: >0.020 in. (>0.508 mm) **diam**, 45 ksi (310 MPa) **TS**, 35 ksi (240 MPa) **YS**, 45–70 **HR30T**; **Hard drawn**: >0.020 in. (>0.508 mm) **diam**, 50 ksi (345 MPa) **TS**, 40 ksi (275 MPa) **YS**, 63 min **HR30T**
ANSI/ASTM B 467	CA 706	US	Pipe, tube	86.5 min **Cu**, 0.05 **Pb**, 1.0 **Zn**, 1.0–1.8 **Fe**, 1.0 **Mn**, 9.0–11.0 **Ni+Co**, 0.02 **P**, 0.02 **S**	**Annealed**: <4.5 in. (<114 mm) **diam**, 40 ksi (275 MPa) **TS**, 15 ksi (105 MPa) **YS**, 25% in 2 in. (50.8 mm) **El**; **Cold rolled**: <4.5 in. (<114 mm) **diam**, 54 ksi (375 MPa) **TS**, 45 ksi (310 MPa) **YS**
MIL-R-19631B	MIL-RBCuZn-D	US	Rod	46.0–50.0 **Cu**, 0.05 **Pb**, rem **Zn**, 0.04–0.25 **Si**, 0.01 **Al**, 9.0–11.0 **Ni+Co**, 0.25 **P**, 0.50 total, others, silver included in copper content	**As milled**: all **diam**, 414 MPa **TS**
MIL-T-16420K	706	US	Tube: seamless and welded	86.5 min **Cu**, 0.02 **Pb**, 0.50 **Zn**, 1.0–1.8 **Fe**, 1.0 **Mn**, 9.0–11.0 **Ni+Co**, 0.02 **P**, 0.02 **S**, 0.05 **C**, silver included in copper content	**Annealed**: ≤4.5 in. **diam**, 262 MPa **TS**, 103 MPa **YS**, 30% in 2 in. (50 mm) **El**; **Light drawn**: 310 MPa **TS**, 241 MPa **YS**, 15% in 2 in. (50 mm) **El**
COPANT 532	Cu Ni 10 Fe 1	COPANT	Bar	86.5 min **Cu**, 0.05 **Pb**, 1.0 **Zn**, 1.0–1.8 **Fe**, 1.0 **Mn**, 9–11 **Ni+Co**, 0.02 **P**, 0.02 **S**, 99.5 min cu and all other specified elements, 0.5 total, others	**Full tempered**: >0.5 mm **diam**, 550 MPa **TS**; **1/4 hard**: 0.5–13 mm **diam**, 410 MPa **TS**
SABS 460	Cu-Ni10FeMn1	South Africa	Tube	rem **Cu**, 0.01 **Pb**, 1.0–2.0 **Fe**, 0.5–1.0 **Mn**, 9.0–11.0 **Ni**, 0.05 **S**, 0.05 **C**, 0.30 total, others	**Annealed**: all **diam**, 300 MPa **TS**, 30% in 2 in. (50 mm) **El**, 100 max **HV**; **As drawn**: all **diam**, 400 MPa **TS**, 130 **HV**
VSM 11557	Cu Ni 10 Fe Mn	Switzerland	Tube	rem **Cu**, 0.5 **Zn**, 1.0–1.7 **Fe**, 0.5–1.0 **Mn**, 9.0–11.0 **Ni**, 0.05 **S**, 0.05 **C**, 0.05 **Pb+Sn**, 0.1 total, others	**Quenched**: all **diam**, 294 MPa **TS**, 98 MPa **YS**, 30% in 2 in. (50 mm) **El**, 75 **HB**

WROUGHT COPPER AND COPPER ALLOYS

specification number	designation	country	product forms	chemical composition	mechanical properties and hardness values
CDA 725		US	Rod, wire, tube, plate, sheet, flatwire, strip	88.2 nom **Cu**, 1.8–2.8 **Sn**, 0.05 **Pb**, 0.50 **Zn**, 0.6 **Fe**, 0.20 **Mn**, 8.5–10.5 **Ni**, 99.8 min **Cu**, + others	**Eighth hard temper (plate, sheet, flatwire, rolled strip)**: 0.040 in. **diam**, 58 ksi (400 MPa) **TS**, 43 ksi (296 MPa) **YS**, 29% in 2 in. (50 mm) **El**, 65 **HRB**; **Half hard temper–45% (wire)**: 0.080 in. **diam**, 90 ksi (621 MPa) **TS**; **0.015 mm temper (wire)**: 0.080 in. **diam**, 60 ksi (414 MPa) **TS**
ANSI/ASTM B 122	C 72500	US	Plate, sheet, strip, bar	99.8 min **Cu**, 1.8–2.8 **Sn**, 0.05 **Pb**, 0.5 **Zn**, 0.6 **Fe**, 0.2 **Mn**, 8.5–10.5 **Ni + Co**, silver included in copper content	**Hard**: >0.012 in. (>0.305 mm) **diam**, 75 ksi (515 MPa) **TS**, 60–75 **HR30T**; **Spring**: >0.012 in. (>0.305mm) **diam**, 85 ksi (585 MPa) **TS**, 72–80 **HR30T**; **Annealed**: >0.015 in. (>0.381 mm) **diam**, 70–81 **HR30T**
COPANT 162	C 725 00	COPANT	Wrought products	85.15–88.15 **Cu**, 1.8–2.8 **Sn**, 0.05 **Pb**, 0.50 **Zn**, 0.6 **Fe**, 0.20 **Mn**, 8.5–10.5 **Ni**	
COPANT 162	C 705 00	COPANT	Wrought products	90 min **Cu**, 0.05 **Pb**, 0.20 **Zn**, 0.10 **Fe**, 0.15 **Mn**, 5.8–7.8 **Ni**	
BS 2870	CN 101	UK	Sheet, strip, foil	rem **Cu**, 0.01 **Sn**, 0.01 **Pb**, 1.05–1.35 **Fe**, 0.01 **Sb**, 0.05 **Si**, 0.30–0.80 **Mn**, 5.0–6.0 **Ni**, 0.03 **P**, 0.05 **S**, 0.05 **C**, 0.05 **As**, 0.30 total, others	**As manufactured**: ≤10.0 mm **diam**, 162 MPa **TS**, 35% in 2 in. (50 mm) **El**
BS 2875	CN 101	UK	Plate	rem **Cu**, 0.01 **Sn**, 0.01 **Pb**, 1.05–1.35 **Fe**, 0.01 **Sb**, 0.05 **Si**, 0.30–0.80 **Mn**, 5.0–6.0 **Ni**, 0.03 **P**, 0.05 **S**, 0.05 **C**, 0.05 **As**, 0.30 total, others	**As manufactured or annealed**: ≥10 mm **diam**, 230 MPa **TS**, 27% in 2 in. (50 mm) **El**
JIS C 2532	Class 10	Japan	Wire, ribbon, and sheet: low resistance	90–93 **Cu**, 0.5 **Fe**, 1.5 **Mn**, 4.0–7.0 **Ni + Co**, 99.0 min **Cu + Ni + Mn**	**Annealed**: <0.2 mm **diam**, 20% in 2 in. (50 mm) **El**
ANSI/ASTM B 111	C 70400	US	Tube: seamless, and stock: ferrule	91.2 min **Cu**, 0.05 **Pb**, 1.0 **Zn**, 1.3–1.7 **Fe**, 0.30–0.8 **Mn**, 4.8–6.2 **Ni + Co**	**Annealed**: all **diam**, 38 ksi (260 MPa) **TS**, 12 ksi (85 MPa) **YS**; **Light drawn**: all **diam**, 40 ksi (275 MPa) **TS**, 30 ksi (205 MPa) **YS**
COPANT 531	Cu Ni 5 Fe 1 Mn	COPANT	Tube	0.05 **Pb**, 1.0 **Zn**, 1.3–1.7 **Fe**, 0.3–0.8 **Mn**, 4.8–6.2 **Ni + Co**, 0.02 **P**, 0.02 **S**, 91.2 min **Cu + Ag**, 0.5 total, others	**Annealed**: 3.18–127 mm **diam**, 255 MPa **TS**, 80 MPa **YS**; **1/4 temper**: 0.406–7.62 mm **diam**, 275 MPa **TS**, 205 MPa **YS**; **Full temper**: all **diam**, 315 MPa **TS**, 245 MPa **YS**

WROUGHT COPPER AND COPPER ALLOYS

specification number	designation	country	product forms	chemical composition	mechanical properties and hardness values
ANSI/ASTM B 466	CA 704	US	Pipe and tube: seamless	91.2 min **Cu**, 0.05 **Pb**, 1.0 **Zn**, 1.3–1.7 **Fe**, 0.30–0.8 **Mn**, 4.8–6.2 **Ni + Co**, 0.02 **P**, 0.02 **S**	**Annealed**: >0.020 in. (>0.508 mm) **diam**, 37 ksi (255 MPa) **TS**, 12 ksi (85 MPa) **YS**, 45 max **HR30T**; **Light drawn**: >0.020 in. (>0.508 mm) **diam**, 40 ksi (275 MPa) **TS**, 30 ksi (205 MPa) **YS**, 41–65 **HR30T**; **Hard drawn**: >0.020 in. (>0.508 mm) **diam**, 45 ksi (310 MPa) **TS**, 35 ksi (240 MPa) **YS**, 60 min **HR30T**
ANSI/ASTM B 543	C 70400	US	Tube: welded	91.2 min **Cu**, 0.05 **Pb**, 1.0 **Zn**, 1.3–1.7 **Fe**, 0.30–0.8 **Mn**, 4.8–6.2 **Ni + Co**	**As welded from annealed strip**: all **diam**, 38 ksi (260 MPa) **TS**, 12 ksi (85 MPa) **YS**; **Welded and annealed**: all **diam**, 38 ksi (260 MPa) **TS**, 12 ksi (85 mPa) **YS**; **Fully finished–annealed**: all **diam**, 38 ksi (260 MPa) **TS**, 12 ksi (85 MPa) **YS**
ANSI/ASTM B 359	C 70400	US	Tube: seamless	91.2 min **Cu**, 0.05 **Pb**, 1.0 **Zn**, 1.3–1.7 **Fe**, 0.30–0.8 **Mn**, 4.8–6.2 **Ni + Co**	**Annealed**: all **diam**, 38 ksi (260 MPa) **TS**, 12 ksi (85 MPa) **YS**
ANSI/ASTM B 395	C 70400	US	Tube: seamless	91.2 min **Cu**, 0.05 **Pb**, 1.0 **Zn**, 1.3–1.7 **Fe**, 0.30–0.8 **Mn**, 4.8–6.2 **Ni + Co**	**Annealed**: <0.048 in. (<1.21 mm) **diam**, 38 ksi (260 MPa) **TS**, 12 ksi (85 MPa) **YS**, 12% in 2 in. (50.8 mm) **El**; **Light drawn**: <0.048 in. (<1.21 mm) **diam**, 40 ksi (275 MPa) **TS**, 30 ksi (205 MPa) **YS**, 12% in 2 in. (50.8 mm) **El**
COPANT 162	C 704 00	COPANT	Wrought products	90–92 **Cu**, 0.05 **Pb**, 1.0 **Zn**, 1.3–1.7 **Fe**, 0.3–0.8 **Mn**, 4.8–6.2 **Ni**	
COPANT 527	Cu Ni 5 Fe 1 Mn	COPANT	Tube	0.05 **Pb**, 1.0 **Zn**, 1.3–1.7 **Fe**, 0.3–0.8 **Mn**, 4.8–6.2 **Ni**, 91.2 min **Cu + Ag**, 99.5 min, total elements	**Annealed**: 0.8–3.4 mm **diam**, 265 MPa **TS**, 80 MPa **YS**; **Tempered**: 0.8–3.4 mm **diam**, 275 MPa **TS**, 205 MPa **YS**
ISO 429	CuNi5Fe1Mn	ISO	Plate, sheet, strip, tube	rem **Cu**, 0.5 **Zn**, 1.0–1.5 **Fe**, 0.3–0.8 **Mn**, 4.5–6.0 **Ni**, 0.05 **S**, 0.05 **C**, 0.05 **Sn + Pb**, 0.1 total, others	
ISO 1635	CuNi5Fe1Mn	ISO	Tube	rem **Cu**, 0.5 **Zn**, 1.0–1.5 **Fe**, 0.5 **Co**, 0.3–0.8 **Mn**, 4.5–6.0 **Ni**, 0.05 **S**, 0.05 **C**, **Co** included in **Ni**; 0.5 **Sn + Pb**; 0.1 total, others	**Annealed**: all **diam**, 30% in 2 in. (50 mm) **El**, 100 max **HV**
ISO 1634	CuNi5Fe1Mn	ISO	Plate, sheet, strip	rem **Cu**, 0.5 **Zn**, 1.0–1.5 **Fe**, 0.3–0.8 **Mn**, 4.5–6.0 **Ni**, 0.05 **S**, 0.05 **C**, 0.05 **Sn + Pb**, total; 0.1 total, others	**Annealed**: all **diam**, 30% in 2 in. (50 mm) **El**, 85 max **HV**

WROUGHT COPPER AND COPPER ALLOYS

specification number	designation	country	product forms	chemical composition	mechanical properties and hardness values
COPANT 162	C 703 00	COPANT	Wrought products	93 min **Cu**, 0.05 **Fe**, 0.50 **Mn**, 4.7–5.7 **Ni**	
COPANT 162	C 701 00	COPANT	Wrought products	95 min **Cu**, 0.25 **Zn**, 0.05 **Fe**, 0.25–0.50 **Mn**, 3–4 **Ni**	
COPANT 162	C 702 00	COPANT	Wrought products	95 min **Cu**, 0.05 **Pb**, 0.10 **Fe**, 0.40 **Mn**, 2–3 **Ni**	
ISO 1637	CuNi2Si	ISO	Solid products: straight lengths	rem **Cu**, 0.5–0.8 **Si**, 1.6–2.5 **Ni**, 0.5 total, others	**Solution heat treated and naturally aged**: 30 max mm **diam**, 450 MPa **TS**, 340 MPa **YS**, 8% in 2 in. (50 mm) **El**, 130 **HV**; **Solution heat treated and artificially aged**: 30 max mm **diam**, 670 MPa **TS**, 590 MPa **YS**, 10% in 2 in. (50 mm) **El**, 180 **HV**
ISO R1187	CuNi2Si	ISO	Strip, rod, bar, wire	rem **Cu**, 0.5–0.8 **Si**, 1.6–2.5 **Ni**, 0.5 total, others	
COPANT 428	Cu Ni 2 Si	COPANT	Wire	rem **Cu**, 0.10 **Pb**, 0.5 **Zn**, 0.10 **Fe**, 0.4–0.8 **Si**, 1.6–2.2 **Ni**, 0.5 total, others	**Solution treated or aged**: 1.02–19 mm **diam**, 615 MPa **TS**, 520 MPa **YS**, 5% in 2 in. (50 mm) **El**
JIS C 2532	Class 5	Japan	Wire, ribbon, and sheet: low resistance	94.0–96.5 **Cu**, 0.5 **Fe**, 1.5 **Mn**, 0.5–3.0 **Ni + Co**, 99.0 min **Cu + Ni + Mn**	**Annealed**: <0.2 mm **diam**, 245 MPa **TS**, 20% in 2 in. (50 mm) **El**
BS 2873	NS109	UK	Wire	55.0–60.0 **Cu**, 0.025 **Pb**, rem **Zn**, 0.30 **Fe**, 0.05–0.75 **Mn**, 24.0–26.0 **Ni**, 0.50 total, others	
BS 2870	NS 109	UK	Sheet, strip, foil	55.0–60.0 **Cu**, 0.025 **Pb**, rem **Zn**, 0.30 **Fe**, 0.05–0.75 **Mn**, 24.0–26.0 **Ni**, 0.50 total, others	**Annealed**: ≤10.0 mm **diam**, 115 max **HV**; **1/2 hard**: ≤10.0 mm **diam**, 150 **HV**; **Hard**: ≤10.0 mm **diam**, 180 **HV**
NF A 51 107	Cu Ni 25 Zn 20	France	Sheet and strip	53–57 **Cu**, 0.05 **Pb**, rem **Zn**, 0.5 **Mn**, 24–26 **Ni**, 0.3 others, total	**1/2 hard**: ≤10 mm **diam**, 500 MPa **TS**, 20% in 2 in. (50 mm) **El**, 145–170 **HV**; **3/4 hard**: ≤10 mm **diam**, 560 MPa **TS**, 10% in 2 in. (50 mm) **El**, 170–190 **HV**; **Hard**: ≤10 mm **diam**, 640 MPa **TS**, 190–210 **HV**
COPANT 162	C 732 00	COPANT	Wrought products	70.0 min **Cu**, 0.05 **Pb**, 3.0–6.0 **Zn**, 0.6 **Fe**, 1.0 **Mn**, 19.0–23 **Ni**	
BS 2873	NS108	UK	Wire	60.0–65.0 **Cu**, 0.025 **Pb**, rem **Zn**, 0.30 **Fe**, 0.05–0.50 **Mn**, 19.0–21.0 **Ni**, 0.50 total, others	

WROUGHT COPPER AND COPPER ALLOYS

specification number	designation	country	product forms	chemical composition	mechanical properties and hardness values
BS 2870	NS 108	UK	Sheet, strip, foil	60.0–65.0 **Cu**, 0.025 **Pb**, rem **Zn**, 0.30 **Fe**, 0.05–0.50 **Mn**, 19.0–21.0 **Ni**, 0.50 total, others	**Annealed**: ≤10.0 mm **diam**, 110 max **HV**; **1/2 hard**: ≤10.0 mm **diam**, 140 **HV**; **Hard**: ≤10.0 mm **diam**, 175 **HV**
AS 1566	770B	Australia	Plate, bar, sheet, strip, foil	54.0–56.0 **Cu**, 0.03 **Pb**, rem **Zn**, 0.30 **Fe**, 0.05–0.35 **Mn**, 17.0–19.0 **Ni**, 0.50 total, others	**Annealed (sheet, strip)**: all **diam**, 115 **HV**; **Strain hardened to 1/2 hard (sheet, strip)**: all **diam**, 170–190 **HV**; **Strain hardened to full hard (sheet, strip)**: all **diam**, 195–200 **HV**
BS 2873	NS106	UK	Wire	60.0–65.0 **Cu**, 0.03 **Pb**, rem **Zn**, 0.30 **Fe**, 0.05–0.50 **Mn**, 17.0–19.0 **Ni**, 0.50 total, others	
BS 2873	NS107	UK	Wire	54.0–56.0 **Cu**, 0.03 **Pb**, rem **Zn**, 0.30 **Fe**, 0.05–0.35 **Mn**, 17.0–19.0 **Ni**, 0.50 total, others	
BS 2874	NS113	UK	Rod, section	60.0–63.0 **Cu**, 0.4–0.8 **Pb**, rem **Zn**, 0.3 **Fe**, 0.1–0.5 **Mn**, 17.0–19.0 **Ni**, 0.10 total, others	
BS 2870	NS 106	UK	Sheet, strip, foil	60.0–65.0 **Cu**, 0.03 **Pb**, rem **Zn**, 0.30 **Fe**, 0.05–0.50 **Mn**, 17.0–19.0 **Ni**, 0.50 total, others	**Annealed**: ≤10.0 mm **diam**, 110 max **HV**; **1/2 hard**: ≤10.0 mm **diam**, 135 **HV**; **Hard**: ≤10.0 mm **diam**, 170 **HV**
BS 2870	NS 107	UK	Sheet, strip, foil	54.0–56.0 **Cu**, 0.03 **Pb**, rem **Zn**, 0.30 **Fe**, 0.05–0.35 **Mn**, 17.0–19.0 **Ni**, 0.50 total, others	
CDA 752		US	Rod, wire, flat wire, strip	63.0–66.5 **Cu**, 0.10 **Pb**, rem **Zn**, 0.25 **Fe**, 0.50 **Mn**, 16.5–19.5 **Ni**, 0.50 total, others	**Quarter hard temper (flat wire, strip)**: 0.040 in. **diam**, 65 ksi (448 MPa) **TS**, 20% in 2 in. (50 mm) **El**, 73 **HRB**; **Half hard–20% (rod)**: 0.500 in. **diam**, 70 ksi (483 MPa) **TS**, 20% in 2 in. (50 mm) **El**, 78 **HRB**; **Hard temper (wire)**: 0.080 in. **diam**, 103 ksi (710 MPa) **TS**, 3% in 2 in. (50 mm) **El**
CDA 770		US	Rod, wire, flat wire, strip	53.5–56.5 **Cu**, 0.10 **Pb**, rem **Zn**, 0.25 **Fe**, 0.50 **Mn**, 16.5–19.5 **Ni**, 0.50 total, others	**Hard temper (flat wire, strip: rolled)**: 0.040 in. **diam**, 100 ksi (690 MPa) **TS**, 3% in 2 in. (50 mm) **El**, 91 **HRB**; **Extra hard temper (flatwire, strip: rolled)**: 0.040 in. **diam**, 108 ksi (745 MPa) **TS**, 2.5% in 2 in. (50 mm) **El**, 96 **HRB**; **Spring temper–68% (wire)**: 0.080 in. **diam**, 145 ksi (1000 MPa) **TS**, 2% in 2 in. (50 mm) **El**

WROUGHT COPPER AND COPPER ALLOYS

specification number	designation	country	product forms	chemical composition	mechanical properties and hardness values
CSA HC 4.4	HC 4.NZ1810	Canada	Bar	70.5–73.5 **Cu**, 0.10 **Pb**, rem **Zn**, 0.25 **Fe**, 0.50 **Mn**, 16.5–19.5 **Ni**, 0.50 others	**Half hard**: all **diam**, 56 ksi (386 MPa) **TS**; **Full hard**: all **diam**, 73 ksi (503 MPa) **TS**; **Extra hard**: all **diam**, 79 ksi (544 MPa) **TS**
CSA HC 4.4	HC 4.NZ1810	Canada	Plate	70.5–73.5 **Cu**, 0.10 **Pb**, rem **Zn**, 0.25 **Fe**, 0.50 **Mn**, 16.5–19.5 **Ni**, 0.50 others	**Quarter hard**: all **diam**, 56 ksi (386 MPa) **TS**; **Half hard**: all **diam**, 63 ksi (434 MPa) **TS**
CSA HC.4.4	HC 4.NZ1810	Canada	Strip	70.5–73.5 **Cu**, 0.10 **Pb**, rem **Zn**, 0.25 **Fe**, 0.50 **Mn**, 16.5–19.5 **Ni**, 0.50 others	**Half hard**: all **diam**, 63 ksi (434 MPa) **TS**; **Full hard**: all **diam**, 73 ksi (503 MPa) **TS**; **Extra hard**: all **diam**, 79 ksi (544 MPa) **TS**
CSA HC.4.4	HC 4.NZ1810	Canada	Sheet	70.5–73.5 **Cu**, 0.40 **Pb**, rem **Zn**, 0.25 **Fe**, 0.50 **Mn**, 16.5–19.5 **Ni**, 0.50 others	**Half hard**: all **diam**, 63 ksi (434 MPa) **TS**; **Full hard**: all **diam**, 73 ksi (503 MPa) **TS**; **Extra hard**: all **diam**, 79 ksi (544 MPa) **TS**
CSA HC.4.4	HC 4.NZ1810	Canada	Strip	70.5–73.5 **Cu**, 0.10 **Pb**, rem **Zn**, 0.25 **Fe**, 0.50 **Mn**, 16.5–19.5 **Ni**, 0.50 others	**Half hard**: all **diam**, 63 ksi (434 MPa) **TS**; **Full hard**: all **diam**, 73 ksi (503 MPa) **TS**; **Extra hard**: all **diam**, 79 ksi (544 MPa) **TS**
CSA HC.4.4	HC 4.NZ1817	Canada	Bar	63.0–66.5 **Cu**, 0.10 **Pb**, rem **Zn**, 0.25 **Fe**, 0.50 **Mn**, 16.5–19.5 **Ni**, 0.50 others	**Half hard**: all **diam**, 66 ksi (455 MPa) **TS**; **Full hard**: all **diam**, 78 ksi (517 MPa) **TS**; **Spring**: all **diam**, 90 ksi (670 MPa) **TS**
CSA HC.4.4	HC 4.NZ1817	Canada	Strip	63.0–66.5 **Cu**, 0.10 **Pb**, rem **Zn**, 0.25 **Fe**, 0.50 **Mn**, 16.5–19.5 **Ni**, 0.50 others	**Half hard**: all **diam**, 66 ksi (455 MPa) **TS**; **Full hard**: all **diam**, 78 ksi (517 MPa) **TS**; **Spring**: all **diam**, 90 ksi (670 MPa) **TS**
CSA HC.4.4	HC 4.NZ1817	Canada	Sheet	63.0–66.5 **Cu**, 0.10 **Pb**, rem **Zn**, 0.25 **Fe**, 0.50 **Mn**, 16.5–19.5 **Ni**, 0.50 others	**Half hard**: all **diam**, 66 ksi (455 MPa) **TS**; **Full hard**: all **diam**, 78 ksi (517 MPa) **TS**; **Spring**: all **diam**, 90 ksi (670 MPa) **TS**
CSA HC.4.4	HC 4.NZ1817	Canada	Plate	63.0–66.5 **Cu**, 0.10 **Pb**, rem **Zn**, 0.25 **Fe**, 0.50 **Mn**, 16.5–19.5 **Ni**, 0.50 others	**Quarter hard**: all **diam**, 58 ksi (399 MPa) **TS**; **Half hard**: all **diam**, 66 ksi (455 MPa) **TS**
CSA HC.4.4	HC 4.ZN 2718	Canada	Sheet	53.5–56.5 **Cu**, 0.10 **Pb**, rem **Zn**, 0.25 **Fe**, 0.50 **Mn**, 16.5–19.5 **Ni**, 0.50 others	**Half hard**: all **diam**, 78 ksi (337 MPa) **TS**; **Full hard**: all **diam**, 92 ksi (634 MPa) **TS**; **Spring**: all **diam**, 108 ksi (744 MPa) **TS**
COPANT 162	C 735 00	COPANT	Wrought products	70.5–73.5 **Cu**, 0.10 **Pb**, rem **Zn**, 0.25 **Fe**, 0.5 **Mn**, 16.5–19.5 **Ni**	

WROUGHT COPPER AND COPPER ALLOYS

specification number	designation	country	product forms	chemical composition	mechanical properties and hardness values
COPANT 162	C 752 00	COPANT	Wrought products	63–66.5 **Cu**, 0.10 **Pb**, rem **Zn**, 0.25 **Fe**, 0.5 **Mn**, 16.5–19.5 **Ni**	
COPANT 162	C 764 00	COPANT	Wrought products	58.5–61.5 **Cu**, 0.05 **Pb**, rem **Zn**, 0.25 **Fe**, 0.5 **Mn**, 16.5–19.5 **Ni**	
COPANT 162	C 770 00	COPANT	Wrought products	53.5–56.5 **Cu**, 0.10 **Pb**, rem **Zn**, 0.25 **Fe**, 0.5 **Mn**, 16.5–19.5 **Ni**	
COPANT 532	Cu Ni 18 Zn 17	COPANT	Bar	63–66.5 **Cu**, 0.05 **Pb**, rem **Zn**, 0.25 **Fe**, 0.5 **Mn**, 16.5–19.5 **Ni**, 0.5 total, others	**Full tempered**: >0.5 mm **diam**, 550 MPa **TS**; **1/4 hard**: 410 MPa **TS**
COPANT 532	Cu Ni 18 Zn 22	COPANT	Bar	58.5–61.5 **Cu**, 0.05 **Pb**, rem **Zn**, 0.25 **Fe**, 0.5 **Mn**, 16.5–19.5 **Ni**, 0.5 total, others	**Full tempered**: >0.5 mm **diam**, 615 MPa **TS**; **1/4 hard**: 0.5–13 mm **diam**, 520 MPa **TS**
COPANT 532	Cu Ni 18 Zn 27	COPANT	Bar	53.5–56.5 **Cu**, 0.05 **Pb**, rem **Zn**, 0.25 **Fe**, 0.5 **Mn**, 16.5–19.5 **Ni**, 0.5 total, others	**Full tempered**: >0.5 mm **diam**, 615 MPa **TS**; **1/4 hard**: 0.5–13 mm **diam**, 520 MPa **TS**
COPANT 532	Cu Ni 18 Zn 18 Pb 1	COPANT	Bar	59–66.5 **Cu**, 0.8–1.2 **Pb**, rem **Zn**, 0.25 **Fe**, 0.5 **Mn**, 16.5–19.5 **Ni**, 0.5 total, others	**Full tempered**: >0.5 mm **diam**, 550 MPa **TS**; **1/4 hard**: 0.5–13 mm **diam**, 410 MPa **TS**
DS 3003	5246	Denmark	Rolled, drawn, extruded, and forged products	60.0–64.0 **Cu**, 0.03 **Pb**, 20 **Zn**, 0.3 **Fe**, 0.7 **Mn**, 17.0–19.0 **Ni**, 0.3 others, total	
QQ-W-321d	794	US	Wire	59.0–66.5 **Cu**, 0.8–1.2 **Pb**, rem **Zn**, 0.25 **Fe**, 0.50 **Mn**, 16.5–19.5 **Ni**, 0.50 total, others; **Co** in **Ni**	**Half hard temper**: 0.02–0.250 in. **diam**, 75 ksi (517 MPa) **TS**
QQ-W-321d	770	US	Wire	53.5–56.5 **Cu**, 0.10 **Pb**, rem **Zn**, 0.25 **Fe**, 0.50 **Mn**, 16.5–19.5 **Ni**, 0.50 total, others; **Co** in **Ni**	**Quarter hard temper**: 0.02–0.250 in. **diam**, 74 ksi (510 MPa) **TS**; **Half hard temper**: 0.02–0.250 in. **diam**, 92 ksi (634 MPa) **TS**; **Hard temper**: 0.02–0.250 in. **diam**, 112 ksi (772 MPa) **TS**
QQ-W-321d	764	US	Wire	58.5–61.5 **Cu**, 0.05 **Pb**, rem **Zn**, 0.25 **Fe**, 0.50 **Mn**, 16.5–19.5 **Ni**, 0.50 total, others; **Co** in **Ni**	**Quarter hard temper**: 0.02–0.250 in. **diam**, 74 ksi (510 MPa) **TS**; **Half hard temper**: 0.02–0.250 in. **diam**, 92 ksi (634 MPa) **TS**; **Hard temper**: 0.02–0.250 in. **diam**, 112 ksi (772 MPa) **TS**
QQ-W-321d	752	US	Wire	63.0–66.5 **Cu**, 0.10 **Pb**, rem **Zn**, 0.25 **Fe**, 0.50 **Mn**, 16.5–19.5 **Ni**, 0.50 total, others; **Co** in **Ni**	**Quarter hard temper**: 0.02–0.250 in. **diam**, 68 ksi (469 MPa) **TS**; **Half hard temper**: 0.02–0.250 in. **diam**, 83 ksi (572 MPa) **TS**; **Hard temper**: 0.02–0.250 in. **diam**, 99 ksi (683 MPa) **TS**

WROUGHT COPPER AND COPPER ALLOYS

specification number	designation	country	product forms	chemical composition	mechanical properties and hardness values
QQ-C-585B	735	US	Plate, sheet, strip, bar	70.5–73.5 **Cu**, 0.10 **Pb**, rem **Zn**, 0.25 **Fe**, 0.50 **Mn**, 16.5–19.5 **Ni**, **Co** in **Ni**content	**Rolled temper–quarter hard**: all **diam**, 56 ksi (386 MPa) **TS**; **Half hard temper**: all **diam**, 63 ksi (434 MPa) **TS**; **Hard temper**: all **diam**, 73 ksi (503 MPa) **TS**
SFS 2935	CuNi18Zn20	Finland	Sheet and strip	60.0–64.0 **Cu**, 0.03 **Pb**, 20 **Zn**, 0.3 **Fe**, 0.7 **Mn**, 17.0–19.0 **Ni**	**Annealed**: 0.2–0.5 mm **diam**, 370 MPa **TS**, 160 MPa **YS**, 15% in 2 in. (50 mm) **El**; **Cold finished**: 0.1–0.5 mm **diam**, 610 MPa **TS**, 540 MPa **YS**, 1% in 2 in. (50 mm) **El**
NF A 51 107	Cu Ni 18 Zn 20	France	Sheet and strip	60–64 **Cu**, 0.05 **Pb**, rem **Zn**, 0.5 **Mn**, 17–19 **Ni**, 0.3 others, total	**1/4 hard**: $\leq$10 mm **diam**, 450 MPa **TS**, 20% in 2 in. (50 mm) **El**, 120–150 **HV**; **1/2 hard**: $\leq$10 mm **diam**, 510 MPa **TS**, 7% in 2 in. (50 mm) **El**, 150–170 **HV**; **3/4 hard**: $\leq$10 mm **diam**, 570 MPa **TS**, 5% in 2 in. (50 mm) **El**, 175–195 **HV**
NF A 51 107	Cu Ni 18 Zn 27	France	Sheet and strip	53–57 **Cu**, 0.05 **Pb**, rem **Zn**, 0.5 **Mn**, 17–19 **Ni**, 0.3 others, total	**1/4 hard**: $\leq$10 mm **diam**, 480 MPa **TS**, 22% in 2 in. (50 mm) **El**, 130–160 **HV**; **1/2 hard**: $\leq$10 mm **diam**, 540 MPa **TS**, 11% in 2 in. (50 mm) **El**, 160–185 **HV**; **3/4 hard**: $\leq$10 mm **diam**, 620 MPa **TS**, 6% in 2 in. (50 mm) **El**, 190–210 **HV**
ISO 1639	CuNi18Zn19Pb1	ISO	Solid product: drawn, on coils or reels	59.0–63.0 **Cu**, 0.5–1.5 **Pb**, rem **Zn**, 0.3 **Fe**, 0–0.7 **Mn**, 17.0–19.0 **Ni**, 0.5 total, others	**As cast**: all **diam**, 450 MPa **TS**, 20% in 2 in. (50 mm) **El**, 160 **HV**
ISO 1638	CuNi18Zn20	ISO	Solid product: drawn, on coils or reels	60.0–64.0 **Cu**, 0.03 **Pb**, rem **Zn**, 0.3 **Fe**, 0–0.7 **Mn**, 17.0–19.0 **Ni**, 0.3 total, others	**Annealed**: 1–5 mm **diam**, 390 MPa **TS**, 35% in 2 in. (50 mm) **El**; **HD–strain hardened**: 1–3 mm **diam**, 640 MPa **TS**
ISO 1637	CuNi18Zn19Pb1	ISO	Solid products: straight lengths	59.0–63.0 **Cu**, 0.5–1.5 **Pb**, rem **Zn**, 0.3 **Fe**, 0–0.7 **Mn**, 17.0–19.0 **Ni**, 0.5 total, others	**HA**: 5–50 mm **diam**, 430 MPa **TS**, 30% in 2 in. (50 mm) **El**, 140 **HV**; **HB–strain hardened**: 5–15 mm **diam**, 490 MPa **TS**, 10% in 2 in. (50 mm) **El**, 170 **HV**
ISO 1637	CuNi18Zn20	ISO	Solid products: straight lengths	60.0–64.0 **Cu**, 0.03 **Pb**, rem **Zn**, 0.3 **Fe**, 0–0.7 **Mn**, 17.0–19.0 **Ni**, 0.3 total, others	**HA**: 5–50 mm **diam**, 470 MPa **TS**, 22% in 2 in. (50 mm) **El**, 150 **HV**; **HB–strain hardened**: 5–15 mm **diam**, 540 MPa **TS**, 8% in 2 in. (50 mm) **El**, 175 **HV**

WROUGHT COPPER AND COPPER ALLOYS

specification number	designation	country	product forms	chemical composition	mechanical properties and hardness values
ISO 1634	CuNi18Zn20	ISO	Plate, sheet, strip	60.0–64.0 **Cu**, 0.03 **Pb**, rem **Zn**, 0.3 **Fe**, 0–0.7 **Mn**, 17.0–19.0 **Ni**, 0.3 total, others	**Specially annealed (OS25)**: 350 max mm **diam**, 30% in 2 in. (50 mm) **El**, 120 max **HV**; **HA**: all **diam**, 460 MPa **TS**, 8% in 2 in. (50 mm) **El**, 160 **HV**; **Strain hardened (HB)**: all **diam**, 560 MPa **TS**, 3% in 2 in. (50 mm) **El**, 190 **HV**
ISO 1634	CuNi18Zn27	ISO	Plate, sheet, strip	53.0–56.0 **Cu**, 0.05 **Pb**, rem **Zn**, 0.3 **Fe**, 0–0.5 **Mn**, 17.0–19.0 **Ni**, 0.3 total, others	**Strain hardened (HD)**: all **diam**, 690 MPa **TS**, 2% in 2 in. (50 mm) **El**, 210 **HV**
ISO 430	CuNi18Zn20	ISO	Plate, sheet, strip, rod, bar, tube, wire	60.0–64.0 **Cu**, 0.03 **Pb**, rem **Zn**, 0.3 **Fe**, 0–0.7 **Mn**, 17.0–19.0 **Ni**, 0.3 total, others	
ISO 430	CuNi18Zn27	ISO	Plate, sheet, strip, rod, bar, wire	53.0–56.0 **Cu**, 0.05 **Pb**, rem **Zn**, 0.3 **Fe**, 0–0.5 **Mn**, 17.0–19.0 **Ni**, 0.3 total, others	
ISO 430	CuNi18Zn19Pb1	ISO	Strip, rod, bar, extruded sections, wire	59.0–63.0 **Cu**, 0.5–1.5 **Pb**, rem **Zn**, 0.3 **Fe**, 0–0.7 **Mn**, 17.0–19.0 **Ni**, 0.5 total, others	
JIS H 3270	C 7701	Japan	Rod and bar	54.0–58.0 **Cu**, 0.10 **Pb**, rem **Zn**, 0.25 **Fe**, 0.50 **Mn**, 16.5–19.5 **Ni**	**1/2 hard**: <6.5 mm **diam**, 520 MPa **TS**, 160 min **HV**; **Hard**: <6.5 mm **diam**, 618 MPa **TS**, 160 min **HV**
JIS H 3270	C 7521	Japan	Rod and bar	61.0–67.0 **Cu**, 0.10 **Pb**, rem **Zn**, 0.25 **Fe**, 0.50 **Mn**, 16.5–19.5 **Ni**	**1/2 hard**: <6.5 mm **diam**, 490 MPa **TS**, 145 min **HV**; **Hard**: <6.5 mm **diam**, 549 MPa **TS**, 145 min **HV**
JIS H 3270	C 7941	Japan	Rod and bar	61.0–67.0 **Cu**, 0.8–1.8 **Pb**, rem **Zn**, 0.25 **Fe**, 0.50 **Mn**, 16.5–19.5 **Ni+Co**	**Hard**: <6.5 mm **diam**, 549 MPa **TS**, 150 min **HV**
QQ-C-585B	752	US	Plate, sheet, strip, bar	63.0–66.5 **Cu**, 0.10 **Pb**, rem **Zn**, 0.25 **Fe**, 0.50 **Mn**, 16.5–19.5 **Ni+Co**	**Rolled–quarter hard**: all **diam**, 58 ksi (400 MPa) **TS**; **Half hard temper**: all **diam**, 66 ksi (455 MPa) **TS**; **Hard temper**: all **diam**, 78 ksi (538 MPa) **TS**
JIS H 3130	C 7701	Japan	Sheet, plate, and strip: for spring	54.0–58.0 **Cu**, 0.10 **Pb**, rem **Zn**, 0.25 **Fe**, 0.50 **Mn**, 16.5–19.5 **Ni+Co**	**1/2 hard**: $\geq$0.15 mm **diam**, 539 MPa **TS**, 8% in 2 in. (50 mm) **El**, 140–200 **HV**; **Hard**: $\geq$0.15 mm **diam**, 628 MPa **TS**, 4% in 2 in. (50 mm) **El**; **Extra hard**: $\geq$0.15 mm **diam**, 706 MPa **TS**, 195–240 **HV**
JIS H 3110	C 7351	Japan	Sheet, plate, and strip	70.0–75.0 **Cu**, 0.10 **Pb**, rem **Zn**, 0.25 **Fe**, 0.50 **Mn**, 16.5–19.5 **Ni+Co**	**Annealed**: 0.3–5 mm **diam**, 324 MPa **TS**, 20% in 2 in. (50 mm) **El**; **1/2 hard**: $\geq$0.15 mm **diam**, 392 MPa **TS**, 5% in 2 in. (50 mm) **El**, 105 min **HV**

WROUGHT COPPER AND COPPER ALLOYS

specification number	designation	country	product forms	chemical composition	mechanical properties and hardness values
JIS H 3110	C 7521	Japan	Sheet, plate, and strip	61.0–67.0 **Cu**, 0.10 **Pb**, rem **Zn**, 0.25 **Fe**, 0.50 **Mn**, 16.5–19.5 **Ni+Co**	**Annealed**: 0.3–5 mm **diam**, 353 MPa **TS**, 20% in 2 in. (50 mm) **El**; **1/2 hard**: ≥0.15 mm **diam**, 441 MPa **TS**, 5% in 2 in. (50 mm) **El**, 120 min **HV**; **Hard**: ≥0.15 mm **diam**, 539 MPa **TS**, 3% in 2 in. (50 mm) **El**, 140 min **HV**
CSA HC.4.4	HC.4.ZN2718	Canada	Bar, plate	53.5–56.5 **Cu**, 0.10 **Pb**, rem **Zn**, 0.25 **Fe**, 0.50 **Mn**, 16.5–19.5 **Ni+Co**, 0.50 total, others	**Quarter hard**: all **diam**, 476 MPa **TS**; **Half hard**: all **diam**, 517 MPa **TS**; **Full hard**: all **diam**, 635 MPa **TS**
QQ-C-586C	794	US	Rod, shapes, flat wire, strip, bar	59.0–66.5 **Cu**, 0.8–1.2 **Pb**, rem **Zn**, 0.25 **Fe**, 0.50 **Mn**, 16.5–19.5 **Ni+Co**, 0.50 total, others	**Quarter hard temper**: all **diam**, 58 ksi (400 MPa) **TS**; **Half hard temper**: all **diam**, 66 ksi (455 MPa) **TS**; **Hard temper**: all **diam**, 78 ksi (538 MPa) **TS**
QQ-C-586C	770	US	Rod, shapes, flatwire, strip, bar	53.5–56.5 **Cu**, 0.05 **Pb**, rem **Zn**, 0.25 **Fe**, 0.50 **Mn**, 16.5–19.5 **Ni+Co**, 0.50 total, others	
QQ-C-586C	764	US	Rod, shapes, flatwire, strip, bar	58.5–61.5 **Cu**, 0.05 **Pb**, rem **Zn**, 0.25 **Fe**, 0.50 **Mn**, 16.5–19.5 **Ni+Co**, 0.50 total, others	
QQ-C-586C	752	US	Rod, shapes, flatwire, strip, bar	63.0–66.5 **Cu**, 0.05 **Pb**, rem **Zn**, 0.25 **Fe**, 0.50 **Mn**, 16.5–19.5 **Ni+Co**, 0.50 total, others	
QQ-C-585B	770	US	Plate, sheet, strip, bar	53.5–56.5 **Cu**, 0.10 **Pb**, rem **Zn**, 0.25 **Fe**, 0.50 **Mn**, 16.5–19.5 **Ni+Co**	**Rolled–quarter hard**: all **diam**, 69 ksi (476 MPa) **TS**; **Half hard temper**: all **diam**, 78 ksi (538 MPa) **TS**; **Hard temper**: all **diam**, 92 ksi (634 MPa) **TS**
ANSI/ASTM B 122	C 77000	US	Plate, sheet, strip, bar	53.5–56.5 **Cu**, 0.10 **Pb**, rem **Zn**, 0.25 **Fe**, 0.50 **Mn**, 16.5–19.5 **Ni+Co**, silver included in copper content	**Hard**: >0.012 in. (>0.305 mm) **diam**, 92 ksi (635 MPa) **TS**, 76–80 **HR30T**; **Spring**: >0.012 in. (>0.305 mm) **diam**, 108 ksi (740 MPa) **TS**, 80 min **HR30T**; **Annealed**: >0.015 in. (>0.381 mm) **diam**, 35–46 **HR30T**
ANSI/ASTM B 206	CA 752	US	Wire	63.0–66.5 **Cu**, 0.05 **Pb**, rem **Zn**, 0.25 **Fe**, 0.50 **Mn**, 16.5–19.5 **Ni+Co**	**Quarter hard**: 0.020–0.250 in. (0.508–6.35 mm) **diam**, 68 ksi (470 MPa) **TS**; **Hard**: 0.020–0.250 in. (0.508–6.35 mm) **diam**, 99 ksi (685 MPa) **TS**

WROUGHT COPPER AND COPPER ALLOYS

specification number	designation	country	product forms	chemical composition	mechanical properties and hardness values
ANSI/ASTM B 206	CA 764	US	Wire	58.5–61.5 **Cu**, 0.05 **Pb**, rem **Zn**, 0.25 **Fe**, 0.50 **Mn**, 16.5–19.5 **Ni + Co**	**Quarter hard**: 0.020–0.250 in. (0.508–6.35 mm) **diam**, 74 ksi (510 MPa) **TS**; **Hard**: 0.020–0.250 in. (0.508–6.35 mm) **diam**, 112 ksi (770 MPa) **TS**; **Spring**: 0.020–0.0253 in. (0.508–0.643 mm) **diam**, 130 ksi (900 MPa) **TS**
ANSI/ASTM B 206	CA 770	US	Wire	53.5–56.5 **Cu**, 0.05 **Pb**, rem **Zn**, 0.25 **Fe**, 0.50 **Mn**, 16.5–19.5 **Ni + Co**	**Quarter hard**: 0.020–0.250 in. (0.508–6.35 mm) **diam**, 74 ksi (510 MPa) **TS**; **Hard**: 0.020–0.250 in. (0.508–6.35 mm) **diam**, 112 ksi (770 MPa) **TS**; **Spring**: 0.020–0.0253 in. (0.508–0.643 mm) **diam**, 130 ksi (900 MPa) **TS**
ANSI/ASTM B 151	CA 764	US	Rod, bar	58.5–61.5 **Cu**, 0.05 **Pb**, rem **Zn**, 0.25 **Fe**, 0.50 **Mn**, 16.5–19.5 **Ni + Co**, silver included in copper content	**Quarter hard**: 0.02–0.50 in. (0.508–12.7 mm) **diam**, 75 ksi (515 MPa) **TS**; **Hard**: all **diam**, 75 ksi (515 MPa) **TS**
ANSI/ASTM B 151	CA 770	US	Rod, bar	53.5–56.5 **Cu**, 0.05 **Pb**, rem **Zn**, 0.25 **Fe**, 0.50 **Mn**, 16.5–19.5 **Ni + Co**, silver included in copper content	**Quarter hard**: 0.02–0.50 in. (0.508–12.7 mm) **diam**, 75 ksi (515 MPa) **TS**; **Hard**: all **diam**, 75 ksi (515 MPa) **TS**
ANSI/ASTM B 151	CA 752	US	Rod, bar	63.0–66.5 **Cu**, 0.05 **Pb**, rem **Zn**, 0.25 **Fe**, 0.50 **Mn**, 16.5–19.5 **Ni + Co**, silver included in copper content	**Quarter hard**: 0.02–0.50 in. (0.508–12.7 mm) **diam**, 60 ksi (415 MPa) **TS**; **Hard**: all **diam**, 60 ksi (415 MPa) **TS**
ANSI/ASTM B 122	C 73500	US	Plate, sheet, strip, bar	70.5–73.5 **Cu**, 0.10 **Pb**, rem **Zn**, 0.25 **Fe**, 0.50 **Mn**, 16.5–19.5 **Ni + Co**, silver included in copper content	**Quarter hard**: >0.012 in. (>0.305 mm) **diam**, 56 ksi (385 MPa) **TS**, 60–70 **HR30T**; **Hard**: >0.012 in. (>0.305 mm) **diam**, 73 ksi (505 MPa) **TS**, 72–75 **HR30T**; **Annealed**: >0.015 in. (>0.381 mm) **diam**, 29–40 **HR30T**
ANSI/ASTM B 122	C 75200	US	Plate, sheet, strip, bar	63.0–66.5 **Cu**, 0.10 **Pb**, rem **Zn**, 0.25 **Fe**, 0.50 **Mn**, 16.5–19.5 **Ni + Co**, silver included in copper content	**Hard**: >0.012 in. (>0.305 mm) **diam**, 78 ksi (540 MPa) **TS**, 70–76 **HR30T**; **Spring**: >0.012 in. (>0.305 mm) **diam**, 90 ksi (620 MPa) **TS**, 75–80 **HR30T**; **Annealed**: >0.015 in. (>0.381 mm) **diam**, 32–43 **HR30T**
NS 16 420	CuNi18Zn20	Norway	Sheet, strip, plate, and bar	62 **Cu**, 20 **Zn**, 0.4 **Mn**, 18 **Ni**	

WROUGHT COPPER AND COPPER ALLOYS

specification number	designation	country	product forms	chemical composition	mechanical properties and hardness values
SIS 14 52 46	SIS Brass 52 46–02	Sweden	Plate, sheet, strip	60.0–64.0 **Cu**, 0.03 **Pb**, 20 **Zn**, 0.3 **Fe**, 0–0.7 **Mn**, 17.0–19.0 **Ni**, 0.3 total, others	**Annealed**: 5 mm **diam**, 370 MPa **TS**, 160 MPa **YS**, 40% in 2 in. (50 mm) **El**, 90–120 **HV**
SIS 14 52 46	SIS Brass 52 46–06	Sweden	Sheet, strip	60.0–64.0 **Cu**, 0.03 **Pb**, 20 **Zn**, 0.3 **Fe**, 0–0.7 **Mn**, 17.0–19.0 **Ni**, 0.3 total, others	**Strain hardened**: 0.1–0.5 mm **diam**, 610 MPa **TS**, 540 MPa **YS**, 1% in 4 in. (100 mm) **El**, 185–225 **HV**
BS 2870	NS 105	UK	Sheet, strip, foil	60.0–65.0 **Cu**, 0.04 **Pb**, rem **Zn**, 0.30 **Fe**, 0.05–0.50 **Mn**, 14.0–16.0 **Ni**, 0.50 total, others	**Annealed**: $\leq$10.0 mm **diam**, 105 max **HV**; **1/2 hard**: $\leq$10.0 mm **diam**, 135 **HV**; **Hard**: $\leq$10.0 mm **diam**, 165 **HV**
BS 2873	NS105	UK	Wire	60.0–65.0 **Cu**, 0.04 **Pb**, rem **Zn**, 0.30 **Fe**, 0.05–0.50 **Mn**, 14.0–16.0 **Ni**, 0.50 total, others	
BS 2874	NS112	UK	Rod, section	60.0–63.0 **Cu**, 0.5–1.0 **Pb**, rem **Zn**, 0.1–0.5 **Mn**, 14.0–16.0 **Ni**, 0.50 total, others	
CDA 754		US	Strip: rolled	63.5–66.5 **Cu**, 0.10 **Pb**, rem **Zn**, 0.25 **Fe**, 0.50 **Mn**, 14.0–16.0 **Ni**, 0.50 total, others	**Half hard temper**: 0.040 in. **diam**, 74 ksi (510 MPa) **TS**, 10% in 2 in. (50 mm) **El**, 80 **HRB**; **Hard temper**: 0.040 in. **diam**, 85 ksi (586 MPa) **TS**, 3% in 2 in. (50 mm) **El**, 87 **HRB**; **Extra hard temper**: 0.040 in. **diam**, 92 ksi (634 MPa) **TS**, 2% in 2 in. (50 mm) **El**, 90 **HRB**
COPANT 162	C 754 00	COPANT	Wrought products	63.5–66.5 **Cu**, 0.10 **Pb**, rem **Zn**, 0.25 **Fe**, 0.5 **Mn**, 14–16 **Ni**	
COPANT 162	C 767 00	COPANT	Wrought products	55–58 **Cu**, rem **Zn**, 0.5 **Mn**, 14–16 **Ni**	
NF A 51 107	Cu Ni 15 Zn 22	France	Sheet and strip	61–65 **Cu**, 0.05 **Pb**, rem **Zn**, 0.5 **Mn**, 14–16 **Ni**, 0.3 others, total	**1/4 hard**: $\leq$10 mm **diam**, 440 MPa **TS**, 22% in 2 in. (50 mm) **El**, 115–145 **HV**; **1/2 hard**: $\leq$10 mm **diam**, 490 MPa **TS**, 9% in 2 in. (50 mm) **El**, 145–170 **HV**; **3/4 hard**: $\leq$10 mm **diam**, 560 MPa **TS**, 5% in 2 in. (50 mm) **El**, 175–195 **HV**
ISO 1638	CuNi15Zn21	ISO	Solid product: drawn, on coils or reels	62.0–66.0 **Cu**, 0.05 **Pb**, rem **Zn**, 0.3 **Fe**, 0–0.5 **Mn**, 14.0–16.0 **Ni**, 0.3 total, others	**Annealed**: 1–5 mm **diam**, 360 MPa **TS**, 35% in 2 in. (50 mm) **El**; **HD–strain hardened**: 1–3 mm **diam**, 590 MPa **TS**, 5% in 2 in. (50 mm) **El**
ISO 1637	CuNi15Zn21	ISO	Solid products: straight lengths	62.0–66.0 **Cu**, 0.05 **Pb**, rem **Zn**, 0.3 **Fe**, 0–0.5 **Mn**, 14.0–16.0 **Ni**, 0.3 total, others	**Annealed**: 5 mm **diam**, 36% in 2 in. (50 mm) **El**, 120 **HV**; **HB–strain hardened**: 5–15 mm **diam**, 440 MPa **TS**, 18% in 2 in. (50 mm) **El**, 140 **HV**

WROUGHT COPPER AND COPPER ALLOYS

specification number	designation	country	product forms	chemical composition	mechanical properties and hardness values
ISO 1634	CuNi15Zn21	ISO	Plate, sheet, strip	62.0–66.0 **Cu**, 0.05 **Pb**, rem **Zn**, 0.3 **Fe**, 0–0.5 **Mn**, 14.0–16.0 **Ni**, 0.3 total, others	**Annealed**: all **diam**, 36% in 2 in. (50 mm) **El**, 120 max **HV**; **Strain hardened (HB)**: all **diam**, 440 MPa **TS**, 18% in 2 in. (50 mm) **El**, 140 **HV**
ISO 430	CuNi15Zn21	ISO	Plate, sheet, strip, rod, bar, tube, wire	62.0–66.0 **Cu**, 0.05 **Pb**, rem **Zn**, 0.3 **Fe**, 0–0.5 **Mn**, 14.0–16.0 **Ni**, 0.3 total, others	
BS 2874	NS102	UK	Rod, section	39.0–42.0 **Cu**, 1.0–2.25 **Pb**, rem **Zn**, 0.3 **Fe**, 1.5–3.0 **Mn**, 13.0–15.0 **Ni**, 0.30 total, others	**As manufactured**: ≥6 mm **diam**, 510 MPa **TS**, 8% in 2 in. (50 mm) **El**
JIS H 3110	C 7541	Japan	Sheet, plate, and strip	59.0–65.0 **Cu**, 0.10 **Pb**, rem **Zn**, 0.25 **Fe**, 0.50 **Mn**, 12.5–15.5 **Ni+Co**	**Annealed**: 0.3–5 mm **diam**, 353 MPa **TS**, 20% in 2 in. (50 mm) **El**; **1/2 hard**: ≥0.15 mm **diam**, 412 MPa **TS**, 5% in 2 in. (50 mm) **El**, 110 min **HV**; **Hard**: ≥0.15 mm **diam**, 490 MPa **TS**, 3% in 2 in. (50 mm) **El**, 135 min **HV**
JIS H 3270	C 7541	Japan	Rod and bar	59.0–65.0 **Cu**, 0.10 **Pb**, rem **Zn**, 0.25 **Fe**, 0.50 **Mn**, 12.5–15.5 **Ni+Co**	**1/2 hard**: <6.5 mm **diam**, 441 MPa **TS**, 135 min **HV**; **Hard**: <6.5 mm **diam**, 569 MPa **TS**, 150 min **HV**
COPANT 162	C 776 00	COPANT	Wrought products	42–45 **Cu**, 0.15 **Sn**, 0.25 **Pb**, rem **Zn**, 0.20 **Fe**, 0.25 **Mn**, 12–14 **Ni**	
COPANT 162	C 762 00	COPANT	Wrought products	57–61 **Cu**, 0.10 **Pb**, rem **Zn**, 0.25 **Fe**, 0.5 **Mn**, 11–13.5 **Ni**	
COPANT 162	C 766 00	COPANT	Wrought products	55–58 **Cu**, 0.10 **Pb**, rem **Zn**, 0.25 **Fe**, 0.5 **Mn**, 11–13.5 **Ni**	
ANSI/ASTM B 122	C 76200	US	Plate, sheet, strip, bar	57.0–61.0 **Cu**, 0.10 **Pb**, rem **Zn**, 0.25 **Fe**, 0.50 **Mn**, 11.0–13.5 **Ni+Co**, silver included in copper content	**Hard**: >0.012 in. (>0.305 mm) **diam**, 90 ksi (620 MPa) **TS**, 76–79 **HR30T**; **Spring**: >0.012 in. (>0.305 mm) **diam**, 107 ksi (740 MPa) **TS**, 80 min **HR30T**
CSA HC.4.4	HC.4.ZN2912	Canada	Bar, sheet, strip, plate	57.0–61.0 **Cu**, 0.10 **Pb**, rem **Zn**, 0.25 **Fe**, 0.50 **Mn**, 11.0–13.5 **Ni+Co**	**Quarter hard**: all **diam**, 448 MPa **TS**; **Half hard**: all **diam**, 517 MPa **TS**; **Full hard**: all **diam**, 621 MPa **TS**
QQ-C-585B	766	US	Plate, sheet, strip, bar	55.0–58.0 **Cu**, 0.10 **Pb**, rem **Zn**, 0.25 **Fe**, 0.50 **Mn**, 11.0–13.5 **Ni+Co**	**Rolled–quarter hard**: all **diam**, 65 ksi (448 MPa) **TS**; **Half hard temper**: all **diam**, 75 ksi (517 MPa) **TS**; **Hard temper**: all **diam**, 90 ksi (621 MPa) **TS**

WROUGHT COPPER AND COPPER ALLOYS

specification number	designation	country	product forms	chemical composition	mechanical properties and hardness values
QQ-C-585B	762	US	Plate, sheet, strip, bar	57.0–61.0 **Cu**, 0.10 **Pb**, rem **Zn**, 0.25 **Fe**, 0.50 **Mn**, 11.0–13.5 **Ni+Co**	**Rolled–quarter hard**: all **diam**, 65 ksi (448 MPa) **TS**; **Half hard temper**: all **diam**, 75 ksi (517 MPa) **TS**; **Hard temper**: all **diam**, 90 ksi (621 MPa) **TS**
AS 1566	757B	Australia	Plate, bar, sheet, strip, foil	60.0–65.0 **Cu**, 0.04 **Pb**, rem **Zn**, 0.25 **Fe**, 0.5–0.30 **Mn**, 11.0–13.0 **Ni**, 0.50 total, others	**Annealed (sheet, strip)**: all **diam**, 100 **HV**; **Strain hardened to 1/2 hard (sheet, strip)**: all **diam**, 140–170 **HV**; **Strain hardened to full hard (sheet, strip)**: all **diam**, 170–190 **HV**
BS 2870	NS 104	UK	Sheet, strip, foil	60.0–65.0 **Cu**, 0.04 **Pb**, rem **Zn**, 0.25 **Fe**, 0.05–0.30 **Mn**, 11.0–13.0 **Ni**, 0.50 total, others	**Annealed**: $\leq$10.0 mm **diam**, 100 max **HV**; **1/2 hard**: $\leq$10.0 mm **diam**, 130 **HV**; **Hard**: $\leq$10.0 mm **diam**, 160 **HV**
BS 2873	NS104	UK	Wire	60.0–65.0 **Cu**, 0.04 **Pb**, rem **Zn**, 0.25 **Fe**, 0.05–0.30 **Mn**, 11.0–13.0 **Ni**, 0.50 total, others	
CDA 757		US	Wire, flat wire, bar, strip	63.5–66.5 **Cu**, 0.05 **Pb**, rem **Zn**, 0.25 **Fe**, 0.50 **Mn**, 11.0–13.0 **Ni**, 0.50 total, others	**Half hard temper (flat wire, bar, strip: rolled)**: 0.040 in. **diam**, 73 ksi (503 MPa) **TS**, 11% in 2 in. (50 mm) **El**, 87 **HRB**; **Hard temper (flat wire, bar, strip: rolled)**: 0.040 in. **diam**, 85 ksi (586 MPa) **TS**, 4% in 2 in. (50 mm) **El**, 89 **HRB**; **Extra hard temper (flat wire, bar, strip: rolled)**: 0.040 in. **diam**, 93 ksi (641 MPa) **TS**, 2% in 2 in. (50 mm) **El**, 92 **HRB**
COPANT 162	C 738 00	COPANT	Wrought products	68.5–71.5 **Cu**, 0.05 **Pb**, rem **Zn**, 0.25 **Fe**, 0.5 **Mn**, 11–13 **Ni**	
COPANT 162	C 757 00	COPANT	Wrought products	63.5–66.5 **Cu**, 0.05 **Pb**, rem **Zn**, 0.25 **Fe**, 0.5 **Mn**, 11–13 **Ni**	
COPANT 162	C 790 00	COPANT	Wrought products	63–67 **Cu**, 1.5–2.2 **Pb**, rem **Zn**, 0.35 **Fe**, 0.50 **Mn**, 11–13 **Ni**	
COPANT 162	C 792 00	COPANT	Wrought products	59–66.5 **Cu**, 0.8–1.4 **Pb**, rem **Zn**, 0.25 **Fe**, 0.50 **Mn**, 11–13 **Ni**	
COPANT 532	Cu Ni 12 Zn 23	COPANT	Bar	63.5–66.5 **Cu**, 0.05 **Pb**, rem **Zn**, 0.25 **Fe**, 0.50 **Mn**, 11.0–13.0 **Ni**, 0.5 total, others	**Full tempered**: >0.5 mm **diam**, 615 MPa **TS**; **1/4 hard**: 0.5–13 mm **diam**, 520 MPa **TS**
COPANT 532	Cu Ni 12 Zn 24 Pb 1	COPANT	Bar	59–66.5 **Cu**, 0.8–1.4 **Pb**, rem **Zn**, 0.25 **Fe**, 0.5 **Mn**, 11–13 **Ni**, 0.5 total, others	**Full tempered**: >0.5 mm **diam**, 550 MPa **TS**; **1/4 hard**: 0.5–13 mm **diam**, 410 MPa **TS**
DS 3003	5243	Denmark	Rolled, drawn, extruded, and forged products	62.0–66.0 **Cu**, 0.05 **Pb**, 24 **Zn**, 0.3 **Fe**, 0.5 **Mn**, 11.0–13.0 **Ni**, 0.3 others, total	

WROUGHT COPPER AND COPPER ALLOYS

specification number	designation	country	product forms	chemical composition	mechanical properties and hardness values
QQ-W-321d	757	US	Wire	63.5–66.5 **Cu**, 0.05 **Pb**, rem **Zn**, 0.25 **Fe**, 0.50 **Mn**, 11.0–13.0 **Ni**, 0.50 total, others; **Co** in **Ni**	**Quarter hard temper**: 0.02–0.250 in. **diam**, 73 ksi (503 MPa) **TS**; **Half hard temper**: 0.02–0.250 in. **diam**, 88 ksi (607 MPa) **TS**; **Hard temper**: 0.02–0.250 in. **diam**, 108 ksi (745 MPa) **TS**
NF A 51 107	Cu Ni 12 Zn 24	France	Sheet and strip	62–66 **Cu**, 0.05 **Pb**, rem **Zn**, 0.5 **Mn**, 11–13 **Ni**, 0.3 others, total	**1/4 hard**: $\leq$10 mm **diam**, 420 MPa **TS**, 20% in 2 in. (50 mm) **El**, 105–135 **HV**; **1/2 hard**: $\leq$10 mm **diam**, 470 MPa **TS**, 8% in 2 in. (50 mm) **El**, 140–160 **HV**; **3/4 hard**: $\leq$10 mm **diam**, 550 MPa **TS**, 5% in 2 in. (50 mm) **El**, 170–190 **HV**
ISO 1638	CuNi12Zn24	ISO	Solid product: drawn, on coils or reels	62.0–66.0 **Cu**, 0.05 **Pb**, rem **Zn**, 0.3 **Fe**, 0–0.5 **Mn**, 11.0–13.0 **Ni**, 0.3 total, others	**Annealed**: 1–5 mm **diam**, 340 MPa **TS**, 38% in 2 in. (50 mm) **El**; **HB–strain hardened**: 1–3 mm **diam**, 490 MPa **TS**, 5% in 2 in. (50 mm) **El**
ISO 1637	CuNi12Zn24	ISO	Solid products: straight lengths	62.0–66.0 **Cu**, 0.05 **Pb**, rem **Zn**, 0.3 **Fe**, 0–0.5 **Mn**, 11.0–13.0 **Ni**, 0.3 total, others	**HA**: 5–50 mm **diam**, 440 MPa **TS**, 22% in 2 in. (50 mm) **El**, 150 **HV**; **HB–strain hardened**: 5–15 mm **diam**, 540 MPa **TS**, 5% in 2 in. (50 mm) **El**, 185 **HV**
ISO 1635	CuNi12Zn24	ISO	Tube	62.0–66.0 **Cu**, 0.05 **Pb**, rem **Zn**, 0.3 **Fe**, 0–0.5 **Mn**, 11.0–13.0 **Ni**, 0.3 total, others	**Annealed**: all **diam**, 38% in 2 in. (50 mm) **El**, 115 max **HV**; **Strain hardened (HB)**: all **diam**, 440 MPa **TS**, 30% in 2 in. (50 mm) **El**, 100 **HV**
ISO 1634	CuNi12Zn24	ISO	Plate, sheet, strip	62.0–66.0 **Cu**, 0.05 **Pb**, rem **Zn**, 0.3 **Fe**, 0–0.5 **Mn**, 11.0–13.0 **Ni**, 0.3 total, others	**Specially annealed (OS25)**: all **diam**, 35% in 2 in. (50 mm) **El**, 110 max **HV**; **Specially annealed (OS35)**: all **diam**, 40% in 2 in. (50 mm) **El**, 100 max **HV**; **HA**: all **diam**, 410 MPa **TS**, 20% in 2 in. (50 mm) **El**, 140 **HV**
ISO 430	CuNi12Zn24	ISO	Plate, sheet, strip, rod, bar, tube, wire	62.0–66.0 **Cu**, 0.05 **Pb**, rem **Zn**, 0.3 **Fe**, 0–0.5 **Mn**, 11.0–13.0 **Ni**, 0.3 total, others	
NS 16 424	CuNi12Zn24	Norway	Sheet, strip, plate, bar, rod, and wire	64 **Cu**, 24 **Zn**, 0.3 **Mn**, 12 **Ni**	
SIS 14 52 43	SIS Brass 52 43–02	Sweden	Plate, sheet, strip, wire	62.0–66.0 **Cu**, 0.05 **Pb**, 24 **Zn**, 0.3 **Fe**, 0–0.5 **Mn**, 11.0–13.0 **Ni**, 0.3 total, others	**Annealed**: 5 mm **diam**, 340 MPa **TS**, 120 MPa **YS**, 45% in 2 in. (50 mm) **El**, 80–110 **HV**

WROUGHT COPPER AND COPPER ALLOYS

specification number	designation	country	product forms	chemical composition	mechanical properties and hardness values
SIS 14 52 43	SIS Brass 52 43–03	Sweden	Plate, sheet, strip	62.0–66.0 **Cu**, 0.05 **Pb**, 24 **Zn**, 0.3 **Fe**, 0–0.5 **Mn**, 11.0–13.0 **Ni**, 0.3 total, others	**Strain hardened**: 5 mm **diam**, 380 MPa **TS**, 200 MPa **YS**, 25% in 2 in. (50 mm) **El**, 105–135 **HV**
SIS 14 52 43	SIS Brass 52 43–04	Sweden	Strip, bar, wire	62.0–66.0 **Cu**, 0.05 **Pb**, 24 **Zn**, 0.3 **Fe**, 0–0.5 **Mn**, 11.0–13.0 **Ni**, 0.3 total, others	**Strain hardened**: 0.5–2.5 mm **diam**, 440 MPa **TS**, 290 MPa **YS**, 6% in 2 in. (50 mm) **El**, 130–170 **HV**
SIS 14 52 43	SIS Brass 52 43–06	Sweden	Sheet, strip	62.0–66.0 **Cu**, 0.05 **Pb**, 24 **Zn**, 0.3 **Fe**, 0–0.5 **Mn**, 11.0–13.0 **Ni**, 0.3 total, others	**Strain hardened**: 0.1–0.5 mm **diam**, 560 MPa **TS**, 510 MPa **YS**, 1%in 4 in. (100 mm) **El**, 175–210 **HV**
SIS 14 52 43	SIS Brass 52 43–07	Sweden	Wire	62.0–66.0 **Cu**, 0.05 **Pb**, 24 **Zn**, 0.3 **Fe**, 0–0.5 **Mn**, 11.0–13.0 **Ni**, 0.3 total, others	**Strain hardened**: 0.1–2.5 mm **diam**, 740 MPa **TS**, 720 MPa **YS**, 1% in 2 in. (50 mm) **El**
QQ–C–586C	792	US	Rod, shapes, flatwire, strip, bar	59.0–66.5 **Cu**, 0.8–1.4 **Pb**, rem **Zn**, 0.25 **Fe**, 0.50 **Mn**, 11.0–13.0 **Ni + Co**, 0.50 total, others	
ANSI/ASTM B 151	CA 792	US	Rod, bar	59.0–66.5 **Cu**, 0.8–1.4 **Pb**, rem **Zn**, 0.25 **Fe**, 0.50 **Mn**, 11.0–13.0 **Ni + Co**, silver included in copper content	**Quarter hard**: 0.02–0.50 in. (0.508–12.7 mm) **diam**, 60 ksi (415 MPa) **TS**; **Hard**: all **diam**, 68 ksi (470 MPa) **TS**
ANSI/ASTM B 151	CA 757	US	Rod, bar	63.5–66.5 **Cu**, 0.05 **Pb**, rem **Zn**, 0.25 **Fe**, 0.50 **Mn**, 11.0–13.0 **Ni + Co**, silver included in copper content	**Quarter hard**: 0.20–0.50 in. (0.508–12.7 mm) **diam**, 75 ksi (515 MPa) **TS**; **Hard**: all **diam**, 75 ksi (515 MPa) **TS**
ANSI/ASTM B 206	CA 757	US	Wire	63.5–66.5 **Cu**, 0.05 **Pb**, rem **Zn**, 0.25 **Fe**, 0.50 **Mn**, 11.0–13.0 **Ni + Co**	**Quarter hard**: 0.020–0.250 in. (0.508–6.35 mm) **diam**, 73 ksi (505 MPa) **TS**; **Hard**: 0.020–0.250 in. (0.508–6.35 mm) **diam**, 108 ksi (740 MPa) **TS**; **Spring**: 0.020–0.0253 in. (0.508–0.643 mm) **diam**, 130 ksi (900 MPa) **TS**
ANSI/ASTM B 206	CA 792	US	Wire	59.0–66.5 **Cu**, 0.8–1.4 **Pb**, rem **Zn**, 0.25 **Fe**, 0.50 **Mn**, 11.0–13.0 **Ni + Co**	**Quarter hard**: 0.020–0.250 in. (0.508–6.35 mm) **diam**, 70 ksi (485 MPa) **TS**; **Hard**: 0.020–0.250 in. (0.508–6.35 mm) **diam**, 104 ksi (715 MPa) **TS**
DS 3003	5282	Denmark	Rolled, drawn, extruded, and forged products	44.0–48.0 **Cu**, 1.0–2.5 **Pb**, 42 **Zn**, 0.5 **Fe**, 0.5 **Mn**, 11.0 **Ni**, 0.5 others, total	
JIS Z 3262	BCuZn–7	Japan	Wire: filler	46–49 **Cu**, rem **Zn**, 0.15 **Si**, 10–11 **Ni**, 0.30–1.00 **Ag**	

WROUGHT COPPER AND COPPER ALLOYS

specification number	designation	country	product forms	chemical composition	mechanical properties and hardness values
AS 1566	761A	Australia	Plate, bar, sheet, strip, foil	59.0–63.0 **Cu**, 0.10 **Pb**, rem **Zn**, 0.25 **Fe**, 0.5 **Mn**, 9.0–11.0 **Ni**, 0.50 total, others	**Strain hardened to 1/2 hard (sheet, strip)**: all **diam**, 125–160 **HV**; **Strain hardened to full hard (sheet, strip)**: all **diam**, 160–180 **HV**
BS 2870	NS 103	UK	Sheet, strip, foil	60.0–65.0 **Cu**, 0.04 **Pb**, rem **Zn**, 0.25 **Fe**, 0.05–0.30 **Mn**, 9.0–11.0 **Ni**, 0.50 total, others	**Annealed**: ≤10.0 mm **diam**, 100 max **HV**; **1/2 hard**: ≤10.0 mm **diam**, 125 **HV**; **Hard**: ≤10.0 mm **diam**, 160 **HV**
BS 2873	NS103	UK	Wire	60.0–65.0 **Cu**, 0.04 **Pb**, rem **Zn**, 0.25 **Fe**, 0.05–0.30 **Mn**, 9.0–11.0 **Ni**, 0.50 total, others	
BS 2874	NS111	UK	Rod, section	58.0–63.0 **Cu**, 1.0–2.0 **Pb**, rem **Zn**, 0.1–0.5 **Mn**, 9.0–11.0 **Ni**, 0.50 total, others	
BS 2874	NS101	UK	Rod, section	44.0–47.0 **Cu**, 1.0–2.5 **Pb**, rem **Zn**, 0.4 **Fe**, 0.2–0.5 **Mn**, 9.0–11.0 **Ni**, 0.30 total, others	**As manufactured**: ≥6 mm **diam**, 466 MPa **TS**, 8% in 2 in. (50 mm) **El**
BS 2872	NS 101	UK	Forging stock, forgings	44.0–47.0 **Cu**, 1.0–2.5 **Pb**, rem **Zn**, 0.40 **Fe**, 0.2–0.5 **Mn**, 9.0–11.0 **Ni**, 0.30 total, others	**As manufactured or annealed**: ≥6 mm **diam**, 466 MPa **YS**, 8% in 2 in. (50 mm) **El**
CDA 745		US	Wire, flat wire, bar, strip	63.5–68.5 **Cu**, 0.10 **Pb**, rem **Zn**, 0.25 **Fe**, 0.50 **Mn**, 9.0–11.0 **Ni**, 0.50 total, others	**Hard temper (flat wire, rolled bar, rolled strip)**: 0.040 in. **diam**, 86 ksi (593 MPa) **TS**, 4% in 2 in. (50 mm) **El**, 89 **HRB**; **Extra hard temper (flat wire, rolled bar, rolled strip)**: 0.040 in. **diam**, 95 ksi (655 MPa) **TS**, 3% in 2 in. (50 mm) **El**, 92 **HRB**; **Spring temper – 84% (wire)**: 0.080 in. **diam**, 130 ksi (896 MPa) **TS**, 1% in 2 in. (50 mm) **El**
CSA HC.4.4	HC 4.ZN2010	Canada	Plate	69.0–73.5 **Cu**, 0.10 **Pb**, rem **Zn**, 0.25 **Fe**, 0.50 **Mn**, 9.0–11.0 **Ni**, 0.50 others	**Quarter hard**: all **diam**, 55 ksi (379 MPa) **TS**; **Half hard**: all **diam**, 63 ksi (434 MPa) **TS**
CSA HC.4.4	HC 4.ZN2010	Canada	Bar	69.0–73.5 **Cu**, 0.10 **Pb**, rem **Zn**, 0.25 **Fe**, 0.50 **Mn**, 9.0–11.0 **Ni**, 0.50 others	**Half hard**: all **diam**, 63 ksi (434 MPa) **TS**; **Full hard**: all **diam**, 73 ksi (503 MPa) **TS**; **Extra hard**: all **diam**, 79 ksi (544 MPa) **TS**
CSA HC.4.4	HC 4.ZN2010	Canada	Strip	69.0–73.5 **Cu**, 0.10 **Pb**, rem **Zn**, 0.25 **Fe**, 0.50 **Mn**, 9.0–11.0 **Ni**, 0.50 others	**Half hard**: all **diam**, 63 ksi (434 MPa) **TS**; **Full hard**: all **diam**, 73 ksi (503 MPa) **TS**; **Extra hard**: all **diam**, 79 ksi (544 MPa) **TS**
CSA HC.4.4	HC 4.ZN2010	Canada	Sheet	69.0–73.5 **Cu**, 0.10 **Pb**, rem **Zn**, 0.25 **Fe**, 0.50 **Mn**, 9.0–11.0 **Ni**, 0.50 others	**Half hard**: all **diam**, 63 ksi (434 MPa) **TS**; **Full hard**: all **diam**, 73 ksi (503 MPa) **TS**; **Extra hard**: all **diam**, 79 ksi (544 MPa) **TS**

WROUGHT COPPER AND COPPER ALLOYS

specification number	designation	country	product forms	chemical composition	mechanical properties and hardness values
CSA HC.4.4	HC 4.ZN2410	Canada	Bar	63.5–68.5 **Cu**, 0.10 **Pb**, rem **Zn**, 0.25 **Fe**, 0.50 **Mn**, 9.0–11.0 **Ni**, 0.50 others	**Half hard**: all **diam**, 67 ksi (461 MPa) **TS**; **Full hard**: all **diam**, 80 ksi (551 MPa) **TS**; **Spring**: all **diam**, 95 ksi (655 MPa) **TS**
CSA HC.4.4	HC 4.ZN2410	Canada	Strip	63.5–68.5 **Cu**, 0.10 **Pb**, rem **Zn**, 0.25 **Fe**, 0.50 **Mn**, 9.0–11.0 **Ni**, 0.50 others	**Half hard**: all **diam**, 67 ksi (461 MPa) **TS**; **Full hard**: all **diam**, 80 ksi (551 MPa) **TS**; **Spring**: all **diam**, 95 ksi (655 MPa) **TS**
CSA HC.4.4	HC 4.ZN2410	Canada	Plate	63.5–68.5 **Cu**, 0.10 **Pb**, rem **Zn**, 0.25 **Fe**, 0.50 **Mn**, 9.0–11.0 **Ni**, 0.50 others	**Quarter hard**: all **diam**, 56 ksi (386 MPa) **TS**; **Half hard**: all **diam**, 67 ksi (461 MPa) **TS**
CSA HC.4.4	HC 4.ZN2410	Canada	Sheet	63.5–68.5 **Cu**, 0.10 **Pb**, rem **Zn**, 0.25 **Fe**, 0.50 **Mn**, 9.0–11.0 **Ni**, 0.50 others	**Half hard**: all **diam**, 67 ksi (461 MPa) **TS**; **Full hard**: all **diam**, 80 ksi (551 MPa) **TS**; **Spring**: all **diam**, 95 ksi (655 MPa) **TS**
COPANT 162	C 740 00	COPANT	Wrought products	69–73.5 **Cu**, 0.10 **Pb**, rem **Zn**, 0.25 **Fe**, 0.5 **Mn**, 9–11 **Ni**	
COPANT 162	C 74500	COPANT	Wrought products	63.5–66.5 **Cu**, 0.10 **Pb**, rem **Zn**, 0.25 **Fe**, 0.5 **Mn**, 9–11 **Ni**	
COPANT 162	C 761 00	COPANT	Wrought products	59–63 **Cu**, 0.10 **Pb**, rem **Zn**, 0.25 **Fe**, 0.5 **Mn**, 9–11 **Ni**	
COPANT 162	C 774 00	COPANT	Wrought products	43–47 **Cu**, 0.20 **Pb**, rem **Zn**, 9–11 **Ni**	
COPANT 162	C 788 00	COPANT	Wrought products	63–67 **Cu**, 1.5–2.0 **Pb**, rem **Zn**, 0.25 **Fe**, 0.50 **Mn**, 9–11 **Ni**	
COPANT 162	C 796 00	COPANT	Wrought products	43.5–46.5 **Cu**, 0.8–1.2 **Pb**, 1.5–2.5 **Mn**, 9–11 **Ni**	
COPANT 162	C 798 00	COPANT	Wrought products	45.5–48.5 **Cu**, 1.5–2.5 **Pb**, rem **Zn**, 0.25 **Fe**, 1.5–2.5 **Mn**, 9–11 **Ni**	
COPANT 532	Cu Ni 10 Zn 25	COPANT	Bar	63.5–66.5 **Cu**, 0.05 **Pb**, rem **Zn**, 0.25 **Fe**, 0.5 **Mn**, 9–11 **Ni**, 0.5 total, others	**Full tempered**: >0.5 mm **diam**, 615 MPa **TS**; **1/4 hard**: 0.5–13 mm **diam**, 520 MPa **TS**
QQ–W–321d	745	US	Wire	63.5–68.5 **Cu**, 0.10 **Pb**, rem **Zn**, 0.25 **Fe**, 0.50 **Mn**, 9.0–11.0 **Ni**, 0.50 total, others; **Co** in **Ni**	**Quarter hard temper**: 0.02–0.250 in. **diam**, 73 ksi (503 MPa) **TS**; **Half hard temper**: 0.02–0.250 in. **diam**, 88 ksi (607 MPa) **TS**; **Hard temper**: 0.02–0.250 in. **diam**, 108 ksi (745 MPa) **TS**
QQ–C–585B	745	US	Plate, sheet, strip, bar	63.5–66.5 **Cu**, 0.10 **Pb**, rem **Zn**, 0.25 **Fe**, 0.50 **Mn**, 9.0–11.0 **Ni**, **Co** in **Ni**	**Rolled–quarter hard**: all **diam**, 56 ksi (386 MPa) **TS**; **Half hard temper**: all **diam**, 67 ksi (462 MPa) **TS**; **Hard temper**: all **diam**, 80 ksi (552 MPa) **TS**

WROUGHT COPPER AND COPPER ALLOYS

specification number	designation	country	product forms	chemical composition	mechanical properties and hardness values
SFS 2938	CuNi10Fe1Mn	Finland	Tube	88 **Cu**, 0.5 **Zn**, 1.0–1.8 **Fe**, 0.5–1.0 **Mn**, 9.0–11.0 **Ni**, 0.05 **S**, 0.05 **C**, 0.05 **Pb+Sn**	**Annealed**: ≤5 mm **diam**, 290 MPa **TS**, 100 MPa **YS**, 35% in 2 in. (50 mm) **El**
IS 6912	Leaded 10% nickel brass	India	Forgings	44.0–47.0 **Cu**, 1.0–2.5 **Pb**, rem **Zn**, 0.40 **Fe**, 0.20 0.50 **Mn**, 9.0–11.0 **Ni**	**As manufactured**: ≥6 mm **diam**, 465 MPa **TS**, 10% in 2 in. (50 mm) **El**
ISO 1640	CuNi10Zn42Pb2	ISO	Forgings	44.0–48.0 **Cu**, 1.0–2.5 **Pb**, rem **Zn**, 0.5 **Fe**, 0–0.5 **Mn**, 9.0–11.0 **Ni**, 0.5 total, others	**As cast**: all **diam**, 490 MPa **TS**, 15% in 2 in. (50 mm) **El**, 170 **HV**
ISO 1639	CuNi10Zn42Pb2	ISO	Solid product: drawn, on coils or reels	44.0–48.0 **Cu**, 1.0–2.5 **Pb**, rem **Zn**, 0.5 **Fe**, 0–0.5 **Mn**, 9.0–11.0 **Ni**, 0.5 total, others	**As cast**: all **diam**, 490 MPa **TS**, 15% in 2 in. (50 mm) **El**, 170 **HV**
ISO 1637	CuNi10Zn28Pb1	ISO	Solid products: straight lengths	59.0–63.0 **Cu**, 1.0–2.0 **Pb**, rem **Zn**, 0.3 **Fe**, 0–0.5 **Mn**, 9.0–11.0 **Ni**, 0.5 total, others	**HA**: 5–50 mm **diam**, 410 MPa **TS**, 15% in 2 in. (50 mm) **El**, 150 **HV**; **HB–strain hardened**: 5–15 mm **diam**, 480 MPa **TS**, 8% in 2 in. (50 mm) **El**, 170 **HV**
ISO 1637	CuNi10Zn42Pb2	ISO	Solid products: straight lengths	44.0–48.0 **Cu**, 1.0–2.5 **Pb**, rem **Zn**, 0.5 **Fe**, 0–0.5 **Mn**, 9.0–11.0 **Ni**, 0.5 total, others	**HA**: 5–50 mm **diam**, 460 MPa **TS**, 15% in 2 in. (50 mm) **El**, 150 **HV**; **HB–strain hardened**: 5–15 mm **diam**, 540 MPa **TS**, 8% in 2 in. (50 mm) **El**, 170 **HV**
ISO 1634	CuNi10Zn27	ISO	Plate, sheet, strip	61.0–65.0 **Cu**, 0.05 **Pb**, rem **Zn**, 0.3 **Fe**, 0–0.5 **Mn**, 9.0–11.0 **Ni**, 0.3 total, others	**Specially annealed (OS35)**: all **diam**, 38% in 2 in. (50 mm) **El**, 100 max **HV**; **HA**: all **diam**, 410 MPa **TS**, 15% in 2 in. (50 mm) **El**, 140 **HV**; **Strain hardened (HB)**: all **diam**, 540 MPa **TS**, 5% in 2 in. (50 mm) **El**, 180 **HV**
ISO 1634	CuNi10Zn28Pb1	ISO	Plate, sheet, strip	59.0–63.0 **Cu**, 1.0–2.0 **Pb**, rem **Zn**, 0.3 **Fe**, 0–0.5 **Mn**, 9.0–11.0 **Ni**, 0.5 total, others	**Annealed**: all **diam**, 35% in 2 in. (50 mm) **El**, 100 max **HV**; **HA**: all **diam**, 410 MPa **TS**, 10% in 2 in. (50 mm) **El**, 150 **HV**; **Strain hardened**: all **diam**, 550 MPa **TS**, 4% in 2 in. (50 mm) **El**, 180 **HV**
ISO 430	CuNi10Zn27	ISO	Plate, sheet, strip	61.0–65.0 **Cu**, 0.05 **Pb**, rem **Zn**, 0.3 **Fe**, 0–0.5 **Mn**, 9.0–11.0 **Ni**, 0.3 total, others	
ISO 430	CuNi10Zn28Pb1	ISO	Strip, rod, bar	59.0–63.0 **Cu**, 1.0–2.0 **Pb**, rem **Zn**, 0.3 **Fe**, 0–0.5 **Mn**, 9.0–11.0 **Ni**, 0.5 total, others	
ISO 430	CuNi10Zn42Pb2	ISO	Rod, bar, extruded sections, forgings	44.0–48.0 **Cu**, 1.0–2.5 **Pb**, rem **Zn**, 0.5 **Fe**, 0.0–0.5 **Mn**, 9.0–11.0 **Ni**, 0.5 total, others	
COPANT 162	C 773 00	COPANT	Wrought products	46–50 **Cu**, 0.05 **Pb**, rem **Zn**, 9–11 **Ni**, 0.01 **Al**	
ANSI/ASTM B 124	C 77400	US	Rod, bar, shapes	43.0–47.0 **Cu**, 0.20 **Pb**, rem **Zn**, 9.0–11.0 **Ni+Co**	

WROUGHT COPPER AND COPPER ALLOYS

specification number	designation	country	product forms	chemical composition	mechanical properties and hardness values
QQ-C-586C	745	US	Rod, shapes, flatwire, strip, bar	63.5–66.5 **Cu**, 0.05 **Pb**, rem **Zn**, 0.25 **Fe**, 0.50 **Mn**, 9.0–11.0 **Ni + Co**, 0.50 total, others	
ANSI/ASTM B 122	C 74500	US	Plate, sheet, strip, bar	63.5–68.5 **Cu**, 0.10 **Pb**, rem **Zn**, 0.25 **Fe**, 0.50 **Mn**, 9.0–11.0 **Ni + Co**, silver included in copper content	**Hard**: >0.012 in. (>0.305 mm) **diam**, 80 ksi (550 MPa) **TS**, 73–78 **HR30T**; **Spring**: >0.012 in. (>0.305 mm) **diam**, 95 ksi (655 MPa) **TS**, 77–80 **HR30T**; **Annealed**: >0.015 in. (>0.381 mm) **diam**, 26–36 **HR30T**
ANSI/ASTM B 122	C 74000	US	Plate, sheet, strip, bar	69.0–73.5 **Cu**, 0.10 **Pb**, rem **Zn**, 0.25 **Fe**, 0.50 **Mn**, 9.0–11.0 **Ni + Co**, silver included in copper content,	**Quarter hard**: >0.020 in. (>0.508 mm) **diam**, 55 ksi (380 MPa) **TS**, 60–80 **HRB**; **Hard**: >0.020 in. (>0.508 mm) **diam**, 73 ksi (505 MPa) **TS**, 79–91 **HRB**; **Annealed**: >0.020 in. (>0.508 mm) **diam**, 5–20 **HRB**
ANSI/ASTM B 151	CA 745	US	Rod, bar	63.5–66.5 **Cu**, 0.05 **Pb**, rem **Zn**, 0.25 **Fe**, 0.50 **Mn**, 9.0–11.0 **Ni + Co**, silver included in copper content	**Quarter hard**: 0.02–0.50 in. (0.508–12.7 mm) **diam**, 75 ksi (515 MPa) **TS**; **Hard**: all **diam**, 75 ksi (515 MPa) **TS**, 95 ksi (650 MPa) **YS**
ANSI/ASTM B 206	CA 745	US	Wire	63.5–66.5 **Cu**, 0.05 **Pb**, rem **Zn**, 0.25 **Fe**, 0.50 **Mn**, 9.0–11.0 **Ni + Co**	**Quarter hard**: 0.020–0.250 in. (0.508–6.35 mm) **diam**, 73 ksi (505 MPa) **TS**; **Hard**: 0.020–0.250 in. (0.508–6.35 mm) **diam**, 108 ksi (740 MPa) **TS**; **Spring**: 0.020–0.0253 in. (0.508–0.643 mm) **diam**, 130 ksi (900 MPa) **TS**
ANSI/ASTM B 283	C 77400	US	Die forgings	43.0–47.0 **Cu**, 0.20 **Pb**, rem **Zn**, 9.0–11.0 **Ni + Co**	
JIS H 3110	C 7451	Japan	Sheet, plate, and strip	62.0–68.0 **Cu**, 0.10 **Pb**, rem **Zn**, 0.25 **Fe**, 0.50 **Mn**, 8.5–11.5 **Ni**, 8.5–11.5 **Ni + Co**	**Annealed**: 0.3–5 mm **diam**, 324 MPa **TS**, 20% in 2 in. (50 mm) **El**; **1/2 hard**: $\geq$0.15 mm **diam**, 392 MPa **TS**, 5% in 2 in. (50 mm) **El**
JIS Z 3262	BCuZn-6	Japan	Wire: filler	46–50 **Cu**, 0.05 **Pb**, rem **Zn**, 0.25 **Si**, 0.02 **Al**, 9–11 **Ni**, 0.25 **P**	
JIS Z 3202	YCuZnNi Nickel silver	Japan	Rod: bare welding	46–50 **Cu**, 0.05 **Pb**, rem **Zn**, 0.25 **Si**, 0.02 **Al**, 9–11 **Ni**, 0.25 **P**	
JIS H 3270	C 7451	Japan	Rod and bar	62.0–68.0 **Cu**, 0.10 **Pb**, rem **Zn**, 0.25 **Fe**, 0.50 **Mn**, 8.5–11.5 **Ni + Co**	

WROUGHT COPPER AND COPPER ALLOYS

specification number	designation	country	product forms	chemical composition	mechanical properties and hardness values
AS 1567	798C	Australia	Rod, bar, section	46.0–48.0 **Cu**, 2.0–3.5 **Pb**, rem **Zn**, 0.5 **Mn**, 8.0–11.0 **Ni**, 0.75 total, others, excl mn	**As manufactured**: >6 mm **diam**, 460 MPa **TS**, 8% in 2 in. (50 mm) **El**
AS 1567	796C	Australia	Rod, bar, section	46.0–48.0 **Cu**, 2.0 **Pb**, rem **Zn**, 0.50 **Mn**, 8.0–11.0 **Ni**, 0.75 total, others, excl pb, mn	**As manufactured**: >6 mm **diam**, 460 MPa **TS**, 8% in 2 in. (50 mm) **El**
AS 1568	796C	Australia	Forgings	46.0–48.0 **Cu**, 2.0 **Pb**, rem **Zn**, 0.50 **Mn**, 8.0–11.0 **Ni**, 0.75 total, others	**As manufactured**: ≥6 mm **diam**, 460 MPa **TS**, 8% in 2 in. (50 mm) **El**
NF A 51 107	Cu Ni 10 Zn 27	France	Sheet and strip	61–65 **Cu**, 0.05 **Pb**, rem **Zn**, 0.5 **Mn**, 8–11 **Ni**, 0.3 others, total	**1/4 hard**: ≤10 mm **diam**, 420 MPa **TS**, 27% in 2 in. (50 mm) **El**, 105–135 **HV**; **1/2 hard**: ≤10 mm **diam**, 470 MPa **TS**, 12% in 2 in. (50 mm) **El**, 140–160 **HV**; **3/4 hard**: ≤10 mm **diam**, 550 MPa **TS**, 7% in 2 in. (50 mm) **El**, 170–190 **HV**
CDA 782		US	Strip: rolled	63.0–67.0 **Cu**, 1.5–2.5 **Pb**, rem **Zn**, 0.35 **Fe**, 0.50 **Mn**, 7.0–9.0 **Ni**, 0.10 total, others	**0.035mm temper**: 0.040 in. **diam**, 53 ksi (365 MPa) **TS**, 40% in 2 in. (50 mm) **El**, 78 **HRF**; **Half hard temper**: 0.040 in. **diam**, 69 ksi (476 MPa) **TS**, 12% in 2 in. (50 mm) **El**, 78 **HRB**; **Extra hard temper**: 0.040 in. **diam**, 91 ksi (627 MPa) **TS**, 3% in 2 in. (50 mm) **El**, 90 **HRB**
COPANT 162	C 760 00	COPANT	Wrought products	60–63 **Cu**, 0.10 **Pb**, rem **Zn**, 0.25 **Fe**, 0.5 **Mn**, 7–9 **Ni**	
COPANT 162	C 782 00	COPANT	Wrought products	63–67 **Cu**, 1.5–2.5 **Pb**, rem **Zn**, 0.35 **Fe**, 0.50 **Mn**, 7–9 **Ni**	
COPANT 162	C 743 00	COPANT	Wrought products	63–66 **Cu**, 0.10 **Pb**, rem **Zn**, 0.25 **Fe**, 0.5 **Mn**, 7.9 **Ni**	
COPANT 162	C 799 00	COPANT	Wrought products	47.5–50.5 **Cu**, 1.0–1.5 **Pb**, rem **Zn**, 0.25 **Fe**, 0.50 **Mn**, 6.5–8.5 **Ni**	

CAST COPPER AND COPPER ALLOYS

CAST COPPER AND COPPER ALLOYS

specification number	designation	country	product forms	chemical composition	mechanical properties and hardness values
CDA C 80100		US	Castings	99.95 min **Cu**, silver included in copper content	**Sand casting**: all **diam**, 19 ksi (131 MPa) **TS**, 6.5 ksi (44.8 MPa) **YS**, 20% in 2 in. (50 mm) **El**, 44 **HB**
COPANT 162	C 801 00	COPANT	Cast products	99.95 min **Cu + Ag**, 0.05 total, others	
AS 1565	801A	Australia	Ingot, castings	99.95 min **Cu**, 0.05 total, others	**Sand cast**: all **diam**, 160 MPa **TS**, 23% in 2 in. (50 mm) **El**, 40 **HB**
COPANT 162	C 803 00	COPANT	Cast products	99.95 min **Cu + Ag**, 0.034 **Ag**, 0.05 total, others	
CDA C 80300		US	Castings	99.95 min **Cu**, 0.034 **Ag**, silver included in copper content	**Sand casting**: all **diam**, 19 ksi (131 MPa) **TS**, 6.5 ksi (44.8 MPa) **YS**, 20% in 2 in. (50 mm) **El**, 44 **HB**
COPANT 671	Cu ETP	COPANT	Ingot	99.90 min **Cu + Ag**, 0.004 **Pb**, 0.003 **Sb**, 0.002 **Ag**, 0.015–0.04 O_2, 0.012 **As**, 0.003 **Bi**, 0.025 **Te**, 0.04 total, impurities	
COPANT 534	Cu QTP	COPANT	Ingot	99.90 min **Cu + Ag**, 0.005 **Pb**, 0.55 O_2, 0.0025 **Bi**	
QQ–C–521d	Copper ingot	US	Ingot	99.90 min **Cu**, 0.005 **Pb**, 0.004 **Fe**, 0.0025 **Sb**, 0.0035 **S**, 0.001 **Bi**, 0.0025 **As**, **Ag** in **Cu** content	
CDA C 80500		US	Castings	99.75 min **Cu**, 0.034 min **Ag**, 0.02 **B**, silver included in copper content	**Sand casting**: all **diam**, 19 ksi (131 MPa) **TS**, 6.5 ksi (44.8 MPa) **YS**, 20% in 2 in. (50 mm) **El**, 44 **HB**
COPANT 162	C 805 00	COPANT	Cast products	99.75 min **Cu + Ag**, 0.034 **Ag**, 0.02 **B**, 0.23 total, others	
CDA C 80700		US	Castings	99.75 min **Cu**, 0.02 **B**, silver included in copper content	**Sand casting**: all **diam**, 19 ksi (131 MPa) **TS**, 6.5 ksi (44.8 MPa) **YS**, 20% in 2 in. (50 mm) **El**, 44 **HB**
COPANT 162	C 807 00	COPANT 162	Cast products	99.75 min **Cu + Ag**, 0.23 total, others	
CDA C 80900		US	Castings	99.70 min **Cu**, 0.034 min **Ag**, silver included in copper content	**Sand casting**: all **diam**, 19 ksi (131 MPa) **TS**, 6.5 ksi (44.8 MPa) **YS**, 20% in 2 in. (50 mm) **El**, 44 **HB**
COPANT 162	C 809 00	COPANT	Cast products	0.034 **Ag**, 99.70 min **Cu + Ag**, 0.30 total, others	
CDA C 81100		US	Castings	99.70 min **Cu**, silver included in copper content	**Sand casting**: all **diam**, 19 ksi (131 MPa) **TS**, 6.5 ksi (44.8 MPa) **YS**, 20% in 2 in. (50 mm) **El**, 44 **HB**
COPANT 162	C 811 00	COPANT	Cast products	99.70 min **Cu + Ag**, 0.30 total, others	

CAST COPPER AND COPPER ALLOYS

specification number	designation	country	product forms	chemical composition	mechanical properties and hardness values
CDA C 81300		US	Castings	98.5 min **Cu**, 0.6–1.0 **Co**, 0.02–0.10 **Be**, 99.5 min **Cu**+**Co**+**Be**	**Heat treated**: all **diam**, 53 ksi (365 MPa) **TS**, 36 ksi (248 MPa) **YS**, 11% in 2 in. (50 mm) **El**, 89 **HB**
COPANT 162	C 813 00	COPANT	Cast products	98.5 min **Cu**, 0.6–1.0 **Co**, 0.02–0.10 **Be**, 99.5 min, total elements	
CDA C 81400		US	Castings	98.5 min **Cu**, 0.6–1.0 **Cr**, 0.02–0.10 **Be**, 99.5 min **Cu**+**Cr**+**Be**	**Heat treated**: all **diam**, 53 ksi (365 MPa) **TS**, 36 ksi (248 MPa) **YS**, 11% in 2 in. (50 mm) **El**, 69 **HRB**
COPANT 162	C 814 00	COPANT	Cast products	98.5 min **Cu**, 0.6–1.0 **Cr**, 0.02–0.10 **Be**, 99.5 min, total elements	
AS 1565	815B	Australia	Ingot, castings	98.79 min **Cu**, 0.6–1.2 **Cr**	**Sand cast**: all **diam**, 270 MPa **TS**, 170 MPa **YS**, 18% in 2 in. (50 mm) **El**, 100 **HB**
CDA C 81500		US	Castings	98.0 min **Cu**, 0.10 **Sn**, 0.02 **Pb**, 0.10 **Zn**, 0.10 **Fe**, 0.15 **Si**, 0.10 **Al**, 0.40–1.5 **Cr**	**Heat treated**: all **diam**, 45 ksi (310 MPa) **TS**, 35 ksi (241 MPa) **YS**, 12% in 2 in. (50 mm) **El**, 89 min **HB**
COPANT 162	C 815 00	COPANT	Cast products	98.0 min **Cu**, 0.4–1.5 **Cr**, 99.5 min, total elements	
SABS 200	Cu–Cr1	South Africa	Castings	98.79 min **Cu**, 0.60–1.2 **Cr**	**Sand cast, chill cast, continuously cast**: all **diam**, 100 **HB**
COPANT 162	C 824 00	COPANT	Cast products	97.2 min **Cu**, 0.20–0.30 **Co**, 1.65–1.75 **Be**, 99.5 min, total elements	
CDA C 82400		US	Castings	96.4 min **Cu**, 0.10 **Sn**, 0.02 **Pb**, 0.10 **Zn**, 0.20 **Fe**, 0.15 **Al**, 0.20–0.40 **Co**, 0.10 **Ni**, 0.10 **Cr**, 1.65–1.75 **Be**	**Sand casting**: all **diam**, 72 ksi (496 MPa) **TS**, 37 ksi (255 MPa) **YS**, 20% in 2 in. (50 mm) **El**, 78 **HRB**; **Heat treated**: all **diam**, 145 ksi (100 MPa) **TS**, 135 ksi (931 MPa) **YS**, 0% **El**, 34 min **HRC**
CDA C 82500		US	Castings	95.5 min **Cu**, 0.10 **Sn**, 0.02 **Pb**, 0.10 **Zn**, 0.25 **Fe**, 0.20–0.35 **Si**, 0.15 **Al**, 0.35–0.7 **Co**, 0.20 **Ni**, 0.10 **Cr**, 1.90–2.15 **Be**	**Sand casting**: all **diam**, 80 ksi (552 MPa) **TS**, 45 ksi (310 MPa) **YS**, 20% in 2 in. (50 mm) **El**, 82 **HRB**; **Heat treated**: all **diam**, 155 ksi (1069 MPa) **TS**, 115 ksi (793 MPa) **YS**, 0% **El**, 38 min **HRC**
CDA C 82600		US	Castings	95.2 min **Cu**, 0.10 **Sn**, 0.02 **Pb**, 0.10 **Zn**, 0.25 **Fe**, 0.20–0.35 **Si**, 0.15 **Al**, 0.35–0.7 **Co**, 0.20 **Ni**, 0.10 **Cr**, 2.25–2.45 **Be**	**Sand casting**: all **diam**, 82 ksi (565 MPa) **TS**, 47 ksi (324 MPa) **YS**, 20% in 2 in. (50 mm) **El**, 83 **HRB**; **Heat treated**: all **diam**, 160 ksi (1103 MPa) **TS**, 150 ksi (1034 MPa) **YS**, 0% **El**, 40 min **HRC**

CAST COPPER AND COPPER ALLOYS

specification number	designation	country	product forms	chemical composition	mechanical properties and hardness values
COPANT 162	C 826 00	COPANT	Cast products	95.6 min **Cu**, 0.2–0.35 **Si**, 0.35–0.7 **Co**, 2.25–2.45 **Be**, 99.5 min, total elements	
CDA C 82700		US	Castings	94.6 min **Cu**, 0.10 **Sn**, 0.02 **Pb**, 0.10 **Zn**, 0.25 **Fe**, 0.15 **Si**, 0.15 **Al**, 1.0–1.5 **Ni**, 0.10 **Cr**, 2.35–2.55 **Be**, 99.5 min **Cu + Sn + Pb + Zn + Fe + Si + Al + Ni + Cr + Be**	**Heat treated**: all **diam**, 155 ksi (1069 MPa) **TS**, 130 ksi (896 MPa) **YS**, 0% **El**, 39 min **HRC**
CDA C 82800		US	Castings	94.8 min **Cu**, 0.10 **Sn**, 0.02 **Pb**, 0.10 **Zn**, 0.25 **Fe**, 0.20–0.35 **Si**, 0.15 **Al**, 0.35–0.7 **Co**, 0.20 **Ni**, 0.10 **Cr**, 2.50–2.75 **Be**	**Sand casting**: all **diam**, 97 ksi (669 MPa) **TS**, 55 ksi (379 MPa) **YS**, 20% in 2 in. (50 mm) **El**, 85 **HRB**; **Heat treated**: all **diam**, 150 ksi (1034 MPa) **TS**, 110 ksi (758 MPa) **YS**, 0% **El**, 42 min **HRC**
CDA C 82200		US	Castings	96.5 min **Cu**, 1.0–2.0 **Ni**, 0.35–0.8 **Be**	**Sand casting**: all **diam**, 57 ksi (393 MPa) **TS**, 30 ksi (207 MPa) **YS**, 20% in 2 in. (50 mm) **El**, 60 **HRB**; **Heat treated**: all **diam**, 90 ksi (621 MPa) **TS**, 70 ksi (483 MPa) **YS**, 5% in 2 in. (50 mm) **El**, 92 min **HRB**
CDA C 82100		US	Castings	95.5 min **Cu**, 0.25–1.5 **Co**, 0.25–1.5 **Ni**, 0.35–0.8 **Be**	**Heat treated**: all **diam**, 85 ksi (586 MPa) **TS**, 62 ksi (427 MPa) **YS**, 5% in 2 in. (50 mm) **El**, 217 **HB**
CDA C 82000		US	Castings	95.0 min **Cu**, 0.10 **Sn**, 0.02 **Pb**, 0.10 **Zn**, 0.10 **Fe**, 0.15 **Si**, 0.10 **Al**, 0.20 **Ni**, 2.4–2.7 **Ni + Co**, 0.10 **Cr**, 0.45–0.8 **Be**	**Sand casting**: all **diam**, 50 ksi (348 MPa) **TS**, 20 ksi (138 MPa) **YS**, 20% in 2 in. (50 mm) **El**, 55 **HRB**; **Heat treated**: all **diam**, 95 ksi (655 MPa) **TS**, 70 ksi (483 MPa) **YS**, 3% in 2 in. (50 mm) **El**, 92 min **HRB**
COPANT 162	C 820 00	COPANT	Cast products	98.13 min **Cu**, 0.10 **Sn**, 0.02 **Pb**, 0.10 **Zn**, 0.10 **Fe**, 0.15 **Si**, 0.10 **Al**, 2.4–2.7 **Co**, 0.10 **Cr**, 0.45–0.8 **Be**, 99.5 min, total elements	
COPANT 162	C 821 00	COPANT	Cast products	95.5 min **Cu**, 0.25–1.5 **Co**, 0.25–1.5 **Ni**, 0.35–0.8 **Be**, 99.5 min, total elements	
COPANT 162	C 822 00	COPANT	Cast products	96.5 min **Cu**, 1.0–2.0 **Ni**, 0.35–0.8 **Be**, 99.5 min, total elements	
COPANT 162	C 825 00	COPANT	Cast products	95.38 min **Cu**, 0.10 **Sn**, 0.02 **Pb**, 0.10 **Zn**, 0.25 **Fe**, 0.20–0.35 **Si**, 0.15 **Al**, 0.35–0.7 **Co**, 0.20 **Ni**, 0.10 **Cr**, 1.9–2.15 **Be**, 99.5 min, total elements	

CAST COPPER AND COPPER ALLOYS

specification number	designation	country	product forms	chemical composition	mechanical properties and hardness values
AMS 4890A		US	Investment castings	95.82 min **Cu**, 0.10 **Sn**, 0.02 **Pb**, 0.10 **Zn**, 0.35 **Fe**, 0.20–0.35 **Si**, 0.15 **Al**, 0.20–0.65 **Co**, 0.20 **Ni**, 0.10 **Cr**, 1.85–2.15 **Be**	**Solution and precipitation heat treated**: all **diam**, 150 ksi (1034 MPa) **TS**, 120 ksi (827 MPa) **YS**, 2% in 1 in. (25 mm) **El**, 36 min **HRC**
COPANT 162	C 827 00	COPANT	Cast products	94.58 min **Cu**, 0.10 **Sn**, 0.02 **Pb**, 0.10 **Zn**, 0.25 **Fe**, 0.15 **Si**, 0.15 **Al**, 1.0–1.5 **Ni**, 0.10 **Cr**, 2.35–2.55 **Be**, 99.5 min, total elements	
COPANT 162	C 828 00	COPANT	Cast products	94.78 min **Cu**, 0.10 **Sn**, 0.02 **Pb**, 0.10 **Zn**, 0.25 **Fe**, 0.20–0.35 **Si**, 0.15 **Al**, 0.35–0.7 **Co**, 0.20 **Ni**, 0.10 **Cr**, 2.5–2.75 **Be**, 99.5 min, total elements	
ANSI/ASTM B 52	Alloy A	US	Ingot	99.75 min **Cu**, 0.15 **Fe**, 14.0 min **P**, phosphorous included in copper content	
ANSI/ASTM B 52	Alloy B	US	Ingot	99.75 min **Cu**, 0.15 **Fe**, 10.0 min **P**, phosphorous included in copper content	
JIS H 2501	Class 2, grade A	Japan	Ingot	89.7 min **Cu**, 0.01 **Sn**, 0.01 **Pb**, 0.05 **Fe**, 10.0 min **P**, 99.70 min **P+Cu**	
JIS H 2501	Class 1, grade A	Japan	Ingot	85.20 min **Cu**, 0.01 **Sn**, 0.01 **Pb**, 0.05 **Fe**, 14.5 min **P**, 99.70 min **P+Cu**	
JIS H 2501	Class 1, grade B	Japan	Ingot	85.50 min **Cu**, 0.15 **Fe**, 14.0 min **P**, 99.50 min **P+Cu**	
JIS H 2501	Class 2, grade B	Japan	Ingot	90.0 min **Cu**, 0.15 **Fe**, 9.5 min **P**, 99.50 min **P+Cu**	
CDA C 81800		US	Castings	95.6 min **Cu**, 1.4–1.7 **Co**, 0.30–0.55 **Be**, 0.8–1.2 **Ag**	**Heat treated**: all **diam**, 90 ksi (621 MPa) **TS**, 70 ksi (483 MPa) **YS**, 3% in 2 in. (50 mm) **El**, 92 min **HRB**
CDA C 81700		US	Castings	94.2 min **Cu**, 0.25–1.5 **Co**, 0.25–1.5 **Ni**, 0.30–0.55 **Be**, 0.8–1.2 **Ag**	**Heat treated**: all **diam**, 85 ksi (586 MPa) **TS**, 62 ksi (427 MPa) **YS**, 5% in 2 in. (50 mm) **El**, 217 **HB**
COPANT 162	C 817 00	COPANT	Cast products	94.2 min **Cu**, 0.25–1.5 **Co**, 0.25–1.5 **Ni**, 0.3–0.55 **Be**, 0.8–1.2 **Ag**, 99.5 min, total elements	
COPANT 162	C 818 00	COPANT	Cast products	95.6 min **Cu**, 1.4–1.7 **Co**, 0.3–0.55 **Be**, 0.8–1.2 **Ag**, 99.5 min, total elements	
QQ-C-390A	820	US	Castings	rem **Cu**, 0.01 **Sn**, 0.01 **Pb**, 0.01 **Zn**, 0.10 **Fe**, 0.15 **Si**, 0.10 **Al**, 2.4–2.7 **Co**, 0.20 **Ni**, 0.01 **Cr**, 0.45–0.8 **Be**, **Ni** in **Co** content	**Solution heat treated and aged**: all **diam**, 95 ksi (655 MPa) **TS**, 3% in 2 in. (50 mm) **El**, 92 **HRB**

CAST COPPER AND COPPER ALLOYS

specification number	designation	country	product forms	chemical composition	mechanical properties and hardness values
QQ-C-390A	824	US	Castings	97.20 min **Cu**, 0.20–0.30 **Co**, 1.7–1.8 **Be**	**Solution heat treated and aged**: all **diam**, 145 ksi (1000 MPa) **TS**, 130 ksi (896 MPa) **YS**, 0% in 2 in. (50 mm) **El**, 35 **HRC**
QQ-C-390A	825	US	Castings	rem **Cu**, 0.10 **Sn**, 0.02 **Pb**, 0.10 **Zn**, 0.25 **Fe**, 0.20–0.35 **Si**, 0.15 **Al**, 0.35–0.7 **Co**, 0.20 **Ni**, 0.10 **Cr**, 1.9–2.2 **Be**, **Ni** in **Co** content	**Solution heat treated and aged**: all **diam**, 155 ksi (1069 MPa) **TS**, 140 ksi (965 MPa) **YS**, 0% in 2 in. (50 mm) **El**, 38 **HRC**
QQ-C-390A	826	US	Castings	95.65 min **Cu**, 0.20–0.35 **Si**, 0.35–0.7 **Co**, 2.3–2.5 **Be**	**Solution heat treated and aged**: all **diam**, 160 ksi (1103 MPa) **TS**, 145 ksi (1000 MPa) **YS**, 0% in 2 in. (50 mm) **El**, 40 **HRC**
QQ-C-390A	827	US	Castings	rem **Cu**, 0.10 **Sn**, 0.02 **Pb**, 0.10 **Zn**, 0.25 **Fe**, 0.15 **Si**, 0.15 **Al**, 1.0–1.5 **Ni**, 0.10 **Cr**, 2.4–2.6 **Be**	**Solution heat treated and aged**: all **diam**, 155 ksi (1069 MPa) **TS**, 140 ksi (965 MPa) **YS**, 0% in 2 in. (50 mm) **El**, 39 **HRC**
QQ-C-390A	828	US	Castings	rem **Cu**, 0.10 **Sn**, 0.02 **Pb**, 0.10 **Zn**, 0.25 **Fe**, 0.20–0.35 **Si**, 0.15 **Al**, 0.35–0.7 **Co**, 0.20 **Ni**, 0.10 **Cr**, 2.5–2.8 **Be**, **Ni** in **Co** content	**Solution heat treated and aged**: all **diam**, 160 ksi (1103 MPa) **TS**, 150 ksi (1034 MPa) **YS**, 0% in 2 in. (50 mm) **El**, 42 **HRC**
QQ-C-571B	Alloy B	US	Ingot, slab, shot	89.75 min **Cu**, 0.15 **Fe**, 10 min **P**, **P** in **Cu** content	
QQ-C-571B	Alloy A	US	Ingot, slab, shot	85.6 min **Cu**, 0.15 **Fe**, 14 min **P**, **P** in **Cu** content	
COPANT 162	C 833 00	COPANT	Cast products	92–94 **Cu**, 1.0–2.0 **Sn**, 1.0–2.0 **Pb**, 2.0–6.0 **Zn**	
CDA C 83300		US	Castings	92.0–94.0 **Cu**, 1.0–2.0 **Sn**, 1.0–2.0 **Pb**, 2.0–6.0 **Zn**	**Sand castings**: all **diam**, 32 ksi (221 MPa) **TS**, 10 ksi (69 MPa) **YS**, 35% in 2 in. (50 mm) **El**, 35 **HB**
COPANT 162	C 834 00	COPANT	Cast products	88–92 **Cu**, 0.2 **Sn**, 0.5 **Pb**, 8–12 **Zn**	
CDA C 83400		US	Castings	88.0–92.0 **Cu**, 0.20 **Sn**, 0.50 **Pb**, 8.0–12.0 **Zn**	**Sand castings**: all **diam**, 35 ksi (241 MPa) **TS**, 10 ksi (69 MPa) **YS**, 30% in 2 in. (50 mm) **El**, 50 **HRF**
AS 1565	835D	Australia	Ingot, castings	88.0–91.0 **Cu**, 1.0–2.0 **Sn**, 0.10 **Pb**, rem **Zn**, 0.30 total, others (castings)	**Sand cast**: all **diam**, 190 MPa **TS**, 20% in 2 in. (50 mm) **El**
JIS H 2204	Class 2	Japan	Ingot: for castings	87.0–91.0 **Cu**, 9.0–12.0 **Sn**, 0.1 **Pb**	
JIS H 2203	Class 3	Japan	Ingot: for castings	86.5–89.5 **Cu**, 9.0–11.0 **Sn**, 1.0–3.0 **Zn**	
JIS H 5111	Class 3	Japan	Castings	86.5–89.5 **Cu**, 9.0–11.0 **Sn**, 1.0 **Pb**, 1.0–3.0 **Zn**	**As cast**: all **diam**, 245 MPa **TS**, 15% in 2 in. (50 mm) **El**
JIS H 5111	Class 2	Japan	Castings	86.0–90.0 **Cu**, 7.0–9.0 **Sn**, 1.0 **Pb**, 3.0–5.0 **Zn**	**As cast**: all **diam**, 245 min MPa **TS**, 20% in 2 in. (50 mm) **El**

CAST COPPER AND COPPER ALLOYS

specification number	designation	country	product forms	chemical composition	mechanical properties and hardness values
JIS H 2203	Class 2	Japan	Ingot: for castings	86.0–89.0 **Cu**, 7.0–9.0 **Sn**, 3.0–5.0 **Zn**	
AMS 4845D		US	Castings	86.0–89.0 **Cu**, 9.0–11.0 **Sn**, 0.30 **Pb**, 1.0–3.0 **Zn**, 0.15 **Fe**, 0.05 **P**	**Sand cast**: all **diam**, 70 min **HRF**; **Chill and centrifugally cast**: all **diam**, 80 min **HRF**
AMS 4846A		US	Castings	84.0–89.0 **Cu**, 9.0–13.5 **Sn**, 1.0 **Pb**, 1.0–3.0 **Zn**, 0.15 **Fe**, 0.05 **P**	**Sand cast**: all **diam**, 70 min **HRF**; **Chill and centrifugally cast**: all **diam**, 80 min **HRF**
JIS H 2204	Class 3	Japan	Ingot: for castings	84.0–88.0 **Cu**, 12.0–15.0 **Sn**, 0.1 **Pb**	
AS 1565	837D	Australia	Ingot, castings	83.0–88.0 **Cu**, 0.5 **Pb**, rem **Zn**, 0.05–0.20 **As**, 1.0 total, others (castings)	**Sand cast**: all **diam**, 170 MPa **TS**, 80 MPa **YS**, 18% in 2 in. (50 mm) **El**, 45 **HB**
DS 3001	Tombac 5456	Denmark	Sand castings	84.5–86.5 **Cu**, 9.0–11.0 **Sn**, 1.0–2.0 **Pb**, 1.5–3.5 **Zn**, 0.30 **Fe**, 0.35 **Sb**, 0.01 **Si**, 0.01 **Al**, 0.20 **Mn**, 2.0 **Ni**, 0.05 **P**, 0.10 **S**	**As cast**: all **diam**, 240 MPa **TS**, 120 MPa **YS**, 12% in 2 in. (50 mm) **El**, 75 min **HB**
JIS H 5111	Class 7	Japan	Castings	86.0–90.0 **Cu**, 5.0–7.0 **Sn**, 1.0–3.0 **Pb**, 3.0–5.0 **Zn**	**As cast**: all **diam**, 216 min MPa **TS**, 18% in 2 in. (50 mm) **El**
JIS H 2203	Class 7	Japan	Ingot: for castings	86.0–90.0 **Cu**, 5.0–7.0 **Sn**, 1.0–3.0 **Pb**, 3.0–5.0 **Zn**	
SFS 2208	CuSn10Zn2	Finland	Sand castings	86.0–89.0 **Cu**, 9.0–11.0 **Sn**, 1.5 **Pb**, 1.0–3.0 **Zn**, 0.25 **Fe**, 0.25 **Sb**, 0.01 **Si**, 0.01 **Al**, 0.2 **Mn**, 2.0 **Ni**, 0.05 **P**, 0.10 **S**	**Sand castings**: all **diam**, 260 MPa **TS**, 120 MPa **YS**, 15% in 2 in. (50 mm) **El**
ANSI/ASTM B 62	CA 836	US	Castings	84.0–86.0 **Cu**, 4.0–6.0 **Sn**, 4.0–6.0 **Pb**, 4.0–6.0 **Zn**, 0.30 **Fe**, 1.0 **Ni**, 0.05 **P**	**As cast**: all **diam**, 30 ksi (205 MPa) **TS**, 14 ksi (95 MPa) **YS**, 20% in 2 in. (51 mm) **El**
ANSI/ASTM B 30	CA 836	US	Ingot	84.0–86.0 **Cu**, 4.3–6.0 **Sn**, 4.0–5.7 **Pb**, 4.3–6.0 **Zn**, 0.25 **Fe**, 0.25 **Sb**, 0.005 **Si**, 0.005 **Al**, 0.8 **Ni**, 0.03 **P**, 0.08 **S**	**As cast**: all **diam**, 30 ksi (205 MPa) **TS**, 14 ksi (95 MPa) **YS**, 20% in 2 in. (51 mm) **El**
ANSI/ASTM B 30	836	US	Ingot: for castings	84.0–86.0 **Cu**, 4.3–6.0 **Sn**, 4.0–5.7 **Pb**, 4.3–6.0 **Zn**, 0.25 **Fe**, 0.25 **Sb**, 0.005 **Si**, 0.005 **Al**, 0.8 **Ni+Co**, 0.03 **P**, 0.08 **S**	
ANSI/ASTM B 584	C 83600	US	Sand castings	84.0–86.0 **Cu**, 4.0–6.0 **Sn**, 4.0–6.0 **Pb**, 4.0–6.0 **Zn**, 0.30 **Fe**, 0.25 **Sb**, 0.005 **Si**, 0.005 **Al**, 1.0 **Ni+Co**, 0.03 **P**, 0.08 **S**, nickel included in copper content	**As cast**: all **diam**, 30 ksi (207 MPa) **TS**, 14 ksi (97 MPa) **YS**, 20% in 2 in. (50 mm) **El**
ANSI/ASTM B 505	CA 836	US	Castings	84.0–86.0 **Cu**, 4.0–6.0 **Sn**, 4.0–6.0 **Pb**, 4.0–6.0 **Zn**, 0.30 **Fe**, 0.25 **Sb**, 0.005 **Si**, 0.005 **Al**, 1.0 **Ni+Co**, 1.5 **P**, 0.08 **S**	**As cast**: all **diam**, 36 ksi (248 MPa) **TS**, 19 ksi (131 MPa) **YS**, 15% in 2 in. (50 mm) **El**
CDA C 83600		US	Castings	84.0–86.0 **Cu**, 4.0–6.0 **Sn**, 4.0–6.0 **Pb**, 4.0–6.0 **Zn**, 0.30 **Fe**, 0.25 **Sb**, 0.005 **Si**, 0.005 **Al**, 1.0 **Ni**, 0.05 **P**, 0.08 **S**	**Sand castings**: all **diam**, 30 ksi (207 MPa) **TS**, 14 ksi (97 MPa) **YS**, 20% in 2 in. (50 mm) **El**, 60 **HB**

CAST COPPER AND COPPER ALLOYS

specification number	designation	country	product forms	chemical composition	mechanical properties and hardness values
COPANT 162	C 836 00	COPANT	Cast products	84–86 **Cu**, 4–6 **Sn**, 4–6 **Pb**, 4–6 **Zn**, 0.30 **Fe**, 0.25 **Sb**, 0.005 **Si**, 0.005 **Al**, 1.0 **Ni**, 0.05 **P**, 0.08 **S**	
COPANT 801	C 836 00	COPANT	Continuous–cast rod, bar and tube	84–86 **Cu**, 4–6 **Sn**, 4–6 **Pb**, 4–6 **Zn**, 0.30 **Fe**, 0.25 **Sb**, 0.005 **Si**, 0.005 **Al**, 1.0 **Ni**, 0.05 **P**, 0.08 **S**	**As manufactured**: all **diam**, 250 MPa **TS**, 130 MPa **YS**, 15% in 2 in. (50 mm) **El**
DS 3001	Tombac 5204	Denmark	Sand castings	84.0–86.0 **Cu**, 4.0–6.0 **Sn**, 4–6 **Pb**, 4.0–6.0 **Zn**, 0.30 **Fe**, 0.25 **Sb**, 0.01 **Si**, 0.01 **Al**, 0.1 **Mn**, 2.0 **Ni**, 0.05 **P**, 0.10 **S**	**As cast**: all **diam**, 230 MPa **TS**, 90 MPa **YS**, 15% in 2 in. (50 mm) **El**, 60 min **HB**
QQ–C–390A	836	US	Castings	84.0–86.0 **Cu**, 4.0–6.0 **Sn**, 4.0–6.0 **Pb**, 4.0–6.0 **Zn**, 0.30 **Fe**, 1.0 **Ni**, 0.05 **P**	**As cast**: all **diam**, 30 ksi (207 MPa) **TS**, 14 ksi (97 MPa) **YS**, 20% in 2 in. (50 mm) **El**
QQ–C–525b	Composition 2	US	Ingot	84.0–86.0 **Cu**, 4.3–6.0 **Sn**, 4.0–5.7 **Pb**, 4.5–6.0 **Zn**, 0.25 **Fe**, 0.25 **Sb**, 0.005 **Si**, 0.005 **Al**, 0.8 **Ni**, 0.03 **P**, 0.08 **S**	
SFS 2209	CuPb5Sn5Zn5	Finland	Sand castings	84.0–86.0 **Cu**, 4.0–6.0 **Sn**, 4.0–6.0 **Pb**, 4.0–6.0 **Zn**, 0.25 **Fe**, 0.25 **Sb**, 0.01 **Si**, 0.01 **Al**, 0.1 **Mn**, 2.0 **Ni**, 0.05 **P**, 0.10 **S**	**As cast**: all **diam**, 230 MPa **TS**, 90 MPa **YS**, 15% in 2 in. (50 mm) **El**
ISO 1338	CuPb5Sn5Zn5	ISO	Castings, ingot	84.0–86.0 **Cu**, 4.0–6.0 **Sn**, 4.0–6.0 **Pb**, 4.0–6.0 **Zn**, 0.30 **Fe**, 0.01 **Si**, 0.01 **Al**, 2.5 **Ni**, 0.05 **P**, 0.10 **S**, **Cu** total includes **Ni**	**Sand cast and permanent mould**: 12–25 mm **diam**, 200 MPa **TS**, 90 MPa **YS**, 13% in 2 in. (50 mm) **El** Centrifugally or continuously cast: 12–25 mm **diam**, 250 MPa **TS**, 100 MPa **YS**, 13% in 2 in. (50 mm) **El**
MIL–C–22229A	836	US	Castings	84.0–86.0 **Cu**, 4.0–6.0 **Sn**, 4.0–6.0 **Pb**, 4.0–6.0 **Zn**, 0.30 **Fe**, 1.0 **Ni**, 0.05 **P**	**As cast**: all **diam**, 207 MPa **TS**, 97 MPa **YS**, 20% in 2 in. (50 mm) **El**
SIS 14 52 04	SIS Red Metal 52 04–03	Sweden	Sand castings	84.0–86.0 **Cu**, 4.0–6.0 **Sn**, 4.0–6.0 **Pb**, 4.0–6.0 **Zn**, 0.30 **Fe**, 0.25 **Sb**, 0.01 **Si**, 0.01 **Al**, 0.1 **Mn**, 2.0 **Ni**, 0.05 **P**, 0.10 **S**	**As cast**: all **diam**, 230 MPa **TS**, 90 MPa **YS**, 15% in 2 in. (50 mm) **El**, 60 **HB**
SIS 14 52 04	SIS Red Metal 52 04–06	Sweden	Chill castings	84.0–86.0 **Cu**, 4.0–6.0 **Sn**, 4.0–6.0 **Pb**, 4.0–6.0 **Zn**, 0.30 **Fe**, 0.25 **Sb**, 0.01 **Si**, 0.01 **Al**, 0.1 **Mn**, 2.0 **Ni**, 0.05 **P**, 0.10 **S**	**As cast**: all **diam**, 200 MPa **TS**, 90 MPa **YS**, 13% in 2 in. (50 mm) **El**, 60 **HB**
SIS 14 52 04	SIS Red Metal 52 04–15	Sweden	Castings: continuous, centrifugal	84.0–86.0 **Cu**, 4.0–6.0 **Sn**, 4.0–6.0 **Pb**, 4.0–6.0 **Zn**, 0.30 **Fe**, 0.25 **Sb**, 0.01 **Si**, 0.01 **Al**, 0.1 **Mn**, 2.0 **Ni**, 0.05 **P**, 0.10 **S**	**As cast**: all **diam**, 250 MPa **TS**, 100 MPa **YS**, 13% in 2 in. (50 mm) **El**, 70 **HB**
AMS 4855B		US	Castings	83.0–86.0 **Cu**, 4.0–6.0 **Sn**, 4.0–6.0 **Pb**, 4.0–6.0 **Zn**, 0.300 **Fe**, 1.0 **Ni**, 0.05 **P**	
DS 3001	Tombac 5445	Denmark	Sand castings	83.5–85.5 **Cu**, 8.0–10.0 **Sn**, 1.5–3.0 **Pb**, 3.0–5.0 **Zn**, 0.30 **Fe**, 0.01 **Si**, 0.01 **Al**, 0.20 **Mn**, 2.0 **Ni**, 0.05 **P**, 0.10 **S**	**As cast**: all **diam**, 230 MPa **TS**, 110 MPa **YS**, 12% in 2 in. (50 mm) **El**, 75 min **HB**
JIS H 5111	Class 6	Japan	Castings	82.0–87.0 **Cu**, 4.0–6.0 **Sn**, 4.0–6.0 **Pb**, 4.0–7.0 **Zn**	**As cast**: all **diam**, 196 min MPa **TS**, 15% in 2 in. (50 mm) **El**

CAST COPPER AND COPPER ALLOYS

specification number	designation	country	product forms	chemical composition	mechanical properties and hardness values
SIS 14 52 04	SIS Red Metal 52 04–00	Sweden	Ingot	83.5–85.5 **Cu**, 4.3–6.0 **Sn**, 4.0–5.7 **Pb**, 4.5–6.0 **Zn**, 0.25 **Fe**, 0.25 **Sb**, 0.01 **Si**, 0.01 **Al**, 0.1 **Mn**, 2.0 **Ni**, 0.3 **P**, 0.06 **S**	
DS 3001	Tombac 5426	Denmark	Sand castings	81.0–85.0 **Cu**, 6.0–8.0 **Sn**, 5.0–7.0 **Pb**, 2.0–5.0 **Zn**, 0.20 **Fe**, 0.35 **Sb**, 0.01 **Si**, 0.01 **Al**, 2.0 **Ni**, 0.10 **P**, 0.10 **S**	**As cast**: all **diam**, 240 MPa **TS**, 100 MPa **YS**, 12% in 2 in. (50 mm) **El**, 65 min **HB**
SFS 2207	CuSn7Pb6Zn3	Finland	Sand castings	82.0–84.0 **Cu**, 6.3–8.0 **Sn**, 5.3–7.0 **Pb**, 2.3–5.0 **Zn**, 0.20 **Fe**, 0.25 **Sb**, 0.01 **Si**, 0.01 **Al**, 2.0 **Ni**, 0.03 **P**, 0.06 **S**	**As cast**: all **diam**, 240 MPa **TS**, 100 MPa **YS**, 12% in 2 in. (50 mm) **El**
ANSI/ASTM B 584	C 83800	US	Sand castings	82.0–83.8 **Cu**, 3.3–4.2 **Sn**, 5.0–7.0 **Pb**, 5.0–8.0 **Zn**, 0.30 **Fe**, 0.25 **Sb**, 0.005 **Si**, 0.005 **Al**, 1.0 **Ni + Co**, 0.005 **P**, 0.08 **S**, nickel included in copper content	**As cast**: all **diam**, 30 ksi (207 MPa) **TS**, 13 ksi (90 MPa) **YS**, 20% in 2 in. (50 mm) **El**
ANSI/ASTM B 505	CA 838	US	Castings	82.0–83.8 **Cu**, 3.3–4.2 **Sn**, 5.0–7.0 **Pb**, 5.0–8.0 **Zn**, 0.30 **Fe**, 0.25 **Sb**, 0.005 **Si**, 0.005 **Al**, 1.0 **Ni + Co**, 1.5 **P**, 0.08 **S**	**As cast**: all **diam**, 30 ksi (207 MPa) **TS**, 15 ksi (97 MPa) **YS**, 16% in 2 in. (50 mm) **El**
CDA C 83800		US	Castings	82.0–83.8 **Cu**, 3.3–4.2 **Sn**, 5.0–7.0 **Pb**, 5.0–8.0 **Zn**, 0.30 **Fe**, 0.25 **Sb**, 0.005 **Si**, 0.005 **Al**, 1.0 **Ni**, 0.03 **P**, 0.08 **S**	**Sand castings**: all **diam**, 30 ksi (207 MPa) **TS**, 13 ksi (90 MPa) **YS**, 20% in 2 in. (50 mm) **El**, 60 **HB**
COPANT 162	C 838 00	COPANT	Cast products	82–83.8 **Cu**, 3.3–4.2 **Sn**, 5–7 **Pb**, 5–8 **Zn**, 0.30 **Fe**, 0.25 **Sb**, 0.005 **Si**, 0.005 **Al**, 1.0 **Ni**, 0.03 **P**, 0.08 **S**	
COPANT 801	C 838 00	COPANT	Continuous–cast rod, bar and tube	82–83.8 **Cu**, 3.3–4.2 **Sn**, 5–7 **Pb**, 5–8 **Zn**, 0.30 **Fe**, 0.25 **Sb**, 0.005 **Si**, 0.005 **Al**, 1.0 **Ni**, 0.03 **P**, 0.08 **S**	**As manufactured**: all **diam**, 210 MPa **TS**, 110 MPa **YS**, 16% in 2 in. (50 mm) **El**
QQ–C–390A	838	US	Castings	82.0–83.7 **Cu**, 3.3–4.2 **Sn**, 5.0–7.0 **Pb**, 5.0–8.0 **Zn**, 0.30 **Fe**, 1.0 **Ni**, 0.03 **P**	**As cast**: all **diam**, 29 ksi (230 MPa) **TS**, 12 ksi (83 MPa) **YS**, 15% in 2 in. (50 mm) **El**
ANSI/ASTM B 30	CA 838	US	Ingot	82.0–83.5 **Cu**, 3.5–4.2 **Sn**, 5.8–6.8 **Pb**, 5.5–8.0 **Zn**, 0.25 **Fe**, 0.25 **Sb**, 0.005 **Si**, 0.005 **Al**, 0.8 **Ni**, 0.02 **P**, 0.08 **S**	
ANSI/ASTM B 30	838	US	Ingot: for castings	82.0–83.5 **Cu**, 3.5–4.2 **Sn**, 5.8–6.8 **Pb**, 5.5–8.0 **Zn**, 0.25 **Fe**, 0.25 **Sb**, 0.005 **Si**, 0.005 **Al**, 0.8 **Ni + Co**, 0.02 **P**, 0.08 **S**	
DS 3001	Tombac 5214	Denmark	Sand castings	79.5–82.5 **Cu**, 2.5–3.5 **Sn**, 6.0–8.0 **Pb**, 6.5–10.0 **Zn**, 0.3 **Fe**, 0.3 **Sb**, 0.01 **Si**, 0.01 **Al**, 0.2 **Mn**, 2.0 **Ni**, 0.05 **P**, 0.10 **S**	**As cast**: all **diam**, 190 MPa **TS**, 80 MPa **YS**, 18% in 2 in. (50 mm) **El**, 50 min **HB**
JIS H 5111	Class 1	Japan	Castings	79.0–83.0 **Cu**, 2.0–4.0 **Sn**, 3.0–7.0 **Pb**, 8.0–12.0 **Zn**	**As cast**: all **diam**, 167 min MPa **TS**, 15% in 2 in. (50 mm) **El**
JIS H 2203	Class 1	Japan	Ingot: for castings	79.0–83.0 **Cu**, 2.0–4.0 **Sn**, 3.0–7.0 **Pb**, 8.0–12.0 **Zn**	

CAST COPPER AND COPPER ALLOYS

specification number	designation	country	product forms	chemical composition	mechanical properties and hardness values
ANSI/ASTM B 30	CA 844	US	Ingot	79.0–82.0 **Cu**, 2.5–3.5 **Sn**, 6.3–7.7 **Pb**, 7.0–10.0 **Zn**, 0.35 **Fe**, 0.25 **Sb**, 0.005 **Si**, 0.005 **Al**, 0.8 **Ni**, 0.02 **P**, 0.08 **S**	
ANSI/ASTM B 30	844	US	Ingot: for castings	79.0–82.0 **Cu**, 2.5–3.5 **Sn**, 6.3–7.7 **Pb**, 7.0–10.0 **Zn**, 0.35 **Fe**, 0.25 **Sb**, 0.005 **Si**, 0.005 **Al**, 0.8 **Ni+Co**, 0.02 **P**	
ANSI/ASTM B 30	CA 842	US	Ingot	78.0–82.0 **Cu**, 4.3–6.0 **Sn**, 2.0–2.8 **Pb**, 10.0–16.0 **Zn**, 0.35 **Fe**, 0.25 **Sb**, 0.005 **Si**, 0.005 **Al**, 0.8 **Ni**, 0.02 **P**, 0.08 **S**	
ANSI/ASTM B 30	842	US	Ingot: for castings	78.0–82.0 **Cu**, 4.3–6.0 **Sn**, 2.0–2.8 **Pb**, 10.0–16.0 **Zn**, 0.35 **Fe**, 0.25 **Sb**, 0.005 **Si**, 0.005 **Al**, 0.8 **Ni+Co**, 0.02 **P**, 0.08 **S**	
ANSI/ASTM B 584	C 84400	US	Sand castings	78.0–82.0 **Cu**, 2.3–3.5 **Sn**, 6.0–8.0 **Pb**, 7.0–10.0 **Zn**, 0.40 **Fe**, 0.25 **Sb**, 0.005 **Si**, 0.005 **Al**, 1.0 **Ni+Co**, 0.02 **P**, 0.08 **S**, nickel included in copper content	**As cast**: all **diam**, 29 ksi (200 MPa) **TS**, 13 ksi (90 MPa) **YS**, 18% in 2 in. (50 mm) **El**
ANSI/ASTM B 505	CA 842	US	Castings	78.0–82.0 **Cu**, 4.0–6.0 **Sn**, 2.0–3.0 **Pb**, 10.0–16.0 **Zn**, 0.40 **Fe**, 0.25 **Sb**, 0.005 **Si**, 0.005 **Al**, 1.8 **Ni+Co**, 1.5 **P**, 0.08 **S**	**As cast**: all **diam**, 32 ksi (221 MPa) **TS**, 16 ksi (110 MPa) **YS**, 13% in 2 in. (50 mm) **El**
ANSI/ASTM B 505	CA 844	US	Castings	78.0–82.0 **Cu**, 2.3–3.5 **Sn**, 6.0–8.0 **Pb**, 7.0–10.0 **Zn**, 0.40 **Fe**, 0.25 **Sb**, 0.005 **Si**, 0.005 **Al**, 1.0 **Ni+Co**, 1.5 **P**, 0.08 **S**	**As cast**: all **diam**, 30 ksi (207 MPa) **TS**, 15 ksi (97 MPa) **YS**, 16% in 2 in. (50 mm) **El**
CDA C 84200		US	Castings	78.0–82.0 **Cu**, 4.0–6.0 **Sn**, 2.0–3.0 **Pb**, 10.0–16.0 **Zn**, 0.40 **Fe**, 0.25 **Sb**, 0.005 **Si**, 0.005 **Al**, 0.8 **Ni**, 0.08 **S**	**Sand castings**: all **diam**, 28 ksi (193 MPa) **TS**, 14 ksi (97 MPa) **YS**, 15% in 2 in. (50 mm) **El**, 60 **HB**
CDA C 84400		US	Castings	78.0–82.0 **Cu**, 2.3–3.5 **Sn**, 6.0–8.0 **Pb**, 7.0–10.0 **Zn**, 0.40 **Fe**, 0.25 **Sb**, 0.005 **Si**, 0.005 **Al**, 1.0 **Ni**, 0.02 **P**, 0.08 **S**	**Sand castings**: all **diam**, 29 ksi (200 MPa) **TS**, 13 ksi (90 MPa) **YS**, 18% in 2 in. (50 mm) **El**, 55 **HB**
COPANT 162	C 842 00	COPANT	Cast products	78–82 **Cu**, 4–6 **Sn**, 2–3 **Pb**, 10–16 **Zn**	
COPANT 162	C 844 00	COPANT	Cast products	78–82 **Cu**, 2.3–3.5 **Sn**, 6–8 **Pb**, 7–10 **Zn**, 0.40 **Fe**, 0.25 **Sb**, 0.005 **Si**, 0.005 **Al**, 1.0 **Ni**, 0.02 **P**, 0.08 **S**	
COPANT 801	C 844 00	COPANT	Continuous-cast rod, bar and tube	78–82 **Cu**, 2.3–3.5 **Sn**, 6–8 **Pb**, 7–10 **Zn**, 0.40 **Fe**, 0.25 **Sb**, 0.005 **Si**, 0.005 **Al**, 1.0 **Ni**, 0.02 **P**, 0.08 **S**	**As manufactured**: all **diam**, 210 MPa **TS**, 110 MPa **YS**, 16% in 2 in. (50 mm) **El**
COPANT 801	C 842 00	COPANT	Continuous-cast rod, bar and tube	78–82 **Cu**, 4–6 **Sn**, 2–3 **Pb**, 10–16 **Zn**	**As manufactured**: all **diam**, 230 MPa **TS**, 110 MPa **YS**, 13% in 2 in. (50 mm) **El**

CAST COPPER AND COPPER ALLOYS

specification number	designation	country	product forms	chemical composition	mechanical properties and hardness values
QQ-C-390A	844	US	Castings	78.0–82.0 **Cu**, 2.3–3.5 **Sn**, 6.0–8.0 **Pb**, 7.0–10.0 **Zn**, 0.40 **Fe**, 1.0 **Ni**, 0.02 **P**	**As cast**: all **diam**, 29 ksi (200 MPa) **TS**, 13 ksi (90 MPa) **YS**, 18% in 2 in. (50 mm) **El**
QQ-C-525b	Composition 3	US	Ingot	78.0–82.0 **Cu**, 4.3–6.0 **Sn**, 2.0–2.8 **Pb**, 12.0–16.0 **Zn**, 0.35 **Fe**, 0.25 **Sb**, 0.005 **Si**, 0.005 **Al**, 0.8 **Ni**, 0.03 **P**, 0.05 **S**	
QQ-C-390A	842	US	Castings	78.0–82.0 **Cu**, 4.0–6.0 **Sn**, 2.0–3.0 **Pb**, 10.0–16.0 **Zn**, 0.35 **Fe**, 0.7 **Ni**, 0.05 **P**, 0.50 total, others	**As cast**: all **diam**, 32 ksi (221 MPa) **TS**, 16 ksi (110 MPa) **YS**, 13% in 2 in. (50 mm) **El**
SABS 200	Cu-Sn3Zn9Pb5	South Africa	Castings	79.99 min **Cu**, 2.0–3.5 **Sn**, 4.0–6.0 **Pb**, 7.0–9.5 **Zn**, 0.02 **Si**, 0.01 **Al**, 2.0 **Ni**, 0.10 **Bi**, 0.75 **As + Fe + Sb**, 0.12 total, others	**Sand cast**: all **diam**, 185 MPa **TS**, 11% in 2 in. (50 mm) **El**; **Chill cast**: all **diam**, 180 MPa **TS**, 2% in 2 in. (50 mm) **El**
SABS 200	Cu-Sn5Zn5Pb5	South Africa	Castings	79.19 min **Cu**, 4.0–6.0 **Sn**, 4.0–6.0 **Pb**, 4.0–6.0 **Zn**, 0.02 **Si**, 0.01 **Al**, 2.0 **Ni**, 0.05 **Bi**, 0.50 **As + Fe + Sb**, 0.22 total, others	**Continuously cast**: all **diam**, 265 MPa **TS**, 13% in 2 in. (50 mm) **El**
AS 1565	836B	Australia	Ingot, castings	78.58 min **Cu**, 4.0–6.0 **Sn**, 4.0–6.0 **Pb**, 4.0–6.0 **Zn**, 0.05 **Si**, 0.01 **Al**, 2.0 **Ni**, 0.05 **Bi**, 0.50 **Fe + As + Sb**, 0.80 total, others (castings)	**Sand cast**: all **diam**, 200 MPa **TS**, 100 MPa **YS**, 13% in 2 in. (50 mm) **El**, 66 **HB**; **Chill cast**: all **diam**, 200 MPa **TS**, 110 MPa **YS**, 6% **El**, 80 **HB**; **Continuously cast**: all **diam**, 270 MPa **TS**, 100 MPa **YS**, 13% in 2 in. (50 mm) **El**, 75 **HB**
CDA C 84500		US	Castings	77.0–79.0 **Cu**, 2.0–4.0 **Sn**, 6.0–7.5 **Pb**, 10.0–14.0 **Zn**, 0.40 **Fe**, 0.25 **Sb**, 0.005 **Si**, 0.005 **Al**, 1.0 **Ni**, 0.02 **P**, 0.08 **S**	**Sand castings**: all **diam**, 29 ksi (200 MPa) **TS**, 13 ksi (90 MPa) **YS**, 16% in 2 in. (50 mm) **El**, 55 **HB**
COPANT 162	C 845 00	COPANT	Cast products	77–79 **Cu**, 2–4 **Sn**, 6–7.5 **Pb**, 10–14 **Zn**, 0.40 **Fe**, 1.0 **Ni**, 0.02 **P**	
AS 1565	838C	Australia	Ingot, castings	77.11 min **Cu**, 2.0–3.5 **Sn**, 4.0–6.0 **Pb**, 7.0–9.5 **Zn**, 0.02 **Si**, 0.01 **Al**, 2.0 **Ni**, 0.10 **Bi**, 0.75 **Fe + As + Sb**, 1.0 total, others (castings)	**Sand cast**: all **diam**, 180 MPa **TS**, 80 MPa **YS**, 11% in 2 in. (50 mm) **El**, 55 **HB**; **Chill cast**: all **diam**, 180 MPa **TS**, 80 MPa **YS**, 2% in 2 in. (50 mm) **El**, 65 **HB**
ANSI/ASTM B 584	C 84800	US	Sand castings	75.0–77.0 **Cu**, 2.0–3.0 **Sn**, 5.5–7.0 **Pb**, 13.0–17.0 **Zn**, 0.40 **Fe**, 0.25 **Sb**, 0.005 **Si**, 0.005 **Al**, 1.0 **Ni + Co**, 0.02 **P**, 0.08 **S**, nickel included in copper content	**As cast**: all **diam**, 24 ksi (193 MPa) **TS**, 12 ksi (83 MPa) **YS**, 16% in 2 in. (50 mm) **El**
ANSI/ASTM B 505	CA 848	US	Castings	75.0–77.0 **Cu**, 2.0–3.0 **Sn**, 5.5–7.0 **Pb**, 13.0–17.0 **Zn**, 0.40 **Fe**, 0.25 **Sb**, 0.005 **Si**, 0.005 **Al**, 1.0 **Ni + Co**, 1.5 **P**, 0.08 **S**	**As cast**: all **diam**, 30 ksi (207 MPa) **TS**, 15 ksi (97 MPa) **YS**, 16% in 2 in. (50 mm) **El**

CAST COPPER AND COPPER ALLOYS

specification number	designation	country	product forms	chemical composition	mechanical properties and hardness values
CDA C 84800		US	Castings	75.0–77.0 **Cu**, 2.0–3.0 **Sn**, 5.5–7.0 **Pb**, 13.0–17.0 **Zn**, 0.40 **Fe**, 0.25 **Sb**, 0.005 **Si**, 0.005 **Al**, 1.0 **Ni**, 0.02 **P**, 0.08 **S**	**Sand castings**: all **diam**, 28 ksi (193 MPa) **TS**, 12 ksi (83 MPa) **YS**, 16% in 2 in. (50 mm) **El**, 55 **HB**
COPANT 162	C 848 00	COPANT	Cast products	75–77 **Cu**, 2–3 **Sn**, 5.5–7 **Pb**, 13–17 **Zn**, 0.40 **Fe**, 0.25 **Sb**, 0.005 **Si**, 0.005 **Al**, 1.0 **Ni**, 0.02 **P**, 0.08 **S**	
COPANT 801	C 848 00	COPANT	Continuous–cast rod, bar and tube	75–77 **Cu**, 2–3 **Sn**, 5.5–7 **Pb**, 13–17 **Zn**, 0.40 **Fe**, 0.25 **Sb**, 0.005 **Si**, 0.005 **Al**, 1.0 **Ni**, 0.02 **P**, 0.08 **S**	**As manufactured**: all **diam**, 210 MPa **TS**, 110 MPa **YS**, 16% in 2 in. (50 mm) **El**
QQ-C-390A	848	US	Castings	75.0–76.8 **Cu**, 2.0–3.0 **Sn**, 5.3–6.7 **Pb**, 13.0–17.0 **Zn**, 0.40 **Fe**, 1.0 **Ni**, 0.02 **P**	**As cast**: all **diam**, 25 ksi (172 MPa) **TS**, 12 ksi (83 MPa) **YS**, 15% in 2 in. (50 mm) **El**
ANSI/ASTM B 30	CA 848	US	Ingot	75.0–76.7 **Cu**, 2.3–3.0 **Sn**, 5.5–6.7 **Pb**, 13.0–16.0 **Zn**, 0.35 **Fe**, 0.25 **Sb**, 0.005 **Si**, 0.005 **Al**, 0.8 **Ni**, 0.02 **P**, 0.08 **S**	
ANSI/ASTM B 30	848	US	Ingot: for castings	75.0–76.7 **Cu**, 2.3–3.0 **Sn**, 5.5–6.7 **Pb**, 13.0–16.0 **Zn**, 0.35 **Fe**, 0.25 **Sb**, 0.005 **Si**, 0.005 **Al**, 0.8 **Ni + Co**, 0.02 **P**, 0.08 **S**	
AS 1565	859D	Australia	Ingot, castings	60.0–63.0 **Cu**, 1.0–1.5 **Sn**, 0.5 **Pb**, rem **Zn**, 0.1 **Al**, 0.75 total, others	**Sand cast**: all **diam**, 250 MPa **TS**, 70 MPa **YS**, 18% in 2 in. (50 mm) **El**, 50 min **HB**
CDA C 85500		US	Castings	59.0–63.0 **Cu**, 0.20 **Sn**, 0.20 **Pb**, rem **Zn** 0.20 **Fe**, 0.8 **Al**, 0.20 **Mn**, 0.20 **Ni**, 1.0 **Sn + Pb + Fe + Mn + Ni**	**Sand castings**: all **diam**, 55 ksi (379 MPa) **TS**, 23 ksi (159 MPa) **YS**, 25% in 2 in. (50 mm) **El**, 55 **HRB**
COPANT 162	C 855 00	COPANT	Cast products	59–63 **Cu**, rem **Zn**, 1.0 **Sn + Pb + Ni + Mn + Fe** total, 0.20 total, others	
COPANT 162	C 856 00	COPANT	Cast products	59–63 **Cu**, rem **Zn**, 1.0 **Sn + Pb + Ni + Mn + Fe** total, 0.20 total, others	
SABS 200	Cu-Zn40 4A	South Africa	Castings	59.0–63.0 **Cu**, 0.25 **Pb**, rem **Zn**, 0.05 **Si**, 0.5 **Al**, 0.5 **Mn**, 0.20 total, others	**Gravity die cast**: all **diam**, 280 MPa **TS**, 25% in 2 in. (50 mm) **El**
AS 1565	855B	Australia	Ingot, castings	59.0–62.0 **Cu**, 0.25 **Pb**, rem **Zn**, 0.25–0.50 **Al**, 0.75 total, others (castings)	**Chill cast**: all **diam**, 280 MPa **TS**, 90 MPa **YS**, 23% in 2 in. (50 mm) **El**, 60 min **HB**
SABS 200	Cu-Zn14	South Africa	Castings	83.0–88.0 **Cu**, 0.5 **Pb**, rem **Zn**, 0.05–0.20 **As**, 1.0 total, others	**Sand cast**: all **diam**, 170 MPa **TS**, 20% in 2 in. (50 mm) **El**
JIS H 5101	Class 1	Japan	Castings	83.0–88.0 **Cu**, 0.5 **Pb**, rem **Zn**, 1.0 **Sn + Al + Fe**	**As cast**: all **diam**, 147 min MPa **TS**, 25% in 2 in. (50 mm) **El**
JIS H 2202	Class 1	Japan	Ingot: for castings	83.0–88.0 **Cu**, 0.5 **Pb**, rem **Zn**, 1.0 **Sn + Al + Fe**	

CAST COPPER AND COPPER ALLOYS

specification number	designation	country	product forms	chemical composition	mechanical properties and hardness values
JIS H 2203	Class 6	Japan	Ingot: for castings	83.0–87.0 **Cu**, 4.0–6.0 **Sn**, 3.0–6.0 **Pb**, 4.0–6.0 **Zn**	
IS 292	Grade 1	India	Castings	71.0–81.0 **Cu**, 1.0–3.5 **Sn**, 2.0–5.0 **Pb**, rem **Zn**, 0.75 **Fe**, 0.01 **Al**, nickel included in copper	**Sand cast**: all **diam**, 172 MPa **TS**, 12% in 2 in. (50 mm) **El**
AS 1565	852C	Australia	Ingot, castings	70.0–80.0 **Cu**, 1.0–3.0 **Sn**, 2.0–5.0 **Pb**, rem **Zn**, 0.75 **Fe**, 0.01 **Al**, 1.0 **Ni**, 1.0 total, others (castings)	**Sand cast**: all **diam**, 170 MPa **TS**, 80 MPa **YS**, 18% in 2 in. (50 mm) **El**, 45 **HB**
IS 292	Grade 1	India	Ingot	71.0–78.0 **Cu**, 1.0–3.5 **Sn**, 2.0–5.0 **Pb**, rem **Zn**, 0.5 **Fe**, 0.01 **Al**, nickel included in copper	**Sand cast**: all **diam**, 172 MPa **TS**, 12% in 2 in. (50 mm) **El**
ANSI/ASTM B 584	C 85200	US	Sand castings	70.0–74.0 **Cu**, 0.7–2.0 **Sn**, 1.5–3.8 **Pb**, 20.0–27.0 **Zn**, 0.6 **Fe**, 0.20 **Sb**, 0.05 **Si**, 0.005 **Al**, 1.0 **Ni+Co**, 0.02 **P**, 0.05 **S**	**As cast**: all **diam**, 35 ksi (241 MPa) **TS**, 12 ksi (83 MPa) **YS**, 25% in 2 in. (50 mm) **El**
CDA C 85200		US	Castings	70.0–74.0 **Cu**, 0.7–2.0 **Sn**, 1.5–3.8 **Pb**, 20.0–27.0 **Zn**, 0.6 **Fe**, 0.20 **Sb**, 0.05 **Si**, 0.005 **Al**, 1.0 **Ni**, 0.02 **P**, 0.05 **S**	**Sand castings**: all **diam**, 35 ksi (241 MPa) **TS**, 12 ksi (83 MPa) **YS**, 25% in 2 in. (50 mm) **El**, 45 **HB**
COPANT 162	C 852 00	COPANT	Cast products	70–74 **Cu**, 0.7–2.0 **Sn**, 1.5–3.8 **Pb**, 20–27 **Zn**, 0.6 **Fe**, 0.20 **Sb**, 0.05 **Si**, 0.005 **Al**, 1.0 **Ni**, 0.02 **P**, 0.05 **S**	
QQ–C–390A	852	US	Castings	70.0–74.0 **Cu**, 0.8–2.0 **Sn**, 1.5–3.7 **Pb**, rem **Zn**, 0.60 **Fe**	**As cast**: all **diam**, 35 ksi (241 MPa) **TS**, 12 ksi (83 MPa) **YS**, 25% in 2 in. (50 mm) **El**
ANSI/ASTM B 30	852	US	Ingot: for castings	70.0–73.0 **Cu**, 0.8–1.7 **Sn**, 1.5–3.5 **Pb**, 21.0–27.0 **Zn**, 0.50 **Fe**, 0.20 **Sb**, 0.05 **Si**, 0.005 **Al**, 0.8 **Ni+Co**, 0.01 **P**, 0.05 **S**	
ANSI/ASTM B 30	CA 852	US	Ingot	70.0–73.0 **Cu**, 0.8–1.7 **Sn**, 1.5–3.5 **Pb**, 21.0–27.0 **Zn**, 0.50 **Fe**, 0.20 **Sb**, 0.05 **Si**, 0.005 **Al**, 0.8 **Ni**, 0.01 **P**, 0.05 **S**	
IS 292	Grade 2	India	Castings	67.0–74.0 **Cu**, 1.5 **Sn**, 2.0–5.0 **Pb**, rem **Zn**, 0.75 **Fe**, 0.01 **Al**, nickel included in copper	**Sand cast**: all **diam**, 172 MPa **TS**, 12% in 2 in. (50 mm) **El**
CDA C 85300		US	Castings	68.0–72.0 **Cu**, 0.50 **Sn**, 0.50 **Pb**, rem **Zn**, 1.0 **Pb+Sn+Mn**	**Sand castings**: all **diam**, 35 ksi (241 MPa) **TS**, 11 ksi (76 MPa) **YS**, 40% in 2 in. (50 mm) **El**, 54 **HRF**
COPANT 162	C 853 00	COPANT	Cast products	68–72 **Cu**, 0.50 **Sn**, 0.50 **Pb**, rem **Zn**, 1.0 total, others	
AS 1565	851D	Australia	Ingot, castings	66.0–73.0 **Cu**, 1.5 **Sn**, 2.0–5.0 **Pb**, rem **Zn**, 0.75 **Fe**, 0.01 **Al**, 1.0 **Ni**, 1.0 total, others (castings)	**Sand cast**: all **diam**, 170 MPa **TS**, 12% in 2 in. (50 mm) **El**

CAST COPPER AND COPPER ALLOYS

specification number	designation	country	product forms	chemical composition	mechanical properties and hardness values
IS 292	Grade 2	India	Ingot	67.0–72.0 **Cu**, 1.5 **Sn**, 2.0–5.0 **Pb**, rem **Zn**, 0.5 **Fe**, 0.01 **Al**, nickel included in copper	**Sand cast**: all **diam**, 172 MPa **TS**, 12% in 2 in. (50 mm) **El**
ANSI/ASTM B 30	854	US	Ingot: for castings	66.0–69.0 **Cu**, 0.50–1.5 **Sn**, 1.5–3.5 **Pb**, 25.0–31.0 **Zn**, 0.50 **Fe**, 0.05 **Si**, 0.005 **Al**, 0.8 **Ni+Co**	
ANSI/ASTM B 584	C 85400	US	Sand castings	65.0–70.0 **Cu**, 0.50–1.5 **Sn**, 1.5–3.8 **Pb**, 24.0–32.0 **Zn**, 0.7 **Fe**, 0.05 **Si**, 0.35 **Al**, 1.0 **Ni+Co**	**As cast**: all **diam**, 30 ksi (207 MPa) **TS**, 11 ksi (76 MPa) **YS**, 20% in 2 in. (50 mm) **El**
ANSI/ASTM B 30	CA 854	US	Ingot	66.0–69.0 **Cu**, 0.50–1.5 **Sn**, 1.5–3.5 **Pb**, 25.0–31.0 **Zn**, 0.50 **Fe**, 0.05 **Si**, 0.005 **Al**, 0.8 **Ni**	
CDA C 85400		US	Castings	65.0–70.0 **Cu**, 0.50–1.5 **Sn**, 1.5–3.8 **Pb**, 24.0–32.0 **Zn**, 0.7 **Fe**, 0.05 **Si**, 0.35 **Al**, 1.0 **Ni**	**Sand castings**: all **diam**, 30 ksi (207 MPa) **TS**, 11 ksi (76 MPa) **YS**, 20% in 2 in. (50 mm) **El**, 50 **HB**
COPANT 162	C 854 00	COPANT	Cast products	65–70 **Cu**, 0.5–1.5 **Sn**, 1.5–3.5 **Pb**, 24–32 **Zn**, 0.7 **Fe**, 0.05 **Si**, 0.35 **Al**, 1.0 **Ni**	
QQ-C-390A	854	US	Castings	65.0–70.0 **Cu**, 1.5 **Sn**, 1.5–3.7 **Pb**, rem **Zn**, 0.75 **Fe**, 0.30 **Al**	**As cast**: all **diam**, 30 ksi (207 MPa) **TS**, 11 ksi (76 MPa) **YS**, 20% in 2 in. (50 mm) **El**
IS 292	Grade 3	India	Castings	64.0–71.0 **Cu**, 1.5 **Sn**, 1.0–3.0 **Pb**, rem **Zn**, 0.75 **Fe**, 0.01 **Al**, nickel included in copper	**Sand cast**: all **diam**, 186 MPa **TS**, 12% in 2 in. (50 mm) **El**
JIS H 2202	Class 2	Japan	Ingot: for castings	65.0–70.0 **Cu**, 1.0 **Sn**, 0.5 **Pb**, rem **Zn**, 0.8 **Fe**, 0.5 **Al**	
AS 1565	854C	Australia	Castings	63.0–70.0 **Cu**, 1.5 **Sn**, 1.0–3.0 **Pb**, rem **Zn**, 0.75 **Fe**, 0.10 **Al**, 1.0 **Ni**, 1.0 total, others (castings)	**Sand cast**: all **diam**, 190 MPa **TS**, 70 MPa **YS**, 11% in 2 in. (50 mm) **El**, 45 **HB**
IS 292	Grade 3	India	Ingot	64.0–68.0 **Cu**, 1.5 **Sn**, 1.0–3.0 **Pb**, rem **Zn**, 0.5 **Fe**, 0.01 **Al**, nickel included in copper	**Sand cast**: all **diam**, 186 MPa **TS**, 12% in 2 in. (50 mm) **El**
NF A 53-703	U-Z 35-Y 20	France	Castings	63.0–68.0 **Cu**, 1.50 **Sn**, 1.0–3.0 **Pb**, rem **Zn**, 0.80 **Fe**, 0.05 **Al**, 0.20 **Mn**	**As cast**: all **diam**, 180 MPa **TS**, 12% in 2 in. (50 mm) **El**
DS 3001	Brass 5144	Denmark	Sand castings	63.0–67.0 **Cu**, 1.5 **Sn**, 1.0–3.0 **Pb**, rem **Zn**, 0.8 **Fe**, 0.10 **Sb**, 0.05 **Si**, 0.10 **Al**, 0.2 **Mn**, 1.0 **Ni**, 0.05 **P**, 0.10 **As**	**As cast**: all **diam**, 180 MPa **TS**, 70 MPa **YS**, 12% in 2 in. (50 mm) **El**, 45 min **HB**
SFS 2203	CuZn33Pb2	Finland	Sand castings	63.0–67.0 **Cu**, 1.5 **Sn**, 1.0–3.0 **Pb**, rem **Zn**, 0.8 **Fe**, 0.10 **Sb**, 0.05 **Si**, 0.10 **Al**, 0.2 **Mn**, 1.0 **Ni**, 0.05 **P**, 0.10 **As**	**As cast**: all **diam**, 180 MPa **TS**, 70 MPa **YS**, 12% in 2 in. (50 mm) **El**
ISO 1338	CuZn33Pb2	ISO	Sand castings, ingot	63.0–67.0 **Cu**, 1.5 **Sn**, 1.0–3.0 **Pb**, rem **Zn**, 0.8 **Fe**, 0.05 **Si**, 0.1 **Al**, 0.2 **Mn**, 1.0 **Ni**, 0.05 **P**, **Cu** total includes **Ni**	**Sand cast**: 12–25 mm **diam**, 180 MPa **TS**, 70 MPa **YS**, 12% in 2 in. (50 mm) **El**

CAST COPPER AND COPPER ALLOYS

specification number	designation	country	product forms	chemical composition	mechanical properties and hardness values
SIS 14 51 44	SIS Brass 51 44–03	Sweden	Sand castings	63.0–67.0 **Cu**, 1.5 **Sn**, 1.0–3.0 **Pb**, rem **Zn**, 0.8 **Fe**, 0.10 **Sb**, 0.05 **Si**, 0.10 **Al**, 0.2 **Mn**, 1.0 **Ni**, 0.05 **P**, 0.10 **As**	**As cast**: all **diam**, 180 MPa **TS**, 70 MPa **YS**, 12% in 2 in. (50 mm) **El**, 45 **HB**
SIS 14 51 44	SIS Brass 51 44–00	Sweden	Ingot	63.0–66.0 **Cu**, 1.5 **Sn**, 1.0–2.8 **Pb**, rem **Zn**, 0.7 **Fe**, 0.05 **Sb**, 0.03 **Si**, 0.03 **Al**, 0.2 **Mn**, 1.0 **Ni**, 0.02 **P**, 0.08 **As**	
COPANT 801	C 863 00	COPANT	Continuous cast rod, bar and tube	60–66 **Cu**, 0.20 **Sn**, 0.20 **Pb**, 22–28 **Zn**, 2–4 **Fe**, 5.0–7.5 **Al**, 2.5–5.0 **Mn**, 1.0 **Ni**	**As manufactured**: 140 mm **diam**, 760 MPa **TS**, 430 MPa **YS**, 14% in 2 in. (50 mm) **El**
COPANT 801	C 862 00	COPANT	Continuous cast rod, bar and tube	60–66 **Cu**, 0.20 **Sn**, 0.20 **Pb**, 22–28 **Zn**, 2–4 **Fe**, 3–4.9 **Al**, 2.5–5.0 **Mn**, 1.0 **Ni**	**As manufactured**: 140 mm **diam**, 620 MPa **TS**, 310 MPa **YS**, 18% in 2 in. (50 mm) **El**
QQ–C–390A	857	US	Castings	60.0–65.0 **Cu**, 0.5–1.5 **Sn**, 0.8–1.5 **Pb**, rem **Zn**, 0.75 **Fe**, 0.50 **Al**	**As cast**: all **diam**, 40 ksi (276 MPa) **TS**, 14 ksi (97 MPa) **YS**, 15% in 2 in. (50 mm) **El**
JIS H 2202	Class 3	Japan	Ingot: for castings	60.0–65.0 **Cu**, 1.0 **Sn**, 0.5 **Pb**, rem **Zn**, 0.8 **Fe**, 0.5 **Al**	
DS 3001	Brass 5241	Denmark	Chill castings	60.0–64.0 **Cu**, 0.3 **Sn**, 1.0–2.0 **Pb**, rem **Zn**, 0.3 **Fe**, 0.10 **Si**, 0.1–0.8 **Al**, 0.5 **Mn**, 0.8 **Ni**	**As cast**: all **diam**, 280 MPa **TS**, 120 MPa **YS**, 15% in 2 in. (50 mm) **El**, 70 min **HB**
ANSI/ASTM B 584	C 85700	US	Sand castings	58.0–64.0 **Cu**, 0.50–1.5 **Sn**, 0.8–1.5 **Pb**, 32.0–40.0 **Zn**, 0.7 **Fe**, 0.05 **Si**, 0.55 **Al**, 1.0 **Ni + Co**	**As cast**: all **diam**, 40 ksi (276 MPa) **TS**, 14 ksi (97 MPa) **YS**, 15% in 2 in. (50 mm) **El**
CDA C 85700		US	Castings	58.0–64.0 **Cu**, 0.50–1.5 **Sn**, 0.8–1.5 **Pb**, 32.0–40.0 **Zn**, 0.7 **Fe**, 0.05 **Si**, 0.55 **Al**, 1.0 **Ni**	**Sand castings**: all **diam**, 40 ksi (276 MPa) **TS**, 14 ksi (97 MPa) **YS**, 15% in 2 in. (50 mm) **El**, 75 **HB**
DS 3001	Manganese Brass 5256	Denmark	Sand castings	57.0–65.0 **Cu**, 1.0 **Sn**, 0.5 **Pb**, rem **Zn**, 0.5–2.0 **Fe**, 0.10 **Si**, 0.5–2.5 **Al**, 0.5–3.0 **Mn**, 3.0 **Ni**, 0.40 **As + P + Sb**	**As cast**: all **diam**, 500 MPa **TS**, 170 MPa **YS**, 20% in 2 in. (50 mm) **El**, 110 min **HB**
SFS 2205	CuZn35AlFeMn	Finland	Sand castings	57.0–65.0 **Cu**, 1.0 **Sn**, 0.5 **Pb**, rem **Zn**, 0.5–2.0 **Fe**, 0.10 **Si**, 0.5–2.5 **Al**, 0.5–3.0 **Mn**, 3.0 **Ni**, 0.40 **As + P + Sb**	**As cast**: all **diam**, 500 MPa **TS**, 170 MPa **YS**, 20% in 2 in. (50 mm) **El**
NF A 53–703	U–Z 40–Y 30	France	Castings	58.0–64.0 **Cu**, 0.80 **Sn**, 0–2.1 **Pb**, rem **Zn**, 0.60 **Fe**, 0–1.0 **Al**	**As cast**: all **diam**, 340 MPa **TS**, 8% in 2 in. (50 mm) **El**
COPANT 162	C 857 00	COPANT	Cast products	58–64 **Cu**, 0.5–1.5 **Sn**, 0.5–1.5 **Pb**, 32–40 **Zn**, 0.7 **Fe**, 0.05 **Si**, 0.55 **Al**, 1.0 **Ni**	
QQ–C–390A	855	US	Castings	59.0–63.0 **Cu**, 0.50–1.0 **Pb**, rem **Zn**, 1.0 **Pb + Fe + Sn + Mn**; 0.2 total, others	**As cast**: all **diam**, 55 ksi (379 MPa) **TS**, 25% in 2 in. (50 mm) **El**
SIS 14 52 53	SIS Brass 52 53–06	Sweden	Chill castings	60.0–62.0 **Cu**, 1.0 **Sn**, 1.0–1.8 **Pb**, rem **Zn**, 0.5 **Fe**, 0.10 **Si**, 0.3–0.5 **Al**, 0.5 **Mn**, 1.0 **Ni**	**As cast**: all **diam**, 280 MPa **TS**, 120 MPa **YS**, 15% in 2 in. (50 mm) **El**, 70 **HB**

CAST COPPER AND COPPER ALLOYS

specification number	designation	country	product forms	chemical composition	mechanical properties and hardness values
ANSI/ASTM B 30	857	US	Ingot: for castings	58.0–63.0 **Cu**, 0.50–1.5 **Sn**, 0.8–1.5 **Pb**, 33.0–40.0 **Zn**, 0.50 **Fe**, 0.05 **Si**, 0.50 **Al**, 0.8 **Ni+Co**	
ANSI/ASTM B 30	CA 857	US	Ingot	58.0–63.0 **Cu**, 0.50–1.5 **Sn**, 0.8–1.5 **Pb**, 33.0–40.0 **Zn**, 0.50 **Fe**, 0.05 **Si**, 0.50 **Al**, 0.8 **Ni**	
AS 1565	857B	Australia	Ingot, castings	58.0–63.0 **Cu**, 1.0 **Sn**, 0.5–2.5 **Pb**, rem **Zn**, 0.8 **Fe**, 0.05 **Si**, 0.2–0.8 **Al**, 0.5 **Mn**, 1.0 **Ni**, 2.0 total, others (castings)	**Chill cast**: all **diam**, 300 MPa **TS**, 90 MPa **YS**, 13% in 2 in. (50 mm) **El**, 60 **HB**
ISO 1338	CuZn40Pb	ISO	Castings, ingot	58.0–63.0 **Cu**, 1.0 **Sn**, 0.5–2.5 **Pb**, rem **Zn**, 0.8 **Fe**, 0.05 **Si**, 0.2–0.8 **Al**, 0.5 **Mn**, 1.0 **Ni**, **Cu** total includes **Ni**	**Sand cast**: 12–25 mm **diam**, 220 MPa **TS**, 15% in 2 in. (50 mm) **El**; **Pressure die cast or permanent mould**: 12–25 mm **diam**, 280 MPa **TS**, 120 MPa **YS**, 15% in 2 in. (50 mm) **El**
DS 3001	Brass 5253	Denmark	Pressure die castings	58.0–61.0 **Cu**, 1.0 **Sn**, 1.5–2.5 **Pb**, rem **Zn**, 0.5 **Fe**, 0.10 **Si**, 0.3–0.5 **Al**, 0.5 **Mn**, 1.0 **Ni**	**As cast**: all **diam**, 280 MPa **TS**, 120 MPa **YS**, 5% in 2 in. (50 mm) **El**, 75 min **HB**
SFS 2204	CuZn40Pb	Finland	Die castings	58.0–61.0 **Cu**, 1.0 **Sn**, 1.5–2.5 **Pb**, rem **Zn**, 0.5 **Fe**, 0.10 **Si**, 0.3–0.5 **Al**, 0.5 **Mn**, 1.0 **Ni**	**As cast**: all **diam**, 280 MPa **TS**, 120 MPa **YS**, 15% in 2 in. (50 mm) **El**
SIS 14 52 52	SIS Brass 52 52	Sweden	Ingot	58.0–61.0 **Cu**, 1.0 **Sn**, 1.5–2.5 **Pb**, rem **Zn**, 0.5 **Fe**, 0.10–0.3 **Si**, 0.10 **Al**, 0.5 **Mn**, 1.0 **Ni**	
SIS 14 52 52	SIS Brass 52 52–10	Sweden	Pressure die castings	58.0–61.0 **Cu**, 1.0 **Sn**, 1.5–2.5 **Pb**, 0.5 **Fe**, 0.10–0.3 **Si**, 0.10 **Al**, 0.5 **Mn**, 1.0 **Ni**	**As cast**: all **diam**, 280 MPa **TS**, 120 MPa **YS**, 5% in 2 in. (50 mm) **El**, 75 **HB**
SIS 14 52 53	SIS Brass 52 53–00	Sweden	Ingot	58.0–61.0 **Cu**, 1.0 **Sn**, 1.5–2.5 **Pb**, rem **Zn**, 0.5 **Fe**, 0.10 **Si**, 0.3–0.5 **Al**, 0.5 **Mn**, 1.0 **Ni**	
SIS 14 52 53	SIS Brass 52 53–10	Sweden	Pressure die castings	58.0–61.0 **Cu**, 1.0 **Sn**, 1.5–2.5 **Pb**, rem **Zn**, 0.5 **Fe**, 0.10 **Si**, 0.3–0.5 **Al**, 0.5 **Mn**, 1.0 **Ni**	**As cast**: all **diam**, 280 MPa **TS**, 120 MPa **YS**, 5% in 2 in. (50 mm) **El**, 75 **HB**
JIS H 2205	Class 1	Japan	Ingot: for castings	55.0–60.0 **Cu**, 0.5–1.5 **Sn**, 0.4 **Pb**, rem **Zn**, 0.5–1.5 **Fe**, 0.1 **Si**, 0.5–1.5 **Al**, 1.5 **Mn**, 1.0 **Ni**	
ANSI/ASTM B 30	858	US	Ingot: for castings	57.0 min **Cu**, 1.5 **Sn**, 1.5 **Pb**, 31.0–41.0 **Zn**, 0.50 **Fe**, 0.05 **Sb**, 0.25 **Si**, 0.50 **Al**, 0.25 **Mn**, 0.50 **Ni+Co**, 0.01 **P**, 0.05 **S**, 0.05 **As**	
ANSI/ASTM B 176	UNS C85800 (858)	US	Die castings	57.0 min **Cu**, 1.50 **Sn**, 1.50 **Pb**, 31–41 **Zn**, 0.50 **Fe**, 0.25 **Si**, 0.25 **Al**, 0.25 **Mn**	**As cast**: all **diam**, 55 ksi (379 MPa) **TS**, 30 ksi (207 MPa) **YS**, 15% in 2 in. (50 mm) **El**, 55–60 **HB**

CAST COPPER AND COPPER ALLOYS

specification number	designation	country	product forms	chemical composition	mechanical properties and hardness values
ANSI/ASTM B 176	C 85800	US	Die castings	57.0 min **Cu**, 1.50 **Sn**, 1.50 **Pb**, 31.0–41.0 **Zn**, 0.50 **Fe**, 0.25 **Si**, 0.25 **Al**, 0.25 **Mn**	**As cast**: all **diam**, 55 ksi (379 MPa) **TS**, 30 ksi (207 MPa) **YS**, 15% in 2 in. (50 mm) **El**, 60 max **HRB**
ANSI/ASTM B 30	CA 858	US	Ingot	57.0 min **Cu**, 1.5 **Sn**, 1.5 **Pb**, 31.0–41.0 **Zn**, 0.50 **Fe**, 0.05 **Sb**, 0.25 **Si**, 0.50 **Al**, 0.25 **Mn**, 0.05 **Ni**, 0.01 **P**, 0.05 **S**	**As cast**: all **diam**, 55 ksi (379 MPa) **TS**, 30 ksi (207 MPa) **YS**, 15% in 2 in. (50 mm) **El**, 55 **HB**
CDA C 85800		US	Castings	57.0 min **Cu**, 1.5 **Sn**, 1.5 **Pb**, 31.0–41.0 **Zn**, 0.50 **Fe**, 0.05 **Sb**, 0.25 **Si**, 0.50 **Al**, 0.25 **Mn**, 0.50 **Ni**, 0.01 **P**. 0.05 **S**, 0.05 **As**	**Die castings**: all **diam**, 55 ksi (689 MPa) **TS**, 30 ksi (207 MPa) **YS**, 15% in 2 in. (50 mm) **El**, 55 **HRB**
COPANT 162	C 858 00	COPANT	Cast products	57 min **Cu**, 1.5 **Sn**, 1.5 **Pb**, 31–41 **Zn**, 0.5 **Fe**, 0.05 **Sb**, 0.25 **Si**, 0.5 **Al**, 0.25 **Mn**, 0.5 **Ni**, 0.01 **P**, 0.05 **S**, 0.05 **As**	
MIL-B-15894B	Class 1	US	Castings	57.0 min **Cu**, 1.50 **Sn**, 1.50 **Pb**, 30.0 min **Zn**, 0.25 **Fe**, 0.25 **Si**, 0.25 **Al**, 0.25 **Mn**, 0.50 total, others	**As cast**: all **diam**, 379 MPa **TS**, 207 MPa **YS**, 15% in 2 in. (50 mm) **El**, 55–60 **HRB**
CDA C 86100		US	Castings	66.0–68.0 **Cu**, 0.20 **Sn**, 0.20 **Pb**, rem **Zn**, 2.0–4.0 **Fe**, 4.5–5.5 **Al**, 2.5–5.0 **Mn**	**Sand castings**: all **diam**, 90 ksi (621 MPa) **TS**, 45 ksi (310 MPa) **YS**, 18% in 2 in. (50 mm) **El**, 180 **HB**
COPANT 162	C 861 00	COPANT	Cast products	66–68 **Cu**, 0.20 **Sn**, 0.20 **Pb**, rem **Zn**, 2–4 **Fe**, 4.5–5.5 **Al**, 2.5–5.0 **Mn**	
QQ-C-523A	Alloy E	US	Ingot	66–68 **Cu**, 0.10 **Sn**, 0.10 **Pb**, rem **Zn**, 2.0–4.0 **Fe**, 4.5–6.0 **Al**, 3.0–5.0 **Mn**, 1.0 **Ni**	**Remelted and sand cast**: all **diam**, 90 ksi (621 MPa) **TS**, 45 ksi (310 MPa) **YS**, 18% in 2 in. (50 mm) **El**
QQ-C-390A	861	US	Castings	66.0–68.0 **Cu**, 0.20 **Sn**, 0.20 **Pb**, rem **Zn**, 2.0–4.0 **Fe**, 4.5–5.5 **Al**, 2.5–5.0 **Mn**	**As cast**: all **diam**, 90 ksi (621 MPa) **TS**, 45 ksi (310 MPa) **YS**, 18% in 2 in. (50 mm) **El**
MIL-C-22229A	861	US	Castings	66.0–68.0 **Cu**, 0.20 **Sn**, 0.20 **Pb**, rem **Zn**, 2.0–4.0 **Fe**, 4.5–5.5 **Al**, 5.0 **Mn**	**As cast**: all **diam**, 621 MPa **TS**, 310 MPa **YS**, 18% in 2 in. (50 mm) **El**
ANSI/ASTM B 30	863	US	Ingot: for castings	60.0–68.0 **Cu**, 0.10 **Sn**, 0.10 **Pb**, 22.0–28.0 **Zn**, 2.0–4.0 **Fe**, 5.0–7.5 **Al**, 2.5–5.0 **Mn**, 0.8 **Ni + Co**	
QQ-C-523A	Alloy C	US	Ingot	60–68 **Cu**, 0.10 **Sn**, 0.10 **Pb**, rem **Zn**, 2.0–4.0 **Fe**, 3.0–7.5 **Al**, 2.5–5.0 **Mn**, 1.0 **Ni**	**Remelted and sand cast**: all **diam**, 110 ksi (758 MPa) **TS**, 60 ksi (414 MPa) **YS**, 12% in 2 in. (50 mm) **El**
QQ-C-523A	Alloy B	US	Ingot	60–68 **Cu**, 0.10 **Sn**, 0.10 **Pb**, rem **Zn**, 2.0–4.0 **Fe**, 3.0–7.5 **Al**, 2.5–5.0 **Mn**, 1.0 **Ni**	**Remelted and sand cast**: all **diam**, 90 ksi (621 MPa) **TS**, 45 ksi (310 MPa) **YS**, 18% in 2 in. (50 mm) **El**

CAST COPPER AND COPPER ALLOYS

specification number	designation	country	product forms	chemical composition	mechanical properties and hardness values
ANSI/ASTM B 30	862	US	Ingot: for castings	60.0–66.0 **Cu**, 0.10 **Sn**, 0.10 **Pb**, 22.0–28.0 **Zn**, 2.0–4.0 **Fe**, 3.0–4.9 **Al**, 2.5–5.0 **Mn**, 0.8 **Ni+Co**	
ANSI/ASTM B 584	C 86300	US	Sand castings	60.0–66.0 **Cu**, 0.20 **Sn**, 0.20 **Pb**, 22.0–28.0 **Zn**, 2.0–4.0 **Fe**, 5.0–7.5 **Al**, 2.5–5.0 **Mn**, 1.0 **Ni+Co**	**As cast**: all **diam**, 110 ksi (758 MPa) **TS**, 60 ksi (414 MPa) **YS**, 12% in 2 in. (50 mm) **El**
ANSI/ASTM B 505	CA 862	US	Castings	60.0–66.0 **Cu**, 0.20 **Sn**, 0.20 **Pb**, 22.0–28.0 **Zn**, 2.0–4.0 **Fe**, 3.0–4.9 **Al**, 2.5–5.0 **Mn**, 1.0 **Ni+Co**	**As cast**: all **diam**, 90 ksi (621 MPa) **TS**, 45 ksi (310 MPa) **YS**, 18% in 2 in. (50 mm) **El**
ANSI/ASTM B 505	CA 863	US	Castings	60.0–66.0 **Cu**, 0.20 **Sn**, 0.20 **Pb**, 22.0–28.0 **Zn**, 2.0–4.0 **Fe**, 5.0–7.5 **Al**, 2.5–5.0 **Mn**, 1.0 **Ni+Co**	**As cast**: all **diam**, 110 ksi (758 MPa) **TS**, 62 ksi (427 MPa) **YS**, 14% in 2 in. (50 mm) **El**
ANSI/ASTM B 30	CA 862	US	Ingot	60.0–66.0 **Cu**, 0.10 **Sn**, 0.10 **Pb**, 22.0–28.0 **Zn**, 3.0–4.9 **Al**, 2.5–5.0 **Mn**, 0.8 **Ni**	
ASTM B 22	CA 863	US	Castings	60.0–66.0 **Cu**, 0.20 **Sn**, 0.20 **Pb**, 22.0–28.0 **Zn**, 2.0–4.0 **Fe**, 5.0–7.5 **Al**, 2.5–5.0 **Mn**, 1.0 **Ni**	**As cast**: all **diam**, 110 ksi (760 MPa) **TS**, 60 ksi (415 MPa) **YS**, 12% in 2 in. (51 mm) **El**, 223 **HB**
ANSI/ASTM B 30	CA 863	US	Ingot	60.0–66.0 **Cu**, 0.10 **Sn**, 0.10 **Pb**, 22.0–28.0 **Zn**, 2.0–4.0 **Fe**, 5.0–7.5 **Al**, 2.5–5.0 **Mn**, 0.8 **Ni**	**As cast**: all **diam**, 110 ksi (760 MPa) **TS**, 60 ksi (415 MPa) **YS**, 12% in 2 in. (51 mm) **El**, 223 **HB**
CDA C 86200		US	Castings	60.0–66.0 **Cu**, 0.20 **Sn**, 0.20 **Pb**, 22.0–28.0 **Zn**, 2.0–4.0 **Fe**, 3.0–4.9 **Al**, 2.5–5.0 **Mn**, 1.0 **Ni**	**Sand castings**: all **diam**, 90 ksi (621 MPa) **TS**, 45 ksi (310 MPa) **YS**, 18% in 2 in. (50 mm) **El**, 180 **HB**
CDA C 86300		US	Castings	60.0–66.0 **Cu**, 0.20 **Sn**, 0.20 **Pb**, 22.0–28.0 **Zn**, 2.0–4.0 **Fe**, 5.0–7.5 **Al**, 2.5–5.0 **Mn**, 1.0 **Ni**	**Sand castings**: all **diam**, 110 ksi (758 MPa) **TS**, 60 ksi (414 MPa) **YS**, 12% in 2 in. (50 mm) **El**, 225 **HB**
COPANT 162	C 862 00	COPANT	Cast products	60–66 **Cu**, 0.20 **Sn**, 0.20 **Pb**, 22–28 **Zn**, 2–4 **Fe**, 3–4.9 **Al**, 2.5–5.0 **Mn**, 1.0 **Ni**	
COPANT 162	C 863 00	COPANT	Cast products	60–66 **Cu**, 0.20 **Sn**, 0.20 **Pb**, 22–28 **Zn**, 2–4 **Fe**, 5.0–7.5 **Al**, 2.5–5.0 **Mn**, 1.0 **Ni**	
ANSI/ASTM B 584	C 86200	US	Sand castings	60.0–66.0 **Cu**, 0.20 **Sn**, 0.20 **Pb**, 22.0–28.0 **Zn**, 2.0–4.0 **Fe**, 3.0–4.9 **Al**, 2.5–5.0 **Mn**, 1.0 **Ni+Co**	**As cast**: all **diam**, 90 ksi (621 MPa) **TS**, 45 ksi (310 MPa) **YS**, 18% in 2 in. (50 mm) **El**
QQ-C-390A	863	US	Castings	60.0–66.0 **Cu**, 0.20 **Sn**, 0.20 **Pb**, rem **Zn**, 2.0–4.0 **Fe**, 5.0–7.5 **Al**, 2.5–5.0 **Mn**	**As cast**: all **diam**, 110 ksi (758 MPa) **TS**, 60 ksi (414 MPa) **YS**, 12% in 2 in. (50 mm) **El**
QQ-C-390A	862	US	Castings	60.0–66.0 **Cu**, 0.20 **Sn**, 0.20 **Pb**, rem **Zn**, 2.0–4.0 **Fe**, 3.0–4.9 **Al**, 2.5–5.0 **Mn**	**As cast**: all **diam**, 90 ksi (621 MPa) **TS**, 45 ksi (310 MPa) **YS**, 18% in 2 in. (50 mm) **El**

CAST COPPER AND COPPER ALLOYS

specification number	designation	country	product forms	chemical composition	mechanical properties and hardness values
ISO 1338	CuZn26Al4Fe3Mn3	ISO	Ingot, castings	60.0–66.0 **Cu**, 0.20 **Sn**, 0.20 **Pb**, rem **Zn**, 1.5–4.0 **Fe**, 0.10 **Si**, 2.5–5.0 **Al**, 1.5–4.0 **Mn**, 3.0 **Ni**	**Sand cast**: 12–25 mm **diam**, 600 MPa **TS**, 300 MPa **YS**, 18% in 2 in. (50 mm) **El**; Continuously or centrifugally cast: 12–25 mm **diam**, 600 MPa **TS**, 300 MPa **YS**, 18% in 2 in. (50 mm) **El**
ISO 1338	CuZn25Al6Fe3Mn3	ISO	Ingot, castings	60.0–66.0 **Cu**, 0.20 **Sn**, 0.20 **Pb**, rem **Zn**, 2.0–4.0 **Fe**, 0.10 **Si**, 4.5–7.0 **Al**, 1.5–4.0 **Mn**, 3.0 **Ni**	**Sand cast**: 12–25 mm **diam**, 725 MPa **TS**, 400 MPa **YS**, 10% in 2 in. (50 mm) **El**; Continuously or centrifugally cast: 12–25 mm **diam**, 740 MPa **TS**, 400 MPa **YS**, 10% in 2 in. (50 mm) **El**
MIL-C-22229A	863	US	Castings	60.0–66.0 **Cu**, 0.20 **Sn**, 0.20 **Pb**, rem **Zn**, 2.0–4.0 **Fe**, 5.0–7.5 **Al**, 2.5–5.0 **Mn**	**As cast**: all **diam**, 758 MPa **TS**, 414 MPa **YS**, 12% in 2 in. (50 mm) **El**
MIL-C-22229A	862	US	Castings	60.0–66.0 **Cu**, 0.20 **Sn**, 0.20 **Pb**, rem **Zn**, 2.0–4.0 **Fe**, 3.0–4.9 **Al**, 2.5–5.0 **Mn**	**As cast**: all **diam**, 621 MPa **TS**, 310 MPa **YS**, 18% in 2 in. (50 mm) **El**
QQ-C-390A	864	US	Castings	56.0–62.0 **Cu**, 1.5 **Sn**, 0.5–1.5 **Pb**, rem **Zn**, 2.0 **Fe**, 1.5 **Al**, 1.5 **Mn**	**As cast**: all **diam**, 60 ksi (414 MPa) **TS**, 20 ksi (138 MPa) **YS**, 15% in 2 in. (50 mm) **El**
QQ-C-523A	Alloy F	US	Ingot	56–61 **Cu**, 1.00 **Sn**, 0.30 **Pb**, 38–40 **Zn**, 0.4–2.0 **Fe**, 0.7–1.0 **Al**, 1.50 **Mn**, 0.50 **Ni**	**Remelted and sand cast**: all **diam**, 65 ksi (448 MPa) **TS**, 20% in 2 in. (50 mm) **El**
ANSI/ASTM B 30	865	US	Ingot: for castings	55.0–60.0 **Cu**, 1.0 **Sn**, 0.30 **Pb**, 36.0–42.0 **Zn**, 0.40–2.0 **Fe**, 0.50–1.5 **Al**, 0.10–1.5 **Mn**, 0.8 **Ni+Co**	
ANSI/ASTM B 30	CA 865	US	Ingot	55.0–60.0 **Cu**, 1.0 **Sn**, 0.30 **Pb**, 36.0–42.0 **Zn**, 0.40–2.0 **Fe**, 0.50–1.5 **Al**, 0.10–1.5 **Mn**, 0.8 **Ni**	
QQ-C-523A	Alloy A	US	Ingot	55–60 **Cu**, 1.00 **Sn**, 0.30 **Pb**, rem **Zn**, 0.4–2.0 **Fe**, 0.5–1.5 **Al**, 1.50 **Mn**, 0.50 **Ni**	**Remelted and sand cast**: all **diam**, 65 ksi (448 MPa) **TS**, 20% in 2 in. (50 mm) **El**
QQ-C-390A	865	US	Castings	55.0–60.0 **Cu**, 1.0 **Sn**, 0.40 **Pb**, rem **Zn**, 0.40–2.0 **Fe**, 0.50–1.5 **Al**, 1.5 **Mn**, 0.50 **Ni**	**As cast**: all **diam**, 65 ksi (448 MPa) **TS**, 25 ksi (172 MPa) **YS**, 20% in 2 in. (50 mm) **El**
IS 304	Grade 1	India	Ingot and castings	56.0 **Cu**, 0.2 **Sn**, 0.2 **Pb**, rem **Zn**, 1.0–2.5 **Fe**, 0.10 **Si**, 3.0–6.0 **Al**, 4.0 **Mn**, 0.2 total, others; nickel included in copper	**Sand cast (separately cast)**: all **diam**, 745 MPa **TS**, 402 MPa **YS**, 12% in 2 in. (50 mm) **El**
CDA C 86800		US	Castings	53.5–57.0 **Cu**, 1.0 **Sn**, 0.20 **Pb**, rem **Zn**, 1.0–2.5 **Fe**, 2.0 **Al**, 2.5–4.0 **Mn**, 2.5–4.0 **Ni**	**Sand castings**: all **diam**, 78 ksi (538 MPa) **TS**, 35 ksi (241 MPa) **YS**, 18% in 2 in. (50 mm) **El**, 80 **HB**
COPANT 162	C 868 00	COPANT	Cast products	53.5–57 **Cu**, 1.0 **Sn**, 0.2 **Pb**, rem **Zn**, 1–2.5 **Fe**, 2.0 **Al**, 2.5–4.0 **Mn**, 2.5–4 **Ni**	
QQ-C-390A	868	US	Castings	53.5–57.0 **Cu**, 1.0 **Sn**, 0.20 **Pb**, rem **Zn**, 1.0–2.5 **Fe**, 2.0 **Al**, 2.5–4.0 **Mn**, 2.5–4.0 **Ni**	**As cast**: all **diam**, 78 ksi (538 MPa) **TS**, 35 ksi (241 MPa) **YS**, 18% in 2 in. (50 mm) **El**

CAST COPPER AND COPPER ALLOYS

specification number	designation	country	product forms	chemical composition	mechanical properties and hardness values
AS 1565	869D	Australia	Ingot, castings	55.0 min **Cu**, 0.20 **Sn**, 0.20 **Pb**, rem **Zn**, 1.5–3.25 **Fe**, 0.10 **Si**, 3.0–6.0 **Al**, 4.0 **Mn**, 1.0 **Ni**, 0.20 total, others (castings)	**Sand cast**: all **diam**, 740 MPa **TS**, 400 MPa **YS**, 11% in 2 in. (50 mm) **El**, 150 **HB**; **Centrifugal cast**: all **diam**, 740 MPa **TS**, 400 MPa **YS**, 13% in 2 in. (50 mm) **El**, 150 **HB**
AS 1565	866D	Australia	Ingot, castings	55.0 min **Cu**, 0.50 **Sn**, 0.50 **Pb**, rem **Zn**, 0.5–2.5 **Fe**, 0.10 **Si**, 5.0 **Al**, 3.0 **Mn**, 2.0 **Ni**, 0.20 total, others (castings)	**Sand cast**: all **diam**, 590 MPa **TS**, 280 MPa **YS**, 15% in 2 in. (50 mm) **El**
AS 1565	865C	Australia	Ingot, castings	55.0 min **Cu**, 1.0 **Sn**, 0.50 **Pb**, rem **Zn**, 0.7–2.0 **Fe**, 0.10 **Si**, 0.5–2.5 **Al**, 3.0 **Mn**, 1.0 **Ni**, 0.20 total, others (castings)	**Sand cast**: all **diam**, 470 MPa **TS**, 170 MPa **YS**, 18% in 2 in. (50 mm) **El**, 100 **HB**; **Chill cast**: all **diam**, 500 MPa **TS**, 210 MPa **YS**, 18% in 2 in. (50 mm) **El**; **Centrifugal cast**: all **diam**, 500 MPa **TS**, 210 MPa **YS**, 20% in 2 in. (50 mm) **El**, 100 **HB**
SABS 200	Cu–Zn18Pb4Sn2	South Africa	Castings	70.0–80.0 **Cu**, 1.0–3.0 **Sn**, 2.0–5.0 **Pb**, rem **Zn**, 0.75 **Fe**, 0.1 **Al**, 1.0 **Ni**, 0.99 total, others	**Sand cast**: all **diam**, 170 MPa **TS**, 20% in 2 in. (50 mm) **El**
SABS 200	Cu–Zn26Pb4	South Africa	Castings	66.0–73.0 **Cu**, 1.5 **Sn**, 2.0–5.0 **Pb**, rem **Zn**, 0.75 **Fe**, 0.01 **Al**, 1.0 **Ni**, 0.99 total, others	**Sand cast**: all **diam**, 170 MPa **TS**, 12% in 2 in. (50 mm) **El**
SABS 200	CuZn32Pb2	South Africa	Castings	63.0–70.0 **Cu**, 1.5 **Sn**, 1.0–3.0 **Pb**, rem **Zn**, 0.75 **Fe**, 0.1 **Al**, 1.0 **Ni**, 0.99 total, others	**Sand cast**: all **diam**, 185 MPa **TS**, 12% in 2 in. (50 mm) **El**
SABS 200	Cu–Zn34Pb2	South Africa	Castings	62.0–68.0 **Cu**, 1.0 **Sn**, 1.0–2.5 **Pb**, rem **Zn**, 0.5 **Fe**, 0.5 **Al**, 1.0 total, others	**Gravity die cast**: all **diam**, 240 MPa **TS**, 16% in 2 in. (50 mm) **El**
SABS 200	Cu–Zn36Sn1	South Africa	Castings	60.0–64.0 **Cu**, 1.0–1.5 **Sn**, 0.5 **Pb**, rem **Zn**, 0.01 **Al**, 0.74 total, others	**Sand cast**: all **diam**, 245 MPa **TS**, 20% in 2 in. (50 mm) **El**
SIS 14 52 56	SIS Brass 52 56–03	Sweden	Sand castings	57.0–65.0 **Cu**, 1.0 **Sn**, 1.5–2.5 **Pb**, rem **Zn**, 0.5 **Fe**, 0.10 **Si**, 0.5–2.5 **Al**, 0.5 **Mn**, 1.0 **Ni**	**As cast**: all **diam**, 500 MPa **TS**, 170 MPa **YS**, 20% in 2 in. (50 mm) **El**, 110 **HB**
SIS 14 52 56	SIS Brass 52 56–06	Sweden	Chill castings	57.0–65.0 **Cu**, 1.0 **Sn**, 1.5–2.5 **Pb**, rem **Zn**, 0.5 **Fe**, 0.10 **Si**, 0.5–2.5 **Al**, 0.5 **Mn**, 1.0 **Ni**	**As cast**: all **diam**, 500 MPa **TS**, 200 MPa **YS**, 18% in 2 in. (50 mm) **El**, 120 **HB**
SIS 14 52 56	SIS Brass 52 56–10	Sweden	Pressure die castings	57.0–65.0 **Cu**, 1.0 **Sn**, 1.5–2.5 **Pb**, rem **Zn**, 0.5 **Fe**, 0.10 **Si**, 0.5–2.5 **Al**, 0.5 **Mn**, 1.0 **Ni**	**As cast**: all **diam**, 500 MPa **TS**, 200 MPa **YS**, 18% in 2 in. (50 mm) **El**, 120 **HB**
SIS 14 52 56	SIS Brass 52 56–15	Sweden	Castings: continuous & centrifugal	57.0–65.0 **Cu**, 1.0 **Sn**, 1.5–2.5 **Pb**, rem **Zn**, 0.5 **Fe**, 0.10 **Si**, 0.5–2.5 **Al**, 0.5 **Mn**, 1.0 **Ni**	**As cast**: all **diam**, 500 MPa **TS**, 200 MPa **YS**, 18% in 2 in. (50 mm) **El**, 120 **HB**
SIS 14 52 56	SIS Brass 52 56–00	Sweden	Ingot	57.0–65.0 **Cu**, 1.0 **Sn**, 0.5 **Pb**, rem **Zn**, 0.5–2.0 **Fe**, 0.10 **Si**, 0.5–2.5 **Al**, 0.5–3.0 **Mn**, 3.0 **Ni**, 0.40 **Al + P + As**, total	

CAST COPPER AND COPPER ALLOYS

specification number	designation	country	product forms	chemical composition	mechanical properties and hardness values
SABS 200	Cu–Zn40 3A	South Africa	Castings	58.0–63.0 **Cu**, 1.0 **Sn**, 0.5–2.5 **Pb**, rem **Zn**, 0.5 **Fe**, 0.2–0.8 **Al**, 1.0 **Ni**, 0.5 total, others	**Gravity die cast**: all **diam**, 295 MPa **TS**, 15% in 2 in. (50 mm) **El**
JIS H 5102	Class 3B	Japan	Castings	60.0 min **Cu**, 0.5 **Sn**, rem **Zn**, 0.1 **Si**, 3.0–7.5 **Al**, 2.5–5.0 **Mn**, 0.2 **P**	**As cast**: all **diam**, 755 min MPa **TS**, 12% in 2 in. (50 mm) **El**
JIS H 5102	Class 3A	Japan	Castings	60.0 min **Cu**, 0.5 **Sn**, rem **Zn**, 0.1 **Si**, 3.0–7.5 **Al**, 2.5–5.0 **Mn**, 0.2 **P**	**As cast**: all **diam**, 637 min MPa **TS**, 15% in 2 in. (50 mm) **El**
ANSI/ASTM B 30	864	US	Ingot: for castings	56.0–62.0 **Cu**, 0.50–1.0 **Sn**, 0.50–1.3 **Pb**, 34.0–42.0 **Zn**, 0.40–2.0 **Fe**, 0.50–1.5 **Al**, 0.10–1.0 **Mn**, 0.8 **Ni+Co**	
CDA C 86400		US	Castings	56.0–62.0 **Cu**, 0.50–1.5 **Sn**, 0.50–1.5 **Pb**, 34.0–42.0 **Zn**, 0.40–2.0 **Fe**, 0.50–1.5 **Al**, 0.10–1.0 **Mn**, 1.0 **Ni**	**Sand castings**: all **diam**, 60 ksi (414 MPa) **TS**, 20 ksi (138 MPa) **YS**, 15% in 2 in. (50 mm) **El**, 90 **HB**
SABS 200	Cu–Zn41Pb2	South Africa	Castings	57.0–61.0 **Cu**, 0.8 **Sn**, 0.5–2.5 **Pb**, rem **Zn**, 0.3 **Fe**, 0.5 **Al**	
AMS 4860		US	Castings	55.0–60.0 **Cu**, 1.0 **Sn**, 0.40 **Pb**, rem **Zn**, 0.40–2.0 **Fe**, 0.50–1.5 **Al**, 1.5 **Mn**, 0.50 **Ni**, 0.20 others, total	**As cast**: all **diam**, 65 ksi (448 MPa) **TS**, 25 ksi (172 MPa) **YS**, 25% in 2 in. (50 mm) **El**
ANSI/ASTM B 30	867	US	Ingot: for castings	55.0–60.0 **Cu**, 1.5 **Sn**, 0.50–1.5 **Pb**, 30.0–38.0 **Zn**, 1.0–3.0 **Fe**, 1.0–3.0 **Al**, 1.0–3.5 **Mn**, 0.8 **Ni+Co**	
CDA C 86500		US	Castings	55.0–60.0 **Cu**, 1.0 **Sn**, 0.40 **Pb**, 36.0–42.0 **Zn**, 0.40–2.0 **Fe**, 0.50–1.5 **Al**, 0.10–1.5 **Mn**, 1.0 **Ni**	**Sand castings**: all **diam**, 65 ksi (448 MPa) **TS**, 25 ksi (172 MPa) **YS**, 20% in 2 in. (50 mm) **El**, 100 **HB**
CDA C 86700		US	Castings	55.0–60.0 **Cu**, 1.5 **Sn**, 0.5–1.5 **Pb**, 30.0–38.0 **Zn**, 1.0–3.0 **Fe**, 1.0–3.0 **Al**, 1.0–3.5 **Mn**, 1.0 **Ni**	**Sand castings**: all **diam**, 80 ksi (552 MPa) **TS**, 32 ksi (221 MPa) **YS**, 15% in 2 in. (50 mm) **El**, 80 **HRB**
JIS H 2205	Class 2	Japan	Ingot: for castings	55.0–60.0 **Cu**, 1.0 **Sn**, 0.4 **Pb**, rem **Zn**, 0.5–2.0 **Fe**, 0.1 **Si**, 0.5–2.0 **Al**, 3.5 **Mn**, 1.0 **Ni**	
MIL–C–22229A	865	US	Castings	55.0–60.0 **Cu**, 1.0 **Sn**, 0.40 **Pb**, rem **Zn**, 0.40–2.0 **Fe**, 0.50–1.5 **Al**, 1.5 **Mn**, 0.50 **Ni**	**As cast**: all **diam**, 448 MPa **TS**, 172 MPa **YS**, 20% in 2 in. (50 mm) **El**
SABS 200	Cu–Zn39Fe1(AlMn)	South Africa	Castings	55.0 min **Cu**, 1.0 **Sn**, 0.50 **Pb**, rem **Zn**, 0.70–2.0 **Fe**, 0.10 **Si**, 0.5–2.5 **Al**, 3.0 **Mn**, 1.0 **Ni**, 0.10 total, others	**Sand cast**: all **diam**, 470 MPa **TS**, 170 MPa **YS**, 18% in 2 in. (50 mm) **El**; **Gravity die cast**: all **diam**, 500 MPa **TS**, 210 MPa **YS**, 18% in 2 in. (50 mm) **El**
SABS 200	Cu–Zn37Fe2(AlMnNi)	South Africa	Castings	55.0 min **Cu**, 0.50 **Sn**, 0.50 **Pb**, rem **Zn**, 0.50–2.5 **Fe**, 0.10 **Si**, 5.0 **Al**, 3.0 **Mn**, 2.0 **Ni**, 0.10 total, others	**Sand cast**: all **diam**, 585 MPa **TS**, 280 MPa **YS**, 14% in 2 in. (50 mm) **El**
JIS H 5102	Class 2	Japan	Castings	55.0 min **Cu**, 1.0 **Sn**, 0.4 **Pb**, rem **Zn**, 0.5–1.5 **Fe**, 0.1 **Si**, 0.5–1.5 **Al**, 1.5 **Mn**, 1.0 **Ni**	**As cast**: all **diam**, 431 min MPa **TS**, 20% in 2 in. (50 mm) **El**

CAST COPPER AND COPPER ALLOYS

specification number	designation	country	product forms	chemical composition	mechanical properties and hardness values
JIS H 5102	Class 1	Japan	Castings	55.0 min **Cu**, 1.0 **Sn**, 0.4 **Pb**, rem **Zn**, 0.5–1.5 **Fe**, 0.1 **Si**, 0.5–1.5 **Al**, 1.5 **Mn**, 1.0 **Ni**	**As cast**: all **diam**, 431 min MPa **TS**, 20% in 2 in. (50 mm) **El**
COPANT 162	C 879 00	COPANT	Cast products	63 min **Cu**, 0.25 **Sn**, 0.25 **Pb**, 30–36 **Zn**, 0.4 **Fe**, 0.05 **Sb**, 8–12 **Si**, 0.15 **Al**, 0.15 **Mn**, 0.5 **Ni**, 0.01 **P**, 0.05 **S**, 0.05 **As**	
COPANT 162	C 865 00	COPANT	Cast products	55–60 **Cu**, 1.0 **Sn**, 0.4 **Pb**, 36–42 **Zn**, 0.4–2.0 **Fe**, 0.5–1.5 **Al**, 0.10–1.5 **Mn**, 1.0 **Ni**	
ANSI/ASTM B 505	CA 865	US	Castings	55.0–60.0 **Cu**, 1.0 **Sn**, 0.40 **Pb**, 36.0–42.0 **Zn**, 0.40–2.0 **Fe**, 0.50–1.5 **Al**, 1.5 **Mn**, 1.0 **Ni+Co**	**As cast**: all **diam**, 70 ksi (483 MPa) **TS**, 25 ksi (172 MPa) **YS**, 25% in 2 in. (50 mm) **El**
ANSI/ASTM B 584	C 86500	US	Sand castings	55.0–60.0 **Cu**, 1.0 **Sn**, 0.40 **Pb**, 36.0–42.0 **Zn**, 0.40–2.0 **Fe**, 0.50–1.5 **Al**, 0.10–1.5 **Mn**, 1.0 **Ni+Co**	**As cast**: all **diam**, 65 ksi (448 MPa) **TS**, 25 ksi (172 MPa) **YS**, 20% in 2 in. (50 mm) **El**
COPANT 801	C 865 00	COPANT	Continuous–cast rod, bar and tube	55–60 **Cu**, 1.0 **Sn**, 0.4 **Pb**, 36–42 **Zn**, 0.4–2.0 **Fe**, 0.5–1.5 **Al**, 0.10–1.5 **Mn**, 1.0 **Ni**	**As manufactured**: 140 mm **diam**, 480 MPa **TS**, 180 MPa **YS**, 25% in 2 in. (50 mm) **El**
COPANT 162	C 864 00	COPANT	Cast products	56–62 **Cu**, 0.5–1.5 **Sn**, 0.5–1.5 **Pb**, 34–42 **Zn**, 0.4–2.0 **Fe**, 0.5–1.5 **Al**, 0.10–1.0 **Mn**, 1.0 **Ni**	
ANSI/ASTM B 584	C 86400	US	Sand castings	56.0–62.0 **Cu**, 0.50–1.5 **Sn**, 0.50–1.5 **Pb**, 34.0–42.0 **Zn**, 0.40–2.0 **Fe**, 0.50–1.5 **Al**, 0.10–1.0 **Mn**, 1.0 **Ni+Co**	**As cast**: all **diam**, 60 ksi (414 MPa) **TS**, 20 ksi (138 MPa) **YS**, 15% in 2 in. (50 mm) **El**
ANSI/ASTM B 30	CA 864	US	Ingot	56.0–62.0 **Cu**, 0.50–1.0 **Sn**, 0.50–1.3 **Pb**, 34.0–42.0 **Zn**, 0.40–2.0 **Fe**, 0.50–1.5 **Al**, 0.10–1.0 **Mn**, 0.8 **Ni**	
QQ-C-523A	Alloy D	US	Ingot	56–62 **Cu**, 0.5–1.0 **Sn**, 0.5–1.0 **Pb**, rem **Zn**, 0.75–1.5 **Fe**, 0.25–1.0 **Al**, 0.1–0.5 **Mn**, 0.50 **Ni**	**Remelted and sand cast**: all **diam**, 60 ksi (414 MPa) **TS**, 20 ksi (138 MPa) **YS**, 15% in 2 in. (50 mm) **El**
IS 304	Grade 3	India	Ingot and castings	56.0 **Cu**, 1.5 **Sn**, 0.5 **Pb**, 34.19 min **Zn**, 0.5–2.0 **Fe**, 0.10 **Si**, 2.5 **Al**, 3.0 **Mn**, 0.2 total, others, nickel included in copper	**Sand cast (separately cast)**: all **diam**, 461 MPa **TS**, 167 MPa **YS**, 20% in 2 in. (50 mm) **El**
ANSI/ASTM B 30	879	US	Ingot: for castings	63.0 min **Cu**, 0.25 **Sn**, 30.0–36.0 **Zn**, 0.40 **Fe**, 0.05 **Sb**, 0.8–1.2 **Si**, 0.15 **Al**, 0.15 **Mn**, 0.50 **Ni+Co**, 0.01 **P**, 0.05 **S**, 0.05 **As**	
ANSI/ASTM B 179	UNS C87900 (879)	US	Die castings	63.0–67.0 **Cu**, 0.25 **Sn**, 0.25 **Pb**, 30–36 **Zn**, 0.15 **Fe**, 0.75–1.25 **Si**, 0.15 **Al**, 0.15 **Mn**	**As cast**: all **diam**, 70 ksi (483 MPa) **TS**, 35 ksi (241 MPa) **YS**, 25% in 2 in. (50 mm) **El**, 68–72 **HB**
ANSI/ASTM B 30	CA 879	US	Ingot	63.0 min **Cu**, 0.25 **Sn**, 0.25 **Pb**, 30.0–36.0 **Zn**, 0.40 **Fe**, 0.05 **Sb**, 0.8–1.2 **Si**, 0.15 **Al**, 0.15 **Mn**, 0.50 **Ni**, 0.01 **P**, 0.05 **S**, 0.05 **As**	**As cast**: all **diam**, 55 ksi (379 MPa) **TS**, 30 ksi (207 MPa) **YS**, 15% in 2 in. (50 mm) **El**, 55 **HB**

CAST COPPER AND COPPER ALLOYS

specification number	designation	country	product forms	chemical composition	mechanical properties and hardness values
ANSI/ASTM B 176	C 87900	US	Plate	63.0–67.0 **Cu**, 0.25 **Sn**, 0.25 **Pb**, 30.0–36.0 **Zn**, 0.15 **Fe**, 0.75–1.25 **Si**, 0.15 **Al**, 0.15 **Mn**, 99.5 min total named elements	**As cast**: all **diam**, 70 ksi (483 MPa) **TS**, 35 ksi (241 MPa) **YS**, 25% in 2 in. (50 mm) **El**, 72 max **HRB**
COPANT 162	C 867 00	COPANT	Cast products	55–60 **Cu**, 1.5 **Sn**, 0.5–1.5 **Pb**, 30–38 **Zn**, 1–3 **Fe**, 1–3 **Al**, 1.0–3.5 **Mn**, 1.0 **Ni**	
ANSI/ASTM B 584	C 86700	US	Sand castings	55.0–60.0 **Cu**, 1.5 **Sn**, 0.50–1.5 **Pb**, 30.0–38.0 **Zn**, 1.0–3.0 **Fe**, 1.0–3.0 **Al**, 1.0–3.5 **Mn**	**As cast**: all **diam**, 80 ksi (552 MPa) **TS**, 32 ksi (221 MPa) **YS**, 15% in 2 in. (50 mm) **El**
ANSI/ASTM B 30	CA 867	US	Ingot	55.0–60.0 **Cu**, 1.5 **Sn**, 0.50–1.5 **Pb**, 30.0–38.0 **Zn**, 1.0–3.0 **Fe**, 1.0–3.0 **Al**, 1.0–3.5 **Mn**, 0.8 **Ni**	
CDA C 87900		US	Castings	63.0 min **Cu**, 0.25 **Sn**, 0.25 **Pb**, 30.0–36.0 **Zn**, 0.40 **Fe**, 0.05 **Sb**, 0.8–1.2 **Si**, 0.15 **Al**, 0.15 **Mn**, 0.50 **Ni**, 0.01 **P**, 0.05 **S**, 0.05 **As**	**Die castings**: all **diam**, 70 ksi (483 MPa) **TS**, 35 ksi (241 MPa) **YS**, 25% in 2 in. (50 mm) **El**, 70 **HRB**
IS 304	Grade 2	India	Ingot and castings	56.0 **Cu**, 0.5 **Sn**, 0.5 **Pb**, 32.19 min **Zn**, 0.5–2.5 **Fe**, 0.10 **Si**, 5.0 **Al**, 3.0 **Mn**, 0.2 total, others, nickel included in copper	**Sand cast (separately cast)**: all **diam**, 588 MPa **TS**, 274 MPa **YS**, 15% in 2 in. (50 mm) **El**
MIL–B–15894B	Class 2	US	Castings	63.0–67.0 **Cu**, 0.25 **Sn**, 0.25 **Pb**, 30.29 min **Zn**, 0.15 **Fe**, 0.75–1.25 **Si**, 0.15 **Al**, 0.15 **Mn**, 0.50 total, others	**As cast**: all **diam**, 483 MPa **TS**, 241 MPa **YS**, 25% in 2 in. (50 mm) **El**, 68–72 **HRB**
QQ–C–581a	Alloy C	US	Ingot	rem **Cu**, 0.25 **Sn**, 0.25 **Zn**, 0.50 **Fe**, 28.5–31.5 **Si**, 0.25 **Al**, 0.15 **Ca**, 99.4 min **Cu + Si + Fe**	
ISO 1338	CuZn35AlFeMn	ISO	Ingot, castings	57–65.0 **Cu**, 1.0 **Sn**, 0.5 **Pb**, 25.49 min **Zn**, 0.5–2.0 **Fe**, 0.10 **Si**, 0.5–2.5 **Al**, 0.1–3.0 **Mn**, 3.0 **Ni**, 0.40 **Sb + P + As**, total	**Sand cast**: 12–25 mm **diam**, 450 MPa **TS**, 170 MPa **YS**, 20% in 2 in. (50 mm) **El**; **Pressure die cast or permanent mould**: 12–25 mm **diam**, 475 MPa **TS**, 200 MPa **YS**, 18% in 2 in. (50 mm) **El**; **Continuously or centrifugally cast**: 12–25 mm **diam**, 475 MPa **TS**, 200 MPa **YS**, 18% in 2 in. (50 mm) **El**
QQ–C–581a	Alloy B	US	Ingot	rem **Cu**, 0.25 **Sn**, 0.50 **Zn**, 18.5–21.5 **Si**, 0.25 **Al**, 0.15 **C**, 99.4 min **Cu + Si + Fe**	
JIS H 5112	Class 2	Japan	Castings	78.5–82.5 **Cu**, 14.0–16.0 **Zn**, 4.0–5.0 **Si**	**As cast**: all **diam**, 441 min MPa **TS**, 12% in 2 in. (50 mm) **El**
ANSI/ASTM B 30	875	US	Ingot: for castings	79.0 min **Cu**, 0.50 **Pb**, 12.0–16.0 **Zn**, 3.0–5.0 **Si**, 0.5 **Al**	

CAST COPPER AND COPPER ALLOYS

specification number	designation	country	product forms	chemical composition	mechanical properties and hardness values
ANSI/ASTM B 30	878	US	Ingot: for castings	80.0 min **Cu**, 0.25 **Sn**, 0.15 **Pb**, 12.0–16.0 **Zn**, 0.15 **Fe**, 0.05 **Sb**, 3.8–4.2 **Si**, 0.15 **Al**, 0.15 **Mn**, 0.20 **Ni+Co**, 0.01 **P**, 0.05 **S**, 0.05 **As**, 0.01 **Mg**	
ANSI/ASTM B 179	UNS C87800 (878)	US	Die castings	80.0–83.0 **Cu**, 0.25 **Sn**, 0.15 **Pb**, 12–16 **Zn**, 0.15 **Fe**, 3.75–4.25 **Si**, 0.15 **Al**, 0.15 **Mn**, 0.01 **Mg**	**As cast**: all **diam**, 85 ksi (586 MPa) **TS**, 50 ksi (345 MPa) **YS**, 25% in 2 in. (50 mm) **El**, 85–90 **HB**
ANSI/ASTM B 584	C 87500	US	Sand castings	79.0 min **Cu**, 0.50 **Pb**, 12.0–16.0 **Zn**, 3.0–5.0 **Si**, 0.50 **Al**	**As cast**: all **diam**, 60 ksi (414 MPa) **TS**, 24 ksi (165 MPa) **YS**, 16% in 2 in. (50 mm) **El**
ANSI/ASTM B 30	CA 875	US	Ingot	79.0 min **Cu**, 0.50 **Pb**, 12.0–16.0 **Zn**, 3.0–5.0 **Si**, 0.5 **Al**	
ANSI/ASTM B 30	CA 878	US	Ingot	80.0 min **Cu**, 0.25 **Sn**, 0.15 **Pb**, 12.0–16.0 **Zn**, 0.15 **Fe**, 0.05 **Sb**, 3.8–4.2 **Si**, 0.15 **Al**, 0.15 **Mn**, 0.20 **Ni**, 0.01 **P**, 0.05 **S**, 0.05 **As**, 0.01 **Mg**	**As cast**: all **diam**, 55 ksi (379 MPa) **TS**, 30 ksi (207 MPa) **YS**, 15% in 2 in. (50 mm) **El**, 55 **HB**
COPANT 162	C 875 00	COPANT	Cast products	79 min **Cu**, 0.5 **Pb**, 12–16 **Zn**, 3–5 **Si**, 0.5 **Al**	
ANSI/ASTM B 176	C 87800	US	Die castings	80.0–83.0 **Cu**, 0.25 **Sn**, 0.15 **Pb**, 12.0–16.0 **Zn**, 0.15 **Fe**, 3.75–4.25 **Si**, 0.15 **Al**, 0.15 **Mn**, 0.01 **Mg**, 99.8 min total named elements	**As cast**: all **diam**, 85 ksi (586 MPa) **TS**, 50 ksi (345 MPa) **YS**, 25% in 2 in. (50 mm) **El**, 90 max **HRB**
CDA C 87500		US	Castings	79.0 min **Cu**, 0.50 **Pb**, 12.0–16.0 **Zn**, 3.0–5.0 **Si**, 0.50 **Al**	**Sand castings**: all **diam**, 60 ksi (414 MPa) **TS**, 24 ksi (165 MPa) **YS**, 16% in 2 in. (50 mm) **El**, 115 **HB**
CDA C 87800		US	Castings	80.0 min **Cu**, 0.25 **Sn**, 0.15 **Pb**, 12.0–16.0 **Zn**, 0.15 **Fe**, 0.05 **Sb**, 3.8–4.2 **Si**, 0.15 **Al**, 0.15 **Mn**, 0.20 **Ni**, 0.01 **P**, 0.05 **S**, 0.05 **As**, 0.01 **Mg**	**Die casting**: all **diam**, 85 ksi (586 MPa) **TS**, 50 ksi (345 MPa) **YS**, 25% in 2 in. (50 mm) **El**, 85 **HRB**
COPANT 162	C 878 00	COPANT	Cast products	80 min **Cu**, 0.25 **Sn**, 0.15 **Pb**, 12–16 **Zn**, 0.15 **Fe**, 3.8–4.2 **Si**, 0.15 **Al**, 0.15 **Mn**, 0.2 **Ni**, 0.01 **Mg**, 0.05 **As**	
JIS H 5112	Class 3	Japan	Castings	80.0–84.0 **Cu**, 13.0–15.0 **Zn**, 3.2–4.2 **Si**, 0.5 **Mn+Fe**	**As cast**: all **diam**, 20% in 2 in. (50 mm) **El**
ANSI/ASTM B 30	874	US	Ingot: for castings	79.0 **Cu**, 1.0 **Pb**, 12.0–16.0 **Zn**, 2.5–4.0 **Si**, 0.5 **Al**	
ANSI/ASTM B 584	C 87400	US	Sand castings	79.0 min **Cu**, 1.0 **Pb**, 12.0–16.0 **Zn**, 2.5–4.0 **Si**, 0.8 **Al**	**As cast**: all **diam**, 50 ksi (345 MPa) **TS**, 21 ksi (145 MPa) **YS**, 18% in 2 in. (50 mm) **El**
ANSI/ASTM B 30	CA 874	US	Ingot	79.0 min **Cu**, 1.0 **Pb**, 12.0–16.0 **Zn**, 2.5–4.0 **Si**, 0.5 **Al**	

CAST COPPER AND COPPER ALLOYS

specification number	designation	country	product forms	chemical composition	mechanical properties and hardness values
COPANT 162	C 87400	COPANT	Cast products	79 min **Cu**, 1.0 **Pb**, 12–16 **Zn**, 2.5–4.0 **Si**, 0.8 **Al**, 99.5 total elements	
CDA C 87400		US	Castings	79.0 min **Cu**, 1.0 **Pb**, 12.0–16.0 **Zn**, 2.5–4.0 **Si**, 0.8 **Al**, 99.5 min total named elements	**Sand castings**: all **diam**, 50 ksi (345 MPa) **TS**, 21 ksi (145 MPa) **YS**, 18% in 2 in. (50 mm) **El**, 70 **HB**
QQ–C–390A	874	US	Castings	78.5 min **Cu**, 1.0 **Pb**, 12.0–16.0 **Zn**, 2.5–4.0 **Si**	**As cast**: all **diam**, 50 ksi (345 MPa) **TS**, 21 ksi (145 MPa) **YS**, 18% in 2 in. (50 mm) **El**
MIL–B–15894B	Class 3	US	Castings	80.0–83.0 **Cu**, 0.25 **Sn**, 0.15 **Pb**, 11.63 min **Zn**, 0.15 **Fe**, 3.75–4.25 **Si**, 0.15 **Al**, 0.15 **Mn**, 0.01 **Mg**, 0.25 total, others	**As cast**: all **diam**, 586 MPa **TS**, 345 MPa **YS**, 25% in 2 in. (50 mm) **El**, 85–90 **HRB**
JIS H 5112	Class 1	Japan	Castings	84.0–88.0 **Cu**, 9.0–11.0 **Zn**, 3.5–4.5 **Si**	**As cast**: all **diam**, 25% in 2 in. (50 mm) **El**
QQ–C–581a	Alloy A	US	Ingot	rem **Cu**, 0.25 **Sn**, 0.25 **Zn**, 0.50 **Fe**, 10.0–12.0 **Si**, 0.25 **Al**, 0.15 **Ca**, 99.4 min **Cu+Si+Fe**	
ANSI/ASTM B 584	C 87600	US	Sand castings	88.0 min **Cu**, 0.50 **Pb**, 4.7–7.0 **Zn**, 3.5–5.5 **Si**	**As cast**: all **diam**, 60 ksi (414 MPa) **TS**, 30 ksi (207 MPa) **YS**, 16% in 2 in. (50 mm) **El**
ANSI/ASTM B 30	876	US	Ingot: for castings	88.0 min **Cu**, 0.50 **Pb**, 4.0–7.0 **Zn**, 3.5–5.5 **Si**	
ANSI/ASTM B 30	CA 876	US	Ingot	88.0 min **Cu**, 0.50 **Pb**, 4.0–7.0 **Zn**, 3.5–5.5 **Si**	
COPANT 162	C 876 00	COPANT	Cast products	88 min **Cu**, 0.5 **Pb**, 4–7 **Zn**, 3.5–5.5 **Si**	
CDA C 87600		US	Castings	88.0 min **Cu**, 0.50 **Pb**, 4.0–7.0 **Zn**, 3.5–5.5 **Si**	**Sand castings**: all **diam**, 60 ksi (414 MPa) **TS**, 30 ksi (207 MPa) **YS**, 16% in 2 in. (50 mm) **El**, 76 **HRB**
ANSI/ASTM B 30	872	US	Ingot: for castings	89.0 min **Cu**, 1.0 **Sn**, 0.50 **Pb**, 5.0 **Zn**, 2.5 **Fe**, 1.0–5.0 **Si**, 1.5 **Al**, 1.5 **Mn**	
ANSI/ASTM B 30	CA 872	US	Ingot	89.0 min **Cu**, 1.0 **Sn**, 0.50 **Pb**, 5.0 **Zn**, 2.5 **Fe**, 1.0–5.0 **Si**, 1.5 **Al**, 1.5 **Mn**	
COPANT 162	C 872 00	COPANT	Cast products	89 min **Cu**, 1.0 **Sn**, 0.5 **Pb**, 5.0 **Zn**, 2.5 **Fe**, 1–5.0 **Si**, 1.5 **Al**, 1.5 **Mn**, 99.5 total elements	
CDA C 87200		US	Castings	89.0 min **Cu**, 1.0 **Sn**, 0.50 **Pb**, 5.0 **Zn**, 2.5 **Fe**, 1.0–5.0 **Si**, 1.5 **Al**, 1.5 **Mn**, 99.5 min total named elements	**Sand castings**: all **diam**, 45 ksi (310 MPa) **TS**, 18 ksi (124 MPa) **YS**, 20% in 2 in. (50 mm) **El**, 85 **HB**
QQ–C–390A	872	US	Castings	83.98 min **Cu**, 1.0 **Sn**, 0.50 **Pb**, 5.0 **Zn**, 2.5 **Fe**, 1.0–5.0 **Si**, 1.5 **Al**, 1.5 **Mn**	**As cast**: all **diam**, 45 ksi (310 MPa) **TS**, 18 ksi (124 MPa) **YS**, 20% in 2 in. (50 mm) **El**
IS 1028		India	Ingot and castings	82.5–86.5 **Cu**, 1.0 **Sn**, 0.5 **Pb**, 5.0 **Zn**, 2.5 **Fe**, 1.0–5.0 **Si**, 1.5 **Al**, 1.5 **Mn**	

CAST COPPER AND COPPER ALLOYS

specification number	designation	country	product forms	chemical composition	mechanical properties and hardness values
MIL-C-22229A	872	US	Castings	82.5 min **Cu**, 1.0 **Sn**, 0.50 **Pb**, 5.0 **Zn**, 2.5 **Fe**, 1.0–5.0 **Si**, 1.5 **Al**, 1.5 **Mn**	**As cast**: all **diam**, 310 MPa **TS**, 124 MPa **YS**, 20% in 2 in. (50 mm) **El**
ANSI/ASTM B 584	C 87200	US	Sand castings	89.0 min **Cu**, 1.0 **Sn**, 0.50 **Pb**, 5.0 **Zn**, 2.5 **Fe**, 1.0–1.5 **Si**, 1.5 **Al**, 1.5 **Mn**	**As cast**: all **diam**, 45 ksi (310 MPa) **TS**, 18 ksi (124 MPa) **YS**, 20% in 2 in. (50 mm) **El**
ANSI/ASTM B 30	913	US	Ingot: for castings	79.0–82.0 **Cu**, 18.3–20.0 **Sn**, 0.25 **Pb**, 0.25 **Zn**, 0.15 **Fe**, 0.20 **Sb**, 0.005 **Al**, 0.50 **Ni+Co**, 1.0 **P**, 0.05 **S**	
ANSI/ASTM B 30	CA 913	US	Ingot	79.0–82.0 **Cu**, 18.3–20.0 **Sn**, 0.25 **Pb**, 0.25 **Zn**, 0.15 **Fe**, 0.20 **Sb**, 0.005 **Si**, 0.005 **Al**, 0.50 **Ni**, 1.0 **P**, 0.05 **S**	**As cast**: all **diam**, 40 ksi (275 MPa) **TS**, 18 ksi (125 MPa) **YS**, 20% in 2 in. (51 mm) **El**
QQ-C-525b	Composition 10	US	Ingot	80.0–82.0 **Cu**, 18.3–20.0 **Sn**, 0.40 **Pb**, 0.25 **Zn**, 0.10 **Fe**, 0.15 **Sb**, 0.005 **Si**, 0.005 **Al**, 0.8 **Ni**, 0.40–0.60 **P**, 0.05 **S**	
ASTM B 22	CA 913	US	Castings	79.0–82.0 **Cu**, 18.0–20.0 **Sn**, 0.25 **Pb**, 0.25 **Zn**, 0.25 **Fe**, 1.0 **P**	
ANSI/ASTM B 505	CA 913	US	Castings	79.0–82.0 **Cu**, 18.0–20.0 **Sn**, 0.50 **Pb**, 0.25 **Zn**, 0.25 **Fe**, 0.20 **Sb**, 0.005 **Si**, 0.005 **Al**, 0.50 **Ni+Co**, 1.0 **P**, 0.05 **S**	**As cast**: all **diam**, 160 min **HB**
CDA C 91300		US	Castings	79.0–82.0 **Cu**, 18.0–20.0 **Sn**, 0.25 **Pb**, 0.25 **Zn**, 0.25 **Fe**, 0.20 **Sb**, 0.005 **Si**, 0.005 **Al**, 0.50 **Ni**, 1.0 **P**, 0.05 **S**	**Sand castings**: all **diam**, 35 ksi (241 MPa) **TS**, 30 ksi (207 MPa) **YS**, .5% in **El**, 160 **HB**
COPANT 162	C 913 00	COPANT	Cast products	79–82 **Cu**, 18–20 **Sn**, 0.25 **Pb**, 0.25 **Zn**, 0.25 **Fe**, 1.0 **P**	
COPANT 801	C 913 00	COPANT	Continuous-cast rod, bar and tube	79–82 **Cu**, 18–20 **Sn**, 0.25 **Pb**, 0.25 **Zn**, 0.25 **Fe**, 1.0 **P**	**As manufactured**: 127 mm **diam**, 86–96 **HRB**
QQ-C-390A	913	US	Castings	79.0–82.0 **Cu**, 18.0–20.0 **Sn**, 0.25 **Pb**, 0.25 **Zn**, 0.25 **Fe**, 1.0 **P**	**As cast**: all **diam**, 160 **HB**
ANSI/ASTM B 30	911	US	Ingot: for castings	82.0–85.0 **Cu**, 15.3–17.0 **Sn**, 0.25 **Pb**, 0.25 **Zn**, 0.15 **Fe**, 0.20 **Sb**, 0.005 **Si**, 0.005 **Al**, 0.50 **Ni+Co**, 1.0 **P**, 0.05 **S**	
ANSI/ASTM B 30	CA 911	US	Ingot	82.0–85.0 **Cu**, 15.3–17.0 **Sn**, 0.25 **Pb**, 0.25 **Zn**, 0.15 **Fe**, 0.20 **Sb**, 0.005 **Si**, 0.005 **Al**, 0.50 **Ni**, 1.0 **P**, 0.05 **S**	**As cast**: all **diam**, 40 ksi (275 MPa) **TS**, 18 ksi (125 MPa) **YS**, 20% in 2 in. (51 mm) **El**
ASTM B 22	CA 911	US	Castings	82.0–85.0 **Cu**, 15.0–17.0 **Sn**, 0.25 **Pb**, 0.25 **Zn**, 0.25 **Fe**, 1.0 **P**	
CDA C 91100		US	Castings	82.0–85.0 **Cu**, 15.0–17.0 **Sn**, 0.25 **Pb**, 0.25 **Zn**, 0.25 **Fe**, 0.20 **Sb**, 0.005 **Si**, 0.005 **Al**, 0.50 **Ni**, 1.0 **P**, 0.05 **S**	**Sand castings**: all **diam**, 35 ksi (241 MPa) **TS**, 25 ksi (172 MPa) **YS**, 2% in 2 in. (50 mm) **El**, 135 **HB**
COPANT 162	C 911 00	COPANT	Cast products	82–85 **Cu**, 15–17 **Sn**, 0.25 **Pb**, 0.25 **Zn**, 0.25 **Fe**, 1.0 **P**	

CAST COPPER AND COPPER ALLOYS

specification number	designation	country	product forms	chemical composition	mechanical properties and hardness values
ANSI/ASTM B 30	CA 910	US	Ingot	84.0–86.0 **Cu**, 14.3–16.0 **Sn**, 0.20 **Pb**, 1.5 **Zn**, 0.10 **Fe**, 0.10 **Sb**, 0.005 **Si**, 0.005 **Al**, 0.8 **Ni**, 0.03 **P**, 0.05 **S**	
ANSI/ASTM B 30	910	US	Ingot: for castings	84.0–86.0 **Cu**, 14.3–16.0 **Sn**, 0.20 **Pb**, 1.5 **Zn**, 0.10 **Fe**, 0.10 **Sb**, 0.005 **Si**, 0.005 **Al**, 0.8 **Ni+Co**, 0.03 **P**, 0.05 **S**	
ANSI/ASTM B 505	CA 910	US	Castings	84.0–86.0 **Cu**, 14.0–16.0 **Sn**, 0.20 **Pb**, 1.5 **Zn**, 0.10 **Fe**, 0.20 **Sb**, 0.005 **Si**, 0.005 **Al**, 0.8 **Ni+Co**, 0.10 **P**, 0.05 **S**	**As cast**: all **diam**, 30 ksi (207 MPa) **TS**
CDA C 91000		US	Castings	84.0–86.0 **Cu**, 14.0–16.0 **Sn**, 0.20 **Pb**, 1.5 **Zn**, 0.10 **Fe**, 0.20 **Sb**, 0.005 **Si**, 0.005 **Al**, 0.8 **Ni**, 0.05 **P**, 0.05 **S**	**Sand castings**: all **diam**, 30 ksi (207 MPa) **TS**, 25 ksi (172 MPa) **YS**, 1% in 2 in. (50 mm) **El**, 105 **HB**
QQ-C-525b	Composition 9	US	Ingot	84.0–86.0 **Cu**, 13.3–15.0 **Sn**, 0.15 **Pb**, 1.5 **Zn**, 0.10 **Fe**, 0.15 **Sb**, 0.005 **Si**, 0.005 **Al**, 0.8 **Ni**, 0.03 **P**, 0.05 **S**	
COPANT 162	C 910 00	COPANT	Cast products	84–86 **Cu**, 13–15 **Sn**, 0.20 **Pb**, 1.5 **Zn**, 0.10 **Fe**, 0.7 **Ni**, 0.05 **P**	
COPANT 801	C 910 00	COPANT	Continuous-cast rod, bar and tube	84–86 **Cu**, 13–15 **Sn**, 0.20 **Pb**, 1.5 **Zn**, 0.10 **Fe**, 0.7 **Ni**, 0.05 **P**	**As manufactured**: 127 mm **diam**, 210 MPa **TS**
QQ-C-390A	910	US	Castings	84.0–86.0 **Cu**, 13.0–15.0 **Sn**, 0.20 **Pb**, 1.5 **Zn**, 0.10 **Fe**, 0.7 **Ni**, 0.05 **P**	**As cast**: all **diam**, 30 ksi (207 MPa) **TS**, 1% in 2 in. (50 mm) **El**
JIS H 5113	Class 3	Japan	Castings	84.0–88.0 **Cu**, 12.0–15.0 **Sn**, 0.15–0.50 **P**	**As cast**: all **diam**, 90 min **HB**
CDA C 90900		US	Castings	86.0–89.0 **Cu**, 12.0–14.0 **Sn**, 0.25 **Pb**, 0.25 **Zn**, 0.15 **Fe**, 0.20 **Sb**, 0.005 **Si**, 0.005 **Al**, 0.50 **Ni**, 0.05 **P**, 0.05 **S**	**Sand castings**: all **diam**, 40 ksi (276 MPa) **TS**, 20 ksi (138 MPa) **YS**, 15% in 2 in. (50 mm) **El**, 90 **HB**
COPANT 162	C 909 00	COPANT	Cast products	86–89 **Cu**, 12–14 **Sn**, 99.5 min, total elements	
ANSI/ASTM B 30	CA 908	US	Ingot	85.0–89.0 **Cu**, 11.3–13.0 **Sn**, 0.25 **Pb**, 0.25 **Zn**, 0.15 **Fe**, 0.10 **Sb**, 0.005 **Si**, 0.005 **Al**, 0.50 **Ni**, 0.30 **P**, 0.05 **S**	**As cast**: all **diam**, 50 ksi (345 MPa) **TS**, 28 ksi (193 MPa) **YS**, 12% in 2 in. (50.8 mm) **El**, 95 **HB**
ANSI/ASTM B 427	CA 908	US	Castings	85.0–89.0 **Cu**, 11.0–13.0 **Sn**, 0.25 **Pb**, 0.25 **Zn**, 0.15 **Fe**, 0.20 **Sb**, 0.005 **Si**, 0.005 **Al**, 0.50 **Ni**, 0.30 **P**, 0.05 **S**, 99.5 min **Cu+Sn+Ni+Pb+P**	**Chill cast**: all **diam**, 50 ksi (345 MPa) **TS**, 28 ksi (193 MPa) **YS**, 12% in 2 in. (50.8 mm) **El**, 95 **HB**; **Sand cast**: all **diam**, 35 ksi (241 MPa) **TS**, 17 ksi (117 MPa) **YS**, 10% in 2 in. (50.8 mm) **El**, 65 **HB**
ANSI/ASTM B 30	CA 917	US	Ingot	84.0–87.0 **Cu**, 11.5–12.5 **Sn**, 0.25 **Pb**, 0.25 **Zn**, 0.15 **Fe**, 0.10 **Sb**, 0.005 **Si**, 0.005 **Al**, 1.2–2.0 **Ni**, 0.30 **P**, 0.05 **S**	

CAST COPPER AND COPPER ALLOYS

specification number	designation	country	product forms	chemical composition	mechanical properties and hardness values
ANSI/ASTM B 30	917	US	Ingot: for castings	84.0–87.0 **Cu**, 11.5–12.5 **Sn**, 0.25 **Pb**, 0.25 **Zn**, 0.15 **Fe**, 0.10 **Sb**, 0.005 **Si**, 0.005 **Al**, 1.2–2.0 **Ni+Co**, 0.30 **P**, 0.05 **S**	
ANSI/ASTM B 30	908	US	Ingot: for castings	85.0–89.0 **Cu**, 11.0–13.0 **Sn**, 0.25 **Pb**, 0.25 **Zn**, 0.15 **Fe**, 0.10 **Sb**, 0.005 **Si**, 0.005 **Al**, 0.50 **Ni+Co**, 0.30 **P**, 0.05 **S**	
COPANT 162	C 908 00	COPANT	Cast products	85–89 **Cu**, 11–13 **Sn**, 0.25 **Pb**, 0.25 **Zn**, 0.15 **Fe**, 0.10 **Sb**, 0.005 **Si**, 0.005 **Al**, 0.50 **Ni**, 0.30 **P**, 0.05 **S**, 99.5 total elements	
ANSI/ASTM B 427	CA 917	US	Castings	84.0–87.0 **Cu**, 11.3–12.5 **Sn**, 0.25 **Pb**, 0.25 **Zn**, 0.20 **Fe**, 0.20 **Sb**, 0.005 **Si**, 0.005 **Al**, 1.2–2.0 **Ni**, 0.30 **P**, 0.05 **S**, 99.5 min **Cu+Sn+Ni+Pb+P**	**Chill cast**: all **diam**, 50 ksi (345 MPa) **TS**, 28 ksi (193 MPa) **YS**, 12% in 2 in. (50.8 mm) **El**, 95 **HB**; **Sand cast**: all **diam**, 35 ksi (241 MPa) **TS**, 17 ksi (117 MPa) **YS**, 10% in 2 in. (50.8 mm) **El**, 65 **HB**
CDA C 91700		US	Castings	84.0–87.0 **Cu**, 11.3–12.5 **Sn**, 0.25 **Pb**, 0.25 **Zn**, 0.15 **Fe**, 0.20 **Sb**, 0.005 **Si**, 0.005 **Al**, 1.2–2.0 **Ni**, 0.30 **P**, 0.05 **S**	**Sand castings**: all **diam**, 35 ksi (241 MPa) **TS**, 17 ksi (117 MPa) **YS**, 10% in 2 in. (50 mm) **El**, 65 **HB**
COPANT 162	C 917 00	COPANT	Cast products	84–87 **Cu**, 11.3–12.5 **Sn**, 0.25 **Pb**, 0.25 **Zn**, 0.15 **Fe**, 0.10 **Sb**, 0.005 **Si**, 0.005 **Al**, 1.2–2.0 **Ni**, 0.30 **P**, 0.05 **S**	
ANSI/ASTM B 30	CA 907	US	Ingot	88.0–90.0 **Cu**, 10.3–12.0 **Sn**, 0.50 **Pb**, 0.50 **Zn**, 0.15 **Fe**, 0.10 **Sb**, 0.005 **Si**, 0.005 **Al**, 0.50 **Ni**, 0.30 **P**, 0.05 **S**	
ANSI/ASTM B 30	907	US	Ingot: for castings	88.0–90.0 **Cu**, 10.3–12.0 **Sn**, 0.50 **Pb**, 0.50 **Zn**, 0.15 **Fe**, 0.10 **Sb**, 0.005 **Si**, 0.005 **Al**, 0.50 **Ni+Co**, 0.30 **P**, 0.05 **S**	
ANSI/ASTM B 427	CA 907	US	Castings	88.0–90.0 **Cu**, 10.0–12.0 **Sn**, 0.50 **Pb**, 0.50 **Zn**, 0.15 **Fe**, 0.20 **Sb**, 0.005 **Si**, 0.005 **Al**, 0.50 **Ni**, 0.30 **P**, 0.05 **S**, 99.5 min **Cu+Sn+Ni+Pb+P**	**Chill cast**: all **diam**, 50 ksi (345 MPa) **TS**, 28 ksi (193 MPa) **YS**, 12% in 2 in. (50.8 mm) **El**, 95 **HB**; **Sand cast**: all **diam**, 35 ksi (241 MPa) **TS**, 17 ksi (117 MPa) **YS**, 10% in 2 in. (50.8 mm) **El**, 65 **HB**
ANSI/ASTM B 505	CA 907	US	Castings	88.0–90.0 **Cu**, 10.0–12.0 **Sn**, 0.50 **Pb**, 0.50 **Zn**, 0.15 **Fe**, 0.10 **Sb**, 0.005 **Si**, 0.005 **Al**, 0.50 **Ni+Co**, 1.5 **P**, 0.05 **S**	**As cast**: all **diam**, 40 ksi (276 MPa) **TS**, 25 ksi (172 MPa) **YS**, 10% in 2 in. (50 mm) **El**
CDA C 90700		US	Castings	88.0–90.0 **Cu**, 10.0–12.0 **Sn**, 0.50 **Pb**, 0.50 **Zn**, 0.15 **Fe**, 0.20 **Sb**, 0.005 **Si**, 0.005 **Al**, 0.50 **Ni**, 0.30 **P**, 0.05 **S**	**Sand castings**: all **diam**, 35 ksi (241 MPa) **TS**, 18 ksi (124 MPa) **YS**, 10% in 2 in. (50 mm) **El**, 80 **HB**
COPANT 162	C 907 00	COPANT	Cast products	88–90 **Cu**, 10–12 **Sn**, 0.5 **Pb**, 0.5 **Zn**, 0.15 **Fe**, 0.005 **Al**, 0.10–0.30 **P**, 1.0 **Pb+Zn+Ni**	

CAST COPPER AND COPPER ALLOYS

specification number	designation	country	product forms	chemical composition	mechanical properties and hardness values
COPANT 801	C 907 00	COPANT	Continuous–cast rod, bar and tube	88–90 **Cu**, 10–12 **Sn**, 0.5 **Pb**, 0.5 **Zn**, 0.15 **Fe**, 0.005 **Al**, 0.10–0.30 **P**, 1.0 **Pb+Zn+Ni**	**As manufactured**: 127 mm **diam**, 280 MPa **TS**, 180 MPa **YS**, 10% in 2 in. (50 mm) **El**
QQ–C–390A	907	US	Castings	88.0–90.0 **Cu**, 10.0–12.0 **Sn**, 0.50 **Pb**, 0.050 **Zn**, 0.15 **Fe**, 0.005 **Al**, 0.10–0.30 **P**, 1.0 **Pb+Zn+Ni**	**As cast**: all **diam**, 40 ksi (276 MPa) **TS**, 25 ksi (172 MPa) **YS**, 10% in 2 in. (50 mm) **El**
ISO 1338	CuSn10P	ISO	Castings, ingot	87–89.5 **Cu**, 10.0–11.5 **Sn**, 0.25 **Pb**, 0.05 **Zn**, 0.10 **Fe**, 0.05 **Sb**, 0.02 **Si**, 0.01 **Al**, 0.05 **Mn**, 0.10 **Ni**, 0.50–1.0 **P**, 0.05 **S**	**Sand cast**: 12–25 mm **diam**, 220 MPa **TS**, 130 MPa **YS**, 3% in 2 in. (50 mm) **El**; **Permanent mould cast**: 12–25 mm **diam**, 310 MPa **TS**, 170 MPa **YS**, 2% in 2 in. (50 mm) **El**; **Continuously cast**: 12–25 mm **diam**, 360 MPa **TS**, 170 MPa **YS**, 6% in 2 in. (50 mm) **El**
ANSI/ASTM B 30	916	US	Ingot: for castings	86.0–89.0 **Cu**, 10.0–10.8 **Sn**, 0.25 **Pb**, 0.25 **Zn**, 0.15 **Fe**, 0.10 **Sb**, 0.005 **Si**, 0.005 **Al**, 1.2–2.0 **Ni+Co**, 0.25 **P**, 0.05 **S**	
ANSI/ASTM B 30	CA 916	US	Ingot	86.0–89.0 **Cu**, 10.0–10.8 **Sn**, 0.25 **Pb**, 0.25 **Zn**, 0.15 **Fe**, 0.10 **Sb**, 0.005 **Si**, 0.005 **Al**, 1.2–2.0 **Ni**, 0.25 **P**, 0.05 **S**	**As cast**: all **diam**, 50 ksi (345 MPa) **TS**, 28 ksi (193 MPa) **YS**, 12% in 2 in. (50.8 mm) **El**
ANSI/ASTM B 427	CA 916	US	Castings	86.0–89.0 **Cu**, 9.7–10.8 **Sn**, 0.25 **Pb**, 0.25 **Zn**, 0.20 **Fe**, 0.20 **Sb**, 0.005 **Si**, 0.005 **Al**, 1.2–2.0 **Ni**, 0.30 **P**, 0.05 **S**, 99.5 min **Cu+Sn+Ni+Pb+P**	**Chill cast**: all **diam**, 45 ksi (310 MPa) **TS**, 25 ksi (172 MPa) **YS**, 10% in 2 in. (50.8 mm) **El**, 85 **HB**; **Sand cast**: all **diam**, 35 ksi (241 MPa) **TS**, 17 ksi (117 MPa) **YS**, 10% in 2 in. (50.8 mm) **El**, 65 **HB**
CDA C 91600		US	Castings	86.0–89.0 **Cu**, 9.7–10.8 **Sn**, 0.25 **Pb**, 0.25 **Zn**, 0.20 **Fe**, 0.20 **Sb**, 0.005 **Si**, 0.005 **Al**, 1.2–2.0 **Ni**, 0.30 **P**, 0.05 **S**	**Sand castings**: all **diam**, 35 ksi (241 MPa) **TS**, 17 ksi (117 MPa) **YS**, 10% in 2 in. (50 mm) **El**, 65 **HB**
COPANT 162	C 916 00	COPANT	Cast products	86–89 **Cu**, 9.7–10.8 **Sn**, 0.25 **Pb**, 0.25 **Zn**, 0.15 **Fe**, 0.10 **Sb**, 0.005 **Si**, 0.005 **Al**, 1.2–2.0 **Ni**, 0.25 **P**, 0.05 **S**, 99.5 total elements	
COPANT 162	C 916 00	COPANT	Cast products	86–89 **Cu**, 9.7–10.8 **Sn**, 0.25 **Pb**, 0.25 **Zn**, 0.15 **Fe**, 0.10 **Sb**, 0.005 **Si**, 0.005 **Al**, 1.2–2.0 **Ni**, 0.25 **P**, 0.05 **S**, 99.5 total elements	
QQ–C–390A	916	US	Castings	86.3 min **Cu**, 9.8–10.7 **Sn**, 0.25 **Pb**, 1.3–2.0 **Ni**, 0.25 **P**, 0.50 total, others	**As cast**: all **diam**, 35 ksi (241 MPa) **TS**, 17 ksi (117 MPa) **YS**, 10% in 2 in. (50 mm) **El**; **Chill cast**: all **diam**, 45 ksi (310 MPa) **TS**, 25 ksi (172 MPa) **YS**, 14% in 2 in. (50 mm) **El**

CAST COPPER AND COPPER ALLOYS

specification number	designation	country	product forms	chemical composition	mechanical properties and hardness values
ANSI/ASTM B 30	915	US	Ingot: for castings	82.0–86.0 **Cu**, 9.3–11.0 **Sn**, 2.0–3.0 **Pb**, 0.25 **Zn**, 0.15 **Fe**, 0.10 **Sb**, 0.005 **Si**, 0.005 **Al**, 2.8–4.0 **Ni + Co**, 0.50 **P**, 0.05 **S**	
ANSI/ASTM B 30	CA 915	US	Ingot	82.0–86.0 **Cu**, 9.3–11.0 **Sn**, 2.0–3.0 **Pb**, 0.25 **Zn**, 0.15 **Fe**, 0.10 **Sb**, 0.005 **Si**, 0.005 **Al**, 2.8–4.0 **Ni**, 0.50 **P**, 0.05 **S**	**As cast**: all **diam**, 50 ksi (345 MPa) **TS**, 28 ksi (193 MPa) **YS**, 12% in 2 in. (50.8 mm) **El**, 95 **HB**
ANSI/ASTM B 584	C 90500	US	Sand castings	86.0–89.0 **Cu**, 9.0–11.0 **Sn**, 0.30 **Pb**, 1.0–3.0 **Zn**, 0.20 **Fe**, 0.20 **Sb**, 0.005 **Si**, 0.005 **Al**, 1.0 **Ni + Co**, 0.05 **P**, 0.05 **S**	**As cast**: all **diam**, 40 ksi (276 MPa) **TS**, 18 ksi (124 MPa) **YS**, 20% in 2 in. (50 mm) **El**
ANSI/ASTM B 505	CA 905	US	Castings	86.0–89.0 **Cu**, 9.0–11.0 **Sn**, 0.30 **Pb**, 1.0–3.0 **Zn**, 0.20 **Fe**, 0.20 **Sb**, 0.005 **Si**, 0.005 **Al**, 1.0 **Ni + Co**, 1.5 **P**, 0.05 **S**	**As cast**: all **diam**, 44 ksi (303 MPa) **TS**, 25 ksi (172 MPa) **YS**, 10% in 2 in. (50 mm) **El**
ASTM B 22	CA 905	US	Castings	86.0–89.0 **Cu**, 9.0–11.0 **Sn**, 0.30 **Pb**, 1.0–3.0 **Zn**, 0.15 **Fe**, 1.0 **Ni**, 0.05 **P**	**As cast**: all **diam**, 40 ksi (275 MPa) **TS**, 18 ksi (125 MPa) **YS**, 20% in 2 in. (51 mm) **El**
ANSI/ASTM B 30	905	US	Ingot: for castings	86.0–89.0 **Cu**, 9.5–10.5 **Sn**, 0.25 **Pb**, 1.5–3.0 **Zn**, 0.15 **Fe**, 0.20 **Sb**, 0.005 **Si**, 0.005 **Al**, 0.8 **Ni + Co**, 0.03 **P**, 0.05 **S**	
ANSI/ASTM B 30	CA 905	US	Ingot	86.0–89.0 **Cu**, 9.5–10.5 **Sn**, 0.25 **Pb**, 1.5–3.0 **Zn**, 0.15 **Fe**, 0.20 **Sb**, 0.005 **Si**, 0.005 **Al**, 0.8 **Ni**, 0.03 **P**, 0.05 **S**	**As cast**: all **diam**, 40 ksi (275 MPa) **TS**, 18 ksi (125 MPa) **YS**, 20% in 2 in. (51 mm) **El**
AS 1565	906D	Australia	Ingot, castings	rem **Cu**, 9.0–11.0 **Sn**, 0.25 **Pb**, 0.05 **Zn**, 0.25 **Ni**, 0.15 **P**, 0.80 total, others (applies to castings only)	**Sand cast**: all **diam**, 230 MPa **TS**, 130 MPa **YS**, 6% in 2 in. (50 mm) **El**, 70 **HB**; **Chill cast**: all **diam**, 270 MPa **TS**, 140 MPa **YS**, 5% in 2 in. (50 mm) **El**, 90 **HB**; **Continuous cast**: all **diam**, 310 MPa **TS**, 160 MPa **YS**, 9% in 2 in. (50 mm) **El**, 90 **HB**
AS 1565	904D	Australia	Ingot, castings	rem **Cu**, 10.0 **Sn**, 0.25 **Pb**, 0.05 **Zn**, 0.10 **Fe**, 0.02 **Si**, 0.01 **Al**, 0.10 **Ni**, 0.60 **P**, 0.60 total, others (applies to castings only)	**Sand cast**: all **diam**, 220 MPa **TS**, 130 MPa **YS**, 3% in 2 in. (50 mm) **El**, 70 **HB**; **Chill cast**: all **diam**, 310 MPa **TS**, 170 MPa **YS**, 2% in 2 in. (50 mm) **El**, 95 **HB**; **Continuous cast**: all **diam**, 360 MPa **TS**, 170 MPa **YS**, 6% in 2 in. (50 mm) **El**, 100 **HB**
CDA C 90500		US	Castings	86.0–89.0 **Cu**, 9.0–11.0 **Sn**, 0.30 **Pb**, 1.0–3.0 **Zn**, 0.20 **Fe**, 0.20 **Sb**, 0.005 **Si**, 0.005 **Al**, 1.0 **Ni**, 0.05 **P**, 0.05 **S**	**Sand castings**: all **diam**, 40 ksi (276 MPa) **TS**, 18 ksi (124 MPa) **YS**, 20% in 2 in. (50 mm) **El**, 75 **HB**
COPANT 162	C 905 00	COPANT	Cast products,	86–89 **Cu**, 9–11 **Sn**, 0.3 **Pb**, 1–3 **Zn**, 0.2 **Fe**, 0.2 **Sb**, 0.005 **Si**, 0.005 **Al**, 1.0 **Ni**, 0.05 **P**, 0.05 **S**	
COPANT 801	C 905 00	COPANT	Continuous–cast rod, bar and tube	86–89 **Cu**, 9–11 **Sn**, 0.3 **Pb**, 1–3 **Zn**, 0.2 **Fe**, 0.2 **Sb**, 0.005 **Si**, 0.005 **Al**, 1.0 **Ni**, 0.05 **P**, 0.05 **S**	**As manufactured**: 127 mm **diam**, 300 MPa **TS**, 180 MPa **YS**, 10% in 2 in. (50 mm) **El**

CAST COPPER AND COPPER ALLOYS

specification number	designation	country	product forms	chemical composition	mechanical properties and hardness values
QQ-C-390A	915	US	Castings	81.3 min **Cu**, 9.0–11.0 **Sn**, 2.0–3.2 **Pb**, 2.8–4.0 **Ni**, 0.50 **P**, 0.50 total, others	**As cast**: all **diam**, 45 ksi (310 MPa) **TS**, 25 ksi (172 MPa) **YS**, 8% in 2 in. (50 mm) **El**
QQ-C-390A	905	US	Castings	86.0–89.0 **Cu**, 9.0–11.0 **Sn**, 0.30 **Pb**, 1.0–3.0 **Zn**, 0.15 **Fe**, 1.0 **Ni**, 0.5 **P**	**As cast**: all **diam**, 40 ksi (276 MPa) **TS**, 18 ksi (124 MPa) **YS**, 20% in 2 in. (50 mm) **El**
SABS 200	Cu-Sn10P 1C	South Africa	Castings	rem **Cu**, 10.0 min **Sn**, 0.25 **Pb**, 0.05 **Zn**, 0.10 **Fe**, 0.02 **Si**, 0.01 **Al**, 0.1 **Ni**, 0.50 min **P**, 0.47 total, others	**Continuously cast**: all **diam**, 355 MPa **TS**, 6% in 2 in. (50 mm) **El**
ANSI/ASTM B 30	CA 903	US	Ingot	86.0–89.0 **Cu**, 7.8–9.0 **Sn**, 0.25 **Pb**, 3.5–5.0 **Zn**, 0.15 **Fe**, 0.20 **Sb**, 0.005 **Si**, 0.005 **Al**, 0.8 **Ni**, 0.03 **P**, 0.05 **S**	
ANSI/ASTM B 30	903	US	Ingot: for castings	86.0–89.0 **Cu**, 7.8–9.0 **Sn**, 0.25 **Pb**, 3.5–5.0 **Zn**, 0.15 **Fe**, 0.20 **Sb**, 0.005 **Si**, 0.005 **Al**, 0.8 **Ni+Co**, 0.03 **P**, 0.05 **S**	
QQ-C-525b	Composition 6 and 6X	US	Ingot	85.0–89.0 **Cu**, 7.8–9.0 **Sn**, 0.90 **Pb**, 3.0–5.0 **Zn**, 0.20 **Fe**, 0.25 **Sb**, 0.005 **Si**, 0.005 **Al**, 0.8 **Ni**, 0.03 **P**, 0.05 **S**	
QQ-C-525b	Composition 5	US	Ingot	86.0–89.0 **Cu**, 7.8–9.0 **Sn**, 0.25 **Pb**, 3.5–5.0 **Zn**, 0.20 **Fe**, 0.25 **Sb**, 0.005 **Si**, 0.005 **Al**, 0.8 **Ni**, 0.03 **P**, 0.05 **S**	
ANSI/ASTM E 310		US	Castings	86.0–89.0 **Cu**, 7.5–9.0 **Sn**, 0.03 **Pb**, 3.0–5.0 **Zn**, 0.15 **Fe**, 1.0 **Ni**, 0.05 **P**	**As cast**: all **diam**, 40 ksi (275 MPa) **TS**, 20% in 2 in. (51 mm) **El**
ANSI/ASTM B 584	C 90300	US	Sand castings	86.0–89.0 **Cu**, 7.5–9.0 **Sn**, 0.30 **Pb**, 3.0–5.0 **Zn**, 0.20 **Fe**, 0.20 **Sb**, 0.005 **Si**, 0.005 **Al**, 1.0 **Ni+Co**, 0.05 **P**, 0.05 **S**	**As cast**: all **diam**, 40 ksi (276 MPa) **TS**, 18 ksi (124 MPa) **YS**, 20% in 2 in. (50 mm) **El**
ANSI/ASTM B 505	CA 903	US	Castings	86.0–89.0 **Cu**, 7.5–9.0 **Sn**, 0.30 **Pb**, 3.0–5.0 **Zn**, 0.20 **Fe**, 0.20 **Sb**, 0.005 **Si**, 0.005 **Al**, 1.0 **Ni+Co**, 1.5 **P**, 0.05 **S**	**As cast**: all **diam**, 44 ksi (303 MPa) **TS**, 22 ksi (152 MPa) **YS**, 18% in 2 in. (50 mm) **El**
CDA C 90300		US	Castings	86.0–89.0 **Cu**, 7.5–9.0 **Sn**, 0.30 **Pb**, 3.0–5.0 **Zn**, 0.20 **Fe**, 0.20 **Sb**, 0.005 **Si**, 0.005 **Al**, 1.0 **Ni**, 0.05 **P**, 0.05 **S**	**Sand castings**: all **diam**, 40 ksi (276 MPa) **TS**, 18 ksi (124 MPa) **YS**, 20% in 2 in. (50 mm) **El**, 70 **HB**
COPANT 162	C 903 00	COPANT	Cast products	86–89 **Cu**, 7.5–9 **Sn**, 0.3 **Pb**, 3–5 **Zn**, 0.2 **Fe**, 0.2 **Sb**, 0.005 **Si**, 0.005 **Al**, 1.0 **Ni**, 0.05 **P**, 0.05 **S**	
COPANT 801	C 903 00	COPANT	Continuous-cast rod, bar, and tube	86–89 **Cu**, 7.5–9 **Sn**, 0.3 **Pb**, 3–5 **Zn**, 0.2 **Fe**, 0.2 **Sb**, 0.005 **Si**, 0.005 **Al**, 1.0 **Ni**, 0.05 **P**, 0.05 **S**	**As manufactured**: 127 mm **diam**, 300 MPa **TS**, 160 MPa **YS**, 18% in 2 in. (50 mm) **El**
QQ-C-390A	903	US	Castings	86.0–89.0 **Cu**, 7.5–9.0 **Sn**, 0.30 **Pb**, 3.0–5.0 **Zn**, 0.15 **Fe**, 1.0 **Ni**, 0.05 **P**	**As cast**: all **diam**, 40 ksi (276 MPa) **TS**, 18 ksi (124 MPa) **YS**, 20% in 2 in. (50 mm) **El**

CAST COPPER AND COPPER ALLOYS

specification number	designation	country	product forms	chemical composition	mechanical properties and hardness values
MIL-C-22229A	903	US	Castings	86.0–89.0 **Cu**, 7.5–9.0 **Sn**, 0.30 **Pb**, 3.0–5.0 **Zn**, 0.15 **Fe**, 1.0 **Ni**, 0.50 **P**	**As cast**: all **diam**, 276 MPa **TS**, 124 MPa **YS**, 20% in 2 in. (50 mm) **El**
CDA C 90200		US	Castings	91.0–94.0 **Cu**, 6.0–8.0 **Sn**, 0.30 **Pb**, 0.50 **Zn**, 0.20 **Fe**, 0.20 **Sb**, 0.005 **Si**, 0.005 **Al**, 0.50 **Ni**, 0.05 **P**, 0.05 **S**	
COPANT 162	C 902 00	COPANT	Cast products	91–94 **Cu**, 6–8 **Sn**, 0.50 **Zn**, 99.0 min, total elements	
ANSI/ASTM B 505	CA 928	US	Castings	78.0–82.0 **Cu**, 15.0–17.0 **Sn**, 4.0–6.0 **Pb**, 0.50 **Zn**, 0.20 **Fe**, 0.25 **Sb**, 0.005 **Si**, 0.005 **Al**, 0.50 **Ni + Co**, 1.5 **P**, 0.05 **S**	**As cast**: all **diam**, 72–82 **HRB**
AS 1565	930D	Australia	Ingot, castings	rem **Cu**, 9.0–11.0 **Sn**, 4.0–6.0 **Pb**, 1.0 **Zn**, 2.0 **Ni**, 0.10 **P**, 0.50 total, others (applies to castings only)	**Sand cast**: all **diam**, 190 MPa **TS**, 80 MPa **YS**, 5% in 2 in. (50 mm) **El**; **Chill cast**: all **diam**, 200 MPa **TS**, 140 MPa **YS**, 3% in 2 in. (50 mm) **El**; **Continuous cast**: all **diam**, 280 MPa **TS**, 160 MPa **YS**, 9% in 2 in. (50 mm) **El**
CDA C 92800		US	Castings	78.0–82.0 **Cu**, 15.0–17.0 **Sn**, 4.0–6.0 **Pb**, 0.50 **Zn**, 0.20 **Fe**, 0.25 **Sb**, 0.005 **Si**, 0.005 **Al**, 0.50 **Ni**, 0.05 **P**, 0.05 **S**	**Sand castings**: all **diam**, 40 ksi (276 MPa) **TS**, 30 ksi (207 MPa) **YS**, 1% in 2 in. (50 mm) **El**, 80 **HRB**
COPANT 162	C 928 00	COPANT	Cast products	78–82 **Cu**, 15–17 **Sn**, 4–6 **Pb**, 0.50 **Zn**, 0.50 **Ni**	
COPANT 801	C 928 00	COPANT	Continuous-cast rod, bar and tube	78–82 **Cu**, 15–17 **Sn**, 4–6 **Pb**, 0.50 **Zn**, 0.50 **Ni**	**As manufactured**: 127 mm **diam**, 72–82 **HRB**
IS 318	Grade 2	India	Ingot and castings	rem **Cu**, 4.0–6.0 **Sn**, 4.0–6.0 **Pb**, 4.0–6.0 **Zn**, 0.35 **Fe**, 0.3 **Sb**, 0.15 total, others, nickel included in copper	**Sand cast (separately cast)**: all **diam**, 186 MPa **TS**, 108 MPa **YS**, 12% in 2 in. (50 mm) **El**; **Sand cast (cast on)**: all **diam**, 186 MPa **TS**, 8% in 2 in. (50 mm) **El**; **Chill cast**: all **diam**, 201 MPa **TS**, 108 MPa **YS**, 7% in 2 in. (50 mm) **El**
IS 1458	Class V	India	Ingot and casting	4.0–6.0 **Sn**, 4.0–6.0 **Pb**, 4.0–6.0 **Zn**, 0.3 **Fe**, 0.3 **Sb**, 0.01 **Al**, 0.05 **P**, 0.5 **Fe + Sb**, rem **Cu + Ni**	**Sand cast (cast-on)**: all **diam**, 186 MPa **TS**, 8% in 2 in. (50 mm) **El**; **Sand cast (separately cast)**: all **diam**, 206 MPa **TS**, 12% in 2 in. (50 mm) **El**
JIS H 5115	Class 2	Japan	Castings	82.0–86.0 **Cu**, 9.0–11.0 **Sn**, 4.0–6.0 **Pb**, 1.0 **Zn**, 0.3 **Fe**, 1.0 **Ni**	**As cast**: all **diam**, 65 min **HB**
NF A 53 707	Cu Pb5 Sn5 Zn5	France	Castings	rem **Cu**, 4.0–6.0 **Sn**, 4.0–6.0 **Pb**, 1.5 **Ni**, 1.0 total, others	**As cast**: all **diam**, 200 MPa **TS**, 12% in 2 in. (50 mm) **El**

CAST COPPER AND COPPER ALLOYS

specification number	designation	country	product forms	chemical composition	mechanical properties and hardness values
SABS 200	Cu–Sn10Pb5	South Africa	Castings	rem **Cu**, 9.0–11.0 **Sn**, 4.0–6.0 **Pb**, 1.0 **Zn**, 2.0 **Ni**, 0.10 **P**, 0.50 total, others	**Continuously cast**: all **diam**, 280 MPa **TS**, 9% in 2 in. (50 mm) **El**
JIS H 2207	Class 2	Japan	Ingot: for castings	rem **Cu**, 9.0–11.0 **Sn**, 4.0–6.0 **Pb**, 1.0 **Zn**, 0.15 **Fe**, 0.1 **P**	
ANSI/ASTM B 30	928	US	Ingot: for castings	78.0–82.0 **Cu**, 15.3–17.0 **Sn**, 4.0–5.7 **Pb**, 0.8 **Zn**, 0.15 **Fe**, 0.20 **Sb**, 0.005 **Si**, 0.005 **Al**, 0.8 **Ni + Co**, 0.30 **P**, 0.05 **S**	
ANSI/ASTM B 30	CA 928	US	Ingot	78.0–82.0 **Cu**, 15.3–17.0 **Sn**, 4.0–5.7 **Pb**, 0.8 **Zn**, 0.15 **Fe**, 0.20 **Sb**, 0.005 **Si**, 0.005 **Al**, 0.8 **Ni**, 0.30 **P**, 0.05 **S**	
SIS 14 54 44	SIS TOMBAC 5444–03	Sweden	Sand castings	85–87 **Cu**, 8.3–10.0 **Sn**, 4 **Pb**, 1.0–2.5 **Zn**, 0.3 **Fe**, 0.3 **Sb**, 0.01 **Si**, 0.01 **Al**, 0.1 **Mn**, 0.75 **Ni**, 0.10 **S**, 0.15 **As**	**As cast**: all **diam**, 250 MPa **TS**, 150 MPa **YS**, 16% in 2 in. (50 mm) **El**, 75 **HB**
AS 1565	931D	Australia	Ingot, castings	rem **Cu**, 6.5–8.5 **Sn**, 2.0–5.0 **Pb**, 2.0 **Zn**, 1.0 **Ni**, 0.30 min **P**, 0.50 total, others (applies to castings only)	**Sand cast**: all **diam**, 190 MPa **TS**, 80 MPa **YS**, 3% in 2 in. (50 mm) **El**, 60 **HB**; **Chill cast**: all **diam**, 220 MPa **TS**, 130 MPa **YS**, 2% in 2 in. (50 mm) **El**, 85 **HB**; **Continuous cast**: all **diam**, 270 MPa **TS**, 130 MPa **YS**, 5% in 2 in. (50 mm) **El**, 85 **HB**
SABS 200	Cu–Sn8Pb4P	South Africa	Castings	rem **Cu**, 6.5–8.5 **Sn**, 2.0–5.0 **Pb**, 2.0 **Zn**, 1.0 **Ni**, 0.30 min **P**, 0.50 total, others	**Continuously cast**: all **diam**, 265 MPa **TS**, 5% in 2 in. (50 mm) **El**
AS 1565	924B	Australia	Ingot, castings	rem **Cu**, 6.0–8.0 **Sn**, 2.5–3.5 **Pb**, 1.5–3.0 **Zn**, 0.20 **Fe**, 0.25 **Sb**, 0.01 **Si**, 0.01 **Al**, 2.0 **Ni**, 0.05 **Bi**, 0.40 **Fe + As + Sb**, 0.70 total, others (castings)	**Sand cast**: all **diam**, 250 MPa **TS**, 130 MPa **YS**, 16% in 2 in. (50 mm) **El**, 70 **HB**; **Chill cast**: all **diam**, 250 MPa **TS**, 130 MPa **YS**, 5% in 2 in. (50 mm) **El**, 80 **HB**; **Continuous cast**: all **diam**, 300 MPa **TS**, 130 MPa **YS**, 13% in 2 in. (50 mm) **El**, 80 **HB**
SABS 200	Cu–Sn7Zn3Pb3	South Africa	Castings	rem **Cu**, 6.0–8.0 **Sn**, 2.5–3.5 **Pb**, 1.5–3.0 **Zn**, 0.20 **Fe**, 0.25 **Sb**, 0.01 **Si**, 0.01 **Al**, 2.0 **Ni**, 0.05 **Bi**, 0.18 total, others	**Continuously cast**: all **diam**, 300 MPa **TS**, 13% in 2 in. (50 mm) **El**
SIS 14 54 44	SIS TOMBAC 54 44–00	Sweden	Ingot	85–87 **Cu**, 8.5–10.0 **Sn**, 2–4 **Pb**, 1.5–2.5 **Zn**, 0.2 **Fe**, 0.3 **Sb**, 0.01 **Si**, 0.01 **Al**, 0.1 **Mn**, 0.75 **Ni**, 0.03 **P**, 0.06 **S**, 0.15 **As**, 0.01–1 **Al + Si**, total	
ANSI/ASTM B 505	CA 929	US	Castings	82.0–86.0 **Cu**, 9.0–11.0 **Sn**, 2.0–3.3 **Pb**, 0.15 **Fe**, 0.15 **Sb**, 0.005 **Si**, 0.005 **Al**, 2.8–4.0 **Ni + Co**, 1.5 **P**, 0.05 **S**	**As cast**: all **diam**, 45 ksi (316 MPa) **TS**, 25 ksi (176 MPa) **YS**, 8% in 2 in. (50 mm) **El**
COPANT 801	C 915 00	COPANT	Continuous–cast rod, bar and tube	82.0–86.0 **Cu**, 9.0–11.0 **Sn**, 2.0–3.3 **Pb**, 0.15 **Fe**, 0.15 **Sb**, 2.8–4.0 **Ni**, 1.5 **P**, 0.50 total, others	**As manufactured**: 127 mm **diam**, 310 MPa **TS**, 180 MPa **YS**, 8% in 2 in. (50 mm) **El**

CAST COPPER AND COPPER ALLOYS

specification number	designation	country	product forms	chemical composition	mechanical properties and hardness values
ANSI/ASTM B 427	CA 929	US	Castings	82.0–86.0 **Cu**, 9.0–11.0 **Sn**, 2.0–3.2 **Pb**, 0.25 **Zn**, 0.20 **Fe**, 0.25 **Sb**, 0.005 **Si**, 0.005 **Al**, 2.8–4.0 **Ni**, 0.50 **P**, 0.05 **S**, 99.5 min **Cu + Sn + Ni + Pb + P**	**Sand or chill cast**: all **diam**, 45 ksi (310 MPa) **TS**, 25 ksi (172 MPa) **YS**, 8% in 2 in. (50.8 mm) **El**, 75 **HB**
CDA C 92900		US	Castings	82.0–86.0 **Cu**, 9.0–11.0 **Sn**, 2.0–3.2 **Pb**, 0.25 **Zn**, 0.20 **Fe**, 0.25 **Sb**, 0.005 **Si**, 0.005 **Al**, 2.8–4.0 **Ni**, 0.50 **P**, 0.05 **S**	**Sand castings**: all **diam**, 45 ksi (310 MPa) **TS**, 25 ksi (172 MPa) **YS**, 8% in 2 in. (50 mm) **El**, 75 **HB**
COPANT 162	C 929 00	COPANT	Cast products	82–86 **Cu**, 9–11 **Sn**, 2–3.2 **Pb**, 0.25 **Zn**, 0.15 **Fe**, 0.10 **Sb**, 0.005 **Si**, 0.005 **Al**, 2.8–4.0 **Ni**, 0.5 **P**, 0.05 **S**	
NF A 53 707	Cu Sn12	France	Castings	rem **Cu**, 10.5–13.0 **Sn**, 2.5 **Pb**, 2.0 **Zn**, 2.0 **Ni**, 0.30 **P**, 0.5 total, others	**As cast**: 240 MPa **TS**, 5% in 2 in. (50 mm) **El**
ISO 1338	CuSn8Pb2	ISO	Castings, ingot	82.0–91.0 **Cu**, 6.0–9.0 **Sn**, 0.5–4.0 **Pb**, 3.0 **Zn**, 0.2 **Fe**, 0.25 **Sb**, 0.01 **Si**, 0.01 **Al**, 2.5 **Ni**, 0.05 **P**, 0.10 **S**, **Cu**, total includes **Ni**	**Sand cast**: 12–25 mm **diam**, 250 MPa **TS**, 130 MPa **YS**, 16% in 2 in. (50 mm) **El**; **Permanent mould cast**: 12–25 mm **diam**, 220 MPa **TS**, 130 MPa **YS**, 2% in 2 in. (50 mm) **El**; **Centrifugally**: 12–25 mm **diam**, 230 MPa **TS**, 130 MPa **YS**, 4% in 2 in. (50 mm) **El**
AS 1565	922C	Australia	Ingot, castings	rem **Cu**, 6.0–8.0 **Sn**, 1.0–3.0 **Pb**, 3.0–5.0 **Zn**, 0.02 **Si**, 0.01 **Al**, 2.0 **Ni**, 0.05 **Bi**, 0.50 **Fe + As + Sb** (castings)	**Sand cast**: all **diam**, 220 MPa **TS**, 100 MPa **YS**, 12% in 2 in. (50 mm) **El**; **Chill cast**: all **diam**, 230 MPa **TS**, 115 MPa **YS**, 5% in 2 in. (50 mm) **El**; **Continuous cast**: all **diam**, 280 MPa **TS**, 115 MPa **YS**, 12% in 2 in. (50 mm) **El**
IS 318	Grade 1	India	Ingot and castings	rem **Cu**, 6.0–8.0 **Sn**, 1.0–3.0 **Pb**, 4.0–6.0 **Zn**, 0.35 **Fe**, 0.3 **Sb**, 0.15 total, others, nickel included in copper	**Sand cast (separately cast)**: all **diam**, 216 MPa **TS**, 113 MPa **YS**, 12% in 2 in. (50 mm) **El**; **Sand cast (cast on)**: all **diam**, 216 MPa **TS**, 8% in 2 in. (50 mm) **El**; **Chill cast**: all **diam**, 230 MPa **TS**, 113 MPa **YS**, 5% in 2 in. (50 mm) **El**
IS 1458	Class II	India	Ingot and casting	5.0–7.0 **Sn**, 1.0–3.0 **Pb**, 2.0–3.0 **Zn**, 0.3 **Fe**, 0.1 **Sb**, 0.01 **Al**, 0.05 **P**, rem **Cu + Ni**, 0.2 total, others	**Sand cast (cast–on)**: all **diam**, 196 MPa **TS**, 8% in 2 in. (50 mm) **El**; **Sand cast (separately cast)**: all **diam**, 216 MPa **TS**, 12% in 2 in. (50 mm) **El**
ANSI/ASTM B 505	CA 927	US	Castings	86.0–89.0 **Cu**, 9.0–11.0 **Sn**, 1.0–2.5 **Pb**, 0.7 **Zn**, 0.20 **Fe**, 0.25 **Sb**, 0.005 **Si**, 0.005 **Al**, 1.0 **Ni + Co**, 1.5 **P**, 0.05 **S**	**As cast**: all **diam**, 38 ksi (252 MPa) **TS**, 20 ksi (138 MPa) **YS**, 8% in 2 in. (50 mm) **El**
CDA C 92700		US	Castings	86.0–89.0 **Cu**, 9.0–11.0 **Sn**, 1.0–2.5 **Pb**, 0.7 **Zn**, 0.20 **Fe**, 0.25 **Sb**, 0.005 **Si**, 0.005 **Al**, 1.0 **Ni**, 0.25 **P**, 0.05 **S**	**Sand castings**: all **diam**, 35 ksi (241 MPa) **TS**, 21 ksi (145 MPa) **YS**, 10% in 2 in. (50 mm) **El**, 77 **HB**

CAST COPPER AND COPPER ALLOYS

specification number	designation	country	product forms	chemical composition	mechanical properties and hardness values
COPANT 162	C 924 00	COPANT	Cast products	86–89 **Cu**, 9–11 **Sn**, 1–2.5 **Pb**, 1–3 **Zn**	
COPANT 162	C 927 00	COPANT	Cast products	86–89 **Cu**, 9–11 **Sn**, 1–2.5 **Pb**, 0.7 **Zn**, 0.15 **Fe**, 0.005 **Al**, 1.0 **Ni**, 0.25 **P**	
COPANT 801	C 927 00	COPANT	Continuous–cast rod, bar and tube	86–89 **Cu**, 9–11 **Sn**, 1–2.5 **Pb**, 0.7 **Zn**, 0.15 **Fe**, 0.005 **Al**, 1.0 **Ni**, 0.25 **P**	**As manufactured**: 127 mm **diam**, 270 MPa **TS**, 140 MPa **YS**, 8% in 2 in. (50 mm) **El**
QQ–C–390A	927	US	Castings	86.0–89.0 **Cu**, 9.0–11.0 **Sn**, 1.0–2.5 **Pb**, 0.7 **Zn**, 0.15 **Fe**, 0.005 **Al**, 1.0 **Ni**, 0.25 **P**	**As cast**: all **diam**, 38 ksi (262 MPa) **TS**, 20 ksi (138 MPa) **YS**, 8% in 2 in. (50 mm) **El**
NF A 53 707	Cu Sn8	France	Castings	rem **Cu**, 7.0–9.0 **Sn**, 0.5–3.0 **Pb**, 3.0 **Zn**, 1.5 **Ni**, 1.0 total, others	**As cast**: all **diam**, 250 MPa **TS**, 16% in 2 in. (50 mm) **El**
ISO 1338	CuSn12Pb2	ISO	Castings, ingot	84.0–87.5 **Cu**, 11.0–13.0 **Sn**, 1.0–2.5 **Pb**, 2.0 **Zn**, 0.20 **Fe**, 0.2 **Sb**, 0.01 **Si**, 0.01 **Al**, 0.2 **Mn**, 2.0 **Ni**, 0.05–0.40 **P**, 0.05 **S**, **Cu** total includes **Ni**	**Sand cast**: 12–25 mm **diam**, 240 MPa **TS**, 130 MPa **YS**, 7% in 2 in. (50 mm) **El**; **Centrifugally cast**: 12–25 mm **diam**, 280 MPa **TS**, 150 MPa **YS**, 5% in 2 in. (50 mm) **El**; **Continuously cast**: 12–25 mm **diam**, 280 MPa **TS**, 150 MPa **YS**, 7% in 2 in. (50 mm) **El**
ANSI/ASTM B 30	927	US	Ingot: for castings	86.0–89.0 **Cu**, 9.3–11.0 **Sn**, 1.0–2.3 **Pb**, 0.8 **Zn**, 0.15 **Fe**, 0.20 **Sb**, 0.005 **Si**, 0.005 **Al**, 0.8 **Ni+Co**, 0.30 **P**, 0.05 **S**	
ANSI/ASTM B 30	CA 927	US	Ingot	86.0–89.0 **Cu**, 9.3–11.0 **Sn**, 1.0–2.3 **Pb**, 0.8 **Zn**, 0.15 **Fe**, 0.20 **Sb**, 0.005 **Si**, 0.005 **Al**, 0.8 **Ni**, 0.30 **P**, 0.05 **S**	
ANSI/ASTM B 584	C 92200	US	Sand castings	86.0–90.0 **Cu**, 5.5–6.5 **Sn**, 1.0–2.0 **Pb**, 3.0–5.0 **Zn**, 0.25 **Fe**, 0.25 **Sb**, 0.005 **Si**, 0.005 **Al**, 1.0 **Ni+Co**, 0.05 **P**, 0.05 **S**	**As cast**: all **diam**, 34 ksi (234 MPa) **TS**, 16 ksi (110 MPa) **YS**, 24% in 2 in. (50 mm) **El**
ANSI/ASTM B 505	CA 922	US	Castings	86.0–90.0 **Cu**, 5.5–6.5 **Sn**, 1.0–2.0 **Pb**, 3.0–5.0 **Zn**, 0.25 **Fe**, 0.25 **Sb**, 0.005 **Si**, 0.005 **Al**, 1.0 **Ni+Co**, 1.5 **P**, 0.05 **S**	**As cast**: all **diam**, 38 ksi (252 MPa) **TS**, 19 ksi (131 MPa) **YS**, 18% in 2 in. (50 mm) **El**
ANSI/ASTM B 61	CA 922	US	Castings	86.0–90.0 **Cu**, 5.5–6.5 **Sn**, 1.0–2.0 **Pb**, 3.0–5.0 **Zn**, 0.25 **Fe**, 1.0 **Ni**, 0.05 **P**	**As cast**: all **diam**, 34 ksi (235 MPa) **TS**, 16 ksi (110 MPa) **YS**, 22% in 2 in. (50.8 mm) **El**
AS 1565	903C	Australia	Ingot, castings	rem **Cu**, 7.5–8.5 **Sn**, 1.5 **Pb**, 3.50–4.50 **Zn**, 0.02 **Si**, 0.01 **Al**, 1.0 **Ni**, 0.03 **Bi**, 0.20 **Fe+As+Sb** (castings)	**Sand cast**: all **diam**, 250 MPa **TS**, 115 MPa **YS**, 15% in 2 in. (50 mm) **El**; **Chill cast**: all **diam**, 220 MPa **TS**, 120 MPa **YS**, 3% in 2 in. (50 mm) **El**; **Continuous cast**: all **diam**, 280 MPa **TS**, 140 MPa **YS**, 10% in 2 in. (50 mm) **El**

CAST COPPER AND COPPER ALLOYS

specification number	designation	country	product forms	chemical composition	mechanical properties and hardness values
AS 1565	905C	Australia	Ingot, castings	rem **Cu**, 9.5–10.5 **Sn**, 1.5 **Pb**, 1.75–2.75 **Zn**, 0.15 **Fe**, 0.02 **Si**, 0.01 **Al**, 1.0 **Ni**, 0.03 **Bi**, 0.20 **Fe**+**As**+**Sb**, (castings)	**Sand cast**: all **diam**, 270 MPa **TS**, 130 MPa **YS**, 13% in 2 in. (50 mm) **El**, 70 **HB**; **Chill cast**: all **diam**, 230 MPa **TS**, 130 MPa **YS**, 3% in 2 in. (50 mm) **El**, 85 **HB**; **Continuous**: all **diam**, 300 MPa **TS**, 140 MPa **YS**, 9% in 2 in. (50 mm) **El**, 90 **HB**
CDA C 92200		US	Castings	86.0–90.0 **Cu**, 5.5–6.5 **Sn**, 1.0–2.0 **Pb**, 3.0–5.0 **Zn**, 0.25 **Fe**, 0.25 **Sb**, 0.005 **Si**, 0.005 **Al**, 1.0 **Ni**, 0.05 **P**, 0.05 **S**	**Sand castings**: all **diam**, 34 ksi (234 MPa) **TS**, 16 ksi (110 MPa) **YS**, 24% in 2 in. (50 mm) **El**, 65 **HB**
COPANT 162	C 922 00	COPANT	Cast products	86–90 **Cu**, 5.5–6.5 **Sn**, 1–2 **Pb**, 3–5 **Zn**, 0.25 **Fe**, 0.20 **Sb**, 0.005 **Si**, 0.005 **Al**, 1.0 **Ni**, 0.05 **P**, 0.05 **S**	
COPANT 801	C 922 00	COPANT	Continuous cast rod, bar and tube	86–90 **Cu**, 5.5–6.5 **Sn**, 1–2 **Pb**, 3–5 **Zn**, 0.25 **Fe**, 0.20 **Sb**, 0.005 **Si**, 0.005 **Al**, 1.0 **Ni**, 0.05 **P**, 0.05 **S**	**As manufactured**: 127 mm **diam**, 270 MPa **TS**, 130 MPa **YS**, 18% in 2 in. (50 mm) **El**
DS 3001	Tin Bronze 5458	Denmark	Sand castings	86.0–89.0 **Cu**, 9.0–11.0 **Sn**, 1.5 **Pb**, 1.0–3.0 **Zn**, 0.25 **Fe**, 0.3 **Sb**, 0.01 **Si**, 0.01 **Al**, 0.2 **Mn**, 2.0 **Ni**, 0.05 **P**, 0.10 **S**	**As cast**: all **diam**, 260 MPa **TS**, 120 MPa **YS**, 15% in 2 in. (50 mm) **El**, 75 min **HB**
QQ-C-390A	922	US	Castings	86.0–90.0 **Cu**, 5.5–6.5 **Sn**, 1.0–2.0 **Pb**, 3.0–5.0 **Zn**, 0.25 **Fe**, 1.0 **Ni**, 0.05 **P**	**As cast**: all **diam**, 34 ksi (234 MPa) **TS**, 16 ksi (110 MPa) **YS**, 22% in 2 in. (50 mm) **El**
IS 306		India	Ingot and castings	85.23 min **Cu**, 7.5–8.5 **Sn**, 1.5 **Pb**, 3.5–4.5 **Zn**, 0.02 **Si**, 0.01 **Al**, 0.03 **Bi**, 0.20 total, others, nickel included in copper content	**Sand cast (cast-on)**: all **diam**, 216 MPa **TS**, 8% in 2 in. (50 mm) **El**; **Sand cast (separately-cast)**: all **diam**, 225 MPa **TS**, 12% in 2 in. (50 mm) **El**; **Chill cast**: all **diam**, 245 MPa **TS**
ISO 1338	CuSn10Zn2	ISO	Castings, ingot	86.0–89.0 **Cu**, 9.0–11.0 **Sn**, 1.5 **Pb**, 1.0–3.0 **Zn**, 0.25 **Fe**, 0.3 **Sb**, 0.01 **Si**, 0.01 **Al**, 0.2 **Mn**, 2.0 **Ni**, 0.05 **P**, 0.10 **S**, **Cu**, total includes iron	**Sand cast**: 12–25 mm **diam**, 240 MPa **TS**, 120 MPa **YS**, 12% in 2 in. (50 mm) **El**; **Centrifugally or continuously cast**: 12–25 mm **diam**, 270 MPa **TS**, 140 MPa **YS**, 7% in 2 in. (50 mm) **El**
SABS 200	Cu-Sn10Zn2	South Africa	Castings	rem **Cu**, 9.5–10.5 **Sn**, 1.5 **Pb**, 1.5–2.5 **Zn**, 0.15 **Fe**, 0.02 **Si**, 0.01 **Al**, 1.0 **Ni**, 0.02 **P**, 0.03 **Bi**, 0.05 **As**+**Sb**	**Continuously cast**: all **diam**, 295 MPa **TS**, 9% in 2 in. (50 mm) **El**
SIS 14 54 58	SIS Bronze 54 58-03	Sweden	Sand castings	86.0–89.0 **Cu**, 9.0–11.0 **Sn**, 1.5 **Pb**, 1.0–3.0 **Zn**, 0.25 **Fe**, 0.3 **Sb**, 0.01 **Si**, 0.01 **Al**, 0.2 **Mn**, 2.0 **Ni**, 0.05 **P**, 0.10 **S**	**As cast**: all **diam**, 260 MPa **TS**, 120 MPa **YS**, 15% in 2 in. (50 mm) **El**, 75 **HB**
SIS 14 54 58	SIS Bronze 54 58-15	Sweden	Castings: centrifugal and continuous	86.0–89.0 **Cu**, 9.0–11.0 **Sn**, 1.5 **Pb**, 1.0–3.0 **Zn**, 0.25 **Fe**, 0.3 **Sb**, 0.01 **Si**, 0.01 **Al**, 0.2 **Mn**, 2.0 **Ni**, 0.05 **P**, 0.10 **S**	**As cast**: all **diam**, 270 MPa **TS**, 140 MPa **YS**, 7% in 2 in. (50 mm) **El**, 80 **HB**

CAST COPPER AND COPPER ALLOYS

specification number	designation	country	product forms	chemical composition	mechanical properties and hardness values
MIL-B-16541B		US	Castings	86.0–89.0 **Cu**, 5.5–6.5 **Sn**, 1.0–2.0 **Pb**, 3.0–5.0 **Zn**, 0.25 **Fe**, 1.00 **Ni**, 0.05 **P**	**As cast**: all **diam**, 234 MPa **TS**, 22% in 2 in. (50 mm) **El**
ANSI/ASTM B 30	922	US	Ingot: for castings	86.0–89.0 **Cu**, 5.8–6.5 **Sn**, 1.0–1.8 **Pb**, 3.5–5.0 **Zn**, 0.20 **Fe**, 0.20 **Sb**, 0.005 **Si**, 0.005 **Al**, 0.8 **Ni + Co**, 0.03 **P**, 0.05 **S**	
ANSI/ASTM B 30	CA 922	US	Ingot	86.0–89.0 **Cu**, 5.8–6.5 **Sn**, 1.0–1.8 **Pb**, 3.5–5.0 **Zn**, 0.20 **Fe**, 0.20 **Sb**, 0.005 **Si**, 0.005 **Al**, 0.8 **Ni**, 0.03 **P**, 0.05 **S**	**As cast**: all **diam**, 34 ksi (235 MPa) **TS**, 16 ksi (110 MPa) **YS**, 22% in 2 in. (50.8 mm) **El**
QQ-C-525b	Composition 1	US	Ingot	86.0–89.0 **Cu**, 5.8–6.5 **Sn**, 1.0–1.8 **Pb**, 3.5–5.0 **Zn**, 0.20 **Fe**, 0.20 **Sb**, 0.005 **Si**, 0.005 **Al**, 0.8 **Ni**, 0.03 **P**, 0.05 **S**	
SIS 14 54 58	SIS Bronze 54 58–00	Sweden	Ingot	86.0–88.5 **Cu**, 9.3–11.0 **Sn**, 1.3 **Pb**, 1.0–3.0 **Zn**, 0.20 **Fe**, 0.3 **Sb**, 0.01 **Si**, 0.01 **Al**, 0.2 **Mn**, 2.0 **Ni**, 0.03 **P**, 0.10 **S**	
ANSI/ASTM B 30	925	US	Ingot: for castings	85.0–88.0 **Cu**, 10.3–12.0 **Sn**, 1.0–1.5 **Pb**, 0.50 **Zn**, 0.20 **Fe**, 0.20 **Sb**, 0.005 **Si**, 0.005 **Al**, 0.8–1.5 **Ni + Co**, 0.30 **P**, 0.05 **S**	
ANSI/ASTM B 505	CA 925	US	Castings	85.0–88.0 **Cu**, 10.0–12.0 **Sn**, 1.0–1.5 **Pb**, 0.50 **Zn**, 0.30 **Fe**, 0.25 **Sb**, 0.005 **Si**, 0.005 **Al**, 0.8–1.5 **Ni + Co**, 1.5 **P**, 0.05 **S**	**As cast**: all **diam**, 40 ksi (270 MPa) **TS**, 24 ksi (165 MPa) **YS**, 10% in 2 in. (50 mm) **El**
ANSI/ASTM B 30	CA 925	US	Ingot	85.0–88.0 **Cu**, 10.3–12.0 **Sn**, 1.0–1.5 **Pb**, 0.50 **Zn**, 0.20 **Fe**, 0.20 **Sb**, 0.005 **Si**, 0.005 **Al**, 0.8–1.5 **Ni**, 0.30 **P**, 0.05 **S**	
CDA C 92500		US	Castings	85.0–88.0 **Cu**, 10.0–12.0 **Sn**, 1.0–1.5 **Pb**, 0.50 **Zn**, 0.30 **Fe**, 0.25 **Sb**, 0.005 **Si**, 0.005 **Al**, 0.8–1.5 **Ni**, 0.30 **P**, 0.05 **S**	**Sand castings**: all **diam**, 35 ksi (241 MPa) **TS**, 20 ksi (138 MPa) **YS**, 10% in 2 in. (50 mm) **El**, 80 **HB**
COPANT 162	C 925 00	COPANT	Cast products	85–88 **Cu**, 10–12 **Sn**, 1.0–1.5 **Pb**, 0.5 **Zn**, 0.30 **Fe**, 0.005 **Al**, 0.8–1.5 **Ni**, 0.20–0.30 **P**	
COPANT 801	C 925 00	COPANT	Continuous-cast rod, bar and tube	85–88 **Cu**, 10–12 **Sn**, 1.0–1.5 **Pb**, 0.5 **Zn**, 0.30 **Fe**, 0.005 **Al**, 0.8–1.5 **Ni**, 0.20–0.30 **P**	**As manufactured**: 127 mm **diam**, 280 MPa **TS**, 170 MPa **YS**, 10% in 2 in. (50 mm) **El**
QQ-C-390A	925	US	Castings	85.0–88.0 **Cu**, 10.0–12.0 **Sn**, 1.0–1.5 **Pb**, 0.50 **Zn**, 0.30 **Fe**, 0.005 **Al**, 0.8–1.5 **Ni**, 0.20–0.30 **P**	**As cast**: all **diam**, 40 ksi (276 MPa) **TS**, 24 ksi (165 MPa) **YS**, 10% in 2 in. (50 mm) **El**
CDA C 92600		US	Castings	86.0–88.5 **Cu**, 9.3–10.5 **Sn**, 0.8–1.2 **Pb**, 1.3–2.5 **Zn**, 0.20 **Fe**, 0.25 **Sb**, 0.005 **Si**, 0.005 **Al**, 0.7 **Ni**, 0.03 **P**, 0.05 **S**	**Sand castings**: all **diam**, 40 ksi (276 MPa) **TS**, 18 ksi (124 MPa) **YS**, 20% in 2 in. (50 mm) **El**, 78 **HRF**
COPANT 162	C 926 00	COPANT	Cast products	86–88.5 **Cu**, 9.3–10.5 **Sn**, 0.8–1.2 **Pb**, 1.3–2.5 **Zn**, 0.15 **Fe**, 0.7 **Ni**, 0.03 **P**	

CAST COPPER AND COPPER ALLOYS

specification number	designation	country	product forms	chemical composition	mechanical properties and hardness values
DS 3001	Tin Bronze 5475	Denmark	Sand castings	84.0–86.5 **Cu**, 13.0–15.0 **Sn**, 1.0 **Pb**, 0.20 **Fe**, 0.2 **Sb**, 0.01 **Si**, 0.01 **Al**, 0.2 **Mn**, 1.0 **Ni**, 0.4 **P**, 0.05 **S**	**As cast**: all **diam**, 250 MPa **TS**, 170 MPa **YS**, 5% in 2 in. (50 mm) **El**, 115 min **HB**
DS 3001	Tin bronze 5465	Denmark	Sand castings	85.0–88.5 **Cu**, 11.0–13.0 **Sn**, 1.0 **Pb**, 0.5 **Zn**, 0.25 **Fe**, 0.2 **Sb**, 0.01 **Si**, 0.01 **Al**, 0.2 **Mn**, 2.0 **Ni**, 0.40 **P**, 0.05 **S**	**As cast**: all **diam**, 280 MPa **TS**, 160 MPa **YS**, 12% in 2 in. (50 mm) **El**, 95 min **HB**
QQ-C-390A	923	US	Castings	85.0–89.0 **Cu**, 7.5–9.0 **Sn**, 1.0 **Pb**, 2.5–5.0 **Zn**, 0.25 **Fe**, 1.0 **Ni**, 0.05 **P**	**As cast**: all **diam**, 36 ksi (248 MPa) **TS**, 16 ksi (110 MPa) **YS**, 18% in 2 in. (50 mm) **El**
SFS 2214	CuSn12	Finland	Sand castings	85.0–88.5 **Cu**, 11.0–13.0 **Sn**, 1.0 **Pb**, 0.25 **Fe**, 0.2 **Sb**, 0.01 **Si**, 0.01 **Al**, 0.2 **Mn**, 2.0 **Ni**, 0.05–0.40 **P**, 0.05 **S**	**As cast**: all **diam**, 280 MPa **TS**, 160 MPa **YS**, 12% in 2 in. (50 mm) **El**
ISO 1338	CuSn10	ISO	Castings, ingot	88.0–91.0 **Cu**, 9.0–11.0 **Sn**, 1.0 **Pb**, 0.5 **Zn**, 0.20 **Fe**, 0.2 **Sb**, 0.01 **Si**, 0.01 **Al**, 0.2 **Mn**, 2.0 **Ni**, 0.20 **P**, 0.05 **S**, **Cu**, total includes **Ni**	**Sand cast**: 12–25 mm **diam**, 240 MPa **TS**, 130 MPa **YS**, 12% in 2 in. (50 mm) **El**
ISO 1338	CuSn12	ISO	Castings, ingot	85.0–88.5 **Cu**, 10.5–13.0 **Sn**, 1.0 **Pb**, 2.0 **Zn**, 0.25 **Fe**, 0.2 **Sb**, 0.01 **Si**, 0.01 **Al**, 0.2 **Mn**, 2.0 **Ni**, 0.05–0.40 **P**, 0.05 **S**, **Cu**, total includes **Ni**	**Sand cast**: 12–25 mm **diam**, 240 MPa **TS**, 130 MPa **YS**, 7% in 2 in. (50 mm) **El**; **Permanent mould cast**: 12–25 mm **diam**, 270 MPa **TS**, 150 MPa **YS**, 5% in 2 in. (50 mm) **El**; **Continuously or centrifugally cast**: 12–25 mm **diam**, 270 MPa **TS**, 150 MPa **YS**, 5% in 2 in. (50 mm) **El**
SIS 14 54 75	SIS Bronze 54–75–03	Sweden	Sand castings	86 **Cu**, 13.0–15.0 **Sn**, 1.0 **Pb**, 0.5 **Zn**, 0.20 **Fe**, 0.2 **Sb**, 0.01 **Si**, 0.01 **Al**, 0.2 **Mn**, 1.0 **Ni**, 0.4 **P**, 0.05 **S**, 0.15 **As**, 99.0 **Cu+Sn+P**	**As cast**: all **diam**, 200 MPa **TS**, 140 MPa **YS**, 3% in 2 in. (50 mm) **El**, 100 **HB**
SIS 14 54 65	SIS Bronze 54 65–15	Sweden	Castings: centrifugal or continuous	85.0–88.5 **Cu**, 11.0–13.0 **Sn**, 1.0 **Pb**, 0.5 **Zn**, 0.25 **Fe**, 0.2 **Sb**, 0.01 **Si**, 0.01 **Al**, 0.2 **Mn**, 2.0 **Ni**, 0.05–0.40 **P**, 0.05 **S**	**As cast**: all **diam**, 300 MPa **TS**, 180 MPa **YS**, 8% in 2 in. (50 mm) **El**, 105 **HB**
SIS 14 54 65	SIS Bronze 54 65–06	Sweden	Chill castings	85.0–88.5 **Cu**, 11.0–13.0 **Sn**, 1.0 **Pb**, 0.5 **Zn**, 0.25 **Fe**, 0.2 **Sb**, 0.01 **Si**, 0.01 **Al**, 0.2 **Mn**, 0.05–0.40 **P**, 0.05 **S**	**As cast**: all **diam**, 280 MPa **TS**, 160 MPa **YS**, 12% in 2 in. (50 mm) **El**, 95 **HB**
SIS 14 54 65	SIS Bronze 54 65–03	Sweden	Sand castings	85.0–88.5 **Cu**, 11.0–13.0 **Sn**, 1.0 **Pb**, 0.5 **Zn**, 0.25 **Fe**, 0.2 **Sb**, 0.01 **Si**, 0.01 **Al**, 0.2 **Mn**, 2.0 **Ni**, 0.05–0.40 **P**, 0.05 **S**	**As cast**: all **diam**, 280 MPa **TS**, 160 MPa **YS**, 12% in 2 in. (50 mm) **El**, 95 **HB**
DS 3001	Tin Bronze 5443	Denmark	Sand castings	88.0–91.0 **Cu**, 9.0–11.0 **Sn**, 0.8 **Pb**, 0.5 **Zn**, 0.20 **Fe**, 0.2 **Sb**, 0.01 **Si**, 0.01 **Al**, 0.2 **Mn**, 2.0 **Ni**, 0.20 **P**, 0.05 **S**	**As cast**: all **diam**, 270 MPa **TS**, 130 MPa **YS**, 18% in 2 in. (50 mm) **El**, 70 min **HB**
SIS 14 54 75	SIS Bronze 54 75–00	Sweden	Ingot	86 **Cu**, 13.5–15.0 **Sn**, 0.8 **Pb**, 0.5 **Zn**, 0.15 **Fe**, 0.2 **Sb**, 0.01 **Si**, 0.01 **Al**, 0.2 **Mn**, 0.8 **Ni**, 0.05 **P**, 0.04 **S**, 0.10 **As**, 99.0 min **Cu+Sn+P**	

CAST COPPER AND COPPER ALLOYS

specification number	designation	country	product forms	chemical composition	mechanical properties and hardness values
SIS 14 54 65	SIS Bronze 54 65–00	Sweden	Ingot	85.5–88.3 **Cu**, 11.3–13.0 **Sn**, 0.8 **Pb**, 0.5 **Zn**, 0.15 **Fe**, 0.2 **Sb**, 0.01 **Si**, 0.01 **Al**, 0.2 **Mn**, 1.8 **Ni**, 0.05 **P**, 0.05 **S**	
SIS 14 54 43	SIS Bronze 54 43–00	Sweden	Ingot	88.0–90.0 **Cu**, 9.3–11.0 **Sn**, 0.8 **Pb**, 0.5 **Zn**, 0.15 **Fe**, 0.2 **Sb**, 0.01 **Si**, 0.01 **Al**, 0.2 **Mn**, 2.0 **Ni**, 0.05 **P**, 0.05 **S**	
SIS 14 54 43	SIS Bronze 54 43–03	Sweden	Sand castings	88.0–91.0 **Cu**, 9.0–11.0 **Sn**, 0.8 **Pb**, 0.5 **Zn**, 0.20 **Fe**, 0.2 **Sb**, 0.01 **Si**, 0.01 **Al**, 0.2 **Mn**, 2.0 **Ni**, 0.20 **P**, 0.05 **S**	**As cast**: all **diam**, 270 MPa **TS**, 130 MPa **YS**, 18% in 2 in. (50 mm) **El**, 70 **HB**
SIS 14 54 43	SIS Bronze 54 43–15	Sweden	Castings: continuous and centrifugal	88.0–91.0 **Cu**, 9.0–11.0 **Sn**, 0.8 **Pb**, 0.5 **Zn**, 0.20 **Fe**, 0.2 **Sb**, 0.01 **Si**, 0.01 **Al**, 0.2 **Mn**, 2.0 **Ni**, 0.20 **P**, 0.05 **S**	**As cast**: all **diam**, 280 MPa **TS**, 130 MPa **YS**, 15% in 2 in. (50 mm) **El**, 80 **HB**
AS 1565	908C	Australia	Ingot, castings	rem **Cu**, 9.5 **Sn**, 0.75 **Pb**, 0.50 **Zn**, 0.50 **Ni**, 0.40 **P**, 0.50 total, others (applies to castings only)	**Sand cast**: all **diam**, 190 MPa **TS**, 100 MPa **YS**, 3% in 2 in. (50 mm) **El**, 70 **HB**; **Chill cast**: all **diam**, 270 MPa **TS**, 140 MPa **YS**, 2% in 2 in. (50 mm) **El**, 95 **HB**; **Continuous cast**: all **diam**, 330 MPa **TS**, 160 MPa **YS**, 7% in 2 in. (50 mm) **El**, 95 **HB**
SABS 200	CuSn10P 2C	South Africa	Castings	rem **Cu**, 9.5 min **Sn**, 0.75 **Pb**, 0.5 **Zn**, 0.5 **Ni**, 0.4 min **P**, 0.50 total, others	**Continuously cast**: all **diam**, 325 MPa **TS**, 7% in 2 in. (50 mm) **El**
ANSI/ASTM B 584	C 92300	US	Sand castings	85.0–89.0 **Cu**, 7.5–9.0 **Sn**, 0.30–1.0 **Pb**, 2.5–5.0 **Zn**, 0.25 **Fe**, 0.25 **Sb**, 0.005 **Si**, 0.005 **Al**, 1.0 **Ni+Co**, 0.05 **P**, 0.05 **S**	**As cast**: all **diam**, 36 ksi (248 MPa) **TS**, 16 ksi (110 MPa) **YS**, 18% in 2 in. (50 mm) **El**
ANSI/ASTM B 505	CA 923	US	Castings	85.0–89.0 **Cu**, 7.5–9.0 **Sn**, 0.3–1.0 **Pb**, 2.5–5.0 **Zn**, 0.25 **Fe**, 0.25 **Sb**, 0.005 **Si**, 0.005 **Al**, 1.0 **Ni+Co**, 1.5 **P**, 0.05 **S**	**As cast**: all **diam**, 40 ksi (270 MPa) **TS**, 19 ksi (131 MPa) **YS**, 16% in 2 in. (50 mm) **El**
CDA C 92300		US	Castings	85.0–89.0 **Cu**, 7.5–9.0 **Sn**, 0.30–1.0 **Pb**, 2.5–5.0 **Zn**, 0.25 **Fe**, 0.25 **Sb**, 0.005 **Si**, 0.005 **Al**, 1.0 **Ni**, 0.05 **P**, 0.05 **S**	**Sand castings**: all **diam**, 36 ksi (248 MPa) **TS**, 16 ksi (110 MPa) **YS**, 18% in 2 in. (50 mm) **El**, 70 **HB**
COPANT 162	C 923 00	COPANT	Cast products	85–89 **Cu**, 7.5–9.0 **Sn**, 0.3–1.0 **Pb**, 2.5–5.0 **Zn**, 0.25 **Fe**, 0.20 **Sb**, 0.005 **Si**, 0.005 **Al**, 1.0 **Ni**, 0.05 **P**, 0.05 **S**	
COPANT 801	C 923 00	COPANT	Continuous–cast rod, bar and tube	85–89 **Cu**, 7.5–9.0 **Sn**, 0.3–1.0 **Pb**, 2.5–5.0 **Zn**, 0.25 **Fe**, 0.20 **Sb**, 0.005 **Si**, 0.005 **Al**, 1.0 **Ni**, 0.05 **P**, 0.05 **S**	**As manufactured**: 127 mm **diam**, 280 MPa **TS**, 130 MPa **YS**, 16% in 2 in. (50 mm) **El**
ANSI/ASTM B 30	923	US	Ingot: for castings	85.0–89.0 **Cu**, 7.8–9.0 **Sn**, 0.30–0.9 **Pb**, 3.0–5.0 **Zn**, 0.20 **Fe**, 0.20 **Sb**, 0.005 **Si**, 0.005 **Al**, 0.8 **Ni+Co**, 0.03 **P**, 0.05 **S**	

CAST COPPER AND COPPER ALLOYS

specification number	designation	country	product forms	chemical composition	mechanical properties and hardness values
ANSI/ASTM B 30	CA 923	US	Ingot	85.0–89.0 **Cu**, 7.8–9.0 **Sn**, 0.30–0.9 **Pb**, 3.0–5.0 **Zn**, 0.20 **Fe**, 0.20 **Sb**, 0.005 **Si**, 0.005 **Al**, 0.8 **Ni**, 0.03 **P**, 0.05 **S**	
AS 1565	907C	Australia	Ingot, castings	rem **Cu**, 11.0–13.0 **Sn**, 0.50 **Pb**, 0.30 **Zn**, 0.15 **Fe**, 0.02 **Si**, 0.01 **Al**, 0.50 **Ni**, 0.15 **P**, 0.20 total, others, (applies to castings only)	**Sand cast**: all **diam**, 220 MPa **TS**, 130 MPa **YS**, 5% in 2 in. (50 mm) **El**, 75 **HB**; **Chill cast**: all **diam**, 270 MPa **TS**, 170 MPa **YS**, 3% in 2 in. (50 mm) **El**, 100 **HB**; **Continuous cast**: all **diam**, 270 MPa **TS**, 170 MPa **YS**, 5% in 2 in. (50 mm) **El**, 100 **HB**
IS 1458	Class I	India	Ingot and casting	6.0–8.0 **Sn**, 0.5 **Pb**, 0.5 **Zn**, 0.3 **Fe**, 0.1 **Sb**, 0.01 **Al**, 0.5–0.6 **P**, rem **Cu + Ni**, 0.2 total, others	**Sand cast (cast–on)**: all **diam**, 186 MPa **TS**, 3% in 2 in. (50 mm) **El**, 60 **HB**; **Sand cast (separately cast)**: all **diam**, 206 MPa **TS**, 5% in 2 in. (50 mm) **El**, 65 **HB**
ISO 1338	CuSn11P	ISO	Castings, ingot	86.0–89.5 **Cu**, 10.0–12.0 **Sn**, 0.5 **Pb**, 0.5 **Zn**, 0.10 **Fe**, 0.02 **Si**, 0.01 **Al**, 0.2 **Ni**, 0.15–1.5 **P**, **Cu**, total includes **Ni**	**Sand cast**: 12–25 mm **diam**, 220 MPa **TS**, 3% in 2 in. (50 mm) **El**; **Permanent mould cast**: 12–25 mm **diam**, 270 MPa **TS**, 2% in 2 in. (50 mm) **El**; **Continuously cast**: 12–25 mm **diam**, 320 MPa **TS**, 6% in 2 in. (50 mm) **El**
SABS 200	Cu–Sn12P	South Africa	Castings	rem **Cu**, 11.0–13.0 **Sn**, 0.50 **Pb**, 0.30 **Zn**, 0.15 **Fe**, 0.02 **Si**, 0.01 **Al**, 0.50 **Ni**, 0.15 min **P**, 0.02 total, others	**Continuously cast**: all **diam**, 310 MPa **TS**, 5% in 2 in. (50 mm) **El**
MIL–B–16540B	A	US	Castings	85.0–89.0 **Cu**, 7.5–9.0 **Sn**, 0–1.0 **Pb**, 3.0–5.0 **Zn**, 0.25 **Fe**, 1.00 **Ni**, 0.50 **P**, 0.50 total, others	**As cast**: all **diam**, 241 MPa **TS**, 18% in 2 in. (50 mm) **El**
JIS H 5113	Class 2B	Japan	Castings	87.0–91.0 **Cu**, 9.0–12.0 **Sn**, 0.15–0.50 **Pb**	**As cast**: all **diam**, 294 MPa **TS**, 5% in 2 in. (50 mm) **El**, 80 min **HB**
SABS 200	Cu–Sn7Zn3Ni5–WP	South Africa	Castings	rem **Cu**, 6.5–7.5 **Sn**, 0.10–0.50 **Pb**, 1.0–3.0 **Zn**, 0.20 **Mn**, 5.25–5.75 **Ni**, 0.02 **P**, 0.01 **S**, 0.02 **Bi**, 0.20 **As + Fe + Sb**, 0.04 total, others	**Continuously cast**: all **diam**, 430 MPa **TS**, 3% in 2 in. (50 mm) **El**
SABS 200	Cu–Sn7Zn3N:5–M	South Africa	Castings	rem **Cu**, 6.5–7.5 **Sn**, 0.10–0.50 **Pb**, 1.0–3.0 **Zn**, 0.01 **Al**, 0.20 **Mn**, 5.25–5.75 **Ni**, 0.02 **P**, 0.01 **S**, 0.02 **Bi**, 0.20 **As + Fe + Sb**, 0.04 total, others	**Continuously cast**: all **diam**, 340 MPa **TS**, 18% in 2 in. (50 mm) **El**
SABS 200	Cu–Sn10	South Africa	Castings	rem **Cu**, 9.0–11.0 **Sn**, 0.25 **Pb**, 0.05 **Zn**, 0.25 **Ni**, 0.15 **P**, 0.80 total, others	**Continuously cast**: all **diam**, 310 MPa **TS**, 10% in 2 in. (50 mm) **El**
JIS H 5113	Class 2A	Japan	Castings	87.0–91.0 **Cu**, 9.0–12.0 **Sn**, 0.05–0.20 **Pb**	**As cast**: all **diam**, 196 MPa **TS**, 5% in 2 in. (50 mm) **El**, 60 min **HB**

CAST COPPER AND COPPER ALLOYS

specification number	designation	country	product forms	chemical composition	mechanical properties and hardness values
AMS 4840A		US	Castings	67.5–72.5 **Cu**, 4.5–6.0 **Sn**, 23.0–26.0 **Pb**, 0.50 **Zn**, 0.30 **Fe**, 0.01 **P**, 0.25 others, total	**As cast**: all **diam**, 35–50 **HB**
IS 318	Grade 5	India	Ingot and castings	rem **Cu**, 4.0–6.0 **Sn**, 22.0–26.0 **Pb**, 0.5 **Zn**, 0.35 **Fe**, 0.5 **Sb**, 0.7 **Fe + Sb**, 0.45 total, others, nickel included in copper	**Sand cast (separately cast)**: all **diam**, 137 MPa **TS**, 59 MPa **YS**, 4% in 2 in. (50 mm) **El**; **Sand cast (cast on)**: all **diam**, 118 MPa **TS**, 2% in 2 in. (50 mm) **El**; **Chill cast**: all **diam**, 152 MPa **TS**, 59 MPa **YS**, 5% in 2 in. (50 mm) **El**
ANSI/ASTM B 584	C 94300	US	Sand castings	68.5–73.5 **Cu**, 4.5–6.0 **Sn**, 22.0–25.0 **Pb**, 0.8 **Zn**, 0.15 **Fe**, 0.8 **Sb**, 0.005 **Si**, 0.005 **Al**, 1.0 **Ni + Co**, 0.05 **P**, 0.08 **S**	**As cast**: all **diam**, 24 ksi (165 MPa) **TS**, 10% in 2 in. (50 mm) **El**
ANSI/ASTM B 505	CA 943	US	Castings	68.5–73.5 **Cu**, 4.5–6.0 **Sn**, 22.0–25.0 **Pb**, 0.8 **Zn**, 0.15 **Fe**, 0.8 **Sb**, 0.005 **Si**, 0.005 **Al**, 0.8 **Ni + Co**, 1.5 **P**, 0.08 **S**	**As cast**: all **diam**, 21 ksi (145 MPa) **TS**, 15 ksi (97 MPa) **YS**, 7% in 2 in. (50 mm) **El**
ANSI/ASTM B 66	CA 943	US	Castings	68.5–73.5 **Cu**, 4.5–6.0 **Sn**, 22.0–25.0 **Pb**, 0.8 **Zn**, 0.15 **Fe**, 0.8 **Sb**, 0.005 **Si**, 0.005 **Al**, 1.0 **Ni**, 0.05 **P**, 0.08 **S**	
CDA C94300		US	Castings	68.5–73.5 **Cu**, 4.5–6.0 **Sn**, 22.0–25.0 **Pb**, 0.8 **Zn**, 1.5 **Fe**, 0.8 **Sb**, 0.005 **Si**, 0.005 **Al**, 1.0 **Ni**, 0.05 **P**, 0.08 **S**	**Sand casting**: all **diam**, 24 ksi **TS**, 13 ksi **YS**, 10% in 2 in. (50 mm) **El**, 48 **HB**
COPANT 162	C 943 00	COPANT	Cast products	68.5–73.5 **Cu**, 4.5–6.0 **Sn**, 22–25 **Pb**, 0.8 **Zn**, 0.15 **Fe**, 0.8 **Sb**, 0.005 **Si**, 0.005 **Al**, 1.0 **Ni**, 0.05 **P**, 0.08 **S**	
QQ–C–390A	943	US	Castings	68.5–73.5 **Cu**, 4.5–6.0 **Sn**, 22.0–25.0 **Pb**, 0.50 **Zn**, 0.15 **Fe**, 0.7 **Sb**, 0.7 **Ni**, 0.05 **P**	**As cast**: all **diam**, 21 ksi (145 MPa) **TS**, 7% in 2 in. (50 mm) **El**, 38 **HB**
COPANT 801	C 943 00	COPANT	Continuous–cast rod, bar and tube	68.5–73.5 **Cu**, 4.5–6.0 **Sn**, 22–25 **Pb**, 0.8 **Zn**, 0.15 **Fe**, 0.8 **Sb**, 0.005 **Si**, 0.005 **Al**, 1.0 **Ni**, 0.05 **P**, 0.08 **S**	**As manufactured**: 127 mm **diam**, 150 MPa **TS**, 110 MPa **YS**, 7% in 2 in. (50 mm) **El**
ANSI/ASTM B 30	943	US	Ingot: for castings	69.0–73.0 **Cu**, 4.7–5.8 **Sn**, 22.0–24.5 **Pb**, 0.8 **Zn**, 0.10 **Fe**, 0.7 **Sb**, 0.005 **Si**, 0.005 **Al**, 0.8 **Ni + Co**, 0.05 **P**, 0.08 **S**	
ANSI/ASTM B 30	CA 943	US	Ingot	69.0–73.0 **Cu**, 4.7–5.8 **Sn**, 22.0–24.5 **Pb**, 0.8 **Zn**, 0.10 **Fe**, 0.7 **Sb**, 0.005 **Si**, 0.005 **Al**, 0.8 **Ni**, 0.05 **P**, 0.08 **S**	
COPANT 162	C 942 00	COPANT	Cast products	68.5–75.5 **Cu**, 3–4 **Sn**, 21–25 **Pb**, 3.0 **Zn**, 0.35 **Fe**, 0.50 **Sb**, 0.50 **Ni**, 0.40 total, others	

CAST COPPER AND COPPER ALLOYS

specification number	designation	country	product forms	chemical composition	mechanical properties and hardness values
AS 1565	941C	Australia	Ingot, castings	rem **Cu**, 4.0–6.0 **Sn**, 18.0–23.0 **Pb**, 1.0 **Zn**, 0.50 **Sb**, 0.01 **Si**, 2.0 **Ni**, 0.10 **P**, 0.30 total, others (applies to castings only)	**Sand cast**: all **diam**, 160 MPa **TS**, 60 MPa **YS**, 5% in 2 in. (50 mm) **El**, 45 **HB**; **Chill cast**: all **diam**, 170 MPa **TS**, 80 MPa **YS**, 5% in 2 in. (50 mm) **El**, 50 **HB**; **Continuous cast**: all **diam**, 190 MPa **TS**, 100 MPa **YS**, 8% in 2 in. (50 mm) **El**, 50 **HB**
SABS 200	Cu-Sn5Pb20	South Africa	Castings	rem **Cu**, 4.0–6.0 **Sn**, 18.0–23.0 **Pb**, 1.0 **Zn**, 0.50 **Sb**, 0.01 **Si**, 2.0 **Ni**, 0.10 **P**	**Continuously cast**: all **diam**, 185 MPa **TS**, 8% in 2 in. (50 mm) **El**
NF A 53-707	CuPb20 Sn5	France	Castings	rem **Cu**, 4.0–6.0 **Sn**, 18–23 **Pb**, 2.0 **Zn**, 0.25 **Fe**, 0.01 **Si**, 0.01 **Al**, 2.5 **Ni**, 0.30 **P**	**As cast**: all **diam**, 150 MPa **TS**, 5% in 2 in. (50 mm) **El**
ISO 1338	CuPb20Sn5	ISO	Castings, ingot	70.0–78.0 **Cu**, 4.0–6.0 **Sn**, 18.0–23.0 **Pb**, 2.0 **Zn**, 0.25 **Fe**, 0.75 **Sb**, 0.01 **Si**, 0.01 **Al**, 0.2 **Mn**, 2.5 **Ni**, 0.10 **P**, 0.10 **S**, **Cu** total includes **Ni**	**Sand cast**: 12–25 mm **diam**, 150 MPa **TS**, 60 MPa **YS**, 5% in 2 in. (50 mm) **El**; **Continuously cast**: 12–25 mm **diam**, 180 MPa **TS**, 80 MPa **YS**, 7% in 2 in. (50 mm) **El**
JIS H 2207	Class 5	Japan	Ingot: for castings	rem **Cu**, 6.5–8.0 **Sn**, 16.5–22.0 **Pb**, 1.0 **Zn**, 0.15 **Fe**, 0.1 **P**	
ANSI/ASTM B 66	CA 945	US	Castings	69.0–75.0 **Cu**, 6.0–8.0 **Sn**, 16.0–22.0 **Pb**, 1.2 **Zn**, 0.15 **Fe**, 0.3 **Sb**, 0.005 **Si**, 0.005 **Al**, 1.0 **Ni**, 0.05 **P**, 0.08 **S**	
CDA C94500		US	Castings	rem **Cu**, 6.0–8.0 **Sn**, 16.0–22.0 **Pb**, 1.2 **Zn**, 0.15 **Fe**, 0.8 **Sb**, 0.005 **Si**, 0.005 **Al**, 1.0 **Ni**, 0.05 **P**, 0.08 **S**	**Sand castings**: all **diam**, 25 ksi (172 MPa) **TS**, 12 ksi (83 MPa) **YS**, 12% in 2 in. (50 mm) **El**, 50 **HB**
COPANT 162	C 945 00	COPANT	Cast products	rem **Cu**, 6–8 **Sn**, 16–22 **Pb**, 1.2 **Zn**, 0.70 total, others	
JIS H 5115	Class 5	Japan	Castings	70.0–76.0 **Cu**, 6.0–8.0 **Sn**, 16.0–22.0 **Pb**, 1.0 **Zn**, 0.3 **Fe**, 1.0 **Ni**	**As cast**: all **diam**, 65 min **HB**
ANSI/ASTM B 30	945	US	Ingot: for castings	70.0–75.0 **Cu**, 6.3–8.0 **Sn**, 16.0–21.5 **Pb**, 1.0 **Zn**, 0.10 **Fe**, 0.7 **Sb**, 0.005 **Si**, 0.005 **Al**, 0.8 **Ni+Co**, 0.05 **P**, 0.08 **S**	
ANSI/ASTM B 30	CA 945	US	Ingot	70.0–75.0 **Cu**, 6.3–8.0 **Sn**, 16.0–21.5 **Pb**, 1.0 **Zn**, 0.10 **Fe**, 0.7 **Sb**, 0.005 **Si**, 0.005 **Al**, 0.8 **Ni**, 0.05 **P**, 0.08 **S**	
ANSI/ASTM B 505	CA 941	US	Castings	65.0–75.0 **Cu**, 4.5–6.5 **Sn**, 15.0–22.0 **Pb**, 3.0 **Zn**, 0.20 **Fe**, 0.7 **Sb**, 0.005 **Si**, 0.005 **Al**, 0.50 **Ni+Co**, 1.5 **P**, 0.08 **S**	**As cast**: all **diam**, 25 ksi (172 MPa) **TS**, 17 ksi (117 MPa) **YS**, 7% in 2 in. (50 mm) **El**
ANSI/ASTM B 67	CA 941	US	Journal bearings: lined	65.0–75.0 **Cu**, 4.5–6.5 **Sn**, 15.0–22.0 **Pb**, 3.0 **Zn**, 0.25 **Fe**, 0.8 **Sb**, 0.005 **Si**, 0.005 **Al**, 0.80 **Ni**, 0.05 **P**, 0.08 **S**	

CAST COPPER AND COPPER ALLOYS

specification number	designation	country	product forms	chemical composition	mechanical properties and hardness values
ANSI/ASTM B 67	C 94100	US	Bearings: backing for tin liner	65.0–75.0 **Cu**, 4.5–6.5 **Sn**, 15.0–22.0 **Pb**, 3.0 **Zn**, 0.25 **Fe**, 0.8 **Sb**, 0.005 **Si**, 0.005 **Al**, 0.80 **Ni**, 0.05 **P**, 0.08 **S**	
COPANT 162	C 941 00	COPANT	Cast products	65–75 **Cu**, 4.5–6.5 **Sn**, 15–22 **Pb**, 3.0 **Zn**, 1.0 total, others	
QQ–C–390A	941	US	Castings	65.0–75.0 **Cu**, 4.5–6.5 **Sn**, 15.0–22.0 **Pb**, 3.0 **Zn**, 1.0 total, others	**As cast**: all **diam**, 25 ksi (172 MPa) **TS**, 10% in 2 in. (50 mm) **El**
COPANT 801	C 941 00	COPANT	Continuous–cast rod, bar and tube	65–75 **Cu**, 4.5–6.5 **Sn**, 15–22 **Pb**, 3.0 **Zn**	**As manufactured**: 127 mm **diam**, 180 MPa **TS**, 120 MPa **YS**, 7% in 2 in. (50 mm) **El**
ANSI/ASTM B 30	941	US	Ingot: for castings	65.0–75.0 **Cu**, 4.7–6.5 **Sn**, 15.0–21.7 **Pb**, 3.0 **Zn**, 0.10 **Fe**, 0.7 **Sb**, 0.005 **Si**, 0.005 **Al**, 0.8 **Ni+Co**, 0.05 **P**, 0.08 **S**	
ANSI/ASTM B 30	CA 941	US	Ingot	65.0–75.0 **Cu**, 4.7–6.5 **Sn**, 15.0–21.7 **Pb**, 3.0 **Zn**, 0.10 **Fe**, 0.7 **Sb**, 0.005 **Si**, 0.005 **Al**, 0.8 **Ni**, 0.05 **P**, 0.08 **S**	
QQ–C–525b	Composition 7	US	Ingot	74.0–79.0 **Cu**, 5.3–7.0 **Sn**, 15.3–19.0 **Pb**, 1.3 **Zn**, 0.20 **Fe**, 0.7 **Sb**, 0.005 **Si**, 0.005 **Al**, 0.8 **Ni**, 0.03 **P**, 0.08 **S**	
ANSI/ASTM B 505	CA 939	US	Castings	76.5–79.5 **Cu**, 5.0–7.0 **Sn**, 14.0–18.0 **Pb**, 1.5 **Zn**, 0.40 **Fe**, 0.50 **Sb**, 0.005 **Si**, 0.005 **Al**, 0.8 **Ni+Co**, 1.5 **P**, 0.08 **S**	**As cast**: all **diam**, 25 ksi (172 MPa) **TS**, 16 ksi (110 MPa) **YS**, 5% in 2 in. (50 mm) **El**
CDA C 93900		US	Castings	76.5–79.5 **Cu**, 5.0–7.0 **Sn**, 14.0–18.0 **Pb**, 1.5 **Zn**, 0.40 **Fe**, 0.50 **Sb**, 0.005 **Si**, 0.005 **Al**, 0.8 **Ni**, 1.5 **P**, 0.08 **S**	**Continuous casting**: all **diam**, 25 ksi (172 MPa) **TS**, 16 ksi (110 MPa) **YS**, 5% in 2 in. (50 mm) **El**, 63 **HB**
COPANT 162	C 939 00	COPANT	Cast products	76.5–79.5 **Cu**, 5–7 **Sn**, 14–18 **Pb**, 1.5 **Zn**, 0.40 **Fe**, 0.8 **Ni**, 1.5 **P**	
QQ–C–390A	939	US	Castings	76.5–79.5 **Cu**, 5.0–7.0 **Sn**, 14.0–18.0 **Pb**, 1.5 **Zn**, 0.40 **Fe**, 0.35 **Sb**, 0.8 **Ni**, 0.50 **P**, **Ni** in **Cu** content	**As cast**: all **diam**, 25 ksi (172 MPa) **TS**, 16 ksi (110 MPa) **YS**, 5% in 2 in. (50 mm) **El**
COPANT 801	C 939 00	COPANT	Continuous–cast rod, bar and tube	76.5–79.5 **Cu**, 5–7 **Sn**, 14–18 **Pb**, 1.5 **Zn**, 0.40 **Fe**, 0.8 **Ni**, 1.5 **P**	**As manufactured**: 127 mm **diam**, 180 MPa **TS**, 110 MPa **YS**, 5% in 2 in. (50 mm) **El**
ANSI/ASTM B 30	939	US	Ingot: for castings	76.5–79.5 **Cu**, 5.3–7.0 **Sn**, 14.0–17.7 **Pb**, 1.5 **Zn**, 0.35 **Fe**, 0.50 **Sb**, 0.005 **Si**, 0.005 **Al**, 0.8 **Ni+Co**, 0.05 **P**, 0.08 **S**	
ANSI/ASTM B 30	CA 939	US	Ingot	76.5–79.5 **Cu**, 5.3–7.0 **Sn**, 14.0–17.7 **Pb**, 1.5 **Zn**, 0.35 **Fe**, 0.50 **Sb**, 0.005 **Si**, 0.005 **Al**, 0.8 **Ni**, 0.05 **P**, 0.08 **S**	
JIS H 2207	Class 4	Japan	Ingot: for castings	rem **Cu**, 7.5–9.0 **Sn**, 14.5–16.0 **Pb**, 1.0 **Zn**, 0.15 **Fe**, 0.1 **P**	

CAST COPPER AND COPPER ALLOYS

specification number	designation	country	product forms	chemical composition	mechanical properties and hardness values
ANSI/ASTM B 30	938	US	Ingot: for castings	76.0–79.0 **Cu**, 6.5–7.5 **Sn**, 14.0–16.0 **Pb**, 0.8 **Zn**, 0.10 **Fe**, 0.50 **Sb**, 0.005 **Si**, 0.005 **Al**, 0.8 **Ni+Co**, 0.05 **P**, 0.08 **S**	
ANSI/ASTM B 505	CA 940	US	Castings	69.0–72.0 **Cu**, 12.0–14.0 **Sn**, 14.0–16.0 **Pb**, 0.50 **Zn**, 0.25 **Fe**, 0.50 **Sb**, 0.005 **Si**, 0.005 **Al**, 0.50–1.0 **Ni+Co**, 1.5 **P**, 0.08 **S**	**As cast**: all **diam**, 80 **HB**
ANSI/ASTM B 30	CA 938	US	Ingot	76.0–79.0 **Cu**, 6.5–7.5 **Sn**, 14.0–16.0 **Pb**, 0.8 **Zn**, 0.10 **Fe**, 0.50 **Sb**, 0.005 **Si**, 0.005 **Al**, 0.8 **Ni**, 0.05 **P**, 0.08 **S**	
AS 1565	946D	Australia	Ingot, castings	rem **Cu**, 8.0–10.0 **Sn**, 13.0–17.0 **Pb**, 1.0 **Zn**, 0.50 **Sb**, 0.02 **Si**, 2.0 **Ni**, 0.10 **P**, 0.30 total, others (applies to castings only)	**Sand cast**: all **diam**, 170 MPa **TS**, 80 MPa **YS**, 4% in 2 in. (50 mm) **El**; **Chill cast**: all **diam**, 200 MPa **TS**, 130 MPa **YS**, 3% in 2 in. (50 mm) **El**, 70 **HB**; **Continuous cast**: all **diam**, 230 MPa **TS**, 130 MPa **YS**, 9% in 2 in. (50 mm) **El**, 70 **HB**
COPANT 162	C 940 00	COPANT	Cast products	69–72 **Cu**, 12–14 **Sn**, 14–16 **Pb**, 0.50 **Zn**, 0.25 **Fe**, 0.35 **Sb**, 0.5–1.0 **Ni**, 0.05 **P**, 0.25 **S**, 0.35 total, others	
QQ-C-390A	940	US	Castings	69.0–72.0 **Cu**, 12.0–14.0 **Sn**, 14.0–16.0 **Pb**, 0.50 **Zn**, 0.25 **Fe**, 0.35 **Sb**, 0.50–1.0 **Ni**, 0.05 **P**	**As cast**: all **diam**, 80 **HB**
SABS 200	Cu-Sn9Pb15	South Africa	Castings	rem **Cu**, 8.0–10.0 **Sn**, 13.0–17.0 **Pb**, 1.0 **Zn**, 0.5 **Sb**, 2.0 **Ni**, 0.10 **P**, 0.25 total, others	**Continuously cast**: all **diam**, 230 MPa **TS**, 9% in 2 in. (50 mm) **El**
COPANT 801	C 940 00	COPANT	Continuous-cast rod, bar and tube	69–72 **Cu**, 12–14 **Sn**, 14–16 **Pb**, 0.50 **Zn**, 0.25 **Fe**, 0.35 **Sb**, 0.5–1.0 **Ni**, 0.05 **P**	
QQ-C-525b	Composition 13	US	Ingot	69.0–72.0 **Cu**, 13.0–14.0 **Sn**, 14.0–16.0 **Pb**, 0.50 **Zn**, 0.25 **Fe**, 0.50 **Sb**, 0.005 **Al**, 0.50–1.0 **Ni**, 0.05 **P**, 0.08 **S**, 0.35 total, others	
SFS 2216	CuPb15Sn8	Finland	Sand castings	75.0–79.0 **Cu**, 7.0–9.0 **Sn**, 13.0–17.0 **Pb**, 2.0 **Zn**, 0.25 **Fe**, 0.5 **Sb**, 0.01 **Si**, 0.01 **Al**, 0.2 **Mn**, 2.0 **Ni**, 0.10 **P**, 0.10 **S**	**As cast**: all **diam**, 170 MPa **TS**, 80 MPa **YS**, 5% in 2 in. (50 mm) **El**
IS 318	Grade 4	India	Ingot and castings	rem **Cu**, 6.0–8.0 **Sn**, 14.0–16.0 **Pb**, 0.5 **Zn**, 0.35 **Fe**, 0.5 **Sb**, 0.7 **Fe+Sb**, 0.45 total, others, nickel included in copper	**Sand cast (separately cast)**: all **diam**, 157 MPa **TS**, 69 MPa **YS**, 4% in 2 in. (50 mm) **El**; **Sand cast (cast on)**: all **diam**, 137 MPa **TS**, 2% in 2 in. (50 mm) **El**; **Chill cast**: all **diam**, 186 MPa **TS**, 69 MPa **YS**, 5% in 2 in. (50 mm) **El**

CAST COPPER AND COPPER ALLOYS

specification number	designation	country	product forms	chemical composition	mechanical properties and hardness values
IS 1458	Class III	India	Ingot and casting	6.0–8.0 **Sn**, 14.0–16.0 **Pb**, 0.5 **Zn**, 0.3 **Fe**, 0.4 **Sb**, 0.01 **Al**, 0.05 **P**, 0.5 **Fe + Sb**, rem **Cu + Ni**	**Sand cast (cast-on)**: all **diam**, 137 MPa **TS**, 2% in 2 in. (50 mm) **El**; **Sand cast (separately cast)**: all **diam**, 157 MPa **TS**, 4% in 2 in. (50 mm) **El**
ISO 1338	CuPb15Sn8	ISO	Castings, ingot	75.0–79.0 **Cu**, 7.0–9.0 **Sn**, 13.0–17.0 **Pb**, 2.0 **Zn**, 0.25 **Fe**, 0.5 **Sb**, 0.01 **Si**, 0.01 **Al**, 0.2 **Mn**, 2.0 **Ni**, 0.10 **P**, 0.10 **S**, **Cu** total includes **Ni**	**Sand cast**: 12–25 mm **diam**, 170 MPa **TS**, 80 MPa **YS**, 5% in 2 in. (50 mm) **El**; **Centrifugally or continuously cast**: 12–25 mm **diam**, 220 MPa **TS**, 100 MPa **YS**, 8% in 2 in. (50 mm) **El**
JIS H 5115	Class 4	Japan	Castings	74.0–78.0 **Cu**, 7.0–9.0 **Sn**, 14.0–16.0 **Pb**, 1.0 **Zn**, 0.3 **Fe**, 1.0 **Ni**	**As cast**: all **diam**, 55 min **HB**
ANSI/ASTM B 30	940	US	Ingot: for castings	69.0–72.0 **Cu**, 12.3–14.0 **Sn**, 14.0–15.7 **Pb**, 0.50 **Zn**, 0.25 **Fe**, 0.50 **Sb**, 0.005 **Si**, 0.005 **Al**, 0.50–1.0 **Ni + Co**, 0.05 **P**, 0.08 **S**	
ANSI/ASTM B 30	CA 940	US	Ingot	69.0–72.0 **Cu**, 12.3–14.0 **Sn**, 14.0–15.7 **Pb**, 0.50 **Zn**, 0.25 **Fe**, 0.50 **Sb**, 0.005 **Si**, 0.005 **Al**, 0.50–1.0 **Ni**, 0.05 **P**, 0.08 **S**	
ANSI/ASTM B 584	C 93800	US	Sand castings	75.0–79.0 **Cu**, 6.3–7.5 **Sn**, 13.0–16.0 **Pb**, 0.8 **Zn**, 0.15 **Fe**, 0.8 **Sb**, 0.005 **Si**, 0.005 **Al**, 1.0 **Ni + Co**, 0.05 **P**, 0.08 **S**	**As cast**: all **diam**, 26 ksi (179 MPa) **TS**, 14 ksi (97 MPa) **YS**, 12% in 2 in. (50 mm) **El**
ANSI/ASTM B 505	CA 938	US	Castings	75.0–79.0 **Cu**, 6.3–7.5 **Sn**, 13.0–16.0 **Pb**, 0.8 **Zn**, 0.15 **Fe**, 0.8 **Sb**, 0.005 **Si**, 0.005 **Al**, 1.0 **Ni + Co**, 1.5 **P**, 0.08 **S**	**As cast**: all **diam**, 25 ksi (172 MPa) **TS**, 16 ksi (110 MPa) **YS**, 5% in 2 in. (50 mm) **El**
CDA C93800		US	Castings	75.0–79.0 **Cu**, 6.3–7.5 **Sn**, 13.0–16.0 **Pb**, 0.8 **Zn**, 0.15 **Fe**, 0.8 **Sb**, 0.005 **Si**, 0.005 **Al**, 1.0 **Ni**, 0.05 **P**, 0.08 **S**	**Sand castings**: all **diam**, 26 ksi (179 MPa) **TS**, 14 ksi (97 MPa) **YS**, 12% in 2 in. (50 mm) **El**, 55 **HB**
COPANT 162	C 938 00	COPANT	Cast products	75–79 **Cu**, 6.3–7.5 **Sn**, 13–16 **Pb**, 0.8 **Zn**, 0.15 **Fe**, 0.8 **Sb**, 0.005 **Si**, 0.005 **Al**, 1.0 **Ni**, 0.05 **P**, 0.08 **S**	
QQ-C-390A	938	US	Castings	75.0–79.0 **Cu**, 6.3–7.5 **Sn**, 13.0–16.0 **Pb**, 0.7 **Zn**, 0.15 **Fe**, 0.7 **Sb**, 0.7 **Ni**, 0.05 **P**	**As cast**: all **diam**, 25 ksi (172 MPa) **TS**, 14 ksi (97 MPa) **YS**, 10% in 2 in. (50 mm) **El**
COPANT 801	C 938 00	COPANT	Continuous-cast rod, bar and tube	75–79 **Cu**, 6.3–7.5 **Sn**, 13–16 **Pb**, 0.8 **Zn**, 0.15 **Fe**, 0.8 **Sb**, 0.005 **Si**, 0.005 **Al**, 1.0 **Ni**, 0.05 **P**, 0.08 **S**	**As manufactured**: 127 mm **diam**, 180 MPa **TS**, 110 MPa **YS**, 5% in 2 in. (50 mm) **El**
ANSI/ASTM B 66	CA 938	US	Castings	75.0–79.0 **Cu**, 6.3–7.5 **Sn**, 9.0–12.0 **Pb**, 0.8 **Zn**, 0.15 **Fe**, 0.8 **Sb**, 0.005 **Si**, 0.005 **Al**, 1.0 **Ni**, 0.20–0.50 **P**, 0.08 **S**	
ANSI/ASTM B 66	CA 944	US	Castings	78.0–82.0 **Cu**, 7.0–9.0 **Sn**, 9.0–12.0 **Pb**, 0.8 **Zn**, 0.15 **Fe**, 0.8 **Sb**, 0.005 **Si**, 0.005 **Al**, 1.0 **Ni**, 0.20–0.50 **P**, 0.08 **S**	

CAST COPPER AND COPPER ALLOYS

specification number	designation	country	product forms	chemical composition	mechanical properties and hardness values
CDA C 94400		US	Castings	rem **Cu**, 9.0–12.0 **Pb**, 0.8 **Zn**, 0.15 **Fe**, 0.8 **Sb**, 0.005 **Si**, 0.005 **Al**, 1.0 **Ni**, 0.50 **P**, 0.08 **S**	**Sand castings**: all **diam**, 32 ksi (221 MPa) **TS**, 16 ksi (110 MPa) **YS**, 18% in 2 in. (50 mm) **El**, 55 **HB**
COPANT 162	C 944 00	COPANT	Cast products	rem **Cu**, 7–9 **Sn**, 9–12 **Pb**, 0.7 **Zn**, 0.15 **Fe**, 0.8 **Sb**, 1.0 **Ni**, 0.2–0.50 **P**, 99.5 total elements	
ANSI/ASTM B 30	944	US	Ingot: for castings	78.0–82.0 **Cu**, 7.3–9.0 **Sn**, 9.0–11.7 **Pb**, 0.8 **Zn**, 0.10 **Fe**, 0.7 **Sb**, 0.005 **Si**, 0.005 **Al**, 0.8 **Ni + Co**, 0.05 **P**, 0.08 **S**	
ANSI/ASTM B 30	CA 944	US	Ingot	78.0–82.0 **Cu**, 7.3–9.0 **Sn**, 9.0–11.7 **Pb**, 0.8 **Zn**, 0.10 **Fe**, 0.7 **Sb**, 0.005 **Si**, 0.005 **Al**, 0.8 **Ni**, 0.05 **P**, 0.08 **S**	
JIS H 2207	Class 3	Japan	Ingot: for castings	rem **Cu**, 9.5–11.0 **Sn**, 9.5–11.0 **Pb**, 1.0 **Zn**, 0.15 **Fe**, 0.1 **P**	
IS 318	Grade 3	India	Ingot and castings	rem **Cu**, 6.0–8.0 **Sn**, 9.0–11.0 **Pb**, 0.5 **Zn**, 0.35 **Fe**, 0.5 **Sb**, 0.7 **Fe + Sb**, 0.45 total, others, nickel included in copper	**Sand cast (separately cast)**: all **diam**, 172 MPa **TS**, 74 MPa **YS**, 4% in 2 in. (50 mm) **El**; **Sand cast (cast on)**: all **diam**, 157 MPa **TS**, 2% in 2 in. (50 mm) **El**; **Chill cast**: all **diam**, 201 MPa **TS**, 74 MPa **YS**, 5% in 2 in. (50 mm) **El**
IS 1458	Class IV	India	Ingot and casting	6.0–8.0 **Sn**, 9.0–11.0 **Pb**, 0.5 **Zn**, 0.3 **Fe**, 0.4 **Sb**, 0.01 **Al**, 0.05 **P**, 0.5 **Fe + Sb**, rem **Cu + Ni**	**Sand cast (cast–on)**: all **diam**, 157 MPa **TS**, 2% in 2 in. (50 mm) **El**; **Sand cast (separately cast)**: all **diam**, 177 MPa **TS**, 4% in 2 in. (50 mm) **El**
JIS H 5115	Class 3	Japan	Castings	77.0–81.0 **Cu**, 9.0–11.0 **Sn**, 9.0–11.0 **Pb**, 1.0 **Zn**, 0.3 **Fe**, 1.0 **Ni**	**As cast**: all **diam**, 65 min **HB**
AS 1565	937B	Australia	Ingot, castings	rem **Cu**, 9.0–11.0 **Sn**, 8.5–11.0 **Pb**, 1.0 **Zn**, 0.15 **Fe**, 0.5 **Sb**, 0.02 **Si**, 0.01 **Al**, 2.0 **Ni**, 0.10 **P**, 0.50 total, others (applies to castings only)	**Sand cast**: all **diam**, 190 MPa **TS**, 80 MPa **YS**, 5% in 2 in. (50 mm) **El**, 65 **HB**; **Chill cast**: all **diam**, 220 MPa **TS**, 140 MPa **YS**, 3% in 2 in. (50 mm) **El**, 80 **HB**; **Continuous cast**: all **diam**, 280 MPa **TS**, 160 MPa **YS**, 6% in 2 in. (50 mm) **El**, 80 **HB**
SABS 200	Cu–Sn10Pb10	South Africa	Castings	rem **Cu**, 9.0–11.0 **Sn**, 8.5–11.0 **Pb**, 1.0 **Zn**, 0.15 **Fe**, 0.5 **Sb**, 0.02 **Si**, 0.1 **Al**, 0.10 **P**	**Continuously cast**: all **diam**, 280 MPa **TS**, 6% in 2 in. (50 mm) **El**
AMS 4842A		US	Castings	77.0–81.0 **Cu**, 9.0–11.0 **Sn**, 8.0–11.0 **Pb**, 0.75 **Zn**, 0.15 **Fe**, 0.20 **Sb**, 0.005 **Al**, 0.50 **Ni**, 0.05 **P**, 0.35 total, others	**Sand cast**: all **diam**, 75 min **HRF**; **Chill and centrifugally cast**: all **diam**, 75 min **HRF**
ASTM B 22	CA 937	US	Castings	78.0–82.0 **Cu**, 9.0–11.0 **Sn**, 8.0–11.0 **Pb**, 0.75 **Zn**, 0.15 **Fe**, 0.75 **Sb**, 1.0 **Ni**, 0.15 **P**, 98.50 min **Cu + Sn + Ni**	

CAST COPPER AND COPPER ALLOYS

specification number	designation	country	product forms	chemical composition	mechanical properties and hardness values
ANSI/ASTM B 584	C 93700	US	Sand castings	78.0–82.0 **Cu**, 9.0–11.0 **Sn**, 8.0–11.0 **Pb**, 0.8 **Zn**, 0.15 **Fe**, 0.55 **Sb**, 0.005 **Si**, 0.005 **Al**, 1.0 **Ni+Co**, 0.15 **P**, 0.08 **S**	**As cast**: all **diam**, 30 ksi (207 MPa) **TS**, 12 ksi (83 MPa) **YS**, 15% in 2 in. (50 mm) **El**
ANSI/ASTM B 505	CA 937	US	Castings	78.0–82.0 **Cu**, 9.0–11.0 **Sn**, 8.0–11.0 **Pb**, 0.8 **Zn**, 0.15 **Fe**, 0.55 **Sb**, 0.005 **Si**, 0.005 **Al**, 1.0 **Ni+Co**, 1.5 **P**, 0.08 **S**	**As cast**: all **diam**, 35 ksi (241 MPa) **TS**, 20 ksi (138 MPa) **YS**, 6% in 2 in. (50 mm) **El**
ANSI/ASTM B 30	937	US	Ingot: for castings	78.0–81.0 **Cu**, 9.3–10.7 **Sn**, 8.3–10.7 **Pb**, 0.8 **Zn**, 0.10 **Fe**, 0.50 **Sb**, 0.005 **Si**, 0.005 **Al**, 0.8 **Ni+Co**, 0.05 **P**, 0.08 **S**	
ANSI/ASTM B 30	CA 937	US	Ingot	78.0–81.0 **Cu**, 9.3–10.7 **Sn**, 8.3–10.7 **Pb**, 0.8 **Zn**, 0.10 **Fe**, 0.50 **Sb**, 0.005 **Si**, 0.005 **Al**, 0.8 **Ni**, 0.05 **P**, 0.08 **S**	**As cast**: all **diam**, 40 ksi (275 MPa) **TS**, 18 ksi (125 MPa) **YS**, 20% in 2 in. (51 mm) **El**
CDA C 93700		US	Castings	78.0–82.0 **Cu**, 9.0–11.0 **Sn**, 8.0–11.0 **Pb**, 0.8 **Zn**, 0.15 **Fe**, 0.55 **Sb**, 0.005 **Si**, 0.005 **Al**, 1.0 **Ni**, 0.15 **P**, 0.8 **S**	**Sand castings**: all **diam**, 30 ksi (207 MPa) **TS**, 12 ksi (83 MPa) **YS**, 15% in 2 in. (50 mm) **El**, 60 **HB**
COPANT 162	C 937 00	COPANT	Cast products	78–82 **Cu**, 9–11 **Sn**, 8–11 **Pb**, 0.8 **Zn**, 0.15 **Fe**, 0.55 **Sb**, 0.005 **Si**, 0.005 **Al**, 1.0 **Ni**, 0.15 **P**, 0.08 **S**	
QQ-C-390A	937	US	Castings	78.0–82.0 **Cu**, 9.0–11.0 **Sn**, 8.0–11.0 **Pb**, 0.7 **Zn**, 0.15 **Fe**, 0.50 **Sb**, 0.7 **Ni**, 0.05 **P**, **Ni** in **Cu** content	**As cast**: all **diam**, 25 ksi (172 MPa) **TS**, 12 ksi (83 MPa) **YS**, 8% in 2 in. (50 mm) **El**
COPANT 801	C 937 00	COPANT	Continuous-cast rod, bar and tube	78–82 **Cu**, 9–11 **Sn**, 8–11 **Pb**, 0.8 **Zn**, 0.15 **Fe**, 0.55 **Sb**, 0.005 **Si**, 0.005 **Al**, 1.0 **Ni**, 0.15 **P**, 0.08 **S**	**As manufactured**: 127 mm **diam**, 250 MPa **TS**, 140 MPa **YS**, 6% in 2 in. (50 mm) **El**
DS 3001	Lead Bronze 5640	Denmark	Sand castings	78.0–82.0 **Cu**, 9.0–11.0 **Sn**, 8.0–11.0 **Pb**, 2.0 **Zn**, 0.25 **Fe**, 0.5 **Sb**, 0.01 **Si**, 0.01 **Al**, 0.2 **Mn**, 2.0 **Ni**, 0.05 **P**, 0.10 **S**	**As cast**: all **diam**, 180 MPa **TS**, 80 MPa **YS**, 7% in 2 in. (50 mm) **El**, 65 min **HB**
SFS 2215	CuPb10Sn10	Finland	Sand castings	78.0–82.0 **Cu**, 9.0–11.0 **Sn**, 8.0–11.0 **Pb**, 2.0 **Zn**, 0.25 **Fe**, 0.5 **Sb**, 0.01 **Si**, 0.01 **Al**, 0.2 **Mn**, 2.0 **Ni**, 0.05 **P**, 0.10 **S**	**As cast**: all **diam**, 180 MPa **TS**, 80 MPa **YS**, 7% in 2 in. (50 mm) **El**
NF A 53 707	CuPb10Sn10	France	Castings	rem **Cu**, 9.0–11.0 **Sn**, 8.0–11.0 **Pb**, 2.0 **Zn**, 0.25 **Fe**, 0.01 **Si**, 0.01 **Al**, 0.30 **P**, 1.0 total, others	**As cast**: all **diam**, 180 MPa **TS**, 7% in 2 in. (50 mm) **El**
ISO 1338	CuPb10Sn10	ISO	Castings, ingot	78.0–82.0 **Cu**, 9.0–11.0 **Sn**, 8.0–11.0 **Pb**, 2.0 **Zn**, 0.25 **Fe**, 0.5 **Sb**, 0.01 **Si**, 0.01 **Al**, 0.2 **Mn**, 2.0 **Ni**, 0.05 **P**, 0.10 **S**, **Cu** total includes **Ni**	**Sand cast**: 12–25 mm **diam**, 180 MPa **TS**, 80 MPa **YS**, 7% in 2 in. (50 mm) **El**; **Permanent mould cast**: 12–25 mm **diam**, 220 MPa **TS**, 140 MPa **YS**, 3% in 2 in. (50 mm) **El**; **Centrifugally or continuously cast**: 12–25 mm **diam**, 110 MPa **YS**

CAST COPPER AND COPPER ALLOYS

specification number	designation	country	product forms	chemical composition	mechanical properties and hardness values
SIS 14 56 40	SIS Bronze 56 40–00	Sweden	Ingot	78.0–81.0 **Cu**, 9.3–11.0 **Sn**, 8.5–10.5 **Pb**, 1.0 **Zn**, 0.15 **Fe**, 0.5 **Sb**, 0.01 **Si**, 0.01 **Al**, 0.2 **Mn**, 2.0 **Ni**, 0.05 **P**, 0.10 **S**	
SIS 14 56 40	SIS Bronze 56 40–03	Sweden	Sand castings	78.0–82.0 **Cu**, 9.0–11.0 **Sn**, 8.0–11.0 **Pb**, 2.0 **Zn**, 0.25 **Fe**, 0.5 **Sb**, 0.01 **Si**, 0.01 **Al**, 0.2 **Mn**, 2.0 **Ni**, 0.05 **P**, 0.10 **S**	**As cast**: all **diam**, 180 MPa **TS**, 80 MPa **YS**, 7% in 2 in. (50 mm) **El**, 65 **HB**
SIS 14 56 40	SIS Bronze 56 40–15	Sweden	Castings: continuous and centrifugal	78.0–82.0 **Cu**, 9.0–11.0 **Sn**, 8.0–11.0 **Pb**, 2.0 **Zn**, 0.25 **Fe**, 0.5 **Sb**, 0.01 **Si**, 0.01 **Al**, 0.2 **Mn**, 2.0 **Ni**, 0.05 **P**, 0.10 **S**	**As cast**: all **diam**, 220 MPa **TS**, 110 MPa **YS**, 6% in 2 in. (50 mm) **El**, 70 **HB**
ANSI/ASTM B 30	935	US	Ingot: for castings	83.0–85.0 **Cu**, 4.5–5.5 **Sn**, 8.5–9.7 **Pb**, 0.50–1.5 **Zn**, 0.10 **Fe**, 0.30 **Sb**, 0.005 **Si**, 0.005 **Al**, 0.8 **Ni+Co**, 0.04 **P**, 0.08 **S**	
ANSI/ASTM B 30	CA 935	US	Ingot	83.0–85.0 **Cu**, 4.5–5.5 **Sn**, 8.5–9.7 **Pb**, 0.50–1.5 **Zn**, 0.10 **Fe**, 0.30 **Sb**, 0.005 **Si**, 0.005 **Al**, 0.8 **Ni**, 0.04 **P**, 0.08 **S**	
ANSI/ASTM B 584	C 93500	US	Sand castings	83.0–86.0 **Cu**, 4.3–6.0 **Sn**, 8.0–10.0 **Pb**, 2.0 **Zn**, 0.20 **Fe**, 0.30 **Sb**, 0.005 **Si**, 0.005 **Al**, 1.0 **Ni+Co**, 0.05 **P**, 0.08 **S**	**As cast**: all **diam**, 28 ksi (193 MPa) **TS**, 12 ksi (83 MPa) **YS**, 15% in 2 in. (50 mm) **El**
ANSI/ASTM B 505	CA 935	US	Castings	83.0–86.0 **Cu**, 4.3–6.0 **Sn**, 8.0–10.0 **Pb**, 2.0 **Zn**, 0.20 **Fe**, 0.30 **Sb**, 0.005 **Si**, 0.005 **Al**, 1.0 **Ni+Co**, 1.5 **P**, 0.08 **S**	**As cast**: all **diam**, 30 ksi (207 MPa) **TS**, 16 ksi (110 MPa) **YS**, 12% in 2 in. (50 mm) **El**
AS 1565	935B	Australia	Ingot, castings	rem **Cu**, 4.0–6.0 **Sn**, 8.0–10.0 **Pb**, 2.0 **Zn**, 0.50 **Sb**, 0.02 **Si**, 2.0 **Ni**, 0.10 **P**, 0.50 total, others (applies to castings only)	**Sand cast**: all **diam**, 160 MPa **TS**, 60 MPa **YS**, 7% in 2 in. (50 mm) **El**, 55 **HB**; **Chill cast**: all **diam**, 200 MPa **TS**, 80 MPa **YS**, 5% in 2 in. (50 mm) **El**, 60 **HB**; **Continuous cast**: all **diam**, 230 MPa **TS**, 130 MPa **YS**, 9% in 2 in. (50 mm) **El**, 60 **HB**
CDA C 93500		US	Castings	83.0–86.0 **Cu**, 4.3–6.0 **Sn**, 8.0–10.0 **Pb**, 2.0 **Zn**, 0.20 **Fe**, 0.30 **Sb**, 0.005 **Si**, 0.005 **Al**, 1.0 **Ni**, 0.05 **P**, 0.08 **S**	**Sand castings**: all **diam**, 28 ksi (193 MPa) **TS**, 12 ksi (83 MPa) **YS**, 15% in 2 in. (50 mm) **El**, 60 **HB**
COPANT 162	C 935 00	COPANT	Cast products	83–86 **Cu**, 4.3–6.0 **Sn**, 8–10 **Pb**, 2.0 **Zn**, 0.20 **Fe**, 0.30 **Sb**, 0.005 **Si**, 0.005 **Al**, 1.0 **Ni**, 0.05 **P**, 0.08 **S**	
SABS 200	Cu–Sn5Pb10	South Africa	Castings	rem **Cu**, 4.0–6.0 **Sn**, 8.0–10.0 **Pb**, 2.0 **Zn**, 0.5 **Sb**, 0.02 **Si**, 2.0 **Ni**, 0.10 **P**	**Continuously cast**: all **diam**, 230 MPa **TS**, 9% in 2 in. (50 mm) **El**
COPANT 801	C 935 00	COPANT	Continuous–cast rod, bar and tube	83–86 **Cu**, 4.3–6.0 **Sn**, 8–10 **Pb**, 2.0 **Zn**, 0.20 **Fe**, 0.30 **Sb**, 0.005 **Si**, 0.005 **Al**, 1.0 **Ni**, 0.05 **P**, 0.08 **S**	**As manufactured**: 127 mm **diam**, 210 MPa **TS**, 110 MPa **YS**, 12% in 2 in. (50 mm) **El**

CAST COPPER AND COPPER ALLOYS

specification number	designation	country	product forms	chemical composition	mechanical properties and hardness values
ISO 1338	CuPb9Sn5	ISO	Castings, ingot	80.0–87.0 **Cu**, 4.0–6.0 **Sn**, 8.0–10.0 **Pb**, 2.0 **Zn**, 0.25 **Fe**, 0.5 **Sb**, 0.01 **Si**, 0.01 **Al**, 0.2 **Mn**, 2.0 **Ni**, 0.10 **P**, 0.10 **S**, **Cu** total includes **Ni**	**Sand cast**: 12–25 mm **diam**, 160 MPa **TS**, 60 MPa **YS**, 7% in 2 in. (50 mm) **El**; **Centrifugally cast**: 12–25 mm **diam**, 220 MPa **TS**, 80 MPa **YS**, 6% in 2 in. (50 mm) **El**; **Continuously cast**: 12–25 mm **diam**, 230 MPa **TS**, 130 MPa **YS**, 9% in 2 in. (50 mm) **El**
ANSI/ASTM B 505	CA 934	US	Castings	82.0–85.0 **Cu**, 7.0–9.0 **Sn**, 7.0–9.0 **Pb**, 0.8 **Zn**, 0.15 **Fe**, 0.50 **Sb**, 0.005 **Si**, 0.005 **Al**, 1.0 **Ni+Co**, 1.5 **P**, 0.08 **S**	**As cast**: all **diam**, 34 ksi (234 MPa) **TS**, 20 ksi (138 MPa) **YS**, 8% in 2 in. (50 mm) **El**
CDA C 93400		US	Castings	82.0–85.0 **Cu**, 7.0–9.0 **Sn**, 7.0–9.0 **Pb**, 0.8 **Zn**, 0.15 **Fe**, 0.50 **Sb**, 0.005 **Si**, 0.005 **Al**, 1.0 **Ni**, 0.50 **P**, 0.08 **S**	**Sand castings**: all **diam**, 25 ksi (172 MPa) **TS**, 12 ksi (83 MPa) **YS**, 8% in 2 in. (50 mm) **El**, 60 **HB**
COPANT 162	C 934 00	COPANT	Cast products	82–85 **Cu**, 7–9 **Sn**, 7–9 **Pb**, 0.7 **Zn**, 0.15 **Fe**, 0.5 **Sb**, 1.0 **Ni**, 0.5 **P**	
COPANT 162	C 936 00	COPANT	Cast products	83 min **Cu**, 3.5–4.5 **Sn**, 7–9 **Pb**, 4.0 **Zn**, 0.35 **Fe**, 0.50 **Sb**, 1.5 **Ni**, 0.30 total, others	
QQ-C-390A	935	US	Castings	83.0–86.0 **Cu**, 7.0–9.0 **Sn**, 7.0–9.0 **Pb**, 0.7 **Zn**, 0.15 **Fe**, 0.50 **Sb**, 1.0 **Ni**, 0.50 **P**, **Ni** in **Cu** content	**As cast**: all **diam**, 25 ksi (172 MPa) **TS**, 12 ksi (83 MPa) **YS**, 8% in 2 in. (50 mm) **El**
QQ-C-390A	934	US	Castings	82.0–85.0 **Cu**, 7.0–9.0 **Sn**, 7.0–9.0 **Pb**, 0.7 **Zn**, 0.15 **Fe**, 0.50 **Sb**, 1.0 **Ni**, 0.50 **P**	**As cast**: all **diam**, 25 ksi (172 MPa) **TS**, 12 ksi (83 MPa) **YS**, 8% in 2 in. (50 mm) **El**
COPANT 801	C 934 00	COPANT	Continuous-cast rod, bar and tube	82–85 **Cu**, 7–9 **Sn**, 7–9 **Pb**, 0.7 **Zn**, 0.15 **Fe**, 0.5 **Sb**, 1.0 **Ni**, 0.5 **P**	**As manufactured**: 127 mm **diam**, 240 MPa **TS**, 140 MPa **YS**, 8% in 2 in. (50 mm) **El**
QQ-C-525b	Composition 8	US	Ingot	83.0–85.0 **Cu**, 7.3–8.7 **Sn**, 7.3–8.7 **Pb**, 0.8 **Zn**, 0.15 **Fe**, 0.35 **Sb**, 0.005 **Si**, 0.005 **Al**, 1.0 **Ni**, 0.03 **P**, 0.08 **S**	
MIL-C-22229A	934	US	Castings	82.0–85.0 **Cu**, 7.0–9.0 **Sn**, 7.0–9.0 **Pb**, 0.7 **Zn**, 0.15 **Fe**, 0.50 **Sb**, 1.0 **Ni**, 0.50 **P**	**As cast**: all **diam**, 172 MPa **TS**, 83 MPa **YS**, 8% in 2 in. (50 mm) **El**
ANSI/ASTM B 30	934	US	Ingot: for castings	82.0–85.0 **Cu**, 7.3–9.0 **Sn**, 7.0–8.7 **Pb**, 0.8 **Zn**, 0.20 **Fe**, 0.30 **Sb**, 0.005 **Si**, 0.005 **Al**, 0.8 **Ni+Co**, 0.03 **P**, 0.08 **S**	
ANSI/ASTM B 30	CA 934	US	Ingot	82.0–85.0 **Cu**, 7.3–9.0 **Sn**, 7.0–8.7 **Pb**, 0.8 **Zn**, 0.20 **Fe**, 0.30 **Sb**, 0.005 **Si**, 0.005 **Al**, 0.8 **Ni**, 0.03 **P**, 0.08 **S**	
ANSI/ASTM B 30	932	US	Ingot: for castings	82.0–84.0 **Cu**, 6.5–7.5 **Sn**, 6.5–7.7 **Pb**, 2.5–4.0 **Zn**, 0.20 **Fe**, 0.30 **Sb**, 0.005 **Si**, 0.005 **Al**, 0.8 **Ni+Co**, 0.03 **P**, 0.08 **S**	

CAST COPPER AND COPPER ALLOYS

specification number	designation	country	product forms	chemical composition	mechanical properties and hardness values
ANSI/ASTM B 30	CA 932	US	Ingot	82.0–84.0 **Cu**, 6.5–7.5 **Sn**, 6.5–7.7 **Pb**, 2.5–4.0 **Zn**, 0.20 **Fe**, 0.30 **Sb**, 0.005 **Si**, 0.005 **Al**, 0.8 **Ni**, 0.03 **P**, 0.08 **S**	
QQ-C-525b	Composition 12	US	Ingot	82.0–84.0 **Cu**, 6.5–7.5 **Sn**, 6.5–7.7 **Pb**, 2.5–4.0 **Zn**, 0.20 **Fe**, 0.30 **Sb**, 0.005 **Si**, 0.005 **Al**, 0.8 **Ni**, 0.03 **P**, 0.08 **S**	
ANSI/ASTM B 584	C 93200	US	Sand castings	81.0–85.0 **Cu**, 6.3–7.5 **Sn**, 6.0–8.0 **Pb**, 2.0–4.0 **Zn**, 0.20 **Fe**, 0.35 **Sb**, 0.005 **Si**, 0.005 **Al**, 1.0 **Ni+Co**, 0.15 **P**, 0.08 **S**	**As cast**: all **diam**, 30 ksi (207 MPa) **TS**, 14 ksi (97 MPa) **YS**, 15% in 2 in. (50 mm) **El**
ANSI/ASTM B 505	CA 932	US	Castings	81.0–85.0 **Cu**, 6.3–7.5 **Sn**, 6.0–8.0 **Pb**, 2.0–4.0 **Zn**, 0.20 **Fe**, 0.30 **Sb**, 0.005 **Si**, 0.005 **Al**, 1.0 **Ni+Co**, 1.5 **P**, 0.08 **S**	**As cast**: all **diam**, 35 ksi (241 MPa) **TS**, 20 ksi (138 MPa) **YS**, 10% in 2 in. (50 mm) **El**
CDA C 93200		US	Castings	81.0–85.0 **Cu**, 6.3–7.5 **Sn**, 6.0–8.0 **Pb**, 2.0–4.0 **Zn**, 0.20 **Fe**, 0.35 **Sb**, 0.005 **Si**, 0.005 **Al**, 1.0 **Ni**, 0.15 **P**, 0.08 **S**	**Sand castings**: all **diam**, 30 ksi (207 MPa) **TS**, 14 ksi (97 MPa) **YS**, 15% in 2 in. (50 mm) **El**, 65 **HB**
COPANT 162	C 932 00	COPANT	Cast products	81–85 **Cu**, 6.3–7.5 **Sn**, 6–8 **Pb**, 2–4 **Zn**, 0.20 **Fe**, 0.35 **Sb**, 0.005 **Si**, 0.005 **Al**, 1.0 **Ni**, 0.15 **P**, 0.08 **S**	
QQ-C-390A	932	US	Castings	81.0–85.0 **Cu**, 6.3–7.5 **Sn**, 6.0–8.0 **Pb**, 2.0–4.0 **Zn**, 0.20 **Fe**, 0.35 **Sb**, 0.50 **Ni**, 0.15 **P**, **Ni** in **Cu** content	**As cast**: all **diam**, 30 ksi (207 MPa) **TS**, 14 ksi (97 MPa) **YS**, 12% in 2 in. (50 mm) **El**
COPANT 801	C 932 00	COPANT	Continuous-cast rod, bar and tube	81–85 **Cu**, 6.3–7.5 **Sn**, 6–8 **Pb**, 2–4 **Zn**, 0.20 **Fe**, 0.35 **Sb**, 0.005 **Si**, 0.005 **Al**, 1.0 **Ni**, 0.15 **P**, 0.08 **S**	**As manufactured**: 127 mm **diam**, 250 MPa **TS**, 140 MPa **YS**, 10% in 2 in. (50 mm) **El**
QQ-C-525b	Composition 11	US	Ingot	79.0–82.0 **Cu**, 2.5–3.5 **Sn**, 6.3–7.7 **Pb**, 7.0–10.0 **Zn**, 0.35 **Fe**, 0.25 **Sb**, 0.005 **Si**, 0.005 **Al**, 0.8 **Ni**, 0.02 **P**, 0.08 **S**	
ISO1338	CuSn7Pb7Zn3	ISO	Castings, ingot	81.0–85.0 **Cu**, 6.0–8.0 **Sn**, 5.0–8.0 **Pb**, 2.0–5.0 **Zn**, 0.20 **Fe**, 0.35 **Sb**, 0.01 **Si**, 0.01 **Al**, 2.0 **Ni**, 0.10 **P**, 0.10 **S**, **Cu** total includes **Ni**	**Sand cast and permanent mould**: 12–25 mm **diam**, 210 MPa **TS**, 100 MPa **YS**, 12% in 2 in. (50 mm) **El**; **Centrifugally or continuously cast**: 12–25 mm **diam**, 260 MPa **TS**, 120 MPa **YS**, 12% in 2 in. (50 mm) **El**
NF A 53-707	CuSn7Pb6Zn4	France	Castings	rem **Cu**, 6.0–8.0 **Sn**, 5.0–7.0 **Pb**, 2.0–5.0 **Zn**, 1.5 **Ni**, 1.0 total, others	**As cast**: all **diam**, 250 MPa **TS**, 16% in 2 in. (50 mm) **El**
QQ-C-390A	948	US	Castings	85.0–89.0 **Cu**, 4.7–6.0 **Sn**, 0.30–0.90 **Pb**, 1.3–2.5 **Zn**, 0.20 **Fe**, 0.10 **Sb**, 0.20 **Mn**, 4.5–6.0 **Ni**, 0.05 **P**, 0.25 total, others	**As cast**: all **diam**, 45 ksi (310 MPa) **TS**, 20 ksi (138 MPa) **YS**, 20% in 2 in. (50 mm) **El**

CAST COPPER AND COPPER ALLOYS

specification number	designation	country	product forms	chemical composition	mechanical properties and hardness values
QQ-C-390A	947	US	Castings	86.0–89.0 **Cu**, 4.7–6.0 **Sn**, 0.08 **Pb**, 1.3–2.5 **Zn**, 0.20 **Fe**, 0.10 **Sb**, 0.20 **Mn**, 4.5–6.0 **Ni**, 0.5 **P**, 0.25 total, others	**As cast**: all **diam**, 45 ksi (310 MPa) **TS**, 20 ksi (138 MPa) **YS**, 25% in 2 in. (50 mm) **El**; **Heat treated**: all **diam**, 75 ksi (517 MPa) **TS**, 50 ksi (345 MPa) **YS**, 5% in 2 in. (50 mm) **El**
CDA C 94700		US	Castings	85.0–90.0 **Cu**, 4.5–6.0 **Sn**, 0.10 **Pb**, 1.0–2.5 **Zn**, 0.25 **Fe**, 0.15 **Sb**, 0.005 **Si**, 0.005 **Al**, 0.20 **Mn**, 4.5–6.0 **Ni**, 0.05 **P**, 0.05 **S**	**Sand castings**: all **diam**, 45 ksi (310 MPa) **TS**, 20 ksi (138 MPa) **YS**, 25% in 2 in. (50 mm) **El**, 85 **HB**
ANSI/ASTM B 584	C 94700	US	Sand castings	85.0–90.0 **Cu**, 4.5–6.0 **Sn**, 0.10 **Pb**, 1.0–2.5 **Zn**, 0.25 **Fe**, 0.15 **Sb**, 0.005 **Si**, 0.005 **Al**, 0.20 **Mn**, 4.5–6.0 **Ni+Co**, 0.05 **P**, 0.05 **S**	**As cast**: all **diam**, 45 ksi (310 MPa) **TS**, 20 ksi (138 MPa) **YS**, 25% in 2 in. (50 mm) **El**; **Heat treated**: all **diam**, 75 ksi (517 MPa) **TS**, 50 ksi (345 MPa) **YS**, 5% in 2 in. (50 mm) **El**
ANSI/ASTM B 30	CA 947	US	Ingot	86.0–89.0 **Cu**, 4.7–6.0 **Sn**, 0.08 **Pb**, 1.3–2.5 **Zn**, 0.20 **Fe**, 0.10 **Sb**, 0.005 **Si**, 0.005 **Al**, 4.5–6.0 **Ni+Co**, 0.05 **P**, 0.05 **S**	
ANSI/ASTM B 30	947	US	Ingot: for castings	86.0–89.0 **Cu**, 4.7–6.0 **Sn**, 0.08 **Pb**, 1.3–2.5 **Zn**, 0.20 **Fe**, 0.10 **Sb**, 0.005 **Al**, 4.5–6.0 **Ni+Co**, 0.05 **P**, 0.05 **S**	
COPANT 162	C 947 00	COPANT	Cast products	85–89 **Cu**, 4.5–6.0 **Sn**, 0.10 **Pb**, 1.0–2.5 **Zn**, 0.25 **Fe**, 0.15 **Sb**, 0.005 **Si**, 0.005 **Al**, 0.20 **Mn**, 4.5–6.0 **Ni**, 0.05 **P**, 0.05 **S**	
ANSI/ASTM B 505	CA 947	US	Castings	85.0–89.0 **Cu**, 4.5–6.0 **Sn**, 0.10 **Pb**, 1.0–3.0 **Zn**, 0.25 **Fe**, 0.15 **Sb**, 0.005 **Si**, 0.005 **Al**, 0.20 **Mn**, 4.5–6.0 **Ni+Co**, 0.050 **P**, 0.05 **S**	**As cast**: all **diam**, 45 ksi (310 MPa) **TS**, 20 ksi (138 MPa) **YS**, 25% in 2 in. (50 mm) **El**; **Heat treated**: all **diam**, 75 ksi (517 MPa) **TS**, 50 ksi (345 MPa) **YS**, 5% in 2 in. (50 mm) **El**
COPANT 801	C 947 00 (HT)	COPANT	Continuous-cast rod, bar and tube	85–89 **Cu**, 4.5–6.0 **Sn**, 0.10 **Pb**, 1.0–2.5 **Zn**, 0.25 **Fe**, 0.15 **Sb**, 0.005 **Si**, 0.20 **Mn**, 4.5–6.0 **Ni**, 0.05 **P**, 0.05 **S**	**Heat treated**: 127 mm **diam**, 310 MPa **TS**, 140 MPa **YS**, 25% in 2 in. (50 mm) **El**
ANSI/ASTM B 30	CA 948	US	Ingot	85.0–89.0 **Cu**, 4.7–6.0 **Sn**, 0.30–0.9 **Pb**, 1.3–2.5 **Zn**, 0.20 **Fe**, 0.10 **Sb**, 0.005 **Si**, 0.005 **Al**, 4.5–6.0 **Ni**, 0.05 **P**, 0.05 **S**	
ANSI/ASTM B 30	948	US	Ingot: for castings	85.0–89.0 **Cu**, 4.7–6.0 **Sn**, 0.30–0.9 **Pb**, 1.3–2.5 **Zn**, 0.20 **Fe**, 0.10 **Sb**, 0.005 **Si**, 0.005 **Al**, 4.5–6.0 **Ni+Co**, 0.05 **P**, 0.05 **S**	

CAST COPPER AND COPPER ALLOYS

specification number	designation	country	product forms	chemical composition	mechanical properties and hardness values
ISO 1338	CuSn12Ni2	ISO	Castings, ingot	84.5–87.5 **Cu**, 11.0–13.0 **Sn**, 0.3 **Pb**, 0.4 **Zn**, 0.20 **Fe**, 0.1 **Sb**, 0.1 **Si**, 0.1 **Al**, 0.2 **Mn**, 1.5–2.5 **Ni**, 0.05–0.40 **P**, 0.05 **S**	**Sand cast**: 12–25 mm **diam**, 280 MPa **TS**, 160 MPa **YS**, 12% in 2 in. (50 mm) **El**; **Centrifugally cast**: 12–25 mm **diam**, 300 MPa **TS**, 180 MPa **YS**, 8% in 2 in. (50 mm) **El**; **Continuously cast**: 12–25 mm **diam**, 300 MPa **TS**, 180 MPa **YS**, 10% in 2 in. (50 mm) **El**
ANSI/ASTM B 505	CA 948	US	Castings	84.0–89.0 **Cu**, 4.5–6.0 **Sn**, 0.3–1.0 **Pb**, 2.5 **Zn**, 0.25 **Fe**, 0.15 **Sb**, 0.005 **Si**, 0.005 **Al**, 0.20 **Mn**, 4.5–6.0 **Ni+Co**, 0.050 **P**, 0.05 **S**	**As cast**: all **diam**, 40 ksi (276 MPa) **TS**, 20 ksi (138 MPa) **YS**, 20% in 2 in. (50 mm) **El**
COPANT 162	C 948 00	COPANT	Cast products	84–89 **Cu**, 4.5–6.0 **Sn**, 0.30–1.0 **Pb**, 1.0–2.5 **Zn**, 0.25 **Fe**, 0.15 **Sb**, 0.005 **Si**, 0.005 **Al**, 0.20 **Mn**, 4.5–6.0 **Ni**, 0.05 **P**, 0.05 **S**	
COPANT 801	C 948 00	COPANT	Continuous–cast rod, bar and tube	84–89 **Cu**, 4.5–6.0 **Sn**, 0.30–1.0 **Pb**, 1.0–2.5 **Zn**, 0.25 **Fe**, 0.15 **Sb**, 0.005 **Si**, 0.005 **Al**, 0.20 **Mn**, 4.5–6.0 **Ni**, 0.05 **P**, 0.05 **S**	**As manufactured**: 127 mm **diam**, 280 MPa **TS**, 140 MPa **YS**, 20% in 2 in. (50 mm) **El**
CDA C 94800		US	Castings	84.0–89.0 **Cu**, 4.5–6.0 **Sn**, 0.30–1.0 **Pb**, 1.0–2.5 **Zn**, 0.25 **Fe**, 0.15 **Sb**, 0.005 **Si**, 0.005 **Al**, 0.20 **Mn**, 4.5–6.0 **Ni**, 0.05 **P**, 0.05 **S**	**Sand castings**: all **diam**, 40 ksi (276 MPa) **TS**, 20 ksi (138 MPa) **YS**, 20% in 2 in. (50 mm) **El**, 80 **HB**
ANSI/ASTM B 584	C 94800	US	Sand castings	84.0–89.0 **Cu**, 4.5–6.0 **Sn**, 0.30–1.0 **Pb**, 1.0–2.5 **Zn**, 0.25 **Fe**, 0.15 **Sb**, 0.005 **Si**, 0.005 **Al**, 0.20 **Mn**, 4.5–6.0 **Ni+Co**, 0.05 **P**, 0.05 **S**	**As cast**: all **diam**, 40 ksi (276 MPa) **TS**, 20 ksi (138 MPa) **YS**, 20% in 2 in. (50 mm) **El**
AS 1565	948C	Australia	Ingot, castings	83.16 min **Cu**, 6.5–7.5 **Sn**, 0.10–0.50 **Pb**, 1.50–3.0 **Zn**, 0.01 **Si**, 0.01 **Al**, 0.02 **Mn**, 5.25–5.75 **Ni**, 0.02 **P**, 0.02 **Bi**, 0.20 **Fe+As+Sb**, 0.50 total, others (castings)	**As cast**: all **diam**, 280 MPa **TS**, 140 MPa **YS**, 16% in 2 in. (50 mm) **El**, 70 **HB**; **Heat treated**: all **diam**, 430 MPa **TS**, 280 MPa **YS**, 3% in 2 in. (50 mm) **El**, 160 **HB**
ANSI/ASTM B 30	CA 949	US	Ingot	79.0–81.0 **Cu**, 4.3–6.0 **Sn**, 4.0–5.7 **Pb**, 4.3–6.0 **Zn**, 0.25 **Fe**, 0.25 **Sb**, 0.005 **Si**, 0.005 **Al**, 4.5–6.0 **Ni**, 0.05 **P**, 0.08 **S**	
COPANT 162	C 949 00	COPANT	Cast products	79–81 **Cu**, 4–6 **Sn**, 4–6 **Pb**, 4–6 **Zn**, 0.30 **Fe**, 0.25 **Sb**, 0.005 **Si**, 0.005 **Al**, 0.10 **Mn**, 4–6 **Ni**, 0.05 **P**, 0.08 **S**	
ANSI/ASTM B 30	949	US	Ingot: for castings	79.0–81.0 **Cu**, 4.3–6.0 **Sn**, 4.0–5.7 **Pb**, 4.3–6.0 **Zn**, 0.25 **Fe**, 0.25 **Sb**, 0.005 **Si**, 0.005 **Al**, 4.5–6.0 **Ni+Co**, 0.05 **P**, 0.08 **S**	
ANSI/ASTM B 584	C 94900	US	Sand castings	79.0–81.0 **Cu**, 4.0–6.0 **Sn**, 4.0–6.0 **Pb**, 4.0–6.0 **Zn**, 0.30 **Fe**, 0.25 **Sb**, 0.005 **Si**, 0.005 **Al**, 0.10 **Mn**, 4.0–6.0 **Ni+Co**, 0.05 **P**, 0.08 **S**	**As cast**: all **diam**, 38 ksi (262 MPa) **TS**, 15 ksi (97 MPa) **YS**, 15% in 2 in. (50 mm) **El**

CAST COPPER AND COPPER ALLOYS

specification number	designation	country	product forms	chemical composition	mechanical properties and hardness values
AMS 4870B		US	Centrifugal and chill castings	83.5 min **Cu**, 3.0–4.25 **Fe**, 10.5–12.0 **Al**, 0.50 **Mn**, 0.50 **Ni**	**As cast**: ≤1 in. **diam**, 90 ksi (621 MPa) **TS**, 36 ksi (248 MPa) **YS**, 179–235 **HB**
AMS 4872B		US	Sand castings	83.5 min **Cu**, 3.0–4.25 **Fe**, 10.5–12.0 **Al**, 0.50 **Mn**, 0.50 **Ni**	**As cast**: >1 in. **diam**, 72 ksi (496 MPa) **TS**, 28 ksi (193 MPa) **YS**, 5% in 2 in. (50 mm) **El**, 150 min **HB**
AMS 4881		US	Sand and centrifugal castings: martensitic	74.5 min **Cu**, 0.25 **Sn**, 0.03 **Pb**, 0.30 **Zn**, 4.0–5.5 **Fe**, 0.15 **Si**, 10.5–11.5 **Al**, 0.20 **Co**, 1.5 **Mn**, 4.2–6.0 **Ni**, 0.05 **Cr**	**Sand cast**: >1.00 in. (>25.4 mm) **diam**, 120 ksi (827 MPa) **TS**, 85 ksi (586 MPa) **YS**, 2% in 2 in. (50 mm) **El**, 25 **HRC**; **Centrifugally cast**: >1.00 in. (>25.4 mm) **diam**, 125 ksi (862 MPa) **TS**, 90 ksi (621 MPa) **YS**, 2% in 2 in. (50 mm) **El**, 28 **HRC**
NF A 53-709	U-All N5 Fe Y200	France	Sand castings	rem **Cu**, 0.2 **Sn**, 0.2 **Pb**, 0.5 **Zn**, 3–6 **Fe**, 10–12 **Al**, 3.0 **Mn**, 4.6–5 **Ni**	**As cast**: all **diam**, 6% in 2 in. (50 mm) **El**, 190 **HB**
AMS 4871B		US	Centrifugal and chill castings	83.5 min **Cu**, 3.0–4.25 **Fe**, 10.3–11.5 **Al**, 0.50 **Mn**, 0.50 **Ni**	**Quenched and tempered**: all **diam**, 90 ksi (621 MPa) **TS**, 45 ksi (310 MPa) **YS**, 5% in 2 in. (50 mm) **El**, 200–235 **HB**
AMS 4873A		US	Sand castings	83.5 min **Cu**, 3.0–4.25 **Fe**, 10.3–11.5 **Al**, 0.50 **Mn**, 0.50 **Ni**	**Quenched and tempered**: ≤1 in. **diam**, 85 ksi (586 MPa) **TS**, 45 ksi (310 MPa) **YS**, 5% in 2 in. (50 mm) **El**, 200–235 **HB**
ANSI/ASTM B 148	CA 954	US	Sand castings	83.0 min **Cu**, 3.0–5.0 **Fe**, 10.0–11.5 **Al**, 0.50 **Mn**, 2.5 **Ni+Co**	**As cast**: all **diam**, 75 ksi (515 MPa) **TS**, 30 ksi (205 MPa) **YS**, 12% in 2 in. (50.8 mm) **El**, 150 **HB**; **Heat treated**: all **diam**, 90 ksi (620 MPa) **TS**, 45 ksi (310 MPa) **YS**, 6% in 2 in. (50.8 mm) **El**, 190 **HB**
ANSI/ASTM B 148	CA 955	US	Sand castings	78.0 min **Cu**, 3.0–5.0 **Fe**, 10.0–11.5 **Al**, 3.5 **Mn**, 3.0–5.5 **Ni+Co**	**As cast**: all **diam**, 90 ksi (620 MPa) **TS**, 40 ksi (275 MPa) **YS**, 6% in 2 in. (50.8 mm) **El**, 190 **HB**; **Heat treated**: all **diam**, 110 ksi (760 MPa) **TS**, 60 ksi (415 MPa) **YS**, 5% in 2 in. (50.8 mm) **El**, 200 **HB**
ANSI/ASTM B 505	CA 954	US	Castings	83.0 min **Cu**, 3.0–5.0 **Fe**, 10.0–11.5 **Al**, 0.50 **Mn**, 2.5 **Ni+Co**	**As cast**: all **diam**, 85 ksi (586 MPa) **TS**, 32 ksi (221 MPa) **YS**, 12% in 2 in. (50 mm) **El**; **Heat treated**: all **diam**, 95 ksi (655 MPa) **TS**, 45 ksi (310 MPa) **YS**, 10% in 2 in. (50 mm) **El**

CAST COPPER AND COPPER ALLOYS

specification number	designation	country	product forms	chemical composition	mechanical properties and hardness values
ANSI/ASTM B 505	CA 955	US	Castings	78.0 min **Cu**, 3.0–5.0 **Fe**, 10.0–11.5 **Al**, 3.5 **Mn**, 3.0–5.5 **Ni** + **Co**	**As cast**: all **diam**, 95 ksi (655 MPa) **TS**, 42 ksi (290 MPa) **YS**, 10% in 2 in. (50 mm) **El**; **Heat treated**: all **diam**, 110 ksi (758 MPa) **TS**, 62 ksi (427 MPa) **YS**, 8% in 2 in. (50 mm) **El**
ANSI/ASTM B 30	954	US	Ingot: for castings	83.0 min **Cu**, 3.0–5.0 **Fe**, 10.0–11.5 **Al**, 0.5 **Mn**, 2.5 **Ni** + **Co**	
ANSI/ASTM B 30	955	US	Ingot: for castings	78.0 min **Cu**, 3.0–5.0 **Fe**, 10.0–11.5 **Al**, 3.5 **Mn**, 3.0–5.5 **Ni** + **Co**	
ANSI/ASTM B 30	CA 954	US	Ingot	83.0 min **Cu**, 3.0–5.0 **Fe**, 10.0–11.5 **Al**, 0.5 **Mn**, 2.5 **Ni**	**As cast**: all **diam**, 75 ksi (515 MPa) **TS**, 30 ksi (205 MPa) **YS**, 12% in 2 in. (50.8 mm) **El**, 150 **HB**
ANSI/ASTM B 30	CA 955	US	Ingot	78.0 min **Cu**, 3.0–5.0 **Fe**, 10.0–11.5 **Al**, 3.5 **Mn**, 3.0–5.0 **Ni**	**As cast**: all **diam**, 90 ksi (620 MPa) **TS**, 40 ksi (275 MPa) **YS**, 6% in 2 in. (50.8 mm) **El**, 150 **HB**
CDA C 95400		US	Castings	83.0 min **Cu**, 3.0–5.0 **Fe**, 10.0–11.5 **Al**, 0.50 **Mn**, 2.5 **Ni**	**Sand castings**: all **diam**, 75 ksi (517 MPa) **TS**, 30 ksi (207 MPa) **YS**, 12% in 2 in. (50 mm) **El**, 170 **HB**; **Heat treated**: all **diam**, 90 ksi (621 MPa) **TS**, 45 ksi (310 MPa) **YS**, 6% in 2 in. (50 mm) **El**, 195 **HB**
CDA C 954100		US	Castings	83.0 min **Cu**, 3.0–5.0 **Fe**, 10.0–11.5 **Al**, 0.50 **Mn**, 1.5–2.5 **Ni**	**Sand castings**: all **diam**, 75 ksi (517 MPa) **TS**, 30 ksi (207 MPa) **YS**, 12% in 2 in. (50 mm) **El**, 170 **HB**; **Heat treated**: all **diam**, 90 ksi (621 MPa) **TS**, 45 ksi (310 MPa) **YS**, 6% in 2 in. (50 mm) **El**, 195 **HB**
CDA C 95500		US	Castings	78.0 min **Cu**, 3.0–5.0 **Fe**, 10.0–11.5 **Al**, 3.5 **Mn**, 3.0–5.5 **Ni**	**Sand castings**: all **diam**, 90 ksi (621 MPa) **TS**, 40 ksi (276 MPa) **YS**, 6% in 2 in. (50 mm) **El**, 195 **HB**; **Heat treated**: all **diam**, 110 ksi (758 MPa) **TS**, 60 ksi (414 MPa) **YS**, 5% in 2 in. (50 mm) **El**, 230 **HB**
COPANT 162	C 955 00	COPANT	Cast products	78 min **Cu**, 3–5 **Fe**, 10–11.5 **Al**, 3.5 **Mn**, 3–5 **Ni**, 99.5 total elements	
QQ-C-390A	955	US	Castings	78.0 min **Cu**, 3.0–5.0 **Fe**, 10.0–11.5 **Al**, 3.5 **Mn**, 3.0–5.5 **Ni**, 0.50 total, others	**As cast**: all **diam**, 90 ksi (621 MPa) **TS**, 40 ksi (276 MPa) **YS**, 6% in 2 in. (50 mm) **El**, 170 **HB**; **Heat treated**: all **diam**, 110 ksi (758 MPa) **TS**, 60 ksi (414 MPa) **YS**, 5% in 2 in. (50 mm) **El**

CAST COPPER AND COPPER ALLOYS

specification number	designation	country	product forms	chemical composition	mechanical properties and hardness values
QQ–C–390A	954	US	Castings	83.0 min **Cu**, 3.0–5.0 **Fe**, 10.0–11.5 **Al**, 0.50 **Mn**, 2.5 **Ni**, 0.50 total, others	**As cast**: all **diam**, 75 ksi (517 MPa) **TS**, 30 ksi (207 MPa) **YS**, 12% in 2 in. (50 mm) **El**; **Heat treated**: all **diam**, 90 ksi (621 MPa) **TS**, 45 ksi (310 MPa) **YS**, 6% in 2 in. (50 mm) **El**
QQ–B–675B	955	US	Ingot	78.0 min **Cu**, 3.0–5.0 **Fe**, 10.0–11.5 **Al**, 3.5 **Mn**, 3.0–5.5 **Ni**, 0.5 total, others	
QQ–B–675B	954	US	Ingot	83.0 min **Cu**, 3.0–5.0 **Fe**, 10.0–11.5 **Al**, 0.5 **Mn**, 2.5 **Ni**, 0.5 total, others	
COPANT 801	C 955 00 (HT)	COPANT	Continuous–cast rod, bar and tube	78 min **Cu**, 3–5 **Fe**, 10–11.5 **Al**, 3.5 **Mn**, 3–5 **Ni**, 99.5 min total elements	**Heat treated**: 140 mm **diam**, 660 MPa **TS**, 290 MPa **YS**, 10% in 2 in. (50 mm) **El**
COPANT 801	C 954 00 (HT)	COPANT	Continuous–cast rod, bar and tube	83 min **Cu**, 3–5 **Fe**, 10–11.5 **Al**, 0.50 **Mn**, 2.5 **Ni**, 99.5 min total elements	**Heat treated**: 140 mm **diam**, 590 MPa **TS**, 230 MPa **YS**, 12% in 2 in. (50 mm) **El**
NF A 53–709	U–All Fe3 Y200	France	Sand castings	rem **Cu**, 0.2 **Sn**, 0.2 **Pb**, 0.5 **Zn**, 2.0–4.0 **Fe**, 0.2 **Si**, 10.0–11.5 **Al**, 3.0 **Mn**, 1.5 **Ni**	**As cast**: all **diam**, 5% in 2 in. (50 mm) **El**, 180 **HB**
MIL–C–22229A	955	US	Castings	78.0 min **Cu**, 3.0–5.0 **Fe**, 10.0–11.5 **Al**, 3.5 **Mn**, 3.0–5.5 **Ni**	**As cast**: all **diam**, 621 MPa **TS**, 276 MPa **YS**, 6% in 2 in. (50 mm) **El**; **Heat treated**: all **diam**, 758 MPa **TS**, 414 MPa **YS**, 5% in 2 in. (50 mm) **El**
AMS 4880		US	Centrifugal castings	78.0 min **Cu**, 0.20 **Sn**, 0.30 **Zn**, 2.0–3.5 **Fe**, 9.7–10.9 **Al**, 1.5 **Mn**, 4.5–5.5 **Ni**	**Quenched and tempered**: ≤1 in. **diam**, 105 ksi (724 MPa) **TS**, 63 ksi (434 MPa) **YS**, 9% in 2 in. (50 mm) **El**, 192–248 **HB**
ANSI/ASTM B 148	CA 953	US	Sand castings	86.0 min **Cu**, 0.75–1.5 **Fe**, 9.0–11.0 **Al**, 99.0 min **Cu**+**Fe**+**Al**	**As cast**: all **diam**, 65 ksi (450 MPa) **TS**, 25 ksi (170 MPa) **YS**, 20% in 2 in. (50.8 mm) **El**, 110 **HB**; **Heat treated**: all **diam**, 80 ksi (550 MPa) **TS**, 40 ksi (275 MPa) **YS**, 12% in 2 in. (50.8 mm) **El**, 160 **HB**
ANSI/ASTM B 505	CA 953	US	Castings	86.0 min **Cu**, 0.8–1.5 **Fe**, 9.0–11.0 **Al**	**As cast**: all **diam**, 70 ksi (483 MPa) **TS**, 26 ksi (179 MPa) **YS**, 25% in 2 in. (50 mm) **El**; **Heat treated**: all **diam**, 80 ksi (552 MPa) **TS**, 40 ksi (276 MPa) **YS**, 12% in 2 in. (50 mm) **El**
ANSI/ASTM B 30	953	US	Ingot: for castings	86.0 min **Cu**, 0.8–1.5 **Fe**, 9.0–11.0 **Al**	

CAST COPPER AND COPPER ALLOYS

specification number	designation	country	product forms	chemical composition	mechanical properties and hardness values
ANSI/ASTM B 30	CA 953	US	Ingot	86.0 min **Cu**, 0.8–1.5 **Fe**, 9.0–11.0 **Al**	**As cast**: all **diam**, 65 ksi (450 MPa) **TS**, 25 ksi (170 MPa) **YS**, 20% in 2 in. (50.8 mm) **El**, 110 **HB**
CDA C 95300		US	Castings	86.0 min **Cu**, 0.8–1.5 **Fe**, 9.0–11.0 **Al**, 99.0 min **Cu**+**Fe**+**Al**	**Sand castings**: all **diam**, 65 ksi (448 MPa) **TS**, 25 ksi (172 MPa) **YS**, 20% in 2 in. (50 mm) **El**, 140 **HB**
COPANT 162	C 953 00	COPANT	Cast products	86 min **Cu**, 0.8–1.5 **Fe**, 9–11 **Al**	
QQ-C-390A	953	US	Castings	86.0 min **Cu**, 0.8–1.5 **Fe**, 9.0–11.0 **Al**, 1.0 total, others	**As cast**: all **diam**, 65 ksi (448 MPa) **TS**, 25 ksi (172 MPa) **YS**, 20% in 2 in. (50 mm) **El**; **Heat treated**: all **diam**, 80 ksi (552 MPa) **TS**, 40 ksi (276 MPa) **YS**, 12% in 2 in. (50 mm) **El**
QQ-B-675B	953	US	Ingot	86.0 min **Cu**, 0.8–1.5 **Fe**, 9.0–11.0 **Al**, 1.0 total, others	
COPANT 801	C 953 00 (HT)	COPANT	Continuous-cast rod, bar and tube	86 min **Cu**, 0.8–1.5 **Fe**, 9–11 **Al**	**Heat treated**: 140 mm **diam**, 480 MPa **TS**, 100 MPa **YS**, 25% in 2 in. (50 mm) **El**
IS 305	Grade 3	India	Ingot and castings	rem **Cu**, 0.75–1.5 **Fe**, 9–11 **Al**	**Sand cast**: all **diam**, 446 MPa **TS**, 20% in 2 in. (50 mm) **El**
ISO 1338	CuAl10Fe3	ISO	Castings, ingot	83.0–89.5 **Cu**, 0.30 **Sn**, 0.20 **Pb**, 0.40 **Zn**, 2.0–5.0 **Fe**, 0.20 **Si**, 8.5–11.0 **Al**, 1.0 **Mn**, 3.0 **Ni**	**Sand cast**: 12–25 mm **diam**, 500 MPa **TS**, 180 MPa **YS**, 13% in 2 in. (50 mm) **El**; **Permanent mould cast**: 12–25 mm **diam**, 550 MPa **TS**, 200 MPa **YS**, 15% in 2 in. (50 mm) **El**; **Continuously or centrifugally cast**: 12–25 mm **diam**, 550 MPa **TS**, 200 MPa **YS**, 15% in 2 in. (50 mm) **El**
MIL-B-21230A	Alloy 1	US	Castings	78 min **Cu**, 0.03 **Pb**, 3.0–5.0 **Fe**, 8.5–11 **Al**, 3.5 **Mn**, 3.0–5.5 **Ni**, 0.50 total, others	**As cast**: all **diam**, 586 MPa **TS**, 241 MPa **YS**, 15% in 2 in. (50 mm) **El**
SIS 14 57 10	SIS Bronze 57 10-00	Sweden	Ingot	83.0–83.3 **Cu**, 0.20 **Sn**, 0.10 **Pb**, 0.40 **Zn**, 2.0–3.5 **Fe**, 0.10 **Si**, 8.7–10.5 **Al**, 1.0 **Mn**, 3.0 **Ni**	
AS 1565	952C	Australia	Ingot, castings	rem **Cu**, 0.10 **Sn**, 0.05 **Pb**, 0.50 **Zn**, 1.5–3.5 **Fe**, 0.25 **Si**, 8.5–10.5 **Al**, 1.0 **Mn**, 1.0 **Ni**, 0.05 **Mg**, 0.30 total, others	**Sand cast**: all **diam**, 500 MPa **TS**, 170 MPa **YS**, 18% in 2 in. (50 mm) **El**, 90 **HB**; **Chill cast**: all **diam**, 540 MPa **TS**, 200 MPa **YS**, 18% in 2 in. (50 mm) **El**, 130 **HB**; **Centrifugal cast**: all **diam**, 560 MPa **TS**, 200 MPa **YS**, 20% in 2 in. (50 mm) **El**, 120 **HB**
JIS H 2206	Class 3	Japan	Ingot: for castings	78.0 min **Cu**, 3.0–6.0 **Fe**, 8.5–10.5 **Al**, 1.5 **Mn**, 3.0–6.0 **Ni**	
SFS 2211	CuAl10Fe3	Finland	Sand castings	83.0–89.5 **Cu**, 0.30 **Sn**, 0.20 **Pb**, 0.40 **Zn**, 2.0–4.0 **Fe**, 0.20 **Si**, 8.5–10.5 **Al**, 1.0 **Mn**, 3.0 **Ni**	**As cast**: all **diam**, 500 MPa **TS**, 180 MPa **YS**, 13% in 2 in. (50 mm) **El**

CAST COPPER AND COPPER ALLOYS

specification number	designation	country	product forms	chemical composition	mechanical properties and hardness values
NF A 53-709	U-A9 Y300	France	Die castings	rem **Cu**, 2.0 **Fe**, 8.5-10.5 **Al**	**As cast**: all **diam**, 500 MPa **TS**, 20% in 2 in. (50 mm) **El**
NF A 53-709	U-A9 Fe3 Y200	France	Sand castings	rem **Cu**, 0.2 **Sn**, 0.2 **Pb**, 0.5 **Zn**, 2-4 **Fe**, 0.2 **Si**, 8.5-10.5 **Al**, 3.0 **Mn**, 1.5 **Ni**, 3.0 **Mn+Ni**	**As cast**: all **diam**, 500 MPa **TS**, 180 MPa **YS**, 15% in 2 in. (50 mm) **El**
IS 305	Grade 1	India	Ingot and castings	rem **Cu**, 0.05 **Pb**, 0.5 **Zn**, 3.5-5.5 **Fe**, 0.25 **Si**, 8.5-10.5 **Al**, 1.5 **Mn**, 4.5-6.5 **Ni**, 0.05 **Mg**, 0.30 **Sn+Pb+Si+Mg**	**Sand cast**: all **diam**, 647 MPa **TS**, 245 MPa **YS**, 15% in 2 in. (50 mm) **El**; **Chill cast**: all **diam**, 647 MPa **TS**, 245 MPa **YS**, 12% in 2 in. (50 mm) **El**
IS 305	Grade 2	India	Ingot and castings	rem **Cu**, 0.10 **Sn**, 0.05 **Pb**, 0.5 **Zn**, 1.5-3.5 **Fe**, 0.25 **Si**, 8.5-10.5 **Al**, 1.0 **Mn**, 1.0 **Ni**, 0.05 **Mg**, 0.30 **Sn+Pb+Si+Mg**	**Sand cast**: all **diam**, 345 MPa **TS**, 167 MPa **YS**, 20% in 2 in. (50 mm) **El**; **Chill cast**: all **diam**, 379 MPa **TS**, 197 MPa **YS**, 20% in 2 in. (50 mm) **El**
ISO 1338	CuAl10Fe5Ni5	ISO	Castings, ingot	76.0 min **Cu**, 0.20 **Sn**, 0.10 **Pb**, 0.50 **Zn**, 3.5-5.5 **Fe**, 0.10 **Si**, 8.0-11.0 **Al**, 3.0 **Mn**, 3.5-6.5 **Ni**, 99.2 min **Cu+Fe+Ni+Al+Mn**	**Sand cast**: 12-25 mm **diam**, 600 MPa **TS**, 250 MPa **YS**, 10% in 2 in. (50 mm) **El**; **Continuously or centrifugally cast**: 12-25 mm **diam**, 680 MPa **TS**, 280 MPa **YS**, 12% in 2 in. (50 mm) **El**
JIS H 5114	Class 3	Japan	Castings	78 min **Cu**, 3.0-6.0 **Fe**, 8.5-10.5 **Al**, 1.5 **Mn**, 3.0-6.0 **Ni**	**As cast**: all **diam**, 588 min MPa **TS**, 15% in 2 in. (50 mm) **El**, 150 min **HB**
SABS 200	CuAl10Fe2	South Africa	Castings	rem **Cu**, 0.1 **Sn**, 0.05 **Pb**, 0.50 **Zn**, 1.5-3.5 **Fe**, 0.25 **Si**, 8.5-10.5 **Al**, 1.0 **Mn**, 1.0 **Ni**, 0.05 **Mg**	**Sand cast**: all **diam**, 500 MPa **TS**, 18% in 2 in. (50 mm) **El**; **Gravity die cast**: all **diam**, 540 MPa **TS**, 18% in 2 in. (50 mm) **El**
SIS 14 57 10	SIS Bronze 57 10-03	Sweden	Sand castings	83.0-89.5 **Cu**, 0.30 **Sn**, 0.20 **Pb**, 0.40 **Zn**, 2.0-4.0 **Fe**, 0.20 **Si**, 8.5-10.5 **Al**, 1.0 **Mn**, 3.0 **Ni**	**As cast**: all **diam**, 500 MPa **TS**, 180 MPa **YS**, 13% in 2 in. (50 mm) **El**, 115 **HB**
SIS 14 57 10	SIS Bronze 57 10-06	Sweden	Chill castings	83.0-89.5 **Cu**, 0.30 **Sn**, 0.20 **Pb**, 0.40 **Zn**, 2.0-4.0 **Fe**, 0.20 **Si**, 8.5-10.5 **Al**, 1.0 **Mn**, 3.0 **Ni**	**As cast**: all **diam**, 550 MPa **TS**, 200 MPa **YS**, 15% in 2 in. (50 mm) **El**, 115 **HB**
SIS 14 57 10	SIS Bronze 57 10-15	Sweden	Castings: centrifugal or continuous	83.0-89.5 **Cu**, 0.30 **Sn**, 0.20 **Pb**, 0.40 **Zn**, 2.0-4.0 **Fe**, 0.20 **Si**, 8.5-10.5 **Al**, 1.0 **Mn**, 3.0 **Ni**	**As cast**: all **diam**, 550 MPa **TS**, 200 MPa **YS**, 15% in 2 in. (50 mm) **El**, 115 **HB**
SIS 14 57 16	SIS Bronze 57 16-00	Sweden	Ingot	77.0-82.0 **Cu**, 0.10 **Sn**, 0.02 **Pb**, 0.20 **Zn**, 3.5-5.3 **Fe**, 0.07 **Si**, 9.0-10.0 **Al**, 2.3 **Mn**, 4.5-6.3 **Ni**, 0.005 **Cr**, 0.04 **Mg**	
AS 1565	958C	Australia	Ingot, castings	rem **Cu**, 0.10 **Sn**, 0.05 **Pb**, 0.50 **Zn**, 4.0-5.5 **Fe**, 0.10 **Si**, 8.8-10.0 **Al**, 1.5 **Mn**, 4.0-5.5 **Ni**, 0.05 **Mg**, 0.30 total, others (castings)	**Sand cast**: all **diam**, 640 MPa **TS**, 250 MPa **YS**, 13% in 2 in. (50 mm) **El**, 140 **HB**; **Chill cast**: all **diam**, 650 MPa **TS**, 250 MPa **YS**, 13% in 2 in. (50 mm) **El**, 160 **HB**; **Centrifugal cast**: all **diam**, 670 MPa **TS**, 250 MPa **YS**, 13% in 2 in. (50 mm) **El**, 140 **HB**

CAST COPPER AND COPPER ALLOYS

specification number	designation	country	product forms	chemical composition	mechanical properties and hardness values
DS 3001	Aluminum bronze 5716	Denmark	Sand castings	77.0–82.0 **Cu**, 0.20 **Sn**, 0.05 **Pb**, 0.20 **Zn**, 3.5–5.5 **Fe**, 0.10 **Si**, 8.8–10.0 **Al**, 0.3–2.5 **Mn**, 4.5–6.5 **Ni**, 0.005 **Cr**	**As cast**: all **diam**, 640 MPa **TS**, 250 MPa **YS**, 13% in 2 in. (50 mm) **El**, 140 min **HB**
SFS 2212	CuAl10Fe5Ni5	Finland	Sand castings	77.0–82.0 **Cu**, 0.20 **Sn**, 0.05 **Pb**, 0.20 **Zn**, 3.5–5.5 **Fe**, 0.10 **Si**, 8.8–10.0 **Al**, 0.3–2.5 **Mn**, 4.5–6.5 **Ni**, 0.005 **Cr**, 0.05 **Mg**	**As cast**: all **diam**, 640 MPa **TS**, 250 MPa **YS**, 13% in 2 in. (50 mm) **El**
SABS 200	CuAl10Ni6Fe4	South Africa	Castings	rem **Cu**, 0.1 **Sn**, 0.05 **Pb**, 0.50 **Zn**, 4.0–5.5 **Fe**, 0.10–0.1 **Si**, 8.8–10.0 **Al**, 1.5 **Mn**, 4.0–5.5 **Ni**, 0.05 **Mg**	**Sand cast**: all **diam**, 640 MPa **TS**, 250 MPa **YS**, 13% in 2 in. (50 mm) **El**; **Gravity die cast**: 650 MPa **TS**, 250 MPa **YS**, 13% in 2 in. (50 mm) **El**
SIS 14 57 16	SIS Bronze 57 16–03	Sweden	Sand castings	77.0–82.0 **Cu**, 0.20 **Sn**, 0.05 **Pb**, 0.20 **Zn**, 3.5–5.5 **Fe**, 0.10 **Si**, 8.8–10.0 **Al**, 2.5 **Mn**, 4.5–6.5 **Ni**, 0.005 **Cr**, 0.05 **Mg**	**As cast**: all **diam**, 640 MPa **TS**, 250 MPa **YS**, 13% in 2 in. (50 mm) **El**, 140 **HB**
SIS 14 57 16	SIS Bronze 57 16–06	Sweden	Chill castings	77.0–82.0 **Cu**, 0.20 **Sn**, 0.05 **Pb**, 0.20 **Zn**, 3.5–5.5 **Fe**, 0.10 **Si**, 8.8–10.0 **Al**, 2.5 **Mn**, 4.5–6.5 **Ni**, 0.005 **Cr**, 0.05 **Mg**	**As cast**: all **diam**, 650 MPa **TS**, 250 MPa **YS**, 13% in 2 in. (50 mm) **El**, 150 **HB**
SIS 14 57 16	SIS Bronze 57 16–15	Sweden	Castings: centrifugal or continuous	77.0–85.0 **Cu**, 0.20 **Sn**, 0.05 **Pb**, 0.20 **Zn**, 3.5–5.5 **Fe**, 0.10 **Si**, 8.8–10.0 **Al**, 2.5 **Mn**, 4.5–6.5 **Ni**, 0.005 **Cr**, 0.05 **Mg**	**As cast**: all **diam**, 670 MPa **TS**, 250 MPa **YS**, 13% in 2 in. (50 mm) **El**, 160 **HB**
NF A 53–709	U–A9 N5 Fe Y200	France	Sand castings	rem **Cu**, 0.2 **Sn**, 0.2 **Pb**, 0.5 **Zn**, 3–6 **Fe**, 8.2–10.5 **Al**, 3.0 **Mn**, 4.6–5 **Ni**	**As cast**: all **diam**, 630 MPa **TS**, 240 MPa **YS**, 12% in 2 in. (50 mm) **El**
JIS H 2206	Class 2	Japan	Ingot: for castings	80.0 min **Cu**, 2.5–5.0 **Fe**, 8.0–10.5 **Al**, 1.5 **Mn**, 1.0–3.0 **Ni**	
NF A 53–709	U–A9 Fe3 Y300	France	Die castings	rem **Cu**, 2–4 **Fe**, 8.5–10.0 **Al**	**As cast**: all **diam**, 600 MPa **TS**, 20% in 2 in. (50 mm) **El**
NF A 53–709	U–A9 N3 Fe Y300	France	Die castings	rem **Cu**, 2–3.0 **Fe**, 8.5–10.0 **Al**, 2.5–4.0 **Ni**	**As cast**: all **diam**, 600 MPa **TS**, 20% in 2 in. (50 mm) **El**
ISO 1338	CuAl9	ISO	Castings, ingot	88.0–92.0 **Cu**, 0.30 **Sn**, 0.30 **Pb**, 0.50 **Zn**, 1.2 **Fe**, 0.20 **Si**, 8.0–10.5 **Al**, 0.50 **Mn**, 1.0 **Ni**, **Cu** total includes **Ni**	**Permanent mould cast**: 12–25 mm **diam**, 450 MPa **TS**, 15% in 2 in. (50 mm) **El**
JIS H 5114	Class 2	Japan	Castings	78 min **Cu**, 2.5–5.0 **Fe**, 8.0–10.5 **Al**, 1.5 **Mn**, 1.0–3.0 **Ni**	**As cast**: all **diam**, 490 min MPa **TS**, 20% in 2 in. (50 mm) **El**, 120 min **HB**
NF A 53–709	U–A9 N3 Fe Y200	France	Sand castings	rem **Cu**, 0.2 **Sn**, 0.2 **Pb**, 0.5 **Zn**, 1.5–3.0 **Fe**, 0.2 **Si**, 8.2–10.0 **Al**, 3.0 **Mn**, 1.5–4.0 **Ni**	**As cast**: all **diam**, 500 MPa **TS**, 180 MPa **YS**, 18% in 2 in. (50 mm) **El**
ANSI/ASTM B 148	CA 952	US	Sand castings	86.0 min **Cu**, 2.5–4.0 **Fe**, 8.5–9.5 **Al**	**As cast**: all **diam**, 65 ksi (450 MPa) **TS**, 25 ksi (170 MPa) **YS**, 20% in 2 in. (50.8 mm) **El**, 110 **HB**
ANSI/ASTM B 148	CA 958	US	Sand castings	79.0 min **Cu**, 3.5–4.5 **Fe**, 0.10 **Si**, 8.5–9.5 **Al**, 0.8–1.5 **Mn**, 4.0–5.0 **Ni**+**Co**, 0.03 **P**	**As cast**: all **diam**, 85 ksi (585 MPa) **TS**, 35 ksi (240 MPa) **YS**, 15% in 2 in. (50.8 mm) **El**

CAST COPPER AND COPPER ALLOYS

specification number	designation	country	product forms	chemical composition	mechanical properties and hardness values
ANSI/ASTM B 505	CA 952	US	Castings	86.0 min **Cu**, 2.5–4.0 **Fe**, 8.5–9.5 **Al**	**As cast**: all **diam**, 68 ksi (469 MPa) **TS**, 26 ksi (179 MPa) **YS**, 20% in 2 in. (50 mm) **El**
ANSI/ASTM B 505	CA 958	US	Castings	79.0 min **Cu**, 0.03 **Pb**, 3.5–4.5 **Fe**, 0.10 **Si**, 8.5–9.5 **Al**, 0.8–1.5 **Mn**, 4.0–5.0 **Ni+Co**	**As cast**: all **diam**, 85 ksi (598 MPa) **TS**, 35 ksi (246 MPa) **YS**, 18% in 2 in. (50 mm) **El**
ANSI/ASTM B 30	952	US	Ingot: for castings	86.0 min **Cu**, 2.5–4.0 **Fe**, 8.5–9.5 **Al**	
ANSI/ASTM B 30	958	US	Ingot: for castings	78.0 min **Cu**, 0.02 **Pb**, 3.0–5.0 **Fe**, 0.05 **Si**, 8.5–9.5 **Al**, 3.5 **Mn**, 4.0–5.5 **Ni+Co**	
ANSI/ASTM B 30	CA 952	US	Ingot	86.0 min **Cu**, 2.5–4.0 **Fe**, 8.5–9.5 **Al**	**As cast**: all **diam**, 65 ksi (450 MPa) **TS**, 25 ksi (170 MPa) **YS**, 20% in 2 in. (50.8 mm) **El**, 110 **HB**
ANSI/ASTM B 30	CA 958	US	Ingot	78.0 min **Cu**, 0.02 **Pb**, 3.0–5.0 **Fe**, 0.05 **Si**, 8.5–9.5 **Al**, 3.5 **Mn**, 4.0–5.5 **Ni**	**As cast**: all **diam**, 85 ksi (585 MPa) **TS**, 35 ksi (240 MPa) **YS**, 15% in 2 in. (50.8 mm) **El**
CDA C 95200		US	Castings	86.0 min **Cu**, 2.5–4.0 **Fe**, 8.5–9.5 **Al**	**Sand castings**: all **diam**, 65 ksi (448 MPa) **TS**, 25 ksi (172 MPa) **YS**, 20% in 2 in. (50 mm) **El**, 125 **HB**
CDA C 95800		US	Castings	79.0 min **Cu**, 0.03 **Pb**, 3.5–4.5 **Fe**, 0.10 **Si**, 8.5–9.5 **Al**, 0.8–1.5 **Mn**, 4.0–5.0 **Ni**	**Sand castings**: all **diam**, 85 ksi (586 MPa) **TS**, 35 ksi (241 MPa) **YS**, 15% in 2 in. (50 mm) **El**, 159 **HB**
COPANT 162	C 958 00	COPANT	Cast products	79 min **Cu**, 0.03 **Pb**, 3.5–4.5 **Fe**, 0.10 **Si**, 8.5–9.5 **Al**, 0.8–1.5 **Mn**, 4.0–5.0 **Ni**, 99.5 total elements	
COPANT 162	C 952 00	COPANT	Cast products	86 min **Cu**, 2.5–4.0 **Fe**, 8.5–9.5 **Al**, 99.0 total elements	
QQ–C–390A	958	US	Castings	78.0 min **Cu**, 0.02 **Pb**, 3.0–5.0 **Fe**, 0.10 **Si**, 8.5–9.5 **Al**, 3.5 **Mn**, 4.0–5.5 **Ni**, 0.50 total, others	**As cast or annealed**: all **diam**, 85 ksi (586 MPa) **TS**, 35 ksi (241 MPa) **YS**, 18% in 2 in. (50 mm) **El**
QQ–C–390A	952	US	Castings	86.0 min **Cu**, 2.5–4.0 **Fe**, 8.5–9.5 **Al**, 1.0 total, others	**As cast**: all **diam**, 65 ksi (448 MPa) **TS**, 25 ksi (172 MPa) **YS**, 20% in 2 in. (50 mm) **El**
QQ–B–675B	958	US	Ingot	78.0 min **Cu**, 0.02 **Pb**, 3.0–5.0 **Fe**, 0.05 **Si**, 8.5–9.5 **Al**, 0.8–1.5 **Mn**, 4.0–5.5 **Ni**, 0.5 total, others	
QQ–B–675B	952	US	Ingot	86.0 min **Cu**, 2.5–4.0 **Fe**, 8.5–9.5 **Al**, 1.0 total, others	
JIS H 2206	Class 1	Japan	Ingot: for castings	85.0 min **Cu**, 1.0–4.0 **Fe**, 8.0–10.0 **Al**, 1.0 **Mn**, 1.0 **Ni**	

CAST COPPER AND COPPER ALLOYS

specification number	designation	country	product forms	chemical composition	mechanical properties and hardness values
COPANT 801	C 952 00	COPANT	Continuous–cast rod, bar and tube	86 min **Cu**, 2.5–4.0 **Fe**, 8.5–9.5 **Al**, 99.0 min, total, others	**As manufactured**: 140 mm **diam**, 470 MPa **TS**, 180 MPa **YS**, 20% in 2 in. (50 mm) **El**
JIS H 5114	Class 1	Japan	Castings	85 min **Cu**, 1.0–4.0 **Fe**, 8.0–10.0 **Al**, 1.0 **Mn**, 1.0 **Ni**	**As cast**: all **diam**, 441 min MPa **TS**, 20% in 2 in. (50 mm) **El**, 90 min **HB**
MIL–C–22229A	952	US	Castings	86.0 min **Cu**, 2.5–4.0 **Fe**, 8.5–9.5 **Al**	**As cast**: all **diam**, 448 MPa **TS**, 172 MPa **YS**, 20% in 2 in. (50 mm) **El**
MIL–C–22229A	958	US	Castings	78.0 min **Cu**, 0.02 **Pb**, 3.0–5.0 **Fe**, 0.10 **Si**, 8.5–9.5 **Al**, 3.5 **Mn**, 4.0–5.5 **Ni**	**As cast**: all **diam**, 586 MPa **TS**, 241 MPa **YS**, 18% in 2 in. (50 mm) **El**
MIL–B–24480		US	Castings	79.0 min **Cu**, 0.03 **Pb**, 3.5–4.5 **Fe**, 0.10 **Si**, 8.5–9.5 **Al**, 0.75–1.5 **Mn**, 4.0–5.0 **Ni**	**Annealed and heat treated**: all **diam**, 586 MPa **TS**, 241 MPa **YS**, 15% in 2 in. (50 mm) **El**
AS 1565	959D	Australia	Ingot, castings	rem **Cu**, 1.0 **Sn**, 0.05 **Pb**, 0.50 **Zn**, 2.0–4.0 **Fe**, 0.15 **Si**, 8.5–9.0 **Al**, 11.0–15.0 **Mn**, 1.5–4.5 **Ni**, 0.05 **P**, 0.30 total, others	**Sand cast**: all **diam**, 740 MPa **TS**, 380 MPa **YS**, 9% in 2 in. (50 mm) **El**, 200 **HB**
AS 1565	957B	Australia	Ingot, castings	rem **Cu**, 1.0 **Sn**, 0.05 **Pb**, 0.50 **Zn**, 2.0–4.0 **Fe**, 0.15 **Si**, 7.5–8.5 **Al**, 11.0–15.0 **Mn**, 1.5–4.5 **Ni**, 0.05 **P**, 0.30 total, others	**Sand cast**: all **diam**, 650 MPa **TS**, 280 MPa **YS**, 18% in 2 in. (50 mm) **El**, 160 **HB**; **Chill cast**: all **diam**, 670 MPa **TS**, 310 MPa **YS**, 27% in 2 in. (50 mm) **El**
ANSI/ASTM B 148	CA 957	US	Sand castings	71.0 min **Cu**, 2.0–4.0 **Fe**, 0.10 **Si**, 7.0–8.5 **Al**, 11.0–14.0 **Mn**, 1.5–3.0 **Ni + Co**, 0.03 **P**	**As cast**: all **diam**, 90 ksi (620 MPa) **TS**, 40 ksi (275 MPa) **YS**, 20% in 2 in. (50.8 mm) **El**
ANSI/ASTM B 30	957	US	Ingot: for castings	71.0 min **Cu**, 0.03 **Pb**, 2.0–4.0 **Fe**, 0.10 **Si**, 7.0–8.5 **Al**, 11.0–14.0 **Mn**, 1.5–3.0 **Ni + Co**	
ANSI/ASTM B 30	CA 957	US	Ingot	71.0 min **Cu**, 0.03 **Pb**, 2.0–4.0 **Fe**, 0.10 **Si**, 7.0–8.5 **Al**, 11.0–14.0 **Mn**, 1.5–3.0 **Ni**	**As cast**: all **diam**, 90 ksi (620 MPa) **TS**, 40 ksi (275 MPa) **YS**, 20% in 2 in. (50.8 mm) **El**
COPANT 162	C 957 00	COPANT	Cast products	71 min **Cu**, 0.03 **Pb**, 2.0–4.0 **Fe**, 0.10 **Si**, 7–8.5 **Al**, 11–14 **Mn**, 1.5–3.0 **Ni**, 99.0 total elements	
QQ–C–390A	957	US	Castings	71.0 min **Cu**, 0.03 **Pb**, 2.0–4.0 **Fe**, 0.10 **Si**, 7.0–8.5 **Al**, 11.0–14.0 **Mn**, 1.5–3.0 **Ni**, 0.50 total, others	**As cast**: all **diam**, 90 ksi (621 MPa) **TS**, 40 ksi (776 MPa) **YS**, 20% in 2 in. (50 mm) **El**
QQ–B–675B	957	US	Ingot	71.0 min **Cu**, 0.03 **Pb**, 2.0–4.0 **Fe**, 0.10 **Si**, 7.0–8.5 **Al**, 11.0–14.0 **Mn**, 1.5–3.0 **Ni**, 0.5 total, others	
MIL–B–21230A	Alloy 2	US	Castings	71 min **Cu**, 0.03 **Pb**, 2.0–4.0 **Fe**, 0.10 **Si**, 7.0–8.5 **Al**, 11–14 **Mn**, 1.5–3.0 **Ni**, 0.50 total, others	**As cast**: all **diam**, 621 MPa **TS**, 276 MPa **YS**, 20% in 2 in. (50 mm) **El**

CAST COPPER AND COPPER ALLOYS

specification number	designation	country	product forms	chemical composition	mechanical properties and hardness values
ANSI/ASTM E 272		US	Castings	71.0 min **Cu**, 0.03 **Pb**, 2.0–4.0 **Fe**, 0.10 **Si**, 7.0–8.5 **Al**, 11.0–14.0 **Mn**, 1.5–3.0 **Ni**	**As cast**: all **diam**, 90 ksi (620 MPa) **TS**, 40 ksi (275 MPa) **YS**, 20% in 2 in. (51 mm) **El**
JIS H 5114	Class 4	Japan	Castings	71 min **Cu**, 2.0–5.0 **Fe**, 6.0–9.0 **Al**, 7.0–15.0 **Mn**, 1.0–4.0 **Ni**	**As cast**: all **diam**, 588 min MPa **TS**, 15% in 2 in. (50 mm) **El**, 150 min **HB**
ANSI/ASTM B 148	CA 956	US	Sand castings	88.0 min **Cu**, 1.75–3.25 **Si**, 6.0–8.0 **Al**, 0.25 **Ni+Co**,	**99.0 min Cu+Si+Al+Ni+Co**
ANSI/ASTM B 30	956	US	Ingot: for castings	88.0 min **Cu**, 1.8–3.3 **Si**, 6.0–8.0 **Al**, 0.25 **Ni+Co**	
ANSI/ASTM B 30	CA 956	US	Ingot	88.0 min **Cu**, 1.8–3.3 **Si**, 6.0–8.0 **Al**, 0.25 **Ni**	**As cast**: all **diam**, 60 ksi (415 MPa) **TS**, 28 ksi (195 MPa) **YS**, 10% in 2 in. (50.8 mm) **El**
CDA C 95600		US	Castings	88.0 min **Cu**, 1.8–3.3 **Si**, 6.0–8.0 **Al**, 0.25 **Ni**	**Sand castings**: all **diam**, 60 ksi (414 MPa) **TS**, 28 ksi (193 MPa) **YS**, 10% in 2 in. (50 mm) **El**, 140 **HB**
COPANT 162	C 956 00	COPANT	Cast products	88 min **Cu**, 1.8–3.3 **Si**, 6–8 **Al**, 0.25 **Ni**, 99.0 total elements	
QQ–B–675B	956	US	Ingot	88.0 min **Cu**, 1.8–3.3 **Si**, 6.0–8.0 **Al**, 0.25 **Ni**, 1.0 total, others	
SFS 2213	CuSn10	Finland	Sand castings	88.0–91.0 **Cu**, 9.0–11.0 **Sn**, 1.0 **Pb**, 0.5 **Zn**, 0.20 **Fe**, 0.2 **Sb**, 0.01 **Si**, 0.01 **Al**, 0.2 **Mn**, 2.0 **Ni**, 0.20 **P**, 0.05 **S**	**As cast**: all **diam**, 270 MPa **TS**, 130 MPa **YS**, 18% in 2 in. (50 mm) **El**
SIS 14 56 82	SIS Cu 56 82–12	Sweden	Tube	67 **Cu**, 0.05 **Pb**, 0.5 **Zn**, 0.4–1.0 **Fe**, 0.5–1.5 **Mn**, 30.0–32.0 **Ni**, 0.05 **S**, 0.06 **C**, **Sn** in **Pb** content; 0.2 total, others	**Annealed**: 5 max mm **diam**, 370 MPa **TS**, 140 MPa **YS**, 30% in 2 in. (50 mm) **El**, 90–120 **HV**
SIS 14 56 82	SIS Cu 56 82–02	Sweden	Tube	67 **Cu**, 0.05 **Pb**, 0.5 **Zn**, 0.4–1.0 **Fe**, 0.5–1.5 **Mn**, 30.0–32.0 **Ni**, 0.05 **S**, 0.06 **C**, **Sn** in **Pb** content; 0.2 total, others	**Annealed**: 3 max mm **diam**, 380 MPa **TS**, 150 MPa **YS**, 30% in 2 in. (50 mm) **El**, 85–120 **HV**
CDA C 96600		US	Castings	rem **Cu**, 0.01 **Pb**, 0.8–1.1 **Fe**, 0.15 **Si**, 1.0 **Mn**, 29.0–33.0 **Ni**, 0.40–0.7 **Be**	**Heat treated**: all **diam**, 110 ksi (758 MPa) **TS**, 70 ksi (483 MPa) **YS**, 7 in 2 in. (50 mm) **El**, 230 **HB**
COPANT 162	C 966 00	COPANT	Cast products	rem **Cu**, 0.01 **Pb**, 0.8–1.1 **Fe**, 0.15 **Si**, 1.0 **Mn**, 29–33 **Ni**, 0.4–0.7 **Be**	
ANSI/ASTM B 30	CA 964	US	Ingot	65.0–67.0 **Cu**, 0.005 **Pb**, 0.25–1.0 **Fe**, 0.30–0.50 **Si**, 0.005 **Al**, 0.8–1.5 **Mn**, 29.5–31.5 **Ni**, 0.02 **P**, 0.02 **S**, 0.7–1.5 **Nb**, 0.05 **C**	**As cast**: all **diam**, 60 ksi (415 MPa) **TS**, 32 ksi (220 MPa) **YS**, 20 in 2 in. (50.8 mm) **El**
MIL–C–20159B	Type I	US	Castings	rem **Cu**, 0.01 **Pb**, 0.25–1.00 **Fe**, 0.70 **Si**, 1.50 **Mn**, 28.0–32.0 **Ni**, 1.50 **Nb**, 0.15 **C**	**As cast**: all **diam**, 414 MPa **TS**, 221 MPa **YS**, 15% in 2 in. (50 mm) **El**

CAST COPPER AND COPPER ALLOYS

specification number	designation	country	product forms	chemical composition	mechanical properties and hardness values
QQ-C-390A	964	US	Castings	rem **Cu**, 0.03 **Pb**, 0.25–1.5 **Fe**, 0.7 **Si**, 1.5 **Mn**, 28.0–32.0 **Ni**, 1.5 **Nb**	**As cast**: all **diam**, 60 ksi (414 MPa) **TS**, 32 ksi (221 MPa) **YS**, 20 in 2 in. (50 mm) **El**
CDA C 96400		US	Castings	65.0–69.0 **Cu**, 0.03 **Pb**, 0.25–1.5 **Fe**, 0.50 **Si**, 1.5 **Mn**, 28.0–32.0 **Ni**, 0.50–1.5 **Nb**, 0.15 **C**	**Sand castings**: all **diam**, 60 ksi (414 MPa) **TS**, 32 ksi (221 MPa) **YS**, 20 in 2 in. (50 mm) **El**, 140 **HB**
COPANT 162	C 964 00	COPANT	Cast products	65–69 **Cu**, 0.03 **Pb**, 0.25–1.5 **Fe**, 0.5 **Si**, 1.5 **Mn**, 28–32 **Ni**, 0.15 **C**, 0.5–1.5 **Cb**	
COPANT 801	C 964 00	COPANT	Continuous-cast rod, bar and tube	65–69 **Cu**, 0.03 **Pb**, 0.25–1.5 **Fe**, 0.5 **Si**, 1.5 **Mn**, 28–32 **Ni**, 0.15 **C**, 0.5–1.5 **Cb**	**As manufactured**: 127 mm **diam**, 450 MPa **TS**, 250 MPa **YS**, 25% in 2 in. (50 mm) **El**
CDA C 96300		US	Castings	rem **Cu**, 0.03 **Pb**, 0.40–1.0 **Fe**, 0.7 **Si**, 1.0 **Mn**, 18.0–22.0 **Ni**, 1.0 **Nb**	**Sand castings**: all **diam**, 75 ksi (517 MPa) **TS**, 55 ksi (379 MPa) **YS**, 10 in 2 in. (50 mm) **El**, 150 **HB**
COPANT 162	C 963 00	COPANT	Cast products	rem **Cu**, 0.03 **Pb**, 0.4–1.0 **Fe**, 0.7 **Si**, 1.0 **Mn**, 18–22 **Ni**, 1.0 **Cb**	
QQ-C-390A	962	US	Castings	rem **Cu**, 0.03 **Pb**, 1.0–1.8 **Fe**, 0.25 **Si**, 1.5 **Mn**, 9.0–11.0 **Ni**, 1.0 **Nb**	**As cast**: all **diam**, 45 ksi (310 MPa) **TS**, 25 ksi (172 MPa) **YS**, 20 in 2 in. (50 mm) **El**
MIL-C-20159B	Type II	US	Castings	rem **Cu**, 0.1 **Pb**, 1.00–1.80 **Fe**, 0.50 **Si**, 1.50 **Mn**, 9.0–11.0 **Ni**, 1.00 **Nb**, 0.10 **C**	**As cast**: all **diam**, 310 MPa **TS**, 172 MPa **YS**, 20% in 2 in. (50 mm) **El**
SIS 14 56 67	SIS Bronze 56 67–02	Sweden	Tube	88 **Cu**, 0.05 **Pb**, 0.5 **Zn**, 1.0–1.8 **Fe**, 0.5–1.0 **Mn**, 9.0–11.0 **Ni**, 0.05 **S**, 0.05 **C**, **Sn** in **Pb** content; 0.1 total, others	**Annealed**: 5 max mm **diam**, 280 MPa **TS**, 80 MPa **YS**, 30% in 2 in. (50 mm) **El**, 60–85 **HV**
SIS 14 56 67	SIS Bronze 56 67–12	Sweden	Tube	88 **Cu**, 0.05 **Pb**, 0.5 **Zn**, 1.0–1.8 **Fe**, 0.5–1.0 **Mn**, 9.0–11.0 **Ni**, 0.05 **S**, 0.05 **C**, **Sn** in **Pb** content; 0.1 total, others	**Annealed**: 3 max mm **diam**, 290 MPa **TS**, 100 MPa **YS**, 35% in 2 in. (50 mm) **El**, 65–85 **HV**
SIS 14 56 67	SIS Bronze 56 67–03	Sweden	Tube	88 **Cu**, 0.05 **Pb**, 0.5 **Zn**, 1.0–1.8 **Fe**, 0.5–1.0 **Mn**, 9.0–11.0 **Ni**, 0.05 **S**, 0.05 **C**, **Sn** in **Pb** content; 0.1 total, others	**Strain hardened**: 5 max mm **diam**, 340 MPa **TS**, 290 MPa **YS**, 20% in 2 in. (50 mm) **El**, 110–135 **HV**
ANSI/ASTM B 30	CA 962	US	Ingot	84.5–87.0 **Cu**, 0.005 **Pb**, 1.0–1.8 **Fe**, 0.25 **Si**, 0.005 **Al**, 0.8–1.5 **Mn**, 9.0–11.0 **Ni**, 0.02 **P**, 0.02 **S**, 1.0 **Nb**, 0.05 **C**	**As cast**: all **diam**, 45 ksi (310 MPa) **TS**, 25 ksi (170 MPa) **YS**, 20 in 2 in. (50.8 mm) **El**
CDA C 96200		US	Castings	84.5–87.0 **Cu**, 0.03 **Pb**, 1.0–1.8 **Fe**, 0.30 **Si**, 1.5 **Mn**, 9.0–11.0 **Ni**, 1.0 **Nb**, 0.15 **C**	**Sand castings**: all **diam**, 45 ksi (310 MPa) **TS**, 25 ksi (172 MPa) **YS**, 20 in 2 in. (50 mm) **El**
COPANT 162	C 962 00	COPANT	Cast products	84.5–87.0 **Cu**, 0.03 **Pb**, 1.0–1.8 **Fe**, 0.30 **Si**, 1.5 **Mn**, 9.0–11.0 **Ni**, 0.15 **C**, 1.0 **Cb**	
ANSI/ASTM B 369	C 96200	US	Castings	84.5 min **Cu**, 0.03 **Pb**, 1.0–1.8 **Fe**, 0.30 **Si**, 1.5 **Mn**, 9.0–11.0 **Ni+Co**, 1.0 **Nb**, 0.10 **C**	**As cast**: all **diam**, 45 ksi (310 MPa) **TS**, 25 ksi (170 MPa) **YS**, 20 in 2 in. (50.8 mm) **El**

CAST COPPER AND COPPER ALLOYS

specification number	designation	country	product forms	chemical composition	mechanical properties and hardness values
ANSI/ASTM B 369	C 96400	US	Castings	65.0 min **Cu**, 0.03 **Pb**, 0.30–1.5 **Fe**, 0.50 **Si**, 1.5 **Mn**, 28.0–32.0 **Ni+Co**, 1.5 **Nb**, 0.15 **C**	**As cast**: all **diam**, 60 ksi (415 MPa) **TS**, 32 ksi (220 MPa) **YS**, 20 in 2 in. (50.8 mm) **El**
ANSI/ASTM B 30	962	US	Ingot: for castings	84.5–87.0 **Cu**, 0.05 **C**, 0.005 **Pb**, 1.0 **Cb**, 1.0–1.8 **Fe**, 0.25 **Si**, 0.005 **Al**, 0.8–1.5 **Mn**, 9.0–11.0 **Ni+Co**, 0.02 **P**, 0.02 **S**	
ANSI/ASTM B 505	CA 964	US	Castings	65.0–69.0 **Cu**, 0.03 **Pb**, 0.25–1.50 **Fe**, 0.5 **Si**, 1.5 **Mn**, 28.0–32.0 **Ni+Co**, 0.02 **P**, 0.02 **S**, 1.5 **Nb**, 0.15 **C**	**As cast**: all **diam**, 65 ksi (448 MPa) **TS**, 35 ksi (241 MPa) **YS**, 25 in 2 in. (50 mm) **El**
ANSI/ASTM B 492		US	Sleeves	74.0–80.0 **Cu**, 0.01 **Pb**, 0.1 **Zn**, 0.5–1.5 **Fe**, 0.03–0.8 **Si**, 0.05 **Al**, 0.25–1.5 **Mn**, 18.0–22.0 **Ni+Co**, 0.02 **P**	**As cast**: all **diam**, 70 ksi (483 MPa) **TS**, 50 ksi (345 MPa) **YS**, 8% in 2 in. (50.8 mm) **El**, 150 **HB**
CDA C 97800		US	Castings	64.0–67.0 **Cu**, 4.0–5.5 **Sn**, 1.0–2.5 **Pb**, 1.0–4.0 **Zn**, 1.5 **Fe**, 0.20 **Sb**, 0.15 **Si**, 0.005 **Al**, 1.0 **Mn**, 24.0–27.0 **Ni**, 0.05 **P**, 0.08 **S**	**Sand castings**: all **diam**, 50 ksi (345 MPa) **TS**, 22 ksi (152 MPa) **YS**, 10 in 2 in. (50 mm) **El**, 130 **HB**
COPANT 162	C 978 00	COPANT	Cast products	64–67 **Cu**, 4–5.5 **Sn**, 1–2.5 **Pb**, 1.0–4.0 **Zn**, 1.5 **Fe**, 0.20 **Sb**, 0.15 **Si**, 0.005 **Al**, 1.0 **Mn**, 24–27 **Ni**, 0.05 **P**, 0.08 **S**	
ANSI/ASTM B 30	CA 978	US	Ingot	64.0–67.0 **Cu**, 4.5–5.5 **Sn**, 1.0–2.0 **Pb**, 1.0–4.0 **Zn**, 1.0 **Fe**, 0.20 **Sb**, 0.005 **Si**, 0.005 **Al**, 1.0 **Mn**, 24.0–26.0 **Ni**, 0.05 **P**, 0.08 **S**	
MIL-C-17112A		US	Castings	63.0–67.0 **Cu**, 3.5–4.5 **Sn**, 3.0–5.0 **Pb**, rem **Zn**, 1.5 **Fe**, 1.0 **Mn**, 19.5–21.5 **Ni**	**As cast**: all **diam**, 207 MPa **TS**, 117 MPa **YS**, 8 in 2 in. (50 mm) **El**
ANSI/ASTM B 30	CA 976	US	Ingot	63.0–66.0 **Cu**, 3.5–4.5 **Sn**, 3.5–5.0 **Pb**, 3.0–9.0 **Zn**, 1.0 **Fe**, 0.25 **Sb**, 0.05 **Si**, 0.005 **Al**, 1.0 **Mn**, 19.5–21.0 **Ni**, 0.05 **P**, 0.08 **S**	
CDA C 97600		US	Castings	63.0–67.0 **Cu**, 3.5–4.5 **Sn**, 3.0–5.0 **Pb**, 3.0–9.0 **Zn**, 1.5 **Fe**, 0.25 **Sb**, 0.15 **Si**, 0.005 **Al**, 1.0 **Mn**, 19.0–21.5 **Ni**, 0.05 **P**, 0.08 **S**	**Sand castings**: all **diam**, 40 ksi (276 MPa) **TS**, 17 ksi (117 MPa) **YS**, 10 in 2 in. (50 mm) **El**, 80 **HB**
COPANT 162	C 976 00	COPANT	Cast products	63–67 **Cu**, 3.5–4.5 **Sn**, 3–5 **Pb**, 3–9 **Zn**, 1.5 **Fe**, 0.25 **Sb**, 0.15 **Si**, 1.0 **Mn**, 19–21.5 **Ni**, 0.05 **P**, 0.08 **S**	
CDA C 97400		US	Castings	58.0–61.0 **Cu**, 2.5–3.5 **Sn**, 4.5–5.5 **Pb**, rem **Zn**, 1.5 **Fe**, 0.50 **Mn**, 15.5–17.0 **Ni**	**Sand castings**: all **diam**, 30 ksi (207 MPa) **TS**, 16 ksi (110 MPa) **YS**, 8 in 2 in. (50 mm) **El**, 70 **HB**
COPANT 162	C 974 00	COPANT	Cast products	58–61 **Cu**, 2.5–3.5 **Sn**, 4.5–5.5 **Pb**, rem **Zn**, 1.5 **Fe**, 0.5 **Mn**, 15.5–17.0 **Ni**	

CAST COPPER AND COPPER ALLOYS

specification number	designation	country	product forms	chemical composition	mechanical properties and hardness values
ANSI/ASTM B 30	CA 973	US	Ingot	53.0–58.0 **Cu**, 1.5–3.0 **Sn**, 8.0–11.0 **Pb**, 17.0–25.0 **Zn**, 1.0 **Fe**, 0.35 **Sb**, 0.05 **Si**, 0.005 **Al**, 0.5 **Mn**, 11.0–14.0 **Ni**, 0.05 **P**, 0.08 **S**	
CDA C 97300		US	Castings	53.0–58.0 **Cu**, 1.5–3.0 **Sn**, 8.0–11.0 **Pb**, 17.0–25.0 **Zn**, 1.5 **Fe**, 0.35 **Sb**, 0.15 **Si**, 0.005 **Al**, 0.50 **Mn**, 11.0–14.0 **Ni**, 0.05 **P**, 0.08 **S**	**Sand castings**: all **diam**, 30 ksi (207 MPa) **TS**, 15 ksi (103 MPa) **YS**, 8 in 2 in. (50 mm) **El**, 55 **HB**
COPANT 162	C 973 00	COPANT	Cast products	53–58 **Cu**, 1.5–3.0 **Sn**, 8–11 **Pb**, 17–25 **Zn**, 1.5 **Fe**, 0.35 **Sb**, 0.15 **Si**, 0.005 **Al**, 0.50 **Mn**, 11–14 **Ni**, 0.05 **P**, 0.08 **S**	
ANSI/ASTM B 30	978	US	Ingot: for castings	64.0–67.0 **Cu**, 4.5–5.5 **Sn**, 1.0–2.0 **Pb**, 1.0–4.0 **Zn**, 1.0 **Fe**, 0.20 **Sb**, 0.05 **Si**, 0.005 **Al**, 1.0 **Mn**, 24.0–26.0 **Ni + Co**, 0.05 **P**, 0.08 **S**	
ANSI/ASTM B 30	976	US	Ingot: for castings	63.0–66.0 **Cu**, 3.5–4.5 **Sn**, 3.5–5.0 **Pb**, 3.0–9.0 **Zn**, 1.0 **Fe**, 0.25 **Sb**, 0.05 **Si**, 0.005 **Al**, 1.0 **Mn**, 19.5–21.0 **Ni + Co**, 0.05 **P**, 0.08 **S**	
ANSI/ASTM B 30	973	US	Ingot: for castings	53.0–58.0 **Cu**, 1.5–3.0 **Sn**, 8.0–11.0 **Pb**, 17.0–25.0 **Zn**, 1.0 **Fe**, 0.35 **Sb**, 0.05 **Si**, 0.005 **Al**, 0.5 **Mn**, 11.0–14.0 **Ni + Co**, 0.05 **P**, 0.08 **S**	
ANSI/ASTM B 584	C 97300	US	Sand castings	53.0–58.0 **Cu**, 1.5–3.0 **Sn**, 8.0–11.0 **Pb**, 17.0–25.0 **Zn**, 1.5 **Fe**, 0.35 **Sb**, 0.15 **Si**, 0.005 **Al**, 0.50 **Mn**, 11.0–14.0 **Ni + Co**, 0.05 **P**, 0.08 **S**	**As cast**: all **diam**, 30 ksi (207 MPa) **TS**, 15 ksi (97 MPa) **YS**, 8 in 2 in. (50 mm) **El**
ANSI/ASTM B 584	C 97600	US	Sand castings	63.0–67.0 **Cu**, 3.5–4.5 **Sn**, 3.0–5.0 **Pb**, 3.0–9.0 **Zn**, 1.5 **Fe**, 0.25 **Sb**, 0.15 **Si**, 0.005 **Al**, 1.0 **Mn**, 19.0–21.5 **Ni + Co**, 0.05 **P**, 0.08 **S**	**As cast**: all **diam**, 40 ksi (276 MPa) **TS**, 17 ksi (117 MPa) **YS**, 10 in 2 in. (50 mm) **El**
ANSI/ASTM B 584	C 97800	US	Sand castings	64.0–67.0 **Cu**, 4.0–5.5 **Sn**, 1.0–2.5 **Pb**, 1.0–4.0 **Zn**, 1.5 **Fe**, 0.20 **Sb**, 0.15 **Si**, 0.005 **Al**, 1.0 **Mn**, 24.0–27.0 **Ni + Co**, 0.05 **P**, 0.08 **S**	**As cast**: all **diam**, 50 ksi (345 MPa) **TS**, 22 ksi (152 MPa) **YS**, 10 in 2 in. (50 mm) **El**
ANSI/ASTM B 505	CA 973	US	Castings	53.0–58.0 **Cu**, 1.5–3.0 **Sn**, 8.0–11.0 **Pb**, 17.0–25.0 **Zn**, 1.5 **Fe**, 0.35 **Sb**, 0.15 **Si**, 0.005 **Al**, 0.50 **Mn**, 11.0–14.0 **Ni + Co**, 0.05 **P**, 0.08 **S**	**As cast**: all **diam**, 30 ksi (211 MPa) **TS**, 15 ksi (105 MPa) **YS**, 8 in 2 in. (50 mm) **El**
ANSI/ASTM B 505	CA 976	US	Castings	63.0–67.0 **Cu**, 3.5–4.5 **Sn**, 3.0–5.0 **Pb**, 3.0–9.0 **Zn**, 1.5 **Fe**, 0.20 **Sb**, 0.15 **Si**, 0.005 **Al**, 1.0 **Mn**, 19.0–25.0 **Ni + Co**, 0.05 **P**, 0.08 **S**	**As cast**: all **diam**, 40 ksi (281 MPa) **TS**, 20 ksi (141 MPa) **YS**, 10 in 2 in. (50 mm) **El**

CAST COPPER AND COPPER ALLOYS

specification number	designation	country	product forms	chemical composition	mechanical properties and hardness values
ANSI/ASTM B 505	CA 978	US	Castings	64.0–67.0 **Cu**, 4.0–5.5 **Sn**, 1.0–2.5 **Pb**, 1.0–4.0 **Zn**, 1.5 **Fe**, 0.20 **Sb**, 0.15 **Si**, 0.005 **Al**, 1.0 **Mn**, 24.0–27.0 **Ni + Co**, 0.05 **P**, 0.08 **S**	**As cast**: all **diam**, 45 ksi (316 MPa) **TS**, 22 ksi (155 MPa) **YS**, 8 in 2 in. (50 mm) **El**
COPANT 162	C 988 00	COPANT	Cast products	56.5–62.5 **Cu**, 0.25 **Sn**, 37.5–42.5 **Pb**, 0.10 **Zn**, 0.35 **Fe**, 0.02 **P**, 5.5 **Ag**	
JIS H 5403	Class 1	Japan	Castings: for bearings	rem **Cu**, 1.0 **Sn**, 38–42 **Pb**, 2.0 **Ag**	
COPANT 162	C 986 00	COPANT	Cast products	60–70 **Cu**, 0.5 **Sn**, 30–40 **Pb**, 0.35 **Fe**, 1.5 **Ag**, 0.30 total, others	
JIS H 5403	Class 2	Japan	Castings: for bearings	rem **Cu**, 1.0 **Sn**, 33–37 **Pb**, 0.80 **Fe**, 2.0 **Ag**	
JIS H 5403	Class 3	Japan	Castings: for bearings	rem **Cu**, 1.0 **Sn**, 28–32 **Pb**, 0.80 **Fe**, 2.0 **Ag**	
COPANT 162	C 984 00	COPANT	Cast products	67–74 **Cu**, 0.25 **Sn**, 25–32 **Pb**, 0.10 **Zn**, 0.35 **Fe**, 1.5 **Ag**, 0.15 total, others	
COPANT 162	C 982 00	COPANT	Cast products	73–79 **Cu**, 0.50 **Sn**, 21–27 **Pb**, 0.35 **Fe**, 0.45 total, others	
JIS H 5403	Class 4	Japan	Castings: for bearings	rem **Cu**, 1.0 **Sn**, 23 **Pb**, 0.80 **Fe**, 2.0 **Ag**, 1.0 total, others	**As cast**: all **diam**, 45 max **HV**
COPANT 162	C 993 00	COPANT	Cast products	rem **Cu**, 0.05 **Sn**, 0.02 **Pb**, 0.40–1.0 **Fe**, 0.02 **Si**, 10.7–11.5 **Al**, 1.0–2.0 **Co**, 13.5–16.5 **Ni**, 0.25 total, others	
CDA C 99300		US	Castings	rem **Cu**, 0.05 **Sn**, 0.02 **Pb**, 0.40–1.0 **Fe**, 0.02 **Si**, 10.7–11.5 **Al**, 1.0–2.0 **Co**, 13.5–16.5 **Ni**	**Sand castings**: all **diam**, 95 ksi (655 MPa) **TS**, 55 ksi (379 MPa) **YS**, 2% in 2 in. (50 mm) **El**, 200 **HB**
COPANT 162	C 994 00	COPANT	Cast products	rem **Cu**, 0.25 **Pb**, 0.5–5.0 **Zn**, 1–3 **Fe**, 0.5–2.0 **Si**, 0.5–2.0 **Al**, 0.5 **Mn**, 1.0–3.5 **Ni**	
CDA 994		US	Centrifugal, continuous and sand castings	rem **Cu**, 0.25 **Pb**, 0.5–2.0 **Zn**, 1.0–3.0 **Fe**, 0.5–2.0 **Si**, 0.50–2.0 **Al**, 0.50 **Mn**, 1.0–3.5 **Ni**	**Sand cast**: all **diam**, 60 ksi (414 MPa) **TS**, 30 ksi (207 MPa) **YS**, 20% in 2 in. (50 mm) **El**, 125 **HB** Solution and precipitation heat treated: all **diam**, 79 ksi (546 MPa) **TS**, 54 ksi (372 MPa) **YS**, 170 **HB**
COPANT 162	C 995 00	COPANT	Cast products	rem **Cu**, 0.25 **Pb**, 0.5–2.0 **Zn**, 3–5 **Fe**, 0.5–2.0 **Si**, 0.5–2.0 **Al**, 0.5 **Mn**, 3.5–5.5 **Ni**	
CDA 995		US	Centrifugal, continuous and sand castings	rem **Cu**, 0.25 **Pb**, 0.5–2.0 **Zn**, 3.0–5.0 **Fe**, 0.5–2.0 **Si**, 0.5–2.0 **Al**, 0.50 **Mn**, 3.5–5.5 **Ni**	**Sand cast**: all **diam**, 70 ksi (483 MPa) **TS**, 40 ksi (276 MPa) **YS**, 12% in 2 in. (50 mm) **El**, 145 **HB** Solution and precipitation heat treated: all **diam**, 86 ksi (593 MPa) **TS**, 62 ksi (427 MPa) **YS**, 8% in 2 in. (50 mm) **El**, 196 **HB**

CAST COPPER AND COPPER ALLOYS

specification number	designation	country	product forms	chemical composition	mechanical properties and hardness values
COPANT 162	C 996 00	COPANT	Cast products	rem **Cu**, 0.10 **Sn**, 0.02 **Pb**, 0.20 **Zn**, 0.20 **Fe**, 0.10 **Si**, 1–2.8 **Al**, 0.20 **Co**, 39–45 **Mn**, 0.20 **Ni**, 0.05 **C**, 99.5 total elements	
CDA C 99700		US	Centrifugal, die, investment, permanent mold, plaster and sand castings	54.0 min **Cu**, 1.0 **Sn**, 2.0 **Pb**, 19.0–25.0 **Zn**, 1.0 **Fe**, 0.5–3.0 **Al**, 11.0–15.0 **Mn**, 4.0–6.0 **Ni**	**Sand cast**: all **diam**, 55 ksi (379 MPa) **TS**, 25 ksi (172 MPa) **YS**, 25% in 2 in. (50 mm) **El**, 110 **HB**
CDA C 99750		US	Castings	55.0–61.0 **Cu**, 0.50–2.5 **Pb**, 17.0–23.0 **Zn**, 1.0 **Fe**, 0.25–3.0 **Al**, 17.0–23.0 **Mn**, 5.0 **Ni**	**Sand castings**: all **diam**, 65 ksi (448 MPa) **TS**, 32 ksi (220 MPa) **YS**, 30% in 2 in. (50 mm) **El**, 77 **HRB**

WROUGHT MAGNESIUM ALLOYS

WROUGHT MAGNESIUM ALLOYS

specification number	designation	country	product forms	chemical composition	mechanical properties and hardness values
MIL-R-6944B	AZ101A	US	Rod and wire	9.5–10.5 **Al**, 0.13 min **Mn**, 0.75–1.25 **Zn**, 0.05 **Cu**, 0.005 **Ni**, 0.005 **Fe**, 0.05 **Si**, 0.0002–0.0008 **Be**, rem **Mg**, 0.30 total, others	
ANSI/ASTM B 93	AZ92A	US	Ingot: for sand, permanent mold and investment castings	8.5–9.5 **Al**, 0.13 min **Mn**, 1.7–2.3 **Zn**, 0.20 **Cu**, rem **Mg**, 0.01 **Ni**, 0.20 **Si**, 0.30 total, others	
ANSI/ASTM B 93	AZ91A	US	Ingot: for die castings	8.5–9.5 **Al**, 0.15 min **Mn**, 0.45–0.9 **Zn**, 0.08 **Cu**, rem **Mg**, 0.01 **Ni**, 0.20 **Si**, 0.30 total, others	
ANSI/ASTM B 93	AZ91B	US	Ingot: for die castings	8.5–9.5 **Al**, 0.15 min **Mn**, 0.45–0.9 **Zn**, 0.25 **Cu**, rem **Mg**, 0.01 **Ni**, 0.20 **Si**, 0.30 total, others	
ANSI/ASTM B 80	AZ92C	US	Sand castings	8.3–9.7 **Al**, 0.10 **Mn**, 1.6–2.4 **Zn**, 0.25 **Cu**, rem **Mg**, 0.01 **Ni**, 0.30 **Si**, 0.30 total, others	**Solution heat treated and naturally aged**: 34 ksi (234 MPa) **TS**, 11 ksi (76 MPa) **YS**, 6% in 2 in. (50 mm) **El**, 63 **HB**; **Cooled and artificially aged**: 23 ksi (158 MPa) **TS**, 12 ksi (83 MPa) **YS**, 69 **HB**; **Solution heat treated and artificially aged**: 34 ksi (234 MPa) **TS**, 18 ksi (124 MPa) **YS**, 1% in 2 in. (50 mm) **El**, 81 **HB**
ANSI/ASTM B 94	UNS M11912 (AZ91B)	US	Die castings	8.3–9.7 **Al**, 0.13 min **Mn**, 0.35–1.0 **Zn**, 0.35 **Cu**, rem **Mg**	
ANSI/ASTM B 94	UNS M11910 (AZ91A)	US	Die castings	8.3–9.7 **Al**, 0.13 min **Mn**, 0.35–1.0 **Zn**, 0.35 **Cu**, rem **Mg**, 0.03 **Ni**, 0.50 **Si**	**As cast**: all diam, 34 ksi (230 MPa) **TS**, 23 ksi (160 MPa) **YS**, 3% in 2 in. (50 mm) **El**, 63 **HB**
ANSI/ASTM B 199	AZ92A	US	Permanent mold castings	8.3–9.7 **Al**, 0.10 min **Mn**, 1.6–2.4 **Zn**, 0.25 **Cu**, 0.01 **Ni**, 0.30 **Si**, 0.30 total, others	**As fabricated**: 23 ksi (159 MPa) **TS**, 11 ksi (76 MPa) **YS**, 65 **HB**; **Solution heat treated and naturally aged**: 34 ksi (234 MPa) **TS**, 11 ksi (76 MPa) **YS**, 6% in 2 in. (50 mm) **El**, 63 **HB**; **Cooled and artificially aged**: 23 ksi (159 MPa) **TS**, 12 ksi (83 MPa) **YS**, 69 **HB**
MIL-R-6944B	AZ92A	US	Rod and wire	8.3–9.7 **Al**, 0.15 min **Mn**, 1.7–2.3 **Zn**, 0.05 **Cu**, 0.005 **Ni**, 0.005 **Fe**, 0.05 **Si**, 0.0002–0.0008 **Be**, rem **Mg**, 0.30 total, others	
ANSI/ASTM B 93	AZ91C	US	Ingot: for sand, permanent mold, and investment castings	8.3–9.2 **Al**, 0.15 min **Mn**, 0.45–0.9 **Zn**, 0.08 **Cu**, rem **Mg**, 0.01 **Ni**, 0.20 **Si**, 0.30 total, others	

WROUGHT MAGNESIUM ALLOYS

specification number	designation	country	product forms	chemical composition	mechanical properties and hardness values
ANSI/ASTM B 199	AZ91C	US	Permanent mold castings	8.1–9.3 **Al**, 0.13 min **Mn**, 0.40–1.0 **Zn**, 0.10 **Cu**, rem **Mg**, 0.01 **Ni**, 0.30 **Si**, 0.30 total, others	**Solution heat treated and naturally aged**: 34 ksi (234 MPa) **TS**, 11 ksi (76 MPa) **YS**, 7% in 2 in. (50 mm) **El**, 55 **HB**; **Cooled and artificially aged**: 23 ksi (159 MPa) **TS**, 12 ksi (83 MPa) **YS**, 2% in 2 in. (50 mm) **El**, 62 **HB**; **Solution heat treated and artificially aged**: 34 ksi (234 MPa) **TS**, 16 ksi (110 MPa) **YS**, 3% in 2 in. (50 mm) **El**
ANSI/ASTM B 91	AZ80A	US	Forgings	7.8–9.2 **Al**, 0.12 min **Mn**, 0.20–0.8 **Zn**, 0.05 **Cu**, 0.005 **Fe**, rem **Mg**, 0.005 **Ni**, 0.10 **Si**, 0.30 total, others	**As fabricated**: 42 ksi (290 MPa) **TS**, 26 ksi (179 MPa) **YS**, 5% in 2 in. (50 mm) **El**; **Cooled and artificially aged**: all **diam**, 42 ksi (290 MPa) **TS**, 28 ksi (193 MPa) **YS**, 2% in 2 in. (50 mm) **El**
ANSI/ASTM B 107	AZ80A	US	Bar, tube, rod, wire, and shapes: extruded	7.8–9.2 **Al**, 0.12 min **Mn**, 0.20–0.8 **Zn**, 0.05 **Cu**, 0.005 **Fe**, rem **Mg**, 0.005 **Ni**, 0.10 **Si**, 0.30 total, others	**As fabricated (except tube)**: 2.500–4.999 in. **diam**, 42 ksi (290 MPa) **TS**, 27 ksi (186 MPa) **YS**, 4% in 2 in. (50 mm) **El**; **Cooled and artificially aged (except tube)**: 2.500–4.999 in. **diam**, 45 ksi (310 MPa) **TS**, 30 ksi (207 MPa) **YS**, 2% in 2 in. (50 mm) **El**
NF A 65–717	G–A8Z	France	Wire	7.5–9.2 **Al**, 0.1–0.4 **Mn**, 0.2–1.0 **Zn**, 0.05 **Cu**, 0.005 **Fe**, rem **Mg**, 0.005 **Ni**, 0.1 **Si**	**As manufactured**: all **diam**, 300 MPa **TS**, 220 MPa **YS**, 8% in 2 in. (50 mm) **El**
ISO 3116	Mg–Al8 Zn (cc Alloy No. 23)	ISO	Bar, full profiles	7.5–9.2 **Al**, 0.12 min **Mn**, 0.2–1.0 **Zn**, 0.1 **Si**, 0.05 **Cu**, 0.005 **Fe**, 0.005 **Ni**	**M**: all **diam**, 290 MPa **TS**, 190 MPa **YS**, 5% in 2 in. (50 mm) **El**
ISO 3116	Mg–Al8 Zn	ISO	Forged products	7.5–9.2 **Al**, 0.12 min **Mn**, 0.2–1.0 **Zn**, 0.1 **Si**, 0.05 **Cu**, 0.005 **Fe**, 0.005 **Ni**	**M**: all **diam**, 290 MPa **TS**, 190 MPa **YS**, 5% in 2 in. (50 mm) **El**; **TE**: all **diam**, 290 MPa **TS**, 200 MPa **YS**, 4% in 2 in. (50 mm) **El**
ANSI/ASTM B 93	AZ81A	US	Ingot: for sand, permanent mold and investment castings	7.2–8.0 **Al**, 0.15 min **Mn**, 0.50–0.9 **Zn**, 0.08 **Cu**, rem **Mg**, 0.01 **Ni**, 0.20 **Si**, 0.30 total, others	
ANSI/ASTM B 199	AZ81A	US	Permanent mold castings	7.0–8.1 **Al**, 0.13 min **Mn**, 0.40–1.0 **Zn**, 0.10 **Cu**, rem **Mg**, 0.01 **Ni**, 0.30 **Si**, 0.30 total, others	**Solution heat treated and naturally aged**: 34 ksi (234 MPa) **TS**, 11 ksi (76 MPa) **YS**, 7% in 2 in. (50 mm) **El**, 55 **HB**
ANSI/ASTM B 91	AZ61A	US	Forgings	5.8–7.2 **Al**, 0.15 min **Mn**, 0.40–1.5 **Zn**, 0.05 **Cu**, 0.005 **Fe**, rem **Mg**, 0.005 **Ni**, 0.10 **Si**, 0.30 total, others	**As fabricated**: 38 ksi (262 MPa) **TS**, 22 ksi (152 MPa) **YS**, 6% in 2 in. (50 mm) **El**

WROUGHT MAGNESIUM ALLOYS

specification number	designation	country	product forms	chemical composition	mechanical properties and hardness values
ANSI/ASTM B 107	AZ61A	US	Bar, tube, rod, wire, and shapes: extruded	5.8–7.2 **Al**, 0.15 min **Mn**, 0.40–1.5 **Zn**, 0.05 **Cu**, 0.005 **Fe**, rem **Mg**, 0.005 **Ni**, 0.10 **Si**, 0.30 total, others	**As fabricated (except tube)**: 2.500–4.999 in. **diam**, 40 ksi (276 MPa) **TS**, 22 ksi (152 MPa) **YS**, 7% in 2 in. (50 mm) **El**; **As fabricated (tube)**: 0.028–0.750 in. **diam**, 36 ksi (248 MPa) **TS**, 16 ksi (110 MPa) **YS**, 7% in 2 in. (50 mm) **El**
MIL-R-6944B	AZ61A	US	Rod and wire	5.8–7.2 **Al**, 0.15 min **Mn**, 0.40–1.5 **Zn**, 0.05 **Cu**, 0.005 **Ni**, 0.005 **Fe**, 0.05 **Si**, 0.0002–0.0008 **Be**, rem **Mg**, 0.30 total, others	
WW-T-825B	AZ61A	US	Tube	5.8–7.2 **Al**, 0.15 **Mn**, 0.40–1.5 **Zn**, 0.10 **Si**, 0.05 **Cu**, 0.005 **Ni**, 0.005 **Fe**, rem **Mg**, 0.30 total, others	**As cast**: .028–0.750 in. **diam**, 36 ksi (248 MPa) **TS**, 16 ksi (110 MPa) **YS**, 7% in 2 in. (50 mm) **El**
JIS H 4202	Class 2	Japan	Tube: seamless	5.8–7.2 **Al**, 0.15 min **Mn**, 0.4–1.5 **Zn**, 0.10 **Cu**, 0.01 **Fe**, rem **Mg**, 0.1 **Si**	**As extruded**: all **diam**, 245 MPa **TS**, 108 MPa **YS**, 7% in (50 mm) **El**
JIS H 4203	Class 2	Japan	Bar	5.8–7.2 **Al**, 0.15 min **Mn**, 0.4–1.5 **Zn**, 0.10 **Cu**, 0.01 **Fe**, rem **Mg**, 0.1 **Si**	
JIS H 4204	Class 2	Japan	Shapes	5.8–7.2 **Al**, 0.15 min **Mn**, 0.4–1.5 **Zn**, 0.10 **Cu**, 0.01 **Fe**, rem **Mg**, 0.1 **Si**	**As extruded**: 245 MPa **TS**, 108 MPa **YS**, 7% in (50 mm) **El**
ISO 3116	Mg-Al6Zn1 (cc.Alloy 22)	ISO	Bar, full profiles	5.5–7.2 **Al**, 0.15 min **Mn**, 0.5–1.5 **Zn**, 0.1 **Si**, 0.05 **Cu**, 0.005 **Fe**, 0.005 **Ni**	**M**: all **diam**, 270 MPa **TS**, 180 MPa **YS**, 6% in 2 in. (50 mm) **El**
ISO 3116	Mg-Al6 Zn1 (cc. Alloy 26)	ISO	Bar, full profiles	5.4–7.3 **Al**, 0.15–0.4 **Mn**, 0.5–1.5 **Zn**, 0.1 **Si**, 0.1 **Cu**, 0.03 **Fe**, 0.005 **Ni**	**As cast**: all **diam**, 270 MPa **TS**, 180 MPa **YS**, 6% in 2 in. (50 mm) **El**
ISO 3116	Mg-Al6 Zn1 (cc Alloy No. 22)	ISO	Tube, thin profiles	5.5–7.2 **Al**, 0.15 min **Mn**, 0.5–1.5 **Zn**, 0.1 **Si**, 0.05 **Cu**, 0.005 **Fe**, 0.005 **Ni**	**M**: all **diam**, 260 MPa **TS**, 150 MPa **YS**, 6% in 2 in. (50 mm) **El**
ISO 3116	Mg-Al6Zn1 (cc Alloy No. 26)	ISO	Tube, thin profiles	5.4–7.3 **Al**, 0.15–10.4 **Mn**, 0.5–1.5 **Zn**, 0.1 **Si**, 0.1 **Cu**, 0.03 **Fe**, 0.005 **Ni**	**M**: all **diam**, 260 MPa **TS**, 150 MPa **YS**, 6% in 2 in. (50 mm) **El**
ISO 3116	Mg-Al6 Zn1 (cc Alloy No. 22)	ISO	Forged products	5.5–7.2 **Al**, 0.15 min **Mn**, 0.5–1.5 **Zn**, 0.1 **Si**, 0.05 **Cu**, 0.005 **Fe**, 0.005 **Ni**	**M**: all **diam**, 270 MPa **TS**, 150 MPa **YS**, 5% in 2 in. (50 mm) **El**
ISO 3116	Mg-Al6 Zn1 (cc Alloy No.26)	ISO	Forged products	5.4–7.3 **Al**, 0.15–0.4 **Mn**, 0.5–1.5 **Zn**, 0.1 **Si**, 0.1 **Cu**, 0.03 **Fe**, 0.005 **Ni**	**M**: all **diam**, 270 MPa **TS**, 150 MPa **YS**, 5% in 2 in. (50 mm) **El**
ANSI/ASTM B 93	AZ63A	US	Ingot: for sand, permanent mold, and investment castings	5.5–6.5 **Al**, 0.18 min **Mn**, 2.7–3.3 **Zn**, 0.20 **Cu**, rem **Mg**, 0.01 **Ni**, 0.20 **Si**, 0.30 total, others	
BS 3373	Mg-A16Zn1Mn	UK	Bar, section, tube, forging stock	5.5–6.5 **Al**, 0.15–0.40 **Mn**, 0.5–1.5 **Zn**, 0.1 **Cu**, 0.1 **Si**, 0.03 **Fe**, 0.005 **Ni**, rem **Mg**	**As manufactured**: 75–150 mm **diam**, 250 MPa **TS**, 160 MPa **YS**, 6% in 2 in. (50 mm) **El**

WROUGHT MAGNESIUM ALLOYS

specification number	designation	country	product forms	chemical composition	mechanical properties and hardness values
BS 2L 503		UK	Tube	5.5–6.5 **Al**, 0.2–0.4 **Mn**, 0.5–1.5 **Zn**, 0.1 **Cu**, 0.1 **Si**, 0.03 **Fe**, 0.005 **Ni**, rem **Mg**	**As extruded and straightened**: all **diam**, 260 MPa **TS**, 150 MPa **YS**, 7% in 2 in. (50 mm) **El**
BS L 513		UK	Forging stock and forgings	5.5–6.5 **Al**, 0.2–0.4 **Mn**, 0.5–1.5 **Zn**, 0.1 **Cu**, 0.1 **Si**, 0.03 **Fe**, 0.005 **Ni**, rem **Mg**	**As extruded or as extruded and straightened (forging stock)**: ≤75 mm **diam**, 270 MPa **TS**, 180 MPa **YS**, 8% in 2 in. (50 mm) **El**; **As forged, or forged and stress relieved (forgings)**: all **diam**, 275 MPa **TS**, 160 MPa **YS**, 7% in 2 in. (50 mm) **El**
BS L 512		UK	Bar and section	5.5–6.5 **Al**, 0.2–0.4 **Mn**, 0.5–1.5 **Zn**, 0.1 **Cu**, 0.1 **Si**, 0.03 **Fe**, 0.005 **Ni**, rem **Mg**	**As extruded and straightened**: ≤75 mm **diam**, 270 MPa **TS**, 180 MPa **YS**, 8% in 2 in. (50 mm) **El**
BS 3372	Mg–Al6Zn1Mn	UK	Forgings and cast forging stock	5.5–6.5 **Al**, 0.15–0.40 **Mn**, 0.5–1.5 **Zn**, 0.1 **Cu**, 0.1 **Si**, 0.03 **Fe**, 0.005 **Ni**, rem **Mg**	**As manufactured**: all **diam**, 270 MPa **TS**, 160 MPa **YS**, 7% in 2 in. (50 mm) **El**
NF A 65–717	G–A6Z1	France	Wire and shapes	5.5–6.5 **Al**, 0.15–0.4 **Mn**, 0.5–1.5 **Zn**, 0.1 **Cu**, 0.03 **Fe**, rem **Mg**, 0.005 **Ni**, 0.1 **Si**	**As manufactured (wire)**: all **diam**, 280 MPa **TS**, 180 MPa **YS**, 8% in 2 in. (50 mm) **El**
ANSI/ASTM B 91	AZ31B	US	Forgings	2.5–3.5 **Al**, 0.20 min **Mn**, 0.6–1.4 **Zn**, 0.04 **Ca**, 0.05 **Cu**, 0.005 **Fe**, rem **Mg**, 0.005 **Ni**, 0.10 **Si**, 0.30 total, others	**As fabricated**: 34 ksi (234 MPa) **TS**, 19 ksi (131 MPa) **YS**, 6% in 2 in. (50 mm) **El**
ANSI/ASTM B 90	AZ31B	US	Sheet and plate	2.5–3.5 **Al**, 0.20 min **Mn**, 0.6–1.3 **Zn**, 0.04 **Ca**, 0.05 **Cu**, 0.005 **Fe**, rem **Mg**, 0.005 **Ni**, 0.10 **Si**, 0.30 total, others	**Annealed**: 0.501–2.000 in. **diam**, 32 ksi (221 MPa) **TS**, 10% in 2 in. (50 mm) **El**; **H24**: 2.001–3.000 in. **diam**, 34 ksi (234 MPa) **TS**, 18 ksi (124 MPa) **YS**, 8% in 2 in. (50 mm) **El**; **H26**: 1.501–2.000 in. **diam**, 35 ksi (241 MPa) **TS**, 21 ksi (145 MPa) **YS**, 6% in 2 in (50 mm) **El**
ANSI/ASTM B 90	AZ31C	US	Sheet and plate	2.4–3.6 **Al**, 0.15 min **Mn**, 0.50–1.5 **Zn**, 0.10 **Cu**, 0.005 **Fe**, rem **Mg**, 0.03 **Ni**, 0.10 **Si**, 0.30 total, others	
ANSI/ASTM B 107	AZ31B	US	Bar, tube, rod, wire, and shapes: extruded	2.5–3.5 **Al**, 0.20 min **Mn**, 0.6–1.4 **Zn**, 0.04 **Ca**, 0.05 **Cu**, 0.005 **Fe**, rem **Mg**, 0.005 **Ni**, 0.10 **Si**, 0.30 total, others	**As fabricated (except tube)**: 2.500–4.999 in. **diam**, 32 ksi (221 MPa) **TS**, 20 ksi (138 MPa) **YS**, 7% in 2 in. (50 mm) **El**; **As fabricated (tube)**: 0.028–0.500 in. **diam**, 32 ksi (221 MPa) **TS**, 16 ksi (110 MPa) **YS**, 8% in 2 in. (50 mm) **El**
ANSI/ASTM B 107	AZ31C	US	Bar, tube, rod, wire, and shapes: extruded	2.4–3.6 **Al**, 0.15 min **Mn**, 0.50–1.5 **Zn**, 0.10 **Cu**, rem **Mg**, 0.03 **Ni**, 0.10 **Si**, 0.30 total, others	

WROUGHT MAGNESIUM ALLOYS

specification number	designation	country	product forms	chemical composition	mechanical properties and hardness values
BS 3373	Mg–A13Zn1Mn	UK	Bar, section, tube, forging stock	2.5–3.5 **Al**, 0.15–0.40 **Mn**, 0.6–1.4 **Zn**, 0.04 **Ca**, 0.1 **Cu**, 0.1 **Si**, 0.03 **Fe**, 0.005 **Ni**, rem **Mg**	**As manufactured**: 10–75 mm **diam**, 245 MPa **TS**, 160 MPa **YS**, 10% in 2 in. (50 mm) **El**
BS 3370	Mg–Al3Zn1Mn	UK	Plate, sheet, strip	2.5–3.5 **Al**, 0.15–0.40 **Mn**, 0.6–1.4 **Zn**, 0.04 **Ca**, 0.1 **Cu**, 0.1 **Si**, 0.03 **Fe**, 0.005 **Ni**, rem **Mg**	**As manufactured**: 0.5–6 mm **diam**, 250 MPa **TS**, 160 MPa **YS**; **Annealed**: 0.5–6 mm **diam**, 220 MPa **TS**, 120 MPa **YS**
WW–T–825B	AZ31B	US	Tube	2.5–3.5 **Al**, 0.20 **Mn**, 0.60–1.4 **Zn**, 0.04 **Ca**, 0.10 **Si**, 0.05 **Cu**, 0.005 **Ni**, 0.005 **Fe**, rem **Mg**, 0.30 total, others	**As cast**: 0.028–0.250 in. **diam**, 32 ksi (221 MPa) **TS**, 16 ksi (110 MPa) **YS**, 8% in 2 in. (50 mm) **El**
JIS H 4201	Class 1	Japan	Sheet and plate	2.4–3.6 **Al**, 0.15 min **Mn**, 0.5–1.5 **Zn**, 0.10 **Cu**, 0.01 **Fe**, rem **Mg**, 0.1 **Si**	**Soft**: ≤0.5 mm **diam**, 216 MPa **TS**, 12% in (50 mm) **El**; **1/2 hard**: (≤6 mm) **diam**, 245 MPa **TS**, 137 MPa **YS**, 4% in (50 mm) **El**
JIS H 4202	Class 1	Japan	Tube: seamless	2.4–3.6 **Al**, 0.15 min **Mn**, 0.5–1.5 **Zn**, 0.10 **Cu**, 0.01 **Fe**, rem **Mg**, 0.1 **Si**	**As extruded**: all **diam**, 216 MPa **TS**, 108 MPa **YS**, 7% in (50 mm) **El**
JIS H 4203	Class 1	Japan	Bar	2.4–3.6 **Al**, 0.15 min **Mn**, 0.5–1.5 **Zn**, 0.10 **Cu**, 0.01 **Fe**, rem **Mg**, 0.1 **Si**	**As extruded**: all **diam**, 216 MPa **TS**, 108 MPa **YS**, 7% in (50 mm) **El**
JIS H 4204	Class 1	Japan	Shapes	2.4–3.6 **Al**, 0.15 min **Mn**, 0.5–1.5 **Zn**, 0.10 **Cu**, 0.01 **Fe**, rem **Mg**, 0.1 **Si**	**As extruded**: 216 MPa **TS**, 108 MPa **YS**, 7% in (50 mm) **El**
ISO 3116	Mg–Al3Zn1 (cc Alloy No.21)	ISO	Bar, full profiles	2.5–3.5 **Al**, 0.2 min **Mn**, 0.5–1.5 **Zn**, 0.04 **Ca**, 0.1 **Si**, 0.05 **Cu**, 0.005 **Fe**, 0.005 **Ni**	**M**: all **diam**, 240 MPa **TS**, 150 MPa **YS**, 6% in 2 in. (50 mm) **El**
ISO 3116	Mg–Al3Zn1 (cc Alloy 25)	ISO	Bar, full profiles	2.4–3.6 **Al**, 0.15–0.4 **Mn**, 0.5–1.5 **Zn**, 0.04 **Ca**, 0.1 **Si**, 0.1 **Cu**, 0.03 **Fe**, 0.005 **Ni**	**M**: all **diam**, 240 MPa **TS**, 150 MPa **YS**, 6% in 2 in. (50 mm) **El**
ISO 3116	Mg–Al13Zn1 (cc Alloy No.21)	ISO	Tube, thin profiles	2.5–3.5 **Al**, 0.2 min **Mn**, 0.5–1.5 **Zn**, 0.04 **Ca**, 0.1 **Si**, 0.05 **Cu**, 0.005 **Fe**, 0.005 **Ni**	**M**: all **diam**, 230 MPa **TS**, 150 MPa **YS**, 6% in 2 in. (50 mm) **El**
ISO 3116	Mg–Al13Zn1 (cc Alloy No. 25)	ISO	Tube, thin profiles	2.4–3.6 **Al**, 0.15–0.4 **Mn**, 0.5–1.5 **Zn**, 0.04 **Ca**, 0.1 **Si**, 0.1 **Cu**, 0.03 **Fe**, 0.005 **Ni**	**M**: all **diam**, 230 MPa **TS**, 150 MPa **YS**, 6% in 2 in. (50 mm) **El**
ISO 3116	Mg–Al3 Zn1 (cc Alloy No. 21)	ISO	Rolled products	2.5–3.5 **Al**, 0.2 min **Mn**, 0.5–1.5 **Zn**, 0.1 **Si**, 0.05 **Cu**,	**Annealed**: all **diam**, 220 MPa **TS**, 105 MPa **YS**, 11% in 2 in. (50 mm) **El**; **Strain hardened**: all **diam**, 250 MPa **TS**, 160 MPa **YS**, 5% in 2 in. (50 mm) **El**
ISO 3116	Mg–Al3 Zn1 (cc Alloy No. 25)	ISO	Rolled products	2.4–3.6 **Al**, 0.15–0.4 **Mn**, 0.5–1.5 **Zn**, 0.1 **Si**, 0.1 **Cu**, 0.003 **Fe**, 0.005 **Ni**	**Annealed**: all **diam**, 220 MPa **TS**, 105 MPa **YS**, 11% in 2 in. (50 mm) **El**; **Strain hardened**: all **diam**, 250 MPa **TS**, 160 MPa **YS**, 5% in 2 in. (50 mm) **El**
ISO 3116	Mg–Al3 Zn1 (cc Alloy No. 21)	ISO	Forged products	2.5–3.5 **Al**, 0.2 min **Mn**, 0.5–1.5 **Zn**, 0.04 **Ca**, 0.1 **Si**, 0.05 **Cu**, 0.005 **Fe**, 0.005 **Ni**	**M**: all **diam**, 240 MPa **TS**, 130 MPa **YS**, 6% in 2 in. (50 mm) **El**

WROUGHT MAGNESIUM ALLOYS

specification number	designation	country	product forms	chemical composition	mechanical properties and hardness values
ISO 3116	Mg–Al3 Zn1 (cc Alloy No. 25)	ISO	Forged products	2.4–3.6 **Al**, 0.15–0.4 **Mn**, 0.5–1.5 **Zn**, 0.04 **Ca**, 0.1 **Si**, 0.1 **Cu**, 0.03 **Fe**, 0.005 **Ni**	**M**: all **diam**, 240 MPa **TS**, 130 MPa **YS**, 6% in 2 in. (50 mm) **El**
NF A 65–717	G–A3Z1	France	Sheet, strip, wire, and shapes	2.3–3.5 **Al**, 0.2 min **Mn**, 0.5–1.5 **Zn**, 0.04 **Ca**, 0.1 **Cu**, 0.03 **Fe**, rem **Mg**, 0.005 **Ni**, 0.1 **Si**	**Annealed (sheet and strip)**: 0.5–6 mm **diam**, 230 MPa **TS**, 130 MPa **YS**, 12% in 2 in. (50 mm) **El**; **As manufactured (wire and shapes)**: all **diam**, 240 MPa **TS**, 160 MPa **YS**, 10% in 2 in. (50 mm) **El**
MIL–R–6944B	EZ33A	US	Rod and wire	2.0–3.1 **Zn**, 2.5–4.0 **RE**, 0.45–1.0 **Zr**, rem **Mg**, 0.30 total, others	
ANSI/ASTM B 80	HK31A	US	Sand castings	0.30 **Zn**, 2.5–4.0 **Th**, 0.10 **Cu**, rem **Mg**, 0.01 **Ni**, 0.40–1.0 **Zr**, 0.30 total, others	**Solution heat treated and artificially aged**: all **diam**, 27 ksi (186 MPa) **TS**, 13 ksi (89 MPa) **YS**, 4% in 2 in. (50 mm) **El**, 66 **HB**
ANSI/ASTM B 90	HK31A	US	Sheet and plate	0.30 **Zn**, 2.5–4.0 **Th**, 0.10 **Cu**, rem **Mg**, 0.01 **Ni**, 0.40–1.0 **Zr**, 0.30 total, others	**Annealed**: 1.001–3.000 in. **diam**, 29 ksi (200 MPa) **TS**, 14 ksi (97 MPa) **YS**, 12% in 2 in. (50 mm) **El**; **H24**: 1.001–3.000 in. **diam**, 33 ksi (228 MPa) **TS**, 25 ksi (172 MPa) **YS**, 4% in 2 in. (50 mm) **El**
ANSI/ASTM B 199	HK31A	US	Permanent mold castings	0.30 **Zn**, 2.5–4.0 **Th**, 0.10 **Cu**, rem **Mg**, 0.01 **Ni**, 0.40–1.0 **Zr**, 0.30 total, others	**Solution heat treated and artificially aged**: all **diam**, 27 ksi (186 MPa) **TS**, 13 ksi (90 MPa) **YS**, 4% in 2 in. (50 mm) **El**, 66 **HB**
MIL–M–8916A		US	Bar, rod, sections	1.2 min **Mn**, 2.5–3.5 **Th**, rem **Mg**, 0.10 each, 0.30 total, others	**Artificially aged**: $<$4.000 sq. in. **diam**, 255 MPa **TS**, 179 MPa **YS**, 4% in 2 in. (50 mm) **El**
ANSI/ASTM B 91	HM21A	US	Forgings	0.45–1.1 **Mn**, 1.5–2.5 **Th**, rem **Mg**, 0.30 total, others	**Cooled and artificially aged**: $\leq$4 in. ($\leq$102 mm) **diam**, 33 ksi (228 MPa) **TS**, 25 ksi (172 MPa) **YS**, 3% in 2 in. (50 mm) **El**
ANSI/ASTM B 90	HM21A	US	Sheet and plate	0.45–1.1 min **Mn**, 1.5–2.5 **Th**, rem **Mg**, 0.30 total, others	**T8**: 0.501–3.000 in. **diam**, 30 ksi (207 MPa) **TS**, 21 ksi (145 MPa) **YS**, 6% in 2 in. (50 mm) **El**
MIL–M–8917A	HM21A	US	Sheet and plate	0.45–1.1 **Mn**, 1.5–2.5 **Th**, rem **Mg**, 0.10 each, 0.30 total, others	**T8**: 0.016–0.250 in. **diam**, 228 MPa **TS**, 124 MPa **YS**, 6% in 2 in. (50 mm) **El**; **T81**: 0.125–0.312 in. **diam**, 234 MPa **TS**, 172 MPa **YS**, 4% in 2 in. (50 mm) **El**
BS 3373	Mg–Mn1.5	UK	Bar, section, tube, forging stock	0.05 **Al**, 1.0–2.0 **Mn**, 0.03 **Zn**, 0.02 **Ca**, 0.02 **Cu**, 0.02 **Si**, 0.03 **Fe**, 0.005 **Ni**, rem **Mg**	**As manufactured**: 50–100 mm **diam**, 200 MPa **TS**, 120 MPa **YS**, 3% in 2 in. (50 mm) **El**
BS 3370	Mg–Mn1.5	UK	Plate, sheet, strip	0.05 **Al**, 1.0–2.0 **Mn**, 0.03 **Zn**, 0.02 **Ca**, 0.02 **Cu**, 0.02 **Si**, 0.03 **Fe**, 0.005 **Ni**, rem **Mg**	**As manufactured**: 0.5–6 mm **diam**, 200 MPa **TS**, 70 MPa **YS**

WROUGHT MAGNESIUM ALLOYS

specification number	designation	country	product forms	chemical composition	mechanical properties and hardness values
BS 3372	Mg–Mn1.5	UK	Forgings and cast forging stock	0.05 **Al**, 1.0–2.0 **Mn**, 0.03 **Zn**, 0.02 **Cu**, 0.02 **Si**, 0.03 **Fe**, 0.005 **Ni**, 0.02 **Ca**, rem **Mg**	**As manufactured**: all **diam**, 200 MPa **TS**, 105 MPa **YS**, 4% in 2 in. (50 mm) **El**
ANSI/ASTM B 107	M1A	US	Bar, tube, rod, wire, and shapes: extruded	1.2 min **Mn**, 0.30 **Ca**, 0.05 **Cu**, rem **Mg**, 0.01 **Ni**, 0.10 **Si**, 0.30 total, others	**As fabricated (except tube)**: 2.500–4.999 in. **diam**, 29 ksi (200 MPa) **TS**, 2% in 2 in. (50 mm) **El**; **As fabricated (tube)**: 0.028–0.750 in. **diam**, 28 ksi (193 MPa) **TS**, 2% in 2 in. (50 mm) **El**
WW–T–825B	MIA	US	Tube	1.20 **Mn**, 0.30 **Ca**, 0.10 **Si**, 0.05 **Cu**, 0.010 **Ni**, rem **Mg**, 0.30 total, others	**As cast**: .028–0.750 in. **diam**, 28 ksi (193 MPa) **TS**, 2% in 2 in. (50 mm) **El**
BS 3372	Mg–Zn2Mn1	UK	Forgings and cast forging stock	0.20 **Al**, 0.6–1.3 **Mn**, 1.5–2.3 **Zn**, 0.1 **Cu**, 0.10 **Si**, 0.06 **Fe**, 0.005 **Ni**, rem **Mg**	**As manufactured**: all **diam**, 200 MPa **TS**, 125 MPa **YS**, 9% in 2 in. (50 mm) **El**
BS 3370	Mg–Zn2Mn1	UK	Plate, sheet, strip	0.20 **Al**, 0.6–1.3 **Mn**, 1.5–2.3 **Zn**, 0.1 **Cu**, 0.10 **Si**, 0.06 **Fe**, 0.005 **Ni**, rem **Mg**	**As manufactured**: 6–25 mm **diam**, 220 MPa **TS**, 120 MPa **YS**, 8% in 2 in. (50 mm) **El**; **Annealed**: 0.5–6 mm **diam**, 220 MPa **TS**, 120 MPa **YS**
BS 3373	Mg–Zn2Mn1	UK	Bar, section, tube, forging stock	0.20 **Al**, 0.6–1.3 **Mn**, 1.5–2.3 **Zn**, 0.1 **Cu**, 0.10 **Si**, 0.06 **Fe**, 0.005 **Ni**, rem **Mg**	**As manufactured**: 10–75 mm **diam**, 245 MPa **TS**, 160 MPa **YS**, 10% in 2 in. (50 mm) **El**
ANSI/ASTM B 90	ZE10A	US	Sheet and plate	1.0–1.5 **Zn**, 0.12–0.22 **RE**, rem **Mg**, 0.30 total, others	**Annealed**: 0.251–0.500 in. **diam**, 29 ksi (200 MPa) **TS**, 12 ksi (83 MPa) **YS**, 12% in 2 in. (50 mm) **El**; **H24**: 0.016–0.125 in. **diam**, 36 ksi (248 MPa) **TS**, 25 ksi (172 MPa) **YS**, 4% in 2 in. (50 mm) **El**
MIL–M–46037B		US	Plate and sheet	1.0–1.5 **Zn**, 0.12–0.22 **RE**, rem **Mg**, 0.30 total, others	**Annealed**: 0.016–0.060 in. **diam**, 207 MPa **TS**, 124 MPa **YS**, 15% in 2 in. (50 mm) **El**; **Strain hardened and partially annealed**: 0.016–0.125 in. **diam**, 248 MPa **TS**, 172 MPa **YS**, 6% in 2 in. (50 mm) **El**
MIL–M–26696		US	Bar, rod, sections	4.8–6.8 **Zn**, 0.45 min **Zr**, rem **Mg**, 0.30 total, others	**Artificially aged**: <2.999 in. **diam**, 310 MPa **TS**, 241 MPa **YS**, 4% in 2 in. (50 mm) **El**
BS 3373	Mg–Zn6Zr	UK	Bar, section, tube, forging stock	0.02 **Al**, 0.15 **Mn**, 4.8–6.2 **Zn**, 0.45–0.8 **Zr**, 0.03 **Cu**, 0.01 **Si**, 0.01 **Fe**, 0.005 **Ni**, rem **Mg**	**Cooled and artificially aged**: 10–50 mm **diam**, 315 MPa **TS**, 230 MPa **YS**, 8% in 2 in. (50 mm) **El**
BS 3372	Mg–Zn6Zr	UK	Forgings and cast forging stock	0.02 **Al**, 0.15 **Mn**, 4.8–6.2 **Zn**, 0.45–0.8 **Zr**, 0.03 **Cu**, 0.01 **Si**, 0.01 **Fe**, 0.005 **Ni**, rem **Mg**	**Cooled and artificially aged**: all **diam**, 280 MPa **TS**, 180 MPa **YS**, 7% in 2 in. (50 mm) **El**

WROUGHT MAGNESIUM ALLOYS

specification number	designation	country	product forms	chemical composition	mechanical properties and hardness values
ANSI/ASTM B 91	ZK60A	US	Die forgings	4.8–6.2 **Zn**, rem **Mg**, 0.45 **Zr**, 0.30 total, others	**Cooled and artificially aged**: ≤3 in. (≤76 mm) **diam**, 42 ksi (290 MPa) **TS**, 26 ksi (179 MPa) **YS**, 7% in 2 in. (50 mm) **El**
ANSI/ASTM B 107	ZK60A	US	Bar, tube, rod, wire, and shapes: extruded	4.8–6.2 **Zn**, rem **Mg**, 0.45 min **Zr**, 0.30 total, others	**As fabricated (except tube)**: 5.000–39.999 in. **diam**, 43 ksi (296 MPa) **TS**, 31 ksi (214 MPa) **YS**, 4% in 2 in. (50 mm) **El**; **As fabricated (tube)**: 0.028–0.750 in. **diam**, 40 ksi (276 MPa) **TS**, 28 ksi (193 MPa) **YS**, 5% in 2 in. (50 mm) **El**; **Cooled and artificially aged (except tube)**: all **diam**, 45 ksi (310 MPa) **TS**, 36 ksi (248 MPa) **YS**, 4% in 2 in. (50 mm) **El**
NF A 65-717	G-Z55Zr	France	Wire and shapes	0.02 **Al**, 0.15 **Mn**, 4.8–6.2 **Zn**, 0.01 **Fe**, 0.005 **Ni**, rem **Mg**, 0.01 **Si**	**As manufactured (wire)**: all **diam**, 300 MPa **TS**, 210 MPa **YS**, 6% in (50 mm) **El**; **T5**: 1–10 mm **diam**, 300 MPa **TS**, 250 MPa **YS**, 5% in (50 mm) **El**
WW-T-825B	ZK60A	US	Tube	4.8–6.2 **Zn**, 0.45 **Zr**, rem **Mg**, 0.30 total, others	**Solution heat treated and artificially aged**: 0.028–0.250 in. **diam**, 46 ksi (317 MPa) **TS**, 38 ksi (262 MPa) **YS**, 4% in 2 in. (50 mm) **El**
ANSI/ASTM B 107	ZK40A	US	Bar, tube, rod, wire, and shapes: extruded	3.5–4.5 **Zn**, rem **Mg**, 0.45 min **Zr**, 0.30 total, others	**Cooled and artificially aged (except tube)**: all **diam**, 276 MPa **TS**, 255 MPa **YS**, 4% in (50 mm) **El**; **Cooled and artificially aged (tube)**: 0.062–0.500 in. **diam**, 276 MPa **TS**, 248 MPa **YS**, 4% in (50 mm) **El**
BS L 514		UK	Forging stock and forgings	0.02 **Al**, 0.15 **Mn**, 2.5–4.0 **Zn**, 0.5–1.0 **Zr**, 0.03 **Cu**, 0.01 **Si**, 0.01 **Fe**, 0.005 **Ni**, rem **Mg**	**As extruded or as extruded and straightened (forging stock)**: <10 mm **diam**, 280 MPa **TS**, 195 MPa **YS**, 8% in 2 in. (50 mm) **El**; **As forged or forged and stress relieved (forgings)**: all **diam**, 290 MPa **TS**, 205 MPa **YS**, 7% in 2 in. (50 mm) **El**
BS 3373	Mg-Zn3Zr	UK	Bar, section, tube, forging stock	0.02 **Al**, 0.15 **Mn**, 2.5–4.0 **Zn**, 0.4–0.8 **Zr**, 0.03 **Cu**, 0.01 **Si**, 0.01 **Fe**, 0.005 **Ni**, rem **Mg**	**As manufactured**: 10–100 mm **diam**, 305 MPa **TS**, 225 MPa **YS**, 8% in 2 in. (50 mm) **El**
BS 3370	Mg-Zn3Zr	UK	Plate, sheet, strip	0.02 **Al**, 0.15 **Mn**, 2.5–4.0 **Zn**, 0.4–0.8 **Zr**, 0.03 **Cu**, 0.01 **Si**, 0.01 **Fe**, 0.005 **Ni**, rem **Mg**	**As manufactured**: 6–50 mm **diam**, 250 MPa **TS**, 150 MPa **YS**, 8% in 2 in. (50 mm) **El**
BS 3372	Mg-Zn3Zr	UK	Forgings and cast forging stock	0.02 **Al**, 0.15 **Mn**, 2.5–4.0 **Zn**, 0.4–0.8 **Zr**, 0.03 **Cu**, 0.01 **Si**, 0.01 **Fe**, 0.005 **Ni**, rem **Mg**	**As manufactured**: all **diam**, 270 MPa **TS**, 180 MPa **YS**, 7% in 2 in. (50 mm) **El**

WROUGHT MAGNESIUM ALLOYS

specification number	designation	country	product forms	chemical composition	mechanical properties and hardness values
MIL-M-46039A		US	Bar, rod, sections	2.0–2.6 **Zn**, 0.48–0.8 **Zr**, rem **Mg**, 0.30 total, others	**As extruded**: <5.000 in. **diam**, 262 MPa **TS**, 193 MPa **YS**, 4% in 2 in. (50 mm) **El**
BS L 515		UK	Sheet and strip	0.02 **Al**, 0.15 **Mn**, 0.75–1.5 **Zn**, 0.4–0.8 **Zr**, 0.03 **Cu**, 0.01 **Si**, 0.01 **Fe**, 0.005 **Ni**, rem **Mg**	**As rolled or as rolled flattened**: 0.5–1.2 mm **diam**, 240 MPa **TS**, 160 MPa **YS**, 5% in 2 in. (50 mm) **El**
BS 3373	Mg–Zn1Zr	UK	Bar, section, tube, forging stock	0.02 **Al**, 0.15 **Mn**, 0.75–1.5 **Zn**, 0.4–0.8 **Zr**, 0.03 **Cu**, 0.01 **Si**, 0.01 **Fe**, 0.005 **Ni**, rem **Mg**	**As manufactured**: 10–75 mm **diam**, 260 MPa **TS**, 185 MPa **YS**, 8% in 2 in. (50 mm) **El**
BS 3370	Mg–Zn1Zr	UK	Plate, sheet, strip	0.02 **Al**, 0.15 **Mn**, 0.75–1.5 **Zn**, 0.4–0.8 **Zr**, 0.03 **Cu**, 0.01 **Si**, 0.01 **Fe**, 0.005 **Ni**, rem **Mg**	**As manufactured**: 25–50 mm **diam**, 220 MPa **TS**, 120 MPa **YS**, 8% in 2 in. (50 mm) **El**
BS 3372	Mg–Zn1Zr	UK	Forgings and cast forging stock	0.02 **Al**, 0.15 **Mn**, 0.75–1.5 **Zn**, 0.4–0.8 **Zr**, 0.03 **Cu**, 0.01 **Si**, 0.01 **Fe**, 0.005 **Ni**, rem **Mg**	**As manufactured**: all **diam**, 200 MPa **TS**, 125 MPa **YS**, 7% in 2 in. (50 mm) **El**
ANSI/ASTM B 90	LA141A	US	Sheet and plate	1.0–1.5 **Al**, 0.15 min **Mn**, 0.04 **Cu**, 0.005 **Fe**, 13.0–15.0 **Li**, rem **Mg**, 0.005 **Ni**, 0.10 **Si**, 0.005 **Na**, 0.30 total, others	
MIL-R-6944B	LA141A	US	Rod and wire	1.0–1.5 **Al**, 0.15 **Mn**, 13.0–15.0 **Li**, 0.04 **Cu**, 0.005 **Ni**, 0.005 **Fe**, 0.10 **Si**, 0.005 **Na**, rem **Mg**, 0.30 total, others	
MIL-M-46130	LA141A	US	Plate, sheet, forgings	1.0–1.5 **Al**, 0.15 **Mn**, 13.0–15.0 **Li**, 0.10 **Si**, 0.005 **Na**, 0.005 **Fe**, 0.005 **Ni**, 0.04 **Cu**, rem **Mg**, 0.30 total, others	**Stabilized**: 0.010–0.090 in. **diam**, 131 MPa **TS**, 103 MPa **YS**, 10% in 2 in. (50 mm) **El**
MIL-M-46130	LS141A	US	Plate, sheet, forgings	0.05 **Al**, 0.15 **Mn**, 12.0–15.0 **Li**, 0.5–0.6 **Si**, 0.005 **Na**, 0.005 **Fe**, 0.005 **Ni**, 0.05 **Cu**, rem **Mg**, 0.20 total, others	**As rolled**: 0.020–2.00 in. **diam**, 124 MPa **TS**, 90 MPa **YS**, 30% in 2 in. (50 mm) **El**
MIL-M-46130	LZ145A	US	Plate, sheet, forgings	0.05 **Al**, 0.15 **Mn**, 12.0–15.0 **Li**, 4.5–5.0 **Zn**, 2.0–3.0 **Ag**, 1.5–2.0 **Si**, 0.005 **Na**, 0.005 **Fe**, 0.005 **Ni**, 0.05 **Cu**, rem **Mg**, 0.20 total, others	**Stabilized**: 0.020–2.00 in. **diam**, 193 MPa **TS**, 165 MPa **YS**, 20% in 2 in. (50 mm) **El**

CAST MAGNESIUM AND MAGNESIUM ALLOYS

CAST MAGNESIUM AND MAGNESIUM ALLOYS

specification number	designation	country	product forms	chemical composition	mechanical properties and hardness values
ANSI/ASTM B 92	9998A	US	Ingot and stick: for remelting	0.004 **Al**, 0.002 **Mn**, 0.0005 **Cu**, 0.002 **Fe**, 99.98 min **Mg**, 0.0005 **Ni**, 0.002 **Pb**, 0.003 **Si**, 0.001 **Ti**, 0.005 each, others	
ANSI/ASTM B 92	9995A	US	Ingot and stick: for remelting	0.01 **Al**, 0.004 **Mn**, 0.003 **Fe**, 99.95 min **Mg**, 0.001 **Ni**, 0.005 **Si**, 0.01 **Ti**, 0.005 each, others	
SIS 14 46 02	SIS Mg 46 02–00	Sweden	Ingot	0.01 **Al**, 0.01 **Mn**, 0.01 **Zn**, 0.01 **Si**, 0.001 **Ni**, 0.002 **Cu**, 0.003 **Fe**, 0.005 **Pb**, 0.001 **Sn**, 99.95 **Mg**, 0.005 **Cu+Fe+Ni**, 0.01 each, others	
SIS 14 46 02	SIS Mg 46 02–00	Sweden	Pigs	0.01 **Al**, 0.01 **Mn**, 0.01 **Zn**, 0.01 **Si**, 0.001 **Ni**, 0.002 **Cu**, 0.003 **Fe**, 0.005 **Pb**, 0.001 **Sn**, 99.95 **Mg**, 0.005 **Cu+Fe+Ni**, 0.01 each, others	
ANSI/ASTM B 92	9990A	US	Ingot and stick: for remelting	0.003 **Al**, 0.004 **Mn**, 0.04 **Fe**, 99.90 min **Mg**, 0.001 **Ni**, 0.005 **Si**	
JIS H 2150	Class 1	Japan	Ingot	0.01 **Al**, 0.01 **Mn**, 0.05 **Zn**, 0.005 **Cu**, 0.01 **Fe**, 99.90 min **Mg**, 0.0001 **Ni**, 0.01 **Si**	
ANSI/ASTM B 92	9980A	US	Ingot and stick: for remelting	0.10 **Mn**, 0.02 **Cu**, 99.80 min **Mg**, 0.001 **Ni**, 0.01 **Pb**, 0.01 **Sn**, 0.05 each, others	
JIS H 2150	Class 2	Japan	Ingot	0.05 **Al**, 0.10 **Mn**, 0.05 **Zn**, 0.02 **Cu**, 0.05 **Fe**, 99.8 min **Mg**, 0.001 **Ni**, 0.01 **Si**	
SIS 14 46 04	SIS Mg 46 04–00	Sweden	Ingot	0.05 **Al**, 0.01 **Mn**, 0.02 **Cu**, 0.05 **Fe**, 0.05 **Si**, 0.002 **Ni**, 99.8 **Mg**, 0.05 each, others	
SIS 14 46 04	SIS Mg 46 04–00	Sweden	Pigs	0.05 **Al**, 0.01 **Mn**, 0.02 **Cu**, 0.05 **Fe**, 0.05 **Si**, 0.002 **Ni**, 99.8 **Mg**, 0.05 each, others	
MIL-M-20161B		US	Ingot and stick	99.80 min **Al**, 0.15 **Mn**, 0.02 **Cu**, 0.01 **Pb**, 0.001 **Ni**, 0.01 **Sn**, 0.05 each, 0.20 total, others	
ISO/R144	Mg 99.8	ISO	Ingot	0.05 **Al**, 0.01 **Mn**, 0.05 **Si**, 0.02 **Cu**, 0.05 **Fe**, 0.002 **Ni**, 0.20 **Al+Mn+Si+Cu+Fe+Ni**; total; 0.05 each, other	
ISO/R207	Mg 99.95	ISO	Ingot	0.01 **Al**, 0.01 **Mn**, 0.01 **Zn**, 0.01 **Si**, 0.002 **Cu**, 0.003 **Fe**, 0.001 **Ni**, 0.005 **Pb**, 0.001 **Sn**, 0.05 **Al+Mn+Zn+Si+Cu+Fe+Ni+Pb+Sn**	
ANSI/ASTM B 80	AM100A	US	Sand castings	9.3–10.7 **Al**, 0.10 **Mn**, 0.30 **Zn**, 0.10 **Cu**, rem **Mg**, 0.01 **Ni**, 0.30 **Si**, 0.30 others, total	**Solution heat treated and artificially aged**: all **diam**, 35 ksi (241 MPa) **TS**, 17 ksi (117 MPa) **YS**, 69 **HB**

CAST MAGNESIUM AND MAGNESIUM ALLOYS

specification number	designation	country	product forms	chemical composition	mechanical properties and hardness values
ANSI/ASTM B 403	AM100A	US	Investment castings	9.3–10.7 **Al**, 0.10 min **Mn**, 0.30 **Zn**, 0.10 **Cu**, rem **Mg**, 0.01 **Ni**, 0.30 **Si**, 0.30 others, total	**As fabricated**: all **diam**, 20 ksi (138 MPa) **TS**, 10 ksi (69 MPa) **YS**; **Solution heat treated and naturally aged**: all **diam**, 34 ksi (234 MPa) **TS**, 10 ksi (69 MPa) **YS**, 6% in 2 in. (50 mm) **El**; **Solution heat treated and artificially aged**: all **diam**, 34 ksi (234 MPa) **TS**, 15 ksi (103 MPa) **YS**, 2% in 2 in. (50 mm) **El**
MIL-M-46062	AM100A	US	Castings	9.3–10.7 **Al**, 0.10 **Mn**, 0.30 **Zn**, rem **Mg**, 0.30 **Si**, 0.10 **Cu**, 0.01 **Ni**, 0.30 total, others	**T6**: all **diam**, 262 MPa **TS**, 138 MPa **YS**, 3% in 2 in. (50 mm) **El**
ANSI/ASTM B 199	AM100A	US	Permanent mold castings	9.3–10.7 **Al**, 0.10 min **Mn**, 0.30 **Zn**, 0.10 **Cu**, rem **Mg**, 0.01 **Ni**, 0.30 **Si**, 0.30 others, total	**As fabricated**: all **diam**, 20 ksi (138 MPa) **TS**, 10 ksi (69 MPa) **YS**, 53 **HB**; **Solution heat treated and naturally aged**: all **diam**, 34 ksi (234 MPa) **TS**, 10 ksi (69 MPa) **YS**, 6% in 2 in. (50 mm) **El**; **Solution heat treated and artificially aged**: all **diam**, 34 ksi (234 MPa) **TS**, 15 ksi (103 MPa) **YS**, 2% in 2 in. (50 mm) **El**
ANSI/ASTM B 93	AM100A	US	Ingot: for sand, permanent mold, and investment castings	9.4–10.6 **Al**, 0.13 min **Mn**, 0.20 **Zn**, 0.08 **Cu**, rem **Mg**, 0.01 **Ni**, 0.20 **Si**, 0.30 others, total	
JIS H 2221	Class 5	Japan	Ingot: for castings	9.3–10.7 **Al**, 0.10–0.5 **Mn**, 0.10 **Zn**, 0.08 **Cu**, rem **Mg**, 0.01 **Ni**, 0.20 **Si**	
JIS H 5203	Class 5	Japan	Castings	9.3–10.7 **Al**, 0.10–0.5 **Mn**, 0.30 **Zn**, 0.10 **Cu**, rem **Mg**, 0.01 **Ni**, 0.30 **Si**	**As cast**: all **diam**, 137 MPa **TS**, 69 MPa **YS**; **Solution heat treated**: all **diam**, 235 MPa **TS**, 69 MPa **YS**, 6% in 2 in. (50 mm) **El**; **Solution heat treated and artificially aged**: all **diam**, 235 MPa **TS**, 108 MPa **YS**, 2% in 2 in. (50 mm) **El**
ANSI/ASTM B 94	UNS M10600 (AM60A)	US	Die castings	5.5–6.5 **Al**, 0.13 min **Mn**, 0.22 **Zn**, 0.35 **Cu**, rem **Mg**, 0.03 **Ni**, 0.50 **Si**	**As cast**: all **diam**, 32 ksi (220 MPa) **TS**, 19 ksi (130 MPa) **YS**, 8% in 2 in. (50 mm) **El**
ANSI/ASTM B 93	AM60A	US	Ingot: for die castings	5.7–6.3 **Al**, 0.15 min **Mn**, 0.20 **Zn**, 0.25 **Cu**, rem **Mg**, 0.01 **Ni**, 0.20 **Si**, 0.30 others, total	
JIS H 2221	Class 1	Japan	Ingot: for castings	5.3–6.7 **Al**, 0.15–0.6 **Mn**, 0.08 **Cu**, rem **Mg**, 0.01 **Ni**, 0.20 **Si**	

CAST MAGNESIUM AND MAGNESIUM ALLOYS

specification number	designation	country	product forms	chemical composition	mechanical properties and hardness values
ANSI/ASTM B 94	UNS M10410 (AS41A)	US	Die castings	3.5–5.0 **Al**, 0.20–0.50 **Mn**, 0.12 **Zn**, 0.06 **Cu**, rem **Mg**, 0.03 **Ni**, 0.50–1.5 **Si**	**As cast**: all **diam**, 31 ksi (210 MPa) **TS**, 20 ksi (140 MPa) **YS**, 6% in 2 in. (50 mm) **El**
ANSI/ASTM B 93	AS41A	US	Ingot: for die castings	3.7–4.8 **Al**, 0.22–0.48 **Mn**, 0.10 **Zn**, 0.04 **Cu**, rem **Mg**, 0.01 **Ni**, 0.60–1.4 **Si**, 0.30 total, others	
BS 3L. 125		US	Castings	9.0–10.5 **Al**, 0.15–0.4 **Mn**, 0.3–1.0 **Zn**, 0.15 **Cu**, 0.3 **Si**, 0.05 **Fe**, 0.01 **Ni**, 0.1 **Sn**, rem **Mg**, 0.40 total, others	**Heat treated and quenched**: all **diam**, 215 MPa **TS**, 120 MPa **YS**, 2% in 2 in. (50 mm) **El**
BS 3L. 125		US	Ingot	9.0–10.5 **Al**, 0.2–0.4 **Mn**, 0.3–1.0 **Zn**, 0.15 **Cu**, 0.2 **Si**, 0.03 **Fe**, 0.01 **Ni**, 0.1 **Sn**, rem **Mg**, 0.35 total, others	**Heat treated and quenched**: all **diam**, 215 MPa **TS**, 120 MPa **YS**, 2% in 2 in. (50 mm) **El**
BS 3L. 124		US	Castings	9.0–10.5 **Al**, 0.15–0.4 **Mn**, 0.3–1.0 **Zn**, 0.15 **Cu**, 0.3 **Si**, 0.05 **Fe**, 0.01 **Ni**, 0.1 **Sn**, rem **Mg**, 0.40 total, others	**Heat treated and quenched**: all **diam**, 200 MPa **TS**, 80 MPa **YS**, 4% in 2 in. (50 mm) **El**
BS 3L. 124		US	Ingot	9.0–10.5 **Al**, 0.2–0.4 **Mn**, 0.3–1.0 **Zn**, 0.15 **Cu**, 0.2 **Si**, 0.03 **Fe**, 0.01 **Ni**, 0.1 **Sn**, rem **Mg**, 0.35 total, others	**Heat treated and quenched**: all **diam**, 200 MPa **TS**, 80 MPa **YS**, 4% in 2 in. (50 mm) **El**
SIS 14 46 35	SIS Mg 46 35–10	Sweden	Pressure die castings	8.3–10.3 **Al**, 0.15–0.6 **Mn**, 0.2–1.0 **Zn**, 0.3 **Si**, 0.01 **Ni**, 0.005–0.0015 **Be**, 0.2 **Cu**, 0.05 **Fe**	**As cast**: all **diam**, 230 MPa **TS**, 160 MPa **YS**, 1% in 2 in. (50 mm) **El**, 70 **HB**
ISO/R121	Mg–Al9Zn	ISO	Sand castings, bar	8.3–10.3 **Al**, 0.15–0.6 **Mn**, 0.2–1.0 **Zn**, 0.3 **Si**, 0.05 **Fe**, 0.2 **Cu**, 0.01 **Ni**	**As cast**: 13 mm **diam**, 140 MPa **TS**, 75 MPa **YS**, 1% in 2 in. (50 mm) **El**; **Solution treated**: 13 mm **diam**, 230 MPa **TS**, 75 MPa **YS**, 6% in 2 in. (50 mm) **El**; **Fully heat treated**: 13 mm **diam**, 235 MPa **TS**, 110 MPa **YS**, 1% in 2 in. (50 mm) **El**
SIS 14 46 35	SIS Mg 46 35–00	Sweden	Ingot	8.3–9.8 **Al**, 0.12–0.6 **Mn**, 0.3–0.8 **Zn**, 0.2 **Si**, 0.01 **Ni**, 0.0005–0.0015 **Be**, 0.15 **Cu**, 0.03 **Fe**, 90 **Mg**	
ISO/R122	Alloy No. 2	ISO	Ingot: for castings	8.3–9.8 **Al**, 0.20–0.6 **Mn**, 0.3–0.8 **Zn**, 0.2 **Si**, 0.03 **Fe**, 0.15 **Cu**, 0.01 **Ni**	
NS 17 709	NS 17 709–32	Norway	Chill castings	9 **Al**, 0.2 **Mn**, 0.7 **Zn**, 90 **Mg**	**Artificially aged**: all **diam**, 170 MPa **TS**, 110 MPa **YS**, 2% in 2 in. (50 mm) **El**
NS 17 709	NS 17 709–31	Norway	Sand castings	9 **Al**, 0.2 **Mn**, 0.7 **Zn**, 90 **Mg**	**Artificially aged**: all **diam**, 170 MPa **TS**, 130 MPa **YS**, 2% in 2 in. (50 mm) **El**
NS 17 709	NS 17 709–22	Norway	Chill castings	9 **Al**, 0.2 **Mn**, 0.7 **Zn**, 90 **Mg**	**Homogenized**: all **diam**, 160 MPa **TS**, 100 MPa **YS**, 3% in 2 in. (50 mm) **El**
NS 17 709	NS 17 709–21	Norway	Sand castings	9 **Al**, 0.2 **Mn**, 0.7 **Zn**, 90 **Mg**	**Homogenized**: all **diam**, 170 MPa **TS**, 90 MPa **YS**, 3% in 2 in. (50 mm) **El**

CAST MAGNESIUM AND MAGNESIUM ALLOYS

specification number	designation	country	product forms	chemical composition	mechanical properties and hardness values
NS 17 709	NS 17 709–05	Norway	Pressure die castings	9 **Al**, 0.2 **Mn**, 0.7 **Zn**, 90 **Mg**	**As cast**: all **diam**, 200 MPa **TS**, 150 MPa **YS**, 1% in 2 in. (50 mm) **El**
NS 17 709	NS 17 709–02	Norway	Chill castings	9 **Al**, 0.2 **Mn**, 0.7 **Zn**, 90 **Mg**	**As cast**: all **diam**, 120 MPa **TS**, 90 MPa **YS**, 1% in 2 in. (50 mm) **El**
NS 17 709	NS 17 709–01	Norway	Sand castings	9 **Al**, 0.2 **Mn**, 0.7 **Zn**, 90 **Mg**	**As cast**: all **diam**, 130 MPa **TS**, 80 MPa **YS**, 1% in 2 in. (50 mm) **El**
MIL-M-46062	AZ92A	US	Castings	8.3–9.7 **Al**, 0.10 **Mn**, 1.6–2.4 **Zn**, rem **Mg**, 0.30 **Si**, 0.10 **Cu**, 0.01 **Ni**, 0.30 total, others	**T6**: all **diam**, 276 MPa **TS**, 172 MPa **YS**, 3% in 2 in. (50 mm) **El**
ANSI/ASTM B 403	AZ92A	US	Investment castings	8.3–9.7 **Al**, 0.10 **Mn**, 1.6–2.4 **Zn**, 0.10 **Cu**, rem **Mg**, 0.01 **Ni**, 0.30 **Si**, 0.30 total, others	**As fabricated**: all **diam**, 20 ksi (138 MPa) **TS**, 10 ksi (69 MPa) **YS**; **Solution heat treated and naturally aged**: all **diam**, 34 ksi (234 MPa) **TS**, 10 ksi (69 MPa) **YS**, 6% in 2 in. (50 mm) **El**; **Cooled and artificially aged**: all **diam**, 20 ksi (138 MPa) **TS**, 11 ksi (76 MPa) **YS**
JIS H 2221	Class 3	Japan	Ingot: for castings	8.3–9.7 **Al**, 0.10–0.5 **Mn**, 1.6–2.4 **Zn**, 0.08 **Cu**, rem **Mg**, 0.01 **Ni**, 0.20 **Si**	
JIS H 2222	Class 1, Grade B	Japan	Ingot: for die castings	8.5–9.5 **Al**, 0.15 **Mn**, 0.45–0.9 **Zn**, 0.25 **Cu**, rem **Mg**, 0.01 **Ni**, 0.30 **Si**	
JIS H 2222	Class 1, Grade A	Japan	Ingot: for die castings	8.5–9.5 **Al**, 0.15 **Mn**, 0.45–0.9 **Zn**, 0.08 **Cu**, rem **Mg**, 0.01 **Ni**, 0.20 **Si**	
JIS H 5303	Class 1A	Japan	Die castings	8.3–9.7 **Al**, 0.15 **Mn**, 0.35–1.0 **Zn**, 0.10 **Cu**, rem **Mg**, 0.03 **Ni**, 0.50 **Si**	
JIS H 5303	Class 1B	Japan	Die castings	8.3–9.7 **Al**, 0.15 **Mn**, 0.35–1.0 **Zn**, 0.35 **Cu**, rem **Mg**, 0.03 **Ni**, 0.50 **Si**	
ISO/R121	Mg-Al9Zn2	ISO	Sand castings, bar	8.0–10.0 **Al**, 0.10–0.5 **Mn**, 1.5–2.5 **Zn**, 0.3 **Si**, 0.05 **Fe**, 0.2 **Cu**, 0.01 **Ni**	**As cast**: 13 mm **diam**, 140 MPa **TS**, 75 MPa **YS**, 1% in 2 in. (50 mm) **El**; **Solution treated**: 13 mm **diam**, 230 MPa **TS**, 75 MPa **YS**, 5% in 2 in. (50 mm) **El**; **Fully heat treated**: 13 mm **diam**, 235 MPa **TS**, 110 MPa **YS**, 1% in 2 in. (50 mm) **El**
JIS H 5203	Class 3	Japan	Castings	8.3–9.7 **Al**, 0.10–0.5 **Mn**, 1.6–2.4 **Zn**, 0.10 **Cu**, rem **Mg**, 0.01 **Ni**, 0.30 **Si**	**As cast**: all **diam**, 157 MPa **TS**, 69 MPa **YS**; **Solution heat treated**: all **diam**, 235 MPa **TS**, 69 MPa **YS**, 6% in (50 mm) **El**; **Artificially aged**: all **diam**, 157 MPa **TS**, 78 MPa **YS**
ISO/R122	Alloy No. 3	ISO	Ingot: for castings	8.0–9.5 **Al**, 0.13–0.5 **Mn**, 1.7–2.3 **Zn**, 0.2 **Si**, 0.03 **Fe**, 0.15 **Cu**, 0.01 **Ni**	

CAST MAGNESIUM AND MAGNESIUM ALLOYS

specification number	designation	country	product forms	chemical composition	mechanical properties and hardness values
ANSI/ASTM B 80	AZ91C	US	Sand castings	8.1–9.3 **Al**, 0.13 **Mn**, 0.40–1.0 **Zn**, 0.10 **Cu**, rem **Mg**, 0.01 **Ni**, 0.30 **Si**, 0.30 total, others	**Solution heat treated and naturally aged**: all **diam**, 34 ksi (234 MPa) **TS**, 11 ksi (76 MPa) **YS**, 7% in 2 in. (50 mm) **El**, 55 **HB**; **Cooled and artificially aged**: all **diam**, 23 ksi (158 MPa) **TS**, 12 ksi (83 MPa) **YS**, 2% in 2 in. (50 mm) **El** 62 **HB**; **Solution heat treated and artificially aged**: all **diam**, 34 ksi (234 MPa) **TS**, 16 ksi (110 MPa) **YS**, 3% in 2 in. (50 mm) **El**, 70 **HB**
MIL–M–46062	AZ91C	US	Castings	8.1–9.3 **Al**, 0.13 **Mn**, 0.40–1.0 **Zn**, rem **Mg**, 0.30 **Si**, 0.10 **Cu**, 0.01 **Ni**, 0.30 total, others	**T6**: all **diam**, 241 MPa **TS**, 124 MPa **YS**, 4% in 2 in. (50 mm) **El**
ANSI/ASTM B 403	AZ91C	US	Investment castings	8.1–9.3 **Al**, 0.13 **Mn**, 0.40–1.0 **Zn**, 0.10 **Cu**, rem **Mg**, 0.01 **Ni**, 0.30 **Si**, 0.30 total, others	**As fabricated**: all **diam**, 18 ksi (124 MPa) **TS**, 10 ksi (69 MPa) **YS**; **Solution heat treated and naturally aged**: all **diam**, 34 ksi (234 MPa) **TS**, 10 ksi (69 MPa) **YS**, 7% in 2 in. (50 mm) **El**; **Cooled and artificially aged**: all **diam**, 20 ksi (138 MPa) **TS**, 11 ksi (76 MPa) **YS**, 2% in 2 in. (50 mm) **El**
JIS H 2221	Class 2	Japan	Ingot: for castings	8.1–9.3 **Al**, 0.13–0.5 **Mn**, 0.40–1.0 **Zn**, 0.08 **Cu**, rem **Mg**, 0.01 **Ni**, 0.20 **Si**	
JIS H 5203	Class 2	Japan	Castings	8.1–9.3 **Al**, 0.13–0.5 **Mn**, 0.40–1.0 **Zn**, 0.10 **Cu**, rem **Mg**, 0.01 **Ni**, 0.30 **Si**	**As cast**: all **diam**, 157 MPa **TS**, 69 MPa **YS**; **Solution heat treated**: all **diam**, 235 MPa **TS**, 69 MPa **YS**, 7% in (50 mm) **El**; **Artificially aged**: all **diam**, 235 MPa **TS**, 108 MPa **YS**, 2% in (50 mm) **El**
ISO/R503	Alloy No. 23	ISO	Ingot, billet, slab: for wrought products	7.5–9.2 **Al**, 0.12 min **Mn**, 0.2–1.0 **Zn**, 0.1 **Si**, 0.05 **Cu**, 0.005 **Fe**, 0.005 **Ni**	
BS 3L. 122		US	Castings	7.5–9.0 **Al**, 0.15–0.4 **Mn**, 0.3–1.0 **Zn**, 0.15 **Cu**, 0.3 **Si**, 0.05 **Fe**, 0.01 **Ni**, 0.1 **Sn**, rem **Mg**, 0.40 total, others	**Heat treated and quenched**: all **diam**, 200 MPa **TS**, 80 MPa **YS**, 7% in 2 in. (50 mm) **El**
SIS 14 16 40	SIS Mg 46 40–10	Sweden	Pressure die castings	7.0–9.5 **Al**, 0.15 min **Mn**, 0.3–2.0 **Zn**, 0.5 **Si**, 0.02 **Ni**, 0.35 **Cu**, 0.05 **Fe**, 91 **Mg**	**As cast**: all **diam**, 200 MPa **TS**, 140 MPa **YS**, 2% in 2 in. (50 mm) **El**, 60 **HB**
SIS 14 46 37	SIS Mg 46 37–03	Sweden	Sand castings	7.5–9.0 **Al**, 0.15–0.6 **Mn**, 0.2–1.0 **Zn**, 0.3 **Si**, 0.01 **Ni**, 0.2 **Cu**, 0.05 **Fe**, 91 **Mg**	**As cast**: all **diam**, 130 MPa **TS**, 90 MPa **YS**, 1% in 2 in. (50 mm) **El**, 60 **HB**
SIS 14 46 37	SIS Mg 46 37–04	Sweden	Sand castings	7.5–9.0 **Al**, 0.15–0.6 **Mn**, 0.2–1.0 **Zn**, 0.3 **Si**, 0.01 **Ni**, 0.2 **Cu**, 0.05 **Fe**, 91 **Mg**	**Solution treated**: all **diam**, 170 MPa **TS**, 90 MPa **YS**, 3% in 2 in. (50 mm) **El**, 60 **HB**

CAST MAGNESIUM AND MAGNESIUM ALLOYS

specification number	designation	country	product forms	chemical composition	mechanical properties and hardness values
SIS 14 46 40	SIS Mg 46 40–03	Sweden	Sand castings	7.0–9.5 **Al**, 0.15 min **Mn**, 0.3–2.0 **Zn**, 0.5 **Si**, 0.02 **Ni**, 0.35 **Cu**, 0.05 **Fe**, 91 **Mg**	**As casat**: all **diam**, 130 MPa **TS**, 90 MPa **YS**, 1% in 2 in. (50 mm) **El**, 60 **HB**
SIS 14 46 40	SIS Mg 46 40–04	Sweden	Sand castings	7.0–9.5 **Al**, 0.15 min **Mn**, 0.3–2.0 **Zn**, 0.5 **Si**, 0.02 **Ni**, 0.35 **Cu**, 0.05 **Fe**, 91 **Mg**	**Solution treated**: all **diam**, 170 MPa **TS**, 90 MPa **YS**, 3% in 2 in. (50 mm) **El**, 60 **HB**
SIS 14 46 40	SIS Mg 46 40–06	Sweden	Chill castings	7.0–9.5 **Al**, 0.15 min **Mn**, 0.3–2.0 **Zn**, 0.5 **Si**, 0.02 **Ni**, 0.35 **Cu**, 0.05 **Fe**, 91 **Mg**	**As cast**: all **diam**, 140 MPa **TS**, 100 MPa **YS**, 1% in 2 in. (50 mm) **El**, 60 **HB**
SIS 14 46 40	SIS Mg 46 40–07	Sweden	Chill castings	7.0–9.5 **Al**, 0.15 min **Mn**, 0.3–2.0 **Zn**, 0.5 **Si**, 0.02 **Ni**, 0.35 **Cu**, 0.05 **Fe**, 91 **Mg**	**Solution treated**: all **diam**, 180 MPa **TS**, 100 MPa **YS**, 3% in 2 in. (50 mm) **El**, 60 **HB**
ISO/R121	Mg–Al8Zn	ISO	Sand castings, bar	7.5–9.0 **Al**, 0.15–0.6 **Mn**, 0.2–1.0 **Zn**, 0.3 **Si**, 0.05 **Fe**, 0.2 **Cu**, 0.01 **Ni**	**As cast**: 13 mm **diam**, 140 MPa **TS**, 75 MPa **YS**, 1% in 2 in. (50 mm) **El**; **Solution treated**: 13 mm **diam**, 230 MPa **TS**, 75 MPa **YS**, 6% in 2 in. (50 mm) **El**; **Fully heat treated**: 13 mm **diam**, 235 MPa **TS**, 95 MPa **YS**, 2% in 2 in. (50 mm) **El**
ISO/R121	Mg–Al8Zn1	ISO	Sand castings, bar	7.0–9.5 **Al**, 0.15 min **Mn**, 0.3–2.0 **Zn**, 0.5 **Si**, 0.05 **Fe**, 0.35 **Cu**, 0.02 **Ni**	**As cast**: 13 mm **diam**, 140 MPa **TS**, 75 MPa **YS**
SIS 14 46 40	SIS Mg 46 40–00	Sweden	Ingot	7.0–9.2 **Al**, 0.2 **Mn**, 0.4–1.8 **Zn**, 0.3 **Si**, 0.02 **Ni**, 0.3 **Cu**, 0.05 **Fe**, 91 **Mg**	
ISO/R122	Alloy No. 5	ISO	Ingot: for castings	7.0–9.2 **Al**, 0.2 min **Mn**, 0.4–1.8 **Zn**, 0.3 **Si**, 0.05 **Fe**, 0.3 **Cu**, 0.02 **Ni**	
BS 3L. 122		US	Ingot	7.5–8.5 **Al**, 0.2–0.4 **Mn**, 0.3–1.0 **Zn**, 0.15 **Cu**, 0.2 **Si**, 0.03 **Fe**, 0.01 **Ni**, 0.1 **Sn**, rem **Mg**, 0.35 total, others	**Heat treated and quenched**: all **diam**, 200 MPa **TS**, 80 MPa **YS**, 7% in 2 in. (50 mm) **El**
NS 17 708	NS 17 708–22	Norway	Chill castings	8 **Al**, 0.2 **Mn**, 0.7 **Zn**, 91 **Mg**	**Homogenized**: all **diam**, 170 MPa **TS**, 80 MPa **YS**, 4% in 2 in. (50 mm) **El**
NS 17 708	NS 17 708–21	Norway	Sand castings	8 **Al**, 0.2 **Mn**, 0.7 **Zn**, 91 **Mg**	**Homogenized**: all **diam**, 170 MPa **TS**, 80 MPa **YS**, 4% in 2 in. (50 mm) **El**
NS 17 708	NS 17 708–05	Norway	Pressure die castings	8 **Al**, 0.2 **Mn**, 0.7 **Zn**, 91 **Mg**	**As cast**: all **diam**, 200 MPa **TS**, 140 MPa **YS**, 1% in 2 in. (50 mm) **El**
NS 17 708	NS 17 708–02	Norway	Chill castings	8 **Al**, 0.2 **Mn**, 0.7 **Zn**, 91 **Mg**	
NS 17 708	NS 17 708–01	Norway	Sand castings	8 **Al**, 0.2 **Mn**, 0.7 **Zn**, 91 **Mg**	
SIS 14 46 37	SIS Mg 46 37–00	Sweden	Ingot	7.5–8.5 **Al**, 0.20–0.6 **Mn**, 0.3–0.8 **Zn**, 0.2 **Si**, 0.01 **Ni**, 0.15 **Cu**, 0.03 **Fe**, 91 **Mg**	
ISO/R122	Alloy No. 1	ISO	Ingot: for castings	7.5–8.5 **Al**, 0.20–0.6 **Mn**, 0.3–0.8 **Zn**, 0.2 **Si**, 0.03 **Fe**, 0.15 **Cu**, 0.01 **Ni**	

CAST MAGNESIUM AND MAGNESIUM ALLOYS

specification number	designation	country	product forms	chemical composition	mechanical properties and hardness values
ANSI/ASTM B 80	AZ81A	US	Sand castings	7.0–8.1 **Al**, 0.13 **Mn**, 0.40–1.0 **Zn**, 0.10 **Cu**, rem **Mg**, 0.01 **Ni**, 0.30 **Si**, 0.30 total, others	**Solution heat treated and naturally aged**: all **diam**, 34 ksi (234 MPa) **TS**, 11 ksi (76 MPa) **YS**, 7% in 2 in. (50 mm) **El**, 55 **HB**
ANSI/ASTM B 403	AZ81A	US	Investment castings	7.0–8.1 **Al**, 0.13 **Mn**, 0.40–1.0 **Zn**, 0.10 **Cu**, rem **Mg**, 0.01 **Ni**, 0.30 **Si**, 0.30 total, others	**Solution heat treated and naturally aged**: all **diam**, 34 ksi (234 MPa) **TS**, 10 ksi (69 MPa) **YS**, 7% in 2 in. (50 mm) **El**
ISO/R503	Alloy No. 22	ISO	Ingot, billet, slab: for wrought products	5.5–7.2 **Al**, 0.15 min **Mn**, 0.5–1.5 **Zn**, 0.1 **Si**, 0.05 **Cu**, 0.005 **Fe**, 0.005 **Ni**	
ISO/R503	Alloy No. 26	ISO	Ingot, billet, slab: for wrought products	5.4–7.3 **Al**, 0.15–0.4 **Mn**, 0.5–1.5 **Zn**, 0.1 **Si**, 0.01 **Cu**, 0.03 **Fe**, 0.005 **Ni**	
ANSI/ASTM B 80	AZ63A	US	Sand castings	5.3–6.7 **Al**, 0.15 **Mn**, 2.5–3.5 **Zn**, 0.25 **Cu**, rem **Mg**, 0.01 **Ni**, 0.30 **Si**, 0.30 total, others	**As fabricated**: all **diam**, 26 ksi (179 MPa) **TS**, 11 ksi (76 MPa) **YS**, 4% in 2 in. (50 mm) **El**, 50 **HB**; **Solution heat treated and naturally aged**: all **diam**, 34 ksi (234 MPa) **TS**, 11 ksi (76 MPa) **YS**, 7% in 2 in. (50 mm) **El**, 55 **HB**; **Cooled and artificially aged**: all **diam**, 26 ksi (179 MPa) **TS**, 12 ksi (83 MPa) **YS**, 2% in 2 in. (50 mm) **El** 55 **HB**
NS 17 706	NS 17 706–05	Norway	Pressure die castings	6.0 **Al**, 0.2 **Mn**, 0.6 **Zn**, 93 **Mg**	**As cast**: all **diam**, 200 MPa **TS**, 130 MPa **YS**, 3% in 2 in. (50 mm) **El**
ISO/R121	Mg-Al6Zn3	ISO	Sand castings, bar	5.0–7.0 **Al**, 0.10–0.5 **Mn**, 2.0–3.5 **Zn**, 0.3 **Si**, 0.05 **Fe**, 0.2 **Cu**, 0.01 **Ni**	**As cast**: 13 mm **diam**, 160 MPa **TS**, 75 MPa **YS**, 3% in 2 in. (50 mm) **El**
JIS H 5203	Class 1	Japan	Castings	5.3–6.7 **Al**, 0.15–0.6 **Mn**, 2.5–3.5 **Zn**, 0.10 **Cu**, rem **Mg**, 0.01 **Ni**, 0.30 **Si**	**As cast**: all **diam**, 177 MPa **TS**, 69 MPa **YS**, 4% in (50 mm) **El**; **Solution heat treated**: all **diam**, 235 MPa **TS**, 69 MPa **YS**, 7% in (50 mm) **El**; **Artificially aged**: all **diam**, 235 MPa **TS**, 108 MPa **YS**, 2% in (50 mm) **El**
ISO/R122	Alloy No. 4	ISO	Ingot: for castings	5.0–6.5 **Al**, 0.15–0.5 **Mn**, 2.3–3.3 **Zn**, 0.2 **Si**, 0.03 **Fe**, 0.15 **Cu**, 0.01 **Ni**	
ISO/R503	Alloy No. 21	ISO	Ingot, billet, slab: for wrought products	2.5–3.5 **Al**, 0.2 min **Mn**, 0.5–1.5 **Zn**, 0.04 **Ca**, 0.1 **Si**, 0.05 **Cu**, 0.005 **Fe**, 0.005 **Ni**	
ISO/R503	Alloy No. 25	ISO	Ingot, billet, slab: for wrought products	2.4–3.6 **Al**, 0.15–0.4 **Mn**, 0.5–1.5 **Zn**, 0.1 **Si**, 0.01 **Cu**, 0.03 **Fe**, 0.005 **Ni**	

CAST MAGNESIUM AND MAGNESIUM ALLOYS

specification number	designation	country	product forms	chemical composition	mechanical properties and hardness values
ANSI/ASTM B 80	EZ33A	US	Sand castings	2.0–3.1 **Zn**, 2.5–4.0 **RE**, 0.10 **Cu**, rem **Mg**, 0.01 **Ni**, 0.50–1.0 **Zr**, 0.30 total, others	**Cooled and artificially aged**: all **diam**, 20 ksi (138 MPa) **TS**, 14 ksi (96 MPa) **YS**, 2% in 2 in. (50 mm) **El**, 50 **HB**
ANSI/ASTM B 199	EZ33A	US	Permanent mold castings	2.0–3.1 **Zn**, 2.5–4.0 **RE**, 0.10 **Cu**, rem **Mg**, 0.01 **Ni**, 0.50–1.0 **Zr**	**Cooled and artificially aged**: all **diam**, 20 ksi (138 MPa) **TS**, 14 ksi (97 MPa) **YS**, 2% in 2 in. (50 mm) **El**, 50 **HB**
ANSI/ASTM B 403	EZ33A	US	Investment castings	2.0–3.1 **Zn**, 2.5–4.0 **RE**, 0.10 **Cu**, rem **Mg**, 0.01 **Ni**, 0.50–1.0 **Zr**, 0.30 total, others	**Cooled and artificially aged**: all **diam**, 20 ksi (138 MPa) **TS**, 14 ksi (97 MPa) **YS**, 2% in 2 in. (50 mm) **El**
ANSI/ASTM B 403	HK31A	US	Investment castings	0.30 **Zn**, 2.5–4.0 **Th**, 0.10 **Cu**, rem **Mg**, 0.03 **Ni**, 0.40–1.0 **Zr**, 0.30 total, others	**Solution heat treated and artificially aged**: all **diam**, 27 ksi (186 MPa) **TS**, 13 ksi (90 MPa) **YS**, 4% in 2 in. (50 mm) **El**
MIL–M–46062	HK31A	US	Castings	0.30 **Zn**, 2.5–4.0 **Th**, rem **Mg**, 0.10 **Cu**, 0.01 **Ni**, 0.50–1.0 **Zr**, 0.30 total, others	**T6**: all **diam**, 227 MPa **TS**, 110 MPa **YS**, 6% in 2 in. (50 mm) **El**
ANSI/ASTM B 80	HZ32A	US	Sand castings	1.7–2.5 **Zn**, 0.10 **RE**, 2.5–4.0 **Th**, 0.10 **Cu**, rem **Mg**, 0.01 **Ni**, 0.50–1.0 **Zr**, 0.30 total, others	**Cooled and artificially aged**: all **diam**, 27 ksi (186 MPa) **TS**, 13 ksi (89 MPa) **YS**, 4% in 2 in. (50 mm) **El**, 55 **HB**
ISO 2119	Mg–Th3Zn2Zr	ISO	Castings	1.7–2.5 **Zn**, 0.10 **RE**, 2.5–4.0 **Th**, 0.40–1.0 **Zr**, 0.10 **Cu**, 0.01 **Ni**	
ISO 3115	Mg–Th3Zn2Zr	ISO	Sand castings	1.7–2.5 **Zn**, 0.10 **RE**, 2.5–4.0 **Th**, 0.40–1.0 **Zr**, 0.10 **Cu**, 0.01 **Ni**	**TE**: all **diam**, 185 MPa **TS**, 90 MPa **YS**, 3% in 2 in. (50 mm) **El**
ANSI/ASTM B 403	K1A	US	Investment castings	rem **Mg**, 0.40–1.0 **Zr**, 0.30 total, others	**As fabricated**: all **diam**, 22 ksi (152 MPa) **TS**, 7 ksi (48 MPa) **YS**, 14% in 2 in. (50 mm) **El**
ANSI/ASTM B 80	K1A	US	Sand castings	rem **Mg**, 0.40 **Zr**, 0.30 total, others	**As fabricated**: all **diam**, 24 ksi (165 MPa) **TS**, 6 ksi (41 MPa) **YS**, 14% in 2 in. (50 mm) **El**
ANSI/ASTM B 403	QE22A	US	Investment castings	1.8–2.5 **RE**, 2.0–3.0 **Ag**, 0.10 **Cu**, rem **Mg**, 0.01 **Ni**, 0.40–1.0 **Zr**, 0.30 total, others	**Solution heat treated and artificially aged**: all **diam**, 35 ksi (241 MPa) **TS**, 25 ksi (172 MPa) **YS**, 2% in 2 in. (50 mm) **El**
ANSI/ASTM B 80	QE22A	US	Sand castings	1.8–2.5 **RE**, 2.0–3.0 **Ag**, 0.10 **Cu**, rem **Mg**, 0.01 **Ni**, 0.40–1.0 **Zr**, 0.30 total, others	**Solution heat treated and artificially aged**: all **diam**, 35 ksi (241 MPa) **TS**, 25 ksi (172 MPa) **YS**, 2% in 2 in. (50 mm) **El**, 78 **HB**
ANSI/ASTM B 199	QE22A	US	Permanent mold castings	1.8–2.5 **RE**, 2.0–3.0 **Ag**, 0.10 **Cu**, rem **Mg**, 0.01 **Ni**, 0.40–1.0 **Zr**, 0.30 total, others	**Solution heat treated and artificially aged**: all **diam**, 35 ksi (241 MPa) **TS**, 25 ksi (172 MPa) **YS**, 2% in 2 in. (50 mm) **El**, 78 **HB**
MIL–M–46062	QE22A	US	Castings	1.8–2.5 **RE**, 2.0–3.0 **Ag**, rem **Mg**, 0.10 **Cu**, 0.01 **Ni**, 0.40–1.0 **Zr**, 0.30 total, others	**T6**: all **diam**, 276 MPa **TS**, 193 MPa **YS**, 4% in 2 in. (50 mm) **El**

CAST MAGNESIUM AND MAGNESIUM ALLOYS

specification number	designation	country	product forms	chemical composition	mechanical properties and hardness values
ANSI/ASTM B 80	ZE63A	US	Sand castings	5.5–6.0 **Zn**, 2.1–3.0 **RE**, 0.10 **Cu**, rem **Mg**, 0.01 **Ni**, 0.40–1.0 **Zr**, 0.30 total, others	**Solution heat treated and artificially aged**: all **diam**, 40 ksi (276 MPa) **TS**, 27 ksi (186 MPa) **YS**, 5% in 2 in. (50 mm) **El**
MIL–M–46062	ZE63A	US	Castings	5.5–6.0 **Zn**, 2.0–3.0 **RE**, rem **Mg**, 0.10 **Cu**, 0.01 **Ni**, 0.40–1.0 **Zr**, 0.30 total, others	**T6**: all **diam**, 290 MPa **TS**, 193 MPa **YS**, 6% in 2 in. (50 mm) **El**
ANSI/ASTM B 80	ZE41A	US	Sand castings	0.15 **Mn**, 3.5–5.0 **Zn**, 0.75–1.75 **RE**, 0.10 **Cu**, rem **Mg**, 0.01 **Ni**, 0.40–1.0 **Zr**, 0.30 total, others	**Cooled and artificially aged**: all **diam**, 29 ksi (200 MPa) **TS**, 20 ksi (133 MPa) **YS**, 3% in 2 in. (50 mm) **El**, 62 **HB**
ISO 3115	Mg–Zn4REZr	ISO	Sand castings	3.5–5.0 **Zn**, 0.75–1.75 **RE**, 0.40–1.0 **Zr**, 0.10 **Cu**, 0.01 **Ni**	**TE**: all **diam**, 200 MPa **TS**, 135 MPa **YS**, 2% in 2 in. (50 mm) **El**
ISO 2119	Mg–Zn4REZr	ISO	Castings	3.5–5.0 **Zn**, 0.75–1.75 **RE**, 0.40–1.0 **Zr**, 0.10 **Cu**, 0.01 **Ni**,	
JIS H 5203	Class 8	Japan	Castings	2.0–3.1 **Zn**, 2.5–4.0 **RE**, 0.10 **Cu**, rem **Mg**, 0.01 **Ni**, 0.50–1.0 **Zr**	**Artificially aged**: all **diam**, 137 MPa **TS**, 98 MPa **YS**, 2% in 2 in. (50 mm) **El**
ISO 3115	Mg–RE3Zn2Zr	ISO	Sand castings	0.8–3.0 **Zn**, 2.5–4.0 **RE**, 0.40–1.0 **Zr**, 0.10 **Cu**, 0.01 **Ni**	**TE**: all **diam**, 140 MPa **TS**, 95 MPa **YS**, 2% in 2 in. (50 mm) **El**
ISO 2119	Mg–RE3Zn2Zr	ISO	Castings	0.8–3.0 **Zn**, 2.5–4.0 **RE**, 0.40–1.0 **Zr**, 0.10 **Cu**, 0.01 **Ni**	
ISO 2119	Mg–Zn6Th2Zr	ISO	Castings	5.0–6.2 **Zn**, 1.5–2.3 **Th**, 0.40–1.0 **Zr**, 0.10 **Cu**, 0.01 **Ni**	
MIL–M–46062	ZH62A	US	Castings	5.2–6.2 **Zn**, 1.4–2.2 **Th**, rem **Mg**, 0.10 **Cu**, 0.01 **Ni**, 0.50–1.0 **Zr**, 0.30 total, others	**T5**: all **diam**, 262 MPa **TS**, 159 MPa **YS**, 5% in 2 in. (50 mm) **El**
ANSI/ASTM B 80	ZH62A	US	Sand castings	5.2–6.2 **Zn**, 1.4–2.2 **Th**, 0.10 **Cu**, rem **Mg**, 0.01 **Ni**, 0.50–1.0 **Zr**, 0.30 total, others	**Cooled and artificially aged**: all **diam**, 35 ksi (241 MPa) **TS**, 22 ksi (152 MPa) **YS**, 5% in 2 in. (50 mm) **El**, 70 **HB**
ISO 3115	Mg–Zn6Th2Zr	ISO	Sand castings	5.0–6.0 **Zn**, 1.5–2.3 **Th**, 0.40–1.0 **Zr**, 0.10 **Cu**, 0.01 **Ni**	**TE**: all **diam**, 240 MPa **TS**, 150 MPa **YS**, 4% in 2 in. (50 mm) **El**
ISO 2119	Mg–Th3Zn2Zr	ISO	Castings	1.7–2.5 **Zn**, 0.10 **RE**, 2.5–4.0 **Th**, 0.40–1.0 **Zr**, 0.10 **Cu**, 0.01 **Ni**	
ISO 3115	Mg–Th3Zn2Zr	ISO	Sand castings	1.7–2.5 **Zn**, 0.10 **RE**, 2.5–4.0 **Th**, 0.40–1.0 **Zr**, 0.10 **Cu**, 0.01 **Ni**	**TE**: all **diam**, 185 MPa **TS**, 90 MPa **YS**, 3% in 2 in. (50 mm) **El**
ANSI/ASTM B 403	ZK61A	US	Investment castings	5.5–6.5 **Zn**, 0.10 **Cu**, rem **Mg**, 0.01 **Ni**, 0.6–1.0 **Zr**, 0.30 total, others	**Solution heat treated and artificially aged**: all **diam**, 40 ksi (276 MPa) **TS**, 25 ksi (172 MPa) **YS**, 5% in 2 in. (50 mm) **El**
MIL–M–46062	ZK61A	US	Castings	5.5–6.5 **Zn**, rem **Mg**, 0.10 **Cu**, 0.01 **Ni**, 0.60–1.0 **Zr**, 0.30 total, others	**T6**: all **diam**, 290 MPa **TS**, 200 MPa **YS**, 6% in 2 in. (50 mm) **El**
ANSI/ASTM B 80	ZK61A	US	Sand castings	5.5–6.5 **Zn**, 0.10 **Cu**, rem **Mg**, 0.01 **Ni**, 0.6–1.0 **Zr**, 0.30 total, others	**Solution heat treated and artificially aged**: all **diam**, 40 ksi (276 MPa) **TS**, 26 ksi (179 MPa) **YS**, 70 **HB**

CAST MAGNESIUM AND MAGNESIUM ALLOYS

specification number	designation	country	product forms	chemical composition	mechanical properties and hardness values
JIS H 5203	Class 7	Japan	Castings	5.5–6.5 **Zn**, 0.10 **Cu**, rem **Mg**, 0.01 **Ni**, 0.60–1.0 **Zr**	**Artificially aged**: all **diam**, 265 MPa **TS**, 177 MPa **YS**, 5% in 2 in. (50 mm) **El**; **Solution heat treated and artificially aged**: all **diam**, 265 MPa **TS**, 177 MPa **YS**, 5% in 2 in. (50 mm) **El**
ISO 2119	Mg–Zn6Zr	ISO	Castings	5.5–6.5 **Zn**, 0.60–1.0 **Zr**, 0.10 **Cu**, 0.01 **Ni**	
ISO 3115	Mg–Zn6Zr	ISO	Sand castings	5.5–6.5 **Zn**, 0.60–1.0 **Zr**, 0.10 **Cu**, 0.01 **Ni**	**Solution and precipitation treated**: all **diam**, 275 MPa **TS**, 180 MPa **YS**, 4% in 2 in. (50 mm) **El**
MIL–M–46062	ZK51A	US	Castings	3.6–5.5 **Zn**, rem **Mg**, 0.10 **Cu**, 0.01 **Ni**, 0.50–1.0 **Zr**, 0.30 total, others	**T5**: all **diam**, 248 MPa **TS**, 145 MPa **YS**, 6% in 2 in. (50 mm) **El**
ANSI/ASTM B 80	ZK51A	US	Sand castings	3.6–5.5 **Zn**, 0.10 **Cu**, rem **Mg**, 0.01 **Ni**, 0.50–1.0 **Zr**, 0.30 total, others	**Cooled and artificially aged**: all **diam**, 34 ksi (234 MPa) **TS**, 20 ksi (138 MPa) **YS**, 5% in 2 in. (50 mm) **El**, 65 **HB**
JIS H 5203	Class 6	Japan	Castings	3.6–5.5 **Zn**, 0.10 **Cu**, rem **Mg**, 0.01 **Ni**, 0.50–1.0 **Zr**	**Artificially aged**: all **diam**, 235 MPa **TS**, 137 MPa **YS**, 5% in 2 in. (50 mm) **El**
ISO 2119	Mg–Zn5Zr	ISO	Castings	3.5–5.5 **Zn**, 0.40–1.0 **Zr**, 0.10 **Cu**, 0.01 **Ni**	
ISO 3115	Mg–Zn5Zr	ISO	Sand castings	3.5–5.5 **Zn**, 0.40–1.0 **Zr**, 0.10 **Cu**, 0.01 **Ni**	**TE**: 235 MPa **TS**, 140 MPa **YS**, 4% in 2 in. (50 mm) **El**
BS 2L. 126		UK	Castings	0.15 **Mn**, 0.8–3.0 **Zn**, 2.5–4.0 **RE**, 0.4–1.0 **Zr**, 0.03 **Cu**, 0.01 **Si**, 0.01 **Fe**, 0.005 **Ni**, rem **Mg**	**Heat treated and quenched**: all **diam**, 140 MPa **TS**, 95 MPa **YS**, 3% in 2 in. (50 mm) **El**
BS 2L. 126		UK	Ingot	0.15 **Mn**, 0.8–3.0 **Zn**, 2.5–4.0 **RE**, 0.1–1.0 **Zr**, 0.03 **Cu**, 0.01 **Si**, 0.01 **Fe**, 0.005 **Ni**, rem **Mg**	**Heat treated and quenched**: all **diam**, 140 MPa **TS**, 95 MPa **YS**, 3% in 2 in. (50 mm) **El**
JIS H 2502	Class 1, Grade A	Japan	Ingot	0.01 **Zn**, 0.05 **Fe**, 47–53 **Mg**, 0.01 **Pb** min **Mg**+ **Ni**, cobalt included in nickel	
JIS H 2502	Class 1, Grade B	Japan	Ingot	0.15 **Fe**, 46–54 **Mg**, 99.0 min **Mg**+ **Ni**, cobalt included in nickel	
JIS H 2502	Class 2, Grade A	Japan	Ingot	0.01 **Zn**, 0.05 **Fe**, 18–22 **Mg**, 0.01 **Pb**, 99.0 min **Mg**+ **Ni**, cobalt included in nickel	
JIS H 2502	Class 2, Grade B	Japan	Ingot	0.15 **Fe**, 17–23 **Mg**, 99.0 min **Mg**+ **Ni**, cobalt included in nickel	

WROUGHT AND CAST TIN AND TIN ALLOYS

WROUGHT AND CAST TIN AND TIN ALLOYS

specification number	designation	country	product forms	chemical composition	mechanical properties and hardness values
ANSI/ASTM B 339	Grade AAA	US	Ingot	99.98 min **Sn**, 0.008 **Sb**, 0.002 **Cu**, 0.010 **Pb**, 0.0005 **As**, 0.005 **Fe**, 0.001 **Zn**, 0.001 **Cd**, 0.001 **Bi**, 0.005 **Ni**+ **Co**, 0.002 **S**	
ANSI/ASTM B 339	Grade AA	US	Ingot	99.95 min **Sn**, 0.02 **Sb**, 0.02 **Cu**, 0.02 **Pb**, 0.01 **As**, 0.01 **Fe**, 0.001 **Zn**, 0.001 **Cd**, 0.01 **Bi**, 0.01 **Ni**+ **Co**, 0.01 **S**	
ANSI/ASTM B 339	Grade A	US	Ingot	99.80 min **Sn**, 0.04 **Sb**, 0.04 **Cu**, 0.05 **Pb**, 0.05 **As**, 0.015 **Fe**, 0.005 **Zn**, 0.001 **Cd**, 0.015 **Bi**, 0.01 **Ni**+ **Co**, 0.01 **S**	
ANSI/ASTM B 339	Grade B	US	Ingot	99.80 min **Sn**, 0.05 **As**	
ANSI/ASTM B 339	Grade C	US	Ingot	99.65 min **Sn**	
ANSI/ASTM B 339	Grade D	US	Ingot	99.50 min **Sn**	
ANSI/ASTM B 339	Grade E	US	Ingot	99.00 min **Sn**	
ANSI/ASTM B 32	Grade 95TA	US	Wire, strip, bar, ingot	95.0 **Sn**, 4.5–5.5 **Sb**, 0.08 **Cu**, 0.20 **Pb**, 0.05 **As**, 0.04 **Fe**, 0.005 **Al**, 0.005 **Zn**, 0.15 **Bi**	
ANSI/ASTM B 32	Grade 96TS	US	Wire, strip, bar, ingot	96.0 **Sn**, 0.20–0.50 **Sb**, 0.08 **Cu**, 0.20 **Pb**, 3.6–4.4 **Ag**, 0.05 **As**, 0.02 **Fe**, 0.005 **Al**, 0.005 **Zn**, 0.15 **Bi**	
JIS Z 3282	H 95 A	Japan	Bar, ribbon and wire	94–96 **Sn**, 0.30 **Sb**, 0.05 **Cu**, rem **Pb**, 0.03 **As**, 0.03 **Fe**, 0.005 **Zn**, 0.005 **Al**, 0.05 **Bi**	
JIS Z 3282	H 95 B	Japan	Bar, ribbon and wire	93–97 **Sn**, 1.0 **Sb**, 0.08 **Cu**, rem **Pb**, 0.35 **Bi**+ **Zn**+ **Fe**+ **Al**+ **As**	
ANSI/ASTM B 32	Grade 70B	US	Wire, strip, bar, ingot	70.0 **Sn**, 0.20–0.50 **Sb**, 0.08 **Cu**, 30.0 **Pb**, 0.03 **As**, 0.02 **Fe**, 0.005 **Zn**, 0.005 **Al**, 0.25 **Bi**	
ANSI/ASTM B 32	Grade 70A	US	Wire, strip, bar, ingot	70.0 **Sn**, 0.12 **Sb**, 0.08 **Cu**, 30.0 **Pb**, 0.03 **As**, 0.02 **Fe**, 0.005 **Zn**, 0.005 **Al**, 0.25 **Bi**	
JIS Z 3282	H 65 S	Japan	Bar, ribbon and wire	64–66 **Sn**, 0.10 **Sb**, 0.03 **Cu**, rem **Pb**, 0.03 **As**, 0.02 **Fe**, 0.005 **Al**, 0.005 **Zn**, 0.03 **Bi**	
ANSI/ASTM B 32	Grade 63B	US	Wire, strip, bar, ingot	63.0 **Sn**, 0.20–0.50 **Sb**, 0.08 **Cu**, 37.0 **Pb**, 0.03 **As**, 0.02 **Fe**, 0.005 **Zn**, 0.005 **Al**, 0.25 **Bi**	
JIS Z 3282	H 63 A	Japan	Bar, ribbon and wire	62–64 **Sn**, 0.30 **Sb**, 0.05 **Cu**, rem **Pb**, 0.03 **As**, 0.03 **Fe**, 0.005 **Zn**, 0.005 **Al**, 0.05 **Bi**	
JIS Z 3282	H 63 S	Japan	Bar, ribbon and wire	62–64 **Sn**, 0.10 **Sb**, 0.03 **Cu**, rem **Pb**, 0.03 **As**, 0.02 **Fe**, 0.005 **Zn**, 0.005 **Al**, 0.03 **Bi**	
JIS Z 3282	H 63 B	Japan	Bar, ribbon and wire	61–65 **Sn**, 1.0 **Sb**, rem **Pb**, 0.08 **Al**, 0.35 **Bi**+ **Zn**+ **Fe**+ **Al**+ **As**	

WROUGHT AND CAST TIN AND TIN ALLOYS

specification number	designation	country	product forms	chemical composition	mechanical properties and hardness values
ANSI/ASTM B 32	Grade 63A	US	Wire, strip, bar, ingot	63 **Sn**, 0.12 **Sb**, 0.08 **Cu**, 37.0 **Pb**, 0.03 **As**, 0.02 **Fe**, 0.005 **Zn**, 0.005 **Al**, 0.25 **Bi**	
ANSI/ASTM B 32	Grade 60A	US	Wire, strip, bar, ingot	60.0 **Sn**, 0.12 **Sb**, 0.08 **Cu**, 40.0 **Pb**, 0.03 **As**, 0.02 **Fe**, 0.005 **Zn**, 0.005 **Al**, 0.25 **Bi**	
ANSI/ASTM B 32	Grade 60B	US	Wire, strip, bar, ingot	60.0 **Sn**, 0.20–0.50 **Sb**, 0.08 **Cu**, 40.0 **Pb**, 0.03 **As**, 0.02 **Fe**, 0.005 **Zn**, 0.005 **Al**, 0.25 **Bi**	
JIS Z 3282	H 60 A	Japan	Bar, ribbon and wire	59–61 **Sn**, 0.30 **Sb**, 0.05 **Cu**, rem **Pb**, 0.03 **As**, 0.03 **Fe**, 0.005 **Zn**, 0.005 **Al**, 0.05 **Bi**	
JIS Z 3282	H 60 S	Japan	Bar, ribbon and wire	59–61 **Sn**, 0.10 **Sb**, 0.03 **Cu**, rem **Pb**, 0.03 **As**, 0.02 **Fe**, 0.005 **Zn**, 0.005 **Al**	
JIS Z 3282	H 60 B	Japan	Bar, ribbon and wire	58–62 **Sn**, 1.0 **Sb**, rem **Pb**, 0.08 **Al**, 0.35 **Bi**+ **Zn**+ **Fe**+ **Al**+ **As**	
JIS Z 3282	H 55 A	Japan	Bar, ribbon and wire	54–56 **Sn**, 0.3 **Sb**, 0.05 **Cu**, rem **Pb**, 0.03 **As**, 0.03 **Fe**, 0.005 **Zn**, 0.005 **Al**, 0.05 **Bi**	
JIS Z 3282	H 55 S	Japan	Bar, ribbon and wire	54–56 **Sn**, 0.10 **Sb**, 0.03 **Cu**, rem **Pb**, 0.03 **As**, 0.02 **Fe**, 0.005 **Zn**, 0.005 **Al**, 0.03 **Bi**	
JIS Z 3282	H 55 B	Japan	Bar, ribbon and wire	53–57 **Sn**, 1.0 **Sb**, rem **Pb**, 0.08 **Al**, 0.35 **Bi**+ **Zn**+ **Fe**+ **Al**+ **As**	
DIN 8512	L-SnPbZn/2.3852	Germany	Soft solder	40–60 **Sn**, 30–55 **Pb**, 2–20 **Zn**+ **Cd**	
JIS Z 3282	H 50 B	Japan	Bar, ribbon and wire	48–52 **Sn**, 1.0 **Sb**, 46.56 min **Pb**, 0.08 **Al**, 0.35 **Bi**+ **Zn**+ **Fe**+ **Al**+ **As**	
JIS Z 3282	H 50 S	Japan	Bar, ribbon and wire	49–51 **Sn**, 0.10 **Sb**, 0.03 **Cu**, 48.79 min **Pb**, 0.03 **As**, 0.02 **Fe**, 0.005 **Zn**, 0.005 **Al**, 0.03 **Bi**	
JIS Z 3282	H 50 A	Japan	Bar, ribbon and wire	49–51 **Sn**, 0.30 **Sb**, 0.05 **Cu**, 48.52 min **Pb**, 0.03 **As**, 0.03 **Fe**, 0.005 **Zn**, 0.005 **Al**, 0.05 **Bi**	
ANSI/ASTM B 23	Alloy 1	US	Bar, ingot	90.0–92.0 **Sn**, 4.0–5.0 **Sb**, 4.0–5.0 **Cu**, 0.35 **Pb**, 0.10 **As**, 0.08 **Fe**, 0.005 **Zn**, 0.05 **Cd**, 0.005 **Al**, 0.08 **Bi**, 99.80 min **Sn**+ **Sb**+ **Cu**+ **Pb**+ **As**+ **Fe**+ **Cd**+ **Al**+ **Zn**+ **Bi**	
ANSI/ASTM B 102	CY44A	US	Die castings	90–92 **Sn**, 4–5 **Sb**, 4–5 **Cu**, 0.35 **Pb**, 0.08 **As**, 0.08 **Fe**, 0.01 **Al**, 0.01 **Zn**	
SABS 695	WM 90	South Africa	Ingot, bar	88.5–91.5 **Sn**, 6.0–8.0 **Sb**, 2.5–3.5 **Cu**, 0.3 **Pb**, 0.05 **Fe**, 0.05 **Cd**, 0.01 **Al**, 0.01 **Zn**, 0.05 **Ni**, 0.3 total, others	

WROUGHT AND CAST TIN AND TIN ALLOYS

specification number	designation	country	product forms	chemical composition	mechanical properties and hardness values
ANSI/ASTM B 23	Alloy 2	US	Bar, ingot	88.0–90.0 **Sn**, 7.0–8.0 **Sb**, 3.0–4.0 **Cu**, 0.35 **Pb**, 0.10 **As**, 0.08 **Fe**, 0.005 **Zn**, 0.05 **Cd**, 0.005 **Al**, 0.08 **Bi**, 99.80 min **Sn**+ **Sb**+ **Cu**+ **Pb**+ **As**+ **Fe**+ **Zn**+ **Cd**+ **Al**+ **Bi**	
ANSI/ASTM B 23	Alloy 11	US	Bar, ingot	86.0–89.0 **Sn**, 6.0–7.5 **Sb**, 5.0–6.5 **Cu**, 0.50 **Pb**, 0.10 **As**, 0.08 **Fe**, 0.005 **Zn**, 0.05 **Cd**, 0.005 **Al**, 0.08 **Bi**, 99.80 min **Sn**+ **Sb**+ **Cu**+ **Pb**+ **As**+ **Fe**+ **Zn**+ **Cd**+ **Al**+ **Bi**	
JIS H 5401	Class 1	Japan	Ingot	87.39 min **Sn**, 5.0–7.0 **Sb**, 3.0–5.0 **Cu**, 0.50 **Pb**, 0.10 **As**, 0.08 **Fe**, 0.01 **Zn**, 0.01 **Al**, 0.08 **Bi**	
ANSI/ASTM B 23	Alloy 3	US	Bar, ingot	83.0–85.0 **Sn**, 7.5–8.5 **Sb**, 7.5–8.5 **Cu**, 0.35 **Pb**, 0.10 **As**, 0.08 **Fe**, 0.005 **Zn**, 0.05 **Cd**, 0.005 **Al**, 0.08 **Bi**, 99.80 min **Sn**+ **Sb**+ **Cu**+ **Pb**+ **As**+ **Fe**+ **Zn**+ **Cd**+ **Al**+ **Bi**	
SABS 696	WM 84	South Africa	Ingot, bar	82.0–86.0 **Sn**, 9.0–11.0 **Sb**, 5.0–7.0 **Cu**, 0.3 **Pb**, 0.05 **Fe**, 0.01 **Zn**, 0.3 total, others	
JIS H 5401	Class 2	Japan	Ingot	83.39 min **Sn**, 8.0–10.0 **Sb**, 5.0–6.0 **Cu**, 0.50 **Pb**, 0.10 **As**, 0.08 **Fe**, 0.01 **Zn**, 0.01 **Al**, 0.08 **Bi**	
JIS H 5401	Class 2B	Japan	Ingot	81.39 min **Sn**, 7.5–9.5 **Sb**, 7.5–8.5 **Cu**, 0.50 **Pb**, 0.10 **As**, 0.08 **Fe**, 0.01 **Al**, 0.01 **Zn**, 0.08 **Bi**	
ANSI/ASTM B 102	YC135A	US	Die castings	80–85 **Sn**, 12–14 **Sb**, 4–6 **Cu**, 0.35 **Pb**, 0.08 **As**, 0.01 **Zn**, 0.01 **Al**	
JIS H 5401	Class 3	Japan	Ingot	82.0–80.0 **Sn**, 11.0–12.0 **Sb**, 4.0–5.0 **Cu**, 3.0 **Pb**, 0.10 **As**, 0.10 **Fe**, 0.01 **Zn**, 0.01 **Al**, 0.08 **Bi**	
SABS 697	WM 80	South Africa	Ingot, bar	78.0–82.0 **Sn**, 8.0–10.0 **Sb**, 5.0–7.0 **Cu**, 5.0–8.0 **Pb**, 0.05 **Fe**, 0.01 **Zn**, 0.3 total, others	
SABS 698	WM 72	South Africa	Ingot, bar	70.0–74.0 **Sn**, 9.0–11.0 **Sb**, 5.0–7.0 **Cu**, 10.0–14.0 **Pb**, 0.05 **Fe**, 0.01 **Zn**, 0.03 total, others	
JIS H 5401	Class 4	Japan	Ingot	67.0–73.0 **Sn**, 11.0–13.0 **Sb**, 3.0–5.0 **Cu**, 13.0–15.0 **Pb**, 0.10 **As**, 0.10 **Fe**, 0.01 **Zn**, 0.01 **Al**, 0.08 **Bi**	
ANSI/ASTM B 102	PY1815A	US	Die castings	64–66 **Sn**, 14–16 **Sb**, 1.5–2.5 **Cu**, 17–19 **Pb**, 0.15 **As**	

WROUGHT AND CAST TIN AND TIN ALLOYS

specification number	designation	country	product forms	chemical composition	mechanical properties and hardness values
JIS H 5401	Class 6	Japan	Ingot	44.0–46.0 **Sn**, 11.0–13.0 **Sb**, rem **Pb**, 0.20 **As**, 0.10 **Fe**, 0.05 **Zn**, 1.0–3.0 **Al**, 0.05 **Zn**	
DIN 8512	L-SnZn10/2.3820	Germany	Soft solder	85–92 **Sn**, 8–15 **Zn**	
JIS H 5401	Class 5	Japan	Ingot	67.89 min **Sn**, 2.0–3.0 **Cu**, 0.10 **Fe**, 28.0–29.0 **Zn**, 0.05 **Al**	
SABS 699	WM 66	South Africa	Ingot, bar	65.0–67.0 **Sn**, 0.5 **Sb**, 3.0–4.0 **Cu**, 0.5 **Pb**, 0.01 **As**, 0.1 **Fe**, 30.0–32.0 **Zn**	
DIN 8512	L-SnZn40/2.3830	Germany	Soft solder	50–70 **Sn**, 30–50 **Zn**	
ASTM B 560	Type 1	US	Bar, ingot, sheet	90.0–93.0 **Sn**, 6.0–8.0 **Sb**, 0.25–2.0 **Cu**, 0.05 **Pb**, 0.05 **As**, 0.015 **Fe**, 0.005 **Zn**	
SABS 700	WM 40	South Africa	Ingot, bar	39.0–41.0 **Sn**, 14.0–16.0 **Sb**, 2.0–3.0 **Cu**, rem **Pb**, 0.05 **Fe**, 0.01 **Zn**, 0.3 total, others	
ASTM B 560	Type 3	US	Bar, ingot, sheet	95.0–98.0 **Sn**, 1.0–3.0 **Sb**, 1.0–2.0 **Cu**, 0.05 **Pb**, 0.05 **As**, 0.015 **Fe**, 0.005 **Zn**	
ASTM B 560	Type 2	US	Bar, ingot, sheet	90.0–93.0 **Sn**, 5.0–7.5 **Sb**, 1.5–3.0 **Cu**, 0.05 **Pb**, 0.05 **As**, 0.015 **Fe**, 0.005 **Zn**	

WROUGHT AND CAST LEAD AND LEAD ALLOYS

WROUGHT AND CAST LEAD AND LEAD ALLOYS

specification number	designation	country	product forms	chemical composition	mechanical properties and hardness values
ANSI/ASTM B 32	Grade 1.5 S	US	Wire, strip, bar, ingot	97.5 **Pb**, 0.02 **As**, 0.25 **Bi**, 1.3–1.7 **Ag**, 0.75–1.25 **Sn**, 0.40 **Sb**, 0.08 **Cu**, 0.02 **Fe**, 0.005 **Al**, 0.005 **Zn**	
ANSI/ASTM B 32	Grade 2.5 S	US	Wire, strip, bar, ingot	97.5 **Pb**, 0.02 **As**, 0.25 **Bi**, 2.3–2.7 **Ag**, 0.25 **Sn**, 0.40 **Sb**, 0.08 **Cu**, 0.02 **Fe**, 0.005 **Al**, 0.005 **Zn**	
ANSI/ASTM B 32	Grade 2A	US	Wire, strip, bar, ingot	98.0 **Pb**, 0.02 **As**, 0.25 **Bi**, 1.5–2.5 **Sn**, 0.12 **Sb**, 0.08 **Cu**, 0.02 **Fe**, 0.005 **Al**, 0.005 **Zn**	
ANSI/ASTM B 32	Grade 2B	US	Wire, strip, bar, ingot	98.0 **Pb**, 0.02 **As**, 0.25 **Bi**, 1.5–2.5 **Sn**, 0.20–0.50 **Sb**, 0.08 **Cu**, 0.02 **Fe**, 0.005 **Al**, 0.005 **Zn**	
JIS Z 3282	H 2 A	Japan	Bar, ribbon, and wire	97.02 min **Pb**, 1.5–2.5 **Sn**, 0.30 **Sb**, 0.005 **Al**, 0.05 **Cu**, 0.03 **Fe**, 0.005 **Zn**	
ANSI/ASTM B 32	Grade 5A	US	Wire, strip, bar, ingot	95.0 **Pb**, 0.02 **As**, 0.25 **Bi**, 4.5–5.5 **Sn**, 0.12 **Sb**, 0.08 **Cu**, 0.02 **Fe**, 0.005 **Al**, 0.005 **Zn**	
ANSI/ASTM B 32	Grade 5B	US	Wire, strip, bar, ingot	95.0 **Pb**, 0.02 **As**, 0.25 **Bi**, 4.5–5.5 **Sn**, 0.20–0.50 **Sb**, 0.08 **Cu**, 0.02 **Fe**, 0.005 **Al**, 0.005 **Zn**	
JIS Z 3282	H 5 A	Japan	Bar, ribbon, and wire	93.52 min **Pb**, 0.03 **As**, 0.05 **Bi**, 4–6 **Sn**, 0.30 **Sb**, 0.005 **Al**, 0.05 **Cu**, 0.03 **Fe**, 0.005 **Zn**	
JIS Z 3282	H 5 B	Japan	Bar, ribbon, and wire	91.6 min **Pb**, 3–7 **Sn**, 1.0 **Sb**, 0.08 **Al**, 0.35 **Bi** + **Zn** + **Fe** + **Al** + **As**	
ANSI/ASTM B 32	Grade 10B	US	Wire, strip, bar, ingot	90.0 **Pb**, 0.02 **As**, 0.25 **Bi**, 10.0 **Sn**, 0.20–0.50 **Sb**, 0.08 **Cu**, 0.02 **Fe**, 0.005 **Al**, 0.005 **Zn**	
JIS Z 3282	H 10 A	Japan	Bar, ribbon, and wire	88.52 min **Pb**, 0.03 **As**, 0.05 **Bi**, 9–11 **Sn**, 0.30 **Sb**, 0.005 **Al**, 0.05 **Cu**, 0.03 **Fe**, 0.005 **Zn**	
JIS Z 3282	H 10 B	Japan	Bar, ribbon, and wire	86.6 min **Pb**, 8–12 **Sn**, 1.0 **Sb**, 0.08 **Al**, 0.35 **Bi** + **Zn** + **Fe** + **Al** + **As**	
ANSI/ASTM B 32	Grade 15B	US	Wire, strip, bar, ingot	85.0 **Pb**, 0.02 **As**, 0.25 **Bi**, 15.0 **Sn**, 0.20–0.50 **Sb**, 0.08 **Cu**, 0.02 **Fe**, 0.005 **Al**, 0.005 **Zn**	
ANSI/ASTM B 32	Grade 20B	US	Wire, strip, bar, ingot	80.0 **Pb**, 0.02 **As**, 0.25 **Bi**, 20.0 **Sn**, 0.20–0.50 **Sb**, 0.08 **Cu**, 0.02 **Fe**, 0.005 **Al**, 0.005 **Zn**	
ANSI/ASTM B 32	Grade 20C	US	Wire, strip, bar, ingot	79.0 **Pb**, 0.02 **As**, 0.25 **Bi**, 20.0 **Sn**, 0.8–1.2 **Sb**, 0.08 **Cu**, 0.02 **Fe**, 0.005 **Al**, 0.005 **Zn**	
JIS Z 3282	H 20 A	Japan	Bar, ribbon, and wire	78.52 min **Pb**, 0.03 **As**, 0.05 **Bi**, 19–21 **Sn**, 0.30 **Sb**, 0.005 **Al**, 0.05 **Cu**, 0.03 **Fe**, 0.005 **Zn**	

WROUGHT AND CAST LEAD AND LEAD ALLOYS

specification number	designation	country	product forms	chemical composition	mechanical properties and hardness values
JIS H 3282	H 20 B	Japan	Bar, ribbon, and wire	76.6 min **Pb**, 18–22 **Sn**, 1.0 **Sb**, 0.08 **Al**, 0.35 **Bi** + **Zn** + **Fe** + **Al** + **As**	
ANSI/ASTM B 32	Grade 25A	US	Wire, strip, bar, ingot	75.0 **Pb**, 0.02 **As**, 0.25 **Bi**, 25.0 **Sn**, 0.25 **Sb**, 0.08 **Cu**, 0.02 **Fe**, 0.005 **Al**, 0.005 **Zn**	
ANSI/ASTM B 32	Grade 25B	US	Wire, strip, bar, ingot	75.0 **Pb**, 0.02 **As**, 0.25 **Bi**, 25.0 **Sn**, 0.20–0.50 **Sb**, 0.08 **Cu**, 0.02 **Fe**, 0.005 **Al**, 0.005 **Zn**	
ANSI/ASTM B 32	Grade 25C	US	Wire, strip, bar, ingot	73.7 **Pb**, 0.02 **As**, 0.25 **Bi**, 25.0 **Sn**, 1.1–1.5 **Sb**, 0.08 **Cu**, 0.02 **Fe**, 0.005 **Al**, 0.005 **Zn**	
ANSI/ASTM B 32	Grade 30B	US	Wire, strip, bar, ingot	70.0 **Pb**, 0.02 **As**, 0.25 **Bi**, 30.0 **Sn**, 0.20–0.50 **Sb**, 0.08 **Cu**, 0.02 **Fe**, 0.005 **Al**, 0.005 **Zn**	
ANSI/ASTM B 32	Grade 30A	US	Wire, strip, bar, ingot	70.0 **Pb**, 0.02 **As**, 0.25 **Bi**, 30.0 **Sn**, 0.25 **Sb**, 0.08 **Cu**, 0.02 **Fe**, 0.005 **Al**, 0.005 **Zn**	
JIS Z 3282	H 30 A	Japan	Bar, ribbon, and wire	68.52 min **Pb**, 0.03 **As**, 0.05 **Bi**, 29–31 **Sn**, 0.30 **Sb**, 0.005 **Al**, 0.05 **Cu**, 0.03 **Fe**, 0.005 **Zn**	
ANSI/ASTM B 32	Grade 30C	US	Wire, strip, bar, ingot	68.4 **Pb**, 0.02 **As**, 0.25 **Bi**, 30.0 **Sn**, 1.4–1.8 **Sb**, 0.08 **Cu**, 0.02 **Fe**, 0.005 **Al**, 0.005 **Zn**	
JIS Z 3282	H 30 B	Japan	Bar, ribbon, and wire	66.6 min **Pb**, 28–32 **Sn**, 1.0 **Sb**, 0.08 **Al**, 0.35 **Bi** + **Zn** + **Fe** + **Al** + **As**	
ANSI/ASTM B 32	Grade 35B	US	Wire, strip, bar, ingot	65.0 **Pb**, 0.02 **As**, 0.25 **Bi**, 35.0 **Sn**, 0.20–0.50 **Sb**, 0.08 **Cu**, 0.02 **Fe**, 0.005 **Al**, 0.005 **Zn**	
ANSI/ASTM B 32	Grade 35A	US	Wire, strip, bar, ingot	65.0 **Pb**, 0.02 **As**, 0.25 **Bi**, 35.0 **Sn**, 0.25 **Sb**, 0.08 **Cu**, 0.02 **Fe**, 0.005 **Al**, 0.005 **Zn**	
JIS Z 3282	H 35 A	Japan	Bar, ribbon, and wire	63.52 min **Pb**, 0.03 **As**, 0.05 **Bi**, 34–36 **Sn**, 0.30 **Sb**, 0.005 **Al**, 0.05 **Cu**, 0.03 **Fe**, 0.005 **Zn**	
ANSI/ASTM B 32	Grade 35C	US	Wire, strip, bar, ingot	63.2 **Pb**, 0.02 **As**, 0.25 **Bi**, 35.0 **Sn**, 1.6–2.0 **Sb**, 0.08 **Cu**, 0.02 **Fe**, 0.005 **Al**, 0.005 **Zn**	
JIS Z 3282	H 35 B	Japan	Bar, ribbon, and wire	61.6 min **Pb**, 33–37 **Sn**, 1.0 **Sb**, 0.08 **Al**, 0.35 **Bi** + **Zn** + **Fe** + **Al** + **As**	
JIS Z 3282	H 38 A	Japan	Bar, ribbon, and wire	60.52 min **Pb**, 0.03 **As**, 0.05 **Bi**, 37–39 **Sn**, 0.30 **Sb**, 0.005 **Al**, 0.05 **Cu**, 0.03 **Fe**, 0.005 **Zn**	
ANSI/ASTM B 32	Grade 40B	US	Wire, strip, bar, ingot	60.0 **Pb**, 0.02 **As**, 0.25 **Bi**, 40.0 **Sn**, 0.20–0.50 **Sb**, 0.08 **Cu**, 0.02 **Fe**, 0.005 **Al**, 0.005 **Zn**	
ANSI/ASTM B 32	Grade 40A	US	Wire, strip, bar, ingot	60.0 **Pb**, 0.02 **As**, 0.25 **Bi**, 40.0 **Sn**, 0.12 **Sb**, 0.08 **Cu**, 0.02 **Fe**, 0.005 **Al**, 0.005 **Zn**	
JIS Z 3282	H 40 S	Japan	Bar, ribbon, and wire	58.77 min **Pb**, 0.03 **As**, 0.03 **Bi**, 39–41 **Sn**, 0.10 **Sb**, 0.005 **Al**, 0.03 **Cu**, 0.02 **Fe**, 0.005 **Zn**	

WROUGHT AND CAST LEAD AND LEAD ALLOYS

specification number	designation	country	product forms	chemical composition	mechanical properties and hardness values
JIS Z 3282	H 40 A	Japan	Bar, ribbon, and wire	58.52 min **Pb**, 0.03 **As**, 0.05 **Bi**, 39–41 **Sn**, 0.30 **Sb**, 0.005 **Al**, 0.05 **Cu**, 0.03 **Fe**, 0.005 **Zn**	
ANSI/ASTM B 32	Grade 40C	US	Wire, strip, bar, ingot	58.0 **Pb**, 0.02 **As**, 0.25 **Bi**, 40.0 **Sn**, 1.8–2.4 **Sb**, 0.08 **Cu**, 0.02 **Fe**, 0.005 **Al**, 0.005 **Zn**	
JIS Z 3282	H 40 B	Japan	Bar, ribbon, and wire	56.6 min **Pb**, 38–42 **Sn**, 1.0 **Sb**, 0.08 **Al**, 0.35 **Bi**+ **Zn**+ **Fe**+ **Al**+ **As**	
ANSI/ASTM B 32	Grade 45B	US	Wire, strip, bar, ingot	55.0 **Pb**, 0.03 **As**, 0.25 **Bi**, 45.0 **Sn**, 0.20–0.50 **Sb**, 0.08 **Cu**, 0.02 **Fe**, 0.005 **Al**, 0.005 **Zn**	
ANSI/ASTM B 32	Grade 45A	US	Wire, strip, bar, ingot	55.0 **Pb**, 0.03 **As**, 0.25 **Bi**, 45.0 **Sn**, 0.20–0.50 **Sb**, 0.08 **Cu**, 0.02 **Fe**, 0.005 **Al**, 0.005 **Zn**	
JIS Z 3282	H 45 S	Japan	Bar, ribbon, and wire	53.77 min **Pb**, 0.03 **As**, 0.03 **Bi**, 44–46 **Sn**, 0.10 **Sb**, 0.005 **Al**, 0.03 **Cu**, 0.02 **Fe**, 0.005 **Zn**	
JIS Z 3282	H 45 A	Japan	Bar, ribbon, and wire	53.52 min **Pb**, 0.03 **As**, 0.05 **Bi**, 44–46 **Sn**, 0.30 **Sb**, 0.005 **Al**, 0.05 **Cu**, 0.03 **Fe**, 0.005 **Zn**	
JIS Z 3282	H 45 B	Japan	Bar, ribbon, and wire	50.84 min **Pb**, 43–47 **Sn**, 1.0 **Sb**, 0.08 **Al**, 0.35 **Bi**+ **Zn**+ **Fe**+ **Al**+ **As**	
ANSI/ASTM B 32	Grade 50B	US	Wire, strip, bar, ingot	50.0 **Pb**, 0.03 **As**, 0.25 **Bi**, 50.0 **Sn**, 0.20–0.50 **Sb**, 0.08 **Cu**, 0.02 **Fe**, 0.005 **Al**, 0.005 **Zn**	
ANSI/ASTM B 32	Grade 50A	US	Wire, strip, bar, ingot	50.0 **Pb**, 0.03 **As**, 0.25 **Bi**, 50.0 **Sn**, 0.12 **Sb**, 0.08 **Cu**, 0.02 **Fe**, 0.005 **Al**, 0.005 **Zn**	
COPANT 448	Cab Pb Sb 0.75	COPANT	Wrought product: for electrical cable sheathing	99.07 **Pb**, 0.001 **As**, 0.05 **Bi**, 0.001 **Ag**, 0.005 **Sn**, 0.65–0.85 **Sb**, 0.001 **Cu**, 0.001 **Fe**, 0.001 **Mg**, 0.001 **Zn**	
SABS 250	LEAD ALLOY B	South Africa	Ingot	98.9 min **Pb**, 0.05 **Bi**, 0.005 **Ag**, 0.01 **Sn**, 0.85–0.95 **Sb**, 0.06 **Cu**, 0.005 **Te**, 0.002 **Zn**, 0.1 total, others	
COPANT 447	Cab Pb Sb 1	COPANT	Wrought product: for electrical cable sheathing	98.82 min **Pb**, 0.001 **As**, 0.05 **Bi**, 0.001 **Ag**, 0.005 **Sn**, 0.9–1.1 **Sb**, 0.001 **Cu**, 0.001 **Fe**, 0.001 **Mg**, 0.001 **Zn**	
COPANT 447	Pb Sb (As)	COPANT	Wrought products	98.67 min **Pb**, 0.02–0.05 **As**, 0.75–1.25 **Sb**	
JIS H 4302	Class 4	Japan	Plate	94.69 min **Pb**, 0.10 **Bi**, 0.50 **Sn**, 3.5–4.5 **Sb**, 0.20 **Cu**	
JIS H 4313	Class 4	Japan	Pipe	94.69 min **Pb**, 0.10 **Bi**, 0.50 **Sn**, 3.5–4.5 **Sb**, 0.20 **Cu**	
COPANT 447	Pb Sb As	COPANT	Wrought products	94.49 min **Pb**, 1.2–1.7 **As**, 2.0–3.8 **Sb**	
COPANT 447	Pb Sb 5	COPANT	Wrought products	92.99 min **Pb**, 5–7 **Sb**	

WROUGHT AND CAST LEAD AND LEAD ALLOYS

specification number	designation	country	product forms	chemical composition	mechanical properties and hardness values
JIS H 4302	Class 6	Japan	Plate	92.69 min **Pb**, 0.10 **Bi**, 0.50 **Sn**, 5.5–6.5 **Sb**, 0.20 **Cu**	
JIS H 4313	Class 6	Japan	Pipe	92.69 min **Pb**, 0.10 **Bi**, 0.50 **Sn**, 5.5–6.5 **Sb**, 0.20 **Cu**	
AMS 7720		US	Castings	91.9 min **Pb**, 0.10 **As**, 0.10 **Bi**, 0.25–0.75 **Sn**, 6.0–7.0 **Sb**, 0.077 others, total	**As cast**: all **diam**, 6 ksi (41 MPa) **TS**, 18% in 2 in. (50 mm) **El**, 11–14 **HB**
COPANT 447	Pb Sb 8	COPANT	Wrought products	91.49 min **Pb**, 7.5–8.5 **Sb**	
COPANT 447	Pb Sb 9	COPANT	Wrought products	90.89 min **Pb**, 0.01 **As**, 0.02 **Bi**, 0.005 **Ag**, 0.01 **Sn**, 8.7–9.0 **Sb**, 0.01 **Cu**, 0.01 **Fe**, 0.001 **Zn**	
COPANT 447	Pb Sb 9X	COPANT	Wrought products	90.87 min **Pb**, 0.015 **As**, 0.04 **Bi**, 0.008 **Ag**, 0.02 **Sn**, 8.7–9.0 **Sb**, 0.02 **Cu**, 0.01 **Fe**, 0.001 **Zn**	
JIS H 4302	Class 8	Japan	Plate	90.69 min **Pb**, 0.10 **Bi**, 0.50 **Sn**, 7.5–8.5 **Sb**, 0.20 **Cu**	
JIS H 4313	Class 8	Japan	Pipe	90.69 min **Pb**, 0.10 **Bi**, 0.50 **Sn**, 7.5–8.5 **Sb**, 0.20 **Cu**	
JIS H 5403	Class 8	Japan	Castings: for bearing	90.69 min **Pb**, 0.10 **Bi**, 0.50 **Sn**, 7.5–8.5 **Sb**, 0.20 **Cu**	**As cast**: all **diam**, 49 MPa **TS**, 20% in 50 mm **El**, 14.0 min **HB**
ANSI/ASTM B 102	Y10A	US	Die castings	89–91 **Pb**, 9.25–10.75 **Sb**, 0.50 **Cu**, 0.01 **Zn**	
JIS H 5403	Class 10	Japan	Castings: for bearing	88.69 min **Pb**, 0.10 **Bi**, 0.50 **Sn**, 9.5–10.5 **Sb**, 0.20 **Cu**	**As cast**: 50 MPa **TS**, 19% in 50 mm **El**, 14.5 min **HB**
COPANT 447	Pb Sb 12	COPANT	Wrought products	86.99 min **Pb**, 12–13 **Sb**	
SABS 12	LINO	South Africa	Ingot	83.34 min **Pb**, 3.5–5.5 **Sn**, 10.5–11.5 **Sb**, 0.05 **Cu**, 0.005 **Zn**, 0.005 **Al**, 0.02 **Fe**+**Ni**, 0.07 total, others	
SIS 14 74 10	SIS Pb 74 10–00	Sweden	Ingot	82.96 min **Pb**, 0.15 **As**, 3.8–4.3 **Sn**, 11.5–12.5 **Sb**, 0.005 **Ni**, 0.05 **Cu**, 0.001 **Zn**, 0.001 **Al**, 0.02 **Fe**	
ANSI/ASTM B 23	Alloy 13	US	Bar, ingot	81.98 min **Pb**, 0.25 **As**, 0.10 **Bi**, 5.5–6.5 **Sn**, 9.5–10.5 **Sb**, 0.50 **Cu**, 0.10 **Fe**, 0.005 **Zn**, 0.005 **Al**, 0.05 **Cd**	
JIS H 5401	Class 9	Japan	Ingot	81.53 min **Pb**, 5.0–7.0 **Sn**, 9.0–11.0 **Sb**, 0.01 **Al**, 0.30 **Cu**, 0.10 **Fe**, 0.05 **Zn**	
JIS H 5401	Class 10	Japan	Ingot	81.38 min **Pb**, 0.75–1.25 **As**, 0.8–1.2 **Sn**, 14.0–15.5 **Sb**, 0.01 **Al**, 0.1–0.5 **Cu**, 0.10 **Fe**, 0.05 **Zn**	
ANSI/ASTM B 102	YT155A	US	Die castings	79–81 **Pb**, 0.15 **As**, 4–6 **Sn**, 14–16 **Sb**, 0.01 **Al**, 0.50 **Cu**, 0.01 **Zn**	

WROUGHT AND CAST LEAD AND LEAD ALLOYS

specification number	designation	country	product forms	chemical composition	mechanical properties and hardness values
SIS 14 74 16	SIS Pb 74 16–00	Sweden	Ingot	79.91 min **Pb**, 0.2 **As**, 4.8–5.3 **Sn**, 13.5–14.5 **Sb**, 0.005 **Ni**, 0.05 **Cu**, 0.001 **Zn**, 0.001 **Al**, 0.02 **Fe**	
ANSI/ASTM B 23	Alloy 15	US	Bar, ingot	79.1 min **Pb**, 0.8–1.4 **As**, 0.10 **Bi**, 0.8–1.2 **Sn**, 14.5–17.5 **Sb**, 0.6 **Cu**, 0.10 **Fe**, 0.005 **Zn**, 0.005 **Al**, 0.05 **Cd**	
SIS 14 74 17	SIS Pb 74 17–00	Sweden	Ingot	77.7 min **Pb**, 0.2 **As**, 6.5–7.5 **Sn**, 13.5–14.5 **Sb**, 0.005 **Ni**, 0.05 **Cu**, 0.001 **Zn**, 0.001 **Al**, 0.02 **Fe**	
SABS 703	WM 6	South Africa	Ingot, bar	77.33 min **Pb**, 5.0–7.0 **Sn**, 13.0–15.0 **Sb**, 0.03 **Cu**, 0.05 **Fe**, 0.01 **Zn**, 0.03 total, others	
ANSI/ASTM B 23	Alloy 8	US	Bar, ingot	77.13 min **Pb**, 0.30–0.60 **As**, 0.10 **Bi**, 4.5–5.5 **Sn**, 14.0–16.0 **Sb**, 0.50 **Cu**, 0.10 **Fe**, 0.005 **Zn**, 0.005 **Al**, 0.05 **Cd**	
SABS 12	MONO A	South Africa	Ingot	74.8 min **Pb**, 6.5–7.5 **Sn**, 16.5–17.5 **Sb**, 0.05 **Cu**, 0.005 **Zn**, 0.005 **Al**, 0.02 **Fe**+ **Ni**, 0.07 total, others	
SABS 12	STEREO	South Africa	Ingot	73.84 min **Pb**, 5.0–10.0 **Sn**, 14.0–16.0 **Sb**, 0.05 **Cu**, 0.005 **Zn**, 0.005 **Al**, 0.02 **Fe**+ **Ni**, 0.07 total, others	
SIS 14 74 21	SIS Pb 74 21–00	Sweden	Ingot	73.7 min **Pb**, 0.2 **As**, 9.5–10.5 **Sn**, 14.5–15.5 **Sb**, 0.005 **Ni**, 0.05 **Cu**, 0.001 **Zn**, 0.001 **Al**, 0.02 **Fe**	
SABS 12	MONO D	South Africa	Ingot	72.8 min **Pb**, 9.5–10.5 **Sn**, 15.5–16.5 **Sb**, 0.05 **Cu**, 0.005 **Zn**, 0.005 **Al**, 0.02 **Fe**+ **Ni**, 0.07 total, others	
SIS 14 74 28	SIS Pb 74 28–00	Sweden	Ingot	72.7 min **Pb**, 0.2 **As**, 9.5–10.5 **Sn**, 15.5–16.5 **Sb**, 0.005 **Ni**, 0.05 **Cu**, 0.001 **Zn**, 0.001 **Al**, 0.02 **Fe**	
JIS H 5401	Class 8	Japan	Ingot	72.63 min **Pb**, 0.20 **As**, 6.0–8.0 **Sn**, 16.0–18.0 **Sb**, 0.01 **Al**, 1.0 **Cu**, 0.10 **Fe**, 0.05 **Zn**	
ANSI/ASTM B 23	Alloy 7	US	Bar, ingot	71.93 min **Pb**, 0.30–0.60 **As**, 0.10 **Bi**, 9.3–10.7 **Sn**, 14.0–16.0 **Sb**, 0.50 **Cu**, 0.10 **Fe**, 0.005 **Zn**, 0.005 **Al**, 0.05 **Cd**	
SABS 12	MONO B	South Africa	Ingot	70.8 min **Pb**, 8.5–9.5 **Sn**, 18.5–19.5 **Sb**, 0.05 **Cu**, 0.005 **Zn**, 0.005 **Al**, 0.02 **Fe**+ **Ni**, 0.07 total, others	
JIS H 5401	Class 7	Japan	Ingot	70.63 min **Pb**, 0.20 **As**, 11.0–13.0 **Sn**, 13.0–15.0 **Sb**, 0.01 **Al**, 1.0 **Cu**, 0.10 **Fe**, 0.05 **Zn**	

WROUGHT AND CAST LEAD AND LEAD ALLOYS

specification number	designation	country	product forms	chemical composition	mechanical properties and hardness values
SABS 702	WM 15	South Africa	Ingot, bar	67.88 min **Pb**, 14.0–16.0 **Sn**, 13.0–15.0 **Sb**, 0.25–0.75 **Cu**, 0.05 **Fe**, 0.01 **Zn**, 0.3 total, others	
SABS 12	MONO C	South Africa	Ingot	62.8 min **Pb**, 11.5–12.5 **Sn**, 23.5–24.5 **Sb**, 0.05 **Cu**, 0.005 **Zn**, 0.005 **Al**, 0.02 **Fe** + **Ni**, 0.07 total, others	
SABS 701	WM 30	South Africa	Ingot, bar	50.13 min **Pb**, 29.0–31.0 **Sn**, 15.0–17.0 **Sb**, 0.5–1.5 **Cu**, 0.05 **Fe**, 0.01 **Zn**, 0.3 total, others	

WROUGHT AND CAST ZINC AND ZINC ALLOYS

WROUGHT AND CAST ZINC AND ZINC ALLOYS

specification number	designation	country	product forms	chemical composition	mechanical properties and hardness values
NOM-W-34	Types A and B	Mexico	Pipe: seamless	0.002–0.003 **Cu**, 0.0005–0.001 **Al**, 0.0003–0.0025 **Mg**, 0.050–0.3 **Pb**, 0.007–0.020 **Cd**, 0.001–0.020 **Fe**, 0.001–0.002 **Sb**, rem **Zn**	**Hot or cold finished:** 3.2–101.6 mm **diam**, 108 MPa **TS**
DIN 8512	L–ZnCd40/2.2360	Germany	Soft solder	4 **Al**, 35–45 **Cd**, 55–65 **Zn**	
JIS H 4321	Class 3	Japan	Plate and sheet: for boilers	1.30 **Pb**, 0.40 **Cd**, 0.09 **Fe**, 0.01 **Ni**, 98.5 min **Zn**	
JIS H 4321	Class 1	Japan	Plate and sheet: for general use	0.01 **Cu**, 1.30 **Pb**, 0.40 **Cd**, 0.09 **Fe**, 98.5 min **Zn**	
JIS H 4321	Class 1	Japan	Plate and sheet: for dry battery	0.005 **Cu**, 0.60 **Pb**, 0.60 **Cd**, 0.025 **Fe**, 98.8 min **Zn**	
JIS H 4321	Class 2	Japan	Plate and sheet: for relief printing	0.02 **Mg**, 0.50 **Pb**, 0.50 **Cd**, 0.25 **Fe**, 0.20 **Ni**, 99.0 min **Zn**	**As manufactured**: all **diam**, 40 min **HV**
JIS H 4321	Class 2	Japan	Plate and sheet: for lithograph	0.005 **Cu** 0.40 **Pb**, 0.40 **Cd**, 0.02 **Fe**, 99.0 min **Zn**	**As manufactured**: all **diam**, 15 MPa **TS**, 12% (50 mm) **El**
COPANT 442	Zn 99.995	COPANT	Ingot, slab	0.001 **Cu**, 0.003 **Pb**, 0.003 **Cd**, 0.001 **Sn**, 0.002 **Fe**, 99.995 min **Zn**, 0.04 **Pb+Cd**, 0.05 total, others	
SABS 20	Zn 99.995	South Africa	Ingot	0.001 **Cu**, 0.0015 **Pb**, 0.0015 **Cd**, 0.0010 **Sn**, 0.0010 **Fe**, 0.001 **Tl**, 0.005 **In**, 99.995 min **Zn**	
COPANT 442	Zn 99.99	COPANT	Ingot, slab	0.002 **Cu**, 0.003 **Pb**, 0.003 **Cd**, 0.001 **Sn**, 0.003 **Fe**, 99.99 min **Zn**, 0.006 **Pb+Cd**, 0.05 total, others	
SABS 20	Zn 99.99	South Africa	Ingot	0.0020 **Cu**, 0.005 **Al**, 0.030 **Pb**, 0.0030 **Cd**, 0.0010 **Sn**, 0.0020 **Fe**, 0.001 **Tl**, 0.0005 **In**, 99.99 min **Zn**	
ANSI/ASTM B 6	Special High Grade	US	Slab and shapes	0.003 **Pb**, 0.003 **Cd**, 0.001 **Sn**, 0.003 **Fe**, 99.990 min **Zn**	
ANSI/ASTM B 6	Special High Grade	US	Slab	0.003 **Pb**, 0.003 **Cd**, 0.003 **Fe**, 99.990 min **Zn**	
ISO/R752	Zn 99.995	ISO	Ingot	0.001 **Cu**, 0.003 **Pb**, 0.003 **Cd**, 0.001 **Sn**, 0.002 **Fe**, 99.983 min **Zn**, 0.004 **Pb+Cd**, 0.0050 **Pb+Cd+Fe+Sn+Cu**, total	
ISO/R752	Zn 99.99	ISO	Ingot	0.002 **Cu**, 0.003 **Pb**, 0.003 **Cd**, 0.001 **Sn**, 0.003 **Fe**, 99.981 min **Zn**, 0.006 **Pb+Cd**, 0.0010 **Cu+Pb+Cd+Sn+Fe**, total	
COPANT 442	Zn 99.95	COPANT	Ingot, slab	0.002 **Cu**, 0.03 **Pb**, 0.02 **Cd**, 0.001 **Sn**, 0.02 **Fe**, 99.95 min **Zn**, 0.05 total, others	

WROUGHT AND CAST ZINC AND ZINC ALLOYS

specification number	designation	country	product forms	chemical composition	mechanical properties and hardness values
SABS 20	Zn 99.95	South Africa	Ingot	0.002 **Cu**, 0.005 **Al**, 0.03 **Pb**, 0.02 **Cd**, 0.001 **Sn**, 0.001 **Fe**, 0.001 **Tl**, 0.0005 **In**, 99.95 min **Zn**	
ANSI/ASTM B 6	High Grade	US	Slab and shapes	0.03 **Pb**, 0.02 **Cd**, 0.02 **Fe**, 99.90 min **Zn**	
ANSI/ASTM B 6	High Grade	US	Slab	0.03 **Pb**, 0.02 **Cd**, 0.02 **Fe**, 99.90 min **Zn**	
ISO/R752	Zn 99.95	ISO	Ingot	0.002 **Cu**, 0.03 **Pb**, 0.02 **Cd**, 0.001 **Sn**, 0.02 **Fe**, 99.877 min **Zn**, 0.050 **Cu** + **Pb** + **Cd** + **Sn** + **Fe**, total	
COPANT 442	Zn 99.5	COPANT	Ingot, slab	0.45 **Pb**, 0.15 **Cd**, 0.005 **Sn**, 0.03 **Fe**, 99.5 min **Zn**, 0.5 total, others	
SABS 20	Zn 99.5	South Africa	Ingot	0.005 **Al**, 0.35–0.45 **Pb**, 0.05 **Cd**, 0.001 **Sn**, 0.03 **Fe**, 0.001 **Tl**, 0.0005 **In**, 99.5 min **Zn**	
COPANT 443	Zn Al Cd	COPANT	Ingot	0.0002 **Cu**, 0.10–0.30 **Al**, 0.001 **Pb**, 0.04–0.06 **Cd**, 0.001 **Fe**, 99. m min **Zn**	
ISO/R752	Zn 99.5	ISO	Ingot	0.45 **Pb**, 0.15 **Cd**, 0.005 **Sn**, 0.03 **Fe**, 98.865 min **Zn**, 0.50 **Pb** + **Cd** + **Fe** + **Sn**, total	
ANSI/ASTM B 6	Prime Western	US	Slab	1.4 **Pb**, 0.20 **Cd**, 0.05 **Fe**, 98.0 min **Zn**	
BS 1004	Alloy A	UK	Ingot	3.9–4.3 **Al**, 0.04–0.06 **Mg**, rem **Zn**, 0.0895 total, others	
SABS 25	ZnAl4	South Africa	Castings	0.030 **Cu**, 3.9–4.3 **Al**, 0.04–0.06 **Mg**, 0.003 **Pb**, 0.003 **Cd**, 0.001 **Sn**, 0.050 **Fe**, 0.001 **Ni**, 0.001 **Tl**, rem **Zn**	**Die–cast**: all **diam**, 286 MPa **TS**, 15% in 2 in. (50 mm) **El**, 83 **HB**
COPANT 443	Zn Al 4	COPANT	Ingot	0.03 **Cu**, 3.9–4.3 **Al**, 0.03–0.06 **Mg**, 0.003 **Pb**, 0.003 **Cd**, 0.001 **Sn**, 0.01 **Fe**, rem **Zn**, 0.04 **Pb** + **Cd**, 0.05 total, others	
JIS H 2201	Class 2	Japan	Ingot: for die castings	3.9–4.3 **Al**, 0.03–0.06 **Mg**, 0.003 **Pb**, 0.002 **Cd**, 0.001 **Sn**, 0.075 **Fe**, 0.03 **Cu**, rem **Zn**	
NF A 55–010	Z–A4 G	France	Pressure die castings	0.10 **Cu**, 3.90–4.30 **Al**, 0.03–0.06 **Mg**, 0.10 **Fe**, rem **Zn**, 0.008 **Pb** + **Cd** + **Sn**	**As cast**: all **diam**, 240 MPa **TS**, 2% in 2 in. (50 mm) **El**
ISO/R301	ZnAl4	ISO	Ingot	0.03 **Cu**, 3.9–4.3 **Al**, 0.03–0.06 **Mg**, 0.003 **Pb**, 0.003 **Cd**, 0.001 **Sn**, 0.05 **Fe**	
BS 1004	Alloy A	UK	Castings	3.8–4.3 **Al**, 0.03–0.06 **Mg**, rem **Zn**, 0.2195 total, others	
SABS 26	ZBD1	South Africa	Castings	0.10 **Cu**, 3.8–4.3 **Al**, 0.03–0.06 **Mg**, 0.005 **Pb**, 0.005 **Cd**, 0.002 **Sn**, 0.10 **Fe**, 0.006 **Ni**, 0.001 **Tl**, rem **Zn**	

WROUGHT AND CAST ZINC AND ZINC ALLOYS

specification number	designation	country	product forms	chemical composition	mechanical properties and hardness values
JIS H 5301	Class 2	Japan	Die castings	0.25 **Cu**, 3.5–4.3 **Al**, 0.020–0.06 **Mg**, 0.005 **Pb**, 0.004 **Cd**, 0.003 **Sn**, 0.10 **Fe**, rem **Zn**	**As cast**: all **diam**, 284 MPa **TS**, 10% in 2 in. (50 mm) **El**, 82 **HB**
DS 3013	7020–11	Denmark	Pressure die castings	0.1 **Cu**, 3.5–4.3 **Al**, 0.020–0.06 **Mg**, 0.005 **Pb**, 0.005 **Cd**, 0.002 **Sn**, 0.10 **Fe**, rem **Zn**	**Aged at 20°C, 5 weeks**: all **diam**, 270 MPa **TS**, 17% in 2 in. (50 mm) **El**, 61 **HB**; **Aged at 20°C, 1 year**: all **diam**, 264 MPa **TS**, 24% in 2 in. (50 mm) **El**, 56 **HB**; **Aged at 20°C, 2 years**: all **diam**, 235 MPa **TS**, 29% in 2 in. (50 mm) **El**, 50 **HB**
DS 3013	7020–10	Denmark	Pressure die castings	0.1 **Cu**, 3.5–4.3 **Al**, 0.020–0.06 **Mg**, 0.005 **Pb**, 0.005 **Cd**, 0.002 **Sn**, 0.10 **Fe**, rem **Zn**	**Aged at 20°C, 5 weeks**: all **diam**, 270 MPa **TS**, 17% in 2 in. (50 mm) **El**, 61 **HB**; **Aged at 20°C, 1 year**: all **diam**, 264 MPa **TS**, 24% in 2 in. (50 mm) **El**, 56 **HB**; **Aged at 20°C, 2 years**: all **diam**, 235 MPa **TS**, 29% in 2 in. (50 mm) **El**, 50 **HB**
SFS 3091	ZnAl4	Finland	Castings	0.1 **Cu**, 3.5–4.3 **Al**, 0.02–0.06 **Mg**, 0.005 **Pb**, 0.005 **Cd**, 0.02 **Sn**, 0.10 **Fe**	
ANSI/ASTM B 86	UNS Z33520 (AG40A)	US	Die castings	0.25 **Cu**, 3.5–4.3 **Al**, 0.020–0.05 **Mg**, 0.005 **Pb**, 0.004 **Cd**, 0.003 **Sn**, 0.100 **Fe**, rem **Zn**	**As cast**: all **diam**, 41 ksi (286 MPa) **TS**, 10% in 2 in. (50 mm) **El**, 82 **HB**
MIL-Z-7068	Class I	US	Ingot	2.5–3.5 **Cu**, 3.5–4.5 **Al**, 0.02–0.10 **Mg**, 0.007 **Pb**, 0.005 **Cd**, 0.005 **Sn**, 0.100 **Fe**, rem **Zn**	
MIL-Z-7068	Class II	US	Ingot	2.5–2.9 **Cu**, 3.9–4.3 **Al**, 0.02–0.05 **Mg**, 0.003 **Pb**, 0.003 **Cd**, 0.001 **Sn**, 0.075 **Fe**, rem **Zn**	
ANSI/ASTM B 86	UNS Z35530 (AC41)	US	Die castings	0.75–1.25 **Cu**, 3.5–4.3 **Al**, 0.03–0.08 **Mg**, 0.005 **Pb**, 0.004 **Cd**, 0.003 **Sn**, 0.100 **Fe**, rem **Zn**	**As cast**: all **diam**, 47 ksi (335 MPa) **TS**, 7% **El**, 91 **HB**
JIS H 5301	Class 1	Japan	Die castings	0.75–1.25 **Cu**, 3.5–4.3 **Al**, 0.03–0.08 **Mg**, 0.007 **Pb**, 0.005 **Cd**, 0.005 **Sn**, 0.10 **Fe**, rem **Zn**	**As cast**: all **diam**, 324 MPa **TS**, 7% in 50 mm **El**, 91 **HB**
BS 1004	Alloy B	UK	Ingot	0.75–1.25 **Cu**, 3.9–4.3 **Al**, 0.04–0.06 **Mg**, rem **Zn**, 0.0595 total, others	
SABS 25	ZnAl4Cu1	South Africa	Castings	0.75–1.25 **Cu**, 3.9–4.3 **Al**, 0.04–0.06 **Mg**, 0.003 **Pb**, 0.003 **Cd**, 0.001 **Sn**, 0.050 **Fe**, 0.001 **Ni**, 0.001 **Tl**, rem **Zn**	**Die cast**: all **diam**, 335 MPa **TS**, 9% in 2 in. (50 mm) **El**, 92 **HB**
BS 1004	Alloy B	UK	Castings	0.75–1.25 **Cu**, 3.8–4.3 **Al**, 0.03–0.06 **Mg**, rem **Zn**, 0.1195 total, others	

WROUGHT AND CAST ZINC AND ZINC ALLOYS

specification number	designation	country	product forms	chemical composition	mechanical properties and hardness values
COPANT 443	Zn Al 4 Cu 1	COPANT	Ingot	0.75–1.25 **Cu**, 3.9–4.3 **Al**, 0.03–0.06 **Mg**, 0.003 **Pb**, 0.003 **Cd**, 0.001 **Sn**, 0.01 **Fe**, rem **Zn**	
SABS 26	ZBD2	South Africa	Castings	0.75–1.25 **Cu**, 3.8–4.3 **Al**, 0.03–0.06 **Mg**, 0.005 **Pb**, 0.005 **Cd**, 0.002 **Sn**, 0.10 **Fe**, 0.006 **Ni**, 0.001 **Tl**, rem **Zn**	
JIS H 2201	Class 1	Japan	Ingot: for die castings	0.75–1.25 **Cu**, 3.9–4.3 **Al**, 0.03–0.06 **Mg**, 0.003 **Pb**, 0.002 **Cd**, 0.001 **Sn**, 0.075 **Fe**, rem **Zn**	
NF A 55–010	Z–A4 U1 G	France	Pressure die castings	0.75–1.25 **Cu**, 3.90–4.30 **Al**, 0.03–0.06 **Mg**, rem **Zn**, 0.008 **Pb + Cd + Sn**	
ANSI/ASTM B 234	AC41A	US	Ingot: for die castings	0.75–1.25 **Cu**, 3.9–4.3 **Al**, 0.03–0.06 **Mg**, 0.004 **Pb**, 0.003 **Cd**, 0.002 **Sn**, 0.075 **Fe**, rem **Zn**	
ISO/R301	ZnAl4Cu1	ISO	Ingot	0.75–1.25 **Cu**, 3.9–4.3 **Al**, 0.03–0.06 **Mg**, 0.003 **Pb**, 0.003 **Cd**, 0.001 **Sn**, 0.05 **Fe**	
DS 3013	7030–10	Denmark	Pressure die castings	0.75–1.25 **Cu**, 3.5–4.3 **Al**, 0.020–0.06 **Mg**, 0.005 **Pb**, 0.005 **Cd**, 0.002 **Sn**, 0.10 **Fe**, rem **Zn**	**Aged at 20°C, 5 weeks**: all **diam**, 330 MPa **TS**, 9% in 2 in. (50 mm) **El**, 9 **HB**; **Aged at 20°C, 1 year**: all **diam**, 320 MPa **TS**, 12% in 2 in. (50 mm) **El**, 12 **HB**; **Aged at 20°C, 2 years**: all **diam**, 255 MPa **TS**, 23% in 2 in. (50 mm) **El**, 23 **HB**
DS 3013	7030–11	Denmark	Pressure die castings	0.75–1.25 **Cu**, 3.5–4.3 **Al**, 0.020–0.06 **Mg**, 0.005 **Pb**, 0.005 **Cd**, 0.002 **Sn**, 0.10 **Fe**, rem **Zn**	**Aged at 20°C, 5 weeks**: all **diam**, 310 MPa **TS**, 10% in 2 in. (50 mm) **El**, 83 **HB**; **Aged at 20°C, 1 year**: all **diam**, 290 MPa **TS**, 14% in 2 in. (50 mm) **El**, 72 **HB**; **Aged at 20°C, 2 years**: all **diam**, 255 MPa **TS**, 23% in 2 in. (50 mm) **El**, 64 **HB**
SFS 3092	ZnAl4Cu1	Finland	Castings	0.75–1.25 **Cu**, 3.5–4.3 **Al**, 0.02–0.06 **Mg**, 0.005 **Pb**, 0.005 **Cd**, 0.002 **Sn**, 0.10 **Fe**	
ANSI/ASTM B 240	AG40A	US	Ingot: for die castings	0.10 **Cu**, 3.9–4.3 **Al**, 0.025–0.05 **Mg**, 0.004 **Pb**, 0.003 **Cd**, 0.002 **Sn**, 0.075 **Fe**, rem **Zn**	
COPANT 442	Zn 98	COPANT	Ingot, slab	1.8 **Pb**, 0.08 **Fe**, 98 min **Zn**, 2.0 total, others	
ISO/752	An98	ISO	Ingot	1.8 **Pb**, 0.08 **Fe**, 2.0 **Pb + Fe**, total	
ISO/R752	Zn 98.5	ISO	Ingot	1.4 **Pb**, 0.20 **Cd**, 0.05 **Fe**, 1.50 **Pb + Cd + Fe**, total	
ANSI/ASTM B6	Prime Western	US	Slab and shapes	0.05 **Al**, 1.4 **Pb**, 0.20 **Cd**, 0.05 **Fe**, 98.0 min **Zn**	

WROUGHT AND CAST ZINC AND ZINC ALLOYS

specification number	designation	country	product forms	chemical composition	mechanical properties and hardness values
COPANT 442	Zn 98.5	COPANT	Ingot, slab	1.4 **Pb**, 0.70 **Cd**, 0.05 **Fe**, 98.5 min **Zn**, 1.50 total, others	
SABS 20	Zn 98.5	South Africa	Ingot	0.005 **Al**, 0.95–1.35 **Pb**, 0.15 **Cd**, 0.02 **Sn**, 0.04 **Fe**, 98.5 min **Zn**	

WROUGHT AND CAST NICKEL AND NICKEL ALLOYS

WROUGHT AND CAST NICKEL AND NICKEL ALLOYS

specification number	designation	country	product forms	chemical composition	mechanical properties and hardness values
ANSI/ASTM B 39		US	Cathode, briquette, pellet	0.02 **Cu**, 0.005 **Mn**, 0.02 **Fe**, 0.005 **Si**, 0.03 **C**, 0.01 **S**, 99.80 min **Ni**, 0.15 **Co**, 0.005 **P**, 0.005 **As**, 0.005 **Pb**, 0.005 **Sb**, 0.005 **Bi**, 0.005 **Sn**, 0.005 **Zn**,	
JIS H 4502	Class 3	Japan	Sheet and strip: for cathode of electronic tube	99.8 min **Ni+Co**, 0.005 **Ti**, 0.05 **Cu**, 0.02 **Mn**, 0.07 **Fe**, 0.01 **Si**, 0.04 **C**, 0.005 **S**, 0.01 **Mg**	**As agreed**: ≥0.5 mm **diam**, 490 MPa **TS**, 35% in 2 in. (50 mm) **El**
JIS H 4522	Class 3	Japan	Tube: seamless, for cathode of vacuum tube	99.7 min **Ni+Co**, 0.01 **Ti**, 0.05 **Cu**, 0.05 **Mn**, 0.05 **Fe**, 0.02 **Si**, 0.05 **C**, 0.008 **S**, 0.02 **Mg**	
NF A 54–101	Ni–02	France	Sheet and strip	99.5 **Ni+Co**, 0.10 **Cu**, 0.30 **Mn**, 0.10 **Fe**, 0.10 **Si**, 0.02 **C**, 0.007 **S**, 0.05 **Mg**	
NF A 54–101	Ni–01	France	Sheet and strip	99.5 **Ni+Co**, 0.10 **Cu**, 0.25 **Mn**, 0.15 **Fe**, 0.10 **Si**, 0.12 **C**, 0.007 **S**, 0.05 **Mg**	**1/2 hard**: ≥0.3 mm **diam**, 490 MPa **TS**, 15% in 2 in. (50 mm) **El**; **3/4 hard**: ≥0.3 mm **diam**, 540 MPa **TS**, 5% in 2 in. (50 mm) **El**; **Hard**: ≥0.3 mm **diam**, 590 MPa **TS**, 3% in 2 in. (50 mm) **El**
JIS H 4502	Class 2C	Japan	Sheet and strip: for cathode of electronic tube	99.2 min **Ni+Co**, 0.05 **Cu**, 0.15 **Mn**, 0.07 **Fe**, 0.05 **Si**, 0.04 **C**, 0.005 **S**, 0.06–0.09 **Mg**	**As agreed**: ≥0.5 mm **diam**, 490 MPa **TS**, 35% in 2 in. (50 mm) **El**
JIS H 4502	Class 2B	Japan	Sheet and strip: for cathode of electronic tube	99.2 min **Ni+Co**, 0.05 **Cu**, 0.15 **Mn**, 0.07 **Fe**, 0.05 **Si**, 0.04 **C**, 0.005 **S**, 0.03–0.06 **Mg**	**As agreed**: ≥0.5 mm **diam**, 490 MPa **TS**, 35% in 2 in. (50 mm) **El**
JIS H 4502	Class 2A	Japan	Sheet and strip: for cathode of electronic tube	99.2 min **Ni+Co**, 0.50 **Cu**, 0.15 **Mn**, 0.07 **Fe**, 0.05 **Si**, 0.04 **C**, 0.005 **S**, 0.01 **Mg**	**As agreed**: ≥0.5 mm **diam**, 490 MPa **TS**, 35% in 2 in. (50 mm) **El**
JIS H 4502	Class 1	Japan	Sheet and strip: for cathode of electronic tube	99.2 min **Ni+Co**, 0.05 **Cu**, 0.15 **Mn**, 0.07 **Fe**, 0.05–0.25 **Si**, 0.04 **C**, 0.005 **S**, 0.001 **Mg**	**As agreed**: ≥0.5 mm **diam**, 490 MPa **TS**, 35% in 2 in. (50 mm) **El**
JIS H 4522	Class 2	Japan	Tube: seamless, for cathode of vacuum tube	99.2 min **Ni+Co**, 0.10 **Cu**, 0.20 **Mn**, 0.20 **Fe**, 0.01–0.05 **Si**, 0.10 **C**, 0.008 **S**, 0.01–0.10 **Mg**	
JIS H 4522	Class 1B	Japan	Tube: seamless, for cathode of vacuum tube	99.2 min **Ni+Co**, 0.10 **Cu**, 0.20 **Mn**, 0.20 **Fe**, 0.05–0.25 **Si**, 0.10 **C**, 0.008 **S**, 0.01–0.15 **Mg**	
JIS H 4522	Class 1A	Japan	Tube: seamless, for cathode of vacuum tube	99.2 min **Ni+Co**, 0.10 **Cu**, 0.20 **Mn**, 0.20 **Fe**, 0.05–0.25 **Si**, 0.10 **C**, 0.008 **S**	
JIS H 4511		Japan	Bar and wire: for electronic tube	99.00 min **Ni+Co**, 0.10 **Cu**, 0.30 **Mn**, 0.20 **Fe**, 0.20 **Si**, 0.10 **C**, 0.008 **S**, 0.10 **Mg**	**As agreed**: ≥1 mm **diam**, 539 MPa **TS**, 30% in 2 in. (50 mm) **El**
JIS H 4501		Japan	Sheet and strip: for electronic use	99.00 min **Ni+Co**, 0.10 **Cu**, 0.30 **Mn**, 0.20 **Fe**, 0.20 **Si**, 0.10 **C**, 0.008 **S**, 0.10 **Mg**	**As agreed**: ≥0.5 mm **diam**, 490 MPa **TS**, 35% in 2 in. (50 mm) **El**
MIL–N–46026A		US	Rod and wire	99.0 min **Ni+Co**, 0.20 **Cu**, 0.35 **Mn**, 0.30 **Fe**, 0.20 **Si**, 0.15 **C**, 0.008 **S**	**Annealed**: all **diam**, 517 MPa **TS**; **As drawn**: all **diam**, 552 MPa **TS**

WROUGHT AND CAST NICKEL AND NICKEL ALLOYS

specification number	designation	country	product forms	chemical composition	mechanical properties and hardness values
MIL–N–46025B		US	Bar, wire, strip	99.0 min **Ni+Co**, 0.20 **Cu**, 0.35 **Mn**, 0.30 **Fe**, 0.20 **Si**, 0.15 **C**, 0.008 **S**	**Annealed (bar)**: all **diam**, 379 MPa **TS**; **As drawn (bar)**: all **diam**, 448 MPa **TS**; **As drawn (wire)**: all **diam**, 552 MPa **TS**
BS 3073	NA12	UK	Strip	99.0 min **Ni+Co**, 0.10 **Ti**, 0.25 **Cu**, 0.35 **Mn**, 0.40 **Fe**, 0.35 **Si**, 0.02 **C**, 0.01 **S**, 0.20 **Mg**	**Cold rolled and material**: 0.25–0.5 mm **diam**, 350 MPa **TS**, 30% in 2 in. (50 mm) **El**
BS 3073	NA11	UK	Strip	99.0 min **Ni+Co**, 0.10 **Ti**, 0.25 **Cu**, 0.35 **Mn**, 0.40 **Fe**, 0.35 **Si**, 0.15 **C**, 0.01 **S**, 0.20 **Mg**	**Cold rolled to 1/4 hard**: 125–150 **HV**; **Cold rolled to 1/2 hard**: 145–175 **HV**; **Cold rolled to hard**: 185–215 **HV**
BS 3074	NA 12	UK	Tube: seamless	99.0 min **Ni+Co**, 0.10 **Ti**, 0.25 **Cu**, 0.35 **Mn**, 0.40 **Fe**, 0.35 **Si**, 0.02 **C**, 0.01 **S**, 0.20 **Mg**	**Cold rolled and annealed**: ≤115 mm **diam**, 350 MPa **TS**, 85 MPa **YS**, 40% in 2 in. (50 mm) **El**; **Cold worked and stress relieved**: ≤115 mm **diam**, 410 MPa **TS**, 205 MPa **YS**, 15% in 2 in. (50 mm) **El**; **Hot worked and annealed**: ≤125 mm **diam**, 350 MPa **TS**, 85 MPa **YS**, 40% in 2 in. (50 mm) **El**
BS 3074	NA 11	UK	Tube: seamless	99.0 min **Ni+Co**, 0.10 **Ti**, 0.25 **Cu**, 0.35 **Mn**, 0.40 **Fe**, 0.35 **Si**, 0.15 **C**, 0.01 **S**, 0.20 **Mg**	**Cold rolled and annealed**: ≤115 mm **diam**, 380 MPa **TS**, 105 MPa **YS**, 40% in 2 in. (50 mm) **El**; **Cold worked and stress relieved**: ≤115 mm **diam**, 450 MPa **TS**, 275 MPa **YS**, 15% in 2 in. (50 mm) **El**; **Hot worked and annealed**: ≤125 mm **diam**, 380 MPa **TS**, 105 MPa **YS**, 40% in 2 in. (50 mm) **El**
BS 3075	NA 11	UK	Wire	99.0 min **Ni+Co**, 0.10 **Ti**, 0.25 **Cu**, 0.35 **Mn**, 0.40 **Fe**, 0.35 **Si**, 0.15 **C**, 0.01 **S**, 0.20 **Mg**	**Cold drawn**: ≤3.20 mm **diam**, 540 MPa **TS**; **Cold drawn and annealed**: ≤0.45 mm **diam**, 380 MPa **TS**, 20% in 2 in. (50 mm) **El**
BS 3076	NA12	UK	Bar	99.0 min **Ni+Co**, 0.10 **Ti**, 0.25 **Cu**, 0.35 **Mn**, 0.40 **Fe**, 0.35 **Si**, 0.02 **C**, 0.01 **S**, 0.20 **Mg**	**Cold worked and annealed**: all **diam**, 340 MPa **TS**, 70 MPa **YS**, 35% in 2 in. (50 mm) **El**; **Hot worked and annealed**: all **diam**, 340 MPa **TS**, 70 MPa **YS**, 35% in 2 in. (50 mm) **El**

WROUGHT AND CAST NICKEL AND NICKEL ALLOYS

specification number	designation	country	product forms	chemical composition	mechanical properties and hardness values
BS 3076	NA11	UK	Bar	99.0 min **Ni + Co**, 0.10 **Ti**, 0.25 **Cu**, 0.35 **Mn**, 0.40 **Fe**, 0.35 **Si**, 0.15 **C**, 0.01 **S**, 0.20 **Mg**	**Cold worked**: 25–55 mm **diam**, 520 MPa **TS**, 345 MPa **YS**, 14% in 2 in. (50 mm) **El**; **Cold worked and annealed**: all **diam**, 380 MPa **TS**, 105 MPa **YS**, 35% in 2 in. (50 mm) **El**; **Hot worked and annealed**: all **diam**, 380 MPa **TS**, 105 MPa **YS**, 35% in 2 in. (50 mm) **El**
BS 3072	NA11	UK	Sheet and plate	99.0 min **Ni + Co**, 0.10 **Ti**, 0.25 **Cu**, 0.35 **Mn**, 0.40 **Fe**, 0.35 **Si**, 0.15 **C**, 0.01 **S**, 0.20 **Mg**	**Cold rolled and annealed**: 0.5–1.5 mm **diam**, 380 MPa **TS**, 105 MPa **YS**, 35% in 2 in. (50 mm) **El**, 110 max **HV**; **Hot rolled**: all **diam**, 380 MPa **TS**, 130 MPa **YS**, 30% in 2 in. (50 mm) **El**, 110 max **HV**; **Hot rolled and annealed**: all **diam**, 380 MPa **TS**, 105 MPa **YS**, 40% in 2 in. (50 mm) **El**, 110 max **HV**
BS 3072	NA12	UK	Sheet and plate	99.0 min **Ni + Co**, 0.10 **Ti**, 0.25 **Cu**, 0.35 **Mn**, 0.40 **Fe**, 0.35 **Si**, 0.02 **C**, 0.01 **S**, 0.20 **Mg**	**Cold rolled and annealed**: 1.5–4 mm **diam**, 350 MPa **TS**, 85 MPa **YS**, 40% in 2 in. (50 mm) **El**, 110 max **HV**; **Hot rolled**: all **diam**, 350 MPa **TS**, 85 MPa **YS**, 30% in 2 in. (50 mm) **El**, 110 max **HV**; **Hot rolled and annealed**: all **diam**, 350 MPa **TS**, 85 MPa **YS**, 40% in 2 in. (50 mm) **El**, 110 max **HV**
AMS 5553B		US	Sheet and strip	99.0 min **Ni + Co**, 0.25 **Cu**, 0.35 **Mn**, 0.40 **Fe**, 0.35 **Si**, 0.02 **C**, 0.010 **S**	**Cold rolled and annealed**: >0.025 in. (>0.64 mm **diam**, 50 ksi (345 MPa) **TS**, 12 ksi (83 MPa) **YS**, 40% in 2 in. (50 mm) **El**, 66 max **HRB**
ANSI/ASTM B 163	N 02201	US	Tube: seamless	99.0 min **Ni + Co**, 0.25 **Cu**, 0.35 **Mn**, 0.40 **Fe**, 0.35 **Si**, 0.02 **C**, 0.01 **S**	**Annealed**: all **diam**, 50 ksi (345 MPa) **TS**, 12 ksi (83 MPa) **YS**, 40% in 2 in. (50 mm) **El**; **Stress relieved**: all **diam**, 60 ksi (414 MPa) **TS**, 30 ksi (207 MPa) **YS**, 15% in 2 in. (50 mm) **El**, 62 max **HRB**

WROUGHT AND CAST NICKEL AND NICKEL ALLOYS

specification number	designation	country	product forms	chemical composition	mechanical properties and hardness values
ANSI/ASTM B 163	N 02200	US	Tube: seamless	99.0 min **Ni+Co**, 0.25 **Cu**, 0.35 **Mn**, 0.40 **Fe**, 0.35 **Si**, 0.15 **C**, 0.01 **S**	**Annealed**: all **diam**, 55 ksi (379 MPa) **TS**, 15 ksi (103 MPa) **YS**, 40% in 2 in. (50 mm) **El**; **Stress relieved**: all **diam**, 65 ksi (448 MPa) **TS**, 40 ksi (276 MPa) **YS**, 15% in 2 in. (50 mm) **El**, 65 max **HRB**
ANSI/ASTM B 162	N 02201	US	Plate, sheet, strip	99.0 min **Ni+Co**, 0.25 **Cu**, 0.35 **Mn**, 0.40 **Fe**, 0.35 **Si**, 0.02 **C**, 0.01 **S**	**Hot rolled, annealed plate**: all **diam**, 50 ksi (345 MPa) **TS**, 12 ksi (80 MPa) **YS**, 40% in 2 in. (50 mm) **El**, 75–135 **HRB**; **Hot rolled, annealed sheet**: all **diam**, 50 ksi (345 MPa) **TS**, 12 ksi (80 MPa) **YS**, 40% in 2 in. (50 mm) **El**, 62 max **HRB**; **Cold rolled, annealed strip**: all **diam**, 50 ksi (345 MPa) **TS**, 12 ksi (80 MPa) **YS**, 40% in 2 in. (50 mm) **El**, 55 max **HRB**
ANSI/ASTM B 162	N 02200	US	Plate, sheet, strip	99.0 min **Ni+Co**, 0.25 **Cu**, 0.35 **Mn**, 0.40 **Fe**, 0.35 **Si**, 0.15 **C**, 0.01 **S**	**Hot rolled, annealed plate**: all **diam**, 55 ksi (380 MPa) **TS**, 15 ksi (100 MPa) **YS**, 40% in 2 in. (50 mm) **El**, 90–140 **HRB**; **Hot rolled, annealed sheet**: all **diam**, 55 ksi (380 MPa) **TS**, 15 ksi (100 MPa) **YS**, 40% in 2 in. (50 mm) **El**, 70 max **HRB**; **Cold rolled, annealed strip**: all **diam**, 55 ksi (380 MPa) **TS**, 15 ksi (100 MPa) **YS**, 40% in 2 in. (50 mm) **El**, 64 max **HRB**
ANSI/ASTM B 161	N 02201	US	Pipe and tube: seamless	99.0 min **Ni+Co**, 0.25 **Cu**, 0.35 **Mn**, 0.40 **Fe**, 0.35 **Si**, 0.02 **C**, 0.01 **S**	**Annealed**: <5.0 in. (<127.0 mm) **diam**, 50 ksi (380 MPa) **TS**, 12 ksi (80 MPa) **YS**, 40% in 2 in. (50 mm) **El**; **Stress relieved**: all **diam**, 60 ksi (413 MPa) **TS**, 30 ksi (205 MPa) **YS**, 15% in 2 in. (50 mm) **El**
ANSI/ASTM B 161	N 02200	US	Pipe and tube: seamless	99.0 min **Ni+Co**, 0.25 **Cu**, 0.35 **Mn**, 0.40 **Fe**, 0.35 **Si**, 0.15 **C**, 0.01 **S**	**Annealed**: <5.0 in. (<127.0 mm) **diam**, 50 ksi (380 MPa) **TS**, 15 ksi (105 MPa) **YS**, 40% in 2 in. (50 mm) **El**; **Stress relieved**: all **diam**, 65 ksi (450 MPa) **TS**, 40 ksi (275 MPa) **YS**, 15% in 2 in. (50 mm) **El**

WROUGHT AND CAST NICKEL AND NICKEL ALLOYS

specification number	designation	country	product forms	chemical composition	mechanical properties and hardness values
ANSI/ASTM B 160	N 02201	US	Rod, bar	99.0 min **Ni+Co**, 0.25 **Cu**, 0.35 **Mn**, 0.40 **Fe**, 0.35 **Si**, 0.02 **C**, 0.01 **S**	**Hot finished**: all **diam**, 50 ksi (345 MPa) **TS**, 10 ksi (70 MPa) **YS**, 40% in 2 in. (50 mm) **El**; **Annealed**: all **diam**, 50 ksi (345 MPa) **TS**, 10 ksi (70 MPa) **YS**, 40% in 2 in. (50 mm) **El**
ANSI/ASTM B 160	N 02200	US	Rod, bar	99.0 min **Ni+Co**, 0.25 **Cu**, 0.35 **Mn**, 0.40 **Fe**, 0.35 **Si**, 0.15 **C**, 0.01 **S**	**Cold drawn (rounds)**: <1.0 in. (<25.4 mm) **diam**, 80 ksi (550 MPa) **TS**, 60 ksi (415 MPa) **YS**; **Hot finished**: all **diam**, 60 ksi (415 MPa) **TS**, 15 ksi (105 MPa) **YS**, 35% in 2 in. (50 mm) **El**; **Annealed**: all **diam**, 55 ksi (380 MPa) **TS**, 15 ksi (105 MPa) **YS**, 40% in 2 in. (50 mm) **El**
MIL-W-19487A	Grade B	US	Wire	97.0 min **Ni+Co**, 0.2–0.6 **Ti**, 0.25 **Cu**, 0.50 **Mn**, 0.60 **Fe**, 0.35 **Si**, 0.40 **C**, 0.010 **S**, 0.20–0.50 **Mg**	**Cold drawn, spring, spring tempered, as drawn**: ≤0.057 in. **diam**, 1172 MPa **TS**; **Cold drawn, spring tempered, age hardened**: ≤0.057 in. **diam**
AMS 5890		US	Bar, forgings, and extrusions: corrosion and heat resistant	96.87 min **Ni+Co**, 0.05 **Ti**, 0.15 **Cu**, 0.05 **Fe**, 0.02 **C**, 0.0025 **S**, 0.20 **Co**, 0.05 **Cr**, 1.80–2.60 **Th**	
AMS 5865		US	Sheet and strip: corrosion and heat resistant	96.87 min **Ni+Co**, 0.05 **Ti**, 0.15 **Cu**, 0.05 **Fe**, 0.02 **C**, 0.0025 **S**, 0.20 **Co**, 0.05 **Cr**, 1.80–2.60 **Th**	
JIS H 4502	Class 4	Japan	Sheet and strip: for cathode of electronic tube	94.5 min **Ni+Co**, 0.005 **Ti**, 0.05 **Cu**, 0.15 **Mn**, 0.07 **Fe**, 0.06 **Si**, 0.04 **C**, 0.005 **S**, 0.01–0.08 **Mg**, 3.50–4.50 **W**	**As agreed**: ≥0.5 mm **diam**, 588 MPa **TS**, 35% in 2 in. (50 mm) **El**
JIS H 4522	Class 4	Japan	Tube: seamless, for cathode of vacuum tube	94.50 min **Ni+Co**, 0.02 **Ti**, 0.20 **Cu**, 0.20 **Mn**, 0.20 **Fe**, 0.02–0.06 **Si**, 0.10 **C**, 0.008 **S**, 0.01–0.10 **Mg**, 3.50–4.50 **W**	
MIL-W-19487A	Grade A	US	Wire	93.0 min **Ni+Co**, 4.0–4.75 **Al**, 0.25–1.00 **Ti**, 0.25 **Cu**, 0.50 **Mn**, 0.60 **Fe**, 1.0 **Si**, 0.30 **C**, 0.010 **S**	**Cold drawn, spring, spring tempered, as drawn**: ≤0.057 in. **diam**, 1172 MPa **TS**; **Cold drawn, spring tempered, age hardened**: ≤0.057 in. **diam**, 1413 MPa **TS**
JIS Z 3265	BNi-4	Japan	Wire and ribbon: brazing filler	92.23 min **Ni+Co**, 1.5 **Fe**, 3.0–4.0 **Si**, 0.06 **C**, 1.0–2.2 **B**, 0.50 total, others	
JIS Z 3265	BNi-3	Japan	Wire and ribbon: brazing filler	89.43 min **Ni+Co**, 1.5 **Fe**, 4.0–5.0 **Si**, 0.06 **C**, 2.75–3.5 **B**, 0.50 total, others	
JIS Z 3265	BNi-6	Japan	Wire and ribbon: brazing filler	79.84 min **Ni+Co**, 0.15 **C**, 10.0–20.0 **P**, 0.50 total, others	

WROUGHT AND CAST NICKEL AND NICKEL ALLOYS

specification number	designation	country	product forms	chemical composition	mechanical properties and hardness values
JIS Z 3265	BNi-2	Japan	Wire and ribbon: brazing filler	78.84 min **Ni + Co**, 2.0–4.0 **Fe**, 4.0–5.0 **Si**, 0.15 **C**, 2.75–3.5 **B**, 6.0–8.0 **Cr**, 0.50 total, others	
JIS H 4552		Japan	Pipe and tube: seamless	63.0–70.0 **Ni + Co**, rem **Cu**, 2.0 **Mn**, 2.5 **Fe**, 0.50 **Si**, 0.30 **C**, 0.024 **S**	**Annealed**: 15–51 mm **diam**, 481 MPa **TS**, 35% in 2 in. (50 mm) **El**; **Stress relieved**: 15–51 mm **diam**, 588 MPa **TS**, 15% in 2 in. (50 mm) **El**
JIS H 4551		Japan	Sheet and plate	63.0–70.0 **Ni + Co**, rem **Cu**, 2.0 **Mn**, 2.5 **Fe**, 0.50 **Si**, 0.30 **C**, 0.024 **S**	**Annealed**: 0.5–50 mm **diam**, 481 MPa **TS**, 196 MPa **YS**, 35% in 2 in. (50 mm) **El**
ASTM F 467	N 04400	US	Nut	63.0–70.0 **Ni + Co**, rem **Cu**, 2.0 **Mn**, 2.5 **Fe**, 0.5 **Si**, 0.024 **S**	**As agreed**: all **diam**, 75 min **HRB**
ASTM F 467	N 04405	US	Nut	63.0–70.0 **Ni + Co**, rem **Cu**, 2.0 **Mn**, 2.5 **Fe**, 0.5 **Si**, 0.3 **C**, 0.025–0.060 **S**	**As agreed**: all **diam**, 60 min **HRB**
ASTM F 467	N 05500	US	Nut	63.0–70.0 **Ni + Co**, 2.30–3.15 **Al**, 0.35–0.85 **Ti**, rem **Cu**, 1.5 **Mn**, 2.0 **Fe**, 0.5 **Si**, 0.25 **C**	**As agreed**: all **diam**, 24 min **HRC**
ASTM F 468	N 04400	US	Bolt, screw, stud	63.0–70.0 **Ni + Co**, rem **Cu**, 2.0 **Mn**, 2.5 **Fe**, 0.5 **Si**, 0.3 **C**, 0.024 **S**	**As manaufactured**: 0.250–0.750 in. (0.635–1.905 mm) **diam**, 80 ksi (550 MPa) **TS**, 40 ksi (275 MPa) **YS**, 25 min **HRC**; **Hot formed**: all **diam**, 70 ksi (480 MPa) **TS**, 30 ksi (205 MPa) **YS**, 60–95 **HRB**
ASTM F 468	N 05500	US	Bolt, screw, stud	63.0–70.0 **Ni + Co**, 2.30–3.15 **Al**, 0.35–0.85 **Ti**, rem **Cu**, 1.5 **Mn**, 2.0 **Fe**, 0.5 **Si**, 0.25 **C**, 0.01 **S**	**As manaufactured**: 0.250–0.875 in. (6.35–22.2 mm) **diam**, 130 ksi (900 MPa) **TS**, 90 ksi (620 MPa) **YS**, 24–27 **HRC**
AMS 4544C		US	Sheet, strip and plate	63.0–70.0 **Ni + Co**, rem **Cu**, 2.00 **Mn**, 2.50 **Fe**, 0.50 **Si**, 0.30 **C**, 0.024 **S**, 1.00 **Co**	**Cold rolled and annealed (sheet and strip)**: ≤0.250 in. (≤6.35 mm) **diam**, 70 ksi (483 MPa) **TS**, 35% in 2 in. (50 mm) **El**, 73 max **HRB**; **Hot rolled and annealed (plate**: ≤0.250 in. (≤6.35 mm) **diam**, 70 ksi (483 MPa) **TS**, 35% in 2 in. (50 mm) **El**, 73 max **HRB**
AMS 4574B		US	Tube: seamless	63.0–70.0 **Ni + Co**, rem **Cu**, 2.0 **Mn**, 2.5 **Fe**, 0.5 **Si**, 0.30 **C**, 0.024 **S**, 1.0 **Co**	**Cold finished and annealed**: all **diam**, 85 ksi (586 MPa) **TS**, 32% in 2 in. (50 mm) **El**
AMS 4677		US	Bar and forgings: corrosion resistant	63.0–70.0 **Ni + Co**, 2.3–3.5 **Al**, 0.5 **Ti**, rem **Cu**, 1.5 **Mn**, 2.0 **Fe**, 0.5 **Si**, 0.10 **C**, 0.010 **S**, 1.0 **Co**, 0.02 **P**, 0.006 **Pb**, 0.006 **Sn**, 0.02 **Zn**	**Hot finished, annealed, and precipitation heat treated**: ≤4.500 in. **diam**, 130 ksi (896 MPa) **TS**, 80 ksi (552 MPa) **YS**, 20% in 2 in. (50 mm) **El**, 233 min **HB**

WROUGHT AND CAST NICKEL AND NICKEL ALLOYS

specification number	designation	country	product forms	chemical composition	mechanical properties and hardness values
AMS 4676A		US	Bar and forgings: corrosion resistant	63.0–70.0 **Ni+Co**, 2.00–4.00 **Al**, 0.25–1.00 **Ti**, rem **Cu**, 1.50 **Mn**, 2.00 **Fe**, 1.00 **Si**, 0.25 **C**, 0.010 **S**, 1.0 **Co**, 0.02 **P**, 0.006 **Pb**, 0.02 **Zn**	**Hot finished, quenched, and solution heat treated**: all **diam**, 248 max **HB**; **Hot finished, quenched, and precipitation heat treated:** >4.25 in. (108 mm) **diam**, 140 ksi (965 MPa) **TS**, 100 ksi (690 MPa) **YS**, 17% in 2 in. (50 mm) **El**, 262 min **HB**
AMS 4675A		US	Bar and forgings: corrosion resistant	63.00–70.00 **Ni+Co**, rem **Cu**, 2.00 **Mn**, 2.50 **Fe**, 0.50 **Si**, 0.30 **C**, 0.024 **S**, 1.0 **Co**	**Cold drawn (bar)**: 0.093–0.500 in. (2.36–12.70 mm) **diam**, 84 ksi (579 MPa) **TS**, 50 ksi (345 MPa) **YS**, 10% in 2 in. (50 mm) **El**, 80 min **HB**; **As forged (forgings)**: all **diam**, 78–96 **HRB**
AMS 4674D		US	Bar and forgings: corrosion resistant	63.00–70.00 **Ni+Co**, rem **Cu**, 2.00 **Mn**, 2.50 **Fe**, 0.50 **Si**, 0.30 **C**, 0.025–0.06 **S**, 1.0 **Co**	**Cold finished (bar)**: 0.50–1.00 in. (12.7–25.4 mm) **diam**, 85 ksi (586 MPa) **TS**, 50 ksi (345 MPa) **YS**, 15% in 2 in. (50 mm) **El**, 80 min **HB**; **As forged (forgings)**: all **diam**, 78–96 **HRB**
AMS 4577A		US	Tube: seamless, corrosion resistant	63.0–70.0 **Ni+Co**, rem **Cu**, 2.00 **Mn**, 2.5 **Fe**, 0.5 **Si**, 0.30 **C**, 0.024 **S**, 1.0 **Co**	**Cold drawn and annealed**: all **diam**, 85 ksi (586 MPa) **TS**, 32% in 2 in. (50 mm) **El**
AMS 4575B		US	Tube: brazed, corrosion resistant	63.0–70.0 **Ni+Co**, rem **Cu**, 2.0 **Mn**, 2.5 **Fe**, 0.5 **Si**, 0.30 **C**, 0.024 **S**, 1.0 **Co**	**Cold drawn and annealed**: all **diam**, 85 ksi (586 MPa) **TS**, 32% in 2 in. (50 mm) **El**
AMS 4731		US	Wire and ribbon: for electronic applications	63.0–70.0 **Ni+Co**, 0.50 **Al**, rem **Cu**, 2.0 **Mn**, 2.5 **Fe**, 0.50 **Si**, 0.20 **C**, 0.015 **S**, 1.0 **Co**, 0.02 **P**, 0.006 **Pb**, 0.006 **Sn**, 0.02 **Zn**	**Cold drawn or cold rolled and annealed**: all **diam**, 85 ksi (586 MPa) **TS**
AMS 4730D		US	Wire: corrosion resistant	63.0–70.0 **Ni+Co**, rem **Cu**, 2.0 **Mn**, 2.5 **Fe**, 0.5 **Si**, 0.3 **C**, 0.024 **S**, 1.0 **Co**	**Cold drawn and annealed**: >0.040 in. (>1.02 mm) **diam**, 90 ksi (621 MPa) **TS**
QQ-N-286D	Class B	US	Bar, rod, plate	63.0–70.0 **Ni+Co**, 2.50–3.50 **Al**, 0.50 **Ti**, rem **Cu**, 1.5 **Mn**, 2.0 **Fe**, 0.50 **Si**, 0.10 **C**, 0.010 **S**, 0.02 **Zn**, 0.02 **P**, 0.006 **Sn**, 0.006 **Pb**	**Hot finished and age hardened**: 8 in. **diam**, 130 ksi (896 MPa) **TS**, 85 ksi (586 MPa) **YS**, 20% in 2 in. (50 mm) **El**, 245 **HB**, 23 **HRC**; **Hot finished, annealed and age hardened**: 4.5 in. **diam**, 130 ksi (896 MPa) **TS**, 80 ksi (552 MPa) **YS**, 20% in 2 in. (50 mm) **El**, 233 **HB**, 21 **HRC**; **Cold drawn and age hardened**: 2 in. **diam**, 130 ksi (896 MPa) **TS**, 90 ksi (621 MPa) **YS**, 17% in 2 in. (50 mm) **El**, 255 **HB**, 23 **HRC**

WROUGHT AND CAST NICKEL AND NICKEL ALLOYS

specification number	designation	country	product forms	chemical composition	mechanical properties and hardness values
QQ-N-281D	Class B	US	Bar, rod, plate, sheet, strip, wire, forgings	63.0–70.0 **Ni + Co**, 0.50 **Al**, rem **Cu**, 2.0 **Mn**, 2.50 **Fe**, 0.5 **Si**, 0.3 **C**, 0.025–0.060 **S**, 0.02 **Zn**, 0.02 **P**, 0.006 **Sn**, 0.006 **Pb**	**Cold drawn**: all **diam**, 85 ksi (586 MPa) **TS**, 50 ksi (344 MPa) **YS**, 8% in 2 in. (50 mm) **El**; **Hot finished**: all **diam**, 75 ksi (517 MPa) **TS**, 35 ksi (241 MPa) **YS**, 30% in 2 in. (50 mm) **El**; **Annealed**: all **diam**, 70 ksi (483 MPa) **TS**, 25 ksi (172 MPa) **YS**, 35% in 2 in. (50 mm) **El**, 60–75 **HRB**
QQ-N-286D	Class A	US	Bar, rod, forgings, sheet, strip, wire	63.0–70.0 **Ni + Co**, 2.30–3.15 **Al**, 0.35–0.85 **Ti**, rem **Cu**, 1.5 **Mn**, 2.0 **Fe**, 0.50 **Si**, 0.25 **C**, 0.010 **S**, 0.02 **Zn**, 0.02 **P**, 0.006 **Sn**, 0.006 **Pb**	**Hot finished and age hardened**: all **diam**, 140 ksi (965 MPa) **TS**, 100 ksi (689 MPa) **YS**, 20% in 2 in. (50 mm) **El**, 265 **HB**, 27 **HRC**; **Cold drawn and hardened**: 0.25–1.0 in. **diam**, 145 ksi (1000 MPa) **TS**, 110 ksi (758 MPa) **YS**, 15% in 2 in. (50 mm) **El**, 290 **HB**, 31 **HRC**; **Annealed and age hardened**: 1 in. **diam**, 130 ksi (896 MPa) **TS**, 90 ksi (621 MPa) **YS**, 20% in 2 in. (50 mm) **El**, 250 **HB**, 24 **HRC**
QQ-N-281D	Class A	US	Bar, rod, plate, sheet, strip, wire, forgings	63.0–70.0 **Ni + Co**, 0.5 **Al**, rem **Cu**, 2.0 **Mn**, 2.50 **Fe**, 0.5 **Si**, 0.2 **C**, 0.015 **S**, 0.02 **P**, 0.02 **Zn**, 0.006 **Sn**, 0.006 **Pb**	**Cold drawn (stress relieved)**: all **diam**, 84 ksi (579 MPa) **TS**, 50 ksi (345 MPa) **YS**, 10% in 2 in. (50 mm) **El**; **Hot finished**: all **diam**, 80 ksi (552 MPa) **TS**, 40 ksi (276 MPa) **YS**, 30% in 2 in. (50 mm) **El**; **Annealed**: all **diam**, 70 ksi (483 MPa) **TS**, 25 ksi (172 MPa) **YS**, 35% in 2 in. (50 mm) **El**
ASTM F 468	N 04405	US	Bolt, screw, stud	63.0–70.0 **Ni + Co**, rem **Cu**, 2.5 **Mn**, 2.5 **Fe**, 0.5 **Si**, 0.3 **C**, 0.025–0.060 **S**	**As manufactured**: all **diam**, 70 ksi (480 MPa) **TS**, 30 ksi (205 MPa) **YS**, 20 min **HRC**
ANSI/ASTM B 127	N 04400	US	Plate, sheet, strip	63.0–70.0 **Ni + Co**, rem **Cu**, 2.0 **Mn**, 2.5 **Fe**, 0.5 **Si**, 0.3 **C**, 0.024 **S**	**Hot rolled, annealed plate**: all **diam**, 70 ksi (485 MPa) **TS**, 28 ksi (195 MPa) **YS**, 35% in 2 in. (50 mm) **El**, 110–140 **HRB**; **Hot rolled, annealed and pickled sheet**: all **diam**, 70 ksi (485 MPa) **TS**, 28 ksi (195 MPa) **YS**, 35% in 2 in. (50 mm) **El**, 73 max **HRB**; **Cold rolled, annealed strip**: all **diam**, 70 ksi (485 MPa) **TS**, 28 ksi (195 MPa) **YS**, 35% in 2 in. (50 mm) **El**, 68 max **HRB**

WROUGHT AND CAST NICKEL AND NICKEL ALLOYS

specification number	designation	country	product forms	chemical composition	mechanical properties and hardness values
QQ-N-288	Composition A	US	Castings	62–68 **Ni + Co**, 0.50 **Al**, 26–33 **Cu**, 1.5 **Mn**, 2.5 **Fe**, 2.0 **Si**, 0.35 **C**	**Solution heat treated and age hardened**: all **diam**, 65 ksi (448 MPa) **TS**, 32 ksi (221 MPa) **YS**, 25% in 2 in. (50 mm) **El**, 120–150 **HB**
QQ-N-288	Composition B	US	Castings	61–68 **Ni + Co**, 0.50 **Al**, 27–33 **Cu**, 1.5 **Mn**, 2.5 **Fe**, 2.7–3.7 **Si**, 0.30 **C**	**Solution heat treated and age hardened**: all **diam**, 100 ksi (689 MPa) **TS**, 60 ksi (414 MPa) **YS**, 10% in 2 in. (50 mm) **El**, 240–290 **HB**
ANSI/ASTM B 163	N 04400	US	Tube: seamless	63.0 min **Ni + Co**, 28.0–34.0 **Cu**, 2.0 **Mn**, 2.5 **Fe**, 0.5 **Si**, 0.3 **C**, 0.024 **S**	**Annealed**: all **diam**, 70 ksi (483 MPa) **TS**, 28 ksi (193 MPa) **YS**, 35% in 2 in. (50 mm) **El**; **Stress relieved**: all **diam**, 85 ksi (586 MPa) **TS**, 55 ksi (379 MPa) **YS**, 15% in 2 in. (50 mm) **El**, 75 max **HRB**
ANSI/ASTM B 165	N 04400	US	Pipe and tube: seamless	63.0 min **Ni + Co**, 28.0–34.0 **Cu**, 2.0 **Mn**, 2.5 **Fe**, 0.5 **Si**, 0.3 **C**, 0.024 **S**	**Annealed**: <5.0 in. (<127.0 mm) **diam**, 70 ksi (480 MPa) **TS**, 28 ksi (195 MPa) **YS**, 35% in 2 in. (50 mm) **El**; **Stress relieved**: all **diam**, 85 ksi (585 MPa) **TS**, 55 ksi (380 MPa) **YS**, 15% in 2 in. (50 mm) **El**
ANSI/ASTM B 164	N 04405	US	Rod, bar	63.0 min **Ni + Co**, 28.0–34.0 **Cu**, 2.0 **Mn**, 2.5 **Fe**, 0.5 **Si**, 0.3 **C**, 0.025–0.060 **S**	**Cold drawn rounds**: <0.50 in. (<12.70 mm) **diam**, 85 ksi (585 MPa) **TS**, 50 ksi (345 MPa) **YS**, 8% in 2 in. (50 mm) **El**, 84–96 **HRB**; **Hot finished rounds**: <3.0 in. (<76.2 mm) **diam**, 75 ksi (515 MPa) **TS**, 35 ksi (240 MPa) **YS**, 30% in 2 in. (50 mm) **El**; **Annealed rods and bars**: all **diam**, 70 ksi (480 MPa) **TS**, 25 ksi (170 MPa) **YS**, 35% in 2 in. (50 mm) **El**, 60–75 **HRB**
ANSI/ASTM B 164	N 04400	US	Rod, bar	63.0 min **Ni + Co**, 28.0–34.0 **Cu**, 2.0 **Mn**, 2.5 **Fe**, 0.5 **Si**, 0.3 **C**, 0.024 **S**	**Cold drawn rounds**: <0.50 in. (<12.70 mm) **diam**, 110 ksi (760 MPa) **TS**, 85 ksi (585 MPa) **YS**, 8% in 2 in. (50 mm) **El**; **Hot finished rounds, squares, rectangles**: all **diam**, 80 ksi (550 MPa) **TS**, 40 ksi (275 MPa) **YS**, 30% in 2 in. (50 mm) **El**; **Annealed rods and bars**: all **diam**, 70 ksi (480 MPa) **TS**, 25 ksi (170 MPa) **YS**, 35% in 2 in. (50 mm) **El**

WROUGHT AND CAST NICKEL AND NICKEL ALLOYS

specification number	designation	country	product forms	chemical composition	mechanical properties and hardness values
BS 3073	NA18	UK	Strip	63.0 min **Ni+Co**, 2.3–3.2 **Al**, 0.35–0.85 **Ti**, 27.0–33.0 **Cu**, 1.5 **Mn**, 2.0 **Fe**, 0.5 **Si**, 0.25 **C**, 0.01 **S**	**Cold rolled, solution treated and precipitation treated**: 0.50–4.0 mm **diam**, 900 MPa **TS**, 620 MPa **YS**, 15% in 2 in. (50 mm) **El**, 200 max **HV**; **Cold rolled to 1/2 hard and precipitation treated**: 0.50–4.0 mm **diam**, 1000 MPa **TS**, 760 MPa **YS**, 8% in 2 in. (50 mm) **El**, 230–280 **HV**; **Cold rolled to hard and precipitation treated**: 0.50–4.0 mm **diam**, 1170 MPa **TS**, 900 MPa **YS**, 5% in 2 in. (50 mm) **El**, 280–330 **HV**
BS 3073	NA13	UK	Strip	63.0 min **Ni+Co**, 28.0–34.0 **Cu**, 2.0 **Mn**, 2.5 **Fe**, 0.5 **Si**, 0.3 **C**, 0.02 **S**	**Cold rolled to 1/4 hard**: 0.5–4 mm **diam**, 480 MPa **TS**, 195 MPa **YS**, 35% in 2 in. (50 mm) **El**, 140–170 **HV**; **Cold rolled to 1/2 hard**: 0.5–4 mm **diam**, 480 MPa **TS**, 195 MPa **YS**, 35% in 2 in. (50 mm) **El**, 165–195 **HV**; **Cold rolled to hard**: 0.5–4 mm **diam**, 480 MPa **TS**, 195 MPa **YS**, 35% in 2 in. (50 mm) **El**, 205–235 **HV**
BS 3074	NA18	UK	Tube: seamless	63.0 min **Ni+Co**, 2.3–3.2 **Al**, 0.35–0.85 **Ti**, 27.0–33.0 **Cu**, 1.5 **Mn**, 2.0 **Fe**, 0.5 **Si**, 0.25 **C**, 0.01 **S**	**Cold worked, solution and precipitation treated**: all **diam**, 900 MPa **TS**, 620 MPa **YS**, 15% in 2 in. (50 mm) **El**
BS 3074	NA13	UK	Tube: seamless	63.0 min **Ni+Co**, 28.0–34.0 **Cu**, 2.0 **Mn**, 2.5 **Fe**, 0.5 **Si**, 0.3 **C**, 0.024 **S**	**Cold worked and annealed**: $\leq$115 mm **diam**, 480 MPa **TS**, 195 MPa **YS**, 35% in 2 in. (50 mm) **El**; **Cold worked and stress relieved**: $\leq$115 mm **diam**, 590 MPa **TS**, 380 MPa **YS**, 15% in 2 in. (50 mm) **El**; **Hot worked and annealed**: $\leq$125 mm **diam**, 480 MPa **TS**, 195 MPa **YS**, 35% in 2 in. (50 mm) **El**
BS 3075	NA18	UK	Wire	63.0 min **Ni+Co**, 2.3–3.2 **Al**, 0.35–0.85 **Ti**, 27.0–33.0 **Cu**, 1.5 **Mn**, 2.0 **Fe**, 1.0 **Si**, 0.25 **C**, 0.01 **S**	**Cold**: $\leq$10 mm **diam**, 760 MPa **TS**; **Cold drawn and solution treated**: $\leq$10 mm **diam**, 760 MPa **TS**; **Cold drawn and precipitation treated**: $\leq$10 mm **diam**, 1070 MPa **TS**
BS 3075	NA13	UK	Wire	63.0 min **Ni+Co**, 28.0–34.0 **Cu**, 2.0 **Mn**, 2.5 **Fe**, 0.5 **Si**, 0.3 **C**, 0.024 **S**	**Cold drawn**: $\leq$3.20 mm **diam**, 770 MPa **TS**; **Cold drawn and annealed**: $\leq$0.45 mm **diam**, 480 MPa **TS**, 20% in 2 in. (50 mm) **El**

WROUGHT AND CAST NICKEL AND NICKEL ALLOYS

specification number	designation	country	product forms	chemical composition	mechanical properties and hardness values
BS 3076	NA18	UK	Bar	63.0 min **Ni + Co**, 2.3–3.2 **Al**, 0.35–0.85 **Ti**, 27.0–33.0 **Cu**, 1.5 **Mn**, 2.0 **Fe**, 0.5 **Si**, 0.25 **C**, 0.01 **S**	**Cold worked and precipitation treated**: all **diam**, 970 MPa **TS**, 690 MPa **YS**, 16% in 2 in. (50 mm) **El**; **Cold worked, solution and precipitation treataed**: all **diam**, 900 MPa **TS**, 585 MPa **YS**, 20% in 2 in. (50 mm) **El**; **Hot worked and precipitation treated**: all **diam**, 830 MPa **TS**, 550 MPa **YS**, 15% in 2 in. (50 mm) **El**
BS 3076	NA13	UK	Bar	63.0 min **Ni + Co**, 28.0–34.0 **Cu**, 2.0 **Mn**, 2.5 **Fe**, 0.5 **Si**, 0.3 **C**, 0.024 **S**	**Cold worked and annealed**: all **diam**, 480 MPa **TS**, 170 MPa **YS**, 35% in 2 in. (50 mm) **El**; **Cold worked and stress relieved**: ≤40 mm **diam**, 600 MPa **TS**, 415 MPa **YS**, 20% in 2 in. (50 mm) **El**; **Hot worked and annealed**: all **diam**, 480 MPa **TS**, 170 MPa **YS**, 35% in 2 in. (50 mm) **El**
BS 3072	NA18	UK	Sheet and plate	63.0 min **Ni + Co**, 2.3–3.2 **Al**, 0.35–0.85 **Ti**, 27.0–33.0 **Cu**, 1.5 **Mn**, 2.0 **Fe**, 0.5 **Si**, 0.25 **C**, 0.01 **S**	**Cold rolled, solution and precipitation treated**: 0.5–4.0 mm **diam**, 900 MPa **TS**, 620 MPa **YS**, 15% in 2 in. (50 mm) **El**, 200 max **HV**; **Hot rolled and precipitation treated**: all **diam**, 970 MPa **TS**, 690 MPa **YS**, 15% in 2 in. (50 mm) **El**, 270 max **HV**; **Hot rolled, solution and precipitation treated**: all **diam**, 900 MPa **TS**, 620 MPa **YS**, 15% in 2 in. (50 mm) **El**, 200 max **HV**
BS 3072	NA13	UK	Sheet and plate	63.0 min **Ni + Co**, 28.0–34.0 **Cu**, 2.0 **Mn**, 2.5 **Fe**, 0.5 **Si**, 0.3 **C**, 0.02 **S**	**Cold rolled and annealed**: 0.5–4 mm **diam**, 480 MPa **TS**, 195 MPa **YS**, 35% in 2 in. (50 mm) **El**, 140 max **HV**; **Hot rolled**: all **diam**, 510 MPa **TS**, 275 MPa **YS**, 25% in 2 in. (50 mm) **El**, 140 max **HV**; **Hot rolled and annealed**: all **diam**, 480 MPa **TS**, 195 MPa **YS**, 35% in 2 in. (50 mm) **El**, 140 max **HV**
ANSI/ASTM F 96	N 04400	US	Wrought products	62.0 min **Ni + Co**, 28.0–34.0 **Cu**, 2.0 **Mn**, 2.5 **Fe**, 0.5 **Si**, 0.015 **S**, 0.01 **Pb**, 0.02 **Zn**, 0.02 **P**	
BS 3071	NA 1	UK	Castings	61.57 min **Ni + Co**, 28.0–32.0 **Cu**, 0.5–1.5 **Mn**, 3.0 **Fe**, 0.5–1.5 **Si**, 0.1–0.3 **C**, 0.05 **S**, 0.08–0.12 **Mg**, 0.005 **Pb**	**As cast**: 386 MPa **TS**, 138 MPa **YS**, 16% in 2 in. (50 mm) **El**

WROUGHT AND CAST NICKEL AND NICKEL ALLOYS

specification number	designation	country	product forms	chemical composition	mechanical properties and hardness values
BS 3071	NA 2	UK	Castings	60.22 min **Ni + Co**, 28.0–32.0 **Cu**, 0.5–1.5 **Mn**, 3.0 **Fe**, 2.5–3.0 **Si**, 0.15 **C**, 0.05 **S**, 0.08–0.12 **Mg**, 0.005 **Pb**	**As cast**: all **diam**, 483 MPa **TS**, 207 MPa **YS**, 10% in 2 in. (50 mm) **El**
QQ–N–288	Composition E	US	Castings	60 min **Ni + Co**, 0.50 **Al**, 26–33 **Cu**, 1.5 **Mn**, 3.5 **Fe**, 1.0–2.0 **Si**, 0.30 **C**, 1.0–3.0 **Nb + Ta**	**Solution treated and age hardened**: all **diam**, 65 ksi (448 MPa) **TS**, 32 ksi (221 MPa) **YS**, 25% in 2 in. (50 mm) **El**, 125–150 **HB**
QQ–N–288	Composition D	US	Castings	60 min **Ni + Co**, 0.50 **Al**, 27–31 **Cu**, 1.5 **Mn**, 2.5 **Fe**, 3.5–4.5 **Si**, 0.25 **C**	**Solution heat treated and age hardened**: all **diam**, 300 **HB**
QQ–N–288	Composition C	US	Castings	60 min **Ni + Co**, 0.50 **Al**, 27–31 **Cu**, 1.5 **Mn**, 2.5 **Fe**, 3.3–4.3 **Si**, 0.20 **C**	**Solution heat treated and age hardened**: all **diam**, 120 ksi (827 MPa) **TS**, 80 ksi (552 MPa) **YS**, 10% in 2 in. (50 mm) **El**, 250–300 **HB**
ANSI/ASTM A 494	M–35	US	Rod, bar, wire, plate	55–60 **Ni + Co**, rem **Cu**, 1.25–2.25 **Mn**, 1.00 **Fe**, 0.75 **Si**, 0.30 **C**, 0.024 **S**	**Cold worked (rod, bar)**: <0.5 in. **diam**, 552 MPa **TS**, 379 MPa **YS**, 10% in 2 in. (50 mm) **El**; **Cold drawn (wire)**: <0.5 in. **diam**, 517 MPa **TS**; **Annealed (plate)**: 448 MPa **TS**, 159 MPa **YS**, 35% in 2 in. (50 mm) **El**, 100–150 **HRB**
ANSI/ASTM F 96	N 04404	US	Wrought products	52.0–57.0 **Ni + Co**, 0.05 **Al**, rem **Cu**, 0.1 **Mn**, 0.5 **Fe**, 0.1 **Si**, 0.15 **C**, 0.015 **S**, 0.01 **Pb**, 0.02 **Zn**, 0.03 **P**	
QQ–C–557B	Alloy B	US	Shot	48.5–51.5 **Ni + Co**, 47 min **Cu**, 0.30 **Mn**, 1.0 **Fe**, 0.05 **Si**, 0.05 **C**, 0.05 **S**, 0.05 **Sn**, 0.05 **Pb**, 0.05 **Zn**	
JIS C 2521		Japan	Wire: resistance	40.0–50.0 **Ni + Co**, 46.5–56.5 **Cu**, 2.5 **Mn**	**As drawn**: $\geq$0.2 mm **diam**, 415 MPa **TS**, 20% in 2 in. (50 mm) **El**
ISO 429	CuNi44Mn1	ISO	Strip, wire	43.0–45.0 **Ni + Co**, rem **Cu**, 0.5–2.0 **Mn**, 0.50 **Fe**, 0.05 **C**, 0.05 **S**, 0.02 **Sn + Pb**, 0.1 total, others	
ISO 1638	CuNi44Mn1	ISO	Solid product: drawn, on coils or reels	43.0–45.0 **Ni + Co**, rem **Cu**, 0.5–2.0 **Mn**, 0.50 **Fe**, 0.05 **C**, 0.05 **S**, 0.02 **Sn + Pb**, 0.1 total, others	**Annealed**: 1–5 mm **diam**, 410 MPa **TS**, 30% in 2 in. (50 mm) **El**
ISO 1634	CuNi44Mn1	ISO	Plate, sheet, strip	43.0–45.0 **Ni + Co**, rem **Cu**, 0.5–2.0 **Mn**, 0.50 **Fe**, 0.05 **C**, 0.05 **S**, 0.2 **Zn**, 0.02 **Sn + P**, 0.1 total, others	

WROUGHT AND CAST TITANIUM AND TITANIUM ALLOYS

WROUGHT AND CAST TITANIUM AND TITANIUM ALLOYS

specification number	designation	country	product forms	chemical composition	mechanical properties and hardness values
BS 2 TA.8		UK	Forging stock	99.79 min **Ti**, 0.20 **Fe**, 0.010 **H**	**Heat treated**: all **diam**, 540 MPa **TS**, 430 MPa **YS**, 16% in 2 in. (50 mm) **El**
BS 2 TA.4		UK	Forging stock	99.79 min **Ti**, 0.20 **Fe**, 0.010 **H**	**Heat treated**: all **diam**, 390 MPa **TS**, 290 MPa **YS**, 20% in 2 in. (50 mm) **El**
BS 2 TA.9		UK	Forgings	99.78 min **Ti**, 0.20 **Fe**, 0.15 **H**	**Heat treated**: all **diam**, 540 MPa **TS**, 430 MPa **YS**, 16% in 2 in. (50 mm) **El**
BS 2 TA.7		UK	Bar and section	99.78 min **Ti**, 0.20 **Fe**, 0.0125 **H**	**Heat treated**: all **diam**, 540 MPa **TS**, 430 MPa **YS**, 16% in 2 in. (50 mm) **El**
BS 2 TA.6		UK	Sheet and strip	99.78 min **Ti**, 0.20 **Fe**, 0.0125 **H**	**Heat treated**: all **diam**, 570 MPa **TS**, 460 MPa **YS**, 15% in 2 in. (50 mm) **El**
BS 2 TA.5		UK	Forgings	99.78 min **Ti**, 0.20 **Fe**, 0.015 **H**	**Heat treated**: all **diam**, 390 MPa **TS**, 290 MPa **YS**, 20% in 2 in. (50 mm) **El**
BS 2 TA.3		UK	Bar and section	99.78 min **Ti**, 0.20 **Fe**, 0.0125 **H**	**Heat treated**: all **diam**, 390 MPa **TS**, 290 MPa **YS**, 20% in 2 in. (50 mm) **El**
BS 2 TA.2		UK	Sheet and strip	99.78 min **Ti**, 0.20 **Fe**, 0.0125 **H**	**Heat treated**: all **diam**, 390 MPa **TS**, 290 MPa **YS**, 22% in 2 in. (50 mm) **El**
BS 2 TA.1		UK	Sheet and strip	99.78 min **Ti**, 0.20 **Fe**, 0.0125 **H**	**Heat treated**: all **diam**, 290 MPa **TS**, 200 MPa **YS**, 25% in 2 in. (50 mm) **El**
MIL-T-46038B	ELI	US	Rod, bar, billet	99.76 min **Ti**, 0.02 **N**, 0.10 **C**, 0.0125 **H**, 0.10 **O**	**Annealed or heat treated**: all **diam**, 827 MPa **YS**, 13% in 2 in. (50 mm) **El**
MIL-T-46035A	EL 1	US	Wrought products	99.76 min **Ti**, 0.02 **N**, 0.10 **C**, 0.0125 **H**, 0.10 O_2	**Annealed or heat treated**: 0.25–2.5 in. **diam**, 827 MPa **YS**, 13% in 2 in. (50 mm) **El**
MIL-T-46038B	LI	US	Rod, bar, billet	99.73 min **Ti**, 0.03 **N**, 0.10 **C**, 0.0125 **H**, 0.12 O_2	**Annealed or heat treated**: all **diam**, 827 MPa **YS**, 13% in 2 in. (50 mm) **El**
MIL-T-46035A	L 1	US	Wrought products	99.73 min **Ti**, 0.03 **N**, 0.10 **C**, 0.0125 **H**, 0.12 O_2	**Annealed or heat treated**: 0.25–2.5 in. **diam**, 827 MPa **YS**, 13% in 2 in. (50 mm) **El**
MIL-T-9047F	Unalloyed	US	Bar and forging stock	99.7 min **Ti**, 0.50 **Fe**, 0.050 **N**, 0.08 **C**, 0.0125 **H**, 0.40 **O**, 0.30 total, others	
MIL-T-9047E	Unalloyed	US	Bar and forging stock	99.7 min **Ti**, 0.50 **Fe**, 0.050 **N**, 0.08 **C**, 0.0125 **H**, 0.40 **O**, 0.30 total, others	**Annealed or solution treated**: all **diam**, 552 MPa **TS**, 483 MPa **YS**, 15% in 2 in. (50 mm) **El**
AIR 9182	T-35	France	Sheet	99.69 min **Ti**, 0.12 **Fe**, 0.05 **N**, 0.08 **C**, 0.015 **H**, 0.04 **Si**	**Annealed and cold finished**: all **diam**, 440 max MPa **TS**, 395 MPa **YS**, 30% in 2 in. (50 mm) **El**
AIR 9182	T-40	France	Sheet	99.69 min **Ti**, 0.12 **Fe**, 0.05 **N**, 0.08 **C**, 0.015 **H**, 0.04 **Si**	**Annealed and cold finished**: <2 mm **diam**, 390 MPa **TS**, 295 MPa **YS**, 28% in 2 in. (50 mm) **El**

WROUGHT AND CAST TITANIUM AND TITANIUM ALLOYS

specification number	designation	country	product forms	chemical composition	mechanical properties and hardness values
JIS H 4631	Class 1	Japan	Pipe and tube: seamless, for heat exchanger	99.67 min **Ti**, 0.25 **Fe**, 0.05 **N**, 0.015 **H**, 0.15 **O**	**Cold drawn and annealed**: 10–60 mm **diam**, 275 MPa **TS**, 27% in 50 mm **El**
MIL–T–46038B	NI	US	Rod, bar, billet	99.66 min **Ti**, 0.04 **N**, 0.10 **C**, 0.0125 **H**, 0.18 O_2	**Annealed or heat treated**: all **diam**, 827 MPa **YS**, 13% in 2 in. (50 mm) **El**
MIL–T–46035A	NI	US	Wrought products	99.66 min **Ti**, 0.04 **N**, 0.10 **C**, 0.0125 **H**, 0.18 O_2	**Annealed or heat treated**: 0.25–2.5 in. **diam**, 827 MPa **YS**, 13% in 2 in. (50 mm) **El**
JIS H 2151	Class 1	Japan	Sponge	99.6 min **Ti**, 0.10 **Fe**, 0.01 **Mn**, 0.02 **N**, 0.03 **C**, 0.005 **H**, 0.10 **Cl**, 0.06 **Mg**, 0.08 **O**, 0.03 **Si**	**As manufactured**: all **diam**, 105 max **HB**
JIS H 4600	Class 1	Japan	Sheet, plate, and strip	99.58 min **Ti**, 0.20 **Fe**, 0.05 **N**, 0.013 **H**, 0.15 **O**	**Hot or cold rolled**: 0.5–15 mm **diam**, 275 MPa **TS**, 167 MPa **YS**, 27% in 50 mm **El**
JIS H 4630	Class 1	Japan	Pipe and tube: seamless	99.58 min **Ti**, 0.20 **Fe**, 0.05 **N**, 0.015 **H**, 0.15 **O**	**Heat extruded and cold drawn**: 10–80 mm **diam**, 275 MPa **TS**, 27% in 50 mm **El**
JIS H 4650	Class 1	Japan	Rod and bar	99.58 min **Ti**, 0.20 **Fe**, 0.05 **N**, 0.015 **H**, 0.15 **O**	**Hot worked or cold drawn**: 8–100 mm **diam**, 275 MPa **TS**, 27% in 50 mm **El**, 100 min **HB**
JIS H 4670	Class 1	Japan	Wire	99.58 min **Ti**, 0.20 **Fe**, 0.05 **N**, 0.015 **H**, 0.15 **O**	**As drawn**: 1–8 mm **diam**, 275 MPa **TS**, 15% in 50 mm **El**
AIR 9182	T–60	France	Sheet	99.56 min **Ti**, 0.30 **Fe**, 0.08 **N**, 0.08 **C**, 0.015 **H**, 0.04 **Si**	**Annealed and cold finished**: <2 mm **diam**, 590 MPa **TS**, 470 MPa **YS**, 20% in 2 in. (50 mm) **El**
AIR 9182	T–50	France	Sheet	99.54 min **Ti**, 0.25 **Fe**, 0.07 **N**, 0.08 **C**, 0.015 **H**, 0.04 **Si**	**Annealed and cold finished**: <2 mm **diam**, 490 MPa **TS**, 390 MPa **YS**, 24% in 2 in. (50 mm) **El**
JIS H 4600	Class 2	Japan	Sheet, plate, and strip	99.53 min **Ti**, 0.20 **Fe**, 0.05 **N**, 0.013 **H**, 0.20 **O**	**Hot or cold rolled**: 0.5–15 mm **diam**, 343 MPa **TS**, 216 MPa **YS**, 23% in 50 mm **El**
JIS H 4650	Class 2	Japan	Rod and bar	99.5 min **Ti**, 0.25 **Fe**, 0.05 **N**, 0.015 **H**, 0.20 **O**	**Hot worked or cold drawn**: 8–100 mm **diam**, 343 MPa **TS**, 23% in 50 mm **El**, 110 min **HB**
JIS H 4630	Class 2	Japan	Pipe and tube: seamless	99.48 min **Ti**, 0.25 **Fe**, 0.05 **N**, 0.015 **H**, 0.20 **O**	**Heat extruded and cold drawn**: 10–80 mm **diam**, 343 MPa **TS**, 23% in 50 mm **El**
JIS H 4631	Class 2	Japan	Pipe and tube: seamless, for heat exchanger	99.48 min **Ti**, 0.25 **Fe**, 0.05 **N**, 0.015 **H**, 0.20 **O**	**Cold drawn and annealed**: 10–60 mm **diam**, 343 MPa **TS**, 23% in 50 mm **El**
JIS H 4670	Class 2	Japan	Wire	99.48 min **Ti**, 0.25 **Fe**, 0.05 **N**, 0.015 **H**, 0.20 **O**	**As drawn**: 343 MPa **TS**, 13% in 50 mm **El**
ASTM F 467	Grade 1	US	Nut	99.45 min **Ti**, 0.20 **Fe**, 0.05 **N**, 0.10 **C**, 0.0125 **H**, 0.18 **O**	**As agreed**: all **diam**, 140 min **HV**
ASTM F 468	Grade 1	US	Bolt, screw, stud	99.45 min **Ti**, 0.20 **Fe**, 0.05 **N**, 0.10 **C**, 0.0125 **H**, 0.18 **O**	**As manufactured**: all **diam**, 40 ksi (280 MPa) **TS**, 30 ksi (205 MPa) **YS**, 140–160 **HV**

WROUGHT AND CAST TITANIUM AND TITANIUM ALLOYS

specification number	designation	country	product forms	chemical composition	mechanical properties and hardness values
JIS H 2151	Class 2	Japan	Sponge	99.4 min **Ti**, 0.10 **Fe**, 0.01 **Mn**, 0.02 **N**, 0.03 **C**, 0.005 **H**, 0.12 **Cl**, 0.07 **Mg**, 0.12 **O**, 0.03 **Si**	**As manufactured**: all **diam**, 106–120 **HB**
MIL-T-81915	A	US	Castings	99.4 min **Ti**, 0.20 **Fe**, 0.05 **N**, 0.08 **C**, 0.015 **H**, 0.20 **O**, 0.60 total, others	**Annealed**: all **diam**, 241 MPa **TS**, 172 MPa **YS**, 24% in 2 in. (50 mm) **El**
MIL-T-9046H	A(40 KSI-YS)	US	Sheet, strip, plate	99.4 min **Ti**, 0.50 **Fe**, 0.05 **N**, 0.08 **C**, 0.015 **H**, 0.20 **O**, 0.60 total, others	**Annealed**: all **diam**, 345 MPa **TS**, 276 MPa **YS**, 20% in 2 in. (50 mm) **El**
MIL-T-9046H	C(55 KSI-YS)	US	Sheet, strip, plate	99.4 min **Ti**, 0.50 **Fe**, 0.05 **N**, 0.08 **C**, 0.015 **H**, 0.30 **O**, 0.60 total, others	**Annealed**: all **diam**, 448 MPa **TS**, 379 MPa **YS**, 18% in 2 in. (50 mm) **El**
JIS H 4600	Class 3	Japan	Sheet, plate, and strip	99.31 min **Ti**, 0.30 **Fe**, 0.07 **N**, 0.013 **H**, 0.30 **O**	**Hot or cold rolled**: 0.5–15 mm **diam**, 481 MPa **TS**, 343 MPa **YS**, 18% in 50 mm **El**
JIS H 4630	Class 3	Japan	Pipe and tube: seamless	99.31 min **Ti**, 0.30 **Fe**, 0.07 **N**, 0.015 **H**, 0.30 **O**	**Heat extruded and cold drawn**: 10–80 mm **diam**, 481 MPa **TS**, 18% in 50 mm **El**
JIS H 4650	Class 3	Japan	Rod and bar	99.31 min **Ti**, 0.30 **Fe**, 0.07 **N**, 0.015 **H**, 0.30 **O**	**Hot worked or cold drawn**: 8–100 mm **diam**, 481 MPa **TS**, 18% in 50 mm **El**, 150 min **HB**
JIS H 4670	Class 3	Japan	Wire	99.31 min **Ti**, 0.30 **Fe**, 0.07 **N**, 0.015 **H**, 0.30 **O**	**As drawn**: <8 mm **diam**, 481 MPa **TS**, 11% in 50 mm **El**
JIS H 2151	Class 3	Japan	Sponge	99.3 min **Ti**, 0.20 **Fe**, 0.05 **Mn**, 0.03 **N**, 0.03 **C**, 0.005 **H**, 0.15 **Cl**, 0.08 **Mg**, 0.25 **O**, 0.03 **Si**	**As manufactured**: all **diam**, 121–140 **HB**
JIS H 4631	Class 3	Japan	Pipe and tube: seamless, for heat exchanger	99.3 min **Ti**, 0.30 **Fe**, 0.07 **N**, 0.015 **H**, 0.30 **O**	**Cold drawn and annealed**: 10–60 mm **diam**, 481 MPa **TS**, 18% in 50 mm **El**
ASTM F 467	Grade 2	US	Nut	99.28 min **Ti**, 0.30 **Fe**, 0.05 **N**, 0.10 **C**, 0.0125 **H**, 0.25 **O**	**As agreed**: all **diam**, 150 min **HV**
ASTM F 468	Grade 2	US	Bolt, screw, stud	99.28 min **Ti**, 0.30 **Fe**, 0.05 **N**, 0.10 **C**, 0.0125 **H**, 0.25 **O**	**As manufactured**: 55 ksi (380 MPa) **TS**, 45 ksi (310 MPa) **YS**, 160–180 **HV**
JIS H 2151	Class 4	Japan	Sponge	99.2 min **Ti**, 0.20 **Fe**, 0.05 **Mn**, 0.03 **N**, 0.03 **C**, 0.005 **H**, 0.15 **Cl**, 0.08 **Mg**, 0.25 **O**, 0.03 **Si**	**As manufactured**: all **diam**, 141–160 **HB**
MIL-T-9046H	B(70 KSI-YS)	US	Sheet, strip, plate	99.2 min **Ti**, 0.50 **Fe**, 0.05 **N**, 0.08 **C**, 0.015 **H**, 0.40 **O**, 0.80 total, others	**Annealed**: all **diam**, 552 MPa **TS**, 483 MPa **YS**, 15% in 2 in. (50 mm) **El**
ASTM F 467	Grade 7	US	Nut	99.03 min **Ti**, 0.30 **Fe**, 0.05 **N**, 0.10 **C**, 0.0125 **H**, 0.25 **O**, 0.12–0.25 **Pd**	**As agreed**: all **diam**, 160 min **HV**
ASTM F 468	Grade 4	US	Bolt, screw, stud	99.03 min **Ti**, 0.50 **Fe**, 0.07 **N**, 0.10 **C**, 0.0125 **H**, 0.40 **O**	**As manufactured**: all **diam**, 85 ksi (590 MPa) **TS**, 75 ksi (515 MPa) **YS**, 200–220 **HV**
ASTM F 468	Grade 7	US	Bolt, screw, stud	99.03 min **Ti**, 0.30 **Fe**, 0.05 **N**, 0.10 **C**, 0.0125 **H**, 0.25 **O**, 0.12–0.25 **Pd**	**As manufactured**: all **diam**, 55 ksi (380 MPa) **TS**, 45 ksi (310 MPa) **YS**, 160–180 **HV**

WROUGHT AND CAST TITANIUM AND TITANIUM ALLOYS

specification number	designation	country	product forms	chemical composition	mechanical properties and hardness values
JIS H 2152	Class 1	Japan	Sponge: compressed	99.0 min **Ti**, 0.60 **Fe**, 0.03 **Mn**, 0.03 **N**, 0.05 **C**, 0.005 **H**, 0.15 **Cl**, 0.10 **Mg**, 0.04 **Si**	
ASTM F 467	Grade 4	US	Nut	98.91 min **Ti**, 0.50 **Fe**, 0.07 **N**, 0.10 **C**, 0.0125 **H**, 0.40 **O**	**As agreed**: all **diam**, 200 min **HV**
JIS H 2152	Class 2	Japan	Sponge: compressed	97.0 min **Ti**, 2.0 **Fe**, 0.05 **Mn**, 0.10 **N**, 0.10 **C**, 0.005 **H**, 0.15 **Cl**, 0.50 **Mg**, 0.10 **Si**	
MIL-T-9047E	5Al-2.5Sn ELI	US	Bar and forging stock	90.83 min **Ti**, 4.70-5.60 **Al**, 2.00-3.00 **Sn**, 0.25 **Fe**, 0.10 **Mn**, 0.035 **N**, 0.05 **C**, 0.0125 **H**, 0.12 **O**, 0.30 total, others	**Annealed or solution treated**: $\leq$2.00 in. **diam**, 689 MPa **TS**, 621 MPa **YS**, 10% in 2 in. (50 mm) **El**
MIL-T-009047F	5Al-2.5Sn ELI	US	Bar and forging stock	90.83 min **Ti**, 4.70-5.60 **Al**, 2.00-3.00 **Sn**, 0.25 **Fe**, 0.10 **Mn**, 0.035 **N**, 0.05 **C**, 0.0125 **H**, 0.12 **O**, 0.30 total, others	
MIL-T-9046H	B(5Al-2.5Sn ELI)	US	Sheet, strip, plate	90.78 min **Ti**, 4.5-5.75 **Al**, 2.0-3.0 **Sn**, 0.25 **Fe**, 0.035 **N**, 0.05 **C**, 0.0125 **H**, 0.12 **O**, 0.30 total, others	**Annealed**: 689 MPa **TS**, 655 MPa **YS**, 10% in 2 in. (50 mm) **El**
MIL-T-9046H	A(5Al-2.5Sn)	US	Sheet, strip, plate	90.4 min **Ti**, 4.5-5.75 **Al**, 2.0-3.0 **Sn**, 0.50 **Fe**, 0.05 **N**, 0.08 **C**, 0.020 **H**, 0.20 **O**, 0.40 total, others	**Annealed**: all **diam**, 827 MPa **TS**, 779 MPa **YS**, 10% in 2 in. (50 mm) **El**
MIL-T-9047E	5Al-5.2Sn	US	Bar and forging stock	90.4 min **Ti**, 4.50-5.75 **Al**, 2.00-3.00 **Sn**, 0.50 **Fe**, 0.05 **N**, 0.08 **C**, 0.020 **H**, 0.20 **O**, 0.40 total, others	**Annealed or solution treated**: all **diam**, 793 MPa **TS**, 758 MPa **YS**, 10% in 2 in. (50 mm) **El**
MIL-T-009047F	5Al-5.2Sn	US	Bar and forging stock	90.4 min **Ti**, 4.50-5.75 **Al**, 2.00-3.00 **Sn**, 0.50 **Fe**, 0.05 **N**, 0.08 **C**, 0.020 **H**, 0.20 **O**, 0.40 total, others	
MIL-T-81915	5Al-2.5Sn	US	Castings	90.0 min **Ti**, 4.50-5.75 **Al**, 2.0-3.0 **Sn**, 0.50 **Fe**, 0.05 **N**, 0.08 **C**, 0.020 **H**, 0.20 **O**, 0.40 total, others	**Annealed**: all **diam**, 758 MPa **TS**, 724 MPa **YS**, 10% in 2 in. (50 mm) **El**
AIR 9183	T-A4M	France	Bar, shapes, and forgings	89.64 min **Ti**, 3.5-5.0 **Al**, 0.15 **Fe**, 3.5-5.0 **Mn**, 0.05 N_2, 0.08 **C**, 0.0125 H_2, 0.2 O_2, 0.04 **Si**	**As manufactured**: all **diam**, 930 MPa **TS**, 830 MPa **YS**, 10% in 2 in. (50 mm) **El**
MIL-T-9046H	F(8Al-1Mo-1V)	US	Sheet, strip, plate	88.6 min **Ti**, 7.3-8.3 **Al**, 0.75-1.25 **V**, 0.30 **Fe**, 0.75-1.25 **Mo**, 0.05 **N**, 0.08 **C**, 0.015 **H**, 0.15 **O**, 0.40 total, others	**Annealed (sheet, strip)**: all **diam**, 1000 MPa **TS**, 931 MPa **YS**, 10% in 2 in. (50 mm) **El**; **Annealed (plate)**: 3/16-1/4 in. **diam**, 896 MPa **TS**, 827 MPa **YS**, 10% in 2 in. (50 mm) **El**
MIL-T-9047E	8Al-1Mo-1V	US	Bar and forging stock	88.58 min **Ti**, 7.35-8.35 **Al**, 0.75-1.25 **V**, 0.30 **Fe**, 0.75-1.25 **Mo**, 0.050 **N**, 0.08 **C**, 0.0120 **H**, 0.12 **O**, 0.40 total, others	**Annealed or solution treated**: $\leq$2.5 in. **diam**, 896 MPa **TS**, 827 MPa **YS**, 10% in 2 in. (50 mm) **El**

WROUGHT AND CAST TITANIUM AND TITANIUM ALLOYS

specification number	designation	country	product forms	chemical composition	mechanical properties and hardness values
MIL–T–009047F	8Al–1Mo–1V	US	Bar and forging stock	88.5 min **Ti**, 7.35–8.35 **Al**, 0.75–1.25 **V**, 0.30 **Fe**, 0.75–1.25 **Mo**, 0.050 **N**, 0.08 **C**, 0.0120 **H**, 0.12 **O**, 0.40 total, others	
MIL–T–9046H	G(6Al–2Cb–1Ta–0.8Mo)	US	Sheet, strip, plate	88.05 min **Ti**, 5.5–6.5 **Al**, 0.25 **Fe**, 0.5–1.0 **Mo**, 0.03 **N**, 0.05 **C**, 0.0125 **H**, 1.5–2.5 **Cd**, 0.5–1.5 **Ta**, 0.10 **O**, 0.40 total, others	**Annealed (plate)**: all **diam**, 710 MPa **TS**, 655 MPa **YS**, 10% in 2 in. (50 mm) **El**
MIL–T–9047E	6Al–Sn–4Zr–2Mo	US	Bar and forging stock	84.18 min **Ti**, 5.50–6.50 **Al**, 1.80–2.20 **Sn**, 0.25 **Fe**, 1.80–2.20 **Mo**, 0.05 **N**, 0.05 **C**, 0.0125 **H**, 1.60–4.40 **Zr**, 0.15 **O**	**Annealed or solution treated**: all **diam**, 896 MPa **TS**, 827 MPa **YS**, 10% in 2 in. (50 mm) **El**; **Heat treated**: ≤1 in. **diam**, 1034 MPa **TS**, 951 MPa **YS**, 10% in 2 in. (50 mm) **El**
MIL–T–009047F	6Al–2Sn–4Zr–2Mo	US	Bar and forging stock	83.58 min **Ti**, 5.50–6.50 **Al**, 1.75–2.25 **Sn**, 0.25 **Fe**, 1.80–2.20 **Mo**, 0.05 **N**, 0.05 **C**, 0.0125 **H**, 0.25 **Si**, 3.50–4.50 **Zr**, 0.15 **O**, 0.10 each, 0.30 total, others	
MIL–T–9046H	G(6Al–2Sn–4Zr–2Mo)	US	Sheet, strip, plate	83.48 min **Ti**, 5.5–6.5 **Al**, 1.5–2.5 **Sn**, 0.35 **Fe**, 1.5–2.5 **Mo**, 0.05 **N**, 0.08 **C**, 0.015 **H**, 3.6–4.4 **Zr**, 0.12 **O**, 0.30 total, others	**Annealed**: all **diam**, 931 MPa **TS**, 862 MPa **YS**, 8% in 2 in. (50 mm) **El**
MIL–T–81915	B(6Al–2Sn–4Zr–2Mo)	US	Castings	78.98 min **Ti**, 5.5–6.5 **Al**, 1.5–2.5 **Sn**, 0.35 **Fe**, 1.5–2.5 **Mo**, 0.05 **N**, 0.08 **C**, 0.015 **H**, 3.6–4.4 **Zr**, 0.12 **O**, 0.40 total, others	**Annealed**: all **diam**, 862 MPa **TS**, 793 MPa **YS**, 8% in 2 in. (50 mm) **El**
MIL–T–9047E	11Sn–5Zr–2Al–1Mo	US	Bar and forging stock	78.22 min **Ti**, 2.00–2.50 **Al**, 10.50–11.50 **Sn**, 0.12 **Fe**, 0.80–1.20 **Mo**, 0.050 **N**, 0.04 **C**, 0.0125 **H**, 0.15–0.20 **Si**, 4.00–6.00 **Zr**, 0.15 **O**, 0.40 total, others	**Annealed or solution treated**: all **diam**, 965 MPa **TS**, 896 MPa **YS**, 10% in 2 in. (50 mm) **El**; **Heat treated**: ≤1 in. **diam**, 1000 MPa **TS**, 931 MPa **YS**, 12% in 2 in. (50 mm) **El**
MIL–T–009047F	11Sn–5Zr–2Al–1Mo	US	Bar and forging stock	78.15 min **Ti**, 2.00–2.50 **Al**, 10.50–11.50 **Sn**, 0.12 **Fe**, 0.80–1.20 **Mo**, 0.050 **N**, 0.04 **C**, 0.0125 **H**, 0.15–0.27 **Si**, 4.00–6.00 **Zr**, 0.15 **O**, 0.40 total, others	
BS TA.18		UK	Bar	78.08 min **Ti**, 2.0–2.5 **Al**, 10.5–11.5 **Sn**, 0.20 **Fe**, 0.8–1.2 **Mo**, 0.0125 **H**, 4.0–6.0 **Zr**, 0.10–0.50 **Si**	**Heat treated and quenched**: all **diam**, 992 MPa **TS**, 866 MPa **YS**, 8% in 2 in. (50 mm) **El**
BS TA.19		UK	Forging stock	78.08 min **Ti**, 2.0–2.5 **Al**, 10.5–11.5 **Sn**, 0.20 **Fe**, 0.8–1.2 **Mo**, 0.0125 **H**, 4.0–6.0 **Zr**, 0.10–0.50 **Si**	**Heat treated and quenched**: all **diam**, 992 MPa **TS**, 866 MPa **YS**, 8% in 2 in. (50 mm) **El**
BS TA.25		UK	Bar	78.08 min **Ti**, 2.0–2.5 **Al**, 10.5–11.5 **Sn**, 0.20 **Fe**, 0.8–1.2 **Mo**, 0.0125 **H**, 4.0–6.0 **Zr**, 0.10–0.50 **Si**	**Heat treated**: all **diam**, 920 MPa **TS**, 786 MPa **YS**, 8% in 2 in. (50 mm) **El**

WROUGHT AND CAST TITANIUM AND TITANIUM ALLOYS

specification number	designation	country	product forms	chemical composition	mechanical properties and hardness values
BS TA.20		UK	Forgings	78.08 min **Ti**, 2.0–2.5 **Al**, 10.5–11.5 **Sn**, 0.20 **Fe**, 0.8–1.2 **Mo**, 0.015 **H**, 4.0–6.0 **Zr**, 0.10–0.50 **Si**	**Heat treated and quenched**: all **diam**, 992 MPa **TS**, 866 MPa **YS**, 8% in 2 in. (50 mm) **El**
BS TA.26		UK	Forging stock	78.08 min **Ti**, 2.0–2.5 **Al**, 10.5–11.5 **Sn**, 0.20 **Fe**, 0.8–1.2 **Mo**, 0.0125 **H**, 4.0–6.0 **Zr**, 0.10–0.50 **Si**	**Heat treated**: all **diam**, 920 MPa **TS**, 786 MPa **YS**, 8% in 2 in. (50 mm) **El**
BS TA.27		UK	Forgings	78.08 min **Ti**, 2.0–2.5 **Al**, 10.5–11.5 **Sn**, 0.20 **Fe**, 0.8–1.2 **Mo**, 0.015 **H**, 4.0–6.0 **Zr**, 0.10–0.50 **Si**	**Heat treated**: all **diam**, 920 MPa **TS**, 786 MPa **YS**, 8% in 2 in. (50 mm) **El**
BS TA.58		UK	Plate	96.79 min **Ti**, 0.20 **Fe**, 0.010 **H**, 2.0–3.0 **Cu**	**Heat treated**: all **diam**, 520 MPa **TS**, 420 MPa **YS**, 20% in 2 in. (50 mm) **El**
BS 2TA.21		UK	Sheet and strip	96.79 min **Ti**, 0.20 **Fe**, 0.010 **H**, 2.0–3.0 **Cu**	**Heat treated**: all **diam**, 540 MPa **TS**, 460 MPa **YS**, 15% in 2 in. (50 mm) **El**
BS 2TA.22		UK	Bar and section	96.79 min **Ti**, 0.20 **Fe**, 0.010 **H**, 2.0–3.0 **Cu**	**Heat treated**: all **diam**, 540 MPa **TS**, 400 MPa **YS**, 16% in 2 in. (50 mm) **El**
BS 2TA.23		UK	Forging stock	96.79 min **Ti**, 0.20 **Fe**, 0.010 **H**, 2.0–3.0 **Cu**	**Heat treated**: all **diam**, 540 MPa **TS**, 400 MPa **YS**, 16% in 2 in. (50 mm) **El**
BS TA.54		UK	Forging stock	96.79 min **Ti**, 0.20 **Fe**, 0.010 **H**, 2.0–3.0 **Cu**	**Heat treated**: all **diam**, 650 MPa **TS**, 525 MPa **YS**, 10% in 2 in. (50 mm) **El**
BS TA.53		UK	Bar and section	96.79 min **Ti**, 0.20 **Fe**, 0.010 **H**, 2.0–3.0 **Cu**	**Heat treated**: all **diam**, 650 MPa **TS**, 525 MPa **YS**, 10% in 2 in. (50 mm) **El**
BS TA.52		UK	Sheet and strip	96.79 min **Ti**, 0.20 **Fe**, 0.010 **H**, 2.0–3.0 **Cu**	**Heat treated**: all **diam**, 690 MPa **TS**, 550 MPa **YS**, 10% in 2 in. (50 mm) **El**
BS 2TA.24		UK	Forgings	96.78 min **Ti**, 0.20 **Fe**, 0.015 **H**, 2.0–3.0 **Cu**	**Heat treated**: all **diam**, 540 MPa **TS**, 400 MPa **YS**, 16% in 2 in. (50 mm) **El**
BS TA.55		UK	Forgings	96.78 min **Ti**, 0.20 **Fe**, 0.015 **H**, 2.0–3.0 **Cu**	**Heat treated**: all **diam**, 650 MPa **TS**, 525 MPa **YS**, 10% in 2 in. (50 mm) **El**
MIL–T–9046H	C(6Al–4V)	US	Sheet, strip, plate	88.35 min **Ti**, 5.5–6.5 **Al**, 3.5–4.5 **V**, 0.30 **Fe**, 0.05 **N**, 0.08 **C**, 0.015 **H**, 0.20 **O**, 0.40 total, others	**Annealed (sheet, strip)**: all **diam**, 924 MPa **TS**, 869 MPa **YS**, 8% in 2 in. (50 mm) **El**; **Solution treated (sheet, strip)**: all **diam**, 1034 MPa **YS**, 6% in 2 in. (50 mm) **El**; **Annealed (plate)**: all **diam**, 896 MPa **TS**, 827 MPa **YS**, 10% in 2 in. (50 mm) **El**
MIL–T–9046H	H(6Al–4V–SPL)	US	Sheet, strip, plate	88.23 min **Ti**, 5.5–6.75 **Al**, 3.5–4.5 **V**, 0.25 **Fe**, 0.05 **N**, 0.08 **C**, 0.005 **H**, 0.13 **O**, 0.30 total, others	**Annealed**: all **diam**, 896 MPa **TS**, 827 MPa **YS**, 10% in 2 in. (50 mm) **El**

WROUGHT AND CAST TITANIUM AND TITANIUM ALLOYS

specification number	designation	country	product forms	chemical composition	mechanical properties and hardness values
MIL-T-9046H	D(6Al-4V ELI)	US	Sheet, strip, plate	88.22 min **Ti**, 5.5-6.75 **Al**, 3.5-4.5 **V**, 0.25 **Fe**, 0.05 **N**, 0.08 **C**, 0.0125 **H**, 0.13 **O**, 0.30 total, others	**Annealed**: all **diam**, 896 MPa **TS**, 827 MPa **YS**, 10% in 2 in. (50 mm) **El**
BS TA.56		UK	Plate	88.2 min **Ti**, 5.5-6.75 **Al**, 3.5-4.5 **V**, 0.30 **Fe**, 0.25 **O+ N**	**Heat treated**: 5-10 mm **diam**, 895 MPa **TS**, 825 MPa **YS**, 10% in 2 in. (50 mm) **El**
BS 2TA.12		UK	Forging stock	88.19 min **Ti**, 5.5-6.75 **Al**, 3.5-4.5 **V**, 0.30 **Fe**, 0.05 **N**, 0.010 **H**, 0.20 **O**	**Heat treated**: all **diam**, 900 MPa **TS**, 830 MPa **YS**, 8% in 2 in. (50 mm) **El**
BS 2TA.28		UK	Forging stock and wire	88.19 min **Ti**, 5.5-6.75 **Al**, 3.5-4.5 **V**, 0.30 **Fe**, 0.05 **N**, 0.0125 **H**, 0.20 **O**	**Heat treated and quenched**: all **diam**, 1100 MPa **TS**, 970 MPa **YS**, 8% in 2 in. (50 mm) **El**
BS 2TA.10		UK	Sheet and strip	88.18 min **Ti**, 5.5-6.75 **Al**, 3.5-4.5 **V**, 0.30 **Fe**, 0.0125 **H**, 0.25 **O+ N**	**Heat treated**: all **diam**, 960 MPa **TS**, 900 MPa **YS**, 8% in 2 in. (50 mm) **El**
BS 2TA.11		UK	Bar and section	88.18 min **Ti**, 5.5-6.75 **Al**, 3.5-4.5 **V**, 0.30 **Fe**, 0.05 **N**, 0.0125 **H**, 0.20 **O**	**Heat treated**: all **diam**, 900 MPa **TS**, 830 MPa **YS**, 8% in 2 in. (50 mm) **El**
BS 2TA.13		UK	Forgings	88.18 min **Ti**, 5.5-6.75 **Al**, 3.5-4.5 **V**, 0.30 **Fe**, 0.05 **Ne**, 0.015 **H**, 0.20 **O**	**Heat treated**: all **diam**, 900 MPa **TS**, 830 MPa **YS**, 8% in 2 in. (50 mm) **El**
MIL-T-9047E	6Al-4V	US	Bar and forging stock	88.1 min **Ti**, 5.50-6.75 **Al**, 3.50-4.50 **V**, 0.30 **Fe**, 0.05 **N**, 0.08 **C**, 0.0125 **H**, 0.20 **O**, 0.40 total, others	**Annealed or solution treated**: all **diam**, 896 MPa **TS**, 827 MPa **YS**, 10% in 2 in. (50 mm) **El**; **Heat treated**: $\leq$0.5 **diam**, 1103 MPa **TS**, 1034 MPa **YS**, 10% in 2 in. (50 mm) **El**
ASTM F 467	Grade 5	US	Nut	87.98 min **Ti**, 5.5-6.75 **Al**, 3.5-4.5 **V**, 0.40 **Fe**, 0.05 **N**, 0.10 **C**, 0.0125 **H**, 0.20 **O**	**As agreed**: all **diam**, 30 min **HV**
ASTM F 468	Grade 5	US	Bolt, screw, stud	87.98 min **Ti**, 5.5-6.75 **Al**, 3.5-4.5 **V**, 0.40 **Fe**, 0.05 **N**, 0.10 **C**, 0.0125 **H**, 0.20 **O**	**As manufactured**: all **diam**, 135 ksi (930 MPa) **TS**, 125 ksi (860 MPa) **YS**, 30-36 **HRC**
AIR 9183	T-A6V	France	Bar, shapes, and forgings	87.88 min **Ti**, 5.5-7.0 **Al**, 3.5-4.5 **V**, 0.25 **Fe**, 0.07 N_2, 0.08 **C**, 0.0125 H_2, 0.20 O_2	**As manufactured**: all **diam**, 880 MPa **TS**, 820 MPa **YS**, 10% in 2 in. (50 mm) **El**
MIL-T-9047F	6Al-4V ELI	US	Bar and forging stock	87.82 min **Ti**, 6.50-6.75 **Al**, 3.50-4.50 **V**, 0.10 **Sn**, 0.15 **Fe**, 0.10 **Mo**, 0.10 **Mn**, 0.05 **N**, 0.08 **C**, 0.0125 **H**, 0.10 **Zr**, 0.10 **Cu**, 0.13 **O**, 0.10 each, 0.30 total, others	
MIL-T-81915	A(6Al-4V)	US	Castings	87.7 min **Ti**, 5.5-6.75 **Al**, 3.5-4.5 **V**, 0.30 **Fe**, 0.05 **N**, 0.08 **C**, 0.015 **H**, 0.20 **O**, 0.40 total, others	**Annealed**: all **diam**, 862 MPa **TS**, 793 MPa **YS**, 8% in 2 in. (50 mm) **El**
MIL-T-9047F	6Al-4V	US	Bar and forging stock	87.60 min **Ti**, 6.50-6.75 **Al**, 3.50-4.50 **V**, 0.10 **Sn**, 0.30 **Fe**, 0.10 **Mo**, 0.10 **Mn**, 0.05 **N**, 0.08 **C**, 0.0125 **H**, 0.10 **Zr**, 0.10 **Cu**, 0.20 **O**, 0.10 each, 0.40 total, others	

WROUGHT AND CAST TITANIUM AND TITANIUM ALLOYS

specification number	designation	country	product forms	chemical composition	mechanical properties and hardness values
MIL-T-9047E	7Al-4Mo	US	Bar and forging stock	87.53 min **Ti**, 6.50–7.30 **Al**, 0.30 **Fe**, 3.50–4.50 **Mo**, 0.05 **N**, 0.10 **C**, 0.013 **H**, 0.20 **O**, 0.40 total, others	**Annealed or solution treated**: <2.0 **diam**, 1000 MPa **TS**, 931 MPa **YS**, 10% in 2 in. (50 mm) **El**; **Heat treated**: ≤1 **diam**, 1172 MPa **TS**, 1103 MPa **YS**, 8% in 2 in. (50 mm) **El**
MIL-T-9047E	6Al-4V-ELI	US	Bar and forging stock	86.83 min **Ti**, 5.50–6.75 **Al**, 5.00–6.00 **V**, 0.15 **Fe**, 0.04 **N**, 0.08 **C**, 0.015 **H**, 0.13 **O**, 0.30 total, others	**Annealed or solution treated**: ≤1.50 **diam**, 827 MPa **TS**, 758 MPa **YS**, 10% in 2 in. (50 mm) **El**; **Heat treated**: ≤0.5 **diam**, 1034 MPa **TS**, 965 MPa **YS**, 12% in 2 in. (50 mm) **El**
MIL-T-9047F	7Al-4Mo	US	Bar and forging stock	86.63 min **Ti**, 6.5–7.30 **Al**, 0.30 **Fe**, 3.50–4.50 **Mo**, 0.05 **N**, 0.10 **C**, 0.013 **H**, 0.20 **O**, 0.40 total, others	
BS TA.51		UK	Forgings	86.33 min **Ti**, 3.0–5.0 **Al**, 1.5–2.5 **Sn**, 0.20 **Fe**, 3.0–5.0 **Mo**, 0.015 **H**, 0.3–0.7 **Si**, 0.25 **O**	**Heat treated**: all **diam**, 1000 MPa **TS**, 870 MPa **YS**, 9% in 2 in. (50 mm) **El**
BS TA.47		UK	Forging stock	86.29 min **Ti**, 3.0–5.0 **Al**, 1.5–2.5 **Sn**, 0.20 **Fe**, 3.0–5.0 **Mo**, 0.05 **N**, 0.010 **H**, 0.3–0.7 **Si**, 0.25 **O**	**Heat treated**: 25–100 mm **diam**, 1050 MPa **TS**, 920 MPa **YS**, 9% in 2 in. (50 mm) **El**
BS TA.50		UK	Forging stock	86.29 min **Ti**, 3.0–5.0 **Al**, 1.5–2.5 **Sn**, 0.20 **Fe**, 3.0–5.0 **Mo**, 0.05 **N**, 0.010 **H**, 0.3–0.7 **Si**, 0.25 **O**	**Heat treated**: 100–150 mm **diam**, 1000 MPa **TS**, 870 MPa **YS**, 9% in 2 in. (50 mm) **El**
BS TA.57		UK	Plate	86.28 min **Ti**, 3.0–5.0 **Al**, 1.5–2.5 **Sn**, 0.20 **Fe**, 3.0–5.0 **Mo**, 0.05 **N**, 0.0125 **H**, 0.3–0.7 **Si**, 0.25 **O**	**Heat treated**: 5–10 mm **diam**, 1030 MPa **TS**, 900 MPa **YS**, 9% in 2 in. (50 mm) **El**
BS TA.45		UK	Bar and section	86.28 min **Ti**, 3.0–5.0 **Al**, 1.5–2.5 **Sn**, 0.20 **Fe**, 3.0–5.0 **Mo**, 0.05 **N**, 0.0125 **H**, 0.3–0.7 **Si**, 0.25 **O**	**Heat treated**: all **diam**, 1100 MPa **TS**, 960 MPa **YS**, 9% in 2 in. (50 mm) **El**
BS TA.46		UK	Bar and section	86.28 min **Ti**, 3.0–5.0 **Al**, 1.5–2.5 **Sn**, 0.20 **Fe**, 3.0–5.0 **Mo**, 0.05 **N**, 0.0125 **H**, 0.3–0.7 **Si**, 0.25 **O**	**Heat treated**: all **diam**, 1050 MPa **TS**, 920 MPa **YS**, 9% in 2 in. (50 mm) **El**
BS TA.48		UK	Forgings	86.28 min **Ti**, 3.0–5.0 **Al**, 1.5–2.5 **Sn**, 0.20 **Fe**, 3.0–5.0 **Mo**, 0.05 **N**, 0.015 **H**, 0.3–0.7 **Si**, 0.25 **O**	**Heat treated**: all **diam**, 1050 MPa **TS**, 920 MPa **YS**, 9% in 2 in. (50 mm) **El**
BS TA.49		UK	Bar and section	86.28 min **Ti**, 3.0–5.0 **Al**, 1.5–2.5 **Sn**, 0.20 **Fe**, 3.0–5.0 **Mo**, 0.05 **N**, 0.0125 **H**, 0.3–0.7 **Si**, 0.25 **O**	**Heat treated**: all **diam**, 1000 MPa **TS**, 870 MPa **YS**, 9% in 2 in. (50 mm) **El**
BS TA.43		UK	Forging stock	86.04 min **Ti**, 5.7–6.3 **Al**, 0.25–0.75 **Mo**, 0.05 **N**, 0.006 **H**, 4.0–6.0 **Zr**, 0.10–0.40 **Si**, 0.20 **Fe**, 0.25 **O**	**Heat treated and quenched**: all **diam**, 990 MPa **TS**, 850 MPa **YS**, 6% in 2 in. (50 mm) **El**

WROUGHT AND CAST TITANIUM AND TITANIUM ALLOYS

specification number	designation	country	product forms	chemical composition	mechanical properties and hardness values
BS TA.44		UK	Forgings	86.04 min **Ti**, 5.7–6.3 **Al**, 0.20 **Fe**, 0.25–0.75 **Mo**, 0.05 **N**, 0.010 **H**, 4.0–6.0 **Zr**, 0.10–0.40 **Si**, 0.25 **O**	**Heat treated and quenched**: all **diam**, 990 MPa **TS**, 850 MPa **YS**, 6% in 2 in. (50 mm) **El**
BS TA.38		UK	Bar	83.78 min **Ti**, 3.0–5.0 **Al**, 3.0–5.0 **Sn**, 0.20 **Fe**, 3.0–5.0 **Mo**, 0.05 **N**, 0.05–0.20 **C**, 0.0125 **H**, 0.3–0.7 **Si**, 0.25 **O**	**Heat treated**: all **diam**, 1250 MPa **TS**, 1095 MPa **YS**, 8% in 2 in. (50 mm) **El**
BS TA.39		UK	Forging stock	83.78 min **Ti**, 3.0–5.0 **Al**, 3.0–5.0 **Sn**, 0.20 **Fe**, 3.0–5.0 **Mo**, 0.05 **N**, 0.05–0.20 **C**, 0.0125 **H**, 0.3–0.7 **Si**, 0.25 **O**	**Heat treated**: all **diam**, 1250 MPa **TS**, 1095 MPa **YS**, 8% in 2 in. (50 mm) **El**
BS TA.40		UK	Bar	83.58 min **Ti**, 3.0–5.0 **Al**, 3.0–5.0 **Sn**, 0.20 **Fe**, 3.0–5.0 **Mo**, 0.05 **N**, 0.05–0.20 **C**, 0.0125 **H**, 0.3–0.7 **Si**, 0.25 **O**	**Heat treated**: all **diam**, 1205 MPa **TS**, 1065 MPa **YS**, 8% in 2 in. (50 mm) **El**
BS TA.41		UK	Forging stock	83.58 min **Ti**, 3.0–5.0 **Al**, 3.0–5.0 **Sn**, 0.20 **Fe**, 3.0–5.0 **Mo**, 0.05 **N**, 0.05–0.20 **C**, 0.0125 **H**, 0.3–0.7 **Si**, 0.25 **O**	**Heat treated**: all **diam**, 1205 MPa **TS**, 1065 MPa **YS**, 8% in 2 in. (50 mm) **El**
BS TA.42		UK	Forgings	83.58 min **Ti**, 3.0–5.0 **Al**, 3.0–5.0 **Sn**, 0.20 **Fe**, 3.0–5.0 **Mo**, 0.05 **N**, 0.05–0.20 **C**, 0.0125 **H**, 0.3–0.7 **Si**, 0.25 **O**	**Heat treated**: all **diam**, 1205 MPa **TS**, 1065 MPa **YS**, 8% in 2 in. (50 mm) **El**
MIL–T–9047F	6Al–6V–2Sn	US	Bar and forging stock	83.19 min **Ti**, 5.00–6.00 **Al**, 5.00–6.00 **V**, 1.50–2.50 **Sn**, 0.35–1.00 **Fe**, 0.04 **N**, 0.05 **C**, 0.015 **H**, 0.35–1.00 **Cu**, 0.20 **O**, 0.30 total, others	
MIL–T–9046H	E(6Al–6V–2Sn)	US	Sheet, strip, plate	83.18 min **Ti**, 5.0–6.0 **Al**, 5.0–6.0 **V**, 1.5–2.5 **Sn**, 0.35–1.00 **Fe**, 0.05 **N**, 0.05 **C**, 0.015 **H**, 0.35–1.0 **Cu**, 0.20 **O**, 0.30 total, others	**Annealed (sheet, strip)**: all **diam**, 1069 MPa **TS**, 1000 MPa **YS**, 10% in 2 in. (50 mm) **El**; **Solution treated (sheet, strip)**: all **diam**, 1103 MPa **YS**, 10% in 2 in. (50 mm) **El**; **Annealed (plate)**: all **diam**, 1034 MPa **TS**, 965 MPa **YS**, 10% in 2 in. (50 mm) **El**
MIL–T–9047E	6Al–6V–2Sn	US	Bar and forging stock	83.16 min **Ti**, 5.00–6.00 **Al**, 5.00–6.00 **V**, 1.50–2.50 **Sn**, 0.35–1.00 **Fe**, 0.04 **N**, 0.05 **C**, 0.015 **H**, 0.35–1.00 **Cu**, 0.20 **O**, 0.30 total, others	**Annealed or solution treated**: 1034 MPa **TS**, 965 MPa **YS**, 8% in 2 in. (50 mm) **El**; **Heat treated**: ≤ 1 **diam**, 1207 MPa **TS**, 1103 MPa **YS**, 6% in 2 in. (50 mm) **El**
MIL–T–9047F	6Al–2Sn–4Zr–6Mo	US	Bar and forging stock	80.35 min **Ti**, 5.50–6.00 **Al**, 1.75–2.25 **Sn**, 0.15 **Fe**, 5.50–6.50 **Mo**, 0.04 **N**, 0.04 **C**, 0.0125 **H**, 3.50–4.50 **Zr**, 0.15 **O**, 0.10 each, 0.40 total, others	

WROUGHT AND CAST TITANIUM AND TITANIUM ALLOYS

specification number	designation	country	product forms	chemical composition	mechanical properties and hardness values
MIL-T-9046H	D(8Mo-8V-2Fe-3Al)	US	Sheet, strip, plate	77.06 min **Ti**, 2.6-3.4 **Al**, 7.5-8.5 **V**, 1.6-2.4 **Fe**, 7.5-8.5 **Mo**, 0.05 **N**, 0.05 **C**, 0.015 **H**, 0.016 **O**, 0.40 total, others	**Solution treated (sheet, strip)**: all **diam**, 862 MPa **TS**, 827 MPa **YS**, 8% in 2 in. (50 mm) **El**; **Annealed (plate)**: all **diam**, 862 MPa **TS**, 827 MPa **YS**, 10% in 2 in. (50 mm) **El**
MIL-T-9047E	11.5Mo-6.0Zr-4.5Sn	US	Bar and forging stock	74.55 min **Ti**, 3.75-5.25 **Sn**, 0.35 **Fe**, 10.0-12.0 **Mo**, 0.05 **N**, 0.10 **C**, 0.015 **H**, 4.5-7.5 **Zr**, 0.18 **O**, 0.40 total, others	**Annealed or solution treated**: ≤1.675 in. **diam**, 689 MPa **TS**, 671 MPa **YS**, 15% in 2 in. (50 mm) **El**; **Heat treated**: ≤1.58 in. **diam**, 1241 MPa **TS**, 1207 MPa **YS**, 8% in 2 in. (50 mm) **El**
MIL-T-9046H	B(11.5Mo-6Zr-4.5Sn	US	Sheet, strip, plate	73.5 min **Ti**, 3.75-5.25 **Sn**, 0.35 **Fe**, 10.0-13.0 **Mo**, 0.05 **N**, 0.10 **C**, 0.015 **H**, 4.5-7.5 **Zr**, 0.18 **O**, 0.40 total, others	**Solution treated (sheet, strip)**: all **diam**, 689 MPa **TS**, 621 MPa **YS**, 12% in 2 in. (50 mm) **El**; **Annealed (plate)**: all **diam**, 689 MPa **TS**, 621 MPa **YS**, 10% in 2 in. (50 mm) **El**
MIL-T-9046H	C(3Al-8V-6Cr-4Mo-4Zr)	US	Sheet, strip, plate	71.48 min **Ti**, 3.0-4.0 **Al**, 7.5-8.5 **V**, 0.30 **Fe**, 5.5-6.5 **Cr**, 3.5-4.5 **Mo**, 0.03 **N**, 0.05 **C**, 0.020 **H**, 3.5-4.5 **Zr**, 0.12 **O**, 0.40 total, others	**Solution treated (sheet, strip)**: all **diam**, 862 MPa **TS**, 827 MPa **YS**, 8% in 2 in. (50 mm) **El**; **Annealed (plate)**: all **diam**, 862 MPa **TS**, 827 MPa **YS**, 10% in 2 in. (50 mm) **El**
MIL-T-9047E	13V-11Cr-3Al	US	Bar and forging stock	69.43 min **Ti**, 2.50-3.50 **Al**, 12.50-14.50 **V**, 0.30 **Fe**, 10.00-12.00 **Cr**, 0.05 **C**, 0.015 **H**, 0.20 **O**, 0.40 total, others	**Annealed or solution treated**: all **diam**, 862 MPa **TS**, 827 MPa **YS**, 10% in 2 in. (50 mm) **El**; **Heat treated**: ≤2 in. **diam**, 1172 MPa **TS**, 1103 MPa **YS**, 4% in 2 in. (50 mm) **El**
MIL-T-9046H	A(13V-11Cr-3Al)	US	Sheet, strip, plate	69.37 min **Ti**, 2.5-3.5 **Al**, 12.5-14.5 **V**, 0.15-0.30 **Fe**, 10.0-12.0 **Cr**, 0.05 **N**, 0.05 **C**, 0.025 **H**, 0.20 **O**, 0.40 total, others	**Solution treated (sheet, strip)**: all **diam**, 862 MPa **TS**, 827 MPa **YS**, 10% in 2 in. (50 mm) **El**; **Annealed (plate)**: all **diam**, 862 MPa **TS**, 827 MPa **YS**, 10% in 2 in. (50 mm) **El**

SPECIFICATIONS INDEX

SPECIFICATIONS INDEX

specification number . . . . page

specification number . . . . page

specification number . . . . page

SPECIFICATIONS INDEX

specification number . . . . page

SPECIFICATIONS INDEX

specification number . . . . page

specification number . . . . page

specification number . . . . page

SPECIFICATIONS INDEX

SPECIFICATIONS INDEX

specification number . . . . page

specification number . . . . page

specification number . . . . page

SPECIFICATIONS INDEX

specification number . . . . page

specification number . . . . page

specification number . . . . page

SPECIFICATIONS INDEX

specification number . . . . page

specification number . . . . page

specification number . . . . page

SPECIFICATIONS INDEX

specification number . . . . page

specification number . . . . page

specification number . . . . page

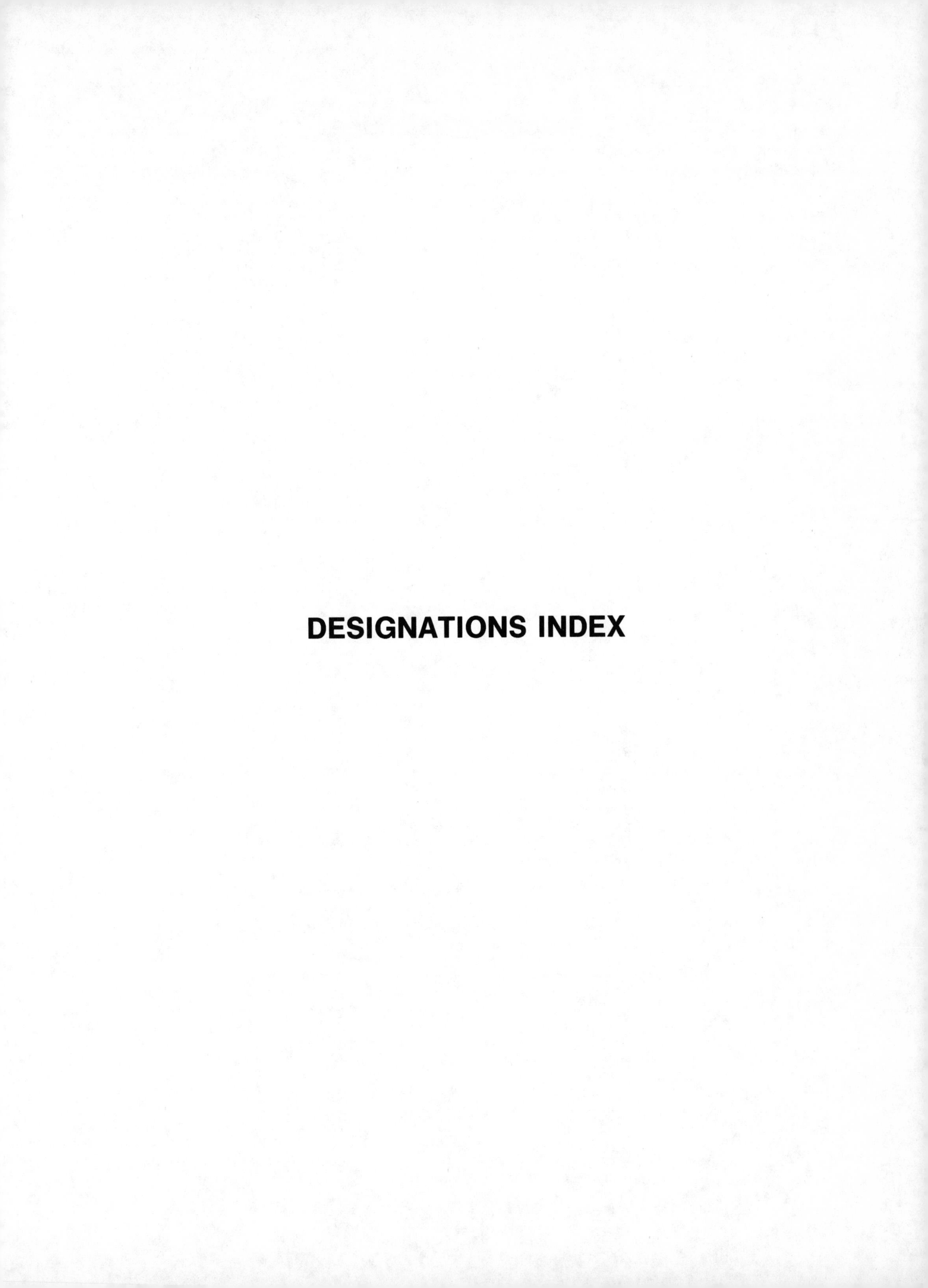

DESIGNATIONS INDEX

DESIGNATIONS INDEX

designation . . . . page

designation . . . . page

designation . . . . page

DESIGNATIONS INDEX

DESIGNATIONS INDEX

designation . . . . page

designation . . . . page

designation . . . . page

DESIGNATIONS INDEX

designation . . . . page

designation . . . . page

designation . . . . page

DESIGNATIONS INDEX

designation . . . . page

DESIGNATIONS INDEX

DESIGNATIONS INDEX

DESIGNATIONS INDEX

DESIGNATIONS INDEX

designation . . . . page

DESIGNATIONS INDEX

designation . . . . page

designation . . . . page

designation . . . . page

DESIGNATIONS INDEX

designation . . . . page

designation . . . . page

designation . . . . page

DESIGNATIONS INDEX

designation . . . . page

DESIGNATIONS INDEX

designation . . . . page

designation . . . . page

designation . . . . page

DESIGNATIONS INDEX

designation . . . . page